MATRIX Book Series 5

MATRIX is Australia's international and residential mathematical research institute. It facilitates new collaborations and mathematical advances through intensive residential research programs, each lasting 1–4 weeks.

David R. Wood (Editor-in-Chief) · Jan de Gier ·
Cheryl E. Praeger
Editors

2021-2022 MATRIX Annals

MATRIX
Australian National University
MONASH University
THE UNIVERSITY OF MELBOURNE
THE UNIVERSITY OF QUEENSLAND AUSTRALIA
ACEMS
Springer

Editors
David R. Wood *(Editor-in-Chief)*
School of Mathematics
Monash University
Melbourne, VIC, Australia

Jan de Gier
School of Mathematics and Statistics
University of Melbourne
Melbourne, VIC, Australia

Cheryl E. Praeger
School of Mathematics and Statistics
University of Western Australia
Perth, WA, Australia

ISSN 2523-3041 ISSN 2523-305X (electronic)
MATRIX Book Series
ISBN 978-3-031-47416-3 ISBN 978-3-031-47417-0 (eBook)
https://doi.org/10.1007/978-3-031-47417-0

This Springer imprint is published by the registered company Springer Nature Switzerland AG
The registered company address is: Gewerbestrasse 11, 6330 Cham, Switzerland

Preface

MATRIX is Australia's residential mathematical research institute. It facilitates new collaborations and mathematical advances through intensive residential research programs, which are held in Creswick, a small town nestled in the beautiful forests of the Macedon Ranges. MATRIX was established in 2015, and is a joint partnership between The Australian National University, Monash University, and The University of Melbourne, with The University of Queensland as an associate member. The ARC Centre of Excellence for Mathematical and Statistical Frontiers (ACEMS) provided seed funding for MATRIX and was an associate member until the end of 2021.

The closure of Australian borders and numerous lock-downs during 2020–2022 meant it was often difficult to hold in-person research programs at Creswick. MATRIX was forced to innovate. First, MATRIX introduced an online seminar series featuring high-profile speakers from around the world, including several Fields Medalists, a Nobel Prize winner, and many others.

MATRIX also ran the following tandem workshops in partnership with the Mathematisches Forschungsinstitut Oberwolfach (MFO):

- *Invariants and Structures in Low-Dimensional Topology*
  (Guest editors: Stefan Friedl, Jessica Purcell, Arunima Ray, Stephan Tillmann)
- *Rough Wave Equations*

Similarly, MATRIX ran the following tandem workshop in partnership with the Research Institute for Mathematical Sciences Kyoto University (RIMS):

- *Integrable Probability, Combinatorics and Representation Theory*

In these workshops, researchers meet simultaneously in Australia and in Germany or Japan to work on a common mathematical topic. The tandem model overcame the border closures, and will be continued into the future, especially in light of many people's desire to reduce long-haul flying.

MATRIX also co-organised the following workshops with the Sydney Mathematical Research Institute (SMRI):

- *Singularities in Geometric Flows: An Ancient Perspective*
- *Nijenhuis Geometry and Integrable Systems*

The following 'regular' research programs were held at MATRIX in 2021–2022. Each program lasted 1–2 weeks, and included ample unstructured time to encourage collaborative research. Some of the longer programs had an embedded lecture series. Because of the pandemic, some of the programs were held online.

- *Integrability and Combinatorics at Finite Temperature*
- *Structured Random Matrices in Down Under*
- *Women in Geometry, Analysis and Topology*
- *Isoperimetric Inequalities in Geometric Partial Differential Equations*
- *Groups and Geometries* (Guest editor: Anne Thomas)
- *Quantum Curves, Integrability and Cluster Algebras* (Guest editor: Norman Do)
- *2D Supersymmetric Theories and Related Topics*
  (Guest editor: Emanuel Scheidegger)

- *Dynamics, Foliations, and Geometry II*
- *Frontiers in Representation Theory*
- *Structural Graph Theory Downunder II*
  (Guest editors: Tony Huynh and Liana Yepremyan)
- *Hyperbolic Differential Equations in Geometry and Physics*
  (Guest editor: Volker Schlue)
- *Symmetry for Group Actions in Differential Geometry*
- *Mathematics of Tissue Dynamics*
  (Guest editors: Ryan Murphy, Domenic Germano and Adriana Zanca)
- *Cell Motility in Dynamic Environments*
- *Theory and Applications of Stable Polynomials* (Guest editor: James Saunderson)
- *Mathematics of Risk 2022* (Guest editor: Kais Hamza)
- *Mathematics of the Interactions between Brain Structure and Brain Functions*
  (Guest editor: Sergiy Shelyag)
- *Convexity, Geometry and Computation*
- *IBS-CGP & MATRIX Workshop on Symplectic Topology*
  (Guest editor: Marco Yat-Hin Suen)

The MATRIX Scientific Committee selected these programs based on scientific excellence and the participation rate of high-profile international participants. The following people served on the MATRIX Scientific Committee during 2021-22: Jan de Gier (Melbourne), Ben Andrews (ANU), Peter Bouwknegt (ANU), Santiago Badia (Monash), Peter Bühlmann (ETH Zurich), Alison Etheridge (Oxford), Yanan Fan (UNSW), Gary Froyland (UNSW), Cecilia González Tokman (Queensland), Joshua Ross (Adelaide), Terence Tao (UCLA), Ngamta Thamwattana (Newcastle), Ole Warnaar (Queensland), Geordie Williamson (Sydney), and David Wood (Monash, Chair). The MFO and RIMS events were selected by sub-committees of the MATRIX, MFO and RIMS scientific committees.

These programs involved organisers from a variety of Australian universities, including Australian National University, Deakin University, Macquarie University, Monash University, Swinburne University, Queensland University of Technology, University of Adelaide, University of Melbourne, University of Newcastle, University of Queensland, University of Sydney, University of Technology Sydney, and University of Western Australia, along with numerous international organisers.

This volume is a scientific record of all of the above events. All participants were encouraged to submit articles to the MATRIX Annals. The articles are grouped into refereed contributions and other contributions. Refereed articles contain original results or surveys on a topic related to the MATRIX program. The other contributions are typically lecture notes or short articles based on talks or activities at MATRIX. A guest editor organised appropriate refereeing and ensured the scientific quality of submitted articles arising from each program. The Editors (Jan de Gier, Cheryl E. Praeger and myself) finally evaluated and approved the papers.

Many thanks to the authors and to the guest editors for their wonderful work.

The final innovation, online PhD Student Symposia, were introduced in partnership with the Australian Mathematical Sciences Institute (AMSI). These events were organised by PhD students for PhD students. Students gave talks about their work, and collaborated together on open problems. This scheme proved very popular, and will be continued into the future. The following MATRIX–AMSI PhD Student Symposia took place in 2020–2022:

- *Australasian Graduate Symposium in Combinatorics* (16–18 Dec 2020). Organisers: Ajani De Vas Gunasekara (Monash), Georgina Liversidge (Auckland), Robert Hickingbotham (Monash), Shasha Zheng (Melbourne), Subhrajyoti Saha (Monash), Timothy Chan (Monash)
- *Spatial and Temporal Statistics* (17–19 Feb 2021). Organisers: Laura Cartwright (Wollongong), Quan Vu (Wollongong)
- *(GT)^2 Graduate Talks in Geometry and Topology* (2 Mar – 9 Nov 2021). Organisers: Andrew Cook (Monash), Ellena Moskovsky (Monash), Em Thompson (Monash), Grace Garden (Sydney), James Morgan (Sydney), Yossi Bokor (ANU)
- *Dynamical Systems in Topology and Triangulated Categories* (7 Jun – 2 Jul 2021). Organisers: Edmund Heng (ANU), KieSeng Nge (ANU), Sinead Wilson (ANU)
- *Categories and Companions* (8 – 12 Jun 2021). Organisers: Bryce Clarke (Macquarie), Matthew Di Meglio (UNSW), Nicola Di Vittorio (Macquarie), Eli Hazel (Macquarie), Yuki Maehara (Kyushu), Adrian Miranda (Macquarie), Giacomo Tendas (Macquarie), Paula Verdugo (Macquarie), Giulian Wiggins (Sydney)
- *Data-Driven Modelling in Mathematical Biology* (28 Jun – 9 Jul 2021). Organisers: Alexander Browning (QUT), Maud El-Hachem (QUT), Ryan John Murphy (QUT)
- *Machine Learning and Decision Making: Theory, Algorithms and Applications* (17 – 18 Mar 2022). Organisers: Jun Ju (UQ), Yeming Lei (UQ), David Maine (UQ)
- *Combinatorics Symposium* (30 May – 3 Jun 2022). Organisers: Afsane Ghafari Baghestani (Monash), Yudhistira Bunjamin (UNSW), Jodie Lee (UNSW), Thomas Lesgourgues (UNSW), Jack Neubecker (UQ), Angus Southwell (Monash)
- *(GT)^3 Graduate Talks in Geometry and Topology Get-Together* (25 – 29 Jul 2022). Organisers: Em Thompson (Monash), Grace Garden (Sydney), Yossi Bokor (Sydney), James Morgan (Sydney)
- *Categories and Companions Symposium* (19 – 23 Sep 2022). Organisers: Matthew Di Meglio (Macquarie), Nicola Di Vittorio (Macquarie), David Farrell (Macquarie), Eli Hazel (Macquarie), James Koussas (La Trobe), Yao Tang (Macquarie), Ivo Vekemans (ANU), Paula Verdugo (Macquarie)

MATRIX is hosting 14 programs in 2023, with more to come in 2024; see `www.matrix-inst.org.au`. Our goal is to facilitate collaboration between researchers in universities and industry, and increase the international impact of Australian research in the mathematical sciences.

David R. Wood
MATRIX Annals Editor-in-Chief

MATRIX Program
Integrability and Combinatorics at Finite Temperature

7 – 18 June 2021

Organisers

Dan Betea
KU Leuven

Greta Panova
University of Southern California

Leonid Petrov
University of Virginia

Tomohiro Sasamoto
Tokyo Institute of Technology

Michael Wheeler
University of Melbourne

In-Person Participants

Jan de Gier (Melbourne), Alexandr Garbali (Melbourne), Will Mead (Melbourne), Weiying Guo (Melbourne), Paul Zinn-Justin (Melbourne), Ajeeth Gunna (Melbourne), Jules Lamers (Melbourne), Peter McNamara (Melbourne), Johanna Knapp (Melbourne)

Online Participants

Valentin Buciumaș (Alberta), Evgeni Dimitrov (Columbia), Andrew Gitlin (Berkeley), Alisa Knizel (Chicago), Alexandre Krajenbrink (SISSA), Alejandro Morales (Amherst), Matteo Mucciconi (Tokyo), GaYee Park (Amherst), Marianna Russkikh (MIT), Travis Scrimshaw (Queensland), Sofia Tarricone (d'Angers & Concordia), Harriet Walsh (ENS de Lyon), Nikos Zygouras (Warwick), Vadim Gorin (Madison), Jérémie Bouttier (Paris-Saclay), Grégory Schehr (Paris-Saclay)

The overall goal of the program was to advance the methods of integrability and combinatorics towards structural and asymptotic results across a wide range of finite temperature models of statistical physics. Here the notion of "finite temperature" might have different meanings in different situations, and a complete understanding of the range of finite temperature models and related asymptotic phenomena is work in progress. Previously, "infinite temperature" models (based on, e.g., sums of independent random variables) in classical and modern probability theory, and of "zero temperature" models (e.g., last passage percolation) have been very successful. The latter were analysed using the methods of integrability and combinatorics, and currently these methods are being extended by a number of groups towards more complicated "finite temperature" settings, in order to capture new universal phenomena. Our aim was to bring together members of these groups, and produce further advances in this direction. New finite-temperature universal phenomena are expected to apply to a wide range of real-world settings, such as to the structure of ice and other condensed matter models, magnetism, quantum spin systems, thermodynamics, and polymers.

The first week of the program was held partly online via Zoom, while the second week was hybrid in-person and online. The schedule featured the following series of mini-courses (with exercise sessions):

- Vadim Gorin "General beta random matrix theory"
- Jérémie Bouttier and Grégory Schehr "Some aspects of noninteracting trapped fermions at finite temperature"
- Michael Wheeler "Integrable vertex models on the cylinder"

Contributed talks were given by: Valentin Buciumaș, Evgeni Dimitrov, Alexandr Garbali, Andrew Gitlin, Alisa Knizel, Alexandre Krajenbrink, William Mead, Alejandro Morales, Matteo Mucciconi, GaYee Park, Marianna Russkikh, Travis Scrimshaw, Sofia Tarricone, Harriet Walsh, and Nikos Zygouras. Recordings of the talks are available on youtube. The second week also included a series of hybrid discussion and collaboration sessions.

The Organisers

Structured Random Matrices in Down Under

26 July – 6 August 2021

Organisers

Giusi Alfano
Politecnico di Torino, Italy

Mario Kieburg
University of Melbourne

Dang-Zheng Liu
Hefei University of Science and Technology, China

Emanuele Viterbo
Monash University

Speakers

Matthew McKay (Hong Kong University of Science and Technology), Wenjian Liu (Queensborough), Thomas Guhr (Duisburg-Essen), Jiyuan Zhang (Melbourne), Laszlo Erdos (IST Austria), Rishabh Dudeja (Columbia), Stephane Dartois (Bordeaux), Fan Yang (Pennsylvania), Zdzislaw Burda (AGH University of Science and Technology, Poland), Arno Kuijlaars (KU Leuven, Belgium), Sergey Berezin (Hebrew University Jerusalem), Jac Verbaarschot (SUNY Stony Brook), Karol Zyczkowski (Jagiellonian), Qiang Wu (Illinois Urbana-Champaign), Benoit Collins (Kyoto), Thomas Seligman (UNAM Mexico), Shurong Zheng (Northeast Normal University, China), Shinsuke Nishigaki (Shimane University, Japan), Peter Forrester (Melbourne), Manan Vyas (UNAM Mexico), Anthony Mays (Melbourne), Nicholas Witte (Massey), Hirdesh Pharasi (Duisburg-Essen), Jesper Ipsen (Melbourne), Lucas Hackl (Melbourne)

The program focused on Random Matrix Theory, its applications, and mathematical developments. Due to the lockdown in Victoria, this program was held online only with reduced research activity. We held two one-hour-talks each evening (Melbourne time). The talks centred around recent applications to quantum information theory (6 talks), time series analysis (9 talks), and many-body systems (4 talks). Of particular interest were those applications which involve structures of random matrices such as tensors (quantum field theory and many-body systems) and block structures (time series analysis). In total 27 presentations were given over two weeks. To realise some research component, we scheduled a daily afternoon discussion talk. These were open-ended, typically lasted two hours, with 10 – 20 participants. The discussions were lively and led to some new collaborations.

The Organisers

MATRIX-MFO Tandem Workshop
Invariants and Structures in Low-Dimensional Topology

6 – 11 September 2021

Organisers

Stefan Friedl
University of Regensburg

Jessica Purcell
Monash University

Arunima Ray
Max-Planck-Institut Bonn

Stephan Tillmann
University of Sydney

MATRIX Participants

Jessica Purcell (Monash), Stephan Tillmann (Sydney), Adele Jackson (Oxford), Jack Brand (ANU), Thiago de Paiva Souza (Monash), Urs Fuchs (Monash), Yossi Bokor (ANU/Sydney), Alex He (UQ), Ben Burton (UQ), Bea Bleile (UNE), Carol Badre (Sydney), Caba Nagy (Melbourne), Dan Mathews (Monash), Diarmuid Crowley (Melbourne), Em Thompson (Monash), Grace Garden (Sydney), Tamara Hogan (Melbourne/Sydney), James Morgan (ANU), Joan Licata (ANU), Jonathan Spreer (Sydney), Marcy Robertson (Melbourne), Zsuzsanna Dancso (Sydney)

MFO Participants

Jonathan Bowden (Regensburg), Rachael Jane Boyd (MPIM Bonn), Rima Chatterjee (Köln), Arnaud de Mesmay (CNRS, LIGM), Peter Feller (ETH Zürich), Stefan Friedl (Regensburg), Holger Kammeyer (Düsseldorf), Daniel Kasprowski (Bonn), Marc Kegel (HU Berlin), Markus Land (Kopenhagen), Delphine Moussard (Aix-Marseille), Boris L. Okun (Wisconsin-Milwaukee), Mark A. Powell (Durham), José Pedro Quintanilha (Regensburg), Katherine Raoux (MPIM Bonn), Arunima Ray (MPIM Bonn), David Reutter (MPIM Bonn), Benjamin Matthias Ruppik (MPIM Bonn), Saul Schleimer (Warwick), Paula Truöl (ETH Zürich), Claudius Zibrowius (Regensburg)

When we initially planned this workshop, we intended for 20 mathematicians to meet at the MATRIX Institute in Creswick, and another 20 to meet at MFO in Germany. In the end 19 mathematicians did indeed make it to MFO. However the ongoing Covid-19 pandemic and travel restrictions within Australia prevented the Australia based mathematicians from travelling to the MATRIX institute. Instead the Australian participants joined the workshop virtually from their respective homes. An unexpected positive outcome of this was that more Australian mathematicians were able to participate than initially planned.

To accommodate the hybrid setting and the time difference, the workshop had an unusual format, consisting primarily of discussion sessions and with very few formal talks. Our first goal was to give young mathematicians an opportunity to introduce themselves and their work. The second goal was to start conversations among all the participants and to try to make the workshop as interactive as possible. To achieve our second goal we wrote to all participants before the meeting, soliciting questions or topics to be discussed at the meeting, and put these into a shared online document. Each participant also got a chance to indicate which problems they would be interested in discussing.

Among the suggestions from the participants, we picked four problems which had enough traction at MFO as well as on the Australian side:

1. Twisted intersection forms of spin 4-manifolds
2. Knots in 3-manifolds
3. Profinite rigidity of 3-manifold groups
4. Cobordisms up to stable diffeomorphism

Each topic had two discussion leaders, one at MFO and one in Australia.

The setup resulted in an a rather unusual schedule. A common time worked for most participants, between 17:00 to 20:00 Australian Eastern Standard Time (AEST), which was 9:00 to 12:00 German time. Participants at MFO were in a standard workshop environment (albeit with certain social distancing obligations), and the common time was bookended by meals provided at MFO. Meanwhile, Australian participants in the privacy of their own homes had different experiences. Some relished the human interactions brought about by the online meetings, whilst others were at times a little distracted and occupied with tasks other than research.

In Australia, the scheduled events of the tandem workshop began at 16:30 AEST on Monday, in an online "MATRIX house" that used the software Sococo. Participants were able to move an avatar around the house to interact with other participants via video and chat. This allowed social interaction for remote Australian participants. Working groups also had access to shared libraries and notes in different rooms of the house. After a brief meeting in the online MATRIX house, participants moved to Zoom to meet with the European participants at 17:00 AEST.

On Monday, every participant on both continents spent one minute introducing themselves to the full group. Afterwards the discussion leaders gave short introductions to the four problems. Finally, the working groups met separately to get started on their projects. The Australians moved into Zoom breakout rooms, and they were joined by German participants who moved into separate rooms at the MFO.

At the end of the Australian day, the MATRIX groups organised times to meet up separately during the following day, and most signed off.

Meanwhile, during their afternoon, the MFO participants met for one scheduled talk each day (by Peter Feller, Holger Kammeyer, Markus Land and Delphine Moussard). These talks took place at 22:15 AEST, and despite the late hour a few Australian mathematicians joined via Zoom. The talks were also recorded and made available to the Australian participants. Next, MFO working groups met for two more hours to work on the problems while Australia slept. After discussions, those at the MFO added notes to the common online documents to let the Aussies know what they had covered; this ensured some continuity for the following day.

During the Australian morning and afternoon, MATRIX participants met up at scheduled times via Zoom or at the Sococo MATRIX house, to go over MFO progress, and to discuss problems separately. The workshop reconvened each day at 17:00 AEST in tandem with MFO. The tandem portion of each day began with short (10 minute) talks by young researchers. There were four such talks from from MFO on Tuesday, and ten from Australia from Wednesday to Friday. The talks were followed by brief summaries from the working groups, and occasionally an indication of the plans for the coming day. The bulk of the shared time was left to the working groups to discuss and collaborate, with more detailed reports from the groups on either continent. Over the course of the week the problems being discussed also evolved – for example the twisted intersection forms group merged with the cobordism group and the knot group split into smaller groups. Participants were encouraged to switch between groups during the week as they wished, but most remained with their original choices.

Although the Australians were unable to meet in person, by setting up regular meetings with small groups, by focussing on particular problems, and by interspersing discussions with short talks, many participants were able to spend real time thinking about mathematical research in collaborative groups. For some, this was the first time they had been able to do so since lockdowns began. Meanwhile, the 19 mathematicians at MFO relished the opportunity to talk in person about mathematics again, and they report that the discussions on mathematical and academic issues were extremely lively and a welcome relief after spending many months in front of a laptop screen. The collaboration across time zones and continents worked out better than expected. Indeed, the seven hour time difference turned out to be a blessing since it gave all participants enough time to meet together, but it also gave each continent time to work on the problem separately and in smaller groups.

The already published MFO Report No. 42/2021 details much of the progress made by the working groups during the workshop and contains extended abstracts of most talks. Preprints and talks on joint projects directly arising as a result of interaction at the workshop and involving groups of people from both continents bear testament to progress that can be made over long distances in a hybrid setting. The organisers feel that a workshop as initially planned, with 20 participants together in Creswick and 20 at MFO with a three hour window for interaction, is a sound model for sustainable workshops in the future, with the caveat that intensive research progress and knowledge transfer is mainly achieved within each of the two nodes.

Stefan Friedl, Jessica Purcell, Arunima Ray, Stephan Tillmann
Guest editors

Ben
Jonathan Bowden

MATRIX-MFO Tandem Workshop
Rough Wave Equations

13 –18 September 2021

Organisers

Zihua Guo
Monash University

Andrew Hassell
Australian National University

Pierre Portal
Australian National University

Po Lam Yung
Australian National University

Dorothee Frey
Karlsruhe Institute of Technology

Jan Rozendaal
Institute of Mathematics of the Polish Academy of Sciences

MATRIX Participants

Pierre Portal (ANU), Zihua Guo (Monash), Andrew Hassell (ANU), Po Lam Yung (ANU), Timothy Candy (Otago), Michael Cowling (UNSW), Xuan Duong (Macquarie), Volker Schlue (Melbourne), Adam Sikora (Macquarie), Melissa Tacy (Auckland), Himani Sharma (ANU), Sean Gomes (ANU), Sean Harris (ANU), Antoine Gansemer (ANU)

Oberwolfach Participants

Dorothee Frey (Karlsruher), Sebastian Herr (Bielefeld), Marina Iliopoulou (Kent), Alessio Martini (Birmingham), Yonas Mesfun (Karlsruher), Detlef Müller (Kiel), Jan Rozendaal (Polish Academy), David Rule (Linköping), Robert Schippa (Karlsruher)

This tandem workshop provided a platform for early career researchers and leading experts in harmonic analysis, microlocal analysis and spectral theory. The workshop focused on collaboration and technical knowledge exchange on topics such as local smoothing, spectral multipliers, restriction estimates, Hardy spaces for Fourier integral operators, and nonlinear partial differential equations.

The workshop brought together 23 early career researchers and leading experts working on these topics, 8 of which were physically present at MFO and 15 of which participated virtually from Australia and New Zealand. This hybrid format led to new interactions between different areas, both geographically and scientifically.

The following talks by the participants were pre-recorded made available online one week before the meeting. We had a one hour discussion session per day between MFO and the MATRIX participants, where we discussed the talks. We scheduled in advance which 3 or 4 talks would be discussed on a particular day. That allowed participants to view, or to review, the talk just beforehand so the content was fresh in their memory. These discussions were lively and conducted in much greater depth than usually happens in the 5–10 minute discussion after a standard talk.

- Pierre Portal "Fixed time L^p estimates for certain wave equations with rough coefficients"
- Po Lam Yung "The local smoothing conjecture for the wave equation: a survey"
- Jan Rozendaal "Local smoothing and Hardy spaces for Fourier integral operators"
- Zihua Guo "Decay of solutions to the energy critical nonlinear Schrodinger and wave equation"
- Andrew Hassell "Wave equations with rough coefficients and regularity in Hardy spaces for Fourier integral operators"
- Melissa Tacy "Intersections between mircolocal, semiclassical and harmonic analysis"
- Alessio Martini "Spectral multipliers and wave equation for sub-Laplacian"
- Sebastian Herr "Nonlinear Dirac Equations"
- Himani Sharma "Spectral multiplier estimates for abstract differential operators"
- Robert Schippa "Quasilinear Maxwell equations"
- David Rule "Boundedness of multilinear oscillatory integral operators in local Hardy spaces"
- Adam Sikora "Quantum separation effect"
- Michael Cowling "Characterising Hardy spaces"
- Antoine Gansemer "Possible exact Egorov theorem on hyperbolic surfaces"
- Xuan Duong "BMO spaces on non-doubling manifolds with ends"
- Timothy Candy "The transference principle and bilinear restriction estimates"
- Volker Schlue "Scattering from infinity for semi-linear wave equations with weak null condition"
- Marina Iliopoulou "Sharp L^p estimates for oscillatory integral operators of according to signature"

In addition, time was reserved for collaboration in smaller groups, both in person and virtually. We had a Google spreadsheet that could be edited by everyone. This was, in the first instance, so that participants could provide a link to their talk, and also schedule meetings on particular topics so that anyone interested could join in. These were the online equivalent of a small group of mathematicians standing around a blackboard discussing some specific problem.

Zihua Guo

Audience MFO L2
Organizer 2137a
Speaker MFO L2
Adam Sikora
Po Lam Yung
Marina Iliopoulou
Pierre Portal (he/him/his)
Himani Sharma
Antoine Gansemer
Zihua Guo
Tim Candy
Andrew Hassell
Sean Gomes
Sean Harris
Melissa Tacy
Michael
Michael

MATRIX Program
Women in Geometry, Analysis and Topology

15 – 19 November 2021

Organisers

Julie Clutterbuck
Monash University

Melissa Tacy
University of Auckland

Vanessa Robins
Australian National Unviersity

Participants

Julie Clutterbuck (Monash), Valentina Wheeler (Wollongong), Zsuzsanna Dancso (Sydney), Marcy Robertson (Melbourne), Tamara Hogan (Sydney), Olivia Borghi (Melbourne), Sophie Raynor (Macquarie)

In 2021, we decided to adapt to the pandemic's international and state travel restrictions by running a small workshop focused on fostering local collaborations between women. Our goal:

This program is a week-long meeting to foster new local collaborations, deepen existing ones, and above all create a space for women to get research done. The research themes are very broad, ranging from analysis, geometry, and topology. We anticipate that substantial progress will be made on open questions. In a time where international travel is curtailed, local collaborations will be strengthened. Here local means anyone able to be in Australia at the time of the meeting: we expect this to be mathematicians based in New Zealand and Australia.

As it turned out, even inviting local mathematicians was somewhat ambitious: by the time of the workshop, travel to and from New Zealand was still difficult, and travel between states was seen as risky, with some jurisdictions having Melbourne listed as a "high risk geographic area". There were seven participants, including two PhD students and one post-doc.

The schedule featured introductory talks on the opening day, and one contributed talk every other day. The remainder of the time was used for collaboration.

Julie Clutterbuck and Valentina Wheeler investigated quasi-concavity of solutions to elliptic equations for domains in the sphere. With Olivia Borghi and Marcy Robertson, Sophie Raynor looked at questions related to algebraic infinity operads. Motivated by Olivia Borghi's work on operads of cactus groups—which emerge in the study of the real points of the moduli spaces of genus 0 surfaces—they considered appropriate definitions for infinity (cyclic) operads that are equivariant with respect to group actions. They also discussed future directions for studying algebraic infinity modular operads (to describe higher genus structures). Zsuzsanna Dansco, Tamara Hogan and Marcy Robertson discussed combinatorial invariants of knotted objects.

Julie Clutterbuck

MATRIX Program
Isoperimetric Inequalities in Geometric Partial Differential Equations

22 – 26 November 2021

Organisers

Paul Bryan
Macquarie University

Julie Clutterbuck
Monash University

Daniel Hauer
University of Sydney

Lei Ni
UC San Diego

Guofang Wang
University of Freiburg

Guofang Wei
UC Santa Barbara

Participants

Julie Clutterbuck (Monash), Valentina Wheeler (Wollongong), Peter Olanipekun (Monash), Cale Rankin (ANU), Yuri Nikolayevsky (La Trobe), Mat Langford (Newcastle & Knoxville), Jiakun Liu (Wollongong)

Isoperimetric comparisons may be used to uncover new properties of curved spaces, and to estimate eigenvalues of operators of physical significance. Isoperimetric estimates have an ancient history, comparing the area and perimeter of regions. Only recently have the complex geometric situations seen in nature begun to be understood. These include excitation energies in quantum mechanics, and the geometry of soap bubbles. The field brings together the theory of PDE, differential geometry and non-smooth metric geometry and has continued to produce novel and exciting mathematics since antiquity.

This was a much-postponed and much-reduced version of the original meeting. Even though we were expecting international visitors to be unable to come, even interstate visitors were prevented from attending by university requirements. We had seven participants, including two students.

The schedule featured introductory talks on the opening day, and one contributed talk every other day. The remainder of the time was used for collaboration.

Langford made progress on free boundary curve shortening and mean curvature flow. Olanipekun worked on Willmore surfaces, including writing up some of his PhD results for a paper. Rankin and Liu both worked on Monge–Ampere problems. Liu gave a talk on the connection between the isoperimetric problem for curves, and the Monge–Ampere problem. Clutterbuck and Wheeler investigated quasi-concavity of solutions to elliptic equations for domains in the sphere.

Julie Clutterbuck

MATRIX Program
Groups and Geometries

29 November – 3 December 2021

Organisers

Alice Devillers
University of Western Australia

James Parkinson
University of Sydney

Jeroen Schillewaert
University of Auckland

Anne Thomas
University of Sydney

Participants
John Bamberg (Western Australia), Anton Baykalov (Auckland), Alex Bishop (Sydney), Max Carter (Newcastle), Marston Conder (Auckland), Matthew Conder (Auckland), Alice Devillers (Western Australia), Heiko Dietrich (Monash), Alex Elzenaar (Auckland), Michael Giudici (Western Australia), Melissa Lee (Auckland), Ari Markowitz (Auckland), Gaven Martin (Massey), Eamonn O'Brien (Auckland), James Parkinson (Sydney), Adam Piggott (ANU), Tomasz Popiel (Auckland), Cheryl Praeger (Western Australia), Youming Qiao (UTS), Arun Ram (Melbourne), Jeroen Schillewaert (Auckland), Anne Thomas (Sydney), Stephan Tornier (Newcastle), Geertrui Van De Voorde (Canterbury), Gabriel Verret (Auckland), George Willis (Newcastle), Yeeka Yau (Sydney)

In the 1950's Jacques Tits provided a converse to Klein's Erlangen program showing that algebraic groups could be understood by the study of geometric objects (buildings) associated to them. In fact his theory is so all-encompassing it also allows us to study groups over local fields, which are of crucial importance in number theory and representation theory. Moreover, the geometry of buildings is very rich in itself and buildings are a key family of examples in geometric group theory, as they form spaces of non-positive curvature (CAT(0) spaces). Lastly the development of buildings played a crucial role in the classification of the finite simple groups.

The original aim of the meeting was to gather world leading experts and early career researchers working on the theory of buildings and its connections to adjacent fields to investigate the research avenues that have opened up in recent years by means of some spectacular breakthroughs in the field.

Due to Covid only researchers based in Australia and New Zealand were able to attend, hence the scope of the workshop was broadened to a wide range of topics in groups and geometry. Despite the obvious drawback of an online meeting engagement was high. The meeting consisted of a mixture of contributed talks, open problem sessions and free discussions. Here is the list of talks:

- Adam Piggott "Rewriting systems, triangles in Cayley graphs, and the isomorphism problem for plain groups"
- Youming Qiao "Optimisation meets CAT(0) spaces"
- Stephan Tornier: "A permutation group problem relating to groups acting on trees"
- Yeeka Yau "Cone Types, Automata and Regular Partitions of Coxeter Groups"
- Alex Bishop "Kazhdan-Lusztig Polynomials and Bruhat Posets"
- Matthew Conder "Discrete two-generated subgroups of PSL$(2, Q_p)$"
- Alex Elzenaar "Approximating the Riley slice exterior"
- Michael Giudici "Orbits of Sylow subgroups of finite permutation groups"

Good progress was made on the open problems. See the workshop website for more details.

The article in the MATRIX Annals is an introduction for the non-expert to the Riley slice and much of the related background material, assuming only graduate level complex analysis and topology. It reviews the history of and literature surrounding the Riley slice; and announces some of the authors' recent results, which extends the work of Keen and Series to the one complex dimensional moduli spaces of Kleinian groups isomorphic to $\mathbb{Z}/a\mathbb{Z} * \mathbb{Z}/b\mathbb{Z}$ acting on the Riemann sphere.

We hope to organise an in-person event in Creswick in coming years.

Anne Thomas
Guest editor

MATRIX Program
Quantum Curves, Integrability and Cluster Algebras

13 – 18 December 2021

Organisers

Leonid Chekhov
Michigan State University

Paul Norbury
University of Melbourne

Michael Shapiro
Michigan State University

Norman Do
Monash University

Elba Garcia-Failde
IRIF, Paris Diderot University

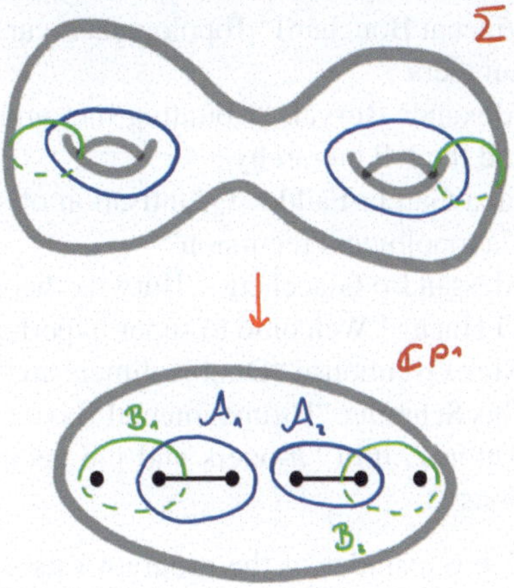

Participants

Anand Deopurkar (ANU), Norman Do (Monash), Paul Norbury (Melbourne), Alexander Alexandrov (IBS Korea), Lukas Anagnostou (Melbourne), Kowshik Bettadapura (Melbourne), Vincent Bouchard (Alberta), Alexandr Buryak (HSE Uni.), Anupam Chaudhuri (Monash), Xavier Coulter (Auckland), Anand Deopurkar (ANU), Elba Garcia-Failde (Sorbonne), Alessandro Giacchetto (Inst. de Physique Théorique), Pedram Hekmati (Auckland), Yi Huang (Tsinghua), Miles Koumouris (Melbourne), Ian Le (ANU), Ellena Moskovsky (Monash), Nick Ovenhouse (Minnesota), Brett Parker (ANU), Lucia Stockdale (Melbourne), Gus Schrader (Columbia), Michael Swaddle (Melbourne), Campbell Wheeler (Max Planck), Anton Zeitlin (Louisiana)

Recent decades have seen a proliferation of ideas surrounding enumerative geometry and mathematical physics. These have led to the discovery of rich mathematical structures and unexpected connections to diverse parts of mathematics. In particular, recent advances in the field indicate that the three topics in the title of the program provide fruitful avenues for current exploration. The general theory of topological recursion provides an underlying theme that ties these topics together.

This one-week program assembled researchers working on quantum curves, integrability and cluster algebras to address recent advances and progress new connections between these topics. Participants came from a diverse range of backgrounds and included leading experts, early career researchers, and research students. Due to the ongoing COVID-19 pandemic and the resulting restrictions on travel, the workshop was held in a hybrid mode, with international participants taking part online and local participants hosted at MATRIX.

Most days of the program were bookended by a research talk in the morning and another in the evening. These talks, listed below, were all delivered by international speakers via a video conferencing platform.

- Alexander Alexandrov "KP integrability of triple Hodge integrals"
- Vincent Bouchard "Topological recursion, quantum curves, and atlantes Hurwitz numbers"
- Alexandr Buryak "Counting meromorphic differentials on the Riemann sphere and the KP hierarchy"
- Elba Garcia-Failde "Quantisation of spectral curves of arbitrary rank and genus via topological recursion"
- Alessandro Giacchetto "Hurwitz theory, with (a) spin"
- Yi Huang "Welcome to super hyperbolic surface theory"
- Nick Ovenhouse "Double dimers and the super Ptolemy relation"
- Gus Schrader "Bifundamental Baxter operators"
- Anton Zeitlin "$\hbar$-opers and the geometric approach to the Bethe ansatz equations"

The remainder of the program was structured in a much more organic way, allowing participants to dictate the flow of events on a day-to-day basis. Academic activities included:

- impromptu lectures on fundamental topics (such as KP integrability);
- large group informal discussions on topics of current interest (such as superhyperbolic geometry);
- presentations of unsolved problems (such as the Kazarian–Norbury conjecture); and
- a great deal of small group research collaboration.

It is worth remarking that the Kazarian–Norbury conjecture, concerning tautological relations on moduli spaces of curves, was presented at the workshop and eventually proven by a team involving two of the program participants.

Overall, the workshop was entirely successful and provided an intense, productive, collegiate atmosphere for all participants. It stimulated mathematical progress and incubated ideas that will guide research in the field for years to come.

Norman Do
Guest editor

MATRIX
QUANTUM CURVES,
INTEGRABILITY and
CLUSTER ALGEBRAS
13 - 18 DECEMBER 2021
CRESWICK
MATRIX
ANU

MATRIX-SMRI Symposium
Singularities in Geometric Flows: An Ancient Perspective

4 – 8 January 2022

WILLIAM P. MINICOZZI II
Massachusetts Institute of Technology

NATAŠA ŠEŠUM
Rutgers University

Organisers

Jonathan Zhu
Australian National University &
Princeton University

Mat Langford
University of Newcastle &
University of Tennessee Knoxville

Participants

William P. Minicozzi II (MIT), Nataša Šešum (Rutgers), Ben Andrews (ANU), Theodora Bourni (Knoxville), Paul Bryan (Macquarie), Tim Buttsworth (Queensland), Xuzhong Chen (Hunan), Otis Chodosh (Stanford), Kyeongsu Choi (KAIST), Sam Cuthbertson (Wollongong), Nick Edelen (Notre Dame), Ben Goldys (Sydney), Andrey Glubokov (Perdue), Ian Hamilton, Seongtag Kim (Inha), Robert Haslhofer (Toronto), Brett Kotschwar (Arizona), Michael Law (ANU), Christos Mantoulidis (Brown), Ramiro Lafuente (Queensland), Tang-Kai Lee (MIT), Jiakun Liu (Wollongong), Stephen Lynch (Tübingen), James McCoy (Newcastle), Vishnu Mangalath (Sydney), Makoto Narita (Okinawa), Bruce Ni (Monash), Huy The Nguyen (Queen Mary), Laurentiu Paunescu (Sydney), Jyotshana Prajapat (Mumbai), Peter Price, Felix Schulze (Warwick), David Shellard (ANU), Lu Wang (CalTech), Glen Wheeler (Wollongong), Haotian Wu (Sydney), Zhenyang Xia, Zhou Zhang (Sydney), Tailong Zhou (Tsinghua)

The joint MATRIX-SMRI symposium focussed on significant developments in the understanding of geometric flows via the analysis of soliton and ancient solutions, particularly in the context of mean curvature flow. It centred around the pioneering work of the joint chairs Nataša Šešum and William P. Minicozzi II.

Nataša Šešum (Rutgers University) has made a number of groundbreaking contributions to the analysis of singularities in geometric evolution equations. Her remarkable work with Angenent, Daskalopoulos and others provide the first general classification results for ancient solutions. Nataša was an invited speaker at the International Congress of Mathematicians in 2014, and was elected as a Fellow of the American Mathematical Society in 2015. Nataša Šešum gave a lecture series on “Geometric flows and ancient solutions” in the symposium.

William P. Minicozzi II is the Singer Professor of Mathematics at the Massachusetts Institute of Technology. Throughout an enduring collaboration with Tobias H. Colding, he has resolved a number of major open problems in several areas of geometric analysis. Colding and Minicozzi received jointly the AMS Oswald Veblen Prize in Geometry in 2010 for major breakthroughs on the structure of embedded minimal surfaces. He gave an invited address at the International Congress of Mathematicians in 2006. William P. Minicozzi II gave a lecture series on "Singularities and diffeomorphisms" in the symposium.

Due to the CoViD-19 pandemic, the symposium was held in a hybrid format, with the first week held both on-campus at MATRIX and online, and the second week entirely online. The first week consisted of thematic talks, discussion sessions and a series of open problem sessions with the chairs, with online participants joining via video link for the talks and open problem sessions. The second week consisted of online lecture series by the chairs, supplemented with the following talks by prominent international researchers.

- Lu Wang "Relative expander entropy and its applications"
- Kyeongsu Choi "Translating flows by sub-affine-critical powers of Gauss curvature"
- Ramiro Lafuente "Noncompact Einstein manifolds with symmetry"
- Brett Kotschwar "Propagation of geometric structure under the Ricci flow and applications"
- Otis Chodosh "Ancient flows and generic regularity"
- Ben Andrews "Limiting shapes of fully nonlinear flows of convex hypersurfaces"
- Felix Schulze "A relative entropy and a unique continuation result for Ricci expanders"
- Huy The Nguyen "Brakke Regularity for the Allen-Cahn Flow"
- Theodora Bourni "Ancient solutions of mean curvature flow"
- Robert Haslhofer "Classification of noncollapsed translators in $\mathbb{R}^4$"
- Stephen Lynch "Collapsing and noncollapsing convex ancient solutions"
- Christos Mantoulidis "Variational aspects of ancient flows"
- Nick Edelen "Degeneration of 7-dimensional minimal hypersurfaces with bounded index"

Recordings of the lecture series and talks are available through SMRI's YouTube page.

The workshop was hit with some initial setbacks, with interstate travel restrictions and some unavoidable late withdrawals resulting in lower than hoped-for in-person numbers, followed by once-a-decade storms hitting the central Victoria region, resulting in significant flooding of Creswick Creek, lengthy power outages, and costly damage to the roof of the reception building (which also contains the lecture theatres). Nonetheless, we were able to continue the programme relatively uninterrupted through the use of mobile internet, back-up power and creative choices of venues, and kept ourselves entertained with expeditions to observe the flooded Creek and games of black-out snooker.

We would like to express our gratitude to Nataša for preparing and presenting her outstanding lectures and open problem sessions despite suffering through a bout of CoViD, and to Adam Crutchfield for his heroic efforts keeping us well-fed despite the dangerous conditions and power outages, and the worry of flood waters threatening his own property.

Mat Langford

MATRIX Program
2D Supersymmetric Theories and Related Topics

17 – 28 January 2022 (online)
13 – 17 June 2022 (in person)

Organisers

Johanna Knapp
University of Melbourne

Jock McOrist
University of New England

Gabriele Tartaglino-Mazzucchelli
University of Queensland

Emanuel Scheidegger
Peking University

Participants

Shabeeb Alalawi (Bonn), Leighton Arnold (Queensland), Deepan Betal (Bhopal), Daniele Bielli (Surrey & Milano), Jean-Emile Bourgine (Melbourne & Sogang), Peter Bouwknegt (ANU), Max Carter (Newcastle), Ashish Chandra (IIT Kanpur), Huailiang Chang (Hong Kong), Jaklyn Crilly (ANU), Ariunzul Davgadorj (Masaryk), Andrea Dei (Harvard), Xenia del la Ossa (Oxford), Zhihao Duan (KIAS), Stephen Ebert (UCLA), Johannes Faller (Queensland), Fotis Farakos U. Padova, Zachary Fehily (Melbourne), Christian Ferko (UC Davis), Urs Fuchs (Monash), Ethan Fursman (ANU), Masoud Ganji (New England), Moji Ghadimi (Queensland), Dongwook Ghim (KIAS), Nicholas Godfrey (Queensland), Gregory Gold (Queensland), Daniel Grumiller (TU Wien), Wei Gu (Virginia Tech), Bin Gui (Tsinghua), Bolin Han (ANU), Zezhuang Hao (Southampton), Adam Harris (New England), Kentaro Hori Kavli (IPMU), Daniel Hutchings (Western Australia), Jessica Hutomo (Queensland), Hongliang Jiang (Queen Mary), Christian Kennedy (Queensland) Saurish Khandelwal (Queensland), Sungjoon Kim (POSTECH), Alyshah Ladhu (Melbourne) Xiaobin Li (Jiaotong), Ulf Lindström (Uppsala & Ankara), Nicholas Macleod (Melbourne), Taniya Mandal (Witwatersrand), John Mashford (Melbourne), Jock McOrist (New England), Davide Morgante (Milan), Suresh Nampuri (Lisboa), Usman Naseer (Uppsala), Satoshi Nawata (Fudan), Takashi Otofuji (Nihon), Silvia Penati (Milano-Bicocca), Michael Ponds (Western Australia), Peter Price (DSTG), Robert Pryor (Melbourne), Thomas Quella (Melbourne), Jorgen Rasmussen (Queensland), Christopher Raymond (ANU), Mauricio Romo (Tsinghua), anik rudra (Witwatersrand), Emanuel Scheidegger (Peking), Thorsten Schimannek (CNRS), Gerd Schmalz (New England), Rak-Kyeong Seong (Ulsan), Mohamed

Ali Seridi (Constantine Mentouri), Savdeep Sethi (Chicago), Alessandro Sfondrini (Princeton & Padova), Eric Sharpe (Virginia Tech), Liam Smith (Queensland), Martin Sticka (New England), Tamera Summerill (Queensland), Kaiwen Sun (KIAS), Haoyu Sun (UT Austin), Gabriele Tartaglino Mazzucchelli (Queensland), James Tener (ANU), Mathai Varghese (Adelaide), Rikard von Unge (Masaryk), Jun Wang (Peking), Junxiao Wang (BICMR), Bryan Wang (ANU), Jesse Woods (Queensland), Ben Wootton (New England), Xian-Jin Xie (THU), Yaping Yang (Melbourne), Jinghao Yu (Tsinghua), Grace Yuan (Melbourne), Gufang Zhao (Melbourne), Zijun Zhou Kavli (IPMU)

The program consisted of three parts, divided into two events.

The first event took place online and lasted for two weeks. The first week was primarily research oriented with sessions of overview talks on the subject areas of the program with plenty of time for discussions and research activities. There was one 90-minute overview talk per day that introduced the topics of the program. This was very well-received with 30-40 participants joining the live Zoom sessions.

In the second week we held a research conference entitled "2D Supersymmetric Theories and Related Topics". The conference schedule consisted of four research talks per day, plus one or two moderated discussion sessions.

After the CoViD-19 restrictions in Australia were lifted we were given the opportunity to have a second event, a one-week in-person extension of the first one. Due to the short notice, all participants were members of the Australian research community working on the topics of the program. Since 15 of the participants were early career researchers, the workshop schedule consisted of 25-minute presentations by them in the mornings, supplemented by a few presentations of senior researchers. The afternoons were left for discussions and collaborations.

Week 1 Lectures

Five online lectures were given in the first week of the program. Since they were given online by various experts from universities in Europe, Japan, the United States and Australia, they were distributed all over the day. In the remaining time, they were supported by online discussion sessions that provided ample opportunity to interact with the speakers. All of the speakers prepared the ground for their (and others') research talks in the second week. The lectures covered five distinct, but interrelated approaches to two–dimensional (2D) supersymmetric field theories.

- Christian Ferko "$T\overline{T}$ and Supercurrent-Squared".

  This lecture considered general 2D quantum field theories and gave an introduction to a particular universal irrelevant deformation of 2D QFTs provided by the $T\overline{T}$ operator. The theories produced by this deformation are especially interesting because they are not local QFTs but rather share some properties of string theories. The lecture focused on cases where the deformation can be presented in a manifestly supersymmetric form known as supercurrent-squared.

- Daniel Grumiller "Introduction to 2D gravity"

  This lecture focused on a very interesting 2D quantum field theory known as 2D gravity. The lecture provided an introduction to 2D gravity with focus on holographic aspects of 2D dilaton gravity.

- Kentaro Hori "Introduction to 2D (2,2) supersymmetric gauged linear sigma models"

  This lecture considered another class of highly interesting 2D quantum field theories known as 2D (2,2) supersymmetric gauged linear sigma models. The lecture provided an introduction to these theories. One key aspect is that the vacuum manifolds can be Calabi–Yau manifolds and hence these theories describe properties of string theories compactified on such manifolds.

- Ulf Lindström (Uppsala) "Supersymmetry and Complex Geometry"

  This lecture focused on 2D supersymmetric nonlinear sigma models. There is a close relation between such theories and geometry. The number of supersymmetries and their representation determine the geometry of their target space. E.g., sigma models with (2,2) chiral superfields lead to Kähler geometry on the target space.

- Jock McOrist "An introduction to string compactifications"

  This lecture provided an overview of string compactification realising four-dimensional Minkowski space with minimal supersymmetry. The lecture explained the relation to the (2,2) supersymmetric gauged linear sigma models and supersymmetric nonlinear sigma models of the other lectures. Then it explained which supergravity theories arise from the heterotic string.

Week 2 Research talks

The second week saw a research conference entitled "2D Supersymmetric Theories and Related Topics". It featured speakers from universities in Europe, Japan, the United States and Australia who originally all had agreed to attend the conference in person had there not been travel restrictions due to CoViD-19. Their background ranged from young postdocs to well–established researchers of international renown.

- Daniele Bielli "Super non-Abelian T-duality"
- Jean-Emile Bourgine "Vortex partition function of 2D $N = (2,2)$ Super Yang-Mills theories and quantum groups"
- Andrea Dei "String correlators on AdS3"
- Xenia de la Ossa "Black holes, periods and arithmetic of Calabi–Yau manifolds"
- Christian Ferko "Holography and irrelevant operators"
- Daniel Grumiller (TU Wien) "Generalized dilaton gravity in 2D"
- Wei Gu "On phases of 3d $N = 2$ Chern-Simons matters theories"
- Kentaro Hori "Grade restriction rule and applications"
- Hongliang Jiang "Infinite symmetries and Ward identities in celestial CFT"
- Christopher Raymond "Logarithmic $N = 2$ superconformal minimal models"

- Mauricio Romo "GLSM, homological projective duality and nc resolutions"
- Savdeep Sethi "Toward a construction of non-classical string solutions"
- Alessandro Sfondrini "A gentle introduction to the AdS3/CFT2 mirror TBA"
- Eric Sharpe "An introduction to decomposition"
- Kaiwen Sun "Twisted elliptic genera"
- Eirik Eik Svanes "Heterotic moduli without field redefinitions"
- Rikard von Unge "$T\overline{T}$ and nonlinearly realized symmetries"

Week 3 Research Talks

The third week featured research talks by invited speakers from the Australian research community, ranging from graduate students to established experts, and covering all aspects of this program.

- Leighton Arnold "SUSY gauge theories, branes, and W-algebras"
- Peter Bouwknegt "A surjective submersion primer"
- Jaklyn Crilly "T-duality and discrete torsion"
- Gregory Gold "A functional approach to cosmological particle production"
- Bolin Han "Parafermionic CFT"
- Daniel Hutchings "AdS superprojectors"
- Jessica Hutomo "Higher spin supercurrents in 4D N=1 AdS superspace"
- Saurish Khandelwal "Dilaton-Weyl multiplet of conformal supergravity"
- James Lafontaine "An effective action approach for superconformal higher-spin theories"
- Michael Ponds "Conformal interactions between matter and higher-spin fields"
- Robert Pryor "Localisation in $N = (2,2)$ GLSMs"
- Emmanouil Raptakis "Duality invariant SCHS models"
- Christpher Raymond "Linking vertex algebras and axiomatic CFT"
- David Ridout "Inverting quantum hamiltonian reduction"
- Liam Smith "$T\overline{T}$-like flows and non-linear electrodynamics"
- Martin Sticka "On heterotic moduli"
- Benjamin Stone "Correlation functions of conserved currents in 3D CFT"

Contributions

Six articles that arose from the program are included in this volume:

- In "On Super Non-Abelian T-Duality of Symmetric and Semi-Symmetric Coset Sigma Models" Daniele Bielli reviews the algebraic approach to super non-Abelian T-Duality that he and his collaborators had introduced.
- In "Minkowski and (A)dS ground states in general 2d dilaton gravity" Daniel Grumiller, Martin Laihartinger and Romain Ruzziconi describe a three-parameter family of 2d dilaton gravity models with integrable mass function.
- In "Vacuum structures revisited" Wei Gu revisits the relationship between higher form symmetries and the dynamical decomposition in supersymmetric gauge theories in various dimensions.

- In "2d Sigma Models and Geometry" Ulf Lindström reviews, based on his lecture in the first week, the relation of 2d supersymmetry to Riemannian, complex, bihermitian, (p, q) hermitean, Kähler, hyperkähler and generalised geometry.
- In "An introduction to decomposition" Eric Sharpe reviews his recent work on 'decomposition,' a property of two-dimensional theories with 1-form symmetries which is based on the observation that such quantum field theories are equivalent to disjoint unions of other QFTs.
- Kaiwen Sun contributed "On Twisted Elliptic Genera" where he studies several aspects of the twisted elliptic genera associated to the BPS strings in the twisted circle compactification of 6d (1, 0) SCFTs.

Emanuel Scheidegger
Guest Editor

MATRIX Program
Dynamics, Foliations, and Geometry II

31 January – 11 February 2022

Organisers

Jonathan Bowden
University of Regensburg

Sergio Fenley
Florida State University

Andy Hammerlindl
Monash University

Rafael Potrie
Universidad de la República

Jessica Purcell
Monash University

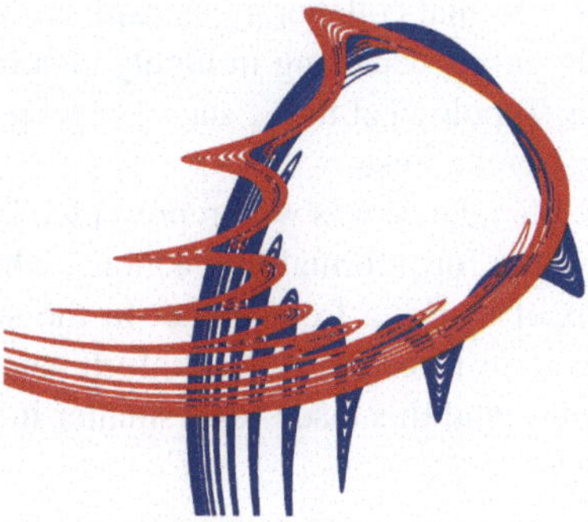

Creswick Participants

Andy Hammerlindl (Monash), Jessica Purcell (Monash), Warwick Tucker (Monash), Layne Hall (Monash), Andrés Bonilla (Monash), Andrew Cook (Monash), Thiago de Paiva Souza (Monash), Em Thompson (Monash), Urs Fuchs (Monash), Grace Shanti Garden (Sydney), Holger Dullin (Sydney), Connie Hui (Monash)

Regensburg Participants

Pierre Dehornoy (Grenoble), Vincent Colin (Nantes), Ana Rechtmann (Strasbourg), Robert Cardona Aguilar (Strasbourg), Katie Mann (Cornell), Helene Eynard (Grenoble), Ioannis Iakovoglou (Bourgogne), Theo Marty (Bonn), Davide Ravotti (Wien), Thierry Barbot (d'Avignon), Paolo Ghiggini (Nantes), Marcelo Alves (Antwerp), Andres Rodriguez Migueles (LMU München), Sebastian Hensel (LMU München), Marc Kegel (HU Berlin), Lukas Böke (LMU München), Jonathan Bowden (Regensburg), Raphael Zentner (Regensburg), Filip Misev (Regensburg)

This 2-week workshop focused on topics bridging the areas of dynamics, geometry, and topology. It was held as a hybrid event with hubs at both the MATRIX Institute in Creswick, Australia and at the University of Regensburg in Germany. Talks were streamed via video-conferencing between the two campuses and many other mathematicians in other parts of the world were able to attend the talks virtually.

There were two mini-courses delivered at this workshop. In Creswick, Andy Hammerlindl delivered a four part mini-course on branching foliations, detailing the proof of their existence, which has not been well understood in the dynamics community. In Regensburg, Pierre Dehornoy delivered a three part mini-course on Birkhoff sections for 3-dimensional vector fields, detailing examples of Birkhoff sections, and presenting some results on their existence.

In addition to the mini-courses, there was a series of talks on recent advances in the field. Speakers included Warwick Tucker, Thierry Barbot, Ioannis Iakovoglou, Théo Marty, Holger Dullin, Andrés Migueles, Vincent Colin, Kathryn Mann, Ana Rechtman, Davide Ravotti, and Marcelo Alves.

The informal collaborations explored applications of knot theory to the study of periodic orbits occurring in highly chaotic dynamical systems, such as the Lorenz system. Topological tools, such as Dehn filling and triangulations were applied to simple Lorenz links.

The attendance was well represented by students and the workshop was a timely opportunity for informal interactions, which was a privilege to treasure after many months of working from home. In the second week, each student gave a talk on their current research. It was unfortunate that international travel to Australia was infeasible, and this made for a smaller in-person group than there would have been otherwise.

Andy Hammerlindl

MATRIX-SMRI Symposium
Nijenhuis Geometry and Integrable Systems

14 – 18 February 2022

Organisers

Emma Carberry
University of Sydney

Holger Dullin
University of Sydney

Vladimir Matveev
Universität Jena

Participants

Reinout Quispel (LaTrobe), Peter Van der Kamp (LaTrobe), Yuri Nikolayevsky (LaTrobe), Jonathan Kress (UNSW), Holger Dullin (Sydney), Nathan Duignan (Sydney), Sean Gasiorek (Sydney)

Running an international workshop under COVID restrictions, we were happy to assemble a small group of Australian researchers in person in Creswick together with our European experts (Alexey Bolsinov, Andrey Konyaev, Vladimir Matveev) who joined via Zoom.

Recent progress in the foundations on Nijenhuis geometry by Bolsinov, Konyaev, and Matveev described in a series of papers have shown that many problems in integrable systems when described in this new language become much more transparent, and thus open up the possibility of generalisation and application in new areas.

During the program there were four local talks (Quispel, Kress, van der Kamp, Carberry) and two ECR talks (Duignan and Gasiorek) all of which were related to various flavours of integrable systems. In the absence of our European experts on Nijenhuis geometry during the day we embarked on connecting their work to "our" integrable systems. We focused on two particular problems: The theory of separation of variables, and the theory of recursion operators. For both we had some local experts, and by each evening we were joined by the European experts on Nijenhuis geometry. Considering that the aim of the workshop was to bring these two groups together, this was a success. However, we were hindered by the COVID situation, in that by the evening the local participants were tired, and to have collaborative zoom problem solving sessions from 8-9pm every evening was ambitious, to say the least. That is not to say we did not try, but we would have been much more effective and much faster had we all been meeting in person.

Jonathan Kress gave a local talk about the "classical" theory of separation of variables, which for geodesic flows is essentially a theory of systems of Killing-2-tensors. In the classical approach due to Kalnins and Miller the Nijenhuis geometry was not apparent for many years. However, it turns out that all Killing-2-tensors can be constructed from a single conformal Nijenhuis tensor. This solid geometrical foundation allowed a recent generalisation of separation of variables in spaces of constant curvature with arbitrary signature by Bolsinov, Konyaev, and Matveev [arXiv:2212.01605].

Peter van der Kamp gave a local talk about integrable PDEs that arise from curvature flows in various geometries, and in this setting recursion operators are an essential tool to establish integrability. This connected very much to the paper [arXiv:2206.12942] by Bolsinov, Konyaev, and Matveev where multi-component KdV, Camassa-Holm, and Dullin-Gottwald-Holm equations are constructed, and in particular we studied how the Nijenhuis approach produces the Lax pair and the recursion operator by considering the theory of compatible Poisson brackets.

In the end the perception was that we all did learn a lot from each other. We established new links between the integrable systems group in LaTrobe and in Sydney (USyd and UNSW). These links are expected to flourish when our European collaborators will be present for the second workshop on Nijenhuis geometry and integrable systems under non-COVID conditions in February 2024.

Holger Dullin

MATRIX Program
Frontiers in Representation Theory

14 – 25 February 2022

Organisers

Nora Ganter
University of Melbourne

Masoud Kamgarpour
University of Queensland

Peter McNamara
University of Melbourne

Peng Shan
Tsinghua University

Yaping Yang
University of Melbourne

Gufang Zhao
University of Melbourne

Participants

Yaping Yang (Melbourne), Gufang Zhao (Melbourne), Peter McNamara (Melbourne), Paul Zinn-Justin (Melbourne), Bregje Pauwels (Sydney), Christopher Hone (Sydney), Davood Nejaty (Melbourne), Weiying Guo (Melbourne), Gaston Burrull (Sydney), Mengfan Lyu (Sydney), Chenyan Wu (Melbourne), Johanna Knapp (Melbourne), David Ridout (Melbourne), Sasha Garbali (Melbourne)

This program focused on the connection between representation theory, homotopy theory and mathematical physics, which are central branches of mathematics. In recent years, new exciting and similar structures have been found in different areas. This workshop bought together leading mathematicians interested in these those important areas of mathematics, with the aim of encouraging and supporting the interaction and collaboration. The workshop lasted for two weeks consisting of research talks by the participants on cutting-edge topics.

Each day there were one or two research talks by the participants followed by the fruitful discussion and collaboration at MATRIX House. Each talk lasted for 1 hour and 40 minutes, consisting of 1-hour background accessible to PhD students, and 40 minutes on the speaker's recent research.

Peter McNamara and Christopher Hone discussed sheaves in modular representation theory, while Peter focussed on the application to Schubert calculus and Christopher talked about the existence of an indecomposable summand in characteristic p, which is motivated by the decomposition theorem in characteristic 0.

Paul Zinn-Justin discussed puzzles in Schubert calculus and relations to quantum integrable systems. Gaston Burrull talked about combinatorics of Schubert varieties focusing on the top-heaviness for intervals in affine Weyl groups.

Yaping Yang talked about the quantum group at roots of unity, and Sasha Garbali discussed about quatum toroidal algebras and R-matrices. Weiying Guo explained double affine Hecke algebra, and Macdonal polynomials. Mengfan Lyu talked about generalized Temperly-Lieb algebras, which is closely related to the Iwahori Hecke algebra. David Ridout discussed about vertex operator algebra and modularity.

Gufang Zhao discussed the moduli of sheaves on Calabi–Yau 3-fold and computations towards the quantum cohomology and Davood Nejaty talked about the Hitchin integrable systems for Langland dual groups. Bregje Pauwels talked about Tannakian theory in category theory in characteristic p. Johanna Knapp talked about the categorical correspondence between Pfaffian and Grassmannian.

Chenyan Wu discussed about the automorphic forms and L-functions in number theory which play an important role in local Langlands correspondence. Masoud Kamgarpour gave his talk on "Arithmetic geometry of character variety of reductive groups" on zoom.

The Organisers

MATRIX Program
Structural Graph Theory Downunder II

20 – 27 March 2022

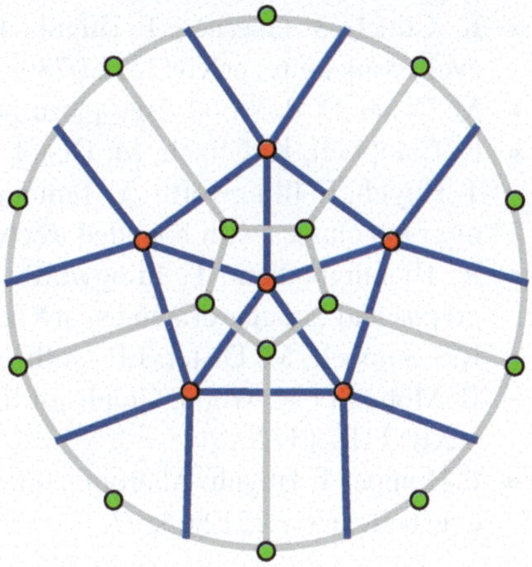

Organisers

David Wood
Monash University

Tony Huynh
Monash University

Sang-il Oum
KAIST & Institute for Basic Science, Korea

Alex Scott
University of Oxford, UK

Liana Yepremyan
Emory University, USA

Participants

Rutger Campbell (IBS, Korea), Katie Clinch (Melbourne), Marc Distel (Monash), Jackson Goerner (Monash), Pascal Gollin (IBS, Korea), Daniel Harvey (Ballarat), Kevin Hendrey (IBS, Korea), Robert Hickingbotham (Monash), Tony Huynh (Monash), Thomas Lesgourges (UNSW Sydney), Freddie Illingworth (Oxford), Bojan Mohar (Simon Fraser), Youri Tamitegama (Oxford), Jane Tan (Oxford), David Wood (Monash)

Graph theory, which is the mathematics of networks, has sprung to prominence as a model of real-world processes, with deep connections to many mathematical branches such as number theory, probability, logic and knot theory, as well as in applied disciplines especially computer science. The subfield of structural graph theory incorporates graph minor theory, graphs embedded in surfaces, graph classes defined by excluded induced subgraphs, generalisations of perfect graphs, graph colourings, etc.

This program consisted of a 1-week intensive research workshop, where mathematicians from Australia, United Kingdom, Canada and Korea came together to work on open problems in structural graph theory. The participants featured a mix of PhD students, early-career, mid-career and senior researchers. The goal was to create an environment where mathematicians at all career stages worked side-by-side. This goal was certainly achieved. Many participants commented on how conducive MATRIX House was for doing collaborative research.

The organisers prepared open problems to work on, and the workshop started with an open problems session where any participant could propose an open problem. We had progress sessions every morning of the workshop. The rest of the time was allocated to collaborative research. People then naturally formed groups working on problems of common interest. The focus became Aharoni's rainbow cycle conjecture, and graph product structure theory. On both of these topics, significant progress was made during the program, which led to the following papers:

- K. Clinch, J. Goerner, T. Huynh, F. Illingworth, Notes on Aharoni's rainbow cycle conjecture, arXiv:2211.07897.
- M. Distel, D. R. Wood. Tree-partitions with bounded degree trees, arXiv:2210.12577.
- R. Campbell, K. Clinch, M. Distel, J.P. Gollin, K. Hendrey, R. Hickingbotham, T. Huynh, F. Illingworth, Y. Tamitegama, J. Tan, D. R. Wood, Product structure of graph classes with bounded treewidth, arXiv:2206.02395.
- R. Hickingbotham, F. Illingworth, B. Mohar, D. R. Wood. Treewidth, circle graphs and circular drawings, arXiv:2212.11436.
- R. Campbell, M. Distel, J.P. Gollin, D. Harvey, K. Hendrey, R. Hickingbotham, B. Mohar, D. R. Wood, Graphs of linear growth have bounded treewidth, arXiv:2210.13720.
- P. Hompe, T. Huynh. Aharoni's rainbow cycle conjecture holds up to an additive constant, arXiv:2212.05697.

The first two papers above are published in full in this volume, along with extended abstracts of the next two papers.

All in all, the program was a great success. For many participants it was their first post-covid mathematics meeting. It was wonderful to be doing mathematics in-person again.

Tony Huynh and Liana Yepremyan
Guest editors

MATRIX Program
Hyperbolic Differential Equations in Geometry and Physics

4 – 8 April 2022

Organisers

Jesse Gell-Redman
University of Melbourne

Andrew Hassell
Australian National University

Todd Oliynyk
Monash University

Volker Schlue
University of Melbourne

Participants

Volker Schlue (Melbourne), Jesse Gell-Redman (Melbourne), Andrew Hassell (ANU), Todd Oliynyk (Monash), Kiril Datchev (Purdue), Dean Baskin (Texas A&M), Allen Fang (Sorbonne), Arick Shao (Queen Mary), Jan Sbierski (Edinburgh), Sean Gomes (Sydney), Zihua Guo (Monash), Trang Nguyen (Monash), Brian Krummel (Melbourne), William Trad (Sydney), Arturo Olvera (Monash), Cal Robertson (Monash), Antoine Gansemer (ANU), Louie Bernhardt (Melbourne), Zijun Chen (Monash)

This workshop focused on hyperbolic partial differential equations, which appear in the mathematical description of many phenomena in science where wave propagation plays a role. The study of hyperbolic PDE is an enormously broad subject, in particular in combination with the geometric and analytic methods developed to study many aspects of their solutions. Moreover, in this workshop we saw many touchpoints with areas such as spectral theory, scattering theory, quantum mechanics, and general relativity.

The inspiration was in part a response to the exciting new results in Hyperbolic PDEs coming from new applications of microlocal analysis to mathematical relativity. These are two major topics in the international PDE community, and our goal was to bring to Australia some of the main representatives of these fields, to foster new research collaborations.

We highlight several nascent research collaborations that began at the workshop. Zihua Guo (dispersive PDEs) began talking fruitfully with Andrew Hassell and postdoc Sean Gomes about the relationship between known Strichartz estimates and the propagation of singularities estimates that come from microlocal techniques for the Schrödinger equation. Dean Baskin and Kiril Datchev gave a minicourse in an evening session to the group on wave propagation, and in particular on the topic of propagation of singularities. They gave a beautiful extended version of this discussion in “Propagation of Singularities for the Wave Equation” in this volume.

Todd Oliynyk led a discussion group on the free boundary problem in fluid mechanics, and elasticity. These were attended in particular by Jan Sbierski, and Arick Shao, and led to collaborations among them on these topics. Allen Fang contributed to many interactions among participants with research interests in general relativity, and participants working primarily in the areas of spectral and microlocal analysis. Arick Shao wrote another paper arising from this program, "Bulk-boundary correspondences and unique continuation in asymptotically Anti-de Sitter spacetimes".

This workshop was both an effective tool for boosting collaboration between two synergistic subfields of PDE and an exciting entree to potential future collaborations.

Volker Schlue
Guest editor

MATRIX Program
Symmetry for Group Actions in Differential Geometry

16 – 27 May 2022

Organisers

Emma Carberry
University of Sydney

Michael Eastwood
University of Adelaide

Carolyn Gordon
Dartmouth College

Joseph Wolf
University of California, Berkeley

The hyperbolic disc in wycinanka łowicka style by Katarzyna Nurowska

Participants

Rod Gover (Auckland), Xavier Coulter (Auckland), Yang Shi (Flinders), Yuri Nikolayevsky (La Trobe), Artem Pulemotov (Queensland), Ian Marquette (Queensland), Peter Van Der Kamp (La Trobe), Tim Buttsworth (Queensland), Emma Carberry (Sydney), Michael Eastwood (Adelaide), Thomas Leistner (Adelaide), Federico Costanza (Adelaide), Joseph Wolf (Berkeley), Holger Dullin (Sydney), Sean Gasiorek (Sydney), Danilo Latini (Queensland)

This research program brought together differential geometers with a particular emphasis on symmetry. On the one hand, geometry gives rise to symmetry, for example as isometry groups, controlled by curvature and topology. Conversely, Lie groups can be used to construct examples of highly symmetrical geometries. We have in mind both classical Riemannian geometry with global invariants, especially the spectrum of the Laplacian and harmonic maps, but also other natural geometric structures including Lorentzian, Finsler, projective, parabolic, and CR geometry where there are many local classification results and problems concerning geometries with a high degree of symmetry. Many of these structures have strong links with physics, optimisation, control theory, and integrable systems. The program featured the following talks, and included a diverse group of participants, having in common a strong interest in symmetry.

- Rod Gover "Conformal submanifolds, distinguished submanifolds, and integrability"
- Yuri Nikolayevsky "Einstein hypersurfaces in symmetric spaces"

- Artem Pulemotov "Mountain pass approach to the prescribed Ricci curvature problem"
- Tim Buttsworth "U(2)-invariant Ricci flows on four-dimensional manifolds"
- Holger Dullin "Momentum Polytopes related to geodesic flows on spheres"
- Ian Marquette "Superintegrable Hamiltonians with higher order symmetries and Painlevé transcendents"
- Sean Gasiorek "Minkowski billiards on the hyperboloid of one sheet"
- Peter van der Kamp "Integrable curvature evolution"
- Yang Shi "Coxeter groups and connections between integrable systems"

The Organisers

MATRIX Program
Mathematics of Tissue Dynamics

30 May – 3 June 2022

Organisers

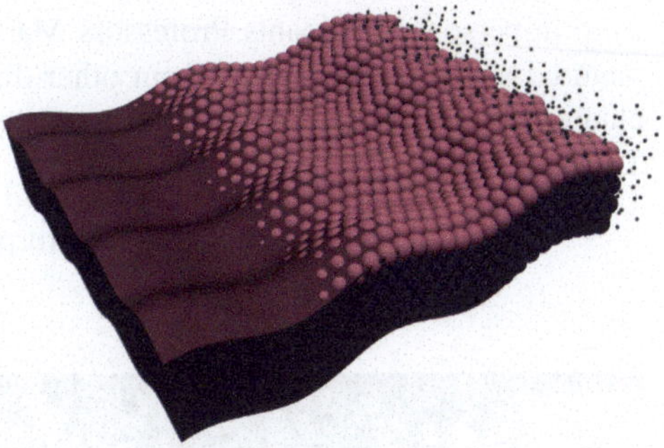

Domenic Germano
University of Melbourne

Adriana Zanca
University of Melbourne

Ryan Murphy
Queensland University of Technology

Emine Endes
Heriot-Watt University

James Osborne
University of Melbourne

Participants

Emine Atici-Endes (Heriot-Watt University), Mat Simpson (QUT), Nizhum Rahman (Queensland), Hasitha Weerasinghe (QUT), Adrianne Jenner (QUT), Ryan Murphy (QUT), Mary Myerscough (Sydney), Joshua Won (Sydney), Domenic Germano (Melbourne), Adriana Zanca (Melbourne), James Osborne (Melbourne), Jennifer Flegg (Melbourne), Ricardo Ruiz Baier (Monash), Claire Miller (Amsterdam), Mark Flegg (Monash), Stephanie Khuu (Auckland), Stuart Johnston (Melbourne), Alexander Browning (QUT), Ishraq Ahmed (Sydney)

Tissue modelling is key to understanding many current biological challenges including cancer, wound healing and tissue disease. Mathematical and computational models offer a means to investigate tissues and their associated challenges. These models complement biological experimental results and may offer suggestions for new avenues of experimental research. Models can be formulated across a range of spatial and temporal scales, from the subcellular through to the tissue spatial scale and from seconds through to years in the temporal scale. Together, these models across scales help complete an understanding of tissues and the various biological problems that occur within them. The Mathematics of Tissue Dynamics program brought together leading mathematical modellers, early career researchers and graduate research students to investigate three key problems:

- modelling diffusion on growing tissues,
- multiphase models of tissues,
- cell-based modelling of tissue rheology.

During the program there were three main groups of collaborators working on each of the problems, with discussion between groups encouraged. Graduate research student and early career researcher leads for each group presented the outcomes for each group at the end of the program. Subsequent to the program, the multiphase models of tissues group submitted a peer-reviewed article to the MATRIX Annals, and the other two groups are preparing publications for other peer-reviewed journals. The program, in addition to group work and key presentations from in-person participants Professors Mary Myerscough and Mat Simpson, also featured virtual presentations from other distinguished researchers in the field including Professors Philip Maini, Hans Othmer, Alys Clark, Qing Tie and Xavier Trepat.

Ryan Murphy, Domenic Germano, Adriana Zanca
Guest editors

MATRIX Program
Cell Motility in Dynamic Environments

6 – 10 June 2022

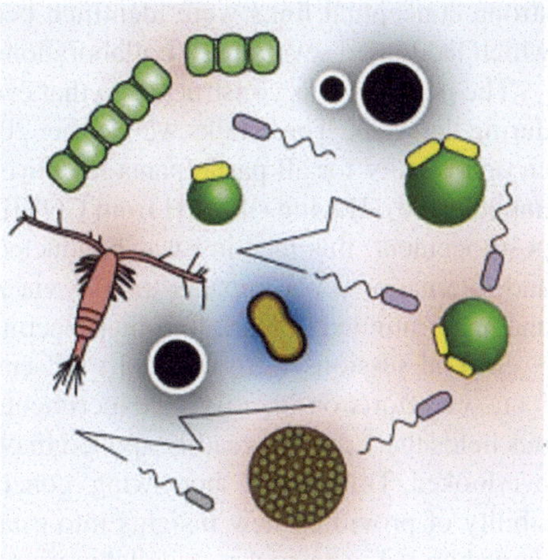

Organisers

Douglas Brumley
University of Melbourne

Jennifer Flegg
University of Melbourne

Jean-Baptiste Raina
University of Technology Sydney

Arezoo Ardekani
Purdue University

Lars Behrendt
Uppsala University

Participants

Douglas Brumley (Melbourne), Jennifer Flegg (Melbourne), Lars Behrendt (Uppsala), James Osborne (Melbourne), Adriana Zanca (Melbourne), Domenic Germano (Melbourne), Allison Mertin (Melbourne), Pranali Deore (Melbourne), Peijing Li (Melbourne), Prabhakar Ranganathan (Monash), Reza Nosrati (Monash), Vijay Rajagopal (Melbourne), Rebecca Yin (Melbourne), Shibani Veeraragavan (Monash), Bhuwan Ghimire (UTS), Jim Mitchell (Flinders University), Susie Grigson (Flinders University), Yvonne Stokes (Adelaide), Arezoo Ardekani (Purdue), Robyn Araujo (QUT), Jean-Baptiste Raina (UTS), Rachel Bennett (Bristol), Yoshiaki Kinosita (Gakushuin), Lyndon Koens (Macquarie)

This research workshop was jointly supported by Matrix, and the Special Emphasis Year in Biological Dynamics in the School of Mathematics and Statistics at Melbourne University. The workshop examined the intricate coupling between fluid mechanics, cell motility, growth and active navigation, ubiquitous features of the microbial world. The participants presented novel mathematical and experimental approaches for studying cell dynamics in carefully controlled microenvironments, discussing mechanistic modelling frameworks which can extrapolate these features to larger scales, and ultimately, experimental platforms for validating bulk properties. This program brought together researchers at all career stages (MSc students, PhD students, ECRs and senior academics) across various branches of mathematics (e.g., fluid dynamics, stochastic theory, continuum modelling, dynamical systems) and experimental sciences (e.g., microbiology, microbial ecology, micro-engineering).

The program emphasised bringing together complementary disciplines, to facilitate collaboration, and the transfer of modelling frameworks between different systems. Topics included, but were not limited to, the navigation of single-cell organisms in marine and wastewater systems; cell growth and the development of tissues; movement of particles in confined fluid flows; the physical interactions between plant seeds and bacteria; and the response of algae to fluctuating environments. Very strong conceptual links were identified between these seemingly disparate topics, which facilitated several new collaborations.

The program was constructed so that every participant could give a presentation during the week. These talks were either 20 min or 40 min in duration and provided an opportunity for all participants to convey their latest work to the group (usually unpublished). Having suffered from COVID interruptions in the past, and an earlier postponement, this meeting was conducted in hybrid form, with several domestic and international participants joining via Zoom. This format was very successful and has continued to work for other meetings and seminars hosted since then.

Several sessions were devoted to discussion and collaboration. The full group focussed efforts on identifying experimental and theoretical approaches from various fields that could be readily applied in other domains, but which are traditionally overlooked. This idea of 'borrowing' concepts from related disciplines has the possibility of providing new insights into established problems. This approach at the workshop led to several new collaborations, as well as a perspective journal article co-authored by the participants (soon to be submitted).

Douglas Brumley

MATRIX Program
Theory and Applications of Stable Polynomials

29 August – 2 September 2022

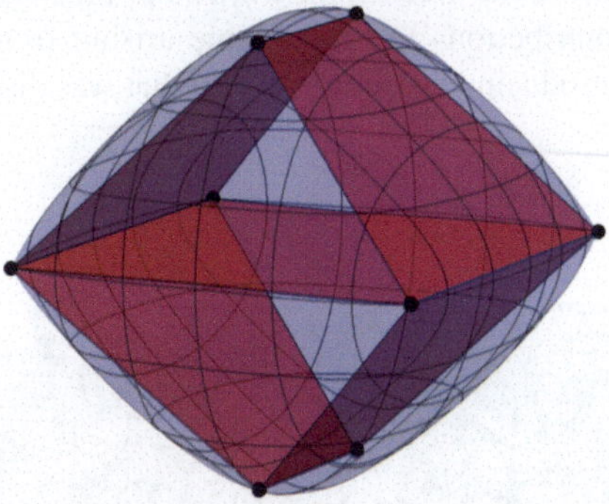

Organisers

James Saunderson
Monash University

Ngoc Mai Tran
University of Texas

Participants

Jonathan Leake (Waterloo), Jacob Matherne (Bonn & Max Planck), Brian Ng (Monash), Vera Roshchina (UNSW), James Saunderson (Monash), Cynthia Vinzant (Washington), Jayden Wang (Austin), Minh Dao (Federation), Graham Farr (Monash), Kerri Morgan (Deakin), Sebastian Avila Molina (UNSW), Norman Do (Monash)

The program focused on real stable polynomials (and the closely related hyperbolic polynomials and Lorentzian polynomials) and their applications in convex optimization, theoretical computer science, and combinatorics. These objects have featured in the resolution of major problems such as the Kadison-Singer problem and Mason's conjecture. The program brought together researchers from areas including optimization, algorithms and complexity, real algebraic geometry, algebraic combinatorics, and graph/matroid theory, all of whom had an interest in developing and applying techniques based on real stable polynomials to their areas of interest.

The program included a mixture of introductory lectures to build common ground among participants.

- Cynthia Vinzant gave an introductory talk: "Introduction to stable and log-concave polynomials" as well as a more research-oriented follow-up talk on "Log-concavity in matroids and expanders"
- Jonathan Leake gave two talks on "Approximate Counting using Stable and Lorentzian Polynomials", the first presenting the foundational ideas and the second covering recent advances in the area;
- James Saunderson gave an introductory talk on "Hyperbolic polynomials and hyperbolicity cones", discussing the convex (algebraic) geometry of hyperbolicity cones.

In addition, Jacob Matherne gave a research talk related to log-concavity properties of symmetric polynomials and open problem presentations were given by Norm Do and Jayden Wang.

The schedule left ample time for informal collaboration among the participants during the week-long workshop, much of it stimulated by the morning talks. The contributions to this volume arising from the program are four brief notes, each introducing an open problem that was discussed at the workshop.

James Saunderson
Guest editor

MATRIX-RIMS Tandem Workshop
Integrable Probability, Combinatorics and Representation Theory

26 – 30 September 2022

Organisers

Sasha Garbali
University of Melbourne

Chihiro Matsui
University of Tokyo

Masato Okado
Osaka Metropolitan University

Tomohiro Sasamoto
Tokyo Institute of Technology

Michael Wheeler
University of Melbourne

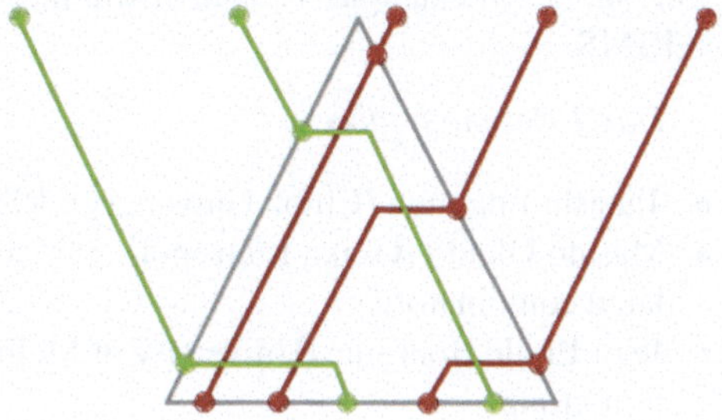

Participants

Arvind Ayyer (Indian Institute of Science), Alexandr Garbali (Melbourne), Jan de Gier (Melbourne), Weiying Guo (Melbourne), Peter McNamara (Melbourne), William Mead (Melbourne), Arun Ram (Melbourne), Michael Wheeler (Melbourne), Paul Zinn-Justin (Melbourne), Jean-Emile Bourgine (Melbourne)

The aim of this tandem workshop was to bring together groups of specialists in integrability, combinatorics and representation theory at MATRIX (Creswick) and RIMS (Kyoto). Talks were streamed between these two institutes and for those attending online were accessible via Zoom. The talks were split between the morning and afternoon sessions with typically two speakers presenting from each of the two institutes. In the morning sessions between the talks a subset of participants from both sides were discussing online a range of selected topics based on the talks which were given prior. In the afternoons the participants were mostly discussing and collaborating with attendees from their location. The main topics included RSK approaches and integrable probability, Macdonald theory and related symmetric functions, Crystals and box-ball systems, and Quantum groups.

Particular attention was given to the topic of the skew RSK dynamics and box-ball systems. This is a deterministic dynamics on the space of pairs of semi-standard tableaux (P,Q) which was recently introduced by Imamura, Sasamoto and Mucciconi who were all present at the workshop. Like the box-ball system it displays solitonic behavior and its symmetries can be described in terms of a novel bi-crystal structure on the set of tableaux (P,Q). Linearized dynamics and implementation of an inverse scattering method were discussed.

A number of discussions and on integrable probability were conducted between A. Ayyer, A. Garbali and J. de Gier. The discussions aimed at technical questions related to computations of stationary states in a totally asymmetric exclusion process with particles of different types.

Some particularly fruitful conversations were conducted between J. de Gier, W. Mead and M. Wheeler, leading to the explicit computation of a new type of partition function in the six-vertex model.

Some participants met over Zoom, taking advantage of the tandem nature of the workshop. For example, P. Zinn-Justin met in Zoom with a number of participants at RIMS.

Day 1 Research talks:

- Takashi Imamura (Chiba University): RSK approaches to integrable probability
- Masato Okado (Osaka Metropolitan University) Introduction to crystals and related combinatorics
- Jean-Emile Bourgine (University of Melbourne) An introduction to shifted quantum groups
- Alexandr Garbali (University of Melbourne) Shuffle algebras and integrability

Day 2 Research talks:

- Matteo Mucciconi (University of Warwick) Combinatorial inverse scattering method for the skew RSK dynamics
- Arvind Ayyer (Indian Institute of Science) A multispecies totally asymmetric zero range process and Macdonald polynomials
- Vladimir Bazhanov (Australian National University) An Ising-type formulation of the six-vertex model
- Taichiro Takagi (National Defense Academy) Crystal interpretation of box-ball systems

Day 3 Research talks:

- Satoshi Naito (Tokyo Institute of Technology) Specializations of symmetric and nonsymmetric Macdonald polynomials at t = zero and t = infinity
- Masatoshi Noumi (Rikkyo University) Elliptic van Diejen difference operators and elliptic hypergeometric integrals of Selberg type
- Arun Ram (University of Melbourne) c-functions and Macdonald polynomials
- Peter McNamara (University of Melbourne) Quantum groups and their bases

Day 4 Research talks:

- Paul Zinn-Justin (University of Melbourne) Schubert puzzles: (almost) separated descents
- Weiying Guo (University of Melbourne) Cauchy identity and symmetrization formula for Izergin-Korepin model
- Soichi Okada (Nagoya University) Symplectic Q-functions
- Takeshi Ikeda (Waseda University) Geometric variants of Schur's Q-functions

Day 5 Research talks:

- Amol Aggarwal (Columbia University) Fusion and LLT Polynomials
- Hayate Suda (Keio University) Relationships between two linearizations of the box-ball system : rigged configuration and slot decomposition
- Anatol Kirillov (Beijing Institute of Mathematical Sciences and Applications) About RSK
- Atsuo Kuniba (University of Tokyo) New solutions to the tetrahedron equation associated with quantized six-vertex models

Sasha Garbali

MATRIX Program
Mathematics of Risk 2022

31 October – 11 November 2022

Organisers

Alex Novikov
University of Technology, Sydney

Kostya Borokov
University of Melbourne

Kais Hamza
Monash University

Yulia Mishura
Taras Shevchenko National
University of Kyiv

Nino Kordzakhia
Macquarie University

Kazutoshi Yamazaki
Kansai University &
University of Queensland

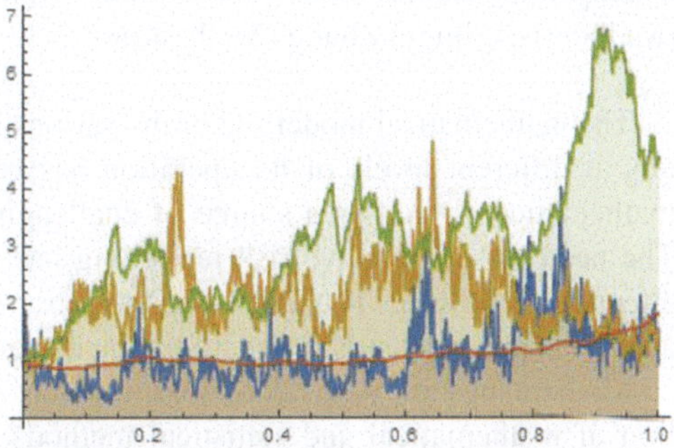

Participants

Alex Novikov (UTS), Kostya Borovkov (Melbourne), Kais Hamza (Monash), Nino Kordzakhia (Macquarie), Kazutoshi Yamazaki (Queensland), Erik Schlogl (UTS), Fima Klebaner (Monash), Artem Prokhorov (Sydney), Robert James (Sydney), Eckhard Platen (UT), Masaaki Fukasawa (Osaka), Alex Tartakovsky (AGT Stat-Consult), George V. Moustakides (Patras), Miryana Grigorova (Warwick), Arturo Kohatsu-Higa (Ritsumeikan), Anna Aksamit (Sydney), Aihua Xia (Melbourne), Peter Spreij (Amsterdam), Jie Yen Fan (Sydney), Jiro Akahori (Ritsumeikan), Robert Elliott (South Australia & Calgary), Ivan Guo (Monash), Kostiantyn Ralchenko (Kyiv), Travis Monk (Western Sydney), Kei Noba (Institute of Statistical Mathematics), Yuri Imamura (Kanazawa), Yushi Hamaguchi (Osaka), Vladka Seckarova (Monash), Alexander Buryak (National Australia Bank), Anton Skrobotov (Moscow), Peter Taylor (Melbourne), Ken Palmer (Taiwan), Peter Morrison (UTS), Pavel Shevchenko (Macquarie), Jessica Leung (Monash), Peilun He (Macquarie), Vincent Liang (Melbourne), Phat Nguyen (Melbourne), Jun Seok Han (Macquarie), Alexander Semenov (Florida), Svitlana Ralchenko (Kyiv)

Online Participants

Vladimir Dragalin (Johnson and Johnson), Anthony Brockwell (Wellhan), Amine Lkabous (Southampton), Oscar Peralta Gutierrez (Lausanne), Jie Yen Fan (Sydney),

Anton Yurchenko-Tytarenko (Oslo) Dai Taguchi (Okayama), Mitsuki Kobayashi (Waseda) Phillip Yam (CUHK), Xiaowen Zhou (Concordia), Kevin Lu (Washington), Oscar Peralta (Lausanne), Rustam Ibragimov (Imperial College) Xiang Yu (Hong Kong Poly), José Luis Pérez Garmendia (CIMAT), Kihun Nam (Monash), Dmitry Muravey (ADIA), Juri Hinz (National Australian Bank), Yuliya Mishura (Kyiv), Anton Yurchenko-Tytarenko (Oslo), Xiaowen Zhou (Concordia), Oscar Tian (Monash), Michael Baron (American), Yajun Mei (Georgia Tech), Christina Nikitopoulos (UTS), Cheng-Der Fuh (Zhejiang), Mikhail Zhitlukhin (Steklov), Jun Han (Macquarie), Yuri Kabanov (Franche-Comte), Georgios Fellouris (Illinois, Urbana-Champaign), Nathan Ross (Melbourne), Timothy Ling (UTS), Duy Phat Nguyen (Melbourne), Jingyi Zhang (Melbourne)

The mathematical modelling of the various types of risk modern society encounters at different levels of its operation has become an important part of applied mathematics as well as a source of challenging problems of a theoretical nature. The need for quantitative risk modelling has, in recent years, attracted enormous attention worldwide. The risk related to extreme events is generating a vast research activity, a pursuit that is international by its very nature.

Mathematics of Risk 2022 was the second MATRIX event on the application of mathematical and statistical methods to model and subsequently analyse risks mostly related to activities in the financial industry and, more generally, economics. The first Mathematics of Risk MATRIX program was successfully run in November–December 2017.

In the 2022 program, there were 41 registered in-person participants from Australia and overseas representing Canada, Greece, Holland, Japan, Russia, Taiwan, UK, Ukraine and USA. Additionally, 38 other researchers and PhD students from Australia, Canada, China, France, Hong-Kong, Japan, Mexico, Norway, Russia, Switzerland, UK, Ukraine and USA, registered for the online (Zoom) participation during the mini-symposia and lecture courses that were part of the program.

The key themes of the present program were:

- modelling of uncertainty and risk events using the theory of stochastic processes;
- new methods for and approaches to computing the prices of financial derivatives;
- systemic risk, including the stability of national and world financial systems;
- risk modelling and quantification, including risk measures;
- analysis of model risk; i.e., the type of risk which arises due to using inappropriate mathematical models for asset price dynamics;
- mathematical modelling of extreme events due to various factors, such as natural disasters, human errors, infrastructure and computer control system failures.

The program featured two embedded mini-symposia:

- Analysis of structural changes in Finance, Economics and Medicine.
- Recent advances in stochastic modelling: Rough volatility, adaptive finance, and machine learning in finance.

Four short lecture courses were given on topics aligned with the main themes:

- D. Muravey, "Generalized Integral Transforms in Mathematical Finance"
- E. Schlogl, "Mathematical Modelling of the Term Structure of Interest Rates – From Fundamentals to a Post-LIBOR World"
- I. Guo "Machine Learning Algorithms in Finance"
- G. Moustakides, "Machine Learning Methods for Statistical Decision Making"

These lectures were also broadcast via Zoom and attracted dozens of researchers and PhD students from both Australia and overseas.

The program included several forums where researchers attending the program in person and their colleagues and collaborators from all over the world discussed during Zoom sessions topics from the following areas:

- First passage problems for long-range dependent processes;
- Modern topics in Financial Econometrics;
- Rough volatility models;
- Stochastic and Statistics analysis of Jump processes;
- Mathematical Modelling of the Term Structure of Interest Rates.

For the present volume, participants of the program (both those who attended in person and those who took part in online sessions) prepared interesting and instructive papers presenting their research on the topics in the focus of the program. Our thanks go to all of them for taking the time to contribute to this volume. Finally, we would like to thank the MATRIX and the Creswick campus staff for facilitating and hosting this event. The participants enjoyed it tremendously. We had an excellent opportunity to engage in joint research work and exchange ideas.

Kais Hamza
Guest editor

MATRIX Program
Mathematics of the interactions between brain structure and brain functions

14 – 18 November 2022

Organisers

Kerri Morgan
Deakin University

Mikail Rubinov
Vanderbilt University

Svetlana Postnova
University of Sydney

Nadezda Sukhorukova
Swinburne University

Julien Ugon
Deakin University

Geoffrey Goodhill
Washington University St Louis

Mark Holmes
University of Melbourne

Sergiy Shelyag
Deakin University

Karen Caeyenberghs
Deakin University

Andrea Collevecchio
Monash University

MATHEMATICS OF THE INTERACTIONS BETWEEN
BRAIN STRUCTURE AND FUNCTION

Participants

Julien Ugon (Deakin), Nadezda Sukhorukova (Swinburne), Mark Holmes (Melbourne), Catie Chang (Vanderbilt), Victor Solo (UNSW Sydney), Sveta Postnova (Sydney), Kerri Morgan (Deakin), Andrea Collevecchio (Monash), Sergiy Shelyag (Deakin), Mikail Rubinov (Vanderbilt), Karen Caeyenberghs (Deakin), Michael Breakspear (Newcastle Unjiversity), Govinda Poudel (Australian Catholic University), Nicholas Parsons (Deakin), Alex Hocking (Deakin), Sue Chan (Deakin), Paula Sanz-Leon (QIMR Berghofer), Stuart Knock (Sydney), Zari Dzalilov (Federation), Adil Bagirov (Federation), Sina Mansour (Melbourne), Lena Vogelsang (Deakin), Maithilee Kunda (Vanderbilt), Will Woods (Swinburne)

The program focussed on fostering collaboration between mathematicians and neuroscientists to develop mathematical models to explain the relationships between the brain's structure and function, and connectivity between and within brain regions.

The first week of the program consisted of introductory tutorials on neuroscience, graph theory and mathematical modelling to provide background for the diverse audience. Three introductory presentations were given based on the three themes: static (graph theory) and dynamic (through dynamical systems) approaches for brain modelling, and multiscale processes in the brain. The presentations were in a format of open discussion. A list of problems and possible solution approaches for mathematical analysis of the structure-function relationship in the human brain were collected and formed the basis of the collaborative groups for the week.

The second week of the program began with discussion and refinement of the research problems and approaches. Three presentations were given based on the three themes in an open discussion format. Most of this week was devoted to collaborative group work based around the following areas:

1. Development of methods for generation of random graphs with properties similar to the human brain. Although, existing random graph models for generating small world networks are used in neuroscience, they may not preserve some of the important brain network properties.
2. Applications of the recently developed graph metrics, such as generalised rich club, *k*-core and *k*-nub for weighted and directed graphs representing brain connectomes.
3. Analysis of the dynamics (at the neuron population or single neuron levels) in small subgraphs and understanding the relevance of the graph metrics to the dynamics.
4. Development of a weighted extension to the existing unweighted graph metrics has been initiated. Apart from theoretical interest, these metrics would be useful for characterisation of the brain connectivity at functional and structural levels. During the workshop, one of the collaborative groups planned to perform this analysis. This is ongoing work.
5. Investigation of preservation of mathematical properties of the brain network across different parcellations (spatial resolutions at which brain connectomes are measured).

A valuable part of this workshop was gaining a better understanding of the limitations of dealing with noisy data particularly the problem of reconciling data taken at different resolutions and/or using different methods. Mathematical modelling based on this data needs to be accurate across multiple scales. The program has led to the emergence of several new collaborative groups, including:

1. Investigation into the relationship between functional and structural connectivity. A new model was proposed which used the weight of structural connection and likelihood of functional connection to determine appropriate thresholding of MRI and fMRI brain data.

2. Identifying graph properties/metrics that can be used to obtain meaningful information about brain networks. Collaborative work on a new metric was initiated.
3. Modelling the dynamics of small groups of neurons was investigated as a bridge between the dynamics of microscale and macroscale levels. The metric in (2) may be used to identify small subgraphs of interest that could be modelled.

Participants reported newly established links between neuroscience and dynamic and static mathematical models of the brain. One participant wrote, "MATRIX is a great opportunity to develop new research directions and collaborations. The informality of most discussions is a very important part of this, and giving people time to talk without pressure is great. We found that most keynote speakers spent more time than allocated, due to many questions from the audience, and having informal times after the keynotes helped give us flexibility to extend the talk time. Both speakers and audience appreciated that."

This volume includes the following two papers arising from the program:

- "Mathematical Foundations for Measurement of Communication Efficiency in the Human Brain" by Alex Hocking, Kerri Morgan, Nicholas Parsons, Govinda Poudel, Sergiy Shelyag, Julien Ugon.
- "*k*-Means clustering in EEG (brain waves) timeseries" by Nadezda Sukhorukova and Liam Kelly.

Sergiy Shelyag
Guest editor

MATRIX Program
Convexity, Geometry and Computation

28 November – 2 December 2022

Organisers

Martin Helmer
North Carolina State University

Cordian Riener
Arctic University of Norway

Vera Roschina
UNSW Sydney

James Saunderson
Monash University

Participants

Martin Helmer (North Carolina), Vera Roshchina (UNSW), James Saunderson (Monash), Kathlen Kohn (KTH Royal Institute of Technology), Bruno Lourenco (Institute of Statistical Mathematics), Mareike Dressler (UNSW), Hoa Bui (Curtin), Hongzhi Liao (UNSW), Vinesha Peiris (Deakin), Guillermo Pineda-Villavicencio (Deakin), Gregorio Malajovich (Universidade Federal do Rio de Janeiro), David Yost (Federation), Jie Wang (Federation)

The research program focused on topics related to convex geometry and algebraic computation, with an emphasis on open problems and collaborative discussions. There were several groups of collaborators who worked on various research projects during the program. Some of the topics discussed included infinite-dimensional polytopes, proofs of polyhedrality, the structure of the cone of SONC (Sum of Non-negative Circuit) polynomials, the Wachpress conjecture, reconstruction of convex bodies from projections, nonlinear versions of the knapsack problem, the structure of faces of convex cones, higher order tangent sets of hyperbolicity cones, mixed volumes of polytopes, and maximizing a potential on the sphere.

Participants had productive research discussions with each other and formed new collaborations, working on open problems in their respective fields as well as subjects outside of their mathematical comfort zones. The program structure included talks by participants to introduce themselves mathematically and stimulate discussions, followed by collaborative problem-solving sessions in the afternoons.

- Hoa Bui "Single-Projection Procedure for Infinite Dimensional Convex Optimization Problems"

- Vinesha Peiris "Approximation of functions by neural networks and rational functions"
- Bruno Lourenço "Koans of cones: self-dual polyhedra"
- James Saunderson "Convex forms and sums of squares"
- David Yost "Decomposable convex bodies"
- Jie Wang "Some conditional decomposable polytopes"
- Guillermo Pineda Villavicencio "Problems on graphs of polytopes"
- Mareike Dressler "A tale of cones: subcones of the nonnegativity cone"
- Kathlén Kohn "Wachspress conjecture"
- Martin Helmer "Conormal Spaces and Whitney Stratifications"
- Gregorio Malajovich "Mixed volume, toric varieties and sparse polynomial solving"

The program provided ample time for research interactions and was organized in a way that facilitated collaboration among researchers with different interests and backgrounds.

Martin Helmer

IBS-CGP & MATRIX Workshop
Symplectic Topology

5 – 16 December 2022

Organisers

Yong-Geun Oh
IBS Center for Geometry and Physics, Korea

Brett Parker
Australian National University

River Chiang
National Cheng Kung University, Taiwan

Participants

Mohammad Abouzaid (Columbia & Stanford), Hanwool Bae (Hong Kong), River Chiang (Taiwan), Mark Gross (Cambridge), Jongmyeong Kim (IBS Korea), Taesu Kim (IBS Korea), Yoosik Kim (Pusan), Ian Le (ANU), Conan Nai Chung Leung (Hong Kong), Grigory Mikhalkin (Geneva), Paul Norbury (Melbourne), Yong-Geun Oh (IBS Korea), Kaoru Ono (RIMS Kyoto), Brett Parker (ANU), Bernd Siebert (Austin), Yat-Hin Suen (IBS Korea), Sara Tukachinsky (Tel Aviv), Bai-Ling Wang (ANU), Yuto Yamamuto (IBS Korea), Ke Zhu (Minnesota State)

The program focused on symplectic topology, with emphasis in the areas of tropical geometry, the Fukaya category, open Gromov–Witten invariants and singular symplectic geometry. The first week featured a series of talks on log Gromov–Witten invariants by Mark Gross (Cambridge) and Bernd Siebert (UT Austin). The second week featured a series of talks on the local Fukaya category by Mohammed Abouzaid (Columbia and Stanford). Each day featured the following talks by other participants, with the rest of the time spent in informal discussions and active collaboration.

- Ian Le "Cluster algebras in representation theory"
- Kaoru Ono "Floer Novikov cohomology"
- Bernd Seibert and Mark Gross "Introduction to logarithmic Gromov–Witten invariants"
- Yat-Hin Suen "Lagrangian multi-sections mirror to holomorphic vector bundles"
- Jongmyeong Kim "Derived Fukaya category and Calabi–Yau pairs"
- Hanwook Bae "Derived Fukaya category and Calabi–Yau pairs 2"
- Sara Tukachinsky "Relative quantum cohomology for complete intersections"
- Mohammed Abouzaid "Local Floer homology"

- River Chiang "Circle actions on symplectic 4-manifolds"
- Grigory Mikhalkin "Breaking to pieces in tropical correspondence"
- Yoosik Kim "Infinitely many lagrangian tori in flag varieties"
- Yong-Geun Oh "Contact instantons and entanglement of Legendrian links"
- Conan Nai Chung Leung "de Rham approach to Gross-Siebert program"
- Ian Le "Braid varieties and mirror symmetry"

Participants particularly liked the structure of the program, where most of the time was reserved for less structured collaboration. A typical comment was, "there was ample collaboration time at MATRIX House, which provided enjoyable settings for discussions. Even though the program looked relaxed at first glance, it was actually intense as it's intellectually stimulating. It should be the protocol for the future workshops."

There were several groups of collaborators, including Ke Zhu and Yong-Geun Oh, Bai-Ling Wang and Kaoru Ono, Mark Gross and Bernd Siebert, and Jongmyeong Kim and Hanwool Bae. That said, groups during collaboration time were fluid, with a lot of mixing and many new connections were made.

Participants reported newly established links between mirror symmetry and the group of diffeomorphisms, and between log geometry and holomorphic curve counting problems involving jet bundles.

This volume includes the following two papers arising from the workshop:

- Ke Zhu "High-jet Relations of the Heat Kernel Embedding Map and Applications"
- Yong-geun Oh, Taesu Kim "Analysis of Pseudoholomorphic Curves on Symplectization: Revisit via Contact Instantons"

Marco Yat-Hin Suen
Guest editor

Contents

Part I

Refereed Articles

Chapter 1

MATRIX–MFO Tandem Workshop: Invariants and Structures in Low-Dimensional Topology

A new family of minimal ideal triangulations of cusped hyperbolic 3–manifolds

J. Hyam Rubinstein, Jonathan Spreer and Stephan Tillmann

Abstract Previous work of the authors with Bus Jaco determined a lower bound on the complexity of cusped hyperbolic 3–manifolds and showed that it is attained by the monodromy ideal triangulations of once-punctured torus bundles. This paper exhibits an infinite family of minimal ideal triangulations of Dehn fillings on the link 8^3_9 that also attain this lower bound on complexity.

1 Introduction

We refer to [10] for background and precise definitions used in this paper and only give a quick summary here. For a cusped orientable hyperbolic 3–manifold M of finite volume, the **norm** $|| \alpha ||$ of a non-trivial class $\alpha \in H_2(M, \mathbb{Z}_2)$ is the negative of the maximal Euler characteristic of a properly embedded surface S (no component of which is a sphere) representing the class. Any surface S with $[S] = \alpha$ and $-\chi(S) = || \alpha ||$ is **taut** if no component of S is a sphere or a torus. A class $\alpha \in H_2(M, \mathbb{Z}_2)$ determines a labelling of the ideal edges of an ideal triangulation of M by elements in $\mathbb{Z}_2 = \{0, 1\}$. These can be interpreted as edge weights of a normal surface, called the **canonical normal representative** of α.

J. Hyam Rubinstein
School of Mathematics and Statistics, The University of Melbourne, VIC 3010, Australia, e-mail: joachim@unimelb.edu.au

Jonathan Spreer
School of Mathematics and Statistics F07, The University of Sydney, NSW 2006 Australia e-mail: jonathan.spreer@sydney.edu.au

Stephan Tillmann
School of Mathematics and Statistics F07, The University of Sydney, NSW 2006 Australia e-mail: stephan.tillmann@sydney.edu.au

D. R. Wood et al. (eds.), *2021-2022 MATRIX Annals*, MATRIX Book Series 5,
https://doi.org/10.1007/978-3-031-47417-0_1

The **complexity** $c(M)$ is the minimal number of ideal tetrahedra in an ideal triangulation of M. In [10, Theorem 1], the authors and Bus Jaco obtain the following lower bound on the complexity of a cusped hyperbolic 3-manifold:

Theorem 1[10] *Let $\mathscr{T}$ be an ideal triangulation of the cusped orientable hyperbolic 3–manifold M of finite volume. If $H \leq H_2(M,\mathbb{Z}_2)$ is a rank 2 subgroup, then the number of ideal tetrahedra $|\mathscr{T}|$ satisfies*

$$|\mathscr{T}| \geq c(M) \geq \sum_{0 \neq \alpha \in H} || \, \alpha \, ||.$$

Moreover, in the case of equality the triangulation is minimal and each canonical normal representative of a non-zero element in $H \leq H_2(M,\mathbb{Z}_2)$ is taut and meets each tetrahedron in a quadrilateral disc.

It is shown in [10] that the above bound implies that the monodromy ideal triangulations of the hyperbolic once-punctured torus bundles are minimal. This note presents another infinite family of once-cusped hyperbolic 3-manifolds $M_{k,n}$, admitting triangulations $\mathscr{T}_{k,n}$ which achieve the lower bound in Theorem 1. These manifolds are shown to arise from Dehn surgery on the link 8^3_9. What makes these triangulations interesting is that their structure is different from the structure of the monodromy ideal triangulations, and that they do not feature layered solid tori as Dehn fillings (as, for instance, the *generic* canonical triangulations exhibited by Guéritaud and Schleimer [7, Theorem 1]). We thank Saul Schleimer and an anonymous referee for pointing out that our triangulations are closely related to the canonical decompositions of hyperbolic Dehn fillings on one cusp of the Whitehead link complement described in [7, §5]. Whether our new family of minimal triangulations is canonical is left as an open problem for further investigation. We now state the main theorem of this paper (the relevant definitions and an outline of the proof are given in Section 2):

Theorem 2. *Let $k,n \geq 3$ be odd integers. The identification space $M_{k,n}$ of the triangulation $\mathscr{T}_{k,n}$ is hyperbolic and homeomorphic to the manifold obtained from the complement of link 8^3_9 (see Figure 1) by Dehn filling the symmetric cusp with slope $(-k,1)$ and one of the other cusps with slope $(n+1,1)$. Here, the natural geometric framing is used for the cusps. The triangulation $\mathscr{T}_{k,n}$ achieves equality in Theorem 1 with $H = H_2(M_{k,n},\mathbb{Z}_2) \cong \mathbb{Z}_2 \oplus \mathbb{Z}_2$, and, in particular, $M_{k,n}$ has complexity $k+n$.*

This theorem, and its proof, have the following consequences.

First, our approach to compute the $\mathbb{Z}_2$–norm in Section 6 introduces new techniques that extend the tool kits from [9, 10, 11] and is also applicable to the computation of the Thurston norm of Dehn fillings.

Second, the construction of the minimal triangulations generalises. We do not provide full details, and simply list a further family of triangulations that we conjecture to be minimal in Section 7.

Third, it remains an open problem to classify all ideal triangulations of cusped hyperbolic 3–manifolds that achieve the lower bound in Theorem 1, and to provide a complete list of all underlying cusped hyperbolic 3–manifolds. In contrast, an analogous bound for the complexity of closed 3–manifolds and a complete characterisation of all tight examples are given in [9].

Acknowledgements We thank Saul Schleimer and an anonymous referee for helpful comments on an earlier draft of this paper. Research of the authors is supported in part under the Australian Research Council's Discovery funding scheme (project number DP190102259).

2 The new infinite family

The first member, $\mathscr{T}_{3,3}$, of the infinite family $\mathscr{T}_{k,n}$ was found by experimentation. Of the 9596 minimal triangulations of 4815 cusped hyperbolic 3–manifolds contained in the orientable cusped hyperbolic census (shipped, for instance, with `Regina` [3]), there are exactly nine manifolds for which the lower bound in Theorem 1 is attained. Eight of these are monodromy ideal triangulations of once-punctured torus bundles, but the remaining one, a triangulation of manifold `s781`, is not.

One can check that this minimal ideal triangulation $\mathscr{T}_{3,3}$ of `s781` achieves the lower bound as follows. It has isomorphism signature `gLLMQbeefffehhqxhqq` and six ideal tetrahedra. The underlying hyperbolic manifold $M_{3,3}$ has one cusp and $H_1(M_{3,3},\mathbb{Z}) \cong \mathbb{Z} \oplus \mathbb{Z}_2 \oplus \mathbb{Z}_4$. The space of all closed normal surfaces has 11 embedded fundamental surfaces consisting of one vertex linking torus, seven orientable surfaces of Euler characteristic at most -2 and three non-orientable surfaces of Euler characteristic equal to -2. The non-orientable surfaces represent distinct non-trivial classes in $H_2(M_{3,3},\mathbb{Z}_2)$. Since each even torsion class has a taut representative that is isotopic to a normal surface, and all non-trivial fundamental normal surfaces in $\mathscr{T}_{3,3}$ have Euler characteristic ≤ -2, it follows that $\mathscr{T}_{3,3}$ attains the lower bound in Theorem 1.

We now briefly describe how this example is extended to the infinite family of triangulations $\mathscr{T}_{k,n}$, $k,n \geq 3$ odd.

The ideal triangulation $\mathscr{T}_{k,n}$ is build from a k–tetrahedron solid torus T_k and an n–tetrahedron solid torus T_n, $k,n \geq 3$ odd, each with two punctures on the boundary, by identifying the eight boundary triangles (four per solid torus) in pairs. The identification space $M_{k,n}$ has the required torsion classes in homology, namely $H_1(M_{k,n},\mathbb{Z}) \cong \mathbb{Z} \oplus \mathbb{Z}_2 \oplus \mathbb{Z}_{k+1}$ (note that $H_1(M_{k,n},\mathbb{Z})$ does not depend on n). Hence we have $H_2(M_{k,n},\mathbb{Z}_2) \cong \mathbb{Z}_2 \oplus \mathbb{Z}_2$. We remark that we do not consider homology relative to the boundary. The triangulations of the solid tori are described in Section 3, and the triangulations $\mathscr{T}_{k,n}$ in Section 4.

In order to apply Theorem 1, we need to show that each $M_{k,n}$ is hyperbolic, i.e. has a complete hyperbolic structure of finite volume. To this end, we show in Section 5 that the manifolds $M_{k,n}$ are obtained by Dehn surgery on the link 8^3_9 and determine the surgery coefficients. The complement of the three component link 8^3_9, shown in Figure 1, is hyperbolic and admits a symmetry which fixes one of the cusps and interchanges the other two. We refer to the former as the *symmetric* cusp, and it is shown in red in Figure 1. The manifolds $M_{k,n}$ are all obtained from the complement of 8^3_9 by filling two of its cusps, where one of these must be the symmetric cusp. An application of Lackenby's combinatorial word hyperbolic Dehn surgery theorem [14, Theorem 4.9] shows that $M_{k,n}$ is hyperbolic for all $k,n \geq 17$. A computation with Regina [3] extends this to all $k,n \geq 3$.

We next determine the norm of the non-trivial classes in $H_2(M_{k,n}, \mathbb{Z}_2)$. This is achieved in Section 6 by using a different triangulation of $M_{k,n}$, namely a triangulation $\mathscr{T}'_{k,n}$ that uses the description of $M_{k,n}$ as Dehn surgery on the link 8^3_9. This allows us to analyse the space of *all* normal surfaces representing the homology classes. Namely, it reduces the argument to a calculation of all normal surfaces with a specific boundary pattern in the link complement, together with an analysis of how these surfaces extend to the surgery tori.

This completes the outline of the proof of Theorem 2. Below is a quick guide to the notation used for the various triangulations in this paper:

$\mathscr{T}$ ideal triangulation of the complement of the link 8^3_9 in $\mathbb{S}^3$

$\mathscr{T}'$ triangulation of the exterior of the link 8^3_9 in $\mathbb{S}^3$
with one boundary component removed

$M_{k,n} \cong$ Manifold('8^3_9')(∞ , $(n+1,-1),(k+1,1)$)

$\mathscr{T}_{k,n} = T_k \cup T_n$ ideal triangulation of $M_{k,n}$

$\mathscr{T}'_{k,n} = \mathscr{T}' \cup \mathrm{LST}(1,m) \cup \mathrm{LST}(1,k-1)$ triangulation of $M_{k,n}$
with one ideal and two material vertices

We remark that the complement of the link 8^3_9 can be identified as a double cover of the complement of the Whitehead link. However, the deck transformation pre-

Fig. 1: Three component hyperbolic (non-alternating) link 8^3_9 (Alexander-Briggs (Rolfsen) Number [17, Appendix C]), L8a16 (Thistlethwaite Number [2]). Picture from [2].

serves the symmetric cusp and interchanges the other two cusps. In particular, the triangulations $\mathscr{T}_{k,n}$ are not obtained as lifts of the triangulations described in [7, Theorem 2].

3 An infinite family of solid tori

The construction of our infinite families of minimal triangulations of once-cusped hyperbolic manifolds is based on a family of solid torus triangulations T_m. Start with a 2–triangle standard triangulation of the annulus. The two boundary curves of the annulus are edges in this triangulation, and each of the remaining three edges is incident with both boundary edges. One now iteratively layers tetrahedra on one side of this annulus, such that the initial triangulation of the annulus is in the boundary of the solid torus, and each interior edge has degree four. This completely characterises these triangulations combinatorially. In what follows, we mainly introduce notation to make this explicit and we highlight some features that will be relevant later.

Let $m \geq 1$ and $\Delta_1, \ldots, \Delta_m$ be a collection of Euclidean 3–simplices. The vertices of each Δ_j are labelled $0,1,2,3$, and for each subset $\Lambda \subseteq \{0,1,2,3\}$, $\Delta_j(\Lambda)$ is the subsimplex spanned by Λ. The triangulation T_1 is obtained by making the edge identification $\Delta_1(03) = \Delta_1(12)$. Note that this is not a triangulation, much less a triangulation of a solid torus. For $m > 1$, T_m is obtained from T_{m-1} by adding Δ_m via the face pairings:

$$\begin{aligned}
\Delta_m(012) &\mapsto \Delta_{m-1}(103), \\
\Delta_m(123) &\mapsto \Delta_{m-1}(123), \text{ if } m \text{ is even}, \\
\Delta_m(023) &\mapsto \Delta_{m-1}(023), \text{ if } m \text{ is odd}.
\end{aligned}$$

If m is even, this is a layering of Δ_m onto the edge $\Delta_{m-1}(12) = \Delta_{m-1}(03)$; and if m is odd, this is a layering of Δ_m onto the edge $\Delta_{m-1}(02) = \Delta_{m-1}(13)$.

Remark 1. Note that Regina [3] starts counting tetrahedra at 0 rather than 1. In particular, in a Regina gluing table of T_m, the parity condition is reversed. Also note that, while the above set of gluings looks very simple, it requires the orientation of the tetrahedra Δ_j to oscillate with their index j.

It follows from the construction that the faces $\Delta_1(012)$ and $\Delta_1(023)$ form an annulus in the boundary of T_m, and that for $m > 1$, T_m is a triangulation of a solid torus, obtained from T_{m-1} by layering a tetrahedron on an edge not contained in the boundary faces $\Delta_1(012)$ and $\Delta_1(023)$. The different pairings for m odd or even ensure that each interior edge of T_m has degree four.

The two boundary edges of T_m corresponding to a longitude of the solid torus are $\Delta_1(01) = \ldots = \Delta_m(01)$ and $\Delta_1(23) = \ldots = \Delta_m(23)$. We denote $\Delta_1(01) = \ldots = \Delta_m(01)$ by λ, see Figure 2.

We note that T_m has two vertices $\Delta_1(0) = \Delta_1(1) = \ldots = \Delta_m(0) = \Delta_m(1)$ and $\Delta_1(2) = \Delta_1(3) = \ldots = \Delta_m(2) = \Delta_m(3)$.

The boundary of the solid torus T_m is a 4–triangle, 2–vertex torus. For m odd this results in boundary faces $\Delta_m(013)$, $\Delta_m(123)$, $\Delta_1(012)$, and $\Delta_1(023)$, see also Figure 2.

In order to determine the meridian μ of T_m we trace the meridian disc of T_m through the layerings:

- In the 2–triangle annulus the seed of the meridian disc is a horizontal arc connecting the two boundary components (Figure 2 top row).
- Layering Δ_1 to the annulus adds a single horizontal quad of type $q_{03/12}$ in Δ_1 to the seed of the meridian disc (Figure 2, second row).
- Layering Δ_2 on edge $\Delta_1(03) = \Delta_1(12)$ adds two normal triangles in Δ_2 (one of type t_0 and one of type t_3) to the meridian disc (Figure 2, third row).
- Layering Δ_3 on edge $\Delta_2(02) = \Delta_2(13)$, adds two normal triangles (of types t_1 and t_3 respectively) and one normal quadrilateral (of type $q_{01/23}$) in Δ_3 to the meridian disc (Figure 2, bottom row).
- Layering Δ_j, $j \geq 4$, adds two more normal triangles (of types as before) and $m-2$ normal quadrilaterals of type $q_{01/23}$ to the meridian disc. See Figure 3 for the meridian disc in T_m and its boundary curve.

The triangulation T_m admits three normal surfaces F_1, F_2, and F_3 with exactly one normal quadrilateral per tetrahedron and every normal quadrilateral type features in exactly one of these three surfaces:

- Starting with quadrilateral $q_{01/23}$ in Δ_1 we obtain a separating annulus F_1 intersecting the boundary of T_m, $m \geq 2$ arbitrary, in two longitudes, denoted by γ_1.
- Starting with quadrilateral $q_{02/13}$ in Δ_1 we obtain a one-sided (hence non-orientable) surface F_2 intersecting the boundary of T_m, $m \geq 3$ and odd, horizontally in curve γ_2.
- Starting with quadrilateral $q_{03/12}$ in Δ_1 we obtain another one-sided (hence non-orientable) surface F_3 intersecting the boundary of T_m, $m \geq 3$ and odd, diagonally in curve γ_3.

See Figure 4 for an illustration.

Due to the following result, we know when and how a curve on the boundary of a solid torus extends to a one-sided incompressible properly embedded (connected) surface in its interior.

Proposition 1 (Corollary 2.2 in [5]). *Suppose T is a solid torus, μ a standard meridian and λ a longitude for T. A one-sided connected incompressible surface $S \subset T$ has boundary a single curve homotopic with $2p\lambda + q\mu$, for some $p, q \in \mathbb{Z}$ satisfying $p \neq 0$, $2p$ and q co-prime. Conversely, every simple closed curve on ∂T homotopic with $2p\lambda + q\mu$, where $p \neq 0$, $2p$ and q co-prime, is the boundary of a one-sided incompressible surface in T.*

Moreover, the negative Euler characteristic of S is determined by $[\partial S] = 2p\lambda + q\mu$ and given by the number $N(2p,q)$ defined by $N(2p,q) = N(2(p-Q), q-2m) + 1$, where Q is the smallest positive integer so that there exists an integer m such that $Qq = 2pm \pm 1$ and $N(2p,1) = p-1$.

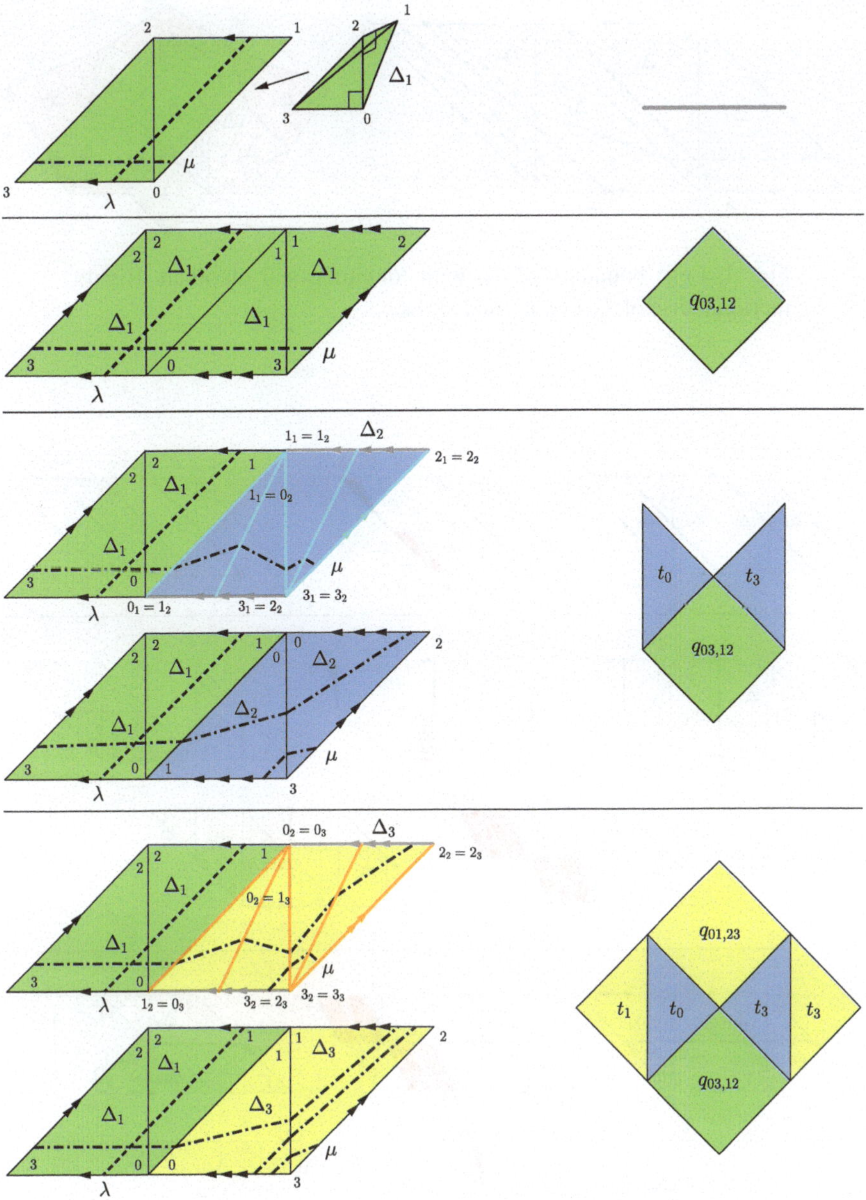

Fig. 2: Construction of T_m, longitude, meridian, and meridian disc. Top: the base annulus. Second, third, and fourth row: T_1, T_2, and T_3, respectively. In each step, another tetrahedron is layered onto the existing triangulation. Right column: the normal pieces of the corresponding meridian discs.

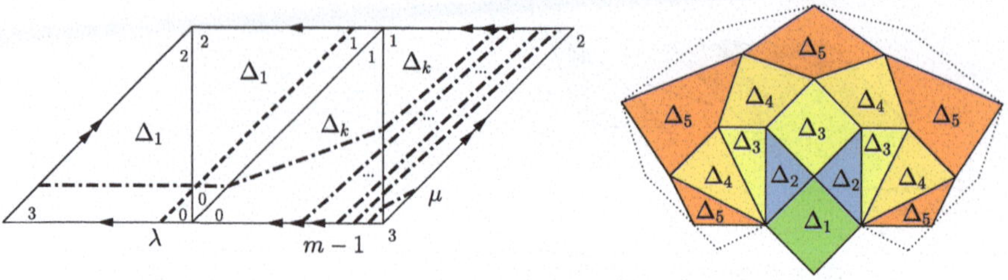

Fig. 3: Left: boundary of T_m with longitude and meridian. Right: meridian disc of T_m as a normal surface.

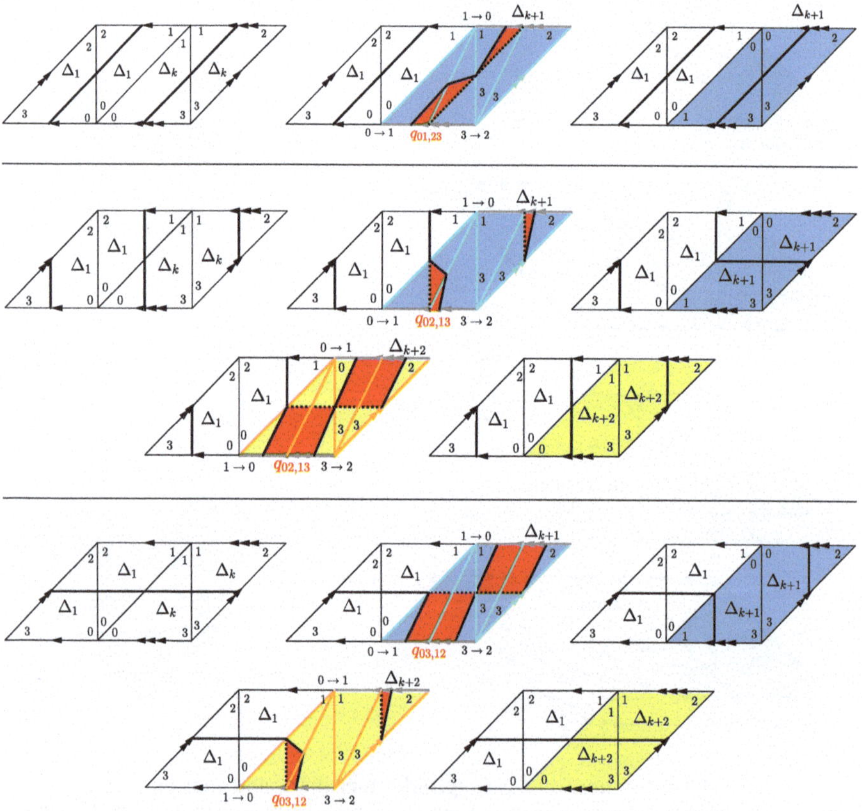

Fig. 4: Boundary patterns of normal surfaces F_1, F_2, and F_3 in T_m. The boundary pattern of F_1 does not depend on m, whereas the boundary patterns of F_2 and F_3 depend on the parity of m.

Corollary 1. *Let $m \geq 3$ be odd. The surface F_2 with boundary γ_2 is an incompressible one-sided surface with negative Euler characteristic $(m-1)/2$ properly embedded in T_m, and the surface F_3 with boundary γ_3 is an incompressible one-sided surface with negative Euler characteristic $(m-3)/2$ properly embedded in T_m.*

Proof. We have
$$\gamma_i \simeq \langle \gamma_i, \mu \rangle \lambda + \langle \gamma_i, \lambda \rangle \mu.$$
This results in $\gamma_2 = (m+1)\lambda + \mu$ and $\gamma_3 = (m-1)\lambda + \mu$ as can be deduced from Figure 4.

Proposition 1 applied to γ_2 and γ_3 shows that these curves bound incompressible surfaces of negative Euler characteristic $(m-1)/2$ and $(m-3)/2$ respectively. Now F_2 and F_3 have precisely this negative Euler characteristic. Hence they must be incompressible.

4 The family of triangulations

Suppose $k, n \geq 3$ odd. Let $T^1 = T^1_k = T_k$ and $T^2 = T^2_n = T_n$ be triangulations of the solid torus from Section 3. Denote their tetrahedra by Δ^i_j, $i = 1, 2$. Glue the boundary of T^1 to the boundary of T^2 in the following way:

$$\begin{aligned} \Delta^1_1(012) &\mapsto \Delta^2_1(120) \\ \Delta^1_k(013) &\mapsto \Delta^2_1(032) \\ \Delta^1_1(023) &\mapsto \Delta^2_n(321) \\ \Delta^1_k(123) &\mapsto \Delta^2_n(130) \end{aligned} \tag{1}$$

We denote the set the face pairings in Equation (1) by Φ.

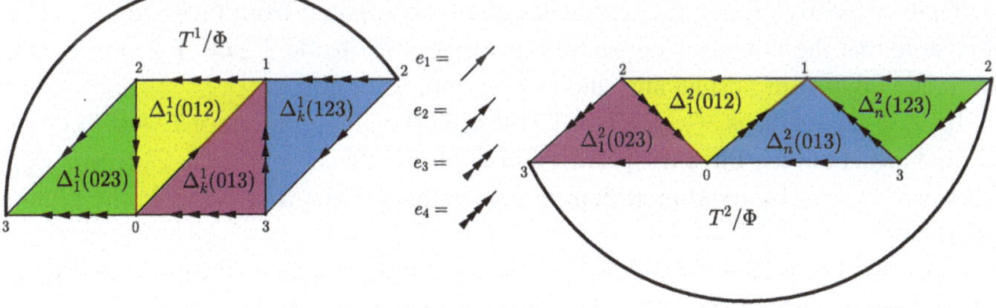

Fig. 5: Gluing two solid tori, $T^1 = T^1_k = T_k$ and $T^2 = T^2_n = T_n$, to obtain $\mathscr{T}_{k,n}$. T^1 is meant to sit on top of the paper (just like in the figures in Section 3), while T^2 is thought of sitting below the paper. Note the different shape of the two tori to keep edge lengths consistant.

This identifies all vertices from both boundary tori, and the link of this single vertex is a torus. The identification space of the resulting triangulation $\mathscr{T}_{k,n}$ is an orientable pseudo-manifold having the single vertex as the only non-manifold point, and the complement of the vertex is denoted $M_{k,n}$. These facts do not depend on k or n and can thus be checked by a straightforward calculation. See Figure 5 for the two torus boundaries of T^1 and T^2 and how they are glued together to form $\mathscr{T}_{k,n}$.

We refer to the 2-complex that is the image of the boundaries of T^1 and T^2 in $\mathscr{T}_{k,n}$ as the *gluing interface*. As observed, this has four triangles, one vertex, and we claim that there are four edges. For the twenty edges involved in the identifications on the boundaries of the tori, we have identifications $\Delta_1^1(01) = \Delta_k^1(01)$, $\Delta_1^1(23) = \Delta_k^1(23)$, $\Delta_1^1(03) = \Delta_1^1(12)$ and $\Delta_k^1(03) = \Delta_k^1(12)$ coming from the gluings inside T^1; and similarly $\Delta_1^2(01) = \Delta_n^2(01)$, $\Delta_1^2(23) = \Delta_n^2(23)$, $\Delta_1^2(03) = \Delta_1^2(12)$ and $\Delta_n^2(03) = \Delta_n^2(12)$ coming from the gluings inside T^2. Combining this with the identifications of the boundaries yields four edge classes in the gluing interface:

$$\begin{aligned}
e_1 &= \Delta_1^1(01) = \Delta_k^1(01) = \Delta_1^2(03) = \Delta_1^2(12),\\
e_2 &= \Delta_1^1(23) = \Delta_k^1(23) = \Delta_n^2(30) = \Delta_n^2(21),\\
e_3 &= \Delta_1^2(01) = \Delta_n^2(01) = \Delta_1^2(23) = \Delta_n^2(23) = \Delta_1^1(20) = \Delta_k^1(31),\\
e_4 &= \Delta_1^1(03) = \Delta_1^1(12) = \Delta_k^1(30) = \Delta_k^1(21) = \Delta_1^2(20) = \Delta_n^2(31).
\end{aligned}$$

Denote the quadrilateral normal surfaces in the solid tori T^i described in Section 3 by F_j^i, $i = 1,2$, $j = 1,2,3$. Their edge weights $(w(e_1), w(e_2), w(e_3), w(e_4))$ in ∂T^i, $i = 1,2$ are

- $(1,1,0,1)$ for F_2^1 and F_1^2;
- $(0,0,1,1)$ for F_1^1 and F_3^2;
- $(1,1,1,0)$ for F_3^1 and F_2^2.

It follows that $\mathscr{T}_{k,n}$ admits three normal surfaces $\mathscr{S}_1 = F_2^1 \cup F_1^2$, $\mathscr{S}_2 = F_1^1 \cup F_3^2$, and $\mathscr{S}_3 = F_3^1 \cup F_2^2$, each consisting of a single quadrilateral per tetrahedron such that every normal quadrilateral type of $\mathscr{T}_{k,n}$ occurs in exactly one of these surfaces.

Their respective Euler characteristics can be computed from Proposition 1. For this, note that the boundary curves of the surfaces F_j^i inside T^i, $i = 1,2$, $j = 1,2,3$, are made up of four normal arcs and four midpoints of edges.

In $\mathscr{T}_{k,n}$, the surfaces $\mathscr{S}_j$, $j = 1,2,3$, still meet the gluing interface $T^1 \cap T^2$ in four normal arcs, but only three (resp. two) midpoints of edges for $\mathscr{S}_3$ and $\mathscr{S}_1$ (resp. $\mathscr{S}_2$). This can be seen by looking at their edge weights for the four edges in the gluing interface.

Thus, the Euler characteristic of each surface pieces is the value given in Proposition 1 minus 1 (resp. minus 2). Altogether we have

$$\chi(\mathscr{S}_1) = (\chi(F_2^1)-1)+(\chi(F_1^2)-1)+1 = -\frac{k-1}{2}+0-1 = -\frac{k+1}{2} \tag{2}$$

$$\chi(\mathscr{S}_2) = (\chi(F_1^1)-2)+(\chi(F_3^2)-2)+2 = 0-\frac{n-3}{2}-2 = -\frac{n+1}{2} \tag{3}$$

$$\chi(\mathscr{S}_3) = (\chi(F_3^1)-1)+(\chi(F_2^2)-1)+1 = -\frac{k-3}{2}-\frac{n-1}{2}-1 = -\frac{k+n-2}{2} \tag{4}$$

Each of the $\mathscr{S}_j$, $j=1,2,3$ is the canonical representative of one of the non-zero $\mathbb{Z}_2$–homology classes and we have

$$\chi(\mathscr{S}_1)+\chi(\mathscr{S}_2)+\chi(\mathscr{S}_3) = -(k+n).$$

In order to apply Theorem 1 to conclude that $\mathscr{T}_{k,n}$ is minimal, we need to show that each $M_{k,n}$ is hyperbolic (see Section 5) and that each $\mathscr{S}_j$ is taut (see Section 6).

5 Hyperbolicity and algebraic topology of $M_{k,n}$

We start by building a 3–cusped manifold using a variant of the construction of $\mathscr{T}_{k,n}$ described above. For this, take a cone $\widehat{C}^1$ over ∂T^1 and a cone $\widehat{C}^2$ over ∂T^2, where $T^1 = T^1_k$ and $T^2 = T^2_n$ are the solid torus triangulations from Section 4 and $k,n \geq 3$ both odd. Now we glue them together using the same face pairings Φ from Equation (1) to obtain a triangulation $\mathscr{T}$ with eight tetrahedra of a pseudo-manifold $\widehat{N}$ with three vertices having torus links. This is defined by the gluing table given in Table 1 and shown in Figure 6.

Using Regina [3] we can check that the complement $N = \widehat{N} \setminus \widehat{N}^{(0)}$ of the three vertices in $\widehat{N}$ is a 3–manifold homeomorphic to the complement of the link 8^3_9, and that this is a minimal triangulation of N. We refer to $C^1 = \widehat{C}^1 \setminus \widehat{N}^{(0)}$ and $C^2 = \widehat{C}^2 \setminus \widehat{N}^{(0)}$ respectively as the *(red) cusp 1* and the *(blue) cusp 2* of N. The red cusp is the symmetric cusp of the link complement, and the blue cusp therefore one of the non-symmetric cusps.

It follows from the construction that the *interface* between the red and blue cusps in N is *identical* to the interface between the solid tori in $M_{k,n}$, see Figures 5 and 6 and their captions. We now truncate the red and blue cusps along vertex linking surfaces (as indicated together with their peripheral curves in Figure 6). These surfaces are naturally combinatorially isomorphic with ∂T^1 and ∂T^2. The cusps are then filled with the solid tori T^1_k and T^2_n respectively, such that the gluing agrees with the natural combinatorial isomorphism. With respect to the framing indicated in Figure 6, and the meridian curves of the solid tori given in Figure 3 on the left, this results in the manifold

$$N_{k,n} = N(\infty,(-k,1),(n+1,1)).$$

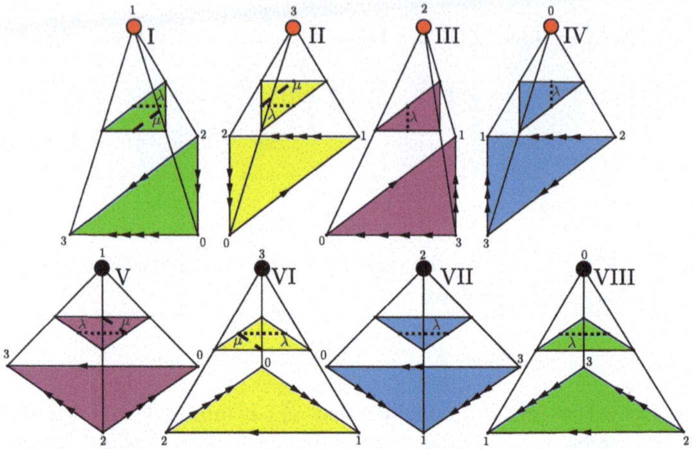

Fig. 6: The ideal triangulation of N, the complement of the link 8^3_9, together with vertex linking surfaces and their peripheral curves. See Figure 5 for how the glued-up interface is seen from the red cusp (on the left), or when looking into the blue cusp (on the right). Between both figures, vertex and edge labels, as well as colours are identical.

Tet	(012)	(013)	(023)	(123)
I	II (032)	II (132)	**VIII (321)**	IV (023)
II	**VI (120)**	III (012)	I (021)	I (031)
III	II (013)	**V (032)**	IV (102)	IV (103)
IV	III (203)	III (213)	I (123)	**VII (130)**
V	VI (032)	VI (132)	**III (031)**	VIII (023)
VI	**II (201)**	VII (012)	V (021)	V (031)
VII	VI (013)	**IV (312)**	VIII (102)	VIII (103)
VIII	VII (203)	VII (213)	V (123)	**I (320)**

Table 1: Gluing table for 3-cusped manifold N, an ideal triangulation homeomorphic to the complement of 8^3_9. Bold entries are gluings across the interface (labelings coincide with Equation (1)). Tetrahedra I, III, V, and VII have the opposite orientations to tetrahedra II, IV, VI, VIII.

For the interested reader who wishes to use SnapPy to study these manifolds, we remark that

$$N_{k,n} = \text{Manifold('8\^{}3\_9')}(\infty, (n+1,-1), (k+1,1)). \tag{5}$$

The *Euclidean* equilateral triangles shown in Figure 5 in fact give the Euclidean structures (up to scale) on the cusp cross sections with respect to the complete hyperbolic structure on N. Here, each ideal tetrahedron has shape i and each triangle therefore has angles $\frac{\pi}{2}, \frac{\pi}{4}, \frac{\pi}{4}$. The framing used in this paper is thus the natural geometric framing of the cusps.

Casson observed that if an ideal triangulation admits an angle structure, then the manifold must admit a complete hyperbolic metric of finite volume (see [16, Theorem 10.2]). Using this, computation with Regina [3] confirms that $N_{k,n}$ is hyperbolic for all $17 \geq k,n \geq 3$. The same result could also be obtained with SnapPy [4].

To show that $N_{k,n}$ also has a complete hyperbolic metric of finite volume for all $k,n \geq 17$, we note that due to Thurston's hyperbolisation theorem for Haken manifolds [13, 15, 18], it suffices to show that $N_{k,n}$ is irreducible, atoroidal, and not Seifert fibred. This conclusion is achieved by an application of Lackenby's combinatorial version of the Gromov-Thurston 2π theorem [14, Theorem 4.9] (see also the sentence after its proof for the case when not all cusps are filled). We refer to [14] for the definitions and statements required to understand the next paragraph.

The hypothesis of [14, Theorem 4.9] is achieved by applying [14, Proposition 4.10], which gives a criterion using angle structures. We use the angle structure on N determined by the complete hyperbolic structure, where each triangle in the vertex link has angles $\frac{\pi}{2}, \frac{\pi}{4}, \frac{\pi}{4}$. In particular, Lackenby's combinatorial length of any curve γ on a vertex link is estimated from below by the shortest simplicial length $|\gamma|$ of a representative of $[\gamma]$ times $\frac{\pi}{8}$. The theorem thus applies to any set of filling curves γ with $|\gamma|\frac{\pi}{8} > 2\pi$, and hence $|\gamma| \geq 17$. Examination of the vertex links gives the simple bounds $|\mu_1^{-k}\lambda_1| \geq k$ and $|\mu_2^{n+1}\lambda_2| \geq n+1$. Therefore $k,n \geq 17$ are sufficient to ensure hyperbolicity.

Next, we observe that by construction the manifold $N_{k,n}$ is homotopy equivalent with $M_{k,n}$, and that we may choose the homotopy to be the identity in a neighbourhood of the single cusp (corresponding to the vertex on the interface). Now take a sufficiently large compact core of each manifold, and double it along its boundary. Then we have two closed Haken 3–manifolds that are homotopy equivalent, and hence homeomorphic by Waldhausen [19]. Since $N_{k,n}$ is hyperbolic and homeomorphism preserves the JSJ decomposition, it follows that $N_{k,n}$ is homeomorphic with $M_{k,n}$. In particular, $M_{k,n}$ is also hyperbolic.

From the description of the normal surfaces in the previous section we can deduce that $M_{k,n} = N_{k,n}$ has homology amenable to an application of Theorem 1 ([10, Theorem 1]). This can be verified using a homology computation using the chosen peripheral framings. Using the triangulation, one computes

$$\pi_1(N) = \langle\, a,b,c \mid abcbc^{-1}a^{-1}cb^{-1}c^{-1}b^{-1},\ abc^{-1}ab^{-1}a^{-1}bca^{-1}b^{-1} \,\rangle$$

with the two peripheral subgroups of interest to us (in the geometric framing) generated by:

$$\mu_1 = c^{-1}a, \quad \lambda_1 = b^{-1}a^{-1}bc, \qquad \mu_2 = a, \quad \lambda_2 = bcbc^{-1}$$

This implies:

$$H_1(M_{k,n}) \cong \langle\, a,b,c \mid (a^{-1}c)^{k+1},\ (a^{(n+1)/2}b)^2 \,\rangle \cong \mathbb{Z} \oplus \mathbb{Z}_2 \oplus \mathbb{Z}_{k+1}$$

In particular, this highlights the requirement for both n and k to be odd.

6 The norm of the homology classes of $M_{k,n}$

In this section we determine the norm of the non-trivial classes in $H_2(M_{k,n}, \mathbb{Z}_2)$. We first provide an overview. Our proofs of correctness of the calculation rely on technical results by Jaco and Sedgwick [12] for normal surfaces with boundary on a single 2–triangle torus and Bachman, Derby-Talbot and Sedgwick [1] for normal surfaces with boundary on two 2–triangle tori.

We consider a triangulation $\mathscr{T}'$ of the 3–manifold obtained by removing open neighbourhoods of the blue and red cusps of the complement of 8^3_9 (Section 6.1). This has one cusp and two boundary components, ∂_1 (the boundary of the blue cusp) and ∂_2 (the boundary of the red cusp). The interior of the identification space $\overline{\mathscr{T}}$ of $\mathscr{T}'$ is homeomorphic to the complement of 8^3_9. Using the natural geometric framing on ∂_1 and ∂_2 and the surgery description of $M_{k,n}$, we determine how to complete $\mathscr{T}'$ to a triangulation $\mathscr{T}'_{k,n}$ of $M_{k,n}$ by gluing suitable layered solid tori to ∂_1 and ∂_2 (Section 6.5).

Each non-trivial class $\alpha \in H_2(M_{k,n}, \mathbb{Z}_2)$ associates with each edge in ∂_1 and ∂_2 a parity, which we call its **boundary pattern** (Section 6.2). This boundary pattern encodes the parities of the number of points in the intersection of the edges with a surface that *(a)* represents the class α, and *(b)* is transverse to the triangulation. There is a taut normal surface S in $\mathscr{T}'_{k,n}$ that represents the class α and its intersection with each of ∂_1 and ∂_2 is a connected and essential curve (Section 6.3).

The norm of α is the maximum Euler characteristic of any normal surface in $\mathscr{T}'_{k,n}$ representing α. The computation of the norm of α thus splits into a computation of all normal surfaces of $\mathscr{T}'$ that have the correct boundary pattern (Section 6.3), and determining how these normal surfaces extend into the layered solid tori attached to $\mathscr{T}'$ along ∂_1 and ∂_2 (Section 6.6). Normal surfaces in layered solid tori are well-known (Section 6.4). Moreover, their Euler characteristics are determined by their boundary slopes via the Bredon-Wood formula (see Proposition 1 or [5, Corollary 2.2]). The conclusion of these computations (see Section 6.6) is that:

$$|| \alpha_1 || = \frac{k+1}{2}, \qquad || \alpha_2 || = \frac{n+1}{2}, \qquad || \alpha_3 || = \frac{n+k-2}{2}$$

In particular, this confirms that the surfaces $\mathscr{S}_1, \mathscr{S}_2$ and $\mathscr{S}_3$ described in Section 4 are taut. We now supply the details.

6.1 A triangulation with two boundary components and one ideal vertex

Starting with the ideal 3-cusped triangulation N of the complement of 8^3_9, see Figure 6 and Table 1, we cut along the vertex links of the blue and red vertex respectively to obtain a 17-tetrahedra triangulation with two boundary components and one ideal vertex. By construction, the interior of the identification space is homeomorphic with the complement of 8^3_9. We denote this triangulation by $\mathscr{T}'$. Its isomorphism signature is

`rfLLHMzLPMwQcddghghjnklomqopqrwgrrgfxrvqdabxs`

and its gluing table is given in Table 2.

Tet	(012)	(013)	(023)	(123)
Δ_0	3(012)	∂_1	2(023)	1(123)
Δ_1	5(012)	3(230)	4(023)	0(123)
Δ_2	7(012)	3(321)	0(023)	6(123)
Δ_3	0(012)	∂_1	1(301)	2(310)
Δ_4	8(012)	7(230)	1(023)	6(023)
Δ_5	1(012)	7(103)	9(023)	6(310)
Δ_6	10(012)	5(321)	4(123)	2(123)
Δ_7	2(012)	5(103)	4(301)	11(123)
Δ_8	4(012)	13(013)	9(021)	12(123)
Δ_9	8(032)	10(230)	5(023)	13(210)
Δ_{10}	6(012)	14(013)	9(301)	11(120)
Δ_{11}	10(312)	12(021)	14(201)	7(123)
Δ_{12}	11(031)	16(013)	15(023)	8(123)
Δ_{13}	9(321)	8(013)	14(032)	16(123)
Δ_{14}	11(230)	10(013)	13(032)	15(210)
Δ_{15}	14(321)	∂_2	12(023)	16(012)
Δ_{16}	15(123)	12(013)	∂_2	13(123)

Table 2: Gluing table for triangulation $\mathscr{T}'$ of the complement of 8^3_9 with two real boundary components and one ideal vertex.

In particular, the two boundary components ∂_1 (bounding a neighbourhood of the blue cusp in Figure 6 – one of the non-symmetric components of 8^3_9) and ∂_2 (bounding a neighbourhood of the red cusp in Figure 6 – the symmetric component of 8^3_9) are two 1-vertex tori made up of triangles $0(013)$ and $3(013)$, and $15(013)$ and $16(023)$ respectively. We have edge classes

$$\begin{aligned} e_0 &= \Delta_3(01) = \Delta_0(01), \\ e_2 &= \Delta_0(03) = \Delta_2(03) = \Delta_3(31), \\ e_4 &= \Delta_3(30) = \Delta_1(13) = \Delta_0(13) \end{aligned}$$

for ∂_1 and

$$e_{18} = \Delta_{16}(03) = \Delta_{12}(03) = \Delta_{15}(03),$$
$$e_{19} = \Delta_{15}(01) = \Delta_{14}(32) = \Delta_{13}(23) = \Delta_{16}(23),$$
$$e_{20} = \Delta_{16}(02) = \Delta_{15}(13)$$

for ∂_2.

6.2 Boundary patterns

By construction, a triangulation $\mathscr{T}'_{k,n}$ of $M_{k,n}$ can be obtained from $\mathscr{T}'$ by gluing two suitable layered solid tori to ∂_1 and ∂_2. Let $\alpha_1, \alpha_2, \alpha_3 \in H_2(M_{k,n}, \mathbb{Z}_2)$ be the three non-zero $\mathbb{Z}_2$– homology classes corresponding to normal surfaces $\mathscr{S}_1$, $\mathscr{S}_2$, and $\mathscr{S}_3$ in $\mathscr{T}_{k,n}$ (see Section 4).

The fundamental group of the complement of 8^3_9 is generated by three meridians. We choose the base point for the fundamental group of $\overline{\mathscr{T}}$ at the vertex of ∂_1. Since $\mathbb{Z}_2$ is abelian, a homomorphism $\pi_1(\overline{\mathscr{T}}) \to \mathbb{Z}_2$ uniquely associates labels in $\mathbb{Z}_2$ to the edges of ∂_1 and ∂_2. Via the inclusion $\overline{\mathscr{T}} \subset \overline{\mathscr{T}'_{k,n}}$, a class in $H_2(M_{k,n}, \mathbb{Z}_2)$ determines a homomorphism $\pi_1(\overline{\mathscr{T}}) \to \mathbb{Z}_2$ mapping the meridian of the cusp of $\overline{\mathscr{T}}$ to zero.

Conversely, any homomorphism $\pi_1(\overline{\mathscr{T}}) \to \mathbb{Z}_2$ mapping the meridian of the cusp of $\overline{\mathscr{T}}$ to zero uniquely determines a class in $H_2(M_{k,n}, \mathbb{Z}_2)$ due to the surgery description.

Hence each α_i associates labels in $\mathbb{Z}_2$ to the edges in ∂_1 and ∂_2 (and in turn is uniquely determined by these). We call these labels the **boundary pattern** of α_i.

6.3 The normal surfaces of $\mathscr{T}'$

Let $U_i \subset M_{k,n}$, $[U_i] = \alpha_i$, $i = 1,2,3$, be a taut representative of α_i in $M_{k,n}$. Since U_i is incompressible and $M_{k,n}$ is irreducible, U_i is isotopic to a normal surface. Hence we may assume that U_i is normal. We further assume that amongst all taut normal surfaces representing α_i, U_i minimises the total edge weight on ∂_1 and ∂_2.

We claim that this implies that U_1 meets ∂_1 in one essential curve and is disjoint from ∂_2; U_2 meets ∂_2 in one essential curve and is disjoint from ∂_1; and U_3 meets both ∂_1 and ∂_2 in one essential curve. Indeed, if there are trivial curves in the intersection with ∂_1 or ∂_2, then one may perform isotopies across balls to reduce the total weight on ∂_1 and ∂_2. The resulting surface may not be normal, but normalises to a normal surface that is still taut and has smaller total edge weight on ∂_1 and ∂_2 than the initial surface. Similarly, if there are two parallel essential curves in the intersection with ∂_1 or ∂_2, then one may perform an annular compression reducing the total edge weight on ∂_1 and ∂_2. The resulting surface has the same Euler characteristic, and hence still has a taut component that therefore normalises to a normal surfaces of smaller total weight on ∂_1 and ∂_2 than the initial surface. It follows that there is at most one essential curve in the intersection with each ∂_1 and ∂_2, and the claimed numbers now follow from the boundary pattern.

Let $S_i = U_i \cap \mathscr{T}'$. We now show that we may assume that S_1 and S_2 are fundamental surfaces in $\mathscr{T}$.

Since S_1 and S_2 have one essential curve on a 1–vertex torus boundary component, it follows from [12, Proposition 3.7], that there exists a fundamental normal surface with the same boundary curve and with equal or larger Euler characteristic. Using Regina [3] we find that $\mathscr{T}'$ has 900 fundamental surfaces in standard coordinates.

First, note that neither $\mathscr{T}'$ nor the filling tori contain closed normal surfaces of positive Euler characteristic. Hence, to find the norm of α_1 (resp. α_2) it is sufficient to take the minimum over all fundamental normal surfaces F in $\mathscr{T}'$ with boundary a single essential curve in ∂_1 (resp. ∂_2) of the sum of

(a) the negative Euler characteristic of F, and
(b) the negative Euler characteristic of the incompressible surface F' in the filling torus with $\partial F' = \partial F$.

There are only a few candidate surfaces F to consider.

There are 10 fundamental surfaces matching the boundary pattern of α_1 and having intersection with ∂_1 one essential curve and empty intersection with ∂_2. There are two distinct boundary curves amongst these surfaces, and it suffices to pick a fundamental surface of maximal Euler characteristic for each of these curves.

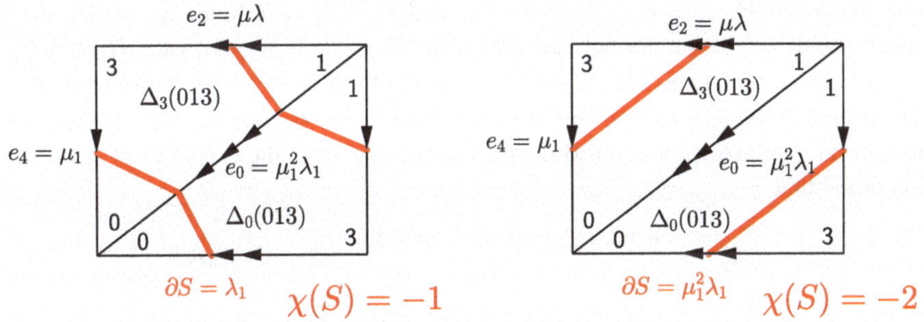

Fig. 7: Boundary curves and Euler characteristics of candidate surfaces for S_1.

Similarly, there are 25 fundamental surfaces matching the boundary pattern of α_2 and having intersection with ∂_2 one essential curve and empty intersection with ∂_1. There are three distinct boundary curves amongst these surfaces, and it suffices to pick a fundamental surface of maximal Euler characteristic for each of these curves; see Figure 8.

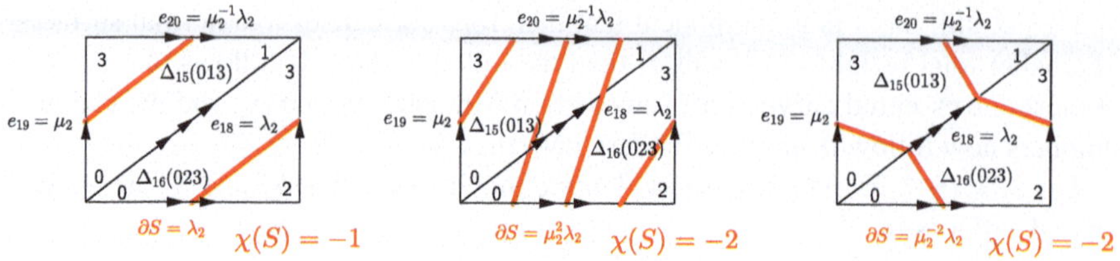

Fig. 8: Boundary curves and Euler characteristics of candidate surfaces for S_2.

The second term is determined by $\partial F' = \partial F$ via the framing and the Bredon-Wood formula, see Proposition 1 or [5, Corollary 2.2]. However, we take an equivalent approach by analysing the space of all normal surfaces of the layered solid torus realising the filling of the respective cusp.

We show in Section 6.6 that this forces

$$|| \alpha_1 || = \frac{k+1}{2} \qquad \text{and} \qquad || \alpha_2 || = \frac{n+1}{2}$$

We use these norms to determine the norm of α_3. The surface S_3 has one essential boundary curve in each of ∂_1 and ∂_2. Since $S_3 \cap \partial_j$ is a single essential curve, its Haken summands cannot have any trivial boundary components. First suppose a Haken sum of fundamental surfaces in $\mathscr{T}'$ giving S_3 contains a surface that is disjoint from one of ∂_1 or ∂_2. It then follows from [1, Lemma 2.10] that there must be exactly one summand H_1 with $H_1 \cap \partial_1 = S_3 \cap \partial_1$ and $H_1 \cap \partial_2 = \emptyset$; and exactly one summand H_2 with $H_2 \cap \partial_2 = S_3 \cap \partial_2$ and $H_2 \cap \partial_1 = \emptyset$. Since S_3 is the intersection with $\mathscr{T}'$ of the normal surface U_3 in $\mathscr{T}'_{k,n}$, it follows that each H_1 and H_2 extends to a normal surface V_1 and V_2 respectively in $\mathscr{T}'_{k,n}$ via a subsurface of U_2. The boundary pattern implies that $[V_1] = \alpha_1$ and $[V_2] = \alpha_2$, and since U_3 is taut, we have $|| \alpha_3 || \geq || \alpha_1 || + || \alpha_2 ||$. But we already know from the existence of $\mathscr{S}_3$ from Section 4 that

$$|| \alpha_3 || \leq \frac{n+k-2}{2} < \frac{k+1}{2} + \frac{n+1}{2} = || \alpha_1 || + || \alpha_2 ||$$

Hence every Haken summand H with non-empty boundary satisfies $H \cap \partial_1 = S_3 \cap \partial_1$ and $H \cap \partial_2 = S_3 \cap \partial_2$. Since there is no closed normal surface with positive Euler characteristic, we may assume that every Haken summand in a Haken sum giving S_3 meets each boundary component of $\mathscr{T}'$ in non-trivial essential curves.

After sorting out duplicates and non-maximal Euler characteristic examples, there are 7 compatibility classes of normal surfaces in $\mathscr{T}'$ representing α_3. In particular, the first class can be reduced to a single fundamental surface, and each of the remaining six classes can be reduced to Haken sums of two fundamental surfaces, i.e. Haken sums of the form $w_1 H_1 + w_2 H_2$. For all pairs H_1, H_2 and their possible

Haken sums $w_1H_1 + w_2H_2$, we list in Table 3 their edge weights on ∂_1 and ∂_2, their Euler characteristics, and weights w_1 and w_2.

#	$w(e_0)$	$w(e_2)$	$w(e_4)$	$w(e_{18})$	$w(e_{19})$	$w(e_{20})$	χ	weight
1	2	1	1	0	1	1	-2	1
2	2	0	2	2	1	1	-2	$2k+1$
	2	1	3	2	2	0	-3	$2l+1$
	$4(k+l+1)$	$2l+1$	$4k+6l+5$	$4(k+l+1)$	$2k+4l+3$	$2k+1$	$-4k-6l-5$	
3	2	0	2	2	1	1	-2	$2k+1$
	2	1	1	2	0	2	-1	$2l+1$
	$4(k+l+1)$	$2l+1$	$4k+2l+3$	$4(k+l+1)$	$2k+1$	$2k+4l+3$	$-4k-2l-3$	
4	0	1	1	0	1	1	-1	$2k+1$
	2	2	0	2	1	3	-2	$2l$
	$4l$	$2k+4l+1$	$2k+1$	$4l$	$2k+2l+1$	$2k+6l+1$	$-2k-4l-1$	
5	2	1	1	2	0	2	-1	$2k+1$
	2	2	0	2	1	3	-2	$2l+1$
	$4(k+l+1)$	$2k+4l+3$	$2k+1$	$4(k+l+1)$	$2l+1$	$4k+6l+5$	$-2k-4l-3$	
6	0	1	1	0	1	1	-1	$2k+1$
	2	1	3	2	2	0	-3	$2l$
	$4l$	$2k+2l+1$	$2k+6l+1$	$4l$	$2k+4l+1$	$2k+1$	$-2k-6l-1$	
7	0	1	1	0	1	1	-1	$2k+1$
	2	3	1	2	2	4	-3	$2l$
	$4l$	$2k+6l+1$	$2k+2l+1$	$4l$	$2k+4l+1$	$2k+8l+1$	$-2k-6l-1$	

Table 3: Compatibility classes of candidates for a norm minimiser for α_3.

6.4 Some basic facts on layered solid tori

Layered solid tori denote a two parameter family $\mathrm{LST}(j,k)$ of 1-vertex triangulations of the solid torus, see [8] for details. Given $\mathrm{LST}(j,k)$ (j and k co-prime), the boundary of the meridian disc has edge weights j, k, and $j+k$ on the three boundary edges.

In this section we are only interested in $\mathrm{LST}(0,1)$, and the family $\mathrm{LST}(1,m)$, $m \geq 2$. The layered solid torus $\mathrm{LST}(0,1)$ has three tetrahedra. One of its edges is parallel to the boundary of its meridian disc (the unique edge with edge weight 0). We call this edge the **meridional edge** of the layered solid torus. The layered solid torus $\mathrm{LST}(1,m)$ is obtained inductively from the 1-tetrahedron layered solid torus $\mathrm{LST}(1,2)$ by layering a tetrahedron on the edge of edge weight m of $\mathrm{LST}(1,m-1)$. In particular, for $m \geq 2$ there is a unique boundary edge with edge weight 1, and we call this the **longitudinal edge**.

It follows from [8, Theorem 5.3] that the only normal surfaces in $\mathrm{LST}(1,m)$ with connected, essential boundary are the meridian disc (with edge weights 1, m, and $m+1$ in the boundary, case (3) in the theorem), and one-sided incompressible normal surfaces with boundary slope what is called the slope of an even ordered edge in $\mathrm{LST}(1,m)$ (this is an edge in $\mathrm{LST}(1,m)$ that represents an even multiple of a generator of the fundamental group of the torus, case (6) in the theorem).

For $\mathrm{LST}(1,m)$ the slopes of even ordered edges have edge weights 1, $m-2$, and $m-1$ (a Möbius strip), 1, $m-4$, and $m-3$ (a surface of Euler characteristic -1), all the way down to 1, 1, and 0 for m odd (a surface of Euler characteristic $\frac{1-m}{2}$) and all the way down to 1, 0, and 1 for m even (a surface of Euler characteristic $\frac{2-m}{2}$). In particular, we have the following:

Corollary 2 (Theorem 5.3 from [8]). *Let S be a normal surface in* $\mathrm{LST}(1,m)$, $m \geq 2$, *with connected essential boundary. Then S has edge weight* 1 *on the longitudinal edge of* $\mathrm{LST}(1,m)$.

If the layered solid torus triangulation is generated using the standard Regina [3] function, then the surface S in Corollary 2 has edge weight 1 on the edge $e_2 = \Delta_0(03) = \Delta_0(21)$.

6.5 Determining the natural geometric framing on ∂_1 and ∂_2

Just as in Section 5, we denote the geometric framing of ∂_i by (μ_i, λ_i), $i = 1,2$.

Gluing $\mathrm{LST}(0,1)$ to ∂_2 with the meridian edge along e_{19} produces a manifold with fundamental group free of rank 2 and hence the 2-component unlink. Hence, e_{19} runs parallel to the meridian μ_2 in the geometric framing of ∂_2. Gluing in the same layered solid torus along the other two edges demonstrates that e_{18} runs parallel to the geometric longitude λ_2 and that e_{20} runs parallel to $\mu_2^{-1}\lambda_2$.

Gluing $\mathrm{LST}(0,1)$ with the meridian edge onto the edges e_4 and e_2 yields the fundamental group of the link complement of the remaining two link components (a double twisted Hopf link). Gluing the meridian edge onto e_0 shows that this edge runs parallel to $\mu_1^2\lambda_1$ in the geometric framing. This implies that e_4 runs parallel to μ_1^{-1} and e_2 runs parallel to $\mu_1\lambda_1$. This computation is confirmed by noting that the homological longitude has edge weight 2 on e_0 and edge weight 1 on e_2 and e_4 (as observed by an orientable surface with this boundary curve, see Figure 7 on the left) and the fact that this homological longitude coincides with the geometric longitude.

The above computations were verified using SnapPy [4] and SnapPy in sage for the homological longitude. See Figures 7 and 8 for pictures of ∂_1 and ∂_2 with their respective framings.

The information about the natural geometric framing on ∂_1 and ∂_2 combined with Equation (5) determine which layered solid tori to glue to ∂_1 and ∂_2 to recover $M_{k,n}$.

In detail, $\mathrm{LST}(1,k-1)$ must be glued to ∂_1 with the edge weight-1 boundary edge of the layered solid torus glued to e_2. Similarly, $\mathrm{LST}(1,n)$ must be glued to ∂_2

with the edge weight-1 boundary edge of the layered solid torus glued to edge e_{19}. See Figure 9 for a detailed picture of the gluings.

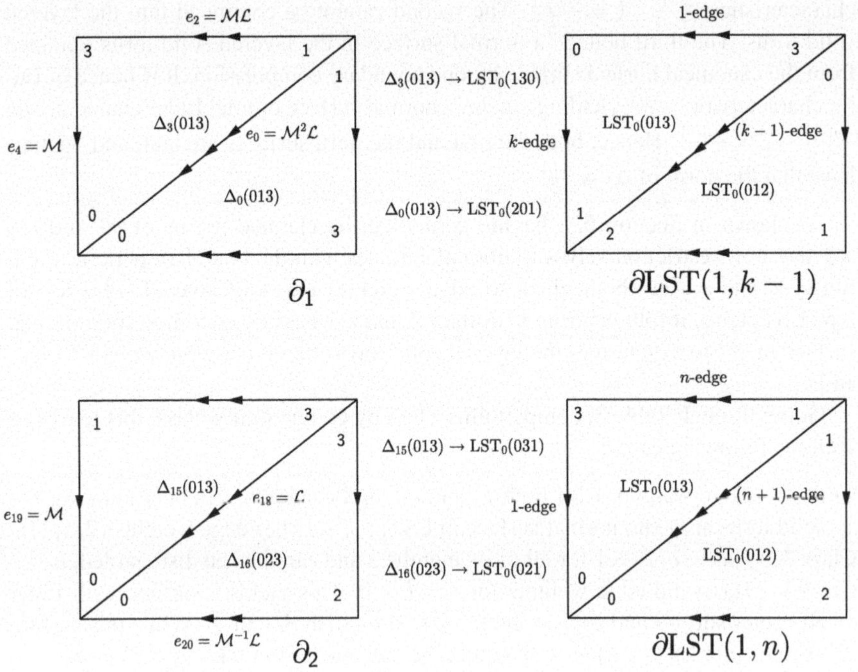

Fig. 9: Filling $\mathscr{T}'$ with suitable layered solid tori.

6.6 Euler characteristic of the completion of S_i, $1 \leq i \leq 3$.

We are now in a position to determine the norms of α_i, $1 \leq i \leq 3$.

The triple (a,b,c) denotes the (parities of the) edge weights of e_0, e_2, and e_4 in that order for ∂_1 and (d,e,f) denotes the (parities of the) edge weights of e_{18}, e_{19}, and e_{20} in that order for ∂_2.

The boundary pattern of α_1 is $(0,1,1)$ with respect to the three edges on ∂_1 and $(0,0,0)$ with respect to ∂_2. These parities are realised by the following edge weights of the fundamental surfaces. We have $(2,1,1)$ ($\chi = -1$), and $(0,1,1)$ ($\chi = -2$). The former cannot be continued into the layered solid torus, the latter continues along the canonical surface of Euler characteristic $\frac{-k+3}{2}$. In total, this yields a surface of Euler characteristic $\frac{-k+3}{2} - 2 = \frac{-k-1}{2}$, hence the norm of α_1 is $\frac{k+1}{2}$.

The boundary pattern of α_2 is $(0,1,1)$ with respect to the three edges on ∂_2 and $(0,0,0)$ with respect to ∂_1. These parities are realised by the following edge weights of the fundamental surfaces. We have $(0,1,1)$ $(\chi=-1)$, $(2,1,3)$ $(\chi=-2)$, and $(2,1,1)$ $(\chi=-2)$. The first can be completed with canonical surface of Euler characteristic $\frac{-n+1}{2}$. In total, this produces a surface representing α_2 with Euler characteristic $\frac{-n+1}{2}-1=\frac{-n-1}{2}$. The second cannot be continued into the layered solid torus. The third bounds a normal surface in the layered solid torus obtained from the canonical normal surface by one boundary compression. It is hence of Euler characteristic $\frac{-n+3}{2}$ yielding, again, a normal surface of total Euler characteristic $\frac{-n+3}{2}-2=\frac{-n-1}{2}$. Hence, both the first and the third surfaces are taut, and we thus have that the norm of α_2 is $\frac{n+1}{2}$.

As shown in Section 6.3, having computed the claimed norms of α_1 and α_2, we may now restrict ourselves to normal surfaces with the boundary pattern of α_3. Since e_2 and e_{19} are both glued to edge weight 1-edges of layered solid tori of type $\mathrm{LST}(1,m)$, it follows from Corollary 2 that a necessary condition for a normal surface in $\mathscr{T}'$ to extend into the layered solid tori is that it has edge weight 1 on e_2 and e_{19}.

Going through Table 3, compatibility class by compatibility class, this leaves us with the following cases:

Class 1 Edge weights 1 for both e_2 and e_{19}, but cannot be extended into layered solid torus at ∂_1 (no normal surface in $\mathrm{LST}(1,k-1)$ has edge weights $(2,1,1)$).

Class 2 Has $w(e_{19})>1$ for all of its members and can thus be disregarded.

Class 3 Has valid edge weights for $k=l=0$. This yields a surface with Euler characteristic -3 and edge weights $(4,1,3)$ on both ∂_1 and ∂_2. While this surface extends into $\mathrm{LST}(1,n)$ at ∂_2, it cannot be extended into $\mathrm{LST}(1,k-1)$ at ∂_1.

Class 4 Has valid edge weights for $k=l=0$. This yields a surface with Euler characteristic -1 and edge weights $(0,1,1)$ on both ∂_1 and ∂_2. This surface extends via the canonical 0/1-surface into both $\mathrm{LST}(1,n)$ and $\mathrm{LST}(1,k-1)$ and produces a closed 1-sided surface of Euler characteristic

$$-1+\frac{1-n}{2}+\frac{3-k}{2}=-\frac{n+k-2}{2}.$$

Class 5 Has $w(e_2)>1$ for all of its members and can thus be disregarded.

Class 6 Has valid edge weights for $k=l=0$. Gives same Euler characteristic and edge weights as Class 4.

Class 7 Has valid edge weights for $k=l=0$. Gives same Euler characteristic and edge weights as Class 4.

Altogether, it follows that the norm of α_3 equals $\frac{n+k-2}{2}$.

7 Generalisations

As mentioned in the introduction, the triangulations $\mathscr{T}_{k,n}$ can be generalised to further conjecturally minimal triangulations. In this section we present one such family that was again found using experimentation with the census.

Consider the triangulation $\mathscr{U}_{3,3}$ with Regina isomorphism signature

`iLLwQPcbeefgehhhhhqhhqhqx`,

and given by the gluing table

Tet	(012)	(013)	(023)	(123)
0	3(012)	1(102)	2(023)	1(123)
1	0(103)	4(102)	4(023)	0(123)
2	6(012)	5(013)	0(023)	4(301)
3	0(012)	4(231)	6(132)	5(032)
4	1(103)	2(231)	1(023)	3(301)
5	7(103)	2(013)	3(132)	7(123)
6	2(012)	7(320)	7(201)	3(032)
7	6(230)	5(102)	6(310)	5(123)

This triangulation coincides with census triangulation `t12546` #1 of a once-cusped hyperbolic 3-manifold. Its first homology group is $\mathbb{Z} \oplus \mathbb{Z}_2 \oplus \mathbb{Z}_4$, and its hyperbolic volume is approximately 7.555.

The triangulation contains two copies of solid torus T_3 (tetrahedra 0, 1, 4, and tetrahedra 5, 6, 7). But unlike in $\mathscr{T}_{3,3}$, in $\mathscr{U}_{3,3}$ they are not identified along their boundaries. Instead, they attach to a central complex consisting of tetrahedra 2, and 3.

We replace the two copies of T_3 in $\mathscr{U}_{3,3}$ by T_k and T_n, $k,n \geq 3$ odd, to obtain a family of triangulations $\mathscr{U}_{k,n}$ with $n+k+2$ tetrahedra, and three non-trivial $\mathbb{Z}_2$–torsion classes. The three normal surfaces F_1, F_2, and F_3 of each solid torus pair up symmetrically through the central 2–tetrahedra interface. This produces three normal surfaces, each consisting entirely of quadrilaterals. Because of the symmetric pairing of surfaces from the solid tori, $\mathscr{U}_{k,n}$ is isomorphic to $\mathscr{U}_{n,k}$. The three quadrilateral normal surfaces have Euler characteristic $-\frac{n+k}{2}$, $-\frac{n+k}{2}$, and -2 respectively. Their sum, hence, equals the negative of the number of tetrahedra in $\mathscr{U}_{k,n}$. We conjecture that these surfaces are taut, and hence these triangulations are minimal due to Theorem 1.

As with the main examples of this paper, the family $\mathscr{U}_{k,n}$ arises by Dehn surgery on a thrice-cusped hyperbolic 3–manifold. Indeed, replacing the two solid tori with a cusp each yields a 3-cusped ideal triangulation with Regina isomorphism signature `kLLPwLQkceefeijijijiiapuuxptxl`. According to SnapPy's identify function [4] this triangulation is decomposed into ten regular ideal hyperbolic tetrahedra and gives the thrice-cusped Platonic manifold `otet10_00015` (see [6]). It follows that the triangulation of the Platonic manifold is minimal.

It is an interesting open problem to determine further generalisations of this construction, using other ways of identifying the two solid torus triangulations T_k and T_n along a subcomplex.

References

1. Bachman, D., Derby-Talbot, R., Sedgwick, E.: Heegaard structure respects complicated JSJ decompositions. Mathematische Annalen **365**, 1137–1154 (2016) 14, 18
2. Bar-Natan, D., Morrison, S., et al.: The Knot Atlas. URL http://katlas.org 4
3. Burton, B.A., Budney, R., Pettersson, W., et al.: Regina: Software for low-dimensional topology. http://regina-normal.github.io/ (1999–2021) 3, 4, 5, 11, 13, 17, 20
4. Culler, M., Dunfield, N.M., Goerner, M., Weeks, J.R.: SnapPy, a computer program for studying the geometry and topology of 3-manifolds. Available at https://snappy.math.uic.edu (Nov 2021) 13, 20, 23
5. Frohman, C.: One-sided incompressible surfaces in Seifert fibered spaces. Topology Appl. **23**(2), 103–116 (1986). URL https://doi.org/10.1016/0166-8641(86)90032-5 6, 14, 18
6. Görner, M.: A census of hyperbolic platonic manifolds and augmented knotted trivalent graphs. New York J. Math. **23**, 527–553 (2017) 23
7. Guéritaud, F., Schleimer, S.: Canonical triangulations of Dehn fillings. Geom. Topol. **14**(1), 193–242 (2010). URL https://doi.org/10.2140/gt.2010.14.193 2, 5
8. Jaco, W., Rubinstein, J.H.: Layered-triangulations of 3-manifolds. ArXiv Mathematics e-prints (2006) 19, 20
9. Jaco, W., Rubinstein, J.H., Spreer, J., Tillmann, S.: $\mathbb{Z}_2$-Thurston norm and complexity of 3-manifolds, II. Algebr. Geom. Topol. **20**(1), 503–529 (2020). DOI 10.2140/agt.2020.20.503. URL https://doi.org/10.2140/agt.2020.20.503 2, 3
10. Jaco, W., Rubinstein, J.H., Spreer, J., Tillmann, S.: On minimal ideal triangulations of cusped hyperbolic 3-manifolds. J. Topol. **13**(1), 308–342 (2020). DOI 10.1112/topo.12127. URL https://doi.org/10.1112/topo.12127 1, 2, 13
11. Jaco, W., Rubinstein, J.H., Tillmann, S.: Coverings and minimal triangulations of 3-manifolds. Algebr. Geom. Topol. **11**(3), 1257–1265 (2011). URL https://doi.org/10.2140/agt.2011.11.1257 2
12. Jaco, W., Sedgwick, E.: Decision problems in the space of Dehn fillings. Topology **42**(4), 845–906 (2003). DOI 10.1016/S0040-9383(02)00083-6. URL https://doi.org/10.1016/S0040-9383(02)00083-6 14, 17
13. Kapovich, M.: Hyperbolic manifolds and discrete groups, *Progress in Mathematics*, vol. 183. Birkhäuser Boston, Inc., Boston, MA (2001) 13
14. Lackenby, M.: Word hyperbolic Dehn surgery. Invent. Math. **140**(2), 243–282 (2000). DOI 10.1007/s002220000047. URL https://doi.org/10.1007/s002220000047 4, 13
15. Otal, J.P.: Thurston's hyperbolization of Haken manifolds. In: Surveys in differential geometry, Vol. III (Cambridge, MA, 1996), pp. 77–194. Int. Press, Boston, MA (1998) 13
16. Rivin, I.: Combinatorial optimization in geometry. Adv. in Appl. Math. **31**(1), 242–271 (2003). DOI 10.1016/S0196-8858(03)00093-9. URL https://doi.org/10.1016/S0196-8858(03)00093-9 13
17. Rolfsen, D.: Knots and Links. AMS Chelsea Publishing Series. AMS Chelsea Pub. (2003). URL https://books.google.com.au/books?id=s4eGEecSgHYC 4
18. Thurston, W.P.: Hyperbolic structures on 3-manifolds. I. Deformation of acylindrical manifolds. Ann. of Math. (2) **124**(2), 203–246 (1986). DOI 10.2307/1971277. URL https://doi.org/10.2307/1971277 13
19. Waldhausen, F.: On irreducible 3-manifolds which are sufficiently large. Ann. of Math. (2) **87**, 56–88 (1968). DOI 10.2307/1970594. URL https://doi.org/10.2307/1970594 13

Chapter 2

MATRIX Program: Groups and Geometries

Concrete one complex dimensional moduli spaces of hyperbolic manifolds and orbifolds: Kleinian groups and the generalised Riley slices

Alex Elzenaar, Gaven Martin, and Jeroen Schillewaert

Abstract The Riley slice is arguably the simplest example of a moduli space of Kleinian groups; it is naturally embedded in $\mathbb{C}$, and has a natural coordinate system (introduced by Linda Keen and Caroline Series in the early 1990s) which reflects the geometry of the underlying 3-manifold deformations. The Riley slice arises in the study of arithmetic Kleinian groups, the theory of two-bridge knots, the theory of Schottky groups, and the theory of hyperbolic 3-manifolds; because of its simplicity it provides an easy source of examples and deep questions related to these subjects. We give an introduction for the non-expert to the Riley slice and much of the related background material, assuming only graduate level complex analysis and topology; we review the history of and literature surrounding the Riley slice; and we announce some results of our own, extending the work of Keen and Series to the one complex dimensional moduli spaces of Kleinian groups isomorphic to $\mathbb{Z}/a\mathbb{Z} * \mathbb{Z}/b\mathbb{Z}$ acting on the Riemann sphere, $2 \leq a, b \leq \infty$. The Riley slice is the case of two parabolic generators where $a = b = \infty$.

Key words: Kleinian groups, Schottky groups, hyperbolic geometry, Teichmüller theory, two-bridge links.
MSC classification (2020): 20H10, 30F40, 30F60, 57K32, 57R18.

Alex Elzenaar
The University of Auckland, New Zealand,
e-mail: aelz176@aucklanduni.ac.nz

Gaven Martin
Institute for Advanced Study, Massey University, Auckland, New Zealand,
e-mail: g.j.martin@massey.ac.nz

Jeroen Schillewaert
The University of Auckland, New Zealand,
e-mail: j.schillewaert@auckland.ac.nz

D. R. Wood et al. (eds.), *2021-2022 MATRIX Annals*, MATRIX Book Series 5,
https://doi.org/10.1007/978-3-031-47417-0_2

1 Introduction

The Riley slice is a one complex dimensional moduli space of Kleinian groups which can be embedded naturally in $\mathbb{C}$ and so is particularly amenable to study. In this article, we will define these words (Kleinian groups and moduli spaces), outline the history of the Riley slice, and finally briefly mention some of our own results which are in the process of appearing in full.

We have written this article in order to be accessible to the non-expert: we assume only a good understanding of areas which are plausibly familiar to most beginning graduate students (in particular, topology and complex analysis) and define almost everything else which we need from scratch. Though many of the results require only a little work to state, the study of the Riley slice has required a great deal of high-powered machinery developed in complex analysis and hyperbolic geometry in the late 20th and early 21st centuries (for instance: the theory of quasiconformal mappings and holomorphic motions, Ahlfors' null-measure conjecture, the theory of measured laminations, hyperbolic knot theory, and so forth). This is both an advantage (in that studying this rather concrete object gives an excuse to explore many of the great achievements of 20th century geometry) and a disadvantage (in that it is essentially impossible to give all detail in a short format like this article); as such we will give copious references (so that the reader inclined to explore has a variety of trails to follow) but sparse details (though we will always point towards a place where the more expert reader can look them up).

1.1 Agenda

Broadly speaking, each section of this article is split into two parts (which may correspond to one or more subsections): the first part of each section will be a general introduction to some broad area which is important to the study of the Riley slice (for example, hyperbolic knot theory). Then in the second part of the section, we will explain how this area is related to the Riley slice, giving some literature references for the details.

- The remainder of the introduction, consisting of Section 1.2, is dedicated to motivation for the study of the Riley slice from other areas of mathematics: computational group theory, arithmetic groups, and 3-dimensional geometry. It requires more background than the rest of the article and may be skipped by readers not familiar with the fields discussed.
- Section 2 defines the notion of a Kleinian group as a discrete group of Möbius transformations (2.1) and gives the definition of the closed Riley slice $\overline{\mathcal{R}}$ as a subsection of $\mathbb{C}$ together with the classical bounds on it (2.2).
- Section 3 describes how Kleinian groups act on hyperbolic space (3.1) and then gives a very fast introduction to hyperbolic knot theory (3.2) before explaining how the Riley slice appears naturally in the study of two-bridge knots (3.3).

- Section 4 first runs quickly through the classical theory of Riemann moduli space (4.1) before explaining the definition of a moduli space —or 'quasiconformal deformation space'—of Kleinian groups motivated by the theory of holomorphic motions (4.2) and then checking that the set $\overline{\mathcal{R}}$ is in fact the closure of the Riley slice as defined in Section 3 (4.3).
- Section 5 explains the philosophy of the Thurston boundary of quasi-Fuchsian space and measured laminations (5.1) before briefly sketching the Keen–Series coordinate system of the Riley slice (5.2).
- Finally, Section 6 indicates some of our recent work which is currently appearing in detail in preprint form: we have extended the Keen–Series coordinate system to the so-called *elliptic* Riley slices (6.1); we have found open neighbourhoods of pleating rays in the Riley slice (6.2) and the elliptic Riley slices; and we have established combinatorial and dynamical results concerning pleating rays and their defining polynomials (6.3).

This arrangement gives some flexibility in terms of the way in which the reader might approach this paper. Experts may wish to skim the sections introducing the Riley slice (2.2, 3.3, 4.3, and 5.2) and then read Section 6 in detail; students just learning the theory of Kleinian groups may want to restrict themselves to the sections introducing general aspects of the theory and only read some of the Riley slice material for context.

1.2 Motivation and applications

We discuss in this section three different motivations for this work, in order to contextualise it. In the following loose discussion we avoid many technical details and definitions; those that are important for the understanding of the Riley slice and associated work can be found in the body of this article.

The discreteness problem

Our first motivation is the decision problem for discrete groups, and in particular subgroups of $\mathrm{SL}(2,\mathbb{C})$ or $\mathrm{PSL}(2,\mathbb{C})$; the latter being isomorphic as a topological group to the orientation preserving isometry group of hyperbolic 3-space and hence containing the fundamental groups of the most interesting class of three-dimensional geometric objects, namely hyperbolic 3-manifolds and orbifolds. The problem is this:

Decision problem. *Given matrices* $A_1, A_2, \ldots, A_n \in \mathrm{PSL}(2,\mathbb{C})$*, decide if the group* $\langle A_1, A_2, \ldots, A_n \rangle$ *is discrete, and if so give a presentation for it.*

We skip a discussion of what "decide" means (see for example Chapter 9 of [136]), but basically we seek a practical way to answer this question.

If there is simply one generator, then the answer is "easy"—$\langle A\rangle$ is discrete if and only if $\mathrm{tr}A \notin (-2,2)$ or if $\mathrm{tr}A \in (-2,2)$ (in the latter case, $\mathrm{tr}A = 2\cos(r\pi/s)$ for some $r,s \in \mathbb{Z}$. If there are three or more generators, then the problem is likely completely intractable. This leaves the case of two-generateds. The decision problem for two-generated subgroups of $\mathrm{PSL}(2,\mathbb{R})$ has been completely resolved by the work of Gilman [56, 58], and the classification of discrete two-generated Fuchsian groups into isomorphism classes can be found in earlier work of Purzitsky and Rosenberger, [121].

This leaves the more general question of groups generated by two matrices in $\mathrm{PSL}(2,\mathbb{C})$. This problem goes back to to the 1950s, and we discuss its history in some detail in the main body of the article. For our discussion here, it suffices to note that it started with the case that both generators are parabolic. There is a good geometric reason for examining this case first as it is associated with the moduli space of the four-times punctured sphere, the simplest Riemann surface with a non-trivial Teichmüller space. A theorem of Jørgensen [61] asserts that any subgroup of $\mathrm{PSL}(2,\mathbb{C})$ is discrete if and only if all of its two-generated subgroups are discrete (a result true in much more generality [91]); this gives utility in an effective way to determine discreteness of two-generated groups.

In a real sense the special case of two parabolic generators is now resolved. Up to conjugacy—which preserves both discreteness and isomorphism type—there is a single complex parameter which determines the group. The closure of the Riley slice, $\overline{\mathcal{R}}$, consists of those parameters where the group is discrete and free (there is a lot of theory behind this statement). The boundary of $\overline{\mathcal{R}}$ is a closed Jordan curve [109], but not much is known beyond this. In particular, we ask if we can determine (in an effective, or algorithmic, way) whether a given complex number lies in $\overline{\mathcal{R}}$. This set has a lot of additional structure: in the mid 1990s Keen and Series [67] gave geometrically relevant coordinates for the interior of this set, primarily through "pleating rays"—a dense set of algebraically defined curves in $\mathbb{C}$ which "land" at a dense set of the boundary $\partial\mathcal{R}$ (the so-called "cusp" groups, which are dense by a famous result of McMullen [101]). Our results which we announce here give canonical open neighbourhoods of these rays which fill $\mathcal{R}$. Using these one can construct an effective process to decide if a group is in $\mathcal{R}$ or not. A group on the boundary $\partial\overline{\mathcal{R}}$ is either a cusp group (and therefore the associated parameter is a root of a countable list of "Farey polynomials" defined below) or is geometrically infinite with limit set the entire complex plane (and seemingly intractable computationally). In $\mathbb{C}\setminus\overline{\mathcal{R}}$ the question is resolved by the recent works [12, 10] where Aimi, Lee, Sakai, and Sakuma, and also Akiyoshi, Sakuma, Wada, and Yamashita, resolve a conjecture of Agol. They identify with topological methods descriptions of all the quotient orbifolds (from which one may write down explicit presentations for the holonomy groups) for the parameters in $\mathbb{C}\setminus\overline{\mathcal{R}}$ which represent discrete groups. These associated groups cannot be free and all these groups lie on "extended pleating rays"—where a Farey word is the identity (giving knot and link groups) or elliptic (giving a so-called Heckoid group). Away from these parameters, the groups are not discrete and their structure is quite wild [92].

Our work seeks to extend the programme of understanding the Riley slice to groups generated by two elements of finite order. Such groups include all generalised triangle groups, groups with a presentation of the form

$$\langle x,y : x^a = y^b = R(x,y)^r = 1\rangle$$

and many groups beside. A reasonable conjecture seems to be the following.

Conjecture 1. Let Γ be a discrete representation of $\langle x,y : x^a = y^b = 1\rangle \simeq \mathbb{Z}/a\mathbb{Z} * \mathbb{Z}/b\mathbb{Z}$ in $\mathrm{PSL}(2,\mathbb{C})$ and suppose Γ is not faithful (i.e. not isomorphic to $\mathbb{Z}/a\mathbb{Z} * \mathbb{Z}/b\mathbb{Z}$). Then there is a Farey word $w_{r/s}(x,y) \in \Gamma$ which is either the identity or elliptic.

In the case that the Farey word is the identity, one would expect that the group is the orbifold fundamental group of $(a,0),(b,0)$ Dehn surgery on a two-bridge knot or link. Otherwise, the group should have a Heckoid-like structure analogous to that outlined in [12, 10].

Arithmetic groups

This brings us to another application. Since the very beginnings of the theory of discrete groups, mathematicians have observed deep connections between arithmetic and geometry: for instance, through special functions and the modular group $\mathrm{PSL}(2,\mathbb{C})$; the Selberg trace formula (generally an expression for the character of the unitary representation of a Lie group G on the space $L^2(G/\Gamma)$ for a discrete lattice in G, but initially for the traces of powers of a Laplacian defining the Selberg zeta function and the strong connections with explicit formulas in prime number theory, [128]); and Margulis super-rigidity, which is a partial converse to the Borel and Harish-Chandra theorem that an arithmetic subgroup in a semisimple Lie group is of finite covolume—Margulis showed that any irreducible lattice in a semisimple Lie group of real rank larger than two is arithmetic.

Roughly, an arithmetic group is a group obtained as the integer points of an algebraic group such as $\mathrm{SL}(n,\mathbb{Z})$. A subgroup of $\mathrm{PSL}(2,\mathbb{C})$ is arithmetic if it is commensurable with another subgroup of $\mathrm{PSL}(2,\mathbb{C})$ isomorphic to such a group. We refer to the book of Machlachlan and Reid [83] for a comprehensive account of the theory of arithmetic subgroups of $\mathrm{PSL}(2,\mathbb{C})$. Expanding on results of Borel [25] a necessary and sufficient criteria has been given to decide if a two-generated subgroup of $\mathrm{PSL}(2,\mathbb{C})$ is a subgroup of an arithmetic group [53]. From this in [80] it was shown that there are only finitely many subgroups of $\mathrm{PSL}(2,\mathbb{C})$ which are either arithmetic or a subgroup of an arithmetic group which is not free, and which is generated by two elliptic or parabolic elements. More generally we have the following conjecture.

Conjecture 2. There are only finitely many two-generated arithmetic subgroups of $\mathrm{PSL}(2,\mathbb{C})$.

This might not seem too surprising as there are, for instance, only finitely many arithmetic subgroups of $\mathrm{PSL}(2,\mathbb{C})$ which lie in groups generated by reflections, [4,

5]. It is not difficult to see that every two-generated group $\mathrm{PSL}(2,\mathbb{C})$ is a subgroup of a group generated by three involutions (elements of order two). There seems a big difference between reflections and involutions though, and so the conjecture remains.

There are infinitely many arithmetic groups generated by three elements. One may see this, for instance, by noting the figure 8 knot group is generated by two parabolic elements, is arithmetic, and fibres over the circle (these sorts of things were first observed by Riley [115, 116] and Jørgensen [62]) and unwrapping the fibre to give a sequence of higher, but finite,index subgroups generated by three elements and these which are obviously arithmetic.

Rather surprisingly, there are only four arithmetic groups generated by two parabolic elements [52], and they are all fundamental groups of knot or link complements. A programme was instigated to identify all the arithmetic groups generated by two elliptic or parabolic elements: for instance, there are 40 such non-uniform lattices (non-compact quotient) in $\mathrm{PSL}(2,\mathbb{C})$ [82]. This programme will soon be completed using the work announced in this paper. Using very recent discriminant bounds on fields with one complex place, the arithmetic criterion [53] identifies finitely many (though sometimes many thousands) parameters. The issue is then deciding which of these possibilities are free groups, and this is the primary question which we give an effective way to decide. Previous efforts in this direction tried to construct a finite volume fundamental domain for the group [82], or show the group in question has infinite index in a Bianchi group or other known maximal arithmetic group [36].

Geodesic neighbourhoods in hyperbolic 3-orbifolds

Finally we discuss one more potential application. In an important work [50], Gabai gave the following very strong generalisation of the Mostow rigity theorem in three dimensions, provided a certain condition holds. This is the concept of existence of "non-coalescable insulator families"—structures associated to an embedded geodesic loop in a hyperbolic 3-manifold M. If M has this property, if M is a closed irreducible 3-manifold, and if $f : M \to N$ is a homotopy equivalence, then f is homotopic to a homeomorphism. If f and g are homotopic homeomorphisms of N, then they are isotopic and finally the space of hyperbolic metrics on N is path connected. Gabai constructs a non-coalescable insulator family for a simple closed geodesic whenever it has an embedded tubular neighbourhood in M of radius larger than $\log(3)/2$. Not every shortest geodesic has a tubular neighborhood of this radius; for example, the 3-manifold conjectured to have the third-smallest volume obtained by $(-3,2)$, $(-6,1)$-surgery on the left-handed Whitehead link. It is believed that only six hyperbolic 3-manifolds have shortest geodesics which do not satisfy this condition, and for them other methods produce non-coalescable insulator families.

The question of $\log(3)/2$ embedded neighbourhoods is really a question about two-generated groups, with one generator of order two. To see this, observe that a geodesic will have such an embedded neighbourhood if its nearest translate (when we look at the group acting on hyperbolic 3-space) is further away than $\log 3$. Consider the group generated by the short geodesic g and its nearest translate h. As g and h are conjugate, there is an involution ϕ so that $\phi g \phi = h$, and $\langle g, \phi \rangle$ contains $\langle g, h \rangle$ with index at most two. The space of all such groups $\langle g, \phi \rangle$ up to conjugacy, discrete or not, has two complex dimensions. Further, if the shortest geodesic in M is quite short, then there is always an embedded $\log(3)/2$ neighbourhood (this is an elementary consequence of Jørgensen's inequality). On the other hand, if the shortest geodesic is long, then its nearest translate is obviously far away. Thus one should explore a compact part of the moduli space of discrete groups generated by two elements, one of which has order two, and whose axes (the invariant hyperbolic line projecting to the geodesic representative in M) are relatively close. We expect there are only finitely many such groups and very few manifold groups among them. This is very roughly the setup for the computational exploration [51] which establishes the results mentioned above in complete generality. If there are only finitely many such exceptions in this region, then the remaining groups cannot be discrete. A contradiction to discreteness can be found by identifying a word in the group sufficiently close to the identity. Distance to the identity is a continuous function of the parameters, so a word sufficiently close to the identity persists, and rules out nearby groups as possibilities as well, removing an open region from the search space. But how should one find these words certifying non-discreteness? These words are called "killer words" in [51] and are obtained via an ad-hoc process (really a search itself). In [54, 89] a similar search was carried out to identify the minimal covolume lattice in $\mathrm{PSL}(2,\mathbb{C})$ using "good words". Given that Farey words are likely relators in such discrete groups (as per Conjecture 1) and that, as we will describe below, the set of roots of Farey words is dense in the complement of the space of discrete free and geometrically finite groups, concrete descriptions of these root sets may well facilitate these searches. In addition, clearing up the relationships between killer words, good words, Farey words and even the palindromic words (see [57]) used in discreteness criteria might yield useful insights into questions of discreteness.

2 Some free groups of matrices

The title of this section is taken from the title of the 1955 paper of Brenner [26], who (to our knowledge) was the first to explicitly study the family of matrices which we are interested in here in the way that we are considering. We begin in Section 2.1 by defining Kleinian groups as Poincaré defined them; then in Section 2.2 we define the particular set of groups which we will study in the sequel.

2.1 Kleinian groups

The notion of a Kleinian group was first defined by Poincaré in 1883 [111], as a group of fractional linear transformations which acts discontinuously on an open subset of the Riemann sphere; the name 'Kleinian' was given by Poincaré after Klein objected to the name 'Fuchsian' which Poincaré had previously given to discrete subgroups of $\mathrm{Isom}(\mathbb{H}^2)$ [87, p.xvi]. We will take a slightly more general view than Poincaré, but before giving our definition we recall some facts from complex analysis and define all of the terms which we have just used. Two nice introductory texts which take this classical angle are [16] and [98]. More modern surveys of the field from the view of Kleinian groups (rather than, say, 3-manifolds) include those by Matsuzaki and Taniguchi [100], by Marden [87], by Kapovich [63], and the very nice paper by Series [130].

We use $\hat{\mathbb{C}}$ to denote the Riemann sphere $\mathbb{C}\cup\{\infty\}$; it is well-known that if $f:\hat{\mathbb{C}}\to\hat{\mathbb{C}}$ is an entire homeomorphism then there exist $a,b,c,d\in\mathbb{C}$ satisfying $ad-bc\neq 0$ such that

$$f(z)=\frac{az+b}{cz+d}$$

for all $z\in\hat{\mathbb{C}}$ (with the obvious conventions $f(-d/c):=\infty$ and $f(\infty):=a/c$). Maps of this form are called *fractional linear transformations*, or *Möbius transformations* (c.f. Chapter 3 of [9]). We denote the group of all fractional linear transformations by $\mathbb{M}$.

Observe next that $\hat{\mathbb{C}}$ is identified with $\mathbb{PC}^1$ (one-dimensional projective space over $\mathbb{C}$). Consider the action of the projective transformation

$$A=\begin{bmatrix} a & b \\ c & d \end{bmatrix}\in \mathrm{PSL}(2,\mathbb{C})$$

on $\xi=(u,v)^t\in\mathbb{PC}^1$ (where we use homogenous coordinates):

$$A\xi=\begin{bmatrix} a & b \\ c & d \end{bmatrix}\begin{bmatrix} u \\ v \end{bmatrix}=\begin{bmatrix} au+bv \\ cu+dv \end{bmatrix};$$

assuming $v\neq 0$ (so $\xi\neq\infty$) we have that A acts on $\mathbb{C}$ as

$$A(u/v)=\begin{bmatrix} a(u/v)+b \\ c(u/v)+d \end{bmatrix}=\frac{a(u/v)+b}{c(u/v)+d}$$

and so the action of $\mathrm{PSL}(2,\mathbb{C})$ on $\mathbb{PC}^1$ agrees with the action of $\mathbb{M}$ on $\hat{\mathbb{C}}$; just as we identify the two spheres, we identify the two groups. The matrix group $\mathrm{PSL}(2,\mathbb{C})$ has an obvious topology (identify $\mathrm{SL}(2,\mathbb{C})$ with a subset of $\mathbb{C}^4$ and take the Euclidean norm; then this norm behaves well when passing to the projective quotient), and we use this topology also for $\mathbb{M}$.

Definition 1. A *Kleinian group* is a discrete subgroup of $\mathbb{M}$.

We now define the term 'discontinuous' which we used in giving Poincaré's definition of a Kleinian group, and explain why our definition is slightly more general. Let G be a group acting by homeomorphisms on a topological space X. We can view this action as a dynamical object, and ask for the orbits of points under the group action. We say that $x \in X$ is a *limit point* of G if there is some $u \in X$ such that x is an accumulation point of the orbit Gu; the set of all limit points is denoted by $\Lambda(G,X)$. In our case, the set X will always be $\hat{\mathbb{C}}$ and so we simply write $\Lambda(G)$ for the limit set of a group of Möbius transformations G. Directly from its definition the limit set is always a closed set. We exhibit an example as Fig. 1.

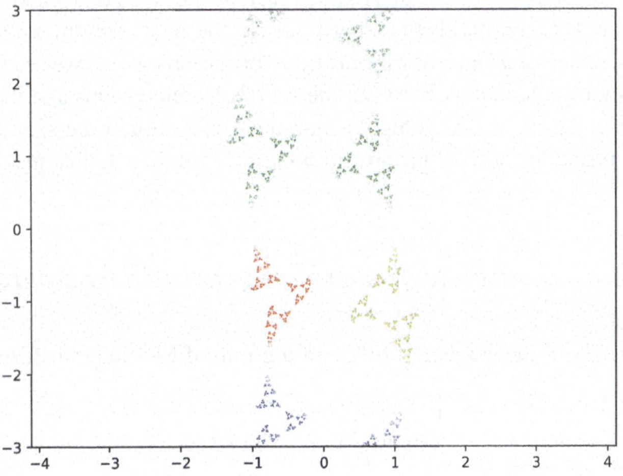

Fig. 1: The limit set of a Kleinian group, generated by two matrices with respective traces $2 \pm 0.2i$ (c.f. Figure 8.5(iv) of [106]).

For a proof of the following proposition, see Section II.E of [98].

Proposition 1. *Suppose that $G \leq \mathbb{M}$. The following two conditions are equivalent for $x \in \hat{\mathbb{C}}$:*

1. $x \in \hat{\mathbb{C}} \setminus \Lambda(G)$
2. *There exists a neighbourhood U of x such that $U \cap gU \neq \emptyset$ for only finitely many $g \in G$.* □

If x satisfies condition (2), then we say that G acts *discontinuously* at x; the set of all points at which G acts discontinuously, which is identical to the open set $\hat{\mathbb{C}} \setminus \Lambda(G)$ by the proposition, is denoted by $\Omega(G)$ and is called the *ordinary set* of G. If G is a Kleinian group and $\Omega(G)$ is non-empty, then $\Omega(G)/G$ naturally has a complex structure; if G is finitely generated, by Alhfors' finiteness theorem the space $\Omega(G)/G$ is a finite union of Riemann surfaces, each with finitely many punctures and no other boundary components [7, 8] (for a modern proof and references to several other proofs see [63, §8.14]).

Poincaré's definition of a Kleinian group was a group G such that $\Omega(G) \neq \emptyset$; it is not too hard to prove that if $G \leq \mathbb{M}$ acts discontinuously at a point $x \in \hat{\mathbb{C}}$, then G is discrete (one uses the fact that a subgroup of $\mathbb{M}$ is discrete iff there are no nontrivial sequences in the group converging to the identity, see [98, II.C.3]); on the other hand, there are discrete groups of Möbius transformations which act discontinuously nowhere [98, VIII.G], so our definition is strictly more general than Poincaré's. (The advantage of our definition is that it includes groups which act discontinuously on the half-space model of hyperbolic space whose sphere at infinity is $\hat{\mathbb{C}}$, even if they do not act discontinuously at infinity; we will take this story up again in Section 3.1.)

As we outlined in the introduction, in general, the question of deciding whether an arbitrary subgroup $G \leq \mathbb{M}$ is discrete (as well as the related problems of computing $\Omega(G)$ and $\Lambda(G)$ explicitly) is a hard one. In the next section, we will describe the particular space of groups of Möbius transformations which we are interested in; we will see that the boundary between the set of discrete groups and the set of non-discrete groups in this space is highly intricate, even though the space is arguably the simplest possible space of groups for which we might ask this question.

2.2 The space of discrete, free groups generated by 2*-parabolics*

We return briefly to the subject of orbits of a group of Möbius transformations.

Proposition 2. *For $\lambda \in \mathbb{C} \setminus \{0\}$, define the function f_{λ^2} by $f_{\lambda^2}(z) := \lambda^2 z$. Every Möbius transformation is conjugate by an element of $\mathbb{M}$ either to $z \mapsto z+1$, or to exactly one of the functions f_{λ^2}.* □

This gives a taxonomy of Möbious transformations. Let $g \in \mathbb{M}$ be non-trivial, that is g is not the identify; then g is called

- *parabolic*, if it is conjugate to the translation $z \mapsto z+1$ (equivalently, $\mathrm{tr}^2 g = 4$);
- *elliptic*, if it is conjugate to some rotation $z \mapsto \lambda^2 z$ with $|\lambda| = 1$ (equivalently, $\mathrm{tr}^2 g \in [0,4)$);
- *hyperbolic*, if it is conjugate to some dilatation $z \mapsto \lambda^2 z$ with $\lambda \in \mathbb{R}$ and $|\lambda| \neq 1$ (equivalently, $\mathrm{tr}^2 g \in (4,\infty)$);
- *strictly loxodromic*, if it is conjugate to $z \mapsto \lambda^2 z$ with $\lambda \in \mathbb{C} \setminus \mathbb{R}$ and $|\lambda| \neq 1$ (equivalently, $\mathrm{tr}^2 g \in \mathbb{C} \setminus [0,\infty)$).

Parabolic elements have exactly one fixed point on the Riemann sphere $\hat{\mathbb{C}}$; all other non-identity elements have exactly two fixed points, and in the case of hyperbolic and strictly loxodromic elements (which together form the class of *loxodromic* elements) these fixed points form a source-sink pair (Fig. 2).

In 1947, Sanov [125] gave the following example of a free Kleinian group:

$$\left\langle \begin{bmatrix} 1 & 2 \\ 0 & 1 \end{bmatrix}, \begin{bmatrix} 1 & 0 \\ 2 & 1 \end{bmatrix} \right\rangle.$$

Fig. 2: The flow of points on $\hat{\mathbb{C}}$ under the four kinds of elements of $\mathbb{M}$.

This group was of interest because it was a free subgroup of $\mathrm{PSL}(2,\mathbb{Z})$; previous work had found free subgroups of $\mathbb{M}$, but all of these had transcendental entries. In 1955, Brenner [26] extended this result, showing that whenever $|m| \geq 2$,

$$G_m = \left\langle A_m = \begin{bmatrix} 1 & m \\ 0 & 1 \end{bmatrix}, B_m = \begin{bmatrix} 1 & 0 \\ m & 1 \end{bmatrix} \right\rangle.$$

is free. Observe that both generators of this group are parabolic (since they both have trace 2). Let $T \in \mathrm{PSL}(2,\mathbb{C})$ be a matrix conjugating A_m to the matrix

$$\begin{bmatrix} 1 & 1 \\ 0 & 1 \end{bmatrix}$$

which represents $z \mapsto z+1$; for example, we can choose

$$T = \begin{bmatrix} \sqrt{m} & 0 \\ 0 & \frac{1}{\sqrt{m}} \end{bmatrix}$$

and then compute that

$$T^{-1} G_m T = \left\langle \begin{bmatrix} 1 & 1 \\ 0 & 1 \end{bmatrix}, \begin{bmatrix} 1 & 0 \\ m^2 & 1 \end{bmatrix} \right\rangle.$$

There is a famous inequality due to Shimizu [132] and Leutbecher [75] (see also [98, II.C.5]) which gives a necessary condition for a group of this form to be discrete:

Lemma 1 (Shimizu–Leutbecher lemma). *Suppose $A, B \in \mathrm{PSL}(2,\mathbb{C})$ are of the form*

$$A = \begin{bmatrix} 1 & 1 \\ 0 & 1 \end{bmatrix}, B = \begin{bmatrix} a & b \\ c & d \end{bmatrix}.$$

If $\langle A, B \rangle$ is discrete, then either $c = 0$ or $|c| \geq 1$. □

Combining this with the result of Brenner, we obtain bounds on the set of $\rho \in \mathbb{C}$ such that

$$\Gamma_\rho := \left\langle X = \begin{bmatrix} 1 & 1 \\ 0 & 1 \end{bmatrix}, Y_\rho = \begin{bmatrix} 1 & 0 \\ \rho & 1 \end{bmatrix} \right\rangle$$

is free and discrete. Let $\overline{\mathcal{R}}$ be this set:

Lemma 2. *Let $\rho \in \mathbb{C}$. If $\rho \in \overline{\mathcal{R}}$, then $|\rho| \geq 1$. On the other hand, if $|\rho| \geq 4$, then $\rho \in \overline{\mathcal{R}}$.* □

We will call the set $\overline{\mathcal{R}}$ the *closed Riley slice* (we will explain this terminology in the next section). Using techniques which we will reference in Section 5 and Section 6, we are able to draw a picture of this set which shows that it has a very intricate structure which is not captured by the crude bounds of Lemma 2; the closed Riley slice is the region of $\mathbb{C}$ which is outside the shaded 'eye' of Fig. 3, and the bounds of Lemma 2 are the two red circles.

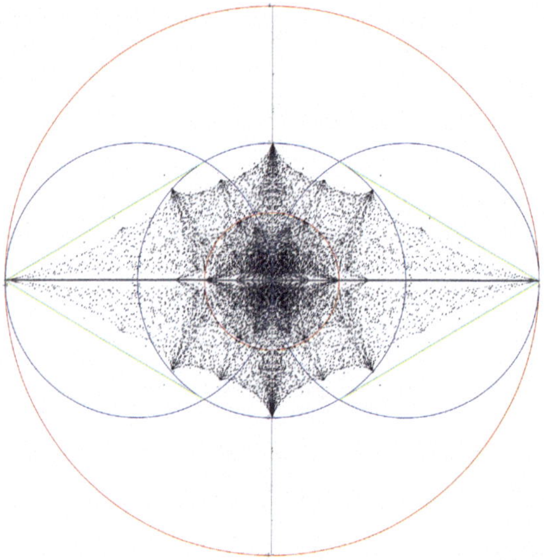

Fig. 3: Crude bounds on the Riley slice.

In 1958, Chang, Jenning, and Ree [35] improved the bound on the interior of the Riley slice from '$|\rho| \geq 4$' to 'ρ lies outside the three circles of radius 2 centred at -2, 0, and 2 respectively'; this is shown in blue in Fig. 3. A couple of years later, Ree [114] showed various density results for the complement of the Riley slice: for instance, that non-free groups are dense in the closed unit disc $\{\rho \in \mathbb{C} : |\rho| \leq 1\}$; and that the two open intervals $(-4,4)$ and $(-2i,2i)$ are contained in open sets in which non-free groups are dense.

Finally, in 1969, Lyndon and Ullman [78] improved the interior bound once more; this bound is shown in green on Fig. 3.

Theorem 1 (Lyndon and Ullman). *Let* LU *be the convex hull in $\mathbb{C}$ of the circle $\{z \in \mathbb{C} : |z| = 2\}$ and the two points ± 4. Then $\overline{\mathcal{R}}$ is contained in $\overline{\mathbb{C} \setminus \mathrm{LU}}$.* □

For some further study of the structure of the Riley slice boundary using Lyndon and Ullman's result, see Section 4.2 of [44]. Beardon has given improvements of some of the other results of Lyndon and Ullman on the set of free groups of the form Γ_ρ [17].

Some more modern bounds on sets intimately related to the Riley slice as defined here (and the elliptic Riley slice which we will describe in Section 6) are given in [93]. The non-discrete groups of the form Γ_ρ have also been studied, for example by Beardon [18] and Martin [92]; the non-free groups have been studied by Bamberg [15]; and the arithmetic groups (whose definition would take us too far afield, see the elementary book by Voight [145] and the more difficult book by Maclachlan and Reid [83]) have been studied by Gehring, Maclachlan, and Martin [52], see also the discussion and citations in the introduction to [44]. Work on parabolic conjugacy classes in the groups Γ_ρ, whether free or not, was done by Gilman [55] in an attempt to explain the patterns of density seen in pictures of the Riley slice exterior like those of Fig. 3 above (which was drawn, following Keen and Series [67], by drawing the set of groups where certain *Farey words* become accidentally parabolic—we discuss this more in Section 5 below).

3 Riley's study of knot groups

In this section we study hyperbolic geometry, and show that the Riley slice appears naturally in association with representations of the fundamental groups of two-bridge knot groups.

3.1 Hyperbolic geometry

We do not give an account here of the interesting history of 'non-Euclidean geometry'; some reading on this matter includes the first chapter of [39] as well as the monograph [122]. We simply note that hyperbolic geometry was first studied as the geometry in which Euclid's *Postulate V*

> For every point A, and any line r not through A, there is not more than one line through A, in the plane spanned by A and r, which does not meet r

(here formulated as in [40, §16.1]) is replaced by the postulate

> For every point A, and any line r not through A, there is more than one line through A, in the plane spanned by A and r, which does not meet r,

and in this form the first people to study it seriously were Lobačevskiĭ [77] and J. Bolyai [23].

For us, *hyperbolic n-space* is the upper half-plane $\mathbb{H}^n := \{(x_1,\dots,x_n) \in \mathbb{R}^n : x_n > 0\}$ endowed with the metric given by the differential

$$ds^2 = \frac{dx_1^2 + \cdots + dx_n^2}{x_n^2};$$

geodesics in this metric are the intersections with $\mathbb{H}^n$ of lines and circles in $\mathbb{R}^n$ orthogonal to the plane $x_n = 0$. This space is isometric (after taking a sphere inversion through a sphere tangent to $x_n = 0$, translating, and scaling) to the space consisting of the interior of the n-disc $\mathbb{B}^n := \{x \in \mathbb{R}^n : |x| < 1\}$ with the metric

$$ds^2 = \frac{2dx_1^2 + \cdots + 2dx_n^2}{(1 - dx_1^2 - \cdots - dx_n^2)^2},$$

in which geodesics are circle arcs orthogonal to the boundary S^{n-1} (see Fig. 4).

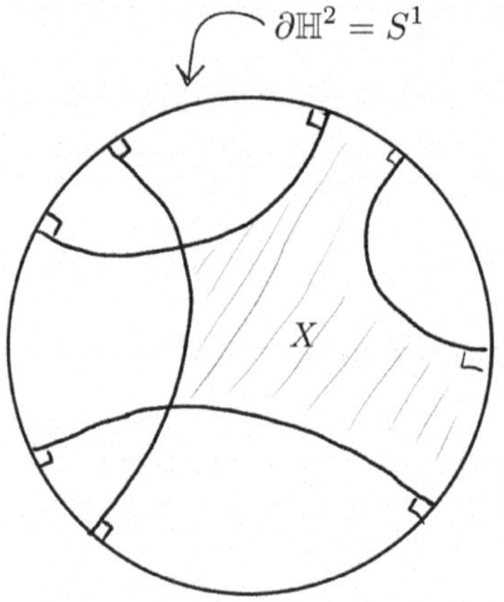

Fig. 4: A polygon X in the ball model of $\mathbb{H}^2$, with four edges given by hyperbolic geodesic segments and two edges on the sphere S^1 at infinity.

We call $\mathbb{H}^n$ the *half-space model* and $\mathbb{B}^n$ the *ball model* of hyperbolic n-space; these are two views of the unique Riemann manifold of constant curvature -1. Some introductory texts on hyperbolic geometry include [40, Chapter 16] (which takes an axiomatic view), [16, Chapter 7] (which takes an analytic view like that we take here), and [113, Chapters 1–5] (which starts with the point of view of Lorentzian space and proceeds to discuss both axiomatic and geometric views).

Clearly both models of hyperbolic space have some kind of 'boundary': this is a common feature of metric spaces of non-positive curvature [27, Chapters 8 and 9] and in this case, the boundary of hyperbolic space is a sphere (either $\mathbb{R}^{n-1} \cup \{\infty\}$ for the half-plane model, or the usual unit sphere S^{n-1} for the ball model). We will primarily be interested in hyperbolic 3-space, in which case we can naturally identify $\mathbb{R}^2$ with $\mathbb{C}$ and hence the sphere at infinity with the Riemann sphere $\hat{\mathbb{C}}$.

Let ϕ be an (orientation-preserving) isometry of $\mathbb{H}^3$. Then ϕ extends naturally to a conformal homeomorphism (hence Möbius transformation) on the boundary $\partial\mathbb{H}^3 = \hat{\mathbb{C}}$—this is a non-trivial observation but isn't too hard to check, see [98, IV.A]—and in fact this map $\mathrm{Isom}(\mathbb{H}^3) \to \mathbb{M}$ is invertible (which is not a feature of all metric spaces of non-positive curvature). The inverse map is the so-called *Poincaré*

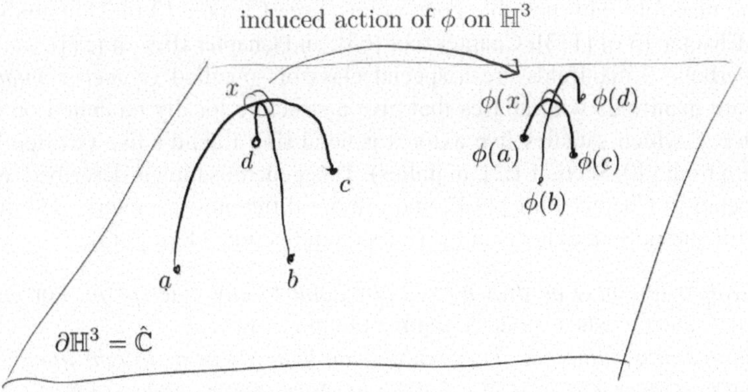

Fig. 5: A conformal map $\phi : \hat{\mathbb{C}} \to \hat{\mathbb{C}}$ extends uniquely to an hyperbolic isometry on $\mathbb{H}^3$: the map ϕ, in sending the points a, b, c, d to their images, induces a map on the hyperbolic segments $[a, c]$ and $[b, d]$; if the original segments intersect at some $x \in \mathbb{H}^3$, then their images intersect at some point which is defined to be $\phi(x)$. The difficulty is in checking that, for any $x \in \mathbb{H}^3$, different choices of segments intersecting at x induce the same image $\phi(x)$.

extension, where to define the action of a conformal map ϕ on the boundary $\hat{\mathbb{C}}$ on a point $x \in \mathbb{H}^n$ one first finds a pair of intersecting geodesics through x, computes the images of the endpoints at infinity of these geodesics under ϕ, and uses these to draw two new geodesics which will intersect at the image of x under the extension of ϕ to $\mathbb{H}^3$; this process is depicted in Fig. 5 (see also the final paragraph of [98, A.13], or for more detail [16, §3.3] and [113, §4.4]). In any case, once we accept that there is a natural geometry-respecting bijection between $\mathbb{M}$ and the group of isometries of $\mathbb{H}^3$ (which in fact is compatible with the compact-open topology on the isometry group) then we may give an alternative definition of a Kleinian group: a discrete group of isometries of $\mathbb{H}^3$.

We now follow Thurston [143] (see also [113, Chapter 8] and [19]) to define *hyperbolic 3-manifolds*. A hyperbolic 3-manifold is a Hausdorff second-countable topological space M equipped with a *hyperbolic atlas*: an open cover U_α with topological embeddings $\phi_\alpha : U_\alpha \to \mathbb{H}^3$ (α in some index set A) where every transition map $\phi_\beta \phi_\alpha^{-1}$ ($\alpha, \beta \in A$) is the restriction of some isometry of $\mathbb{H}^3$ to the domain of definition. The holonomy group of M is a structure which detects the tiling that is induced when M is rolled out onto its universal cover $\mathbb{H}^3$ in a way that respects the geometry (this process is called *developing*; see [143, §3.4] for an intuitive description, much more detail may be found in [113, §8.4]). If Γ is the holonomy group of a connected and metrically complete hyperbolic 3-manifold M, then it is a Kleinian group and M is isometric to $\mathbb{H}^3/\Gamma$ [113, Theorem 8.5.9]. Conversely, if Γ is a Kleinian group then $\mathbb{H}^3/\Gamma$ is a hyperbolic manifold if Γ is torsion-free; if Γ is allowed to have torsion then the quotient is a hyperbolic orbifold (i.e. a hyperbolic manifold with possible cone points—see Chapter 13 of Thurston's notes [140], Chapter 13 of [113], Chapter 6 of [63], and Chapter III.$\mathcal{G}$ of [27]).

Hyperbolic 3-manifolds are a special class of so-called *geometric manifolds*. These are manifolds with atlases that give a structure locally modelled on a metric space X which satisfies five axioms generalising Euclid's five (written below, verbatim from [40, Section 1.2], in italics). This generalisation is described in much more detail in Chapter 8 of [113], and a more differentio-geometric definition of geometric metric spaces is given by Thurston in Section 3.8 of [143].

1. *A straight line may be drawn from any point to any other point.* For any pair $x, y \in X$ there exists a geodesic segment joining x to y.
2. *A finite straight line may be extended continuously in a straight line.* Given a geodesic segment $\gamma : [0,1] \to X$ there exists a unique geodesic line $\hat{\gamma} : \mathbb{R} \to X$ with $\hat{\gamma} \restriction_{[0,1]} = \gamma$.
3. *A circle may be described with any centre and any radius.* There exists a continuous function $\sigma : \mathbb{R}^n \to X$ and a real number $\varepsilon > 0$ such that σ is a homeomorphism of $B(0,\varepsilon)$ onto $B(\sigma(0),\varepsilon)$, such that for any $x \in S^{n-1}$ the map $\gamma : \mathbb{R} \to X$ defined by $\gamma(t) = \sigma(tx)$ is a geodesic line such that $\gamma \restriction_{[-\varepsilon,\varepsilon]}$ is a geodesic segment.
4. *All right angles are equal to each other.* X is homogeneous (i.e. for all $x, y \in X$ there exist isometric neighbourhoods U of x and V of y).

There are eight such geometries in three dimensions: Euclidean space $\mathbb{E}^3$, spherical space S^3, and hyperbolic space $\mathbb{H}^3$; the two product geometries $S^2 \times \mathbb{E}^1$ and $\mathbb{H}^2 \times \mathbb{E}^1$; the universal cover of $\mathrm{SL}(2,\mathbb{R})$; solvegeometry Sol; and nilgeometry Nil. Some excellent visualisation work has been done with these geometries; for instance, the work of Coulon, Matsumoto, Segerman, and Trettel [37, 38].

It is a great insight of Thurston that 'most' geometric manifolds are hyperbolic. He established this in many interesting cases, particularly the Haken case (a *Haken manifold* is a compact, orientable, irreducible 3-manifold that contains an orientable, incompressible surface;[1] see [63] for the details) and promoted the *geometrisation conjecture* which was completely proved by Grigori Perelman using ideas of Richard S. Hamilton who developed the the Ricci flow (see the book [105]).

3.2 Knots

The theory of knots was found to be related to hyperbolic geometry by Robert Riley and William Thurston in the mid 1970s. In this section, we define some of the basic concepts from geometric knot theory and explain exactly how hyperbolic geometry comes into it all. The three most useful references for the parts of knot theory which

[1] If S is a properly embedded surface in a 3-manifold M, then a *compression disc* D (for S) is an embedded disk in M such that $D \cap S = \partial D$ and such that the loop $\partial D \subset S$ does not bound a disc in S. If no such compression disk exists, then S is is said to be *incompressible*.

are of interest to our story are (in rough order of expected prior knowledge) the books by Purcell [112], Rolfsen [120], and Burde and Zieschang [32]. Unfortunately there seems not to be a detailed monograph on the history of knot theory; [32] has some historical notes at the end of each chapter, and the book chapter by Epple [47], though focused primarily on the history of knot invariants, does have some discussion of the history of knot geometry. In addition, Chapter 3 of *In the tradition of Thurston* [110] has a detailed description of Thurston's contributions as well as some earlier history.

Definition 2. A *link* is a topological embedding of a collection of disjoint circles into S^3; the image of each circle is called a *component* of the link, and a *knot* is a link with a single component. We restrict ourselves to links which are *tame*: every component must be ambient isotopic in $\mathbb{R}^3$ to a simple closed piecewise linear curve. In general, we view two links as equivalent if they are ambient isotopic in S^3.

Clearly the topological type of the complement of a knot k in S^3 (formally, $\overline{S^3 \setminus V}$ where V is an open tubular neighbourhood of k) is an isotopy invariant of the knot. That the converse is also true is a famous result of Gorden and Luecke [41]. It follows that studying the topology of knot complements is equivalent to studying the structure of knots. (The corresponding result is not true for links.) It is an insight of Thurston's programme of geometry that one can study the topology of spaces by studying their geometry.

In 1982, Thurston announced the following pair of results, Theorems 2 and 3 in his seminal paper [141].

Theorem 2 (Corollary 2.5 of [141]). *If $k \subseteq S^3$ is a knot, $S^3 \setminus k$ has a geometric structure (i.e. admits a geometric atlas) iff k is not a satellite knot. It has a hyperbolic structure iff, in addition, k is not a torus knot.* □

Thurston conjectured Theorem 2, which essentially shows that most knot complements admit a hyperbolic structure, based on the work of Riley which we discuss in the next section. The history surrounding this discovery is very interesting; various accounts beyond [141] include [119] and the accompanying commentary [28], and the additional references given in the historical notes to Section 10.3 on p.504 of [113].

If L is a link in a 3-manifold M, then a *Dehn surgery* along L is effected by cutting out a tubular neighborhood of L, performing a diffeomorphism on each torus component of that neighbourhood, and gluing it back into M (see Section 1.5 of [113], Chapter 6 of [112], Section E.4 of [19]).

Theorem 3 (Theorem 2.6 of [141]). *Suppose L is a link in a 3-manifold M such that $M \setminus L$ has a hyperbolic structure. Then most manifolds obtained from M by Dehn surgery along L have hyperbolic structures.*

Every closed 3-manifold is obtained from S^3 by Dehn surgery along some link whose complement is hyperbolic; this follows from the Lickorish-Wallace theorem (which shows that every manifold comes from Dehn surgery along some link) [146, 76] (see also [120, §9I] and Theorem 6.5 of [112]) together with the observation that every link in the 3-sphere is a sublink of a hyperbolic link [107]: just do the trivial Dehn surgery on the added components. Thus Theorem 3 shows that most closed 3-manifolds admit hyperbolic structures.

3.3 *Riley's study of two-bridge knots and links*

The Riley slice arises naturally in the study of knots. In particular, a dense subset of the groups in the Riley slice can be viewed as degenerations of holonomy groups of two-bridge knot complements. In this section we briefly discuss Riley's study of these groups, and explain how to obtain the Riley slice from them. Much more detail may be found in Section 2 of our upcoming preprint, [45], as well as in the papers of Akiyoshi, Ohshika, Parker, Sakuma, and Yoshida [12] and Aimi, Lee, Sakai, and Sakuma [10].

We first define two-bridge links, following the treatment of [112, Chapter 10]. A *rational tangle* is obtained by drawing two disjoint arcs on the boundary of a 3-ball B and then pushing the interiors of the arcs slightly into the 3-ball, forming a pair of disjoint arcs embedded in B with endpoints on ∂B so the boundary is a sphere with four marked points. A *two-bridge link* is a link obtained by joining these marked points up with a pair of disjoint arcs on the boundary (the *bridge arcs*). There is a natural bijection due to Schubert [126], described in detail as Algorithm 2.1 of [45], which sets up a bijection between the set of two-bridge links (up to ambient isotopy) and $\hat{\mathbb{Q}} := \mathbb{Q} \cup \{\infty\}$; the rational number associated to a link is called the *slope* of the link.[2]

One of the simplest examples of a two-bridge link is the figure 8 knot k (which has slope $3/5$). It is important historically as it was the first known example of a link whose complement manifold admits a hyperbolic structure; it is Riley's proof of this which (according to Thurston [141]) inspired the conjecture of Theorem 2.

Riley's proof of the hyperbolicity is described in [117]; it is a consequence of his algebraic theory of parabolic representations of two-bridge link groups developed in [115, 116]. The main idea is to consider a representation $\theta : G \to \mathrm{PSL}(2,\mathbb{C})$ (where $G = \langle x, y : x^{-1}yxy^{-1}x = yx^{-1}yxy^{-1} \rangle$ is the fundamental group of the knot complement) defined by

$$\theta x := A = \begin{bmatrix} 1 & 1 \\ 0 & 1 \end{bmatrix} \qquad \theta y := B = \begin{bmatrix} 1 & 0 \\ -\omega & 1 \end{bmatrix}$$

where ω is a primitive cube root of unity. Then:

Theorem 4 (Theorem 1 of [117]). *If $\Gamma := \langle \theta x, \theta y \rangle$ then Γ has presentation*

$$\langle A, B : WAW^{-1} = B \rangle \tag{1}$$

where $W = A^{-1}BAB^{-1}$, and so θ gives an isomorphism $G \simeq \Gamma$. □

[2] Our 'slope' is actually the *reciprocal* of Schubert's normal form.

Hyperbolicity of the complement manifold (Corollary on p.284 of [117]) now follows from this theorem and some standard results on 3-manifold topology.

Every two-bridge link has a fundamental group with an presentation on two-generateds like (1) for the figure 8 knot (such a representation is called an *over-presentation*—the two-generateds correspond to loops about the two-bridge arcs), and Riley showed that every such group admits a non-abelian representation with the images of the two-generateds being parabolic (Theorem 2 of [115]). In fact, in 2002 Agol [2] announced a full classification of the non-free discrete groups generated by two non-parabolic elements (later given a formal and correct presentation in [12] and [10], see also Section 2.2 of [45]).

If X and Y are two parabolics which do not share fixed points then, up to conjugacy, they are represented by matrices

$$X = \begin{bmatrix} 1 & 1 \\ 0 & 1 \end{bmatrix}, Y_\rho = \begin{bmatrix} 1 & 0 \\ \rho & 1 \end{bmatrix}$$

where $\rho \in \mathbb{C}$. This normalisation associates a complex number ρ with each two-bridge link group representation. Riley devoted much effort to finding the set of $\rho \in \mathbb{C}$ such that

$$\Gamma_\rho := \langle X, Y_\rho \rangle$$

is a two-bridge link group, producing the famous image in Figure 6.

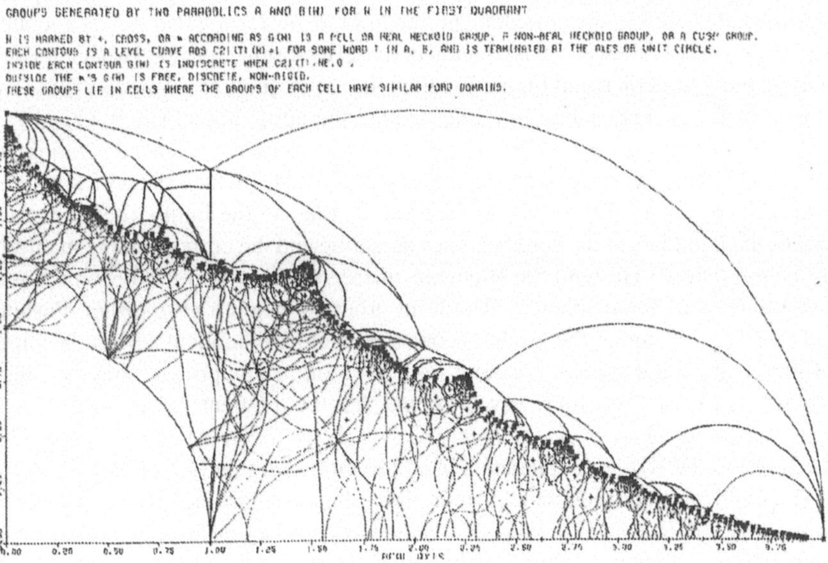

Fig. 6: Riley's plot of two-bridge link groups in the $(+,+)$-quadrant of $\mathbb{C}$.

The most obvious feature of Figure 6 is the fractal curve running diagonally across the centre of the frame. This is the set of so-called *cusp groups*. The half of the plot on the left (closer to 0) contains (marked with $+$ or $\times$) various *Heckoid groups* (introduced by Riley in [118] and formalised by Lee and Sakuma in [73]), which are the non-free Kleinian groups mentioned above in association to Agol's 2002 announcement. Observe that these points appear to lie on curves (called *extended pleating rays*) which end on the diagonal boundary curve.

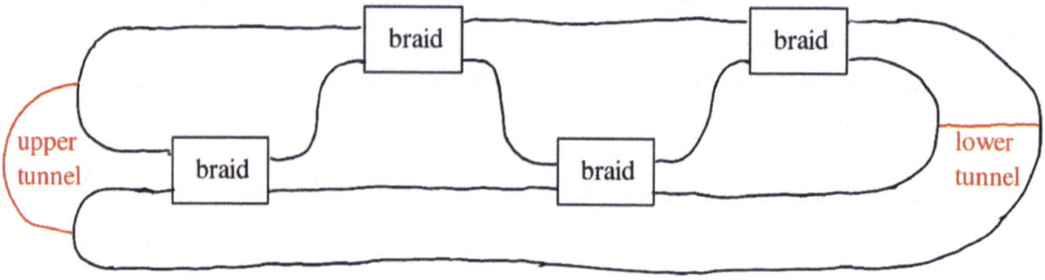

Fig. 7: A two-bridge knot with four braid regions (in black) together with upper and lower unknotting tunnels (in red).

Let $\Gamma_{\rho_0} = \langle X, Y_{\rho_0} : WX = Y_{\rho_0} W \rangle$ be a faithful representation of the two-bridge link group of slope p/q (where W is a word in X and Y_{ρ_0} depending on the slope; compare (1) and [115, Proposition 1]). An *unknotting tunnel* for a link L is a properly embedded arc t in the complement manifold M for L such that if N is a tubular neighbourhood of t then $M \setminus N$ is a handlebody; a two-bridge link has two natural unknotting tunnels, depicted in Figure 7, called the *upper* and *lower tunnels* [124]. The element in Γ_{ρ_0} corresponding to a loop around the upper tunnel t is $WXY_{\rho_0}^{-1}W^{-1}$; in Γ_{ρ_0}, this element is the identity. The discrete groups on the extended pleating ray which contains Γ_{ρ_0} correspond to the values of $\rho \in \mathbb{C}$ for which the corresponding word in Γ_ρ becomes elliptic of ever-increasing order; in the limit (when the curve reaches the boundary) this word becomes parabolic and the corresponding manifold has a third deleted arc (and the Riemann surface of the group has become a pair of disjoint 3-punctured spheres). The limit group is called a *cusp group*. Pushing further along this curve, the group becomes free and the quotient surface becomes a 4-times punctured sphere. (A more in-depth look at this process may be found in [45, §2.2].) An alternative way of seeing that the quotient, for ρ large, is a 4-times punctured sphere is via the *Poincaré polyhedron theorem* (for which see [113, §11.2] or [98, IV.H]); we explain this in a little more detail in [45, §5.1].

The quotient manifolds which occur along the curve are sketched in Figure 8. Looking back at Riley's plot, Figure 6, the fractal boundary is the set of cusp groups, and the empty space in the upper right half of the plot is the moduli space of four-times punctured spheres.

Definition 3. The *parabolic Riley slice*, which we will often shorten to *Riley slice* and denote by $\mathcal{R}$, is the set of $\rho \in \mathbb{C}$ such that Γ_ρ is free, discrete, and the Riemann surface $\Omega(\Gamma_\rho)/\Gamma_\rho$ is homeomorphic to a 4-times punctured sphere.

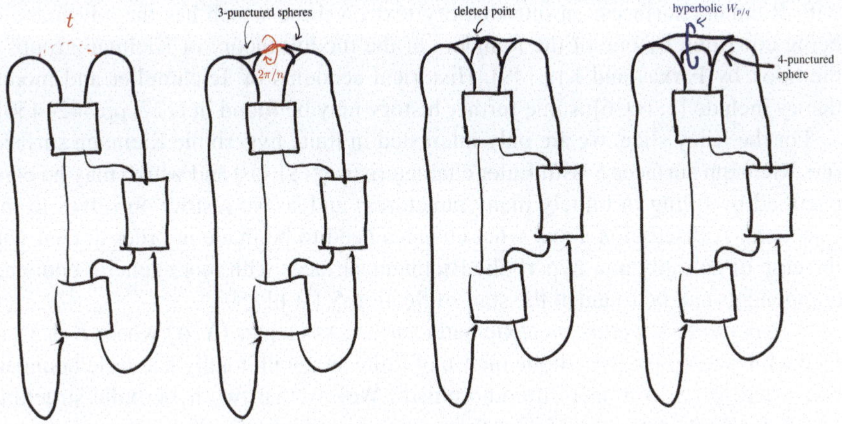

Fig. 8: The four kinds of orbifolds found along an extended pleating ray.

Indeed, $\overline{\mathcal{R}}$ as we defined it in Section 2.2 is the closure in $\mathbb{C}$ of this set, and $\mathcal{R}$ is the interior of $\overline{\mathcal{R}}$ in $\mathbb{C}$; this is a 2010 result of Ohshika and Miyahi [109] (see also Section 3.3 of [45] for a thorough discussion of the topology of the boundary).

Finally, we remark that of course not every group Γ_ρ for $\rho \in \mathcal{R}$ is obtained by twisting out from a two-bridge link group; the ones that can be obtained in this way are the groups lying on the so-called *rational pleating rays* of Keen and Series [67] (see Section 5 below). The remainder of the groups in $\mathcal{R}$ are obtained by pushing out from geometrically infinite groups, and can be traced back to the fundamental groups of *wild* knots, c.f. [64].

4 Quasiconformal deformation spaces

In this section we define formally what we mean by a moduli space of Kleinian groups; the first two sections (4.1 and 4.2) give a fast introduction to the general theory, and the third section (4.3) explains what this theory means for the Riley slice.

4.1 Riemann moduli theory

We begin by quickly recapping the classical theory of Riemann moduli space; some nice introductory books on this subject include those by Farb and Margalit [48], by Imayoshi and Taniguchi [59], and by Lehto [74]. These books all assume comfort with Riemann surfaces; an introductory text on these which has the advantage of being cowritten by one of the founders of the moduli theory of Kleinian groups is the book by Farkas and Kra [49]. Historical accounts of Teichmüller and moduli theory include [1, 60, 6]; some further history may be found in [127, pp.442–450].

For the Riley slice, we are only interested in finite hyperbolic Riemann surfaces (i.e. Riemann surfaces S with Euler characteristic $\chi(S) < 0$ and which may be compactified by filling in finitely many punctures) and so we restrict ourselves to this case here. A discussion about what changes need to be made in order to deal with the case of possibly non-hyperbolic Riemann surfaces with more general boundary components can be found at the start of Section 5.1.1 of [59].

A *hyperbolic structure* on a Riemann surface S is a pair (X, ϕ) where X is a surface with a complete hyperbolic metric of finite area with totally geodesic boundary and where $\phi : S \to X$ is a diffeomorphism. We say that two hyperbolic structures (X_1, ϕ_1), (X_2, ϕ_2) are *equivalent* if there exists an isometry $\iota : X_1 \to X_2$ such that the diagram

$$\begin{array}{ccc} & S & \\ {}^{\phi_1}\swarrow & & \searrow^{\phi_2} \\ X_1 & \xrightarrow{\quad \iota \quad} & X_2 \end{array}$$

commutes up to homotopy. The Teichmüller space of S, $\mathrm{Teich}(S)$, is the set of all hyperbolic structures on S modulo equivalence. It has a natural metric space structure, but defining the metric requires quasiconformal machinery and so we put off the explanation until Section 4.2.

The *mapping class group* of a Riemann surface S, denoted $\mathrm{Mod}(S)$ (also called the *modular group* of S), is the group of orientation-preserving homeomorphisms of S, moduli isotopies. (Observe that these homeomorphisms are allowed to permute the punctures. In the general case, we define the mapping class group in such a way that punctures are permuted, but all other boundary components are fixed pointwise; see [48, §2.1] and [59, §6.3.1].) This group has a natural action on $\mathrm{Teich}(S)$: if f is a representative of an element of the mapping class group, and (X, ϕ) is a representative of an element of the Teichmüller space, then

$$f \cdot (X, \phi) := (X, \phi \circ f^{-1}).$$

(The use of f^{-1} rather than f is necessary for the axioms for a group action to hold.)

The mapping class group acts disconti fundamental group nuously as a group of isometries on $\mathrm{Teich}(S)$ [59, Theorem 6.18], and the quotient $\mathrm{Teich}(S)/\mathrm{Mod}(S)$ is called the *(Riemann) moduli space* and denoted by $\mathcal{M}(S)$. The Riemann moduli space parameterises hyperbolic structures but now doesn't care about the position of punctures in the metric; quoting Thurston [143, p.259]:

> Informally, in Teichmüller space, we pay attention not just to what metric a surface is wearing, but also to how it is worn. In moduli space, all surfaces wearing the same metric are equivalent. The importance of the distinction will be clear to anybody who, after putting a pajama suit on an infant, has found one leg to be twisted.

We give two examples.

Example 1 (Tori). The canonical first example of a moduli space is that of genus 1 Riemann surfaces (i.e. elliptic curves). A complex structure on the torus is determined by the lengths of the meridian and longitude; equivalently, every complex torus is of the form $\mathbb{C}/G_\tau$ where $\tau \in \mathbb{H}^2$ and

$$G_\tau := \{m + n\tau : m, n \in \mathbb{Z}\}.$$

Further, the tori corresponding to τ_1 and τ_2 are biholomorphic iff $\tau' = A\tau$ for some $A \in \mathrm{PSL}(2,\mathbb{Z})$ (acting on the upper half-plane as a fractional linear transformation). We therefore get the following for a complex torus T:

$$\mathrm{Teich}(T) = \mathbb{H}^2, \quad \mathrm{Mod}(T) = \mathrm{PSL}(2,\mathbb{Z}),$$
$$\text{and } \mathcal{M}(T) = \mathbb{H}^2/\mathrm{PSL}(2,\mathbb{Z}) \simeq_{\text{bihol.}} S_{(2,3,\infty)}.$$

Here the space $S_{(2,3,\infty)}$ is topologically the complex plane; however, the natural geometric structure it inherits as a quotient by the group action is hyperbolic: $S_{(2,3,\infty)}$ is the finite area (in fact, area $\pi/3$) Riemann surface obtained as the 2-sphere with two cone points of order 2 and 3 and a single puncture. There is a unique hyperbolic metric on this space.

For further detail see [59, §1.2] and [48, §10.2, §12.2]. As we alluded to, this is very closely related to the arithmetic study of elliptic curves and the theory of modular forms: see for instance [133, C.12].

Example 2 (Four-punctured spheres). The Teichmüller theory of the 4-times punctured sphere $S_{0,4}$ is intimately related to that of the once punctured torus T_*, since there is a natural involution (the *hyperelliptic involution*) which swaps the homotopy classes of non-boundary-parallel simple closed curves on T_* and $S_{0,4}$ (see [48, §2.2.5] and [11, Chapter 2]). In fact T_* can be obtained much as in the previous example,

$$T_* = (\mathbb{C} \setminus \{G_\tau(0)\})/G_\tau, \qquad \tau \in \mathbb{H}^2$$

and if we put the complete hyperbolic metric on $\mathbb{C} \setminus \{G_\tau(0)\}$, the action of G_τ will be by isometry. This construction gives all the complex structures on T_*. In any case, it can be shown that

$$\mathrm{Teich}(S_{0,4}) = \mathbb{H}^2 \quad \text{and} \quad \mathrm{Mod}(S_{0,4}) = \mathrm{PSL}(2,\mathbb{Z}) \ltimes (\mathbb{Z}/2\mathbb{Z} \times \mathbb{Z}/2\mathbb{Z});$$

one can also show that $\mathcal{M}(S_{0,4})$ is topologically a thrice-punctured sphere.

Moduli spaces of Riemann surfaces appear naturally in many interesting areas of mathematics, through to the study of string interactions in physics [150, §22.6–22.7]. We recommend the survey article by Alex Wright [147] of Mirzakhani's work on the moduli spaces of Riemann surfaces for an accessible update on the most recent advances, ideas, and conjectures.

4.2 Holomorphic motions and quasiconformal deformations

The analogue of the Riemann moduli space for hyperbolic 3-manifolds (moving from parameterising the conformal boundary to the interior) requires the notion of quasiconformal mappings. The goal of this section is to give an intuitive reason for the occurence of this class of deformations in particular, via the theory of holomorphic motions; a more detailed exposition of this route may be found in Chapter 12 of the book of Astala, Iwaniec, and Martin [14]; it is known, following Sullivan and McMullen [137, 138, 103, 102], that the entire theory is basically analogous to the theory of complex dynamics and so the reader should try to focus on the deformations occuring in $\hat{\mathbb{C}}$ (in this case, deformations of limit sets) as well as deformations of the hyperbolic 3-manifolds: we find it much easier anyway to visualise points moving around in $\hat{\mathbb{C}}$ compared with trying to think about deforming large hyperbolic manifolds. The idea is that, although the Riemann surface cannot detect all of the Kleinian structure (the Riemann moduli space is a quotient of the Kleinian moduli space which we will define), the limit set *can* (as it should, since it is a global reflection of the group orbits).

The basic definition is due to Mañé, Sad, and Sullivan [84]; it encodes what we mean when we say that a parameter 'varies holomorphically' in a set.

Definition 4 (Holomorphic motions). Let $A \subseteq \hat{\mathbb{C}}$. A **holomorphic motion** of A is a map $\Phi : \mathbb{B}^2 \times A \to \hat{\mathbb{C}}$ (where $\mathbb{B}^2$ is the unit disc in $\mathbb{C}$) such that

1. For each $a \in A$, the map $\mathbb{B}^2 \ni \lambda \mapsto \Phi(\lambda, a) \in \hat{\mathbb{C}}$ is holomorphic;
2. For each $\lambda \in \mathbb{B}^2$, the map $A \ni a \mapsto \Phi(\lambda, a) \in \hat{\mathbb{C}}$ is injective;
3. The mapping $A \ni a \mapsto \Phi(0, a) \in \hat{\mathbb{C}}$ is the identity on A.

See the schematic in Fig. 9.

Mañé, Sad, and Sullivan also proved the so-called 'λ-lemma'; we need an extended version (Theorem 5), due to Słodkowski. The result shows that holomorphic motions are 'rigid': they are determined everywhere on $\hat{\mathbb{C}}$ even when only known on 'small' sets, in such a way that the extension is (remarkably) almost conformal in a. More precisely, its action on the tangent bundle of $\mathbb{C}$ maps circles to ellipses of bounded eccentricity—this is the germ of the notion of quasiconformality, which we pause to define following Section 1.4 and Chapter 4 of [59]. An extended treatment can be found in [14], but this is perhaps too technical for bedtime reading.

Definition 5 (Quasiconformality). Let D, D' be domains in $\mathbb{C}$, and let $f : D \to D'$ be an orientation-preserving diffeomorphism. Define the two differential operators

$$\partial_z := (1/2)(\partial_x - i\partial_y) \quad \text{and} \quad \partial_{\bar{z}} := (1/2)(\partial_x + i\partial_y)$$

where $x = \Re z$ and $y = \Im z$. The function

$$\mu = \frac{\partial_{\bar{z}} f}{\partial_z f}$$

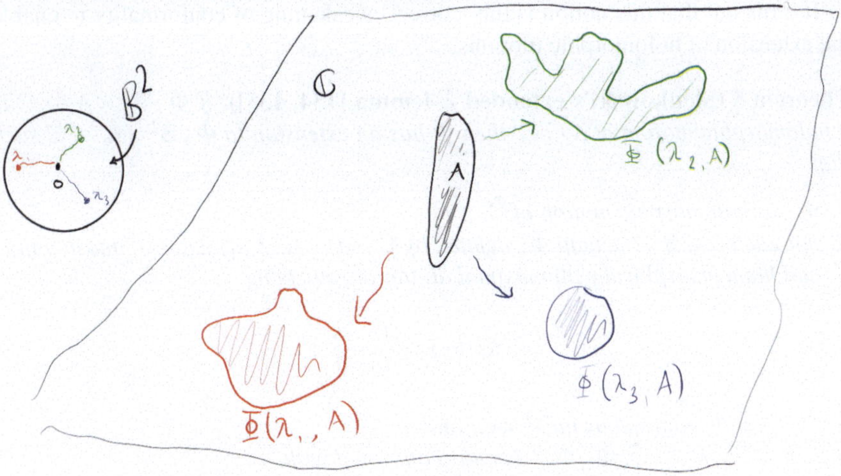

Fig. 9: Images of $A \subset \mathbb{C}$ under a holomorphic motion Φ.

is a smooth complex-valued function on D, and $\|\mu\|_\infty < 1$. The function μ is called the *Beltrami coefficient* of f, and the function f is called a *quasiconformal mapping* if

$$K(f) := \sup_{z \in D} \frac{1+|\mu(z)|}{1-|\mu(z)|} < \infty$$

in which case $K(f)$ is called the *maximal dilatation* of f.

This definition relates to the action on tangent bundle circles in the sense that $\mu(z)$ measures the eccentricity of ellipses which are the images of circles in $T_z\mathbb{C}$ under the differential df and $K(f)$ measures the maximal ratio of one axis of these ellipses to the other, see [59, p.18] and Fig. 10.

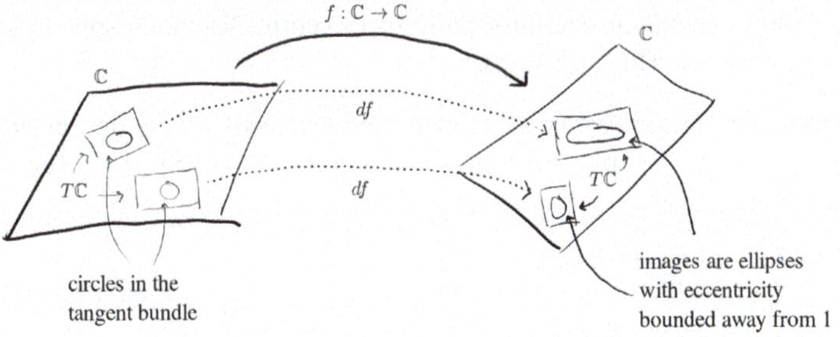

Fig. 10: The image of circles in the tangent bundle to $\mathbb{C}$ under the differential of a quasiconformal map $f : \mathbb{C} \to \mathbb{C}$.

It turns out that this notion is the 'correct' weakening of conformality to enable the extension of holomorphic motions.

Theorem 5 (Słodkowski's extended λ-lemma [134, 135]). *If $\Phi : \mathbb{B}^2 \times A \to \hat{\mathbb{C}}$ is a holomorphic motion of $A \subseteq \mathbb{C}$, then Φ has an extension to $\tilde{\Phi} : \mathbb{B}^2 \times \hat{\mathbb{C}} \to \hat{\mathbb{C}}$ such that*

1. *$\tilde{\Phi}$ is a holomorphic motion of $\hat{\mathbb{C}}$;*
2. *For each $\lambda \in \mathbb{B}^2$, the map $\tilde{\Phi}_\lambda$ defined by $\hat{\mathbb{C}} \ni a \mapsto \tilde{\Phi}(\lambda, a) \in \hat{\mathbb{C}}$ is a quasiconformal homeomorphism with maximal dilatation satisfying*

$$K(\tilde{\Phi}_\lambda) \leq \frac{1+|\lambda|}{1-|\lambda|};$$

3. *$\tilde{\Phi}$ is jointly continuous in $\mathbb{B}^2 \times \hat{\mathbb{C}}$; and*
4. *For all $\lambda_1, \lambda_2 \in \mathbb{B}^2$, $\tilde{\Phi}_{\lambda_1} \tilde{\Phi}_{\lambda_2}^{-1}$ is quasiconformal with*

$$\log K\left(\tilde{\Phi}_{\lambda_1} \tilde{\Phi}_{\lambda_2}^{-1}\right) \leq \rho(\lambda_1, \lambda_2)$$

(where ρ is the hyperbolic metric on $\mathbb{B}^2$). □

There is also an equivarant version due to Earle, Kra, and Krushkal', appearing as Theorem 1 of their paper [42]:

Theorem 6 (Equivariant λ-lemma). *Let $A \subseteq \hat{\mathbb{C}}$ have at least three points, and let Γ be a group of conformal motions preserving A. Let $\Phi : \mathbb{B}^2 \times A \to \hat{\mathbb{C}}$ be a holomorphic motion on A, and suppose that for each $\gamma \in \Gamma$ and each $\lambda \in \mathbb{B}^2$ there is a conformal map $\theta_\lambda(\gamma)$ such that*

$$\Phi(\lambda, \gamma(z)) = \theta_\lambda(\gamma)(\Phi(\lambda, z)) \tag{2}$$

for all $z \in A$. Then Φ can be extended to a holomorphic motion on $\hat{\mathbb{C}}$ such that (2) holds for all $z \in \hat{\mathbb{C}}$. □

Observe that, by the λ-lemmas, a holomorphic motion of the limit set of a Kleinian group extends quasiconformally to the entire Riemann sphere; in particular, the ordinary set. This motivates:

Definition 6. The *quasiconformal deformation space* of a Kleinian group Γ, denoted by $\mathrm{QH}(\Gamma)$, is the space of representations $\theta : \Gamma \to \mathrm{PSL}(2, \mathbb{C})$ (up to conjugacy) such that

1. θ is faithful and $\theta\Gamma$ is discrete;
2. θ is type-preserving, that is if $\gamma \in \Gamma$ is parabolic (resp. elliptic of order n) then $\theta\gamma$ is parabolic (resp. elliptic of order n); and
3. the groups $\theta\Gamma$ are all quasiconformally conjugate (i.e. there exists some quasiconformal $f : \hat{\mathbb{C}} \to \hat{\mathbb{C}}$ depending on θ such that, as functions, $\theta\Gamma = f\Gamma f^{-1}$).

The space is equipped with a natural metric defined in roughly the same way as the classical Teichmüller metric (the distance between two deformations is defined to be the log of the maximal dilatation of the composition of the two quasiconformal homeomorphisms).

This definition was studied first by Bers [20, 22], Kra [71, 72] and Maskit [94, 95]; modern textbooks and monographs which discuss this theory include [100] (essentially the whole book), [87, Chapter 5], and [63, Chapter 8]

The exact relationship between $\mathrm{QH}(\Gamma)$ and $\mathrm{Teich}(\Omega(\Gamma)/\Gamma)$ is subtle to define (see the references following the theorem), so we give only a rough version here. This result is attibuted by Bers [22, §2.4] to Bers and Greenberg [21] and Marden [85].

Theorem 7. *Let Γ be a finitely generated non-elementary*[3] *Kleinian group with $\Omega(\Gamma) \neq \emptyset$; let $S = \Omega(\Gamma)/\Gamma$. Then there is a well-defined holomorphic surjection $p : \mathrm{Teich}(S) \to \mathrm{QH}(\Gamma)$. Further, there is a discrete subgroup $\widehat{\mathrm{Mod}}(S) \leq \mathrm{Mod}(S)$ and a natural bijection $\mathrm{QH}(\Gamma) \approx \mathrm{Teich}(S)/\widehat{\mathrm{Mod}}(S)$ such that the two projection maps agree.* □

In the case that Γ is *geometrically finite* (i.e. its quotient manifold arises as the projection of a tiling of $\mathbb{H}^3$ by finite-sided hyperbolic polyhedra [98, VI]) , the group $\widehat{\mathrm{Mod}}(S)$ is the subgroup generated by Dehn twists along simple closed curves which bound compression discs.

The torsion-free version of this theorem appears in [100] as Theorem 5.27 and the following discussion. The proof given there actually works for the torsion case as well, as long as we are careful to replace all manifold theorems and definitions with the corresponding 'orbi-theorems' and 'orbi-definitions'. Such a version of this theorem is found as Theorem 5.1.3 of [87] (the proof is sketched as Exercise 5-35, p.367ff).

In the case that a Kleinian group G is *Fuchsian* (that is it leaves a round disc Δ in $\Omega(G)$ invariant—and hence acts as a group of hyperbolic isometries on Δ, with $\Lambda(G) \subseteq \partial\Delta$, see the latter chapters of [16] and [98, V.G, IX, and X]) then the quasiconformal deformation space $\mathrm{QH}(G)$ is given the special name of *quasi-Fuchsian deformation space*; a Kleinian group Γ is called *quasi-Fuchsian* if it is a quasiconformal conjugate of some Fuchsian group. The limit sets of quasi-Fuchsian groups are contained in quasiconformal images of circles (called *quasicircles*) and the general theory of these deformation spaces is fairly well-understood: for instance, see [87, Chapter 4] and the (very accessible) [106]. The limit set of Fig. 1 is the limit set of a quasi-Fuchsian group, and shows how intricate these quasicircles can become; controling their large-scale geometry is one of the themes of our preprint [44], in which we study how small quasi-deformations of limit sets modify the quotient surface (see also Section 6.2 below).

[3] An 'elementary' group is a Kleinian group G which is particularly simple, in the sense that $|\Lambda(G)| < \infty$; basically this happens when most of the fixed points of group elements collide. We do not need to worry so much since it turns out that Γ_ρ is elementary iff $\rho = 0$, and this is excluded from the Riley slice by Lemma 2. For more on elementary groups see [98, Chapter V] or [113, §5.5].

4.3 The Riley slice as a quasiconformal deformation space

In Section 2 and Section 3, we introduced the Riley slice $\mathcal{R}$ as a subset of the complex plane $\mathbb{C}$ with no additional structure. However, this set was defined in a way that seems at first glance to be compatible with holomorphic motions: if Γ_ρ is free and discrete with quotient surface a 4-times punctured sphere, then surely so is $\Gamma_{\tilde{\rho}}$ for a small holomorphic motion $\tilde{\rho}$ of ρ? That this is the case is actually surprisingly nontrivial to prove; we give full details in our preprint [45], and in this section we give some of the history of the study of the Riley slice from a quasi-point of view.

The first mention that $\mathcal{R}$ is a quasiconformal deformation space which we can find in the literature is in the proof of Theorem 1 of Lyubich and Suvorov's 1988 paper [79], though the proof is only sketched; a more detailed proof along the same lines can be found in Section 3.3 of our preprint [45]. Using this equivalence, they prove that:

- $\mathcal{R}$ is connected and has connected complement—note that their Γ is our $\mathbb{C}\setminus\mathcal{R}$ [79, Theorem 1];
- There are only two maximal parabolic conjugacy classes in Γ_ρ for $\rho \in \mathcal{R}$ [79, Lemma 1];
- The Riley slice is invariant under a particular dynamical system [79, Lemma 2], c.f. [90], [92], [13], [44, Section 3], and our upcoming preprint [46];
- The set of ρ such that Γ_ρ is non-discrete is dense in $\mathbb{C}\setminus\mathcal{R}$ [79, Theorem 2].

The same proof (again, only sketched) was given by Maskit and Swarup in 1989 [99, §1]; in this paper, they show that if $\rho \in \overline{\mathcal{R}}$ and $\Omega(\Gamma_\rho) \neq \emptyset$ then the Riemann surface $\Omega(\Gamma_\rho)/\Gamma_\rho$ is either a 4-times punctured sphere, or a disjoint union of 3-punctured spheres. Thus, since quasiconformally conjugate groups have homeomorphic quotient surfaces, the boundary $\partial\mathcal{R}$ contains either groups with empty ordinary set, or groups with surface a pair of 3-punctured spheres. The groups in the latter class are called the *cusp groups*; they are exactly the *maximal cusps* of the Bers deformation theory [20, 94, 97, 108]. The 1989 paper also shows that every group in $\mathcal{R}$ is geometrically finite; thus we can apply the theorem relating the quasiconformal deformation space with the Teichmüller space (Theorem 7), along with Example 2, to see that $\mathcal{R}$ is homeomorphic to an (open) annulus—this follows from the observation that $\widehat{\mathrm{Mod}}(\Gamma_\rho)$ is generated by a single parabolic element [45, Corollary 3.9].

Finally we note that the Riley slice is not just a quasiconformal deformation space, it is a *quasi-Fuchsian* deformation space: when $\rho \in \mathbb{R}\cap\mathcal{R}$, the group Γ_ρ is Fuchsian. (A detailed discussion of this point may be found in Section 6.3.1 of [45].)

5 A geometric coordinate system

In this section we motivate and describe the Bers-Thurston deformation theory of quasi-Fuchsian groups in the geometrically finite case (5.1), and then describe how the Riley slice is a concrete example of this theory via the results of Keen and Series (5.2).

5.1 Measured laminations and cusps

Given a quasi-Fuchsian space $\mathcal{F}$, there are two natural definitions for the boundary $\partial\mathcal{F}$.

1. The *Bers boundary*, which is defined analytically as the boundary of a particular embedding of $\mathcal{F}$ into a particular Banach space of Γ-automorphic forms (originally defined by Bers [20], see also [100, Section 4.2] and [87, §5.11.1]); and
2. the *Thurston boundary*, which is defined geometrically in terms of measured laminations; this appears in [144], see also [100, Chapter 6] and [87, §5.11.2].

In this paper, we are interested in the Thurston boundary; and in this section we give a vague intuitive description which will be sufficient for motivating the Keen–Series theory in the next subsection. For motivation of our description of this boundary, we will use one of the major conjectures of hyperbolic 3-manifold theory posed by Thurston [141, Problem 11 in §6] as a very strong generalisation of Mostow rigidity; namely, the *ending lamination conjecture/theorem*. This conjecture was ultimately proved by Brock, Canary and Minsky [29], and see earlier work of Minsky [104], in the tame case; that this is sufficient follows from the resolution of Marden's tameness conjecture (posed in [86], proved independently by Agol [3] and Calegari and Gabai [33]).

Our exposition follows [87] (though see also [63, Chapter 14], [130, Section 4], and [140]); we will explain all of the words used in the theorem statement after giving it.

Theorem 8 (Ending lamination theorem). *A hyperbolic 3-manifold whose fundamental group is finitely generated is uniquely determined by its topological type and its end invariants.*

Here, 'topological type' just means homeomorphism class. Thus the only thing we need to explain is what we mean by 'end invariant'. This consists of two things: an *end* of the manifold, and a *measured lamination* on that end. An end of a 3-manifold, roughly speaking, is an ideal boundary component of the manifold; in the Riley slice interior (the main case which we are interested in), all groups are geometrically finite and in this case all ideal ends actually 'exist' in the sense that they are just boundary components in the usual sense (the components of the quotient Riemann surface).

The definition of a measured lamination requires more work. A *geodesic lamination*—or simply *lamination*—on a surface S is a compact subset of $\operatorname{int} S$ which is a union of complete disjoint geodesics (each geodesic is called a *leaf* of the lamination); an example of a lamination with infinitely many windings around the surface is given in Fig. 11, and many other pictures are found in [100, Chapter 6]. A *measured lamination* is a pair (Λ, μ) where Λ is a lamination, and μ is a *transverse measure* supported on Λ. The reader is not expected to be familiar with the ins-and-outs of measure theory; we only need the general idea that a measure is 'a density which accumulates on curves when integrating'. (A good reference, though, is Rudin's second textbook [123].) In this case we wish to assign to every curve γ which is not tangent to any leaf of Λ ('transverse') a number $\int_\gamma d\mu$ in a consistent way ('measure'). The set of measured laminations on S admits a natural topology (the weak convergence topology) and the resulting topological space is denoted by $\mathcal{ML}(S)$.

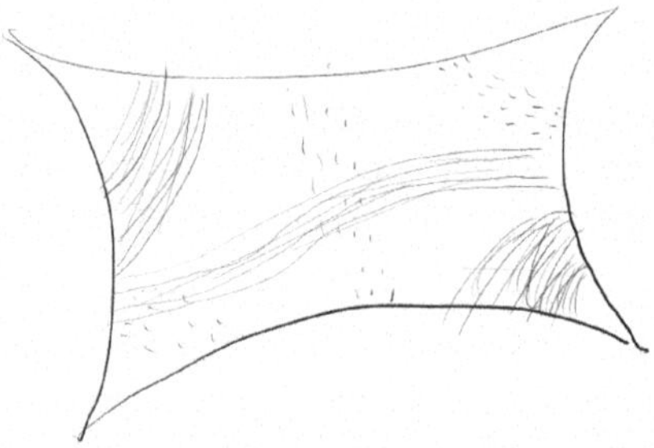

Fig. 11: A lamination of infinite length on the 4-times punctured sphere.

Example 3 (Pleated surfaces). Take a surface S and fold it along some geodesics so that along each geodesic the folding angle is constant. If the union of the geodesics forms a lamination Λ, then the *pleating measure* on Λ is the measure which assigns to each curve on S (...which is not tangent to any of the folds...) the sum of the folding angles of the folds which it crosses.[4] We draw a Euclidean example with three folds in Fig. 12. A very common way that one runs into pleated surface is by considering the hyperbolic convex hull of the limit set of a Kleinian group; two very nice visualisations of this have been produced by Brock and Dumas [31, 30].

An ending invariant of a Kleinian group G in the sense of Theorem 8 (in the geometrically finite case...) is just a connected component Ω^* of $\Omega(G)/G$ together with a measured lamination on Ω^*.

We now proceed to describe the *Thurston boundary* of quasi-Fuchsian space. A measured lamination (Λ, μ) on a hyperbolic surface S has a natural *lamination length*, defined to be the measure on S which is locally the product of μ (in the direction locally transverse to leaves of Λ) and the usual hyperbolic length measure (in the direction locally parallel to Λ). If we change the complex structure on S by shrinking the length of (Λ, μ) then (intuitively) the topological type of the surface does not change until the length is pinched down to 0; neighbourhoods of the leaves of Λ will turn into neighbourhoods of deleted points on the surface (if the leaves

[4] That all this is well-defined is really non-trivial, see Thurston's description [140, §8.6] and the paper by Canary, Epstein, and Green [34].

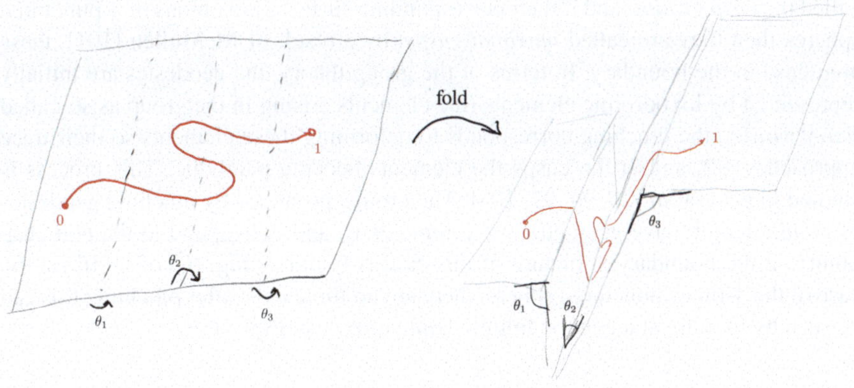

Fig. 12: A (Euclidean) pleated surface, with pleating locus (set of folds) consisting of three geodesics. The red curve γ (oriented from 0 to 1) has measure $\theta_1 + \theta_2 - \theta_2 + \theta_2 + \theta_3 = \theta_1 + \theta_2 + \theta_3$ with respect to the induced measured lamination.

are discrete), and we have hit the boundary of the Teichmüller space of the surface. It turns out that this process gives *quasiconformal* deformations of the underlying group, and (if everything is well-behaved) the union of all the surfaces obtained in this way forms a boundary of the deformation space (in the sense that these boundary surfaces, together with the deformation space itself, form a compact set).

Now observe that all of the surfaces which we have produced on the boundary are essentially different (they have different complex structures); thus, we have a natural bijection between this boundary and the set of measured-laminations-up-to-length: this latter set is usually called the space of *projective classes of measured laminations*, $\mathcal{PML}(S)$.

All of this (the geometrically finite ending lamination theorem and the Thurston compactification theory) gives the following 'picture' of the quasiconformal deformation space $\mathcal{F}$: it consists of a set of groups representing a family of surfaces, where the surfaces are all of the same topological type and where deforming the group slightly either changes the length of a distinguished set of geodesics on the surface (this is a 'radial' deformation, towards the boundary) or changes from one distinguished set of geodesics to another which is 'close' in the sense both of the closeness of the physical set of geodesics and of the measure on the geodesics. The boundary is the set of groups which parameterise those surfaces in which these geodesics have been pinched to deleted points. There is a subtlety here, in the sense that this procedure *a priori* will only produce the geometrically finite groups on the boundary since a measured lamination might correspond to a geodesic of infinite length, and so some normalisation needs to take place to make sense of 'decreasing the length'; in the case of the deformation space of a once-punctured torus, this is explained in [66, §6] and in the case of the Riley slice we explain the process in [45, §6.3–6.4]. The groups obtained by pinching down geodesics of finite length are called the *cusp groups*, and if their corresponding surfaces are unions of 3-punctured spheres then they are called *maximal cusps*; by a result of McMullen [101], these are dense in the boundary. In terms of the group theory, the geodesics are initially represented by loxodromic elements (the elements arising in the group as so-called *Farey words*), the pinching corresponds to deforming these elements so their trace approaches -2, and at the cusps the elements become parabolic. This process is studied in general in [20, 94, 97, 108]. The groups produced by pinching geodesics of *infinite* length have geometrically infinite ends, and correspond to the non-cusp points on the boundary; a picture of this case is found as Fig. 6.5 of [100] (in the case of the 4-times punctured sphere, there are no finite ends after pinching: this can be seen by looking at a generic infinite lamination, like Fig. 11).

5.2 The Keen–Series coordinate system

We now discuss the measured lamination theory in the specific case of the Riley slice; this was developed by Keen and Series [67] with minor corrections by Komori and Series [68]; some of the theory was not developed in detail in those papers as it is similar to the case of the once-punctured torus moduli space (the *Maskit slice*) which Keen and Series had studied earlier [66, 129]. A full account of the theory may be found in our preprint [45], though we develop the elliptic case at the same time which adds some difficulty for the reader.

In any case, the first main theorem is the following.

Theorem 9 ([67, Theorem 5.4]). *There is a natural homeomorphism*

$$\Pi : \mathcal{R} \to \mathbb{R}/2\mathbb{Z} \times \mathbb{R}_{>0}.$$

To prove this theorem, one first shows that $\mathcal{PML}(S_{0,4})$ is naturally homeomorphic to $\mathbb{R}/2\mathbb{Z}$ (see e.g. Section 2 of [67] and Section 5 of [45]—each measured lamination on the surface can be assigned a 'sloping angle' $\pi\theta$ and the homeomorphism sends $\Lambda \mapsto \theta(\Lambda)$) and then defines Π to be the map which sends $\rho \in \mathcal{R}$ to the pair $(\theta(\Lambda), l)$ where Λ is the lamination on $\mathbb{R}$ coming from the ending lamination theorem (which can be found very concretely in this case: if H is the hyperbolic convex hull of the limit set $\Lambda(\Gamma_\rho)$ in $\mathbb{H}^3$, then H/Γ_ρ is a deformation retract of $\mathbb{H}^3/\Gamma_\rho$ whose boundary is a pleated surface; the lamination Λ is the corresponding measured lamination), and l is its lamination length. The continuity then follows from various known results about continuity of functionals associated to measured laminations, discussed for example in [65].

In the process of showing Theorem 9, Keen and Series show that the fibres of Π in $\mathcal{R}$ corresponding to fixed angles in $\mathbb{R}/2\mathbb{Z}$ are analytic curves in $\mathbb{C}$ (the *pleating rays*, corresponding to pinching a single measured lamination) and in the case that it is a lamination of finite hyperbolic length that is being pinched (equivalently, deforming along a fibre above $\mathbb{Q}/2\mathbb{Z}$) these curves are branches of the inverse image of $(-\infty, -2)$ under a certain polynomial in one complex variable. We call these polynomials the *Farey polynomials* since they are very closely related to the sequences in number theory named after Farey; see the paper [12], the monograph [11], and our preprint [46].

A rough map of the Riley slice may be found in Fig. 13. Various people have produced computer-generated images of the coordinate system and the rational pleating rays; for example Wright [55, Fig. 1] and Yamashita [11, Fig. 0.2b], [44, Fig. 1].

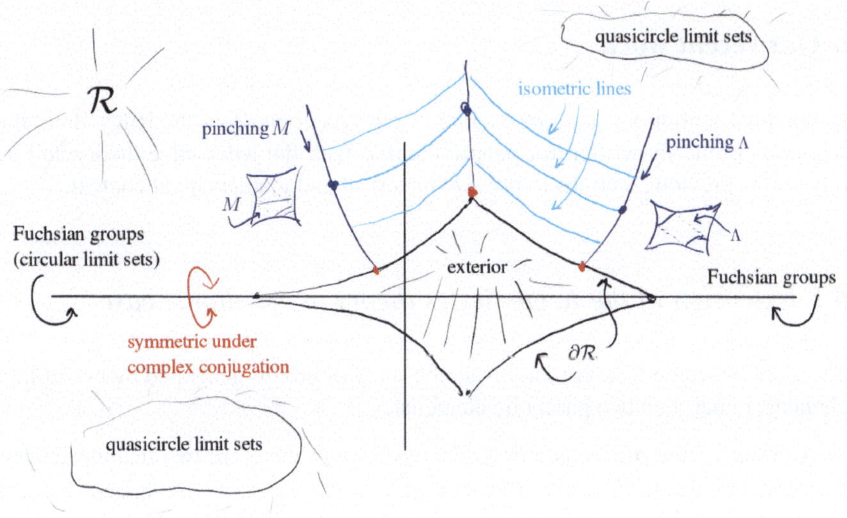

Fig. 13: A map of the Keen–Series coordinate system of the Riley slice.

We believe that the following problem is open.

Problem 1. Prove or disprove that Π is conformal.

If it is true that Π is conformal, then the coordinate lines of the Keen–Series coordinate system are Teichmüller geodesics (see [43, §10.2] for an argument). We think that this is probably not the case, however we make a specific closely related conjecture. Precise definitions and various facts concerning quasigeodesics can be found in [27]. Roughly, α is a *quasigeodesic* if it is the image of a geodesic β under a map f which which admits constants λ and ε so that

$$\frac{1}{\lambda}d(z,w) - \varepsilon \leq d(f(z), f(w)) \leq \lambda d(z,w) + \varepsilon \qquad \text{for all } z, w \in \beta;$$

the map f is called a *quasiisometry*. Notice such maps need not even be continuous, but they are injective and bilipschitz "at large scales".

Conjecture 3. The rational pleating rays are quasigeodesics in the Teichmüller metric.

Remark 1. We note that similar structures and coordinates have been worked out in detail for many other moduli spaces beyond the Maskit and Riley slices we have mentioned above; for instance, the so-called *diagonal slice of Schottky space* [131], the *Earle embedding* [69], and various other linear slices through complex Fenchel-Nielsen coordinates for the deformation space of quasi-Fuchsian once-punctured torus groups [70].

6 Our recent work

In this final section we announce some of our recent work on the Riley slice; this section is primarily written for people familiar with the Riley slice theory and we rely on the previous sections to provide historical and mathematical context.

6.1 Extension of the Keen–Series theory to the elliptic case

There are two natural ways to motivate the study of groups generated by two elliptic elements rather than two parabolic elements.

1. A group Γ_ρ for $\rho \in \mathcal{R}$ is a *Schottky type group*: there are two disjoint tangent pairs of circles in $\hat{\mathbb{C}}$, say C_1, C_1' and C_2, C_2' (where C_i and C_i' are tangent for each i) such that $X(C_1) = C_1'$, $Y_\rho(C_2) = C_2'$, and X and Y_ρ send the inside of the 'domain circle' to the outside of the 'target circle' (see e.g. [106, Chapter 4], [98, Exercises IV.J.21 and VII.F.9], and [43, §3.5] for some properties of Schottky groups). These groups arise naturally from the study of (classical) Schottky groups, where the paired circles are assumed disjoint and the quotient is a handlebody, by pushing the circles together and hence deforming the corresponding groups (in fact, by some work of Maskit [96], the Riley slice is a slice through the boundary of the space of Schottky groups—hence the name). If this deformation is continued and the circles are pushed through each other to form two pairs of transversely intersecting circles, then the pairing elements become elliptic.
2. If one prefers to motivate the Riley slice via knot theory, it is natural to consider knot orbifolds rather than knot complement manifolds; more precisely, given a two-bridge knot k we may produce a manifold M which is a branched cover of the knot complement where the knot outlines 'portals' through which one may step between the different covering sheets; since $\pi_1(S^3 \setminus k)$ is two-generated, we can choose the orders of two of these portals (and then the rest are determined); the covering induces an orbifold structure on the knot complement, where the orbifold has a knotted singular locus in the shape of k. This is visualised in Thurston's talk *Knots to Narnia* [142], and more modern visualisation work has been done by Brakke [24] and Sümmerman [139].

The generalised setting is now the following (where $\mathbb{Z}/n\mathbb{Z}$ denotes the cyclic group of order n):

Definition 7. Let $a,b \in \hat{\mathbb{N}} := \mathbb{N} \cup \{\infty\}$ with $\max\{a,b\} \geq 3$. Define $\Gamma_\rho^{a,b}$ to be the subgroup of $\mathrm{PSL}(2,\mathbb{C})$ generated by

$$X = \begin{bmatrix} \exp(\pi i/a) & 1 \\ 0 & \exp(-\pi i/a) \end{bmatrix}, Y_\rho = \begin{bmatrix} \exp(\pi i/b) & 0 \\ \rho & \exp(-\pi i/b) \end{bmatrix}.$$

The *Riley slice*, which we will denote by $\mathcal{R}^{a,b}$, is the set of $\rho \in \mathbb{C}$ such that $\Gamma_\rho^{a,b}$ is discrete, isomorphic to $\mathbb{Z}/a\mathbb{Z} * \mathbb{Z}/b\mathbb{Z}$, and such that $\Omega(\Gamma_\rho^{a,b})/\Gamma_\rho^{a,b}$ is a sphere with four cone points—two of order a and two of order b. When a or b is equal to ∞ the cone points are punctures, or geometrically cusps, in the quotient surface.

By convention, we take $\pi i/\infty := 0$; with this definition, $\mathcal{R}^{\infty,\infty}$ is the classical parabolic Riley slice. We draw pictures of various Riley slices (analogous to the parabolic slice drawn in Fig. 3 above) in Fig. 14.

Only limited study of the elliptic case has been carried out to this point; for example, the PhD thesis [149], and the study of arithmetic groups in [80]. We remedy this in our preprint [45], where we justify why all of the topological theory mentioned above in Section 4.3 carries through in the elliptic case, and carefully generalise the Keen–Series coordinate system of Theorem 9 to the elliptic case. The main difficulty is in replacing topological arguments with orbi-analogues, and generalising various hyperbolic arguments from ideal polyhedra to polyhedra with vertices contained in hyperbolic space. We also take the opportunity to give more detail than Keen and Series gave in [67] (who gave only sketches of some proofs which were similar to those found in their earlier paper on the Maskit slice [66]) and give more detailed motivation via knot theory in order to make the theory more accessible to non-experts.

Observing Figure 14 one is struck by the apparent continuous deformations of these slices as the 'variables' a,b move. One is tempted to conjecture that that there are natural quasiconformal deformations which transition between the various slices. As these regions are defined dynamically or analytically (assuming the density of roots of Farey polynomials, which seems true based on our computer calculations but which we have been unable to verify) this seems likely but we have not been able to establish this.

6.2 Open neighbourhoods of pleating rays

The theory of Keen and Series depends on deforming the limit sets of Kleinian groups in a conformal way. More precisely, and as they note, the rational pleating rays are curves along which the combinatorics of certain patterns of circles in the limit set are preserved. Allowing small deformations off the rational pleating rays also preserves these combinatorics; however, the circles now become quasicircles.

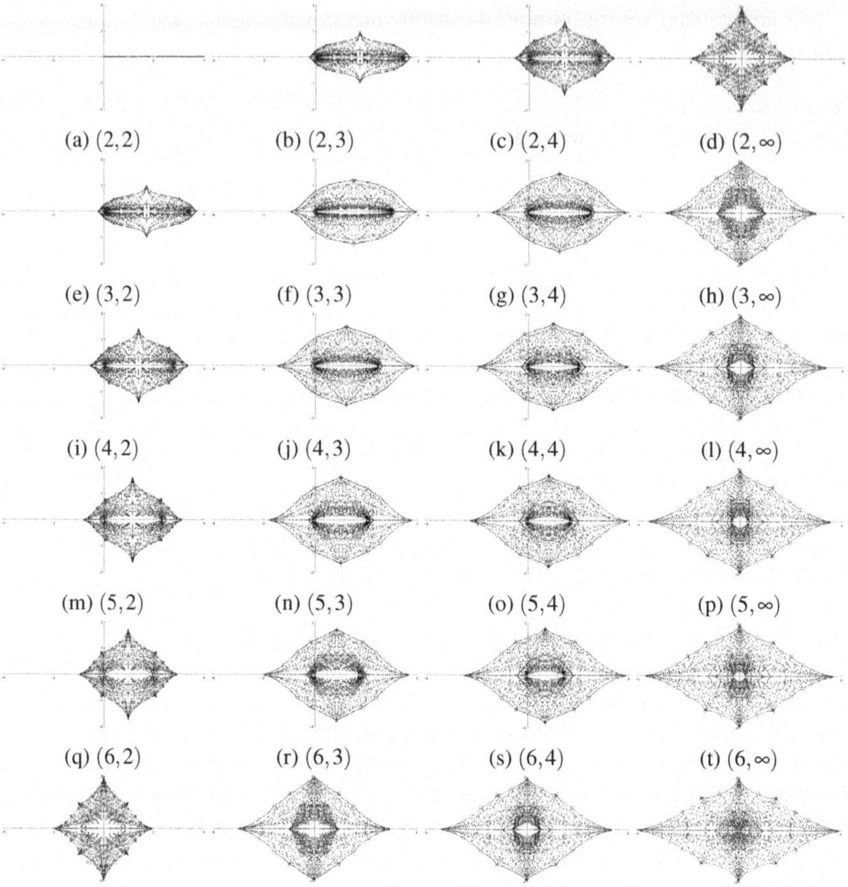

(a) $(2,2)$ (b) $(2,3)$ (c) $(2,4)$ (d) $(2,\infty)$

(e) $(3,2)$ (f) $(3,3)$ (g) $(3,4)$ (h) $(3,\infty)$

(i) $(4,2)$ (j) $(4,3)$ (k) $(4,4)$ (l) $(4,\infty)$

(m) $(5,2)$ (n) $(5,3)$ (o) $(5,4)$ (p) $(5,\infty)$

(q) $(6,2)$ (r) $(6,3)$ (s) $(6,4)$ (t) $(6,\infty)$

(u) $(\infty,2)$ (v) $(\infty,3)$ (w) $(\infty,4)$ (x) (∞,∞)

Fig. 14: Riley slices for various cone point orders (a,b).
The $(2,2)$ picture is simply an interval as any group generated by two elements of order 2 is virtually abelian and can be deformed until it becomes elliptic.

In our preprint [44], we show that the large-scale geometry of these quasicircles is controllable and that the resulting groups still lie in the Riley slice; thus we can construct open neighbourhoods of the pleating rays in the Riley slice, using the Farey polynomials that are used to construct the rays themselves; these open neighbourhoods have well-behaved boundary, which gives some control of the geometry of the Riley slice boundary. More precisely, it is a result of Keen and Series that if α is a rational pleating ray in $\mathcal{R}^{a,b}$, then there is a rational slope r/s and a Farey polynomial $p_{r/s}$ such that for a suitable choice of branch of $p_{r/s}^{-1}$, the curve α is given by $p_{r/s}^{-1}((-\infty,-2])$ and $p_{r/s}^{-1}(-2)$ represents a cusp group in $\partial\mathcal{R}^{a,b}$. Choosing the same branch of $p_{r/s}^{-1}$, we prove that

$$p_{r/s}^{-1}(H) \subset \mathcal{R}^{a,b}, \text{ where } H = \{x+iy : x \leq -2, \, y \in \mathbb{R} \setminus \{0\}\},$$

which gives a neighbourhood of the pleating ray that lies inside the Riley slice and is in some sense tangent to the boundary—we display some of these neighbourhoods in Fig. 15. This gives some precise information about regularity of the boundary of the Riley slice.

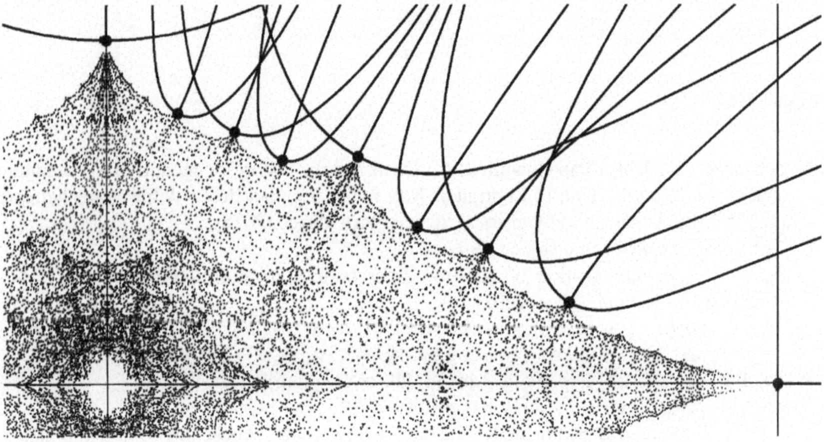

Fig. 15: Neighbourhoods of pleating rays "tangent" to $\mathcal{R}^{a,b}$.

Because of the useful coarse description of $\mathcal{R}^{\infty,\infty}$ given by Lyndon and Ullman (see Fig. 3 and [78]) a somewhat stronger result is true in the parabolic case [44]. We have been unable to completely replicate their result in the case of groups generated by two elliptics (we discuss this further in the paper).

In the same preprint we also give some dynamical properties of the Farey polynomials, both for use in studying these neighborhoods and for their own interest; we note that many of these properties were known to Lyubich and Suvorov [79], but were not studied by them in detail. This theory extends to the elliptic case in a similar way to the extension of the 'round-circle' Keen–Series theory, and the details will appear in the final part of [45].

6.3 Farey polynomials

The Farey polynomials introduced by Keen and Series in [67] also extend to the elliptic case, and we study their combinatorics in [46]. More precisely, we derive a recursion for these polynomials (using results similar to those developed by Mumford, Series, and Wright [106, p.277], see also [148]), show formal analogies with the Chebyshev polynomials and other combinatorial objects, and give some possible applications to the approximation of the Riley slice boundary.

Acknowledgements The authors are grateful to the organisers of the MATRIX Workshop on Groups and Geometries in Nov–Dec 2021 and the subsequent workshop on Waiheke for the opportunity to speak on the Riley slice theory and our results. Some of the material in this article is based on sections of the first author's MSc thesis [43]. Work of authors partially supported by the New Zealand Marsden Fund and by a University of Auckland FRDF grant.

References

1. A'Campo, N., Ji, L., Papadopoulos, A.: On the early history of moduli and Teichmüller spaces. In: Keen, L., Kra, I., Rodríguez, R.E. (eds.), Lipman Bers—A life in mathematics, pp.175–262. American Mathematical Society, Providence (2015) `https://arxiv.org/abs/1602.07208`
2. Agol, I.: The classification of non-free 2-parabolic generator Kleinian groups. Talk at Budapest Bolyai conference (2002) Talk notes: `https://web.archive.org/web/20100729151803/http://www.math.uic.edu/~agol/parabolic/parabolic01.html`
3. Agol, I.: Tameness of hyperbolic 3-manifolds. arXiv preprint (2004) `https://arxiv.org/abs/math/0405568`
4. Agol, I.: Finiteness of arithmetic Kleinian reflection groups, Proceedings of the International Congress of Mathematicians, Madrid, 2006, Vol. 2, 951–960.
5. Agol, I., Belolipetsky, M., Storm, P. and Whyte, K.; Finiteness of arithmetic hyperbolic reflection groups. Groups Geom. Dyn., 2, (2008), no. 4, 481–498.
6. Ahlfors, L.V.: Development of the theory of conformal mapping and Riemann surfaces through a century. In: Ahlfors, L.V., Calabi, E., Morse, M., Sario, L., Spencer, D.C.: Contributions to the theory of Riemann surfaces, pp.3–14. Princeton University Press, Princeton (1953)
7. Ahlfors, L.V.: Finitely generated Kleinian groups. Amer. J. Math. **86**(2), 413–429 (1964).
8. Ahlfors, L.V.: Correction to "Finitely generated Kleinian groups". Amer. J. Math. **87**(3), 759 (1965).
9. Ahlfors, L.V.: Complex analysis. Second edition. McGraw-Hill, New York (1966)
10. Aimi, S., Lee, D., Sakai, S., Sakuma, M.: Classification of parabolic generating pairs of Kleinian groups with two parabolic generators. arXiv preprint (2020) `https://arxiv.org/abs/2001.11662`
11. Akiyoshi, H., Sakuma, M., Wada, M., Yamashita, Y.: Punctured torus groups and two-bridge knot groups I. Springer-Verlag, Berlin (2007)
12. Akiyoshi, H., Ohshika, K., Parker, J., Sakuma, M., Yoshida, H.: Classification of non-free Kleinian groups generated by two parabolic transformations. arXiv preprint (2020) `https://arxiv.org/abs/2001.09564`
13. Alaqad, H., Gong, J., Martin, G.J.: Chebyshev polynomials and inequalities for Kleinian groups. arXiv preprint (2021) `https://arxiv.org/abs/2103.05696`
14. Astala, K., Iwaniec, T., Martin, G.: Elliptic partial differential equations and quasiconformal mappings in the plane. Princeton University Press, Princeton (2009)
15. Bamberg, J.: Non-free points for groups generated by a pair of 2×2 matrices. J. Lond. Math. Soc. (2). **62**(3), 795–801 (2000)
16. Beardon, A.F.: The geometry of discrete groups. Springer-Verlag, New York (1983)

17. Beardon, A.F.: Pell's equation and two-generated free Möbius groups. Bull. Amer. Math. Soc. **25**(6), 527–532 (1993)
18. Beardon, A.F.: Some remarks on non-discrete Möbius groups. Ann. Acad. Sci. Fenn. Math. **21**, 69–79 (1996) http://www.acadsci.fi/mathematica/Vol21/beardon.html
19. Benedetti, R., Petronio, C.: Lectures on hyperbolic geometry. Springer-Verlag, Berlin (1992)
20. Bers, L.: On boundaries of Teichmüller spaces and on Kleinian groups I. Ann. Math. **91**(3), 570–600 (1970)
21. Bers, L., Greenberg, L.: Isomorphisms between Teichmüller spaces. In: Ahlfors, L.V., Bers, L. (eds.), Advances in the theory of Riemann surfaces—proceedings of the 1969 Stony Brook conference, pp.53–79. Princeton University Press, Princeton (1971)
22. Bers, L.: Uniformization, moduli, and Kleinian groups. Bull. Lond. Math. Soc. **4**, 257–300 (1972)
23. Bolyai, J.: Appendix a tér tudománya.–Szerkesetette, beveretéssel, magyarázatakkal Kárteszi, F. Akademiai Kiado, Budapest (1937)
24. Brakke, K. Polycut—Connecting Multiple Universes. Software. https://facstaff.susqu.edu/brakke/polycut/polycut.htm
25. Borel, A.: Commensurability classes and volumes of hyperbolic 3-manifolds, Ann. Scuola Norm. Sup. Pisa, **8**, (1981), 1–33.
26. Brenner, J.L.: Quelques groupes libres de matrices. C. R. Math. Acad. Sci. Paris. **248**, (1955), 1689–1691
27. Bridson, M.R., Haefliger, A.: Metric spaces of non-positive curvature. Springer, Berlin (1999)
28. Brin, M.G., Jones, G.A., Singerman, D.: Commentary on Robert Riley's article "A personal account of the discovery of hyperbolic structures on some knot complements". Expo. Math. **31**(2), 99–103 (2013)
29. Brock, J.F., Canary, R.D., Minsky, Y.D.: The classification of Kleinian surface groups, II: The ending lamination conjecture. Ann. Math. **176**(1), pp.1–149 (2012) https://arxiv.org/abs/math/0412006
30. Brock, J., Dumas, D.: Bug on notes of Thurston. Computer render (2006). https://www.dumas.io/poster/
31. Brock, J., Dumas, D.: Thurston's jewel. In: Marden, A., Hyperbolic manifolds, frontispiece. Cambridge University Press, Cambridge (2015) https://www.dumas.io/convex/
32. Burde, G., Zieschang, H.: Knots. Walter de Gruyter, New York (1985)
33. Calegari, D., Gabai, D.: Shrinkwrapping and the taming of hyperbolic 3-manifolds. J. Amer. Math. Soc. **19**(2), pp.385–446 (2006) https://arxiv.org/abs/math/0407161
34. Canary, R.D., Epstein, D., Green, P.L.: Notes on notes of Thurston. In: Canary, R.D., Epstein, D., Marden, A. (eds.), Fundamentals of hyperbolic geometry—Selected expositions. Cambridge University Press, Cambridge (2006)
35. Chang, B., Jenning, S.A., Ree, R.: On certain pairs of matrices which generate free groups. Canad. J. Math. **10**, 279–284 (1958)
36. Conder, M. D. E., Maclachlan, C., Martin, G. J., and O'Brien, E. A.; 2-generator arithmetic Kleinian groups. III. Math. Scand. 90 (2002), no. 2, 161–179.
37. Coulon, R., Matsumoto, S., Segerman, H., Trettel, S.: Ray-marching in Thurston geometries. GitHub repository. https://github.com/henryseg/non-euclidean_VR
38. Coulon, R., Matsumoto, S., Segerman, H., Trettel, S.: Ray-marching Thurston geometries. arXiv preprint (2020–2022). https://arxiv.org/abs/2010.15801
39. Coxeter, H.S.M.: Non-Euclidean geometry. Second edition. The University of Toronto Press. Toronto (1947)
40. Coxeter, H.S.M.: Introduction to geometry. Second edition. John Wiley & Sons. Toronto (1980)

41. Gordon, C., Luecke, J.: Knots are determined by their complements. J. Amer. Math. Soc. **2**, 371-–415 (1989)
42. Earle, C.J., Kra, I., Krushkal', S.L.: Holomorphic motions and Teichmüller spaces. Trans. Amer. Math. Soc. **343**(2), 927–948 (1994)
43. Elzenaar, A.: Deformation Spaces of Kleinian Groups. Master's thesis, The University of Auckland, New Zealand (2022)
`https://aelzenaar.github.io/msc_thesis.pdf`
44. Elzenaar, A., Martin, G., Schillewaert, J.: Approximations of the Riley slice. arXiv preprint (2021)
`https://arxiv.org/abs/2111.03230`
45. Elzenaar, A., Martin, G., Schillewaert, J.: The elliptic cousins of the Riley slice. In preparation.
46. Elzenaar, A., Martin, G., Schillewaert, J.: The combinatorics of Farey words and their traces. arXiv preprint (2022)
`https://arxiv.org/abs/2204.08076`
47. Epple, M.: Geometric aspects in the development of knot theory. In: James, IM. (ed.), History of topology, pp.301–357. North-Holland, Amsterdam (1999)
48. Farb, B., Margalit, D.: A primer on mapping class groups. Princeton University Press, Princeton (2012)
49. Farkas, H.M., Kra, I.: Riemann surfaces. Second edition. Springer-Verlag, New York (1992)
50. Gabai, D.; On the geometric and topological rigidity of hyperbolic 3-manifolds. J. Amer. Math. Soc. 10 (1997), no. 1, 37–74.
51. Gabai, D., Meyerhoff, G. R.,and Thurston, N.; Homotopy hyperbolic 3-manifolds are hyperbolic. Ann. of Math. (2) 157 (2003), no. 2, 335–431.
52. Gehring, F.W., Maclachlan, C., Martin, G.J.: Two-generator arithmetic Kleinian groups II. Bull. Lond. Math. Soc. **30**(3), 258–266 (1998)
53. Gehring, F. W., Maclachlan, C., Martin, G. J. and Reid, A. W.: Arithmeticity, discreteness and volume. Trans. Amer. Math. Soc. 349 (1997), no. 9, 3611–3643
54. Gehring, F. W. and Martin, G. J.: Minimal co-volume hyperbolic lattices. I. The spherical points of a Kleinian group. Ann. of Math. (2) 170 (2009), no. 1, 123–161.
55. Gilman, J.: The structure of two-parabolic space—Parabolic dust and iteration. Geom. Dedicata. **131**, 27–48 (2008)
56. Gilman, J.: Two-generator discrete subgroups of PSL(2,R). Mem. Amer. Math. Soc. 117 (1995), no. 561, x+204 pp.
57. Gilman, J. and Keen, L.: Discreteness criteria and the hyperbolic geometry of palindromes. Conform. Geom. Dyn. 13 (2009), 76–90.
58. Gilman, J.; Maskit, B. An algorithm for 2-generator Fuchsian groups. Michigan Math. J. 38 (1991), no. 1, 13–32.
59. Imayoshi, Y., Taniguchi, M.: An introduction to Teichmüller spaces. Springer-Verlag, Tokyo (1987)
60. Ji, L., Papadopoulos, A.: Historical development of Teichmüller theory. Arch. Hist. Exact Sci. **67**(2), 119–147 (2013)
61. Jørgensen, T.; On discrete groups of Möbius transformations. Amer. J. Math. 98 (1976), no. 3, 739–749.
62. Jørgensen, T.; Compact 3-manifolds of constant negative curvature fibering over the circle, Ann. of Math. (2) 106 (1977), no. 1, 61–72.
63. Kapovich, M.: Hyperbolic manifolds and discrete groups. Birkhaüser, Boston (2001)
64. Kauffman, L.H., Lambropoulou, S.: On the classification of rational tangles. Adv. Appl. Math. **33** 199-237 (2004)
`https://arxiv.org/abs/math/0311499v2`
65. Keen, L., Series, C.: Continuity of convex hull boundaries. Pacific J. Math. **168**(1), pp.183–206 (1995)
66. Keen, L., Series, C.: Pleating coordinates for the Teichmüller space of a punctured torus. Topology **32** 4, 719–749 (1993)

67. Keen, L., Series, C.: The Riley slice of Schottky space. Proc. Lond. Math. Soc. (3). **69**, 72-90 (1994)
68. Komori, Y., Series, C.: The Riley slice revisited. In: Rivin, I., Rourke, C., Series, C. (eds.) The Epstein birthday schrift, pp. 303–316. Geometry & Topology Monographs 1 (1998)
69. Komori, Y., Series, C.: Pleating coordinates for the Earle embedding. Ann. Fac. Sci. Toulouse Math. (6). **6**(10) 1, 69–105 (2001)
70. Komori, Y., Yamashita, Y.: Linear slices of the quasi-Fuchsian space of punctured tori. Conform. Geom. Dyn. **16**(5), 89—102 (2012) `https://arxiv.org/abs/1111.3407`
71. Kra, I.: On spaces of Kleinian groups. Comment. Math. Helv. **47**, 53–69 (1972)
72. Kra, I.: Deformation spaces. In: Bers, L., Kra, I. (eds.), A crash course on Kleinian groups—Lectures given at a special session at the January 1974 meeting of the American Mathematical Society at San Francisco, pp.48–70. Springer-Verlag, New York (1974)
73. Lee, D. and Sakuma, M.: Epimorphisms from two-bridge link groups onto Heckoid groups (I). Hiroshima Math. J. **43**(2), 239–264 (2013)
74. Lehto, O.: Univalent functions and Teichmüller spaces. Springer-Verlag, New York (1987)
75. Leutbecher, A.: Über Spitzen diskontinuierlicher Gruppen von linear gebrochenen Transformationen. Math. Z. **100**, 183–200 (1967)
76. Lickorish, W.B.R.: A representation of orientable combinatorial 3-manifolds. Ann. of Math. (2) **76**, 531–540 (1962)
77. Lobačevskiĭ, N.I.: O načalag geometrii [On the principles of geometry]. In: The complete works of Lobačevskiĭ, Vol. 1, pp.185–261. Gostehizdat, Moscow (1946)
78. Lyndon, R.C., Ullman, J.L.: Groups Generated by Two Parabolic Linear Fractional Transformations. Canad. J. Math. **21**, 1388–1403 (1969)
79. Lyubich, M.Y., Suvorov, V.V.: Free subgroups of $SL_2(\mathbb{C})$ with two parabolic generators. J. Soviet Math. **41**(2), 976–979 (1988)
80. Maclachlan, C., Martin, G.J.: 2-generator arithmetic Kleinian groups. J. Reine Angew. Math. **511**, 95–117 (1999)
81. Maclachlan, C., Martin, G. J.: The non-compact arithmetic generalised triangle groups. Topology 40 (2001), no. 5, 927–944.
82. Maclachlan, C., Martin, G. J.: The (6,p)-arithmetic hyperbolic lattices in dimension 3. Pure Appl. Math. Q. 7 (2011), no. 2, Special Issue: In honor of Frederick W. Gehring, Part 2, 365–382.
83. Maclachlan, C., Reid, A.W.: The arithmetic of hyperbolic 3-manifolds. Springer-Verlag, New York (2003)
84. Mañé, R., Sad, P., Sullivan, D.: On the dynamics of rational maps. Ann. Sci. Éc. Norm. Supér. **16**(2), 193–217 (1983)
85. Marden, A.: An inequality for Kleinian groups. In: Ahlfors, L.V., Bers, L. (eds.), Advances in the theory of Riemann Surfaces—Proceedings of the 1969 Stony Brook conference, pp.295–296. Princeton University Press, Princeton (1971)
86. Marden, A.: The geometry of finitely generated Kleinian groups. Ann. Math. **99**, pp.383–462 (1974)
87. Marden, A.: Hyperbolic manifolds. Cambridge University Press, Cambridge (2016)
88. Margulis, G. A.: Discrete subgroups of semisimple Lie groups. Ergebnisse der Mathematik und ihrer Grenzgebiete (3) 17. Springer-Verlag, Berlin, 1991.
89. Marshall, T.H., Martin, G.J.: Minimal co-volume hyperbolic lattices, II: Simple torsion in a Kleinian group. Ann. of Math. (2) 176 (2012), no. 1, 261–301.
90. Marshall, T.H., Martin, G.J.: Polynomial trace identities in $SL(2,\mathbb{C})$, quaternion algebras, and two-generated Kleinian groups. arXiv preprint (2019) `https://arxiv.org/abs/1911.11643`
91. Martin, G. J.: On discrete Möbius groups in all dimensions: a generalization of Jørgensen's inequality. Acta Math. 163 (1989), no. 3-4, 253–289.
92. Martin, G.J.: Nondiscrete parabolic characters of the free group F_2: supergroup density and Nielsen classes in the complement of the Riley slice. J. Amer. Math. Soc. **103**(4), 1402–1414 (2020) `https://arxiv.org/abs/2001.10077`

93. Martin, G.J.: The continuous part of the axial distance spectrum for Kleinian groups. arXiv preprint (2020)
https://arxiv.org/abs/2007.00867
94. Maskit, B.: On boundaries of Teichmüller spaces and on Kleinian groups II. Ann. Math. **91**(3), 607–639 (1970)
95. Maskit, B.: Self-maps of Kleinian groups. Amer. J. Math. **93**, 840–856 (1971)
96. Maskit, B.: On free Kleinian groups. Duke Math. J. **48**, 755–765 (1981)
97. Maskit, B.: Parabolic elements in Kleinian groups. Ann. Math. **117**(3), 659–668 (1983)
98. Maskit, B.: Kleinian Groups. Springer-Verlag, Berlin (1988)
99. Maskit, B., Swarup, G.: Two parabolic generator Kleinian groups. Israel J. Math. **64**(3), 257–266 (1989)
100. Matsuzaki, K., Taniguchi, M.: Hyperbolic Manifolds and Kleinian Groups. Oxford University Press, Oxford (1998)
101. McMullen, C.: Cusps are Dense. Ann. of Math. **133**, 217–247 (1991)
102. McMullen, C.T.: Renormalization and 3-manifolds which fiber over the circle. Princeton University Press, Princeton (1996)
103. McMullen, C.T., Sullivan, D.P.: Quasiconformal homeomorphisms and dynamics. III. The Teichmüller space of a holomorphic dynamical system. Adv. Math. **135**(2), 351–395 (1998)
104. Minsky, Y.N.: The classification of Kleinian surface groups, I: Models and bounds. Ann. Math. **171**(1), pp.1–107 (2010)
https://arxiv.org/abs/math/0302208
105. Morgan, J. W., Tian, G.: The geometrization conjecture, Clay Mathematics Monographs, vol. 5, American Mathematical Society (2014)
106. Mumford, D., Series, C., Wright, D.: Indra's pearls. Cambridge University Press, Cambridge (2002)
107. Myers, R.: Simple knots in compact, orientable 3-manifolds. Trans. Amer. Math. Soc. **273**(1), 75–91 (1982)
108. Ohshika, K.: Geometrically finite Kleinian groups and parabolic elements. Proc. Edinb. Math. Soc. **41**(1), 141–159 (1998)
109. Ohshika, K., Miyachi, H.: Uniform models for the closure of the Riley slice. Contemporary Math. **510**, 249–306 (2010)
110. Ohshika, K., Papadopoulos, A. (eds.): In the tradition of Thurston—Geometry and topology. Springer Nature, Switzerland (2020)
111. Poincaré, H.: Mémoire sur les groupes kleinéens. Acta Math. **3**, 49–92 (1883).
English translation: Memoir on Kleinian groups. In: Stillwell, J. (trans.), Papers on Fuchsian functions, pp.255–304. Springer-Verlag, New York (1985)
112. Purcell, J.S.: Hyperbolic knot theory. American Mathematical Society, Providence (2020)
113. Ratcliffe, J.G.: Foundations of hyperbolic manifolds. Third edition. Springer Nature, Switzerland (2019)
114. Ree, R.: On certain pairs of matrices which do not generate a free group. Canad. Math. Bull. **4**(1), 49–52 (1961)
115. Riley, R.: Parabolic representations of knot groups, I. Proc. Lond. Math. Soc. (3). **24**, 217–242 (1972)
116. Riley, R.: Parabolic representations of knot groups, II. Proc. Lond. Math. Soc. (3). **31**, 495–512 (1975)
117. Riley, R.: A quadratic parabolic group. Math. Proc. Cambridge Philos. Soc. **77**, 281–288 (1975)
118. Riley, R.: Algebra for Heckoid groups. Trans. Amer. Math. Soc. **334**(1), 389–409 (1992).
119. Riley, R.: A personal account of the discovery of hyperbolic structures on some knot complements. Expo. Math. **31**(2), 104–115 (2013)
120. Rolfsen, D.: Knots and links. Publish or Perish, Wilmington, U.S.A. (1976)
121. , Rosenberger, G.: All generating pairs of all two-generated Fuchsian groups. Arch. Math. (Basel) 46 (1986), no. 3, 198–204.
122. Rosenfeld, B.A.: A history of non-Euclidean geometry. Translated by Shenitzer, A. Springer-Verlag, New York (1988)

123. Rudin, W.: Real and complex analysis. Third edition. McGraw-Hill, New York (1986)
124. Sakuma, M. The topology, geometry and algebra of unknotting tunnels. Chaos Solitons Fractals. **9**(4-5), 739–748 (1998).
125. Sanov, I.N.: A property of a representation of a free group. Doklady Akad. Nauk SSSR (N. S.). **57**, 657–659 (1947)
126. Schubert, H.: Knoten mit zwei Brücken. Math. Z. **65**, 113–170 (1956)
127. Segal, S.L.: Mathematicians under the Nazis. Princeton University Press, Princeton (2003)
128. Selberg, A. Harmonic analysis and discontinuous groups in weakly symmetric Riemannian spaces with applications to Dirichlet series, J. Indian Math. Soc., 20, 47–87 (1956)
129. Series, C: Lectures on pleating coordinates for once punctured tori. Departmental Bulletin Paper on Hyperbolic Spaces and Related Topics, Kyoto University (1999) http://hdl.handle.net/2433/63221
130. Series, C.: A crash course on Kleinian groups. Rend. Istit. Mat. Univ. Trieste. **37**, pp.1–38 (2005) https://rendiconti.dmi.units.it/volumi/37/01.pdf
131. Series, C., Tan, S., Yamashita, Y.: The diagonal slice of Schottky space. Alg. Geom. Top. **17**(4) 2239—2282 (2017). https://arxiv.org/abs/1409.6863
132. Shimizu, H.: On discontinuous groups operating on the product of the upper half-planes. Ann. of Math. (2). **77**, 33–71 (1963)
133. Silverman, J.H.: The arithmetic of elliptic curves. Second printing of second edition. Springer, New York (2016)
134. Słodkowski, Z.: Holomorphic motions and polynomial hulls. Proc. Amer. Math. Soc. **111**, 347–355 (1991)
135. Słodkowski, Z.: Natural extensions of holomorphic motions. J. Geom. Anal. **7**(4), 637–651 (1997)
136. Stillwell, J.: Classical topology and combinatorial group theory. Second edition. Springer, New York (1983)
137. Sullivan, D.: Quasiconformal homeomorphisms and dynamics. I. Solution of the Fatou-Julia problem on wandering domains. Ann. of Math. (2). **122**(3), 401–418 (1985)
138. Sullivan, D.: Quasiconformal homeomorphisms and dynamics. II. Structural stability implies hyperbolicity for Kleinian groups. Acta Math. **155**, 243–260 (1985)
139. Sümmerman, M.L. Knotted Portals in Virtual Reality. Math. Intelligencer. **43**(1), 55–63 (2021).
140. Thurston, W.: Geometry and Topology of 3-Manifolds. Unpublished lecture notes, Princeton University (1980)
141. Thurston, W.: Three dimensional manifolds, Kleinian groups, and hyperbolic geometry. Bull. Amer. Math. Soc. (N.S.) **6**(3), 357–381 (1982)
142. Thurston, W.: Knots to Narnia. Recorded lecture at UC Berkeley (1992). Published to YouTube by Anthony Phillips. https://www.youtube.com/watch?v=IKSrBt2kFD4
143. Thurston, W.: Three-dimensional geometry and topology. Vol. 1. Princeton University Press, Princeton (1997)
144. Thurston, W.: Hyperbolic Structures on 3-manifolds, II: Surface groups and 3-manifolds which fiber over the circle. arXiv preprint (1998) https://arxiv.org/abs/math/9801045
145. Voight, J.: Quaternion algebras. Springer International Publishing (2021) https://link.springer.com/book/10.1007/978-3-030-56694-4
146. Wallace, A.H.: Modifications and cobounding manifolds. Canad. J. Math. **12**, 503–528 (1960)
147. Wright, A.: A tour through Mirzakhani's work on moduli spaces of Riemann surfaces, https://arxiv.org/pdf/1905.01753.pdf
148. Wright, D.: Searching for the cusp. In: Minsky, Y.N., Sakuma, M., Series, C. (eds.), Spaces of Kleinian Groups, pp.301–336. Cambridge University Press, Cambridge (2005)

149. Zhang, Q.: Two elliptic generator Kleinian groups. PhD thesis. Massey University, Auckland (2010)
150. Zwiebach, B.: A first course in string theory. Cambridge University Press, Cambridge (2004)

Chapter 3

MATRIX Program: Quantum Curves, Integrability and Cluster Algebras

First properties of supermanifolds, their functor of points and the DeWitt topology

Kowshik Bettadapura

Abstract In this article we show how the classification of supermanifolds to first order can be recovered through their functor of points. Subsequently, we consider the notion of L-points of a supermanifold. In analogy with the complex topology on schemes, we describe a topology on the locus of L-points of a superscheme in line with the DeWitt approach to supermanifold theory.

1 Introduction

Mathematically, supermanifolds form a class of spaces representing supercommutative rings. They originally arose as a foundational framework for new ideas in particle physics and were developed along similar lines to contemporary algebraic geometry. In algebraic geometry, nilpotent elements in a commutative ring are essential in forming infinitesimal deformations. In supergeometry, nilpotent elements are part of the foundation. They represent coordinates on 'superspace' in their own right rather than deformations of coordinate rings.

There are a number of approaches to the construction of supermanifolds, many of which are detailed in [1]. In this article we study supermanifolds from two approaches: firstly as local, supercommutative ringed spaces (LSRS) developed by Berezin [2], Kostant [3] and Leites [4]; secondly, in line with the approach by DeWitt in [5].[1] Among our main results in this article is Theorem 5.14 relating the two approaches in the category of superschemes.

Kowshik Bettadapura
University of Melbourne, e-mail: k.bettad@gmail.com

[1] Supermanifolds in this approach could also be attributed to Rogers, e.g., [6], and so might be referred to as 'Rogers-DeWitt' supermanifolds. We will refer to them as 'DeWitt supermanifolds' however since the approach by Rogers is substantially more general, see e.g., [7].

D. R. Wood et al. (eds.), *2021-2022 MATRIX Annals*, MATRIX Book Series 5,
https://doi.org/10.1007/978-3-031-47417-0_3

1.1 Article summary and outline

We begin in §2 with an introduction to supermanifolds in the LSRS approach and present their classification to first order. Subsequently we give a general characterization of submanifolds in supermanifolds. In §3 we illustrate this classification for affine superspace, projective superspace and super Riemann surfaces.

A deep viewpoint propounded by Grothendieck [22] is to consider the functor of points represented by a space to be the more natural object to study, rather than the space itself. A number of authors have adapted this viewpoint to supermanifold theory, yielding significant results. For instance, Vaintrob in [28] used this viewpoint to study supersymplectic structures; Stolz and Teichner et. al. in [29, 30] relied heavily on this viewpoint in deriving links between supergeometry and topological invariants. For a more introductory account of the functor-of-points viewpoint to supermanifold theory, see Manin in [26], [27, Ch. 4] and Deligne and Morgan in [8].

In this article we will also consider the functor-of-points viewpoint to supermanifolds. In §4 we show how to recover this aforementioned classification of supermanifolds to first order through the functorial viewpoint. This is the content in Theorem 4.5 and Theorem 4.6. The example of a quadratic form in dimension $(1|1)$ is then given to illustrate the distinction between L-points for $L = 0, 1, 2$.

One of our motivations to study the functor of points of a supermanifold is to arrive at a framework to unify other approaches in supermanifold theory in [1]. We propose that such a unification amounts to understanding these approaches as Grothendieck topologies on the category of superschemes. This in line with a similar program outlined by Molotkov in [20] and furthered by Sachse [21]. In Definition 4.4 we establish the notion of 'L-points of a supermanifold'. In §5 we present a brief introduction to the DeWitt approach to supermanifold theory, following [5]. Subsequently, we revisit the locus of L-points of a superscheme. Our main result is Theorem 5.14 where it is shown how this locus can be given the structure of a DeWitt supermanifold. This leads to Question 5.15 and 5.16 on understanding supermanifold topologies as Grothendieck topologies.

1.2 Further remarks

Through a unification proposed as above we can hope to arrive at a deeper understanding into superspace constructions and relations among them. As an illustration of a typical problem, consider that of relating constructions in super Teichmüller theory with supermoduli theory. Penner and Zeitlin in [23] have shown, through DeWitt's approach, that super Teichmüller space is supersymplectic; Sachse in [21] give models of super Teichmüller space from both the DeWitt and LSRS approaches; Rothstein in [24] gives a classification of supersymplectic structures on supermanifolds as LSRS; Donagi and Witten in [25] have shown, using the LSRS approach, that the supermoduli space of curves admits a natural gauge pairing between certain cohomology classes and deformation parameters. In light of these results, some natural questions are:

- does Penner and Zeitlin's supersymplectic structure translate to a supersymplectic structure on super Teichmüller space as LSRS?
- can we apply Rothstein's classification to study Penner and Zeitlin's supersymplectic structure?
- can Penner and Zeitlin's supersymplectic form be expressed in terms of Donagi and Witten's deformation gauge pairing?

Progress on the above questions will invariably involve a unification of different approaches to supermanifold theory. As such, it would lead to a wealth of insight into the nature of supermanifolds more broadly.

Note: throughout this article, by 'rational points' it is meant $\mathbb{K}$-rational points for a field $\mathbb{K}$.

2 Supermanifolds

Supermanifolds are, in analogy with complex analytic spaces, a geometric object whose functions under multiplication form a *supercommutative ring*.

2.1 Supercommutative rings

Fix a field $\mathbb{K}$. A $\mathbb{K}$-algebra A is said to be a *superalgebra* if it is equipped with a morphism $A \xrightarrow{p} \mathbb{Z}_2$. The *even subalgebra* is then the preimage $A_{\bar{0}} := p^{-1}\bar{0}$ and the *fermionic part* is the preimage $A_{\bar{1}} := p^{-1}\bar{1}$. See that $A_{\bar{0}} \subset A$ is indeed subalgebra while $A_{\bar{1}}$ is an $A_{\bar{0}}$-module. This gives a $\mathbb{Z}_2$-decomposition of A over $A_{\bar{0}}$, $(A, p) \cong A_{\bar{0}} \oplus A_{\bar{1}}$.

The $\mathbb{K}$-superalgebra (A, p) is said to be a *supercommutative ring* if, for homogeneous $a, b \in A$, that

$$a \cdot b = (-1)^{p(a)p(b)} b \cdot a. \tag{1}$$

From (1) we can immediately deduce the existence of nilpotent elements: if $a \in A_{\bar{1}}$ then $a \cdot a = -a \cdot a$ giving $a^2 = 0$.

Remark 2.1 *While a supercommutative ring A is not* strictly *commutative, it is weakly commutative in the sense that* (1) *gives rise to an equivalence between the categories of left- and right A-modules.* [2] □

[2] c.f., the commutativity isomorphism in [8, p. 62].

2.1.1 First properties

Recall that any superalgebra (A, p) admits a *canonical* $\mathbb{Z}_2$-decomposition $(A, p) \cong A_{\bar{0}} \oplus A_{\bar{1}}$ with $A_{\bar{1}}$ the 'fermionic part'. It lends itself to the construction of the left- and right-*fermionic ideals*: $J_{left} := A_{\bar{1}} \cdot A$ and $J_{right} := A \cdot A_{\bar{1}}$. By Remark 2.1, if A is supercommutative then $J_{left} \cong J_{right}$ and we will just refer to $J_{left} = J_{right} = J \subset A$ as the fermionic ideal. This leads now to two characterizing structures derived from A:

I) the *reduced ring* $R := A/J$;
II) the *fermionic module* $F := J/J^2$.

Note that R here is commutative and F is an R-module. Despite the prefix 'reduced', R may be non-reduced as a commutative ring.

Example 2.2 Over a field $\mathbb{K}$ let V be a $\mathbb{K}$-vector space and $\wedge^{\bullet}_{\mathbb{K}}V$ its exterior algebra. Then $\wedge^{\bullet}_{\mathbb{K}}V$ is a supercommutative ring with: I) $\mathbb{K}$ as its reduced ring; II) V as its fermionic module. □

The prototypical example above of a supercommutative ring motivates the following generalization of affineness from commutative algebra.

Definition 2.3 *An* affine supercommutative ring over $\mathbb{K}$ *is a supercommutative ring over* $\mathbb{K}$ *for which:*

i) its reduced ring R *is affine;*
ii) it is isomorphic to $\wedge^{\bullet}_{R}V/I$ *for* V *a free* R*-module and* $I \subset \wedge^{\bullet}_{R}V$ *an ideal.* □

Accordingly, the exterior algebra from Example 2.2 is an example of an affine, supercommutative ring.

2.2 Local supercommutative ringed spaces

2.2.1 Local supercommutative rings

Recall that a ring is said to be *local* if it contains a unique, maximal ideal $\mathfrak{m} \subset R$. Similarly, a supercommutative algebra A is local if it contains a unique, maximal ideal $\mathfrak{m} \subset A$.

Lemma 2.4 Let A be a supercommutative ring and $J \subset A$ the fermionic ideal. Then A is local if and only if A/J is local. □

Proof Let A be a local, supercommutative ring and $J \subset A$ the fermionic ideal. Recall that the reduced ring $R = A/J$ is commutative. Let $A \xrightarrow{p} A/J = R$ be the quotient morphism. Suppose $\mathfrak{m} \subset A$ is maximal and let $I \subset R$ be an ideal. Since p is an epimorphism (onto), $p(p^{-1}I) = I$. Moreover, p preserves inclusions giving $I = p(p^{-1}I) \subset p(\mathfrak{m})$. As this holds for any ideal $I \subset R$, it follows that $p(\mathfrak{m}) \subset R$ is maximal.

Conversely, let $m \subset R$ be a maximal ideal. Set $\mathfrak{m} = p^{-1}m$ and let $J' \subset A$ be an ideal. Since p is a morphism, $p(J') \subset R$ is an ideal and so contained in m. Thus $J' \subset p^{-1}(p(J')) \subset p^{-1}m = \mathfrak{m}$, i.e., that $J' \subset \mathfrak{m}$, so $\mathfrak{m} \subset A$ must be maximal. □

Example 2.5 Continuing on from Example 2.2, the maximal ideal in a field $\mathbb{K}$ is the zero ideal $\{0\}$. By Lemma 2.4, the maximal ideal in the supercommutative ring $\wedge^\bullet_{\mathbb{K}} V$ is then the preimage of $\{0\}$ under the quotient $\wedge^\bullet_{\mathbb{K}} V \stackrel{p}{\to} \mathbb{K}$. Equivalently, it is the kernel $J = \ker\{\wedge^\bullet_{\mathbb{K}} V \stackrel{p}{\to} \mathbb{K}\}$. Since $\mathbb{K}$ is the reduced ring of $\wedge^\bullet_{\mathbb{K}} V$, the maximal ideal $J \subset \wedge^\bullet_{\mathbb{K}} V$ is precisely the fermionic ideal. □

The above example leads to a more general result pertaining to maximal ideals.

Lemma 2.6 Let A be a supercommutative ring and suppose its reduced ring is also reduced as a commutative ring. Then the fermionic ideal $J \subset A$ is maximal.

2.2.2 Manifolds

Let X be a topological space. A *presheaf* of rings $\mathcal{O}$ on X assigns to each open set $U \subset X$ a ring $\mathcal{O}(U)$. Elements of $\mathcal{O}$ are referred to as functions. To inclusions of open sets $U \subset V$ the presheaf $\mathcal{O}$ comes with a ring morphism, referred to as restriction $res_{V,U} : \mathcal{O}(V) \to \mathcal{O}(U)$. If $U = V$, $res_{U,V}$ is the identity. The presheaf $\mathcal{O}$ is a *sheaf* if it satisfies the *locality* and *glueing* axioms. Locality requires, for functions $s, t \in \mathcal{O}(V)$ and $U \subset V$, if $res_{U,V}s = res_{U,V}t$, then $s = t$. Glueing requires, for functions $s_U \in \mathcal{O}(U)$, $s_V \in \mathcal{O}(V)$, if $res_{U,U\cap V}s_U = res_{V,V\cap U}s_V$, then there exists a function $s \in \mathcal{O}(U \cup V)$ such that $s_U = res_{U\cup V,U}s$ and $s_V = res_{U\cup V,V}s$. The *stalk* of the sheaf $\mathcal{O}$ on X at a point $P \in X$ is the pullback of $\mathcal{O}$ along the inclusion of the point $\iota_P : \{P\} \hookrightarrow X$ and denoted $\mathcal{O}_P$. The pullback morphism of sheaves on X, $\mathcal{O} \to \iota_{P*}\mathcal{O}_P$ given by $f \mapsto f(P)$ is referred to as *evaluation at* P. Note that for different points $P, P' \in X$ the evaluation is valued in the rings $\mathcal{O}_P$ resp., $\mathcal{O}_{P'}$. As such, it doesn't make sense to generally ask whether functions can separate points. This question does make sense if the rings $\mathcal{O}_P$ are local for each $P \in X$ however, since then one can instead ask whether functions vanish modulo the unique, maximal ideal.

Definition 2.7 *A pair $(X, \mathcal{O}_X)$, for X a topological space and $\mathcal{O}_X$ a sheaf of rings is said to be a* locally ringed space *if, for each point $p \in X$, the stalk $\mathcal{O}_{X,p}$ is a local ring. In this case $\mathcal{O}_X$ is referred to as the* structure sheaf. □

Example 2.8 The pair $(\{pt\}, \mathbb{K})$ for $\{pt\}$ the one-point, topological space and $\mathbb{K}$ a field defines a locally ringed space. More generally, for any commutative ring R, its spectrum $\operatorname{Spec} R = (R^{primes}, \widetilde{R})$, for R^{primes} the set of prime ideals in R equipped with the Zariski topology and $\widetilde{R}$ the sheafification of the presheaf assigning R to the open sets in R^{primes} defines a locally ringed space. See e.g., [9, Ch. II] or [10, §3] for a detailed introduction bypassing the language of sheaves. □

Example 2.9 Let $X = \mathbb{R}^m$. A smooth structure on X is a identified with the sheaf of smooth, $\mathbb{R}$-valued functions C^∞. To each open set $U \subset X$, $C^\infty(U) = C^\infty(U, \mathbb{R})$ is the ring of smooth functions $U \xrightarrow{f} \mathbb{R}$. At each point $P \in X$ the stalk C^∞_P is a local ring with maximal ideal $\mathfrak{m}_P \subset C^\infty_P$ comprising those (germs of) functions vanishing at P, i.e., $\mathfrak{m}_P = \{f \in C^\infty \mid f(P) = 0\}$. Smooth, m-dimensional Euclidean space is then the locally ringed space $(\mathbb{R}^m, C^\infty)$. □

Example 2.10 Replacing $\mathbb{R}^m$ with $\mathbb{C}^m$ and the sheaf C^∞ in Example 2.9 with the sheaf of complex-valued, analytic functions O defines the m-dimensional, analytic Euclidean space $(\mathbb{C}^m, O)$. □

Definition 2.11 *A morphism of locally ringed spaces* $\phi : (X, O_X) \to (Y, O_Y)$ *consists of a pair* $\phi = (f, h^\sharp)$ *where* $f : X \to Y$ *is a continuous map of topological spaces and* $h^\sharp : f^*O_Y \to O_X$ *is a morphism of sheaves of local rings on* X. □

Remark 2.12 *Adjunction states* $\mathrm{Hom}_{\mathsf{Sh}(X)}(f^*O_Y, O_X) \cong \mathrm{Hom}_{\mathsf{Sh}(Y)}(O_Y, f_*O_X)$. *As such, W.L.O.G.,* $h^\sharp$ *in Definition 2.11 can be understood equivalently as a morphism of sheaves of local rings* $O_Y \to f_*O_X$ *on* Y. □

With Definition 2.7 and 2.11, locally ringed spaces form a category. It makes sense then to ask whether two such spaces are isomorphic. Accordingly, through Example 2.9 and 2.10 we can define smooth and analytic manifolds, circumventing the more familiar definition involving charts and atlases.[3]

Definition 2.13 *A smooth resp., complex analytic, m-dimensional manifold is a locally ringed space* (X, O_X)*, locally isomorphic to* $(\mathbb{R}^m, C^\infty)$ *resp.,* $(\mathbb{C}^m, O)$. □

2.2.3 Supermanifolds

Definition 2.13 captures the familiar intuition that the real and analytic Euclidean spaces in Example 2.9 and 2.10 are model spaces on which manifolds are built. The notion of a supermanifold can be arrived at in parallel.

Definition 2.14 A supercommutative space $\mathfrak{X}$ *is a pair* $\mathfrak{X} = (X, O_\mathfrak{X})$ *where* X *is a topological space and* $O_\mathfrak{X}$ *is a sheaf of local, supercommutative rings.* □

Following Definition 2.11, a *morphism* of supermanifolds $\mathfrak{X} = (X, O_\mathfrak{X})$ and $\mathfrak{Y} = (Y, O_\mathfrak{Y})$ is defined by a pair $(f, h^\sharp) : \mathfrak{X} \to \mathfrak{Y}$ where $f : X \to Y$ is continuous and $h^\sharp : f^*O_\mathfrak{Y} \to O_\mathfrak{X}$ is a morphism of sheaves of local, supercommutative rings on X (or equivalently on Y by Remark 2.12). The analogue of smooth and analytic Euclidean space in the present context is obtained by reference to the prototypical, supercommutative ring from Example 2.2.

Example 2.15 Fix an n-dimensional $\mathbb{R}$- or $\mathbb{C}$-vector space V and take as the structure sheaf $\wedge^\bullet_{C^\infty} V$ resp., $\wedge^\bullet_O V$. Note that these are indeed local, supercommutative rings by Lemma 2.6. The pair $(\mathbb{R}^m, \wedge^\bullet_{C^\infty} V)$ resp., $(\mathbb{C}^m, \wedge^\bullet_O V)$ defines $m|n$-dimensional real, resp., analytic Euclidean superspace. They are denoted $\mathbb{R}^{m|n}$ and $\mathbb{C}^{m|n}$ respectively.□

[3] see e.g., [10, §2.2].

Remark 2.16 *Throughout this article we will denote by $\mathbb{K}$ a field, which we intend to mean either $\mathbb{R}$ or $\mathbb{C}$. Then $\mathbb{K}^{m|n}$ is taken to mean either $\mathbb{R}^{m|n}$ or $\mathbb{C}^{m|n}$ from Example 2.15.* □

Following Definition 2.13 we arrive at the notion of a supermanifold.[4]

Definition 2.17 *An $m|n$-dimensional real resp., complex supermanifold $\mathfrak{X}$ is a supercommutative space $(X, \mathcal{O}_{\mathfrak{X}})$ which is locally isomorphic to the Euclidean superspace $\mathbb{R}^{m|n}$ resp., $\mathbb{C}^{m|n}$.* □

2.3 First properties of supermanifolds

We denote by $\mathfrak{X}$ a supermanifold and $\mathcal{O}_{\mathfrak{X}}$ its structure sheaf. As $\mathcal{O}_{\mathfrak{X}}$ is a sheaf of supercommutative rings it will imply $\mathfrak{X}$ can be classified to first order through analogous properties I) and II) for supercommutative rings in §2.1.1. Denote by $|\mathfrak{X}|$ the underlying topological space so that, as a supercommutative space, $\mathfrak{X} = (|\mathfrak{X}|, \mathcal{O}_{\mathfrak{X}})$. Over each open set $U \subset |\mathfrak{X}|$, $\mathcal{O}_{\mathfrak{X}}(U)$ is a supercommutative ring and so decomposes into a (commutative) subring and a module over it, $\mathcal{O}_{\mathfrak{X}}(U) \cong A_{\bar{0}} \oplus A_{\bar{1}}$. Projecting onto $A_{\bar{0}}$ for each U defines then a presheaf of commutative rings $U \overset{\mathcal{O}_{\mathfrak{X},\bar{0}}}{\mapsto} A_{\bar{0}}$ and a presheaf of $\mathcal{O}_{\mathfrak{X},\bar{0}}$-modules $U \overset{\mathcal{O}_{\mathfrak{X},\bar{1}}}{\mapsto} A_{\bar{1}}$. Since $\mathcal{O}_{\mathfrak{X}}$ is a sheaf we have:

Lemma 2.18 The presheaves $U \overset{\mathcal{O}_{\mathfrak{X},\bar{0}}}{\mapsto} A_{\bar{0}}$ and $U \overset{\mathcal{O}_{\mathfrak{X},\bar{1}}}{\mapsto} A_{\bar{1}}$ define a sheaf of commutative rings $\mathcal{O}_{\mathfrak{X},\bar{0}}$ and a sheaf of $\mathcal{O}_{\mathfrak{X},\bar{0}}$-modules $\mathcal{O}_{\mathfrak{X},\bar{1}}$ on $|\mathfrak{X}|$ respectively.

As a consequence of Lemma 2.18 we obtain an isomorphism of *sheaves on* $|\mathfrak{X}|$, $\mathcal{O}_{\mathfrak{X}} \overset{\cong}{\to} \mathcal{O}_{\mathfrak{X},\bar{0}} \oplus \mathcal{O}_{\mathfrak{X},\bar{1}}$. The sheaf of fermionic ideals is $\mathcal{J}_{\mathfrak{X}} := \mathcal{O}_{\mathfrak{X}} \cdot \mathcal{O}_{\mathfrak{X},\bar{1}}$ and, modulo $\mathcal{J}_{\mathfrak{X}}$ we obtain a commutative sheaf of rings $C = \mathcal{O}_{\mathfrak{X}}/\mathcal{J}_{\mathfrak{X}}$. By Lemma 2.4, C will be a sheaf of local, commutative rings on $|\mathfrak{X}|$. Hence by Definition 2.13, the pair $M = (|\mathfrak{X}|, C)$ will define a differentiable manifold. This is the analogue of property I) from §2.1.1 for supermanifolds.

Definition 2.19 *The differentiable manifold $M = (|\mathfrak{X}|, C)$ associated to any supermanifold $\mathfrak{X}$ is referred to as the* body *of $\mathfrak{X}$.* □

For each $U \subset |\mathfrak{X}|$ we have the fermonic ideal $\mathcal{J}_{\mathfrak{X}}(U) \subset \mathcal{O}_{\mathfrak{X}}(U)$ giving, over $C(U)$, the fermionic modules $\mathcal{J}_{\mathfrak{X}}(U)/\mathcal{J}_{\mathfrak{X}}(U)^2$. For $\mathfrak{X}$ a supermanifold, it is locally isomorphic to $\mathbb{K}^{m|n}$. Therefore, there exists a covering $(U_i)_i$ such that $\mathcal{J}_{\mathfrak{X}}(U_i)/\mathcal{J}_{\mathfrak{X}}(U_i)^2$ is a free $C(U_i)$-module for all i. Hence $\mathcal{J}/\mathcal{J}^2_{\mathfrak{X}}$ is a locally free sheaf of C-modules on $|\mathfrak{X}|$ and so will be the sheaf of sections of a vector bundle $\mathbf{V}_{\mathfrak{X}} \to M$ over the body of $\mathfrak{X}$. This is the analogue of property II) in §2.1.1 for supermanifolds.

Definition 2.20 *The vector bundle $\mathbf{V}_{\mathfrak{X}} \to M$ is referred to as the* soul *of the supermanifold $\mathfrak{X}$.* □

[4] This viewpoint was developed by the authors [2, 4, 3] and is sometimes referred to as the 'BLK-approach' to supermanifold theory.

The body and soul of $\mathfrak{X}$ admit the following 'geometric relation' to $\mathfrak{X}$.

Proposition 2.21 The body of a supermanifold $\mathfrak{X}$ is naturally embedded in $\mathfrak{X}$; the soul of $\mathfrak{X}$ is the conormal bundle to the embedding of its body. □

Proof Recall that the body of $\mathfrak{X}$ is the locally ringed space $M = (|\mathfrak{X}|, C)$ where $C = O_{\mathfrak{X}}/\mathcal{J}_{\mathfrak{X}}$. An embedding is a morphism of supermanifolds $(f, h^\sharp)$ where f is an embedding of the underlying, topological space and $h^\sharp$ is an epimorphism on structure sheaves. Denote by $h^\sharp$ the quotient $O_{\mathfrak{X}} \xrightarrow{h^\sharp} O_{\mathfrak{X}}/\mathcal{J}_{\mathfrak{X}} = C$. Then we obtain a morphism $\iota = (\mathbf{1}, h^\sharp) : M \to \mathfrak{X}$, where $\mathbf{1} : |\mathfrak{X}| \to |\mathfrak{X}|$ is the identity map. The morphism ι is an embedding of supermanifolds.[5] The ideal sheaf of the body embedding $\iota : M \hookrightarrow \mathfrak{X}$ is $\ker h^\sharp = \mathcal{J}_{\mathfrak{X}}$. The conormal bundle to ι is a vector bundle over the body M whose sheaf of sections is given by $\mathcal{J}_{\mathfrak{X}}/\mathcal{J}^2_{\mathfrak{X}}$. This is precisely the sheaf of fermonic modules in $\mathfrak{X}$. Hence the soul of $\mathfrak{X}$ *is* the conormal bundle to the body embedding ι. □

Lemma 2.22 Let $\mathfrak{X}$ be a supermanifold and $\mathcal{J}_{\mathfrak{X}} \subset O_{\mathfrak{X}}$ the fermionic ideal. Then the soul $\mathbf{V}_{\mathfrak{X}}$ of $\mathfrak{X}$ is given by $\mathbf{V}_{\mathfrak{X}} = \operatorname{Spec} \Gamma(M, \operatorname{Sym}^\bullet(\mathcal{J}_{\mathfrak{X}}/\mathcal{J}^2_{\mathfrak{X}}))$. □

Proof This follows from the characterization of vector bundles as locally ringed spaces from algebraic geometry (see e.g., [11, Tag 01M1]) along with the identification of the sheaf of sections of $\mathbf{V}_{\mathfrak{X}}$ in the proof of Proposition 2.21. □

2.4 First properties of supermanifold embeddings

The following result characterizes the body and soul of submanifolds[6] of a supermanifold $\mathfrak{X}$ in terms of the body and soul of $\mathfrak{X}$.[7]

Theorem 2.23 Let $\mathfrak{X}$ be a supermanifold, $\mathfrak{Y} \subset \mathfrak{X}$ a submanifold and suppose that $\mathcal{I}_{\mathfrak{Y}} \cap \mathcal{J}_{\mathfrak{X}} = (0)$, for $\mathcal{I}_{\mathfrak{Y}}$ the ideal sheaf of $\mathfrak{Y}$ and $\mathcal{J}_{\mathfrak{X}} \subset O_{\mathfrak{X}}$ the fermionic ideal. Then:

(i) the body N of $\mathfrak{Y}$ is a submanifold of the body M of $\mathfrak{X}$;
(ii) the ideal sheaf of $N \subset M$ is given by the image of $\mathcal{I}_{\mathfrak{Y}}$ in O_M;
(iii) $N \cong \mathfrak{Y} \cap M$;
(iv) $\mathbf{V}_{\mathfrak{Y}} \cong \mathbf{V}_{\mathfrak{X}}|_{|\mathfrak{Y}|}$

where $\mathbf{V}_{\mathfrak{Y}}$ resp., $\mathbf{V}_{\mathfrak{X}}$ denotes the soul of $\mathfrak{Y}$ resp., $\mathfrak{X}$. □

Proof *(i)* Let $(i, h^\sharp) : \mathfrak{Y} \subset \mathfrak{X}$ denote the embedding of supermanifolds where $i : |\mathfrak{Y}| \subset |\mathfrak{X}|$ the embedding of topological spaces and $h^\sharp : O_{\mathfrak{X}} \to i_* O_{\mathfrak{Y}}$ an epimorphism. The body of $\mathfrak{Y}$ resp., $\mathfrak{X}$ is the locally ringed space $N = (|\mathfrak{Y}|, O_{\mathfrak{Y}}/\mathcal{J}_{\mathfrak{Y}})$

[5] Scheme-theoretically, ι is a *closed immersion*.

[6] By a submanifold of a supermanifold, it is meant a subspace which is itself a supermanifold.

[7] Theorem 2.23 is adapted from work in [12]. We refer here for more details on the nature of supermanifold embeddings.

resp., $M = (|\mathfrak{X}|, \mathcal{O}_{\mathfrak{X}}/\mathcal{J}_{\mathfrak{X}})$. Since $h^\sharp$ preserves the fermionic ideal it gives a commuting diagram,

$$\begin{array}{ccccc} \mathcal{O}_{\mathfrak{X}} & \xrightarrow{h^\sharp} & i_*\mathcal{O}_{\mathfrak{Y}} & \longrightarrow & 0 \\ \downarrow & & \downarrow & & \\ \mathcal{O}_M & \xrightarrow{\bar{h}^\sharp} & \mathcal{O}_N & \longrightarrow & 0 \end{array}$$

Hence we have an embedding $(i, \bar{h}^\sharp) : N \subset M$.

(ii) Let $\mathcal{I}_{\mathfrak{Y}}$ denote the ideal sheaf of $\mathfrak{Y} \subset \mathfrak{X}$. There exist short exact sequences,

$$0 \to \mathcal{I}_{\mathfrak{Y}} \cap \mathcal{J}_{\mathfrak{X}} \to \mathcal{I}_{\mathfrak{Y}} \to \mathcal{O}_M \to 0 \quad \text{and} \quad 0 \to \mathcal{I}_{\mathfrak{Y}} \cap \mathcal{J}_{\mathfrak{X}} \to \mathcal{J}_{\mathfrak{X}} \to \mathcal{O}_{\mathfrak{Y}} \to 0 \qquad (2)$$

where $\mathcal{I}_{\mathfrak{Y}} \to \mathcal{O}_M$ is the composition $\mathcal{I}_{\mathfrak{Y}} \to \mathcal{O}_{\mathfrak{X}} \to \mathcal{O}_M$ and $\mathcal{J}_{\mathfrak{X}} \to \mathcal{O}_{\mathfrak{Y}}$ is the composition $\mathcal{J}_{\mathfrak{X}} \to \mathcal{O}_{\mathfrak{X}} \to \mathcal{O}_{\mathfrak{Y}}$. Assuming $\mathcal{I}_{\mathfrak{Y}} \cap \mathcal{J}_{\mathfrak{X}} = (0)$, the ideal $\mathcal{I}_{\mathfrak{Y}}$ can be identified with its image in $\mathcal{O}_M$ and $\mathcal{J}_{\mathfrak{X}}$ with its image in $\mathcal{O}_{\mathfrak{Y}}$. The support of $\mathcal{O}_M/\mathcal{I}_{\mathfrak{Y}}$ is $|\mathfrak{X}| \setminus (|\mathfrak{X}| \setminus |\mathfrak{Y}|) = |\mathfrak{Y}|$. Set $N' := (|\mathfrak{Y}|, \mathcal{O}_M/\mathcal{I}_{\mathfrak{Y}})$. We need to show $N' \cong N$, for N the body of $\mathfrak{Y}$. To show this, firstly note $\mathcal{I}_{\mathfrak{Y}} = \ker\{\mathcal{O}_{\mathfrak{X}} \stackrel{h^\sharp}{\to} i_*\mathcal{O}_{\mathfrak{Y}}\}$. Therefore $\mathcal{J}_{\mathfrak{Y}} = \mathcal{J}_{\mathfrak{X}}/\mathcal{I}_{\mathfrak{Y}}\mathcal{J}_{\mathfrak{X}}$. Since the morphism $\mathcal{J}_{\mathfrak{X}} \to \mathcal{O}_{\mathfrak{Y}}$ from (2) is injective we have the commuting diagram,

$$\begin{array}{ccccccccc} 0 & \longrightarrow & \mathcal{I}_{\mathfrak{Y}}\mathcal{J}_{\mathfrak{X}} & \longrightarrow & \mathcal{J}_{\mathfrak{X}} & \longrightarrow & \mathcal{J}_{\mathfrak{Y}} & \longrightarrow & 0 \\ & & & & \downarrow & & \downarrow & & \\ & & & & \mathcal{O}_{\mathfrak{Y}} & = & \mathcal{O}_{\mathfrak{Y}} & & \end{array}$$

Hence the fermionic ideal for $\mathfrak{Y}$ satisfies $\mathcal{J}_{\mathfrak{Y}} = \mathrm{im}\{\mathcal{J}_{\mathfrak{X}} \to \mathcal{O}_{\mathfrak{Y}}\}$. Therefore,

$$\frac{\mathcal{O}_{\mathfrak{Y}}}{\mathcal{J}_{\mathfrak{Y}}} = \frac{\mathcal{O}_{\mathfrak{Y}}}{\mathcal{J}_{\mathfrak{X}}} \cong \frac{(\mathcal{O}_{\mathfrak{X}}/\mathcal{I}_{\mathfrak{Y}})}{\mathcal{J}_{\mathfrak{X}}} \cong \frac{(\mathcal{O}_{\mathfrak{X}}/\mathcal{J}_{\mathfrak{X}})}{\mathcal{I}_{\mathfrak{Y}}} = \frac{\mathcal{O}_M}{\mathcal{I}_{\mathfrak{Y}}}. \qquad (3)$$

Denoting by φ the isomorphism in (3) we have an isomorphism of locally ringed spaces $(\mathbf{1}, \varphi) : N \cong N'$.

(iii) Note that $\mathfrak{Y} \cap M \cong \mathfrak{Y} \times_{\mathfrak{X}} M$, i.e., that the intersection can be identified with the fiber product of embeddings $\mathfrak{Y} \subset \mathfrak{X}$ and $M \subset \mathfrak{X}$. This leads to $i_*\mathcal{O}_{\mathfrak{Y}\cap M} = i_*\mathcal{O}_{\mathfrak{Y}} \otimes_{\mathcal{O}_{\mathfrak{X}}} \mathcal{O}_M$. With $\mathcal{O}_M = \mathcal{O}_{\mathfrak{X}}/\mathcal{J}_{\mathfrak{X}}$ we have,

$$i_*\mathcal{O}_{\mathfrak{Y}\cap M} = i_*\mathcal{O}_{\mathfrak{Y}} \otimes_{\mathcal{O}_{\mathfrak{X}}} \mathcal{O}_M = i_*\mathcal{O}_{\mathfrak{Y}} \otimes_{\mathcal{O}_{\mathfrak{X}}} (\mathcal{O}_{\mathfrak{X}}/\mathcal{J}_{\mathfrak{X}}) \cong \frac{i_*\mathcal{O}_{\mathfrak{Y}}}{\mathcal{J}_{\mathfrak{X}} \cdot i_*\mathcal{O}_{\mathfrak{Y}}} \cong \frac{i_*\mathcal{O}_{\mathfrak{Y}}}{i_*\mathcal{J}_{\mathfrak{Y}}} = i_*\mathcal{O}_N$$

for $N = (|\mathfrak{Y}|, \mathcal{O}_{\mathfrak{Y}}/\mathcal{J}_{\mathfrak{Y}})$ the body of $\mathfrak{Y}$. In identifying $|\mathfrak{Y}| = |\mathfrak{Y}| \cap |\mathfrak{X}|$ see that, as a locally ringed space, $\mathfrak{Y} \cap M = (|\mathfrak{Y}|, \mathcal{O}_{\mathfrak{Y}\cap M})$. Now since i is an embedding, the isomorphism $i_*\mathcal{O}_{\mathfrak{Y}\cap M} \cong i_*\mathcal{O}_N$ implies $\mathcal{O}_{\mathfrak{Y}\cap M} \cong \mathcal{O}_N$ as sheaves on $|\mathfrak{Y}|$. Hence we obtain the desired isomorphism $N \cong \mathfrak{Y} \cap M$.

(iv) Recall that $\mathcal{I}_{\mathfrak{Y}} = \ker\{\mathcal{O}_{\mathfrak{X}} \stackrel{h^\sharp}{\to} i_*\mathcal{O}_{\mathfrak{Y}}\}$ is the ideal sheaf defining the embedding $\mathfrak{Y} \subset \mathfrak{X}$. Accordingly we have a relation between fermonic ideals, $\mathcal{J}_{\mathfrak{Y}} = \mathcal{J}_{\mathfrak{X}}/\mathcal{I}_{\mathfrak{Y}}\mathcal{J}_{\mathfrak{X}}$. This leads to the following morphism of short exact sequences,

$$\begin{array}{ccccccccc} 0 & \longrightarrow & \mathcal{I}_{\mathfrak{Y}}\mathcal{J}_{\mathfrak{X}}^2 & \longrightarrow & \mathcal{J}_{\mathfrak{X}}^2 & \longrightarrow & i_*\mathcal{J}_{\mathfrak{Y}}^2 & \longrightarrow & 0 \\ & & \downarrow & & \downarrow & & \downarrow & & \\ 0 & \longrightarrow & \mathcal{I}_{\mathfrak{Y}}\mathcal{J}_{\mathfrak{X}} & \longrightarrow & \mathcal{J}_{\mathfrak{X}} & \longrightarrow & i_*\mathcal{J}_{\mathfrak{Y}} & \longrightarrow & 0 \end{array}$$

giving the right-exact sequence of $\mathcal{O}_M$-modules,

$$\frac{\mathcal{I}_{\mathfrak{Y}}\mathcal{J}_{\mathfrak{X}}}{\mathcal{I}_{\mathfrak{Y}}\mathcal{J}_{\mathfrak{X}}^2} \longrightarrow \frac{\mathcal{J}_{\mathfrak{X}}}{\mathcal{J}_{\mathfrak{X}}^2} \longrightarrow i_*\frac{\mathcal{J}_{\mathfrak{Y}}}{\mathcal{J}_{\mathfrak{Y}}^2} \longrightarrow 0 \tag{4}$$

With $\mathcal{O}_M = \mathcal{O}_{\mathfrak{X}}/\mathcal{J}_{\mathfrak{X}}$ we have,

$$\frac{\mathcal{I}_{\mathfrak{Y}}\mathcal{J}_{\mathfrak{X}}}{\mathcal{I}_{\mathfrak{Y}}\mathcal{J}_{\mathfrak{X}}^2} \cong (\mathcal{I}_{\mathfrak{Y}}\mathcal{J}_{\mathfrak{X}}) \otimes_{\mathcal{O}_{\mathfrak{X}}} \mathcal{O}_M \cong \mathcal{I}_{\mathfrak{Y}} \otimes_{\mathcal{O}_{\mathfrak{X}}} \mathcal{J}_{\mathfrak{X}} \otimes_{\mathcal{O}_{\mathfrak{X}}} \mathcal{O}_M \cong \mathcal{I}_{\mathfrak{Y}} \otimes_{\mathcal{O}_{\mathfrak{X}}} \frac{\mathcal{J}_{\mathfrak{X}}}{\mathcal{J}_{\mathfrak{X}}^2} \cong \mathcal{I}_N \otimes_{\mathcal{O}_M} \frac{\mathcal{J}_{\mathfrak{X}}}{\mathcal{J}_{\mathfrak{X}}^2}$$

where $\mathcal{I}_N = \operatorname{im}\{\mathcal{I}_{\mathfrak{Y}} \to \mathcal{O}_M\}$ is the ideal sheaf of $N \subset M$ by part *(ii)*. In pulling back the sequence (4) to N now, the morphism $i^*(\mathcal{I}_{\mathfrak{Y}}\mathcal{J}_{\mathfrak{X}}/\mathcal{I}_{\mathfrak{Y}}\mathcal{J}_{\mathfrak{X}}^2) \to i^*(\mathcal{J}_{\mathfrak{X}}/\mathcal{J}_{\mathfrak{X}}^2)$ necessarily vanishes. Hence that $i^*(\mathcal{J}_{\mathfrak{X}}/\mathcal{J}_{\mathfrak{X}}^2) \cong \mathcal{J}_{\mathfrak{Y}}/\mathcal{J}_{\mathfrak{Y}}^2$. As these are (respectively) the sheaves of sections of souls $\mathbf{V}_{\mathfrak{Y}}$ and $\mathbf{V}_{\mathfrak{X}}$, it follows that $\mathbf{V}_{\mathfrak{Y}} \cong \mathbf{V}_{\mathfrak{X}}|_N$. □

3 Examples

3.1 Affine algebraic superspaces

Let $A = \mathbb{K}[x^1, \ldots, x^m|\theta_1, \ldots, \theta_n]$ be the free algebra on $(m+n)$-generators subject to the relations $x^ix^j = x^jx^i, x^i\theta_a = \theta_ax^i$ and $\theta_a\theta_b = -\theta_b\theta_a$. Then A defines a supercommutative ring. The reduced ring of A is the polynomial ring $\mathbb{K}[x^1, \ldots, x^n]$. If we denote by V a free $\mathbb{K}[x^1, \ldots, x^m]$-module of rank n, then $A \cong \wedge^\bullet_{\mathbb{K}[x^1,\ldots,x^m]} V$. Hence by Definition 2.3, A defines an *affine*, supercommutative ring. The spectrum construction from algebraic geometry generalizes to the supergeometry setting and establishes a duality between affine algebraic superspaces and supercommutative rings.[8] The $m|n$-dimensional, affine algebraic superspace over $\mathbb{K}$, denoted $\mathbb{A}^{m|n}_{\mathbb{K}}$, is then the spectrum of A, i.e., $\mathbb{A}^{m|n}_{\mathbb{K}} = \operatorname{Spec} \mathbb{K}[x^1, \ldots, x^m|\theta_1, \ldots, \theta_n]$.

(i) the body of $\mathbb{A}^{m|n}_{\mathbb{K}}$ is $\operatorname{Spec} \mathbb{K}[x^1, \ldots, x^m] = \mathbb{A}^m_{\mathbb{K}}$;

[8] See e.g., [13].

(ii) the soul of $\mathbb{A}_{\mathbb{K}}^{m|n}$ is the free $\mathbb{K}[x^1, \ldots, x^m]$-module V of rank n such that $\mathcal{O}(\mathbb{A}_{\mathbb{K}}^{m|n}) \cong \wedge^{\bullet}_{\mathbb{K}[x^1,\ldots,x^m]} V$.

3.1.1 Affine subvarieties

As in algebraic geometry, affine subvarieties are subsets $S \subset \mathbb{A}_{\mathbb{K}}^{m|n}$ defined by a loci of polynomials. With $\mathbb{A}_{\mathbb{K}}^{m|n} = \operatorname{Spec} \mathbb{K}[x^1, \ldots, x^m|\theta_1, \ldots, \theta_n]$, the coordinate ring $\mathbb{K}[S]$ of S is given by $\mathbb{K}[S] = \mathbb{K}[x^1, \ldots, x^m|\theta_1, \ldots, \theta_n]/I$ for I an ideal. Note by Definition 2.3, since $\mathbb{K}[x^1, \ldots, x^m|\theta_1, \ldots, \theta_n]$ is affine, then so is the coordinate ring $\mathbb{K}[S]$. Suppose $I = (f_i)_i$ for $f_i \in \mathbb{K}[x^1, \ldots, x^m|\theta_1, \ldots, \theta_n]$. With the characterization of the body and soul of submanifolds in Theorem 2.23 we have:

(i) the body of S is given by the intersection of S with the body of $\mathbb{A}_{\mathbb{K}}^{m|n}$. This is precisely the locus,

$$f_i(x^1, \ldots, x^m|\theta_1, \ldots, \theta_n) = 0 \qquad \text{and} \qquad \theta_1 = \cdots = \theta_n = 0 \tag{5}$$

for all i.

(ii) The soul of S is given by restricting the soul of $\mathbb{A}_{\mathbb{K}}^{m|n}$ to the body of S. With $\mathcal{O}(\mathbb{A}_{\mathbb{K}}^{m|n}) \cong \wedge^{\bullet}_{\mathbb{K}[x^1,\ldots,x^m]} V$, the soul of S is then $V|_{|S|}$.

3.1.2 Superschemes

With affine superspaces over a field $\mathbb{K}$ now established, we can define superschemes following Definition 2.17 and in line with Leites in [13].

Definition 3.1 *An* algebraic superscheme *over a field $\mathbb{K}$ is supercommutative space which is locally isomorphic to an affine superspace over $\mathbb{K}$.* □

3.2 Projective superspaces

The multiplicative group $\mathbb{G}_m$ acts on the affine superspace $\mathbb{A}_{\mathbb{K}}^{m|n}$ by scalar multiplication. On generators $x|\theta$ of the coordinate ring of $\mathbb{A}_{\mathbb{K}}^{m|n}$ and $\lambda \in \mathbb{G}_m$ we have the scaling $x|\theta \mapsto \lambda x|\lambda\theta$. The Euclidean superspace $\mathbb{K}^{m|n}$ was formed as a local, supercommutative ringed space in Example 2.15. Projective superspace over $\mathbb{K}$ is then formed as the quotient $\mathbb{P}_{\mathbb{K}}^{m-1|n} := (\mathbb{K}^{m|n} \setminus \{(0|0)\})/\mathbb{G}_m$.

(i) the body of $\mathbb{P}_{\mathbb{K}}^{m-1|n}$ is $\mathbb{P}_{\mathbb{K}}^{m-1}$;

(ii) the soul of $\mathbb{P}_{\mathbb{K}}^{m-1|n}$ is the n-fold sum of the tautological line bundle over $\mathbb{P}_{\mathbb{K}}^{m-1}$. Its sheaf of sections is $\oplus^n \mathcal{O}_{\mathbb{P}_{\mathbb{K}}^m}(-1)$.

An atlas on projective superspace $\mathbb{P}^{m-1|n}_{\mathbb{K}}$ can be described in a manner analogous to the case of $\mathbb{P}^{m-1}_{\mathbb{K}}$. An open set in this atlas can be described as follows. Denoting by $A = \mathbb{K}[X^1, \ldots, X^m|\Theta_1, \ldots, \Theta_n]$ the homogeneous coordinate ring, let U_i be the localization of A at X^i. Set $x^j = \frac{X^j}{X^i}$ and $\theta_a = \frac{\Theta_a}{x^i}$. Then $O(U_i) = \mathbb{K}[x^1, \ldots, x^{i-1}, x^{i+1}, \ldots, x^m|\theta_1, \ldots, \theta_n] \cong A_{X^i}$ is the coordinate ring of the open set $U_i \subset \mathbb{P}^{m-1|n}_{\mathbb{K}}$. From here it ought to be clear why the projective superspace is $m-1|n$-dimensional.

3.2.1 Projective superspace varieties

Homogeneous ideals in the coordinate ring $\mathbb{K}[X^1, \ldots, X^m|\Theta_1, \ldots, \Theta_n]$ with respect to the $\mathbb{G}_m$-action define subvarieties of projective superspace. As an illustration over $\mathbb{K} = \mathbb{C}$, the superspace quadric $Q \subset \mathbb{P}^{2|2}_{\mathbb{C}}$ is given by the ideal $((X^1)^2+(X^2)^2+(X^3)^2+\Theta_2\Theta_2) \subset \mathbb{C}[X^1, X^2, X^3|\Theta_1, \Theta_2]$. In the open set U_1 we have the affine subvariety $(1 + (x^2)^2 + (x^3)^2 + \theta_1\theta_2) \subset \mathbb{C}[x^2, x^3|\theta_1, \theta_2]$ where $x^i = X^i/x^1$ and $\theta_a = \Theta_a/x^1$.

(i) by (5), the body of Q is the projective subvariety defined by the ideal $((X^1)^2 + (X^2)^2 + (X^3)^2) \subset \mathbb{C}[X^1, X^2, X^3]$, which describes the degree two embedding $\mathbb{P}^1_{\mathbb{C}} \subset \mathbb{P}^2_{\mathbb{C}}$. Hence the body of Q is isomorphic to $\mathbb{P}^1_{\mathbb{C}}$;

(ii) by Theorem 2.23, the soul of Q is the restriction of the soul of $\mathbb{P}^{2|2}_{\mathbb{C}}$ to the degree two embedding $\mathbb{P}^1_{\mathbb{C}} \subset \mathbb{P}^2_{\mathbb{C}}$. Recall from §3.2 that the soul of $\mathbb{P}^{2|2}_{\mathbb{C}}$ is $\oplus^2 O_{\mathbb{P}^2_{\mathbb{C}}}(-1)$. Hence, the soul of Q is $\oplus^2 O_{\mathbb{P}^1_{\mathbb{C}}}(-2)$.

For more details and theory on projective superspaces and subvarieties, including their weighted analogues, see [14].

3.3 Super Riemann surfaces

Definition 3.2 A super Riemann surface *is a pair $(\mathcal{S}, \mathcal{D})$ where $\mathcal{S}$ is a $(1|1)$-dimensional, complex supermanifold and $\mathcal{D} \subset T_{\mathcal{S}}$ is a totally non-integrable distribution, i.e., defines a short exact sequence,*

$$0 \longrightarrow \mathcal{D} \longrightarrow T_{\mathcal{S}} \longrightarrow \mathcal{D} \otimes \mathcal{D} \longrightarrow 0$$

where $T_{\mathcal{S}} = Der\, O_{\mathcal{S}}$ is the tangent sheaf. □

In coordinates $x|\theta$ on the complex Euclidean superspace $\mathbb{C}^{1|1}$, the vector field $\frac{\partial}{\partial\theta} + \theta\frac{\partial}{\partial x}$ generates a totally non-integrable distribution $D \subset T_{\mathbb{C}^{1|1}}$. As a result $(\mathbb{C}^{1|1}, D)$ defines a (non-compact) super Riemann surface. It is in fact the local model for any super Riemann surface. In more detail, let $U, V \subset \mathcal{S}$ denote open sets, $x|\theta$ resp., $y|\eta$ coordinates on U resp., V and $\rho_{UV} = \rho^+_{UV}|\rho^-_{UV}$ transition functions on $U \cap V$,

$$y = \rho^+(x|\theta) = f(x) \qquad \eta = \rho^-(x|\theta) = g(x)\theta. \tag{6}$$

The system $(U, V, U \cap V, \{\rho_{UV}, \rho_{VU}\})$ form (part of) an atlas for a super Riemann surface $(\mathcal{S}, \mathcal{D})$ iff ρ_{UV} preserves the generators of $\mathcal{D}|_U$ and $\mathcal{D}|_V$. With $\frac{\partial}{\partial\theta} + \theta\frac{\partial}{\partial x}$ and $\frac{\partial}{\partial\eta} + \eta\frac{\partial}{\partial y}$ the generators of $\mathcal{D}|_U$ and $\mathcal{D}|_V$ respectively, that ρ_{UV} preserves $\mathcal{D}$ means $(\rho_{UV})_*(\frac{\partial}{\partial\theta} + \theta\frac{\partial}{\partial x}) \propto \frac{\partial}{\partial\eta} + \eta\frac{\partial}{\partial y}$. Enforcing this necessitates $g(x) = \sqrt{f'(x)}$ in (6). We then have as a consequence:

(i) the body $|\mathcal{S}|$ of a super Riemann surface $(\mathcal{S}, \mathcal{D})$ is a Riemann surface;
(ii) the soul of a super Riemann surface is a spin structure $(T^*_{|\mathcal{S}|})^{1/2}$ on the body.

Evidently,

Theorem 3.3 The set of totally non-integrable distributions $\mathcal{D}$ on compact, $(1|1)$-dimensional supermanifold $\mathcal{S}$ is in bijective correspondence with spin structures on the body $|\mathcal{S}|$ of $\mathcal{S}$.

Remark 3.4 *Authors such as Rabin and Crane in [15] and Witten in [16] take a broader definition of a super Riemann surface. For these authors, the functions in* (6) *are themselves valued in an auxiliary exterior algebra. The author in [17] views these describing infinitesimal deformations however, which is in line with the treatment of super Riemann surfaces and their moduli from [18]. Through the viewpoint in this article, we would understand the super Riemann surfaces from [15, 16] as 'DeWitt structures on a specified locus of points' of super Riemann surfaces from [17, 18].*□

4 Points of a supermanifold

4.1 L-points

We have so far introduced supermanifolds as local, supercommutative ringed spaces. That is, a supermanifold $\mathfrak{X}$ consists principally of a topological space $|\mathfrak{X}|$ and a sheaf of local, supercommutative rings $\mathcal{O}_{\mathfrak{X}}$ on $|\mathfrak{X}|$. It is arguably more natural however to adopt the functor of points approach and view a supermanifold $\mathfrak{X}$ through the representable functor it defines, $\underline{\mathfrak{X}} = \mathrm{Hom}_{\mathsf{SM}}(-, \mathfrak{X})$ for SM the category of supermanifolds. This allows for defining points of a supermanifold in a very general setting. Over a field $\mathbb{K}$ and for an integer L, consider the supercommutative ring $\wedge^\bullet_{\mathbb{K}}\mathbb{K}^L$. By §3.1, its spectrum is the affine superspace $\mathbb{A}^{0|L}_{\mathbb{K}}$.

Definition 4.1 *An L-point of a supermanifold $\mathfrak{X}$ is a morphism $\mathbb{A}^{0|L}_{\mathbb{K}} \to \mathfrak{X}$. If $L = 0$, we refer to the L-point as* reduced. *If $L = 1$ the L-point is referred to as a* superpoint.□

In analogy with schemes we have:[9]

[9] c.f., [11, Lemma 01J6]

Lemma 4.2 An L-point $\mathbb{A}_{\mathbb{K}}^{0|L} \xrightarrow{\phi} \mathfrak{X}$ is equivalent to a homomorphism of (local) supercommutative rings $\mathcal{O}_{\mathfrak{X},P} \to \mathbb{K}[\xi_1, \ldots, \xi_L]$, where P is the image of ϕ and $\mathbb{A}_{\mathbb{K}}^{0|L} = \operatorname{Spec} \mathbb{K}[\xi_1, \ldots, \xi_L]$. □

Proof With $\mathbb{A}_{\mathbb{K}}^{0|L} = \operatorname{Spec} \mathbb{K}[\xi_1, \ldots, \xi_L]$, as a locally ringed space we can write $\mathbb{A}_{\mathbb{K}}^{0|L} = (\{pt\}, \mathbb{K}[\xi_1, \ldots, \xi_L])$ as in Example 2.8. With $\mathbb{A}_{\mathbb{K}}^{0|L} \xrightarrow{\phi} \mathfrak{X}$ an L-point of $\mathfrak{X}$ it is a morphism of loca, supercommutative ringed spaces. Accordingly, we can write $\phi = (i, h^\sharp)$ where $i : \{pt\} \to |\mathfrak{X}|$ and $h^\sharp : i^*\mathcal{O}_{\mathfrak{X}} \to \mathbb{K}[\xi_1, \ldots, \xi_L]$. Writing $P = i(\{pt\}) \in |\mathfrak{X}|$ note that $i^*\mathcal{O}_{\mathfrak{X}} \cong \mathcal{O}_{\mathfrak{X},P}$. As such ϕ is completely determined by the homomorphism of supercommutative rings $\mathcal{O}_{\mathfrak{X},P} \xrightarrow{h^\sharp} \mathbb{K}[\xi_1, \ldots, \xi_L]$. □

To expand further on Lemma 4.2: with the L-point $\mathbb{A}_{\mathbb{K}}^{0|L} \xrightarrow{\phi} \mathfrak{X}$ equivalent to a homomorphism of local, supercommutative rings $\mathcal{O}_{\mathfrak{X},P} \xrightarrow{h^\sharp} \mathbb{K}[\xi_1, \ldots, \xi_L]$, recall from Example 2.5 that the maximal ideal in $\mathbb{K}[\xi_1, \ldots, \xi_L]$ is the fermionic ideal $J = \ker\{\mathbb{K}[\xi_1, \ldots, \xi_L] \to \mathbb{K}\}$. As such $h^\sharp$ sends the unique maximal ideal $\mathfrak{m}_P \subset \mathcal{O}_{\mathfrak{X},P}$ to J. Hence that it induces a morphism of short exact sequences,

$$\begin{array}{ccccccccc} 0 & \longrightarrow & \mathfrak{m}_P & \longrightarrow & \mathcal{O}_{\mathfrak{X},P} & \longrightarrow & \mathbb{K} & \longrightarrow & 0 \\ & & \downarrow & & \downarrow{\scriptstyle h^\sharp} & & \downarrow & & \\ 0 & \longrightarrow & J & \longrightarrow & \mathbb{K}[\xi_1, \ldots, \xi_L] & \longrightarrow & \mathbb{K} & \longrightarrow & 0 \end{array} \tag{7}$$

A function $f \in \mathcal{O}_{\mathfrak{X}}$ *vanishes at the L-point* ϕ if its germ $[f]_P$ vanishes modulo the maximal ideal $\mathfrak{m}_P$. Equivalently by (7), if $h^\sharp([f]_P)$ vanishes modulo the fermionic ideal J.

Remark 4.3 *For convenience we will often conflate the point P with the image of the L-point* $\phi = (i, h^\sharp)$ *on* $\{pt\}$ □

Recall our notation: for any supermanifold $\mathfrak{X}$, $\underline{\mathfrak{X}} := \operatorname{Hom}_{\mathsf{SM}}(-, \mathfrak{X})$.

Definition 4.4 *The* locus of L-points of $\mathfrak{X}$ *is the image* $\underline{\mathfrak{X}}(\mathbb{A}_{\mathbb{K}}^{0|L})$ *in* Set. *For brevity, it is denoted by* $\underline{\mathfrak{X}}(L)$ □

4.2 First properties

Theorem 4.5 The locus of reduced points of a supermanifold is isomorphic to the locus of rational points of its body. □

Proof Recall that a reduced point of $\mathfrak{X}$ is an L-point with $L = 0$. The locus of reduced points is then $\underline{\mathfrak{X}}(0)$. By Lemma 4.2 and the diagram in (7), any reduced point P is determined by a morphism $\mathcal{O}_{\mathfrak{X},P} \to \mathbb{K}$. Since $\mathbb{K}$ is commutative we have an isomorphism of sets[10] $\mathrm{Hom}_{\mathsf{SCR}}(\mathcal{O}_{\mathfrak{X},P}, \mathbb{K}) \cong \mathrm{Hom}_{\mathsf{CR}}(\mathcal{O}_{M,P}, \mathbb{K})$ where M is the body of $\mathfrak{X}$.[11] Therefore $\underline{\mathfrak{X}}(0) \cong \underline{M}(\mathrm{Spec}\,\mathbb{K})$. As $\underline{M}(\mathrm{Spec}\,\mathbb{K})$ is precisely the locus of rational points of M, the theorem follows. □

Theorem 4.6 The locus of superpoints in a supermanifold is isomorphic to the locus of rational points of its soul. □

Proof We need to show there exists an isomorphism $\underline{\mathfrak{X}}(1) \cong \underline{\mathbf{V}}_{\mathfrak{X}}(\mathrm{Spec}\,\mathbb{K})$ where $\mathbf{V}_{\mathfrak{X}}$ is the soul of $\mathfrak{X}$, characterized in Lemma 2.22. We begin by constructing a map $\underline{\mathfrak{X}}(1) \to \underline{\mathbf{V}}_{\mathfrak{X}}(\mathrm{Spec}\,\mathbb{K})$. Let $\mathbb{A}^{0|1}_{\mathbb{K}} \stackrel{\phi}{\to} \mathfrak{X}$ be a superpoint with image P in $|\mathfrak{X}|$. By Lemma 4.2 and the diagram in (7), ϕ is determined by the morphism $h^\sharp$ on local supercommutative rings,

$$\begin{array}{ccccccccc} 0 & \longrightarrow & \mathfrak{m}_P & \longrightarrow & \mathcal{O}_{\mathfrak{X},P} & \longrightarrow & \mathbb{K} & \longrightarrow & 0 \\ & & \downarrow & & \downarrow{\scriptstyle h^\sharp} & & \downarrow & & \\ 0 & \longrightarrow & \mathbb{K} & \stackrel{\xi}{\longrightarrow} & \mathbb{K}[\xi] & \longrightarrow & \mathbb{K} & \longrightarrow & 0 \end{array} \tag{8}$$

where we have identified $\ker\{\mathbb{K}[\xi] \to \mathbb{K}\} \cong \mathbb{K}$ as $\mathbb{K}$-modules and $\mathbb{K} \stackrel{\xi}{\to} \mathbb{K}[\xi]$ denotes multiplication by ξ. For any $[f]_P \in \mathcal{O}_{\mathfrak{X},P}$ we can write $h^\sharp([f]_P) = a([f]_P) + \xi b([f]_P)$. Since $h^\sharp$ is a morphism of supercommutative rings, $a : \mathcal{O}_{\mathfrak{X},p} \to \mathbb{K}$ will be an even homomorphism and $b : \mathcal{O}_{\mathfrak{X},P} \to \mathbb{K}[\xi]$ will be an odd derivation over a, i.e., that

$$b([ff']_P) = a([f]_P)b([f']_P) + b([f]_P)a([f']_P). \tag{9}$$

By Theorem 4.5, a defines a rational point in the body of $\mathfrak{X}$. Now with $\mathcal{J}_{\mathfrak{X}} \subset \mathcal{O}_{\mathfrak{X}}$ the fermionic ideal and $\mathcal{J}_{\mathfrak{X},P} \subset \mathcal{O}_{\mathfrak{X},P}$ the stalk at P, $h^\sharp$ induces, analogously to (8), the morphism of short exact sequences,

$$\begin{array}{ccccccccc} 0 & \longrightarrow & \mathcal{J}_{\mathfrak{X},P} & \longrightarrow & \mathcal{O}_{\mathfrak{X},P} & \longrightarrow & C_{M,P} & \longrightarrow & 0 \\ & & {\scriptstyle h^\sharp}\downarrow & \swarrow{\scriptstyle b} & \downarrow{\scriptstyle h^\sharp} & \searrow{\scriptstyle a} & \downarrow & & \\ 0 & \longrightarrow & \mathbb{K} & \longrightarrow & \mathbb{K}[\xi] & \longrightarrow & \mathbb{K} & \longrightarrow & 0 \end{array} \tag{10}$$

where C_M is the structure sheaf of the body of $\mathfrak{X}$. Since $a : \mathcal{O}_{\mathfrak{X},P} \to \mathbb{K}$ is even and $\mathbb{K}$ contains no non-trivial, nilpotent elements, it follows that $a(j) = 0, \forall j \in \mathcal{J}_{\mathfrak{X},P}$. Then by (9) see that for any element $j'' = jj' \in \mathcal{J}^2_{\mathfrak{X},P}$,

---

[10] Here SCR refers to the category of supercommutative rings while CR is that of commutative rings

[11] To see that we will indeed have an isomorphism of sets note that conversely, given a rational point of M, this is a morphism $\mathrm{Spec}\,\mathbb{K} \stackrel{\psi}{\to} M$. It defines on local rings, $C_{M,P} \stackrel{g^\sharp}{\to} \mathbb{K}$. Using that $C_{M,P} = \mathcal{O}_{\mathfrak{X},P}/\mathcal{J}_{\mathfrak{X},P}$, the composition $h^\sharp : \mathcal{O}_{\mathfrak{X},P} \to C_{M,P} \stackrel{g^\sharp}{\to} \mathbb{K}$ defines a reduced point of $\mathfrak{X}$.

$$b(j'') = b(jj') = b(j)a(j') + a(j)b(j') = 0.$$

Therefore $h^\sharp(j'') = 0$ so it will induce a commuting diagram,

$$\begin{array}{ccc} \mathcal{J}_{\mathfrak{X},P} & \xrightarrow{h^\sharp} & \mathbb{K} \\ \downarrow & \nearrow_{b} & \\ \mathcal{J}_{\mathfrak{X},P}/\mathcal{J}^2_{\mathfrak{X},P} & & \end{array} \tag{11}$$

Recall now the universal property for the symmetric algebra which says: for any commutative algebra A over a commutative R, an R-module V and any R-linear morphism $V \stackrel{b}{\to} A$, there exists a unique, algebra morphism $\widetilde{b} : \mathrm{Sym}^\bullet_R V \to A$ commuting the following diagram,

$$\begin{array}{ccc} V & \longrightarrow & \mathrm{Sym}^\bullet_R V \\ & {\scriptstyle b}\searrow \quad \swarrow{\scriptstyle \widetilde{b}} & \\ & A & \end{array}$$

where $V \subset \mathrm{Sym}^\bullet_R V$ is the inclusion of V as the degree one component of its symmetric algebra. Applying this universal property to (11) yields a unique morphism of algebras $\widetilde{b} : \mathrm{Sym}^\bullet_{\mathcal{O}_{\mathfrak{X},P}}(\mathcal{J}_{\mathfrak{X},P}/\mathcal{J}^2_{\mathfrak{X},P}) \to \mathbb{K}$. With $\mathbf{V}_{\mathfrak{X}} = \mathrm{Spec}\,\mathrm{Sym}^\bullet(\mathcal{J}_{\mathfrak{X},P}/\mathcal{J}^2_{\mathfrak{X},P})$ from Lemma 2.22 see that $\widetilde{b}$ now defines a rational point $\mathrm{Spec}\,\mathbb{K} \to \mathbf{V}_{\mathfrak{X}}$. We have therefore constructed a map $\underline{\mathfrak{X}}(1) \to \underline{\mathbf{V}}_{\mathfrak{X}}(\mathrm{Spec}\,\mathbb{K})$. A inverse map $\underline{\mathbf{V}}_{\mathfrak{X}}(\mathrm{Spec}\,\mathbb{K}) \to \underline{\mathfrak{X}}(1)$ can be formed by suitably reversing the construction given above. We omit the details here. □

4.3 A (1|1)-quadratic form

To illustrate the distinction between L-points for $L = 0, 1, 2$ we present the $(1|1)$-dimensional, quadratic form as follows. Consider a form Q given by

$$\langle (x|\theta), (y|\eta)\rangle = xy + \theta\eta. \tag{12}$$

Here Q defines a function $\mathbb{A}^{2|2}_{\mathbb{K}} \cong \mathbb{A}^{1|1}_{\mathbb{K}} \times \mathbb{A}^{1|1}_{\mathbb{K}} \to \mathbb{A}^{1|1}_{\mathbb{K}}$. As a pairing, we can write

$$Q = \begin{pmatrix} 1 & 0 \\ 0 & 1 \end{pmatrix} \tag{13}$$

so that $\langle (x|\theta), (y|\eta)\rangle = (x|\theta)^T Q(y|\eta)$.

On the locus of superpoints, set $x = z^{1,0}$, $y = z^{2,0}$, $\theta = \vartheta_1{}^1\xi$ and $\eta = \vartheta_2{}^1\xi$. Then Q defines on $\underline{\mathbb{A}}^{2|2}_{\mathbb{K}}(1)$,

$$h^\sharp(\langle(x|\theta),(y|\eta)\rangle) = z^{1,0}z^{2,0} + (\vartheta_1{}^1\xi)(\vartheta_2{}^1\xi) = z^{1,0}z^{2,0}.$$

As a pairing then on $\underline{\mathbb{A}}_{\mathbb{K}}^{2|2}(1) \cong \mathbb{K}^{1+1} = \mathbb{K}^2 \cong \mathbb{K}\times\mathbb{K}$, Q from (13) is represented by the 1×1-matrix, $\underline{Q}(1) = (1)$.[12]

We obtain a more interesting expression on the locus of 0|2-points. Here, write

$$x \overset{h^\sharp}{\longmapsto} z^{1,0} + z^{1,12}\xi_1\xi_2;$$

$$y \overset{h^\sharp}{\longmapsto} z^{2,0} + z^{2,12}\xi_1\xi_2;$$

$$\theta \overset{h^\sharp}{\longmapsto} \vartheta_1{}^1\xi_1 + \vartheta_1{}^2\xi_2;$$

$$\eta \overset{h^\sharp}{\longmapsto} \vartheta_2{}^1\xi_1 + \vartheta_2{}^2\xi_2.$$

Then,

$$h^\sharp(\langle(x|\theta),(y|\eta)\rangle) = z^{1,0}z^{2,0} + \left(z^{1,0}z^{2,12} + z^{2,0}z^{1,12} + \vartheta_1{}^1\vartheta_2{}^2 - \vartheta_1{}^2\vartheta_2{}^1\right)\xi_1\xi_2$$

As a pairing on $\underline{\mathbb{A}}_{\mathbb{K}}^{2|2}(2) \cong \mathbb{K}^{2^{2-1}(2+2)} = \mathbb{K}^8 \cong \mathbb{K}^4\times\mathbb{K}^4$, Q from (13) is represented by the 4×4-matrix,

$$\underline{Q}(2) = \begin{pmatrix} 1 & 1 & & \\ 1 & 0 & & \\ & & 0 & 1 \\ & & -1 & 0 \end{pmatrix}$$

where the blank spaces in the above matrix are zeroes. Note that $\underline{Q}(2)$ not symmetric, in contrast to (13). This is because, due to anti-commutativity of the odd variables, the product (12) itself is not symmetric, i.e., that $\langle(x|\theta),(y|\eta)\rangle \neq \langle(y|\eta),(x|\theta)\rangle$. Indeed, we have the relations $\langle(x|\theta),(y|\eta)\rangle + \langle(y|\eta),(x|\theta)\rangle = 2xy$ and $\langle(x|\theta),(y|\eta)\rangle - \langle(y|\eta),(x|\theta)\rangle = 2\theta\eta$. In a sense, these relations are not 'seen' by the reduced and superpoints.

5 The DeWitt topology

In §2 we presented the local, supercommutative ringed space (LSRS) approach to supermanifold theory (c.f., Definition 2.17). In §4 we recovered its first properties through its functor of points and arrived general class of points in Definition 4.4. In the present section we detail the approach to supermanifold theory instigated by DeWitt in [5], aptly termed the 'DeWitt approach'. Our main result here is Theorem 5.14. In analogy with the complex topology in algebraic geometry, we construct in

[12] Note, this coincides with $\underline{Q}(0)$.

Theorem 5.14 a topology on the locus of L-points of a superscheme, referred to as the *DeWitt topology*.

5.1 The DeWitt approach

5.1.1 DeWitt superspaces

Fix a field $\mathbb{K}$, an integer L and set $B_L := \wedge^\bullet_{\mathbb{K}}\mathbb{K}^L$. It is referred to as the *auxiliary algebra on L generators*. In the vector space topology, $B_L \cong \mathbb{K}^{2^{L-1}(1+1)}$. Let $B_L^{m,n}$ now denote the product,

$$B_L^{m,n} := \prod^m B_{L,\bar{0}} \times \prod^n B_{L,\bar{1}} \tag{14}$$

where $B_{L,\bar{0}}$ resp., $B_{L,\bar{1}}$ is the subring resp., $B_{L,\bar{0}}$ module of even and odd elements of B_L. Evidently $B_L^{1,1} \cong B_L$. In the vector space topology $B_L^{m,n} \cong \mathbb{K}^{2^{L-1}(m+n)}$. Since B_L is an exterior algebra it comes equipped with a projection $\sigma : B_L^{m,n} \to \mathbb{K}^m$ induced by $B_L \to \mathbb{K}$. We refer to $B_L^{m,n} \stackrel{\sigma}{\to} \mathbb{K}^m$ as the *body projection*.

Definition 5.1 *The topology on $B_L^{m,n}$ comprised of open sets of the form $\sigma^{-1}U$ for $U \subset \mathbb{K}^m$ open is referred to as the* DeWitt topology on $B_L^{m,n}$. □

Relating the vector space and DeWitt topologies on $B_L^{m,n}$, we have for any open set $U \subset \mathbb{K}^m$ that $\sigma^{-1}U \cong U \times \mathbb{K}^{2^{L-1}(m+n)-m}$.

Definition 5.2 *The spaces $B_L^{m,n}$ equipped with the DeWitt topology from* (14) *are referred to as* (m,n)-dimensional, DeWitt superspaces over B_L. □

Recall from the LSRS approach (Definition 2.17), a supermanifold is built on model superspaces $(\mathbb{R}^m, \wedge^\bullet_{C^\infty}V)$ and $(\mathbb{C}^m, \wedge^\bullet_{\mathcal{O}}V)$ from Example 2.15. In the DeWitt approach, a *DeWitt supermanifold* is built by patching together DeWitt superspaces over a fixed, auxiliary algebra B_L with respect to a class of smooth functions.

5.1.2 Supersmooth functions

Let $z^\alpha|\vartheta_a$ denote coordinates on $B_L^{m,n}$ where $\alpha = 1,\ldots,m$ and $a = 1,\ldots,n$. Here z^α is even, so valued in $B_{L,\bar{0}}$ while ϑ_a is odd, so valued in $B_{L,\bar{1}}$. With $1,\xi_1,\ldots,\xi_L$ denoting generators for B_L we write,

$$z^\alpha = z^{\alpha,0} + z^{\alpha,ij}\xi_i\xi_j + \cdots \qquad \text{and} \qquad \vartheta_a = \vartheta_a^{\,i}\xi_i + \vartheta_a^{\,ijk}\xi_i\xi_j\xi_k + \cdots \tag{15}$$

If $P \in B_L^{m,n}$ denotes a point so that $z^\alpha(P) \in B_{L,\bar{0}}$ and $\vartheta_a(P) \in B_{L,\bar{1}}$, then $z^{\alpha,0}(P)$ are coordinates for the body $\sigma(P) \in \mathbb{K}^m$, i.e., $z^{\alpha,0}(P) = z^{\alpha,0}(\sigma(P))$.

Remark 5.3 *More generally, in the vector space topology we have the isomorphism* $\varphi : B_L^{m,n} \cong \mathbb{K}^{2^{L-1}(m+n)}$. *On coordinates, this establishes a unique representation of* $z^\alpha|\vartheta_a$ *in* (15) *with the tuple* $z^\alpha \mapsto (z^{\alpha,0}, z^{\alpha,ij}, \ldots)$ *and* $\vartheta_a \mapsto (\vartheta_a^{\,i}, \vartheta_a^{\,ijk}, \ldots)$. *In the case* $m = 1, n = 1$, *for* $P \in B_L^{1,1}$ *a point,* $\varphi(P) \in \mathbb{K}^{2^{L-1}(1+1)}$ *is a point in a classical, Euclidean space. Coordinates of* $\varphi(P)$ *are* $(z^0(P), z^{ij}(P), \ldots, \vartheta^i(P), \vartheta^{ijk}(P), \ldots)$. □

Definition 5.4 *The ring of smooth functions on the DeWitt superspace* $B_L^{m,n}$ *is given by the tensor product,*

$$C^\infty(B_L^{m,n}) := C^\infty(\sigma(B_L^{m,n})) \otimes B_L$$

where $\sigma(B_L^{m,n}) = \mathbb{K}^m$ *is the body projection and* $C^\infty(\sigma(B_L^{m,n})) = C^\infty(\mathbb{K}^m)$ *are smooth functions on* $\mathbb{K}^m$. □

With the above notion of smooth functions on DeWitt superspaces, the topological isomorphism $B_L^{m,n} \cong \mathbb{K}^{2^{L-1}(m+n)}$ does *not* induce an isomorphism of smooth structures. Indeed, the ring of smooth functions on $\mathbb{K}^{2^{L-1}(m+n)}$ is much larger than that of $B_L^{m,n}$.[13] The above notion of smoothness captures the property that any smooth function F on $B_L^{m,n}$ can be written in coordinates $z|\vartheta$,

$$F(z|\vartheta) = F^0(z) + F^a(z)\vartheta_a + F^{aa'}(z)\vartheta_a\vartheta_{a'} + \cdots \tag{16}$$

The $\mathcal{Z}$-expansion introduced by Bruzzo et. al. in [1] is the morphism of rings $C^\infty(\mathbb{K}^m) \xrightarrow{\mathcal{Z}} C^\infty(B_L^{m,0})$ given by

$$\mathcal{Z}(f)(z) = f(z^0) + (z - z^0)f'(z^0) + \frac{1}{2!}(z - z^0)^2 f''(z^0) + \cdots$$

In [1, §2] it is shown:

Lemma 5.5 The $\mathcal{Z}$-expansion $C^\infty(\mathbb{K}^m) \xrightarrow{\mathcal{Z}} C^\infty(B_L^{m,0})$ defines a monomorphism of rings.

Comparing smooth functions on DeWitt superspaces from (16) with the image of $\mathcal{Z}$ leads to the notion of 'supersmoothness'.

Definition 5.6 *A smooth function* $F \in C^\infty(B_L^{m,n})$ *is said to be* supersmooth *if the coefficient functions* F^I *in* (16) *are in the image of* $\mathcal{Z}$. □

Example 5.7 We give here an example of a function which is, and which is not supersmooth. Let $z|\vartheta$ be coordinates on $B_2^{1,1}$ so that $z = z^0 + z^{12}\xi_1\xi_2$ and $\vartheta = \vartheta^1\xi_1 + \vartheta^2\xi_2$. Now let $F \in C^\infty(B_2^{1,1})$ be given by,

$$F(z|\vartheta) = z$$

[13] E.g., the function $\exp(\vartheta^1)$, which is a smooth function in $C^\infty(\mathbb{K}^{2^{L-1}(m+n)})$, will not come from a smooth function on $B_L^{m,n}$.

Then F is supersmooth since,

$$F(z|\vartheta) = z = z^0 + z^{12}\xi_1\xi_2 = \mathcal{Z}(f)(z)$$

where $f \in C^\infty(\sigma(B_2^{1,1})) = C^\infty(\mathbb{K})$ is given by $f(z^0) = z^0$. Indeed, any algebraic function $F(z|\vartheta) = G(z)$ will be supersmooth and so any $F(z|\vartheta) = G_0(z) + G_1(z)\vartheta + \cdots$ where $G_i(z)$ are algebraic functions of z. The following simple function

$$F(z|\vartheta) = z + \xi_1\xi_2$$

is not supersmooth however. □

5.1.3 DeWitt supermanifolds

Definition 5.8 *An* (m,n)-dimensional, DeWitt supermanifold over B_L *is a locally ringed space which admits an open covering by open sets* $(\mathcal{U}_i)_{i\in I}$ *for* $\mathcal{U}_i \subset B_L^{m,n}$ *open in the DeWitt topology and such that over intersections the transition function*

$$\varphi_j^{-1}(\mathcal{U}_i \cap \mathcal{U}_j) \xrightarrow{\varphi_j^{-1}} \mathcal{U}_i \cap \mathcal{U}_j \xrightarrow{\varphi_i} \varphi_i(\mathcal{U}_i \cap \mathcal{U}_j)$$

is supersmooth for all i, j *for which* $\mathcal{U}_i \cap \mathcal{U}_j \neq \emptyset$. □

Remark 5.9 *Recalling that* $B_L^{m,n} \cong \mathbb{K}^{2^{L-1}(m+n)}$, *it follows from Definition 5.8 that underlying any* (m,n)*-dimensional, DeWitt supermanifold over* B_L *is a* $2^{L-1}(m+n)$*-dimensional, differentiable manifold over* $\mathbb{K}$. *In parallel with Definition 5.8 one can construct a wider class of supermanifold where the open sets are open in the vector space topology on* $B_L^{m,n}$, *i.e., open under the isomorphism* $B_L^{m,n} \cong \mathbb{K}^{2^{L-1}(m+n)}$. *The contrast to the DeWitt case was studied by Rabin and Crane in [7] and it was shown by Bruzzo et. al. in [1, Ch. 5] how to distinguish different topologies with an analogue of de Rham cohomology.* □

5.2 Superschemes and DeWitt supermanifolds

5.2.1 The atlas for DeWitt supermanifolds

As for manifolds, an atlas for an (m,n)-dimensional, DeWitt supermanifold over B_L consists of a collection of disjoint, DeWitt open sets $(\mathcal{U}_i)_{i\in I}$, $\mathcal{U}_i \subset B_L^{m,n}$, distinguished DeWitt open subsets $\mathcal{U}_{ij} \subset \mathcal{U}_i$ and supersmooth diffeomorphisms $\mathcal{U}_{ij} \xrightarrow{\cong} \mathcal{U}_{ji}$ which is the identity if $i = j$ and for all i, j, k fits into a commuting diagram,

$$\begin{array}{ccc} \mathcal{U}_{ij} & \xrightarrow{\quad\cong\quad} & \mathcal{U}_{jk} \\ & \searrow^{\cong} \qquad \nearrow_{\cong} & \\ & \mathcal{U}_{ik} & \end{array}$$

The construction $\frac{\bigsqcup_{i\in I}\mathcal{U}_i}{\sim}$, where the equivalence relation is defined though the transition functions $(\mathcal{U}_{ij} \cong \mathcal{U}_{ji})_{ij}$, defines a DeWitt supermanifold up to isomorphism. In the DeWitt topology, recall that each open set is of the form $\mathcal{U}_i = \sigma^{-1}U_i$ for $U_i \subset \mathbb{K}^m$. As such the glueing data above describes a classical, m-dimensional submanifold of the DeWitt supermanifold, referred to as the *body*. From [5, pp. 55-6]:

Theorem 5.10 Underlying any (m, n)-dimensional, DeWitt supermanifold over B_L is a rank-$(2^{L-1}(m+n) - m)$ vector bundle over the body.

Remark 5.11 *It is suggestive to compare Theorem 5.10 with the first properties of supermanifolds as LSRS, described in §2.3 and §4.2.* □

5.2.2 *L*-points of affine superspace

To prepare our argument for superschemes more generally, we establish the case of affine superspaces in what follows. Over a field $\mathbb{K}$, recall the affine superspace $\mathbb{A}^{m|n}_{\mathbb{K}}$ from §3.1. By Theorem 4.5, the rational points ($L = 0$) of $\mathbb{A}^{m|n}_{\mathbb{K}}$ recover the Euclidean space $\mathbb{K}^m$. By Theorem 4.6, the superpoints ($L = 1$) recover a rank n, trivial vector bundle over $\mathbb{K}^m$. Hence that $\underline{\mathbb{A}}^{m|n}_{\mathbb{K}}(0) \cong \mathbb{K}^m$ and $\underline{\mathbb{A}}^{m|n}_{\mathbb{K}}(1) \cong \mathbb{K}^{m+n}$. More generally we have for the L-points:

Lemma 5.12

$$\underline{\mathbb{A}}^{m|n}_{\mathbb{K}}(L) \cong B^{m,n}_L.$$

Proof Firstly recall $B^{m,n}_L \cong \mathbb{K}^{2^{L-1}(m+n)}$. The statement in the lemma now relates the functor of points $\underline{\mathbb{A}}^{m|n}_{\mathbb{K}}$ with that of the affine space $\underline{\mathbb{A}}^{2^{L-1}(m+n)}_{\mathbb{K}}$, where $\mathbb{K}^{2^{L-1}(m+n)} \cong \underline{\mathbb{A}}^{2^{L-1}(m+n)}_{\mathbb{K}}(\text{Spec}\,\mathbb{K})$. That is, we need to show there exists a natural isomorphism between the sets,[14]

$$\text{Hom}_{\text{LSRS}}(\text{Spec}\,\mathbb{K}[\xi_1, \ldots, \xi_L], \mathbb{A}^{m|n}_{\mathbb{K}}) \cong \text{Hom}_{\text{LSRS}}(\text{Spec}\,\mathbb{K}, \mathbb{A}^{2^{L-1}(m+n)}_{\mathbb{K}}) \qquad (17)$$

where $\mathbb{A}^{0|L}_{\mathbb{K}} = \text{Spec}\,\mathbb{K}[\xi_1, \ldots, \xi_L]$. Write, $\mathbb{A}^{m|n}_{\mathbb{K}} = \text{Spec}\,\mathbb{K}[x^1, \ldots, x^m|\theta_1, \ldots, \theta_n]$ and $\mathbb{A}^{2^{L-1}(m+n)}_{\mathbb{K}} = \text{Spec}\,\mathbb{K}[z^I, \vartheta^J]$ where I and even resp., odd J are multi-indices on L-generators. Then $x^\alpha|\theta_a$ resp., z^I, ϑ^J are coordinates on $\mathbb{A}^{m|n}_{\mathbb{K}}$ resp., $\mathbb{A}^{2^{L-1}(m+n)}_{\mathbb{K}}$. The statement in (17) is now equivalent to the following as $\mathbb{K}$-algebras,

[14] here LSRS denotes the category of local, supercommutative ringed spaces

$$\mathrm{Hom}_{\mathbb{K}\text{-ALG}}(\mathbb{K}[x^\alpha|\theta_a], \mathbb{K}[\xi_1, \ldots, \xi_L]) \cong \mathrm{Hom}_{\mathbb{K}\text{-ALG}}(\mathbb{K}[z^{\alpha I}, \vartheta_a^{\,J}], \mathbb{K}). \tag{18}$$

The above isomorphism is a consequence of the familiar tensor-hom adjunction adapted to supercommutative rings. Thus we obtain the isomorphism (17) and so the lemma. □

To elaborate further on the proof of Lemma 5.12, with $\mathbb{A}_{\mathbb{K}}^{m|n} = \mathrm{Spec}\ \mathbb{K}[x^\alpha|\theta_a]$, $\alpha = 1, \ldots, m$ and $a = 1, \ldots, n$; the coordinate ring of $\mathbb{A}_{\mathbb{K}}^{m|n}$ is $\mathbb{K}[x^\alpha|\theta_a]$. An L-point $\phi \in \underline{\mathbb{A}}_{\mathbb{K}}^{m|n}(\mathbb{K}^{0|L})$ is a morphism $\phi = (f, h^\sharp) : \mathbb{K}^{0|L} \to \mathbb{A}_{\mathbb{K}}^{m|n}$ where $f : \{pt\} \to \mathbb{A}^m$ is a rational point of $\mathbb{A}_{\mathbb{K}}^{m|n}$ (c.f., Theorem 4.5) and $h^\sharp : \mathbb{K}[x^i|\theta_a] \to \mathbb{K}[\xi_1, \ldots, \xi_L]$ is an morphism of supercommutative rings. By Lemma 4.2, ϕ is completely determined by $h^\sharp$ which, on generators, is

$$h^\sharp(x^\alpha) = z^{\alpha,0} + z^{\alpha,ij}\xi_i\xi_j + \cdots \quad \text{and} \quad h^\sharp(\theta_a) = \vartheta_a^{\,i}\xi_i + \vartheta_a^{\,ijk}\xi_i\xi_j\xi_k + \cdots \tag{19}$$

The coefficients of $h^\sharp$ generate the algebra $\mathbb{K}[z^{\alpha,I}, \vartheta_a^{\,J}]$ which is the coordinate ring for $\mathbb{A}_{\mathbb{K}}^{2^{L-1}(m+n)}$. The assignment $\xi_j \mapsto 1, \forall j$ mediates the isomorphism (18). As coordinates, let $P = f(\{pt\})$ be the image of f. Lemma 5.12 says: $h^\sharp(x^\alpha(P)) = h^\sharp(x^\alpha)(P)$ and $h^\sharp(\theta_a(P)) = h^\sharp(\theta_a)(P)$ and by (19) we have a correspondence,

$$h^\sharp(x^\alpha(P)) = (z^{\alpha,0}(P), z^{\alpha,ij}(P), \ldots); \quad h^\sharp(\theta_a(P)) = (\vartheta_a^{\,i}(P), \vartheta_a^{\,ijk}(P), \ldots). \tag{20}$$

This shows explicitly how the locus of L-points of $\mathbb{A}_{\mathbb{K}}^{m|n}$ are precisely tuples (rational points) in the Euclidean space $\mathbb{K}^{2^{L-1}(m+n)}$, or equivalently, the superspace $B_L^{m,n}$ by Remark 5.3. Note, we have only considered $B_L^{m,n}$ here with its vector space topology.

Proposition 5.13 $\underline{\mathbb{A}}_{\mathbb{K}}^{m|n}(L)$ can be given the structure of a DeWitt superspace. □

Proof Recall the body map $B_L^{m,n} \stackrel{\sigma_L}{\to} \mathbb{K}^m$ defining the DeWitt topology. Then σ_L defines the pullback morphism on functions $\sigma_L^\sharp : \mathbb{K}[x^1, \ldots, x^m] \to C^\infty(B_L^{m,n})$. We can extend this to a morphism $\widetilde{\sigma}_L^\sharp : \mathbb{K}[x^1, \ldots, x^m|\theta_1, \ldots, \theta_n] \to C^\infty(B_L^{m,n})$ through the assignment $\theta_i \mapsto \vartheta_i$. Setting $h^\sharp = \sigma_L^\sharp$ in (20), the body map σ_L leads thereby to an identification $\underline{\mathbb{A}}_{\mathbb{K}}^{m|n}(L) \cong \sigma_L^{-1}\underline{\mathbb{A}}_{\mathbb{K}}^{m|n}(0)$. This establishes the proposition. □

5.2.3 *L*-points of superschemes

Theorem 5.14 For any $m|n$-dimensional superscheme $\mathfrak{X}$ over $\mathbb{K}$, its locus of L-points $\underline{\mathfrak{X}}(L)$ admits the structure of an (m, n)-dimensional, DeWitt supermanifold over B_L. □

Proof From [19] we can write $\mathfrak{X} = \frac{\bigsqcup_i U_i}{\sim}$ where $U_i \subset \mathbb{K}^{m|n}$ are a collection of disjoint, open sets which cover $\mathfrak{X}$ and the equivalence relation $\sim$ is defined by isomorphisms $\rho_{ij} : U_{ij} \cong U_{ji}$ between distinguished subsets $U_{ij} \subset U_i$ and $U_{ji} \subset U_j$. In the case where $\mathfrak{X}$ is a superscheme over $\mathbb{K}$, the open sets are isomorphic to affine superspaces $U_i \cong \mathbb{A}_{\mathbb{K}}^{m|n}$. By Lemma 5.12, the L-points of each open U_i is isomorphic to $B_L^{m,n}$, so we have isomorphisms $\underline{U}_i(L) \cong \mathcal{U}_i \cong B_L^{m,n}$ and therefore a collection of distinguished subsets $\underline{U}_{ij}(L) = \mathcal{U}_{ij} \subset \mathcal{U}_i \cong B_L^{m,n}$. By Proposition 5.13 these open sets are open in the DeWitt topology on $B_L^{m,n}$. Now recall from Lemma 5.5 that the $\mathcal{Z}$-expansion is a monomorphism. Accordingly, it will preserve isomorphisms. Applying the $\mathcal{Z}$-expansion then to the transition functions $\rho_{ij} : U_{ij} \cong U_{ji}$ leads to the supersmooth isomorphisms $\mathcal{Z}(\rho_{ij}) : \mathcal{U}_{ij} \cong \mathcal{U}_{ji}$ between DeWitt superspaces. Hence from the covering data (U_i, U_{ij}, ρ_{ij}) defining $\mathfrak{X}$, its L-points give covering data $(\mathcal{U}_i, \mathcal{U}_{ij}, \mathcal{Z}(\rho_{ij}))$ which can be compiled as in §5.2.1 to form an atlas for a DeWitt supermanifold. The locus of L-points the superscheme $\mathfrak{X}$ therefore admits the structure of a DeWitt supermanifold. □

From [10, §4], the *complex topology* on a scheme M over $\mathbb{C}$ is a topology on the rational points $\underline{M}(\mathrm{Spec}\ \mathbb{C})$ of M. It inherits this topology from complex Euclidean space in a manner similar to our proof of Theorem 5.14. Now in Theorem 5.14 above we have shown, for any superscheme $\mathfrak{X}$, that its locus of L-points $\underline{\mathfrak{X}}(L)$ can be given the structure of a DeWitt supermanifold for any L. In analogy with the complex topology then, we can understand Theorem 5.14 as describing a topology which by Definition 5.1 we can refer to as: *the DeWitt topology on the superscheme* $\mathfrak{X}$.

Question 5.15 Can one interpret the DeWitt topology from Theorem 5.14 as a Grothendieck topology on the category of superschemes SSCH? □

Recall from the proof of Theorem 5.14 that the DeWitt topology on $\mathfrak{X}$ is a consequence of Proposition 5.13. Note by Lemma 5.12 however that it is feasible to form more general topologies on $\mathfrak{X}$ than the DeWitt topology.

Question 5.16 Can one interpret the other topologies on supermanifolds from [1] as Grothendieck topologies on SSCH? □

References

1. C. Bartocci, U. Bruzzo, D. Hernandez-Ruiperez: Geometry of supermanifolds. Kluwer Academic Publishers, Boston (1991)
2. F. A. Berezin: Introduction to superanalysis. D. Reidel Publishing Company (1991)
3. B. Kostant: Graded manifolds, Lie theory and prequantization. In: Differential geometrical methods in mathematical physics, volume 570 of Lecture notes in Mathematics, pp. 177–306. Springer (1975)
4. D. Leites: Introduction to the theory of supermanifolds. Russian math surveys, **35**(1), pp. 1–64 (1980)
5. B. DeWitt: Supermanifolds. Cambridge University Press (1984)
6. A. Rogers: Supermanifolds: Theory and Applications. World Scientific (2007)
7. J. Rabin, L. Crane: How different are the supermanifolds of Rogers and DeWitt? Comm. Math. Phys. **102**, 123–137 (1985)

8. P. Deligne, J. W. Morgan: Notes on supersymmetry. In: Quantum fields and strings: a course for mathematicians, pp. 41–97. Springer, Heidelberg (1999)
9. R. Hartshorne: Deformation theory. Springer (2010)
10. A. Neeman: Analystic and algebraic geometry. Cambridge University Press (2007)
11. Stacks Project Authors: Stacks Project. http://stacks.math.columbia.edu (2016)
12. K. Bettadapura: Embeddings of complex supermanifolds. Adv. Th. and Math. Phys. **23**(6), 1427–1466 (2019)
13. D. Leites: Spectra of graded-commutative rings. Uspekhi. Mat. Nauk. **29**(3), 209–210 (1974)
14. K. Bettadapura: Projective superspace varieties, superspace quadrics and non-splitting. Documenta Mathematica. 65–91 (2020)
15. J. Rabin, L. Crane: Super Riemann surfaces: uniformization and Teichmueller theory. Comm. Math. Phys. **113** 601–603 (1988)
16. E. Witten: Notes on super Riemann surfaces and their moduli. Pure an applied mathematics quarterly, **15**(1), 57–211 (2019)
17. K. Bettadapura: On the problem of splitting deformations of super Rieman surfaces. Lett. Math. Phys. **109(2)**, 381–402 (2019)
18. R. Donagi, E. Witten: Supermoduli space is not projected. In Proc. Symp. Pure Math., **90**, 19–72 (2015)
19. K. Bettadapura: Obstructed thickenings and supermanifolds. J. Geom. and Phys. **139**, 25–49 (2019)
20. V. Molotkov: Infinite dimensional Z**k(2) dimensional supermanifolds. International center for theoretical physics, Triste, Italy (1984)
21. C. Sachse: Global analytic approach to super Teichmueller spaces. Phd Thesis, Leipzig University (2009)
22. A. Grothendieck: Introduction to functorial algebraic geometry. Notes written by F. Gaetta, SUNY, Buffalo (1973)
23. R. C. Penner, A. Zeitlin: Decorated super Teichmueller space. J. Diff. Geom. **111**(3), (2019)
24. M. Rothstein: The structure of symplectic supermanifolds. In: 19th international conference on differential geometric methods in theoretical physics, **3**, 331–343 (1990)
25. R. Donagi, E. Witten: Super Atiyah classes and obstructions to splitting supermoduli space. Pure and applied mathematics quarterly **9**(4) (2014)
26. Y. Manin: New directions in geometry. Talk at 25th Arbeitstagung, Bonn (1984)
27. Y. Manin: Gauge fields and complex geometry. Springer-Verlag (1988)
28. A. Vaintrob: Darboux theorem and equivariant Morse lemma. J. Geom. Phys. **18**, 59–75 (1996)
29. H. Hohnhold, M. Kreck, S. Stolz, P. Teichner: Differential forms and zero-dimensional supersymmetric field theories. Quantum Topol. **2**, 1–41 (2011)
30. S. Stolz, P. Teichner: Supersymmetric field theories and generalized cohomology. Mathematical foundations of quantum field theory and perturbative string theory, 279–340 (2011)

Chapter 4

MATRIX Program: 2D Supersymmetric Theories and Related Topics

On Super Non-Abelian T-Duality of Symmetric and Semi-Symmetric Coset Sigma Models

Daniele Bielli

Abstract We review the algebraic approach to super non-Abelian T-Duality considered in [1], focusing on symmetric and semi-symmetric coset spaces on G/H. We discuss a potential impediment, appearing in these models when integrating out the gauge fields in favour of the dual variables. This process cannot be performed in general and we isolate the obstruction, highlighting three cases in which a solution can be found. After writing the T-dual action we provide solution for two specific models. The first based on the symmetric space $S^3 \simeq SO(4)/SO(3)$, well-known in the literature, the second on the semi-symmetric coset $OSp(1|2)/SO(1,1)$, a Green-Schwarz-like string sigma model satisfying the supergravity torsion constraints.[1]

1 Introduction

We start by fixing notation on semi-symmetric cosets, as relevant formulae can be reduced to those for symmetric spaces. We consider two-dimensional sigma models involving a generic Lie (super)group G with associated Lie (super)algebra $\mathfrak{g}$ and defined via smooth maps $g \in C^\infty(\Sigma, G)$ from a two-dimensional Lorentzian worldsheet Σ to G. The main building block of these models is the pull-back to Σ, via g, of the Maurer-Cartan form $j := g^{-1}\mathrm{d}g \in \Omega^1(\Sigma, \mathfrak{g})$, invariant under the global left action G_L of the group on itself $g \to g_0^{-1}g$ and satisfying the Maurer-Cartan flatness condition $F_j := \mathrm{d}j + \frac{1}{2}[j, j] = 0$, with d and $\Omega^p(\Sigma, \mathfrak{g})$ respectively denoting exterior derivative and $\mathfrak{g}$-valued p-forms on Σ. Semi-symmetric spaces are cosets on G/H where H, bosonic subgroup of the supergroup G with associated Lie algebra $\mathfrak{h}$, arises as the fixed point set of an automorphism of G of order four [2, 3]. At the Lie algebra level this implies the existence of an automorphism $\sigma : \mathfrak{g} \to \mathfrak{g}$ such that $\sigma^4 = 1$. This allows to identify four subspaces $\mathfrak{g}_k \subset \mathfrak{g}$, $k = 0, 1, 2, 3$, charac-

Daniele Bielli, University of Surrey (UK) and Milano-Bicocca (Italy).
e-mail: d.bielli@surrey.ac.uk

[1] DMUS-MP-22/18.

D. R. Wood et al. (eds.), *2021-2022 MATRIX Annals*, MATRIX Book Series 5,
https://doi.org/10.1007/978-3-031-47417-0_4

terised by $\sigma(\mathfrak{g}_k) = i^k \mathfrak{g}_k$ and $[\mathfrak{g}_k, \mathfrak{g}_l] \subset \mathfrak{g}_{(k+l)\,mod 4}$. These are singled out by projectors $\mathfrak{g}_k = P_k(\mathfrak{g}) = \frac{1}{4}(1 + i^{3k}\sigma + i^{2k}\sigma^2 + i^k\sigma^3)(\mathfrak{g})$, leading to an orthogonal decomposition $\mathfrak{g} \simeq \mathfrak{g}_0 \oplus \mathfrak{g}_1 \oplus \mathfrak{g}_2 \oplus \mathfrak{g}_3$. Noting that $\sigma^2 = (-1)^k$, the subspaces $\mathfrak{g}_0 \equiv \mathfrak{h}$, $\mathfrak{g}_2 \equiv \mathfrak{m}$ and $\mathfrak{g}_1 \equiv \mathfrak{p}$, $\mathfrak{g}_3 \equiv \mathfrak{q}$ are respectively recognised as purely bosonic and purely fermionic. σ is an involution on bosonic generators, hence semi-symmetric models are the superspace analogue of symmetric spaces, which are recovered by removing fermions. In turn, j decomposes as $j = A + p + m + q$, with the components transforming as $A \to h^{-1}Ah + h^{-1}\mathrm{d}h$ and $\{p, m, q\} \to h^{-1}\{p, m, q\}h$ under the local right action H_R of H on G, $g \to gh$. A locally H_R-invariant and globally G_L-invariant action is then constructed out of p, m, q as $S_{SS} = \frac{1}{2}\int_\Sigma \langle m, \star m\rangle + \frac{1}{2}\int_\Sigma \langle p, q\rangle$. Wedge product is understood and $\star$ is the Hodge star operator on Σ with respect to the world-sheet metric, satisfying $\star^2 = 1$ on $\Omega^1(\Sigma, \mathfrak{g})$. Brackets denote an Ad-invariant, non-degenerate graded-symmetric bilinear form on $\mathfrak{g}$ compatible with the decomposition.

T-duality can then be performed, assuming a topologically trivial world-sheet, by gauging a sub-(super)group $K_L \subseteq G_L$ of the isometries with gauge fields $\omega \in \Omega^1(\Sigma, \mathfrak{k}_L)$ and introducing a term enforcing their flatness $F_\omega := \mathrm{d}\omega + \frac{1}{2}[\omega, \omega] = 0$ by means of Lagrange multipliers $\Lambda \in C^\infty(\Sigma, \mathfrak{k}_L)$, which transform as $\Lambda \to h^{-1}\Lambda h$ under local H_R. Integrating out the multipliers, ω is set to be pure gauge and the initial model can be recovered, while integrating out ω a T-dual model is obtained in which the multipliers play the role of dual coordinates. The gauged action, with Lagrange multipliers, reads $S^\omega_{SS} = \frac{1}{2}\int_\Sigma \langle m_\omega, \star m_\omega\rangle + \frac{1}{2}\int_\Sigma \langle p_\omega, q_\omega\rangle + \frac{1}{2}\int_\Sigma \langle D(j_\omega), j_\omega\rangle + \int_\Sigma \langle \tilde{\Lambda}, F_{j_\omega}\rangle$. Where $\tilde{\Lambda} := g^{-1}\Lambda g + g^{-1}D(g)$, while $j_\omega := j + g^{-1}\omega g$ and $A_\omega, p_\omega, m_\omega, q_\omega$ are its projections on the subspaces. $D : \mathfrak{g} \to \mathfrak{g}$ defines a deformation of the initial model introduced in [4, 5, 6], which will be set to zero in our examples. We shall dualise the whole isometry groups and also choose gauge $g = \mathbb{1}$. The EOM for ω reads $\star m_\omega - \frac{1}{2}p_\omega + \frac{1}{2}q_\omega + \nabla_{j_\omega}\tilde{\Lambda} - D(j_\omega) = 0$ and its projections on $\mathfrak{p}, \mathfrak{m}, \mathfrak{q}$ can be solved for $p_\omega, m_\omega, q_\omega$ [1]. The projection on $\mathfrak{h}$ reads

$$[\tilde{\Lambda}_\mathfrak{q}, p_\omega] + D_{\tilde{\Lambda}_\mathfrak{m}}(m_\omega) + [\tilde{\Lambda}_\mathfrak{p}, q_\omega] = \nabla_{A_\omega}\tilde{\Lambda}_\mathfrak{h} \quad \text{with} \quad \begin{cases} D_{\tilde{\Lambda}_\mathfrak{m}} := D + ad_{\tilde{\Lambda}_\mathfrak{m}} \\ \nabla_{j_\omega} := \mathrm{d} + ad_{j_\omega} \end{cases} \tag{1}$$

and cannot be generally solved for A_ω due to the lack of linear terms. This issue is a result of local H_R-invariance and forces a case-by-case study.

2 Solving the EOM - two simple examples

Exploiting the solution for $p_\omega, m_\omega, q_\omega$, equation (1) can be rearranged as

$$W(A_\omega) + Z(\star A_\omega) = \zeta \quad \text{with} \quad W, Z : \mathfrak{h} \to \mathfrak{h}, \quad A_\omega, \zeta \in \Omega^1(\Sigma, \mathfrak{h}) \quad \text{and}$$
$$\begin{cases} W := ad_{\tilde{\Lambda}_\mathfrak{h}} + N + (D_{\tilde{\Lambda}_\mathfrak{m}} - M^\dagger) \circ \sum_{k=0}^\infty S^{2k+1} \circ (D_{\tilde{\Lambda}_\mathfrak{m}} + M) \\ Z := (D_{\tilde{\Lambda}_\mathfrak{m}} - M^\dagger) \circ \sum_{k=0}^\infty S^{2k} \circ (D_{\tilde{\Lambda}_\mathfrak{m}} + M) \\ \zeta := \mathrm{d}\tilde{\Lambda}_\mathfrak{h} + \xi + (D_{\tilde{\Lambda}_\mathfrak{m}} - M^\dagger) \circ (\star + S) \circ \sum_{k=0}^\infty S^{2k} \circ (\mathrm{d}\tilde{\Lambda}_\mathfrak{m} + \chi) \end{cases} \tag{2}$$

Where we defined $S := ad_{\tilde{\Lambda}_{\mathfrak{h}}} + L$, with $L := ad_{\tilde{\Lambda}_{\mathfrak{p}}} \circ \mathscr{O}_1 + ad_{\tilde{\Lambda}_{\mathfrak{q}}} \circ \mathscr{O}_2$, and

$$\begin{cases} M := ad_{\tilde{\Lambda}_{\mathfrak{p}}} \circ \mathscr{O}_3 + ad_{\tilde{\Lambda}_{\mathfrak{q}}} \circ \mathscr{O}_4 \\ N := ad_{\tilde{\Lambda}_{\mathfrak{q}}} \circ \mathscr{O}_3 + ad_{\tilde{\Lambda}_{\mathfrak{p}}} \circ \mathscr{O}_4 \\ \xi := \mathscr{O}_3^\dagger(\mathrm{d}\tilde{\Lambda}_{\mathfrak{q}}) + \mathscr{O}_4^\dagger(\mathrm{d}\tilde{\Lambda}_{\mathfrak{p}}) \\ \chi := \mathscr{O}_1^\dagger(\mathrm{d}\tilde{\Lambda}_{\mathfrak{q}}) + \mathscr{O}_2^\dagger(\mathrm{d}\tilde{\Lambda}_{\mathfrak{p}}) \end{cases} \qquad \begin{cases} \mathscr{O}_1 := R_{11} \circ ad_{\Lambda_{\mathfrak{q}}} + R_{12} \circ ad_{\Lambda_{\mathfrak{p}}} \\ \mathscr{O}_2 := R_{21} \circ ad_{\Lambda_{\mathfrak{q}}} + R_{22} \circ ad_{\Lambda_{\mathfrak{p}}} \\ \mathscr{O}_3 := R_{12} \circ ad_{\Lambda_{\mathfrak{q}}} + R_{11} \circ ad_{\Lambda_{\mathfrak{p}}} \\ \mathscr{O}_4 := R_{22} \circ ad_{\Lambda_{\mathfrak{q}}} + R_{21} \circ ad_{\Lambda_{\mathfrak{p}}} \end{cases} \tag{3}$$

We refer to [1] for the expressions of R_{ij} and clarify that any $\mathscr{O}^\dagger$ is defined, with respect to the inner product $\langle \mathscr{O}^\dagger(X), Y\rangle = \langle X, \mathscr{O}(Y)\rangle$ for any two 1-forms X, Y, using $\langle R_{12}(X), Y\rangle = -\langle X, R_{12}(Y)\rangle$, $\langle R_{21}(X), Y\rangle = -\langle X, R_{21}(Y)\rangle$, $\langle R_{11}(X), Y\rangle = -\langle X, R_{22}(Y)\rangle$. As anticipated, all the above formulae reduce to those for symmetric spaces, in which $\mathfrak{g} \simeq \mathfrak{h} \oplus \mathfrak{m}$, by simply setting to zero any element in $\mathfrak{p}$ and $\mathfrak{q}$. In symmetric supercosets, $\mathfrak{h}$ and $\mathfrak{m}$ contain both bosonic and fermionic generators.

The possibility of solving (2) for A_ω, depends on the invertibility of W and Z and in turn on the structure of the algebra. This potential obstruction has been mentioned in [7, 5, 6] and problems in the presence of fermions have been discussed in [8]. We now highlight three situations in which the equation can be solved: in the first two cases W and $1 \pm ZW^{-1}$ or Z and $1 \pm WZ^{-1}$ are invertible, while in the third one neither W nor Z is invertible, but their sum and difference are

$$A_\omega = \tfrac{1}{2}(\zeta + \star\zeta)B_+ + \tfrac{1}{2}(\zeta - \star\zeta)B_- \quad \text{with} \quad \begin{cases} B_\pm := W^{-1}[(1 \pm ZW^{-1})^{-1}] \\ B_\pm := \pm Z^{-1}[(1 \pm WZ^{-1})^{-1}] \\ B_\pm := (W \pm Z)^{-1} \end{cases} \tag{4}$$

Writing $A_\omega = \star\alpha + \beta$ with $\alpha := \frac{1}{2}(\zeta_+ B_+ - \zeta_- B_-)$, $\beta := \frac{1}{2}(\zeta_+ B_+ + \zeta_- B_-)$ and $\zeta_\pm := \mathrm{d}\tilde{\Lambda}_{\mathfrak{h}} + \xi \pm (D_{\tilde{\Lambda}_{\mathfrak{m}}} - M^\dagger) \circ \sum_{k=0}^\infty (\pm S)^k \circ (\mathrm{d}\tilde{\Lambda}_{\mathfrak{m}} + \chi)$ defined as $(1 \pm \star)\zeta = (1 \pm \star)\zeta_\pm$, one obtains the full T-dual action $\tilde{S}_{SS} = \int_\Sigma \tilde{g} + \tilde{B}$ from the hybrid one in [1]

$$\tilde{g} := \tfrac{1}{2}\langle \lambda_-, \frac{1}{1-S} \star \lambda_+\rangle - \langle \nabla_\beta \tilde{\Lambda}_{\mathfrak{h}} + \mathscr{O}_3^\dagger(\nabla_\beta \tilde{\Lambda}_{\mathfrak{q}}) + \mathscr{O}_4^\dagger(\nabla_\beta \tilde{\Lambda}_{\mathfrak{p}}), \star\alpha\rangle \tag{5}$$
$$\begin{aligned} \tilde{B} :=& \tfrac{1}{2}\langle \lambda_-, \frac{1}{1-S}\lambda_+\rangle + \langle \tilde{\Lambda}_{\mathfrak{h}}, F_\beta - \tfrac{1}{2}[\alpha, \alpha]\rangle + \tfrac{1}{2}\langle \alpha, N(\alpha)\rangle + \\ &+ \tfrac{1}{2}\langle \nabla_\beta \tilde{\Lambda}_{\mathfrak{p}}, R_{21}(\nabla_\beta \tilde{\Lambda})_{\mathfrak{p}} + R_{22}(\nabla_\beta \tilde{\Lambda})_{\mathfrak{q}}\rangle + \tfrac{1}{2}\langle \nabla_\beta \tilde{\Lambda}_{\mathfrak{q}}, R_{11}(\nabla_\beta \tilde{\Lambda})_{\mathfrak{p}} + R_{12}(\nabla_\beta \tilde{\Lambda})_{\mathfrak{q}}\rangle \end{aligned}$$

with $\lambda_\pm := \nabla_{\beta\pm\alpha}\tilde{\Lambda}_{\mathfrak{m}} + \mathscr{O}_1^\dagger(\nabla_{\beta\pm\alpha}\tilde{\Lambda}_{\mathfrak{q}}) + \mathscr{O}_2^\dagger(\nabla_{\beta\pm\alpha}\tilde{\Lambda}_{\mathfrak{p}}) - D(\beta \pm \alpha)$.

We shall now solve (2) in two examples. The first one involves dualisation of the $SO(4)$ isometry of the coset space $SO(4)/SO(3) \simeq S^3$. This model has been studied in the literature [9, 7] and it is thus interesting to understand how the procedure goes through from the purely algebraic point of view. Given the $SO(4)$ algebra

$$[R_{IJ}, R_{KL}] = -\tfrac{i}{2}(\delta_{IK}R_{JL} - \delta_{JK}R_{IL} - \delta_{IL}R_{JK} + \delta_{JL}R_{IK}) \tag{6}$$

one can separate the $\mathfrak{h} = SO(3) = \{H_i := -\frac{1}{2}\varepsilon_i{}^{jk}R_{jk}\}$ subalgebra from the rest $\mathfrak{m} = \{M_i := R_{i4}\}$ by using indices $I, J = \{1,2,3,4\}$ and $i, j = \{1,2,3\}$. Indices are raised and lowered using the Euclidean metric δ_{IJ}. Generators in $\mathfrak{h}$ and $\mathfrak{m}$ satisfy

$$[H_i, H_j] = \tfrac{i}{2}\varepsilon_{ij}{}^k H_k \qquad [M_i, H_j] = \tfrac{i}{2}\varepsilon_{ij}{}^k M_k \qquad [M_i, M_j] = \tfrac{i}{2}\varepsilon_{ij}{}^k H_k \tag{7}$$

Writing the multipliers, i.e. the dual coordinate, as $\Lambda_{\mathfrak{h}} := \tilde{y}^i H_i$ and $\Lambda_{\mathfrak{m}} := \tilde{x}^i M_i$, recalling that $A_\omega := A^i_\omega H_i$ and using the above commutators, one finds

$$\begin{aligned} W(A_\omega) &= v^i \varepsilon_{ij}{}^k A^j_\omega H_k \qquad \text{with} \qquad v^i := \left[\tfrac{i}{2}\tilde{y}^i + \frac{i(\tilde{x}\cdot\tilde{y})}{2(4-\tilde{y}^2)}\tilde{x}^i\right] \\ Z(\star A_\omega) &= \tfrac{1}{4(\tilde{y}^2-4)}\{[(\tilde{y}^2-4)\tilde{x}^2 - (\tilde{y}\cdot\tilde{x})^2]\delta^k_j + \\ &\quad + [(\tilde{y}\cdot\tilde{x})\tilde{y}_j - (\tilde{y}^2-4)\tilde{x}_j]\tilde{x}^k + [(\tilde{y}\cdot\tilde{x})\tilde{x}_j - \tilde{x}^2\tilde{y}_j]\tilde{y}^k\} \star A^j_\omega H_k \end{aligned} \tag{8}$$

It is clear that v_k and $\tilde{x}_k$ respectively lie in the kernels of W^k_j and Z^k_j, but it is also not hard to find that $W \pm Z$ can be inverted, so that $B_\pm := (W \pm Z)^{-1}$ reads

$$\begin{aligned} (B_\pm)_k{}^l &= a_1^\pm \delta^l_k + \tilde{x}_k(a_2^\pm \tilde{x}^l + a_3^\pm \tilde{y}^l) + \tilde{y}_k(a_4^\pm \tilde{x}^l + a_5^\pm \tilde{y}^l) + \varepsilon_{ak}{}^l(a_6^\pm \tilde{x}^a + a_7^\pm \tilde{y}^a) + \\ &+ \tilde{x}^a\tilde{y}^b\varepsilon_{ab}{}^l(a_8^\pm \tilde{x}_k + a_9^\pm \tilde{y}_k) + \tilde{x}^a\tilde{y}^b\varepsilon_{abk}(a_{10}^\pm \tilde{x}^l + a_{11}^\pm \tilde{y}^l) + a_{12}^\pm \tilde{x}^a\tilde{y}^b\varepsilon_{abk}\tilde{x}^c\tilde{y}^d\varepsilon_{cd}{}^l \end{aligned} \tag{9}$$

with $a_1^\pm, \ldots, a_{12}^\pm$ complicated functions of $\tilde{x}^2, \tilde{y}^2, (\tilde{x}\cdot\tilde{y})$ which we omit for brevity.

The second example has to do with super T-dualisation of the $OSp(1|2)$ isometry of the semi-symmetric space $OSp(1|2)/SO(1,1)$. The interest in such a coset, which has also been considered in the context of holography [10], is due to its structure, which is that of a 2d Green-Schwarz string sigma model satisfying the torsion constraints of supergravity. For this reason, dualising such a model would not only represent a natural next step to the super T-dualisation of the Principal Chiral Model on $OSp(1|2)$, for which in [1] it was argued that T-duality breaks the supergravity constraint, but also a concrete example of super T-duality on supercosets, that could be compared to [4, 5, 6]. Given the $OSp(1|2)$ algebra in light-cone notation

$$\begin{aligned} &\{Q_\pm, Q_\pm\} = L_{\pm\pm} && \{Q_+, Q_-\} = L_{+-} && [L_{\pm\pm}, Q_\mp] = \mp i Q_\pm \\ &[L_{+-}, L_{\pm\pm}] = \pm i L_{\pm\pm} && [L_{++}, L_{--}] = -2iL_{+-} && [L_{+-}, Q_\pm] = \pm\tfrac{i}{2}Q_\pm \end{aligned} \tag{10}$$

the four subspaces are $\mathfrak{h} = \{L_{+-}\}$ $\mathfrak{p} = \{Q_+\}$ $\mathfrak{m} = \{L_{++}, L_{--}\}$ $\mathfrak{q} = \{Q_-\}$. Then, writing the Lagrange multipliers, i.e. the dual coordinates, as $\Lambda_{\mathfrak{h}} := \tilde{y}L_{+-}$, $\Lambda_{\mathfrak{p}} := \tilde{\theta}^+ Q_+$, $\Lambda_{\mathfrak{m}} := \tilde{x}^{++}L_{++} + \tilde{x}^{--}L_{--}$, $\Lambda_{\mathfrak{q}} := \tilde{\theta}^- Q_-$ and recalling that $A_\omega := A^{+-}_\omega L_{+-}$, one can exploit the above commutators to compute $W(A_\omega) = 0$ and

$$Z(\star A_\omega) = \frac{4\tilde{x}^{++}\tilde{x}^{--}}{1+\tilde{y}^2}\left[1 + \frac{4i\tilde{\theta}^+\tilde{\theta}^-}{(1-i\tilde{y})[4\tilde{x}^{++}\tilde{x}^{--} + (1+i\tilde{y})^2]}\right] \star A^{+-}_\omega L_{+-} \tag{11}$$

Equation (2) can thus be immediately solved as in (4) with $B_\pm := \pm Z^{-1}$ and

$$Z^{-1} = \frac{1+\tilde{y}^2}{4\tilde{x}^{++}\tilde{x}^{--}}\left[1 - \frac{4i\tilde{\theta}^+\tilde{\theta}^-}{(1-i\tilde{y})[4\tilde{x}^{++}\tilde{x}^{--} + (1+i\tilde{y})^2]}\right] \tag{12}$$

3 Conclusions and Outlook

We focused on a delicate step in the T-dualisation procedure of symmetric and semi-symmetric cosets, namely solving the EOM for the gauge fields ω in the $\mathfrak{h}$ subspace. We highlighted how such EOM is not necessarily solvable due to the lack of a linear term in A_ω: this may heavily affect T-duality, as the removal of A_ω is necessary to write down the full dual action. Rewriting the EOM as in (2), three cases can be recognised in which the solution is of the form (4). This has been used to write the full T-dual action and construct solutions for two explicit examples.

An important step toward a better comprehension of T-dualisation would be represented by a deeper understanding of the EOM (2) and of the constraints its resolution imposes on the underlying algebra. Additionally, given the physical relevance of the coset $OSp(1|2)/SO(1,1)$ and the simple result (12) for its EOM, it would certainly be very interesting to carry out the dualisation of such model in full details, so as to have a concrete example of super non-Abelian T-duality on supercosets, which could hopefully serve as a base model in view of more complicated ones.

Acknowledgements I am extremely grateful to Silvia Penati, Dmitri Sorokin and Martin Wolf for stimulating discussions and useful comments on this contribution. I also thank the organisers of the workshop “2D SUSY@MATRIX” for the opportunity to present my work and the referee for useful comments. This work was partially supported by Università degli studi di Milano-Bicocca, by the Italian Ministero dell'Università e della Ricerca (MUR), and by the Istituto Nazionale di Fisica Nucleare (INFN) through the research project ‘Gauge theories, Strings, Supergravity’ (GSS).

References

1. Bielli, D., Penati, S., Sorokin, D., Wolf, M.: Super Non-Abelian T-Duality. Nucl.Phys.B **983** (2022) 115904
2. Zarembo, K.: Strings on Semisymmetric Superspaces. JHEP 05 (2010) 002
3. Serganova, V.V.: Classification of real simple Lie superalgebras and symmetric superspaces. Funct.Anal.Appl 17 (1983) 200
4. Borsato, R., Wulff, L.: Integrable Deformations of T-Dual σ Models. Phys.Rev.Lett.**117** (2016) 25, 251602
5. Borsato, R., Wulff, L.: On non-abelian T-duality and deformations of supercoset string sigma-models. JHEP 10 (2017) 024
6. Borsato, R., Wulff, L.: Non-abelian T-duality and Yang-Baxter deformations of Green-Schwarz strings. JHEP 08 (2018) 027
7. Lozano, Y., Colgain, E.O., Sfetsos, K., Thompson, D.C.: Non-abelian T-duality, Ramond Fields and Coset Geometries. JHEP 06 (2011) 106
8. Grassi, P.A., Mezzalira, A.: Aspects of Quantum Fermionic T-duality. JHEP 05 (2011) 019
9. de la Ossa, X.C., Quevedo, F.: Duality symmetries from non-Abelian isometries in string theory. Nucl.Phys.B **403** (1993) 377-394
10. Verlinde, H.L.: Superstrings on AdS(2) and superconformal matrix quantum mechanics. (2004) e-Print: hep-th/0403024

On Twisted Elliptic Genera

Kaiwen Sun

Abstract We study the twisted elliptic genera associated to the BPS strings in the twisted circle compactification of 6d $(1,0)$ SCFTs. Such objects can arise when the 6d gauge algebra allows outer automorphism. We study several aspects of twisted elliptic genera including twisted elliptic blowup equations, $\Gamma_1(N)$ modular bootstrap and spectral flow symmetry. This is a short summary of the author's talk at MATRIX Program "2D Supersymmetric Theories and Related Topic", and based on a joint work with Kimyeong Lee and Xin Wang.

1 Introduction

For the last decade, there has been huge progress on 6d $(1,0)$ superconformal field theories (SCFTs) in both classification [1] and computation. These 6d SCFTs are geometrically engineered by F-theory compactified on local elliptic Calabi-Yau (CY) threefolds, and contain BPS strings with worldsheet theories as some highly nontrivial 2d $(0,4)$ gauge theories. Upon circle compactification and deformations, they produce 5d Kaluza-Klein (KK) theories which are conjectured to flow to all 5d SCFTs by decoupling matters. A simple example is the 6d E-string theory, geometrically engineered by local half-K3 Calabi-Yau threefold. The worldsheet theories for k E-strings are certain 2d $(0,4)$ $O(k)$ gauge theories [2]. Upon circle compactification, E-string theory gives the well-known 5d $SU(2)$ theory with 8 fundamentals. On the partition function level, the above correspondence can be summarized as the following relation chain

$$\mathbb{E}^{\text{2d (0,4) SCFT}} = Z^{\text{6d (1,0) SCFT}}_{\mathbb{R}^4\times T^2} = Z^{\text{5d KK}}_{\mathbb{R}^4\times S^1} = Z^{\text{ref. top.}}_{\text{local elliptic CY3}}. \tag{1}$$

Kaiwen Sun
Korea Institute for Advanced Study, 85 Hoegi-ro, Dongdaemun-gun, Seoul, South Korea.
e-mail: ksun@kias.re.kr

D. R. Wood et al. (eds.), *2021-2022 MATRIX Annals*, MATRIX Book Series 5,
https://doi.org/10.1007/978-3-031-47417-0_5

This states that the partition function of a 6d $(1,0)$ SCFT on 6d Omega background should be equal to the elliptic genera of the 2d worldsheet theories, and the Nekrasov partition function of the 5d KK theory, and the refined topological string partition function on the underlying local elliptic Calabi-Yau threefold. The relation chain (1) has been tested extensively for many theories including E-string theory.

When a 6d (1,0) SCFT has a discrete global symmetry, one can do twisted circle compactification to a 5d KK theory [3]. There are two possibilities:

1. gauge algebra allows outer automorphism: folding vector multiplet.
2. quiver structure has discrete symmetry: folding tensor multiplet.

Here we focus on the first kind. In such situations, fractional KK charges naturally appear upon circle compactification and the local elliptic Calabi-Yau threefolds are generalized to local genus-one fibered ones. For partition functions, we expect a twisted version of relation chain (1):

$$\mathbb{E}^{\text{2d (0,4) SCFT}}_{\text{twisted}} = Z^{\text{6d (1,0) SCFT}}_{\mathbb{R}^4\times S^1\times S^1,\text{twisted}} = Z^{\text{5d KK}}_{\mathbb{R}^4\times S^1} = Z^{\text{ref. top.}}_{\text{local genus-one fibered CY3}}. \tag{2}$$

We are interested in the twisted elliptic genera arising here.

Now we show some simple examples of twisted 6d SCFTs. A 6d $(1,0)$ SCFT with rank-one tensor branch is specified by self-intersection number $\mathfrak{n}$, gauge algebra G, flavor symmetry F and matter represention R. The $(\mathfrak{n},G,F,R)$ is highly constrained by the Calabi-Yau condition. The gauge algebras that allow outer automorphism are classified by twisted affine Lie algebras which are $E_6^{(2)}$, $D_{r+4}^{(2)}$, $A_r^{(2)}$ and $D_4^{(3)}$ [4, Chapter 8]. Removing the affine node, we denote the truncated algebra as $\mathring{G}$ which is the 5d low energy gauge algebra. The twisted matter content $\mathring{R}$ are in the representation of $\mathring{G}$ and can have fractional KK charges. The flavor symmetry normally get reduced which we denote as $\mathring{F}$. We summarize some typical twisted 6d $(1,0)$ rank one theories with $\mathfrak{n}\geq 3$ in Table 1. The main goal of this work is to compute the twisted elliptic genera of all these theories and study their properties.

Table 1 Some typical twisted 6d $(1,0)$ rank-one theories with $\mathfrak{n}\geq 3$. The subscripts in $\mathring{R}$ denote the KK charge, and $\mathbf{V},\mathbf{F},\mathbf{S}$ denote vector, fundamental and spinor representation respectively. The constants c are used in the modular bootstrap in Sect 3. There are four pure gauge theories: $\mathfrak{n}=6$ with $E_6^{(2)}$, $\mathfrak{n}=4$ with $D_4^{(2)},D_4^{(3)}$ and $\mathfrak{n}=3$ with $A_2^{(2)}$.

$\mathfrak{n}$	G	$\mathring{G}$	$\mathring{R}$	$\mathring{F}$	c
6	$E_6^{(2)}$	F_4	–	–	5/4
4	$D_4^{(3)}$	G_2	–	–	5/9
4	$D_{r+4}^{(2)}$	B_{r+3}	$2r(\mathbf{V}_0\oplus\mathbf{1}_{1/2})$	$\mathfrak{sp}(2r)$	3/4
4	$E_6^{(2)}$	F_4	$\mathbf{F}_0\oplus\mathbf{F}_{1/2}\oplus\mathbf{1}_0\oplus\mathbf{1}_{1/2}$	$\mathfrak{sp}(1)$	1
3	$A_2^{(2)}$	C_1	–	–	5/16
3	$D_4^{(2)}$	B_3	$\mathbf{V}_0\oplus\mathbf{1}_{1/2}\oplus\mathbf{S}_0\oplus\mathbf{S}_{1/2}$	$\mathfrak{sp}(1)\times\mathfrak{sp}(1)$	1/2
3	$D_4^{(3)}$	G_2	$\mathbf{F}_0\oplus\mathbf{F}_{1/3}\oplus\mathbf{F}_{2/3}\oplus\mathbf{1}_0\oplus\mathbf{1}_{1/3}\oplus\mathbf{1}_{2/3}$	$\mathfrak{sp}(1)$	1/2

2 Twisted Elliptic Blowup Equations

Blowup equations as the functional equations of Nekrasov partition functions [5] were proposed by Nakajima and Yoshioka [6] for 4d $\mathcal{N}=2$ $SU(N)$ gauge theories to prove the Nekrasov conjecture. They originated by comparing the Nekrasov partition functions on the Omega background $\mathbb{C}^2_{\varepsilon_1,\varepsilon_2}$ and on its one-point blown-up. By blowing up and down the exceptional divisor, it can be expected that the two partition functions should be closely connected. Blowup equations are later generalized to various 4d/5d/6d gauge theories and become powerful tools to solve partition functions and refined BPS invariants. For the elliptic genera associated to 6d $(1,0)$ SCFTs, such functional equations were established in [7, 8, 9, 10] and called elliptic blowup equations. Now we want to further generalize them to twisted elliptic blowup equations and solve the twisted elliptic genera for all rank-one twisted theories. Some cases have been recently studied in [11]. For the total partition function Z of refined topological strings on a non-compact Calabi-Yau threefold X, the general form of blowup equations was proposed in [12]:

$$\begin{aligned}&\Lambda(\varepsilon_1,\varepsilon_2,m_i)Z(\varepsilon_1,\varepsilon_2,t_i+\pi\mathrm{i}B_i)\\ &=\sum_{\mathbf{k}\in\mathbb{Z}^{b_4}}(-1)^{|\mathbf{k}|}Z(\varepsilon_1,\varepsilon_2-\varepsilon_1,t_i+(C_{ij}k_j+B_i/2)\varepsilon_1+\pi\mathrm{i}B_i)\\ &\qquad\times Z(\varepsilon_1-\varepsilon_2,\varepsilon_2,t_i+(C_{ij}k_j+B_i/2)\varepsilon_2+\pi\mathrm{i}B_i),\end{aligned} \tag{3}$$

Here b_4 counts the number of compact divisors of X. The t_i is the Kähler parameter and C_{ij} is the intersection numbers of divisors and curves in X. The B_i is the flux for the corresponding Kähler parameter. The $\Lambda(\varepsilon_1,\varepsilon_2,m_i)$ is a function that only depends on the mass parameters m_i of the theory. The $\Lambda(\varepsilon_1,\varepsilon_2,m_i)\equiv 0$ case is called vanishing, otherwise unity. By geometric engineering, the partition function can be written by 6d gauge quantities as

$$Z(\varepsilon_1,\varepsilon_2,t_i)=e^{\mathscr{F}^{\mathrm{cls}}}Z_0(\varepsilon_1,\varepsilon_2,m_{\mathring{G}},m_{\mathring{F}})\left(1+\sum_{d=1}^{\infty}Q^d\mathbb{E}_d(\varepsilon_1,\varepsilon_2,m_{\mathring{G}},m_{\mathring{F}})\right), \tag{4}$$

Here $\mathscr{F}^{\mathrm{cls}}$ is the classical free energy, Z_0 is the one-loop partition function, $\mathbb{E}_d$ is the d-string twisted elliptic genus and Q counts the number of strings. As the classical and one-loop part have known formulas, by substituting (4) into (3), we can derive the functional equations of $\mathbb{E}_d$ which we call twisted elliptic blowup equations.

We find that for all twisted theories in Table 1, there exist unity blowup equations such that the twisted d-string elliptic genera $\mathbb{E}_d(\tau,m_{\mathring{G}},m_{\mathring{F}},\varepsilon_{1,2})$ can be explicitly solved. We are particular interested in the reduced twisted one-string elliptic genus defined by

$$\mathbb{E}_1^{\mathrm{red}}(\tau,\varepsilon_+)=\frac{\theta_1(\tau,\varepsilon_1)\theta_1(\tau,\varepsilon_2)}{\eta(\tau)^2}\mathbb{E}_1(\tau,\varepsilon_1,\varepsilon_2). \tag{5}$$

The simplicity of $\mathbb{E}_1^{\mathrm{red}}$ is that it only depends on $\varepsilon_+=(\varepsilon_1+\varepsilon_2)/2$ rather than both $\varepsilon_1,\varepsilon_2$. We will discuss its modular bootstrap and spectral flow in the later sections.

3 Modular Bootstrap for $\Gamma_1(N)$

For untwisted 6d $(1,0)$ SCFTs, the modular ansatz for reduced one-string elliptic genus was proposed in [13] which involves Jacobi form on $SL(2,\mathbb{Z})$. For twisted cases, it is natural to expect from modularity that the modular ansatz should be established on the principal congruence subgroup $\Gamma_1(N)$. Here N is the twist coefficient given by:

$$N = \begin{cases} 2, A_{2r-1}^{(2)}, D_r^{(2)}, E_6^{(2)}, \\ 3, D_4^{(3)}, \\ 4, A_{2r}^{(2)}. \end{cases} \tag{6}$$

From the aspect of geometry, the modular ansatz for N-section Calabi-Yau threefolds has been proposed in [14], see also [15]. Combining together, we propose the following modular ansatz for the reduced twisted one-string elliptic genus:

$$\mathbb{E}_1^{\rm red}(\tau,\varepsilon_+) = \frac{\mathcal{N}(\tau,\varepsilon_+)}{\eta(\tau)^{12(\mathfrak{n}-2)-4+24\delta_{\mathfrak{n},1}}\Delta_{2N}(\frac{\tau}{N})^s\phi_{-2,1}(\tau,2\varepsilon_+)^{h^\vee_{\mathring{G}}-1}}, \tag{7}$$

where

$$s = \frac{N}{N-1}\left(c - \frac{\mathfrak{n}-2}{2} - \delta_{\mathfrak{n},1}\right). \tag{8}$$

Here c is the constant given in Table 1. The numerator $\mathcal{N}(\tau,\varepsilon_+)$ is a Jacobi form of weight $6(\mathfrak{n}-2)+2Ns+12\delta_{\mathfrak{n},1}-2h^\vee_{\mathring{G}}$ and index $4(h^\vee_{\mathring{G}}-1)+\mathfrak{n}-h^\vee_G$ and

$$\mathcal{N}(N\tau,\varepsilon_+) \in M_*(N)[\phi_{-2,1}(N\tau,\varepsilon_+),\phi_{0,1}(N\tau,\varepsilon_+)]. \tag{9}$$

Here $M_*(N)$ is the ring of holomorphic modular forms of $\Gamma_1(N)$. Besides, the Δ_{2N} are some cusp forms on $\Gamma_0(N)$:

$$\Delta_4(\tau) = \frac{\eta(2\tau)^{16}}{\eta(\tau)^8}, \quad \Delta_6(\tau) = \frac{\eta(3\tau)^{18}}{\eta(\tau)^6}, \quad \Delta_8(\tau) = \frac{\eta(2\tau)^8\eta(4\tau)^{16}}{\eta(\tau)^8},$$

and $\phi_{-2,1}(\tau,z), \phi_{0,1}(\tau,z)$ are Eichler-Zagier's generators for weak Jacobi forms. We successfully fix the numerator for all theories in Table 1, check them against the elliptic genera solved from blowup equations and find perfect agreements.

4 Spectral Flow Symmetry

Spectral flow is a characteristic feature of 2d $\mathcal{N}=2$ SCFTs. The spectral flow for reduced one-string elliptic genera associated to 6d $(1,0)$ SCFTs was studied in [16, 13]. Such flow relates the R-R elliptic genera we have discussed and the NS-R elliptic genera. For the reduced twisted one-string elliptic genera, we find the spectral flow from the R-R sector to NS-R sector is induced by the following transformation:

$$\mathbb{E}^{\mathring{R}_{\mathrm{KK}}}_{\mathrm{NS-R}}\Big(q,v\Big) = \pm\Big(\frac{q^{1/4}}{v}\Big)^{\mathfrak{n}-h^{\vee}_{G}} \mathbb{E}^{\mathring{R}_{\mathrm{KK}}}_{\mathrm{R-R}}\Big(q,\frac{q^{1/2}}{v}\Big). \tag{10}$$

Here $q = e^{2\pi i\tau}, v = e^{2\pi i\epsilon_+}$. On the other hand, the spectral flow shifts the KK charges of hypermultiplets by half. For 6d $(1,0)$ pure gauge $G^{(n)}$ theories, the above two transformations imply the following spectral flow symmetry for the reduced twisted one-string elliptic genera:

$$\mathbb{E}^{G^{(n)}}_{1}\Big(q,\frac{q^{1/2}}{v}\Big) = \Big(-\frac{q^{1/2}}{v^2}\Big)^{\mathfrak{n}-3}\mathbb{E}^{G^{(n)}}_{1}(q,v). \tag{11}$$

We have explicitly checked this symmetry for $A_2^{(2)}, D_4^{(2)}, D_4^{(3)}, E_6^{(2)}$ theories.

Acknowledgements The author would like to thank Kimyeong Lee and Xin Wang for the collaboration on this subject and Hee-Cheol Kim and Haowu Wang for useful discussions. The author is supported by KIAS Grant QP081001.

References

1. J. J. Heckman, D. R. Morrison, T. Rudelius and C. Vafa, "Atomic Classification of 6D SCFTs," Fortsch. Phys. **63** (2015), 468-530.
2. J. Kim, S. Kim, K. Lee, J. Park and C. Vafa, "Elliptic Genus of E-strings," JHEP **09** (2017), 098.
3. L. Bhardwaj, P. Jefferson, H. C. Kim, H. C. Tarazi and C. Vafa, "Twisted Circle Compactifications of 6d SCFTs," JHEP **12** (2020), 151.
4. V. G. Kac, "Infinite dimensional Lie algebras," Cambridge university press, 1990.
5. N. A. Nekrasov, "Seiberg-Witten prepotential from instanton counting," Adv. Theor. Math. Phys. **7** (2003) no.5, 831-864.
6. H. Nakajima and K. Yoshioka, "Instanton counting on blowup. 1.," Invent. Math. **162** (2005), 313-355.
7. J. Gu, B. Haghighat, K. Sun and X. Wang, "Blowup Equations for 6d SCFTs. I," JHEP **03** (2019), 002.
8. J. Gu, A. Klemm, K. Sun and X. Wang, "Elliptic blowup equations for 6d SCFTs. Part II. Exceptional cases," JHEP **12** (2019), 039.
9. J. Gu, B. Haghighat, A. Klemm, K. Sun and X. Wang, "Elliptic blowup equations for 6d SCFTs. Part III. E-strings, M-strings and chains," JHEP **07** (2020), 135.
10. J. Gu, B. Haghighat, A. Klemm, K. Sun and X. Wang, "Elliptic blowup equations for 6d SCFTs. Part IV. Matters," JHEP **11** (2021), 090.
11. H. C. Kim, M. Kim, S. S. Kim and K. H. Lee, "Bootstrapping BPS spectra of 5d/6d field theories," JHEP **04** (2021), 161.
12. M. x. Huang, K. Sun and X. Wang, "Blowup Equations for Refined Topological Strings," JHEP **10** (2018), 196.
13. M. Del Zotto and G. Lockhart, "Universal Features of BPS Strings in Six-dimensional SCFTs," JHEP **08** (2018), 173.
14. C. F. Cota, A. Klemm and T. Schimannek, "Topological strings on genus one fibered Calabi-Yau 3-folds and string dualities," JHEP **11** (2019), 170.

15. J. Knapp, E. Scheidegger and T. Schimannek, "On genus one fibered Calabi-Yau threefolds with 5-sections," [arXiv:2107.05647 [hep-th]].
16. M. Del Zotto and G. Lockhart, "On Exceptional Instanton Strings," JHEP **09** (2017), 081.

Minkowski and (A)dS ground states in general 2d dilaton gravity

Daniel Grumiller, Martin Laihartinger and Romain Ruzziconi

Abstract We reorganize the derivative expansion of general (power-counting non-renormalizable) 2d dilaton gravity such that the mass function is integrable. As an example, we consider a three-parameter family of models and provide conditions on the parameters such that the ground state is either Minkowski, Rindler, or (A)dS.

1 Introduction to general 2d dilaton gravity

Two is the lowest spacetime dimension with enough room for space and time. Consequently, it is the lowest dimension that allows for lightcones, horizons, and black holes. Moreover, two is the lowest dimension permitting non-trivial intrinsic curvature and topology. For all these reasons, two-dimensional (2d) theories of gravity serve as minimal toy models to elucidate conceptual aspects of classical and quantum gravity, including black hole evaporation and holographic correspondences.

Since Einstein gravity does not exist in 2d, one has to resort to some alternative gravity theory. All known approaches[1] lead to the same class of theories (possibly after some field redefinitions and/or integrating in/out auxiliary fields), namely 2d

Daniel Grumiller
Institute for Theoretical Physics, TU Wien, Wiedner Hauptstrasse 8-10, A-1040 Vienna, Austria.
e-mail: grumil@hep.itp.tuwien.ac.at

Martin Laihartinger
Institute for Theoretical Physics, TU Wien, Wiedner Hauptstrasse 8-10, A-1040 Vienna, Austria.
e-mail: e1608104@student.tuwien.ac.at

Romain Ruzziconi
Institute for Theoretical Physics, TU Wien, Wiedner Hauptstrasse 8-10, A-1040 Vienna, Austria.
e-mail: romain.ruzziconi@tuwien.ac.at

[1] This includes intrinsic models [20, 41, 3], higher curvature/torsion models [22], gravity as linear [19, 4, 42] or non-linear gauge theory [18, 38], the target-space action of non-critical strings in 2d [30, 7, 43, 6], a scaling limit of gravity in $2+\varepsilon$ dimensions [31, 27, 8] and various dimensional reductions to 2d, see [9] and additional refs. therein.

D. R. Wood et al. (eds.), *2021-2022 MATRIX Annals*, MATRIX Book Series 5,
https://doi.org/10.1007/978-3-031-47417-0_6

dilaton gravity, a scalar-tensor theory. In its most general version, the bulk action

$$I[g_{\mu\nu}, X] = -\frac{\kappa}{4\pi}\int d^2x\sqrt{-g}\Big(XR - 2\mathscr{V}(X, -(\partial X)^2)\Big) \tag{1}$$

contains an arbitrary function $\mathscr{V}$ of two variables, the dilaton X and its kinetic term $(\partial X)^2$. We adapted signs consistently with [13]. Special cases often considered in the literature are the restriction to power-counting renormalizable models [33, 35] or to potentials that are independent from the kinetic term [28, 44]. In our work, we do not impose such restrictions.

While the action (1) has appeared, e.g., in [39, 9, 14, 26, 40] (also known as "kinetic gravity braiding" [5] or "Horndeski theory" [17]), besides the class of dilaton scale invariant models studied in detail in [13] no charting of the full model space was attempted yet.

In this proceedings contribution, we make the first steps in this direction. In section 2 we review how to obtain all classical solutions, focussing on the linear dilaton sector, up to the integration of a conservation equation. In section 3 we reorganize the way that models are formulated by demanding the conservation equation to be integrable and providing some function therein as input, from which the potential $\mathscr{V}$ is then deduced. In section 4 we focus on a three-parameter family as an example and discuss under which conditions the models exhibit ground states that are maximally symmetric (Minkowski, Rindler, or (A)dS). In section 5 we conclude.

2 Classical solutions

To discuss the classical solutions of general 2d dilaton gravity (1), it is convenient to work in the first order formulation of the theory [24, 25], whose action

$$I[\omega, e_a, X, X^a] = \frac{\kappa}{2\pi}\int \Big(X\,d\omega + X^a\left(de_a + \varepsilon_a{}^b\,\omega\wedge e_b\right) + \varepsilon\,\mathscr{V}(X, X^cX_c)\Big) \tag{2}$$

depends on the (dualized) Lorentz connection ω, the zweibein e_a, the dilaton X, and auxiliary scalar fields X^a. We introduced the volume form $\varepsilon = \frac{1}{2}\varepsilon^{ab}e_a\wedge e_b$. In light-cone coordinates (x^+, x^-) for the Lorentz indices a, b, the Minkowski metric reads as $\eta_{+-} = 1$, $\eta_{\pm\pm} = 0$. Our sign convention for the Levi-Civitá symbol is $\varepsilon^{\pm}{}_{\pm} = \pm 1$. The equations of motion

$$dX + X^-e^+ - X^+e^- = 0 \tag{3a}$$

$$(d\pm\omega)X^\pm \pm e^\pm\mathscr{V} = 0 \tag{3b}$$

$$d\omega + \varepsilon\frac{\partial\mathscr{V}}{\partial X} = 0 \tag{3c}$$

$$(d\pm\omega)e^\pm + \varepsilon\frac{\partial\mathscr{V}}{\partial X^\mp} = 0 \tag{3d}$$

are of first order in derivatives. Solving (3c) and (3a) for ω and X^a and re-injecting these solutions into the action (2) recovers the second order action (1) (see e.g. [13] for details).

The constant dilaton sector of the theory is given by the solutions of (3a)-(3d) satisfying

$$\mathcal{V}(X, X^a X_a) = 0 = X^a \tag{4}$$

These solutions have constant dilaton, constant curvature, and are locally maximally symmetric. They are covered extensively in the literature (see e.g. [16]), so we do not discuss them in this work.

Instead, we focus on the linear dilaton sector of the theory and summarize the solution algorithm discussed in [13]. Combining (3b) and (3a) obtains

$$\mathrm{d}Y - \mathcal{V}(X, 2Y)\,\mathrm{d}X = 0 \qquad\qquad Y := X^+ X^- . \tag{5}$$

As discussed below, this relation implies Casimir conservation $\mathrm{d}C = 0$ and can be integrated to yield the Casimir function $C(X,Y)$. Assuming $X^+ \neq 0$ (essentially without loss of generality) the equation (3b) implies

$$\omega = -\frac{\mathrm{d}X^+}{X^+} - Z\mathcal{V} \tag{6}$$

where $Z = e^+/X^+$. Similarly, the equation (3a) yields

$$e^- = \frac{\mathrm{d}X}{X^+} + X^- Z \tag{7}$$

and the volume element $\varepsilon = -Z \wedge \mathrm{d}X$. Combining the upper signs (3d) and (3b) gives

$$\mathrm{d}Z = (Z \wedge \mathrm{d}X)\,\frac{\partial \mathcal{V}}{\partial Y} . \tag{8}$$

Inserting $Z = \mathrm{d}v\, e^{Q(X)}$ into this equation yields

$$\frac{\mathrm{d}Q}{\mathrm{d}X} = -\frac{\partial \mathcal{V}}{\partial Y} \tag{9}$$

which can be formally integrated as

$$Q(X) = -\int^X \frac{\partial \mathcal{V}}{\partial Y} \tag{10}$$

by virtue of (5). The line element reads as

$$\mathrm{d}s^2 = 2e^+ e^- = 2e^Q\,\mathrm{d}v\,\mathrm{d}X + 2Ye^{2Q}\,\mathrm{d}v^2 \tag{11}$$

where Y is constrained through (5) and $Q(X)$ is given in (10). Introducing the radial coordinate $\mathrm{d}r = e^Q\,\mathrm{d}X$, which can be integrated to give the dilaton as a function of the radius $X(r)$, the line element (11) takes Eddington–Finkelstein form,

$$ds^2 = 2e^{2Q} Y\, dv^2 + 2\, dv\, dr\,. \tag{12}$$

The solution (12) is completely determined once the potential $\mathcal{V}$ is provided, up to the integration constant C in (5) corresponding to the Casimir of the theory.[2] All solutions exhibit at least one Killing vector, namely ∂_v.

We highlight three properties of the Casimir C: 1. Physically, it encodes the mass of the solution. 2. Notably, the function $Q(X)$, in general, depends on the Casimir, except for power-counting renormalizable models. 3. To obtain explicit solutions, we still need to integrate (5). How to achieve this is the subject of the next section.

3 Integrable conservation equation

In practice, given a potential $\mathcal{V}$, it might be difficult to integrate the conservation equation (5) and find an explicit solution for Y. Furthermore, it is not a priori obvious to determine which class of potentials $\mathcal{V}$ will allow us to find explicit solutions and obtain a complete classification of the ground states. Instead, we will use another route and start from a class of exact differential equations for (5). They are of the form

$$p(X,Y)\ \mathrm{d}X + q(X,Y)\ \mathrm{d}Y = 0 \tag{13}$$

where $p(X,Y)$ and $q(X,Y)$ are functions satisfying the integrablity condition

$$\frac{\partial p}{\partial Y} = \frac{\partial q}{\partial X}\,. \tag{14}$$

With this condition, the conservation equation (13) can be rewritten in terms of the Casimir $C(X,Y)$ as

$$\mathrm{d}C = 0 \qquad \text{with} \qquad p = \frac{\partial C}{\partial X} \qquad q = \frac{\partial C}{\partial Y} \tag{15}$$

which readily can be integrated. Comparing with (5), we identify

$$\mathcal{V} = -\frac{p}{q}\,. \tag{16}$$

Our approach is reminiscent of inverse scattering methods: instead of starting with some potential $\mathcal{V}$ and determining the mass function C, we postulate some suitable mass function C, use (15) to get the functions p and q, and obtain the potential $\mathcal{V}$ as output by virtue of (16).[3] In terms of (16), equation (10) leads to

$$Q(X) = \int^X \frac{\partial}{\partial Y}\left(\frac{p}{q}\right) \mathrm{d}X' = \int^X \left(\frac{1}{q}\frac{\partial p}{\partial Y} - \frac{p}{q}\,\frac{1}{q}\frac{\partial q}{\partial Y}\right) \mathrm{d}X'\,. \tag{17}$$

[2] The constant coming from the integration of (9) is trivial since it can be fixed by a choice of units.

[3] The analog of the integrability condition (14) would be difficult to formulate if one started from a generic potential $\mathcal{V}$ instead of starting from equation (13), which justifies our approach.

The first term in the integral on the right can be rewritten with the relation (14).

$$Q(X) = \int^X \left(\frac{1}{q} \frac{\partial q}{\partial X'} + \mathscr{V} \frac{1}{q} \frac{\partial q}{\partial Y} \right) \mathrm{d}X' \tag{18}$$

Performing the change of integration variable $\mathrm{d}X = \mathscr{V}^{-1}\,\mathrm{d}Y$ in the second term, and using $\mathrm{d}q = \frac{\partial q}{\partial Y}\,\mathrm{d}Y + \frac{\partial q}{\partial X}\,\mathrm{d}X$, we finally obtain

$$Q(X) = \int \frac{1}{q}\,\mathrm{d}q = \ln q(X) \tag{19}$$

where, as explained in the previous section, the integration constant does not contain physical information and has therefore been set to zero. In general, $Q(X)$ depends on the Casimir C when q depends on Y, since we use the integrated version of (5) to express Y as function of X and the Casimir C. Power-counting renormalizable models, $\mathscr{V}(X,2Y) = V(X) - YU(X)$, are a notable exception: for these models, the integrating factor $Q(X)$ is independent from the Casimir C.

Up to this stage, given p, q in equation (13) satisfying the integrability condition (14), we have shown how to write explicitly the line element (11) using (19) and the constraint on Y that can be readily obtained from the Casimir conservation deduced from (13). Now we intend to determine the ground states of these families of metrics, which are characterized by solutions with *(i)* constant curvature and *(ii)* vanishing Casimir function. The first condition is by choice — we are mostly interested in models that have a maximally symmetric ground state, like flat space, Rindler, or (A)dS_2. The second condition is without loss of generality: we can always shift the whole spectrum in such a way that the value of the Casimir function for the ground state vanishes.

The Ricci scalar R is expressed in terms of the Killing norm K as[4]

$$R = \partial_r^2 K \qquad\qquad K = 2q^2 Y\,. \tag{20}$$

We distinguish three types of vacua, depending on the values of the curvature and Killing norm:

- **Minkowski ground states** are characterized by vanishing Ricci scalar, $R = 0$, and constant Killing norm, $K(r) = \text{const}$.
- **Rindler ground states** are given by vanishing Ricci scalar, $R = 0$, and linear Killing norm, $K(r) \propto r$.
- **(A)dS ground states** are characterized by positive (negative) Ricci scalar, so that the Killing norm is quadratic. We can distinguish between Poincaré (A)dS with $K(r) \propto (-)r^2$, global (A)dS with $K(r) \propto ((-)r^2 + 1)$ and (A)dS Rindler where $K(r) \propto (-)r^2 + Ar + B$ for some constants A, B. In the following, when discussing (A)dS ground states, we will only refer to Poincaré (A)dS ground states.

[4] Zeros of the Killing norm K in (20) correspond to loci of Killing horizons. They arise either for $q = 0$ or $Y = 0$. The case of $q = 0$ leads to an extremal Killing horizon, as $q = 0$ is a double zero for the Killing norm. The loci of non-extremal Killing horizons are those for which $Y = 0$.

4 Three-parameter family as example

Power-counting renormalizable models involve potentials and Casimirs that are at most linear in Y. The ground states of these models have been extensively discussed in the literature (see e.g. [21]). Here, we focus on the simplest general dilaton gravity that is not power-counting renormalizable by considering a Casimir $C(X,Y)$ quadratic in Y,

$$C(X,Y) = V_0(X) + V_1(X)Y + V_2(X)Y^2 \tag{21}$$

where the potentials V_0, V_1, V_2 are taken to be monomials in X.

$$V_0(X) = A\,X^a \qquad V_1(X) = B\,X^b \qquad V_2(X) = D\,X^c \tag{22}$$

The parameters A, B, D and a, b, c are constant. The condition *(ii)* to obtain ground state solutions requires solving $C(X,Y) = 0$ for Y. This leads to two branches of solutions.

$$Y^\pm = \frac{-V_1 \pm \sqrt{V_1^2 - 4V_2V_0}}{2V_2} \tag{23}$$

Evaluating the function $q(X,Y)$ on these solutions yields

$$q(X,Y^\pm) = V_1 + 2V_2Y_\pm = \pm\sqrt{V_1^2 - 4V_2V_0}\,. \tag{24}$$

Plugging (23) and (24) into the second equation of (20) gives the ground state Killing norm

$$K_0^\pm := K^\pm|_{C=0} = (V_1^2 - 4V_2V_0)\frac{-V_1 \pm \sqrt{V_1^2 - 4V_2V_0}}{V_2}\,. \tag{25}$$

For the three-parameter family (22), this reads as

$$K_0^\pm = \left(B^2X^{2b} - 4AD\,X^{a+c}\right)\left(-\frac{B}{D}X^{b-c} \pm \sqrt{\frac{B^2}{D^2}X^{2b-2c} - 4\frac{A}{D}X^{a-c}}\right). \tag{26}$$

By expanding $Y^\pm$ for small values of D, one can see that the branch Y^- is continuously connected to the linear solution, while the branch Y^+ is not. This constitutes a criterion to select the branch Y^- rather than Y^+, which is what we are going to do.

We now focus on the condition *(i)* defining ground states. Following the classification of vacua provided at the end of the previous section, we treat each case separately. We determine first the conditions on the model parameters a,b,c such that the ground state of the corresponding model has the desired properties, then write down the potential associated with the model and finally display the metric (12) that emerges as solution of the model, for any value of the Casimir C. The relation between dilaton and radial coordinate is displayed in the appendix. As an abbreviation we often use the discriminant $\Delta := B^2 - 4AD$.

4.1 Minkowski ground states

Minkowski ground states are found by requiring the expression of the Killing norm (26) to be constant. Rewriting (26) as

$$K_0^- = X^{3b-c}\left(B^2 - 4AD\,X^{a+c-2b}\right)\left(-\frac{B}{D} - \sqrt{\frac{B^2}{D^2} - 4\frac{A}{D}X^{a+c-2b}}\right) = \text{const}\,, \quad (27)$$

we see that this condition is satisfied for generic A, B and D only if each term in the product is constant, and therefore $a+c-2b=0$ and $3b-c=0$. Minkowski ground state models then have to obey the relations

$$\boxed{\text{Minkowski ground state models:} \qquad a=-b \qquad c=3b} \quad (28)$$

between the parameters, yielding the potential

$$\boxed{\mathscr{V}_{\text{Mink}}\left(X, -(\partial X)^2\right) = aX^{2a-1}\,\frac{4A+2BX^{-2a}(\partial X)^2 - 3DX^{-4a}(\partial X)^4}{-4B+4DX^{-2a}(\partial X)^2}\,.} \quad (29)$$

The metric (12) for Minkowski ground state models simplifies to ($\Delta = B^2 - 4AD$)

$$\mathrm{d}s^2 = 2\,\mathrm{d}v\,\mathrm{d}r - I_0\left(\Delta + 4CDX^{-a}\right)\left(B + \sqrt{\Delta + 4CDX^{-a}}\right)\mathrm{d}v^2 \quad (30)$$

where we labelled by I_0 the freedom of fixing the integration constant in the function $Q(X)$ given in (19). Demanding the terms proportional to the Casimir C to be subleading for large X establishes the convexity condition $a>0$. For positive a and a convenient choice of I_0, the metric (30) has the asymptotically flat expansion

$$\mathrm{d}s^2 = 2\,\mathrm{d}v\,\mathrm{d}r - \left(1 - MX^{-a} + \mathscr{O}\left(M^2X^{-2a}\right)\right)\mathrm{d}v^2 \quad (31)$$

where the mass parameter M is proportional to C. Demanding that large values of the radial coordinate also correspond to large values of the dilaton requires $a \leq 1$. Thus, standard Minkowski ground state models obey $a \in (0,1]$ in addition to (28).

Non-generic cases for specific values of the constants A, B, D exist as well. If $A=0$ and $B \neq 0$, we get the ground state condition $c=3b$ with no further restriction. In the case of $B=0$ and $A \neq 0$, $K_0^- = \text{const}$ leads to $c=-3a$. The case $D=0$ leads to a two-parameter family of power-counting renormalizable models with the conditions on the parameter $K_0 = 2q^2Y = -2AB\,X^{a+b} = \text{const}$, so that $a=-b$. Comparing with the well-known two-parameter family $(a'b')$ (see e.g. [21]) where the potential reads as $\mathscr{V}_{(a'b')}(X,2Y) = \frac{a'}{X}Y + X^{a'+b'}$ and using the dictionary $a'=-b$ and $b'=a-1$, we recover the known result for Minkowski ground state $a'=1+b'$, which constitutes a consistency check of our derivation.

4.2 Rindler ground states

With the radial coordinate defined by $r = \int^X q(X',Y)\,\mathrm{d}X'$, the condition $K_0^- \propto r$ reads as follows

$$\begin{aligned} K_0^- &= \left(B^2X^{2b} - 4AD\,X^{a+c}\right)\left(-\frac{B}{D}X^{b-c} - \sqrt{\frac{B^2}{D^2}X^{2b-2c} - 4\frac{A}{D}X^{a-c}}\right) \\ &\propto X^{b+1}\left[\frac{B}{b+1} + \frac{2D}{c+1}\left(-\frac{B}{D} - \sqrt{\frac{B^2}{D^2} - 4\frac{A}{D}X^{a-2b+c}}\right)\right]. \end{aligned} \tag{32}$$

This can be rewritten as

$$\left(-\frac{B}{D} - \sqrt{\frac{B^2}{D^2} - 4\frac{A}{D}X^{a-2b+c}}\right)\left(-\frac{2D}{c+1} + X^{2b-c-1}\left(B^2 - 4AD\,X^{a-2b+c}\right)\right) \propto \frac{B}{b+1}. \tag{33}$$

For generic A, B and D this can be solved for the parameters

$$\boxed{\text{Rindler ground state models:} \qquad a = 1 \qquad c = 2b-1} \tag{34}$$

yielding the potential

$$\boxed{\mathscr{V}_{\mathrm{R}}(X, -(\partial X)^2) = X^{-b}\frac{4A - 2bBX^{b-1}(\partial X)^2 + (2b-1)DX^{2b-2}(\partial X)^4}{-4B + 4DX^{b-1}(\partial X)^2}.} \tag{35}$$

The metric (12) for Rindler ground state models simplifies to ($\Delta = B^2 - 4AD$)

$$\mathrm{d}s^2 = 2\,\mathrm{d}v\,\mathrm{d}r - I_0X^{1+b}\left(\Delta + \frac{4CD}{X}\right)\left(B + \sqrt{\Delta + \frac{4CD}{X}}\right)\mathrm{d}v^2 \tag{36}$$

where I_0 has the same meaning as before. The terms proprotional to the Casimir C are always subleading at large X. Demanding that large values of the radial coordinate also correspond to large values of the dilaton requires $b \geq -1$. For these standard Rindler ground state models, the metric (36) has the asymptotically Rindler expansion

$$\mathrm{d}s^2 = 2\,\mathrm{d}v\,\mathrm{d}r - \left(r - Mr^{b/(1+b)} + \mathscr{O}\left(M^2r^{(b-1)/(1+b)}\right)\right)\mathrm{d}v^2 \tag{37}$$

where the mass parameter M is proportional to C.

Non-generic cases for specific values of the constants A, B, D exist as well. For $A = 0$, the condition $K_0^- \propto r$ is satisfied provided $c = 2b-1$. For $B = 0$ and $A \neq 0$ we get the solution $a = 1$ satisfying $K_0^- \propto r$. For $D = 0$, the condition $K_0^- \propto r$ yields $a = 1$ with b arbitrary. In terms of the $(a'b')$ family, we get $b' = 0$, which confirms the known result [21].

4.3 (A)dS ground states

The (A)dS condition $K_0^- \propto r^2$ is given by

$$K_0^- = X^{3b-c}\left(B^2 - 4AD\,X^{a-2b+c}\right)\left(-\frac{B}{D} - \sqrt{\frac{B^2}{D^2} - 4\frac{A}{D}X^{a-2b+c}}\right)$$
$$\propto X^{2b+2}\left[\left(\frac{B}{b+1}\right)^2 + \frac{4BD}{(b+1)(c+1)}\left(-\frac{B}{D} - \sqrt{\frac{B^2}{D^2} - 4\frac{A}{D}X^{a-2b+c}}\right) + \right.$$
$$\left. + \left(\frac{2D}{c+1}\right)^2\left(-\frac{B}{D} - \sqrt{\frac{B^2}{D^2} - 4\frac{A}{D}X^{a-2b+c}}\right)^2\right]. \tag{38}$$

For generic A, B and D this is solved for the parameters obeying

$$\boxed{\text{(A)dS ground state models:} \qquad a = b+2 \qquad c = b-2.} \tag{39}$$

yielding the potential

$$\boxed{\mathscr{V}_{\text{(A)dS}}\left(X, -(\partial X)^2\right) = X\frac{4aA - 2(a-2)BX^{-2}(\partial X)^2 + (a-4)DX^{-4}(\partial X)^4}{-4B + 4DX^{-2}(\partial X)^2}.} \tag{40}$$

The metric (12) for (A)dS ground state models simplifies to ($\Delta = B^2 - 4AD$)

$$\mathrm{d}s^2 = 2\,\mathrm{d}v\,\mathrm{d}r + I_0 X^{2a-2}\left(\Delta + \frac{4CD}{X^a}\right)\left(B + \sqrt{\Delta + \frac{4CD}{X^a}}\right)\mathrm{d}v^2 \tag{41}$$

where again I_0 has the same meaning as before. The terms proprotional to the Casimir C are subleading at large values of the dilaton if $a > 0$. If $a > 1$, then large values of the radial coordinate also correspond to large values of the dilaton. For these standard (A)dS ground state models, the metric (41) has the asymptotically (A)dS expansion

$$\mathrm{d}s^2 = 2\,\mathrm{d}v\,\mathrm{d}r + \left(\Lambda r^2 + M r^{(a-2)/(a-1)}\mathscr{O}\left(M^2 r^{-2/(a-1)}\right)\right)\mathrm{d}v^2 \tag{42}$$

where $\Lambda = \Lambda(A,B,D)$ is positive (negative) for (A)dS, and the mass parameter M is proportional to the Casimir C.

Again, some non-generic cases for specific values of the constants A, B, D exist. If $A = 0$, $K^- \propto r^2$ yields $c = b-2$. For $B = 0$ we get $c = a+4b$. For $D = 0$ we get the result $a = b+2$ from the condition $K_0 \propto r^2$. This is consistent with the AdS ground states previously found for the model $\mathscr{V}_{(a')}(X,2Y) = \frac{a'}{X}Y + X$ [21].

5 Generalizations

There are three types of generalizations of the standard Minkowski-, Rindler-, and (A)dS-ground state models discussed in this work.

1. **Beyond standard models.** One can relax the convexity conditions imposed as standard. For instance, one could consider Minkowski ground state models (29) where the parameter a lies outside the interval $(0,1]$. Then either the weak coupling region, $X \to \infty$, no longer corresponds to large values of the radial coordinate, $r \to \infty$, or the Killing norm is dominated asymptotically by the terms containing the Casimir C. In the latter case, generic solutions no longer are asymptotically flat. Similar remarks apply to Rindler and (A)dS ground state models. In the restricted class of power-counting renormalizable models, this issue was addressed in [11, 2].
2. **Beyond maximally symmetric ground states.** One can give up the requirement of having a maximally symmetric ground state and study all members of the family of models defined by (21) with (22). It could be interesting to scan this model space by imposing as constraint reasonable thermodynamics, along the lines of section 4.2 in [10].
3. **Beyond quadratic ansatz.** One can extend the discussion beyond the 3-parameter family considered in the present work. In particular, generalizing the ansatz (21) to $C(X,Y) = \sum_{n=0}^{4} V_n(X)Y^n$ yields explicit (if cumbersome) expressions, and all our results generalize straightforwardly to this case. Generalizations beyond these algebraic examples require new techniques or special choices of $C(X,Y)$.

Besides these generalizations, there are also some evident potential applications of the models studied in the present work. We list three examples for possible follow-up work and expect that some of these points will be addressed in the near future.

1. **Thermodynamics and holographic renormalization.** Presently, it is unclear which boundary terms have to be added to the bulk action (1) to render the variational principle well-defined. For the class of power-counting renormalizable models, these boundary terms were constructed in [10] (using the Hamilton–Jacobi method, see e.g. [34]). They play an important role in thermodynamics since in the Euclidean version of the theory the Helmholtz free energy is determined from the (holographically renormalized) on-shell action. It could be rewarding to extend the thermodynamical discussion of [10] to generic 2d dilaton gravity models (1), like in section 5 of [13].
2. **Asymptotic symmetries and boundary excitations.** Models with maximally symmetric ground states could allow for some standard boundary conditions, using notions of locally asymptotically flat or locally asymptotically (A)dS spacetimes, along the lines of e.g. [12, 13].
3. **SYK-like correspondences.** Given the relation between Schwarzian actions and AdS isometries (see [29] and the reviews [32, 37, 15]) one might expect SYK-like models [23, 36] dual to the AdS ground state models. Similarly, by analogy to [1] Rindler ground state models could lead to a twisted warped boundary action.

Acknowledgements DG thanks the organizers of the MATRIX Event "2D Supersymmetric Theories and Related Topics" for their efforts.

DG and RR were supported by the Austrian Science Fund (FWF), projects P 30822, P 32581, P 33789, and Y 1447.

Appendix — Relation between dilaton and radial coordinate

The relation between dilaton and radial coordinate,

$$\mathrm{d}r = e^{Q(X)}\,\mathrm{d}X \tag{43}$$

can be integrated in closed form in terms of elementary functions and the hypergeometric function ${}_2F_1$ for all models considered in our work. Dropping the integration constant in $Q(X)$, we list now these results, using again the definition $\Delta = B^2 - AD$ and introducing the definition $M = -4CD/\Delta$.

For Minkowski ground state models, we obtain

$$r = \frac{2X\sqrt{\Delta(1-MX^{-a})}}{(a-2)(3a-2)M^2}\Big(\frac{2a}{a+2}X^a\,{}_2F_1(1,1+\tfrac{1}{a},\tfrac{3}{2}+\tfrac{1}{a};X^a/M) + M\big(a+(2-a)MX^{-a}\big)\Big) \tag{44}$$

which for large values of X and $a \in (0,1]$ expands as

$$r \propto X^{1-a} + \dots \tag{45}$$

For Rindler ground state models, we obtain

$$r = \frac{2}{3}\sqrt{\Delta}\left(\frac{1}{M}\right)^{1/2-b}(X-M)^{3/2}\,{}_2F_1(\tfrac{3}{2},\tfrac{1}{2}-b,\tfrac{5}{2};1-X/M) \tag{46}$$

which for large values of X and $b > -1$ expands as

$$r \propto X^{1+b} + \dots \tag{47}$$

For (A)dS ground state models, we obtain

$$r = \frac{a\sqrt{\Delta(1-MX^{-a})}X^{a-1}}{(a-2)(a-1)}\Big({}_2F_1(1,1-\tfrac{1}{a},\tfrac{3}{2}-\tfrac{1}{a};X^a/M) + 1 - \tfrac{2}{a}\Big) \tag{48}$$

which for large values of X and $a > 1$ expands as

$$r \propto X^{-1+a} + \dots \tag{49}$$

In all cases, the omitted proportionality constants are independent from M and can be adjusted conveniently by fixing the integration constant in $Q(X)$.

References

1. Afshar, H., González, H.A., Grumiller, D., Vassilevich, D.: Flat space holography and the complex Sachdev-Ye-Kitaev model. Phys. Rev. D **101**(8), 086,024 (2020). DOI 10.1103/PhysRevD.101.086024
2. Bagchi, A., Grumiller, D., Salzer, J., Sarkar, S., Schöller, F.: Flat space cosmologies in two dimensions - Phase transitions and asymptotic mass-domination (2014)
3. Callan Jr., C.G., Giddings, S.B., Harvey, J.A., Strominger, A.: Evanescent black holes. Phys. Rev. **D45**, 1005–1009 (1992)
4. Chamseddine, A.H., Wyler, D.: Gauge theory of topological gravity in (1+1)-dimensions. Phys. Lett. **B228**, 75 (1989)
5. Deffayet, C., Pujolas, O., Sawicki, I., Vikman, A.: Imperfect Dark Energy from Kinetic Gravity Braiding. JCAP **10**, 026 (2010). DOI 10.1088/1475-7516/2010/10/026
6. Dijkgraaf, R., Verlinde, H., Verlinde, E.: String propagation in a black hole geometry. Nucl. Phys. **B371**, 269–314 (1992)
7. Elitzur, S., Forge, A., Rabinovici, E.: Some global aspects of string compactifications. Nucl. Phys. **B359**, 581–610 (1991)
8. Grumiller, D., Jackiw, R.: Liouville gravity from Einstein gravity. In: S. Gosh, G. Kar (eds.) Recent Developments in Theoretical Physics, pp. 331–343. World Scientific, Singapore (2010)
9. Grumiller, D., Kummer, W., Vassilevich, D.V.: Dilaton gravity in two dimensions. Phys. Rept. **369**, 327–429 (2002)
10. Grumiller, D., McNees, R.: Thermodynamics of black holes in two (and higher) dimensions. JHEP **04**, 074 (2007)
11. Grumiller, D., McNees, R., Salzer, J.: Cosmological constant as confining U(1) charge in two-dimensional dilaton gravity. Phys.Rev. **D90**, 044,032 (2014). DOI 10.1103/PhysRevD.90.044032
12. Grumiller, D., McNees, R., Salzer, J., Valcárcel, C., Vassilevich, D.: Menagerie of AdS_2 boundary conditions. JHEP **10**, 203 (2017). DOI 10.1007/JHEP10(2017)203
13. Grumiller, D., Ruzziconi, R., Zwikel, C.: Generalized dilaton gravity in 2d. SciPost Phys. **12**, 032 (2022). DOI 10.21468/SciPostPhys.12.1.032
14. Grumiller, D., Vassilevich, D.V.: Non-existence of a dilaton gravity action for the exact string black hole. JHEP **11**, 018 (2002)
15. Gu, Y., Kitaev, A., Sachdev, S., Tarnopolsky, G.: Notes on the complex Sachdev-Ye-Kitaev model. JHEP **02**, 157 (2020). DOI 10.1007/JHEP02(2020)157
16. Hartman, T., Strominger, A.: Central Charge for AdS(2) Quantum Gravity. JHEP **0904**, 026 (2009). DOI 10.1088/1126-6708/2009/04/026
17. Horndeski, G.W.: Second-order scalar-tensor field equations in a four-dimensional space. Int. J. Theor. Phys. **10**, 363–384 (1974). DOI 10.1007/BF01807638
18. Ikeda, N.: Two-dimensional gravity and nonlinear gauge theory. Annals Phys. **235**, 435–464 (1994). DOI 10.1006/aphy.1994.1104
19. Isler, K., Trugenberger, C.A.: A gauge theory of two-dimensional quantum gravity. Phys. Rev. Lett. **63**, 834 (1989)
20. Jackiw, R.: Lower dimensional gravity. Nucl. Phys. **B252**, 343–356 (1985)
21. Katanaev, M.O., Kummer, W., Liebl, H.: On the completeness of the black hole singularity in 2-d dilaton theories. Nucl. Phys. B **486**, 353–370 (1997). DOI 10.1016/S0550-3213(96)00624-4
22. Katanaev, M.O., Volovich, I.V.: String model with dynamical geometry and torsion. Phys. Lett. **B175**, 413–416 (1986)
23. Kitaev, A.: A simple model of quantum holography. KITP strings seminars, April/May 2015, http://online.kitp.ucsb.edu/online/entangled15/ and http://online.kitp.ucsb.edu/online/entangled15/kitaev2/
24. Klosch, T., Strobl, T.: Classical and quantum gravity in (1+1)-Dimensions. Part 1: A Unifying approach. Class. Quant. Grav. **13**, 965–984 (1996). DOI 10.1088/0264-9381/13/5/015. [Erratum: Class.Quant.Grav. 14, 825 (1997)]

25. Klosch, T., Strobl, T.: Classical and quantum gravity in (1+1)-dimensions. Part 2: The Universal coverings. Class. Quant. Grav. **13**, 2395–2422 (1996). DOI 10.1088/0264-9381/13/9/007
26. Kunstatter, G., Maeda, H., Taves, T.: New 2D dilaton gravity for nonsingular black holes. Class. Quant. Grav. **33**(10), 105,005 (2016). DOI 10.1088/0264-9381/33/10/105005
27. Lemos, J.P.S., Sa, P.M.: The Black holes of a general two-dimensional dilaton gravity theory. Phys. Rev. D **49**, 2897–2908 (1994). DOI 10.1103/PhysRevD.49.2897. [Erratum: Phys.Rev.D 51, 5967–5968 (1995)]
28. Louis-Martinez, D., Gegenberg, J., Kunstatter, G.: Exact Dirac quantization of all 2-D dilaton gravity theories. Phys. Lett. B **321**, 193–198 (1994). DOI 10.1016/0370-2693(94)90463-4
29. Maldacena, J., Stanford, D.: Remarks on the Sachdev-Ye-Kitaev model. Phys. Rev. D **94**(10), 106,002 (2016). DOI 10.1103/PhysRevD.94.106002
30. Mandal, G., Sengupta, A.M., Wadia, S.R.: Classical solutions of two-dimensional string theory. Mod. Phys. Lett. **A6**, 1685–1692 (1991)
31. Mann, R.B., Ross, S.F.: The D $\to$ 2 limit of general relativity. Class. Quant. Grav. **10**, 345–351 (1993)
32. Mertens, T.G.: The Schwarzian theory — origins. JHEP **05**, 036 (2018). DOI 10.1007/JHEP05(2018)036
33. Odintsov, S.D., Shapiro, I.L.: One loop renormalization of two-dimensional induced quantum gravity. Phys. Lett. **B263**, 183–189 (1991)
34. Papadimitriou, I.: Holographic renormalization as a canonical transformation. JHEP **1011**, 014 (2010). DOI 10.1007/JHEP11(2010)014
35. Russo, J.G., Tseytlin, A.A.: Scalar tensor quantum gravity in two-dimensions. Nucl. Phys. **B382**, 259–275 (1992)
36. Sachdev, S., Ye, J.: Gapless spin fluid ground state in a random, quantum Heisenberg magnet. Phys. Rev. Lett. **70**, 3339 (1993). DOI 10.1103/PhysRevLett.70.3339
37. Sárosi, G.: AdS_2 holography and the SYK model. PoS **Modave2017**, 001 (2018). DOI 10.22323/1.323.0001
38. Schaller, P., Strobl, T.: Poisson structure induced (topological) field theories. Mod. Phys. Lett. **A9**, 3129–3136 (1994)
39. Strobl, T.: Gravity in two spacetime dimensions (1999). Habilitation thesis
40. Takahashi, K., Kobayashi, T.: Generalized 2D dilaton gravity and kinetic gravity braiding. Class. Quant. Grav. **36**(9), 095,003 (2019). DOI 10.1088/1361-6382/ab1355
41. Teitelboim, C.: Gravitation and Hamiltonian structure in two space-time dimensions. Phys. Lett. **B126**, 41 (1983)
42. Verlinde, H.: Black holes and strings in two dimensions. In: Trieste Spring School on Strings and Quantum Gravity, pp. 178–207 (1991). The same lectures were given at MGVI in Japan, June, 1991
43. Witten, E.: On string theory and black holes. Phys. Rev. **D44**, 314–324 (1991)
44. Witten, E.: Deformations of JT Gravity and Phase Transitions (2020)

$2d$ Sigma Models and Geometry

Ulf Lindström

Abstract Supersymmetric nonlinear sigma models have target spaces that carry interesting geometry. The geometry is richer the more supersymmetries the model has. The study of models with two dimensional world sheets is particularly rewarding since they allow for torsionful geometries. In this review I describe and exemplify the relation of $2d$ supersymmetry to Riemannian, complex, bihermitian, (p,q) hermitean, Kähler, hyperkähler, generalised geometry and more.

1 Introduction

Supersymmetric models are closely associated to complex geometry in several different ways. To be able to write some extended models in a manifest form superspaces may be extended with a CP^1 at each point thus relating them to twistors [23], conformal supergravity can be formulated in terms of local twistors [14], [15], and supersymmetic nonlinear sigma models is often best formulated in terms of complex superfields and typically have complex target space geometries [28],[1]. This last property is what shall concern us here, although we have to mainly restrict to two-dimensional models with two left and two right going supersymmetries i,e, to $2d$, $(2,2)$ supersymmetry. Fortunately this case is sufficiently rich to warrant independent scrutiny.

The format of the presentation is to first introduce bosonic $2d$ sigma models and then $(1,1)$ and $(2,2)$ supersymmetry. There are essentially three different superfield representations of the $(2,2)$ supersymmetric sigma models and they are presented one by one and their their target space geometry identified. After the introduction of semichiral superfields and their sigma models, generalised Kähler geometry is defined. This geometry covers all the target space geometries introduced.

Ulf Lindström
METU, Ankara, and Uppsala University. e-mail: ulf.lindstrom@physics.uu.se

D. R. Wood et al. (eds.), *2021-2022 MATRIX Annals*, MATRIX Book Series 5,
https://doi.org/10.1007/978-3-031-47417-0_7

2 Sigma models

For a review of various applications of sigma models see [24].
A non-linear sigma model is a theory of maps from a (super) manifold Σ to a target space $\mathscr{T}$

$$X : \Sigma \to \mathscr{T}$$
$$X(x) \longmapsto X \in \mathscr{T} , \tag{1}$$

with dynamics specified by extremising an action which is, schematically,

$$S = \int_{\Sigma} dx \mathscr{L}(X) . \tag{2}$$

Its precise form depends on the world volume dimension and the number of supersymmetries. That same number then constrains the geometry of $\mathscr{T}$.

2.1 The bosonic sigma model

To be concrete and quickly see the relation to geometry, we consider a bosonic sigma model in $2d$ where the general bosonic sigma model action reads

$$S = \int d^2x \partial_{++} X^{\mu} \left(G_{\mu\nu}(X) + B_{\mu\nu}(X)\right) \partial_{=} X^{\nu} =: \int d^2x \partial_{++} X^{\mu} E_{\mu\nu} \partial_{=} X^{\nu} . \tag{3}$$

The $2d$ light-cone coordinates are

$$x^{++} = x^0 + x^1 , \quad x^{=} = x^0 - x^1 . \tag{4}$$

The G field is a symmetic tensorfield on $\mathscr{T}$ which we identify as a *metric*. The B field is a *gerbe connection*[1] and the action depends only on its field strength

$$H_{\mu\nu\rho} = \partial_{[\rho} B_{\mu\nu]} . \tag{5}$$

as seen directly from the field equations for X^{μ} or from the alternative form of the action

$$S = \int_{\partial V} d^2x \partial_{++} X^{\mu} G_{\mu\nu} \partial_{=} X^{\nu} + \tfrac{1}{3} \int_V d^3x \varepsilon^{ijk} \partial_i X^{\mu} \partial_j X^{\nu} \partial_k X^{\rho} H_{\mu\nu\rho} , \tag{6}$$

where the $2d$ space is a boundary of a contractible $3d$ space V, and X is extended to live on the $3d$ space. In general there may be different inequivalent extensions which can lead to quantisation conditions on H.

---

[1] This refers to its global properties, more specifically the behaviour under change of coordinate patches, See e,g,[18] .

The B (or H) term is called a *Wess-Zumino term*. Since it depends on B only through its field strength H, the model is invariant under a B field gauge transformations;

$$B \to B + d\lambda \ , \iff \ B_{\mu\nu} \to B_{\mu\nu} + \partial_{[\mu}\lambda_{\nu]} \ . \tag{7}$$

The X^μ field equations that follow from either form of the action read

$$G_{\mu\nu}\nabla^{(-)}_{\#}\partial_{=}X^\nu = G_{\mu\nu}\nabla^{(+)}_{=}\partial_{\#}X^\nu = 0 \ , \tag{8}$$

where the connections now have torsion due to the inclusion of the B field;

$$\Gamma^{(\pm)\mu}_{\sigma\rho} = \Gamma^{(0)\mu}_{\sigma\rho} \pm T_{\sigma\rho}{}^{\mu} \ , \ \ T_{\sigma\rho}{}^{\mu} = \tfrac{1}{2}H_{\sigma\rho\nu}G^{\nu\mu} \ . \tag{9}$$

We see that the target space of the bosonic sigma model carries *Riemann geometry with torsion*. This is also the target space geometry of models with one left and one right supersymmetry, a $(1,1)$ sigma model.

2.2 The $(1,1)$ sigma model

A general sigma model in $(1,1)$ superspace is

$$S = \int d^2xD_+D_-\Big(D_+\Phi^i(x,\theta)\big(G_{ij}+B_{ij}\big)(\Phi)D_-\Phi^j(x,\theta)\Big)\Big|$$
$$= \int d^2x\partial_{\#}\phi^iE_{ij}(\phi)\partial_{=}\phi^j + ..., \tag{10}$$

where the vertical bar in the first line denotes the θ independent part. and $\phi = \Phi_|$. It has $(1,1)$ supersymmetry manifest by construction. The $(1,1)$ algebra is[2]

$$D_+^2 = \partial_{\#} \ , \ \ D_-^2 = \partial_{=} \tag{11}$$

and the $\Phi^i(x,\theta)$s are $(1,1)$ superfields. Additional supersymmetries will constrain the geometry further, however. They will have the infinitesimal form [6]

$$\delta\Phi^i = \varepsilon^+J^i_{(+)k}D_+\Phi^k + \varepsilon^-J^i_{(-)k}D_-\Phi^k \tag{12}$$

The conditions on the Js follow from two requirements:

- Closure of the algebra $[\delta_1,\delta_2]\Phi = -2i\varepsilon_1\varepsilon_2\partial\Phi$
- Invariance of the action $\delta S = 0$

[2] In addition to the generators Q of supersymmetry there are spinorial covariant derivatives D such that $\{D,Q\} = 0$, The algebra may be defined in terms of either set.

From closure of the algebra it follows that $J^2_{(\pm)} = -1$ and the vanishing of the Nijenhuis tensors:

$$\mathcal{N}_{J_{(\pm)}}(X,Y) = [X,Y] + J_{(\pm)}[J_{(\pm)}X,Y] + J_{(\pm)}[X,J_{(\pm)}Y] - [J_{(\pm)}X,J_{(\pm)}Y] = 0\,. \quad (13)$$

The latter condition ensures integrability of $J_{(\pm)}$ so that there is a global atlas of complex coordinates with holomorphic transition functions.

From invariance of the action it follows that $J^t_{(\pm)}GJ_{(\pm)} = G$, i.e., hermiticity of the metric with respect to both complex structures, and that $\nabla^{(\pm)}J_{(\pm)} = 0$.

When there are $p-1$ left and $q-1$ right complex structures, we have (p,q) supersymmety. There will then be additional conditions. For example, $(4,4)$ supersymmetry with vanishing torsion gives hyperkähler geometry on $\mathcal{T}$.

It is not always possible to find an off-shell manifest superspace formulation of the extended geometry. It *is* possible for $(2,2)$ in $2d$ however.

2.3 $(2,2)$ *sigma models*

We denote the supersymmetry generators and covariant derivatives by $\mathbb{Q}$ and $\mathbb{D}$ respectively. Their anticommutator is required to vanish $\{\mathbb{Q},\mathbb{D}\} = 0$. In terms of covariant derivatives the $(2,2)$ algebra is,

$$\{\mathbb{D}_\pm,\bar{\mathbb{D}}_\pm\} = 2i\partial_{\substack{+\\=}}\,, \quad (14)$$

all other (anti)commutators are zero. Since we now have four θs, the multiplet contained in a general superfield will consist of 16 fields. Such a multiplet is recducible and not suitable for a sigma model description since the lowest bosonic component is not a scalar field. But since the covariant derivatives anticommute with the supersymmety generators we can use them to impose constraints that will reduce the multiplet.

2.3.1 Representations of $(2,2)$

There are three[3] types of constrained $(2,2)$ superfields and correspondingly three different target space geometries.

[3] Not strictly true. There are also complex linear fields Σ satisfying $\bar{\mathbb{D}}_+\bar{\mathbb{D}}_-\Sigma = 0$ and twisted complex linear fields $\tilde{\Sigma}$ satisfying $\bar{\mathbb{D}}_+\mathbb{D}_-\tilde{\Sigma} = 0$ and their complex conjugate, but these are dual to chirals and twisted chirals, respectively. So there is always an equivalent formulation in terms of the latter fields.

Chiral superfields Φ

$$\bar{\mathbb{D}}_{\pm}\Phi = 0 \tag{15}$$

Twisted Chiral superfields χ

$$\bar{\mathbb{D}}_{+}\chi = 0\,, \quad \mathbb{D}_{-}\chi = 0 \tag{16}$$

Left ℓ *and Right* $\mathfrak{r}$ *Semichiral superfields*

$$\bar{\mathbb{D}}_{+}\ell = 0\,, \quad \bar{\mathbb{D}}_{-}\mathfrak{r} = 0 \tag{17}$$

We will introduce the sigma models corresponding to the three kind of fields above and display their target space geometries. We begin with the chiral fields.

Chiral superfields

The chiral superfields $\bar{\mathbb{D}}_{\pm}\Phi = 0$ are complex and their lowest components will serve as complex coordinates on the target space $\mathscr{T}$, suggesting that the target space geometry will be complex. Their component fields are given by

$$\begin{aligned}
&\Phi = \phi + \theta^{\alpha}\psi_{\alpha} + \theta^{\alpha}\theta_{\alpha}\mathscr{F}\,, \qquad \alpha = (+,-)\\
&\phi(x) = \Phi(x,\theta)| \qquad (18)\\
&\psi_{\alpha}(x) = \mathbb{D}_{\alpha}\Phi(x,\theta)|\,,\\
&\mathscr{F}(x) = \mathbb{D}^{2}\Phi(x,\theta)|\,, \qquad (19)
\end{aligned}$$

where the vertical bar denotes setting $\theta = 0$. This is the preferred way of defining components, whereas the first line represents an expansion in θ. Such an expansion gets cumbersome when there are more supersymmetries and thus more θs.

The most general superspace action for chiral fields reads [28]

$$S = \int d^2x \mathbb{D}^2\bar{\mathbb{D}}^2 K(\Phi,\bar{\Phi})\,. \tag{20}$$

Pushing in the spinorial derivatives and using the definition of the components (19) we find a supersymmetric sigma model with K as a potential for the metric:

$$S=\int d^2x\Big[\partial_{++}\phi^a G_{ab}\partial_{=}\phi^b+i\tfrac{1}{2}(\psi_+^a\nabla_=\psi_+^b+\psi_-^a\nabla_{++}\psi_-^b)G_{ab}$$
$$-\tfrac{1}{4}R_{cdab}\psi_+^a\psi_+^b\psi_-^c\psi_-^d\Big] \tag{21}$$

after eliminating the auxiliary fields $\mathscr{F}^a$. Here $a=(i,\bar{i})$ etc. The geometry is a particular complex geometry called *Kähler* geometry. The metric, Levi-Civita connection and curvature tensor are all expressible in terms of the potential K:

$$G_{i\,\bar{j}}=\partial^2K/\partial\phi^i\partial\bar{\phi}^{\bar{j}}=:K_{,i\bar{j}} \tag{22}$$

$$\Gamma_{ij}{}^{k}=G^{k\bar{s}}\partial_i G_{j\bar{s}}=K^{k\bar{s}}K_{,ij\bar{s}}\ , \tag{23}$$

$$R_{i\,\bar{j}k\bar{s}}=G_{m\bar{j}}\partial_{\bar{s}}(\Gamma_{ik}{}^{m})=K_{,i\,\bar{j}k\bar{s}}-\Gamma_{ik}{}^{m}\Gamma^{\bar{n}}{}_{\bar{j}\bar{s}}K_{,m\bar{n}} \tag{24}$$

Here comma denotes derivative with respect to the fields indicated by the indices and $G^{i\,\bar{j}}=K^{i\,\bar{j}}$ is the inverse metric.

Kähler geometry is the target space geometry of $\mathscr{N}=1$ sigma models in $4d$ and for chiral $(2,2)$ sigma models in $2d$. The relation is $1-1$.

Here is a quick reminder of the definition of Kähler geometry: A manifold carrying a complex structure J and a metric g hermitean with respect to the complex structure

$$J^t gJ=g\ , \tag{25}$$

is called a Kähler manifold if the complex structure is annihilated by the Levi-Civita connection

$$\nabla J=0\ . \tag{26}$$

The metric then has a potential K such that $g_{i\bar{j}}=K_{,i\bar{j}}$ in complex coordinates. Further there is a (globally defined) symplectic form

$$\omega=gJ\ , \tag{27}$$

called the Kähler form.

In (21) the reduction is all the way from $(2,2)$ to components. But if we are only interested in the geometry as defined by the metric and B-field, it is already displayed in the reduction from $(2,2)$ to $(1,1)$ superspace. This is generally true for our $(2,2)$ sigma models and now we demonstrate it for the chiral sigma model.

The $(1,1)$ superfields are $\Phi=\Phi|$, where the vertical bar denotes setting half of the $(1,1)$ Fermi coordinates to zero $\theta-\bar{\theta}=0$. The spinorial derivatives reduce as

$$\mathbb{D}_\pm = D_\pm - iQ_\pm$$
$$\bar{\mathbb{D}}_\pm = D_\pm + iQ_\pm \tag{28}$$
$$\Longrightarrow \mathbb{D}^2\bar{\mathbb{D}}^2 \sim D^2Q^2 \tag{29}$$

where the Ds are the $(1,1)$ derivatives and the Qs generate the non manifest second supersymmetries. The action becomes

$$S = \int d^2x\mathbb{D}^2\bar{\mathbb{D}}^2K(\Phi,\bar{\Phi})| \to \int d^2x D^2Q^2K(\Phi,\bar{\Phi})| \,. \tag{30}$$

We evaluate the action using

$$Q_\pm\Phi^a = J^a_bD_\pm\Phi^b \;, \quad J = \begin{pmatrix} i\delta^i_j & 0 \\ 0 & -i\delta^{\bar{i}}_{\bar{j}} \end{pmatrix} \tag{31}$$

which follows from the reduction of the chirality constraints.

$$\begin{aligned} 0 = \bar{\mathbb{D}}_\pm\Phi^i &= (D_\pm + iQ_\pm)\Phi^i \\ , \quad &\Rightarrow Q_\pm\Phi^i = iD_\pm\Phi^i \\ &\Rightarrow Q_\pm\bar{\Phi}^{\bar{i}} + = -iD_\pm\bar{\Phi}^{\bar{i}} \end{aligned} \tag{32}$$

Here we have again assumed that there are d chiral and their d complex conjugate antichiral fields labeled by $\Phi^a = (\Phi^i, \bar{\Phi}^{\bar{i}})$. Using this and integrating by parts we find the $(1,1)$ action:

$$\int d^2xD^2Q^2K(\Phi,\bar{\Phi})| = \int d^2xD^2\left(D_+\Phi^iK_{,i\bar{j}}(\Phi)D_-\Phi^{\bar{j}}\right)\,, \tag{33}$$

where again comma denotes derivative. Comparing to (10) we see that the second derivative matrix $K_{,i\bar{j}}$ is the complex metric.

Chiral and Twisted Chiral models

We now turn to a sigma model based on both chiral superfields, $\bar{\mathbb{D}}_\pm\Phi = 0$, and twisted chiral ones, $\bar{\mathbb{D}}_+\chi = 0 \,, \; \mathbb{D}_-\chi = 0$, [6].

The $(1,1)$ reduction (31) implies the following form for the chirality constraints:

$$Q_\pm\Phi^a = J^a_bD_\pm\Phi^b \,, \quad Q_\pm\chi^{a'} = \pm J^{a'}_{b'}D_\pm\chi^{b'} \,. \tag{34}$$

Consider a sigma model with both types of fields

$$\int d^2xD^2Q^2K(\Phi,\bar{\Phi},\chi,\bar{\chi})| =: \int d^2xD^2Q^2K(\mathbb{X})|$$
$$= \int d^2xD^2\left(D_+\mathbb{X}^A(G_{AB}+B_{AB})(\mathbb{X})D_-\mathbb{X}^B\right)$$
$$= \int d^2xD^2\left(D_+\mathbb{X}^A E_{AB}(\mathbb{X})D_-\mathbb{X}^B\right) \qquad (35)$$

Where we introduced the notation $\mathbb{X}^A = (\Phi^a,\chi^{a'}) = (\Phi^i,\bar{\Phi}^{\bar{i}},\chi^{i'},\bar{\chi}^{\bar{i}'})$, with the indices ranging over $i = 1,\ldots,d$,$i' = 1,\ldots,d'$. The middle and last line refers to (10). The metric and B-field are then given by

$$E_{AB} = \begin{pmatrix} 0 & K_{,i\bar{j}} & K_{,ij'} & 0 \\ K_{,\bar{i}j} & 0 & 0 & K_{,\bar{i}\bar{j}'} \\ -K_{,i'j} & 0 & 0 & -K_{,i'\bar{j}'} \\ 0 & -K_{,\bar{i}'\bar{j}} & -K_{,\bar{i}'j'} & 0 \end{pmatrix} \qquad (36)$$

which leads to the following field strength

$$H_{i\bar{j}k'} = K_{,i\bar{j}k'} , \qquad H_{i\bar{j}\bar{k}'} = -K_{,i\bar{j}\bar{k}'}$$
$$H_{i'\bar{j}'k} = -K_{,i'\bar{j}'k} , \quad H_{i'\bar{j}'\bar{k}} = K_{,i'\bar{j}'\bar{k}} . \qquad (37)$$

Note that, as in the previous example, the Lagrangian K is a potential for the metric and now also for the B field/torsion.

The two complex structures can be read off from the non manifest transformations of Φ and χ according to (suppressing indices)

$$\delta_\pm\mathbb{X} = \varepsilon^{\alpha}_{(\pm)}\mathbb{J}^{(\pm)}D_\alpha\mathbb{X} \qquad (38)$$

which leads to

$$\mathbb{J}^{(+)} = \begin{pmatrix} J & 0 \\ 0 & J \end{pmatrix} , \quad \mathbb{J}^{(-)} = \begin{pmatrix} J & 0 \\ 0 & -J \end{pmatrix} , \qquad (39)$$

where again $J = diag(i,-i)$. It is easy to see that they commute $[\mathbb{J}^{(+)},\mathbb{J}^{(-)}] = 0$ and a bit more effort shows that

$$\nabla_{(+)}\mathbb{J}^{(+)} = 0 , \quad \nabla_{(-)}\mathbb{J}^{(-)} = 0 , \qquad (40)$$

where $\pm$ refers to $\pm\frac{1}{2}HG^{-1}$ torsion. Finally, a local product structure is seen to be

$$\mathbb{K} = -\mathbb{J}^{(+)}\mathbb{J}^{(-)} = \begin{pmatrix} \mathbf{1} & 0 \\ 0 & -\mathbf{1} \end{pmatrix} . \qquad (41)$$

These structures define bihermitian geometry [6] whose formal definition we now recapitulate:
A manifold endowed with two complex structures J_+ and J_-, a metric g and an antisymmetric B-field B carries a *bihermitian geometry* if g is hermitian with respect to both complex structures

$$J^t_\pm g J_\pm = g \,, \tag{42}$$

and the two complex structures are covariantly constant with respect to two connections with torsion

$$\nabla^{(+)} J_+ = 0 \,, \quad \nabla^{(-)} J_- = 0 \,, \tag{43}$$

where the torsionfull connections are

$$\nabla^{(+)} = \nabla^{(0)} + T \,, \quad \nabla^{(-)} = \nabla^{(0)} - T \,, \quad T_{ij}{}^k = \tfrac{1}{2} H_{ijn} g^{nk} \,, \tag{44}$$

and $\nabla^{(0)}$ is the Levi-Civita connection for g.

There are two distinct cases of this geometry depending on whether the two complex structures commute or not.

When the complex structures commute, $[J_+, J_-] = 0$, they define a third structure, a local product structure $\mathbb{K}$, by

$$\mathbb{K} := -J_+ J_- \,, \quad \Rightarrow \mathbb{K}^2 = 1 \,, \tag{45}$$

This geometry is sometimes called a *BILP geometry* (for bihermitian local product).

BILP geometry is the target space geometry of $(2,2)$ sigma models with B field and commuting complex structures. It becomes manifest when the model is written in terms of chiral and twisted chiral superfields. It is a special case of Generalised Kähler geometry.

There is a generalisation to p left and q right complex structures: (p,q) *Hermitean geometry* [16].

Semichiral superfields

Semichiral superfields [3] obey only half the chirality constraints compared to chirals and twisted chirals. This is mirrored in the reduction to $(1,1)$ superfields in that in addition to the constraints

$$\begin{aligned} \bar{\mathbb{D}}_+ \ell &= (D_+ + iQ_+)\ell = 0 \,, \quad \Rightarrow Q_+ L = J D_+ L \\ \bar{\mathbb{D}}_- \mathfrak{r} &= (D_- + iQ_-)\mathfrak{r} = 0 \,, \quad \Rightarrow Q_- R = J D_- R \,, \end{aligned} \tag{46}$$

where $L = (\ell, \bar{\ell})$ and $R = (\mathfrak{r}, \bar{\mathfrak{r}})$, there arise two unconstrained spinorial $(1,1)$ fields:

$$Q_- L = \Psi_- \,, \quad Q_+ R = \Psi_+ \tag{47}$$

These are auxiliary fields. When they are integrated out of an action they become part of the complex structures:

$$Q_-L = \Psi_-(L,R)\,, \qquad Q_+R = \Psi_+(L,R)\,. \tag{48}$$

A general action is

$$S = \int d^2x\mathbb{D}^2\bar{\mathbb{D}}^2 K(L,R)| \to \int d^2xD^2Q^2K(L,R)| \tag{49}$$

Pushing in the Qs and using the definitions of the $(1,1)$ components gives, after integrating out the auxiliary spinors,

$$\int d^2xD^2\left(D_+\mathbb{X}^A E_{AB}(\mathbb{X})D_-\mathbb{X}^B\right) \tag{50}$$

where $\mathbb{X}^A = (L,R) = (\ell^a, \bar{\ell}^{\bar{a}}, \mathfrak{r}^a, \bar{\mathfrak{r}}^{\bar{a}})$. To integrate out $\Psi_\pm$ the number of left must equal the number of right semichiral fields[4] $a = 1\ldots, d$. Integrating out $\Psi_\pm$ then results in *non-linear* relations [20], [2] .

The metric plus B field are

$$\begin{aligned}
E_{LL} &= [J, K_{LL}]K^{LR}JK_{RL}\\
E_{LR} &= JK_{LR}J + [J, K_{LL}]K^{LR}[J, K_{RR}]\\
E_{RL} &= -K_{RL}JK^{LR}JK_{RL}\\
E_{RR} &= -K_{RL}JK^{LR}[J, K_{RR}]\,.
\end{aligned} \tag{51}$$

The notation here is a bit stenographic. K_{LR} is the matrix of second derivatives of K with respect to left and right fields and K^{RL} is its inverse. So the Lagrangian is now a potential for all the geometry, albeit a nonlinear one. The complex structures involve the commutator $C_{LL} := [J, K_{LL}]$ with J the canonical complex structure in the left sector, et.c. The complex structures read

$$\mathbb{J}_+ = \begin{pmatrix} J & 0\\ K^{RL}C_{LL} & K^{RL}JK_{LR}\end{pmatrix}, \quad \mathbb{J}_- = \begin{pmatrix} K^{LR}JK_{RL} & K^{LR}C_{RR}\\ 0 & J\end{pmatrix} \tag{52}$$

and will not commute in general.

[4] The case of different number of left and right fields can be related to $\beta\gamma$-systems interacting with sigma-models and is treated in [25].

This describes bihermitean geometry for the symplectic case where the complex structures do not commute. The general case involves chiral, twisted chiral and semichiral fields. $K \to K(\Phi, \chi, L, R)$

The general case then covers all the three cases previously discussed and correspondingly there is a geometry that includes all the hitherto discussed complex geometries. We now turn to that geometry.

Generalised Complex Geometry

In *generalised complex geometry* [12], [9], the tangent bundle $T\mathscr{M}$ is replaced by the sum of the tangent and cotangent bundles,

$$\mathbb{T} := T\mathscr{M} \oplus T^*\mathscr{M} , \tag{53}$$

called the *generalised tangent bundle*[5]. Elements of $\mathbb{X} \in \mathbb{T}$ may be written as

$$\mathbb{X} = X + \xi , \tag{54}$$

where

$$X \in T\mathscr{M} , \quad \xi \in T^*\mathscr{M} . \tag{55}$$

Alternatively, it is often useful to write $\mathbb{X}$ as a column vector

$$\mathbb{X} = \begin{pmatrix} X \\ \xi \end{pmatrix} \tag{56}$$

A *generalised almost complex structure* on $\mathscr{M}$ is an endomorphism $\mathscr{J}$ of the tangent bundle which squares to minus one

$$\mathscr{J} : \mathbb{T}\mathscr{M} \longrightarrow \mathbb{T}\mathscr{M} , \quad \mathscr{J}^2 = -\mathbf{1} , \tag{57}$$

and preserves the natural pairing metric

$$\mathscr{J}^t \eta \mathscr{J} = \eta , \quad \eta = \begin{pmatrix} 0 & \mathbf{1} \\ \mathbf{1} & 0 \end{pmatrix}$$

$$\mathbb{Y}^t \eta \mathbb{X} = (Y, \sigma)\eta \begin{pmatrix} X \\ \xi \end{pmatrix} = Y^\mu \xi_\mu + \sigma_\mu X^\mu . \tag{58}$$

The projection operators

$$\Pi_\pm := \tfrac{1}{2}(\mathbf{1} \pm \mathbf{i}\mathscr{J}) \tag{59}$$

[5] See Fig.1 below for the local structure around a regular point.

may be used to split the generalised tangent space in two parts at a point: The $+i$ eigenspace $\mathbb{L}$ and the $-i$ eigenspace $\overline{\mathbb{L}}$

$$\mathbb{T}\otimes\mathbb{C}=\mathbb{L}\oplus\overline{\mathbb{L}}=\Pi\mathbb{L}\oplus\overline{\Pi}\mathbb{L}\,. \tag{60}$$

In complete analogy to

$$T\otimes\mathbb{C}=T^{(0,1)}\oplus T^{(1,0)} \tag{61}$$

for the ordinary complex structure on $\mathscr{M}$.

The *Courant bracket* $[\![\ ,\ ]\!]_C$ on $\mathbb{T}$ is defined by

$$[\![\mathbb{X},\mathbb{Y}]\!]_C=[\![X+\xi,Y+\eta]\!]_C=[X,Y]+\mathscr{L}_X\eta-\mathscr{L}_Y\xi-\frac{1}{2}d(i_X\eta-i_Y\xi), \tag{62}$$

where $\mathbb{X},\mathbb{Y}\in C^\infty(\mathbb{T})$ and $\mathscr{L}_X$ is the Lie derivative with respect to X.

The Courant bracket is antisymmetric in its arguments but does not satisfy the Jacobi identity. The *Dorfman bracket*

$$[\![\mathbb{X},\mathbb{Y}]\!]_D=[X,Y]+\mathscr{L}_Xd\eta-i_Yd\xi\,, \tag{63}$$

does satisfy the Jacobi identity but is not antisymmetric. The relation between the brackets is

$$[\![\mathbb{X},\mathbb{Y}]\!]_D=[\![\mathbb{X},\mathbb{Y}]\!]_C+d\eta(\mathbb{X},\mathbb{Y})\,. \tag{64}$$

Clearly, the brackets are equal when restricted to an isotropic subspace (i.e. a subspace $\mathscr{M}$ for which $\eta(\mathbb{X},\mathbb{Y})=0$ for all $\mathbb{X},\mathbb{Y}\in\mathscr{M}$).

An important feature of the Courant bracket is that its automorphisms include B-transforms by closed two forms B.

$$[\![e^B\mathbb{X},e^B\mathbb{Y}]\!]_C=e^B[\![\mathbb{X},\mathbb{Y}]\!]_C+i_Yi_XdB\,, \tag{65}$$

where the last term vanishes precisely when $dB=0$. In the representation (56) we have

$$e^B=\begin{pmatrix}1&0\\B&1\end{pmatrix} \tag{66}$$

This automorphism corresponds to the B-field gauge transformations in the sigma model [10].

Both the Courant and the Dorfman brackets may be twisted by a closed three form H:

$$[\![\mathbb{X},\mathbb{Y}]\!]\to[\![\mathbb{X},\mathbb{Y}]\!]+i_Yi_XH. \tag{67}$$

Integrability of a generalised almost complex structure $\mathscr{J}$ is defined by requiring that the subspaces defined by the projections are involutive[6] i.e., that

$$\Pi_\mp[\![\Pi_\pm\mathbb{X},\Pi_\pm\mathbb{Y}]\!]_C=0\,,\quad\forall\,\mathbb{X},\mathbb{Y}\in C^\infty(\mathbb{T}). \tag{68}$$

---

[6] This definition of integrability runs parallel to the usual one for a hermitian space $(\mathscr{M},g,J)$ if we take $g\to\eta$ and $J\to\mathscr{J}$ and replace the Lie bracket by the Courant bracket.

This requires the vanishing of the *generalised Nijenhuis tensor* $\mathbb{N}$. In index free notation

$$\mathbb{N}_{\mathscr{J}}(\mathbb{X},\mathbb{Y}) = [\![\mathbb{X},\mathbb{Y}]\!]_C + \mathscr{J}[\![\mathscr{J}\mathbb{X},\mathbb{Y}]\!]_C + \mathscr{J}[\![\mathbb{X},\mathscr{J}\mathbb{Y}]\!]_C - [\![\mathscr{J}\mathbb{X},\mathscr{J}\mathbb{Y}]\!]_C = 0\,. \quad (69)$$

It is sometimes advantageous to define integrability with respect to the H-twisted bracket. An integrable generalised almost complex structure is called a *generalised complex structure*.

Comment: A generalised complex structure $\mathscr{J}$ comes associated with a Poisson structure which sits in the upper right quadrangle when $\mathscr{J}$ is viewed as a matrix, mapping

$$\mathbb{T} \to \mathbb{T}\,. \quad (70)$$

An *irregular point* is a point where this Poisson structure changes rank. A lot of the mathematical interest is focused on these points, but we will not comment on them further in this presentation.

A complex geometry $(\mathscr{M},J)$ or a symplectic geometry $(\mathscr{M},\omega)$ are seen to be special cases of generalised geometry where the generalised complex structures are

$$\mathscr{J}_J = \begin{pmatrix} J & 0 \\ 0 & -J^t \end{pmatrix}\,, \quad (71)$$

and

$$\mathscr{J}_\omega = \begin{pmatrix} 0 & -\omega^{-1} \\ \omega & 0 \end{pmatrix}\,, \quad (72)$$

respectively.

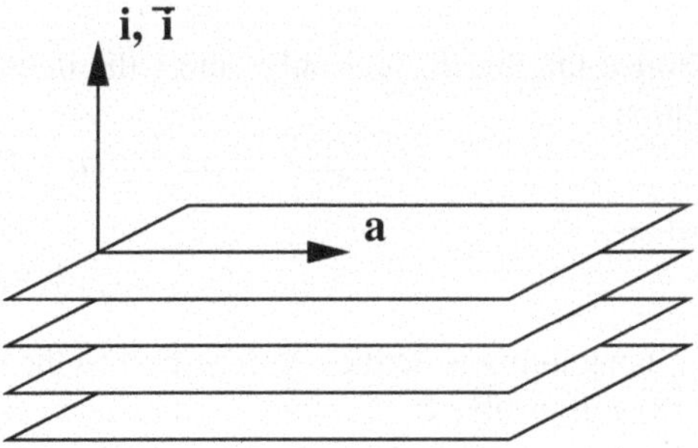

Fig. 1 A generalisation of the Newlander-Nirenberg theorem shows that locally, in the neighbourhood of a regular point. a GC manifold $\mathscr{M}$ looks like a foliation with $z^i, \bar{z}^{\bar{i}}$ complex and x^a Darboux coordinates.

Generalised Kähler Geometry [10], is the form bihermitian geometry takes when lifted to generalised complex geometry. The additional data is that there are *two* generalised complex structures, $\mathscr{J}_1$ and $\mathscr{J}_2$, that *commute*, $[\mathscr{J}_1,\mathscr{J}_2]=0$, and whose product gives an integrable local product structure $\mathscr{G}$

$$\mathscr{G} = -\mathscr{J}_1\mathscr{J}_2\,, \quad \mathscr{G}^2 = \mathbf{1}\,. \tag{73}$$

Integrability is again defined with respect to the Courant bracket or its H-twisted version. Under quite general conditions the existence on $\mathscr{M}$ of a metric g acting on T, with inverse g^{-1} that acts on T^*, ensures that one can find coordinates and B-transforms such that [4]

$$\mathscr{G} = \begin{pmatrix} 0 & g^{-1} \\ g & 0 \end{pmatrix}\,. \tag{74}$$

Using $\mathscr{G}$ we may introduce projection operators

$$P_\pm := \tfrac{1}{2}\left(1 \pm \mathscr{G}\right)\,, \tag{75}$$

that may be used to split the generalised tangent space $\mathbb{T}$ into the ± 1 eigenspaces of $\mathscr{G}$:

$$\mathbb{T} = \mathbb{T}_+ \oplus \mathbb{T}_-\,, \tag{76}$$

where

$$\mathbb{T}_\pm := P_\pm \mathbb{T}\,. \tag{77}$$

Gualtieri's map

In [10] the relation between bihermitian geometry $(\mathscr{M}, G, J_\pm)$ and generalised Kähler geometry was established. Explicitly the map constructs the two generalised complex structures $\mathscr{J}_1$ and $\mathscr{J}_2$ from the bihermitian data as follows:

$$\mathscr{J}_{1/2} = \tfrac{1}{2}\begin{pmatrix} J_+ & -(\omega)_+^{-1} \\ \omega_+ & -J_+^t \end{pmatrix} \pm \tfrac{1}{2}\begin{pmatrix} J_- & -(\omega)_-^{-1} \\ \omega_- & -J_-^t \end{pmatrix}\,, \tag{78}$$

where $\omega_\pm = gJ_\pm$ are the two-forms associated with $J_\pm$.

Notice that the B field only enters the definitions via H in the integrability conditions

$$\Pi_\mp^{1/2}[\![\Pi_\pm^{1/2}\mathbb{X}, \Pi_\pm^{1/2}\mathbb{Y}]\!]_H = 0\,, \quad \forall\, \mathbb{X}, \mathbb{Y} \in C^\infty(\mathbb{T}). \tag{79}$$

where

$$\Pi_\mp^{1/2} := \tfrac{1}{2}\left(1 \pm i\mathscr{J}^{1/2}\right)\,. \tag{80}$$

If integrability is defined with respect to the untwisted Courant bracket the relation (78) will involve B-transforms with a non exact B instead.

3 Concluding comments

Generalised Kähler geometry covers the target space geometry of all three types of $(2,2)$ sigma models described in the previous section, as well as the general one that includes all three types of fields. For different number of left q and right p supersymmetries, i.e., (p,q) supersymmetry, the geometries will differ. Some cases are listed in table 1.

Table 1 Geometries of some 2*d* sigma-models with (p,p) supersymmetries.

Susy	(1,1)	(2,2)	(2,2)	(4,4)	(4,4)
E=G+B	G,B	G	G,B	G	G,B
Geometry	Rieman	Kähler	bihermitian	hyperkähler	bihypercomplex

Other cases such as Stong Kähler with torsion [13] or Strong hyperkähler with torsion can also be described in generalised geometry [19], [4].

These relations between supersymmetric sigma models and geometry are utilised, e.g., in constructing quotients [21], [11] and T-dual models [27], [22] . Both these directions require the gauging of isometries of the models, studied in [17] [5] and, more relevant to generalised geometry, in [22], [26]. Yet another issue concerns quantisation and is partly discussed in [7], [8].

Acknowledgements I thank Martin Roček for reading and commenting on the manuscript. My research is supported in part by the 2236 Co-Funded Scheme2 (CoCirculation2) of TÜBİTAK (Project No:120C067)[7].

References

1. Alvarez-Gaume, L., and Freedman, D.Z., "Geometrical Structure and Ultraviolet Finiteness in the Supersymmetric Sigma Model," Commun. Math. Phys. **80** (1981), 443 doi:10.1007/BF01208280
2. Bogaerts, J., Sevrin, A., van der Loo, S., and Van Gils, S., "Properties of semichiral superfields," Nucl. Phys. B **562** (1999), 277-290 doi:10.1016/S0550-3213(99)00490-3 [arXiv:hep-th/9905141 [hep-th]].
3. Buscher, T., Lindström, U., and Roček, M., "New Supersymmetric σ Models With Wess-Zumino Terms," Phys. Lett. B **202** (1988), 94-98 doi:10.1016/0370-2693(88)90859-3
4. Cavalcanti, G. R., "Hodge theory of SKT manifolds" Adv. Math.,Volume 374, 18 November 2020, 107270. arXiv:1203.0493 [math.DG]
5. de Wit, B. , Roček, M., and Vandoren, S., "Gauging isometries on hyperKahler cones and quaternion Kahler manifolds," Phys. Lett. B **511** (2001), 302-310 doi:10.1016/S0370-2693(01)00636-0 [arXiv:hep-th/0104215 [hep-th]].
6. Gates, Jr. S.J., Hull, C.M., and Roček, M., "Twisted Multiplets and New Supersymmetric Nonlinear Sigma Models," Nucl. Phys. B **248** (1984), 157-186 doi:10.1016/0550-3213(84)90592-3
7. Grisaru, M.T., Massar, M., Sevrin,A. and Troost J., "The Quantum geometry of N=(2,2) non-linear sigma models," Phys. Lett. B **412** (1997), 53-58 doi:10.1016/S0370-2693(97)01053-8 [arXiv:hep-th/9706218 [hep-th]].

[7] However the entire responsibility for the publication is ours. The financial support received from TÜBİTAK does not mean that the content of the publication is approved in a scientific sense by TÜBİTAK.

8. Grisaru, M.T., Massar, M., Sevrin,A. and Troost J., "Some aspects of N=(2,2), D = 2 supersymmetry," Fortsch. Phys. **47** (1999), 301-307 doi:10.1002/(SICI)1521-3978(199901)47:1/3 [arXiv:hep-th/9801080 [hep-th]].
9. Gualtieri, M., "Generalized complex geometry," [arXiv:math/0401221 [math.DG]].
10. Gualtieri, M., "Generalized Kaehler Geometry," Commun. Math. Phys. **331** (2014) no.1, 297-331 doi:10.1007/s00220-014-1926-z
11. Hitchin, N.J., Karlhede, A., Lindström, U. and Roček, M., "Hyperkahler Metrics and Supersymmetry," Commun. Math. Phys. **108** (1987), 535 doi:10.1007/BF01214418
12. N. Hitchin, "Generalized Calabi-Yau manifolds," Quart. J. Math. **54** (2003), 281-308 doi:10.1093/qjmath/54.3.281 [arXiv:math/0209099 [math.DG]].
13. Howe, P.S., and Papadopoulos, G., "Further Remarks on the Geometry of Two-dimensional Nonlinear σ Models," Class. Quant. Grav. **5** (1988), 1647-1661 doi:10.1088/0264-9381/5/12/014
14. Howe, P.S., and Lindström, U., "Local supertwistors and conformal supergravity in six dimensions," Proc. Roy. Soc. Lond. A **476** (2020) no.2243, 20200683 doi:10.1098/rspa.2020.0683 [arXiv:2008.10302 [hep-th]].
15. Howe, P.S., and Lindström, U., "Superconformal geometries and local twistors," JHEP **04** (2021), 140 doi:10.1007/JHEP04(2021)140 [arXiv:2012.03282 [hep-th]].
16. Hull, C.M., and Witten, E., "Supersymmetric Sigma Models and the Heterotic String," Phys. Lett. B **160** (1985), 398-402 doi:10.1016/0370-2693(85)90008-5
17. Hull, C.M., Karlhede, A. Lindström, U., and Roček, M., "Nonlinear σ Models and Their Gauging in and Out of Superspace," Nucl. Phys. B **266** (1986), 1-44 doi:10.1016/0550-3213(86)90175-6
18. Hull, C.M. Lindström, U., Roček, M., von Unge, R. and Zabzine, M., Generalized Kahler geometry and gerbes, JHEP **10** (2009), 062 doi:10.1088/1126-6708/2009/10/062 [arXiv:0811.3615 [hep-th]].
19. Hull, C.M., and Lindström, U., "The Generalised Complex Geometry of (p,q) Hermitian Geometries," Commun. Math. Phys. **375** (2019) no.1, 479-494 doi:10.1007/s00220-019-03488-3 [arXiv:1810.06489 [hep-th]].
20. Lindström, U., Roček, M., von Unge R., and Zabzine, M., "Generalized Kahler manifolds and off-shell supersymmetry," Commun. Math. Phys. **269** (2007), 833-849 doi:10.1007/s00220-006-0149-3 [arXiv:hep-th/0512164 [hep-th]].
21. Lindström, U. and Roček, M., "Scalar Tensor Duality and N=1, N=2 Nonlinear Sigma Models," Nucl. Phys. B **222** (1983), 285-308 doi:10.1016/0550-3213(83)90638-7
22. Lindström, U., Roček, M., Ryb, I, von Unge R., and Zabzine, M., "T-duality and Generalized Kahler Geometry," JHEP **02** (2008), 056 doi:10.1088/1126-6708/2008/02/056 [arXiv:0707.1696 [hep-th]].
23. Lindström, U., and Roček, M., "Properties of hyperkahler manifolds and their twistor spaces," Commun. Math. Phys. **293** (2010), 257-278 doi:10.1007/s00220-009-0923-0 [arXiv:0807.1366 [hep-th]].
24. Lindström, U., "Uses of Sigma Models," PoS **CORFU2017** (2018), 166 doi:10.22323/1.318.0166 [arXiv:1803.08873 [hep-th]].
25. Lindström, U., and Roček, M., "$\beta\gamma$-systems interacting with sigma-models," JHEP **06** (2020), 039 doi:10.1007/JHEP06(2020)039 [arXiv:2004.06544 [hep-th]].
26. Merrell, W. and Vaman, D., 'T-duality, quotients and generalized Kahler geometry," Phys. Lett. B **665** (2008), 401-408 doi:10.1016/j.physletb.2008.06.031 [arXiv:0707.1697 [hep-th]].
27. Roček, M. and Verlinde, E.P., "Duality, quotients, and currents," Nucl. Phys. B **373** (1992), 630-646 doi:10.1016/0550-3213(92)90269-H [arXiv:hep-th/9110053 [hep-th]].
28. Zumino, B. "Supersymmetry and Kahler Manifolds," Phys. Lett. B **87** (1979), 203 doi:10.1016/0370-2693(79)90964-X

An introduction to decomposition

Eric Sharpe

Abstract We review work on 'decomposition,' a property of two-dimensional theories with 1-form symmetries and, more generally, d-dimensional theories with $(d-1)$-form symmetries. Decomposition is the observation that such quantum field theories are equivalent to ('decompose into') disjoint unions of other QFTs, known in this context as "universes." Examples include two-dimensional gauge theories and orbifolds with matter invariant under a subgroup of the gauge group. Decomposition explains and relates several physical properties of these theories – for example, restrictions on allowed instantons arise as a "multiverse interference effect" between contributions from constituent universes. First worked out in 2006 as part of efforts to resolve technical questions in string propagation on stacks, decomposition has been the driver of a number of developments since. We give a general overview of decomposition, describe features of decomposition arising in gauge theories, then dive into specifics for orbifolds. We conclude with a discussion of the recent application to anomaly resolution of Wang-Wen-Witten in two-dimensional orbifolds. This is a contribution to the proceedings of the conference *Two-dimensional supersymmetric theories and related topics* (Matrix Institute, Australia, January 2022), giving an overview of a talk given there and elsewhere.

1 Introduction

Briefly, decomposition is the observation that some QFTs with a local action are secretly equivalent to sums (disjoint unions) of other QFTs, known in this context as 'universes.'

When this happens, we say that the QFT decomposes (into its constitutent univeres). Decomposition of the QFT can be applied to give insight into its properties.

Eric Sharpe
Virginia Tech, Department of Physics MC 0435, 850 West Campus Drive, Blacksburg, VA 24061, e-mail: ersharpe@vt.edu

D. R. Wood et al. (eds.), *2021-2022 MATRIX Annals*, MATRIX Book Series 5,
https://doi.org/10.1007/978-3-031-47417-0_8

Decomposition was first described in 2006 in [26], where it arose as part of efforts to understand string compactifications on generalizations of spaces known as stacks and gerbes, and resolved some of the apparent physical inconsistencies of those theories. It has since been developed in numerous other papers, see for example [4–8, 13, 14, 16–18, 20, 27, 29, 32, 33, 38, 39, 43–46, 51, 53, 54, 57–59, 66] and discussed in review articles including [48–50, 52].

For one QFT to be a sum (or disjoint union[1]) of other QFTs means that, for example, there exist projection operators: a set of topological operators Π_i that commute with all operators in the theory

$$[\Pi_i, \mathcal{O}] = 0, \tag{1}$$

and with the properties that

$$\Pi_i \Pi_j = \delta_{ij} \Pi_j, \ \sum_i \Pi_i = 1. \tag{2}$$

From (1), the projectors are mutually commuting, so they can be simultaneously diagonalized: the Fock space can be diagonalized into eigenmodes of the projectors Π_i, which are the states of the constituent universes. From (2), one can show that correlation functions are a sum of correlation functions in the constituent theories. Using the properties above, we can write

$$\langle \mathcal{O}_1 \cdots \mathcal{O}_m \rangle = \sum_i \langle \Pi_i \mathcal{O}_1 \cdots \mathcal{O}_m \rangle, \tag{3}$$

$$= \sum_i \langle (\Pi_i \mathcal{O}_1) \cdots (\Pi_i \mathcal{O}_m) \rangle, \tag{4}$$

$$= \sum_i \langle \tilde{\mathcal{O}}_1 \cdots \tilde{\mathcal{O}}_m \rangle_i, \tag{5}$$

where the $\tilde{\mathcal{O}}$ are the projections of the operators $\mathcal{O}$ into the universes.

In practice, beyond exhibiting projection operators, another property will be used frequently in this article in describing and checking decompositions, namely that (on a connected spacetime) the partition function of a disjoint union is the sum of the partition functions of the constituent universes:

$$Z = \sum_i Z_i. \tag{6}$$

In essence, this is because the state space of the disjoint union is a sum of the state spaces of the constituent theories. Formally, if we write

$$Z = \sum_{\text{states}} \exp(-\beta H), \tag{7}$$

then using the fact that the whole state (Fock) space is a sum over state spaces of the constituent universes, we have immediately that

[1] We will use the terms 'sum' and 'disjoint union' interchangeably in this article.

$$Z = \sum_i \sum_{\text{states in i}} \exp(-\beta H) = \sum_i Z_i. \tag{8}$$

Let us take a moment to distinguish universes from superselection sectors arising in spontaneous symmetry breaking. Briefly, the idea of a superselection sector is that it is a sector of the theory characterized by a vacuum that cannot be shifted perturbatively by local operators, but can be changed by adding energy (broadly proportional to volume). A prototypical example of superselection sectors is the orientation of magnetization (microscopically, spins) in a bar magnet. In that example, local operators cannot change the direction of magnetization; however, by adding energy (broadly proportional to the volume), one can randomize the spins, then as the magnet cools, the spins may align in a new direction. Here, however, it should be noted that the different superselection sectors (magnetizations) are linked to one another – except in deep IR or infinite volume limits, there is a continuous path in field space connecting them. By contrast, in decomposition, one has disjoint QFTs at every energy scale, meaning that there is no continuous path linking states in different universes. A more detailed discussion can be found in [58].

To be clear, disjoint unions can also emerge as deep IR / infinite volume limits in spontaneous symmetry breaking, and also from limits of KK reductions as in [3]. However, once one has a disjoint union of QFTs, to 'rejoin' the theories at higher energies, one must deform by a (typically non-local) operator that bridges the disjoint union. (At the fixed point, if such an operator were local, it would be an irrelevant operator.) The coupling of that operator vanishes in the IR limit, but at higher energies, could result in a single theory. However, without such a deformation, a disjoint union remains disjoint at all energy scales. In any event, we usually reserve the term 'decomposition' for the special case of a disjoint union of quantum field theories described by a local action.

Later we will work through examples of decomposition in detail, but for the moment, let us outline some known examples.

- Orbifolds. Broadly speaking, orbifolds in which a subgroup of the gauge group acts trivially decompose, see e.g. [26, 43–46]. There is a longer story behind why orbifolds with trivially-acting subgroups differ from ordinary orbifolds, see [40–42] for early work, but in any event, much of this review will be devoted to orbifolds, so we will see detailed examples shortly.
- Two-dimensional gauge theories in which a subgroup of the gauge group acts trivially also decompose. Such gauge theories, and their differences from ordinary gauge theories, were discussed in [40–42]. Examples of their decomposition include the following:
  - A two-dimensional $U(1)$ gauge theory with nonminimal charges is equivalent to a sum of $U(1)$ theories with minimal charges [26],

  - A two-dimensional G gauge theory with center-invariant matter is equivalent to a union of $G/Z(G)$ gauge theories (for $Z(G)$ the center of G) with discrete theta angles [51],
  - Two-dimensional pure Yang-Mills theory for gauge group G is equivalent to a sum of invertible field theories, indexed by irreducible representations of G [38, 39] (see also [13] for the abelian case).

- Four-dimensional Yang-Mills theory with a restriction to instantons of degree divisible by $k > 1$ is equivalent to a disjoint union of k ordinary four-dimensional Yang-Mills theories with different theta angles [58]. (This also, correctly, suggests a subtlety involving cluster decomposition, to which we shall return shortly.)
- Unitary two-dimensional topological field theories (with semisimple local operator algebras). It has been known for many years that these are equivalent to disjoint unions of invertible field theories, see e.g. [15, 37], and by utilizing non-invertible higher-form symmetries, it was argued in [32, 33] that these are also examples of decomposition.
- Finally, sigma models on gerbes. Gerbes are examples of stacks, generalizations of spaces which admit metrics, spinors, gauge fields, and everything else one would require to make sense of a sigma model. Sigma models with target stacks were studied in [40–42], in the hope of discovering new string compactifications, new (2,2) SCFTs, and amongst the technical challenges that arose (construction of an action, presentation dependence, naively inconsistent moduli) was, in the case of gerbes, a violation of cluster decomposition. The original motivation for decomposition [26] was to resolve this problem. The resolution observed that a sigma model on a gerbe is equivalent to a disjoint union of sigma models on spaces, solving the issue with cluster decomposition, but also clarifying that one could not construct new (2,2) SCFTs in this fashion.

So far we have outlined a number of rather diverse-looking examples of decomposition, in both two and four dimensions. The reader may well ask, what do these examples have in common?

Briefly, in d spacetime dimensions, a theory decomposes when it has a $(d-1)$-form symmetry. (In two dimensions, this was the point of [26], and it was generalized to higher dimensions in [13, 58].) Thus, decomposition and higher-form symmetries go hand-in-hand.

In this review, we will primarily focus on the case $d = 2$, for which one will have a decomposition if a $(d-1) = 1$-form symmetry is present.

To that end, let us take a moment to briefly review one-form symmetries. For this review, intuitively, a one-form symmetry group is (something like) a group that exchanges nonperturbative sectors.

For example, consider a G gauge theory or orbifold in which the matter/fields are invariant under a subgroup $K \subset G$. For simplicity, let us assume that K is abelian, and in fact lies within the center of G. Then, in this case, there is a permutation symmetry amongst the nonperturbative sectors. Schematically, the path integral is invariant under

$$(G\text{-bundle}) \mapsto (G\text{-bundle}) \otimes (K\text{-bundle}), \tag{9}$$

or if the reader prefers, in terms of gauge fields,

$$A \mapsto A + A', \tag{10}$$

where A is a G-instanton and A' is a K-instanton.

Instead of an action of elements of groups, we have an action of bundles of groups. This structure is almost a group, except that associativity of multiplication only holds up to multiplication. Technically, this is known as a 2-group.

The 2-group whose elements are K-bundles, is denoted either $K^{(1)}$ (recently in physics) or BK (in math). The latter notation has been used for decades, so we will use the notation BK to denote a 2-group of K bundles.

One-form symmetries can also been seen in the algebra of topological local operators, where they are often realized nonlinearly. This is how the decomposition story connects to two-dimensional topological field theories [32, 33], but is beyond the scope of this article.

There are several descriptions of the two-dimensional quantum field theories which we will discuss in this article:

- A gauge theory or orbifold with a trivially-acting subgroup (i.e. a non-complete charge spectrum),
- A theory with a restriction on instantons,
- Sigma models on gerbes,
- A theory with multiple topological local operators.

Decomposition often relates these different pictures.

- For one example, we will see that restrictions on instantons are implemented as a "multiverse interference effect" between the different universes of a decomposition.
- The one-form symmetry of the quantum field theory can be understood in the sigma model language. A 'gerbe' is a fiber bundle whose fibers are 2-groups BK of one-form symmetries. In any case in which the target space of a sigma model is a fiber bundle, the sigma model possesses a global symmetry corresponding to translations along the fibers. For a sigma model whose target is a gerbe, since the fibers are copies of BK, the sigma model has a BK symmetry.
- This is described by gauge theories with trivially-acting subgroups because $BK = [\text{point}/K]$. (In ordinary geometry, a quotient of a point by any group is the same point back again, but the pertinent mathematics keeps track of automorphisms, and so this is different from a point.) Utlimately, a sigma model whose target is a gerbe involves fibering a trivially-acting K gauge theory over an ordinary theory, hence gauge theories with trivially-acting subgroups.

2 Generalities on gauge theories

Suppose we have a two-dimensional G gauge theory, where G is semisimple, and a subgroup K of the center of G acts trivially on all the matter.

As outlined above, this theory has a global BK one-form symmetry, and so one expects that it should decompose.

The projection operators are, schematically, twist fields / Gukov-Witten operators [23, 24] corresponding to elements of the center of the group algebra $\mathbb{C}[K]$. Existence of projectors (idempotents), forming a basis for the center, is a consequence of Wedderburn's theorem (see e.g. [35, section XVII.3]).

In particular, judging from the projectors of Wedderburn's theorem, universes are in one-to-one correspondence with irreducible representations of K.

Now, abstractly, that is a formal argument for a decomposition, but it does not specify the form of the decomposition. In this particular case, decomposition takes the form (see e.g. [51, section 2])

$$\mathrm{QFT}\,(G-\text{gauge theory}) = \coprod_{\theta\in\hat{K}} \mathrm{QFT}\,(G/K-\text{gauge theory with discrete theta angle }\theta)\,. \tag{11}$$

For example, $SU(2)$ gauge theory with center invariant matter is the disjoint union of a pair of $SO(3)$ theories, schematically

$$SU(2) = SO(3)_+ + SO(3)_-, \tag{12}$$

where the $\pm$ denotes the ($\mathbb{Z}_2$-valued) discrete theta angle coupling to the second Stiefel-Whitney class $w_2 \in H^2(\mathbb{Z}_2)$.

Perturbatively, the $SU(2)$ and $SO(3)_\pm$ theories are identical, but nonperturbatively, they differ. Specifically, there are more $SO(3)$ instantons (bundles) than $SU(2)$ instantons (bundles). The effect of the discrete theta angle is to weight the non-$SU(2)$ $SO(3)$ instantons by a sign, so that when the partition functions for the $SO(3)_\pm$ theories are added, contributions from non-$SU(2)$ $SO(3)$ instantons cancel out between the two theories, leaving only $SU(2)$ instantons – consistent with decomposition.

We can describe this more formally as follows. Write the partition function of the disjoint union as

$$Z = \sum_{\theta\in\hat{K}} \int [DA]\exp(-S)\exp\left(\theta\int w_2(A)\right), \tag{13}$$

where we use $w_2 \in H^2(K)$ to denote the degree-two characteristic class of G/K bundles. Now, moving the summation inside the path integral, we have

$$Z = \int [DA]\exp(-S)\left(\sum_{\theta\in\hat{K}} \exp\left(\theta\int w_2(A)\right)\right). \tag{14}$$

However,

$$\sum_{\theta \in \hat{K}} \exp\left(\theta \int w_2(A)\right) \tag{15}$$

is proportional[2] to a projection operator, projecting out instantons (bundles) for which $w_2 \neq 0$, leaving only those bundles which exist in the G gauge theory.

In effect, an interference effect between the universes of decomposition – a "multiverse interference effect" – has cancelled out some of the nonperturbative sectors.

As a quick consistency check, let us compare to pure $SU(2)$ Yang-Mills theory in two dimensions, for which in essence everything is computable [36,47,64]. We will check a decomposition due to the $B\mathbb{Z}_2$ center symmetry [51, section 2.4]; a more extreme decomposition (to invertible field theories, utilizing noninvertible higher-form symmetries) was discussed in [38, 39].

Consider partition functions in pure Yang-Mills theory in two dimensions. From [36, 47, 64], the partition functions of the pure $SU(2)$ and the pure $SO(3)$ theory without a discrete theta angle (denoted $SO(3)_+$), the partition functions are of the form

$$Z(G) = \sum_R (\dim R)^{2-2g} \exp(-AC_2(R)), \tag{16}$$

where G is the gauge group ($SU(2)$ or $SO(3)$ here), g is the genus of the two-dimensional spacetime, A its area, $C_2(R)$ the second Casimir, and the sum is over all irreducible representations of G. We also need the partition function of the $SO(3)_-$ theory; this was computed in [55], and takes the same form as above, namely

$$Z(SO(3)_-) = \sum_R (\dim R)^{2-2g} \exp(-AC_2(R)), \tag{17}$$

where the sum is now over representations of $SU(2)$ that are not representations of $SO(3)$, a complementary sum to that appearing in the $SO(3)_+$ partition function. Assembling these pieces, we see immediately that adding a sum over $SO(3)$ representations to a sum over $SU(2)$ representations minus $SO(3)$ representations, we get

$$Z(SU(2)) = Z(SO(3)_+) + Z(SO(3)_-), \tag{18}$$

We have discussed a number of theories with properties such as a restriction on instantons, and multiple identity operators, properties which are often taken to signal a violation of cluster decomposition (see e.g. [63, section 23.6]). Especially as cluster decomposition is sometimes used interchangeably with locality, this is often taken to indicate a sickness or inconsistency of the theory.

[2] The proportionality factor reflects that fact that the G gauge theory has more gauge transformations than the G/K gauge theory, so the path integrals have, in principle, slightly different normalization factors.

However, that is an oversimplification. For example, cluster decomposition can also be violated in spontaneous symmetry breaking, in infinite-volume limits. This does not imply an inconsistency of the theory, but rather is just a statement about the vacuum. Perhaps cluster decomposition is best thought of as a property of vacua, and locality a property of the theory, and as these examples illustrate, these are distinct notions, not interchangeable with one another.

In any event, the theories we are describing are manifestly local, in that they have local Lagrangians, and the separate universes are perfectly consistent.

3 Consistency tests and applications

Since 2005, decomposition has been checked in a wide variety of examples, in a wide variety of different kinds of examples, and in many ways. We list a few examples below:

- Gauged linear sigma models (GLSMs): Decomposition has been checked in gauged linear sigma models via mirror symmetry and in quantum cohomology rings (the latter through Coulomb branch computations). For abelian GLSMs, this was described in [40, 41] using Hori-Vafa mirrors [31]; for nonabelian GLSMs, this was checked in the papers describing nonabelian mirror constructions [11, 19, 21, 22].
- In orbifolds, decomposition has been checked extensively [26, 43–46] in, for example, partition functions and massless spectra, as we will outline later in this article.
- Decomposition has also been checked in open strings and K theory [26]. Briefly, in a gauge theory in which a subgroup K of the gauge group acts trivially on bulk degrees of freedom, K can still act nontrivially on boundary degrees of freedom, which therefore organize according to irreducible representations of K, precisely matching the description of universes earlier. In this picture, from gauge invariance, open string states can only exist on open strings connecting the same irreducible representations of K – meaning, that there are no open strings connecting different universes. This also can be understood in terms of K theory [65]. As discussed in [26], K theory on gerbes is equivalent to (twisted) K theory on a disjoint union of spaces, following the same pattern as decomposition.
- In supersymmetric gauge theories in two dimensions, supersymmetric localization can be applied to give further tests of decomposition, as discussed in [51].
- In nonsupersymmetric pure Yang-Mills in two dimensions, decomposition can also be checked. We have previously outlined tests of decomposition along center one-form symmetries, which are described in greater detail in [51]. In addition, there exists a more extreme decomposition of nonsupersymmetric pure

Yang-Mills to a disjoint union of invertible field theories, indexed by irreducible representations of the gauge group [38, 39].
- Decomposition in adjoint QCD_2 was studied in [33].
- Decomposition has been checked numerically in lattice gauge theory [29].
- Finally, lest we give a different impression, decomposition is not restricted to two-dimensional theories, but has also been studied in other dimensions, see e.g. [13, 14, 58].

We should also mention that decomposition has a number of applications:

- The original application was to understand and resolve certain technical issues in making sense of sigma models whose targets are generalized spaces known as 'stacks' [40–42]. This was part of a program of trying to construct new string compactifications, new conformal field theories. After one understands basic issues such as the construction of an action to describe such a sigma model, potential presentation-dependence issues, and moduli mismatches, more subtle issues remain. For example, in the special cases of stacks known as 'gerbes' (fiber bundles whose fibers are one-form symmetry groups), the sigma models violated cluster decomposition. Understanding this issue was the original motivation for work on decomposition, and the resolution was that such sigma models are equivalent to disjoint unions of sigma models on ordinary spaces. As a result, we were not able to construct new (2,2) supersymmetric SCFTs, though we did learn about decomposition. (In the more general case of (0,2) SCFTs for gerbes, it is still an open question of whether new string compactifications exist, see e.g. [4].)
- Decomposition makes predictions for Gromov-Witten theory, specifically the Gromov-Witten theory of stacks and gerbes [1, 9, 10]. From decomposition, the Gromov-Witten theory of a gerbe must match that of a disjoint union of spaces, and this was checked rigorously and discussed in e.g. [5–7, 17, 57, 59].
- In gauged linear sigma models, decomposition was used to provide a novel nonperturbative construction of branched double covers in [8], giving examples of GLSMs with nonbirational phases, realizing examples of Kuznetsov's homological projective duality [34]. This construction has been utilized in the GLSM community in a number of examples since, see e.g. [12, 25, 30] for a few examples.
- Applications to computing elliptic genera of pure gauge theory, studying IR limits of pure supersymmetric gauge theories in two dimensions [2], were discussed in [16].
- Recently decomposition has been applied [44–46] to understand and simplify the Wang-Wen-Witten anomaly resolution proposal [62], as we shall discuss in section 7.

4 Multiverse interference, portals, and wormholes

Next, let us summarize some of the more entertaining features of decomposition:

- *Multiverse interference effects.* We have already seen that summing over universes has the effect of projecting out some nonperturbative contribution, hence a "multiverse interference effect." Our primary exmaples was of a two-dimensional $SU(2)$ gauge theory with center-invariant matter, for which, schematically,

$$\mathrm{QFT}(SU(2)) = \mathrm{QFT}(SO(3)_+) \coprod \mathrm{QFT}(SO(3)_-). \tag{19}$$

- *Fundamentally-charged Wilson lines are defects bridging universes.* Consider for example two-dimensional abelian BF theory at level k. This theory decomposes into a disjoint union of k invertible field theories. The projectors are

$$\Pi_m = \frac{1}{k}\sum_{n=0}^{k-1} \xi^{nm}\mathcal{O}_n, \tag{20}$$

where $\xi = \exp(2\pi i/k)$ and

$$\mathcal{O}_n =: \exp(nB) : . \tag{21}$$

These local operators have clock-shift commutation relations with the Wilson lines (see e.g. [27])

$$\mathcal{O}_p W_q = \xi^{pq} W_q \mathcal{O}_p, \tag{22}$$

which algebraically are equivalent to

$$\Pi_m W_p = W_p \Pi_{m+p \mod k}. \tag{23}$$

Thus, moving a projector past a Wilson line changes the projector, and so Wilson lines in abelian BF theory act as (nondynamical) defects bridging different universes.

- *Wormholes between universes.* This is how GLSMs realize branched double covers nonperturbatively [8]. Consider for example a GLSM with gauge group $U(1)$, two chiral superfields p_a of charge $+2$ and four chiral superfields ϕ_i of charge -1, with a superpotential

$$W = \sum_{ij} \phi_i \phi_j A^{ij}(p). \tag{24}$$

In the phase $r \ll 0$, this describes a branched double cover of $\mathbb{P}^1$, where the sheets of the cover are (approximate) universes, spanned by the p fields (of nonminimal charge), and the branch locus is the region where the mass matrix A^{ij} develops zero eigenvalues, forming a Euclidean wormhole.

5 Specifics on orbifolds

Now, let us turn to examples of decomposition in two-dimensional orbifolds.

Consider an orbifold $[X/\Gamma]$, where $K \subset \Gamma$ acts trivially, and let $G = \Gamma/K$. We can write Γ as

$$1 \longrightarrow K \longrightarrow \Gamma \longrightarrow G \longrightarrow 1. \tag{25}$$

Decomposition is known for general extensions [26], but for simplicity in this overview, let us assume for the moment that this is a central extension, so that K is a subset of the center of Γ. Let $[\omega] \in H^2(G,K)$ denote the element of group cohomology classifying the extension.

In this case, if K acts trivially on X and lies within the center of Γ, then this orbifold decomposes as

$$\mathrm{QFT}\left([X/\Gamma]\right) = \mathrm{QFT}\left(\coprod_{\rho \in \hat{K}} [X/G]_{\hat{\omega}(\rho)} \right), \tag{26}$$

where $\hat{K}$ denotes the set of isomorphism classes of irreducible representations of K, and $\hat{\omega}(\rho)$ represent discrete torsion phases, essentially finite-group analogues of theta angles, which is the image of the extension class under ρ:

$$\begin{aligned} H^2(G,K) &\xrightarrow{\rho} H^2(G,U(1)), \\ \omega &\mapsto \omega \circ \rho = \hat{\omega}(\rho). \end{aligned}$$

This will be a finite-group analogue of the $SU(2)$ decomposition described earlier, with $[X/\Gamma]$ playing the analogue of the $SU(2)$ theory and the $[X/G]_{\hat{\omega}(\rho)}$ theories playing the analogue of the $SO(3)_{\pm}$ theories.

To justify this decomposition, we must first provide projectors. Corresponding to any irreducible representation $R \in \hat{K}$, the projector is [54]

$$\Pi_R = \sum_i \frac{\dim R_i}{|K|} \sum_{k \in K} \chi_{R_i}\left(k^{-1}\right) \tau_k, \tag{27}$$

where τ_k is a twist field for the trivially-acting element $k \in K$, $\chi_R(g)$ denotes the character of g in representation R, and the R_i are a set of representatives of the irreducible representations. (For a detailed examination of why trivially-acting group elements have associated twist fields, and the unitarity violations that ensue if one assumes otherwise, see [40].) It can be shown that these projectors have the expected properties, namely

$$\Pi_R \Pi_S = \delta_{R,S} \Pi_R, \; \sum_R \Pi_R = 1. \tag{28}$$

As the twist fields in this case can be understood formally as the center of the group algebra, this expression for the projector is a formal consequence of Wedderburn's theorem in mathematics (see e.g. [35, section XVII.3]).

To make this more concrete, let us examine all the details in one particular example (taken from [26, section 5.2]). Take $\Gamma = D_4$, the eight-element dihedral group, with center $K = \mathbb{Z}_2$, which we will assume acts trivially on X. In this case, decomposition (26) predicts

$$\mathrm{QFT}\left([X/D_4]\right) = \mathrm{QFT}\left([X/\mathbb{Z}_2 \times \mathbb{Z}_2]_{\mathrm{w/o\,d.t.}}\right) \coprod \mathrm{QFT}\left([X/\mathbb{Z}_2 \times \mathbb{Z}_2]_{\mathrm{d.t.}}\right), \tag{29}$$

a disjoint union of two $\mathbb{Z}_2 \times \mathbb{Z}_2$ orbifolds, one with discrete torsion, and the other without.

In passing, note that this is a very precise analogue of the earlier example of an $SU(2)$ gauge theory (compare $[X/D_4]$) decomposing into a pair of $SO(3)$ theories (compare $[X/\mathbb{Z}_2 \times \mathbb{Z}_2]$).

Now, let us check this statement. First, we consider projection operators. If let let $z \in D_4$ denote the generator of the (trivially-acting) $\mathbb{Z}_2$ center, and $\hat{z}$ the corresponding twist field, then $\hat{z}^2 = 1$, and the projectors are

$$\Pi_{\pm} = \frac{1}{2}\left(1 \pm \hat{z}\right), \tag{30}$$

which obey

$$\Pi_{\pm}^2 = \Pi_{\pm}, \; \Pi_{\pm}\Pi_{\mp} = 0, \; \Pi_{+} + \Pi_{-} = 1. \tag{31}$$

These projectors are in principle the specialization of (27) to this case, but in fact here are sufficiently simple that they can be seen by inspection, as indeed was the case in [26].

So far we have produced a pair of local projection operators, which tell us that the theory breaks into two pieces, but to verify the decomposition 29, we need more information about the pieces. To that end, we will next compute the partition function of this orbifold.

To compute the partition function, we need to describe the dihedral group more explicitly. Briefly, it is generated by elements z (generating the center), a, and b, whose products we list below:

$$D_4 = \{1, z, a, b, az, bz, ab, ba = abz\}. \tag{32}$$

Let us compute the partition function on T^2. For any orbifold $[X/\Gamma]$, the partition function on T^2 is

$$Z_{T^2}\left([X/\Gamma]\right) = \frac{1}{|\Gamma|} \sum_{gh=hg} Z_{g,h}, \tag{33}$$

where each $Z_{g,h}$ represents the path integral contribution from the Γ nonperturbative sector ("twisted sector") defined by the commuting pair $g, h \in G$. (Since Γ is finite, there is no perturbative contribution to the Γ gauge theory, only nonperturbative contributions.) Schematically, each $Z_{g,h}$ is a path integral sum over maps $T^2 \to X$ with branch cuts defined by g, h:

$$Z_{g,h} = \left({}_{g}\underset{h}{\square} \longrightarrow X \right) \tag{34}$$

(In order for the corners of the square to close, one only sums over commuting $g, h \in G$.)

We will argue that

$$Z_{T^2}([X/D_4]) = Z_{T^2}([X/\mathbb{Z}_2 \times \mathbb{Z}_2]) + Z_{T^2}([X/\mathbb{Z}_2 \times \mathbb{Z}_2]_{\text{d.t.}}) \tag{35}$$

verifying decomposition (29).

To that end, first note that since z acts trivially and each $Z_{g,h}$ only depends upon boundary conditions, we immediately have that

$$Z_{g,h} = {}_{g}\underset{h}{\square} = {}_{gz}\underset{h}{\square} = {}_{g}\underset{hz}{\square} = {}_{gz}\underset{hz}{\square} = Z_{gz,hz}. \tag{36}$$

Each square ${}_{g}\square_{h}$ can be associated (modulo automorphisms) with a Γ bundle, so this is a symmetry amongst the nonperturbative sectors of the Γ orbifold. Furthermore, those sectors are related by tensoring in a $B\mathbb{Z}_2$ bundle:

$$Z_{g,h} = {}_{g}\underset{h}{\square} \xrightarrow{{}_{z}\underset{1}{\blacksquare}} {}_{gz}\underset{h}{\square} \xrightarrow{{}_{z}\underset{z}{\blacksquare}} {}_{g}\underset{hz}{\square} \xrightarrow{{}_{z}\underset{1}{\blacksquare}} {}_{gz}\underset{hz}{\square} = Z_{gz,hz} \tag{37}$$

This is the $B\mathbb{Z}_2$ one-form symmetry, explicitly.

Next, to help clarify, write the elements of $\mathbb{Z}_2 \times \mathbb{Z}_2 = D_4/\mathbb{Z}_2$ as

$$\mathbb{Z}_2 \times \mathbb{Z}_2 = \{1, \overline{a}, \overline{b}, \overline{ab}\}, \tag{38}$$

where $\overline{a}$ is the projection of $\{a, az\}$, and $\overline{b}$ is the projection of $\{b, bz\}$. Then, we see that each D_4 twisted sector ($Z_{g,h}$) that appears is the same as a $D_4/\mathbb{Z}_2 = \mathbb{Z}_2 \times \mathbb{Z}_2$ twisted sector, *except* for the sectors

$${}_{\overline{a}}\underset{\overline{b}}{\square}, \quad {}_{\overline{a}}\underset{\overline{ab}}{\square}, \quad {}_{\overline{b}}\underset{\overline{ab}}{\square}, \tag{39}$$

which do not appear, because their lifts do not commute in D_4. (These form a modular orbit – modular invariance is ensured at every step.)

This is a restriction on the nonperturbative sectors.

So far, we have argued that

$$\begin{aligned} Z_{T^2}([X/D_4]) &= \frac{|\mathbb{Z}_2 \times \mathbb{Z}_2|}{|D_4|} |\mathbb{Z}_2|^2 \left(Z_{T^2}([X/\mathbb{Z}_2 \times \mathbb{Z}_2]) - (\text{some twisted sectors}) \right), \\ &= 2 \left(Z_{T^2}([X/\mathbb{Z}_2 \times \mathbb{Z}_2]) - (\text{some twisted sectors}) \right). \end{aligned} \tag{40}$$

In particular, despite the fact that the $\mathbb{Z}_2$ acts trivially, this is a different theory than the $\mathbb{Z}_2 \times \mathbb{Z}_2$ orbifold. *Physics knows when we gauge even a trivially-acting group.*

We can simplify the expression above via the use of discrete torsion [60], which is a set of modular-invariant phases that one can add to partition functions – in essence, theta angles for the finite gauge theory. The new partition function in a G orbifold on T^2, for example, has the form

$$Z_{T^2}([X/G]) = \frac{1}{|G|} \sum_{gh=hg} \varepsilon(g,h) Z_{g,h}, \tag{41}$$

where $\varepsilon(g,h)$ represent the discrete torsion phases.

Now, in a G orbifold, possible choices of discrete torsion phases are classified by $H^2(G,U(1))$. In the case $G = \mathbb{Z}_2 \times \mathbb{Z}_2$,

$$H^2(\mathbb{Z}_2 \times \mathbb{Z}_2, U(1)) = \mathbb{Z}_2, \tag{42}$$

and the twisted sectors that get a phase (specifically, a sign) are

$$\bar{a}\,\underset{\bar{b}}{\square}, \quad \bar{a}\,\underset{\bar{a}\bar{b}}{\square}, \quad \bar{b}\,\underset{\bar{a}\bar{b}}{\square}, \tag{43}$$

the same sectors that are omitted from the $\mathbb{Z}_2 \times \mathbb{Z}_2$ orbifold partition functino in the description of the D_4 orbifold partition function in (40). Thus, we see that the D_4 partition function on T^2 can be rewritten as

$$Z_{T^2}([X/D_4]) = Z_{T^2}([X/\mathbb{Z}_2 \times \mathbb{Z}_2]) + Z_{T^2}([X/\mathbb{Z}_2 \times \mathbb{Z}_2]_{\text{d.t.}}). \tag{44}$$

This matches the prediction of decomposition (29) in this case.

In particular, adding the universes has the effect of cancelling out some of the nonperturbative sectors, namely those listed in (39). This is an example of a *multiverse interference effect.*

So far, we have verified that partition functions on T^2 reproduce decomposition. Analogous computations on higher-genus Riemann surfaces also reproduce decomposition, though the combinatorics is more complex. See [26, section 5.2] for details.

Now, let us turn to massless spectrum computations. (In principle, this is all implicit in the partition function computations, but we find it instructive to explicitly study this particular facet.) For the case $X = T^6$, with a standard $\mathbb{Z}_2 \times \mathbb{Z}_2$ action [61], the massless spectrum of $[T^6/D_4]$ is easily computed and given by

$$\begin{array}{ccccccc} & & & 2 & & & \\ & & 0 & & 0 & & \\ & 0 & & 54 & & 0 & \\ 2 & & 54 & & 54 & & 2 \\ & 0 & & 54 & & 0 & \\ & & 0 & & 0 & & \\ & & & 2 & & & \end{array} \tag{45}$$

This result is problematic – the 2's in the corners signal a violation of cluster decomposition, which ordinarily would be reason to believe that the result is incorrect. However, as we have seen, cluster decomposition arises in any theory describing a disjoint union of QFTs, and so this is not unexpected. Hand-in-hand, the 2's can also be interpreted to mean that the theory has two components. In particular, the massless spectra of the $[T^6/\mathbb{Z}_2 \times \mathbb{Z}_2]$ orbifolds, with and without discrete torsion, are given by [61]

$$\begin{array}{ccccccc} & & & 1 & & & \\ & & 0 & & 0 & & \\ & 0 & & 51 & & 0 & \\ 1 & & 3 & & 3 & & 1 \\ & 0 & & 51 & & 0 & \\ & & 0 & & 0 & & \\ & & & 1 & & & \end{array} + \begin{array}{ccccccc} & & & 1 & & & \\ & & 0 & & 0 & & \\ & 0 & & 3 & & 0 & \\ 1 & & 51 & & 51 & & 1 \\ & 0 & & 3 & & 0 & \\ & & 0 & & 0 & & \\ & & & 1 & & & \end{array} \tag{46}$$

It is easy to see that the sum of the two $[T^6/\mathbb{Z}_2 \times \mathbb{Z}_2]$ orbifold spectra matches that of $[T^6/D_4]$, verifying decomposition (29).

This example was not a one-off, but in fact verifies the general prediction of [26] for orbifolds with trivially-acting subgroups.

In most of this review we have focused on cases of gauge theories and orbifolds in which the trivially-acting subgroup is in the center, but more general examples exist and have been studied. Decomposition in the more general case is the statement [26]

$$\mathrm{QFT}([X/\Gamma]) = \mathrm{QFT}\left(\left[\frac{X \times \hat{K}}{G}\right]_{\hat{\omega}}\right), \tag{47}$$

where the discrete torsion on universes $\hat{\omega}$ is described in [26]. In the general case, where the trivially-acting $K \subset \Gamma$ need not be central, $G = \Gamma/K$ can act nontrivially on the set of isomorphism classes of irreducible representations $\hat{K}$, and the universes are identified with orbits of G in $\hat{K}$. In the special case of central extensions, the G action on $\hat{K}$ is trivial, the orbits are single elements of $\hat{K}$, and the decomposition reduces to a disjoint union of copies of $[X/G]_{\hat{\omega}}$, indexed by $\hat{K}$, as described earlier in (26).

For example, consider the orbifold $[X/\mathbb{H}]$, where $\mathbb{H}$ is the eight-element group of unit quaternions, and where $K = \langle i \rangle \subset \mathbb{H}$ acts trivially. In this case, the center of $\mathbb{H}$ is $\mathbb{Z}_2$, which is contained within $K \cong \mathbb{Z}_4$, but K also has elements that are not central. On the set $\hat{K}$, G leaves two elements invariant but exchanges two elements, so that there are a total of three G orbits, and three universes.

As is discussed in detail in [26, section 5.4], in this case decomposition predicts

$$\mathrm{QFT}\left([X/\mathbb{H}\right) = \mathrm{QFT}\left(X \coprod [X/\mathbb{Z}_2] \coprod [X/\mathbb{Z}_2]\right), \tag{48}$$

which is checked by exhibiting projection operators, computing partition functions at arbitrary genus, and comparing massless spectra. In particular, the universes in this example are not all just the same orbifold with different choices of discrete torsion, but rather they are qualitatively different from one another.

So far we have outlined decomposition in ordinary orbifolds with trivially-acting subgroups. Next we outline decomposition in such orbifolds in the presence of discrete torsion, following [43].

Consider an orbifold $[X/\Gamma]_\omega$, where $K \subset \Gamma$ acts trivially, $\omega \in H^2(\Gamma, U(1))$ is an element of discrete torsion, and we define $G = \Gamma/K$, so that

$$1 \longrightarrow K \stackrel{\iota}{\longrightarrow} \Gamma \stackrel{\pi}{\longrightarrow} G \longrightarrow 1. \tag{49}$$

For simplicity, we assume that this is a central extension (that K maps to a subgroup of the center of Γ). It will be helpful to utilize the maps below:

$$H^2(G, U(1)) \stackrel{\pi^*}{\longrightarrow} \left(\mathrm{Ker}\, \iota^* \subset H^2(\Gamma, U(1))\right) \stackrel{\beta}{\longrightarrow} H^1(G, H^1(K, U(1))) = \mathrm{Hom}(G, \hat{K}). \tag{50}$$

Then, we can describe decomposition of $[X/\Gamma]_\omega$ in terms of the following cases:

1. If $\iota^* \omega \neq 0$, then

$$\mathrm{QFT}\left([X/\Gamma]_\omega\right) = \mathrm{QFT}\left(\left[\frac{X \times \hat{K}_{\iota^*\omega}}{G}\right]_{\hat{\omega}}\right). \tag{51}$$

2. If $\iota^* \omega = 0$ and $\beta(\omega) \neq 0$, then

$$\mathrm{QFT}\left([X/\Gamma]_\omega\right) = \mathrm{QFT}\left(\left[\frac{X \times \widehat{\mathrm{Coker}\, \beta(\omega)}}{\mathrm{Ker}\, \beta(\omega)}\right]_{\hat{\omega}}\right). \tag{52}$$

3. If $\iota^* \omega = 0$ and $\beta(\omega) = 0$, then $\omega = \pi^* \overline{\omega}$ for some $\overline{\omega} \in H^2(G, U(1))$ and

$$\mathrm{QFT}\left([X/\Gamma]_\omega\right) = \mathrm{QFT}\left(\left[\frac{X \times \hat{K}}{G}\right]_{\hat{\omega} + \overline{\omega}}\right), \tag{53}$$

essentially the same decomposition as in the case of no discrete torsion, but with the discrete torsion on the universes shifted by $\overline{\omega}$.

This description was developed more systematically in [43], which also checked the results in numerous examples.

6 Quantum symmetries in noneffective orbifolds

As one final prerequisite before describing the Wang-Wen-Witten anomaly resolution procedure in orbifolds [62], we describe some novel modular-invariant phases that one can add to noneffective orbifolds, which were deemed 'quantum symmetries' in [46] (see also [56]) as they generalize the notion of quantum symmetries in orbifolds from the late 1980s.

Briefly, these phases describe actions on twisted sector states. Consider an orbifold $[X/\Gamma]$ in which a subgroup $K \subset \Gamma$ acts trivially, and define $G = \Gamma/K$, as before. For simplicity, assume that K is central in Γ, so that

$$1 \longrightarrow K \longrightarrow \Gamma \longrightarrow G \longrightarrow 1 \tag{54}$$

is a central extension. In this case, the quantum symmetries (as the term is used in [46]) are classified by $H^1(G,H^1(K,U(1))) = \mathrm{Hom}(G,\hat{K})$, and describe an action of K on G-twisted sector states. Schematically, in terms of path integral data,

$${}_{gz}\underset{h}{\square} = B(\pi(h),z)\left({}_{g}\underset{h}{\square} \right), \tag{55}$$

for $B \in H^1(G,H^1(K,U(1)))$, $z \in K$, and $g,h \in \Gamma$.

It will be useful to note that the gruop classifying quantum symmetries fits into an exact sequence [28]

$$\left(\mathrm{Ker}\,\iota^* \subset H^2(\Gamma,U(1))\right) \stackrel{\beta}{\longrightarrow} H^1(G,H^1(K,U(1))) \stackrel{d_2}{\longrightarrow} H^3(G,U(1)). \tag{56}$$

(Technically, this is part of a seven-term sequence slightly extending the inflation-restriction sequence.) The map d_2 is a differential in the Lyndon-Hochschild-Serre spectral sequence, and for any $\omega \in \mathrm{Ker}\,\iota^*$,

$$\beta(\omega)(\pi(g),z) = \frac{\omega(g,z)}{\omega(z,g)}, \tag{57}$$

for $g \in \Gamma$, $z \in K$. (This is the same map β that appeared in (50).)

The first term in the sequence above can be interpreted in terms of discrete torsion, and as in two-dimensional orbifolds $[X/G]$ gauge anomalies are counted by $H^3(G,U(1))$, the last term can be interpreted in terms of anomalies, so the sequence (56) can be represented schematically as

$$(\text{discrete torsion}) \stackrel{\beta}{\longrightarrow} (\text{quantum symmetries}) \stackrel{d_2}{\longrightarrow} (\text{anomalies}). \tag{58}$$

As this sequence suggests, those quantum symmetries in the image of β are equivalent to choices of discrete torsion, and in fact are equivalent to quantum symmetries in the older sense of the term.

For applications to Wang-Wen-Witten, we will need quantum symmetries B such that $d_2(B) \neq 0$. As the sequence above is exact, such quantum symmetries are necessarily not equivalent to discrete torsion.

Before going on, let us briefly describe decomposition in orbifolds with quantum symmetries. Briefly,

$$\mathrm{QFT}\left([X/\Gamma]_B\right) = \mathrm{QFT}\left(\coprod_{\widehat{\mathrm{Coker}\,B}} [X/\mathrm{Ker}\,B]_{\hat{\omega}} \right), \tag{59}$$

where we interpret the quantum symmetry B as an element of $\mathrm{Hom}(G, \hat{K})$. Decomposition in this case is more or less uniquely dictated by results for decomposition in orbifolds with discrete torsion such that $\iota^* \omega = 0$ and $\beta(\omega) \neq 0$, as described earlier in (52). It was also checked in numerous examples in [46].

7 Wang-Wen-Witten anomaly resolution procedure

In [62], Wang, Wen, and Witten proposed an algorithm to resolve gauge anomalies in anomalous orbifolds $[X/G]$. In this section we will review that procedure, and then observe how decomposition gives an alternative interpretation of the result that clarifies why the procedure removes the anomaly.

We will use the fact that in two dimensions, gauge anomalies in finite G gauge theories are classified by elements of $H^3(G, U(1))$. In the anomalous orbifold $[X/G]$, we will let $\alpha \in H^3(G, U(1))$ denote the anomaly.

The Wang-Wen-Witten procedure has two steps:

1. We replace G by a larger group Γ,

$$1 \longrightarrow K \longrightarrow \Gamma \stackrel{\pi}{\longrightarrow} G \longrightarrow 1, \tag{60}$$

   which we will assume is a central extension, and where K acts trivially.
   From decomposition, if all we did was to replace G by Γ, we would not have resolved the orbifold, as physically $[X/\Gamma]$ is equivalent to copies and covers of $[X/G]$, as we have seen previously.
2. The second step of the Wang-Wen-Witten procedure is to turn on a quantum symmetry phase $B \in H^1(G, H^1(K, U(1)))$, chosen so that $d_2 B = \alpha$. This implies that $\pi^* \alpha \in H^3(\Gamma, U(1))$ is trivial.

These two choices together – an extension Γ plus a choice of suitable quantum symmetry B – resolve the anomaly.

From decomposition, we can see how the anomaly is resolved. Recall from (59 that

$$\mathrm{QFT}\left([X/\Gamma]_B\right) = \mathrm{QFT}\left(\coprod_{\widehat{\mathrm{Coker}\,B}} [X/\mathrm{Ker}\,B]_{\hat{\omega}} \right). \tag{61}$$

As we chose the quantum symmetry B so that $d_2 B = \alpha$, we have immediately that

$$\alpha|_{\mathrm{Ker} B} = 0. \tag{62}$$

Thus, each orbifold $[X/\mathrm{Ker} B]$ is automatically anomaly-free.

Put another way, the result of the Wang-Wen-Witten procedure – replacing $[X/G]$ by a larger orbifold $[X/\Gamma]_B$ – is equivalen to replacing $[X/G]$ by a collection of smaller orbifolds $[X/\mathrm{Ker} B]$, in which $\mathrm{Ker}\, B \subset G$ is non-anomalous.

Let us see this explicitly in examples. We will consider several resolutions of orbifolds of the form $[X/G]$ for $G = \mathbb{Z}_2 \times \mathbb{Z}_2$. For this group, $H^3(G, U(1)) = (\mathbb{Z}_2)^3$, corresponding to the three $\mathbb{Z}_2$ subgroups, so if we write

$$G = \mathbb{Z}_2 \times \mathbb{Z}_2 = \{1, a, b, ab\}, \tag{63}$$

then

$$H^3(G, U(1)) = (\mathbb{Z}_2)^3 = \langle a \rangle \times \langle b \rangle \times \langle ab \rangle. \tag{64}$$

To apply the Wang-Wen-Witten procedure, one must make two choices,

- a choice of larger gauge group Γ, and
- a choice of quantum symmetry.

We will list several examples of larger group Γ, and for each Γ, all possible choices of quantum symmetry and the resulting theories.

For our first resolution, we take $\Gamma = D_4$,

$$1 \longrightarrow \mathbb{Z}_2 \longrightarrow D_4 \longrightarrow \mathbb{Z}_2 \times \mathbb{Z}_2 \longrightarrow 1. \tag{65}$$

The quantum symmetry B is determined by its image on the generators $\{a, b\}$, and we list all possibilites in table 1 (including cases in which we turn on discrete torsion in the $[X/\Gamma]$ orbifold, in addition to the quantum symmetry).

Table 1 A list of all possible quantum symmetries, anomalies resolved, and corresponding orbifolds $[X/\Gamma]_B$ for the case $\Gamma = D_4$. The first two columsn describe the quantum symmetry; the column $d_2(B)$ gives the image of the quantum symmetry, and hence the anomaly that can be resolved; and the last two columns give the physical theory equivalent to $[X/\Gamma]_B$, for either choice of discrete torsion in the $\Gamma = D_4$ orbifold.

$B(a)$	$B(b)$	$d_2(B)$	$[X/\Gamma]_B$ w/o d.t.	$[X/\Gamma]_B$ with d.t.
1	1	–	$[X/G] \coprod [X/G]_{\mathrm{d.t.}}$	$[X/\langle b \rangle]$
-1	1	–	$[X/\langle b \rangle]$	$[X/G] \coprod [X/G]_{\mathrm{d.t.}}$
1	-1	$\langle b \rangle$	$[X/\langle b \rangle]$	$[X/\langle ab \rangle]$
-1	-1	$\langle b \rangle$	$[X/\langle ab \rangle]$	$[X/\langle a \rangle]$

The first row of table 1 describes the case of no quantum symmetry. Nothing is resolved, and the physical theories are (copies of) the G orbifold. The last two rows are more interesting. These describe cases in which an anomaly in the subgroup $\langle b \rangle \subset G$ can be resolved. The resulting physical theories, listed in the last two columns (corresponding to either choice of discrete torsion in the $\Gamma = D_4$ orbifold) are orbifolds by subgroups not containing $\langle b \rangle$, and so by assumption are anomaly-free. In particular, we see that the Wang-Wen-Witten prescription works, explicitly.

For our next resolution of the anomalous $[X/G]$ orbifold, for the same G as before, we take $\Gamma = \mathbb{H}$, the eight-element group of unit quaternions,

$$1 \longrightarrow \mathbb{Z}_2 \longrightarrow \mathbb{H} \longrightarrow \mathbb{Z}_2 \times \mathbb{Z}_2 \longrightarrow 1. \tag{66}$$

We list all possibilites for the quantum symmetry in table 2.

Table 2 A list of all possible quantum symmetries, anomalies resolved, and corresponding orbifolds $[X/\Gamma]_B$ for the case $\Gamma = \mathbb{H}$. The first two columsn describe the quantum symmetry; the column $d_2(B)$ gives the image of the quantum symmetry, and hence the anomaly that can be resolved; and the last column gives the physical theory equivalent to $[X/\Gamma]_B$. (No discrete torsion is possible for this choice of Γ.)

$B(a)$	$B(b)$	$d_2(B)$	$[X/\Gamma]_B$
1	1	–	$[X/G] \coprod [X/G]_{\text{d.t.}}$
-1	1	$\langle a \rangle, \langle ab \rangle$	$[X/\langle b \rangle]$
1	-1	$\langle b \rangle, \langle ab \rangle$	$[X/\langle a \rangle]$
-1	-1	$\langle a \rangle, \langle b \rangle$	$[X/\langle ab \rangle]$

The first row of table 2 describes the case of no quantum symmetry. In this case, all of the nontrivial quantum symmetries can be used to resolve an anomaly, and in each case, the resulting physical theory $[X/\Gamma]_B$ is equivalent to an orbifold which does not intersect an anomalous subgroup. Again, we see that the Wang-Wen-Witten prescription works.

For our next resolution of the anomalous $[X/G]$ orbifold, for the same G as before, we take $\Gamma = \mathbb{Z}_2 \times \mathbb{Z}_4$,

$$1 \longrightarrow \mathbb{Z}_2 \longrightarrow \mathbb{Z}_2 \times \mathbb{Z}_4 \longrightarrow \mathbb{Z}_2 \times \mathbb{Z}_2 \longrightarrow 1. \tag{67}$$

We list all possibilites for the quantum symmetry in table 3.

Table 3 A list of all possible quantum symmetries, anomalies resolved, and corresponding orbifolds $[X/\Gamma]_B$ for the case $\Gamma = \mathbb{Z}_2 \times \mathbb{Z}_4$. The first two columsn describe the quantum symmetry; the column $d_2(B)$ gives the image of the quantum symmetry, and hence the anomaly that can be resolved; and the last two columns give the physical theory equivalent to $[X/\Gamma]_B$, for either choice of discrete torsion in the Γ orbifold.

$B(a)$	$B(b)$	$d_2(B)$	$[X/\Gamma]_B$ w/o d.t.	$[X/\Gamma]_B$ with d.t.
1	1	–	$[X/G] \coprod [X/G]$	$[X/G]_{\text{d.t.}} \coprod [X/G]_{\text{d.t.}}$
-1	1	$\langle ab \rangle$	$[X/\langle b \rangle]$	$[X/\langle b \rangle]$
1	-1	$\langle b \rangle, \langle ab \rangle$	$[X/\langle a \rangle]$	$[X/\langle a \rangle]$
-1	-1	$\langle b \rangle$	$[X/\langle ab \rangle]$	$[X/\langle ab \rangle]$

The details here are different, but we see the same overall pattern. As before, the first row corresponds to the case of no quantum symmetry. In each of the next three rows, we see that it is possible to resolve an anomaly, and if one chooses B such that $d_2(B)$ describes the anomaly, then the resulting physical theory $[X/\Gamma]_B$ is equivalent to an orbifold by a subgroup which does not contain the anomalous subgroup. As before, Wang-Wen-Witten works.

So far we have picked 'minimal' resolutions. For our next resolution of the anomalous $[X/G]$ orbifold, for the same G as before, we take $\Gamma = \mathbb{Z}_2 \times \mathbb{H}$,

$$1 \longrightarrow \mathbb{Z}_2 \times \mathbb{Z}_2 \longrightarrow \mathbb{Z}_2 \times \mathbb{H} \longrightarrow \mathbb{Z}_2 \times \mathbb{Z}_2 \longrightarrow 1, \tag{68}$$

so that in effect there is an extra $\mathbb{Z}_2$. We list all possibilities for the quantum symmetry in table 4.

Table 4 A list of all possible quantum symmetries, anomalies resolved, and corresponding orbifolds $[X/\Gamma]_B$ for the case $\Gamma = \mathbb{Z}_2\mathbb{H}$. The first two columsn describe the quantum symmetry; the column $d_2(B)$ gives the image of the quantum symmetry, and hence the anomaly that can be resolved; and the last column gives the physical theory equivalent to $[X/\Gamma]_B$.

$B(a)$	$B(b)$	$d_2(B)$	$[X/\Gamma]_B$
1	1	$-$	$\coprod_2 ([X/G] \coprod [X/G]_{\text{d.t.}})$
-1	1	$\langle a \rangle, \langle ab \rangle$	$\coprod_2 [X/\langle b \rangle]$
1	-1	$\langle b \rangle, \langle ab \rangle$	$\coprod_2 [X/\langle a \rangle]$
-1	-1	$\langle a \rangle, \langle b \rangle$	$\coprod_2 [X/\langle ab \rangle]$

The details are different, but the pattern is the same: by picking B such that the anomaly is described by $d_2(B)$, the resulting physical theory is non-anomalous, as predicted by Wang-Wen-Witten.

8 Conclusions

In this article we have reviewed work on decomposition, the observation that sometimes one local quantum field theory is equivalent to a disjoint union of other local quantum field theories, known as universes. This arises when a d-dimensional quantum field theory has a $(d-1)$-form symmetry. After reviewing examples of decomposition and its properties (such as multiverse interference effects), we discussed the application to the anomaly-resolution procedure of Wang-Wen-Witten [62].

Acknowledgements We would like to thank the many collaborators we have worked with over the years in making sense of decomposition and its various applications. Two in particular stand out: Tony Pantev, for frequent collaborations on this and related subjects going back roughly twenty years now, and Simeon Hellerman, for early work on resolving technical challenges in making sense of strings on stacks that led to this work. We would also like to specifically thank A. Cherman, S. Hellerman, T. Jacobson, Y. Tanizaki, and M. Ünsal for discussions of locality and cluster decomposition and their role in these theories. We gratefully acknowledge NSF support, most recently via NSF grant PHY-2014086.

References

1. Abramovich, D., Graber, T., Vistoli, A.: Algebraic orbifold quantum products. In: Orbifolds in mathematics and physics (Madison, WI, 2001), *Contemp. Math.*, vol. 310, pp. 1–24. Amer. Math. Soc., Providence, RI (2002). DOI 10.1090/conm/310/05397. URL https://doi.org/10.1090/conm/310/05397
2. Aharony, O., Razamat, S.S., Seiberg, N., Willett, B.: The long flow to freedom. JHEP **02**, 056 (2017). DOI 10.1007/JHEP02(2017)056
3. Aharony, O., Razamat, S.S., Willett, B.: From 3d duality to 2d duality. JHEP **11**, 090 (2017). DOI 10.1007/JHEP11(2017)090
4. Anderson, L.B., Jia, B., Manion, R., Ovrut, B., Sharpe, E.: General aspects of heterotic string compactifications on stacks and gerbes. Adv. Theor. Math. Phys. **19**, 531–611 (2015). DOI 10.4310/ATMP.2015.v19.n3.a2
5. Andreini, E., Jiang, Y., Tseng, H.H.: On Gromov-Witten theory of root gerbes (2008)
6. Andreini, E., Jiang, Y., Tseng, H.H.: Gromov-Witten theory of root gerbes I: Structure of genus 0 moduli spaces. J. Differential Geom. **99**(1), 1–45 (2015). URL http://projecteuclid.org/euclid.jdg/1418345536
7. Andreini, E., Jiang, Y., Tseng, H.H.: Gromov-Witten theory of product stacks. Comm. Anal. Geom. **24**(2), 223–277 (2016). DOI 10.4310/CAG.2016.v24.n2.a1. URL https://doi.org/10.4310/CAG.2016.v24.n2.a1
8. Caldararu, A., Distler, J., Hellerman, S., Pantev, T., Sharpe, E.: Non-birational twisted derived equivalences in abelian GLSMs. Commun. Math. Phys. **294**, 605–645 (2010). DOI 10.1007/s00220-009-0974-2
9. Chen, W.m., Ruan, Y.b.: Orbifold quantum cohomology (2000)
10. Chen, W.m., Ruan, Y.b.: A New cohomology theory for orbifold. Commun. Math. Phys. **248**, 1–31 (2004). DOI 10.1007/s00220-004-1089-4
11. Chen, Z., Gu, W., Parsian, H., Sharpe, E.: Two-dimensional supersymmetric gauge theories with exceptional gauge groups. Adv. Theor. Math. Phys. **24**(1), 67–123 (2020). DOI 10.4310/ATMP.2020.v24.n1.a3
12. Chen, Z., Guo, J., Romo, M.: A GLSM view on Homological Projective Duality (2020)
13. Cherman, A., Jacobson, T.: Lifetimes of near eternal false vacua. Phys. Rev. D **103**(10), 105,012 (2021). DOI 10.1103/PhysRevD.103.105012
14. Cherman, A., Jacobson, T., Neuzil, M.: Universal deformations (2021)
15. Durhuus, B., Jonsson, T.: Classification and construction of unitary topological field theories in two-dimensions. J. Math. Phys. **35**, 5306–5313 (1994). DOI 10.1063/1.530752
16. Eager, R., Sharpe, E.: Elliptic genera of pure gauge theories in two dimensions with semisimple non-simply-connected gauge groups. Commun. Math. Phys. **387**(1), 267–297 (2021). DOI 10.1007/s00220-021-04189-6
17. Gholampour, A., Tseng, H.H.: On Donaldson-Thomas invariants of threefold stacks and gerbes. Proc. Amer. Math. Soc. **141**(1), 191–203 (2013). DOI 10.1090/S0002-9939-2012-11346-2. URL https://doi.org/10.1090/S0002-9939-2012-11346-2
18. Gu, W.: Vacuum structures revisited (2021)
19. Gu, W., Parsian, H., Sharpe, E.: More non-Abelian mirrors and some two-dimensional dualities. Int. J. Mod. Phys. A **34**(30), 1950,181 (2019). DOI 10.1142/S0217751X19501811
20. Gu, W., Pei, D., Zhang, M.: On phases of 3d N=2 Chern-Simons-matter theories. Nucl. Phys. B **973**, 115,604 (2021). DOI 10.1016/j.nuclphysb.2021.115604
21. Gu, W., Sharpe, E.: A proposal for nonabelian mirrors (2018)
22. Gu, W., Sharpe, E., Zou, H.: Notes on two-dimensional pure supersymmetric gauge theories. JHEP **04**, 261 (2021). DOI 10.1007/JHEP04(2021)261

23. Gukov, S., Witten, E.: Gauge Theory, Ramification, And The Geometric Langlands Program (2006)
24. Gukov, S., Witten, E.: Rigid Surface Operators. Adv. Theor. Math. Phys. **14**(1), 87–178 (2010). DOI 10.4310/ATMP.2010.v14.n1.a3
25. Guo, J., Romo, M.: Hybrid models for homological projective duals and noncommutative resolutions (2021)
26. Hellerman, S., Henriques, A., Pantev, T., Sharpe, E., Ando, M.: Cluster decomposition, T-duality, and gerby CFT's. Adv. Theor. Math. Phys. **11**(5), 751–818 (2007). DOI 10.4310/ATMP.2007.v11.n5.a2
27. Hellerman, S., Sharpe, E.: Sums over topological sectors and quantization of Fayet-Iliopoulos parameters. Adv. Theor. Math. Phys. **15**, 1141–1199 (2011). DOI 10.4310/ATMP.2011.v15.n4.a7
28. Hochschild, G.: Basic constructions in group extension theory. In: Contributions to algebra (collection of papers dedicated to Ellis Kolchin), pp. 183–201 (1977)
29. Honda, M., Itou, E., Kikuchi, Y., Tanizaki, Y.: Negative string tension of higher-charge Schwinger model via digital quantum simulation (2021)
30. Hori, K.: Duality In Two-Dimensional (2,2) Supersymmetric Non-Abelian Gauge Theories. JHEP **10**, 121 (2013). DOI 10.1007/JHEP10(2013)121
31. Hori, K., Vafa, C.: Mirror symmetry (2000)
32. Huang, T.C., Lin, Y.H., Seifnashri, S.: Construction of two-dimensional topological field theories with non-invertible symmetries. JHEP **12**, 028 (2021). DOI 10.1007/JHEP12(2021)028
33. Komargodski, Z., Ohmori, K., Roumpedakis, K., Seifnashri, S.: Symmetries and strings of adjoint QCD_2. JHEP **03**, 103 (2021). DOI 10.1007/JHEP03(2021)103
34. Kuznetsov, A.: Derived categories of quadric fibrations and intersections of quadrics. Adv. Math. **218**(5), 1340–1369 (2008). DOI 10.1016/j.aim.2008.03.007. URL https://doi.org/10.1016/j.aim.2008.03.007
35. Lang, S.: Algebra, *Graduate Texts in Mathematics*, vol. 211, third edn. Springer-Verlag, New York (2002). DOI 10.1007/978-1-4613-0041-0. URL https://doi.org/10.1007/978-1-4613-0041-0
36. Migdal, A.A.: Recursion Equations in Gauge Theories. Sov. Phys. JETP **42**, 413–418 (1975)
37. Moore, G.W., Segal, G.: D-branes and K-theory in 2D topological field theory (2006)
38. Nguyen, M., Tanizaki, Y., Ünsal, M.: Noninvertible 1-form symmetry and Casimir scaling in 2D Yang-Mills theory. Phys. Rev. D **104**(6), 065,003 (2021). DOI 10.1103/PhysRevD.104.065003
39. Nguyen, M., Tanizaki, Y., Ünsal, M.: Semi-Abelian gauge theories, non-invertible symmetries, and string tensions beyond N-ality. JHEP **03**, 238 (2021). DOI 10.1007/JHEP03(2021)238
40. Pantev, T., Sharpe, E.: Notes on gauging noneffective group actions (2005)
41. Pantev, T., Sharpe, E.: GLSM's for gerbes (and other toric stacks). Adv. Theor. Math. Phys. **10**(1), 77–121 (2006). DOI 10.4310/ATMP.2006.v10.n1.a4
42. Pantev, T., Sharpe, E.: String compactifications on Calabi-Yau stacks. Nucl. Phys. B **733**, 233–296 (2006). DOI 10.1016/j.nuclphysb.2005.10.035
43. Robbins, D., Sharpe, E., Vandermeulen, T.: A generalization of decomposition in orbifolds. JHEP **21**, 134 (2020). DOI 10.1007/JHEP10(2021)134
44. Robbins, D.G., Sharpe, E., Vandermeulen, T.: Anomalies, extensions, and orbifolds. Phys. Rev. D **104**(8), 085,009 (2021). DOI 10.1103/PhysRevD.104.085009
45. Robbins, D.G., Sharpe, E., Vandermeulen, T.: Anomaly resolution via decomposition. Int. J. Mod. Phys. A **36**(29), 2150,220 (2021). DOI 10.1142/S0217751X21502201
46. Robbins, D.G., Sharpe, E., Vandermeulen, T.: Quantum symmetries in orbifolds and decomposition. JHEP **02**, 108 (2022). DOI 10.1007/JHEP02(2022)108
47. Rusakov, B.E.: Loop averages and partition functions in U(N) gauge theory on two-dimensional manifolds. Mod. Phys. Lett. A **5**, 693–703 (1990). DOI 10.1142/S0217732390000780
48. Sharpe, E.: Derived categories and stacks in physics. Lect. Notes Phys. **757**, 249–272 (2009). DOI 10.1007/978-3-540-68030-7_8

49. Sharpe, E.: Landau-Ginzburg models, gerbes, and Kuznetsov's homological projective duality. Proc. Symp. Pure Math. **81**, 237–249 (2010). DOI 10.1090/pspum/081/2681766
50. Sharpe, E.: GLSM's, gerbes, and Kuznetsov's homological projective duality. J. Phys. Conf. Ser. **462**(1), 012,047 (2013). DOI 10.1088/1742-6596/462/1/012047
51. Sharpe, E.: Decomposition in diverse dimensions. Phys. Rev. D **90**(2), 025,030 (2014). DOI 10.1103/PhysRevD.90.025030
52. Sharpe, E.: Categorical equivalence and the renormalization group. Fortsch. Phys. **67**(8-9), 1910,019 (2019). DOI 10.1002/prop.201910019
53. Sharpe, E.: Undoing decomposition. Int. J. Mod. Phys. A **34**(35), 1950,233 (2020). DOI 10.1142/S0217751X19502336
54. Sharpe, E.: Topological operators, noninvertible symmetries and decomposition (2021)
55. Tachikawa, Y.: On the 6d origin of discrete additional data of 4d gauge theories. JHEP **05**, 020 (2014). DOI 10.1007/JHEP05(2014)020
56. Tachikawa, Y.: On gauging finite subgroups. SciPost Phys. **8**(1), 015 (2020). DOI 10.21468/SciPostPhys.8.1.015
57. Tang, X., Tseng, H.H.: Duality theorems for étale gerbes on orbifolds. Adv. Math. **250**, 496–569 (2014). DOI 10.1016/j.aim.2013.10.002. URL https://doi.org/10.1016/j.aim.2013.10.002
58. Tanizaki, Y., Ünsal, M.: Modified instanton sum in QCD and higher-groups. JHEP **03**, 123 (2020). DOI 10.1007/JHEP03(2020)123
59. Tseng, H.H.: On degree-0 elliptic orbifold Gromov-Witten invariants. Int. Math. Res. Not. IMRN (11), 2444–2468 (2011). DOI 10.1093/imrn/rnq158. URL https://doi.org/10.1093/imrn/rnq158
60. Vafa, C.: Modular Invariance and Discrete Torsion on Orbifolds. Nucl. Phys. B **273**, 592–606 (1986). DOI 10.1016/0550-3213(86)90379-2
61. Vafa, C., Witten, E.: On orbifolds with discrete torsion. J. Geom. Phys. **15**, 189–214 (1995). DOI 10.1016/0393-0440(94)00048-9
62. Wang, J., Wen, X.G., Witten, E.: Symmetric Gapped Interfaces of SPT and SET States: Systematic Constructions. Phys. Rev. X **8**(3), 031,048 (2018). DOI 10.1103/PhysRevX.8.031048
63. Weinberg, S.: The quantum theory of fields. Vol. 2: Modern applications. Cambridge University Press (2013)
64. Witten, E.: On quantum gauge theories in two-dimensions. Commun. Math. Phys. **141**, 153–209 (1991). DOI 10.1007/BF02100009
65. Witten, E.: D-branes and K theory. JHEP **12**, 019 (1998). DOI 10.1088/1126-6708/1998/12/019
66. Yu, M.: Gauging Categorical Symmetries in 3d Topological Orders and Bulk Reconstruction (2021)

Chapter 5

MATRIX Program: Structural Graph Theory Downunder II

Product Structure of Graph Classes with Bounded Treewidth

Rutger Campbell, Katie Clinch, Marc Distel, J. Pascal Gollin, Kevin Hendrey, Robert Hickingbotham, Tony Huynh, Freddie Illingworth, Youri Tamitegama, Jane Tan, and David R. Wood

Abstract We show that many graphs with bounded treewidth can be described as subgraphs of the strong product of a graph with smaller treewidth and a bounded-size complete graph. To this end, define the *underlying treewidth* of a graph class $\mathscr{G}$ to be the minimum non-negative integer c such that, for some function f, for every graph $G \in \mathscr{G}$ there is a graph H with $\text{tw}(H) \leqslant c$ such that G is isomorphic to a subgraph of $H \boxtimes K_{f(\text{tw}(G))}$. We introduce disjointed coverings of graphs and show they determine the underlying treewidth of any graph class. Using this result, we prove that the class of planar graphs has underlying treewidth 3; the class of $K_{s,t}$-minor-free graphs has underlying treewidth s (for $t \geqslant \max\{s,3\}$); and the class of K_t-minor-free graphs has underlying treewidth $t-2$. In general, we prove that a monotone class has bounded underlying treewidth if and only if it excludes some fixed topological minor. We also study the underlying treewidth of graph classes defined by an excluded subgraph or excluded induced subgraph. We show that the class of graphs with no H subgraph has bounded underlying treewidth if and only if every component of H is a subdivided star, and that the class of graphs with no in-

---

Rutger Campbell, J. Pascal Gollin, Kevin Hendrey
Institute for Basic Science, Daejeon, Republic of Korea,
e-mail: {rutger, pascalgollin, kevinhendrey}@ibs.re.kr

Katie Clinch
School of Computer Science and Engineering, UNSW, Sydney, Australia,
e-mail: k.clinch@unsw.edu.au

Marc Distel, Robert Hickingbotham, David Wood
Monash University, Melbourne, Australia,
e-mail: {marc.distel,robert.hickingbotham,david.wood}@monash.edu

Tony Huynh
Sapienza Università di Roma, Rome, Italy,
e-mail: huynh@di.uniroma1.it

Freddie Illingworth
University College London, London, United Kingdom,
e-mail: f.illingworth@ucl.ac.uk

Youri Tamitegama, Jane Tan
University of Oxford, Oxford, United Kingdom,
e-mail: {tamitegama, jane.tan}@maths.ox.ac.uk

D. R. Wood et al. (eds.), *2021-2022 MATRIX Annals*, MATRIX Book Series 5,
https://doi.org/10.1007/978-3-031-47417-0_9

duced H subgraph has bounded underlying treewidth if and only if every component of H is a star.

1 Introduction

Graph product structure theory describes complicated graphs as subgraphs of strong products[1] of simpler building blocks. The building blocks typically have bounded treewidth, which is the standard measure of how similar a graph is to a tree. Examples of graphs classes that can be described this way include planar graphs [14, 29], graphs of bounded Euler genus [10, 14], graphs excluding a fixed minor [14], and various non-minor-closed classes [15, 21]. These results have been the key to solving several open problems regarding queue layouts [14], nonrepetitive colouring [13], p-centered colouring [11], adjacency labelling [12, 18], twin-width [1, 5], and comparable box dimension [16].

This paper shows that graph product structure theory can even be used to describe graphs of bounded treewidth in terms of simpler graphs. Here the building blocks are graphs of smaller treewidth and complete graphs of bounded size. For example, a classical theorem by the referee of [8] can be interpreted as saying that every graph G of treewidth k and maximum degree Δ is contained[2] in $T \boxtimes K_{O(k\Delta)}$ for some tree T.

This result motivates the following definition. The *underlying treewidth* of a graph class $\mathcal{G}$ is the minimum $c \in \mathbb{N}_0$ such that, for some function f, for every graph $G \in \mathcal{G}$ there is a graph H with $\text{tw}(H) \leqslant c$ such that G is contained in $H \boxtimes K_{f(\text{tw}(G))}$. If there is no such c, then $\mathcal{G}$ has *unbounded* underlying treewidth. We call f the *treewidth-binding function*. For example, the above-mentioned result in [8] says that any graph class with bounded degree has underlying treewidth at most 1 with treewidth-binding function $O(k)$.

This paper introduces disjointed coverings of graphs and shows that they are intimately related to underlying treewidth; see Section 3. Indeed, we show that disjointed coverings characterise the underlying treewidth of any graph class (Theorem 7). The remainder of the paper uses disjointed coverings to determine the underlying treewidth of several graph classes of interest, with a small treewidth-binding function as a secondary goal. In this extended abstract, most proofs are omitted; see [6] for all the details.

---

[1] The *strong product* of graphs A and B, denoted by $A \boxtimes B$, is the graph with vertex-set $V(A) \times V(B)$, where distinct vertices $(v,x),(w,y) \in V(A) \times V(B)$ are adjacent if $v = w$ and $xy \in E(B)$, or $x = y$ and $vw \in E(A)$, or $vw \in E(A)$ and $xy \in E(B)$.

[2] A graph G is *contained* in a graph X if G is isomorphic to a subgraph of X.

2 Preliminaries

2.1 Basic Definitions

See [7] for graph-theoretic definitions not given here. We consider simple, finite, undirected graphs G with vertex-set $V(G)$ and edge-set $E(G)$. A *clique* in a graph is a set of pairwise adjacent vertices. Let $\omega(G)$ be the size of the largest clique in a graph G.

A *graph class* is a collection of graphs closed under isomorphism. A graph class is *monotone* if it is closed under taking subgraphs. A graph H is a *minor* of a graph G if H is isomorphic to a graph obtained from a subgraph of G by contracting edges. A graph G is *H-minor-free* if H is not a minor of G. A graph class $\mathcal{G}$ is *minor-closed* if every minor of each graph in $\mathcal{G}$ is also in $\mathcal{G}$.

Let $\mathbb{N} := \{1,2,\dots\}$ and $\mathbb{N}_0 := \{0,1,\dots\}$. All logarithms in this paper are binary.

2.2 Tree-Decompositions

For a tree T, a *T-decomposition* of a graph G is a collection $\mathcal{W} = (W_x : x \in V(T))$ of subsets of $V(G)$ indexed by the nodes of T such that (i) for every edge $vw \in E(G)$, there exists a node $x \in V(T)$ with $v, w \in W_x$; and (ii) for every vertex $v \in V(G)$, the set $\{x \in V(T) : v \in W_x\}$ induces a (connected) subtree of T. Each set W_x in $\mathcal{W}$ is called a *bag*. The *width* of $\mathcal{W}$ is $\max\{|W_x| : x \in V(T)\} - 1$. A *tree-decomposition* is a T-decomposition for any tree T. The *treewidth* $\text{tw}(G)$ of a graph G is the minimum width of a tree-decomposition of G. Treewidth is the standard measure of how similar a graph is to a tree. Indeed, a connected graph has treewidth 1 if and only if it is a tree. Treewidth is of fundamental importance in structural and algorithmic graph theory; see [2, 20, 27] for surveys.

We use the following well-known facts about treewidth. Every minor H of a graph G satisfies $\text{tw}(H) \leqslant \text{tw}(G)$. In every tree-decomposition of a graph G, each clique of G appears in some bag. Thus $\text{tw}(G) \geqslant \omega(G) - 1$ and $\text{tw}(K_n) = n - 1$. If $\{v_1, \dots, v_k\}$ is a clique in a graph G_1 and $\{w_1, \dots, w_k\}$ is a clique in a graph G_2, and G is the graph obtained from the disjoint union of G_1 and G_2 by identifying v_i and w_i for each $i \in \{1, \dots, k\}$, then $\text{tw}(G) = \max\{\text{tw}(G_1), \text{tw}(G_2)\}$. A greedy algorithm shows that $\chi(G) \leqslant \text{tw}(G) + 1$.

2.3 Partitions

To describe our main results in Section 1, it is convenient to use the language of graph products. However, to prove our results, it is convenient to work with the equivalent notion of graph partitions, which we now introduce.

For graphs G and H, an *H-partition* of G is a partition $(V_x: x \in V(H))$ of $V(G)$ indexed by the nodes of H, such that for every edge vw of G, if $v \in V_x$ and $w \in V_y$, then $x = y$ or $xy \in E(H)$. We say that H is the *quotient* of such a partition. The *width* of an H-partition is $\max\{|V_x|: x \in V(H)\}$. For $c \in \mathbb{N}_0$, an H-partition where $\text{tw}(H) \leqslant c$ is called a *c-tree-partition*. The *c-tree-partition-width* of a graph G, denoted $\text{tpw}_c(G)$, is the minimum width of a c-tree-partition of G.

It follows from the definitions that a graph G has an H-partition of width at most ℓ if and only if G is contained in $H \boxtimes K_\ell$. Thus, $\text{tpw}_c(G)$ equals the minimum $\ell \in \mathbb{N}_0$ such that G is contained in $H \boxtimes K_\ell$ for some graph H with $\text{tw}(H) \leqslant c$. Hence, the underlying treewidth of a graph class $\mathscr{G}$ equals the minimum $c \in \mathbb{N}_0$ such that, for some function f, every graph $G \in \mathscr{G}$ has c-tree-partition-width at most $f(\text{tw}(G))$. We henceforth use this as our working definition of underlying treewidth.

Before continuing, we review work on the $c = 1$ case. A *tree-partition* is a T-partition for some tree T. The *tree-partition-width* of G, denoted by $\text{tpw}(G)$, is the minimum width of a tree-partition of G. Thus $\text{tpw}(G) = \text{tpw}_1(G)$, which equals the minimum $\ell \in \mathbb{N}_0$ for which G is contained in $T \boxtimes K_\ell$ for some tree T. Tree-partitions were independently introduced by Seese [28] and Halin [19], and have since been widely investigated [3, 4, 8, 9, 17, 30, 31].

Bounded tree-partition-width implies bounded treewidth, as noted by Seese [28]. This fact easily generalises for c-tree-partition-width:

$$\text{tw}(G) \leqslant (c+1)\,\text{tpw}_c(G) - 1.$$

Of course, $\text{tw}(T) = \text{tpw}(T) = 1$ for every tree T. But in general, $\text{tpw}(G)$ can be much larger than $\text{tw}(G)$. For example, fan graphs on n vertices have treewidth 2 and tree-partition-width $\Omega(\sqrt{n})$; see Lemma 8 below. On the other hand, the referee of [8] showed that if the maximum degree and treewidth are both bounded, then so is the tree-partition-width, which is one of the most useful results about tree-partitions.

Lemma 1 ([8])**.** *For $k, \Delta \in \mathbb{N}$, every graph of treewidth less than k and maximum degree at most Δ has tree-partition-width at most $24k\Delta$.*

This bound is best possible up to the multiplicative constant [31]. Note that bounded maximum degree is not necessary for bounded tree-partition-width (for example, stars). Ding and Oporowki [9] characterised graph classes with bounded tree-partition-width in terms of excluded topological minors. We give an alternative characterisation, which says that graph classes with bounded tree-partition-width are exactly those that have bounded treewidth and satisfy a further 'disjointedness' condition. Furthermore, this result naturally generalises for c-tree-partition-width and thus for underlying treewidth.

3 Disjointed Coverings

This section introduces disjointed coverings and shows that they can be used to characterise bounded c-tree-partition-width and underlying treewidth.

An *ℓ-covering* of a graph G is a set $\beta \subseteq 2^{V(G)}$ such that $|B| \leqslant \ell$ for every $B \in \beta$, and $\bigcup\{B\colon B \in \beta\} = V(G)$.[3] If $B_1 \cap B_2 = \varnothing$ for all distinct $B_1, B_2 \in \beta$, then β is an *ℓ-partition*. As illustrated in Figure 1, an ℓ-covering β of a graph G is *(c,d)-disjointed* if for every c-tuple $(B_1,\dots,B_c) \in \beta^c$ and every component X of $G-(B_1 \cup \dots \cup B_c)$ there exists $Q \subseteq V(X)$ with $|Q| \leqslant d$ such that for each component Y of $X-Q$, for some $i \in \{1,\dots,c\}$ we have $V(Y) \cap N_G(B'_i) = \varnothing$, where $B'_i := B_i \setminus (B_1 \cup \dots \cup B_{i-1})$. Note that we can take $Q = \varnothing$ if some $B'_i = \varnothing$, since $N_G(\varnothing) = \varnothing$.

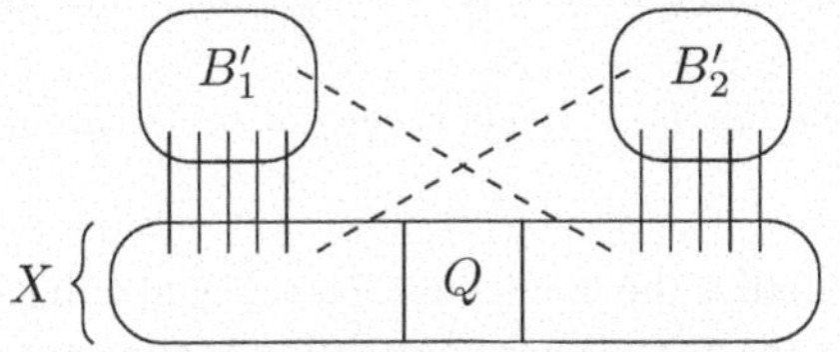

Fig. 1 A disjointed partition with $c = 2$, where non-edges are dashed.

Given positive integers ℓ and t and an ℓ-covering β of a graph G, we define $\beta[t] := \{\bigcup\mathscr{B}\colon \mathscr{B} \subseteq \beta, |\mathscr{B}| \leqslant t\}$. So $\beta[t]$ is a $t\ell$-covering of G. For a function $f\colon \mathbb{N} \to \mathbb{R}^+$ we say that β is *(c,f)-disjointed* if $\beta[t]$ is $(c,f(t))$-disjointed for every $t \in \mathbb{N}$.

While (c,d)-disjointed coverings are conceptually simpler than (c,f)-disjointed coverings, we show they are roughly equivalent (Theorem 3). Moreover, (c,f)-disjointed coverings are essential for the main proof (Lemma 6) and give better bounds on the c-tree-partition-width, leading to smaller treewidth-binding functions when determining the underlying treewidth of several graph classes of interest (for K_t-minor-free graphs for example).

Note that we often consider the singleton partition $\beta := \{\{v\}\colon v \in V(G)\}$ of a graph G, which is (c,f)-disjointed if and only if, for every $t \in \mathbb{N}$, every t-partition of G is $(c,f(t))$-disjointed.

This section characterises c-tree-partition-width in terms of (c,d)-disjointed coverings (or partitions) and (c,f)-disjointed coverings (or partitions). The following observation deals with the $c = 0$ case.

Observation 2. *The following are equivalent for any graph G and $d \in \mathbb{N}$:*

(a) *G has a $(0,d)$-disjointed covering;*
(b) *every covering of G is $(0,d)$-disjointed;*
(c) *each component of G has at most d vertices;*
(d) *G has 0-tree-partition-width at most d.*

Observation 2 implies that a graph class $\mathscr{G}$ has underlying treewidth 0 if and only if there is a function f such that every component of every graph $G \in \mathscr{G}$ has at most $f(\mathrm{tw}(G))$ vertices.

[3] Our definition of ℓ-covering differs from the standard usage where it refers to a covering in which each element of the ground set is covered ℓ times.

We prove the following characterisation of bounded c-tree-partition-width (which is new even in the $c = 1$ case).

Theorem 3. *For fixed $c \in \mathbb{N}_0$, the following are equivalent for a graph class $\mathcal{G}$ with bounded treewidth:*

(a) *$\mathcal{G}$ has bounded c-tree-partition-width;*
(b) *for some $d, \ell \in \mathbb{N}$, every graph in $\mathcal{G}$ has a (c,d)-disjointed ℓ-partition;*
(c) *for some $d, \ell \in \mathbb{N}$, every graph in $\mathcal{G}$ has a (c,d)-disjointed ℓ-covering;*
(d) *for some $\ell \in \mathbb{N}$ and function f, every graph in $\mathcal{G}$ has a (c,f)-disjointed ℓ-partition;*
(e) *for some $\ell \in \mathbb{N}$ and function f, every graph in $\mathcal{G}$ has a (c,f)-disjointed ℓ-covering.*

Proof. Observation 2 handles the $c = 0$ case. Now assume that $c \geqslant 1$. Lemma 5 below says that (a) implies (b). Since every ℓ-partition is an ℓ-covering, (b) implies (c), and (d) implies (e). Lemma 4 below says that (c) implies (d). Finally, Lemma 6 below says that (e) implies (a). □

By definition, every (c,f)-disjointed ℓ-covering is $(c,f(1))$-disjointed. The next lemma gives a qualitative converse to this.

Lemma 4. *Let $\ell, c, d \in \mathbb{N}$, and let β be a (c,d)-disjointed ℓ-covering of a graph G. Then β is (c,f)-disjointed, where $f(t) := dt^c$ for each $t \in \mathbb{N}$.*

Proof. Fix $t \in \mathbb{N}$. Let $B_1, \dots, B_c \in \beta[t]$. Let X be a component of $G - (B_1 \cup \dots \cup B_c)$. For each $i \in \{1, \dots, c\}$, let $\mathcal{B}_i$ be a set of at most t elements of β whose union is B_i. Let $\mathcal{F} := \mathcal{B}_1 \times \dots \times \mathcal{B}_c$, and for each $y = (A_1, \dots, A_c) \in \mathcal{F}$, define Q_y as follows. Let X_y the component of $G - (A_1 \cup \dots \cup A_c)$ containing X. Since β is (c,d)-disjointed, there exists $Q_y \subseteq V(X_y)$ of size at most d such that for every component Y of $X_y - Q_y$ there is some $i \in \{1, \dots, c\}$ such that $V(Y) \cap N_G(A_i \setminus (A_1 \cup \dots \cup A_{i-1}))$ is empty. Now let $Q := \bigcup_{y \in \mathcal{F}} Q_y$, and note that $|Q| \leqslant d|\mathcal{F}| \leqslant dt^c$.

Suppose for contradiction that there exists a component Y of $X - Q$ such that for all $i \in \{1, \dots, c\}$, there is a vertex $b_i \in N_G(Y) \cap B'_i$, where $B'_i := B_i \setminus (B_1 \cup \dots \cup B_{i-1})$. Let $y = (A_1, \dots, A_c) \in \mathcal{F}$ be such that $(b_1, \dots, b_c) \in A_1 \times \dots \times A_c$, and consider that component Y' of $X_y - Q_y$ containing Y. By the definition of Q_y, there is some $i \in \{1, \dots, c\}$ such that Y' contains no neighbour of a vertex in $A_i \setminus (A_1 \cup \dots \cup A_{i-1})$. In particular, all neighbours of vertices of Y are either vertices of Y' or neighbours of vertices of Y', so b_i is not a neighbour of any vertex of Y, a contradiction. □

Having a (c,d)-disjointed partition is necessary for bounded c-tree-partition-width.

Lemma 5. *For all $c, \ell \in \mathbb{N}_0$, every graph G with c-tree-partition-width ℓ has a $(c, c\ell)$-disjointed ℓ-partition.*

Proof. By assumption, G has an H-partition $\beta = (V_h : h \in V(H))$ where H is a graph of treewidth at most c and $|V_h| \leqslant \ell$ for all h. We first show that the singleton partition of H is (c,c)-disjointed. Let $v_1, \dots, v_c \in V(H)$ and let X be a component of $H - \{v_1, \dots, v_c\}$. Let $(W_x : x \in V(T))$ be a tree-decomposition of H where $|W_x| \leqslant c+1$ for all $x \in V(T)$. We may assume that $W_x \neq W_y$ whenever $x \neq y$. For each $i \in \{1, \dots, c\}$, let T_i be the subtree of T induced by $\{x \in V(T) : v_i \in W_x\}$.

First suppose that $V(T_i) \cap V(T_j) = \varnothing$ for some $i, j \in \{1, \dots, c\}$. Let $z \in V(T_i)$ be the closest node (in T) to T_j. Let $Q := W_z \cap X$. Note that $Q \subseteq W_z \setminus \{v_i\}$ so $|Q| \leqslant c$. Any path from v_i to v_j in H passes through W_z, so each component of $X - Q$ is disjoint from $N_H(v_i)$ or $N_H(v_j)$.

Now assume that $V(T_i) \cap V(T_j) \neq \varnothing$ for all $i, j \in \{1, \dots, c\}$. Let T_X be the subgraph of T induced by $\{x \in V(T) : V(X) \cap W_x \neq \varnothing\}$. Since X is connected, T_X is a subtree of T. Suppose that $V(T_i) \cap V(T_X) = \varnothing$ for some i. Since every neighbour of v_i in H is in $\bigcup(W_x : x \in V(T_i))$, it follows that $N_H(v_i) \cap V(X) = \varnothing$ and so we may take $Q := \varnothing$ in this case. Now assume that $V(T_i) \cap V(T_X) \neq \varnothing$ for all $i \in \{1, \dots, c\}$. By the Helly property, $\tilde{T} := T_1 \cap \dots \cap T_c \cap T_X$ is a non-empty subtree of T. For each $x \in V(\tilde{T})$, we have $|W_x| \leqslant c+1$ and so $W_x = \{v_1, \dots, v_c, u\}$ for some $u \in V(X)$. First assume $|V(\tilde{T})| \geqslant 2$. Then there are adjacent $x, y \in V(\tilde{T})$ with $W_x = \{v_1, \dots, v_c, u\}$ and $W_y = \{v_1, \dots, v_c, v\}$ for $u, v \in V(X)$. Since $W_x \neq W_y$, we have $u \neq v$ and thus there is no (u,v)-path in $H - \{v_1, \dots, v_c\}$, contradicting the connectedness of X. Hence $\tilde{T}$ consists of a single vertex z; thus $W_z = \{v_1, \dots, v_c, u\}$ for some $u \in V(X)$. Let $Q := \{u\}$ and consider a component Y of $X - Q$. Let T_Y be the subtree of T induced by $\{y \in V(T) : V(Y) \cap W_y \neq \varnothing\}$. Since T_Y is connected and does not contain z, it is disjoint from some T_i. As above, $N_H(v_i) \cap V(Y) = \varnothing$, as required.

We have shown that the singleton partition of H is (c,c)-disjointed. Now focus on G. By assumption, β is an ℓ-partition of G. Let $V_{v_1}, \dots, V_{v_c}$ be parts in β, and let X be a component of $G - (V_{v_1} \cup \dots \cup V_{v_c})$. Then $X \subseteq \bigcup\{V_h : h \in X'\}$ where X' is a component of $H - \{v_1, \dots, v_c\}$. Since H is (c,c)-disjointed, there exists $Q' \subseteq V(X')$ of size at most c such that each component $X' - Q'$ is disjoint from some $N_H(v_i)$. Let $Q := \bigcup\{V_h : h \in Q'\}$, which has size at most $c\ell$. Each component of $X - Q$ is disjoint from some $N_G(V_{v_i})$. □

The next lemma lies at the heart of the paper.

Lemma 6. *Let $k, c, \ell \in \mathbb{N}$ and $f : \mathbb{N} \to \mathbb{R}^+$. For any graph G, if* $\text{tw}(G) < k$ *and G has a (c,f)-disjointed ℓ-covering, then G has c-tree-partition-width at most* $\max\{12\ell k, 2c\ell f(12k)\}$.

Lemmas 4 and 6 imply the following result:

Corollary 1. *Let $k, c, d, \ell \in \mathbb{N}$. For any graph G, if* $\text{tw}(G) < k$ *and G has a (c,d)-disjointed ℓ-covering, then G has c-tree-partition-width at most* $2cd\ell(12k)^c$.

Observe that the singleton partition of any graph with maximum degree Δ is $(1,\Delta)$-disjointed. So Corollary 1 with $c = \ell = 1$ and $d = \Delta$ implies Lemma 1 (even with the same constant 24). Indeed, the proof of Lemma 6 in the case of graphs with bounded degree is equivalent to the proof of Lemma 1.

To conclude this section, Lemmas 5 and 6 imply the following characterisation of underlying treewidth.

Theorem 7. *The underlying treewidth of a graph class $\mathscr{G}$ is equal to the minimum $c \in \mathbb{N}_0$ such that, for some function $g\colon \mathbb{N} \to \mathbb{N}$, every graph $G \in \mathscr{G}$ has a $(c, g(\text{tw}(G)))$-disjointed $g(\text{tw}(G))$-partition.*

4 Lower Bounds

We now define two graphs that provide lower bounds on the underlying treewidth of various graph classes. For a graph G and $\ell \in \mathbb{N}$, let ℓG be the union of ℓ vertex-disjoint copies of G. Let $\widehat{G}$ be the graph obtained from G by adding one dominant vertex. Observe that $\text{tw}(\widehat{G}) = \text{tw}(G) + 1$ and $\text{tw}(\ell G) = \text{tw}(G)$ for any $\ell \in \mathbb{N}$, implying $\text{tw}(\widehat{\ell G}) = \text{tw}(G) + 1$. For $c, \ell \in \mathbb{N}$, define graphs $G_{c,\ell}$ and $C_{c,\ell}$ recursively as follows. First, $G_{1,\ell} := P_{\ell+1}$ is the path on $\ell + 1$ vertices, and $C_{1,\ell} := K_{1,\ell}$ is the star with ℓ leaves. Further, for $c \geqslant 2$, let $G_{c,\ell} := \widehat{\ell G_{c-1,\ell}}$ and $C_{c,\ell} := \widehat{\ell C_{c-1,\ell}}$.

The next lemma collects together some useful and well-known properties of $G_{c,\ell}$ and $C_{c,\ell}$.

Lemma 8. *For all $c, \ell \in \mathbb{N}$,*

(i) $\text{tw}(G_{c,\ell}) = \text{tw}(C_{c,\ell}) = c$;
(ii) *for any ℓ-partition of $G \in \{G_{c,\ell}, C_{c,\ell}\}$, there is a $(c+1)$-clique in G whose vertices are in distinct parts;*
(iii) *$G_{c,\ell}$ and $C_{c,\ell}$ both have $(c-1)$-tree-partition-width greater than ℓ;*
(iv) *$G_{2,\ell}$ is outerplanar, and $G_{3,\ell}$ is planar;*
(v) *$G_{c,\ell}$ is $K_{c,\max\{c,3\}}$-minor-free;*
(vi) *$C_{c,\ell}$ does not contain P_4 as an induced subgraph;*
(vii) *$C_{c,\ell}$ does not contain P_n as a subgraph for $n \geqslant 2^{c+1}$.*

Proof. Since $\text{tw}(\widehat{\ell G}) = \text{tw}(G) + 1$ for any graph G and $\ell \in \mathbb{N}$, part (i) follows by induction.

We establish (ii) by induction on c. In the case $c = 1$, every ℓ-partition of $P_{\ell+1}$ or $K_{1,\ell}$ contains an edge whose endpoints are in different parts, and we are done. Now assume the claim for $c - 1$ ($c \geqslant 2$) and let $G \in \{G_{c-1,\ell}, C_{c-1,\ell}\}$. Consider an ℓ-partition of $\widehat{\ell G}$. At most $\ell - 1$ copies of G contain a vertex in the same part as the dominant vertex v. Thus, some copy G_0 of G contains no vertices in the same part as v. By induction, G_0 contains a c-clique K whose vertices are in distinct parts. Since v is dominant, $K \cup \{v\}$ satisfies the induction hypothesis.

Let $G \in \{G_{c,\ell}, C_{c,\ell}\}$. Consider an H-partition of G of width at most ℓ. By (ii), G contains a $(c+1)$-clique whose vertices are in distinct parts. So $\omega(H) \geqslant c + 1$, implying $\text{tw}(H) \geqslant c$. This establishes (iii).

Observe that $G_{2,\ell}$ is outerplanar (called a *fan* graph). The disjoint union of outerplanar graphs is outerplanar and the graph obtained from any outerplanar graph by adding a dominant vertex is planar; thus $G_{3,\ell}$ is planar.

We next show that $G_{c,\ell}$ is $K_{c,\max\{c,3\}}$-minor-free. $G_{1,\ell}$ is a path and so has no $K_{1,3}$-minor. $G_{2,\ell}$ is outerplanar and so has no $K_{2,3}$-minor. Let $c \geqslant 3$ and assume the result holds for smaller c. Suppose that $G_{c,\ell}$ contains a $K_{c,c}$-minor. Since $K_{c,c}$ is 2-connected, some copy of $G_{c-1,\ell}$ in $G_{c,\ell}$ contains a $K_{c-1,c}$-minor. This contradiction establishes (v).

We show that $C_{c,\ell}$ contains no induced P_4 by induction on c. First, $C_{1,\ell} = K_{1,\ell}$ does not contain P_4. Next, suppose that $C_{c-1,\ell}$ does not contain an induced P_4. P_4 does not have a dominant vertex and so any induced P_4 in $C_{c,\ell}$ must lie entirely within one copy of $C_{c-1,\ell}$. In particular, $C_{c,\ell}$ does not contain an induced P_4. This proves (vi).

Finally, Nešetřil and Ossona de Mendez [26] proved (vii). □

The underlying treewidth of the class of graphs of treewidth at most k is obviously at most k. Lemma 8 (i) and (iii) imply the following.

Corollary 2. *The underlying treewidth of the class of graphs of treewidth at most k equals k.*

Corollary 3. *The classes $\{G_{c,\ell}\colon c,\ell \in \mathbb{N}\}$ and $\{C_{c,\ell}\colon c,\ell \in \mathbb{N}\}$ both have unbounded underlying treewidth.*

Proof. Suppose that $\{G_{c,\ell}\colon c,\ell \in \mathbb{N}\}$ has underlying treewidth b. Thus, for some function f, for all $c,\ell \in \mathbb{N}$, we have $\text{tpw}_b(G_{c,\ell}) \leqslant f(\text{tw}(G_{c,\ell})) = f(c)$. In particular, with $c := b+1$ and $\ell := f(c)$, we have $\text{tpw}_{c-1}(G_{c,\ell}) \leqslant \ell$, which contradicts Lemma 8 (iii). The proof for $\{C_{c,\ell}\colon c,\ell \in \mathbb{N}\}$ is analogous. □

5 Excluding a Minor

This section uses disjointed partitions to determine the underlying treewidth of several minor-closed classes of interest.

The next definition enables K_t-minor-free graphs and $K_{s,t}$-minor-free graphs to be handled simultaneously. For $s,t \in \mathbb{N}$, let $\mathscr{K}_{s,t}$ be the class of graphs G for which there is a partition $\{A,B\}$ of $V(G)$ such that $|A| = s$ and $|B| = t$; $vw \in E(G)$ for all $v \in A$ and $w \in B$; and $G[B]$ is connected. Obviously, every graph in $\mathscr{K}_{s,t}$ contains $K_{s,t}$. Similarly, we obtain K_t as a minor of any $G \in \mathscr{K}_{t-2,t}$ by contracting a matching between A and B of size $t-2$ whose end-vertices are distinct from the end-vertices of some edge of $G[B]$.

The next lemma is proved by a well-known technique [23, 24].

Lemma 9. *Let G be a graph with no minor in $\mathscr{K}_{s,t}$. Assume $\{A,B\}$ is a partition of $V(G)$ such that $G[B]$ is connected and every vertex in B has at least s neighbours in A. Then $|B| \leqslant \delta|A|$ for some $\delta = \delta(s,t)$.*

The following well-known Erdős–Pósa type result is useful for showing disjointedness.

Lemma 10. *Let $\mathscr{H}$ be a set of connected subgraphs of a graph G. Then, for every non-negative integer ℓ, either there are $\ell+1$ vertex-disjoint graphs in $\mathscr{H}$ or there is a set $Q \subseteq V(G)$ of size at most $\ell(\text{tw}(G)+1)$ such that $G-Q$ contains no graph of $\mathscr{H}$.*

Lemma 11. *For fixed $s,t \in \mathbb{N}$, every graph G with no minor in $\mathscr{K}_{s,t}$ and of treewidth k has s-tree-partition-width $O(k^2)$.*

Proof. By Lemma 6 it suffices to show that the singleton partition of G is (s,f)-disjointed, where $f(n) := \delta sn(k+1)$ and $\delta := \delta(s,t)$ from Lemma 9. Let $S_1,\dots,S_s$ be subsets of $V(G)$ of size at most n, let $S := S_1 \cup \dots \cup S_s$, and for each $i \in \{1,\dots,s\}$ let $S_i' := S_i \setminus (S_1 \cup \dots \cup S_{i-1})$. Let X be a connected component of $G-S$. Let $\mathscr{H}$ be the set of connected subgraphs H of X such that $H \cap N(S_i') \neq \varnothing$ for all $i \in \{1,\dots,s\}$. Say $\mathscr{R}$ is a maximum-sized set of pairwise disjoint subgraphs in $\mathscr{H}$. We may assume that $\bigcup\{V(R)\colon R \in \mathscr{R}\} = V(X)$. Let X' be the graph obtained from $G[S \cup V(X)]$ by contracting each subgraph $R \in \mathscr{R}$ into a vertex v_R. So $V(X') = S \cup \{v_R\colon R \in \mathscr{R}\}$. Since X is connected, $\{v_R\colon R \in \mathscr{R}\}$ induces a connected subgraph of X'. By construction, in X', each vertex v_R has at least s neighbours in S. By Lemma 9, $|\mathscr{R}| \leqslant \delta|S|$. By Lemma 10, there is a set $Q \subseteq V(X)$ of size at most $\delta|S|(k+1) \leqslant f(n)$ such that $X-Q$ contains no graph in $\mathscr{H}$. Thus each component Y of $X-Q$ satisfies $V(Y) \cap N_G(S_i') = \varnothing$ for some $i \in \{1,\dots,s\}$. Hence, the singleton partition of G is (s,f)-disjointed. □

We now determine the underlying treewidth of K_t- and $K_{s,t}$-minor-free graphs.

Theorem 12. *For fixed $t \in \mathbb{N}$ with $t \geqslant 2$, the underlying treewidth of the class of K_t-minor-free graphs equals $t-2$. In particular, every K_t-minor-free graph of treewidth k has $(t-2)$-tree-partition-width $O(k^2)$.*

Proof. Since K_t is a minor of every graph in $\mathscr{K}_{t-2,t}$, Lemma 11 implies that every K_t-minor-free graph of treewidth k has $(t-2)$-tree-partition-width $O(k^2)$. Thus the underlying treewidth of the class of K_t-minor-free graphs is at most $t-2$. Suppose for contradiction that equality does not hold. That is, for some function f, every K_t-minor-free graph G has $(t-3)$-tree-partition-width at most $f(\text{tw}(G))$. Let $\ell := f(t-2)$. The graph $G_{t-2,\ell}$ in Lemma 8 has treewidth $t-2$ and is thus K_t-minor-free. However, by Lemma 8, $\text{tpw}_{t-3}(G_{t-2,\ell}) > \ell = f(t-2) = f(\text{tw}(G_{t-2,\ell}))$, which is the required contradiction. □

A similar proof shows:

Theorem 13. *For fixed $s,t \in \mathbb{N}$ with $t \geqslant \max\{s,3\}$, the underlying treewidth of the class of $K_{s,t}$-minor-free graphs equals s. For $s,t \in \mathbb{N}$ with $s \leqslant t \leqslant 2$, the underlying treewidth of the class of $K_{s,t}$-minor-free graphs equals $s-1$.*

The class of planar graphs is minor-closed. Since planar graphs are K_5- and $K_{3,3}$-minor-free, Theorem 12 or Theorem 13 imply the next result (where the lower bound holds since the graph $G_{3,\ell}$ in Lemma 8 is planar).

Theorem 14. *The underlying treewidth of the class of planar graphs equals* 3. *In particular, every planar graph of treewidth k has* 3*-tree-partition-width* $O(k^2)$.

The class of graphs embeddable on a given surface (that is, a closed compact 2-manifold) is minor-closed. The *Euler genus* of a surface with h handles and c cross-caps is $2h+c$. The *Euler genus* of a graph G is the minimum $g \in \mathbb{N}_0$ such that there is an embedding of G in a surface of Euler genus g; see [25] for more about graph embeddings in surfaces.

It follows from Euler's formula that every graph with Euler genus at most g is $K_{3,2g+3}$-minor-free. Thus Lemma 8 and Theorem 13 imply the following.

Theorem 15. *The underlying treewidth of the class of graphs embeddable on any fixed surface* Σ *equals* 3. *In particular, every graph embeddable in* Σ *and of treewidth k has* 3*-tree-partition-width* $O(k^2)$.

6 Excluding a Topological Minor

A graph $\tilde{G}$ is a *subdivision* of a graph G if $\tilde{G}$ can be obtained from G by replacing each edge vw by a path P_{vw} with endpoints v and w (internally disjoint from the rest of $\tilde{G}$). If each P_{vw} has t internal vertices, then $\tilde{G}$ is the *t-subdivision* of G. If each P_{vw} has at most t internal vertices, then $\tilde{G}$ is a *$(\leqslant t)$-subdivision* of G. A graph H is a *topological minor* of G if a subgraph of G is isomorphic to a subdivision of H. A graph G is *H-topological-minor-free* if H is not a topological minor of G. Using disjointed partitions, we prove:

Theorem 16. *For every fixed multigraph X with p vertices, every X-topological minor-free graph G of treewidth k has p-tree-partition-width* $O(k^2)$.

Theorem 16 implies the upper bound in the next result, which we show is tight.

Theorem 17. *The underlying treewidth of the class of K_t-topological-minor-free graphs equals $t-2$ if $t \in \{2,3,4\}$ and equals t if $t \geqslant 5$.*

Determining the underlying treewidth of the class of $K_{s,t}$-topological-minor-free graphs is an interesting open problem (for $s \geqslant 4$).

We show that c-tree-partition-width is well-behaved (in a certain sense) under subdivisions. The next theorem follows.

Theorem 18. *A monotone graph class $\mathcal{G}$ has bounded underlying treewidth if and only if $\mathcal{G}$ excludes some fixed topological minor.*

7 Excluding a Subgraph or Induced Subgraph

For a graph H, a graph G is *H-free* if G contains no subgraph isomorphic to H. For a finite set of graphs $\mathcal{H}$, we say that G is *$\mathcal{H}$-free* if G is H-free for all $H \in \mathcal{H}$. Let $\mathcal{G}_H$

be the class of H-free graphs and let $\mathcal{G}_{\mathcal{H}}$ be the class of $\mathcal{H}$-free graphs. The next result characterise when $\mathcal{G}_{\mathcal{H}}$ has bounded underlying treewidth, and determines the exact underlying treewidth for several natural classes. The proof is based on disjointed coverings.

Let P_n be the n-vertex path. A *spider* is a subdivision of a star and a *spider-forest* is a subdivision of a star-forest. For $s,t \in \mathbb{N}$ with $s \geqslant 2$, the *(s,t)-spider*, denoted $S_{s,t}$, is the $(t-1)$-subdivision of $K_{1,s}$. If v is the centre of $S_{s,t}$, then each component of $S_{s,t} - v$ is called a *leg*.

Theorem 19. *For all $\ell,n,s,t \in \mathbb{N}$ where $n,s \geqslant 3$ and $\ell \geqslant 2$, and for every finite set $\mathcal{H}$ of graphs,*

(i) *$\mathcal{G}_{\mathcal{H}}$ has bounded underlying treewidth if and only if $\mathcal{H}$ contains a spider-forest;*
(ii) *the underlying treewidth of $\mathcal{G}_{P_n}$ equals $\lfloor \log n \rfloor - 1$;*
(iii) *the underlying treewidth of $\mathcal{G}_{\ell P_n}$ equals $\lfloor \log n \rfloor$;*
(iv) *the underlying treewidth of $\mathcal{G}_{S_{s,t}}$ equals $\lfloor \log t \rfloor + 1$;*
(v) *the underlying treewidth of $\mathcal{G}_{\ell S_{s,t}}$ equals $\lfloor \log t \rfloor + 2$.*

For a graph H, let $\mathcal{I}_H$ be the class of graphs with no induced subgraph isomorphic to H. We characterise the graphs H such that $\mathcal{I}_H$ has bounded underlying treewidth, and determines the precise underlying treewidth for each such H.

Theorem 20. *For any graph H,*

(i) *$\mathcal{I}_H$ has bounded underlying treewidth if and only if H is a star-forest;*
(ii) *if H is a star-forest, then $\mathcal{I}_H$ has underlying treewidth at most 2;*
(iii) *$\mathcal{I}_H$ has underlying treewidth at most 1 if and only if H is a star or each component of H is a path on at most three vertices;*
(iv) *$\mathcal{I}_H$ has underlying treewidth 0 if and only if H is a path on at most three vertices, or $E(H) = \varnothing$.*

8 Graph Drawings

A graph is *k-planar* if it has a drawing in the plane with at most k crossings on each edge, where we assume that no three edges cross at the same point. Of course, the class of 0-planar graphs is the class of planar graphs, which has underlying treewidth 3 (Theorem 14). However, 1-planar graphs behave very differently. It is well-known that every graph has a 1-planar subdivision: take an arbitrary drawing of G and for each edge e add a subdivision vertex between consecutive crossings on e. Since the class of 1-planar graphs is monotone, Theorem 18 implies that the class of 1-planar graphs has unbounded underlying treewidth.

By restricting the type of drawing, we obtain positive results. A *circular drawing* of a graph G positions each vertex on a circle in the plane, and draws each edge as an arc across the circle, such that no two edges cross more than once. A graph is *outer k-planar* if it has a circular drawing such that each edge is involved in at most k

crossings. The outer 0-planar graphs are precisely the outerplanar graphs, which have treewidth 2. We show below that for each $k \in \mathbb{N}$, the class of outer k-planar graphs has underlying treewidth 2. In fact, we prove a slightly more general result. A graph is *weakly outer k-planar* if it has a circular drawing such that whenever two edges e and f cross, e or f crosses at most k edges. Clearly every outer k-planar graph is weakly outer k-planar.

Theorem 21. *Each weakly outer k-planar graph has* 2*-tree-partition-width* $O(k^3)$.

Theorem 21 implies the next result, where the lower bound holds since $G_{2,\ell}$ from Lemma 8 is outerplanar.

Theorem 22. *For every fixed* $k \in \mathbb{N}$*, the underlying treewidth of the class of weakly outer k-planar graphs equals* 2*, with constant treewidth-binding function.*

9 Universal Graphs

A graph U is *universal* for a graph class $\mathcal{G}$ if $U \in \mathcal{G}$ and U contains every graph in $\mathcal{G}$. This definition is only interesting when considering infinite graphs. For each $k \in \mathbb{N}$ there is a universal graph $\mathcal{T}_k$ for the class of countable graphs of treewidth k. Huynh, Mohar, Šámal, Thomassen, and Wood [22] gave an explicit construction for $\mathcal{T}_k$, and showed how product structure theorems for finite graphs lead to universal graphs. Their results imply that for any hereditary class $\mathcal{G}$ of countable graphs, if the class of finite graphs in $\mathcal{G}$ has underlying treewidth c with treewidth-binding function f, then every graph in $\mathcal{G}$ of treewidth at most k is contained in $\mathcal{T}_c \boxtimes K_{f(k)}$. This result is applicable to all minor-closed classes, monotone classes, and hereditary classes. For example, every countable K_t-minor free graph of treewidth k is contained in $\mathcal{T}_{t-2} \boxtimes K_{O(k^2)}$.

Acknowledgements This research was initiated at the *Structural Graph Theory Downunder II* workshop at the Mathematical Research Institute MATRIX (March 2022).

Research of Rutger Campbell and Kevin Hendrey supported by the Institute for Basic Science (IBS-R029-C1). Research of J. Pascal Gollin supported by the Institute for Basic Science (IBS-R029-Y3). Research of Marc Distel and Robert Hickingbotham supported by an Australian Government Research Training Program Scholarship. Research of Tony Huynh supported by the Australian Research Council. Research of Freddie Illingworth supported by EPSRC grant EP/V007327/1. Research of David Wood supported by the Australian Research Council and a Visiting Research Fellowship of Merton College, University of Oxford.

References

1. Bekos, M.A., Da Lozzo, G., Hliněný, P., Kaufmann, M.: Graph product structure for h-framed graphs (2022) *Proc. 33rd Int'l Symp. on Algorithms and Computation (ISAAC '22)*, Vol. 248 of LIPIcs (S. W. Bae and H. Park, eds), Schloss Dagstuhl, pp. 23:1–23:15. arXiv:2204.11495.

2. Bodlaender, H.L.: A partial *k*-arboretum of graphs with bounded treewidth. Theoret. Comput. Sci. **209**(1-2), 1–45 (1998). DOI 10.1016/S0304-3975(97)00228-4
3. Bodlaender, H.L.: A note on domino treewidth. Discrete Math. Theor. Comput. Sci. **3**(4), 141–150 (1999). URL https://dmtcs.episciences.org/256
4. Bodlaender, H.L., Engelfriet, J.: Domino treewidth. J. Algorithms **24**(1), 94–123 (1997). DOI 10.1006/jagm.1996.0854
5. Bonnet, E., Kwon, O., Wood, D.R.: Reduced bandwidth: a qualitative strengthening of twin-width in minor-closed classes (and beyond) (2022) arXiv:2202.11858.
6. Campbell, R., Clinch, K., Distel, M., Gollin, J.P., Hendrey, K., Hickingbotham, R., Huynh, T., Illingworth, F., Tamitegama, Y., Tan, J., Wood, D.R.: Product structure of graph classes with bounded treewidth (2022)
7. Diestel, R.: Graph theory, *Graduate Texts in Mathematics*, vol. 173, 5th edn. Springer (2018). URL http://diestel-graph-theory.com/
8. Ding, G., Oporowski, B.: Some results on tree decomposition of graphs. J. Graph Theory **20**(4), 481–499 (1995). DOI 10.1002/jgt.3190200412
9. Ding, G., Oporowski, B.: On tree-partitions of graphs. Discrete Math. **149**(1–3), 45–58 (1996). DOI 10.1016/0012-365X(94)00337-I
10. Distel, M., Hickingbotham, R., Huynh, T., Wood, D.R.: Improved product structure for graphs on surfaces (2021) Discrete Math. Theor. Comput. Sci. 24(2) 6.
11. Dębski, M., Felsner, S., Micek, P., Schröder, F.: Improved bounds for centered colorings. Adv. Comb. **#8** (2021). DOI 10.19086/aic.27351
12. Dujmović, V., Esperet, L., Gavoille, C., Joret, G., Micek, P., Morin, P.: Adjacency labelling for planar graphs (and beyond). J. ACM **68**(6), 42 (2021). DOI 10.1145/3477542
13. Dujmović, V., Esperet, L., Joret, G., Walczak, B., Wood, D.R.: Planar graphs have bounded nonrepetitive chromatic number. Adv. Comb. **#5** (2020). DOI 10.19086/aic.12100
14. Dujmović, V., Joret, G., Micek, P., Morin, P., Ueckerdt, T., Wood, D.R.: Planar graphs have bounded queue-number. J. ACM **67**(4), #22 (2020). DOI 10.1145/3385731
15. Dujmović, V., Morin, P., Wood, D.R.: Graph product structure for non-minor-closed classes (2023) *J. Combin. Theory Ser. B* **162** 34–67.
16. Dvorák, Z., Gonçalves, D., Lahiri, A., Tan, J., Ueckerdt, T.: On comparable box dimension. In: X. Goaoc, M. Kerber (eds.) Proc. 38th Int'l Symp. on Computat. Geometry (SoCG 2022), *LIPIcs*, vol. 224, pp. 38:1–38:14. Schloss Dagstuhl (2022). DOI 10.4230/LIPIcs.SoCG.2022.38
17. Edenbrandt, A.: Quotient tree partitioning of undirected graphs. BIT **26**(2), 148–155 (1986). DOI 10.1007/BF01933740
18. Esperet, L., Joret, G., Morin, P.: Sparse universal graphs for planarity (2023) *J. Lond. Math. Soc.* **108**(4) 1333–1357.
19. Halin, R.: Tree-partitions of infinite graphs. Discrete Math. **97**, 203–217 (1991). DOI 10.1016/0012-365X(91)90436-6
20. Harvey, D.J., Wood, D.R.: Parameters tied to treewidth. J. Graph Theory **84**(4), 364–385 (2017). DOI 10.1002/jgt.22030
21. Hickingbotham, R., Wood, D.R.: Shallow minors, graph products and beyond planar graphs *SIAM J. DiscreteMath.*, to appear. arXiv:2111.12412.
22. Huynh, T., Mohar, B., Šámal, R., Thomassen, C., Wood, D.R.: Universality in minor-closed graph classes (2021) arXiv:2109.00327.
23. Liu, C.H., Wood, D.R.: Clustered graph coloring and layered treewidth (2019)
24. Ossona de Mendez, P., Oum, S., Wood, D.R.: Defective colouring of graphs excluding a subgraph or minor. Combinatorica **39**(2), 377–410 (2019). DOI 10.1007/s00493-018-3733-1
25. Mohar, B., Thomassen, C.: Graphs on surfaces. Johns Hopkins University Press (2001)
26. Nešetřil, J., Ossona de Mendez, P.: Sparsity, *Algorithms and Combinatorics*, vol. 28. Springer (2012). DOI 10.1007/978-3-642-27875-4
27. Reed, B.A.: Algorithmic aspects of tree width. In: Recent advances in algorithms and combinatorics, vol. 11, pp. 85–107. Springer (2003). DOI 10.1007/0-387-22444-0_4
28. Seese, D.: Tree-partite graphs and the complexity of algorithms. In: L. Budach (ed.) Proc. Int'l Conf. on Fundamentals of Computation Theory, *Lecture Notes in Comput. Sci.*, vol. 199, pp. 412–421. Springer (1985). DOI 10.1007/BFb0028825

29. Ueckerdt, T., Wood, D.R., Yi, W.: An improved planar graph product structure theorem. Electron. J. Combin. **29**, P2.51 (2022). DOI 10.37236/10614
30. Wood, D.R.: Vertex partitions of chordal graphs. J. Graph Theory **53**(2), 167–172 (2006). DOI 10.1002/jgt.20183
31. Wood, D.R.: On tree-partition-width. European J. Combin. **30**(5), 1245–1253 (2009). DOI 10.1016/j.ejc.2008.11.010

Notes on Aharoni's rainbow cycle conjecture

Katie Clinch, Jackson Goerner, Tony Huynh, and Freddie Illingworth

Abstract In 2017, Ron Aharoni made the following conjecture about rainbow cycles in edge-coloured graphs: If G is an n-vertex graph whose edges are coloured with n colours and each colour class has size at least r, then G contains a rainbow cycle of length at most $\lceil \frac{n}{r} \rceil$. One motivation for studying Aharoni's conjecture is that it is a strengthening of the Caccetta-Häggkvist conjecture on digraphs from 1978. In this article, we present a survey of Aharoni's conjecture, including many recent partial results and related conjectures. We also present two new results. Our first result (Theorem 21) provides sharp thresholds for rainbow cycles in edge-coloured graphs with 3 colours. Our second, and main, result (Theorem 35) is for the $r = 3$ case of Aharoni's conjecture. We prove that if G is an n-vertex graph whose edges are coloured with n colours and each colour class has size at least 3, then G contains a rainbow cycle of length at most $\frac{4n}{9} + 7$.

Katie Clinch
School of Computer Science and Engineering, UNSW, Sydney, Australia
e-mail: `k.clinch@unsw.edu.au`

Jackson Goerner
School of Mathematics, Monash University, Melbourne, Australia
e-mail: `jackson.goerner@monash.edu`

Tony Huynh
Dipartimento di Informatica, Sapienza Università di Roma, Italy
Supported by the Australian Research Council
e-mail: `huynh@di.uniroma1.it`

Freddie Illingworth
Mathematical Institute, University of Oxford, Oxford, United Kingdom
Supported by EPSRC grant EP/V007327/1
e-mail: `illingworth@maths.ox.ac.uk`

D. R. Wood et al. (eds.), *2021-2022 MATRIX Annals*, MATRIX Book Series 5,
https://doi.org/10.1007/978-3-031-47417-0_10

1 Introduction

In 1978, Caccetta and Häggkvist made the following conjecture about directed cycles in digraphs.

Conjecture 1 (Caccetta-Häggkvist). For all positive integers n, r, every simple[1] n-vertex digraph with minimum outdegree at least r contains a directed cycle of length at most $\lceil \frac{n}{r} \rceil$.

Despite considerable effort from numerous researchers, the Caccetta-Häggkvist conjecture remains open. A complete summary of the plethora of results related to the Caccetta-Häggkvist is beyond the scope of this survey. We refer the interested reader to Sullivan [23] for a brief synopsis. We instead focus on the following generalisation of Conjecture 1.

Conjecture 2 (Aharoni). Let G be a simple[2] edge-coloured graph with n vertices and n colours, where each colour class has size at least r. Then G contains a rainbow[3] cycle of length at most $\lceil \frac{n}{r} \rceil$.

For completeness, we now give a proof that Conjecture 2 indeed implies Conjecture 1. In fact, as noted in [10], the following weakening of Aharoni's conjecture already implies the Caccetta-Häggkvist conjecture.

Conjecture 3 (DeVos et al. [10]). Let G be a simple edge-coloured graph with n vertices and n colours, where each colour class has size at least r. Then G contains a cycle C of length at most $\lceil \frac{n}{r} \rceil$ such that no two incident edges of C are the same colour.

Proof of Conjecture 1, assuming Conjecture 3. Let D be a simple digraph with n vertices and minimum outdegree at least r. Let G be the graph obtained from D by forgetting the orientations of all arcs. Colour $uv \in E(G)$ with colour u if $(u, v) \in E(D)$. Clearly, this colouring uses $|V(D)| = n$ colours. Moreover, since D has minimum outdegree at least r, each colour class has size at least r. Therefore, by Conjecture 3, G contains a properly edge-coloured cycle C of length at most $\lceil \frac{n}{r} \rceil$. The set of arcs in D corresponding to the edges of C must be a directed cycle; otherwise C is not properly edge-coloured. □

Despite the fact that Aharoni's conjecture implies the Caccetta-Häggkvist conjecture, a proof of Aharoni's conjecture may be easier to find than a proof of the Caccetta-Häggkvist conjecture. Although this might sound counterintuitive, the method of proving a stronger statement is very common in combinatorics. Moreover, generalisation often leads to new techniques and new questions which one would not even consider in the original setting. We will see that this is the case for Aharoni's conjecture in the next section.

---

[1] A digraph is *simple* if for all vertices u and v, there is at most one arc from u to v.

[2] An edge-coloured graph G is *simple* if each colour class does not contain parallel edges.

[3] A subgraph of an edge-coloured graph is *rainbow* if no two of its edges are the same colour.

2 Related Results and Conjectures

In this section, we survey results and conjectures related to rainbow cycles. For a general survey on rainbow sets, we refer the reader to Aharoni and Briggs [2].

2.1 Larger Colour Classes

Much of the research on the Caccetta-Häggkvist conjecture has focused on the directed triangle case ($r = \lceil \frac{n}{3} \rceil$). A natural strategy is to increase the outdegree condition until one can prove the existence of a directed triangle. The best result in this direction is the following result of Hladký, Král', and Norin [13], which uses the flag algebra machinery developed by Razborov [21].

Theorem 4. *Every simple n-vertex digraph with minimum outdegree at least* $0.3465n$ *contains a directed triangle.*

Similarly, for Aharoni's conjecture, one can ask how large must the colour classes be to ensure a rainbow cycle of length at most $\lceil \frac{n}{r} \rceil$? The first non-trivial bound was proven by Hompe, Pelikánová, Pokorná, and Spirkl [17].

Theorem 5. *Let* $r \geqslant 2$, $n \geqslant 1$, *G be a simple edge-coloured graph with n vertices and n colours, where each colour class has size at least* $301 r \log r$. *Then G contains a rainbow cycle of length at most* $\lceil \frac{n}{r} \rceil$.

Theorem 5 was subsequently improved by Hompe and Spirkl [19], who removed the $\log r$ term.

Theorem 6. *Let* $r, n \in \mathbb{N}$, *G be a simple edge-coloured graph with n vertices and n colours, where each colour class has size at least* $10^{11} r$. *Then G contains a rainbow cycle of length at most* $\frac{n}{r}$.

2.2 Number of Colour Classes

One appealing aspect of Aharoni's conjecture is that we can ask what happens when the number of colour classes is different from the number of vertices. Note that this question does not even make sense in the digraph setting. To be precise, we now define a function $f(n,t,r)$ which will be useful to state many of the results that appear in this survey. The *rainbow girth* of an edge-coloured graph G, denoted $\mathrm{rg}(G)$, is the length of a shortest rainbow cycle in G. If G does not contain a rainbow cycle, then $\mathrm{rg}(G) = \infty$. Let

$$f(n,t,r) := \max\{\mathrm{rg}(G)\},$$

where the maximum is taken over all simple edge-coloured graphs G with n vertices and at least t colours, such that each colour class has size at least r.

We can rephrase Aharoni's conjecture via our function $f(n,t,r)$ as follows.

Conjecture 7 (Aharoni). For all $n, r \geqslant 1$,

$$f(n,n,r) \leqslant \lceil \tfrac{n}{r} \rceil.$$

We believe that there is no reason to restrict attention to the case $t = n$, and that the following question is of independent interest.

Question 8. *Obtain good upper and lower bounds for $f(n,t,r)$ for all n,t, and r.*

An important special case of Question 8 is when $r = 1$, which was considered by Bollobás and Szemerédi [7].

Theorem 9. *For all $n \geqslant 4$ and $k \geqslant 2$,*

$$f(n,n+k,1) \leqslant \frac{2(n+k)}{3k}(\log k + \log\log k + 4).$$

In other words, Theorem 9 asserts that every n-vertex graph with at least $n+k$ edges contains a cycle of length at most $\frac{2(n+k)}{3k}(\log k + \log\log k + 4)$. This is a key tool used in many of the results presented in this survey.

Hompe and Spirkl [19] also obtained the following bounds when the number of colours is more than the number of vertices.

Theorem 10. *For all $n \geqslant 1$ and $k \geqslant 2$,*

$$f(n,n+k,10^9 k) \leqslant \max\{6, \frac{n(\log k)^2}{10k^{3/2}} + 14\log k\}.$$

When the number of colours is *less* than the number of vertices, DeVos et al. [10] obtained the following tight bounds for $r = 2$.

Theorem 11. *For all $n \geqslant 3$ and $t \leqslant n$,*

$$f(n,t,2) = \begin{cases} \infty & \text{if } t \leqslant n-2, \\ n-1 & \text{if } t = n-1, \\ \lceil \frac{n}{2} \rceil & \text{if } t = n. \end{cases}$$

2.3 Structured Colour Classes

In the proof of the Caccetta-Häggkvist conjecture (assuming Aharoni's conjecture) given in the Introduction, the colour classes of the derived edge-colouring are all stars. Therefore, it is natural to ask what happens when the colour classes are not stars. Note that if a colour class is not a star, then it must contain a matching of size 2, or it is a triangle.

In the extreme case when all colour classes contain a matching of size 2, Aharoni and Guo [5] proved that there is a much shorter rainbow cycle than the $\lceil \frac{n}{2} \rceil$ bound which follows from [10].

Theorem 12. *There exists an absolute constant C such that if G is a simple edge-coloured graph with n vertices, n colours, and each colour class is a matching of size 2, then G contains a rainbow cycle of length at most $C \log n$.*

Kevin Hendrey (private communication) proved that an $\mathcal{O}(\log n)$ bound also holds in the case when all colour classes are a triangle. The same proof also appears in a recent paper of Aharoni, Berger, Chudnovsky, Guo, and Zerbib [1].

Theorem 13. *There exists an absolute constant C such that if G is a simple edge-coloured graph with n vertices, n colours, and each colour class is a triangle, then G contains a rainbow cycle of length at most $C \log n$.*

Some mixed cases were investigated by Guo [11].

Theorem 14. *There exists an absolute constant C such that if G is a simple edge-coloured graph with n vertices, n colours, and each colour class is a matching of size 2 or a triangle, then G contains a rainbow cycle of length at most $C \log n$.*

Theorem 15. *For any constants $0 \leqslant \alpha < 1$ and $0 \leqslant \beta \leqslant \alpha$ with $\beta < (1-\alpha)/3$, there exists a constant $C(\alpha, \beta)$ such that if G is an n-vertex simple edge-coloured graph containing at least $(\alpha - \beta)n$ colour classes consisting of a single edge and at least $(1-\alpha-\beta)n$ colour classes consisting of a triangle, then G contains a rainbow cycle of length at most $C(\alpha, \beta) \log n$.*

2.4 Rainbow Triangles

The rainbow triangle case (when $r = \lceil n/3 \rceil$) of Aharoni's conjecture states that $f(n,n,\lceil n/3 \rceil) \leqslant 3$. This is of course still open since the directed triangle case of the Caccetta-Häggkvist conjecture is still open. However, there have been partial results which increase the number of colours or sizes of the colour classes. Two such results were obtained by Aharoni, DeVos, and Holzman [4].

Theorem 16. *For all $n \geqslant 1$,*

$$f(n, 9n/8, n/3) \leqslant 3.$$

Theorem 17. *For all $n \geqslant 1$,*

$$f(n, n, 2n/5) \leqslant 3.$$

Both of these results have been subsequently improved by Hompe, Qu, and Spirkl [18].

Theorem 18. *For all* $n \geqslant 1$,

$$f(n, 1.1077n, n/3) \leqslant 3.$$

Theorem 19. *For all* $n \geqslant 1$,

$$f(n, n, 0.3988n) \leqslant 3.$$

Aharoni, DeVos, González Hermosillo de la Maza, Montejano, and Šámal [3] proved the following theorem showing when a simple edge-coloured graph with 3 colours contains a rainbow triangle. Their theorem actually implies Mantel's theorem.

Theorem 20. *Every n-vertex simple edge-coloured graph with 3 colours and each colour class of size at least* $\frac{26-2\sqrt{7}}{81} \cdot n^2$ *contains a rainbow triangle.*

They also prove that the constant $\frac{26-2\sqrt{7}}{81}$ in Theorem 20 is best possible. However, their extremal example has rainbow 2-cycles. Our first new result is the following sharp thresholds for $f(n, 3, r)$.

Theorem 21. *For all* $n \geqslant 100$,

$$f(n,3,r) = \begin{cases} \infty & \text{if } r \leqslant \lfloor \binom{n}{2}/3 \rfloor, \\ 2 & \text{if } \lfloor \binom{n}{2}/3 \rfloor < r \leqslant \binom{n}{2}, \\ 1 & \text{if } r > \binom{n}{2}. \end{cases}$$

Proof. First suppose $r > \lfloor \binom{n}{2}/3 \rfloor$. Let G be a simple edge-coloured graph with n vertices, 3 colours and each colour class of size at least r. Since K_n has only $\binom{n}{2}$ edges, G must contain a loop or a rainbow 2-cycle. Moreover, if $r > \binom{n}{2}$, then G must contain a loop. The n-vertex edge-coloured graph with a red, blue, and green edge between every pair of vertices proves equality.

We now show that $f(n,3,r) = \infty$ if $n \geqslant 100$ and $r \leqslant \lfloor \binom{n}{2}/3 \rfloor$. Partition the vertices of K_n as $X \cup Y \cup Z$ where $|X| = |Y| = \lceil \frac{2n}{5} \rceil$. Colour all edges in $K_n[X]$ or $K_n[Y]$ or $K_n[Z]$ red, all edges between X and Y blue, and all edges between Z and $X \cup Y$ green. Observe that this is an edge-colouring of K_n with no rainbow triangle. The colour classes do not quite have the same size, which we now fix. Since $n \geqslant 100$, we have $|X||Y| \leqslant \binom{n}{2}/3$ and $|Z|(|X|+|Y|) \leqslant \binom{n}{2}/3$. Thus, we may recolour some of the edges in $K_n[X]$ blue, and some of the edges in $K_n[Y]$ green, so that the number of red, blue, and green edges differ by at most 1. This new edge-colouring of K_n still does not contain a rainbow triangle. This of course implies that there are no rainbow cycles since there are only 3 colours. □

2.5 Fixed Values of r

The Caccetta-Häggkvist conjecture is known to hold for small values of r. The $r = 1$ case is trivial, the $r = 2$ case was proven by Caccetta and Häggkvist [8]; the $r = 3$ case was proven by Hamidoune [12]; and the $r \in \{4,5\}$ cases were proven by Hoàng and Reed [14]. For Aharoni's conjecture, the $r = 2$ case was proven by DeVos et al. [10].

Theorem 22. *For all $n \geqslant 1$,*

$$f(n,n,2) \leqslant \lceil \tfrac{n}{2} \rceil.$$

In Section 3, we prove the following result for $r = 3$.

Theorem 23. *For all $n \geqslant 1$,*

$$f(n,n,3) \leqslant \tfrac{4n}{9} + 7.$$

After a previous version of this paper was made publicly available [9], Hompe and Huynh [16] extended our methods to prove the following theorem for all fixed values of r.

Theorem 24. *For all $r \geqslant 1$, there exists a constant c_r such that*

$$f(n,n,r) \leqslant \tfrac{n}{r} + c_r,$$

for all $n \geqslant 1$.

2.6 Non-uniform Versions

For the Caccetta-Häggkvist conjecture, it is natural to ask if there is a version which takes into account *all* the outdegrees rather than just the minimum outdegree. Seymour (see [15]) proposed the following generalization. Given a digraph D with no sink[4], define

$$\psi(D) := \sum_{v \in V(D)} \frac{1}{\deg^+(v)}.$$

Conjecture 25. Every simple digraph D with no sink contains a directed cycle of length at most $\lceil \psi(D) \rceil$.

Note that in the case that all outdegrees are r, then $\psi(D) = \frac{n}{r}$, so Conjecture 25 implies the Caccetta-Häggkvist conjecture. Unfortunately, Conjecture 25 was disproved by Hompe [15]. However, Aharoni et al. [1] proved that "half" of Conjecture 25 holds.

Theorem 26. *Every simple digraph D with no sink contains a directed cycle of length at most $2\psi(D)$.*

[4] A *sink* is a vertex with outdegree zero.

Theorem 26 has a natural generalization in the rainbow setting. Given an edge-coloured graph G, define

$$\psi(G) = \sum_{A} \frac{1}{|A|},$$

where the sum is taken over all colour classes A of G.

Conjecture 27. Every simple edge-coloured graph G with the same number of colours as vertices contains a rainbow cycle of length at most $2\psi(G)$.

In the case that all colour classes have size at most 2, Aharoni et al. [1] proved the following strengthening of Conjecture 27.

Theorem 28. *Let G be a simple edge-coloured graph with the same number of colours as vertices and such that each colour class has size at most 2. Then G contains a rainbow cycle of length at most $\lceil \psi(G) \rceil$.*

Note that Theorem 28 is a strengthening of Theorem 22 since it allows colour classes of size 1.

2.7 Matroids

We can generalise Aharoni's conjecture to any setting in which the notion of 'cycle' makes sense. One natural candidate is that of a *matroid*. For the reader unfamiliar with matroids, we introduce all the necessary definitions now. For a more thorough introduction to matroids, we refer the reader to Oxley [20].

A *matroid* is a pair $M = (E, \mathscr{C})$ where E is a finite set, called the *ground set* of M, and $\mathscr{C}$ is a collection of subsets of E, called *circuits*, satisfying

1. $\emptyset \notin \mathscr{C}$,
2. if C' is a proper subset of $C \in \mathscr{C}$, then $C' \notin \mathscr{C}$,
3. if C_1 and C_2 are distinct members of $\mathscr{C}$ and $e \in C_1 \cap C_2$, then there exists $C_3 \subseteq (C_1 \cup C_2) \setminus \{e\}$.

We now give some examples of matroids. Let G be a graph. We will consider two different matroids with ground set $E(G)$. The circuits of the first matroid are the (edges of) cycles of G. This is the *cycle matroid* of G, denoted $M(G)$. A matroid is *graphic* if it is isomorphic to the cycle matroid of some graph. A *cocycle* of G is an inclusion-wise minimal edge-cut of G. The collection of cocycles of G is also a matroid, called the *cocycle matroid* of G, and is denoted $M(G)^*$. A matroid is *cographic* if it is isomorphic to the cocycle matroid of some graph.

Let $\mathbb{F}$ be a field. An *$\mathbb{F}$-matrix* is a matrix with entries in $\mathbb{F}$. Let A be an $\mathbb{F}$-matrix whose columns are labelled by a finite set E. The *column matroid* of A, denoted $M[A]$, is the matroid with ground set E whose circuits correspond to the minimal (under inclusion) linearly dependent columns of A. A matroid is *representable over $\mathbb{F}$* if it is isomorphic to $M[A]$ for some $\mathbb{F}$-matrix A. A matroid is *binary* if it is representable over the two-element field, and it is *regular* if it is representable over every field.

In order to formulate Aharoni's conjecture for matroids, we need to define simple matroids and how to express the number of vertices of a graph as a matroid parameter. We do this now. A matroid is *simple* it it does not contain any circuits of size 1 or 2. A set $I \subseteq E$ is *independent* if it does not contain a circuit. The *rank* of $X \subseteq E$ is the size of a largest independent set contained in X, and is denoted $r_M(X)$. The *rank* of M is $r(M) := r_M(E)$. Notice that the number of vertices of a connected graph G is $r(M(G)) - 1$. Thus, Aharoni's conjecture can be phrased in the language of matroids as follows.

Conjecture 29 (Aharoni)**.** Let M be a simple rank-$(n-1)$ graphic matroid and c be a colouring of $E(M)$ with n colours, where each colour class has size at least r. Then M contains a rainbow circuit of size at most $\lceil \frac{n}{r} \rceil$.

One way to generalise Conjecture 1 is to replace 'graphic matroid' with some larger superclass of matroids. In [10], it was shown that one cannot replace 'graphic matroid' by 'binary matroid' in Conjecture 29.

Theorem 30. *For all $n \geqslant 6$, there exists a simple rank-$(n-1)$ binary matroid M on $2n$ elements, and a colouring of $E(M)$ where each colour class has size 2, such that all rainbow circuits of M have size strictly greater than $\lceil \frac{n}{2} \rceil$.*

The main result of [10] can be phrased in matroid language as follows.

Theorem 31. *Let M be a simple rank-$(n-1)$ graphic matroid and c be a colouring of $E(M)$ with n colours, where each colour class has size at least 2. Then M contains a rainbow circuit of size at most $\lceil \frac{n}{2} \rceil$.*

In [10], it is also proved that the matroid analogue of Theorem 22 holds for cographic matroids.

Theorem 32. *Let N be a simple rank-$(n-1)$ cographic matroid and c be a colouring of $E(N)$ with n colours, where each colour class has size at least 2. Then N contains a rainbow circuit of size at most $\lceil \frac{n}{2} \rceil$.*

Regular matroids are a well-studied superclass of graphic matroids, and are 'essentially' graphic or cographic via Seymour's regular matroid decomposition theorem [22]. Therefore, by combining Theorem 31, Theorem 32, and Seymour's regular matroid decomposition theorem, it may be possible to prove the following conjecture.

Conjecture 33 (DeVos et al. [10])**.** Let M be a simple rank-$(n-1)$ regular matroid and c be a colouring of $E(M)$ with n colours, where each colour class has size at least 2. Then M contains a rainbow circuit of size at most $\lceil \frac{n}{2} \rceil$.

Of course, we can also consider Question 8 for various classes of matroids. Bérczi and Schwarcz [6] obtained one such result.

Theorem 34. *Let M be an n-element rank-t binary matroid whose ground set is coloured with t colours. Then M either contains a rainbow circuit or a monochromatic cocircuit.*[5]

[5] A *monochromatic cocircuit* is a circuit in the dual matroid whose elements are all the same colour.

Bérczi and Schwarcz [6, Theorem 4] also show that it is possible to characterize binary matroids as exactly those matroids which do not admit a specific type of colouring with no rainbow circuits. They also prove that if a simple graph G has an edge-colouring with no rainbow cycle, where each colour class has size at most 2, then G is independent in the 2-dimensional rigidity matroid.[6]

3 Proof of Main Theorem

In this section, we prove our main theorem.

Theorem 35. *Every simple edge-coloured graph with n vertices, n colours, and each colour class of size at least 3, contains a rainbow cycle of length at most* $\frac{4n}{9}+7$.

3.1 Excess-k Graphs

We begin by establishing some basic properties about graphs which have at least k more edges than vertices, which we call *excess-k graphs*. The first property follows from Theorem 9: for all $n \geqslant 4$ and $k \geqslant 2$, every n-vertex, excess-k graph has a cycle of length at most $\frac{2(n+k)}{3k}(\log k + \log\log k + 4)$. We require the following tighter bounds when $k \leqslant 2$.

Lemma 36. *Every n-vertex, excess-1 graph has a cycle of length at most* $\frac{2n}{3}+1$.

The proof is easy and is omitted (see [10] for a proof of a stronger claim).

Lemma 37. *Every n-vertex, excess-2 graph H has a cycle of length at most* $\frac{n}{2}+1$.

Proof. A *block* of H is a 2-connected subgraph of H which is maximal under the subgraph relation. If H contains at least two blocks, then H contains two cycles which meet in at most one vertex. Hence, one of these cycles has length at most $\frac{n}{2}+1$. So, we may assume H contains exactly one block B. Note that B is excess-2, since H is excess-2. Let $C \cup P_1 \cup \cdots \cup P_k$ be an ear-decomposition of B. Note that $k \geqslant 2$, since B is excess-2. Let $B' = C \cup P_1 \cup P_2$. Either B' contains two cycles which meet in at most two vertices, or B' is a subdivision of K_4. In the first case, one of the two cycles has length at most $\frac{n}{2}+1$. In the second case, B' contains four cycles $C_1,\ldots,C_4$ such that $\sum_{i\in[4]}|V(C_i)| = 2|V(B')|+4$. Thus, one of these four cycles has length at most $\frac{|V(B')|}{2}+1 \leqslant \frac{n}{2}+1$. □

We now prove that the largest stable set of an excess-k graph is roughly half the number of vertices.

[6] Equivalently, for every $X \subseteq V(G)$ with $|X| \geqslant 2$, $|E(G[X])| \leqslant 2|X|-3$.

Lemma 38. *Let H be a simple excess-k graph of minimum degree at least 2. Then a maximum stable set of H has size at most $\frac{|V(H)|+k}{2}$.*

Proof. We proceed by induction on $|V(H)|+|E(H)|$. We may assume that H has exactly k more edges than vertices. If H is the disjoint union of $H_1,\ldots,H_\ell$ and H_i has excess k_i, then $k_1+\cdots+k_\ell=k$. Therefore, we are done by applying induction to each H_i. So, we may assume H is connected. If $k=0$, then H is a cycle, so the lemma clearly holds. Now suppose $k\geqslant 1$. Let X denote the set of vertices of H of degree at least 3. If $|X|=1$, then H consists of $k+1$ cycles which meet at the same vertex. It is easy to see that the lemma holds in this case. So, we may assume $|X|\geqslant 2$. Suppose $u,v\in X$ and $uv\in E(G)$. By induction, a maximum stable set of $H-uv$ has size at most $\frac{|V(H)|+k-1}{2}$. Hence, a maximum stable set of H has size at most $\frac{|V(H)|+k-1}{2}$. So, we may assume that no two vertices of X are adjacent.

Let P be the shortest path between any two vertices of X. By the minimality of P, each internal vertex of P has degree 2. Let $H'=H-I$, where I is the set of internal vertices of P. Note that H' has excess $k-1$ and minimum degree at least 2. Let S be a maximum size stable set in H. Note that $S\cap V(H')$ is a stable set in H'. By induction, $|S\cap V(H')|\leqslant(|V(H')|+k-1)/2$. Also, $|S\cap I|\leqslant(|I|+1)/2$. Thus, $|S|=|S\cap V(H')|+|S\cap I|\leqslant(|V(H)|+k)/2$. □

We finish by establishing the following lemma about 'minimal' excess-k rainbow subgraphs of an edge-coloured graph.

Lemma 39. *Let H be a simple edge-coloured graph where each colour class has size at most r. Let R be an excess-k rainbow subgraph of H such that $V(R)$ is minimal under inclusion. Then H contains a rainbow 2-cycle, or there are at most $\max\{\binom{2k+2}{2},6k(r-1)\}$ chords of R in H.*

Proof. We may assume that H does not contain parallel edges; otherwise H contains a rainbow 2-cycle. By the minimality of R, $\deg_R(v)\geqslant 2$ for all $v\in V(R)$. Let $V(R)=X\cup Y$, where X is the set of vertices of degree at least 3, and Y is the set of degree-2 vertices of R. By the Handshaking Lemma, $|X|\leqslant 2k$.

Say that a chord of R is *novel* if its colour does not appear in R, and *plain* otherwise. Suppose $e=uv$ is a novel chord. If $|Y|\geqslant 3$, then there is a vertex $y\in Y$ such $y\notin\{u,v\}$, and so $(R\cup e)-y$ contradicts the minimality of R. Thus, there are either no novel chords, or the total number of chords is at most $\binom{|X|+|Y|}{2}\leqslant\binom{2k+2}{2}$.

We may thus assume that there are no novel chords. Let $e\in E(H)$ be a plain chord of R. Since e is plain, there is an edge $f\in E(R)$ of the same colour as colour e. If both ends of f are in Y, then $R':=(R\cup e)\setminus f$ contains a degree-1 vertex x. But now $R'-x$ contradicts the minimality of R. Therefore, at least one end of f is in X. Let R_0 be the multigraph on vertex set X obtained by suppressing all degree-2 vertices. Note that R_0 also has excess k and at most $2k$ vertices. Therefore, $|E(R_0)|\leqslant 3k$. There are at most $2|E(R_0)|$ edges of R which are incident to a vertex in X. Each of these edges can correspond to at most $r-1$ plain chords. Hence, there are at most $6k(r-1)$ plain chords. □

3.2 The set-up

Let G be a simple edge-coloured graph with n vertices, n colours, and each colour class of size exactly 3. An *r-star* is a star with r edges. A colour class of G is a *star class* if it is a 3-star. A vertex of G is a *star vertex* if it is the centre of a star class, and is otherwise a *non-star vertex*. Let S denote the set of star vertices of G, and N denote the set of non-star vertices of G. Let N' be the set of non-star classes. Since star classes may be centred at the same vertex, we have $|N'| \leqslant |N|$. We will do a case analysis depending on whether $|N| \geqslant 8$ or $|N| \leqslant 7$.

3.3 Many non-star vertices

Throughout this section we suppose $|N| \geqslant 8$. Let $\{x,y\} \subseteq N$. We say that a colour class A *dominates* $\{x,y\}$ if every edge in A has an end in $\{x,y\}$.

Claim 40. *There exists $\{x,y\} \subseteq N$ such that no colour class dominates $\{x,y\}$.*

Proof. Suppose A is a star class. Then A is a 3-star centred at a vertex not in N. Therefore, for all $\{u,v\} \subseteq N$, A does not dominate $\{u,v\}$ since at least one leaf of A is not in $\{u,v\}$.

Suppose A is a non-star class. Up to isomorphism, $A \in \{K_3, P_3, P_2 \sqcup P_1, P_1 \sqcup P_1 \sqcup P_1\}$, where P_i is a path with i edges, and $\sqcup$ denotes disjoint union. Let $\gamma(A)$ be the number of vertex covers of A of size 2. Observe that the number of pairs $\{u,v\} \subseteq N$ which A dominates is exactly equal to $\gamma(A)$. We have $\gamma(K_3) = 3, \gamma(P_3) = 1, \gamma(P_2 \sqcup P_1) = 2$, and $\gamma(P_1 \sqcup P_1 \sqcup P_1) = 0$. Thus, every non-star class dominates at most 3 pairs of vertices in N. So, the number of pairs dominated by non-star classes is at most $3|N'| \leqslant 3|N| < \binom{N}{2}$, since $|N| \geqslant 8$. Therefore, at least one pair $\{x,y\} \subseteq N$ is not dominated by any colour class. □

Claim 41. *G contains an excess-2 rainbow subgraph R.*

Proof. A *transversal* of G is a subgraph consisting of exactly one edge of each colour. By Claim 40, G has a transversal R such that at least two vertices x, y are not in $V(R)$. Since R has exactly n edges, R is excess-2 (and clearly rainbow). □

We now choose R to be an excess-2 rainbow subgraph of G such that $V(R)$ is minimal under inclusion.

Claim 42. *G contains a rainbow cycle C' such that $E(R) \cap E(C') = \emptyset$.*

Proof. Choose a transversal R' which is edge-disjoint from R. Since R' has n vertices and n edges, R' contains a cycle C'. Clearly, C' is rainbow since R' is rainbow. □

Claim 43. *G contains a rainbow cycle C of length at most $\frac{2n}{5} + 7$.*

Proof. Let $n_1 = |V(R) \setminus V(C')|$, $n_2 = |V(R) \cap V(C')|$, and $n_3 = |V(C') \setminus V(R)|$. First suppose $n_3 \geqslant \frac{n}{5} - 12$. Then, $|V(R)| \leqslant \frac{4n}{5} + 12$. By Lemma 37, R contains a rainbow cycle of length at most $\frac{1}{2} \cdot (\frac{4n}{5} + 12) + 1 = \frac{2n}{5} + 7$.

Thus, we may assume that $n_3 < \frac{n}{5} - 12$. Let A be the subset of edges of C' which are chords of R. Applying Lemma 39 to R in $C' \cup R$ (so $r = 2$), we have $|A| \leqslant 15$. Note that $V(R) \cap V(C')$ is a stable set of $C' \setminus A$. The maximum stable set of C' has size at most $\frac{|V(C')|}{2}$. Deleting one edge from a graph can increase the size of a maximum stable set by at most 1. Therefore, we conclude that $n_2 \leqslant n_3 + 2|A| \leqslant n_3 + 30$. Thus,

$$|V(C')| = n_2 + n_3 \leqslant 2n_3 + 30 < 2(\tfrac{n}{5} - 12) + 30 = \tfrac{2n}{5} + 6,$$

and so we may take $C = C'$ in this case. □

This completes the case $|N| \geqslant 8$, since we have found a rainbow cycle of length at most $\frac{2n}{5} + 7$, which is better than the bound of $\frac{4n}{9} + 7$ required by Theorem 35.

3.4 Few non-star vertices

We complete the proof by considering the case $|N| \leqslant 7$.

Claim 44. *At least one vertex of G is a non-star vertex.*

Proof. Suppose every vertex of G is a star vertex. Since G has the same number of vertices as colours, this implies that at each $v \in V(G)$ there is a star class S_v centred at v. Let D be the digraph obtained from G by orienting the edges of S_v away from v for all $v \in V(G)$. By the $r = 3$ case of the Caccetta-Häggkvist conjecture [12], D contains a directed cycle $\vec{C}$ of length at most $\lceil \frac{n}{3} \rceil$. Note that $\vec{C}$ corresponds to a rainbow cycle C in G. □

Fix a non-star vertex $z \in V(G)$. Since z is a non-star vertex, for every colour a, there is an edge e_a coloured a such that e_a is not incident to z. Thus, there is a transversal R of G such that $z \notin V(R)$. In particular, R is an excess-1 rainbow subgraph of G such that $z \notin V(R)$. Let R_1 be an excess-1 rainbow subgraph of $G - z$ such that $V(R_1)$ is minimal under inclusion.

Claim 45. *There exists a transversal R_2' of G such that $z \notin V(R_2')$ and $|E(R_1) \cap E(R_2')| \leqslant 7$.*

Proof. Let A be a star colour class. Observe that there are at least 2 edges of A which are not incident to z. Thus, there exists an edge $e_A \in A$ such that $e_A \notin E(R_1)$ and e_A is not incident to z. If A is a non-star colour class, there is an edge $e_A \in A$ which is not incident to z. Let $R_2' = \bigcup_A \{e_A\}$, where the union is over all colour classes. Since there are at most $|N'| \leqslant |N| \leqslant 7$ non-star colour classes, $|E(R_1) \cap E(R_2')| \leqslant 7$. □

Since $z \notin V(R_2')$, R_2' is excess-1 (and rainbow). Let $R_2 \subseteq R_2'$ be excess-1 and rainbow such that $V(R_2)$ is minimal under inclusion. In particular, R_2 has minimum degree at least 2.

Claim 46. *G contains a rainbow cycle C of length at most* $\frac{4n}{9}+7$.

Proof. Let $n_1 = |V(R_1)\setminus V(R_2)|$, $n_2 = |V(R_1)\cap V(R_2)|$, and $n_3 = |V(R_2)\setminus V(R_1)|$. First suppose $n_3 \geqslant \frac{n}{3}-9$. Thus, $|V(R_1)| \leqslant \frac{2n}{3}+9$. By Lemma 36, R_1 contains a rainbow cycle of length at most $\frac{2}{3}\cdot(\frac{2n}{3}+9)+1 = \frac{4n}{9}+7$.

Thus, we may assume that $n_3 < \frac{n}{3}-9$. Let A be the subset of edges of R_2 which are chords or edges of R_1. By Lemma 39 applied to R_1 in $R_1 \cup R_2$ (so $r=2$), and Claim 45, $|A| \leqslant 6+7=13$. Note that $V(R_1)\cap V(R_2)$ is a stable set of $R_2 \setminus A$. By Lemma 38, the maximum stable set of R_2 has size at most $\frac{|V(R_2)|+1}{2}$. Deleting one edge from a graph can increase the size of a maximum stable set by at most 1. Therefore, we conclude that $n_2 \leqslant n_3+2|A|+1 \leqslant n_3+27$. By Lemma 36, R_2 contains a rainbow cycle of length at most

$$\frac{2}{3}\cdot(n_2+n_3)+1 \leqslant \frac{2}{3}\cdot(2n_3+27)+1 < \frac{2}{3}\cdot(2(\frac{n}{3}-9)+27)+1 = \frac{4n}{9}+7. \qquad \square$$

Claim 46 completes the proof of Theorem 35.

4 Generalising our Approach

For the $r=3$ case of Aharoni's conjecture, we proved that there is a rainbow cycle of length at most $\frac{4n}{9}+7$. The additive constant can be easily improved since there are stronger versions of Lemma 39 for $k \leqslant 2$. However, we opted to instead prove Lemma 39 for all r and k as we suspected our approach could be generalised. To that end, the previous version of this section in [9] outlined two strategies to achieve this. Combining one of the strategies with some new ideas, Hompe and Huynh [16] subsequently succeeded in generalising our result, by proving the following theorem for all values of r.

Theorem 47. *For all $r \geqslant 1$, there exists a constant c_r such that if G is a simple n-vertex edge-coloured graph with n colour classes of size at least r, then G contains a rainbow cycle of length at most*

$$\frac{n}{r}+c_r.$$

As our previous discussion is now out-dated, we have removed it from this version of the paper. Interested readers can find the old discussion in [9].

Acknowledgements This research was initiated at the Structural Graph Theory Downunder II workshop at the Mathematical Research Institute MATRIX (March 2022). We thank Ron Aharoni and Zach Hunter for carefully reading an earlier version of this paper and making some very helpful suggestions.

References

[1] RON AHARONI, ELI BERGER, MARIA CHUDNOVSKY, HE GUO, AND SHIRA ZERBIB. Non-uniform degrees and rainbow versions of the Caccetta-Häggkvist conjecture. 2021, arXiv:2110.11183.

[2] RON AHARONI AND JOSEPH BRIGGS. Choice functions. *Order*, pp. 1–19, 2022.

[3] RON AHARONI, MATT DEVOS, SEBASTIÁN GONZÁLEZ HERMOSILLO DE LA MAZA, AMANDA MONTEJANO, AND ROBERT ŠÁMAL. A rainbow version of Mantel's theorem. *Adv. Comb.*, pp. Paper No. 2, 12, 2020.

[4] RON AHARONI, MATTHEW DEVOS, AND RON HOLZMAN. Rainbow triangles and the Caccetta-Häggkvist conjecture. *J. Graph Theory*, 92(4):347–360, 2019.

[5] RON AHARONI AND HE GUO. Rainbow cycles for families of matchings. 2021, arXiv:2110.14332.

[6] KRISTÓF BÉRCZI AND TAMÁS SCHWARCZ. Rainbow and monochromatic circuits and cocircuits in binary matroids. *Discrete Math.*, 345(6):Paper No. 112830, 11, 2022.

[7] BÉLA BOLLOBÁS AND ENDRE SZEMERÉDI. Girth of sparse graphs. *J. Graph Theory*, 39(3):194–200, 2002.

[8] LOUIS CACCETTA AND R. HÄGGKVIST. On minimal digraphs with given girth. In *Proceedings of the Ninth Southeastern Conference on Combinatorics, Graph Theory, and Computing (Florida Atlantic Univ., Boca Raton, Fla., 1978)*, Congress. Numer., XXI, pp. 181–187. Utilitas Math., Winnipeg, Man., 1978.

[9] KATIE CLINCH, JACKSON GOERNER, TONY HUYNH, AND FREDDIE ILLINGWORTH. Notes on Aharoni's rainbow cycle conjecture. 2022, arXiv:2211.07897.

[10] MATT DEVOS, MATTHEW DRESCHER, DARYL FUNK, SEBASTIÁN GONZÁLEZ HERMOSILLO DE LA MAZA, KRYSTAL GUO, TONY HUYNH, BOJAN MOHAR, AND AMANDA MONTEJANO. Short rainbow cycles in graphs and matroids. *J. Graph Theory*, 96(2):192–202, 2021.

[11] HE GUO. Rainbow cycles for matchings, single edges, and triangles. 2022, arXiv:2210.12243.

[12] YAHYA OULD HAMIDOUNE. A note on minimal directed graphs with given girth. *J. Combin. Theory Ser. B*, 43(3):343–348, 1987.

[13] JAN HLADKÝ, DANIEL KRÁL', AND SERGEY NORIN. Counting flags in triangle-free digraphs. *Combinatorica*, 37(1):49–76, 2017.

[14] C. T. HOÀNG AND B. REED. A note on short cycles in digraphs. *Discrete Math.*, 66(1-2):103–107, 1987.

[15] PATRICK HOMPE. The girth of digraphs and concatenating bipartite graphs. *Princeton University Undergraduate Senior Thesis*, 2019.

[16] PATRICK HOMPE AND TONY HUYNH. Aharoni's rainbow cycle conjecture holds up to an additive constant. 2022, arXiv:2212.05697.

[17] PATRICK HOMPE, PETRA PELIKÁNOVÁ, ANETA POKORNÁ, AND SOPHIE SPIRKL. On Aharoni's rainbow generalization of the Caccetta-Häggkvist conjecture. *Discrete Math.*, 344(5):Paper No. 112319, 4, 2021.

[18] PATRICK HOMPE, ZISHEN QU, AND SOPHIE SPIRKL. Improved bounds for the triangle case of Aharoni's rainbow generalization of the Caccetta-Häggkvist conjecture. 2022, arXiv:2206.10733.

[19] PATRICK HOMPE AND SOPHIE SPIRKL. Further approximations for Aharoni's rainbow generalization of the Caccetta-Häggkvist conjecture. *Electron. J. Combin.*, 29(1):Paper No. 1.55, 13, 2022.

[20] JAMES OXLEY. Matroid theory, vol. 21 of *Oxford Graduate Texts in Mathematics*. Oxford University Press, Oxford, 2nd edn., 2011.

[21] ALEXANDER A. RAZBOROV. What is...a flag algebra? *Notices Amer. Math. Soc.*, 60(10):1324–1327, 2013.

[22] P. D. SEYMOUR. Decomposition of regular matroids. *J. Combin. Theory Ser. B*, 28(3):305–359, 1980.

[23] BLAIR D. SULLIVAN. A summary of problems and results related to the Caccetta-Häggkvist conjecture. 2006, arXiv:math/0605646.

Tree-Partitions with Bounded Degree Trees

Marc Distel and David R. Wood

Abstract A *tree-partition* of a graph G is a partition of $V(G)$ such that identifying the vertices in each part gives a tree. It is known that every graph with treewidth k and maximum degree Δ has a tree-partition with parts of size $O(k\Delta)$. We prove the same result with the extra property that the underlying tree has maximum degree $O(\Delta)$.

1 Introduction

For a graph G and a tree T, a *T-partition* of G is a partition $(V_x : x \in V(T))$ of $V(G)$ indexed by the nodes of T, such that for every edge vw of G, if $v \in V_x$ and $w \in V_y$, then $x = y$ or $xy \in E(T)$. The *width* of a T-partition is $\max\{|V_x| : x \in V(T)\}$. The *tree-partition-width*[1] of a graph G is the minimum width of a tree-partition of G.

Tree-partitions were independently introduced by Seese [31] and Halin [24], and have since been widely investigated [6–8, 15, 16, 21, 32, 33]. Applications of tree-partitions include graph drawing [12, 14, 19, 20, 34], graphs of linear growth [11], nonrepetitive graph colouring [2], clustered graph colouring [1, 27], monadic second-order logic [26], network emulations [3, 4, 9, 22], statistical learning theory [35], and the edge-Erdős-Pósa property [13, 23, 28]. Tree-partitions are also related to graph product structure theory since a graph G has a T-partition of width at most k if and only if G is isomorphic to a subgraph of $T \boxtimes K_k$ for some tree T; see [10, 17, 18] for example.

Marc Distel, David R. Wood
School of Mathematics, Monash University, Melbourne, Australia
e-mail: {marc.distel, david.wood}@monash.edu

[1] Tree-partition-width has also been called *strong treewidth* [7, 31].

D. R. Wood et al. (eds.), *2021-2022 MATRIX Annals*, MATRIX Book Series 5,
https://doi.org/10.1007/978-3-031-47417-0_11

Bounded tree-partition-width implies bounded treewidth[2], as noted by Seese [31]. In particular, for every graph G,

$$\operatorname{tw}(G) \leqslant 2\operatorname{tpw}(G) - 1.$$

Of course, $\operatorname{tw}(T) = \operatorname{tpw}(T) = 1$ for every tree T. But in general, $\operatorname{tpw}(G)$ can be much larger than $\operatorname{tw}(G)$. For example, fan graphs on n vertices have treewidth 2 and tree-partition-width $\Omega(\sqrt{n})$. On the other hand, the referee of [15] showed that if the maximum degree and treewidth are both bounded, then so is the tree-partition-width, which is one of the most useful results about tree-partitions. A graph G is *trivial* if $E(G) = \varnothing$. Let $\Delta(G)$ be the maximum degree of G.

Theorem 1 ([15]). *For any non-trivial graph G,*

$$\operatorname{tpw}(G) \leqslant 24(\operatorname{tw}(G)+1)\Delta(G).$$

Theorem 1 is stated in [15] with "$\operatorname{tw}(G)$" instead of "$\operatorname{tw}(G)+1$", but a close inspection of the proof shows that "$\operatorname{tw}(G)+1$" is needed. Wood [33] showed that Theorem 1 is best possible up to the multiplicative constant, and also improved the constant 24 to $9+6\sqrt{2} \approx 17.48$.

This paper considers the maximum degree of T in a T-partition. Consider a tree-partition $(B_x : x \in V(T))$ of a graph G with width k. For each node $x \in V(T)$, there are at most $\sum_{v \in B_x} \deg(v)$ edges between B_x and $G - B_x$. Thus we may assume that $\deg_T(x) \leqslant |B_x|\Delta(G) \leqslant k\Delta(G)$, otherwise delete an 'unused' edge of T and add an edge to T between leaf vertices of the resulting component subtrees. It follows that if $\operatorname{tpw}(G) \leqslant k$ then G has a T-partition of width at most k for some tree T with maximum degree at most $\max\{k\Delta(G), 2\}$. By Theorem 1, every graph G has a T-partition of width at most $24(\operatorname{tw}(G)+1)\Delta(G)$ for some tree T with maximum degree at most $24(\operatorname{tw}(G)+1)\Delta(G)^2$. This fact has been used in several applications of Theorem 1 (see [12, 20] for example). The following theorem improves this upper bound on $\Delta(T)$. Indeed, $\Delta(T)$ is independent of $\operatorname{tw}(G)$.

Theorem 2. *Every non-trivial graph G has a T-partition of width at most*

$$18(\operatorname{tw}(G)+1)\Delta(G)$$

for some tree T with $\Delta(T) \leqslant 6\Delta(G)$.

Theorem 2 enables a $\operatorname{tw}(G)\Delta(G)^2$ term to be replaced by a $\Delta(G)$ term in various results [12, 20].

---

[2] A *tree-decomposition* of a graph G is a collection $(B_x \subseteq V(G) : x \in V(T))$ of subsets of $V(G)$ (called *bags*) indexed by the nodes of a tree T, such that: (a) for every edge $uv \in E(G)$, some bag B_x contains both u and v; and (b) for every vertex $v \in V(G)$, the set $\{x \in V(T) : v \in B_x\}$ induces a non-empty subtree of T. The *width* of a tree-decomposition is the size of the largest bag, minus 1. The *treewidth* of a graph G, denoted by $\operatorname{tw}(G)$, is the minimum width of a tree-decomposition of G. Treewidth is the standard measure of how similar a graph is to a tree. Indeed, a connected graph has treewidth 1 if and only if it is a tree. Treewidth is of fundamental importance in structural and algorithmic graph theory; see [25, 29?] for surveys.

As mentioned above, Wood [33] improved the constant 24 to $9+6\sqrt{2}$ in Theorem 1. By tweaking the constants in the proof of Theorem 2, we match this constant with a small increase in the bound on $\Delta(T)$; see Section 3. We choose to first present the proof with integer coefficients for ease of understanding.

Our final result shows that the linear upper bound on $\Delta(T)$ in Theorem 2 is best possible even for trees.

Proposition 3. *For any integer $\Delta \geqslant 3$ there exist $\alpha > 0$ such that there are infinitely many trees X with maximum degree Δ such that for every tree T with maximum degree less than Δ, every T-partition of X has width at least $|V(X)|^{\alpha}$. Moreover, if $\Delta = 3$ then α can be taken to be arbitrarily close to 1.*

2 Main Proofs

Theorem 2 is implied by the following lemma. The proof is identical to the proof of Theorem 1, except that we pay attention to $\Delta(T)$. Let $\mathbb{N} = \{1,2,\dots\}$.

Lemma 4. *Fix $k,d \in \mathbb{N}$. Let G be a graph with $\operatorname{tw}(G) \leqslant k-1$ and $\Delta(G) \leqslant d$. Then G has a tree-partition $(B_x : x \in V(T))$ of width at most $18kd$ such that $\Delta(T) \leqslant 6d$. Moreover, for any set $S \subseteq V(G)$ with $4k \leqslant |S| \leqslant 12kd$, there exists a tree-partition $(B_x : x \in V(T))$ of G with width at most $18kd$, such that $\Delta(T) \leqslant 6d$ and there exists $z \in V(T)$ such that:*

- $S \subseteq B_z$,
- $|B_z| \leqslant \frac{3}{2}|S| - 2k$,
- $\deg_T(z) \leqslant \frac{|S|}{2k} - 1$.

Proof. We proceed by induction on $|V(G)|$.

Case 1. $|V(G)| < 4k$: Then S is not specified. Let T be the 1-vertex tree with $V(T) = \{x\}$, and let $B_x := V(G)$. Then $(B_x : x \in V(T))$ is the desired tree-partition, since $|B_x| = |V(G)| < 4k \leqslant 18kd$ and $\Delta(T) = 0 \leqslant 6d$.

Now assume that $|V(G)| \geqslant 4k$. If S is not specified, then let S be any set of $4k$ vertices in G.

Case 2. $|V(G-S)| \leqslant 18kd$: Let T be the 2-vertex tree with $V(T) = \{y,z\}$ and $E(T) = \{yz\}$. Note that $\Delta(T) = 1 \leqslant 6d$ and $\deg_T(z) = 1 \leqslant \frac{|S|}{2k} - 1$. Let $B_z := S$ and $B_y := V(G-S)$. Thus $|B_z| = |S| \leqslant \frac{3}{2}|S| - 2k \leqslant 18kd$ and $|B_y| \leqslant |V(G-S)| \leqslant 18kd$. Hence $(B_x : x \in V(T))$ is the desired tree-partition of G.

Now assume that $|V(G-S)| \geqslant 18kd$.

Case 3. $4k \leqslant |S| \leqslant 12k$: Let $S' := \bigcup\{N_G(v) \setminus S : v \in S\}$. Thus $|S'| \leqslant d|S| \leqslant 12kd$. If $|S'| < 4k$ then add $4k - |S'|$ vertices from $V(G-S-S')$ to S', so that $|S'| = 4k$. This is well-defined since $|V(G-S)| \geqslant 18kd \geqslant 4k$, implying $|V(G-S-S')| \geqslant 4k - |S'|$. By induction, there exists a tree-partition $(B_x : x \in V(T'))$ of $G-S$ with width at most $18kd$, such that $\Delta(T') \leqslant 6d$ and there exists $z' \in V(T')$ such that:

- $S' \subseteq B_{z'}$,

- $|B_{z'}| \leqslant \frac{3}{2}|S'| - 2k \leqslant 18kd - 2k$,
- $\deg_{T'}(z') \leqslant \frac{|S'|}{2k} - 1 \leqslant 6d - 1$.

Let T be the tree obtained from T' by adding one new node z adjacent to z'. Let $B_z := S$. So $(B_x : x \in V(T))$ is a tree-partition of G with width at most $\max\{18kd, |S|\} \leqslant \max\{18kd, 12k\} = 18kd$. By construction, $\deg_T(z) = 1 \leqslant \frac{|S|}{2k} - 1$ and $\deg_T(z') = \deg_{T'}(z') + 1 \leqslant (6d-1) + 1 = 6d$. Every other vertex in T has the same degree as in T'. Hence $\Delta(T) \leqslant 6d$, as desired. Finally, $S = B_z$ and $|B_z| = |S| \leqslant \frac{3}{2}|S| - 2k$.

Case 4. $12k \leqslant |S| \leqslant 12kd$: By the separator lemma of Robertson and Seymour [30, (2.6)], there are induced subgraphs G_1 and G_2 of G with $G_1 \cup G_2 = G$ and $|V(G_1 \cap G_2)| \leqslant k$, where $|S \cap V(G_i)| \leqslant \frac{2}{3}|S|$ for each $i \in \{1,2\}$. Let $S_i := (S \cap V(G_i)) \cup V(G_1 \cap G_2)$ for each $i \in \{1,2\}$.

We now bound $|S_i|$. For a lower bound, since $|S \cap V(G_1)| \leqslant \frac{2}{3}|S|$, we have $|S_2| \geqslant |S \setminus V(G_1)| \geqslant \frac{1}{3}|S| \geqslant 4k$. By symmetry, $|S_1| \geqslant 4k$. For an upper bound, $|S_i| \leqslant \frac{2}{3}|S| + k \leqslant 8kd + k \leqslant 12kd$. Also note that $|S_1| + |S_2| \leqslant |S| + 2k$.

We have shown that $4k \leqslant |S_i| \leqslant 12kd$ for each $i \in \{1,2\}$. Thus we may apply induction to G_i with S_i the specified set. Hence there exists a tree-partition $(B^i_x : x \in V(T_i))$ of G_i with width at most $18kd$, such that $\Delta(T_i) \leqslant 6d$ and there exists $z_i \in V(T_i)$ such that:

- $S_i \subseteq B_{z_i}$,
- $|B_{z_i}| \leqslant \frac{3}{2}|S_i| - 2k$,
- $\deg_{T_i}(z_i) \leqslant \frac{|S_i|}{2k} - 1$.

Let T be the tree obtained from the disjoint union of T_1 and T_2 by merging z_1 and z_2 into a vertex z. Let $B_z := B^1_{z_1} \cup B^2_{z_2}$. Let $B_x := B^i_x$ for each $x \in V(T_i) \setminus \{z_i\}$. Since $G = G_1 \cup G_2$ and $V(G_1 \cap G_2) \subseteq B^1_{z_1} \cap B^2_{z_2} \subseteq B_z$, we have that $(B_x : x \in V(T))$ is a tree-partition of G. By construction, $S \subseteq B_z$ and since $V(G_1 \cap G_2) \subseteq B^i_{z_i}$ for each i,

$$\begin{aligned}
|B_z| &\leqslant |B^1_{z_1}| + |B^2_{z_2}| - |V(G_1 \cap G_2)| \\
&\leqslant (\tfrac{3}{2}|S_1| - 2k) + (\tfrac{3}{2}|S_2| - 2k) - |V(G_1 \cap G_2)| \\
&= \tfrac{3}{2}(|S_1| + |S_2|) - 4k - |V(G_1 \cap G_2)| \\
&\leqslant \tfrac{3}{2}(|S| + 2|V(G_1 \cap G_2)|) - 4k - |V(G_1 \cap G_2)| \\
&\leqslant \tfrac{3}{2}|S| + 2|V(G_1 \cap G_2)| - 4k \\
&\leqslant \tfrac{3}{2}|S| - 2k \\
&< 18kd.
\end{aligned}$$

Every other part has the same size as in the tree-partition of G_1 or G_2. So this tree-partition of G has width at most $18kd$. Note that

$$\begin{aligned}
\deg_T(z) = \deg_{T_1}(z_1) + \deg_{T_2}(z_2) &\leqslant (\tfrac{|S_1|}{2k} - 1) + (\tfrac{|S_2|}{2k} - 1) \\
&= \tfrac{|S_1| + |S_2|}{2k} - 2 \\
&\leqslant \tfrac{|S| + 2k}{2k} - 2
\end{aligned}$$

$$= \tfrac{|S|}{2k} - 1$$
$$< 6d.$$

Every other node of T has the same degree as in T_1 or T_2. Thus $\Delta(T) \leqslant 6d$. This completes the proof. □

We now prove the lower bound. For $\Delta, d \in \mathbb{N}$ with $\Delta \geqslant 2$, let $X_{\Delta,d}$ be the tree rooted at a vertex r such that every leaf is at distance d from r and every non-leaf vertex has degree Δ. Observe that $X_{\Delta,d}$ has the maximum number of vertices in a tree with maximum degree Δ and radius d, where

$$|V(X_{\Delta,d})| = 1 + \Delta \sum_{i=0}^{d-1} (\Delta-1)^i.$$

Note that $|V(X_{2,d})| = 2d+1$, and if $\Delta \geqslant 3$ then

$$(\Delta-1)^d \leqslant |V(X_{\Delta,d})| = 1 + \tfrac{\Delta}{\Delta-2}\left((\Delta-1)^d - 1\right) \leqslant 3(\Delta-1)^d.$$

Proposition 3. *For any integer $\Delta \geqslant 3$ there exist $\alpha > 0$ such that there are infinitely many trees X with maximum degree Δ such that for every tree T with maximum degree less than Δ, every T-partition of X has width at least $|V(X)|^\alpha$. Moreover, if $\Delta = 3$ then α can be taken to be arbitrarily close to 1.*

Proof. First suppose that $\Delta \geqslant 4$. Let $d_0 \in \mathbb{N}$ be sufficiently large so that $(\frac{\Delta-1}{\Delta-2})^{d_0} > 3$. Let $\alpha := 1 - \log_{\Delta-1}(3^{1/d_0}(\Delta-2))$, which is positive by the choice of d_0. Let $d \in \mathbb{N}$ with $d \geqslant d_0$. It follows that $(\Delta-1)^{(1-\alpha)d} \geqslant 3(\Delta-2)^d$. Consider any tree-partition $(B_u : u \in V(T))$ of $X_{\Delta,d}$, where T is any tree with maximum degree at most $\Delta - 1$. Let z be the vertex of T such that the root $r \in B_z$. Since adjacent vertices in $X_{\Delta,d}$ belong to adjacent parts or the same part in T, every vertex in T is at distance at most d from z. Thus T has radius at most d, and

$$|V(T)| \leqslant |V(X_{\Delta-1,d})| \leqslant 3(\Delta-2)^d \leqslant (\Delta-1)^{(1-\alpha)d} \leqslant |V(X_{\Delta,d})|^{1-\alpha}.$$

By the pigeon-hole principle, there is a vertex $u \in V(T)$ such that $|B_u| \geqslant \frac{|V(X_{\Delta,d})|}{|V(T)|} \geqslant |V(X_{\Delta,d})|^\alpha$.

Now assume that $\Delta = 3$. Let $\alpha \in (0,1)$, let $d_0 \in \mathbb{N}$ be sufficiently large so that $2d_0 + 1 \leqslant 2^{(1-\alpha)d_0}$, and let $d \geqslant d_0$. So $2d+1 \leqslant 2^{(1-\alpha)d}$. Consider any tree-partition $(B_u : u \in V(T))$ of $X_{3,d}$, where T is any tree with maximum degree at most 2. By the argument above, T has radius at most d, implying

$$|V(T)| \leqslant |V(X_{2,d})| = 2d+1 \leqslant 2^{(1-\alpha)d} \leqslant |V(X_{3,d})|^{1-\alpha}.$$

Again, there is a vertex $u \in V(T)$ such that $|B_u| \geqslant \frac{|V(X_{3,d})|}{|V(T)|} \geqslant |V(X_{3,d})|^\alpha$. □

3 Tweaking the Constants

Consider the following generalisation of Lemma 4.

Lemma 5. *Fix $k,d \in \mathbb{N}$ and $\alpha \in \mathbb{R}$ with $\alpha > 2$. Let G be a graph with $\mathrm{tw}(G) \leqslant k-1$ and $\Delta(G) \leqslant d$. Then G has a tree-partition $(B_x : x \in V(T))$ of width at most*

$$\tfrac{3\alpha(\alpha-1)}{\alpha-2}kd - \tfrac{\alpha}{\alpha-2}k$$

such that $\Delta(T) \leqslant \frac{3\alpha}{\alpha-2}d + \frac{\alpha-4}{\alpha-2}$. Moreover, for any set $S \subseteq V(G)$ with $\alpha k \leqslant |S| \leqslant 3\alpha kd$, there exists a tree-partition $(B_x : x \in V(T))$ of G with width at most $\frac{3\alpha(\alpha-1)}{\alpha-2}kd - \frac{\alpha}{\alpha-2}k$, such that $\Delta(T) \leqslant \frac{3\alpha}{\alpha-2}d + \frac{\alpha-4}{\alpha-2}$ and there exists $z \in V(T)$ such that:

- $S \subseteq B_z$,
- $|B_z| \leqslant \frac{\alpha-1}{\alpha-2}|S| - \frac{\alpha}{\alpha-2}k$,
- $\deg_T(z) \leqslant \frac{1}{(\alpha-2)k}|S| - \frac{2}{\alpha-2}$.

Proof. We proceed by induction on $|V(G)|$.

Case 1. $|V(G)| < \alpha k$: Then S is not specified. Let T be the 1-vertex tree with $V(T) = \{x\}$, and let $B_x := V(G)$. Then $(B_x : x \in V(T))$ is the desired tree-partition, since $|B_x| = |V(G)| < \alpha k \leqslant \frac{3\alpha(\alpha-1)}{\alpha-2}kd - \frac{\alpha}{\alpha-2}k$ and $\Delta(T) = 0 \leqslant \frac{3\alpha}{\alpha-2}d + \frac{\alpha-4}{\alpha-2}$.

Now assume that $|V(G)| \geqslant \alpha k$. If S is not specified, then let S be any set of $\lceil \alpha k \rceil$ vertices in G (implying $|S| \leqslant 3\alpha kd$).

Case 2. $|V(G-S)| \leqslant \frac{3\alpha(\alpha-1)}{\alpha-2}kd - \frac{\alpha}{\alpha-2}k$: Let T be the 2-vertex tree with $V(T) = \{y,z\}$ and $E(T) = \{yz\}$. Note that $\Delta(T) = 1 \leqslant \frac{3\alpha}{\alpha-2}d + \frac{\alpha-4}{\alpha-2}$ and $\deg_T(z) = 1 \leqslant \frac{1}{(\alpha-2)k}|S| - \frac{2}{\alpha-2}$. Let $B_z := S$ and $B_y := V(G-S)$. Thus $|B_z| = |S| \leqslant \frac{\alpha-1}{\alpha-2}|S| - \frac{\alpha}{\alpha-2}k \leqslant \frac{3\alpha(\alpha-1)}{\alpha-2}kd - \frac{\alpha}{\alpha-2}k$ and $|B_y| \leqslant |V(G-S)| \leqslant \frac{3\alpha(\alpha-1)}{\alpha-2}kd$. Hence $(B_x : x \in V(T))$ is the desired tree-partition of G.

Now assume that $|V(G-S)| \geqslant \frac{3\alpha(\alpha-1)}{\alpha-2}kd - \frac{\alpha}{\alpha-2}k$.

Case 3. $\alpha k \leqslant |S| \leqslant 3\alpha k$: Let $S' := \bigcup_{v \in S} N_G(v) \setminus S$. Thus $|S'| \leqslant d|S| \leqslant 3\alpha kd$. If $|S'| < \alpha k$ then add $\alpha k - |S'|$ vertices from $V(G-S-S')$ to S', so that $|S'| = \alpha k$. This is well-defined since $|V(G-S)| \geqslant \frac{3\alpha(\alpha-1)}{\alpha-2}kd - \frac{\alpha}{\alpha-2}k \geqslant \alpha k$, implying $|V(G-S-S')| \geqslant \alpha k - |S'|$. By induction, there exists a tree-partition $(B_x : x \in V(T'))$ of $G-S$ with width at most $\frac{3\alpha(\alpha-1)}{\alpha-2}kd - \frac{\alpha}{\alpha-2}k$, such that $\Delta(T') \leqslant \frac{3\alpha}{\alpha-2}d + \frac{\alpha-4}{\alpha-2}$ and there exists $z' \in V(T')$ such that:

- $S' \subseteq B_{z'}$,
- $|B_{z'}| \leqslant \frac{\alpha-1}{\alpha-2}|S'| - \frac{\alpha}{\alpha-2}k \leqslant \frac{3\alpha(\alpha-1)}{\alpha-2}kd - \frac{\alpha}{\alpha-2}k$,
- $\deg_{T'}(z') \leqslant \frac{1}{(\alpha-2)k}|S'| - \frac{2}{\alpha-2} \leqslant \frac{3\alpha}{\alpha-2}d - \frac{2}{\alpha-2}$.

Let T be the tree obtained from T' by adding one new node z adjacent to z'. Let $B_z := S$. So $(B_x : x \in V(T))$ is a tree-partition of G with width at most

$$\max\{\tfrac{3\alpha(\alpha-1)}{\alpha-2}kd - \tfrac{\alpha}{\alpha-2}k, |S|\} \leqslant \max\{\tfrac{3\alpha(\alpha-1)}{\alpha-2}kd - \tfrac{\alpha}{\alpha-2}k, 3\alpha k\} = \tfrac{3\alpha(\alpha-1)}{\alpha-2}kd - \tfrac{\alpha}{\alpha-2}k.$$

By construction,

$$\deg_T(z) = 1 \leqslant \tfrac{1}{(\alpha-2)k}|S| - \tfrac{2}{\alpha-2} \quad \text{and}$$
$$\deg_T(z') = \deg_{T'}(z') + 1 \leqslant \tfrac{3\alpha}{\alpha-2}d - \tfrac{2}{\alpha-2} + 1 = \tfrac{3\alpha}{\alpha-2}d + \tfrac{\alpha-4}{\alpha-2}.$$

Every other vertex in T has the same degree as in T'. Hence $\Delta(T) \leqslant \frac{3\alpha}{\alpha-2}d + \frac{\alpha-4}{\alpha-2}$, as desired. Finally, $S = B_z$ and $|B_z| = |S| \leqslant \frac{\alpha-1}{\alpha-2}|S| - \frac{\alpha}{\alpha-2}k$.

Case 4. $3\alpha k \leqslant |S| \leqslant 3\alpha kd$: By the separator lemma of Robertson and Seymour [30, (2.6)], there are induced subgraphs G_1 and G_2 of G with $G_1 \cup G_2 = G$ and $|V(G_1 \cap G_2)| \leqslant k$, where $|S \cap V(G_i)| \leqslant \frac{2}{3}|S|$ for each $i \in \{1,2\}$. Let $S_i := (S \cap V(G_i)) \cup V(G_1 \cap G_2)$ for each $i \in \{1,2\}$.

We now bound $|S_i|$. For a lower bound, since $|S \cap V(G_1)| \leqslant \frac{2}{3}|S|$, we have $|S_2| \geqslant |S \setminus V(G_1)| \geqslant \frac{1}{3}|S| \geqslant \frac{1}{3}3\alpha k \geqslant \alpha k$. By symmetry, $|S_1| \geqslant \alpha k$. For an upper bound, $|S_i| \leqslant \frac{2}{3}|S| + k \leqslant 2\alpha kd + k \leqslant 3\alpha kd$. Also note that $|S_1| + |S_2| \leqslant |S| + 2|V(G_1 \cap G_2)| \leqslant |S| + 2k$.

We have shown that $\alpha k \leqslant |S_i| \leqslant 3\alpha kd$ for each $i \in \{1,2\}$. Thus we may apply induction to G_i with S_i the specified set. Hence there exists a tree-partition $(B^i_x : x \in V(T_i))$ of G_i with width at most $\frac{3\alpha(\alpha-1)}{\alpha-2}kd - \frac{\alpha}{\alpha-2}k$, such that $\Delta(T_i) \leqslant \frac{3\alpha}{\alpha-2}d + \frac{\alpha-4}{\alpha-2}$ and there exists $z_i \in V(T_i)$ such that:

- $S_i \subseteq B_{z_i}$,
- $|B_{z_i}| \leqslant \frac{\alpha-1}{\alpha-2}|S_i| - \frac{\alpha}{\alpha-2}k$,
- $\deg_{T_i}(z_i) \leqslant \frac{1}{(\alpha-2)k}|S_i| - \frac{2}{\alpha-2}$.

Let T be the tree obtained from the disjoint union of T_1 and T_2 by merging z_1 and z_2 into a vertex z. Let $B_z := B^1_{z_1} \cup B^2_{z_2}$. Let $B_x := B^i_x$ for each $x \in V(T_i) \setminus \{z_i\}$. Since $G = G_1 \cup G_2$ and $V(G_1 \cap G_2) = B^1_{z_1} \cap B^2_{z_2} \subseteq B_z$, we have that $(B_x : x \in V(T))$ is a tree-partition of G. By construction, $S \subseteq B_z$ and since $V(G_1 \cap G_2) \subseteq B^i_{z_i}$ for each i,

$$\begin{aligned}
|B_z| &\leqslant |B^1_{z_1}| + |B^2_{z_2}| - |V(G_1 \cap G_2)| \\
&\leqslant \left(\tfrac{\alpha-1}{\alpha-2}|S_1| - \tfrac{\alpha}{\alpha-2}k\right) + \left(\tfrac{\alpha-1}{\alpha-2}|S_2| - \tfrac{\alpha}{\alpha-2}k\right) - |V(G_1 \cap G_2)| \\
&= \tfrac{\alpha-1}{\alpha-2}(|S_1| + |S_2|) - \tfrac{2\alpha}{\alpha-2}k - |V(G_1 \cap G_2)| \\
&\leqslant \tfrac{\alpha-1}{\alpha-2}(|S| + 2|V(G_1 \cap G_2)|) - \tfrac{2\alpha}{\alpha-2}k - |V(G_1 \cap G_2)| \\
&= \tfrac{\alpha-1}{\alpha-2}|S| - \tfrac{2\alpha}{\alpha-2}k + \tfrac{\alpha}{\alpha-2}|V(G_1 \cap G_2)| \\
&\leqslant \tfrac{\alpha-1}{\alpha-2}|S| - \tfrac{2\alpha}{\alpha-2}k + \tfrac{\alpha}{\alpha-2}k \\
&= \tfrac{\alpha-1}{\alpha-2}|S| - \tfrac{\alpha}{\alpha-2}k \\
&\leqslant \tfrac{3\alpha(\alpha-1)}{\alpha-2}kd - \tfrac{\alpha}{\alpha-2}k.
\end{aligned}$$

Every other part has the same size as in the tree-partition of G_1 or G_2. So this tree-partition of G has width at most $\frac{3\alpha(\alpha-1)}{\alpha-2}kd - \frac{\alpha}{\alpha-2}k$. Note that

$$\deg_T(z) = \deg_{T_1}(z_1) + \deg_{T_2}(z_2)$$

$$
\begin{aligned}
&\leqslant \tfrac{1}{(\alpha-2)k}|S_1| - \tfrac{2}{\alpha-2} + \tfrac{1}{(\alpha-2)k}|S_2| - \tfrac{2}{\alpha-2}\\
&\leqslant \tfrac{1}{(\alpha-2)k}(|S_1|+|S_2|) - \tfrac{4}{\alpha-2}\\
&\leqslant \tfrac{1}{(\alpha-2)k}(|S|+2k) - \tfrac{4}{\alpha-2}\\
&\leqslant \tfrac{1}{(\alpha-2)k}|S| - \tfrac{2}{\alpha-2}\\
&\leqslant \tfrac{3\alpha}{\alpha-2}d - \tfrac{2}{\alpha-2}\\
&< \tfrac{3\alpha}{\alpha-2}d + \tfrac{\alpha-4}{\alpha-2}.
\end{aligned}
$$

Every other node of T has the same degree as in T_1 or T_2. Thus $\Delta(T) \leqslant \frac{3\alpha}{\alpha-2}d + \frac{\alpha-4}{\alpha-2}$. This completes the proof. □

Lemma 5 with $\alpha = 4$ implies the following slight strengthening of Theorem 2.

Theorem 6. *Every non-trivial graph G has a T-partition of width at most*

$$2(\mathrm{tw}(G)+1)(9\Delta(G)-1),$$

for some tree T with $\Delta(T) \leqslant 6\Delta(G)$.

Lemma 5 with $\alpha = 2+\sqrt{2}$ (chosen to minimise $\frac{3\alpha(\alpha-1)}{\alpha-2}$) implies the next result.

Theorem 7. *Every non-trivial graph G has a T-partition of width at*

$$(1+\sqrt{2})(\mathrm{tw}(G)+1)(3(1+\sqrt{2})\Delta(G)-1),$$

for some tree T with $\Delta(T) \leqslant (3+3\sqrt{2})\Delta(G) - 3(\sqrt{2}-1)$.

Acknowledgements Research of Distel is supported by an Australian Government Research Training Program Scholarship. Research of Wood is supported by the Australian Research Council.

References

[1] Noga Alon, Guoli Ding, Bogdan Oporowski, and Dirk Vertigan. Partitioning into graphs with only small components. *J. Combin. Theory Ser. B*, 87(2):231–243, 2003.

[2] János Barát and David R. Wood. Notes on nonrepetitive graph colouring. *Electron. J. Combin.*, 15:R99, 2008.

[3] Hans L. Bodlaender. The complexity of finding uniform emulations on fixed graphs. *Inform. Process. Lett.*, 29(3):137–141, 1988.

[4] Hans L. Bodlaender. The complexity of finding uniform emulations on paths and ring networks. *Inform. and Comput.*, 86(1):87–106, 1990.

[5] Hans L. Bodlaender. A partial k-arboretum of graphs with bounded treewidth. *Theoret. Comput. Sci.*, 209(1-2):1–45, 1998.

[6] HANS L. BODLAENDER. A note on domino treewidth. *Discrete Math. Theor. Comput. Sci.*, 3(4):141–150, 1999.

[7] HANS L. BODLAENDER AND JOOST ENGELFRIET. Domino treewidth. *J. Algorithms*, 24(1):94–123, 1997.

[8] HANS L. BODLAENDER, CARLA GROENLAND, AND HUGO JACOB. On the parameterized complexity of computing tree-partitions. 2022, arXiv:2206.11832.

[9] HANS L. BODLAENDER AND JAN VAN LEEUWEN. Simulation of large networks on smaller networks. *Inform. and Control*, 71(3):143–180, 1986.

[10] RUTGER CAMPBELL, KATIE CLINCH, MARC DISTEL, J. PASCAL GOLLIN, KEVIN HENDREY, ROBERT HICKINGBOTHAM, TONY HUYNH, FREDDIE ILLINGWORTH, YOURI TAMITEGAMA, JANE TAN, AND DAVID R. WOOD. Product structure of graph classes with bounded treewidth. 2022, arXiv:2206.02395.

[11] RUTGER CAMPBELL, MARC DISTEL, J. PASCAL GOLLIN, DANIEL J. HARVEY, KEVIN HENDREY, ROBERT HICKINGBOTHAM, BOJAN MOHAR, AND DAVID R. WOOD. Graphs of linear growth have bounded treewidth. 2022, arXiv:tba.

[12] PAZ CARMI, VIDA DUJMOVIĆ, PAT MORIN, AND DAVID R. WOOD. Distinct distances in graph drawings. *Electron. J. Combin.*, 15:R107, 2008.

[13] DIMITRIS CHATZIDIMITRIOU, JEAN-FLORENT RAYMOND, IGNASI SAU, AND DIMITRIOS M. THILIKOS. An O(log OPT)-approximation for covering and packing minor models of θ_r. *Algorithmica*, 80(4):1330–1356, 2018.

[14] EMILIO DI GIACOMO, GIUSEPPE LIOTTA, AND HENK MEIJER. Computing straight-line 3D grid drawings of graphs in linear volume. *Comput. Geom. Theory Appl.*, 32(1):26–58, 2005.

[15] GUOLI DING AND BOGDAN OPOROWSKI. Some results on tree decomposition of graphs. *J. Graph Theory*, 20(4):481–499, 1995.

[16] GUOLI DING AND BOGDAN OPOROWSKI. On tree-partitions of graphs. *Discrete Math.*, 149(1–3):45–58, 1996.

[17] VIDA DUJMOVIĆ, LOUIS ESPERET, PAT MORIN, BARTOSZ WALCZAK, AND DAVID R. WOOD. Clustered 3-colouring graphs of bounded degree. *Combin. Probab. Comput.*, 31(1):123–135, 2022.

[18] VIDA DUJMOVIĆ, GWENAËL JORET, PIOTR MICEK, PAT MORIN, AND DAVID R. WOOD. Bad news for product structure of bounded-degree graphs. 2022, arXiv:tba.

[19] VIDA DUJMOVIĆ, PAT MORIN, AND DAVID R. WOOD. Layout of graphs with bounded tree-width. *SIAM J. Comput.*, 34(3):553–579, 2005.

[20] VIDA DUJMOVIĆ, MATTHEW SUDERMAN, AND DAVID R. WOOD. Graph drawings with few slopes. *Comput. Geom. Theory Appl.*, 38:181–193, 2007.

[21] ANDERS EDENBRANDT. Quotient tree partitioning of undirected graphs. *BIT*, 26(2):148–155, 1986.

[22] JOHN P. FISHBURN AND RAPHAEL A. FINKEL. Quotient networks. *IEEE Trans. Comput.*, C-31(4):288–295, 1982.

[23] ARCHONTIA C. GIANNOPOULOU, O-JOUNG KWON, JEAN-FLORENT RAYMOND, AND DIMITRIOS M. THILIKOS. Packing and covering immersion models of planar subcubic graphs. In PINAR HEGGERNES, ed., *Proc. 42nd Int'l Workshop on Graph-Theoretic Concepts in Computer Science* (WG 2016), vol. 9941 of *Lecture Notes in Comput. Sci.*, pp. 74–84. 2016.

[24] RUDOLF HALIN. Tree-partitions of infinite graphs. *Discrete Math.*, 97:203–217, 1991.

[25] DANIEL J. HARVEY AND DAVID R. WOOD. Parameters tied to treewidth. *J. Graph Theory*, 84(4):364–385, 2017.

[26] DIETRICH KUSKE AND MARKUS LOHREY. Logical aspects of Cayley-graphs: the group case. *Ann. Pure Appl. Logic*, 131(1–3):263–286, 2005.

[27] CHUN-HUNG LIU AND SANG-IL OUM. Partitioning H-minor free graphs into three subgraphs with no large components. *J. Combin. Theory Ser. B*, 128:114–133, 2018.

[28] JEAN-FLORENT RAYMOND AND DIMITRIOS M. THILIKOS. Recent techniques and results on the Erdős-Pósa property. *Discrete Appl. Math.*, 231:25–43, 2017.

[29] BRUCE A. REED. Algorithmic aspects of tree width. In *Recent advances in algorithms and combinatorics*, vol. 11, pp. 85–107. Springer, 2003.

[30] NEIL ROBERTSON AND PAUL SEYMOUR. Graph minors. II. Algorithmic aspects of tree-width. *J. Algorithms*, 7(3):309–322, 1986.

[31] DETLEF SEESE. Tree-partite graphs and the complexity of algorithms. In LOTHAR BUDACH, ed., *Proc. Int'l Conf. on Fundamentals of Computation Theory*, vol. 199 of *Lecture Notes in Comput. Sci.*, pp. 412–421. Springer, 1985.

[32] DAVID R. WOOD. Vertex partitions of chordal graphs. *J. Graph Theory*, 53(2):167–172, 2006.

[33] DAVID R. WOOD. On tree-partition-width. *European J. Combin.*, 30(5):1245–1253, 2009.

[34] DAVID R. WOOD AND JAN ARNE TELLE. Planar decompositions and the crossing number of graphs with an excluded minor. *New York J. Math.*, 13:117–146, 2007.

[35] RUI-RAY ZHANG AND MASSIH-REZA AMINI. Generalization bounds for learning under graph-dependence: A survey. 2022, arXiv:2203.13534.

Treewidth, Circle Graphs and Circular Drawings

Robert Hickingbotham, Freddie Illingworth, Bojan Mohar, and David R. Wood

Abstract A circle graph is an intersection graph of a set of chords of a circle. We describe the unavoidable induced subgraphs of circle graphs with large treewidth. This includes examples that are far from the 'usual suspects'. Our results imply that treewidth and Hadwiger number are linearly tied on the class of circle graphs, and that the unavoidable induced subgraphs of a vertex-minor-closed class with large treewidth are the usual suspects if and only if the class has bounded rank-width. Using the same tools, we also study the treewidth of graphs G that have a circular drawing whose crossing graph is well-behaved in some way. In this setting, we show that if the crossing graph is K_t-minor-free, then G has treewidth at most $12t-23$ and has no $K_{2,4t}$-topological minor. On the other hand, we show that there are graphs with arbitrarily large Hadwiger number that have circular drawings whose crossing graphs are 2-degenerate.

1 Introduction

This paper studies the treewidth of graphs that are defined by circular drawings. Treewidth is the standard measure of how similar a graph is to a tree, and is of fundamental importance in structural and algorithmic graph theory. The motivation for

Robert Hickingbotham, David Wood
Monash University, Melbourne, Australia,
e-mail: {robert.hickingbotham,david.wood}@monash.edu

Freddie Illingworth
University College London, London, United Kingdom,
e-mail: f.illingworth@ucl.ac.uk

Bojan Mohar
Department of Mathematics, Simon Fraser University, Burnaby, BC, Canada,
e-mail: mohar@sfu.ca

D. R. Wood et al. (eds.), *2021-2022 MATRIX Annals*, MATRIX Book Series 5,
https://doi.org/10.1007/978-3-031-47417-0_12

this study is two-fold. See Section 2 for definitions omitted from this introduction. In this extended abstract, most proofs are omitted; see [28] for all the details.

1.1 Theme #1: Circle Graphs

A *circle graph* is the intersection graph of a set of chords of a circle. Circle graphs form a widely studied graph class [15, 17, 23, 25, 26, 31, 33] and there have been several recent breakthroughs concerning them. In the study of graph colourings, Davies and McCarty [17] showed that circle graphs are quadratically χ-bounded improving upon a previous longstanding exponential upper bound. Davies [15] further improved this bound to $\chi(G) \in O(\omega(G) \log \omega(G))$, which is best possible. Circle graphs are also fundamental to the study of vertex-minors and are conjectured to lie at the heart of a global structure theorem for vertex-minor-closed graph classes (see [36]). To this end, Geelen, Kwon, McCarty, and Wollan [26] recently proved an analogous result to the excluded grid minor theorem for vertex-minors using circle graphs. In particular, they showed that a vertex-minor-closed graph class has bounded rankwidth if and only if it excludes a circle graph as a vertex-minor. For further motivation and background on circle graphs, see [16, 36].

Our first contribution determines when a circle graph has large treewidth.

Theorem 1. *Let $t \in \mathbb{N}$ and let G be a circle graph with treewidth at least $12t+2$. Then G contains an induced subgraph H that consists of t vertex-disjoint cycles $(C_1,\dots,C_t)$ such that for all $i<j$ every vertex of C_i has at least two neighbours in C_j. Moreover, every vertex of G has at most four neighbours in any C_i $(1 \leqslant i \leqslant t)$.*

Observe that in Theorem 1 the subgraph H has a K_t-minor obtained by contracting each of the cycles C_i to a single vertex, implying that H has treewidth at least $t-1$. Moreover, since circle graphs are closed under taking induced subgraphs, H is also a circle graph. We now highlight several consequences of Theorem 1.

First, Theorem 1 describes the unavoidable induced subgraphs of circle graphs with large treewidth. Recently, there has been significant interest in understanding the induced subgraphs of graphs with large treewidth [2, 3, 4, 5, 6, 7, 8, 11, 34, 41, 46]. To date, most of the results in this area have focused on graph classes where the unavoidable induced subgraphs are the following graphs, the *usual suspects*: a complete graph K_t, a complete bipartite graph $K_{t,t}$, a subdivision of the $(t \times t)$-wall, or the line graph of a subdivision of the $(t \times t)$-wall (see [46] for definitions). Circle graphs do not contain subdivisions of large walls nor the line graphs of subdivisions of large walls and there are circle graphs of large treewidth that do not contain large complete graphs nor large complete bipartite graphs (see Theorem 17). To the best of our knowledge this is the first result to describe the unavoidable induced subgraphs of the large treewidth graphs in a natural hereditary class when they are not the usual suspects. Later we show that the unavoidable induced subgraphs of graphs with large treewidth in a vertex-minor-closed class $\mathcal{G}$ are the usual suspects if and only if $\mathcal{G}$ has bounded rankwidth (see Theorem 18).

Second, the subgraph H in Theorem 1 is an explicit witness to the large treewidth of G (with only a multiplicative loss). Circle graphs being χ-bounded says that circle graphs with large chromatic number must contain a large clique witnessing this. Theorem 1 can therefore be considered to be a treewidth analogue to the χ-boundedness of circle graphs.

Third, since the subgraph H has a K_t-minor, it follows that every circle graph contains a complete minor whose order is at least one twelfth of its treewidth. This is in stark contrast to the general setting where there are K_5-minor-free graphs with arbitrarily large treewidth (for example, grids). Theorem 1 also implies the following relationship between the treewidth, Hadwiger number and Hajós number of circle graphs (see Section 5)[1].

Theorem 2. *For the class of circle graphs, the treewidth and Hadwiger number are linearly tied. Moreover, the Hajós number is quadratically tied to both of them. Both 'linear' and 'quadratic' are best possible.*

1.2 Theme #2: Graph Drawing

The second thread of this paper aims to understand the relationship between circular drawings of graphs and their crossing graphs. A *circular drawing* (also called *convex* drawing) of a graph places the vertices on a circle with edges drawn as straight line segments. Circular drawings are well-studied by the graph drawing community. The *crossing graph* of a drawing D of a graph G has vertex-set $E(G)$ where two vertices are adjacent if the corresponding edges cross. Circle graphs are precisely the crossing graphs of circular drawings. If a graph has a circular drawing with a well-behaved crossing graph, must the graph itself also be well-behaved? Graphs that have a circular drawing with no crossings are exactly the outerplanar graphs, which have treewidth at most 2. Put another way, outerplanar graphs are those that have a circular drawing whose crossing graph is K_2-minor-free. Our next result extends this fact, relaxing 'K_2-minor-free' to 'K_t-minor-free'.

Theorem 3. *For every integer $t \geqslant 3$, if a graph G has a circular drawing where the crossing graph has no K_t-minor, then G has treewidth at most $12t - 23$.*

Theorem 3 says that G having large treewidth is sufficient to force a complicated crossing graph in every circular drawing of G. A topological $K_{2,4t}$-minor also suffices.

Theorem 4. *If a graph G has a circular drawing where the crossing graph has no K_t-minor, then G contains no $K_{2,4t}$ as a topological minor.*

Outerplanar graphs are exactly those graphs that have treewidth at most 2 and exclude a topological $K_{2,3}$-minor. As such, Theorems 3 and 4 extends these struc-

---

[1] For a graph class $\mathcal{G}$, two graph parameters α and β are *tied on* $\mathcal{G}$ if there exists a function f such that $\alpha(G) \leqslant f(\beta(G))$ and $\beta(G) \leqslant f(\alpha(G))$ for every graph $G \in \mathcal{G}$. Moreover, α and β are *quadratically/linearly tied on* $\mathcal{G}$ if f may be taken to be quadratic/linear.

tural properties of outerplanar graphs to graphs with circular drawings whose crossing graphs are K_t-minor-free. We also prove a product structure theorem for such graphs, showing that every graph that has a circular drawing whose crossing graph has no K_t-minor is isomorphic to a subgraph of $H \boxtimes K_{O(t^3)}$ where $\text{tw}(H) \leqslant 2$ (see Corollary 11).

In the other direction, we consider sufficient conditions for a graph G to have a circular drawing whose crossing graph has no K_t-minor. By Theorems 3 and 4, G must have bounded treewidth and no $K_{2,4t}$-topological minor. While these conditions are necessary, we show that they are not sufficient, but that bounded treewidth with bounded maximum degree is; see Lemma 14 and Proposition 15 in Section 4.2 for details.

In addition, we show that the assumption in Theorem 3 that the crossing graph has bounded Hadwiger number cannot be weakened to bounded degeneracy. In particular, we construct graphs with arbitrarily large complete graph minors that have a circular drawing whose crossing graph is 2-degenerate (Theorem 16). This result has applications to the study of general (non-circular) graph drawings, and in particular, leads to the solution of an open problem asked by Hickingbotham and Wood [29].

Our proofs of Theorems 1, 2 and 3 are all based on the same core lemmas in Section 3. The results about circle graphs are in Section 5, while the results about graph drawings are in Section 4.

2 Preliminaries

2.1 Graph Basics

We use standard graph-theoretic definitions and notation; see [18].

For a tree T, a *T-decomposition* of a graph G is a collection $\mathscr{W} = (W_x : x \in V(T))$ of subsets of $V(G)$ indexed by the nodes of T such that (i) for every edge $vw \in E(G)$, there exists a node $x \in V(T)$ with $v, w \in W_x$; and (ii) for every vertex $v \in V(G)$, the set $\{x \in V(T) : v \in W_x\}$ induces a (connected) subtree of T. Each set W_x in $\mathscr{W}$ is called a *bag*. The *width* of $\mathscr{W}$ is $\max\{|W_x| : x \in V(T)\} - 1$. A *tree-decomposition* is a T-decomposition for any tree T. The *treewidth* $\text{tw}(G)$ of a graph G is the minimum width of a tree-decomposition of G. When T is a path, $\mathscr{W}$ is a *path-decomposition*. The *pathwidth*, $\text{pw}(G)$, of G is the minimum width of a path-decomposition of G.

A graph H is a *minor* of a graph G if H is isomorphic to a graph obtained from a subgraph of G by contracting edges. The *Hadwiger number*, $h(G)$, of a graph G is the maximum integer t such that K_t is a minor of G.

A graph $\tilde{G}$ is a *subdivision* of a graph G if $\tilde{G}$ can be obtained from G by replacing each edge vw by a path P_{vw} with endpoints v and w (internally disjoint from the rest of $\tilde{G}$). A graph H is a *topological minor* of G if a subgraph of G is isomorphic to a subdivision of H. The *Hajós number*, $h'(G)$, of G is the maximum integer t such that K_t is a topological minor of G. A graph G is *H-topological minor-free* if H is not a topological minor of G.

It is well-known that for every graph G, $h'(G) \leqslant h(G) \leqslant \text{tw}(G)+1$.

A *graph class* is a collection of graphs closed under isomorphism. A *graph parameter* is a real-valued function α defined on all graphs such that $\alpha(G_1) = \alpha(G_2)$ whenever G_1 and G_2 are isomorphic.

2.2 Drawings of Graphs

A *drawing* of a graph G is a function ϕ that maps each vertex $v \in V(G)$ to a point $\phi(v) \in \mathbb{R}^2$ and maps each edge $e = vw \in E(G)$ to a non-self-intersecting curve $\phi(e)$ in $\mathbb{R}^2$ with endpoints $\phi(v)$ and $\phi(w)$, such that:

- $\phi(v) \neq \phi(w)$ for all distinct vertices v and w;
- $\phi(x) \notin \phi(e)$ for each edge $e = vw$ and each vertex $x \in V(G) \setminus \{v, w\}$;
- each pair of edges intersect at a finite number of points: $\phi(e) \cap \phi(f)$ is finite for all distinct edge e, f; and
- no three edges internally intersect at a common point: for distinct edges e, f, g the only possible element of $\phi(e) \cap \phi(f) \cap \phi(g)$ is $\phi(v)$ where v is a vertex incident to all of e, f, g.

A *crossing* of distinct edges $e = uv$ and $f = xy$ is a point in $(\phi(e) \cap \phi(f)) \setminus \{\phi(u), \phi(v), \phi(x), \phi(y)\}$; that is, an internal intersection point. A *plane graph* is a graph G equipped with a drawing of G with no crossings.

The *crossing graph* of a drawing D of a graph G is the graph X_D with vertex set $E(G)$, where for each crossing between edges e and f in D, there is an edge of X_D between the vertices corresponding to e and f. Note that X_D is actually a multigraph, where the multiplicity of ef equals the number of times e and f cross in D. In most drawings that we consider, each pair of edges cross at most once, in which case X_D has no parallel edges.

Numerous papers have studied graphs that have a drawing whose crossing graph is well-behaved in some way; for example, see [9, 20, 21, 24, 40]. A drawing is *circular* if the vertices are positioned on a circle and the edges are straight line segments. A theme of this paper is to study circular drawings D in which X_D is well-behaved in some way. Many papers have considered properties of X_D in this setting; see for example [19, 22, 45].

3 Tools

In this section, we introduce two auxiliary graphs that are useful tools for proving our main theorems. For a drawing D of a graph G, the *planarisation*, P_D, of D is the plane graph obtained by replacing each crossing with a dummy vertex of degree 4.

Note that P_D depends upon the drawing D (and not just upon G). Figure 1 shows a drawing and its planarisation.

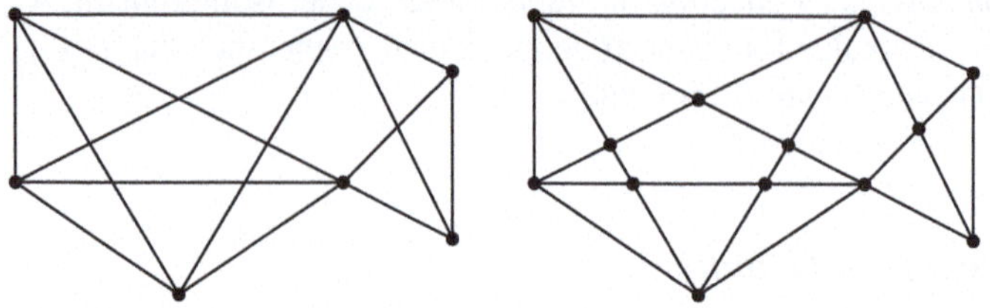

(a) Drawing D of a graph G. (b) Planarisation P_D.

Fig. 1: A drawing and its planarisation.

For a drawing D of a graph G, the *map graph*, M_D, of D is obtained as follows. First let P_D be the planarisation of D. The vertices of M_D are the faces of P_D, where two vertices are adjacent in M_D if the corresponding faces share a vertex. If G is itself a plane graph, then it is already drawn in the plane and so we may talk about the map graph, M_G, of G. Note that all map graphs are connected graphs. Figure 2 shows the map graph M_D for the drawing D in Figure 1.

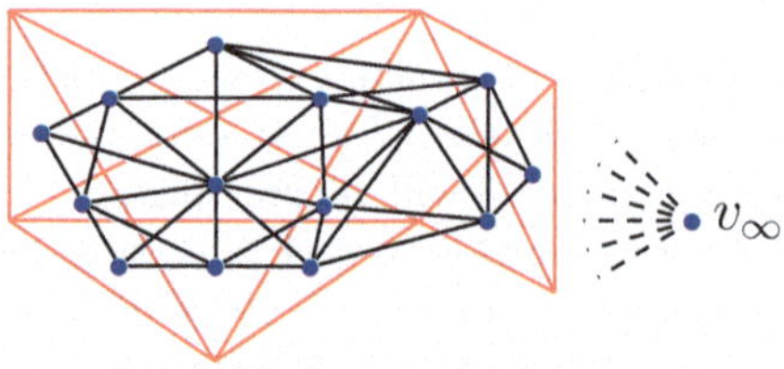

Fig. 2: Map graph M_D. v_∞ is the vertex corresponding to the outer face: it is adjacent to all vertices except the central vertex of degree 10.

The *radius* of a connected graph G, denoted $\text{rad}(G)$, is the minimum non-negative integer r such that for some vertex $v \in V(G)$ and for every vertex $w \in V(G)$ we have $\text{dist}_G(v,w) \leqslant r$.

The following results say that the radius of M_D provides a useful bridge between the treewidth of G, the treewidth of X_D, and the subgraphs of X_D. First, the radius of M_D acts as an upper bound for both the treewidth of G and the treewidth of X_D.

Theorem 5. *For every drawing D of a graph G,*

$$\text{tw}(G) \leqslant 6\,\text{rad}(M_D) + 7 \quad \text{and} \quad \text{tw}(X_D) \leqslant 6\,\text{rad}(M_D) + 7.$$

It is not surprising that treewidth and radius are related for drawings. A classical result of Robertson and Seymour [44, (2.7)] says that $\text{tw}(G) \leqslant 3\,\text{rad}(G) + 1$ for every connected planar graph G. Several authors improved this bound as follows.

Lemma 6 ([10, 21])**.** *For every connected planar graph G,*

$$\operatorname{tw}(G) \leqslant 3\operatorname{rad}(G).$$

The next lemma says that if a planar graph G has large treewidth, then the map graph of any plane drawing of G has large radius. A *triangulation* of a plane G is a plane supergraph of G on the same vertex set and where each face is a triangle.

Lemma 7. *Let G be a plane graph with map graph M_G. Then there is a plane triangulation H of G with* $\operatorname{rad}(H) \leqslant \operatorname{rad}(M_G)+1$. *In particular,*

$$\operatorname{tw}(G) \leqslant 3\operatorname{rad}(M_G)+3.$$

We use the following lemma about planarisations to extend Lemma 7 from plane drawings to arbitrary drawings.

Lemma 8. *For every drawing D of a graph G, the planarisation P_D of D satisfies*

$$\operatorname{tw}(G) \leqslant 2\operatorname{tw}(P_D)+1 \quad \text{and} \quad \operatorname{tw}(X_D) \leqslant 2\operatorname{tw}(P_D)+1.$$

To prove Theorem 5, let P_D be the planarisation of D. By definition, $M_D \cong M_{P_D}$. Lemma 7 implies

$$2\operatorname{tw}(P_D)+1 \leqslant 2(3\operatorname{rad}(M_{P_D})+3)+1 = 6\operatorname{rad}(M_D)+7,$$

and Lemma 8 now gives the required result.

The next lemma is a cornerstone of this paper. It shows that if the map graph of a circular drawing has large radius, then the crossing graph contains a useful substructure.

Lemma 9. *Let D be a circular drawing of a graph G. If the map graph M_D has radius at least $2t$, then the crossing graph X_D contains t vertex-disjoint induced cycles $C_1,\dots,C_t$ such that for all $i<j$ every vertex of C_i has at least two neighbours in C_j. Moreover, every vertex of X_D has at most four neighbours in any C_i $(1 \leqslant i \leqslant t)$.*

4 Structural Properties of Circular Drawings

Theorem 5 says that for any drawing D of a graph G, the radius of M_D provides an upper bound for $\operatorname{tw}(G)$ and $\operatorname{tw}(X_D)$. For a general drawing it is impossible to relate $\operatorname{tw}(X_D)$ to $\operatorname{tw}(G)$. Firstly, planar graphs can have arbitrarily large treewidth (for example, the $(n \times n)$-grid has treewidth n) and admit drawings with no crossings. In the other direction, $K_{3,n}$ has treewidth 3 and crossing number $\Omega(n^2)$, as shown by Kleitman [30]. In particular, the crossing graph of any drawing of $K_{3,n}$ has average degree linear in n and thus has arbitrarily large complete minors [35] and so arbitrarily large treewidth.

Happily, this is not so for circular drawings. Using the tools in Section 3 we show that if a graph G has large treewidth, then the crossing graph of any circular drawing of G has large treewidth. In fact, the crossing graph must contain a large (topological) complete graph minor (see Theorems 3 and 10). In particular, if X_D is K_t-minor-free, then G has small treewidth. We further show that if X_D is K_t-minor-free, then G does not contain a subdivision of $K_{2,4t}$ (Theorem 4). Using these results, we deduce a product structure theorem for G (Corollary 11).

In the other direction, we ask what properties of a graph G guarantee that it has a circular drawing D where X_D has no K_t-minor. Certainly G must have small treewidth. Adding the constraint that G does not contain a subdivision of $K_{2,f(t)}$ is not sufficient (see Lemma 14) but a bounded maximum degree constraint is: we show that if G has bounded maximum degree and bounded treewidth, then G has a circular drawing where the crossing graph has bounded treewidth (Proposition 15).

4.1 Necessary Conditions for K_t-Minor-Free Crossing Graphs

This subsection studies the structure of graphs that have circular drawings whose crossing graph is (topological) K_t-minor-free. Much of our understanding of the structure of these graphs is summarised by Theorems 3 and 4 and the next two results.

Theorem 10. *If a graph G has a circular drawing where the crossing graph has no topological K_t-minor, then G has treewidth at most $6t^2+6t+1$.*

We may deduce a product structure theorem for graphs that have a circular drawing whose crossing graph is K_t-minor-free. For two graphs G and H, the *strong product* $G \boxtimes H$ is the graph with vertex-set $V(G) \times V(H)$, and with an edge between two vertices (v,w) and (v',w') if and only if $v=v'$ and $ww' \in E(H)$, or $w=w'$ and $vv' \in E(G)$, or $vv' \in E(G)$ and $ww' \in E(H)$. Campbell et al. [13, Prop. 55] showed that if a graph G is $K_{2,t}$-topological minor-free and has treewidth at most k, then G is isomorphic to a subgraph of $H \boxtimes K_{O(t^2k)}$ where $\mathrm{tw}(H) \leqslant 2$. Thus Theorems 3 and 4 imply the following product structure result.

Corollary 11. *If a graph G has a circular drawing where the crossing graph has no K_t-minor, then G is isomorphic to a subgraph of $H \boxtimes K_{O(t^3)}$ where $\mathrm{tw}(H) \leqslant 2$.*

En route to proving these results, we use the cycle structure built by Lemma 9 to find (topological) complete minors in the crossing graph of circular drawings. We first show that the treewidth and Hadwiger number of X_D as well as the radius of M_D are all linearly tied.

Lemma 12. *For every circular drawing D,*

$$\mathrm{tw}(X_D) \leqslant 6\,\mathrm{rad}(M_D)+7 \leqslant 12h(X_D)-11 \leqslant 12\,\mathrm{tw}(X_D)+1.$$

Proof. The first inequality is exactly Theorem 5, while the final one is the well-known fact that $h(G) \leqslant \mathrm{tw}(G)+1$ for every graph G. To prove the middle inequality we need to show that for any circular drawing D,

$$\mathrm{rad}(M_D) \leqslant 2h(X_D)-3. \tag{1}$$

Let $t := h(X_D)$ and suppose, for a contradiction, that $\text{rad}(M_D) \geqslant 2t-2$. By Lemma 9, X_D contains $t-1$ vertex disjoint cycles $C_1,\dots,C_{t-1}$ such that for all $i<j$ every vertex of C_i has a neighbour in C_j. Contracting C_1 to a triangle and each C_i $(i \geqslant 2)$ to a vertex gives a K_{t+1}-minor in X_D. This is the required contradiction. □

Clearly the Hajós number of a graph is at most the Hadwiger number. Our next lemma implies that the Hajós number of X_D is quadratically tied to the radius of M_D and to the treewidth and Hadwiger number of X_D.

Lemma 13. *For every circular drawing D,*

$$\text{rad}(M_D) \leqslant h'(X_D)^2 + 3h'(X_D) + 1.$$

Proof. Let $t = h'(X_D)+1$ and suppose, for a contradiction, that $\text{rad}(M_D) \geqslant t^2+t$. By Lemma 9, X_D contains $(t^2+t)/2$ vertex disjoint cycles $C_1,\dots,C_{(t^2+t)/2}$ such that for all $i<j$ every vertex of C_i has a neighbour in C_j. For each $i \in \{1,\dots,t\}$, let $v_i \in V(C_i)$. We assume that $V(K_t) = \{1,\dots,t\}$ and let $\phi\colon E(K_t) \to \{t+1,\dots,(t^2+t)/2\}$ be a bijection. Then for each $ij \in E(K_t)$, there is a (v_i,v_j)-path P_{ij} in X_D whose internal vertices are contained in $V(C_{\phi(ij)})$. Since ϕ is a bijection, it follows that $(P_{ij}\colon ij \in E(K_t))$ defines a topological K_t-minor in X_D, a contradiction. □

We are now ready to prove Theorems 3 and 10.

Proof of Theorem 3. Let D be a circular drawing of G with $h(X_D) \leqslant t-1$. By (1), $\text{rad}(M_D) \leqslant 2t-5$. Finally, by Theorem 5, $\text{tw}(G) \leqslant 12t-23$. □

Proof of Theorem 10. Let D be a circular drawing of G with $h'(X_D) \leqslant t-1$. By Lemma 13, $\text{rad}(M_D) \leqslant t^2+t-1$. Finally, by Theorem 5, $\text{tw}(G) \leqslant 6t^2+6t+1$. □

By considering grid graphs, we show that the bound on $\text{tw}(G)$ in Theorem 3 is within a constant factor of being optimal.

The proof of Theorem 4 is more involved—see [28] for details.

4.2 Sufficient Conditions for K_t-Minor-Free Crossing Graphs

It is natural to consider whether the converse of Theorems 3 and 4 holds. That is, does there exist a function f such that if a $K_{2,t}$-topological minor-free graph G has treewidth at most k, then there is a circular drawing of G whose crossing graph is $K_{f(t,k)}$-minor-free. Our next result shows that this is false in general. A *t-rainbow* in a circular drawing of a graph is a non-crossing matching consisting of t edges between two disjoint arcs in the circle.

Lemma 14. *For every $t \in \mathbb{N}$, there exists a $K_{2,4}$-topological minor-free graph G with $\text{tw}(G) = 2$ such that, for every circular drawing D of G, the crossing graph X_D contains a K_t-minor.*

Proof. Let T be any tree with maximum degree 3 and sufficiently large pathwidth (as a function of t). Such a tree exists as the complete binary tree of height $2h$ has pathwidth h. Let G be obtained from T by adding a dominant vertex v, so G has treewidth 2. Since $G-v$ has maximum degree 3, it follows that G is $K_{2,4}$-topological minor-free.

Let D be a circular drawing of G and let D_T be the induced circular drawing of T. Since T has sufficiently large pathwidth, a result of Pupyrev [43, Thm. 2] implies that X_D has large chromatic number or a $4t$-rainbow[2]. Since the class of circle graphs is χ-bounded [27], it follows that if X_D has large chromatic number, then it contains a large clique and we are done. So we may assume that D_T contains a $4t$-rainbow. By the pigeonhole principle, there is a subset $\{a_1b_1,\dots,a_{2t}b_{2t}\}$ of the rainbow edges such that a_ib_i topologically separates v from a_j and b_j whenever $i<j$. As such, a_ib_i crosses the edges va_j and vb_j in D whenever $i<j$. Therefore X_D contains a $K_{t,2t}$ subgraph with bipartition $(\{a_1b_1,\dots,a_tb_t\},\{va_{t+1},vb_{t+1},\dots,va_{2t},vb_{2t}\})$ and this contains a K_t-minor. □

While $K_{2,4}$-topological minor-free and bounded treewidth is not sufficient to imply that a graph has a circular drawing whose crossing graph is K_t-minor-free, bounded degree and bounded treewidth is sufficient.

Proposition 15. *For $k,\Delta\in\mathbb{N}$, every graph G with treewidth less than k and maximum degree at most Δ has a circular drawing in which the crossing graph X_D has treewidth at most* $(6\Delta+1)(18k\Delta)^2-1$.

4.3 Circular Drawings and Degeneracy

Theorems 3 and 10 say that if a graph G has a circular drawing D where the crossing graph X_D excludes a fixed (topological) minor, then G has bounded treewidth. Graphs excluding a fixed (topological) minor have bounded average degree and degeneracy [35]. Despite this, we now show that X_D having bounded degeneracy is not sufficient to bound the treewidth of G. In fact, it is not even sufficient to bound the Hadwidger number of G.

Theorem 16. *For every $t\in\mathbb{N}$, there is a graph G_t and a circular drawing D of G_t such that:*

- *G_t contains a K_t-minor,*
- *G_t has maximum degree 3, and*
- *X_D is 2-degenerate.*

Proof. We draw G_t with vertices placed on the x-axis (x-coordinate between 1 and t) and edges drawn on or above the x-axis. This can then be wrapped to give a circular drawing of G_t.

[2] The result of Pupyrev [43] is in terms of stacks and queues but is equivalent to our statement.

For real numbers $a_1 < a_2 < \cdots < a_n$, we say a path P is drawn as a *monotone path* with vertices $a_1, \ldots, a_n$ if it is drawn as follows where each vertex has x-coordinate equal to its label:

In all our monotone paths, $a_1, a_2, \ldots, a_n$ will be an arithmetic progression. We construct our drawing of G_t as follows (see Figure 3 for the construction with $t = 4$).

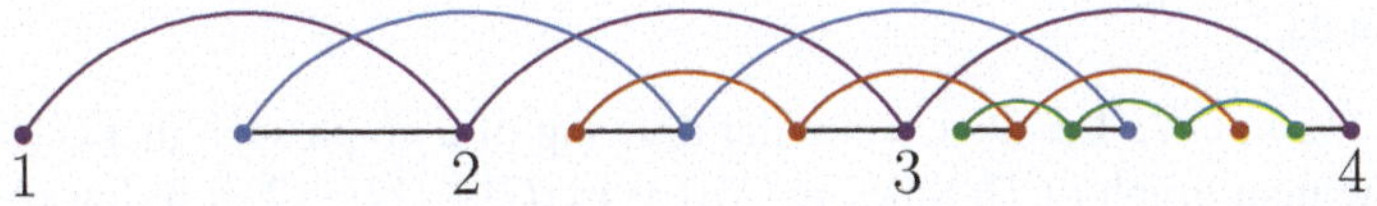

Fig. 3: G_4 where P_0 is purple, P_1 is blue, P_2 is red, P_3 is green, and the $e_{r,s}$ are black.

First let P_0 be the monotone path with vertices $1, 2, \ldots, t$. For $s \in \{1, 2, \ldots, t-1\}$, let P_s be the monotone path with vertices

$$s + 2^{-s}, s + 3 \cdot 2^{-s}, s + 5 \cdot 2^{-s}, \ldots, t - 2^{-s}.$$

Observe that these paths are vertex-disjoint. For $0 \leqslant r < s \leqslant t-1$, let $I_{r,s}$ be the interval

$$[s + 2^{-r} - 2^{-s}, s + 2^{-r}].$$

Note that the lower end-point of $I_{r,s}$ is a vertex in P_s and the upper end-point is a vertex in P_r. Also note that no vertex of any P_i lies in the interior of $I_{r,s}$. Hence for all $r < s$ we may draw a horizontal edge $e_{r,s}$ between the end-points of $I_{r,s}$.

Graph G_t and the drawing D are obtained as a union of the P_s together with all the $e_{r,s}$. The paths P_s are vertex-disjoint and edge $e_{r,s}$ joins P_r to P_s, so G_t contains a K_t-minor. It is simple to check that the $I_{r,s}$ are pairwise disjoint. In particular, any vertex v is the end-point of at most one $e_{r,s}$ and so G_t has maximum degree three.

Each edge $e_{r,s}$ is horizontal and crosses no other edges so has no neighbours in X_D. Next consider an edge aa' of P_s. We have $a' = a + 2 \cdot 2^{-s}$. Exactly one vertex in $V(P_0) \cup V(P_1) \cup \cdots \cup V(P_s)$ lies between a and a': their midpoint, $m = a + 2^{-s}$. Vertex m has at most two non-horizontal edges incident to it and so, in X_D, every $aa' \in E(P_s)$ has at most two neighbours in $E(P_0) \cup E(P_1) \cup \cdots \cup E(P_s)$. Thus X_D is 2-degenerate, as required. □

5 Structural Properties of Circle Graphs

Recall that a circle graph is the intersection graph of a set of chords of a circle. More formally, let C be a circle in $\mathbb{R}^2$. A *chord* of C is a closed line segment with distinct endpoints on C. Two chords of C either cross, are disjoint, or have a common endpoint. Let S be a set of chords of a circle C such that no three chords in S cross at a single point. Let G be the crossing graph of S. Then G is called a *circle graph*. Note that a graph G is a circle graph if and only if $G \cong X_D$ for some circular drawing D of a graph H, and in fact one can take H to be a matching.

We are now ready to prove Theorems 1 and 2. While the treewidth of circle graphs has previously been studied from an algorithmic perspective [31], to the best of our knowledge, these theorems are the first structural results on the treewidth of circle graphs.

Proof of Theorem 1. Let D be a circular drawing of a graph such that $G \cong X_D$. Let M_D be the map graph of D. Since $\text{tw}(X_D) = \text{tw}(G) \geqslant 12t + 2$, it follows by Theorem 5 that M_D has radius at least $2t$. The claim then follows from Lemma 9. □

Proof of Theorem 2. Let G be a circle graph and let D be a circular drawing with $G \cong X_D$. By Lemma 12, $\text{tw}(G) \leqslant 6\text{rad}(M_D) + 7 \leqslant 12h(G) - 11 \leqslant 12\,\text{tw}(G) + 1$. Hence, the Hadwiger number and treewidth are linearly tied for circle graphs. This inequality and Lemma 13 imply

$$h'(G) - 1 \leqslant h(G) - 1 \leqslant \text{tw}(G) \leqslant 6\text{rad}(M_D) + 7 \leqslant 6h'(G)^2 + 18h'(G) + 13.$$

Hence the Hajós number is quadratically tied to both the treewidth and Hadwiger number for circle graphs. Finally, $K_{t,t}$ is a circle graph which has treewidth t, Hadwiger number $t+1$, and Hajós number $\Theta(\sqrt{t})$. Hence, 'quadratic' is best possible. □

We now discuss several noteworthy consequences of Theorems 1 and 2. Say an hereditary class of graphs $\mathcal{G}$ is *induced*-tw-*bounded* if there is a function f such that for every graph $G \in \mathcal{G}$ with $\text{tw}(G) \geqslant f(t)$, G contains at least one of the usual suspects defined in Section 1. While the class of all graphs is not induced-tw-bounded [3, 11, 14, 42, 46], many natural graph classes are. For example, Aboulker, Adler, Kim, Sintiari, Trotignon [1] showed that every proper minor-closed class is induced-tw-bounded and Korhonen [32] recently showed that the class of graphs with bounded maximum degree is induced-tw-bounded. We now show that the class of circle graphs is not induced-tw-bounded.

Theorem 17. *The class of circle graphs is not induced*-tw-*bounded.*

Proof. We first show that for all $t \geqslant 50$, no circle graph contains a subdivision of the $(t \times t)$-wall or a line graph of a subdivision of the $(t \times t)$-wall as an induced subgraph. As the class of circle graphs is hereditary, it suffices to show that for all

$t \geqslant 50$, these two graphs are not circle graphs. These two graphs are planar (so K_5-minor-free) and have treewidth $t \geqslant 50$. However, Lemma 12 implies that every K_5-minor-free circle graph has treewidth at most 49, which is the required contradiction.

Now consider the family of couples of graphs $((G_t, X_t) : t \in \mathbb{N})$ given by Theorem 16 where X_t is the crossing graph of the drawing of G_t. Then $(X_t : t \in \mathbb{N})$ is a family of circle graphs. Since $(G_t : t \in \mathbb{N})$ has unbounded treewidth, Theorem 3 implies that $(X_t : t \in \mathbb{N})$ also has unbounded treewidth. Moreover, since X_t is 2-degenerate for all $t \in \mathbb{N}$, it excludes K_4 and $K_{3,3}$ as (induced) subgraphs, as required. □

We now discuss applications of Theorem 1 to vertex-minor-closed classes. For a vertex v of a graph G, to *locally complement at* v means to replace the induced subgraph on the neighbourhood of v by its complement. A graph H is a *vertex-minor* of a graph G if H can be obtained from G by a sequence of vertex deletions and local complementations. Vertex-minors were first studied by Bouchet [12] under the guise of isotropic systems. The name 'vertex-minor' is due to Oum [37]. Circle graphs are a key example of a vertex-minor-closed class.

We now show that a vertex-minor-closed graph class is induced-tw-bounded if and only if it has bounded rank-width. Rank-width is a graph parameter introduced by Oum and Seymour [39] that describes whether a graph can be decomposed into a tree-like structure by simple cuts. For a formal definition and a survey on this parameter, see [38]. Oum [37] showed that rank-width is closed under vertex-minors.

Theorem 18. *A vertex-minor-closed class $\mathcal{G}$ is induced-*tw*-bounded if and only if it has bounded rankwidth.*

Proof. Suppose $\mathcal{G}$ has bounded rankwidth. By a result of Abrishami, Chudnovsky, Hajebi, and Spirkl [7], there is a function f such that every graph in $\mathcal{G}$ with treewidth at least $f(t)$ contains K_t or $K_{t,t}$ as an induced subgraph. Thus $\mathcal{G}$ is induced-tw-bounded. Now suppose $\mathcal{G}$ has unbounded rank-width. By a result of Gellen, Kwon, McCarty, and Wollan [26], $\mathcal{G}$ contains all circle graphs. It therefore follows by Theorem 17 that $\mathcal{G}$ is not induced-tw-bounded. □

We conclude with the following question:

> Let $\mathcal{G}$ be a vertex-minor-closed class with unbounded rank-width. What are the unavoidable induced subgraphs of graphs in $\mathcal{G}$ with large treewidth?

The cycle structure (or variants thereof) in Theorem 1 must be included in the list of unavoidable induced subgraphs.

Acknowledgements This research was initiated at the *Structural Graph Theory Downunder II* workshop at the Mathematical Research Institute MATRIX (March 2022). Thanks to all the participants, especially Marc Distel, for helpful conversations.

Research of Robert Hickingbotham supported by an Australian Government Research Training Program Scholarship. Research of Freddie Illingworth supported by EPSRC grant EP/V007327/1. Research of Bojan Mohar supported in part by the NSERC Discovery Grant R611450 (Canada). Research of David Wood supported by the Australian Research Council and a Visiting Research Fellowship of Merton College, University of Oxford.

References

1. Aboulker, P., Adler, I., Kim, E.J., Sintiari, N.L.D., Trotignon, N.: On the tree-width of even-hole-free graphs. European J. Combin. **98**, #103,394 (2021). DOI 10.1016/j.ejc.2021.103394
2. Abrishami, T., Alecu, B., Chudnovsky, M., Hajebi, S., Spirkl, S.: Induced subgraphs and tree decompositions VII. Basic obstructions in H-free graphs (2022). URL https://arxiv.org/abs/2207.05538
3. Abrishami, T., Alecu, B., Chudnovsky, M., Hajebi, S., Spirkl, S., Vušković, K.: Induced subgraphs and tree decompositions V. One neighbor in a hole (2022). URL https://arxiv.org/abs/2205.04420
4. Abrishami, T., Chudnovsky, M., Dibek, C., Hajebi, S., Rzążewski, P., Spirkl, S., Vušković, K.: Induced subgraphs and tree decompositions II. Toward walls and their line graphs in graphs of bounded degree (2021). URL https://arxiv.org/abs/2108.01162
5. Abrishami, T., Chudnovsky, M., Hajebi, S., Spirkl, S.: Induced subgraphs and tree decompositions III. Three-path-configurations and logarithmic treewidth. Adv. Comb. **#6** (2022). DOI 10.19086/aic.2022.6
6. Abrishami, T., Chudnovsky, M., Hajebi, S., Spirkl, S.: Induced subgraphs and tree decompositions IV. (Even hole, diamond, pyramid)-free graphs (2022). URL https://arxiv.org/abs/2203.06775
7. Abrishami, T., Chudnovsky, M., Hajebi, S., Spirkl, S.: Induced subgraphs and tree decompositions VI. Graphs with 2-cutsets (2022). URL https://arxiv.org/abs/2207.05538
8. Abrishami, T., Chudnovsky, M., Vušković, K.: Induced subgraphs and tree decompositions I. Even-hole-free graphs of bounded degree. J. Combin. Theory Ser. B **157**, 144–175 (2022). DOI 10.1016/j.jctb.2022.05.009
9. Bae, S.W., Baffier, J., Chun, J., Eades, P., Eickmeyer, K., Grilli, L., Hong, S., Korman, M., Montecchiani, F., Rutter, I., Tóth, C.D.: Gap-planar graphs. Theor. Comput. Sci. **745**, 36–52 (2018). DOI 10.1016/j.tcs.2018.05.029
10. Bodlaender, H.: Planar graphs with bounded treewidth (1988). URL https://dspace.library.uu.nl/handle/1874/16567. Technical Report RUU-CS-88-14, Department of Computer Science, University of Utrecht
11. Bonamy, M., Bonnet, É., Déprés, H., Esperet, L., Geniet, C., Hilaire, C., Thomassé, S., Wesolek, A.: Sparse graphs with bounded induced cycle packing number have logarithmic treewidth (2022). URL https://arxiv.org/abs/2206.00594
12. Bouchet, A.: Isotropic systems. European J. Combin. **8**(3), 231–244 (1987). DOI 10.1016/S0195-6698(87)80027-6
13. Campbell, R., Clinch, K., Distel, M., Gollin, J.P., Hendrey, K., Hickingbotham, R., Huynh, T., Illingworth, F., Tamitegama, Y., Tan, J., Wood, D.R.: Product structure of graph classes with bounded treewidth (2022). URL https://arxiv.org/abs/2206.02395
14. Davies, J.: Counterexample to the conjecture. In: Graph Theory, *Mathematisches Forschungsinstitut Oberwolfach Reports*, vol. 1, pp. 62–63 (2022). DOI 10.14760/OWR-2022-1
15. Davies, J.: Improved bounds for colouring circle graphs. Proc. Amer. Math. Soc. **150**, 5121–5135 (2022). DOI 10.1090/proc/16044
16. Davies, J.: Local properties of graphs with large chromatic number. Ph.D. thesis, University of Waterloo (2022). URL http://hdl.handle.net/10012/18679
17. Davies, J., McCarty, R.: Circle graphs are quadratically χ-bounded. Bull. Lond. Math. Soc. **53**(3), 673–679 (2021). DOI 10.1112/blms.12447
18. Diestel, R.: Graph theory, *Graduate Texts in Mathematics*, vol. 173, 5th edn. Springer (2018)
19. Dujmovic, V., Eppstein, D., Hickingbotham, R., Morin, P., Wood, D.R.: Stack-number is not bounded by queue-number. Combinatorica **42**(2), 151–164 (2022). DOI 10.1007/s00493-021-4585-7
20. Dujmović, V., Eppstein, D., Wood, D.R.: Structure of graphs with locally restricted crossings. SIAM J. Discrete Math. **31**(2), 805–824 (2017). DOI 10.1137/16M1062879

21. Dujmović, V., Morin, P., Wood, D.R.: Layered separators in minor-closed graph classes with applications. J. Combin. Theory Ser. B **127**, 111–147 (2017). DOI 10.1016/j.jctb.2017.05.006
22. Dujmović, V., Wood, D.R.: Graph treewidth and geometric thickness parameters. Discrete Comput. Geom. **37**(4), 641–670 (2007). DOI 10.1007/s00454-007-1318-7
23. Durán, G., Grippo, L.N., Safe, M.D.: Structural results on circular-arc graphs and circle graphs: A survey and the main open problems. Discrete Appl. Math. **164**, 427–443 (2014). DOI 10.1016/j.dam.2012.12.021
24. Eppstein, D., Gupta, S.: Crossing patterns in nonplanar road networks. In: Proc. 25th ACM SIGSPATIAL Int'l Conf. on Advances in Geographic Information Systems, pp. 40:1–9 (2017). DOI 10.1145/3139958.3139999
25. de Fraysseix, H.: A characterization of circle graphs. European J. Combin. **5**(3), 223–238 (1984). DOI 10.1016/S0195-6698(84)80005-0
26. Geelen, J., Kwon, O.j., McCarty, R., Wollan, P.: The grid theorem for vertex-minors. J. Combin. Theory, Ser. B **158**, 93–116 (2023). DOI 10.1016/j.jctb.2020.08.004
27. Gyárfás, A.: On the chromatic number of multiple interval graphs and overlap graphs. Discrete Math. **55**(2), 161–166 (1985). DOI 10.1016/0012-365X(85)90044-5
28. Hickingbotham, R., Illingworth, F., Mohar, B., Wood, D.R.: Treewidth, circle graphs and circular drawings (2022). URL https://arxiv.org/abs/2212.11436
29. Hickingbotham, R., Wood, D.R.: Shallow minors, graph products and beyond planar graphs (2021). URL https://arxiv.org/abs/2111.12412
30. Kleitman, D.J.: The crossing number of $K_{5,n}$. J. Combinatorial Theory **9**, 315–323 (1970). DOI 10.1016/S0021-9800(70)80087-4
31. Kloks, T.: Treewidth of circle graphs. Int. J. Found. Comput. Sci. **7**(2), 111–120 (1996). DOI 10.1142/S0129054196000099
32. Korhonen, T.: Grid induced minor theorem for graphs of small degree (2023). DOI 10.1016/j.jctb.2023.01.002
33. Kostochka, A., Kratochvíl, J.: Covering and coloring polygon-circle graphs. Discrete Math. **163**(1–3), 299–305 (1997). DOI 10.1016/S0012-365X(96)00344-5
34. Lozin, V., Razgon, I.: Tree-width dichotomy. European J. Combin. **103**, # 103,517 (2022). DOI 10.1016/j.ejc.2022.103517
35. Mader, W.: Homomorphieeigenschaften und mittlere Kantendichte von Graphen. Math. Ann. **174**, 265–268 (1967). DOI 10.1007/BF01364272
36. McCarty, R.: Local structure for vertex-minors. Ph.D. thesis, University of Waterloo (2021). URL http://hdl.handle.net/10012/17633
37. Oum, S.i.: Rank-width and vertex-minors. J. Combin. Theory Ser. B **95**(1), 79–100 (2005). DOI 10.1016/j.jctb.2005.03.003
38. Oum, S.i.: Rank-width: algorithmic and structural results. Discrete Appl. Math. **231**, 15–24 (2017). DOI 10.1016/j.dam.2016.08.006
39. Oum, S.i., Seymour, P.: Approximating clique-width and branch-width. J. Combin. Theory Ser. B **96**(4), 514–528 (2006). DOI 10.1016/j.jctb.2005.10.006
40. Pach, J., Tóth, G.: Graphs drawn with few crossings per edge. Combinatorica **17**(3), 427–439 (1997). DOI 10.1007/BF01215922
41. Pilipczuk, M., Sintiari, N.L.D., Thomassé, S., Trotignon, N.: (Theta, triangle)-free and (even hole, K_4)-free graphs. Part 2: Bounds on treewidth. J. Graph Theory **97**(4), 624–641 (2021). DOI 10.1002/jgt.22675
42. Pohoata, C.: Unavoidable induced subgraphs of large graphs. Senior thesis, Department of Mathematics, Princeton University (2014). URL http://arks.princeton.edu/ark:/88435/dsp012514nk67q
43. Pupyrev, S.: Book embeddings of graph products (2020). URL https://arxiv.org/abs/2007.15102
44. Robertson, N., Seymour, P.: Graph minors. III. Planar tree-width. J. Combin. Theory Ser. B **36**(1), 49–64 (1984). DOI 10.1016/0095-8956(84)90013-3
45. Schaefer, M.: The graph crossing number and its variants: A survey. Electron. J. Combin. **#DS21** (2022). URL https://www.combinatorics.org/DS21
46. Sintiari, N.L.D., Trotignon, N.: (Theta, triangle)-free and (even hole, K_4)-free graphs—Part 1: Layered wheels. J. Graph Theory **97**(4), 475–509 (2021). DOI 10.1002/jgt.22666

Chapter 6

MATRIX Program: Hyperbolic Differential Equations in Geometry and Physics

Bulk-boundary correspondences and unique continuation in asymptotically Anti-de Sitter spacetimes

Arick Shao

Abstract This article surveys the research presented by the author at the MATRIX Institute workshop *Hyperbolic Differential Equations in Geometry and Physics* in April 2022. The work is centred about establishing rigorous mathematical statements toward the AdS/CFT correspondence in theoretical physics, in particular in dynamical settings. The contents are mainly based on the recent paper [37] with Gustav Holzegel that proved a unique continuation result for the Einstein-vacuum equations from asymptotically Anti-de Sitter (aAdS) conformal boundaries. We also discuss some preceding results [20, 35, 36, 55, 59], in particular novel Carleman estimates for wave equations on aAdS spacetimes, which laid the foundations toward the main results of [37].

1 Introduction

One of the most influential research directions in theoretical physics over the past three decades is the *AdS/CFT correspondence*. AdS/CFT—first formulated by Maldacena [51, 52]—roughly posits a *correspondence between gravitational dynamics in asymptotically Anti-de Sitter spacetimes and conformal field theories on their boundaries*; see also the works of Gubser, Klebanov, and Polyakov [32] and Witten [68] for further early developments. AdS/CFT is also often associated with the term *holography* [61], as the boundary theory is of one dimension lower than the gravitational theory.

In spite of extensive research and discussions in physics, surprisingly few mathematical statements pertaining to the AdS/CFT correspondence have been rigorously proven, or even formulated. Furthermore, nearly all rigorous results have been

Arick Shao
School of Mathematical Sciences, Queen Mary University of London
London E1 4NS, United Kingdom
e-mail: a.shao@qmul.ac.uk

D. R. Wood et al. (eds.), *2021-2022 MATRIX Annals*, MATRIX Book Series 5,
https://doi.org/10.1007/978-3-031-47417-0_13

within stationary backgrounds (see for instance [2, 12, 13, 16, 69]), whereas dynamical (time-dependent) settings are almost entirely untreated.

The works described in this article represent an ongoing programme to establish rigorous mathematical theory related to the AdS/CFT correspondence. The aim is to rigorously state and prove statements along this direction, as well as to better understand the mechanisms behind such correspondences. In particular, the key results here will be formulated in terms of unique continuation for the Einstein equations. Also, although discussions of AdS/CFT in physics extend well beyond classical relativistic settings, here we will restrict our attention to within general relativity, as this provides us a physical theory with firmly established mathematical foundations.

1.1 Anti-de Sitter Spacetime

In Einstein's theory of general relativity, spacetime is modeled as an $(n+1)$-dimensional Lorentzian manifold $(\mathscr{M}, g)$; as a matter of convention, we assume g has signature $(-,+,\dots,+)$. In this article, we will only consider spacetimes that do not involve matter fields, so that g satisfies the *Einstein-vacuum equations* (*EVE*),

$$\operatorname{Ric}[g] = \frac{2\Lambda}{n-1} g, \tag{1}$$

with $\operatorname{Ric}[g]$ denoting the Ricci curvature associated to g, and $\Lambda \in \mathbb{R}$ being the *cosmological constant*. Furthermore, here we are concerned with the case in which $\Lambda < 0$.

Remark 1. While classical relativity lies in $3+1$ dimensions, here we work in any dimension, since many settings of interest within the AdS/CFT correspondence concern other dimensions.

Anti-de Sitter, or *AdS*, *spacetime* is the analogue of Minkowski spacetime in the case of negative cosmological constant. More specifically, it is the maximally symmetric solution of the EVE

$$\operatorname{Ric}[g] = -ng, \qquad \Lambda := \frac{-n(n-1)}{2} < 0. \tag{2}$$

While there are many ways to describe AdS spacetime, one useful global representation is as the manifold $(\mathbb{R}_t \times \mathbb{R}^n_x, g_{\mathrm{AdS}})$, where the metric is given in polar coordinates by

$$g_{\mathrm{AdS}} := (1+r^2)^{-1} dr^2 - (1+r^2) dt^2 + r^2 \mathring{\gamma}, \tag{3}$$

with $\mathring{\gamma}$ being the unit round metric on $\mathbb{S}^{n-1}$.

Remark 2. One can rescale (3) to obtain a solution of (1) for any $\Lambda < 0$. The particular normalization $\Lambda = \frac{-n(n-1)}{2}$ is convenient for simplifying various constants.

For our purposes, it will be useful to express (3) using an inverted radius ρ, defined by

$$4r := \rho^{-1}(2+\rho)(2-\rho), \qquad \rho \in (0,2]. \tag{4}$$

Observe that $\rho \searrow 0$ corresponds to "infinity" $r \nearrow \infty$, while $\rho \nearrow 2$ corresponds to the centre $r \searrow 0$. A direct computation shows that (4) transforms (3) into the form

$$g_{\text{AdS}} = \frac{1}{\rho^2}\left[d\rho^2 + (-dt^2 + \mathring{\gamma}_{n-1}) - \frac{1}{2}\rho^2(dt^2 + \mathring{\gamma}_{n-1}) + \frac{1}{16}\rho^4(-dt^2 + \mathring{\gamma}_{n-1})\right], \tag{5}$$

with ρ being a conformal boundary-defining function. In anticipation of upcoming terminology, we will refer to (5) as a *Fefferman-Graham* (or *FG*) *gauge* for the AdS metric.

In particular, ignoring for the moment the conformal factor ρ^{-2} in (5), we can then formally associate at "$\rho = 0$" a timelike boundary for AdS spacetime:

$$(\mathscr{I}_{\text{AdS}}, \mathfrak{g}_{\text{AdS}}) := (\mathbb{R}_t \times \mathbb{S}^{n-1}, -dt^2 + \mathring{\gamma}). \tag{6}$$

We refer to (6) as a *conformal boundary*, or *conformal infinity*, of AdS spacetime.

Similar to how one can study asymptotically flat spacetimes that only behave like Minkowski spacetime "at infinity", we can expand our outlook from AdS spacetime to *asymptotically AdS (aAdS) spacetimes*. Very roughly, aAdS spacetimes are those with similar qualitative properties "at infinity" as AdS spacetime, in particular having a timelike conformal boundary. We will provide a more precise description of aAdS spacetimes in the subsequent section.

Remark 3. A more well-known conformal transformation of AdS spacetime embeds AdS spacetime into half the Einstein cylinder, $\mathbb{R}_t \times \mathbb{S}^{n-1}_+$, with metric

$$g_{\text{AdS}} = \frac{1}{\cos^2\theta}[-dt^2 + d\theta^2 + (\sin^2\theta)\,\mathring{\gamma}].$$

However, the FG gauge will be more useful, as it extends directly to the study of aAdS metrics.

1.2 Unique Continuation

Informally, our main objective is to address the following question, representing a purely classical statement in support of the AdS/CFT correspondence:

Problem 1. Is there some one-to-one correspondence between (a) asymptotically AdS solutions $(\mathscr{M}, g)$ of the EVE (2) (representing "gravitational dynamics"), and (b) appropriate data prescribed at the conformal boundary $\mathscr{I}$ (representing the "conformal field theory")?

If we lack additional foresight from partial differential equations (PDEs), then a reasonable attempt at Problem 1 might be to *solve* (2) *given appropriate Cauchy* *(i.e. Dirichlet and Neumann) data on $\mathscr{I}$*. However, this approach immediately runs into fundamental issues, since the EVE is primarily hyperbolic (in the appropriate gauges). As a result, (2) becomes ill-posed when data is prescribed on a timelike hypersurface, such as the conformal boundary $(\mathscr{I}, \mathfrak{g})$. In particular, given Cauchy data on $\mathscr{I}$, a solution to (2) may not exist, much less depend continuously on the data.

Remark 4. In contrast, the appropriate well-posed problem for the EVE in aAdS settings is the initial-boundary value problem, in which Cauchy data is prescribed on a spacelike hypersurface, and *only one* piece of boundary data (e.g. Dirichlet, Neumann, or Robin boundary condition) is imposed on the conformal boundary. For further discussions, see [23, 24, 27, 66, 67].

We are thus forced to attack Problem 1 differently. The key idea is that we instead formulate Problem 1 as a *unique continuation* problem—this is a classical PDE question for ill-posed settings that has an extensive history and literature; see, e.g., [18, 19, 34, 40] for classical results. In short, for unique continuation problems, one avoids solving the PDE altogether and instead asks: *if the solution exists, then must it be uniquely determined by the prescribed data*?

In light of this, we can restate Problem 1 as a unique continuation problem for the EVE:

Problem 2. If two aAdS spacetimes $(\mathscr{M}, g)$ and $(\bar{\mathscr{M}}, \bar{g})$ solving the Einstein-vacuum equations (2) have the same conformal boundary data, then must g and $\bar{g}$ be isometric?

Very informally, the answer to Problem 2—the main result of the joint work [37] with Holzegel—is "*yes, assuming certain conditions on the conformal boundary $(\mathscr{I}, \mathfrak{g})$ hold*". The goal of the upcoming sections is to work toward a precise formulation of Problem 2 and a precise statement of the main result of [37]. Moreover, we will discuss the main ideas behind the proof of this result.

1.3 Organisation of the Article

In Section 2, we give a precise description of aAdS spacetimes, the setting of our analysis, and we introduce the notion of vertical tensor fields, the main quantities that we will work with. Next, Section 3 is dedicated to the precise statements of our main result (see Theorem 3.2) and its corollaries, as well as the geometric assumptions required for these results to hold. In Section 4, we discuss recently established Carleman estimates for tensorial wave equations in aAdS spacetimes, which is the main technical innovation behind our results. Section 5 introduces some of the main ideas behind the proof of our main result. Finally, in Section 6, we briefly discuss some additional questions of interest that are related to the contents of this article.

Acknowledgements The author thanks the MATRIX Institute for holding the workshop, which provided a wonderful environment for discussions. Special thanks go to Jesse Gell-Redman, Andrew Hassell, Todd Oliynik, and Volker Schlue for their efforts organising the workshop, as well as to Sakis Chatzikaleas, Gustav Holzegel, Alex McGill, and Simon Guisset for the collaborations and discussions that made this research possible. Furthermore, a considerable portion of this work was supported by EPSRC grant EP/R011982/1, *Unique continuation for geometric wave equations, and applications to relativity, holography, and controllability.*

2 Asymptotically AdS Spacetimes

While we had previously characterized aAdS spacetimes as "roughly like AdS near and at the conformal boundary", we will need a much more precise definition if we are to state precise results. In this section, we provide the mathematical setup for studying various relevant quantities on aAdS backgrounds, both near and at the conformal boundary.

2.1 The Fefferman-Graham Gauge

Similar to the AdS metric, one can apply a transformation to metrics that are only asymptotically like AdS into a FG gauge, characterized by a boundary defining function ρ that is both normalised and fully decoupled from the other components. As a result, we will, as a matter of convenience, *define* the aAdS spacetimes that we consider in terms of such FG gauges. We refer to these backgrounds as *FG-aAdS segments*, representing an appropriate portion of an aAdS spacetime near the conformal boundary.

Remark 5. See [29] for a treatment of FG gauges in asymptotically hyperbolic manifolds, that is, the Riemannian analogue of our setting. In fact, the argument in [29] for reducing more general metrics into FG gauges extends directly to Lorentzian, aAdS settings.

Definition 1. Let $(\mathscr{I}, \mathfrak{g})$ be a n-dimensional Lorentzian manifold, and let $\rho_0 > 0$. We say that $(\mathscr{M} := (0, \rho_0]_\rho \times \mathscr{I}, g)$ is an *FG-aAdS segment*, with *conformal infinity* $(\mathscr{I}, \mathfrak{g})$, if g has the form

$$g = \rho^{-2}(d\rho^2 + \mathsf{g}), \tag{1}$$

where $\mathsf{g} := \mathsf{g}(\rho)$, $\rho \in (0, \rho_0]$, is a smooth family of Lorentzian metrics on $\mathscr{I}$ such that g has a continuous limit $\mathsf{g}(0) := \mathfrak{g}$ as $\rho \searrow 0$. Furthermore:

- We refer to the form (1) as the *Fefferman-Graham (FG) gauge condition.*
- We say $(\mathscr{M}, g)$ is *vacuum* iff g also satisifes the EVE (2).

Observe in particular that AdS spacetime can itself be formulated as a vacuum FG-aAdS segment, with the standard conformal infinity (6). In particular, the FG gauge (5) for the AdS metric is precisely of the form (1). Similarly, the usual Schwarzschild-AdS and Kerr-AdS spacetimes can also be expressed as vacuum FG-aAdS segments, at least near the conformal boundary. More generally, a large class of vacuum FG-aAdS segments with conformal infinity (6) can be constructed by solving initial-boundary value problems for the EVE; see [24, 27].

Remark 6. Note that Definition 1 allows for very general conformal boundary topologies. In particular, we can consider settings with flat conformal boundaries,

$$\mathscr{I}_{\mathrm{pAdS}} := \mathbb{R}_t \times \mathbb{R}^n, \qquad \mathscr{I}_{t\mathrm{AdS}} := \mathbb{R}_t \times \mathbb{T}^n. \tag{2}$$

These are realized as conformal boundaries for the *planar* and *toric AdS spacetimes*.

Remark 7. We refer to $\mathfrak{g}$, which fully describes the spacetime geometry, as the *vertical metric*. We will discuss vertical tensor fields more generally at the end of this section.

2.2 Fefferman-Graham Expansions

We now turn our attention to vacuum FG-aAdS segments, and can immediately ask the following: *if $(\mathscr{M}, g)$ is also vacuum, then what structure does this impose on $\mathfrak{g}$ at the conformal boundary*? The answer comes from the seminal work of Fefferman and Graham [25, 26] on the ambient construction. While these concerned extending vacuum metrics from null cones, this has since been widely adapted to aAdS settings in the physics literature.

In short, when $(\mathscr{M}, g)$ is vacuum, one can derive a formal series expansion for g near $\rho = 0$:

$$\mathfrak{g}(\rho) = \begin{cases} \mathfrak{g}^{(0)} + \mathfrak{g}^{(2)}\rho^2 + \cdots + \mathfrak{g}^{(n-1)}\rho^{n-1} + \mathfrak{g}^{(n)}\rho^n + \ldots & n \text{ odd}, \\ \mathfrak{g}^{(0)} + \mathfrak{g}^{(2)}\rho^2 + \cdots + \mathfrak{g}^{(n-2)}\rho^{n-2} + \mathfrak{g}^{(\star)}\rho^n \log\rho + \mathfrak{g}^{(n)}\rho^n + \ldots & n \text{ even}. \end{cases} \tag{3}$$

The above is commonly known as the *Fefferman-Graham expansion* for g. The $\mathfrak{g}^{(k)}$'s and $\mathfrak{g}^{(\star)}$ are covariant symmetric rank-2 tensor fields on $\mathscr{I}$, with the anomalous logarithmic coefficient $\mathfrak{g}^{(\star)}$ only appearing when the boundary dimension n is even. Note the leading coefficient $\mathfrak{g}^{(0)} = \mathfrak{g}$ is simply the conformal boundary metric. Moreover, (2) implies that *all the coefficients $\mathfrak{g}^{(k)}$ for $0 < k < n$—as well as $\mathfrak{g}^{(\star)}$ when n is even—are determined locally by $\mathfrak{g}$ and its derivatives.*

Remark 8. We will use the notations $\mathfrak{g}$ and $\mathfrak{g}^{(0)}$ interchangeably, depending on context.

The situation changes once we reach $\mathfrak{g}^{(n)}$. Here, one can derive from (2) that both the divergence and the trace of $\mathfrak{g}^{(n)}$ are determined by $\mathfrak{g}$ and its derivatives. On the other hand, the remaining components of $\mathfrak{g}^{(n)}$ are free, meaning that they are not formally determined by the EVE.

Finally, the formal expansion (3) can be further continued beyond $\mathfrak{g}^{(n)}$, with all subsequent coefficients formally determined by the pair $(\mathfrak{g}^{(0)}, \mathfrak{g}^{(n)})$ alone. Also, when n is even, the expansion remains polyhomogeneous beyond $\mathfrak{g}^{(n)}$, involving additional powers of $\log \rho$.

Now, when $(\mathfrak{g}^{(0)}, \mathfrak{g}^{(n)})$ is real-analytic, Kichenassamy [46] showed, via Fuchsian techniques, that (3) converges near $\rho = 0$ to a vacuum aAdS metric. However, analyticity is a far too restrictive condition, as this is not satisfied by generic aAdS metrics (e.g. arising from an initial-boundary value problem), for which the expansion (3) is merely formal.

In these non-analytic settings, in which (3) fails to converge, one still expects that a *partial FG expansion* holds, up to some finite order. (An apt analogy would be applying Taylor's theorem as opposed to Taylor series.) This was rigorously proved by the author in [59]—more specifically, assuming minimal regularity conditions, g retains the form (3), but only up to n-th order. We state a simplified version of this result below (see [59, Theorem 3.3] for the full statement):

Theorem 2.1 ([59]). *Let $(\mathscr{M}, g)$ be a vacuum FG-aAdS segment, with $n > 1$. Moreover, suppose:*

- *$\mathfrak{g}$ is locally bounded in C^{n+4} in the $\mathscr{I}$-directions, up to the conformal boundary.*
- *$\mathscr{L}_\rho \mathfrak{g}$ is locally bounded in C^0 in the $\mathscr{I}$-directions, up to the conformal boundary.*

Then, $\mathfrak{g}$ satisfies the following partial FG expansion near $\rho = 0$:

$$\mathfrak{g} = \begin{cases} \sum_{k=0}^{\frac{n-1}{2}} \mathfrak{g}^{(2k)} \rho^{2k} + \mathfrak{g}^{(n)} \rho^n + o(\rho^n) & n \text{ odd}, \\ \sum_{k=0}^{\frac{n-2}{2}} \mathfrak{g}^{(2k)} \rho^{2k} + \mathfrak{g}^{(\star)} \rho^n \log \rho + \mathfrak{g}^{(n)} \rho^n + o(\rho^n) & n \text{ even}. \end{cases} \tag{4}$$

Furthermore, the coefficients of the expansion satisfies the following:

- *$\mathfrak{g}^{(2k)}$, for any $0 \leq 2k < n$, is determined by $\mathfrak{g}$ and its derivatives:*

$$\mathfrak{g}^{(2k)} = \mathscr{R}_n^{(2k)}(\partial^{\leq 2k} \mathfrak{g}). \tag{5}$$

- *In particular, $\mathfrak{g}^{(0)} = \mathfrak{g}$, and $-\mathfrak{g}^{(2)}$ is the Schouten tensor for $\mathfrak{g}$ whenever $n \geq 3$:*

$$-\mathfrak{g}^{(2)} = \mathscr{P}[\mathfrak{g}] := \frac{1}{n-2} \left(\mathrm{Ric}[\mathfrak{g}] - \frac{1}{2(n-1)} \mathrm{R}[\mathfrak{g}] \cdot \mathfrak{g} \right). \tag{6}$$

- *When n is even, $\mathfrak{g}^{(\star)}$ is also determined by $\mathfrak{g}$ and its derivatives:*

$$\mathfrak{g}^{(\star)} = \mathscr{R}_n^{(\star)}(\partial^{\leq n} \mathfrak{g}). \tag{7}$$

- *Both the $\mathfrak{g}$-divergence and $\mathfrak{g}$-trace of $\mathfrak{g}^{(n)}$ are determined by $\mathfrak{g}$ and its derivatives:*

$$\mathrm{div}_{\mathfrak{g}}\, \mathfrak{g}^{(n)} = \mathscr{R}_n^{\mathrm{div}}(\partial^{\leq n+1} \mathfrak{g}), \qquad \mathrm{tr}_{\mathfrak{g}}\, \mathfrak{g}^{(n)} = \mathscr{R}_n^{\mathrm{tr}}(\partial^{\leq n} \mathfrak{g}). \tag{8}$$

In fact, when n is odd, both $\mathscr{R}_n^{\mathrm{div}}$ and $\mathscr{R}_n^{\mathrm{tr}}$ are identically zero.

Remark 9. Here and throughout the rest of this article, we use $\mathscr{L}$ to denote the Lie derivative, and $\mathscr{L}_\rho$ for the Lie derivative in the "$(0,\rho_0]$-direction" of $\mathscr{M}$.

Remark 10. The relations $\mathscr{R}_n^{(2k)}$, $\mathscr{R}_n^{(\star)}$, $\mathscr{R}_n^{\mathrm{div}}$, $\mathscr{R}_n^{\mathrm{tr}}$ in (5)–(8) are universal, in that they only depend on the dimension n, and not on $\mathscr{I}$ or $\mathfrak{g}$. Moreover, each of these relations could in principle be computed, although the exact formulas tend to be exceedingly complicated.

Remark 11. The partial FG expansion can be continued beyond $\mathfrak{g}^{(n)}$ to any arbitrary (but finite) order, as long as one assumes sufficient additional regularity on g.

The coefficients $\mathfrak{g}^{(2k)}$ $(0 < 2k < n)$ and $\mathfrak{g}^{(\star)}$ are generated by computing the limits of ρ-derivatives of g at the conformal boundary. These boundary limits are obtained through an inductive process by taking successive ρ-derivatives of (an appropriate form of) the EVE. This induction continues until one reaches $\mathfrak{g}^{(n)}$, which—other than (8)—cannot be obtained from the EVE. Nonetheless, one can still extract existence of the free coefficient $\mathfrak{g}^{(n)}$ in (4), although it is not determined by any of the preceding coefficients. For details, see the discussions in [59].

Remark 12. In particular, the conclusions of [59] show that in practice, one needs not separately assume that $\mathfrak{g}^{(n)}$ and the ρ^n-term in the partial FG expansion exist.

In particular, Theorem 2.1 shows that even though we no longer have a full infinite series expansion, *we can nonetheless still view* $(\mathfrak{g}^{(0)}, \mathfrak{g}^{(n)})$ *as free boundary data* (excepting the constraints (8)) *for the EVE.* Additionally, in the physics literature, $\mathfrak{g}^{(n)}$ is closely connected to the *stress-energy tensor* for the boundary conformal field theory; for further discussions, see [22, 60].

For future convenience, we introduce the following terminology for our boundary data:

Definition 2. In general, we refer to a triple $(\mathscr{I}, \mathfrak{g}^{(0)}, \mathfrak{g}^{(n)})$ as *holographic data* iff $(\mathscr{I}, \mathfrak{g}^{(0)})$ is an n-dimensional Lorentzian manifold, and $\mathfrak{g}^{(n)}$ is a symmetric 2-tensor on $\mathscr{I}$ satisfying (8).

Furthermore, given a vacuum FG-aAdS segment $(\mathscr{M}, g)$, we refer to the $(\mathscr{I}, \mathfrak{g}^{(0)}, \mathfrak{g}^{(n)})$ obtained from Theorem 2.1 as the *holographic data* associated to $(\mathscr{M}, g)$.

Finally, one can explicitly compute the partial FG coefficients for the Schwarzschild-AdS metric with mass $M \in \mathbb{R}$. These are well-known in the physics literature (see, for instance, [15, 65]), but they are also computed in [59]. In particular, when $n > 2$ and $n \neq 4$, we have

$$\mathfrak{g}^{(0)} = -dt^2 + \mathring{\gamma}, \qquad \mathfrak{g}^{(2)} = -\frac{1}{2}(dt^2 + \mathring{\gamma}), \qquad \mathfrak{g}^{(n)} = \frac{M}{n}[(n-1)dt^2 + \mathring{\gamma}]. \quad (9)$$

Note $\mathfrak{g}^{(0)}$ is the same for all Schwarzschild-AdS metrics, and the mass M is encoded in $\mathfrak{g}^{(n)}$.

2.3 Gauge Transformations

Though the partial FG expansions of Theorem 2.1 capture near-boundary geometries of vacuum FG-aAdS segments, they fail to account for one additional gauge freedom in our setting. The issue is that Definition 1 of FG-aAdS segments presupposes a choice of FG gauge. However, there does exist a nontrivial class of transformations $\rho \mapsto \bar{\rho}$ of the boundary defining function that preserve the form (1) of the FG gauge.

The key observation is that (1) can be achieved by any function $\bar{\rho}$ satisfying

$$\bar{\rho} > 0, \qquad \lim_{\rho \searrow 0} \bar{\rho} = 0, \qquad \bar{\rho}^{-2} g(d\bar{\rho}, d\bar{\rho}) = 1. \tag{10}$$

(The critical condition is that the $\bar{\rho}^{-2}g$-gradient of $\bar{\rho}$ is unit, while the other components are then transported along the integral curves of this gradient; see [29] for details.) Adopting the ansatz

$$\bar{\rho} = e^{\mathsf{a}}\rho,$$

then (10) reduces to fully nonlinear initial-value problem

$$(2 + \rho \mathscr{L}_\rho \mathsf{a}) \mathscr{L}_\rho \mathsf{a} + \rho\, \mathsf{g}^{-1}(\mathsf{Da}, \mathsf{Da}) = 0,$$

which can be uniquely solved from $\rho = 0$ using the method of characteristics. This implies that *for any* $\mathfrak{a} \in C^\infty(\mathscr{I})$*, one can find a unique FG gauge, with boundary defining function* $\bar{\rho}$*, such that*

$$\lim_{\rho \searrow 0} \frac{\bar{\rho}}{\rho} = e^{\mathfrak{a}}. \tag{11}$$

Remark 13. Using an appropriate isometry, one can then formulate the FG gauge defined with respect to $\bar{\rho}$ as an FG-aAdS segment (at least near the conformal boundary).

Of course, these different FG gauges represent the same physical object, since they are different representations of the same spacetime. On the other hand, a change of FG gauge $\rho \mapsto \bar{\rho}$ induces a corresponding transformation of the partial FG expansion. For instance, one can show that for the FG change satisfying (11), the boundary metric $\mathfrak{g}^{(0)}$ undergoes a conformal transformation:

$$\mathfrak{g}^{(0)} \mapsto \bar{\mathfrak{g}}^{(0)} = e^{2\mathfrak{a}} \mathfrak{g}^{(0)}. \tag{12}$$

In other words, only the conformal class $[\mathfrak{g}]$ of the boundary metric can be invariantly associated with a given aAdS spacetime; this motivates the term "conformal boundary" for $(\mathscr{I}, \mathfrak{g})$.

The other coefficients in (4) are also transformed via changes of FG gauge (see [22, 41]), though the explicit formulas quickly become complicated. In general, there do exist explicitly computable (given enough effort) universal functions such that the FG coefficients transform as

$$\begin{aligned}
\mathfrak{g}^{(2k)} &\mapsto \bar{\mathfrak{g}}^{(2k)} = \mathscr{G}_n^{2k}(\partial^{\leq 2k}\mathfrak{a}, \partial^{\leq 2k}\mathfrak{g}^{(0)}), \\
\mathfrak{g}^{(\star)} &\mapsto \bar{\mathfrak{g}}^{(\star)} = \mathscr{G}_n^{\star}(\partial^{\leq n}\mathfrak{a}, \partial^{\leq n}\mathfrak{g}^{(0)}), \\
\mathfrak{g}^{(n)} &\mapsto \bar{\mathfrak{g}}^{(n)} = \mathscr{G}_n^{n}(\partial^{\leq n}\mathfrak{a}, \partial^{\leq n}\mathfrak{g}^{(0)}, \mathfrak{g}^{(n)}),
\end{aligned} \tag{13}$$

where $0 \leq 2k < n$. (In particular, $-\mathfrak{g}^{(2)}$ transforms like the Schouten tensor when $n \geq 3$.)

The physical significance, then, is that *pairs* $(\mathfrak{g}^{(0)}, \mathfrak{g}^{(n)})$ *and* $(\bar{\mathfrak{g}}^{(0)}, \bar{\mathfrak{g}}^{(n)})$ *that are related via* (12) *and* (13) *should be viewed as "the same"*, since they arise from the same aAdS spacetime:

Definition 3. Let $(\mathscr{I}, \mathfrak{g}^{(0)}, \mathfrak{g}^{(n)})$, $(\bar{\mathscr{I}}, \bar{\mathfrak{g}}^{(0)}, \bar{\mathfrak{g}}^{(n)})$ denote holographic data, and let $\phi : \mathscr{I} \leftrightarrow \bar{\mathscr{I}}$ be a diffeomorphism. We say that $(\mathscr{I}, \mathfrak{g}^{(0)}, \mathfrak{g}^{(n)})$ and $(\bar{\mathscr{I}}, \bar{\mathfrak{g}}^{(0)}, \bar{\mathfrak{g}}^{(n)})$ are ϕ-*gauge-equivalent* on $\mathscr{D} \subset \mathscr{I}$ iff there exists $\mathfrak{a} \in C^\infty(\mathscr{I})$ such that the following relations hold:

$$\phi_* \bar{\mathfrak{g}}^{(0)}|_{\mathscr{D}} = e^{2\mathfrak{a}} \mathfrak{g}^{(0)}|_{\mathscr{D}}, \qquad \phi_* \bar{\mathfrak{g}}^{(n)}|_{\mathscr{D}} = \mathscr{G}_n^n(\partial^{\leq n}\mathfrak{a}, \partial^{\leq n}\mathfrak{g}^{(0)}, \mathfrak{g}^{(n)})|_{\mathscr{D}}. \tag{14}$$

Remark 14. As usual, ϕ_* in (14) denotes the pullback of ϕ. In practice, by pulling back through ϕ, we can always restrict to the case in which $\bar{\mathscr{I}} = \mathscr{I}$ and ϕ is the identity map.

2.4 The Vertical Tensor Calculus

This subsection is devoted to a more detailed discussion of the main quantities we will analyze—vertical tensor fields. We remark that readers who are only interested in the statement of the main result and its corollaries can skip this part entirely. On the other hand, the contents here will be relevant to the proof of the main result.

We begin with a precise definition of vertical tensor fields:

Definition 4. Let $(\mathscr{M}, g)$ be an FG-aAdS region. Then:

- The *vertical bundle* $\mathsf{V}^k_l \mathscr{M}$ of rank (k,l) over $\mathscr{M}$ is defined to be the manifold consisting of all tensors of rank (k,l) on each level set of ρ in $\mathscr{M}$:

$$\mathsf{V}^k_l \mathscr{M} = \bigcup_{\sigma \in (0, \rho_0]} T^k_l \{\rho = \sigma\}. \tag{15}$$

- A section of $\mathsf{V}^k_l \mathscr{M}$ is called a *vertical tensor field* of rank (k,l).

In short, a vertical tensor field on $\mathscr{M}$ is a tensor field that is entirely tangent in every component to the level sets of ρ. Observe also that vertical tensor fields can be equivalently characterized as a family, parametrized by ρ, of tensor fields on $\mathscr{I}$. In particular, the second interpretation provides a natural notion of limits at the conformal boundary—for a vertical tensor field $\mathsf{A} = \mathsf{A}(\rho)$, its limit as $\rho \searrow 0$ (when exists) is simply a tensor field of the same rank at $\mathscr{I}$. This allows one to easily connect quantities in the bulk spacetime with those on the conformal boundary.

One basic example of a vertical tensor field is the vertical metric $\mathsf{g} = \mathsf{g}(\rho)$ in (1). Note that the boundary limit of g is the conformal boundary metric $\mathfrak{g} = \mathfrak{g}^{(0)}$. Moreover, g induces a notion of *vertical covariant derivative* D; more specifically, since $\mathsf{g}(\rho)$ is a Lorentzian metric for each ρ, we can define D by aggregating the Levi-Civita connections of all the $\mathsf{g}(\rho)$'s.

Analogues of vertical tensor fields have been widely used in mathematical relativity; these have sometimes been referred to as horizontal tensor fields. (Common examples include the connection and curvature components on level spheres in a double null foliation.) However, we will also extend these ideas in some novel ways in order to reach our main results.

Remark 15. We adopt the term "vertical" rather than "horizontal" here, since the hypersurfaces in our foliation—the level sets of ρ—are timelike rather than spacelike.

In particular, since tensorial wave equations play a key role in the proofs of our results, we will want to make sense of a *spacetime wave operator* $\check{\Box}_g$ *applied to vertical tensor fields*. Furthermore, we aim to do this in a covariant manner, so that the usual operations of geometric analysis—e.g. Leibniz rules, integrations by parts—continue to hold. The upshot is that we will be able to treat vertical tensor fields covariantly, in nearly the same manner as we do scalar fields.

We achieve these goals in a series of steps, with the first being to make sense of first spacetime covariant derivatives of vertical tensor fields. While the vertical connection D already makes sense of derivatives in vertical directions, we now want to also extend D to the ρ-direction. There are many ways to accomplish this, the standard approach being to define D_ρ as an appropriate projection of the g-covariant derivative. However, here we elect to do this differently, by instead projecting the $\rho^2 g$-covariant derivative. This is primarily a matter of convenience, since $\rho^2 g$ is well-defined at the conformal boundary, so this definition yield formulas with fewer singular weights. Observe that, crucially, this extended covariant derivative remains torsion-free:

$$\mathsf{D}_\rho \mathsf{g} = 0.$$

However, the key issues arise from making proper sense of *second spacetime* derivatives of vertical tensor fields. To do this covariantly, one requires a small detour—we must extend our vertical tensor calculus to *mixed tensor fields*, containing both spacetime and vertical components:

Definition 5. Let $(\mathscr{M}, g)$ be an FG-aAdS segment. Then:

- We define the *mixed bundle* of ranks $(\kappa, \lambda; k, l)$ over $\mathscr{M}$ to be the tensor product bundle
$$T^\kappa_\lambda \mathsf{V}^k_l \mathscr{M} := T^\kappa_\lambda \mathscr{M} \otimes \mathsf{V}^k_l \mathscr{M}. \tag{16}$$
- We refer to sections of $T^\kappa_\lambda \mathsf{V}^k_l \mathscr{M}$ as *mixed tensor fields* of ranks $(\kappa, \lambda; k, l)$.

The tensor product structure of the mixed bundles induces a natural notion of covariant derivative $\check{\nabla}$ for mixed tensor fields—in short, $\check{\nabla}$ *acts like the spacetime connection* ∇ *on spacetime components and like the vertical connection* D *on vertical components.* This behavior can be generated from the following defining relation, which can be shown to uniquely define $\check{\nabla}$: given any vector field X on $\mathscr{M}$, tensor field G on $\mathscr{M}$, and vertical tensor field B, we set

$$\check{\nabla}_X(G \otimes \mathsf{B}) := \nabla_X G \otimes \mathsf{B} + G \otimes \mathsf{D}_X \mathsf{B}.$$

The torsion-free property extends also to the mixed connection $\check{\nabla}$. As both the spacetime metric g and the vertical metric g are mixed tensor fields, our definition of $\check{\nabla}$ immediately implies

$$\check{\nabla} g = 0, \qquad \check{\nabla} \mathsf{g} = 0.$$

Mixed bundles allow us to now covariantly define a spacetime Hessian of a vertical tensor field A, say of rank (k,l). Recall that the first derivative $\check{\nabla}\mathsf{A}$ is simply the extended vertical connection applied to A. The trick is now to view $\check{\nabla}\mathsf{A}$ as a mixed tensor field of rank $(0,1;k,l)$. The Hessian $\check{\nabla}^2\mathsf{A}$ *is then naturally defined as* $\check{\nabla}$ *applied to the mixed tensor field* $\check{\nabla}\mathsf{A}$.

Finally, for the wave operator, since $\check{\nabla}^2\mathsf{A}$ is a mixed tensor field of rank $(0,2;k,l)$, we can simply define $\check{\Box}_g\mathsf{A}$ as the g-trace of $\check{\nabla}^2\mathsf{A}$ along the two spacetime components.

Remark 16. Since all our connections, by their construction, satisfy the requisite Leibniz rules and are torsion-free, then, with regards to integration by parts, we can treat both vertical and mixed tensor fields in the same way that we treat scalar fields and spacetime tensor fields.

Remark 17. Mixed tensor fields and extended wave operators $\check{\Box}_g$ originated from [58]; for aAdS contexts, an earlier formulation using horizontal tensor fields was used [35, 36]. (Similar notions were also independently developed and used in [44].) The full vertical (and mixed) tensor calculus, in the form shown here, was first constructed in [55, 59] and then used in [20, 37].

3 The Main Result

Having provided, in Section 2, a more complete description of our aAdS spacetimes and their properties, we can now return to the main correspondence question posed in this article. In particular, we can now give a more precise formulation of Problem 2:

Problem 3 (Correspondence). Let $(\mathscr{M}, g)$ and $(\bar{\mathscr{M}}, \bar{g})$ be two vacuum FG-aAdS segments, and suppose that their respective associated holographic data, $(\mathscr{I}, \mathfrak{g}^{(0)}, \mathfrak{g}^{(n)})$ and $(\bar{\mathscr{I}}, \bar{\mathfrak{g}}^{(0)}, \bar{\mathfrak{g}}^{(n)})$, are gauge-equivalent. Then, must g and $\bar{g}$ be isometric to each other?

Compared to Problem 2, we have replaced the vague "aAdS spacetimes" with the precisely defined FG-aAdS segments of Definition 1. Furthermore, the "conformal boundary data" mentioned in Problem 2 have been clarified as holographic data, up to gauge-equivalence.

A closely related question is that of *symmetry extension—are (appropriately defined) symmetries of the holographic data necessarily inherited by the bulk aAdS spacetime*? Using the terminology of Section 2, we can again provide a more precise statement:

Problem 4 (Symmetry Extension). Let $(\mathscr{M}, g)$ be a vacuum FG-aAdS segment, with holographic data $(\mathscr{I}, \mathfrak{g}^{(0)}, \mathfrak{g}^{(n)})$. If ψ is a symmetry of $(\mathscr{I}, \mathfrak{g}^{(0)}, \mathfrak{g}^{(n)})$ (that is, a diffeomorphism $\psi : \mathscr{I} \leftrightarrow \mathscr{I}$ that preserves $\mathfrak{g}^{(0)}$ and $\mathfrak{g}^{(n)}$ up to gauge-equivalence), then must ψ extend to a isometry of g?

Our main result, Theorem 3.2 below, will give a positive resolution to Problem 3. Answers to Problem 4 will then follow as corollaries; see Theorems 3.3 and 3.4.

3.1 Previous Literature

Before stating our main theorems, we first review some of the existing literature related to Problems 3 and 4, starting with stationary settings.

First, Biquard [16] positively answered the Riemannian analogue of Problem 3. More specifically, [16] considered asymptotically hyperbolic Einstein manifolds (N, h), which have both a conformal boundary $(\partial N, \mathfrak{h})$ and an analogous Fefferman-Graham expansion from ∂N. The main result of [16] then showed that *the free coefficients* $(\mathfrak{h}^{(0)}, \mathfrak{h}^{(n)})$ *in the FG expansion uniquely determine the metric* h *on* N. The key technical tool in the proof is a unique continuation result of Mazzeo [53] from the conformal boundary of asymptotically hyperbolic manifolds.

A similar correspondence result, also for (Riemannian) asymptotically hyperbolic Einstein manifolds, was established independently by Anderson and Herzlich; see [12, 13]. Notice that in each of the results [12, 13, 16], the key PDEs are always elliptic in nature.

Next, the methods of [16] were adapted by Chrusciel and Delay [21] to *Lorentzian* aAdS spacetimes, under the specific restriction that the spacetime is *stationary*. More specfically, [21] showed that *if* $(\mathscr{M}, g)$ *is time-independent, then* g *is indeed uniquely determined by its holographic data* $(\mathscr{I}, \mathfrak{g}^{(0)}, \mathfrak{g}^{(n)})$. Furthermore, using similar techniques, [21] positively addressed Problem 4 in stationary settings—*if* $(\mathscr{M}, g)$ *is time-independent, then any Killing vector field on the conformal boundary that also preserves* $\mathfrak{g}^{(n)}$ *necessarily extends into* $\mathscr{M}$ *as a Killing field for* g.

Remark 18. Even though [21] treated Lorentzian settings, the assumption of time-independence meant that all the key PDEs are elliptic. Thus, there remains a fundamental gap between [21] and the questions posed here in general dynamical settings, which instead involve hyperbolic PDEs.

In addition, we mention, again in the Riemannian context, the classical result of Graham and Lee [30], which proved the existence of asymptotically hyperbolic Einstein metrics on the Poincaré ball $\mathbb{B}^n$, with prescribed metric $\mathfrak{h}^{(0)}$ on the conformal boundary $\mathbb{S}^{n-1}$ that is sufficiently close to the round metric. Observe, however, that the above corresponds to solving an elliptic Dirichlet problem, which has no analogue in hyperbolic contexts.

We now turn our attention to dynamical Lorentzian settings. First, Anderson [11] partly showed that if a vacuum aAdS spacetime $(\mathscr{M}, g)$ has a stationary conformal boundary and becomes asymptotically stationary as $t \to \pm\infty$, then $(\mathscr{M}, g)$ must itself be stationary. However, the result in [11] is *conditional*, in that *it assumes that a unique continuation property holds for the linearized Einstein equations from the conformal boundary*. In particular, this unique continuation condition was crucially used to prove a symmetry extension result, in the sense of Problem 4, for timelike Killing vector fields from the conformal boundary into the bulk spacetime.

Remark 19. The main symmetry extension result of this article, Theorem 3.4, can be applied in [11] to prove an improved, *unconditional* version of Anderson's theorem.

The first concrete steps toward resolving Problem 3, as well as a precursor to the main result of this article, was provided by McGill [54], which gave a holographic characterization of locally AdS spacetimes. Very roughly, [54] established the following:

Theorem 3.1 ([54]). *Let $(\mathscr{M}, g)$ be a vacuum FG-aAdS segment with the AdS conformal boundary $(\mathscr{I}_{AdS}, g_{AdS})$. Then, $(\mathscr{M}, g)$ is locally isometric to AdS spacetime (at least close to the conformal boundary) if and only if both $\mathfrak{g}^{(0)}$ is conformally flat and $\mathfrak{g}^{(n)} = 0$ on a sufficiently large timespan of the conformal boundary. Furthermore, the above criteria for $\mathfrak{g}^{(0)}$ and $\mathfrak{g}^{(n)}$ are gauge-invariant, in the sense that they are preserved by gauge transformations.*

McGill's result was proved using the same setting and tools [35, 36, 55, 59] as our main theorems, most notably the Carleman estimates for tensorial waves near aAdS conformal boundaries. One key step is a unique continuation result for the spacetime Weyl curvature, which satisfies a tensorial wave equation, and whose vanishing ensures the spacetime is locally AdS. In particular, [54] represented the first application of the Carleman estimates of [35, 36, 55] to the EVE, and it demonstrated that these estimates are indeed powerful enough to treat the EVE.

More generally, symmetry extension results for Killing vector fields have seen several applications in general relativity, outside of aAdS settings. For instance, Killing extension theorems have played a crucial role in recent black hole rigidity results [3, 4, 5, 28, 43, 47, 70], which showed, in various circumstances, that vacuum (or electrovacuum) stationary spacetimes must be isometric to Kerr (or Kerr-Newman) spacetime. Next, [56] proved a Killing extension theorem from compact Cauchy horizons, which can then be applied to questions in both black hole rigidity and cosmic censorship. Furthermore, the results in [6] proved non-existence of time-periodic vacuum asymptotically flat spacetimes; a key step is a Killing extension theorem from spacelike and null infinity.

3.2 Correspondence Results

The main result of this article provides an *affirmative answer to Problem 3, provided the conformal boundary also satisfies a gauge-invariant geometric condition.* We refer to this extra assumption, which in its current form was first identified and formulated by Chatzikaleas and the author in [20], as the *generalised null convexity criterion*, or *GNCC*.

We will define and further discuss the GNCC (which can seem quite technical at first glance) in the following subsection. But before doing so, let us first state our main result:

Theorem 3.2 (Bulk-boundary correspondence [37]). *Let $n > 2$, and let $(\mathscr{M}, g)$ and $(\bar{\mathscr{M}}, \bar{g})$ denote vacuum FG-aAdS segments, with associated holographic data $(\mathscr{I}, \mathfrak{g}^{(0)}, \mathfrak{g}^{(n)})$ and $(\bar{\mathscr{I}}, \bar{\mathfrak{g}}^{(0)}, \bar{\mathfrak{g}}^{(n)})$, respectively. In addition, fix $\mathscr{D} \subset \mathscr{I}$ with $\bar{\mathscr{D}}$ compact, and assume $(\mathscr{D}, \mathfrak{g}^{(0)})$ satisfies the GNCC.*

Then, if $(\mathscr{I}, \mathfrak{g}^{(0)}, \mathfrak{g}^{(n)})$ and $(\bar{\mathscr{I}}, \bar{\mathfrak{g}}^{(0)}, \bar{\mathfrak{g}}^{(n)})$ are ϕ-gauge-equivalent on $\mathscr{D}$, then there exist two neighborhoods $\mathscr{U} \subset \mathscr{M}$ and $\bar{\mathscr{U}} \subset \bar{\mathscr{M}}$—of $\{0\} \times \mathscr{D}$ and $\{0\} \times \phi(\mathscr{D})$, respectively—as well as an isometry $\Phi : (\mathscr{U}, g) \leftrightarrow (\bar{\mathscr{U}}, \bar{g})$ whose limit at the conformal boundary is ϕ.

In particular, Theorem 3.2 shows that the near-boundary vacuum aAdS geometries (up to isometry) are in one-to-one correspondence with the space of holographic data (up to gauge-equivalence).

Remark 20. That $\mathscr{D}$ in Theorem 3.2 satisfies the GNCC can be interpreted as $\mathscr{D}$ being "sufficiently large" compared to the geometry of $(\mathscr{I}, \mathfrak{g}^{(0)})$. In other words, the correspondence result holds only when the holographic data match (up to gauge) on a sufficiently large boundary region.

Remark 21. As GNCC is a gauge-invariant condition (see Proposition 1 below), then $(\mathscr{D}, \mathfrak{g}^{(0)})$ satisfying the GNCC is equivalent to $(\phi(\mathscr{D}), \bar{\mathfrak{g}}^{(0)})$ satisfying the GNCC. In other words, all assumptions on the conformal boundaries in Theorem 3.2 are gauge-invariant.

Remark 22. Note the correspondence described in Theorem 3.2 is local, as the FG-aAdS segments are only isometric near the subregions $\mathscr{D}$ and $\phi(\mathscr{D})$ of the conformal boundaries. In particular, one needs additional information to prove a global correspondence.

Remark 23. By pulling back through ϕ, we can, without loss of generality, reduce Theorem 3.2 to the case in which $\bar{\mathscr{M}} = \mathscr{M}$, $\bar{\mathscr{I}} = \mathscr{I}$, and ϕ is the identity map on $\mathscr{I}$; see the beginning of Section 5. In fact, the main result of [37] is only formulated in this special case. We state the general case in Theorem 3.2 in order to highlight the geometric and gauge-covariant nature of the result.

Remark 24. The main result Theorem 3.1 of [54] can be viewed as a special case of Theorem 3.2, for spacetimes that are locally isometric to AdS. Furthermore, the assumption in Theorem 3.1 of "sufficiently large timespan of the conformal boundary" is a direct consequence of the GNCC.

Next, one important application of Theorem 3.2 is toward proving symmetry extension results on aAdS spacetimes, which address Problem 4. The first such result deals with extending discrete symmetries from the conformal boundary into the bulk spacetime:

Definition 6. Let $(\mathscr{I}, \mathfrak{g}^{(0)}, \mathfrak{g}^{(n)})$ denote holographic data, and fix $\mathscr{D} \subset \mathscr{I}$ with $\bar{\mathscr{D}}$ compact. We say that $\psi : \mathscr{I}' \to \mathscr{I}$, where $\bar{\mathscr{D}} \subset \mathscr{I}' \subset \mathscr{I}$, is a *holographic gauge symmetry* on $\mathscr{D}$ iff $(\mathscr{I}', \mathfrak{g}^{(0)}, \mathfrak{g}^{(n)})$ and $(\psi(\mathscr{I}'), \mathfrak{g}^{(0)}, \mathfrak{g}^{(n)})$ are ψ-gauge-equivalent on $\mathscr{D}$. If ψ satisfies, in addition,

$$\psi_* \mathfrak{g}^{(0)}|_{\mathscr{D}} = \mathfrak{g}^{(0)}|_{\mathscr{D}}, \qquad \psi_* \mathfrak{g}^{(n)}|_{\mathscr{D}} = \mathfrak{g}^{(n)}|_{\mathscr{D}}. \tag{1}$$

then ψ is called a *holographic symmetry* on $\mathscr{D}$.

Informally, a holographic gauge symmetry is a diffeomorphism within $\mathscr{I}$ that preserves $\mathfrak{g}^{(0)}$ and $\mathfrak{g}^{(n)}$ up to gauge-equivalence, while a holographic symmetry preserves $\mathfrak{g}^{(0)}$ and $\mathfrak{g}^{(n)}$ entirely.

Theorem 3.3 (Extension of symmetries [37]). *Let $n > 2$, and let $(\mathscr{M}, g)$ be a vacuum FG-aAdS segment, with holographic data $(\mathscr{I}, \mathfrak{g}^{(0)}, \mathfrak{g}^{(n)})$. Also, fix $\mathscr{D} \subset \mathscr{I}$ with $\bar{\mathscr{D}}$ compact, and assume $(\mathscr{D}, \mathfrak{g}^{(0)})$ satisfies the GNCC. Then, if ψ is a holographic gauge symmetry on $\mathscr{D}$, then there is a neighborhood $\mathscr{U} \subset \mathscr{M}$ of $\{0\} \times \mathscr{D}$ such that ψ extends to an isometry Ψ between $(\mathscr{U}, g)$ and $(\Psi(\mathscr{U}), g)$.*

Proof. This follows immediately from Theorem 3.2—assuming $\psi : \mathscr{I}' \to \mathscr{I}$, where $\bar{\mathscr{D}} \subset \mathscr{I}' \subset \mathscr{I}$, then we simply apply Theorem 3.2 to the vertical truncations of $\mathscr{M}$ to $\mathscr{I}'$ and $\psi(\mathscr{I}')$. □

Remark 25. If ψ in Theorem 3.3 is, in addition, a holographic symmetry on $\mathscr{D}$, then the desired isometry Ψ can be given explicitly in terms of ψ by

$$\Psi(\rho, p) = (\rho, \psi(p)), \qquad (\rho, p) \in \mathscr{U} \subset \mathscr{M}. \tag{2}$$

Another variant (as well as a direct corollary) of Theorem 3.3 concerns extending a smooth family of symmetries, in the form of a Killing vector field. This is stated in the subsequent theorem, which, on its own, can also be seen as an analogue of the symmetry extension results [3, 6, 28, 43, 56], but in the setting of conformal boundaries of vacuum aAdS spacetimes.

Definition 7. Let $(\mathscr{I}, \mathfrak{g}^{(0)}, \mathfrak{g}^{(n)})$ denote holographic data, and fix $\mathscr{D} \subset \mathscr{I}$ with $\bar{\mathscr{D}}$ compact. We say that a vector field $\mathfrak{K}$ on $\mathscr{I}'$, where $\bar{\mathscr{D}} \subset \mathscr{I}' \subset \mathscr{I}$, is *holographic gauge Killing* on $\mathscr{D}$ iff there exist holographic data $(\bar{\mathscr{I}}, \bar{\mathfrak{g}}^{(0)}, \bar{\mathfrak{g}}^{(n)})$ that is ϕ-gauge-equivalent to $(\mathscr{I}, \mathfrak{g}^{(0)}, \mathfrak{g}^{(n)})$ and such that

$$\mathscr{L}_{\phi^* \mathfrak{K}} \bar{\mathfrak{g}}^{(0)}|_{\phi(\mathscr{D})} = 0, \qquad \mathscr{L}_{\phi^* \mathfrak{K}} \bar{\mathfrak{g}}^{(n)}|_{\phi(\mathscr{D})} = 0. \tag{3}$$

Furthermore, $\mathfrak{K}$ is called *holographic Killing* on $\mathscr{D}$ iff

$$\mathscr{L}_{\mathfrak{K}} \mathfrak{g}^{(0)}|_{\mathscr{D}} = 0, \qquad \mathscr{L}_{\mathfrak{K}} \mathfrak{g}^{(n)}|_{\mathscr{D}} = 0. \tag{4}$$

Remark 26. As is standard, $\phi^*\mathfrak{K}$ denotes the push-forward of $\mathfrak{K}$ through ϕ.

Informally, a holographic Killing vector field $\mathfrak{K}$ preserves $\mathfrak{g}^{(0)}$ and $\mathfrak{g}^{(n)}$ along its integral curves, while a holographic gauge Killing field only preserves $\mathfrak{g}^{(0)}$ and $\mathfrak{g}^{(n)}$ up to gauge-equivalence.

Theorem 3.4 (Extension of Killing vector fields [37]). *Let $n > 2$, and let $(\mathscr{M}, g)$ be a vacuum FG-aAdS segment, with holographic data $(\mathscr{I}, \mathfrak{g}^{(0)}, \mathfrak{g}^{(n)})$. Fix $\mathscr{D} \subset \mathscr{I}$ with $\bar{\mathscr{D}}$ compact, and assume $(\mathscr{D}, \mathfrak{g}^{(0)})$ satisfies the GNCC. Then, if $\mathfrak{K}$ is a holographic gauge Killing vector field on $\mathscr{D}$, then there is a neighborhood $\mathscr{U} \subset \mathscr{M}$ of $\{0\} \times \mathscr{D}$ such that $\mathfrak{K}$ extends to a (g-)Killing vector field K on $\mathscr{U}$.*

Proof. By pushing through the gauge transformation ϕ in Definition 7 and relying on Theorem 3.2, we can reduce to the case in which $\mathfrak{K}$ is holographic Killing on $\mathscr{D}$ and (4) holds. Let ψ_s, for small $s \in \mathbb{R}$, be the family of holographic symmetries generated by transporting along the integral curves of $\mathfrak{K}$. By Theorem 3.3, these extend to a family of g-isometries Ψ_s in the bulk spacetime. The Killing vector field K is then obtained as the generator of the isometries Ψ_s. □

One immediate consequence of Theorem 3.4 and the classical Birkhoff theorem is the following holographic rigidity result for Schwarzschild-AdS spacetimes:

Corollary 1 (Rigidity of Schwarzschild-AdS spacetime). *Let $n > 2$, and let $(\mathscr{M}, g)$ be a vacuum FG-aAdS segment, with holographic data $(\mathscr{I}_{\mathrm{AdS}}, \mathfrak{g}^{(0)}, \mathfrak{g}^{(n)})$. Also, fix $\mathscr{D} \subset \mathscr{I}$ with $\bar{\mathscr{D}}$ compact, and assume $(\mathscr{D}, \mathfrak{g}^{(0)})$ satisfies the GNCC. If $(\mathfrak{g}^{(0)}, \mathfrak{g}^{(n)})$ is spherically symmetric on $\mathscr{D}$, then there is a neighborhood $\mathscr{U}$ of $\{0\} \times \mathscr{D}$ such that $(\mathscr{U}, g)$ is isometric to part of a Schwarzschild-AdS spacetime.*

Remark 27. If one takes the AdS conformal boundary, $\mathfrak{g}^{(0)} = \mathfrak{g}_{\mathrm{AdS}}$, in Corollary 1, then its conclusions hold with $\mathscr{D} := \{t_- < t < t_+\}$ whenever $t_+ - t_- > \pi$; see Proposition 2 below.

Finally, note that Theorem 3.3 also can be applied toward discrete symmetries that are not generated by Killing vector fields. In particular, one consequence of Theorem 3.3 is that time periodicity of the conformal boundary is inherited by the bulk spacetime:

Corollary 2 (Extension of time periodicity). *Let $n > 2$, and let $(\mathscr{M}, g)$ denote a vacuum FG-aAdS segment, with holographic data $(\mathscr{I}, \mathfrak{g}^{(0)}, \mathfrak{g}^{(n)})$. In addition, fix $\mathscr{D} \subset \mathscr{I}$ with $\bar{\mathscr{D}}$ compact, and assume that $(\mathscr{D}, \mathfrak{g}^{(0)})$ satisfies the GNCC. If $(\mathfrak{g}^{(0)}, \mathfrak{g}^{(n)})$ is time-periodic on $\mathscr{D}$, then there exists a neighborhood $\mathscr{U}$ of $\{0\} \times \mathscr{D}$ such that $(\mathscr{U}, g)$ is also time-periodic.*

3.3 Null Convexity Criteria

Having stated our key results, we now turn our attention toward the crucial geometric assumption required for these results—the GNCC of [20]. The first task is to state the GNCC, adapted here to the special case of vacuum aAdS spacetimes:

Definition 8 (GNCC [20]). Let $(\mathscr{I}, \mathfrak{g})$ be an n-dimensional Lorentzian manifold, and fix any Riemannian metric $\mathfrak{h}$ on $\mathscr{I}$. In addition, fix an open subset $\mathscr{D} \subset \mathscr{I}$, with $\bar{\mathscr{D}}$ compact. We say that $(\mathscr{D}, \mathfrak{g})$ satisfies the *generalized null convexity criterion* (abbreviated *GNCC*) iff there is a C^4-function η, defined on a neighborhood of $\bar{\mathscr{D}}$, such that the following hold:

- $\eta > 0$ on $\mathscr{D}$.
- $\eta = 0$ on the boundary $\partial \mathscr{D}$.
- There exists $c > 0$ such that the following holds on $\mathscr{D}$ for any $\mathfrak{g}$-null vector field $\mathfrak{Z}$,
$$(\mathfrak{D}^2 \eta + \eta \cdot \mathscr{P}[\mathfrak{g}])(\mathfrak{Z}, \mathfrak{Z}) > c\eta \cdot \mathfrak{h}(\mathfrak{Z}, \mathfrak{Z}), \tag{5}$$
where $\mathfrak{D}^2$ and $\mathscr{P}[\mathfrak{g}]$ denote the Hessian and Schouten tensor with respect to $\mathfrak{g}$, respectively.

Remark 28. Note Definition 8 is independent of the choice of $\mathfrak{h}$, since $\bar{\mathscr{D}}$ is compact. Moreover, observe that $\mathscr{P}[\mathfrak{g}]$ in (5) can be replaced by $\frac{1}{n-2} \operatorname{Ric}[\mathfrak{g}]$, since their difference is proportional to $\mathfrak{g}$—see (6)—and hence vanishes along all $\mathfrak{g}$-null directions.

Remark 29. The GNCC can be extended to FG-aAdS segments that are not vacuum by replacing the Schouten tensor $\mathscr{P}[\mathfrak{g}]$ in (5) with $-\mathfrak{g}^{(2)}$. (Note the general case reduces to Definition 8 in vacuum settings due to (6).) This is the GNCC that was stated in [20].

One important feature of the GNCC is that it is a conformally invariant property. In particular, [20, Proposition 3.6] showed, in the context of Definition 8, that *if* $(\mathscr{D}, \mathfrak{g})$ *satisfies the GNCC, with associated function* η*, then* $(\mathscr{D}, e^{2\mathfrak{a}} \mathfrak{g})$ *also satisfies the GNCC, with associated function* $\eta' := e^{\mathfrak{a}} \eta$. We can immediately rephrase the above in terms of our holographic gauge-invariance.

Proposition 1. *Let* $(\mathscr{I}, \mathfrak{g}^{(0)}, \mathfrak{g}^{(n)})$, $(\bar{\mathscr{I}}, \bar{\mathfrak{g}}^{(0)}, \bar{\mathfrak{g}}^{(n)})$ *denote holographic boundary data, and suppose* $(\mathscr{I}, \mathfrak{g}^{(0)}, \mathfrak{g}^{(n)})$, $(\bar{\mathscr{I}}, \bar{\mathfrak{g}}^{(0)}, \bar{\mathfrak{g}}^{(n)})$ *are* ϕ*-gauge-equivalent. Moreover, let* $\mathscr{D} \subset \mathscr{I}$ *be open, with* $\bar{\mathscr{D}}$ *compact. Then,* $(\mathscr{D}, \mathfrak{g}^{(0)})$ *satisfies the GNCC if and only if* $(\phi(\mathscr{D}), \bar{\mathfrak{g}}^{(0)})$ *satisfies the GNCC.*

The main significance of the GNCC, demonstrated in [20], is that it provides the crucial conditions on the conformal boundary that lead to pseudoconvexity in the near-boundary aAdS geometry. This observation was a crucial ingredient in proving the Carleman estimates of [20] for tensorial waves on FG-aAdS segments. These Carleman estimates, in turn, serve as a key tool in the proof of Theorem 3.2. See Sections 4 and 5 for further discussions of these points.

Definition 8 may seem, at first glance, rather abstract and technical. Furthermore, the conditions for the GNCC, which involve finding a desirable candidate function η, can be rather difficult to check in practice. Therefore, to further flesh out Definition 8, we consider, in the remainder of this subsection, the GNCC applied to some special cases of interest.

Let us begin with the standard AdS conformal boundary $(\mathscr{I}_{\text{AdS}}, \mathfrak{g}_{\text{AdS}})$ of (6), which satisfies

$$\mathfrak{g}_{\mathrm{AdS}} = -dt^2 + \mathring{\gamma}, \qquad \mathscr{P}[\mathfrak{g}_{\mathrm{AdS}}] = \frac{1}{2}(dt^2 + \mathring{\gamma}). \tag{6}$$

In addition, we take $\mathscr{D} := \mathscr{D}_0$ to be the time slab

$$\mathscr{D}_0 := (t_-, t_+) \times \mathbb{S}^{n-1}, \qquad \partial \mathscr{D}_0 = \{t_-, t_+\} \times \mathbb{S}^{n-1}, \qquad t_- < t_+. \tag{7}$$

The following observation was established in [20, Corollary 3.14]:

Proposition 2 ([20, 35]). *$(\mathscr{D}_0, \mathfrak{g}_{Ads})$ satisfies the GNCC if and only if $t_+ - t_- > \pi$.*

Proof. The key observation is that by taking η to depend only on t, then (5) and (6) yield

$$(\mathfrak{D}^2 \eta + \eta \cdot \mathscr{P}[\mathfrak{g}])(3,3) = (3t)^2 (\ddot{\eta} + \eta). \tag{8}$$

Then, one can directly check that the choice

$$\eta(t) := \sin\left(\pi \cdot \frac{t - t_-}{t_+ - t_-}\right)$$

satisfies the conditions of Definition 8 whenever $t_+ - t_- > \pi$. (Observe $t_+ - t_- > \pi$ is needed for the right-hand side of (8) to be positive when $t_- < t < t_+$.)

For the converse, the basic idea is that one can apply a standard Sturm comparison argument with the function $\sin(\pi t)$ to show that if $\eta := \eta(t)$ satisfies

$$\eta(t_\pm) = 0, \qquad \eta|_{(t_-, t_+)} > 0, \qquad \ddot{\eta} + \eta > 0,$$

then $t_+ - t_- > \pi$. In particular, this, along with (8), shows that no function of the form $\eta = \eta(t)$ can satisfy the conditions of Definition 8 whenever $t_+ - t_- \leq \pi$. Moreover, the above argument can be adapted to general functions η, from which one concludes that $(\mathscr{D}_0, \mathfrak{g}_{\mathrm{AdS}})$ cannot satisfy the GNCC whenever $t_+ - t_- \leq \pi$; see [20, Lemma 3.7] for details. □

Remark 30. Note that Proposition 2 applies to AdS spacetime, as well as every Schwarzschild-AdS and Kerr-AdS spacetime, since these all induce the AdS conformal boundary.

Remark 31. The key consequence of Proposition 2—that unique continuation for wave equations holds from $\mathscr{D}_0$ when $t_+ - t_- > \pi$—was first proved as a special case of the results of [35].

Next, we consider general conformal boundary domains that are foliated by a time function t,

$$\mathscr{I}_* := \mathbb{R}_t \times \mathscr{S}, \qquad \mathscr{D}_* := (t_-, t_+) \times \mathscr{S}, \tag{9}$$

with $\mathscr{S}$ a compact manifold of dimension $n-1$. Previous versions of Carleman estimates and unique continuation results for wave equations were developed in the setting (9), under conditions on the conformal boundary that can be viewed as special cases of the GNCC.

First, Holzegel and the author [35] treated the case in which the metric $\mathfrak{g}$ is static (with time function t), establishing Carleman estimates and unique continuation under a so-called *pseudoconvexity criterion.* In particular, this was the first unique continuation result for wave equations from aAdS conformal boundaries. The result was subsequently extended by Holzegel and the author [36] to some non-static $\mathfrak{g}$, and then by McGill and the author [55] to a wider class of non-static $\mathfrak{g}$ and time foliations and under a slightly weaker conformal boundary criterion.

Let us focus on the key criterion of [55], as well as its relation to the GNCC:

Proposition 3 ([20, 55]). *Assume the setting of* (9), *and let* $\mathfrak{g}$ *be a Lorentzian metric on* $\mathscr{I}_*$. *Suppose there exist* $0 \leq B < C$ *such that the following holds for any* $\mathfrak{g}$*-null vector field* $\mathfrak{Z}$*:*

$$\mathscr{P}[\mathfrak{g}](\mathfrak{Z},\mathfrak{Z}) \geq C^2 \cdot (\mathfrak{Z}t)^2, \qquad |\mathfrak{D}^2 t(\mathfrak{Z},\mathfrak{Z})| \leq 2B \cdot (\mathfrak{Z}t)^2. \tag{10}$$

Then, $(\mathscr{D}_*, \mathfrak{g})$ satisfies the GNCC *as long as* $t_+ - t_-$ *is large enough (depending on* B *and* C*).*

Proof. The process is similar to the proof of Proposition 2, except one now chooses η (still depending only on t) to roughly solve a damped harmonic oscillator:

$$\ddot{\eta} - 2b|\dot{\eta}| + c^2\eta = 0, \qquad B \leq b < c < C. \tag{11}$$

The reader is referred to [20, Proposition 3.13] for details. □

Remark 32. The second condition of (10) can be viewed as a bound on the non-stationarity of $\mathfrak{g}$, since $\mathfrak{D}^2 t$ is proportional to the Lie derivative of $\mathfrak{g}$ along the gradient of t.

Remark 33. The connection between damped harmonic oscillators (11) and unique continuation from $\mathscr{D}_*$ was first illuminated in [36]; see the discussions therein.

Remark 34. Proposition 3 can be used to generalize the conclusions of Proposition 2:

- For instance, if $t_+ - t_- > \pi$, then Proposition 3 implies that $(\mathscr{D}_0, \mathfrak{g})$ satisfies the GNCC whenever $\mathfrak{g}$ is a sufficiently small perturbation of $\mathfrak{g}_{\mathrm{AdS}}$.
- If $\mathfrak{g}$ is static with respect to t, and if the cross-sections $\mathscr{S}$ have positive Ricci curvature, then $(\mathscr{D}_*, \mathfrak{g})$ satisfies the GNCC for sufficiently large $t_+ - t_-$; see [36, Proposition B.2].

The conditions (10) were first identified in [55] and were called the *null convexity criterion* (or *NCC*). In particular, [55] established Carleman estimates and unique continuation results, under the assumption that the NCC is satisfied. Note that Proposition 3 shows that the GNCC indeed generalizes the NCC. Furthermore, the GNCC removes the need for a predetermined time function and allows for a larger class of boundary domains $\mathscr{D}$ to be treated.

One advantage of the NCC (10) is that it is more concrete and easier to check than the GNCC. On the other hand, one shortcoming of the NCC is that (10) fails to be gauge-invariant, since a conformal transformation of $\mathfrak{g}$ can cause (10) to no longer hold. This makes the NCC undesirable for Theorem 3.2, and this served as a key motivation for developing the GNCC.

3.4 Geodesic Return

While the GNCC provides a sufficient condition for Theorem 3.2, due to its connection to pseudoconvexity and to Carleman estimates for wave equations, one can also make a heuristic case that the GNCC may be a necessary condition for Theorem 3.2.

The necessity of some geometric condition in Theorem 3.2 was first conjectured in [35, 36], due to the special properties of AdS geometry near its conformal boundary $\mathscr{I}_{\text{AdS}}$. More specifically, on AdS spacetime, *there exist null geodesics which initiate from* $\mathscr{I}_{\text{AdS}}$ *at time* $t = 0$, *remain arbitrarily close to* $\mathscr{I}_{\text{AdS}}$, *and then terminate at the boundary at time* $t = \pi$; see [35, Section 1.2].

Remark 35. In terms of the standard conformal embedding of AdS spacetime into the Einstein cylinder $\mathbb{R}_t \times \mathbb{S}^n_+$, these null geodesics can be explicitly described as

$$\Lambda : (0, \pi) \to \mathbb{R} \times \mathbb{S}^n_+, \qquad \Lambda(\tau) = (\tau, \lambda(\tau)),$$

where λ denotes unit speed parametrizations of great circles in $\mathbb{S}^n_+$, with the limits $\lambda(0+)$ and $\lambda(\pi-)$ lying on antipodal points of the equator $\partial\mathbb{S}^n_+$. The near-boundary trapping effect caused by these Λ is also connected to the relative lack of decay for waves on AdS spacetime; see, e.g., [38].

From the above, one can construct, via the geometric optics methods of Alinhac and Baouendi [10], local solutions to linear wave equations on AdS spacetime that propagate along these null geodesics. Most notably, these solutions vanish to infinite order toward the portion $(0, \pi) \times \mathbb{S}^{n-1}$ of the conformal boundary, but immediately become nonzero away from this boundary segment. This yields, for AdS spacetime, counterexamples to unique continuation for various wave equations when the data on the conformal boundary is imposed on a timespan of less than π—the return time of the above-mentioned near-boundary null geodesics.

On the other hand, if we consider a slab $\mathscr{D}_0$ as in (7), with timespan $t_+ - t_- > \pi$, then none of the aforementioned null geodesics (translated forward or backward in time as needed) can travel entirely over $\mathscr{D}_0$. In fact, if such a geodesic goes over $\mathscr{D}_0$ at all, then it must either start from or terminate at $\mathscr{D}_0$. From this, one concludes that the Alinhac-Baouendi counterexamples cannot be constructed over the time slab $\mathscr{D}_0$ whenever $t_+ - t_- > \pi$.

Remark 36. There are multiple important caveats regarding the preceding counterexamples. The first is that the methods of [10] only apply directly to wave operators $\Box_g + \sigma$ with $4\sigma := n^2 - 1$. (For other values of σ, the corresponding wave equation in the conformally compactified setting gains a potential that becomes critically singular at the conformal boundary, which significantly complicates the constructions.) The case of general σ was recently treated in joint work with Guisset [33].

Remark 37. Another caveat is that not every linear wave equation on AdS spacetime can have these counterexamples to unique continuation. For example, from Holmgren's theorem [34, 39], if all the coefficients of the wave equation are real-analytic (in particular, this applies to $(\Box_g + \sigma)u = 0$), then counterexamples to unique continuation cannot exist. In particular, the constructions of [10] only show that there exist potentials V (vanishing at the conformal boundary) such that

$$(\Box_g + \sigma + V)u = 0$$

has the above-mentioned counterexamples to unique continuation, but it does not address whether such counterexamples exist for any fixed, chosen potential V.

One can in fact view the GNCC as a generalization of the above intuitions for AdS spacetime to aAdS settings. This was observed by Chatzikaleas and the author in [20, Theorem 4.1], which connected the GNCC to the trajectories of spacetime null geodesics near the conformal boundary. We restate the result here, but slightly adapted to the special case of vacuum FG-aAdS segments:

Theorem 3.5 (Geodesic return [20, 55]). *Let $(\mathscr{M}, g)$ be a vacuum FG-aAdS segment, with holographic data $(\mathscr{I}_{\mathrm{AdS}}, \mathfrak{g}^{(0)}, \mathfrak{g}^{(n)})$, and fix $\mathscr{D} \subset \mathscr{I}$, with $\bar{\mathscr{D}}$ compact, so that $(\mathscr{D}, \mathfrak{g}^{(0)})$ satisfies the GNCC. In addition, let $\Lambda : (s_-, s_+) \longrightarrow \mathscr{M}$ be a complete null geodesic with respect to $\rho^2 g$, written as*

$$\Lambda(s) := (\rho(s), \lambda(s)) \in (0, \rho_0) \times \mathscr{I}, \qquad s \in (s_-, s_+),$$

with s being an affine parameter for Λ. Assume also that

$$0 < \rho(s_0) < \varepsilon_0, \qquad |\dot{\rho}(s_0)| \lesssim \rho(s_0), \qquad \lambda(s_0) \in \mathscr{D}. \tag{12}$$

for some $s_0 \in (s_-, s_+)$, where $\varepsilon_0 > 0$ is sufficiently small (depending on $\mathfrak{g}$ and $\mathscr{D}$). Then,

- *either Λ initiates from the conformal boundary within $\mathscr{D}$,*

$$\lim_{s \searrow s_-} \rho(s) = 0, \qquad \lim_{s \searrow s_-} \lambda(s) \in \mathscr{D}, \tag{13}$$

- *or Λ terminates at the conformal boundary within $\mathscr{D}$,*

$$\lim_{s \nearrow s_+} \rho(s) = 0, \qquad \lim_{s \nearrow s_+} \lambda(s) \in \mathscr{D}. \tag{14}$$

Remark 38. Since g and $\rho^2 g$ have the same null geodesics, we can more usefully interpret Λ in Theorem 3.5 as being a g-null geodesic. However, it is more convenient to parametrize Λ with respect to $\rho^2 g$, since the conformal boundary is finite from the perspective of $\rho^2 g$.

Note that the first and third conditions of (12) mean that Λ is both "hovering over $\mathscr{D}$" and "ε_0-close to $\mathscr{D}$" at parameter s_0. The second part of (12) can be viewed as a necessary condition in order for Λ to remain similarly close to the conformal boundary. Consequently, one can interpret Theorem 3.5 as saying that any spacetime null geodesic Λ that is sufficiently close to the conformal boundary and that travels over $\mathscr{D}$ (which satisfes the GNCC) must return to the conformal boundary within $\mathscr{D}$, in either the future or past direction. In other words, there cannot exist any null geodesics sufficiently near the boundary that travel over $\mathscr{D}$ but do not terminate within $\mathscr{D}$ itself.

In summary, we conclude that *analogues of the Alinhac-Baouendi counterexamples in AdS spacetime cannot be constructed over any $\mathscr{D}$ that satisfies the GNCC.*

Remark 39. The first version of the geodesic return theorem was proven by McGill and the author in [55, Theorem 4.5]. This showed that the NCC (10) being satisfied implies that counterexamples to unique continuation cannot be constructed over the slab $\mathscr{D}_*$ in Proposition 3. One can hence view Theorem 3.5 as an extension of [55, Theorem 4.5] to the GNCC.

To conclude, the preceding discussions provide two justifications for the GNCC being the crucial condition for unique continuation and for Theorem 3.2:

- The GNCC rules out the known counterexamples to unique continuation for waves.
- The GNCC implies pseudoconvexity, leading to positive unique continuation results.

Remark 40. Though the GNCC is crucial to our proof of Theorem 3.2, it is not known whether the methods of [10] extend to the nonlinear EVE. The construction of counterexamples to unique continuation in Theorem 3.2 when the GNCC is violated is a challenging open problem.

Finally, we note that this connection with null geodesics can, in some cases, be exploited to show that some domains fail to satisfy the GNCC. The most notable examples involve flat conformal boundaries, in particular those of the planar and toric (Kerr-)AdS spacetimes:

Proposition 4. *Consider the planar AdS and toric AdS conformal boundaries* (2), *both with the flat Lorentzian metric. Then, no subdomain $\mathscr{D}$ of either boundary can satisfy the GNCC.*

The proof is similar to that of the "negative" half of Proposition 2; see [20, Corollary 3.10]. The key intuition is that on both planar and toric AdS spacetimes, there are spacetime null geodesics that remain arbitrarily close to but never intersect the conformal boundary for all times.

4 Carleman Estimates

As mentioned before, the main technical tool behind the proof of Theorem 3.2 is a novel Carleman estimate for (vertical) tensorial wave equations near conformal boundaries of FG-aAdS segments. Various versions of this Carleman estimate were proved, in increasing levels of generality, in [35], [36], [55], and finally culminating in [20] (which introduced the GNCC). In this section, we discuss the Carleman estimate [20], along with some of the main ideas behind its proof.

4.1 Unique Continuation

Carleman estimates have long played a fundamental role in the theory of unique continuation for PDEs. They have been spectacularly successful as tools for proving robust classes of unique continuation results that, in particular, do not require any assumption of analyticity—in either the solution or the PDE. (This is in direct contrast to the earliest techniques, through the Cauchy-Kovalevskaya and Holmgren theorems.)

Applications of Carleman estimates to unique continuation began with the work of Carleman [19], which proved strong unique continuation for elliptic equations in 2 dimensions. Further breakthroughs extending to elliptic PDEs in all dimensions are attributed to Calderón [18] (for unique continuation from an open set or a hypersurface) and Aronszajn [14] (for strong unique continuation, i.e., from a point). The classical unique continuation theory for more general equations—including wave equations—is largely attributed to Hörmander; see [40].

In this classical theory, the key condition required for unique continuation is *pseudoconvexity*. Roughly speaking, the result is that *if u solves linear PDE $\mathscr{L} u = 0$, and if u has zero Cauchy data on a hypersurface Σ that is pseudoconvex with respect to $\mathscr{L}$ and a given side of Σ, then u identically vanishes near Σ on the side indicated by the pseudoconvexity*. This result is proved by deriving a local Carleman estimate near Σ, from which one then obtains unique continuation via a standard argument. Furthermore, pseudconvexity has been shown to be crucial, as counterexamples to unique continuation in its absence were constructed by Alinhac [9] and Alinhac-Baouendi [10].

The reader is referred to [40, 48] for the general definition of pseudoconvexity (in the language of microlocal analysis). Here, we focus exclusively on wave operators on Lorentzian manifolds. In this case, one can give a geometric characterization of pseudoconvexity:

Definition 9 ([50]). Let (M, g) be a Lorentzian manifold, and consider a hypersurface

$$\Sigma := \{f = 0\} \subset M,$$

which is a level set of a function f on M. We say that Σ is *pseudoconvex* (with respect to the wave operator $\Box_g$ and the direction of increasing f) if for any vector field Z on Σ satisfying

$$Zf = 0, \qquad g(Z,Z) = 0, \tag{1}$$

the following inequality holds (with ∇^2 denoting the Hessian with respect to g):

$$\nabla^2 f(Z,Z) < 0. \tag{2}$$

To summarize, Σ is pseudoconvex if $-f$ is convex on Σ with respect to all null directions that are tangent to Σ. One qualitative interpretation of this is as follows: *if Λ is a null geodesic that goes through $p \in \Sigma$ and is tangent to Σ at p, then Λ lies on the side $\{f < 0\}$ near p.*

With regards to the wave operator $\Box_g$ on M, a Carleman estimate roughly takes the form

$$\lambda \int_\Omega e^{-\lambda F} (|u|^2 + |\nabla u|^2)\, dg \lesssim \int_\Omega e^{-\lambda F} |\Box_g u|^2\, dg. \tag{3}$$

Here, $\Omega \subset M$ is a spacetime domain, and both integrals are with respect to the volume form induced by g. Thus, one can view (3) as a spacetime weighted H^1-bound for solutions u of wave equations. A key feature of (3) is the parameter $\lambda > 0$, which can be freely chosen as long as it is sufficiently large. Finally, regarding the weight $e^{-\lambda F}$, its main component is a function F that is constructed from the function f (in Definition 9) defining the pseudoconvex hypersurface Σ.

The free parameter λ plays a special role, as it allows one to absorb any lower-order terms arising from the right-hand side into the left-hand side. To obtain unique continuation from (3), the very rough idea is to apply the wave equation to reduce the right-hand side of (3) into lower-order quantities, which can then be absorbed into the left. Afterwards, one can then conclude that $u = 0$ by letting $\lambda \nearrow \infty$. See the end of Section 5 for a demonstration of this process.

Remark 41. In contrast, for elliptic equations, every hypersurface Σ is trivially pseudoconvex. Thus, in elliptic settings, one immediately bypasses any difficulties from pseudoconvexity.

The main difficulty in our setting is that *the conformal boundary of an FG-aAdS segment $(\mathscr{M}, g)$ fails to be pseudoconvex*, so the classical unique continuation results no longer apply. To see this, one can observe that there exist null geodesics near the conformal boundary that asymptote toward being everywhere tangent to the boundary. By the above intuition, we can hence view the conformal boundary as being *zero-pseudoconvex*, as it just barely fails to be pseudoconvex.

Remark 42. Note that pseudoconvexity is a conformally invariant property, so we can realize the conformal boundary by working with the rescaled metric $\rho^2 g$.

Some early results that included zero-pseudoconvex settings are found in [45, 50]. However, the modern machinery for deriving Carleman estimates (and hence unique continuation) for geometric wave equations in zero-pseudoconvex settings originated from works of the author with Alexakis and Schlue [7, 8], in the context of unique continuation of waves from null infinities of asymptotically flat spacetimes. Similar results were later (independently) established by Petersen [56] for compact Cauchy horizons. Our aAdS Carleman estimates [20, 35, 36, 55] also fit within this framework.

One encounters a number of additional difficulties when dealing with a zero-pseudoconvex hypersurface Σ. In this case, a much more careful study of the geometry near Σ is needed to find the requisite pseudoconvexity for a Carleman estimate. Moreover, any pseudoconvexity (if exists) must necessarily degenerate toward Σ, and this leads to various vanishing or singular weights in the Carleman estimates, making their proofs far more delicate.

Perhaps the most striking contrast with the classical theory is the size of the region from which one can uniquely continue. The classical results are local, in that one always attains unique continuation on sufficiently small neighbourhoods of any point on the pseudoconvex hypersurface Σ. In contrast, for zero-pseudoconvex Σ, Carleman estimates (and hence unique continuation properties) may only hold along sufficiently large regions in Σ. This feature is exclusive to zero-pseudoconvex settings, and it is particularly relevant to aAdS conformal boundaries.

Remark 43. There do exist stronger unique continuation results, for which one needs far weaker assumptions than pseudoconvexity; see [42, 49, 57, 63, 64]. But, these results require either additional symmetries in the solution or some partial analyticity in the differential operator. Unfortunately, neither of these assumptions is applicable to our aAdS setting.

4.2 The Main Estimate

The difficulties and features mentioned above apply in particular to aAdS settings. A key problem is to find a foliation of pseudoconvex hypersurfaces near the conformal boundary. To find the boundary regions from which one can uniquely continue solutions of wave equations, one must determine which regions can be covered by such pseudoconvex foliations.

One important finding in aAdS settings, which is a hallmark of the Carleman estimates presented here, is that *whether appropriate pseudoconvex foliations exist near the conformal boundary depends on geometric properties of the boundary itself*—namely, the GNCC.

Let us now state the Carleman estimate proved in [20], again restricted to vacuum settings:

Theorem 4.1 (Carleman estimate for wave equations [20]). *Let $(\mathscr{M}, g)$ denote a vacuum FG-aAdS segment, with conformal infinity $(\mathscr{I}, \mathfrak{g})$. Fix $\mathscr{D} \subset \mathscr{I}$, with $\bar{\mathscr{D}}$ compact, and assume $(\mathscr{D}, \mathfrak{g})$ satisfies the GNCC, with the associated function denoted by η (see Definition 8). Moreover, let*

$$f = \frac{\rho}{\eta}, \tag{4}$$

and define, for any $f_\star > 0$, the bulk regions

$$\Omega_{f_\star} := \{f < f_\star\} \subset \mathscr{M}. \tag{5}$$

Then, the following holds for any vertical tensor field Φ *on* $\mathscr{M}$ *with* Φ, $\nabla\Phi$ *vanishing on* $\{f = f_\star\}$,

$$\begin{aligned} &\int_{\Omega_{f_\star}} e^{-\frac{2\lambda f^p}{p}} f^{n-2-p-2\kappa} |(\check{\Box}_g + \sigma)\Phi|^2 \, dg \\ &\quad + \lambda^3 \limsup_{\rho_\star \searrow 0} \int_{\Omega_{f_\star} \cap \{\rho = \rho_\star\}} (|\mathrm{D}_\rho(\rho^{-\kappa}\Phi)|^2 + |\mathrm{D}(\rho^{-\kappa}\Phi)|^2 + |\rho^{-\kappa-1}\Phi|^2) \, dg \\ &\gtrsim \lambda \int_{\Omega_{f_\star}} e^{-\frac{2\lambda f^p}{p}} f^{n-2-2\kappa} (f\rho^3 |\mathrm{D}_\rho \Phi|^2 + f\rho^3 |\mathrm{D}\Phi|^2 + f^{2p}|\Phi|^2) \, dg, \end{aligned} \tag{6}$$

provided κ *and* λ *are sufficiently large (depending on* g, $\mathscr{D}$, σ, *and the rank of* Φ*),* $f_\star$ *is sufficiently small (also depending on* g, $\mathscr{D}$, σ, *and the rank of* Φ*), and* $0 < p < \frac{1}{2}$.

The statement of Theorem 4.1 is quite involved, so a number of remarks are in order:

Remark 44. $\check{\Box}_g$ is the wave operator defined on vertical tensor fields; see Section 2. Moreover, the norm $|\cdot|$ can be defined relative to a given Riemannian metric on the space of vertical tensors.

Remark 45. The role of the Klein-Gordon mass σ in (6) is not superfluous. In particular, if we rewrite the wave operator with respect to $\rho^2 g$, where the conformal boundary is finite, then $\Box_g + \sigma$ is equivalent to a wave operator $\Box_{\rho^2 g} + c_\sigma \rho^{-2} +$l.o.t. with a singular potential, where c_σ is a constant depending on σ. Furthermore, this singular potential is "critical", in the sense that it scales in the same way as $\Box_{\rho^2 g}$. As a result, one must think of σ as a principal part of the wave operator.

The parameter σ also has a significant effect on the boundary asymptotics of solutions u to

$$(\Box_g + \sigma + \text{l.o.t.})u = 0.$$

In particular, u near the conformal boundary will behave like specific powers of ρ, with the exponents determined by σ; for details, see, for instance, discussions in [17, 35].

Remark 46. In applications of Theorem 4.1, the boundary integral over $\Omega_{f_\star} \cap \{\rho = 0+\}$ can be disregarded, as the Carleman estimate is generally applied to quantities that vanish to sufficiently high order at the conformal boundary. Thus, the estimate (6) takes the same qualitative form as (3), save for a few additional weights depending on f and ρ.

Central to Theorem 4.1 and its proof is the function f defined in (4). Note that the level sets $\{f = f_*\}$ of f, for $f_* > 0$ sufficiently small, form timelike hypersurfaces that "hover over" the region $\mathscr{D}$. Furthermore, by the assumptions on η in Definition 8, all these level sets of f terminate at the conformal boundary precisely on $\partial\mathscr{D}$. Qualitatively speaking, each level set starts from the boundary at $\partial\mathscr{D}$ and travels inward into the bulk as one goes into $\mathscr{D}$.

The primary observation behind the proof of Theorem 4.1 is that *the level sets of f which are sufficiently near the conformal boundary are pseudoconvex*; this will be further elaborated in the following subsection. We note here that the pseudoconvexity of these level sets is a consequence of $\mathscr{D}$ being "sufficiently large" as to satisfy the GNCC. Moreover, the level sets $\{f = f_*\}$ asymptote precisely to the boundary region $\mathscr{D}$ as $f_\star \searrow 0$. As a result, the pseudoconvexity of these level sets degenerates toward the conformal boundary, which causes considerable complications.

There is one additional difference between the forms of (3) and (6)—the presence of additional weights in (3), in the form of powers of ρ and f. The most dramatic weight $f\rho^3$, in the first-order terms on the right-hand side of (6), is present because of the degeneration of pseudoconvexity toward the conformal boundary. On the other hand, the relatively mild weight f^{2p} in the zero-order terms come from σ and its effects on the boundary asymptotics of waves.

Together, these weights determine the class of wave equations for which one can obtain unique continuation results. In particular, observe that for the left-hand side of (6) to absorb into the right-hand side, Φ must satisfy a Klein-Gordon equation in which the lower-order terms have weights that at most match those in the right-hand side of (6). More specifically, we can only prove unique continuation for Klein-Gordon equations of the form

$$(\Box_g + \sigma)\Phi = \mathscr{O}(\rho^{2+\varepsilon})(\mathrm{D}_\rho\Phi, \mathrm{D}\Phi) + \mathscr{O}(\rho^\varepsilon)\Phi, \qquad \varepsilon > 0. \tag{7}$$

Thus, another challenge in the proof of Theorem 3.2 is to ensure that the relevant wave equations arising from the EVE are indeed of the above form, so that Theorem 4.1 is applicable.

Remark 47. To obtain (7), we replaced f-weights in (6) by ρ-weights, since (4) implies that ρ is bounded by f, and because ρ is a much more physically meaningful quantity.

Remark 48. The arbitrarily small powers ρ^ε in the right-hand side of (7) can be removed, at the cost of requiring additional orders of vanishing for Φ; see [20, 36, 55] for details.

Remark 49. For the weight $f^{n-2-2\kappa}$ in (6), note f^n counters the singular factor ρ^{-n} hidden in the volume form dg, while $f^{-2\kappa-2}$ matches the ρ-weight applied to Φ in the boundary term.

4.3 Pseudoconvexity

Finally, we close this section with a discussion of some of the main ideas in the proof of Theorem 4.1. Here, we focus primarily on derivation of pseudoconvexity for the level sets of f and its connection to the GNCC. For full details, the reader is referred to [20, Section 5].

According to Definition 9, we must show, in our aAdS setting, that (2) holds whenever (1) holds. Note that this would immediately follow if we find a smooth (scalar) function w such that

$$-(\nabla^2 f + w \cdot g)(X,X) > 0, \tag{8}$$

for any vector field X (not necessarily null) that is tangent to the level sets of f. For this, we look at the leading-order asymptotics of (8), in terms of properties on the conformal boundary.

First, one can naturally relate tangent vectors on the conformal boundary $\mathscr{I}$ to tangent vectors to level sets of f through the following isomorphism:

$$\mathfrak{X} \in T\mathscr{I} \leftrightarrow X := \rho[f(\mathfrak{X}\eta)\partial_\rho + \mathfrak{X}] \in T\{f = f_0\}. \tag{9}$$

(The weight ρ on the right-hand is to ensure that $\mathfrak{X}$ and X have almost the same $\mathfrak{g}$- and g-lengths, respectively.) Then, a direct computation—see [20, Lemma 2.18]—yields that

$$(\nabla^2 f + f \cdot g)(X,X) = \rho f^2 \mathsf{D}^2\eta(\mathfrak{X},\mathfrak{X}) + \frac{1}{2}\rho f \mathscr{L}_\rho \mathsf{g}(\mathfrak{X},\mathfrak{X}) + \rho f^3 \mathfrak{X}\eta \mathscr{L}_\rho \mathsf{g}(\mathsf{D}^\sharp\eta,\mathfrak{X}),$$

where D^2 and $\mathsf{D}^\sharp$ denote the Hessian and gradient with respect to g. The right-hand side can then be expanded using Definition 1; extracting the leading-order term at the boundary yields

$$\begin{aligned} -(\nabla^2 f + f \cdot g)(X,X) &= \rho f^2(\mathfrak{D}^2\eta - \eta \cdot \mathfrak{g}^{(2)})(\mathfrak{X},\mathfrak{X}) + \mathscr{O}(\rho f^3)(\mathfrak{X},\mathfrak{X}) \\ &= \rho f^2(\mathfrak{D}^2\eta + \eta \cdot \mathscr{P}[\mathfrak{g}])(\mathfrak{X},\mathfrak{X}) + \mathscr{O}(\rho f^3)(\mathfrak{X},\mathfrak{X}), \end{aligned}$$

where we applied the property (6) from the partial FG expansion the last step. Choosing

$$w := f + f\rho^2\zeta, \tag{10}$$

for some $\zeta \in C^\infty(\mathscr{I})$ to be determined, we then have

$$-(\nabla^2 f + w \cdot g)(X,X) = \rho f^2(\mathfrak{D}^2\eta + \eta \cdot \mathscr{P}[\mathfrak{g}] - \zeta \cdot \mathfrak{g})(\mathfrak{X},\mathfrak{X}) + \mathscr{O}(\rho f^3)(\mathfrak{X},\mathfrak{X}). \tag{11}$$

The above can be directly linked to the GNCC through the following algebraic property:

Proposition 5. *The following property holds for any* $\mathfrak{g}$*-null* $\mathfrak{Z}$*,*

$$(\mathfrak{D}^2\eta + \eta \cdot \mathscr{P}[\mathfrak{g}])(\mathfrak{Z},\mathfrak{Z}) \gtrsim \mathfrak{h}(\mathfrak{Z},\mathfrak{Z}),$$

if and only if there exists $\zeta \in C^\infty(\mathscr{I})$ *such that the following holds for any vector field* $\mathfrak{X}$ *on* $\mathscr{I}$*:*

$$(\mathfrak{D}^2\eta + \eta \cdot \mathscr{P}[\mathfrak{g}] - \zeta \cdot \mathfrak{g})(\mathfrak{X},\mathfrak{X}) \gtrsim \mathfrak{h}(\mathfrak{X},\mathfrak{X}).$$

Remark 50. See [55, Corollary 3.5] or [62, Lemma 4.3] for proofs of Proposition 5. Note the $n = 2$ case can be proved directly, but when $n \geq 3$, the main (rather nontrivial) observation is that two bilinear forms that do not vanish simultaneously can be simultaneously diagonalized [31].

Combining (11) with Proposition 5, we conclude that *if the GNCC holds, then* (8) *holds sufficiently near the conformal boundary, and the near-boundary level sets of f are pseudoconvex.*

Remark 51. In fact, the above computations give a slightly stronger bound:

$$-(\nabla^2 f + w \cdot g)(X,X) \gtrsim \rho f^2 \cdot \mathfrak{h}(\mathfrak{X},\mathfrak{X}).$$

The specific weights on the right-hand side eventually lead to corresponding weights in (6).

As for the Carleman estimate (6), the process is similar to other proofs of Carleman estimates for wave equations. Conjugating the Klein-Gordon operator $(\Box_g + \sigma)$ with the Carleman weight,

$$\mathscr{L} := e^{-\frac{\lambda f^p}{p}} f^{-\kappa} (\check{\Box}_g + \sigma) e^{\frac{\lambda f^p}{p}} f^{\kappa},$$

we can then roughly view the proof of (6) (modulo some additional weights) as a positive commutator estimate for $\mathscr{L}$. The challenge is to obtain positivity for the H^1-norm of Φ:

- For derivatives of Φ along the normal to the level sets of f, the positivity comes, as usual, for free from the square of the odd part of $\mathscr{L}$.
- For Φ itself, the positivity comes from the choice of Carleman weight—in fact, from choosing the parameter κ to be sufficiently large compared to σ.
- For derivatives of Φ along the level sets of f, we look, as usual, to the commutator between the odd and even parts of $\mathscr{L}$. For this, the principal part is roughly given by
$$-f^{n-3}(\nabla^2 f + w \cdot g)(\check{\nabla}\Phi, \check{\nabla}\Phi),$$
which we can show is positive due to the GNCC and pseudoconvexity.

The proof of (6) is completed (after a substantial amount of computations) by using the above observations, and by noting that all remaining quantities are "lower-order" and can be absorbed.

Remark 52. The proof of (6) in [20] does not use a positive commutator argument, but instead derives positivity directly through geometric computations via integrations by parts. We adopted the language of positive commutators here, since it is widely used in the Carleman estimate literature.

Finally, all the above is significantly complicated by the zero-pseudoconvexity, which introduces various weights that vanish at the conformal boundary. In particular, considerable care is needed to ensure that terms which should be "lower-order" can still be absorbed into the "principal" terms—that is, that the various weights are compatible with each other. For details of these computations, the reader is referred to the full proof of Theorem 4.1 in [20, Section 5].

5 Proof of Theorem 3.2

In this section, we give an outline of the proof of our main result, Theorem 3.2, and we highlight the main ideas behind the proof. The reader is referred to [37] for further details.

5.1 Reduction to Unique Continuation

Let us now assume the hypotheses of Theorem 3.2. As a matter of convention, we use the same notations for quantities defined from both g and $\bar{g}$, except that objects associated with $\bar{g}$ have a bar over their symbols.

The initial step is to reduce Theorem 3.2 to a setting in which one can perform analysis. By pulling $\bar{\mathscr{I}}$ and $\bar{\mathscr{M}} := (0, \bar{\rho}_0] \times \bar{\mathscr{I}}$ back through the gauge transformation ϕ, we are then in the setting in which $\bar{\mathscr{I}} = \mathscr{I}$ and $\bar{\mathscr{M}} = \mathscr{M}$—more specifically, $\bar{\mathscr{M}}$ is pulled back to $\mathscr{M}$ via

$$(\rho, p) \in \mathscr{M} \mapsto (\rho, \phi(p)) \in \bar{\mathscr{M}}.$$

Furthermore, $(\mathfrak{g}^{(0)}, \mathfrak{g}^{(n)})$ and $(\bar{\mathfrak{g}}^{(0)}, \bar{\mathfrak{g}}^{(n)})$ are now related on $\mathscr{D}$ by (12)–(13), for some conformal factor $\mathfrak{a}$. Next, applying a change of FG boundary defining function $\rho \mapsto \bar{\rho}$ satisfying the boundary condition (11), we can then further reduce to the case in which

$$\mathscr{I} = \bar{\mathscr{I}}, \qquad \mathscr{M} = \bar{\mathscr{M}}, \qquad (\mathfrak{g}^{(0)}, \mathfrak{g}^{(n)})|_{\mathscr{D}} = (\bar{\mathfrak{g}}^{(0)}, \bar{\mathfrak{g}}^{(n)})|_{\mathscr{D}}. \tag{1}$$

Now that g and $\bar{g}$—as well as all the quantities defined from them—lie on a common manifold, the proof will be complete if we can show that $g = \bar{g}$, or equivalently, $g - \bar{g} = 0$. By the last part of (1), and by the properties of the partial FG expansion in Theorem 2.1, we have that

$$\mathfrak{g}^{(k)} - \bar{\mathfrak{g}}^{(k)} = 0, \qquad \mathfrak{g}^{(\star)} - \bar{\mathfrak{g}}^{(\star)} = 0, \qquad 0 \leq k \leq n. \tag{2}$$

Thus, thanks to (1), Theorem 3.2 will be proved if we show that (2) implies

$$\mathsf{g} - \bar{\mathsf{g}} = 0. \tag{3}$$

5.2 The Wave-Transport System

The strategy for proving (3) will be to formulate $\mathsf{g} - \bar{\mathsf{g}}$ as an unknown in a closed system of (vertical) tensorial transport and wave equations, and to apply our Carleman estimates to this system to conclude that all its unknowns (including $\mathsf{g} - \bar{\mathsf{g}}$) vanish.

In this subsection, we derive a viable wave-transport system. For this, we also define

$$\mathsf{m}_{ab} := \mathscr{L}_\rho \mathsf{g}_{ab}. \tag{4}$$

In addition, we let W denote the Weyl curvature for g, and we decompose W into vertical quantities:

$$\mathsf{w}^0_{abcd} := \rho^2 W_{abcd}, \qquad \mathsf{w}^1_{abc} := \rho^2 W_{\rho abc}, \qquad \mathsf{w}^2_{ab} := \rho^2 W_{\rho a \rho b}, \tag{5}$$

The unknowns for our system will be constructed from g, (4), and (5).

Remark 53. We use lowercase Latin indices for vertical components, i.e., in directions along $\mathscr{I}$.

From the Gauss-Codazzi equations on level sets of ρ and (2), one derives the following:

$$\mathscr{L}_\rho \mathsf{m}_{ab} = -2\mathsf{w}^2_{ab} + \rho^{-1}\mathsf{m}_{ab} + \frac{1}{2}\mathsf{g}^{cd}\mathsf{m}_{ad}\mathsf{m}_{bc}, \qquad \mathsf{D}_b \mathsf{m}_{ac} - \mathsf{D}_a \mathsf{m}_{bc} = 2\mathsf{w}^1_{cab}. \tag{6}$$

Of course, analogous formulas also hold with respect to quantities constructed from $\bar{\mathsf{g}}$. We wish to couple the equations (6) to wave equations satisfied by the Weyl curvature W:

$$\Box_g W + 2nW = W \cdot W. \tag{7}$$

(The right-hand side is quadratic in W; see [37, Proposition 3.4] or [54, Proposition 4.5] for the exact formula or a derivation.) For this, we define one additional renormalization for w^0:

$$\mathsf{w}^\star_{abcd} := \mathsf{w}^0_{abcd} - \frac{1}{n-2}(\mathsf{g}_{ad}\mathsf{w}^2_{bc} + \mathsf{g}_{bc}\mathsf{w}^2_{ad} - \mathsf{g}_{ac}\mathsf{w}^2_{bd} - \mathsf{g}_{bd}\mathsf{w}^2_{ac}). \tag{8}$$

Then, one obtains, from (7), the following tensorial wave equations for w^2, w^1, and $\mathsf{w}^\star$ (see either [37, Proposition 3.6] or [54, Proposition 4.15] for detailed derivations):

$$\begin{aligned} \check{\Box}_g \mathsf{w}^2 + 2(n-2)\mathsf{w}^2 &= \mathsf{N}^2(\mathsf{g}, \mathsf{m}, \mathsf{Dm}, \mathsf{w}^\star, \mathsf{w}^1, \mathsf{w}^2, \mathsf{Dw}^\star, \mathsf{Dw}^1, \mathsf{Dw}^2), \\ \check{\Box}_g \mathsf{w}^1 + (n-1)\mathsf{w}^2 &= \mathsf{N}^1(\mathsf{g}, \mathsf{m}, \mathsf{Dm}, \mathsf{w}^\star, \mathsf{w}^1, \mathsf{w}^2, \mathsf{Dw}^\star, \mathsf{Dw}^1, \mathsf{Dw}^2), \\ \check{\Box}_g \mathsf{w}^\star &= \mathsf{N}^\star(\mathsf{g}, \mathsf{m}, \mathsf{Dm}, \mathsf{w}^\star, \mathsf{w}^1, \mathsf{w}^2, \mathsf{Dw}^\star, \mathsf{Dw}^1, \mathsf{Dw}^2). \end{aligned} \tag{9}$$

Here, $\mathsf{N}^2(\cdot)$, $\mathsf{N}^1(\cdot)$, $\mathsf{N}^\star(\cdot)$ represent nonlinear terms involving (contractions of) the listed quantities that decay sufficiently quickly toward the conformal boundary.

Remark 54. The power ρ^2 in (5) is crucial, as this is the only power of ρ for which the resulting quantities will satisfy wave equations that are compatible with (6). The renormalization (8) is made for the same reason, as the wave equation for w^0 cannot be used with (6).

Remark 55. That different Klein-Gordon masses appear in (9), at least when $n > 3$, is due to $\mathsf{w}^\star$, w^1, and w^2 having different asymptotic behaviors at the conformal boundary. (The case $n = 3$ is exceptional, since w^1 and w^2 have the same mass, and $w^\star$ is fully determined by w^1 and w^2.) As a result, we must treat the components $w^\star$, w^1, w^2 separately in our analysis.

Subtracting (4), (6), and (9) from their counterparts for $\bar{g}$ yields a closed wave-transport system for the quantities $g - \bar{g}$, $m - \bar{m}$, $w^\star - \bar{w}^\star$, $w^1 - \bar{w}^1$, and $w^2 - \bar{w}^2$. However, *this system fails to close for the purpose of applying our Carleman estimates*. In particular, the wave Carleman estimate (6) will only allow us to control up to one derivative of $w^\star - \bar{w}^\star$, $w^1 - \bar{w}^1$, and $w^2 - \bar{w}^2$. This, in turn, only allows to control—through (4) and (6)—one derivative of $m - \bar{m}$ and $g - \bar{g}$. On the other hand, when we take differences of the wave equations (9), we obtain terms involving the difference of $\check{\Box}$ and $\bar{\check{\Box}}$, which contains second derivatives of $g - \bar{g}$ that we a priori cannot treat.

The resolution, inspired by the symmetry extension result [43] of Ionescu and Klainerman, is to apply a careful renormalization of the system that eliminates the troublesome quantities.

The first crucial observation is that while $D^2(g - \bar{g})$ is off limits, we can obtain some control if one derivative is a curl. In particular, (4) and the second part of (6) yield, schematically,

$$\begin{aligned}\mathscr{L}_\rho[D_{db}(g-\bar{g})_{ac} - D_{da}(g-\bar{g})_{bc}] &\sim D_d[(D_b m_{ac} - D_a m_{bc}) - (\bar{D}_b \bar{m}_{ac} - \bar{D}_a \bar{m}_{bc})] \\ &\sim D_d(w^1 - \bar{w}^1)_{cab}.\end{aligned}$$

The above still does not quite suffice, and we need one more renormalization—this is due to error terms on the right-hand side involving $D(\check{D} - D)$, which again contain the undesirable $D^2(g - \check{g})$. All this leads us to define the following auxiliary quantities:

$$\begin{aligned} B_{cab} &:= D_c(g-\bar{g})_{ab} - D_a(g-\bar{g})_{cb} - D_b Q_{ca}, \\ \mathscr{L}_\rho Q_{ca} &:= g^{de} m_{ce}(g-\bar{g})_{ad} - g^{de} m_{ae}(g-\bar{g})_{cd}, \end{aligned} \tag{10}$$

with $Q \to 0$ as $\rho \to 0$. One then shows DB can indeed be adequately controlled by $D(w^1 - \bar{w}^1)$.

The second crucial observation comes from a detailed examination of the difference $\check{\Box} - \bar{\check{\Box}}$. To appreciate this, let us consider the wave equation for $w^2 - \bar{w}^2$:

$$\check{\Box}(w^2 - \bar{w}^2) = \check{\Box} w^2 - \bar{\check{\Box}} \bar{w}^2 - (\check{\Box} - \bar{\check{\Box}})\bar{w}^2. \tag{11}$$

The dangerous terms arise from the following (rather long) computation,

$$\begin{aligned}(\check{\Box} - \bar{\check{\Box}})\bar{w}^2_{ab} = &-\frac{1}{2}\rho^2 g^{cd} g^{ef} \bar{w}^2_{eb} D_c B_{afd} - \frac{1}{2} g^{ef} \bar{w}^2_{eb} \check{\Box}(g - \bar{g} + Q)_{af} \\ &-\frac{1}{2}\rho^2 g^{cd} g^{ef} \bar{w}^2_{ea} D_c B_{bfd} - \frac{1}{2} g^{ef} \bar{w}^2_{ea} \check{\Box}(g - \bar{g} + Q)_{bf} + \mathrm{Err}_{ab},\end{aligned} \tag{12}$$

where Err consists of (many) terms containing only difference quantities that we can control.

The key point is that the only instances of $\mathsf{D}^2(\mathsf{g}-\check{\mathsf{g}})$ appear either as DB, which we can control, or as $\check{\Box}$ applied to difference quantities. This leads us to the renormalized curvature difference

$$\mathsf{W}^2_{ab} := \mathsf{w}^2_{ab} - \bar{\mathsf{w}}^2_{ab} + \frac{1}{2}\mathsf{g}^{de}\bar{\mathsf{w}}^2_{ad}(\mathsf{g}-\bar{\mathsf{g}}+\mathsf{Q})_{be} + \frac{1}{2}\mathsf{g}^{de}\bar{\mathsf{w}}^2_{db}(\mathsf{g}-\bar{\mathsf{g}}+\mathsf{Q})_{ae}, \tag{13}$$

which in essence shifts the $\check{\Box}$-terms from the right-hand side of (12) into the left; one can also define the remaining W^1 and $\mathsf{W}^\star$ similarly. In light of (11) and (12), we obtain that $\mathsf{W}^\star$, W^1, W^2 satisfy wave equations that do not contain $\mathsf{D}^2(\mathsf{g}-\check{\mathsf{g}})$ as sources.

Finally, the renormalized wave-transport system is obtained by treating the quantities

$$\mathsf{g}-\bar{\mathsf{g}}, \qquad \mathsf{Q}, \qquad \mathsf{m}-\bar{\mathsf{m}}, \qquad \mathsf{B}, \qquad \mathsf{W}^\star, \qquad \mathsf{W}^1, \qquad \mathsf{W}^2 \tag{14}$$

as unknowns. In particular, from the above discussions, and from various asymptotic properties of geometric quantities, we arrive at the (schematic) transport equations

$$\begin{aligned} \mathscr{L}_\rho(\mathsf{g}-\bar{\mathsf{g}}) &= \mathsf{m}-\bar{\mathsf{m}}, \\ \mathscr{L}_\rho\mathsf{Q} &= \mathscr{O}(\rho)\,(\mathsf{g}-\bar{\mathsf{g}},\mathsf{Q}), \\ \mathscr{L}_\rho\mathsf{B} &= 2(\mathsf{w}^1-\bar{\mathsf{w}}^1) + \mathscr{O}(\rho)\,(\mathsf{g}-\bar{\mathsf{g}},\mathsf{Q},\mathsf{B}) + \mathscr{O}(1)\,(\mathsf{m}-\bar{\mathsf{m}}), \\ \mathscr{L}_\rho[\rho^{-1}(\mathsf{m}-\bar{\mathsf{m}})] &= -2\rho^{-1}\mathsf{W}^2 + \mathscr{O}(1)\,(\mathsf{g}-\bar{\mathsf{g}},\mathsf{m}-\bar{\mathsf{m}},\mathsf{Q}), \end{aligned} \tag{15}$$

coupled to the following (schematic) wave equations:

$$\left.\begin{matrix} \check{\Box}\mathsf{W}^2 + 2(n-2)\mathsf{W}^2 \\ \check{\Box}\mathsf{W}^1 + (n-1)\mathsf{W}^1 \\ \check{\Box}\mathsf{W}^\star \end{matrix}\right\} = \sum_{\mathsf{V}\in\{\mathsf{W}^\star,\mathsf{W}^1,\mathsf{W}^2\}} \left[\mathscr{O}(\rho^2)\mathsf{V} + \mathscr{O}(\rho^3)\mathsf{D}\mathsf{V}\right] + \mathscr{O}(\rho)\,(\mathsf{m}-\bar{\mathsf{m}}) \tag{16}$$

$$+\,\mathscr{O}(\rho^2)\,(\mathsf{g}-\bar{\mathsf{g}},\mathsf{Q},\mathsf{D}(\mathsf{g}-\bar{\mathsf{g}}),\mathsf{D}\mathsf{Q},\mathsf{D}(\mathsf{m}-\bar{\mathsf{m}}),\mathsf{D}\mathsf{B}).$$

The $\mathscr{O}(\cdot)$'s in (15)–(16) indicate the asymptotics of various coefficients as $\rho \to 0$.

For more precise formulas and derivations, see [37, Proposition 3.13, Proposition 3.14]. In particular, the wave-transport system (15)–(16) indeed closes from the point of view of derivatives.

Remark 56. The system (15)–(16) could also be used to derive unique continuation results for the EVE near finite timelike hypersurfaces, providing an alternative approach to that of [1].

Remark 57. Note that [16, 21], the Riemannian and stationary analogues of Theorem 3.2, avoid the issues that make the extended system (15)–(16) necessary. In particular, the main equations in [16, 21] are elliptic rather than hyperbolic, and the Carleman estimates in this setting control one extra derivative. Thus, the analogue of $\mathsf{D}^2(\mathsf{g}-\bar{\mathsf{g}})$ does not cause any issues there.

Furthermore, [16, 21] avoid the curvature entirely. Instead, they can work with a much simpler system that is based on second-order elliptic equations for $\mathsf{m} - \bar{\mathsf{m}}$.

Finally, since all hypersurfaces are pseudoconvex in elliptic settings, [16, 21] avoid all the major difficulties of zero-pseudoconvexity and of constructing pseudoconvex hypersurfaces.

Remark 58. As previously mentioned, our system (15)–(16) is heavily inspired by [43], which devised an analogous wave-transport system for proving a Killing field extension result across finite hypersurfaces. In particular, [43] used as unknowns the analogues of $\mathscr{L}_K g$, $\nabla \mathscr{L}_K g$, $\mathscr{L}_K W$, along with a number of renormalizations, where K is the Killing vector field to be extended.

In fact, a modification of the wave-transport system in [43] (that also decomposes the unknowns into vertical tensor fields) can be used to give a direct proof of Theorem 3.4, without appealing to Theorem 3.2 or Theorem 3.3. This modified wave-transport system would have the same qualities as (15)–(16), and we can similarly apply our Carleman estimates to this system.

Furthermore, one can draw a direct parallel between the system in [43] and our wave-transport system (15)–(16). More specifically, one can map the unknowns in [43] to the unknowns here by replacing every $\mathscr{L}_K$ applied to a quantity in [43] by the corresponding difference of that quantity for two metrics here. (To be fully precise, one must also decompose spacetime quantities into vertical tensor fields.) Furthermore, the renormalizations in [43] can be related to our renormalizations in this same manner. For details, see the discussions in [37, Section 1.5].

5.3 Unique Continuation

With the wave-transport system (15)–(16) in place, we now apply our Carleman estimates to this system to derive unique continuation—in particular (3).

The first step is to show that the unknowns of (15)–(16) vanish to high enough order at the conformal boundary. For this, we begin by expressing them in terms of our partial FG expansion (4). Using that the expansions for g and $\bar{\mathsf{g}}$ match up to n-th order by (2), we derive

$$\mathsf{g} - \bar{\mathsf{g}} = o(\rho^n), \qquad \mathsf{m} - \bar{\mathsf{m}}, \mathsf{w}^1 - \bar{\mathsf{w}}^1 = o(\rho^{n-1}), \qquad \mathsf{w}^2 - \bar{\mathsf{w}}^2, \mathsf{w}^0 - \bar{\mathsf{w}}^0 = o(\rho^{n-2}). \tag{17}$$

From here, the key observation is the following system (see [37, Proposition 3.15] for details):

$$\begin{aligned} \mathscr{L}_\rho(\mathsf{g} - \bar{\mathsf{g}}) &= \mathsf{m} - \bar{\mathsf{m}}, \\ \mathscr{L}_\rho[\rho^{-1}(\mathsf{m} - \bar{\mathsf{m}})] &= -2\rho^{-1}(\mathsf{w}^2 - \bar{\mathsf{w}}^2) + \mathscr{O}(\rho)(\mathsf{g} - \bar{\mathsf{g}}) + \mathscr{O}(1)(\mathsf{m} - \bar{\mathsf{m}}), \\ \mathscr{L}_\rho[\rho^{2-n}(\mathsf{w}^2 - \bar{\mathsf{w}}^2)] &= \mathscr{O}(\rho^{2-n})(\mathsf{D}(\mathsf{w}^1 - \bar{\mathsf{w}}^1), \mathsf{m} - \bar{\mathsf{m}}) \\ &\quad + \mathscr{O}(\rho^{3-n})(\mathsf{g} - \bar{\mathsf{g}}, \mathsf{D}(\mathsf{g} - \bar{\mathsf{g}}), \mathsf{w}^0 - \bar{\mathsf{w}}^0, \mathsf{w}^2 - \bar{\mathsf{w}}^2), \end{aligned} \tag{18}$$

$$\mathscr{L}_\rho[\rho^{-1}(\mathrm{w}^1-\bar{\mathrm{w}}^1)] = \mathscr{O}(\rho^{-1})\,(\mathrm{D}(\mathrm{w}^2-\bar{\mathrm{w}}^2), \mathrm{m}-\bar{\mathrm{m}}) \\ + \mathscr{O}(1)\,(\mathrm{g}-\bar{\mathrm{g}}, \mathrm{D}(\mathrm{g}-\bar{\mathrm{g}}), \mathrm{w}-\bar{\mathrm{w}}^1),$$

$$\mathscr{L}_\rho(\mathrm{w}^0-\bar{\mathrm{w}}^0) = \mathscr{O}(\rho^{-1})\,(\mathrm{w}^2-\bar{\mathrm{w}}^2) + \mathscr{O}(1)\,(\mathrm{D}(\mathrm{w}^1-\bar{\mathrm{w}}^1), \mathrm{g}-\bar{\mathrm{g}}, \mathrm{m}-\bar{\mathrm{m}}) \\ + \mathscr{O}(\rho)\,(\mathrm{D}(\mathrm{g}-\bar{\mathrm{g}}), \mathrm{w}^0-\bar{\mathrm{w}}^0).$$

Note that the first two parts of (18) are just (4) and (15) (without renormalizations), while the last three parts of (18) are consequences of the Bianchi equations for W. Integrating (18) from $\rho = 0$, one can improve the vanishing in (17) by two powers of ρ:

$$\mathrm{g}-\bar{\mathrm{g}} = o(\rho^{n+2}), \qquad \mathrm{m}-\bar{\mathrm{m}}, \mathrm{w}^1-\bar{\mathrm{w}}^1 = o(\rho^{n+1}), \qquad \mathrm{w}^2-\bar{\mathrm{w}}^2, \mathrm{w}^0-\bar{\mathrm{w}}^0 = o(\rho^n). \tag{19}$$

Remark 59. The initial vanishing (17) from the partial FG expansions is crucial here. Due to the ρ^{2-n} in the third part of (18), we cannot use (18) to improve our vanishing without (17).

Remark 60. The equations (18) have a hierarchical structure, in that we have to apply them in a specific order to correctly derive (19). In particular, in [37], the quantities are estimated in the following order: $\mathrm{w}^2-\bar{\mathrm{w}}^2$, $\mathrm{w}^0-\bar{\mathrm{w}}^0$, $\mathrm{m}-\bar{\mathrm{m}}$, $\mathrm{g}-\bar{\mathrm{g}}$, $\mathrm{w}^1-\bar{\mathrm{w}}^1$.

The vanishing rates can be further improved by repeating the above argument, so that each iteration improves all vanishing rates by ρ^2. From this, we conclude that all the unknowns (14) vanish to arbitrarily high order at the conformal boundary. As a result, *there will be no boundary terms present when we apply the Carleman estimates.*

Remark 61. More accurately, we can obtain as much vanishing for (14) as we have regularity in the vertical directions for g. As a result, we must also assume sufficient regularity in Theorem 3.2 in order to achieve the vanishing rates needed for (14) so that the Carleman estimates can be applied. However, this is mainly a technical issue, and we avoid dealing with this here.

From here, the process is mostly standard. We fix constants

$$0 < f_i < f_e < f_\star, \qquad \kappa \gg 1, \tag{20}$$

with $f_\star$ sufficiently small and κ large enough, and we fix a smooth cutoff function

$$\chi := \hat{\chi}(f), \qquad \hat{\chi}(s) = \begin{cases} 1 & s < f_i, \\ 0 & s > f_e. \end{cases} \tag{21}$$

Applying the wave Carleman estimate (6) on $\Omega_{f_\star}$ to $\chi\mathrm{W}^\star$, $\chi\mathrm{W}^1$, and $\chi\mathrm{W}^2$, summing the results, and recalling the wave equations (16), we obtain that

$$\lambda \int_{\Omega_i} w_\lambda(f) \sum_{\mathrm{V}\in\{\mathrm{W}^\star,\mathrm{W}^1,\mathrm{W}^2\}} (\rho^{2p}|\mathrm{V}|^2 + \rho^4|\mathrm{DV}|^2)\,dg \tag{22}$$

$$\lesssim \int_{\Omega_i} w_\lambda(f) \sum_{\mathsf{U}\in\{\mathsf{g}-\bar{\mathsf{g}},\mathsf{Q},\mathsf{m}-\bar{\mathsf{m}},\mathsf{B}\}} (\rho^{2-p}|\mathsf{U}|^2 + \rho^{4-p}|\mathsf{DU}|^2)\,dg$$
$$+ \int_{\Omega_i} w_\lambda(f) \sum_{\mathsf{V}\in\{\mathsf{W}^\star,\mathsf{W}^1,\mathsf{W}^2\}} (\rho^{4-p}|\mathsf{V}|^2 + \rho^{6-p}|\mathsf{DV}|^2)\,dg + \int_{\Omega_e} w_\lambda(f)(\dots)\,dg,$$

where

$$w_\lambda(f) := e^{-2\lambda p^{-1} f^p} f^{n-2-2\kappa}, \qquad \Omega_i := \{0 < f_i\}, \qquad \Omega_e := \{f_i \leq f < f_e\}. \quad (23)$$

(The "..." in the Ω_e-integral depends on the unknowns (14), various weights in ρ and f, and the cutoff χ. Its precise contents are irrelevant, as we only require that this integral is finite.)

For the quantities satisfying transport equations, we must apply a separate transport Carleman estimate (see [37, Proposition 4.6] for the precise statement and proof):

Proposition 6 (Carleman estimate for transport equations [37]). *Assume the same setup as Theorem 4.1, and fix $s \geq 0$. Then, the following holds for any vertical tensor field Ψ on $\mathscr{M}$; large enough κ, λ (depending on n and s); small enough $f_\star$ (depending on g and $\mathscr{D}$); and $0 < p < \frac{1}{2}$:*

$$\int_{\Omega_{f_\star}} e^{-\lambda p^{-1} f^p} f^{n-2-p-2\kappa} \rho^{s+2} |\mathscr{L}_\rho \Psi|^2\,dg + \lambda \limsup_{\rho_\star \searrow 0} \int_{\Omega_{f_\star}\cap\{\rho=\rho_\star\}} \rho^s |\rho^{-\kappa-1}\Psi|^2\,dg \quad (24)$$
$$\gtrsim \lambda \int_{\Omega_{f_\star}} e^{-\lambda p^{-1} f^p} f^{n-2-2\kappa} \rho^s |\Psi|^2\,dg.$$

Remark 62. While the wave Carleman estimate, Theorem 4.1, was quite involved and required several new ideas to treat, the transport Carleman estimate is straightforward to prove.

We now apply Proposition 6 to $\mathsf{g}-\bar{\mathsf{g}}$, Q, $\mathsf{m}-\bar{\mathsf{m}}$, B, and their first derivatives (with the same κ and $f_\star$ as before, and with appropriate s). Recalling (15) and summing the results yields

$$\lambda \int_{\Omega_i} w_\lambda(f) \sum_{\mathsf{U}\in\{\mathsf{g}-\bar{\mathsf{g}},\mathsf{Q},\mathsf{m}-\bar{\mathsf{m}},\mathsf{B}\}} (|\mathsf{U}|^2 + \rho^3|\mathsf{DU}|^2)\,dg \quad (25)$$
$$\lesssim \int_{\Omega_i} w_\lambda(f) \sum_{\mathsf{U}\in\{\mathsf{g}-\bar{\mathsf{g}},\mathsf{Q},\mathsf{m}-\bar{\mathsf{m}},\mathsf{B}\}} (\rho^{2-p}|\mathsf{U}|^2 + \rho^{5-p}|\mathsf{DU}|^2)\,dg$$
$$+ \int_{\Omega_i} w_\lambda(f) \sum_{\mathsf{V}\in\{\mathsf{W}^\star,\mathsf{W}^1,\mathsf{W}^2\}} (\rho^{2-p}|\mathsf{V}|^2 + \rho^{5-p}|\mathsf{DV}|^2)\,dg.$$

The key point here is that the ρ-weights on the right-hand sides of (22) and (25) come from the $\mathscr{O}(\cdot)$-coefficients in (15)–(16). The final crucial feature of our system is these ρ-weights are strong enough that, after summing (22) and (25), *the*

Ω_i-integrals on the right-hand side can be absorbed into the left-hand side (once λ is sufficiently large). From the above, we conclude that

$$\lambda \int_{\Omega_i} w_\lambda(f) \sum_{\mathsf{V}\in\{\mathsf{W}^\star,\mathsf{W}^1,\mathsf{W}^2\}} \rho^{2p}|\mathsf{V}|^2 \, dg + \lambda \int_{\Omega_i} w_\lambda(f) \sum_{\mathsf{U}\in\{\mathsf{g}-\bar{\mathsf{g}},\mathsf{Q},\mathsf{m}-\bar{\mathsf{m}},\mathsf{B}\}} |\mathsf{U}|^2 \, dg$$
$$\lesssim \int_{\Omega_e} w_\lambda(f)(\dots)\, dg.$$

Finally, $w_\lambda(f)$ in the above can be removed in the standard fashion by noting that $w_\lambda(f) \leq w_\lambda(f_i)$ on Ω_e and $w_\lambda(f) \geq w_\lambda(f_i)$ on Ω_i; this yields the estimate

$$\lambda \int_{\Omega_i} \sum_{\mathsf{V}\in\{\mathsf{W}^\star,\mathsf{W}^1,\mathsf{W}^2\}} \rho^{2p}|\mathsf{V}|^2 \, dg + \lambda \int_{\Omega_i} \sum_{\mathsf{U}\in\{\mathsf{g}-\bar{\mathsf{g}},\mathsf{Q},\mathsf{m}-\bar{\mathsf{m}},\mathsf{B}\}} |\mathsf{U}|^2 \, dg$$
$$\lesssim \int_{\Omega_e} (\dots)\, dg.$$

Letting $\lambda \nearrow \infty$ in the above, we conclude that (3) holds on Ω_i.

Remark 63. By setting f_i to be arbitrarily close to $f_\star$, we hence show that (3) holds on $\Omega_{f_\star}$.

6 Related Problems

We conclude this article by briefly discussing some other research directions related to Theorem 3.2. Much of this discussion can also be found in [37, Section 1.6].

6.1 Low Dimensions

Recall that Theorems 3.2, 3.3, and 3.4 all assume that the conformal boundary dimension n is strictly greater than 2. This raises the question of whether analogues of these theorems hold in the low-dimensional case $n = 2$.

In fact, the problem simplifies considerably when $n = 2$ due to the rigidity of low-dimensional settings. Since the spacetime Weyl curvature W now vanishes identically, it follows that any vacuum aAdS spacetime when $n = 2$ must be locally isometric to the AdS spacetime. Furthermore, as all curvature terms disappear from the system (6), one can prove unique continuation using only transport equations (and avoiding wave equations). This yields analogues of all our main theorems for $n = 2$—but *from any domain $\mathscr{D}$, without requiring the GNCC.*

6.2 Extensions of the GNCC

An often studied setting in the physics literature is the case when the boundary region $\mathscr{D} \subset \mathscr{I}$ in Theorem 3.2 is a causal diamond, that is, of the form

$$\mathscr{D} := \mathscr{I}^+(p) \cap \mathscr{I}^-(q), \qquad p,q \in \mathscr{I}. \tag{1}$$

($\mathscr{I}^+$ and $\mathscr{I}^-$ denote the chronological future and past, respectively, in $(\mathscr{I}, \mathfrak{g})$.) Unfortunately, causal diamonds (1), regardless of how large they are, generically fail to satisfy the GNCC when $n > 2$; see the argument in [20, Section 3.3]. As a result, Theorem 3.2 fails to apply when $\mathscr{D}$ is as in (1)—in other words, our result cannot establish that vacuum aAdS spacetimes are uniquely determined by their holographic data on any non-pathological causal diamond.

This leads to the question of whether the GNCC can be further refined, so that Theorem 3.2 can be extended to apply to $\mathscr{D}$ as in (1) in some capacity. One observation here is that the failure of the GNCC is due only to the presence of corners in $\partial\mathscr{D}$, where the boundaries of $\mathscr{I}^+(p)$ and $\mathscr{I}^-(q)$ intersect. Near these corners, one can find near-boundary null geodesics "hovering over" but avoiding $\mathscr{D}$. This leads to the following question: *could holographic data on $\mathscr{D}$ uniquely determine the vacuum aAdS spacetime near some proper subset $\mathscr{D}' \subset \mathscr{D}$*—in particular when $\mathscr{D}$ is sufficiently large, and when $\mathscr{D}'$ is sufficiently far from any corners in $\partial\mathscr{I}^+(p) \cap \partial\mathscr{I}^-(q)$?

At the same time, one may ask whether this refined GNCC can also be formulated for more general domains $\mathscr{D} \subset \mathscr{I}$. More specifically, one can formulate the following:

Problem 5. Consider the setting of Theorem 3.2. Show that if $\mathscr{D}' \subseteq \mathscr{D}$ satisfies some (yet to be formulated) "refined GNCC" relative to $\mathscr{D}$, then $(\mathscr{M}, g)$ is uniquely determined (up to isometry) near $\mathscr{D}'$ by the holographic data on $\mathscr{D}$ (up to gauge equivalence).

Keeping with our intuitions, an optimal formulation of such an extended GNCC would be one that directly characterizes null geodesic trajectories near the conformal boundary. Such a criterion would fully confirm the belief that the only impediment to unique continuation for waves from the conformal boundary is the near-boundary geometric optics solutions of Alinhac and Baouendi [10]. However, a proof of such a statement may require novel ideas from microlocal analysis.

6.3 Global Correspondences

Recall that our main result, Theorem 3.2, is "local" in nature, in the sense that the vacuum spacetime is only uniquely determined near a given conformal boundary region $\mathscr{D}$. This is due to our general setup, which does not provide any information on the global spacetime geometry. However, this leaves open the question of *whether a more global unique continuation result can be established if additional assumptions are imposed.*

For example, one can consider aAdS spacetimes $(\mathscr{M}, g)$ that are global perturbations, in some sense, of a Kerr-AdS spacetime. One can then ask *whether any given holographic data* $(\mathscr{I}, \mathfrak{g}^{(0)}, \mathfrak{g}^{(n)})$ *determines the spacetime in the full domain of outer communications, or if additional conditions are needed to rule out bifurcating counterexamples.* Also, if uniqueness holds, then another question of interest is *whether one has a one-to-one correspondence between holographic data on the boundary and some (appropriately conceived) holographic data on the black hole horizon.*

In [35, Section 6], the Carleman estimates of that paper were used to show that the linearized EVE on AdS spacetime (formulated as Bianchi equations for spin-2 fields) are globally characterized by their holographic boundary data over a sufficiently long timespan. Upcoming work by McGill and the second author will extend this to the nonlinear setting—roughly, under additional global topological assumptions, AdS spacetime is globally uniquely determined, as a solution to the EVE, by its holographic data. An interesting next step would be to explore whether these analyses can be extended to black hole aAdS spacetimes, such as Schwarzschild-AdS and Kerr-AdS.

6.4 Nonlinear Counterexamples

As mentioned before, all the intuitions for the GNCC being the crucial criterion for Theorem 3.2 are based on the observation that violations of the GNCC lead to counterexamples to unique continuation for wave equations. Certainly, it would be more preferable to *construct counterexamples to unique continuation for the nonlinear EVE itself when the GNCC fails to hold.* However, this problem is wide open and is expected to be very difficult.

6.5 Nature of the Correspondence

While Theorem 3.2 establishes that there exists a one-to-one correspondence between vacuum aAdS spacetimes (near the conformal boundary) and some space of holographic data, it says little about the nature of the correspondence, leaving open the question of how the gravitational dynamics and the conformal field theories are connected.

One can pose a number of interesting problems here that are both wide open and challenging. For instance, one such problem would be to *determine which holographic data* $(\mathscr{I}, \mathfrak{g}^{(0)}, \mathfrak{g}^{(n)})$ *can be realized as the conformal boundary data for a vacuum aAdS spacetime.* Moreover, given holographic data that is associated to a vacuum aAdS spacetime, another challenge would be to *reconstruct or approximate the aAdS spacetime that induces the holographic data.*

References

1. S. Alexakis, *Unique continuation for the vacuum Einstein equations*, arXiv:0902.1131, 2009.
2. S. Alexakis, T. Balehowsky, and A. Nachman, *Determining a Riemannian metric from minimal areas*, Adv. Math. **366** (2020), 107025.
3. S. Alexakis, A. Ionescu, and S. Klainerman, *Hawking's local rigidity theorem without analyticity*, Geom. Funct. Anal. **20** (2010), no. 4, 845–869.
4. ———, *Uniqueness of smooth stationary black holes in vacuum: small perturbations of the Kerr spaces*, Comm. Math. Phys. **299** (2010), no. 1, 89–127.
5. ———, *Rigidity of stationary black holes with small angular momentum on the horizon*, Duke Math. J. **163** (2014), no. 14, 2603–2615.
6. S. Alexakis and V. Schlue, *Non-existence of time-periodic vacuum spacetimes*, J. Differential Geom. **108** (2018), no. 1, 1–62.
7. S. Alexakis, V. Schlue, and A. Shao, *Unique continuation from infinity for linear waves*, Adv. Math. **286** (2016), 481–544.
8. S. Alexakis and A. Shao, *Global uniqueness theorems for linear and nonlinear waves*, J. Func. Anal. **269** (2015), no. 11, 3458–3499.
9. S. Alinhac, *Non unicité du problème de Cauchy*, Ann. Math. **117** (1983), no. 1, 77–108.
10. S. Alinhac and M. S. Baouendi, *A non uniqueness result for operators of principal type*, Math. Z. **220** (1995), no. 1, 561–568.
11. M. T. Anderson, *On the uniqueness and global dynamics of AdS spacetimes*, Class. Quantum Grav. **23** (2006), 6395.
12. M. T. Anderson and M. Herzlich, *Unique continuation results for Ricci curvature and applications*, J. Geom. Phys. **58** (2008), no. 2, 179–207.
13. ———, *Erratum to "unique continuation results for Ricci curvature and applications"*, J. Geom. Phys. **60** (2010), 1062–1067.
14. N. Aronszajn, *A unique continuation theorem for solutions of elliptic partial differential equations or inequalities of second order*, J. Math. Pures Appl. (9) **36** (1957), 235–249.
15. K. Bautier, F. Englert, M. Rooman, and Ph. Spindel, *The Fefferman–Graham ambiguity and AdS black holes*, Phys. Lett. B **479** (2000), 291–298.
16. O. Biquard, *Continuation unique à partir de l'infini conforme pour les métriques d'Einstein*, Math. Res. Lett. **15** (2008), no. 6, 1091–1099.
17. P. Breitenlohner and D. Z. Freedman, *Stability in gauged extended supergravity*, Annals Phys. **144** (1982), 249–281.
18. A. P. Calderón, *Uniqueness in the Cauchy problem for partial differential equations*, Amer. J. Math. **80** (1958), 16–36.
19. T. Carleman, *Sur un problème d'unicité pour les systèmes déquations aux dérivées partielles à deux variables indépendentes*, Ark. Mat. Astr. Fys. **26** (1939), no. 17, 1–9.
20. A. Chatzikaleas and A. Shao, *A gauge-invariant unique continuation criterion for waves in asymptotically Anti-de Sitter spacetimes*, Commun. Math. Phys. (2022), to appear.
21. P. Chruściel and E. Delay, *Unique continuation and extensions of Killing vectors for stationary vacuum space-times*, J. Geom. Phys. **61** (2011), no. 8, 1249–1257.
22. S. de Haro, K. Skenderis, and S. N. Solodukhin, *Holographic reconstruction of spacetime and renormalization in the AdS/CFT correspondence*, Comm. Math. Phys. **217** (2001), 595–622.
23. A. Enciso and N. Kamran, *A singular initial-boundary value problem for nonlinear wave equations and holography in asymptotically anti-de Sitter spaces*, J. Math. Pures Appl. **103** (2015), no. 4, 1053–1091.
24. ———, *Lorentzian Einstein metrics with prescribed conformal infinity*, J. Differential Geom. **112** (2019), no. 3, 505–554.
25. C. Fefferman and C. R. Graham, *Conformal invariants*, Élie Cartan et les mathématiques d'aujourd'hui - Lyon, 25-29 juin 1984, Astérisque, Société mathématique de France, 1985, pp. 95–116.
26. ———, *The ambient metric*, Princeton University Press, 2011.

27. H. Friedrich, *Einstein equations and conformal structure: Existence of anti-de Sitter-type space-times*, J. Geom. Phys. **17** (1995), no. 2, 125–184.
28. E. Giorgi, *On the local extension of Killing vector fields in electrovacuum spacetimes*, Ann. Henri Poincaré **20** (2019), 2271–2293.
29. C. R. Graham, *Volume and area renormalizations for conformally compact Einstein metrics*, Proceedings of the 19th Winter School "Geometry and Physics", Circolo Matematico di Palermo, 2000, pp. 31–42.
30. C. R. Graham and J. M. Lee, *Einstein metrics with prescribed conformal infinity on the ball*, Adv. Math. **87** (1991), 186–225.
31. W. Greub, *Linear algebra*, Springer-Verlag, 1975.
32. S.S. Gubser, I.R. Klebanov, and A.M. Polyakov, *Gauge theory correlators from non-critical string theory*, Phys. Lett. B **428** (1998), 105–114.
33. S. Guisset and A. Shao, arXiv:2308.03525
34. E. Holmgren, *Über systeme von linearen partiellen differentialgleichungen*, Öfversigt af Kongl. Vetenskaps-Akad. Förh. **58** (1901), 91–105.
35. G. Holzegel and A. Shao, *Unique continuation from infinity in asymptotically Anti-de Sitter spacetimes*, Comm. Math. Phys. **347** (2016), no. 3, 1–53.
36. ———, *Unique continuation from infinity in asymptotically Anti-de Sitter spacetimes II: Non-static boundaries*, Comm. Partial Differential Equations **42** (2017), no. 12, 1871–1922.
37. ———, *The bulk-boundary correspondence for the Einstein equations in asymptotically Anti-de Sitter spacetimes*, arXiv: 2207.14217, 2022.
38. G. Holzegel and J. Smulevici, *Decay properties of Klein-Gordon fields on Kerr-AdS spacetimes*, Comm. Pure Appl. Math. **66** (2013), no. 11, 1751–1802.
39. L. Hörmander, *The analysis of linear partial differential operators II: Differential operators with constant coefficients*, Springer-Verlag, 1985.
40. ———, *The analysis of linear partial differential operators IV: Fourier integral operators*, Springer-Verlag, 1985.
41. C. Imbimbo, A. Schwimmer, S. Theisen, and S. Yankielowicz, *Diffeomorphisms and holographic anomalies*, Class. Quantum Grav. **17** (2000), 1129–1138.
42. A. Ionescu and S. Klainerman, *Uniqueness results for ill-posed characteristic problems in curved space-times*, Commun. Math. Phys. **285** (2009), no. 3, 873–900.
43. A. D. Ionescu and S. Klainerman, *On the local extension of Killing vector-fields in Ricci flat manifolds*, J. Amer. Math. Soc. **26** (2013), 563–593.
44. J. Keir, *The weak null condition and global existence using the p-weighted energy method*, arXiv:1808.09982, 2018.
45. C. E. Kenig, A. Ruiz, and C. D. Sogge, *Uniform Sobolev inequalities and unique continuation for second order constant coefficient differential operators*, Duke Math. J. **55** (1987), no. 2, 329–347.
46. S. Kichenassamy, *On a conjecture of Fefferman and Graham*, Adv. Math. **184** (2004), no. 2, 268–288.
47. L. Lai, J.-Y. Li, and P. Yu, *On the rigidity of stationary charged black holes: small perturbations of the non-extremal Kerr-Newman family*, arXiv:1911.10560, to appear in J. Differential Geom., 2019.
48. N. Lerner, *Carleman Inequalities: An Introduction and More*, Springer, 2019.
49. ———, *Unique continuation through transversal characteristic hypersurfaces*, J. Anal. Math. **138** (2019), 135–156.
50. N. Lerner and L. Robbiano, *Unicité de Cauchy pour des opérateurs de type principal par*, J. Anal. Math. **44** (1984), 32–66.
51. J. M. Maldacena, *The large N limit of superconformal field theories and supergravity*, Adv. Theor. Math. Phys. **2** (1998), 231–252.
52. ———, *The large N limit of superconformal field theories and supergravity*, Int. J. Theor. Phys. **38** (1999), 1113–1133.
53. R. Mazzeo, *Unique continuation at infinity and embedded eigenvalues for asymptotically hyperbolic manifolds*, Amer. J. Math. (1991), no. 113, 25–45.

54. A. McGill, *Holographic characterisation of locally anti-de Sitter spacetimes*, arXiv:2111.11155, 2021.
55. A. McGill and A. Shao, *Null geodesics and improved unique continuation for waves in asymptotically Anti-de Sitter spacetimes*, Class. Quantum Grav. **38** (2020), 054001.
56. O. L. Petersen, *Extension of Killing vector fields beyond compact Cauchy horizons*, arXiv:1903.09135, 2019.
57. L. Robbiano and C. Zuily, *Uniqueness in the Cauchy problem for operators with partially holomorphic coefficients*, Invent. Math. **131** (1998), no. 3, 493–539.
58. A. Shao, *A generalized representation formula for systems of tensor wave equations*, Commun. Math. Phys. **306** (2011), no. 1, 51–82.
59. ———, *The near-boundary geometry of Einstein-vacuum asymptotically Anti-de Sitter spacetimes*, Class. Quantum Grav. **38** (2020), 034001.
60. K. Skenderis, *Asymptotically Anti-de Sitter spacetimes and their stress energy tensor*, Int. J. Mod. Phys. A **16** (2001), no. 5, 740–749.
61. G. 't Hooft, *Dimensional reduction in quantum gravity*, arXiv:gr-qc/9310026, 1993.
62. D. Tataru, *Carleman estimates, unique continuation and applications*, http://math.berkeley.edu/~tataru/papers/ucpnotes.ps.
63. ———, *Unique continuation for solutions to PDEs; Between Hörmander's theorems and Holmgren's theorem*, Commun. Part. Diff. Eq. **20** (1995), no. 5-6, 855–884.
64. ———, *Unique continuation for operators with partially analytic coefficients*, J. Math. Pures Appl. **78** (1999), no. 5, 505–521.
65. N. Tetradis, *Black holes, holography and thermodynamics of gauge theories*, J. Phys. Conf. Ser. **283** (2011), 012038.
66. A. Vasy, *The wave equation on asymptotically anti-de Sitter spaces*, Anal. PDE **5** (2012), no. 1, 81–144.
67. C. M. Warnick, *The massive wave equation in asymptotically AdS spacetimes*, Commun. Math. Phys. **321** (2013), 85–111.
68. E. Witten, *Anti de Sitter space and holography*, Adv. Theor. Math. Phys. **2** (1998), 253–291.
69. E. Witten and S.-T. Yau, *Connectedness of the boundary in the AdS/CFT-correspondence*, Adv. Theor. Math. Phys. **3** (1999), no. 6, 1635–1655.
70. P. Yu, *On Hawking's local rigidity theorems for charged black holes*, Ann. Henri Poincaré **11** (2010), 1–21.

Propagation of Singularities for the Wave Equation

Dean Baskin and Kiril Datchev

Abstract This expository note gives a digest version of Hörmander's propagation of singularities theorem for the wave equation.

1 Introduction

Let $P = \Box = -\partial_{x_0}^2 + \partial_{x_1}^2 + \cdots + \partial_{x_n}^2$ be the flat D'Alembertian on $\mathbb{R}^d$, $d = n+1$. We use coordinates $x = (x_0, \ldots, x_n)$. Hörmander's propagation of singularities result for solutions to $\Box u = 0$ says that if u has a given degree of Sobolev regularity at a given position and direction in spacetime $(x, \hat{\xi}) \in \mathbb{R}^d \times \mathbb{S}^{d-1}$, then it must have the same Sobolev regularity at all $(x', \hat{\xi})$ such that x' is on the light ray emanating from x with direction $\hat{\xi}$.

To make precise the notion of a given degree of Sobolev regularity at a given position and direction, we introduce the following definitions. For $m \in \mathbb{R}$, let $S^m = S^m(\mathbb{R}^d)$ be the *Kohn–Nirenberg* class of symbols, defined by the condition that $a \in S^m$ means $a \in C^\infty(\mathbb{R}^d \times \mathbb{R}^d)$ and the partial derivatives of a obey the bounds

$$|\partial_x^\alpha \partial_\xi^\beta a(x,\xi)| \le C_{\alpha,\beta} \langle \xi \rangle^{m-|\beta|}.$$

To each such symbol we associate a *pseudodifferential operator* A given by

$$Au(x) = \frac{1}{(2\pi)^d} \iint e^{i(x-y)\cdot\xi} a(x,\xi) u(y)\, dy\, d\xi. \tag{1}$$

Dean Baskin
Department of Mathematics, Texas A&M University, Mailstop 3368, College Station, TX 77843, USA
e-mail:

Kiril Datchev
Mathematics Department, Purdue University, West Lafayette, IN 47907, USA

D. R. Wood et al. (eds.), *2021-2022 MATRIX Annals*, MATRIX Book Series 5,
https://doi.org/10.1007/978-3-031-47417-0_14

The set of A corresponding to some $a \in S^m$ is denoted Ψ^m. If $A \in \Psi^m$, then A is a bounded operator from H^k to H^{k-m} for every k. For a proof of this general statement see Corollary 4.32 of [Hi], but note the following simpler cases: if $a(x,\xi) = a(x)$ is independent of ξ then A is the multiplier $u(x) \mapsto a(x)u(x)$, and if $a(x,\xi) = \langle\xi\rangle^m$ then $\|Au\|_{L^2}$ is the usual norm on the Sobolev space H^m given in terms of the Fourier transform.

To each $a \in S^m$ which is compactly supported in x, we associate an *essential support at fiber infinity*[1] (i.e. a support as the frequency ξ tends to infinity), denoted by ess supp(a), and defined as follows: $(x,\hat{\xi})$ is not in the essential support if and only if there is a neighborhood $U \subset \mathbb{R}^d \times \mathbb{S}^{d-1}$ of $(x,\hat{\xi})$, such that the partial derivatives of a obey the bounds

$$|\partial_x^\alpha \partial_\xi^\beta a(x',\xi')| \le C_{\alpha,\beta,N} \langle\xi'\rangle^{-N}$$

for all N when $(x',\xi'/|\xi'|) \in U$.

We say $a \in S^m$ is *elliptic* at $(x,\hat{\xi})$ if there are positive constants C and ε, and a neighborhood $U \subset \mathbb{R}^d \times \mathbb{S}^{d-1}$ of $(x,\hat{\xi})$, such that

$$|a(x',\xi')| \ge \varepsilon \langle\xi'\rangle^m,$$

when $(x',\xi'/|\xi'|) \in U$ and $|\xi| \ge C$. The set of points in $\mathbb{R}^d \times \mathbb{S}^{d-1}$ at which a is elliptic is denoted ell(a).

Given symbols b, e, and g, we say that b is *controlled by e through g* if for each point $(x,\hat{\xi})$ in ess supp(b), there is a light ray contained in ell(g) which starts at some point $(x',\hat{\xi}')$ in ell(e) and ends at $(x,\hat{\xi})$. A *light ray* is an integral curve of the *Hamilton vector field H_P* lying in the *characteristic set* $\Sigma(P) = \{(x,\hat{\xi}) \in \mathbb{R}^d \times \mathbb{S}^{d-1} : \hat{\xi}_0^2 - \sum \hat{\xi}_j^2 = 0\}$. Here the Hamilton vector field is given by

$$H_P = 2\xi_0 \partial_{x_0} - 2\xi_1 \partial_{x_1} - \cdots - 2\xi_n \partial_{x_n}.$$

Note that the Hamilton vector field has no ξ derivatives, so if $(x,\hat{\xi})$ is connected by a light ray to $(x',\hat{\xi}')$, then $\hat{\xi} = \hat{\xi}'$. See Figure 1.

We are now ready to state Hörmander's propagation of singularities theorem.

Theorem 1. *Let $b \in S^k$, $e \in S^k$, $g \in S^{k-1}$ for some real k be compactly supported in x. Suppose $u \in H^{-N}$, $Eu \in L^2$, and $GPu \in L^2$. If b is controlled by e through g, then $Bu \in L^2$ and there is a constant C (independent of u) such that*

$$\|Bu\|_{L^2} \le C(\|Eu\|_{L^2} + \|GPu\|_{L^2} + \|u\|_{H^{-N}}). \tag{2}$$

Since B and E both map $H^k \to L^2$, this says in particular that if we have a solution to $\Box u = 0$ which is in some Sobolev space H^{-N}, and if we happen to know that it is in some better Sobolev space H^k in the positions and directions given by the elliptic

[1] One can extend this definition to symbols which need not be compactly supported in x: see Chapter 6 of [Hi] or Section E.2.1 of [DyZw].

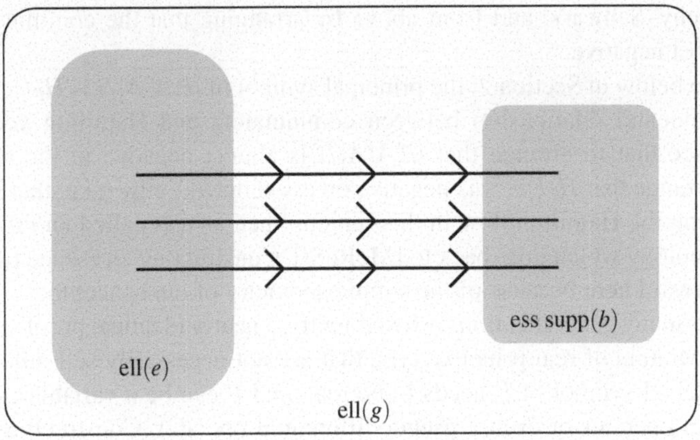

Fig. 1 The horizontal coordinate s is chosen so that the Hamilton vector field is ∂_s, thus the arrows are integral curves of the Hamilton vector field.

set of e, then it is also in this better Sobolev space H^k at points and directions emanating from that set. In this sense (2) is a propagation of regularity estimate. Since a singularity is a place where regularity is lacking, we also call (2) a propagation of singularities estimate.

As b is compactly supported in x, without loss of generality we can take e and g also to have compact support in x and then u need only lie in H^{-N} locally. In particular, if $\chi \in C_c^\infty$ satisfies $\chi \equiv 1$ on the support of g, then

$$\|B_0 u\| \leq C(\|Eu\| + \|GPu\| + \|\chi u\|_{H^{-N}}).$$

For example, let $u = \delta(x_0 - x_1)$. Then $u \in H^s_{\mathrm{loc}}$ if and only if $s < 1/2$. Microlocally, however, one can show that, for almost all $(x, \hat{\xi})$, we have $u \in H^s$ at $(x, \hat{\xi})$ for all s. More precisely, u is in these better Sobolev spaces at $(x, \hat{\xi})$ if and only if $x_0 \neq x_1$ or $|\hat{\xi}_0| \neq 1/\sqrt{2}$ or $\hat{\xi}_1 \neq -\hat{\xi}_0$.

This theorem was first proved by using a Fourier integral operator to microlocally conjugate P to the simpler operator ∂_{x_0}/i, for which the theorem is easy: see [Hö1] and [DuHö], specifically Section 6 of the latter paper. The proof we present here uses a positive commutator argument with a pseudodifferential commutant. It is a special case of Hörmander's proof in Section 3.5 of [Hö2]. Other presentations of this proof in more general settings than the one here can be found in Section E.4 of [DyZw] and Section 8 of [Hi].

The crux of the proof involves bounding the pairing

$$\operatorname{Im}\langle Au, APu\rangle = \frac{1}{2i}(\langle PA^*Au, u\rangle - \langle A^*APu, u\rangle) = \left\langle \frac{1}{2i}[P, A^*Au]u, u \right\rangle. \tag{3}$$

for a well-chosen $A \in \Psi^{k-\frac{1}{2}}$. We bound (3) from below via a straightforward application of Cauchy–Schwarz, and from above by arranging that the commutator $[P, A^*A]/i$ is almost negative.

As we will see below in Section 2, the principal symbol of $[P, A^*A]/i$ is $H_P(|a|^2)$: this is the fundamental relationship between commutators and Hamilton vector fields. We will see that to arrange that $[P, A^*A]/i$ is almost negative in the right sense we must arrange that $H_P(|a|^2)$ is negative on ess supp$(b) \setminus$ ell(e), i.e. that $|a|^2$ is decreasing along the Hamilton flow in this region. Such an a is called an *escape function*, a terminology which goes back to [MoRaSt]. Constructing an escape function is straightforward here because of the simple geometry of our spacetime.

Theorem 1 has many generalizations. Most directly, a nearly identical proof strategy applies to operators of real principal type that are not necessarily self-adjoint: i.e. only the principal symbol of P needs to be real, and P can be a variable coefficient differential operator or even a pseudodifferential operator. Constructing an escape function is then done by locally straightening out the Hamilton flow. The additional error term arising in equation (3) from $P - P^*$ is a lower order term. One can also treat cases where the principal part of P is not real, provided the imaginary part has a sign; this sign dictates the direction in which we can propagate regularity.

A common use for such propagation results is to help obtain global estimates for solutions of wave-like equations. As an example, in the non-trapping setting, these estimates let you propagate regularity from where it is a priori known (e.g., in the distant past for the forward propagator). Closing the estimate, however, typically requires additional estimates, which can be more complicated in settings with trapping.

2 Preliminaries from microlocal analysis

The important calculations take place on the level of symbols. To translate the results into statements about operators, we use the following formula for the symbol of the composition of two pseudodifferential operators. If $A_1 \in \Psi^{m_1}$ and $A_2 \in \Psi^{m_2}$ have symbols a_1 and a_2 respectively, then the composition A_1A_2 belongs to $\Psi^{m_1+m_2}$ and has symbol given by $a_3 \in S^{m_1+m_2}$ such that

$$r := a_3 - a_1a_2 - \frac{1}{i}\partial_\xi a_1 \cdot \partial_x a_2 \in S^{m_1+m_2-2}, \tag{4}$$

and ess supp$(r) \subset$ ess supp$(a_1) \cap$ ess supp(a_2).

We define the *principal symbol* $\sigma_m(A)$ of an operator $A \in \Psi^m$ to be the equivalence class of its symbol a in S^m/S^{m-1}. An immediate consequence of the composition formula above is the observation that

$$\sigma_{m_1+m_2}(A_1A_2) = \sigma_{m_1}(A_1)\sigma_{m_2}(A_2).$$

Similarly, for $A_1 \in \Psi^{m_1}$ and $A_2 \in \Psi^{m_2}$, the commutator $[A_1,A_2] = A_1A_2 - A_2A_1$ lies in $\Psi^{m_1+m_2-1}$ and satisfies

$$\sigma_{m_1+m_2-1}([A_1,A_2]) = \frac{1}{i}\left(\frac{\partial \sigma_{m_1}(A_1)}{\partial \xi}\frac{\partial \sigma_{m_2}(A_2)}{\partial x} - \frac{\partial \sigma_{m_2}(A_2)}{\partial \xi}\frac{\partial \sigma_{m_1}(A_1)}{\partial x}\right)$$

$$= \frac{1}{i} H_{\sigma_{m_1}(A_1)}(\sigma_{m_2}(A_2)),$$

where the Hamilton vector field $H_{\sigma_{m_1}(A_1)}$ is defined by the equation.

Concretely, the principal symbol of $P = \Box$ is

$$\sigma_2(P) = \xi_0^2 - \xi_1^2 - \cdots - \xi_n^2,$$

and its Hamilton vector field is

$$H_P = 2\xi_0\partial_{x_0} - 2\xi_1\partial_{x_1} - \cdots - 2\xi_n\partial_{x_n}.$$

We also use the following adjoint formula: If $A \in \Psi^m$ has symbol a, then the adjoint operator has symbol $a' \in S^m$ obeying

$$a' - \overline{a} \in S^{m-1}.$$

The composition formula can be found in Theorem 4.16 of [Hi] and the adjoint formula can be found in Corollary 4.13 of [Hi]. They are easy to check for differential operators, i.e. in the case that the symbols are polynomials in ξ. In general they may be checked by writing out the definitions and Taylor expanding.[2]

Our first application of the composition formula is to the following elliptic estimate.

Lemma 1. *Let $a \in S^m$ and $a' \in S^{m'}$ be such that a is compactly supported in x and* ess supp$\,a \subset$ ell(a'). *For any k and N, if $u \in H^{-N}$ and $A'u \in H^{k+m-m'}$, then $Au \in H^k$ and there is C (independent of u) such that*

$$\|Au\|_{H^k} \le C(\|A'u\|_{H^{k+m-m'}} + \|u\|_{H^{-N}}). \tag{5}$$

---

[2] Observe that if $A_3 = A_1A_2$ then by definition $A_3u(x)$ is given by (1) with $a(x,\xi)$ replaced by

$$a_3(x,\xi) = \frac{1}{(2\pi)^d}\iint e^{-ix\cdot\xi}e^{ix\cdot\eta}e^{-iz\cdot\eta}e^{iz\cdot\xi}a_1(x,\eta)a_2(z,\xi)\,d\eta\,dz$$

$$= \frac{1}{(2\pi)^d}\iint e^{-iw\cdot\zeta}a_1(x,\xi+\zeta)a_2(x+w,\xi)\,d\zeta\,dw,$$

where we used $e^{-ix\cdot\xi}e^{ix\cdot\eta}e^{-iz\cdot\eta}e^{iz\cdot\xi} = e^{i(x-z)\cdot(\eta-\xi)}$, substituted $w = z - x$ and $\zeta = \eta - \xi$, and ignored issues of convergence. To deduce the expansion (4), Taylor expand

$$a_1(x,\xi+\zeta) = a_1(x,\xi) + \zeta\cdot\partial_\xi a_1(x,\xi) + \cdots, \qquad a_2(x+w,\xi) = a_2(x,\xi) + w\cdot\partial_x a_2(x,\xi) + \cdots,$$

and use the fact that $\frac{1}{(2\pi)^d}\int\int e^{-iw\zeta}w^\alpha\zeta^\beta\,d\zeta\,dw = i^{|\beta|}\int w^\alpha\partial^\beta\delta(w)dw$ which is $(-i)^{|\beta|}$ if $\alpha = \beta$ and 0 otherwise. The convergence issues can be handled by a partition of unity, and the adjoint formula can be proved in the same way: see Chapters 8 and 9 of [Wo] for more.

Proof. Because a' is elliptic on the essential support of a, there exists $g \in S^{m-m'}$ such that $ga' = a$ when ξ is large.[3] By the composition formula (4), there is $R_1 \in \Psi^{m-1}$ such that

$$GA' = A + R_1.$$

Since $G\colon H^{k+m-m'} \to H^k$ is bounded, it follows that

$$\|Au\|_{H^k} \le C(\|A'u\|_{H^{k+m-m'}} + \|R_1 u\|_{H^k}).$$

This gives (5) when $-N = k+m-1$. To get it for larger values of N we apply the same reasoning with A replaced by R_1. That gives $R_2 \in \Psi^{m-2}$ such that

$$\|R_1 u\|_{H^k} \le C(\|A'u\|_{H^{k+m-1-m'}} + \|R_2 u\|_{H^k}),$$

which implies (5) when $-N = k+m-2$. Repeating this argument gives (5) for arbitrary N. □

We further require Gårding's inequality, which states that operators with non-negative principal symbols are non-negative to leading order. In other words, if $A \in \Psi^m$ is compactly supported in x and has $\sigma_m(A) \ge 0$ for $|\xi|$ sufficiently large, then there is some constant C so that

$$\mathrm{Re}\langle Au, u\rangle \ge -C\|u\|^2_{H^{\frac{m-1}{2}}}.$$

The proof of this general form of Gårding's inequality is somewhat involved (see Theorem 18.1.14 of [Hö3]), but is straightforward when the principal symbol of A is a square (or sum of squares). Indeed, if $\sigma_m(A) = b^2$, then by the composition and adjoint formulas, we may write $A = B^*B + R$, where $B \in \Psi^{\frac{m}{2}}$ and $R \in \Psi^{m-1}$. It then follows that

$$\mathrm{Re}\langle Au, u\rangle = \|Bu\|^2 + \langle Ru, u\rangle.$$

The first term on the right side is non-negative. Writing R as the composition of two operators of order $\frac{m-1}{2}$ (such as $R = (I-\Delta)^{\frac{m-1}{4}} \circ (I-\Delta)^{-\frac{m-1}{4}} R$) then shows that this last term is bounded below by $-C\|u\|^2_{H^{\frac{m-1}{2}}}$ for some C. Observe further that combining this proof with the elliptic estimate (5) shows that if $A \in \Psi^m$ is compactly supported in x, has non-negative principal symbol, and $B \in \Psi^0$ has $\operatorname{ess\,supp} a \subset \operatorname{ell}(b)$, then for any N there is a constant C with

$$\mathrm{Re}\langle Au, u\rangle \ge -C\|Bu\|^2_{H^{\frac{m-1}{2}}} - C\|u\|^2_{H^{-N}}. \tag{6}$$

[3] To construct such a g, for C as in the definition of ellipticity, let $g(x,\xi)a'(x,\xi) = (1 - \chi_C(\xi))\psi(x,\hat{\xi})a(x,\xi)$, where $\chi_C \in C_c^\infty(\mathbb{R}^d)$ is identically 1 on the ball of radius C centered at 0, $\psi \in C^\infty(\mathbb{R}^d \times \mathbb{S}^{d-1})$, $\operatorname{supp}\psi \subset \operatorname{ell}(a')$, and $\psi = 1$ near $\operatorname{ess\,supp}(a)$.

3 Proof of theorem

In this section we use the notation (z,ζ) for a variable point of $\mathbb{R}^d \times \mathbb{R}^d$, to avoid a typographical collision with $(x,\hat{\xi})$ and $(x',\hat{\xi}')$ which we use to denote points in the essential support or elliptic set of a particular symbol.

The main lemma in the proof of the theorem is the following similar but weaker statement.

Lemma 2. *Let $b \in S^k$, $e \in S^k$, $g \in S^{k-1}$ for some real k be compactly supported in x. Let $\Lambda_s = (I - \Delta_{\mathbb{R}^d})^{s/2} \in \Psi^s$. If b is controlled by e through g, then for every N there is a constant C such that*

$$\|Bu\|^2 \le C\left(\|\Lambda_{1-k} Gu\|^2_{H^{k-\frac{1}{2}}} + \|Eu\|^2 + \|GPu\|^2 + \|u\|^2_{H^{-N}} \right), \tag{7}$$

for all functions $u \in H^{k+1}$.

Proof. We assume without loss of generality that ess supp$(e) \subset$ ell(g); this can be arranged to hold without changing the hypotheses by shrinking the essential support of e. An application of the elliptic estimate (5) then yields the result with the original E.

Similarly, by a partition of unity argument, it is enough to prove that, for each point $(x,\hat{\xi})$ in ess supp(b), there is $b_1 \in S^k$ which is elliptic at $(x,\hat{\xi})$ such that (7) holds with B replaced by B_1.

For $A \in \Psi^{k-\frac{1}{2}}$ to be specified later, consider the pairing

$$\operatorname{Im}\langle Au, APu\rangle = \frac{1}{2i}\left(\langle PA^*Au, u\rangle - \langle A^*APu, u\rangle\right) = \left\langle \frac{1}{2i}[P, A^*Au]u, u \right\rangle.$$

1. We first bound the pairing below. By Cauchy–Schwarz, for any $\varepsilon > 0$ we have

$$\operatorname{Im}\langle Au, APu\rangle \ge -\frac{1}{4\varepsilon}\|APu\|^2_{H^{-1/2}} - \varepsilon\|Au\|^2_{H^{1/2}}.$$

To write this bound in terms of the L^2 pairing we use $\|\Lambda_s u\|_{L^2} = \|u\|_{H^s}$. That gives

$$\left\langle \frac{1}{2i}[P, A^*Au]u, u \right\rangle \ge -\frac{1}{4\varepsilon}\|\Lambda_{-1/2}APu\|^2_{L^2} - \varepsilon\|\Lambda_{1/2}Au\|^2_{L^2}. \tag{8}$$

2. We next bound the pairing above. Recall that $\frac{1}{2i}[P, A^*A] \in \Psi^{2k}$ and has principal symbol given by $\frac{1}{2}H_P(a^2) = aH_p(a)$. We will construct a in such a way that there exist a compact set $K \subset$ ell(e) and a $\gamma > 0$ such that

$$-aH_p(a) - \gamma\left(\langle\zeta\rangle^{1/2}a\right)^2 \ge 0, \qquad \text{off of } K. \tag{9}$$

Hence there exists C large enough that[4]

$$Ce^2 - aH_p(a) - \gamma\left(\langle\zeta\rangle^{1/2}a\right)^2 \geq 0.$$

In our construction of a we will also obtain

$$\text{ess supp}(a) \subset \text{ell}(g). \tag{10}$$

Using the version of the Gårding inequality stated in (6)[5] then yields

$$\left\langle \frac{1}{2i}[P,A^*A]u,u \right\rangle \leq C\|Eu\|^2 + C\|\Lambda_{1-k}Gu\|^2_{H^{k-\frac{1}{2}}} + C\|u\|_{H^{-N}} - \gamma\|\Lambda_{1/2}Au\|^2. \tag{11}$$

3. We now combine the two bounds on the pairing. Putting together (8) and (11) and taking $\varepsilon = \gamma/2$ yields

$$\|\Lambda_{1/2}Au\|^2 \leq C\left(\|\Lambda_{1-k}Gu\|^2_{H^{k-\frac{1}{2}}} + \|Eu\|^2 + \|\Lambda_{-1/2}APu\|^2 + \|u\|^2_{H^{-N}}\right).$$

To deduce (7) we use the fact that the elliptic estimate (5) and (10) imply $\|\Lambda_{-1/2}APu\|^2 \leq C(\|GPu\|^2 + \|u\|^2_{H^{-N}})$, and moreover we can use $\Lambda_{1/2}A$ as B_1 provided we construct a so that

$$(x,\hat{\xi}) \in \text{ell}(a). \tag{12}$$

Thus is remains to construct $a \in S^{k-\frac{1}{2}}$ such that (9), (10), and (12) all hold.

4. To construct a, use the fact that there is a light ray contained in ell(g) starting at some $(x',\hat{\xi}) \in \text{ell}(e)$ ending at $(x,\hat{\xi})$, and this light ray can be parametrized by

$$t \mapsto \left(x_0 \pm t\hat{\xi}_0, x_j \mp t\hat{\xi}_j, \hat{\xi}\right),$$

with the choice of sign depending on whether the flow from $(x',\hat{\xi})$ to $(x,\hat{\xi})$ is in the direction the flow of the Hamilton vector field or its opposite. We assume for the argument below that it is with the direction of the Hamilton vector field (and hence the top sign is chosen).

---

[4] In our setting, $\gamma > 0$ can be arbitrary, but when dealing with error terms γ must be taken sufficiently large.

[5] If we want to avoid invoking the full strength of the sharp Gårding inequality as in Theorem 18.1.14 of [Hö3], we must check that the quantity $Ce^2 - aH_p(a) - \gamma\langle\zeta\rangle a^2$ is a sum of squares. For that, take $\psi \in S^0$ such that $\psi = 0$ near K and $\psi = 1$ near the complement of ell(e), and use (15) to write $Ce^2 - aH_p(a) - \gamma\langle\zeta\rangle a^2 = a_1^2 + a_2^2$, where

$$a_1 = e\sqrt{C + (1-\psi^2)(-aH_p(a) - \gamma\langle\zeta\rangle a^2)/e^2}, \qquad a_2 = \psi\langle\zeta\rangle^k\chi\sqrt{(-2\varphi' - \gamma\varphi)\varphi}.$$

To see that a_1 is smooth use the fact that the square root of a positive function is smooth. To see that a_2 is smooth, use the explicit formula for φ.

Recall that $(x, \hat{\xi})$ lies on a light ray, and because light rays must lie in the characteristic set of P, $\zeta_0 \neq 0$ on any light ray. Using also the fact that $\mathrm{ell}(g)$ and $\mathrm{ell}(e)$ are open sets, fix $t_0 > 0$, $\delta > 0$, and an open neighborhood $U \subset \mathbb{R}^{d-1} \times \mathbb{S}^{d-1}$ of $(x, \hat{\xi})$ such that $\hat{\zeta}_0$ is non-vanishing on U, such that

$$t \in (-t_0 - \delta, \delta) \text{ and } (y_1, \ldots, y_n, \hat{\zeta}) \in U \implies (x_0 + t\hat{\zeta}_0, y_j - t\hat{\zeta}_j, \hat{\zeta}) \in \mathrm{ell}(g), \tag{13}$$

and such that

$$t \in (-t_0 - \delta, t_0 + \delta) \text{ and } (y_1, \ldots, y_n, \hat{\zeta}) \in U \implies (x_0 + t\hat{\zeta}_0, y_j - t\hat{\zeta}_j, \hat{\zeta}) \in \mathrm{ell}(e). \tag{14}$$

We fix a real function $\chi_1 \in C_c^\infty(U)$ that is identically 1 on a neighborhood of $(x, \hat{\xi})$, and put $\chi = \chi_1^2$. Let

$$\varphi(t) = \exp(-\gamma t + (t - \delta)^{-1} - (t + t_0 + \delta)^{-1}), \qquad \text{when } -t_0 - \delta < t < \delta,$$

and $\varphi(t) = 0$ otherwise.

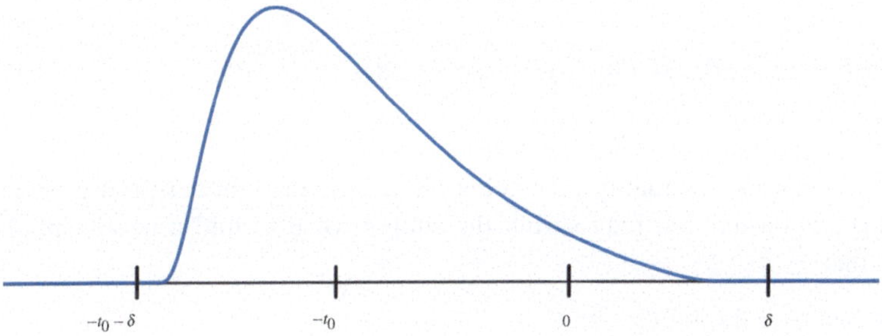

Fig. 2 A graph of φ, from https://www.desmos.com/calculator/tqmurytqe9.

Let

$$a(z, \zeta) = \varphi(t)\chi\left(z_1 + \hat{\zeta}_1 t, \ldots, z_n + \hat{\zeta}_n t, \hat{\zeta}\right) \langle\zeta\rangle^{k-\frac{1}{2}},$$

where $t = t(z, \zeta) = (z_0 - x_0)/\hat{\zeta}_0$. The expression for a is well-defined as $\hat{\zeta}_0 \neq 0$ on the support of χ and it is straightforward to check that $a \in S^{k-\frac{1}{2}}$. We write more shortly $a = \varphi\chi\langle\zeta\rangle^{k-\frac{1}{2}}$.

Now (10) follows from the property (13) of U, and (12) follows from the fact that $a\langle\zeta\rangle^{-k+\frac{1}{2}} \neq 0$ at (x,ξ). It remains to check (9), and for this we compute $aH_P(a)$.

The computation is made straightforward by the fact that we have defined a to be the product of $\varphi(t)$ and a function which is constant along flows of H_P:

$$aH_P(a) = 2\frac{\zeta_0}{\hat{\zeta}_0}\varphi'\varphi\chi^2\langle\zeta\rangle^{2k-1},$$

which has principal symbol

$$2\langle\zeta\rangle^{2k}\varphi'\varphi\chi^2,$$

because $\zeta_0/\hat{\zeta}_0 = |\zeta|$. Thus the principal symbol of $-aH_p(a) - \gamma\langle\zeta\rangle a^2$ is

$$-(2\varphi' + \gamma\varphi)\varphi\chi^2\langle\zeta\rangle^{2k}, \tag{15}$$

and the set where this is negative is contained in the set where $-t_0 - \delta \leq t \leq -t_0$ and in the support of χ, i.e. by the property (14) of U above it is contained in a compact subset of $\mathrm{ell}(e)$. □

With Lemma 2 in hand, we now turn our attention to the proof of Theorem 1. We proceed by removing the first term of the right side of the estimate (7) with an inductive argument and then finally relax the regularity hypothesis with a regularization argument.

We first consider the inductive argument. Indeed, we claim that if b is controlled by e through g, then for any m, there is a constant so that

$$\|Bu\|^2 \leq C\left(\|\Lambda_{1-k}Gu\|^2_{H^{k-\frac{m}{2}}} + \|Eu\|^2 + \|GPu\|^2 + \|u\|^2_{H^{-N}}\right), \tag{16}$$

for all $u \in H^{k+1}$. Lemma 2 provides the base case $m = 1$.

Suppose now that the inequality (16) holds for m. As $\mathrm{ell}(g)$ and $\mathrm{ell}(e)$ are open and $\mathrm{ess\,supp}(b)$ is closed, we may shrink the support of g to find a new symbol $g_1 \in S^{k-1}$ so that

1. b is controlled by e through g_1, and
2. $\langle\zeta\rangle^{1-\frac{m}{2}}g_1$ is controlled by $\langle\zeta\rangle^{-\frac{m}{2}}e$ through $\langle\zeta\rangle^{-\frac{m}{2}}g$.

The inductive hypothesis yields the estimate

$$\|Bu\|^2 \leq C\left(\|\Lambda_{1-k}G_1u\|^2_{H^{k-\frac{m}{2}}} + \|Eu\|^2 + \|G_1Pu\|^2 + \|u\|^2_{H^{-N}}\right).$$

We then apply Lemma 2 to $\langle\zeta\rangle^{1-\frac{m}{2}}g_1 \in S^{k-\frac{m}{2}}$, which is controlled by $\langle\zeta\rangle^{-\frac{m}{2}}e \in S^{k-\frac{m}{2}}$ through $\langle\zeta\rangle^{-\frac{m}{2}}g \in S^{k-1-\frac{m}{2}}$, yielding

$$\|\Lambda_{1-\frac{m}{2}}G_1u\|^2 \leq C\left(\|\Lambda_{1-k}Gu\|^2_{H^{k-\frac{m+1}{2}}} + \|\Lambda_{-\frac{m}{2}}Eu\|^2 + \|\Lambda_{-\frac{m}{2}}GPu\|^2 + \|u\|^2_{H^{-N}}\right).$$

We note first that for any $v \in H^r$, $\|\Lambda_r v\| = \|v\|_{H^r}$. One consequence is the estimate

$$\|\Lambda_{-\frac{1}{2}}v\| = \|v\|_{H^{-\frac{1}{2}}} \leq \|v\|_{L^2}.$$

Similarly,

$$\|\Lambda_{1-k}G_1u\|_{H^{k-\frac{1}{2}}} = \|\Lambda_{\frac{1}{2}}G_1u\|,$$

so that combining the two estimates yields

$$\|Bu\|^2 \leq C\left(\|\Lambda_{1-k}Gu\|^2_{H^{k-\frac{m+1}{2}}} + \|Eu\|^2 + \|GPu\|^2 + \|u\|^2_{H^{-N}}\right),$$

finishing the inductive step.

For $m \geq 2N + 2k$,

$$\|\Lambda_{1-k}Gu\|_{H^{k-\frac{m}{2}}} \leq C\|u\|_{H^{-N}},$$

allowing us to remove this term. Thus we have (2) for all $u \in H^{k+1}$.

We conclude with the regularization argument which proves (2) under the weaker hypotheses of the Theorem. We roughly follow Exercises E.10 and E.31 from [DyZw]. We introduce, for fixed $r > 0$, a family of regularizing operators depending on a parameter $\varepsilon \in (0,1)$:

$$\Lambda_{\varepsilon,-r} = (1 - \varepsilon^2\Delta_{\mathbb{R}^d})^{-r/2}.$$

The inverse of $\Lambda_{\varepsilon,-r}$ is given by

$$\Lambda_{\varepsilon,r} = (1 - \varepsilon^2\Delta)^{r/2}.$$

For each $\varepsilon > 0$, $\Lambda_{\varepsilon,-r} \in \Psi^{-r}$ with symbol $\langle\varepsilon\zeta\rangle^{-r}$; regarded as a subset of S^0, this family of symbols is uniformly bounded in ε. In particular, $\Lambda_{\varepsilon,-r} \in \Psi^{-r}$ is a uniformly bounded family in Ψ^0 converging to the identity map in Ψ^s for any $s > 0$. Similarly, $\Lambda_{\varepsilon,r} \in \Psi^r$ is a uniformly bounded family in Ψ^r.

One useful application of these operators forms the backbone of the regularization argument. The monotone convergence theorem implies that if $u \in H^{s-r}$ and $\|\Lambda_{\varepsilon,-r}u\|_{H^s}$ is uniformly bounded in ε, then in fact $u \in H^s$.

Now let u be as in the statement of the Theorem, let $r = N + k + 1$, and apply (2) with u replaced by $\Lambda_{\varepsilon,-r}u$, B replaced by $\Lambda_{\varepsilon,-r}B\Lambda_{\varepsilon,r}$, E replaced by $\Lambda_{\varepsilon,-r}E\Lambda_{\varepsilon,r}$, and G replaced by $\Lambda_{\varepsilon,-r}G\Lambda_{\varepsilon,r}$; note that $\Lambda_{\varepsilon,r}$ commutes with P. We are using here the fact that the symbol of $\Lambda_{\varepsilon,-r}B\Lambda_{\varepsilon,r}$ is controlled by the symbol of $\Lambda_{\varepsilon,-r}E\Lambda_{\varepsilon,r}$ through the symbol of $\Lambda_{\varepsilon,-r}G\Lambda_{\varepsilon,r}$. Note that the corresponding symbol bounds are uniform in ε, because for every r and β we have

$$|\partial_\zeta^\beta\langle\varepsilon\zeta\rangle^r| \leq C_{r,\beta}\langle\varepsilon\zeta\rangle^{r-|\beta|},$$

uniformly in ε. Hence an inspection of the proof of Lemma 2 and of the inductive argument shows that we obtain

$$\|\Lambda_{\varepsilon,-r}Bu\|^2_{L^2} \leq C(\|\Lambda_{\varepsilon,-r}Eu\|^2_{L^2} + \|\Lambda_{\varepsilon,-r}GPu\|^2_{L^2} + \|\Lambda_{\varepsilon,-r}u\|^2_{H^{-N}}),$$

with a constant C independent of ε. Taking $\varepsilon \to 0$, and using the monotone convergence theorem argument mentioned in the last paragraph, finishes the proof of the Theorem.

Acknowledgements The authors are grateful to the participants of the MATRIX workshop *Hyperbolic Differential Equations in Geometry and Physics* for their interest in this topic and for several lively discussions which were the basis of this note. Thanks also to Semyon Dyatlov for helpful comments and corrections. DB was supported by NSF CAREER grant DMS-1654056. Part of this work was supported by NSF grant DMS-1928930 while DB was in residence at the Mathematical Sciences Research Institute during Summer 2022. KD was supported by NSF grant DMS-1708511.

References

[DuHö] J. J. Duistermaat and L. Hörmander. *Fourier Integral Operators II.* Acta Math., Vol 128, pp. 183–269, 1972.

[DyZw] Semyon Dyatlov and Maciej Zworski. *Mathematical Theory of Scattering Resonances.* Grad. Stud. Math. 200. Amer. Math. Soc., 2019.

[Hi] Peter Hintz. *Introduction to Microlocal Analysis.* https://people.math.ethz.ch/ hintzp/notes/micro.pdf.

[Hö1] Lars Hörmander. *Fourier Integral Operators I.* Acta Math., Vol 127, pp. 79–183, 1971.

[Hö2] Lars Hörmander. *On the Existence and Regularity of Solutions of Linear Pseudo-Differential Equations.* Enseignement mathématique, Vol.17, No. 1, pp. 99–163, 1971.

[Hö3] Lars Hörmander. *The Analysis of Linear Partial Differential Operators III: Pseudo-Differential Operators.* Grundlehren Math. Wiss. 274. Springer, 1994.

[MoRaSt] Cathleen S. Morawetz, James V. Ralston, and Walter A. Strauss. *Decay of Solutions of the Wave Equation Outside Nontrapping Obstacles.* Comm. Pure Appl. Math. Vol 30, pp.447–508, 1977.

[Wo] M. W. Wong. *Introduction To Pseudo-Differential Operators. 3rd Edition.* Ser. Anal. Appl. Comput., Vol. 6, World Scientific, 2014.

Chapter 7

MATRIX Program: Mathematics of Tissue Dynamics

Multiphase models for moving boundary problems in biology

Ishraq U. Ahmed, Jennifer A. Flegg, Claire Miller, Ricardo Ruiz-Baier, Joshua Won, and Adriana Zanca

Abstract We address the modelling of tissue deformation using mixture theory. The governing equations consist of mass and momentum balances for two phases, and are written in Eulerian and Lagrangian frameworks using an active stress approach. We specify the general formulation for relatively simple representative examples in the context of tumour growth and early stages of atherosclerotic plaque development.

1 Introduction

Our current understanding of both the physiological and pathological conditions under which tissue growth and development occurs remains incomplete. In this work we explore the process of tissue growth: the addition of mass to the tissue through

Ishraq U. Ahmed
School of Mathematics and Statistics, University of Sydney, Camperdown 2006 NSW, Australia
e-mail: ishraq.uddin@sydney.edu.au

Jennifer A. Flegg
School of Mathematics and Statistics, University of Melbourne, Parkville 3052 VIC, Australia
e-mail: jennifer.flegg@unimelb.edu.au

Claire Miller
Auckland Bioengineering Institute, The University of Auckland, Auckland 1010, New Zealand
e-mail: claire.miller@auckland.ac.nz

Ricardo Ruiz-Baier
School of Mathematics, Monash University, 9 Rainforest Walk, Clayton 3800 VIC, Australia
e-mail: ricardo.ruizbaier@monash.edu

Joshua Won
School of Mathematics and Statistics, University of Sydney, Camperdown 2006 NSW, Australia
e-mail: joshua.won@sydney.edu.au

Adriana Zanca
School of Mathematics and Statistics, University of Melbourne, Parkville 3052 VIC, Australia
e-mail: adriana.zanca@unimelb.edu.au

D. R. Wood et al. (eds.), *2021-2022 MATRIX Annals*, MATRIX Book Series 5,
https://doi.org/10.1007/978-3-031-47417-0_15

cell growth and rearrangement without changing its mechanical properties. Tissue growth is observed in several biological processes including remodelling and morphogenesis [31]. In many biological contexts, for example avascular tumour growth, external nutrients diffuse into the tumour providing sufficient energy for the tumour cells to proliferate [17]. There are multiple approaches to modelling these phenomena, from cell-based models to large-scale continuum models including high-order phase-field frameworks [26, 27, 35]. The purpose of this research project is to explore a configuration of multiphase models in general geometries, where the specific physical and geometrical characteristics of individual cells are not considered but rather each point in space is regarded as occupied by a particle belonging to each constituent (coexisting continua). As specific biological applications of this framework, we consider two phenomena: tumour growth and atherosclerotic plaque formation. We note that this framework can also be applied to other biological contexts such as chemotactic cell aggregates [22]. We build on the Eulerian two-phase model for avascular tumour growth described in [9] which considers two phases: viscous tumour cells and inviscid interstitial extracellular fluid. We extend this from a single dimension to general geometries then pull-back the set of equations to their Lagrangian counterpart. The active deformation of tissue (which encodes the growth process in the model) is represented using the active stress approach (also known as additive split of forces) [2].

Finding closed-form solutions to the nonlinear coupled problems discussed above is only possible in very restricted scenarios and simplified settings. We opt for solving the governing equations numerically. The numerical framework undertaken here uses the method of lines, adopting a backward Euler scheme for the discretisation in time and finite element methods in space using a total Lagrangian approach. This method has been used extensively for growth modelling (see for example [28, 31]). Here we derive a formulation which is consistent with the multiphase poromechanical description. The specific form of the resulting system of equations (in the Lagrangian setting) has as unknowns the displacement of the first phase, the pressure of the second phase, the nutrient (oxygen) tension, and the volume fraction of the first phase. We adopt a monolithic method that requires stable mixed element pairs (in the sense of the Babuška–Brezzi theory [8]) for the approximation of displacement and volume fraction. For the problem at hand, using the MINI element (piecewise linear displacements enriched with cubic bubble functions and piecewise linear and continuous elements for volume fraction) is efficient and sufficiently accurate, but other techniques including stabilisation of low-order schemes with discontinuous volume fractions [6] or mixed methods based on the Hu–Washizu three-field variational principle [34], could be employed in cases where high gradients of the volume fraction are expected and when the material parameters approach the incompressibility limit.

Understanding the mechanisms of moving boundaries in the applications addressed herein can contribute to gaining insight on tissue characterisation, the design and optimisation of medical devices, and the personalisation of surgical and pharmacological medical treatment.

In Section 2 we present a mixture theory model for tumour growth, addressing its extension to the Lagrangian framework and its finite element discretisation. Section 3 contains the derivation of a new model for the early stages of atherosclerosis (more specifically, regarding plaque formation). We continue the theory of mixtures, but the active mechanisms in the plaque formation are concentrated on interfacial fluxes. We write a simplified model valid in the axisymmetric case. In Section 4 we close with a discussion on the limitations of our study and explore possible extensions.

2 A general multiphase tumour model in general geometries

Tumour growth has long been an area of considerable interest to the mathematical biology community. In addition to the devastating impacts tumours can have on individuals and the healthcare system, tumour growth is a process that is amenable to mathematical studies. Early stage tumour growth - such as that we are considering here - is often avascular, in other words, it does not involve blood vessels. Avascular tumours are simpler to model, since interactions with and development of vasculature are not considered. In the early 1970's, Greenspan developed a continuum, compartmental model for avascular tumour growth that is still expanded upon and used for comparison today [23, 24, 36]. In contrast, Landman and Please [32] developed a mixture model of cells and extracellular water and proposed that the differing pressures in each phase drive tumour behaviour. Subsequent tumour models have utilised the mixture model framework, extending and refining earlier assumptions [9–11,13,33,37]; for comprehensive reviews on modelling of avascular tumour growth, refer to [4,12,38]. One such model is that proposed by Breward *et al.* [9]. In their model, the cellular phase behaves similarly to viscous fluid, and the extracellular phase is assumed to be inviscid. Furthermore, the development of the necrotic core of the tumour depends on oxygen tension. In this section we generalise Breward *et al.*'s model to general geometries, and then numerically solve the system in the two-dimensional case for two different geometries. We note that alternative approaches to solving two and three-dimensional multiphase avascular and vascular tumour models are possible [16–18,25,43].

2.1 Governing equations in Eulerian form

We derive here a set of governing equations starting from the main assumption that a given biological tissue is a medium composed by a solid phase (containing cells of diverse types) and an interstitial fluid completely filling the void spaces. For this we consider the two phase model from [9] and continue the basic assumptions held in their model; that there are no voids such that

$$\alpha+\beta=1, \tag{1}$$

where α and β represent the (dimensionless) volume fractions of the two phases. We also assume that mass and momentum are conserved. We redefine these initial equations to general geometries, where the mass balance equations in each phase become:

$$\frac{\partial \alpha}{\partial t}+\nabla_x\cdot(u_\alpha \alpha)=q_\alpha, \tag{2a}$$

$$\frac{\partial \beta}{\partial t}+\nabla_x\cdot(u_\beta \beta)=q_\beta, \tag{2b}$$

where u_α and u_β denote the velocities of their respective phases, while q_α and q_β denote the respective net production rates of tumour cells and fluid. The x-subindex indicates that the spatial derivatives are taken with respect to the Eulerian coordinates.

The momentum balance equations become:

$$\nabla_x\cdot(\alpha\sigma_\alpha)+f_\alpha=0, \tag{3a}$$

$$\nabla_x\cdot(\beta\sigma_\beta)+f_\beta=0. \tag{3b}$$

Here σ_α and σ_β are the respective Cauchy stresses and f_α and f_β are the respective net sources of momentum. We set

$$\begin{aligned} \sigma_\alpha &= -p_\alpha\mathbf{I}+2\mu_\alpha\varepsilon(u_\alpha)+\lambda_\alpha(\nabla_x\cdot u_\alpha)\mathbf{I}, & \sigma_\beta &= -p_\beta\mathbf{I},\\ f_\alpha &= p_\beta\nabla\alpha+\kappa\alpha\beta(u_\beta-u_\alpha), & f_\beta &= p_\beta\nabla\beta-\kappa\alpha\beta(u_\beta-u_\alpha), \end{aligned} \tag{4}$$

where $\varepsilon(u_\alpha)=\frac{1}{2}(\nabla_x u_\alpha+\nabla_x u_\alpha^T)$ is the strain rate tensor, μ_α and λ_α are the shear and bulk viscosities respectively for the cell phase, and p_α and p_β are the respective average pressures. The net momentum sources contain a drag term accounting for the relative motion of the two phases, and an interfacial force coupling pressure with the gradient of volume fraction. The drag coefficient is denoted by κ. We define Σ_α to encode a capillary pressure (the difference between the pressures in the two phases) $\Sigma=p_\alpha-p_\beta$, and it is assumed to have the following form (depending on α)

$$\Sigma(\alpha)=\begin{cases}\gamma\dfrac{\alpha-\alpha^*}{(1-\alpha)^2} & \text{if } \alpha>\alpha_{\min},\\ 0 & \text{if } \alpha<\alpha_{\min}.\end{cases} \tag{5}$$

Based on our assumptions, it also follows that $q_\alpha=-q_\beta$, reflecting conservation of mass for the cell-fluid mixture. Moreover, $f_\alpha=-f_\beta$, reflecting that no external forces act upon the cell-fluid mixture.

The source/sink terms are assumed to be dependent on the concentration of an externally supplied nutrient, here taken to be oxygen. The functional form of q_α is defined to agree with [40]

$$q_\alpha(\alpha, C) = \frac{(1+s_1)\alpha(1-\alpha)C}{(1+s_1 C)} - \frac{s_2+s_3 C}{1+s_4}\alpha, \tag{6}$$

where s_i, $i \in \{1,2,3,4\}$, are constants and the oxygen tension is denoted by C. The positive and negative terms in the equation for q_α correspond to cell division and death, respectively. The factor $(1-\alpha)$ in the cell division term indicate that the growth rate decreases together with the available space.

Due to the separation of time–scales between the cell dynamics and nutrient transport (the timescale of tumour cell proliferation is typically much longer than that of oxygen diffusion), we simply describe oxygen dynamics by

$$-\nabla_x^2 C = q_C(\alpha, C), \tag{7}$$

where ∇ is the Laplacian operator, and the reactive term assumes the following specification

$$q_C(\alpha, C) = -\frac{Q\alpha C}{1+\hat{Q}_1 C}, \tag{8}$$

where Q represents the scaled oxygen consumption rate (scaled by the oxygen diffusion coefficient), and the positive constant $\hat{Q}_1$ describes how fast the tumour cells consume oxygen as oxygen levels change.

Based on the above considerations, a mass conservation equation and a momentum balance for the mixture can be derived by summing up (2a) and (2b), and (3a) and (3b), respectively. This leads to the following set of equations, all written in an Eulerian frame:

$$\frac{\partial \alpha}{\partial t} + \nabla_x \cdot (u_\alpha \alpha) = q_\alpha(\alpha, C), \qquad \text{continuity of phase } \alpha$$

$$\nabla_x \cdot \left(u_\alpha - \frac{(1-\alpha)}{\kappa\alpha} \nabla p_\beta \right) = 0, \qquad \text{continuity of the mixture}$$

$$-\nabla_x \cdot (\sigma) = 0, \qquad \text{momentum balance}$$

$$\sigma = -\alpha\Sigma(\alpha)\mathbf{I} - p_\beta \mathbf{I} + 2\alpha\mu_\alpha \varepsilon(u_\alpha) + \lambda_\alpha(\nabla_x \cdot u_\alpha)\mathbf{I}, \qquad \text{constitutive equation}$$

$$-\nabla_x^2 C = q_C(\alpha, C). \qquad \text{continuity of oxygen}$$

Note that the conservation of angular momentum reads $\sigma^T = \sigma$, and note also that effective unknowns of the problem are the viscous fluid (cell) volume fraction α, the viscous fluid (cell) velocity u_α, the fluid pressure p_β, and the oxygen tension C.

We have considered the above equations in the current configuration Ω (which depends on t) with coordinates x. This system of equations in general coordinates coincides with the one derived in [17].

2.2 Governing equations in Lagrangian form

To circumvent the need of solving the set of equations in a moving domain, the goal in this section is to re-define the problem in Lagrangian coordinates. The motion of material points in a body that occupies region Ω_0 with boundary $\partial\Omega_0$ in the initial state, moves to region Ω with boundary $\partial\Omega$ in the current state. The motion map can be characterised by a $\mathscr{C}^1$ (and bijective) vector field (the deformation or motion map)

$$\chi : \Omega_0 \to \Omega, \qquad \mathbf{X} \mapsto \chi(\mathbf{X}) = x(\mathbf{X},t) \qquad \text{for all } \mathbf{X} \in \Omega_0,\ t \in \times(0,T],$$

and we have that $\chi^{-1}(x,t) = \mathbf{X}$ for all $x \in \Omega$.

The material time derivative $\frac{D}{Dt}$ (with respect to a fixed material coordinate $\mathbf{X}$) for a scalar function $\phi(x(\mathbf{X},t),t)$, defined by applying the chain rule for differentiation, is:

$$\frac{D\phi}{Dt} = \left.\frac{D\phi}{Dt}\right|_{\mathbf{X}} = \frac{\partial\phi}{\partial t} + \nabla_x\phi \cdot \frac{\partial x}{\partial t} = \frac{\partial\phi}{\partial t} + \nabla_x\phi \cdot u.$$

We consider a displacement field $d : \Omega_0 \to \Omega$. This displacement is associated with the deformation map χ such that

$$\begin{aligned} x &= \chi(\mathbf{X}) \\ &= \mathbf{X} + d \text{ where } d \text{ is the displacement.} \end{aligned} \tag{9}$$

The tensor $\mathbf{F}$ is the gradient of the deformation map such that

$$\mathbf{F} = \nabla_{\mathbf{X}}\chi = \mathbf{I} + \nabla_{\mathbf{X}} d, \tag{10}$$

and the Jacobian of the transformation representing local change of volume is

$$J(\mathbf{X},t) = \det \mathbf{F}(\mathbf{X},t).$$

For the two configurations Ω and Ω_0, we denote the outward unit normals of the respective domains as n and $\mathbf{N}_0$, related through $n = \mathbf{F}^{-T}\mathbf{N}_0$, where we have used the notation $\mathbf{F}^{-T} = (\mathbf{F}^{-1})^T$. Understanding ds and ds_0 as spatial differences in the respective domains and dV and dV_0 as current and initial volume elements, we recall the relation

$$J\,dV_0 = dV, \tag{11}$$

and Nanson's formula

$$J\mathbf{F}^{-T}\mathbf{N}_0\,dS_0 = n\,ds. \tag{12}$$

From the Eulerian momentum balance for the fluid phase (3b) we can obtain the following relation (corresponding to Darcy's law, see, e.g., [15])

$$\frac{1-\alpha}{\alpha\kappa}\nabla_x p_\beta = (1-\alpha)(u_\alpha - u_\beta). \tag{13}$$

Using the Maxwell principle for transport and localisation, (13) can be rewritten in the undeformed domain in the following form

$$J\frac{1-\alpha}{\alpha\kappa}\mathbf{F}^{-1}\mathbf{F}^{-T}\nabla_{\mathbf{X}} p_\beta = J(1-\alpha)\mathbf{F}^{-1}(u_\alpha - u_\beta). \tag{14}$$

The mass balance for α (2a), becomes, in Lagrangian coordinates

$$\frac{D}{Dt}(\alpha J) = Jq_\alpha \qquad \text{in } \Omega_0 \times (0,T], \tag{15}$$

whereas for β we use the generalised Reynolds theorem together with (14) to arrive at

$$\frac{D}{Dt}(\beta J) - \nabla_{\mathbf{X}} \cdot (J\mathbf{F}^{-1}\mathbf{F}^{-T}\frac{1}{\alpha\kappa}\nabla_{\mathbf{X}} p_\beta) = Jq_\beta; \tag{16}$$

using the complementary relation (1), and summing the two equations above, gives us

$$\frac{DJ}{Dt} - \nabla_{\mathbf{X}} \cdot (J\mathbf{F}^{-1}\mathbf{F}^{-T}\frac{1}{\alpha\kappa}\nabla_{\mathbf{X}} p_\beta) = 0 \qquad \text{in } \Omega_0 \times (0,T]. \tag{17}$$

For the Lagrangian balance of linear momentum, following the same steps as in the Eulerian case we obtain

$$-\nabla_{\mathbf{X}} \cdot \mathbf{P} = 0 \qquad \text{in } \Omega_0 \times (0,T],$$

where

$$\mathbf{P} = J\mathbf{T}\mathbf{F}^{-T}, \tag{18}$$

is the first Piola-Kirchhoff stress tensor, and $\mathbf{T}$ is the total Cauchy stress tensor (the stress of the mixture) in the reference configuration.

Note that the use of the Lagrangian setting allows us to consider the cell phase as a solid. Regarding the mechanochemical coupling, we adopt the assumption of active stress. In this case the microscopically generated active stresses of the mixture are considered as an additive contribution to the total stress. In a simplified rheological model, the active stress formalism corresponds to considering a parallel arrangement of a passive element and a newly introduced active element [29]. We decompose the first Piola–Kirchhoff stress tensor into

$$\mathbf{P} = \mathbf{P}_\alpha + \mathbf{P}_\beta + \mathbf{P}_{\text{active}}, \tag{19}$$

where $\mathbf{P}_\alpha$ is the effective solid stress for hyperelastic materials. We have

$$\mathbf{P}_\alpha = \frac{\partial \mathscr{W}}{\partial \mathbf{F}},$$

where $\mathscr{W}$ is a strain energy density depending on $\mathbf{F}$. For the specific case of Neo-Hookean poroelastic solids with material parameters $\mu_\alpha, \lambda_\alpha$, we obtain the following form for the effective first Piola–Kirchhoff stress tensor

$$\mathbf{P}_\alpha = \mu_\alpha(\mathbf{F} - \mathbf{F}^{-T}) + \lambda_\alpha J(J-1)\mathbf{F}^{-T}.$$

The stress due to the fluid phase and to the active mass change are, respectively

$$\mathbf{P}_\beta = -p_\beta J\mathbf{F}^{-T},$$

and

$$\mathbf{P}_{\text{active}} = -\alpha\Sigma(\alpha)J\mathbf{F}^{-T}. \tag{20}$$

In general, the active component of stress can depend on a set of phenomenological or physiologically relevant parameters that describe, for example, the state of a population of cells. This is precisely what is added in (20).

Remark 2.1. A more adequate description of growth in the Lagrangian framework is the multiplicative decomposition of the deformation gradient into a pure growth deformation tensor and an elastic rearrangement deformation $\mathbf{F} = \mathbf{F}_e\mathbf{F}_g$ (see, e.g., [21,26]). In such a case, $\mathbf{F}_g$ is determined from a growth factor $g(\alpha,t)$ whose evolution is dictated to preserve (15). We instead adopt an active stress approach which is much simpler to relate with the constitutive equation used for the linear elastic case in Eulerian coordinates.

Note that the conservation of angular momentum is also included, and it is encoded in the relation $\mathbf{P}\mathbf{F}^T = \mathbf{F}\mathbf{P}^T$.

In summary, we get the following initial-boundary value problem for the multiphase growth model in Lagrangian form

$$\frac{D}{Dt}(\alpha J) = Jq_\alpha(\alpha, C), \qquad \text{continuity of phase } \alpha$$

$$\frac{D}{Dt}(J) - \nabla_{\mathbf{X}}\cdot(\frac{J}{\alpha\kappa}\nabla_{\mathbf{X}} p_\beta) = 0, \qquad \text{continuity of the mixture}$$

$$\nabla_X \cdot \mathbf{P} = 0, \qquad \text{equation of motion for the mixture}$$

$$\mathbf{P} = \frac{\partial \mathscr{W}}{\partial \mathbf{F}} - p_\beta J\mathbf{F}^{-T} - \alpha\Sigma(\alpha)J\mathbf{F}^{-T}, \qquad \text{constitutive equation}$$

$$-\nabla_{\mathbf{X}}\cdot(J\mathbf{F}^{-1}\mathbf{F}^{-T}\nabla_{\mathbf{X}} C) = Jq_C(\alpha, C). \qquad \text{continuity of oxygen}$$

To close the set of equations we need to provide suitable initial data and boundary conditions. We suppose that the system is initially at rest and with a given initial distribution of cells

$$\alpha(0) = \alpha_0, \quad d_\alpha(0) = 0 \quad \text{in } \Omega_0. \tag{21}$$

Boundary conditions depend on the specific case under consideration in the following sections. For the mixture we assume general mixed-loading boundary conditions including displacement essential conditions

$$d_\alpha = \hat{d}_0 \qquad \text{on } \partial\Omega_0^D \times (0,T],$$

and natural boundary conditions in the Lagrangian setting

$$\mathbf{P}\mathbf{N}_0 = \hat{t}_0 \qquad \text{on } \partial\Omega_0^N \times (0,T],$$

where $\partial\Omega_0^N$ and $\partial\Omega_0^D$ are two disjoint parts of the undeformed boundary, $\mathbf{N}_0$ is the unit vector outward normal to the body surface $\partial\Omega_0$ in the referential configuration, and $\hat{d}_0$, $\hat{t}_0$ are the prescribed Lagrangian displacement and traction on the boundary.

On the sub-boundary $\partial\Omega_0^N$ the oxygen concentration is maintained at a given value, whereas a flux is imposed on the remainder of the boundary

$$C = C_{\text{out}} \quad \text{on } \partial\Omega_0^N \times (0,T], \qquad J\mathbf{F}^{-1}\mathbf{F}^{-T}\nabla_{\mathbf{X}} C \cdot \mathbf{N}_0 = g_0 \quad \text{on } \partial\Omega_0^D \times (0,T].$$

2.3 Finite element discretisation for the Lagrangian model

The nonlinear weak form of the multiphase equations of motion collected in Section 2.2 can be established following a standard approach, that is, multiplying each field equation by a suitable test function, integrating over Ω_0, and invoking the divergence theorem (in tensorial and in vectorial form) whenever appropriate. For almost all $t > 0$, we look for $d_\alpha \in \mathbf{H}^1_D(\Omega_0) = \{v \in \mathbf{H}^1(\Omega) : v|_{\partial\Omega_0^D} = 0\}$, $p_\beta, C \in \mathrm{H}^1_N(\Omega_0) = \{s \in \mathrm{H}^1(\Omega) : s|_{\partial\Omega_0^N} = 0\}$, and $\alpha \in \mathrm{L}^2(\Omega_0)$, such that

$$\int_{\Omega_0} (\mathbf{P}_\alpha - [p_\beta + \alpha\Sigma(\alpha)]J\mathbf{F}^{-T}) : \nabla_{\mathbf{X}}\zeta \, dV_0 - \int_{\partial\Omega_0^N} J\mathbf{F}^{-T}\hat{t}_0 \cdot \zeta \, dS_0 = 0 \quad \forall \zeta \in \mathbf{H}^1_D(\Omega_0), \tag{22a}$$

$$\int_{\Omega_0} \frac{D}{Dt}(\alpha J)\eta \, dV_0 - \int_{\Omega_0} J q_\alpha(\alpha, C)\eta \, dV_0 = 0 \quad \forall \eta \in \mathrm{L}^2(\Omega_0), \tag{22b}$$

$$\int_{\Omega_0} \frac{D}{Dt}(J)\psi \, dV_0 + \int_{\Omega_0} \frac{J}{\alpha\kappa}\mathbf{F}^{-1}\mathbf{F}^{-T}\nabla_{\mathbf{X}} p_\beta \cdot \nabla_{\mathbf{X}}\psi \, dV_0 = 0 \quad \forall \psi \in \mathrm{H}^1_N(\Omega_0), \tag{22c}$$

$$\int_{\Omega_0} J\mathbf{F}^{-1}\mathbf{F}^{-T}\nabla_{\mathbf{X}} C \cdot \nabla_{\mathbf{X}}\varphi \, dV_0 - \int_{\Omega_0} J q_C(\alpha, C)\varphi \, dV_0 = 0 \quad \forall \varphi \in \mathrm{H}^1_N(\Omega_0). \tag{22d}$$

In order to define a Galerkin finite element method we denote by $\{\mathscr{T}_h\}_{h>0}$ a shape-regular family of partitions of $\bar{\Omega}_0$, conformed by triangles K of diameter h_K, with mesh size $h := \max\{h_K : K \in \mathscr{T}_h\}$. Given an integer $k \geq 1$ and a subset S of $\mathbb{R}^2$, by $\mathbb{P}_k(S)$ we will denote the space of polynomial functions defined locally in S and being of total degree up to k. Let us also denote by $b_K := \varphi_1\varphi_2\varphi_3$ a $\mathbb{P}_3$ bubble function in K, where $\varphi_1, \varphi_2, \varphi_3$ are the barycentric coordinates of the triangle K. Then the finite-dimensional subspaces for cell displacement $\mathbf{V}_h \subseteq \mathbf{H}^1_D(\Omega_0)$, porous

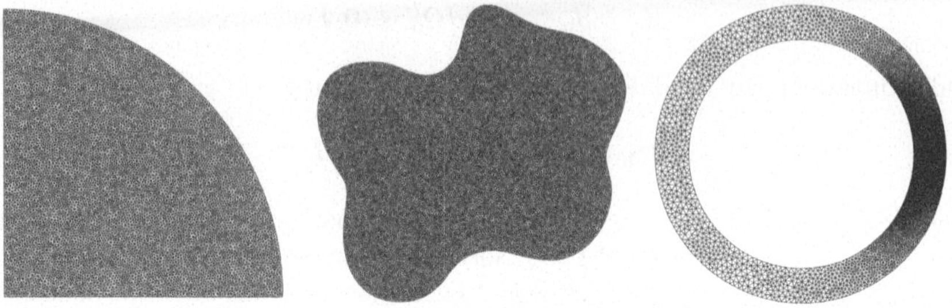

Fig. 1: Samples of unstructured triangular meshes generated for the tumour growth and plaque formation applications.

fluid pressure and oxygen $\mathrm{Q}_h \subseteq \mathrm{H}^1_N(\Omega_0)$, and solid phase volume fraction $\mathrm{R}_h \subseteq \mathrm{L}^2(\Omega_0)$ are defined, respectively, as follows

$$\begin{aligned}
\mathbf{V}_h &:= \{\zeta_h \in C(\overline{\Omega_0}) : \zeta_h|_K \in [\mathbb{P}_1(K) \oplus \mathrm{span}\{b_K\}]^2 \; \forall K \in \mathscr{T}_h,\; \zeta_h|_{\partial\Omega_0^D} = 0\}, \\
\mathrm{Q}_h &:= \{\psi_h \in C(\overline{\Omega_0}) : \psi_h|_K \in \mathbb{P}_1(K),\; \forall K \in \mathscr{T}_h,\; \psi_h|_{\partial\Omega_0^N} = 0\}, \qquad (23) \\
\mathrm{R}_h &:= \{\eta_h \in C(\overline{\Omega}) : \eta_h|_K \in \mathbb{P}_1(K),\; \forall K \in \mathscr{T}_h\}.
\end{aligned}$$

The pair $(\mathbf{V}_h, \mathrm{R}_h)$ is the well-known MINI-element, which is inf-sup stable in the context of saddle-point Stokes equations in their usual velocity-pressure formulation [14].

Next we partition the interval $[0, t_{\text{final}}]$ into N evenly spaced non-overlapping subintervals of fixed length Δt and apply a time semi-discretisation of (22a) using the unconditionally stable, backward Euler's method. Further details on the discretisation and its suitability in solving the equations of motion are omitted here and we simply refer to, e.g., [5, 39].

2.4 Numerical results

In this section we present numerical simulations conducted with algorithms implemented using the open source finite element library `FEniCS` [1]. Sample codes (including meshes) are available from the public repository
`https://github.com/ruizbaier/MATRIX_MultiphaseTissueModelling`.
A fixed tolerance of 10^{-6} is used on the residuals for the convergence criterion of the Newton-Raphson iterative algorithm, and all the linear systems emanating from the discretisation of the tangent systems are solved with the direct method MUMPS. The three meshes used in our studies were generated with the library Gmsh [20] (see Figure 1).

2.4.1 Radial growth on a quarter disk

In our first set of simulations a tumour domain is considered as a unit disk. Only a quarter of the domain is taken into account. On the straight segments of the boundary we impose sliding conditions

$$d_\alpha \cdot \mathbf{N}_0 = 0 \qquad \text{on} \quad \partial\Omega_0^D \times (0,T].$$

For this example the time step is fixed to $\Delta t = 0.5$ and the system is evolved until $t_{\text{end}} = 60$ (adimensional time units). The parameter values are taken to be (see for example, [17])

$$\alpha_{\min} = \alpha_0 = \alpha^* = 0.8, \quad \mu_\alpha = C_{\text{out}} = k = 1, \quad \lambda_\alpha = p_\beta^{\text{out}} = \hat{Q}_1 = 0,$$
$$s_1 = s_4 = 10, \quad s_2 = s_3 = Q = 0.5,$$

where we note that permeability, diffusivity, and active stress are isotropic. In Figure 2 we portray snapshots of the oxygen tension plotted on the deformed configuration (which indicates domain growth). As the tumour grows larger, there is less oxygen available at the tumour 'centre' (the region furthest away from the free boundary). Once the oxygen concentration is below approximately 0.3, a necrotic core forms. From Figure 2, we observe that a necrotic core forms shortly after $t = 40$. In the necrotic core, (viable) tumour cells are replaced with necrotic material. The volume of tumour cells remains lower within the necrotic core (see the lighter region in the α snapshot at $t = 60$ in Figure 2), before rapidly increasing across the boundary between the necrotic core and the remainder of the tumour, then increasing linearly to the free boundary.

2.4.2 Progression of a glioblastoma

We simulate tumour growth in free suspension and therefore for this case we include pure traction boundary conditions with $\hat{t}_0 = 0$. In this situation we are required to remove rigid motions from the set of admissible displacement solutions. This is done using a Lagrange multiplier approach (imposing orthogonality with respect to the space of rigid motions; see for example, [30]). We employ here the same parameters as in the previous subsection.

The top row of Figure 3 depicts the volume fraction of cells on the deformed configuration, showing also the directions of displacement. These arrows confirm a radial growth imposed by the isotropic active stress distribution. In addition, we observe that at early times oxygen levels are everywhere sufficiently high to enable cell proliferation and therefore the tumour cell grows monotonically. The bottom row shows, for the time $t = 100$, the oxygen distribution and the magnitude of the first Piola–Kirchhoff stress tensor (19). From this we note a higher accumulation of stress near the reentrant boundaries of the tumour. We also note that the range of values for α and C vary considerably to the quarter-disk case. Over 100 time units, the

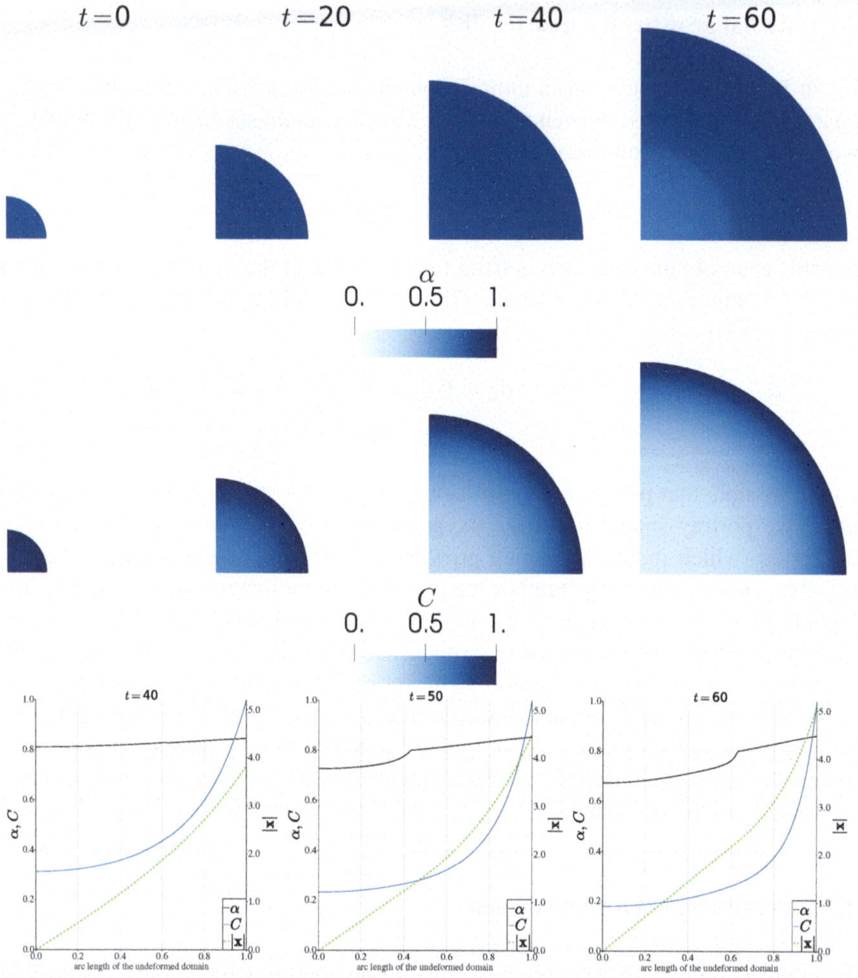

Fig. 2: Snapshots of the time evolution of cell volume fraction (top) and oxygen tension (middle row), plotted on the deformed configuration at times $t = 0, 20, 40, 60$ (adimensional units). The bottom row shows plots of α, C and the magnitude of the current position x along the centre line of the domain (at an angle of 45°), for three times ($t = 40, 50, 60$).

concentration of oxygen never drops below 0.8 anywhere within the glioblastoma, even at the centre, and therefore a necrotic core never forms. Contrast this with the quarter-disk, where a necrotic core forms not long after $t = 40$ time units.

Next steps for this case study are including anisotropy and heterogeneity, not only in the active stress contribution, but also in the permeability parameter κ. For

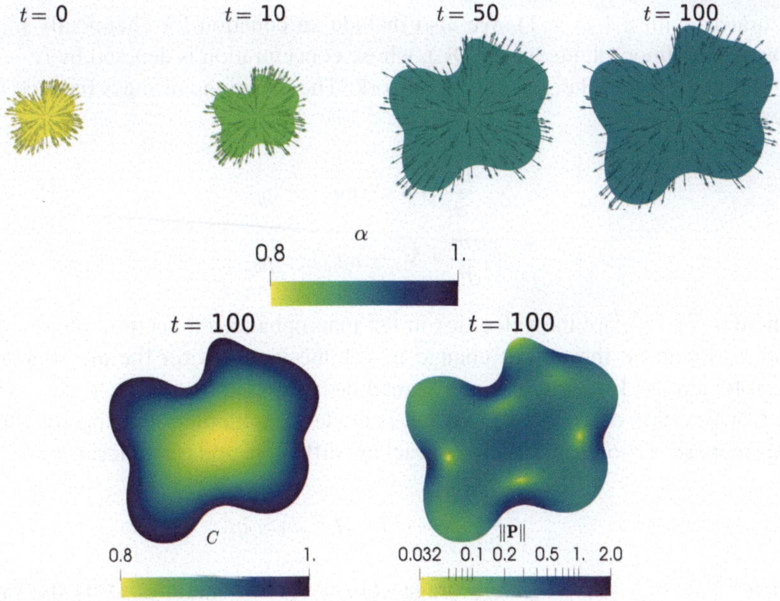

Fig. 3: Top: Snapshots of the time evolution of cell volume fraction, plotted on the deformed configuration at times $t = 0, 10, 50, 100$ (adimensional units). Arrows represent the displacement of the surface. Bottom: concentration of oxygen and magnitude of total stress at the final time $t = 100$. Note the log-scale on the magnitude of total stress. Shape based on [16, 19].

permeability we can use the same medical images employed to generate contours and meshes.

3 Model for plaque development

The mechanisms used in the modelling of tumour growth in Section 2, see also [3, 11], can serve as a starting point for formulating two-phase models for an early stage atherosclerotic plaque. The model that we develop herein follows [41, 42] and assumes that all cells coexist in a mixture medium where proteins diffuse. To capture the evolution of the plaque, we consider the region occupied by the plaque and its surroundings as a moving boundary domain. In the mixture we consider two phases: a macrophage phase (whose quantities are labelled with the subscript m) denoting its volume fraction also m, and a second phase (whose quantities are labelled with the subscript n), with volume fraction n, comprising foam cells, apoptotic and necrotic material, and other material such as extracellular matrix (which we henceforth refer to as necrotic material). We assume that there are no voids (the mixture is fully saturated, with $n+m=1$). We also include an equation for chemically modified low density lipoproteins (modLDL), whose concentration is denoted by l.

We first present the general framework. The equations of mass balance for the two phases gives:

$$\frac{\partial n}{\partial t}+\nabla_x\cdot(nu_n)=q_n, \tag{1a}$$

$$\frac{\partial m}{\partial t}+\nabla_x\cdot(mu_m)=q_m, \tag{1b}$$

where u_n and u_m are the velocities in the macrophage and necrotic phases, and q_n and q_m represent the rate of change in volume fractions for the macrophage and necrotic phases. Both phases are assumed incompressible.

Conservation of mass for modLDL is modelled with the following equation that considers advection in both phases, Fickian diffusion, and a sink term:

$$\frac{\partial l}{\partial t}+\nabla_x\cdot(l\hat{u}-D_l\nabla_x l)=q_l, \tag{2}$$

where $\hat{u}=mu_m+nu_n$ is the average velocity of the mixture, D_l is the intrinsic diffusivity of modLDL (assumed isotropic) and q_l is the rate of removal of modLDL by macrophages. Here we take

$$q_n=-q_m=k_m lm, \quad q_l=-k_l lm, \tag{3}$$

where we adopt a simple mass-action form for the phagocytic consumption of modLDL by macrophages (with k_l, k_m positive constants), although we note that more complicated forms could be considered. We also assume here that macrophages die of necrosis soon after becoming foam cells due to the impairment of normal apoptotic pathways, and that macrophages do not ingest necrotic material [7].

Conservation of momentum is assumed in both phases, neglecting inertial effects (which is justified by the timescales of cell growth and proliferation), leading to:

$$-\nabla_x\cdot(n\sigma_n)=\mathbf{b}_n, \tag{4a}$$

$$-\nabla_x\cdot(m\sigma_m)=\mathbf{b}_m, \tag{4b}$$

where σ_m and σ_n are the Cauchy stress tensors in the macrophage and necrotic phases, and $\mathbf{b}_m$ and $\mathbf{b}_n$ are the net sources of momentum in each phase. Assuming that there are no external momentum sources, we simply take $\mathbf{b}_m=-\mathbf{b}_n$ and consider that interaction forces only include a Darcy-type drag term:

$$\mathbf{b}_n=bn(1-n)(u_m-u_n).$$

Next, we supply the form of stresses in the necrotic and macrophage phases. We simply consider constitutive laws including an hydrostatic and a viscous contribution, giving:

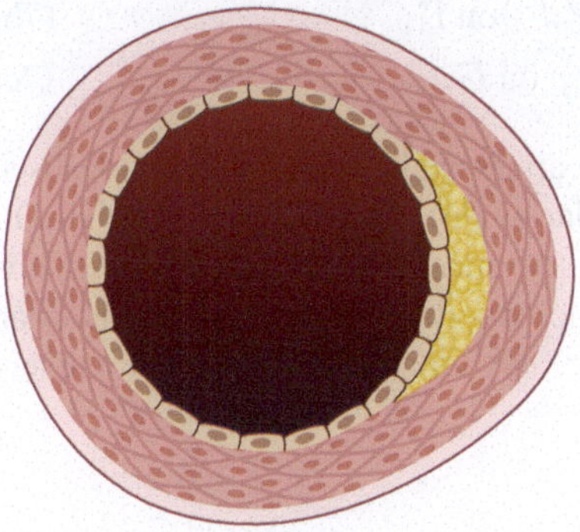

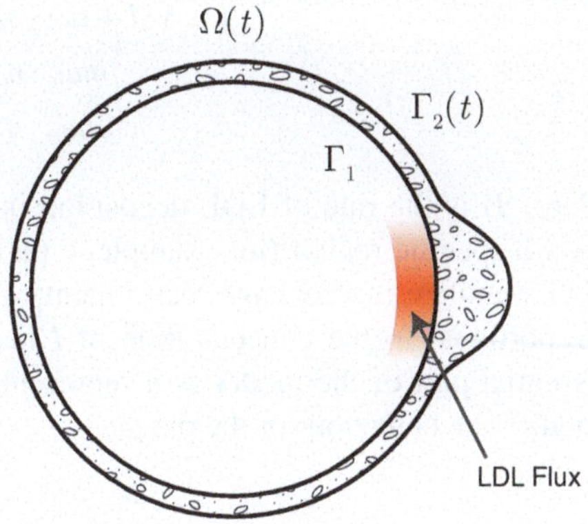

Fig. 4: Left: Cross-section of an artery. The pink cells are smooth muscle cells, the beige cells are endothelial cells, the yellow region shows the necrotic core and the red region is the lumen. Right: Model schematic.

$$\sigma_n = -p_n\mathbf{I} + 2\mu_n\varepsilon(u_n) + \lambda_n(\nabla \cdot u_n)\mathbf{I}, \tag{5a}$$

$$\sigma_m = -p_m\mathbf{I} + 2\mu_m\varepsilon(u_m) + \lambda_m(\nabla \cdot u_m)\mathbf{I}, \tag{5b}$$

where p_n and p_m are the pressures in the nectrotic and macrophage phases, μ_n and μ_m and λ_n and λ_m are the shear and bulk viscosity coefficients in the two phases. We further assume that the pressure difference between the macrophage and necrotic phases contributes to macrophage movement:

$$p_m = p_n + \Sigma,$$

and we take a form for the capillary pressure depending only on the volume fraction of the macrophage phase and the concentration of modLDL. This term is assumed to characterise the chemotaxis of macrophages towards higher levels of modLDL:

$$\Sigma(l,m) = \frac{A}{1+Bl^k}, \tag{6}$$

where A, B and k are positive constants. These terms adopt the structure used in, e.g., [33].

For spatial domain $\Omega(t)$ we consider a cross-section through the artery wall, transverse to the direction of blood flow. Early and mid-stage plaque formation is observed to cause outward remodelling of the artery wall, where the radius of the lumen is preserved. We therefore take the inner endothelial boundary Γ_1 to be fixed. The outer boundary $\Gamma_2(t)$ separates the intima from the layers of muscle cells supporting the artery, and we assume that this boundary may vary with time (see a sketch of the geometry and boundary configuration in Figure 4).

We impose the following flux-type boundary conditions on the inner endothelial boundary:

$$(-\nabla_x l + l\hat{u}) \cdot n = \Psi_l \quad \text{on } \Gamma_1, \tag{7a}$$
$$m u_m \cdot n = \Psi_m l \quad \text{on } \Gamma_1, \tag{7b}$$
$$n u_n \cdot n = 0 \quad \text{on } \Gamma_1. \tag{7c}$$

Here, Ψ_l is the rate of LDL deposition in the artery wall, which will be localised over a specific region (for example, a particular angle). The presence of modLDL will stimulate macrophage recruitment, where the macrophage influx $mu_m|_{\Gamma_1} \cdot n$ is proportional to the concentration of l near the endothelium. Note that this is an essential part of the model as it represents the only contribution that will actively modify the behaviour of the plaque.

3.1 Simplification to one-dimensional geometry

In one dimension the model can be written as follows:

$$\frac{\partial n}{\partial t} + \frac{\partial}{\partial x}(n u_n) = k_n l m, \tag{8a}$$
$$\frac{\partial}{\partial x}(m u_m + n u_n) = 0, \tag{8b}$$
$$\frac{\partial l}{\partial t} + \frac{\partial}{\partial x}\left(l\hat{u} - \frac{\partial l}{\partial x}\right) = k_l l m \text{ where } \hat{u} = n u_n + m u_m, \tag{8c}$$
$$-\frac{\partial}{\partial x}\left(n\left(-p_n + \mu_n \frac{\partial u_n}{\partial x} + \lambda_n \frac{\partial u_n}{\partial x}\right)\right) = b_n n(1-n)(u_m - u_n), \tag{8d}$$
$$-\frac{\partial}{\partial x}\left(m(-p_n - \Sigma(l,m) + \mu_m \frac{\partial u_m}{\partial x} + \lambda_m \frac{\partial u_m}{\partial x}\right)\Bigg) = -(b_n n(1-n)(u_m - u_n)), \tag{8e}$$
$$\left(-\frac{\partial l}{\partial x} + l\hat{u}\right)\Bigg|_{x=0} = \Psi_l, \tag{8f}$$
$$m u_m|_{x=0} = \Psi_m l|_{x=0}, \tag{8g}$$
$$n u_n|_{x=0} = 0, \tag{8h}$$

while the boundary conditions on the outer layer become:

$$m(u_m - L')|_{x=L} = n(u_n - L')|_{x=L} = l(u_l - L')|_{x=L} = 0. \tag{9}$$

For numerical purposes, we scale the spatial domain using $\xi = \frac{x}{L(t)}$ which then gives:

$$\frac{\partial n}{\partial t} - \frac{\xi}{L(t)} L'(t) \frac{\partial n}{\partial \xi} + \frac{1}{L(t)} \frac{\partial}{\partial \xi}(n u_n) = k_m l m, \tag{10a}$$

$$\frac{\partial l}{\partial t} - \frac{\xi}{L(t)} L'(t) \frac{\partial l}{\partial \xi} + \frac{1}{L(t)} \frac{\partial}{\partial \xi}\left(l\hat{u} - \frac{1}{L(t)} \frac{\partial l}{\partial \xi}\right) = -k_l l m. \tag{10b}$$

Implementing this model numerically and comparing to atherosclerotic plaque formation data is left for future work.

4 Discussion

This paper generalises the tumour growth model of Breward *et al.* [9] into general geometries and shows numerical solutions for the two-dimensional case using finite element variational formulations of the system. Our findings are consistent with previous work; however, we have used a different approach that can handle different geometries with relative ease. Given the current assumptions of the model the domain grows uniformly, and therefore the power of this technique is not fully realised. Extensions of this work could consider anisotropic stress and heterogeneous diffusivity and permeability to explore the effects of different geometries.

We additionally hypothesise a two-phase model for early atherosclerotic plaque formation. This work aims to complement the existing literature on late stage plaque behaviour [41, 42]. We leave the exploration of the numerical solution for future work. We note that to better understand early plaque formation it would be beneficial to separate the necrotic material phase further into its component parts. For example, formulating a three (or more) phase model to incorporate the difference between foam cells and apoptotic and necrotic material.

Acknowledgements We thank Professor Mary Myerscough for her advice on atherosclerotic plaques. We also thank Dr Alexander Browning for producing Figure 1 and for the robust discussions on the equations describing early plaque formation. Finally we thank the mathematical research institute MATRIX in Australia where part of this research was performed, as part of the MATRIX Research Program: Mathematics of Tissue Dynamics.

References

1. M. S. Alnæs, J. Blechta, J. Hake, A. Johansson, B. Kehlet, A. Logg, C. Richardson, J. Ring, M. E. Rognes, and G. N. Wells, *The FEniCS Project Version 1.5*, Arch. Numer. Softw., 3 (2015), pp. 9–23.
2. D. Ambrosi and S. Pezzuto, *Active stress vs. active strain in mechanobiology: Constitutive issues*, Journal of Elasticity, 107 (2012), pp. 199–212.
3. D. Ambrosi and L. Preziosi, *On the closure of mass balance models for tumor growth*, Mathematical Models and Methods in Applied Sciences, 12 (2002), pp. 737–754.
4. R. P. Araujo and D. S. McElwain, *A history of the study of solid tumour growth: the contribution of mathematical modelling*, Bulletin of Mathematical Biology, 66 (2004), pp. 1039–1091.
5. N. Barnafi, B. Gómez-Vargas, W. d. J. Lourenço, R. F. Reis, B. M. Rocha, M. Lobosco, R. Ruiz-Baier, and R. Weber dos Santos, *Finite element methods*

for large-strain poroelasticity/chemotaxis models simulating the formation of myocardial oedema, Journal of Scientific Computing, 92 (2022), pp. e92(1–40).
6. L. Berger, R. Bordas, D. Kay, and S. Tavener, *A stabilized finite element method for finite-strain three-field poroelasticity*, Computational Mechanics, 60 (2017), pp. 51–68.
7. Y. V. Bobryshev, E. A. Ivanova, D. A. Chistiakov, N. G. Nikiforov, and A. N. Orekhov, *Macrophages and their role in atherosclerosis: pathophysiology and transcriptome analysis*, BioMed Research International, 2016 (2016).
8. D. Boffi, F. Brezzi, and M. Fortin, *Mixed Finite Element Methods and Applications*, vol. 1, Springer-Verlag, Berlin, 2013.
9. C. Breward, H. Byrne, and C. Lewis, *The role of cell-cell interactions in a two-phase model for avascular tumour growth*, Journal of Mathematical Biology, 45 (2002), pp. 125–152.
10. C. J. Breward, H. M. Byrne, and C. E. Lewis, *A multiphase model describing vascular tumour growth*, Bulletin of Mathematical Biology, 65 (2003), pp. 609–640.
11. H. Byrne and L. Preziosi, *Modelling solid tumour growth using the theory of mixtures*, Mathematical Medicine and Biology: A Journal of the IMA, 20 (2003), pp. 341–366.
12. H. M. Byrne, *12.1 continuum models of avascular tumor growth*, in Mathematics and Life Sciences, De Gruyter, 2012, pp. 279–312.
13. H. M. Byrne, J. R. King, D. S. McElwain, and L. Preziosi, *A two-phase model of solid tumour growth*, Applied Mathematics Letters, 16 (2003), pp. 567–573.
14. A. Cioncolini and D. Boffi, *The MINI mixed finite element for the Stokes problem: An experimental investigation*, Comput. Math. Appl., 77 (2019), pp. 2432–2446.
15. O. Coussy, *Poromechanics*, John Wiley & Sons, 2004.
16. V. Cristini and J. Lowengrub, *Multiscale modeling of cancer: an integrated experimental and mathematical modeling approach*, Cambridge University Press, 2010.
17. J. Droniou, J. A. Flegg, and G. C. Remesan, *Numerical solution of a two dimensional tumour growth model with moving boundary*, Journal of Scientific Computing, 85 (2020), pp. 1–31.
18. H. B. Frieboes, F. Jin, Y.-L. Chuang, S. M. Wise, J. S. Lowengrub, and V. Cristini, *Three-dimensional multispecies nonlinear tumor growth—ii: tumor invasion and angiogenesis*, Journal of Theoretical Biology, 264 (2010), pp. 1254–1278.
19. H. B. Frieboes, J. S. Lowengrub, S. Wise, X. Zheng, P. Macklin, E. L. Bearer, and V. Cristini, *Computer simulation of glioma growth and morphology*, Neuroimage, 37 (2007), pp. S59–S70.
20. C. Geuzaine and J. F. Remacle, *Gmsh: A 3-D finite element mesh generator with built-in pre- and post-processing facilities*, Int. J. Numer. Methods Eng., 79 (2009), pp. 1309–1331.
21. A. Goriely and R. Vandiver, *On the mechanical stability of growing arteries*, IMA J. App. Math., 75 (2010), pp. 549–570.
22. J. Green, J. Whiteley, J. Oliver, H. Byrne, and S. Waters, *Pattern formation in multiphase models of chemotactic cell aggregation*, Mathematical Medicine and Biology: A Journal of the IMA, 35 (2018), pp. 319–346.
23. H. Greenspan, *Models for the growth of a solid tumor by diffusion*, Studies in Applied Mathematics, 51 (1972), pp. 317–340.
24. ———, *On the growth and stability of cell cultures and solid tumors*, Journal of Theoretical Biology, 56 (1976), pp. 229–242.
25. M. Hubbard and H. Byrne, *Multiphase modelling of vascular tumour growth in two spatial dimensions*, Journal of Theoretical Biology, 316 (2013), pp. 70–89.
26. G. W. Jones and S. J. Chapman, *Modeling growth in biological materials*, SIAM Rev., 54 (2012), pp. 52–118.
27. A. Karolak, D. A. Markov, L. J. McCawley, and K. A. Rejniak, *Towards personalized computational oncology: from spatial models of tumour spheroids, to organoids, to tissues*, Journal of The Royal Society Interface, 15 (2018), p. 20170703.
28. N. Kida and Y. Morishita, *Continuum mechanical modeling of developing epithelial tissues with anisotropic surface growth*, Finite Elem. Anal. Des., 144 (2018), pp. 49–60.

29. T. Klotz, C. Bleiler, and O. Röhrle, *A physiology-guided classification of active-stress and active-strain approaches for continuum-mechanical modeling of skeletal muscle tissue*, Frontiers in Physiology, 12 (2021).
30. M. Kuchta, K.-A. Mardal, and M. Mortensen, *On the singular neumann problem in linear elasticity*, Numerical Linear Algebra with Applications, 26 (2019), p. e2212.
31. E. Kuhl, *Growing matter: A review of growth in living systems*, J. Mech. Behav. Biomed. Mater., 29 (2014), pp. 529–543.
32. K. Landman and C. Please, *Tumour dynamics and necrosis: surface tension and stability*, Mathematical Medicine and Biology: A Journal of the IMA, 18 (2001), pp. 131–158.
33. G. Lemon, J. R. King, H. M. Byrne, O. E. Jensen, and K. M. Shakesheff, *Mathematical modelling of engineered tissue growth using a multiphase porous flow mixture theory*, Journal of Mathematical Biology, 52 (2006), pp. 571–594.
34. X. Li, Z. Liu, R. W. Lewis, and K. Suzuki, *Mixed finite element method for saturated poroelastoplastic media at large strains*, International Journal for Numerical Methods in Engineering, 57 (2003), pp. 875–898.
35. J. Metzcar, Y. Wang, R. Heiland, and P. Macklin, *A review of cell-based computational modeling in cancer biology*, JCO Clinical Cancer Informatics, 2 (2019), pp. 1–13.
36. R. J. Murphy, A. P. Browning, G. Gunasingh, N. K. Haass, and M. J. Simpson, *Designing and interpreting 4d tumour spheroid experiments*, Communications Biology, 5 (2022), pp. 1–11.
37. L. Preziosi and A. Tosin, *Multiphase modelling of tumour growth and extracellular matrix interaction: mathematical tools and applications*, Journal of Mathematical Biology, 58 (2009), pp. 625–656.
38. T. Roose, S. J. Chapman, and P. K. Maini, *Mathematical models of avascular tumor growth*, SIAM Review, 49 (2007), pp. 179–208.
39. R. Ruiz-Baier, *Primal-mixed formulations for reaction-diffusion systems on deforming domains*, Journal of Computational Physics, 299 (2015), pp. 320–338.
40. J. P. Ward and J. King, *Mathematical modelling of avascular-tumour growth*, Mathematical Medicine and Biology: A Journal of the IMA, 14 (1997), pp. 39–69.
41. M. G. Watson, H. M. Byrne, C. Macaskill, and M. R. Myerscough, *A two-phase model of early fibrous cap formation in atherosclerosis*, Journal of Theoretical Biology, 456 (2018), pp. 123–136.
42. ———, *A multiphase model of growth factor-regulated atherosclerotic cap formation*, Journal of Mathematical Biology, 81 (2020), pp. 725–767.
43. S. M. Wise, J. S. Lowengrub, H. B. Frieboes, and V. Cristini, *Three-dimensional multispecies nonlinear tumor growth—i: model and numerical method*, Journal of Theoretical Biology, 253 (2008), pp. 524–543.

Chapter 8

MATRIX Program: Mathematics of Risk 2022

Asian Option Pricing via Laguerre Quadrature: A Diffusion Kernel Approach

P. G. Morrison

Abstract This paper will demonstrate some new techniques for developing the theory of Asian (arithmetic average) options pricing. We discuss the basic derivation of the diffusion equations, and how various techniques from potential theory can be applied to solve these complex expressions. The Whittaker-type confluent hypergeometric functions are introduced, and we discuss how these functions are related to other systems including Mehler-Fock and modified Bessel functions. We close with a brief analysis of some index transforms and the kernels related to these integral transforms.

1 Introduction

Asian (arithmetic average) options are a derivative instrument written on the difference between the time average of the stock price over the term, and some nominal strike. Floating strike options have payoff given by the max/min of the difference between the time average and the stock price on the execution date. These derivatives have useful properties when offsetting risk for continuously produced goods such as commodities and electricity. Other use cases could be in order to reward executive performance above some average benchmark, say the average stock price is greater than some nominal value over the term of the option. It is harder to influence in a negative way vs standard European contracts; indeed it is well known that stock volatility increases towards the execution date of European options which leads to inverse incentives, i.e. increases the probability of stock price pumps by executives near the time of expiry.

If one were to think in terms of peak average performance, rather than the performance on any particular one day, it is obvious that such contracts as we shall describe in this paper have great utility in this area. For a commodities producer, the

Peter Morrison
University of Technology, Sydney, Broadway 2000, e-mail: peter.morrison@uts.edu.au

D. R. Wood et al. (eds.), *2021-2022 MATRIX Annals*, MATRIX Book Series 5,
https://doi.org/10.1007/978-3-031-47417-0_16

value in such contracts is that it enables producers to lock in prices for known deposits and stabilise pricing outcomes separate to the spot market, given known fixed production costs at any given time. A minerals producer might for example know that they can produce at an average rate 10,000 tons of ore at a fixed concentration of gold per day. Pricing an option on the average amount of gold produced over a month then becomes of utility when locking in a price for the total amount of gold produced, which in turn can be used to stabilise the costs to market the commodity by using this sort of technique. Indeed, this paper concerns the deep problem of the nature of the average of an exponential Brownian motion.

There is a deep, known link between the diffusion problems we associate with options pricing problems and diffusion theory. It is known that the Black-Scholes equation may be transformed to the heat equation, other results include Feynman-Kac for generalised potentials. Other exotic options, e.g. in CDS pricing equation, one may obtain the diffusion equation related to the CIR process, which has probability function given by a non-central chi-squared distribution. As shall be demonstrated in the following calculations, there is a fundamental relationship between diffusion systems, special functions and spectral theory which is of great use in understanding these types of problems in the theory of stochastic processes.

2 Review

The work contained within this paper pays homage to the pioneers in the field of spectral analysis, principal amongst them Feller [6], who explored the methods of semi-groups and partial differential equations in order to understand stochastic processes and other random variables. These initial researches, carried out in the 1950s, can be seen as the ancestor of the theories of stochastic analysis, especially PDE methods. However, it is a lesser-known topic, with the applications of these types of kernel analysis often obscured by the bright lights of martingales and stochastic calculus. The construction of kernels to solve differential equations is an older question again, and although it is well-known in the fields of physics that an understanding of kernel solutions can be gained through spectral decomposition, this observation has escaped much detailed investigation in the field of quantitative finance.

Spectral decomposition is an elegant way of describing the transition probability density of any system with a spectrum, and the linkage between resolvent or Green's function methods and the probability density of a diffusion is given through Laplace transformation. We note the papers of Albanese et. al [2, 1] examine this problem in general, and have found a number of systems with spectral decompositions that are of finite type. On a deeper level, underlying this whole infrastructure is the effort of Mehler [16], as expressed and popularised through the Feynman kernel, path integral method c. 1948 [7]. This formula is a compact description of a kernel using a set of base states, which in this case is given by the Hermite polynomials. The search for other such formulae is an active area of investigation and we shall touch on some simple topics related to this historical example. As an example of a system related to our investigation, use of the various types of special functions as a result of path integration in the hyperbolic plane may be found in [10]. The authors in this case have succeeded in deriving many formulae that appear in a different context to our research. In many ways, the idea of hyperbolic diffusion is closely allied to the types of hyperbolic random variables that underpin the diffusion problems in this work. Further tabulations of path integrals may be found in [11], in particular the results concerning the resolvent of Whittaker functions and the Morse potential.

The link between special function theory and arithmetic average or Asian option pricing has been explored by a number of authors. The foundational paper in the field is taken to be Geman and Yor [8], who established the connection between the theory of square Bessel processes, stochastic calculus, and exponential Brownian motion. The topic was explored from a different perspective by Linetsky [15], who has demonstrated that spectral theory may be used to calculate Asian options prices. Both of these papers present formidable work, and our research has shown that many of the results which are gained through sheer force of will and brute mathematical force in these works can be found through simpler means, via the techniques of spectral decomposition of the kernel. The stochastic calculus in [8] shows that prima facie the concept of exponential Brownian motion is the important branch of mathematics on which to grasp; one is presented with the idiom of time-averaged exponential Brownian motion. The functional analysis of these types of systems is carried out in detail by Pintoux and Privault [20, 21], where the authors were able to obtain some exact solutions to the Asian option pricing problem by deriving some spectral decompositions of polynomial initial conditions. In particular, the papers in [20, 21] rely on many combinatorical identities and are complicated to verify and replicate. Our methods represent a parallel line of thought which arrives at similar results by a completely different path. A simpler approach has been illustrated through the works of Sousa and Yakubovich [24], who have shown that by understanding the nature of kernel solutions, eigenfunctions and integral transforms one may gain many of the advantages of the PDE method while avoiding necessary complication in the matter. Further, Craddock [5] has shown that analysis of the fundamental solution for Asian options pricing can be understood through the use of special functions and the sine transform.

We shall expand upon this method in this paper and bring this program to fruition by a process of synthesis of the results from these disparate groups. Our aim is to find direct ways of calculating Asian options prices that will be of use to any trader in the same way that European options prices may be found via the Black-Scholes equation in the simplest of cases. Such formulae have proved elusive in the past, and the notational complexity of the field makes our job much more difficult. By stripping away this additional complication, we shall show that the answers to many of the problems involved in pricing these types of exotic options reduce to a small number of simply addressed questions in integration theory.

We must note here the use of special functions known as the Bessel polynomials in this paper. These polynomials have many properties peculiar to the problems in diffusion we shall consider, and it can be hard to find established or tabulated results that one can expect to find for more well-known groups of orthogonal polynomials.

The interested reader is directed to the works of Krall and Frink [14], who along with Grosswald [12] wrote the pioneering works on the topic, see also [13] for further information and references on integration theory of Bessel polynomials. These polynomials, familiar to filter design and signal processing, are less known in modern mathematics. Their many useful properties in describing systems with discrete and continuous spectral components shall be one principal focus explored in this calculation.

Examples of index transforms and related integrals for Whittaker, modified Bessel and Mehler-Fock functions may be found in the tabulated works in [9, 23, 22, 3, 18, 17]. The monumental works of Oberhettinger [18, 17] are of great use in the following sections. Unfortunately, like all sets of integral tables, deficiencies abound, and part of this research is devoted to unpicking the methods by which the known and established results in index transform theory may be derived.

3 Notation and Mathematical Details

This paper shall be primarily concerned with various sets of special functions which we shall now define. There are other notational necessities that must also be set up for the calculation proper. We shall not be focused in this work on the exact nature of how these functions arise, for our purpose it is sufficient to know that they exist and are useful to our aim of analysing diffusion processes of a type that is central to our work.

3.1 Special functions

The important special functions used in this paper are defined as follows:

3.1.1 Whittaker Functions

The Whittaker function, denoted $W_{\alpha,\beta}(z)$, is a type of confluent hypergeometric function. It is the solution of the differential equation:

$$\frac{d^2 f}{dz^2} + \left(-\frac{1}{4} + \frac{\alpha}{z} + \frac{1/4 - \beta^2}{z^2} \right) f = 0 \tag{1}$$

The indices of the function may take any value in the complex plane, however it is important to realise that this function is multi-branched and care must be taken to choose appropriate sets of parameters. For our purposes, as we shall demonstrate, the positive plane defines a continuous solution which may be used to build a set of eigenfunctions. The second solution for the Whittaker differential equation is not used in this proof. Applications of the Whittaker function include the use in the index Whittaker transform [24] where, in a similar way to how functions are mapped to other functions via a Laplace or Fourier transform, in this case we integrate over the index of the function via:

$$F(x) = \frac{1}{\pi^2 x^2} \int_0^\infty |\Gamma(-\mu + 1/2 + ip)|^2 W_{\mu,ip}(x) f(p) p \sinh(2\pi p) dp \tag{2}$$

This is the most generalised form of the index transformation. Note that this is of half-index in the gamma function. An inversion integral of the form:

$$f(p) = \int_0^\infty W_{\mu,ip}(x) F(x) dx \tag{3}$$

is known to exist. We shall rely on the relationships between the confluent hypergeometric function and other special functions to compute a number of different integrals related to the solution of various PDEs. Note the useful hyperbolic identity:

$$\frac{\sinh(2\pi p)}{\cosh(\pi p)} = 2\sinh(\pi p) \tag{4}$$

3.1.2 Modified Bessel Functions

The modified Bessel function $K_\nu(z)$ is another type of hypergeometric function. One may relate it to the restriction of the Whittaker function via:

$$W_{0,\nu}(2z) = \sqrt{\frac{2z}{\pi}} K_\nu(z) \tag{5}$$

The modified Bessel function therefore satisfies the differential equation defined through:

$$\frac{d^2 f}{dz^2} + \left(-\frac{1}{4} + \frac{1/4 - \nu^2}{z^2}\right) f = 0 \tag{6}$$

Using the scaling $x = z/2$, we may transform the differential equation, so in the new variables we have:

$$\frac{1}{4} f_{xx} + \left(-\frac{1}{4} + \frac{1/4 - \nu^2}{4x^2}\right) f = 0 \tag{7}$$

or

$$f_{xx} + \left(-1 + \frac{1/4 - \nu^2}{x^2}\right) f = 0 \tag{8}$$

Substituting $f(x) = \sqrt{x} U(x)$, we find the modified Bessel differential equation:

$$x^2 U_{xx} + x U_x - (\nu^2 + x^2) U(x) = 0 \tag{9}$$

This is an important type of hypergeometric function as shown in the calculation. In an analogous way to the index Whittaker transform, one may define an index transform for the modified Bessel function known as the Kontorovich-Lebedev transform pair. Substituting using the functional relationship, we have:

$$f(p) = \int_0^\infty W_{0,ip}(2x)F(x)dx \tag{10}$$

$$F(x) = \frac{1}{\pi^2 x^2} \int_0^\infty |\Gamma(1/2+ip)|^2 W_{0,ip}(2x) f(p) p \sinh(2\pi p) dp \tag{11}$$

and, using the gamma function identity

$$|\Gamma(ip+1/2)|^2 = \frac{\pi}{\cosh(\pi p)} \tag{12}$$

we find:

$$W_{0,\nu}(2z) = \sqrt{\frac{2z}{\pi}} K_\nu(z) \tag{13}$$

$$f(p) = \int_0^\infty \sqrt{\frac{2x}{\pi}} K_{ip}(x)F(x)dx = \sqrt{\frac{2}{\pi}} \int_0^\infty \sqrt{x} K_{ip}(x)F(x)dx \tag{14}$$

$$F(x) = \frac{1}{\pi^2 x^2} \int_0^\infty \frac{\pi p \sinh(\pi p)}{\cosh(\pi p)} \sqrt{\frac{2x}{\pi}} K_{ip}(x) f(p) dp \tag{15}$$

$$= \sqrt{\frac{2x}{\pi}} \frac{2}{\pi x^2} \int_0^\infty p \sinh(\pi p) K_{ip}(x) f(p) dp \tag{16}$$

As the same numerical factor appears on both sides of the transform and inverse it may be moved to either side, similarly with the square root factor, and we may write:

$$f(p) = \int_0^\infty K_{ip}(x)F(x)dx \tag{17}$$

$$F(x) = \frac{2}{\pi^2 x} \int_0^\infty K_{ip}(x) f(p) p \sinh(\pi p) dp \tag{18}$$

This transform pair is known as the Kontorovich-Lebedev transform. It is a crucial part of our calculation method, and a number of our results depend entirely on its computation using some clever tricks. Note that the integration in the second part of the transform pair is over the index of the function, which requires a lot more deft and tact than integrating over the functional variable. This is a delicate matter requiring some involved mathematics which we shall encounter many times throughout this work.

3.1.3 Associated Legendre Functions

The final special function we shall require relates to both the Kontorovich-Lebedev and Whittaker transforms in the following way. If we look at the differential equation for the modified Bessel function at half-integral values:

$$x^2 f_{xx} + 2(1-\alpha)x f_x + (\alpha^2 - \alpha + 1/4 - \frac{1}{4}(1+2k)^2 - x^2) f(x) = 0 \quad (19)$$

which we write as $\mathscr{H} f(x) = 0$. The solution of this differential equation is given by:

$$f(x) = C x^{\alpha - 1/2} K_{k+1/2}(x) \quad (20)$$

where we neglect the second part of the solution due to boundary conditions. The Laplace transform, which we write as

$$\mathscr{F}[f(x)] = F(u) \quad (21)$$

transforms the differential equation such that we have:

$$\mathscr{F}\left[\mathscr{H} f(x)\right] = 0 \quad (22)$$

where the new differential equation is defined through

$$(u^2-1)\frac{\partial^2 F}{\partial u^2} + 2u(\alpha+1)\frac{\partial F}{\partial u} + (\alpha^2 + \alpha - k(k+1))F(u) = 0 \quad (23)$$

and we have neglected boundary terms due to the initial condition, which is permissible for the purposes of this crude analysis. The solution to this differential equation, which is related to spherical harmonic functions, is given by the generalised Legendre function defined by:

$$F(u) = (u^2-1)^{-\alpha/2} C \mathscr{P}_k^{\alpha}(u) \quad (24)$$

Again, we neglect the second solution to the differential equation as it is not useful in the calculation for our boundary conditions. The function $\mathscr{P}^{\alpha}_{ip-1/2}(u)$, i.e. where we have the special values $k(k+1) = -p^2 + 1/4$ is the Mehler-Fock or conical functions and is well known in many applications in physics. It is important to understand that we are taking the function to be defined for the range $[1^+, \infty)$. In an analogous way, the Mehler-Fock transform may be derived from the Kontorovich-Lebedev transform pair by using the following integral relationship [9]:

$$\int_1^{\infty} e^{-xv}(v^2-1)^{-\mu/2} \mathscr{P}^{\mu}_{ip-1/2}(v)dv = \sqrt{\frac{2}{\pi}} x^{\mu-1/2} K_{ip}(x) \quad (25)$$

This represents the result obtained prior through the Laplace transform of the differential equations. If we then take the Kontorovich-Lebedev transform, we will have:

$$f(p) = \int_0^\infty K_{ip}(x)F(x)dx \tag{26}$$

$$= \int_0^\infty F(x)dx\sqrt{\frac{\pi}{2}}x^{-\mu+1/2}\int_1^\infty e^{-xv}(v^2-1)^{-\mu/2}\mathscr{P}^\mu_{ip-1/2}(v)dv \tag{27}$$

$$= \sqrt{\frac{\pi}{2}}\int_1^\infty dv.\mathscr{P}^\mu_{ip-1/2}(v)(v^2-1)^{-\mu/2}\int_0^\infty dx.F(x)x^{-\mu+1/2}e^{-xv} \tag{28}$$

Denoting

$$S_1(v) = (v^2-1)^{-\mu/2}\int_0^\infty dx.F(x)x^{-\mu+1/2}e^{-xv} \tag{29}$$

which is a Mellin-Laplace composition, we have therefore the forward transform:

$$\sqrt{\frac{\pi}{2}}\int_1^\infty dv.\mathscr{P}^\mu_{ip-1/2}(v)S_1(v) = f(p) \tag{30}$$

On the other hand, the inverse transformation may be easily found using the same method, and we obtain:

$$F(x) = \frac{2}{\pi^2 x}\int_0^\infty K_{ip}(x)f(p)p\sinh(\pi p)dp \tag{31}$$

$$K_{ip}(x) = \sqrt{\frac{\pi}{2}}x^{-\mu+1/2}\int_1^\infty e^{-xv}(v^2-1)^{-\mu/2}\mathscr{P}^\mu_{ip-1/2}(v)dv \tag{32}$$

$$F(x) = \frac{2}{\pi^2 x}\int_0^\infty K_{ip}(x)f(p)p\sinh(\pi p)dp \tag{33}$$

$$= \frac{2}{\pi^2 x}\int_0^\infty dp.\sqrt{\frac{\pi}{2}}x^{-\mu+1/2}\int_1^\infty dv.e^{-xv}(v^2-1)^{-\mu/2}\mathscr{P}^\mu_{ip-1/2}(v)f(p)p\sinh(\pi p) \tag{34}$$

$$= \frac{2x^{-\mu-1/2}}{\pi^2}\sqrt{\frac{\pi}{2}}\int_1^\infty dv.e^{-xv}(v^2-1)^{-\mu/2}\int_0^\infty dp.\mathscr{P}^\mu_{ip-1/2}(v)p\sinh(\pi p)f(p) \tag{35}$$

$$F(x) = \int_1^\infty dv.e^{-xv}(v^2-1)^{-\mu/2}S_2(v) \tag{36}$$

The functional equation appearing in the inner integral is the Mehler-Fock transformation. The standard choice of paired, normalised transforms is given by:

$$F(v) = \frac{1}{\pi}\int_0^\infty dp.|\Gamma(1/2-\mu+ip)|\mathscr{P}^\mu_{ip-1/2}(v)p\sinh(\pi p)f(p) \tag{37}$$

with inverse

$$f(p) = \int_1^\infty dv.\mathscr{P}^\mu_{ip-1/2}(v)F(v) \tag{38}$$

Although these formulae appear forbidding, as we shall show, there is a commonality of structure between these different groups of special functions which renders many problems solvable in similar ways. Tables of these transforms are available in some instances, for example [9, 18, 17, 23, 22], however, there is little to no description of a general method for finding these formulae. This paper shall outline in detail how one goes about deriving these necessary integrals for the solution of various problems in diffusion.

3.1.4 Parabolic Cylinder and Hermite Functions

The final type of special functions necessary for understanding the details of the calculations contained herein is the parabolic cylinder function. These are given by the special case of the Whittaker function via the connection formula:

$$W_{\mu+1/4,1/4}(x) = D_{2\mu}\left(\sqrt{2x}\right)\frac{x^{1/4}}{2^{\mu}} \tag{39}$$

Parabolic cylinder functions satisfy the Weber differential equation:

$$f_{zz} + \left(\nu + \frac{1}{2} - \frac{z^2}{4}\right) f(z) = 0 \tag{40}$$

$$f(z) = cD_{\nu}(z) \tag{41}$$

The parabolic cylinder functions and the Hermite polynomials are linked by the formula:

$$D_n(\sqrt{2}x) = 2^{-\nu/2}e^{-x^2/2}H_n(x) \tag{42}$$

Indeed, the common factor in all these groups of special functions is that they all may be described as a limiting case of the confluent hypergeometric function.

3.1.5 Other Sundries

We shall briefly define all the other necessary objects for the calculation in this paper. Let us begin with the representation of scalars. A single valued object (number) without functional dependence shall be denoted λ. A set of eigenvalues with indicial parameter n shall be given the symbol λ_n, and Latin n represents a discrete set as opposed to a continuous set defined through e.g. λ_u. Functional dependence will be illustrated through the use of both discrete $\psi_n(x)$ and continuous $\Phi_u(x)$.

The kernel solution of a diffusion equation shall be written as $\mathscr{K}(x,y;t)$ for the case of two positions and one time. We indicate that there is a difference between the time and space variables by use of the comma. Wherever possible an effort has been made to ensure that the notation used is consistent for each application. The central focus of the paper is on the calculation of these types of kernels in an effort to understand the nature of the dynamical solutions to the diffusion equations.

We shall make a certain definition in terms of style of writing nested integrals. This paper shall encounter many of these types of formulae and they quickly become entangled without a solid start. We make the operational statement that all integral variables are to be written to the left and associated with their limit. For example, the integral:

$$I = \int_0^\infty dx.f(x) \times \int_1^\infty dy.g(x,y) \tag{43}$$

is to be read as the integral of $g(x,y)$ over y, then the result of this times $f(x)$ integrated over x. Many of the calculations in this work stem from the use of the Fubini-Tonelli theorem and the integrals can become lengthy. For this reason, this convention is one we follow out of necessity so as to avoid confusion.

Finally, we must define the idea of a transform. This paper deals with the Whittaker, Kontorovich-Lebedev and Mehler-Fock transforms in depth, and also obliquely touches upon the Laplace, Fourier, cosine, Mellin and sine transforms. In the interests of book-keeping, we have the following approach to offer. When a function is transformed, if it is a single parameter function, we have automatically no confusion if we write:

$$F(p) = \mathscr{F}[f(x)] \tag{44}$$

and, under the axiom of invertability we have:

$$f(x) = \mathscr{F}^{-1}[F(p)] \tag{45}$$

We shall always use the cursive script in this fashion when referring to a transformation of some type. If we have a function with multiple parameter, such as $\mathscr{K}(x,y) = \mathscr{K}(x,y;0)$, it is necessary to specify the co-ordinate over which the transform is occurring. Often the context will make the integration variable in the transform superfluous and we shall omit it for brevity and refer to the type of transform by name, or explicitly define the integrals whenever possible.

4 Basic Diffusion Theory

We may write a generic diffusion equation in the form:

$$-\frac{\partial u}{\partial \tau} = \frac{\sigma^2}{2}\frac{\partial^2 u}{\partial x^2} + \mu\frac{\partial u}{\partial x} + V(x,\tau)u \tag{46}$$

The basic diffusion problem can be solved using a kernel solution:

$$K(x,y;t) = \int_\Omega d\mathfrak{m}(\lambda).e^{-E_\lambda t}\Phi_\lambda^*(x)\Phi_\lambda(y) + \sum_{n=1}^{n_{max}} |C_n|^2 e^{-E_n t}\Psi_n^*(x)\Psi_n(y) \tag{47}$$

Briefly, the moving parts of this expression are the continuous eigenfunction $\Phi_\lambda(y)$, the discrete eigenfunctions $\Psi_n(y)$, the cutoff for the discrete states n_{max}, which may be some finite number or infinity for a purely discrete system, and the measures for the continuous spectrum $\mathfrak{m}(\lambda)$ and discrete states $|C_n|^2$. Applying the method of separation of variables, we can extract the eigenfunctions by solving the time independent differential equations:

$$u(x,\tau) = \phi(x)\Upsilon(\tau)$$

$$-\frac{1}{\Upsilon(\tau)}\frac{d\Upsilon(\tau)}{d\tau} = \frac{1}{\phi(x)}\left(\frac{\sigma^2}{2}\frac{d^2\phi}{dx^2} + \mu\frac{d\phi}{dx} + V(x,\tau)\phi(x)\right) \tag{48}$$

Assuming a time-independent potential/killing function $V(x,\tau) = V(x)$, it is straightforward to write down the eigenfunction solution for the PDE system, which enables us to determine the kernel.

$$\frac{\sigma^2}{2}\frac{d^2\phi}{dx^2} + \mu\frac{d\phi}{dx} + V(x)\phi(x) = E\phi(x) \tag{49}$$

$$\frac{d\Upsilon(\tau)}{d\tau} = -E\Upsilon(\tau) \tag{50}$$

It is clear that to understand the nature of the solution space, we need to determine the spectrum of the eigenvalues via the constant E.

5 Completeness of Eigenfunction Representations

The fundamental notion which we shall now exploit relates to the completeness of basis sets, and the relationship with orthonormal eigenfunctions. The spectral theory of the kernel states that we can find a set of eigenfunctions with the completeness property expressed through the delta function expansion:

$$\delta(x-y) = \int_\Omega d\mathfrak{m}(\lambda).\Phi_\lambda^*(x)\Phi_\lambda(y) + \sum_{n=1}^{n_{max}} |C_n|^2\Psi_n^*(x)\Psi_n(y) \tag{51}$$

The integral in this case is over the continuous eigenvalue; the sum over the discrete set of states. If we are able to determine the delta function expansion, we can find the kernel in a similar way. If we succeed in this, we will have solved the diffusion problem, as we may then construct the kernel solution which encodes all the measurable information in the system.

The two parts which compose the delta function are the continuous and discrete parts of the state. The continuous state, defined through the eigenfunction $\Phi_\lambda(y)$, has measure $\mathfrak{m}(\lambda)$. In particular, many of the situations considered in this paper have a single continuous eigenfunction defined in a way which is analogous to a scattered state in quantum physics. In the same way that the Born equation gives the solution to the quantum equations for a particle scattered to infinity, the continuous state in our situation will often have an integral measure which is integrated over the positive part of the plane or section thereof. The second part of the completeness relationship is the discrete states; the upper bound n_{max} in this case takes into account the generalised example whereby there may be a finite rather than infinite set of orthogonal polynomials that span the discrete set of eigenfunctions.

6 Potential Theory

If we consider contracts which have positive prices, it is clear that if the process is positive valued, then the average is by necessity a positive valued function. By a similar logic, the potential or killing function $V(x, \tau)$ must necessarily have values defined in the positive plane. Two examples are given in the potential function sketched below:

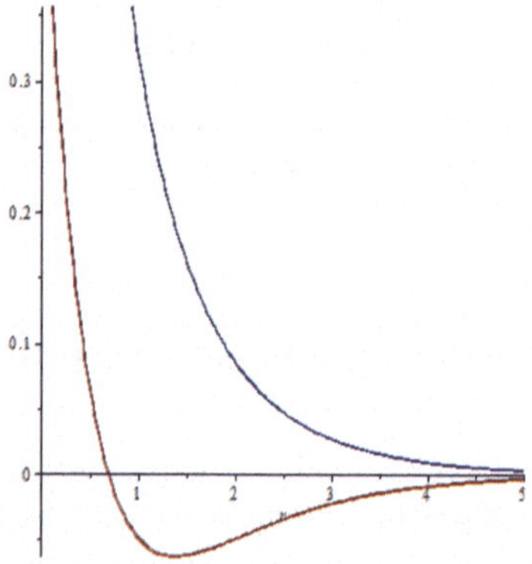

Fig. 1: Potential functions for the positive plane

The upper (blue) line in this case represents a potential which generates a continuous eigenstate. A particle released at any point on this curve will travel to infinity. The lower (red) line gives a different situation. Particles released above the x-axis will travel to infinity, but a particle released below the axis will oscillate. The spectrum in the first case will be a single continuous eigenvalue, and in the latter we will have a spectrum with both a continuous and a discrete part. In many ways we may understand these two different scenarios as analogous to scattered and bound eigenstates in quantum theory.

We shall now show a simple technique that we can use to find simplified, transformed versions of the basic diffusion equations. The fundamental diffusion may be written:

$$a(x)\frac{d^2\phi}{dx^2} + b(x)\frac{d\phi}{dx} + c(x)\phi(x) = 0 \tag{52}$$

In terms of the drift and diffusion, we have:

$$a(x) = \frac{\sigma^2(x)}{2}, b(x) = \mu(x), c(x) = V(x) - E \tag{53}$$

and the Bose invariant, defined through:

$$\frac{d^2\hat{\phi}}{dx^2} + \mathscr{I}(x)\hat{\phi}(x) = 0 \tag{54}$$

$$\mathscr{I}(x) = \frac{4ac - b^2 + 2(ba' - ab')}{4a^2} \tag{55}$$

The basic transformations of the eigenfunction are given by the speed and scale factors, defined through the fundamental symmetry relation:

$$[a(x)]^{-1} h(x)\mathscr{H}(x,\partial_x)\,[h(x)]^{-1} = \partial_x^2 + \mathscr{I}(x) \tag{56}$$

where the generator $\mathscr{H}(x,\partial_x)$ is given by:

$$\mathscr{H}(x,\partial_x) = a(x)\frac{\partial^2}{\partial x^2} + b(x)\frac{\partial}{\partial x} + c(x) \tag{57}$$

and the transformed eigenfunction is defined through the scale factor:

$$\phi(x) = \frac{\hat{\phi}(x)}{h(x)} \tag{58}$$

$$h(x) = \exp\left(\int^x \frac{b(s)}{2a(s)}ds\right) \tag{59}$$

Indeed, it is simple to see that the application of the Bose invariant to any diffusion problem results in a plain Gaussian diffusion, plus a transformed potential or killing function. This is a very powerful method as we shall show; the speed and scale measures of Feller [6] are obtained through:

$$dm(x) = m(x)dx = \frac{e^{B(x)}}{a(x)}dx \tag{60}$$

$$s(x) = e^{-B(x)} \tag{61}$$

$$B(x) = \int^x \frac{b(s)}{a(s)}d\tau \tag{62}$$

and, as we shall show, many of the integration measures for the diffusion processes which are important in this paper are related to these distributions. Finally, the killed measure is given by the formula:

$$k(y) = \frac{c(y)e^{B(y)}}{4a(y)} \tag{63}$$

Note that the killing measure is essentially just the killing potential multiplied by the scale factor. For an excellent discussion of these topics, the interested reader is directed to the works of Kuznetzov and Albanese [1], also Sousa and Yakubovich [24] which contains an excellent discussion of the construction of kernels using the Titchmarsh-Kodaira theorem. Our argument shall differ from theirs in that we shall be examining the case where the continuous and discrete parts of the spectrum must both be taken into account.

7 Asian Options Pricing Formulae

We shall report on some results that are available in the literature. Geman and Yor [8] gave the canonical result for the fixed strike call option by exploiting the scaling property of Brownian motion via:

$$\mathbb{E}_T\left[\left(\frac{1}{T-t_0}\int_{t_0}^{T} S_u du - K\right)^+\right] \tag{64}$$

$$= \frac{4S_t e^{-r(T-t)}}{(T-t_0)\sigma^2}\mathbb{E}^{\mathbb{Q}}_{\sigma^2 s/4}\left[\left(\int_0^{\sigma^2 s/4} \exp\left(2W_{s'} + \frac{4}{\sigma^2}\nu s'\right) ds' - \frac{\mathcal{K}\sigma^2}{4}\right)^+\right] \tag{65}$$

We may price call options for a fixed strike by using the transition probability density:

$$C(T,0;S_0,K) = \frac{4S_0 e^{-rT}}{T\sigma^2}\int_0^{\infty}(u - \frac{\mathcal{K}\sigma^2}{4})^+ f(u,0;\tau)du \tag{66}$$

where the fundamental diffusion equation is defined through:

$$2u^2\frac{\partial^2 f}{\partial u^2} + [(2\zeta+1)u+1]\frac{\partial f}{\partial u} = -\frac{\partial f}{\partial t} \tag{67}$$

and the variable u is the exponential Brownian motion:

$$u = \int_0^{\sigma^2 s/4} \exp\left(2\left[W_{s'} + \zeta s'\right]\right) ds' \tag{68}$$

Other known results include the floating strike call option formula, developed in [5], which can be easily derived via the following means. The price of the option is defined through the payoff function:

$$C_f(S_T, A_T, T) = \max\left(S_T - \frac{A_T}{T}, 0\right) \tag{69}$$

from which it is possible to show that diffusion equation that relates to this system is given through the degenerate parabolic equation:

$$\frac{\partial C_f}{\partial t} + \frac{\sigma^2 S^2}{2}\frac{\partial^2 C_f}{\partial S^2} + S\frac{\partial C_f}{\partial \xi} + rS\frac{\partial C_f}{\partial S} - rC_f = 0 \tag{70}$$

To solve this expression, as in [5], one may employ the ansatz $C_f(S,A,t) = Sf(z,t)$ Again, we find the differential equation which defines the probability density:

$$-\frac{\partial f}{\partial t} + \frac{1}{2}\sigma^2 z^2 \frac{\partial^2 f}{\partial z^2} - \left(\frac{1}{T} + rz\right)\frac{\partial f}{\partial z} = 0 \tag{71}$$

The redundant variable in this case is defined through the time average in a similar way to the Yor system, via:

$$\xi_t = \int_0^t S_u du \tag{72}$$

The change of variables in this case is given by the relationship:

$$z_t = \frac{1}{S}(k - \xi_t) = \frac{1}{S}\left(k - \frac{A_t}{T}\right) \tag{73}$$

8 Application of the Bose Invariant

If we can solve either the system of Craddock [5] or Yor [8], it is possible to find the solution of the other problem. We shall now show how to directly address the Yor problem using the Bose invariant. In an analogous way to the fixed strike, the scale factor for the Yor equation is readily shown to be inverse gamma distributed, via:

$$h(u) = \exp\left(\int^u \frac{[(2\zeta+1)s+1]}{4s^2}ds\right) = e^{-1/(4u)}u^{1/4+\zeta/2} \tag{74}$$

Evaluating the isomorphism of the generator of the diffusion, we find:

$$\frac{1}{2u^2}e^{-1/(4u)}u^{1/4+\zeta/2}\mathscr{L}\left(e^{1/(4u)}u^{-1/4-\zeta/2}f(u)\right) = \partial_u^2 + \mathscr{I}(u) \tag{75}$$

$$2u^2\frac{\partial^2 f}{\partial u^2} + [(2\zeta+1)u+1]\frac{\partial f}{\partial u} = Ef \Leftrightarrow \mathscr{L}f = Ef \tag{76}$$

We recover the Bose invariant potential, either by using the scale factor, or the formula using the PDE coefficients:

$$\mathscr{I}(u) = \frac{1}{16u^4} + \frac{\left(-\frac{\zeta}{4}+\frac{3}{8}\right)}{u^3} + \frac{\left(-\frac{\zeta^2}{4}+\frac{\zeta}{4}-\frac{E}{2}+\frac{3}{16}\right)}{u^2}$$
$$= \frac{C_1}{u^2} + \frac{C_2}{u^3} + \frac{C_3}{u^4} \tag{77}$$

The solution to the PDE system is then given by the Whittaker-type confluent hypergeometric function:

$$\phi(u) = \frac{\hat{\phi}(u)}{h(u)} \sim Ce^{1/(4u)} u^{3/4-\zeta/2} W_{\alpha,ip}\left(\frac{1}{2u}\right) \tag{78}$$

The scalar parameters in this expression are defined through:

$$\alpha = \frac{3}{4} - \frac{\zeta}{2}, ip = \frac{1}{4}\sqrt{4(\zeta^2 - \zeta) + 8E + 1} \tag{79}$$

If we use the inverse variable $x = (2u)^{-1}$ we recover the Whittaker equation directly. Further transformations give the Morse potential, viz.:

$$\frac{\partial^2 F(x)}{\partial x^2} + \left(\frac{C_1}{x^2} + \frac{C_2}{x} + C_3\right) F(x) = 0 \tag{80}$$

where the Whittaker-type solution may be understood through the spherical relation

$$F(x) = uF(u^{-1}) \tag{81}$$

. The Morse potential is given by the substitution $u = e^{-r}$ where the PDE system in the new variables is given by:

$$\left(\partial_r^2 + \mathscr{J}(r)\right)\Phi(r) = 0 \tag{82}$$

$$\mathscr{J}(r) = \frac{\gamma^2}{4}e^{-2r} - \frac{\nu}{2}e^{-r} + \frac{\alpha}{4} - \frac{1}{2} \tag{83}$$

This is, indeed, the potential function described in the previous sections. For various different regimes of the parameters, we are able to tune between a purely scattering potential and a bound potential state. The spectrum is then given by a direct sum over the two contributions from the discrete and continuous parts of the state.

9 Spectral Theory of Whittaker/Morse Potential

The potential derived in the previous section can be related to the spectral theory of the kernel. The standard form of the Whittaker equation can be written as:

$$\left(-\partial_z^2 + \frac{\gamma^2}{4} - \frac{\nu}{2z} + \left(\frac{\alpha}{4} - \frac{1}{4}\right)\frac{1}{z^2}\right)\phi(z) = 0 \tag{84}$$

The potential function in this case is given by:

$$V(z) = \frac{\gamma^2}{4} - \frac{\nu}{2z} + \left(\frac{\alpha}{4} - \frac{1}{4}\right)\frac{1}{z^2} \tag{85}$$

Note the generalised problem defined through the killed process:

$$\frac{\partial f}{\partial t} = \frac{\partial^2 f}{\partial x^2} - V(x)f(x,t) \tag{86}$$

As before, we have the expression for the kernel:

$$K(x,y;t) = \int_\Omega d\mathfrak{m}(\lambda).e^{-E_\lambda t}\Phi_\lambda^*(x)\Phi_\lambda(y) + \sum_{n=1}^{n_{max}} |C_n|^2 e^{-E_n t}\Psi_n^*(x)\Psi_n(y) \tag{87}$$

Now, it is known that the Green's function is given by the Laplace transform of the kernel via:

$$G(x,y;E) = \int_0^\infty e^{-Et}K(x,y;t)dt \tag{88}$$

where, for the Whittaker potential, a known result from [11] gives the Green's function as:

$$\pi^2 \frac{\Gamma(1/2-\kappa+s)}{\Gamma(1+2s)} W_{\kappa,s}(x_>)M_{\kappa,s}(x_<)$$

$$= \int_0^\infty \frac{u\sinh(2\pi u)\Gamma(1/2-\kappa-iu)\Gamma(1/2-\kappa+iu)}{s^2+u^2} W_{\kappa,iu}(x)W_{\kappa,iu}(y)du \tag{89}$$

$$+\pi^2 e^{-(x+y)/2}\sum_{n=0}^{[Re(\kappa)-1/2]} \frac{\alpha n!}{\Gamma(\alpha+n+1)} \frac{(xy)^{(\alpha+1)/2}}{s^2-\alpha^2/4} L_n^{(\alpha)}(x)L_n^{(\alpha)}(y) \tag{90}$$

The first part of this expression gives the continuous eigenstate, and the second part is therefore the discrete component. The eigenvalues are related by:

$$\alpha = 2\kappa - 2n - 1 \tag{91}$$

Concisely, the Green's function may be written as the compact product:

$$G(x,y;E) = Ce^{-(x+y)/2}W_{\kappa,s}(x_<)M_{\kappa,s}(x_>) \tag{92}$$

10 Discrete Green's Function and Bessel Polynomials

We shall now discuss the discrete part of the Green's function as given by the Bessel polynomials [19]. These polynomials differ from the classical orthogonal polynomials, as for a fixed total energy, we have only a finite set of orthogonal polynomials in the positive plane. This is to be compared with e.g. the Hermite polynomials, which possess an infinite dimensional set of orthogonal eigenfunctions. The basic expression for the Bessel polynomials is given in [12, 14] as the Laguerre-type formula:

$$y_n(x;\alpha,\beta) = (-1)^n n! \left(\frac{x}{\beta}\right)^n L_n^{(1-2n-\alpha)}\left(\frac{\beta}{x}\right) \tag{93}$$

One confluent form of the Bessel polynomials is given by:

$$y_n(x;\alpha+1,1/2) = (-1)^n 2^n x^n n! L_n^{(-2n-\alpha)}\left(\frac{1}{2x}\right) \tag{94}$$

which satisfies the differential equation:

$$2x^2\frac{d^2 f}{dx^2} + (2(\alpha+1)x+1)\frac{df}{dx} - 2\left[n(n+\alpha)\right] f = 0 \tag{95}$$

Note that the eigenstates for this differential system are given through:

$$\Psi_n(x) = C_n^{\alpha} y_n(x;\alpha+1/2,1/2) \tag{96}$$

and we see that the differential equation of interest can be easily related to the Asian option diffusion equation:

$$2x^2\frac{d^2 f}{dx^2} + ((2\alpha+1)x+1)\frac{df}{dx} = 2\left[n(n+\alpha)\right] f \tag{97}$$

11 Continuous Part of the Whittaker Kernel

If we examine the pricing diffusion for the Asian option, following the results in the previous sections, we have:

$$2x^2\frac{d^2 f}{dx^2} + ((2\alpha+1)x+1)\frac{df}{dx} = Ef \tag{98}$$

In this case, the energy eigenvalue is given through the relation:

$$E = \frac{1}{2}\alpha(1-\alpha) - \frac{p^2}{2} - \frac{1}{8} \tag{99}$$

with continuous eigenfunction solution:

$$\Phi_p(x) = C(p)e^{1/(4x)}x^{\kappa}W_{\kappa,ip/2}\left(\frac{1}{2x}\right) \tag{100}$$

$$\kappa = -\alpha/2 + 3/4 \tag{101}$$

We shall now show how one may evaluate the measure of the space. The scale measure is given through the relationship [4]:

$$dm(x) = m(x)dx = \frac{e^{B(x)}}{a(x)}dx \tag{102}$$

$$B(x) = \int^{x} \frac{b(s)}{a(s)} ds \tag{103}$$

Evaluating for the diffusion equation, we find the inverse gamma distribution:

$$m(x) = \frac{1}{2} e^{-1/(2x)} x^{-3/2+\alpha} \tag{104}$$

Writing out the continuous part of the kernel, we find the expression:

$$K_c(x,y;\tau) = \int_0^{\infty} e^{-E(p)\tau} \Phi_p^*(x) \Phi_p(y) d\mathfrak{m}(p) \tag{105}$$

Now, using the symmetries of the Whittaker function, it is not too hard to see that we have the relationship:

$$\Phi_p^*(x) = \Phi_{-p}(x) = \Phi_p(x) \tag{106}$$

which may be rewritten for our eigenfunction as:

$$W_{\kappa,ip}^*(z) = W_{\kappa,-ip}(z) = W_{\kappa,ip}(z) \tag{107}$$

The principal aim is then to recover a completeness relationship of the type:

$$K_c(x,y;0) = \lambda \delta(x-y) \tag{108}$$

Following results found in [25], a known formula for orthonormality of Whittaker functions is given by the delta function relation:

$$\int_0^{\infty} C_{\kappa,\mu}^2 \frac{W_{\kappa,i\mu}(x) W_{\kappa,i\mu'}(x)}{x^2} dx = \delta(\mu - \mu') \tag{109}$$

and the measure distribution is then given by:

$$d\mathfrak{m} = C_{\kappa,\mu}^2 \frac{dx}{x^2} = \frac{\mu \sinh(2\pi\mu) \Gamma(1/2 - \kappa + i\mu) \Gamma(1/2 - \kappa - i\mu)}{\pi^2 x^2} \tag{110}$$

Simplification using the properties of the gamma function then gives us the compact relation:

$$d\mathfrak{m}(p) = \frac{1}{\pi^2 x^2} p \sinh(2\pi p) |\Gamma(1/2 - \kappa + ip)|^2 dp \tag{111}$$

from which we can evaluate the continuous part of the kernel as:

$$K_c(x,y;\tau) = C_\kappa \int_0^{\infty} e^{-p^2\tau/2} W_{\kappa,ip}\left(\frac{1}{2x}\right) W_{\kappa,ip}\left(\frac{1}{2y}\right) d\mathfrak{m}(p) \tag{112}$$

Finally, by using inverse coordinate transforms, we find the simplified form of the kernel:

$$\tilde{K}_c(x,y;\tau) = K_c((2x)^{-1}, (2y)^{-1}; \tau) \tag{113}$$

$$= \int_0^\infty e^{-2p^2\tau} W_{\kappa,ip}(x) W_{\kappa,ip}(y)\, d\mathfrak{m}(p) \tag{114}$$

These are the basic properties of the continuous spectral solution. We note that this is the important contribution in the parameter space defined by financial contracts. However, in contrast to say e.g. the results in [24], we take the position that for full understanding of the kernel solution, both the discrete and continuous parts must be taken into account. We shall discuss in the following sections an extension of the Titchmarsh-Kodaira theorem according to this scenario and some ways in which the results of Sousa and Yakubovich may be modified to derive a similar result.

12 Pricing Asian Options via Laguerre Interpolation

We shall now calculate Asian options prices using the kernel formulae developed through this type of analysis. The papers of Linetsky [15], Geman and Yor [8], and Sousa and Yakubovich [24] are of use in developing the necessary stochastic calculus to specify the pricing formula. For completeness, we shall briefly review the basic equations, and then show how one can use techniques from classical numerical analysis to construct various approximations for pricing this option. The basic analysis begins with the exponential Brownian motion:

$$A_\tau^{(\nu)} = \int_0^\tau e^{2(B_u + \nu u)} du \tag{115}$$

Put and call prices are computed using the time-invariant solution to the differential equation [8, 15]:

$$P^{(\nu)}(k,\tau) = \mathbb{E}\left[(k - A_\tau^{(\nu)})^+\right] \tag{116}$$

$$V_p = e^{-rT}\mathbb{E}\left[(K - \mathscr{A}_T)^+\right] = e^{-rT}\left(\frac{4S_0}{\sigma^2 T}\right) P^{(\nu)}(k,\tau) \tag{117}$$

$$V_c = e^{-rT}\mathbb{E}\left[(\mathscr{A}_T - K)^+\right] = V_p + \frac{1 - e^{-rT}}{rT} S_0 - e^{-rT} K \tag{118}$$

Using the results from the previous sections, it is possible to evaluate the price of the Asian option via:

$$P^{(\nu)}(k,\tau) = e^{-\nu^2\tau/2} \frac{(2k)^{-\kappa} e^{-1/(4k)}}{8\pi^2} \int_0^\infty e^{-p^2\tau/2} W_{\kappa,ip/2}\left(\frac{1}{2k}\right) d\mathfrak{m} + .. \tag{119}$$

where the measure is defined by:

$$d\mathfrak{m} = \left|\Gamma\left(\frac{\nu + ip}{2}\right)\right|^2 \sinh \pi p . p dp \tag{120}$$

which we derived previously. Various integrals give interesting relationships between the relevant groups of special functions, viz.:

$$\frac{1}{2\pi i}\int_{-i\infty}^{+i\infty} e^{\xi q} q^{-\nu/2} K_{2ip}\left(\sqrt{8qx}\right) dq = \frac{1}{\sqrt{8e^{x}}}\xi^{(\nu-1)/2} e^{-x/\xi} W_{(1-\nu)/2, ip/2}\left(\frac{2x}{\xi}\right)$$

where the scalar parameters are defined through:

$$\nu = \frac{2r}{\sigma^2} - 1, k = \frac{\tau K}{S_0}, \tau = \frac{T\sigma^2}{4}, \kappa = -\frac{(\nu+3)}{2}$$

We shall now show how these formulae may be computed using Laguerre quadrature of the integrals implied through the expectation value.

$$P^{(\nu)}(k,\tau) = \int_0^\infty f(p)dp \approx \sum_{i=2}^{N-1} f(p_i)\Delta p - \frac{1}{2}[f(p_1) - f(p_N)]\Delta p$$

The integral implied through the option price can be approximated using the trapezoidal rule. However, for Laguerre interpolation we use the set of Lagrange points:

$$\int_a^b f(x)W(x)dx = \sum_{i=1}^{n} w_i(x_i) f(x_i) \tag{121}$$

where the weights give the integral-sum relations via:

$$\int_0^\infty f(x)e^{-x}dx = \sum_{i=1}^{n} w_i f(x_i) \tag{122}$$

and are defined by the Laguerre points:

$$w_i = \frac{1}{(n+1)^2 [L_{n+1}(x_i)]^2} \tag{123}$$

In terms of the exponential geometric Brownian motion, by virtue of the positive nature of the GBM, we know that the average can only be a positive non-zero process. This makes Laguerre or generalised Laguerre quadrature the appropriate choice in this situation, because this is also defined over the positive part of the plane. Other systems [15] rely on a spectral decomposition over the zeroes of the eigenfunction.

$$P_b^{(\nu)}(k,\tau) = \sum_{n=1}^{\infty} \frac{e^{-\lambda_{n,b} t}}{\langle \Psi, \Psi \rangle_{p_{n,b}}} \int_0^k (k-y)\Psi(y;\lambda_{n,b})\mathfrak{m}(y)dy \tag{124}$$

There is the question of convergence of the spectral method especially in the low volatility/low drift case. We expect to recover the heat equation for some limit. This is particularly relevant to markets today in an economy experiencing low interest rates for a prolonged time. A more efficient way of establishing a numerical pricing scheme is by using quadrature of the integral:

$$\int_a^b f(x)W(x)dx = \sum_{i=1}^{n} w_i(x_i)f(x_i) \tag{125}$$

Gauss-Laguerre quadrature has the weighting function:

$$\int_0^\infty f(x)e^{-x}dx = \sum_{i=1}^{n} w_i f(x_i) \tag{126}$$

Weights are readily calculated using the orthogonality property of Laguerre polynomials:

$$w_i = \frac{1}{(n+1)^2[L_{n+1}(x_i)]^2} \tag{127}$$

To compute the integral, it is important to recognise that numerically stable forms of the confluent hypergeometric function are required.

$$U(a,b,x) = \frac{\pi}{\sin \pi b}\left(\frac{M(a,b,x)}{\Gamma(1+a-b)\Gamma(b)} + x^{1-b}\frac{M(1+a-b,2-b,x)}{\Gamma(a)\Gamma(2-b)}\right)$$

$$M(a,b,x) = e^x M(b-a,b,-x) \tag{128}$$

$$\tilde{W}_{\kappa,\mu}(z) = e^{-z/2} z^{b+1/2} \tilde{U}\left(b-a+1/2, 1+2b, z\right) \tag{129}$$

The parameters used in the integral are:

$$z = \frac{1}{2k}, \kappa = -\frac{(\nu+3)}{2}, \mu = \frac{ip}{2} \tag{130}$$

With these formulae for the Whittaker function, we are able to recover prices for Asian options in a variety of regimes. The final ingredients in this mathematical algorithm are the precision requirements of the integral. Using the concept of exponential addition and multiplication for extra precision, it was simple to control any numerical errors in the procedure:

$$P^{(\nu)}(k,\tau) = \int_0^\infty g(p;\nu,\tau,k)dp \tag{131}$$

$$g(p;\nu,\tau,k) = \Re\mathrm{e}\left(\frac{V_4}{8\pi^2}\exp\left(\ln V_1 + \ln V_2 + \ln V_3\right)\right) \tag{132}$$

We were able to demonstrate convergence for a series of well-known test cases [15], however the low volatily case was shown to require extra care. The current R package (fAsianOptions, now defunct) fails on several test cases due to limitations in precision. The cases from [15] are outlined in the table below.

12.0.1 Parameters for Test Cases from [15]

Case	r	σ	T	S_0	ν	τ	k
1	0.02	0.10	1	2.0	3	0.0025	0.0025
2	0.18	0.30	1	2.0	3	0.0225	0.0225
3	0.0125	0.25	2	2.0	−0.6	0.03125	0.03125
4	0.05	0.50	1	1.9	−0.6	0.0625	0.065789
5	0.05	0.50	1	2.0	−0.6	0.0625	0.625
6	0.05	0.50	1	2.1	−0.6	0.0625	0.059524
7	0.05	0.50	2	2.0	−0.6	0.125	0.125

Note that these figures are by no means an exhaustive search over the parameter space. The final three columns, given by the parameters ν, τ, k are derived from the first set of parameters, which define the stochastic system. It isn't intuitively clear what distinguishes the first case as being difficult and less likely to converge, other than the combined situation of both low drift, as expressed through the parameter r, as well as a smaller value for the volatility, as given by σ. The following tables compare the results for various methods of calculating Asian option prices.

12.0.2 Table 1- Asian Option Call Prices

Case	Spectral Method	Trapezoidal Rule	Vecer Method	Monte Carlo
1	14.7694	14.7694	0.0559843109	0.055929(24)
2	0.2183875465	0.2183875465	0.2183869858	0.21704(83)
3	0.1722687410	0.1722687383	0.1722682714	0.173279(89)
4	0.193173790	0.193145214	0.1931716834	0.1922(11)
5	0.246415690	0.246397641	0.2464144912	0.2450(13)
6	0.306220364	0.306208936	0.3062195980	0.3055(14)
7	0.3500952189	0.348084942	0.3498467545	0.3496(20)

12.0.3 Asian Options Prices via Laguerre Quadrature

	Laguerre	
Case	Put	Call
1	9.55*10^5	9.55*10^5
1*(200 points)	0.0362615331589084254661	0.0559968558698675989430
2	0.0585969850989125223683	0.2183875465955682429392
3	0.1476815247399004291038	0.1722687384166216016910
4	0.2423229661816367419105	0.1931459861530763782535
5	0.1980339582371130057155	0.2463981292071246238756
6	0.1603039232499463874824	0.3062092452185299874597
7	0.2545623438519754839573	0.3481391470608648743438

As we can see, with the exception of the pathological case 1, the Laguerre method is highly accurate. This, while not particularly useful for quantitative purposes, may be of relevance to physical situations where the concept of exponential Brownian motion is of interest. To the author's knowledge, these are the most accurate figures produced for these types of options prices and form a useful sandbox test set for validating other models.

The low volatility case is specified by the parameter set:

$$r = 0.02, \sigma = 0.1, T = 1, K = 2.0 \tag{133}$$

$$S_0 = 2.0, \nu = 3, \tau = 0.0025, k = 0.0025 \tag{134}$$

Understanding why this case is difficult when the values seem reasonable is the challenge in deriving formulae for Asian options pricing. The low volatility/low drift approximation of equation can be expected to have some quirks due to limiting behaviour of Whittaker function. We ran some simulations to compare this method against standard techniques, inc. Monte Carlo, spectral theory, discretised PDE, other integral approximations. As the data shows, it seems to hold up well across a variety of situations- with the caveat of an error control term. We can define our convergence region via the difference between a Monte Carlo sim with a known tolerance and the output from our formula. Numerical experiments allow us to probe this region for different parameter sets and to find upper bounds on the number of abscissa points in the quadrature. The chart below shows this convergence for the pathological case as discussed in detail in the previous sections.

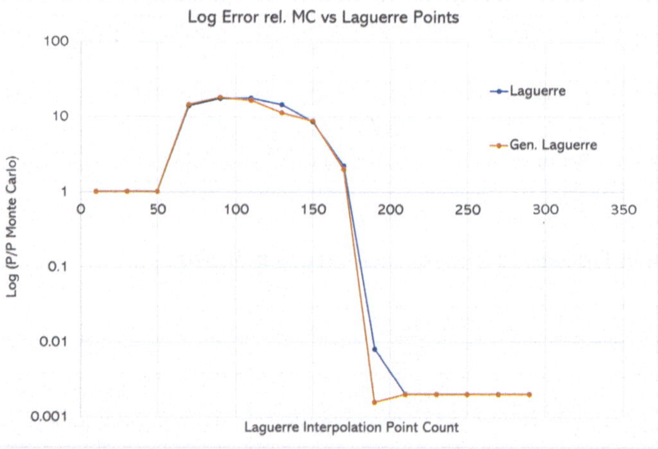

Fig. 2: Convergence for Case 1- Low volatility, low drift

13 Spectral Expansions

We shall now discuss some generalities related to the calculation in the previous sections. Sousa and Yakubovich [24] give the expression for the Titchmarsh-Kodaira theorem via the integral transform pair:

$$(\mathscr{F} f)_i(\lambda) = \tilde{f}_i(\lambda) = \int_a^b f(x) w_i^*(x,\lambda) r(x) dx \tag{135}$$

$$(\mathscr{F}^{-1}\tilde{f})(x) = \int_{0^-}^{\infty} \sum_{i,j=1}^{2} \tilde{f}_i(\lambda) w_j(x,\lambda) d\rho_{ij}(\lambda) \tag{136}$$

The measure in the inverse transform is given by the Titchmarsh-Kodaira theorem:

$$\rho_{ij}(\lambda_1,\lambda_2) = \lim_{\delta\to 0}\lim_{\varepsilon\to 0} \frac{1}{2\pi i}\int_{\lambda_1+\delta}^{\lambda_2-\delta} \left[m_{ij}^{+}(\lambda+i\varepsilon) - m_{ij}^{+}(\lambda-i\varepsilon)\right] d\lambda \tag{137}$$

$$= \lim_{\delta\to 0}\lim_{\varepsilon\to 0} \frac{1}{2\pi i}\int_{\lambda_1+\delta}^{\lambda_2-\delta} \left[m_{ij}^{-}(\lambda+i\varepsilon) - m_{ij}^{-}(\lambda-i\varepsilon)\right] d\lambda \tag{138}$$

where the kernel of the system is defined through the eigenfunction expansion:

$$K(x,y;\lambda) = \sum_{i,j=1}^{2} m_{ij}^{-}(\lambda) w_i(x,\lambda) w_j(y,\lambda^*) \tag{139}$$

where $x < y$

$$K(x,y;\lambda) = \sum_{i,j=1}^{2} m_{ij}^{+}(\lambda) w_i(x,\lambda) w_j^*(y,\lambda^*) \tag{140}$$

analogously, where $x > y$ These different representations give the complex spherical and hyperbolic decompositions for the kernel. The functions $w_i(x,\lambda)$ are naturally then associated with the eigenfunctions of the differential equations which define the system, via the solutions of the equation for the kernel $\mathscr{L}u = \lambda u$. From this definition one can go on to generate various representations for the Green's function, resolvent and other such operators relevant to the study of such differential equations.

13.1 Heat Kernel for Hermite Polynomials

We shall now briefly comment on the natural generalisation of this type of system. Note that this is only strictly true for a system with a single continuous degree of freedom. As in the previous examples we have examined, a simple system is given by the Kontorovich-Lebedev transform, where the continuous eigenvalue represents a continuum of scattered states. However, in the case where we consider the heat equation equivalent of the harmonic oscillator, it is simple to see that the relevant diffusion equation is defined through:

$$-\frac{\partial u}{\partial t} = -\frac{1}{2}\frac{\partial^2 u}{\partial x^2} + \frac{1}{2}x^2 u \tag{141}$$

hence the eigenvalue equation is given by:

$$E\phi(x) = -\frac{1}{2}\frac{d^2\phi}{dx^2} + \frac{1}{2}x^2\phi \tag{142}$$

with solution:

$$\phi(x) = C_1\phi_1(x) + C_2\phi_2(x) \tag{143}$$

$$\phi_1(x) = \frac{M_{E/2,1/4}(x^2)}{\sqrt{x}} \tag{144}$$

$$= -\frac{1}{4}\frac{\Gamma(1/4 - E/2)}{\sqrt{\pi}2^{E/2-1/4}}x^{3/2}\left(D_{E-1/2}(\sqrt{2}x) - D_{E-1/2}(-\sqrt{2}x)\right) \tag{145}$$

$$\phi_2(x) = \frac{W_{E/2,1/4}(x^2)}{\sqrt{x}} = \frac{H_{E-1/2}(x)}{2^{E-1/2}}e^{-x^2/2} = \frac{D_{E-1/2}(\sqrt{2}x)}{2^{E/2-1/4}} \tag{146}$$

If we consider known results from quantum mechanics [7], it is clear that the physical solution is defined through the function $\phi_2(x)$; to quantise the system in the appropriate fashion we must therefore have $H_{E-1/2}(x) = H_n(x)$, hence the energy eigenstates are defined by:

$$E_n = n + \frac{1}{2} \tag{147}$$

which, up to scaling, gives the correct equipartition law for blackbody radiation. If we then calculate the nature of the first solution, we find it must take the form:

$$\phi_1(x) = -\frac{1}{4}\frac{\Gamma(-n/2)}{\sqrt{\pi}2^{n/2}}\left(D_n(\sqrt{2}x) - D_n(-\sqrt{2}x)\right) \tag{148}$$

and it is obvious that this represents a solution that is scattered in a forwards and backwards direction in space. Returning to the physical solution, the Mehler kernel is given by the infinite sum:

$$K(x,y;t) = \sum_{n=0}^{\infty} e^{-E_n t}\psi_n(x)\psi_n^*(y) \tag{149}$$

whence, upon simplification

$$K(x,y;t) = \sum_{n=0}^{\infty} e^{-(n+1/2)t}|c_n|^2 H_n(x)H_n(y)e^{-(x^2+y^2)/2} \tag{150}$$

The normalisation of the eigenfunctions is defined through:

$$\int_0^\infty |\psi_n(x)|^2 dx = |c_n|^2 \int_0^\infty H_n^2(x) e^{-x^2} dx = 1 \tag{151}$$

$$\int_0^\infty H_n^2(x) e^{-x^2} dx = \sqrt{\pi} 2^n n! \tag{152}$$

$$|c_n|^2 = \frac{1}{\sqrt{\pi} 2^n n!} \tag{153}$$

Writing out the kernel, we find:

$$K(x,y;t) = \sum_{n=0}^{\infty} \frac{e^{-(n+1/2)t}}{\sqrt{\pi} 2^n n!} H_n(x) H_n(y) e^{-(x^2+y^2)/2}$$

$$= \frac{e^{-t/2}}{\sqrt{\pi}} \sum_{n=0}^{\infty} \frac{(e^{-t}/2)^n}{n!} H_n(x) H_n(y) e^{-(x^2+y^2)/2} \tag{154}$$

$$\sum_{n=0}^{\infty} \frac{(\rho/2)^n}{n!} H_n(x) H_n(y) e^{-(x^2+y^2)/2} = \frac{1}{\sqrt{1-\rho^2}} \exp\left(\frac{4xy\rho - (1+\rho^2)(x^2+y^2)}{2(1-\rho^2)} \right) \tag{155}$$

Using $\rho = e^{-t}$, the kernel is found via the concise formula:

$$K(x,y;t) = \frac{e^{-t/2}}{\sqrt{\pi}} \frac{1}{\sqrt{1-e^{-2t}}} \exp\left(\frac{4xye^{-t} - (1+e^{-2t})(x^2+y^2)}{2(1-e^{-2t})} \right) \tag{156}$$

$$= \frac{1}{\sqrt{2\pi \sinh(t)}} \exp\left(xy\mathrm{cosech}(t) - (x^2+y^2)\coth(t)/2\right) \tag{157}$$

This is a known result and demonstrates that using appropriate mathematical logic that one can derive compact, succinct formulae for kernels by using the eigenfunction expansions. We shall now show how this relates to the Bessel polynomials as considered earlier in the paper.

13.2 Bessel Polynomials as Transforms and Kernels

Now, let us consider the fundamental differences here. In the example for the Asian option pricing equation, as we have shown, there is a discrete and a continuous component, both of which contribute to the spectral solution given through the kernel. The discrete component in this case has only a finite number of energy states, and we also have a continuous eigenstate which will contribute. The natural generalisation of the Titchmarsh-Kodaira formula will then be given by:

$$K(x,y;t) = \sum_{i=0}^{n_{max}} e^{-E_i t} |c_i|^2 \Phi_i(x,\lambda) \Phi_i^*(y,\lambda) + \int_\Omega e^{-E_\lambda t} \Psi_\lambda(x) \Psi_\lambda^*(y) dm(\lambda) \tag{158}$$

In this case we actually have two solutions to the eigenvalue equation $\mathscr{L}u = Eu$ depending on the values of the parameters which define the differential equation. It is interesting to consider the nature of the integral transform pairs that by necessity underlay the Bessel polynomials, which are the discrete part complementary to the Whittaker transform.

$$(\mathscr{F}f)_i(\lambda) = \tilde{f}_i(\lambda) = \int_a^b f(x)w_i^*(x,\lambda)r(x)dx \tag{159}$$

We shall now briefly discuss the Bessel polynomials and other finite sets of orthogonal functions. For any infinite set of orthogonal polynomials, we may write the following:

$$f(x) = \sum_{n=0}^{\infty} A_n p_n(x) \tag{160}$$

Assume we have orthogonality defined over some interval, then we may extract the coefficients using the orthogonality property:

$$A_n = \int_a^b f(x)p_n(x)dx \tag{161}$$

where

$$\int_a^b p_n(x)p_{n'}(x)dx = \delta_{nn'} \tag{162}$$

Now if we consider orthogonal polynomials of this type, we indeed have the transform pair:

$$A_n = \int_a^b f(x)p_n(x)dx \tag{163}$$

$$f(x) = \sum_{n=0}^{\infty} A_n p_n(x) \tag{164}$$

Although there is an orthogonality implied in the complex plane, we shall not consider this as it is beyond the scope of this simple discussion. Using results from [13], the following is known for the generalised Bessel polynomials $y_n(x;a,b)$ (notation ours) with support in $[0^+,\infty)$:

$$\int_0^{\infty} \left(\frac{x}{b}\right)^{1-a} e^{-x/b} y_m(x;a,b)y_n(x;a,b)dx = b(n!)\Gamma(2-a-n)\delta_{mn} \tag{165}$$

This is the appropriate form for use in the following. The weight function is then:

$$w_b(x) = \left(\frac{x}{b}\right)^{1-a} e^{-x/b} \tag{166}$$

which we recognise as being gamma distributed in x, and inverse gamma in b. Obviously this case differs significantly from the situation with the standard classical orthogonal polynomial series. Let us therefore expand a function defined in the positive plane into the Bessel polynomials via:

$$f(x) = \sum_{n=0}^{\infty} A_n y_n(x;a,b) \tag{167}$$

$$A_n = \frac{1}{b(n!)\Gamma(2-a-n)} \int_0^{\infty} f(x) y_n(x;a,b)dx \tag{168}$$

We shall call this the Bessel polynomial transform pair. Interestingly, there is a result from [20] which is of relevance in determining the correct expansion. The authors of the series of papers [21, 20] report the following expansion equation:

$$x^{-p} = \frac{2^{-p}}{2\pi^2} \int_0^{\infty} u \sinh(\pi u) \left| \Gamma\left(-\frac{p}{2} + \frac{iu}{2}\right) \right|^2 K_{iu}(x)du$$
$$+2^{-p+1} \sum_{k=0}^{\lfloor p/2 \rfloor} \frac{(p-2k)!}{k!(p-k)!} K_{p-2k}(x) \tag{169}$$

which is true for all p. We can see that in this expansion that the first part is given by a Kontorovich-Lebedev transformation, and the second part is given by a Laguerre-type summation. It is this second part that is equivalent to the Bessel polynomial transform as discussed above. Indeed, using the definition of the Bessel polynomial [19], we may write:

$$y_n(x) = \sqrt{\frac{2}{\pi x}} e^{1/x} K_{n+1/2}(x^{-1}) \tag{170}$$

$$K_{p-2k}(x) = \sqrt{\frac{\pi}{2x}} e^{-1/x} y_{p-2k-1/2}(x^{-1}) \tag{171}$$

The finite, discrete polynomial part of the transform is then given by the Bessel polynomial sum:

$$\hat{\mathscr{T}}_{disc.} f(x) = 2^{-p+1} \sum_{k=0}^{\lfloor p/2 \rfloor} \frac{(p-2k)!}{k!(p-k)!} \sqrt{\frac{\pi}{2x}} e^{-x} y_{p-2k-1/2}(x^{-1}) f(x,u) \tag{172}$$

$$= 2^{-p+1} \sqrt{\frac{\pi}{2x}} e^{-x} \sum_{k=0}^{\lfloor p/2 \rfloor} \frac{(p-2k)!}{k!(p-k)!} y_{p-2k-1/2}(x^{-1}) f(x,u) \tag{173}$$

The continuous part of the same transform implied through the formula is, of course, the generalised Kontorovich-Lebedev transform:

$$\hat{\mathscr{T}}_{cont.} f(x,u) = \frac{2^{-p}}{2\pi^2} \int_0^{\infty} u \sinh(\pi u) \left| \Gamma\left(-\frac{p}{2} + \frac{iu}{2}\right) \right|^2 K_{iu}(x) f(x,u)du \tag{174}$$

as required. This completes our discussion of the spectral theory of the Asian option pricing problem. It is clear that the nature of the solutions to the modified Bessel equation requires us, for certain ranges of parameters, to take into account both the discrete and continuous parts which contribute to the solution as a whole. It is easily seen that there exist certain polynomial subspaces of finite dimensions both for this equation, and for the Whittaker equation more generally. In the latter situation, we know that the Bessel polynomials are of importance, however, it seems to have been overlooked that this is necessary in order that the completeness relationship be well defined and analytical in the case of the Bessel equation for all possible values of the parameters.

14 Index Integration

One simple technique that can be used to determine formulae for index transformations are the Whittaker-Bessel-Legendre integrals, which allow the reduction of more complicated integrals to known types involving Legendre equations. We shall briefly show several integrals in which this method of calculation is effective, and indicate some further types of problems that may arise for more complicated scenarios that may be of interest.

14.1 A Mixed Index Integral

For example, if we take the completeness relationship:

$$\pi\delta(x-y) = \int_0^\infty p\sinh(\pi p)|\Gamma(-\mu+1/2+ip)|^2 \mathscr{P}^\mu_{ip-1/2}(x)\mathscr{P}^\mu_{ip-1/2}(y)dp \quad (175)$$

it is possible to show, using some results from integral theory that the following addition formula is available:

$$\frac{2\Gamma(1-\mu)}{\pi(a+b)}W_{\mu,1/2}(2(a+b))$$

$$= \frac{2}{\pi^2}\int_0^\infty p\sinh(\pi p)|\Gamma(-\mu+1/2+ip)|^2 \frac{W_{\mu,ip}(2a)}{a}\frac{W_{\mu,ip}(2b)}{b}dp \quad (176)$$

We shall now show how another different type of index integral may be found using the technique of completeness. We shall "mix" over the basis sets defined by the modified Bessel and Mehler-Fock functions viz.:

$$I = \int_0^\infty p\sinh(\pi p)\left|\Gamma\left(\frac{\mu}{2}+ip\right)\right|^2 K_{ip}(a)\mathscr{P}^\mu_{ip-1/2}(y)dp \quad (177)$$

Using known integral relations between the Macdonald and Mehler-Fock functions, we may write:

$$\int_1^\infty e^{-ax}(x^2-1)^{-\mu/2}\mathscr{P}^\mu_{ip-1/2}(x)dx = \sqrt{\frac{2}{\pi}}a^{\mu-1/2}K_{ip}(a) \quad (178)$$

or:

$$K_{ip}(a) = \sqrt{\frac{\pi}{2}} a^{-\mu+1/2} \int_1^\infty e^{-ax}(x^2-1)^{-\mu/2} \mathscr{P}^\mu_{ip-1/2}(x)dx \tag{179}$$

hence the integral rolls over into the nested formula:

$$I = \sqrt{\frac{\pi}{2}} a^{-\mu+1/2} \int_0^\infty p\sinh(\pi p) \left|\Gamma\left(\frac{\mu}{2}+ip\right)\right|^2 \mathscr{P}^\mu_{ip-1/2}(y)dp \tag{180}$$

$$\times \int_1^\infty e^{-ax}(x^2-1)^{-\mu/2} \mathscr{P}^\mu_{ip-1/2}(x).dx \tag{181}$$

Invoking the Fubini theorem and interchanging the order of integrals, we find:

$$I = \sqrt{\frac{\pi}{2}} a^{-\mu+1/2} \int_1^\infty dx.e^{-ax}(x^2-1)^{-\mu/2} \tag{182}$$

$$\times \int_0^\infty p\sinh(\pi p) \left|\Gamma\left(\frac{\mu}{2}+ip\right)\right|^2 \mathscr{P}^\mu_{ip-1/2}(y)\mathscr{P}^\mu_{ip-1/2}(x)dp \tag{183}$$

$$= \sqrt{\frac{\pi}{2}} a^{-\mu+1/2} \int_1^\infty e^{-ax}(x^2-1)^{-\mu/2} \pi\delta(x-y)dx \tag{184}$$

$$= \sqrt{\frac{\pi^3}{2}} a^{-\mu+1/2} e^{-ay}(y^2-1)^{-\mu/2} \tag{185}$$

The result then states that

$$\int_0^\infty p\sinh(\pi p) \left|\Gamma\left(\frac{\mu}{2}+ip\right)\right|^2 K_{ip}(a) \mathscr{P}^\mu_{ip-1/2}(y)dp \tag{186}$$

$$= \sqrt{\frac{\pi^3}{2}} a^{-\mu+1/2} e^{-ay}(y^2-1)^{-\mu/2} \tag{187}$$

This simple technique of expansion, interchange and resolution using the completeness identity has stark and profound results for a large class of index transforms. As we shall show, many different integrals can be evaluated by this method.

14.2 Second Mixed Integral

We shall now show how to treat another type of mixed integral of a similar kind. In this case we have e.g.:

$$I = \int_0^\infty p\sinh(\pi p) \left|\Gamma\left(\frac{1}{2}-\mu+\frac{ip}{2}\right)\right|^2 W_{\mu,ip}(2x)K_{ip}(y)dp \tag{188}$$

$$= \sqrt{\frac{\pi}{2}} x y^{-\mu+1/2} \int_0^\infty p \sinh(\pi p) \left| \Gamma\left(\frac{1}{2} - \mu + \frac{ip}{2}\right) \right|^2 \sqrt{\frac{2}{\pi}} y^{\mu-1/2} \frac{W_{\mu,ip}(2x)}{x} K_{ip}(y) dp \tag{189}$$

$$= \sqrt{\frac{\pi}{2}} x y^{-\mu+1/2} \int_0^\infty dp.p \sinh(\pi p) \left| \Gamma\left(\frac{1}{2} - \mu + \frac{ip}{2}\right) \right|^2 \tag{190}$$

$$\times \int_1^\infty e^{-xu} \left(\frac{u+1}{u-1}\right)^{-\mu/2} \mathscr{P}^{\mu}_{ip-1/2}(u) du \int_1^\infty e^{-yv} (v^2-1)^{-\mu/2} \mathscr{P}^{\mu}_{ip-1/2}(v) dv \tag{191}$$

It is important at this point to take the correct symmetries for the Whittaker, modified Bessel and Mehler-Fock functions. Indeed, as they are all spherical functions, we may extend the support of the function to take in negative values, via.:

$$\mathscr{P}^{\mu}_{ip-1/2}(u) = \mathscr{P}^{\mu}_{ip-1/2}(-u) \tag{192}$$

$$W_{\mu,ip}(2x) = W_{\mu,ip}(-2x) \tag{193}$$

$$K_{ip}(y) = K_{ip}(-y) \tag{194}$$

The integral is invariant if we take these definitions, hence we must ensure that the final answer respects the same symmetry.

$$I = \int_0^\infty p \sinh(\pi p) \left| \Gamma\left(\frac{1}{2} - \mu + \frac{ip}{2}\right) \right|^2 W_{\mu,ip}(2x) K_{ip}(y) dp \tag{195}$$

Rearrangement via the Fubini-Tonelli theorem yields:

$$I = \sqrt{\frac{\pi}{2}} \pi . x y^{-\mu+1/2} \int_1^\infty \int_1^\infty du dv. e^{-(xu+yv)} \left(\frac{u+1}{u-1}\right)^{-\mu/2} (v^2-1)^{-\mu/2} \tag{196}$$

$$\times \int_0^\infty dp.p \sinh(\pi p) \left| \Gamma\left(\frac{1}{2} - \mu + \frac{ip}{2}\right) \right|^2 \mathscr{P}^{\mu}_{ip-1/2}(u) \mathscr{P}^{\mu}_{ip-1/2}(v) \tag{197}$$

$$= \sqrt{\frac{\pi^3}{2}} x y^{-\mu+1/2} \int_1^\infty \int_1^\infty du dv. e^{-(xu+yv)} \left(\frac{u+1}{u-1}\right)^{-\mu/2} (v^2-1)^{-\mu/2} \delta(u-v) \tag{198}$$

$$= \sqrt{\frac{\pi^3}{2}} x y^{-\mu+1/2} \int_1^\infty du. e^{-(x+y)u} \left(\frac{u+1}{u-1}\right)^{-\mu/2} (u^2-1)^{-\mu/2} \tag{199}$$

$$= \sqrt{\frac{\pi^3}{2}} x y^{-\mu+1/2} \int_1^\infty du. e^{-(x+y)u} \left(\frac{u+1}{u-1}\right)^{-\mu/2} (u-1)^{-\mu/2} (u+1)^{-\mu/2} \tag{200}$$

$$= \sqrt{\frac{\pi^3}{2}} x y^{-\mu+1/2} \int_1^\infty du. e^{-(x+y)u} (u+1)^{-\mu} \tag{201}$$

One way in which to understand this is through the following integral [9] 3.382.2:

$$\int_{u}^{\infty}(x-u)^{\nu}e^{-\mu x}dx=\mu^{-\nu-1}e^{-u\mu}\Gamma(\nu+1) \tag{202}$$

$$\int_{1}^{\infty}(x-1)^{-\nu}e^{-\mu x}dx=\mu^{\nu-1}e^{-\mu}\Gamma(-\nu+1)=\int_{\infty}^{1}(-1)^{-\nu}(1+u)^{-\nu}e^{\mu u}du \tag{203}$$

$$=-(-1)^{-\nu}\int_{1}^{\infty}(1+u)^{-\nu}e^{\mu u}du \tag{204}$$

$$\int_{1}^{\infty}(1+u)^{-\nu}e^{\mu u}du=-(-1)^{\nu}\mu^{\nu-1}e^{-\mu}\Gamma(-\nu+1) \tag{205}$$

The phase factor is therefore cancelled, and we arrive at:

$$\int_{0}^{\infty}p\sinh(\pi p)\left|\Gamma\left(\frac{1}{2}-\mu+\frac{ip}{2}\right)\right|^{2}W_{\mu,ip}(2x)K_{ip}(y)dp \tag{206}$$

$$=\sqrt{\frac{\pi^{3}}{2}xy^{-\mu+1/2}}e^{-(x+y)}(x+y)^{\mu-1}\Gamma(1-\mu,-2(x+y)) \tag{207}$$

In the sector where the parameters are defined through:

$$\mu\leq 1/2 \tag{208}$$

the formula simplifies, yielding that of Oberhettinger [18] pp17, 34

$$\int_{0}^{\infty}p\sinh(\pi p)\left|\Gamma\left(\frac{1}{2}-\mu+\frac{ip}{2}\right)\right|^{2}W_{\mu,ip}(2x)K_{ip}(y)dp \tag{209}$$

$$=\sqrt{\frac{\pi^{3}}{2}xy^{-\mu+1/2}}e^{-(x+y)}(x+y)^{\mu-1}\Gamma(1-\mu) \tag{210}$$

This is a perfect application for complex analysis and residue calculus and shall form the basis of a future paper.

14.3 Cosine Transform

Asian options may be treated in a similar fashion by expanding the Whittaker function into the Legendre state. If we take the formula derived for Asian option pricing and perform this operation, we find:

$$\int_{0}^{\infty}e^{-p^{2}\tau}p\sinh(\pi p)\left|\Gamma\left(\frac{1}{2}-\mu+\frac{ip}{2}\right)\right|^{2}W_{\mu,ip}(2x)dp \tag{211}$$

$$=x\int_{0}^{\infty}e^{-p^{2}\tau}p\sinh(\pi p)\left|\Gamma\left(\frac{1}{2}-\mu+\frac{ip}{2}\right)\right|^{2}\frac{W_{\mu,ip}(2x)}{x}dp \tag{212}$$

$$= x\int_0^\infty dp.e^{-p^2\tau} p\sinh(\pi p)\left|\Gamma\left(\frac{1}{2}-\mu+\frac{ip}{2}\right)\right|^2 \tag{213}$$

$$\times\int_1^\infty e^{-xu}\left(\frac{u+1}{u-1}\right)^{-\mu/2} \mathscr{P}^{\mu}_{ip-1/2}(u)du \tag{214}$$

$$= x\int_1^\infty du.e^{-xu}\left(\frac{u+1}{u-1}\right)^{-\mu/2} \tag{215}$$

$$\times\int_0^\infty dp.e^{-p^2\tau} p\sinh(\pi p)\left|\Gamma\left(\frac{1}{2}-\mu+\frac{ip}{2}\right)\right|^2 \mathscr{P}^{\mu}_{ip-1/2}(u) \tag{216}$$

Note that we have used the multiplication symbol here to indicate a nested integral in the interests of brevity. The inner integral here is of primary interest. We note the connection between integrals of this type and the so-called Yor integrals. We shall now show some methods that can be used to evaluate this integral by relying on various transforms between groups of special functions. Calculating the cosine transform of the Whittaker function:

$$g_c(y) = \int_0^\infty f(x)\cos(xy)dx \tag{217}$$

$$g_c(y) = \int_0^\infty W_{\mu,ip}(2u)\cos(py)dp \tag{218}$$

$$= \int_0^\infty u\int_1^\infty e^{-ux}\left(\frac{x+1}{x-1}\right)^{\mu/2} \mathscr{P}^{\mu}_{ip-1/2}(x)\cos(py).dxdp \tag{219}$$

$$= u\int_1^\infty dx.e^{-ux}\left(\frac{x+1}{x-1}\right)^{\mu/2}\int_0^\infty dp.\mathscr{P}^{\mu}_{ip-1/2}(x)\cos(py) \tag{220}$$

Oberhettinger [17] 12.85 gives the following cosine transform of the Mehler-Fock function:

$$\int_0^\infty dp.\mathscr{P}^{\mu}_{ip-1/2}(x)\cos(py) = \sqrt{\frac{\pi}{2}}\frac{(x^2-1)^{\mu/2}(x-\cosh y)^{-\mu-1/2}}{\Gamma(1/2-\mu)} \tag{221}$$

$$\cosh y < a \tag{222}$$

so we obtain the cosine transform of the Whittaker function via:

$$g_c(y) = \sqrt{\frac{\pi}{2}}\frac{u}{\Gamma(1/2-\mu)}\int_1^\infty dx.e^{-ux}(x+1)^{\mu}(x-\cosh y)^{-\mu-1/2} \tag{223}$$

Using formula from [9] 3.385.3

$$\int_\alpha^\infty dx.e^{-ax}(x+\beta)^{2\nu-1}(x-\alpha)^{2\rho-1} \tag{224}$$

$$= \frac{(\alpha+\beta)^{\nu+\rho-1}}{a^{\nu+\rho}} \exp\left(\frac{(\beta-\alpha)a}{2}\right) \Gamma(2\rho) W_{\nu-\rho,\nu+\rho-1/2}([\alpha+\beta]a) \tag{225}$$

This formula is not quite correct for our application. In the region we need to have the following definition:

$$\int_{\alpha}^{\infty} dx.e^{-ax} (x+\alpha)^{2\rho-1} (x-\beta)^{2\nu-1} \tag{226}$$

$$= \frac{(\alpha+\beta)^{\nu+\rho-1}}{a^{\nu+\rho}} \exp\left(\frac{(\beta-\alpha)a}{2}\right) \Gamma(2\nu) W_{\nu-\rho,\nu+\rho-1/2}([\alpha+\beta]a) \tag{227}$$

Making the substitutions

$$\int_{\alpha}^{\infty} dx.e^{-ax} (x+\alpha)^{2\rho-1} (x-\beta)^{2\nu-1} \tag{228}$$

$$= \frac{(\alpha+\beta)^{\nu+\rho-1}}{a^{\nu+\rho}} \exp\left(\frac{(\beta-\alpha)a}{2}\right) \Gamma(2\nu) W_{\nu-\rho,\nu+\rho-1/2}([\alpha+\beta]a) \tag{229}$$

$$\alpha = 1, \beta = -\cosh y, 2\rho - 1 = \mu, a = u, 2\nu - 1 = -\mu - 1/2 \tag{230}$$

$$\rho = \frac{\mu+1}{2} \tag{231}$$

$$\nu = \frac{1}{2}(-\mu+1/2) = -\frac{\mu}{2} + \frac{1}{4} \tag{232}$$

$$\nu - \rho = -\frac{\mu}{2} + \frac{1}{4} - \frac{\mu}{2} - \frac{1}{2} = -\mu - \frac{1}{4} \tag{233}$$

$$\nu + \rho - 1/2 = -\frac{\mu}{2} + \frac{1}{4} + \frac{\mu+1}{2} - 1/2 = \frac{1}{4} \tag{234}$$

$$\nu + \rho = 3/4, \nu + \rho - 1 = -1/4 \tag{235}$$

$$\int_{1}^{\infty} dx.e^{-ux} (x+1)^{\mu} (x-\cosh y)^{-\mu-1/2} \tag{236}$$

$$= \frac{(1+\cosh y)^{-1/4}}{u^{3/4}} \exp\left(\frac{-(\cosh y - 1)u}{2}\right) \Gamma(1/2-\mu) W_{-\mu-1/4,1/4}([1+\cosh y]u) \tag{237}$$

Using the hyperbolic identities and some simple trigonometry, it is simple to reduce this integral further to:

$$\cosh y = \cosh\left(\frac{y}{2}+\frac{y}{2}\right) = \cosh^2\frac{y}{2} + \sinh^2\frac{y}{2} \tag{238}$$

$$= 2\cosh^2\frac{y}{2} - 1 = 1 + 2\sinh^2\frac{y}{2} \tag{239}$$

$$1 + \cosh y = 2\cosh^2\frac{y}{2} \tag{240}$$

$$1-\cosh y = -2\sinh^2\frac{y}{2} \tag{241}$$

$$\int_1^\infty dx.e^{-ux}(x+1)^\mu (x-\cosh y)^{-\mu-1/2} \tag{242}$$

$$= \frac{(\cosh^2\frac{y}{2})^{-1/4}}{u^{3/4}} \exp\left(-u\sinh^2\frac{y}{2}\right)\Gamma(1/2-\mu)W_{-\mu-1/4,1/4}\left(u\cosh^2\frac{y}{2}\right) \tag{243}$$

Using identities for the Whittaker function and parabolic cylinder functions, we find:

$$W_{-\mu-1/4,1/4}(x) = W_{\mu+1/4,1/4}(x) = D_{2\mu}\left(\sqrt{2x}\right)\frac{x^{1/4}}{2^\mu} \tag{244}$$

$$W_{-\mu-1/4,1/4}\left(u\cosh^2\frac{y}{2}\right) = D_{2\mu}\left(\sqrt{2u}\cosh\frac{y}{2}\right)\frac{u^{1/4}\cosh\frac{y}{2}}{2^\mu} \tag{245}$$

$$\int_1^\infty dx.e^{-ux}(x+1)^\mu (x-\cosh y)^{-\mu-1/2} \tag{246}$$

$$= \frac{(\cosh^2\frac{y}{2})^{-1/4}}{u^{3/4}} \exp\left(-u\sinh^2\frac{y}{2}\right)\Gamma(1/2-\mu)D_{2\mu}\left(\cosh\frac{y}{2}\sqrt{2u}\right)\frac{(u\cosh^2\frac{y}{2})^{1/4}}{2^\mu} \tag{247}$$

$$= \frac{1}{2^\mu u^{1/2}} \exp\left(-u\sinh^2\frac{y}{2}\right)\Gamma(1/2-\mu)D_{2\mu}\left(\cosh\frac{y}{2}\sqrt{2u}\right) \tag{248}$$

Finalising the result for the index transform, we obtain:

$$g_c(y) = \sqrt{\frac{\pi}{2}}\frac{u}{\Gamma(1/2-\mu)}\int_1^\infty dx.e^{-ux}(x+1)^\mu (x-\cosh y)^{-\mu-1/2} \tag{249}$$

$$= \sqrt{\frac{\pi}{2}}\frac{\Gamma(1/2-\mu)}{2^\mu u^{1/2}}\frac{u}{\Gamma(1/2-\mu)}\exp\left(-u\sinh^2\frac{y}{2}\right)D_{2\mu}\left(\cosh\frac{y}{2}\sqrt{2u}\right) \tag{250}$$

$$= \sqrt{\frac{\pi u}{2}}2^{-\mu}\exp\left(-u\sinh^2\frac{y}{2}\right)D_{2\mu}\left(\cosh\frac{y}{2}\sqrt{2u}\right) \tag{251}$$

which is consistent with a tabulated result of Oberhettinger using an unknown method [18] 22.28. We have thus established, with the caveat of the admission of the representation for the shifted confluent hypergeometric function, the following cosine transform:

$$\int_0^\infty W_{\mu,ip}(2u)\cos(py)dp$$

$$= \sqrt{\frac{\pi u}{2}}2^{-\mu}\exp\left(-u\sinh^2\frac{y}{2}\right)D_{2\mu}\left(\cosh\frac{y}{2}\sqrt{2u}\right) = \mathscr{I}_\mu(y,u) \tag{252}$$

15 Conclusions and Future Directions

This paper has demonstrated a number of techniques that can be used to compute Asian options pricing. In particular, we have shown that the method of Gauss-Laguerre quadrature is a highly accurate algorithm that produces answers that are valid to a degree not possible to achieve via known methods, including the spectral method. Difficulties in convergence are easily overcome by increasing the number of abscissa points, which is not simple to achieve using other techniques.

We have also shown that the problem of exponential Brownian motion is deeply associated with various groups of special functions, including those of Bessel, Mehler-Fock and Whittaker. By using spectral decompositions of the kernel over the continuous and discrete eigenfunctions, a large number of formulae related to various integral transforms are readily derived. This is a fruitful area of research that has relevance to the fields of quantitative finance as well as physical implications.

There are some further topics of interest that present themselves that may be amenable to resolution in a similar way to the calculations we have discussed. We shall outline some advanced techniques related to Whittaker functions that are of interest in studying these types of spectral theories and pricing algorithms. This study has skipped over the dynamics associated with the discrete part of the state. It is possible to show that it contributes via a similar integral, a part given by:

$$\sum_{n=0}^{[\nu/2]} e^{-2n(|\nu|-n)\tau} \frac{(-1)^n 2(|\nu|-2n)}{\Gamma(1+|\nu|-n)} (2y)^{n-1-|\nu|} e^{-1/(2y)} L_n^{(|\nu|-2n)}\left(\frac{1}{2y}\right) \tag{253}$$

The origin of this term is the poles of the Whittaker W-function. It is important to realise that the Green's function can be defined using the Cauchy integral formula, so we cannot neglect the discrete components contributing at the poles in the complex plane. We can look at both this expression and the continuous counterpart as being equivalent to the Yor integral which gives the pdf for the particle originating at zero. Research in the integral theory of special functions has turned up the following relationship:

$$\int_0^\infty \nu \sinh(\pi\nu)\Gamma(\lambda+i\nu)\Gamma(\lambda-i\nu)K_{i\nu}(a)K_{i\nu}(b)d\nu \tag{254}$$

$$= \frac{\pi^{3/2}\Gamma(\lambda+1/2)}{2}\left(\frac{a+b}{2ab}\right)^{-\lambda} K_\lambda(a+b) \tag{255}$$

There are relationships between the Whittaker systems and other groups of hyperbolic functions, i.e. associated Legendre, modified Bessel functions. It is an active topic of research aimed at better understanding the interaction between eigenstates, spectral theory and special functions.

Acknowledgements The author wishes to acknowledge the support of the Department of Maths and Physical Sciences, UTS in conducting this research.

References

1. Albanese, C., Kuznetsov, A.: Transformations of Markov processes and classification scheme for solvable driftless diffusions. arXiv preprint arXiv:0710.1596 (2007)
2. Albanese, C., Lawi, S.: Laplace transforms for integrals of Markov processes. arXiv preprint arXiv:0710.1599 (2007)
3. Bateman, H.: Tables of integral transforms. California Institute of Technology Bateman Manuscript Project (1954)
4. Borodin, A.N., Salminen, P.: Handbook of Brownian motion-facts and formulae. Springer Science & Business Media (2015)
5. Craddock, M.: On an integral arising in mathematical finance. In: Nonlinear Economic Dynamics and Financial Modelling, pp. 355–370. Springer (2014)
6. Feller, W.: The parabolic differential equations and the associated semi-groups of transformations. Annals of Mathematics pp. 468–519 (1952)
7. Feynman, R.P., Hibbs, A.R., Styer, D.F.: Quantum mechanics and path integrals (2010)
8. Geman, H., Yor, M.: Bessel processes, Asian options, and perpetuities. Mathematical finance **3**(4), 349–375 (1993)
9. Gradshteyn, I.S., Ryzhik, I.M.: Table of integrals, series, and products. Academic press (2014)
10. Grosche, C.: The path integral on the Poincaré upper half-plane with a magnetic field and for the Morse potential. Annals of Physics **187**(1), 110–134 (1988)
11. Grosche, C., Steiner, F., Steiner, F.: Handbook of Feynman path integrals (1998)
12. Grosswald, E.: On some algebraic properties of the Bessel polynomials. Transactions of the American Mathematical Society **71**(2), 197–210 (1951)
13. Hamza, A.A.: Properties of Bessel polynomials. Ph.D. thesis, Loughborough University (1974)
14. Krall, H.L., Frink, O.: A new class of orthogonal polynomials: The Bessel polynomials. Transactions of the American Mathematical Society **65**(1), 100–115 (1949)
15. Linetsky, V.: Spectral expansions for Asian/average price options. Operations Research **52**(6), 856–867 (2004)
16. Mehler, F.G.: Ueber die Entwicklung einer Function von beliebig vielen Variablen nach Laplaceschen Functionen höherer Ordnung. (1866)
17. Oberhettinger, F.: Tables of Fourier transforms and Fourier transforms of distributions. Springer Science & Business Media (2012)
18. Oberhettinger, F., Higgins, T.P.: Tables of Lebedev, Mehler, and generalized Mehler transforms. Tech. rep., Boeing Scientific Research Labs Seattle WA (1961)
19. Olver, F.W., Lozier, D.W., Boisvert, R.F., Clark, C.W., Dalhuis, A.: Digital library of mathematical functions. National Institute of Standards and Technology from http://dlmf. nist. gov/(release date 2011-07-01), Washington, DC (2010)
20. Pintoux, C., Privault, N.: A direct solution to the Fokker–Planck equation for exponential Brownian functionals. Analysis and Applications **8**(03), 287–304 (2010)
21. Pintoux, C., Privault, N.: The Dothan pricing model revisited. Mathematical Finance: An International Journal of Mathematics, Statistics and Financial Economics **21**(2), 355–363 (2011)
22. Prudnikov, A.P., Brychkov, Y.A., Marichev, O.I., Romer, R.H.: Integrals and series (1988)
23. Prudnikov, H., Brychkov, Y., Marichev, O.: Special Functions, Integrals and Series, Vol. 2 (1986)
24. Sousa, R., Yakubovich, S.: The spectral expansion approach to index transforms and connections with the theory of diffusion processes. arXiv preprint arXiv:1706.05194 (2017)
25. Szmytkowski, R., Bielski, S.: An orthogonality relation for the Whittaker functions of the second kind of imaginary order. Integral transforms and special functions **21**(10), 739–744 (2010)

On Dupire Formula and Diffusion with Given Marginals

Jie Yen Fan and Fima C. Klebaner

Abstract This paper presents relations between the drift and the diffusion coefficients of a diffusion in terms of the prices of a European call option, giving an extension to the Dupire formula. Due to the correspondence between the call option prices and the marginal distributions of the underlying process, a necessary condition for a diffusion with given marginal distributions is obtained. Some specific examples, including the the cases of Normal and Lognormal marginals, are considered. In particular, we construct fake Brownian motion diffusions, a family of diffusions, which all are Gaussian processes with Brownian marginals $N(0,t)$ but not a Brownian motion unless $\sigma^2 = 1$.

1 Introduction

Motivated by Hamza and Klebaner, [9], where a family of processes with Normal marginal distributions was constructed, Mudakkar [15] considered diffusion processes with Normal marginals and obtained a necessary relation between the drift and the diffusion coefficients. In this paper, we generalised Mudakkar's result via a Dupire-like formula. We derive a general version of Dupire formula. Under the assumption of Normal marginal distributions, we recover Mudakkar's result. We can also extend the result to other types of distributions, such as the Lognormal distribution.

The topic of constructing processes with given marginal distributions, also known as mimicking, has been of interest of many in the last two decades. For example,

Jie Yen Fan
School of Mathematics and Statistics, The University of Sydney, NSW 2006, Australia,
e-mail: jieyen.fan@sydney.edu.au

Fima C. Klebaner
School of Mathematics, Monash University, VIC 3800, Australia,
e-mail: fima.klebaner@monash.edu

D. R. Wood et al. (eds.), *2021-2022 MATRIX Annals*, MATRIX Book Series 5,
https://doi.org/10.1007/978-3-031-47417-0_17

[1, 2, 3, 9, 16] gave explicit constructions to mimic Brownian motion, i.e. construct processes of marginals $N(0,t)$. Generalisation from [9] to mimicking self-similar processes was carried out in [7]. For Ito processes, studies include [4, 8, 12]. In the monograph [11] by Hirsch, Profeta, Roynette and Yor on peacocks (processes that are increasing in the convex order), a few methods for constructing processes with given marginals are also discussed.

We give the main results in Section 2 and postpone the proofs to Section 3. In Section 4, we will consider some examples.

2 A general version of Dupire formula

Consider a local martingale $(X_t)_{t\geq 0}$ satisfying the stochastic differential equation $dX_t = \sigma(X_t,t)dB_t$. Let $C(x,t) = \mathbb{E}[(X_t - x)^+]$, the price of a call option with strike x and maturity t. Then, the relation between the functions σ and C is given by the well-known Dupire formula [6]

$$\sigma^2(x,t) = \frac{2\frac{\partial C}{\partial t}(x,t)}{\frac{\partial^2 C}{\partial x^2}(x,t)}, \tag{1}$$

or $\frac{\partial C}{\partial t}(x,t) = \frac{1}{2}\sigma^2(x,t)\frac{\partial^2 C}{\partial x^2}(x,t)$. Note that the probability density function of X_t, if exists, coincides with $\frac{\partial^2 C}{\partial x^2}$. Thus, (1) gives the unique diffusion with given marginal distributions. (In this paper, we assume that the distribution of X_t is absolutely continuous with respect to the Lebesgue measure, i.e. the probability density function exists, and that $\mathbb{E}[(X_t - x)^+]$ is finite.)

This can be generalised to semimartingales as follows.

Theorem 1. *Let $(X_t)_{t\geq 0}$ be a diffusion process satisfying*

$$dX_t = \mu(X_t,t)dt + \sigma(X_t,t)dB_t, \tag{2}$$

with $\int_0^t \sigma^2(X_s,s)ds < \infty$. And, let $C(x,t) = \mathbb{E}[(X_t - x)^+]$. Then,

$$\frac{\partial C}{\partial t}(x,t) = \frac{1}{2}\sigma^2(x,t)\frac{\partial^2 C}{\partial x^2}(x,t) + \int_x^\infty \mu(y,t)\frac{\partial^2 C}{\partial x^2}(y,t)dy, \tag{3}$$

equivalently,

$$\frac{\partial C}{\partial t}(x,t) = \frac{1}{2}\sigma^2(x,t)\frac{\partial^2 C}{\partial x^2}(x,t) + \mathbb{E}[\mu(X_t,t)1_{X_t > x}]. \tag{4}$$

From the put-call parity, we can establish a relation similar to (4) but with put option prices $P(x,t) = \mathbb{E}\left[(x - X_t)^+\right]$. Note that

$$\frac{\partial P}{\partial x}(x,t) = \int_{-\infty}^{x} p_t(y)dy = 1 + \frac{\partial C}{\partial x}(x,t) \tag{5}$$

and

$$\frac{\partial^2 P}{\partial x^2}(x,t) = p_t(x) = \frac{\partial^2 C}{\partial x^2}(x,t). \tag{6}$$

Proposition 1. *Let $(X_t)_{t\geq 0}$ satisfies SDE (2), and let $P(x,t) = \mathbb{E}\left[(x - X_t)^+\right]$. Then,*

$$\frac{\partial P}{\partial t}(x,t) = \frac{1}{2}\sigma^2(x,t)\frac{\partial^2 P}{\partial x^2}(x,t) - \mathbb{E}\left[\mu(X_t,t)1_{X_t<x}\right]. \tag{7}$$

Comparing (4) and (7), we see that

$$\frac{\partial C}{\partial t}(x,t) - \frac{\partial P}{\partial t}(x,t) = \mathbb{E}[\mu(X_t,t)].$$

Remark 1. To get an expression relating $\mu(x,t)$ and $\sigma(x,t)$ without integral, we differentiate (3) with respect to x, resulting in

$$\frac{\partial^2 C}{\partial x \partial t}(x,t) = \frac{1}{2}\sigma^2(x,t)\frac{\partial^3 C}{\partial x^3}(x,t) + \frac{1}{2}\frac{\partial}{\partial x}\{\sigma^2(x,t)\}\frac{\partial^2 C}{\partial x^2}(x,t) - \mu(x,t)\frac{\partial^2 C}{\partial x^2}(x,t). \tag{8}$$

The "put" version follows from (5) and (6),

$$\frac{\partial^2 P}{\partial x \partial t}(x,t) = \frac{1}{2}\sigma^2(x,t)\frac{\partial^3 P}{\partial x^3}(x,t) + \frac{1}{2}\frac{\partial}{\partial x}\{\sigma^2(x,t)\}\frac{\partial^2 P}{\partial x^2}(x,t) - \mu(x,t)\frac{\partial^2 P}{\partial x^2}(x,t),$$

which can also be obtained from (7).

One can easily see that, when $\mu(y,t) = 0$, (3) reduces to the classical Dupire formula (1), and when $\mu(y,t) = \mu(t)$, (3) becomes

$$\frac{\partial C}{\partial t}(x,t) = \frac{1}{2}\sigma^2(x,t)\frac{\partial^2 C}{\partial x^2}(x,t) - \mu(t)\frac{\partial C}{\partial x}(x,t),$$

which is also a known result (see for example [5]). Thus, (3) can be considered as a general version of the Dupire formula.

3 Proofs

There are a few ways to derive (3) and prove Theorem 1. The usual approach is through the forward Kolmogorov equation. Alternatively, it can be obtained through the Meyer-Tanaka formula and local times, like in [10] and [13]. Detailed derivation by the two approaches will be given in Sections 3.1 and 3.2.

3.1 Proof of Theorem 1 by forward Kolmogorov equation

This proof requires further assumptions and is given as a motivation.

Let $(X_t)_{t\geq 0}$ be a process satisfying the SDE (2). Denote by $p(y,t)$ the density function of X_t at point y. Suppose that $\mu(x,t)$ and $\sigma(x,t)$ are bounded and continuous, with $\sigma^2(x,t)$ bounded away from 0; that $\mu(x,t)$ and $\sigma^2(x,t)$ are Holder continuous with respect to x and t; and that $\mu(x,t)$ and $\sigma(x,t)$ have two partial derivatives with respect to x, which are bounded and Holder continuous with respect to x. (See [14, Theorem 5.15], for example.) Then, we have the forward Kolmogorov equation

$$\frac{\partial p}{\partial t}(y,t) = \frac{1}{2}\frac{\partial^2}{\partial y^2}\{\sigma^2(y,t)p(y,t)\} - \frac{\partial}{\partial y}\{\mu(y,t)p(y,t)\}. \tag{9}$$

Before we proceed, we give the derivatives concerning the function

$$C(x,t) = \mathbb{E}[(X_t - x)^+] = \int_x^\infty (y-x)p(y,t)dy.$$

Differentiate with respect to x, we have

$$\frac{\partial C}{\partial x}(x,t) = -\int_x^\infty p(y,t)dy$$

and

$$\frac{\partial^2 C}{\partial x^2}(x,t) = p(x,t).$$

Suppose that for $\frac{\partial p}{\partial t}(y,t)$ exists almost everywhere and that $\int_a^b \int_x^\infty |(y-x)\frac{\partial p}{\partial t}(y,t)|dydt$ is finite for any compact time interval $[a,b]$ and any x. Then, we also have

$$\frac{\partial C}{\partial t}(x,t) = \int_x^\infty (y-x)\frac{\partial p}{\partial t}(y,t)dy.$$

Now, multiply both sides of (9) by $(y-x)^+$ and integrate with respect to y, we obtain

$$\begin{aligned}&\int_x^\infty (y-x)\frac{\partial p}{\partial t}(y,t)dy\\ &= \frac{1}{2}\int_x^\infty (y-x)\frac{\partial^2}{\partial y^2}\{\sigma^2(y,t)p(y,t)\}dy - \int_x^\infty (y-x)\frac{\partial}{\partial y}\{\mu(y,t)p(y,t)\}dy. \end{aligned}\tag{10}$$

By integration by parts, the last integral

$$\begin{aligned}\int_x^\infty (y-x)\frac{\partial}{\partial y}\{\mu(y,t)p(y,t)\}dy &= \Big[(y-x)\mu(y,t)p(y,t)\Big]_x^\infty - \int_x^\infty \mu(y,t)p(y,t)dy\\ &= -\int_x^\infty \mu(y,t)p(y,t)dy,\end{aligned}$$

assuming that $\lim_{y\to\infty}\mu(y,t)p(y,t)=0$. Similarly, the first integral on the right hand side of (10)

$$\begin{aligned}
&\int_x^\infty (y-x)\frac{\partial^2}{\partial y^2}\{\sigma^2(y,t)p(y,t)\}dy \\
&\quad = \left[(y-x)\frac{\partial}{\partial y}\{\sigma^2(y,t)p(y,t)\}\right]_x^\infty - \int_x^\infty \frac{\partial}{\partial y}\{\sigma^2(y,t)p(y,t)\}dy \\
&\quad = -\left[\sigma^2(y,t)p(y,t)\right]_x^\infty = \sigma^2(x,t)p(x,t),
\end{aligned}$$

assuming that $\lim_{y\to\infty}\frac{\partial}{\partial y}\{\sigma^2(y,t)p(y,t)\}=0$ and $\lim_{y\to\infty}\sigma^2(y,t)p(y,t)=0$. Thus, Equation (10) yields the general Dupire formula (3), or (4) equivalently.

Note that in this proof we require $\lim_{y\to\infty}\mu(y,t)p(y,t)$, $\lim_{y\to\infty}\sigma^2(y,t)p(y,t)$ and $\lim_{y\to\infty}\frac{\partial}{\partial y}\{\sigma^2(y,t)p(y,t)\}$ to be zero. We also assume that $\frac{\partial p}{\partial t}(y,t)$ exists and $\int_a^b\int_x^\infty |(y-x)\frac{\partial p}{\partial t}(y,t)|dydt<\infty$ for any time interval $[a,b]$ and any x. These further assumptions are not required in an alternative proof showing in the next section.

3.2 *Proof of Theorem 1 by Meyer-Tanaka formula*

Following [10] and [13], we give an alternative derivation of the general Dupire formula (3) through the local time of X. By the Meyer-Tanaka formula, we have

$$(X_t-x)^+ = (X_0-x)^+ + \int_0^t 1_{X_s>x}dX_s + \frac{1}{2}L_t^x, \tag{11}$$

where L_t^x denotes the local time of X at point x. Take expectation on both sides and use the SDE of X,

$$\mathbb{E}\left[(X_t-x)^+\right] = \mathbb{E}\left[(X_0-x)^+\right] + \int_0^t \mathbb{E}\left[\mu(X_s,s)1_{X_s>x}\right]ds + \frac{1}{2}\mathbb{E}\left[L_t^x\right]. \tag{12}$$

Taking expectation of the occupation times formula, we have, for any positive measurable function g,

$$\begin{aligned}
\int g(x)\mathbb{E}\left[L_t^x\right]dx &= \mathbb{E}\left[\int_0^t g(X_s)d\langle X\rangle_s\right] = \int_0^t \mathbb{E}\left[g(X_s)\frac{d\langle X\rangle_s}{ds}\right]ds \\
&= \int_0^t \mathbb{E}\left[g(X_s)\sigma^2(X_s,s)\right]ds = \int_0^t\int g(x)\sigma^2(x,s)p_s(x)dxds \\
&= \int g(x)\int_0^t \sigma^2(x,s)p_s(x)dsdx.
\end{aligned}$$

Thus, $\mathbb{E}\left[L_t^x\right]=\int_0^t\sigma^2(x,s)p_s(x)ds$. Substituting this into (12), and taking derivative with respect to t, we have

$$\frac{\partial}{\partial t}\mathbb{E}\left[(X_t-x)^+\right]=\mathbb{E}\left[\mu(X_t,t)1_{X_t>x}\right]+\frac{1}{2}\sigma^2(x,t)p_t(x),$$

which is the general Dupire formula (3).

3.3 Proof of Proposition 1

As $(x-X_t)^+=(X_t-x)^+-(X_t-x)$, using (11) we have

$$\begin{aligned}(x-X_t)^+ &= (X_0-x)^+ + \int_0^t 1_{X_s>x}dX_s+\frac{1}{2}L_t^x-(X_t-x)\\ &= (x-X_0)^+ + \int_0^t (1_{X_s>x}-1)dX_s+\frac{1}{2}L_t^x\\ &= (x-X_0)^+ - \int_0^t 1_{X_s<x}dX_s+\frac{1}{2}L_t^x.\end{aligned}$$

Taking expectation, we have

$$\mathbb{E}\left[(x-X_t)^+\right]=\mathbb{E}\left[(x-X_0)^+\right]-\int_0^t\mathbb{E}\left[\mu(X_s,s)1_{X_s<x}\right]ds+\frac{1}{2}\int_0^t\sigma^2(x,s)p_s(x)ds,$$

where $p_s(x)$ is the probability density function of X_s. Differentiate with respect to t, we obtain (7), the "put" version of (3).

4 Some examples on diffusion with given marginals

As the function C uniquely determines the marginal distributions of the process X, (3) and (8) serve as necessary conditions for the diffusion X to match a given marginal distributions. In this section, we derive from Theorem 1 a necessary relation between the drift and diffusion coefficients of X for some specific marginals. For the Normal marginals, the result coincides with Mudakkar [15]. We will also consider the Lognormal marginals and marginals that satisfy selfsimilarity. Some specific examples are also given in this section.

In the following, we will write $\phi(\cdot)$ for the probability density function of $N(0,1)$ and $\Phi(\cdot)$ for the cumulative distribution function of $N(0,1)$.

4.1 Diffusion with Normal marginals

Suppose that $X_t\sim N(a_t,\gamma_t)$, $\gamma_t>0$, then writing Z for a $N(0,1)$ random variable, we have

$$C(x,t) = \mathbb{E}[(X_t - x)^+] = \sqrt{\gamma_t}\phi\Big(\frac{x-a_t}{\sqrt{\gamma_t}}\Big) + (a_t - x)\Big(1 - \Phi\Big(\frac{x-a_t}{\sqrt{\gamma_t}}\Big)\Big).$$

Proposition 2. *Assume that $(X_t)_{t\geq 0}$ has marginals $N(a_t, \gamma_t)$ and satisfies the SDE (2). Then the coefficients satisfy*

$$\mu(x,t) = a_t' + \frac{1}{2}\frac{\partial}{\partial x}\{\sigma^2(x,t)\} + \frac{x-a_t}{2\gamma_t}(\gamma_t' - \sigma^2(x,t)). \tag{13}$$

Proof. Computing the partial derivatives of $C(x,t)$ and using the fact that $\phi'(z) = -z\phi(z)$, we obtain

$$\frac{\partial C}{\partial t} = \frac{\gamma_t'}{2\sqrt{\gamma_t}}\phi\Big(\frac{x-a_t}{\sqrt{\gamma_t}}\Big) + a_t'\Big(1 - \Phi\Big(\frac{x-a_t}{\sqrt{\gamma_t}}\Big)\Big),$$
$$\frac{\partial^2 C}{\partial x^2} = \frac{1}{\sqrt{\gamma_t}}\phi\Big(\frac{x-a_t}{\sqrt{\gamma_t}}\Big).$$

Therefore, Equation (3) gives

$$\frac{\gamma_t}{2\sqrt{\gamma_t}}\phi\Big(\frac{x-a_t}{\sqrt{\gamma_t}}\Big) + a_t'\Big(1 - \Phi\Big(\frac{x-a_t}{\sqrt{\gamma_t}}\Big)\Big) =$$
$$\frac{1}{2}\sigma^2(x,t)\frac{1}{\sqrt{\gamma_t}}\phi\Big(\frac{x-a_t}{\sqrt{\gamma_t}}\Big) + \int_x^{\infty}\mu(y,t)\frac{1}{\sqrt{\gamma_t}}\phi\Big(\frac{y-a_t}{\sqrt{\gamma_t}}\Big)dy. \tag{14}$$

Differentiate with respect to x and simplify, we obtain

$$-\frac{\gamma_t'(x-a_t)}{2\gamma_t} - a_t' = -\frac{1}{2}\sigma^2(x,t)\frac{x-a_t}{\gamma_t} + \frac{1}{2}\frac{\partial}{\partial x}\{\sigma^2(x,t)\} - \mu(x,t).$$

□

Example 1. Suppose $\mu(x,t) = 0$ and let $X_0 = x_0$, i.e. $X_t = x_0 + \int_0^t \sigma(X_s,s)dB_s$. Then, $a_t = x_0$ and (14) gives

$$\sigma^2(x,t) = \gamma_t'$$

as a necessary condition for X to have Normal marginals. In particular, $\sigma^2(x,t)$ does not depend on x.

Example 2. Suppose $\mu(x,t) = \mu_t x$ and $\sigma(x,t) = \sigma_t x$, where μ_t and σ_t are functions of t only, i.e.

$$dX_t = \mu_t X_t dt + \sigma_t X_t dB_t.$$

Since $dX_t = X_t dR_t$, with $dR_t = \mu_t dt + \sigma_t dB_t$, X is the stochastic exponential of R, so that

$$X_t = X_0 \exp\Big(R_t - \frac{1}{2}[R,R]_t\Big) = X_0 \exp\Big(\int_0^t(\mu_s - \frac{1}{2}\sigma_s^2)ds + \int_0^t \sigma_s dB_s\Big).$$

We will show that σ_t must be zero for the process to have a Normal marginals, provided X_0 has a normal distribution. Moreover, the mean and variance a_t and γ_t must be of a certain exponential form of μ_t. Proposition 2 gives

$$\mu_t x = a'_t + x\sigma_t^2 + \frac{x - a_t}{2\gamma_t}(\gamma'_t - \sigma_t^2 x^2)$$

for all x and t. Grouping the terms,

$$\frac{\sigma_t^2}{2\gamma_t}x^3 - \frac{a_t\sigma_t^2}{2\gamma_t}x^2 + \left(\mu_t - \sigma_t^2 - \frac{\gamma'_t}{2\gamma_t}\right)x + \frac{a_t\gamma'_t}{2\gamma_t} - a'_t = 0,$$

thus, we must have

$$\sigma_t^2 = 0\,, \quad \mu_t = \frac{\gamma'_t}{2\gamma_t}\,, \quad a'_t = \mu_t a_t\,.$$

The last two equations give

$$\gamma_t = \gamma_0 \exp\left(2\int_0^t \mu_s ds\right) \qquad \text{and} \qquad a_t = a_0 \exp\left(\int_0^t \mu_s ds\right).$$

Note that the a_t and γ_t are consistent with the property of $X_t = X_0 e^{\int_0^t \mu_s ds}$ when X_0 is Normally distributed.

Example 3. Suppose $\mu(x,t) = \mu x$ and $\sigma(x,t) = \sigma$, where μ and σ are constant, i.e.

$$dX_t = \mu X_t dt + \sigma dB_t$$

and X is the Ornstein Uhlenbeck process. Under the assumption of Normal marginals, Proposition 2 gives

$$\mu x = a'_t + \frac{x - a_t}{2\gamma_t}(\gamma'_t - \sigma^2)\,.$$

Thus, we must have

$$\mu - \frac{\gamma'_t - \sigma^2}{2\gamma_t} = 0 \qquad \text{and} \qquad a'_t - \frac{a_t(\gamma'_t - \sigma^2)}{2\gamma_t} = 0.$$

Solving the two equations, we arrive at

$$a_t = a_0 e^{\mu t} \qquad \text{and} \qquad \gamma_t = e^{2\mu t}\left(\gamma_0 + \frac{\sigma^2}{2\mu}(1 - e^{-2\mu t})\right).$$

In this case equation (13) gives also sufficient conditions, because

$$X_t = e^{\mu t}\left(X_0 + \int_0^t \sigma e^{-\mu s} dB_s\right)$$

is a Gaussian process.

Example 4 (Fake Brownian motion diffusions). Here we give an example of a family of diffusions (indexed by σ) which are all Gaussian processes with Brownian marginals $N(0,t)$ but not a Brownian motion unless $\sigma^2 = 1$.

Let $\sigma > 0$ and $\alpha = \frac{1}{2}(\sigma^2 - 3)$. Define $(X_t)_{t \geq 0}$ by

$$X_t = \sigma t^{-(\alpha+1)} \int_0^t s^{\alpha+1} dB_s, \quad X_0 = 0.$$

Note that the Ito integral is well defined because $\alpha = -\frac{3}{2} + \frac{\sigma^2}{2} > -\frac{3}{2}$. For any $t > 0$, $\mathbb{E}[X_t] = 0$ and

$$Var(X_t) = \sigma^2 t^{-2(\alpha+1)} \int_0^t s^{2(\alpha+1)} ds = t$$

as $2\alpha + 3 = \sigma^2$. It is clear that X is continuous at 0 for $\alpha \leq -1$. From the calculation of variance, it follows that X is continuous at 0 in L^2 also for $\alpha > -1$. Thus, the process $(X_t)_{t \geq 0}$ is continuous at 0 for any $\sigma > 0$.

Note that X is a Gaussian process, with marginals $N(0,t)$. Obtaining the SDE of X, we have

$$dX_t = \frac{1 - \sigma^2}{2t} X_t dt + \sigma dB_t.$$

With

$$a_t = 0, \quad \gamma_t = t, \quad \mu(x,t) = \frac{1 - \sigma^2}{2t} x, \quad \sigma(x,t) = \sigma,$$

we see that equation (13) is indeed satisfied.

The process X is a Brownian motion only for $\sigma = 1$ (or $\alpha = -1$). For $\sigma \neq 1$, X is a martingale multiplied by a monotone function.

Remark that the covariance function, which also defines the process uniquely, is given by, for $u < t$,

$$Cov(X_u, X_t) = \sigma^2 (ut)^{-(\alpha+1)} \int_0^u s^{2\alpha+2} ds = u^{\alpha+2} t^{-\alpha-1}.$$

When $\sigma^2 = 1$ (or $\alpha = -1$), the above covariance function is that of Brownian motion.

4.2 *Diffusion with Lognormal marginals*

Suppose that the marginal distributions of X are $LN(a_t, \gamma_t)$. Let $C(x,t) = \mathbb{E}[(X_t - x)^+]$. The function C and its derivatives can be computed as follows:

$$\begin{aligned}
C(x,t) &= e^{a_t+\frac{1}{2}\gamma_t}\Phi\Big(\frac{a_t+\gamma_t-\ln x}{\sqrt{\gamma_t}}\Big)-x\Phi\Big(\frac{a_t-\ln x}{\sqrt{\gamma_t}}\Big),\\
\frac{\partial C}{\partial x}(x,t) &= -\Phi\Big(\frac{a_t-\ln x}{\sqrt{\gamma_t}}\Big),\\
\frac{\partial^2 C}{\partial x^2}(x,t) &= \frac{1}{x\sqrt{\gamma_t}}\phi\Big(\frac{a_t-\ln x}{\sqrt{\gamma_t}}\Big),\\
\frac{\partial^2 C}{\partial x\partial t}(x,t) &= \frac{(a_t-\ln x)\gamma_t'-2\gamma_t a_t'}{2\gamma_t^{3/2}}\phi\Big(\frac{a_t-\ln x}{\sqrt{\gamma_t}}\Big),\\
\frac{\partial^3 C}{\partial x^3}(x,t) &= \frac{a_t-\gamma_t-\ln x}{x^2\gamma_t^{3/2}}\phi\Big(\frac{a_t-\ln x}{\sqrt{\gamma_t}}\Big).
\end{aligned}$$

Proposition 3. *Assume that $(X_t)_{t\geq 0}$ has marginals $LN(a_t,\gamma_t)$ and satisfies the SDE (2). Then we must have*

$$\mu(x,t)=\frac{1}{2}\frac{\partial}{\partial x}\{\sigma^2(x,t)\}+\frac{a_t-\gamma_t-\ln x}{2x\gamma_t}\sigma^2(x,t)-\frac{x(a_t-\ln x)\gamma_t'}{2\gamma_t}+xa_t'.$$

Proof. This follows from (8). □

Example 5. Suppose that $\mu(x,t)=0$ and $a_t=a_0$. Under the assumption of Lognormal marginals, the diffusion coefficient must satisfy

$$x\gamma_t\frac{\partial}{\partial x}\{\sigma^2(x,t)\}+(a_0-\gamma_t-\ln x)\sigma^2(x,t)-x^2(a_0-\ln x)\gamma_t'=0,$$

due to Proposition 3. Suppose $\sigma(0,t)=0$. Solving for $\sigma^2(x,t)$, we obtain

$$\begin{aligned}
\sigma^2(x,t) &= \gamma_t' x^{2-\ln x/(2\gamma_t)}\exp\Big(\frac{(\ln x)^2}{2\gamma_t}\Big)\\
&\quad -\sqrt{2\pi\gamma_t}\gamma_t' x^{1-a_0/\gamma_t}\exp\Big(\frac{(a_0-\gamma_t)^2+(\ln x)^2}{2\gamma_t}\Big)\Phi\Big(\frac{\ln x-a_0-\gamma_t}{\sqrt{\gamma}}\Big)
\end{aligned}$$

as a necessary condition for matching the marginal distributions. In particular, if $a_0=0$ and $\gamma_t=t$, we must have

$$\sigma^2(x,t)=x^{2-\ln x/(2t)}\exp\Big(\frac{(\ln x)^2}{2t}\Big)-\sqrt{2\pi t}x\exp\Big(\frac{t^2+(\ln x)^2}{2t}\Big)\Phi\Big(\frac{\ln x-t}{\sqrt{t}}\Big)$$

for the SDE $dX_t=\sigma(X_t,t)dB_t$ to have the $LN(0,t)$ marginal distributions.

Example 6. Suppose $\mu(x,t)=\mu_t x$ and $\sigma(x,t)=\sigma_t x$, where μ_t and σ_t are functions of t only, i.e.

$$dX_t=\mu_t X_t dt+\sigma_t X_t dB_t.$$

Under the Lognormal regime, Proposition 3 gives

$$\mu_t x = x\sigma_t^2 + \frac{a_t - \gamma_t - \ln x}{2\gamma_t} x\sigma_t^2 - \frac{x(a_t - \ln x)\gamma_t'}{2\gamma_t} + xa_t',$$

that is,

$$\mu_t - a_t' - \left(\frac{1}{2} + \frac{a_t}{2\gamma_t}\right)\sigma_t^2 + \frac{a_t}{2\gamma_t}\gamma_t' = \left(\frac{\gamma_t'}{2\gamma_t} - \frac{\sigma_t^2}{2\gamma_t}\right)\ln x.$$

Therefore, we must have

$$\sigma_t^2 = \gamma_t' \qquad \text{and} \qquad \mu_t - a_t' - \frac{1}{2}\sigma_t^2 = 0. \tag{15}$$

Checking this with the solution to the SDE,

$$X_t = X_0 e^{\int_0^t (\mu_s - \frac{1}{2}\sigma_s^2)ds + \int_0^t \sigma_s dB_s},$$

which has marginal distribution

$$LN\left(\ln x_0 + \int_0^t \left(\mu_s - \frac{1}{2}\sigma_s^2\right)ds, \int_0^t \sigma_s^2 ds\right),$$

(15) is indeed satisfied.

Example 7 (Fake Lognormal diffusions). Let $Y_t = e^{X_t}$, where X is the process in Example 4. Then Y has marginal distribution $LN(0,t)$ and

$$\begin{aligned} dY_t &= e^{X_t} dX_t + \frac{1}{2} e^{X_t} (dX_t)^2 \\ &= \sigma Y_t dB_t + \left(\frac{1}{2}\sigma^2 - (\alpha+1)\frac{1}{t}\ln Y_t\right) Y_t dt. \end{aligned}$$

That is, $\mu(x,t) = (\frac{1}{2}\sigma^2 - (\alpha+1)\frac{1}{t}\ln x)x$ and $\sigma(x,t) = \sigma x$. We can see that, as $\alpha + 1 = \frac{\sigma^2 - 1}{2}$, we have

$$\begin{aligned} &\frac{1}{2}\frac{\partial}{\partial x}\{\sigma^2(x,t)\} + \frac{a_t - \gamma_t - \ln x}{2x\gamma_t}\sigma^2(x,t) - \frac{x(a_t - \ln x)\gamma_t'}{2\gamma_t} + xa_t' \\ &= \frac{1}{2}\sigma^2 x + (1-\sigma^2)\frac{x\ln x}{2t} \\ &= \mu, \end{aligned}$$

which is consistent with Proposition 3.

4.3 Selfsimilar diffusion

In a more general case, suppose that a diffusion process $(X_t)_{t\geq 0}$, satisfying SDE (2), is selfsimilar. Its marginal densities $(p_t(\cdot))_{t\geq 0}$ satisfies the scaling property $p_t(y) =$

$t^{-\kappa}p_1(yt^{-\kappa})$ for all t, for some $\kappa > 0$. We also have

$$\frac{\partial C}{\partial x} = -\mathbb{P}(X_t \geq x) = \mathbb{P}(X_t < x) - 1 = \mathbb{P}(X_1 < xt^{-\kappa}) - 1$$
$$= \int_{-\infty}^{xt^{-\kappa}} p_1(y)dy = \int_{-\infty}^{t} -\kappa xy^{-\kappa-1}p_1(xy^{-\kappa})dy$$

and

$$\frac{\partial^2 C}{\partial x \partial t} = -\kappa xt^{-\kappa-1}p_1(xt^{-\kappa}).$$

Then, Equation (8) simplifies to

$$\kappa xt^{-1}p_1(xt^{-\kappa}) + \left(\frac{1}{2}\frac{\partial}{\partial x}\{\sigma^2(x,t)\} - \mu(x,t)\right)p_1(xt^{-\kappa}) + \frac{1}{2}\sigma^2(x,t)t^{-\kappa}p_1'(xt^{-\kappa}) = 0,$$

or,

$$\kappa xt^{-1}p_t(x) + \left(\frac{1}{2}\frac{\partial}{\partial x}\{\sigma^2(x,t)\} - \mu(x,t)\right)p_t(x) + \frac{1}{2}\sigma^2(x,t)\frac{\partial}{\partial x}\{p_t(x)\} = 0.$$

Acknowledgements This research was supported by the Australian Research Council grant DP220100973.

References

1. Albin, J.M.P. (2008) A continuous non-Brownian motion martingale with Brownian motion marginal distributions. *Statistics and Probability Letters* **78**, 682–686.
2. Baker, D., Donati-Martin, C., Yor, M. (2011) A sequence of Albin type continuous martingales with Brownian marginals and scaling. *Séminaire de Probabilités XLIII, Lecture Notes in Mathematics* **2006**, 441–449.
3. Beiglböck, M., Lowther, G., Pammer, G., Schachermayer, W. (2021) Faking Brownian motion with continuous Markov martingales. *arXiv preprint arXiv:2109.12927.*
4. Brunick, G., Shreve, S. (2013) Mimicking an Itô process by a solution of a stochastic differential equation. *The Annals of Applied Probability* **23**(4), 1584–1628.
5. Derman, E., Kani, I. (1994) Riding on a smile. *Risk* **7**(2), 32–39.
6. Dupire, B. (1994) Pricing with a smile. *Risk* **7**(1), 18–20.
7. Fan, J.Y., Hamza, K., Klebaner, F.C. (2015) Mimicking self-similar processes. *Bernoulli* **21**(3), 1341–1360.
8. Gyöngy, I. (1986) Mimicking the one-dimensional marginal distributions of processes having an Ito differential. *Probability Theory and Related Fields* **71**, 501–516.
9. Hamza, K., Klebaner, F.C. (2007) A family of non-Gaussian martingales with Gaussian marginals. *Journal of Applied Mathematics and Stochastic Analysis* 2007. Article ID 92723, 19 pages.
10. Hamza, K., Klebaner, F.C. (2022) Expectation of local times and the Dupire formula. *Stochastic Processes and their Aplications* **150**, 782–787.
11. Hirsch, F., Profeta, C., Roynette, B., Yor, M. (2011) *Peacocks and Associated Martingales, with Explicit Constructions.* Springer-Verlag.

12. Hobson, D.G. (2013) Fake exponential Brownian motion. *Statistics and Probability Letters* **83**, 2386–2390.
13. Klebaner, F. (2002) Option price when the stock is a semimartingale. *Electronic Communications in Probability* **7**, 79–83.
14. Klebaner, F.C. (2012) *Introduction to Stochastic Calculus with Applications*. 3rd Edition. Imperial College Press.
15. Mudakkar, S.R. (2012) A necessary characteristic equation of diffusion processes having Gaussian marginals. *Abstract and Applied Analysis* 2012. Article ID 598590, 9 pages.
16. Oleszkiewicz, K. (2008) On fake Brownian motions. *Statistics and Probability Letters* **78**, 1251–1254.

Multi-Factor Polynomial Diffusion Models and Inter-Temporal Futures Dynamics

Peilun He, Nino Kordzakhia, Gareth W. Peters, Pavel V. Shevchenko

Abstract In stochastic multi-factor commodity models, it is often the case that futures prices are explained by two latent state variables which represent the short and long term stochastic factors. In this work, we develop the family of stochastic models using polynomial diffusion to obtain the unobservable spot price to be used for modelling futures curve dynamics. The polynomial family of diffusion models allows one to incorporate a variety of non-linear, higher-order effects, into a multi-factor stochastic model, which is a generalisation of Schwartz and Smith [17] two-factor model. Two filtering methods are used for the parameter and the latent factor estimation to address the non-linearity. We provide a comparative analysis of the performance of the estimation procedures. We discuss the parameter identification problem present in the polynomial diffusion case, regardless, the futures prices can still be estimated accurately. Moreover, we study the effects of different methods of calculating matrix exponential in the polynomial diffusion model. As the polynomial order increases, accurately and efficiently approximating the high-dimensional matrix exponential becomes essential in the polynomial diffusion model.

Peilun He
Macquarie University, Macquarie Park, NSW 2109, Australia, e-mail: peilun.he@mq.edu.au

Nino Kordzakhia
Macquarie University, Macquarie Park, NSW 2109, Australia, e-mail: nino.kordzakhia@mq.edu.au

Gareth W. Peters
University of California Santa Barbara, Santa Barbara, CA 93106, United States, e-mail: gareth-peters@ucsb.edu

Pavel V. Shevchenko
Macquarie University, Macquarie Park, NSW 2109, Australia, e-mail: pavel.shevchenko@mq.edu.au

D. R. Wood et al. (eds.), *2021-2022 MATRIX Annals*, MATRIX Book Series 5,
https://doi.org/10.1007/978-3-031-47417-0_18

1 Introduction

Stochastic models employed in the analysis of commodity futures play an important role in various financial areas, including price forecasting, risk management, portfolio optimisation. In contrast to other commodities that are traded in both spot and futures markets, crude oil is primarily traded in the futures market. As a consequence, it becomes impractical to directly estimate the price of crude oil futures based on the spot price. Instead, people usually model the underlying spot price, denoted as S_t, as a function of certain factors. Under the assumption of an arbitrage-free market, the futures price at current time t, denoted as $F_{t,T}$, is equivalent to the expected spot price at maturity time T:

$$F_{t,T} = \mathbb{E}^*(S_T|\mathscr{F}_t), \tag{1}$$

where $\mathscr{F}_t$ be a natural σ-algebra generated up to time t and $\mathbb{E}^*(\cdot)$ is the expectation taken with respect to the risk-neutral processes. Under this framework, the derivation of a closed-form expression of $F_{t,T}$ necessitates an accurate distribution of S_t. Consequently, this requirement imposes additional constraints on the factors involved in the modelling process.

Over the past few decades, stochastic processes have been employed to model the factors. In 1990, the Ornstein-Uhlenbeck (OU) process was introduced for the modelling of oil futures in a two-factor setup to represent spot price and convenience yield [10]. Building upon this work, Schwartz and Smith [17] modelled the logarithm of the underlying spot price of crude oil futures as the sum of two hidden factors. These factors, assumed to follow the OU process, capture short-term fluctuation and long-term equilibrium price level, respectively. Subsequently, this latent factor model and its extensions became widely utilized in stochastic modelling.

Researchers have further extended this model to enhance its applicability. In the electricity market, a multi-factor model including a deterministic seasonality with additional stochastic factors were modelled by Levy processes in [6]. A time-changed Levy process were commonly used in option pricing to describe the jump behaviour and dynamics of volatility in [3, 7, 11]. Sorensen [18] extended the model by introducing three hidden factors, including an additional deterministic seasonal component, to capture the dynamics of agricultural commodity prices. In [12], the authors focused on direct modelling of the electricity futures prices instead of modelling of electricity spot price first. Ames et al. [1] incorporated time-varying drift and speed of mean reversion parameters in their modelling of crude oil futures. Favetto and Samson [8] applied this model in the field of biology, and used both maximum likelihood and expectation maximization methods for parameter estimations. Further, the performance of the multi-factor model in deriving spot prices was improved by incorporating the analyst's forecasts for futures prices, as demonstrated in [4]. Peters et al. [15] developed a partial Markov Chain Monte Carlo method to deal with the non-linear non-Gaussian multi-factor model. Comparing different models is also of significance. Schwartz [16] compared models with up to three factors, including a hidden factor, convenience yield and interest rate, for copper, oil and gold. Cortazar and Naranjo [5] compared the performance of one- to four-factor models in crude oil futures, and found that the three-factor and four-factor models excelled in explaining the term structure of futures prices, while the four-factor model outperformed others in fitting the volatility term structure.

While this framework has gained popularity in the past two decades, it does possess certain limitations. Firstly, as previously mentioned, obtaining a closed-form expression of futures price necessitates an accurate distribution of spot price S_t. To address this issue, it is common practice to assume that all factors involved follow a Gaussian distribution, and the logarithm of spot price is a linear function of these factors. This assumption ensures a log-normal distribution of the spot price. Secondly, under this framework, it is common to model the logarithm of spot price. In most instances, this poses no significant problems. However, a noticeable event occurred on 20th April 2020 when the front-month May 2020 WTI crude oil futures settled at a unprecedented value of -$37.63 per barrel on the New York Mercantile Exchange. This exceedingly rare phenomenon significantly changes the validity of the entire framework.

This paper aims to address the aforementioned limitations by introducing a polynomial diffusion framework that allows for a more complicated structure of the spot price. The mathematical foundations were introduced in [9]. Under the polynomial diffusion framework, the spot price is represented as a polynomial of any order in terms of the factors. In particular, under certain conditions, it can be proven that the conditional expectation of spot price, which is equivalent to the futures price under the assumption of an arbitrage-free market, is also a polynomial in terms of factors. An application of this framework can be found in the modelling of electricity forwards, where the spot price is represented by a quadratic form of two factors [13]. Additionally, in this study, the Quadratic Kalman Filter was employed to estimate model parameters and unknown state variables. The state space was augmented to include both linear and quadratic terms of the factors. However, as the order of the polynomial increases, the dimension of the state space grows exponentially, making it challenging to derive the explicit form of the augmented state equations. To address this issue, we propose the use of the Extended Kalman Filter (EKF) and Unscented Kalman Filter (UKF). These methods provide effective tools for estimating parameters and the state variables within the context of the polynomial diffusion framework. The EKF and UKF help overcome the computational challenges associated with higher-order polynomial models, enabling the practical implementation of this framework.

This paper is structured as follows. In Sect. 2 and Sect. 3, we present the two aforementioned frameworks for the pricing of commodity futures. The first framework extends the Schwartz-Smith two-factor model [17], while the second framework models the spot price using polynomial forms. In the second framework, polynomial diffusion is employed to price the futures contracts. Sect. 4 introduces the Extended Kalman Filter (EKF) and Unscented Kalman Filter (UKF) as estimation methods for the hidden factors and unknown parameters in the polynomial diffusion models. These filters are specifically designed to handle the non-linear dynamics present in the models. Sect. 5 presents a numerical analysis of the applications of the polynomial diffusion model. Firstly, we compare seven different methods for calculating the matrix exponential, which is required in the polynomial diffusion model. Our results indicate that the eigen-decomposition method provides an efficient and accurate approximation of the matrix exponential. Next, we evaluate the performance of the polynomial diffusion model through a simulation study. While the futures contracts can be estimated accurately, parameter estimation remains challenging, even when separating parameters in the state equation and the coordinate representations in the measurement equation. As a consequence, selecting the order of the polynomial diffusion model proves to be a challenging task, and further study on constraints is required. Finally, Sect. 6 concludes the paper.

2 Schwartz-Smith Two-Factor Model

In this section, we describe a classical approach to modelling commodity futures, which is an extension of [17].

This approach models the logarithm of spot price S_t as the sum of two unobservable factors χ_t and ξ_t,

$$\log(S_t) = \chi_t + \xi_t, \tag{2}$$

where χ_t represents the short-term fluctuation and ξ_t is the long-term equilibrium price level. We assume both χ_t and ξ_t follow an OU process,

$$d\chi_t = -\kappa\chi_t dt + \sigma_\chi dW_t^\chi, \tag{3}$$

and

$$d\xi_t = (\mu_\xi - \gamma\xi_t)dt + \sigma_\xi dW_t^\xi, \tag{4}$$

while in [17] only one factor follows the OU process. We assume the changes in the short-term factor χ_t are temporary and converging to 0 as $t \to \infty$. The processes $(W_t^\chi)_{t\geq 0}$ and $(W_t^\xi)_{t\geq 0}$ are correlated standard Brownian Motions with correlation coefficient ρ. Here, $\kappa, \gamma \in \mathbb{R}^+$ are the speed of mean-reversion parameters; $\mu_\xi \in \mathbb{R}$ is the mean level of the long-term factor; $\sigma_\chi, \sigma_\xi \in \mathbb{R}^+$ are the volatility parameters; and $\lambda_\chi, \lambda_\xi \in \mathbb{R}$ are risk premiums.

By assuming a constant risk premium λ_χ and λ_ξ, the risk-neutral processes of χ_t and ξ_t are given by

$$d\chi_t = (-\kappa\chi_t - \lambda_\chi)dt + \sigma_\chi dW_t^{\chi *}, \tag{5}$$

and

$$d\xi_t = (\mu_\xi - \gamma\xi_t - \lambda_\xi)dt + \sigma_\xi dW_t^{\xi *}, \tag{6}$$

where $W_t^{\chi *}$ and $W_t^{\xi *}$ are correlated standard Brownian Motions with correlation coefficient ρ. This approach stems from the risk-neutral futures pricing theory developed in [2].

Let $\mathscr{F}_t$ be a natural σ-algebra generated up to time t. In discrete time, given the initial values χ_{t_0} and ξ_{t_0}, χ_t and ξ_t are jointly normally distributed with mean

$$\mathbb{E}^*\left(\begin{bmatrix}\chi_t\\ \xi_t\end{bmatrix}\middle|\mathscr{F}_{t_0}\right)=\begin{bmatrix}e^{-\kappa(t-t_0)}\chi_{t_0}-\frac{\lambda_\chi}{\kappa}\left(1-e^{-\kappa(t-t_0)}\right)\\ e^{-\gamma(t-t_0)}\xi_{t_0}+\frac{\mu_\xi-\lambda_\xi}{\gamma}\left(1-e^{-\gamma(t-t_0)}\right)\end{bmatrix}$$

and covariance matrix

$$Cov^*\left(\begin{bmatrix}\chi_t\\ \xi_t\end{bmatrix}\middle|\mathscr{F}_{t_0}\right)=\begin{bmatrix}\frac{1-e^{-2\kappa(t-t_0)}}{2\kappa}\sigma_\chi^2 & \frac{1-e^{-(\kappa+\gamma)(t-t_0)}}{\kappa+\gamma}\sigma_\chi\sigma_\xi\rho\\ \frac{1-e^{-(\kappa+\gamma)(t-t_0)}}{\kappa+\gamma}\sigma_\chi\sigma_\xi\rho & \frac{1-e^{-2\gamma(t-t_0)}}{2\gamma}\sigma_\xi^2\end{bmatrix},$$

where $\mathbb{E}^*(\cdot)$ and $Cov^*(\cdot)$ represent the expectation and covariance taken with respect to the risk-neutral processes. Therefore, the spot price, which is defined in (2), is log-normally distributed with

$$\begin{aligned}\log[\mathbb{E}^*(S_t|\mathscr{F}_{t_0})]&=\mathbb{E}^*[\log(S_t)|\mathscr{F}_{t_0}]+\frac{1}{2}Var^*[\log(S_t)|\mathscr{F}_{t_0}]\\&=e^{-\kappa(t-t_0)}\chi_{t_0}+e^{-\gamma(t-t_0)}\xi_{t_0}+A(t-t_0),\end{aligned}\tag{7}$$

where

$$\begin{aligned}A(t)=&-\frac{\lambda_\chi}{\kappa}(1-e^{-\kappa t})+\frac{\mu_\xi-\lambda_\xi}{\gamma}(1-e^{-\gamma t})\\&+\frac{1}{2}\left(\frac{1-e^{-2\kappa t}}{2\kappa}\sigma_\chi^2+\frac{1-e^{-2\gamma t}}{2\gamma}\sigma_\xi^2+2\frac{1-e^{-(\kappa+\gamma)t}}{\kappa+\gamma}\sigma_\chi\sigma_\xi\rho\right).\end{aligned}\tag{8}$$

Next, we derive the equations for the futures prices. Let $F_{t,T}$ be the market price of a futures contract at time t with maturity time T. For eliminating arbitrage, given all information until time t, the futures price must be equal to the expected spot price at the maturity time T. Therefore, under the risk-neutral measure, we have (assuming the interest rate is not stochastic)

$$\log(F_{t,T})=\log\left[\mathbb{E}^*(S_T|\mathscr{F}_t)\right]=e^{-\kappa(T-t)}\chi_t+e^{-\gamma(T-t)}\xi_t+A(T-t).$$

After discretization, we have the following AR(1) dynamics for bivariate state variable x_t

$$x_t=c+Ex_{t-1}+w_t,\tag{9}$$

where

$$x_t=\begin{bmatrix}\chi_t\\ \xi_t\end{bmatrix},\ c=\begin{bmatrix}0\\ \frac{\mu_\xi}{\gamma}\left(1-e^{-\gamma\Delta t}\right)\end{bmatrix},\ E=\begin{bmatrix}e^{-\kappa\Delta t} & 0\\ 0 & e^{-\gamma\Delta t}\end{bmatrix},$$

and w_t is a column vector of correlated normally distributed noises with $\mathbb{E}(w_t)=\mathbf{0}$ and

$$Cov(w_t)=\Sigma_w=\begin{bmatrix}\frac{1-e^{-2\kappa\Delta t}}{2\kappa}\sigma_\chi^2 & \frac{1-e^{-(\kappa+\gamma)\Delta t}}{\kappa+\gamma}\sigma_\chi\sigma_\xi\rho\\ \frac{1-e^{-(\kappa+\gamma)\Delta t}}{\kappa+\gamma}\sigma_\chi\sigma_\xi\rho & \frac{1-e^{-2\gamma\Delta t}}{2\gamma}\sigma_\xi^2\end{bmatrix},$$

Δt is the time step between $(t-1)$ and t. Moreover, we have the measurement equation

$$y_t = d_t + F_t^\top x_t + v_t, \tag{10}$$

where

$$y_t = (\log(F_{t,T_1}),\ldots,\log(F_{t,T_m}))^\top, d_t = (A(T_1-t),\ldots,A(T_m-t))^\top,$$

$$F_t = \begin{bmatrix} e^{-\kappa(T_1-t)},\ldots,e^{-\kappa(T_m-t)} \\ e^{-\gamma(T_1-t)},\ldots,e^{-\gamma(T_m-t)} \end{bmatrix},$$

and m is the number of futures contracts. v_t is an m-dimensional vector of normally distributed noises with $\mathbb{E}(v_t) = \mathbf{0}$ and

$$Cov(v_t) = \Sigma_v = \begin{bmatrix} \sigma_1^2 & 0 & \ldots & 0 \\ 0 & \sigma_2^2 & \ldots & 0 \\ \vdots & \vdots & \ddots & \vdots \\ 0 & 0 & \ldots & \sigma_m^2 \end{bmatrix}.$$

The prediction error $e_t = y_t - \mathbb{E}(y_t|\mathscr{F}_{t-1})$ are supposed to be multivariate normally distributed. Therefore, the log-likelihood function of $y = (y_1,\ldots,y_n)$ can be written as

$$l(\theta;y) = -\frac{nm\log(2\pi)}{2} - \frac{1}{2}\sum_{t=1}^{n}\left[\log[\det(L_t)] + e_t^\top L_t^{-1} e_t\right], \tag{11}$$

where the set of unknown parameters $\theta = (\kappa,\gamma,\mu_\xi,\sigma_\chi,\sigma_\xi,\rho,\lambda_\chi,\lambda_\xi,\sigma_1,\ldots,\sigma_m)$; n is the number of observations; m is the number of contracts; $L_t = Cov(e_t)$. Given all observations y, the maximum likelihood estimate (MLE) of θ is obtained by maximising the log-likelihood function (11).

3 Polynomial Diffusion Model

In this section, we provide some important theorems of polynomial diffusions and apply these theorems to the two-factor model. The mathematical foundation and applications of polynomial diffusions in finance are provided in [9].

Definition 1. Consider the stochastic differential equation

$$dX_t = b(X_t)dt + \sigma(X_t)dW_t, \tag{12}$$

where W_t is a d-dimensional standard Brownian motion and map $\sigma : \mathbb{R}^d \to \mathbb{R}^{d\times d}$ is continuous. Define $a := \sigma\sigma^\top$. For maps $a : \mathbb{R}^d \to \mathbb{S}^d$ and $b : \mathbb{R}^d \to \mathbb{R}^d$, suppose we have $a_{ij} \in Pol_2$ and $b_i \in Pol_1$. $\mathbb{S}^d$ is the set of all real symmetric $d \times d$ matrices and

Pol_n is the set of all polynomials of degree at most n. Then the solution of (12) is a polynomial diffusion.

Moreover, we define the generator $\mathscr{G}$ associated with the polynomial diffusion X_t as

$$\mathscr{G} f(x) = \frac{1}{2} Tr\left(a(x)\nabla^2 f(x)\right) + b(x)^\top \nabla f(x) \tag{13}$$

for $x \in \mathbb{R}^d$ and any C^2 function f. Let $N = C(d+n, n)$ be the dimension of Pol_n, and $H : \mathbb{R}^d \to \mathbb{R}^N$ be a function whose components form a basis of Pol_n. Then for any $p \in Pol_n$, there exists a unique vector $\vec{p} \in \mathbb{R}^N$ such that

$$p(x) = H(x)^\top \vec{p} \tag{14}$$

and $\vec{p}$ is the coordinate representation of $p(x)$. Moreover, there exists a unique matrix representation $G \in \mathbb{R}^{N\times N}$ of the generator $\mathscr{G}$, such that $G\vec{p}$ is the coordinate vector of $\mathscr{G}p$. So we have

$$\mathscr{G} p(x) = H(x)^\top G\vec{p}. \tag{15}$$

Theorem 1. *Let $p(x) \in Pol_n$ be a polynomial with coordinate representation $\vec{p} \in \mathbb{R}^N$ satisfying* (14)*, $G \in \mathbb{R}^{N\times N}$ be a matrix representation of generator $\mathscr{G}$ satisfying* (15)*, and $X_t \in \mathbb{R}^d$ satisfy* (12)*. Then for $0 \le t \le T$, we have*

$$\mathbb{E}\left[p(X_T)|\mathscr{F}_t\right] = H(X_t)^\top e^{(T-t)G}\vec{p},$$

where $\mathscr{F}_t$ represents all information available until time t.

Proof. The proof is given in [9]. □

Next, we apply this theorem to the two-factor model. Assume the spot price S_t is modelled as

$$S_t = p_n(x_t), \tag{16}$$

where $x_t = (\chi_t, \xi_t)^\top$ is a vector of state variables and $p_n(\cdot)$ is a polynomial function with a degree at most n. χ_t and ξ_t are the short-term and long-term factors defined in (3) and (4) for real processes and (5) and (6) for risk-neutral processes. Obviously, x_t satisfies the stochastic differential equation (12), with

$$b(x_t) = \begin{bmatrix} -\kappa\chi_t - \lambda_\chi \\ \mu_\xi - \gamma\xi_t - \lambda_\xi \end{bmatrix}, \sigma(x_t) = \begin{bmatrix} \sigma_\chi & 0 \\ 0 & \sigma_\xi \end{bmatrix}, a(x_t) = \sigma(x_t)\sigma(x_t)^\top = \begin{bmatrix} \sigma_\chi^2 & 0 \\ 0 & \sigma_\xi^2 \end{bmatrix}.$$

For any basis $H_n(x_t)$, the polynomial $p_n(x_t)$ can be uniquely represented as

$$p_n(x_t) = H_n(x_t)^\top \vec{p}.$$

The generator $\mathscr{G}$ is given by

$$\mathscr{G} f(x) = \frac{1}{2} Tr\left(\begin{bmatrix} \sigma_\chi^2 & 0 \\ 0 & \sigma_\xi^2 \end{bmatrix} \nabla^2 f(x)\right) + \begin{bmatrix} -\kappa\chi_t - \lambda_\chi \\ \mu_\xi - \gamma\xi_t - \lambda_\xi \end{bmatrix}^\top \nabla f(x).$$

By applying $\mathscr{G}$ to each element of $H_n(x_t)$, we get the matrix representation G. Then, by Theorem 1, the futures price $F_{t,T}$ is given by

$$F_{t,T} = \mathbb{E}^*(S_T|\mathscr{F}_t) = H(x_t)^\top e^{(T-t)G}\vec{p}. \tag{17}$$

Therefore, we have the non-linear state-space model

$$x_t = c + Ex_{t-1} + w_t, w_t \sim N(\mathbf{0}, \Sigma_w), \tag{18}$$

and

$$y_t = H_n(x_t)^\top e^{(T-t)G}\vec{p} + v_t, v_t \sim N(\mathbf{0}, \Sigma_v). \tag{19}$$

In this paper, we consider a polynomial with degree 2,

$$S_t = \alpha_1 + \alpha_2\chi_t + \alpha_3\xi_t + \alpha_4\chi_t^2 + \alpha_5\chi_t\xi_t + \alpha_6\xi_t^2.$$

The basis of Pol_2 is

$$H(x_t) = (1, \chi_t, \xi_t, \chi_t^2, \chi_t\xi_t, \xi_t^2)^\top,$$

which has a dimension $N = 6$. The polynomial S_t can be uniquely represented as

$$S_t = H(x_t)^\top \vec{p},$$

where the coordinate representation $\vec{p}$ is given by

$$\vec{p} = (\alpha_1, \alpha_2, \alpha_3, \alpha_4, \alpha_5, \alpha_6)^\top.$$

Then, applying $\mathscr{G}$ to each element of $H(x_t)$, we get

$$G = \begin{bmatrix} 0 & -\lambda_\chi & \mu_\xi - \lambda_\xi & \sigma_\chi^2 & 0 & \sigma_\xi^2 \\ 0 & -\kappa & 0 & -2\lambda_\chi & \mu_\xi - \lambda_\xi & 0 \\ 0 & 0 & -\gamma & 0 & -\lambda_\chi & 2\mu_\xi - 2\lambda_\xi \\ 0 & 0 & 0 & -2\kappa & 0 & 0 \\ 0 & 0 & 0 & 0 & -\kappa-\gamma & 0 \\ 0 & 0 & 0 & 0 & 0 & -2\gamma \end{bmatrix}.$$

4 Non-Linear Filtering Algorithm

Kalman Filter (KF) is the most popular filtering method for estimating the state vector $x_t = (\chi_t, \xi_t)^\top$ based on a filtration $\mathscr{F}_t$. Moreover, the unknown parameters can be estimated jointly by maximising the log-likelihood function which is calculated using the KF. However, the KF can only deal with linear Gaussian state-space models. For dealing with the non-linearity of the polynomial diffusion model, we introduce two extensions of KF, the Extended Kalman Filter (EKF) and Unscented Kalman Filter (UKF).

In this paper, we use the notation

$$a_{t|t-1} := \mathbb{E}(x_t|\mathscr{F}_{t-1}), \qquad P_{t|t-1} := Cov(x_t|\mathscr{F}_{t-1}),$$
$$a_t := \mathbb{E}(x_t|\mathscr{F}_t), \qquad P_t := Cov(x_t|\mathscr{F}_t)$$

to represent the expected values and covariance matrices.

The first non-linear filtering method is EKF. In order to capture the non-linear dynamics in the polynomial diffusion model, EKF linearises the state and measurement equations. Consider the following non-linear dynamic system:

$$x_t = f(x_{t-1}) + w_t, w_t \sim N(\mathbf{0}, \Sigma_w), \tag{20}$$

$$y_t = h(x_t) + v_t, v_t \sim N(\mathbf{0}, \Sigma_v). \tag{21}$$

where $f(\cdot)$ and $h(\cdot)$ are non-linear functions; x_t is the unobservable state vector; y_t is the observation. The main idea of EKF is to linearise the functions $f(\cdot)$ and $h(\cdot)$ by the first-order Taylor series. Let J_f and J_h be the Jacobian of $f(\cdot)$ and $h(\cdot)$ respectively.

Given the prior mean and a new observation at current time t, we calculate the posterior mean $a_{t|t-1}$ and prediction error e_t as

$$a_{t|t-1} = f(a_{t-1}), \tag{22}$$

$$e_t = y_t - h(a_{t|t-1}). \tag{23}$$

However, because of the non-linearity in the state and measurement equations, we cannot get a closed-form of the posterior covariance $P_{t|t-1}$ and the covariance of prediction error L_t. In that case, we approximate $f(\cdot)$ and $h(\cdot)$ by first-order Taylor expansion at point a_{t-1} and $a_{t|t-1}$ respectively:

$$f(x_{t-1}) = f(a_{t-1}) + J_f(a_{t-1})(x_{t-1} - a_{t-1}), \tag{24}$$

$$h(x_t) = h(a_{t|t-1}) + J_h(a_{t|t-1})(x_t - a_{t|t-1}). \tag{25}$$

Therefore, $P_{t|t-1}$ and L_t are calculated as:

$$P_{t|t-1} = J_f(a_{t-1})P_{t-1}J_f^\top(a_{t-1}) + \Sigma_w, \tag{26}$$

$$L_t = J_h(a_{t|t-1})P_{t|t-1}J_h^\top(a_{t|t-1}) + \Sigma_v. \tag{27}$$

Finally, we update the prior mean and covariance at current time t as:

$$a_t = a_{t|t-1} + K_t e_t, \tag{28}$$

$$P_t = \left(I - K_t J_h(a_{t|t-1})\right) P_{t|t-1}, \tag{29}$$

where $K_t = P_{t|t-1}J_h^\top(a_{t|t-1})L_t^{-1}$ is the Kalman gain matrix.

In the EKF, the state distribution is propagated by linearising the non-linear system using the first-order approximation. However, this linearisation process can introduce significant errors in the true state distribution, especially when the system exhibits a strong non-linearity. Additionally, obtaining an analytical Jacobian for complicated state and measurement equations may be impractical. To address these issues, we introduce a derivative-free filtering method called the Unscented Kalman Filter (UKF). Instead of linearising the system, the UKF employs a set of carefully selected points, known as sigma points, to represent the true distributions of the state variables. These sigma points are then propagated through the state equation. The true prior and posterior means, and covariance would be captured by the sigma points.

At previous time $t-1$, the sigma points are defined as

$$\mathscr{X}_{t-1} = \left[a_{t-1}, a_{t-1} \pm \sqrt{(n_x+\lambda)P_{t-1}}\right], \tag{30}$$

where n_x is the number of state variables. In this paper, we set the scaling parameter $\lambda = 0$. A detailed description is available in [19].

Next, these sigma points are propagated through the non-linear state and measurement equations:

$$\mathscr{X}_{t|t-1} = f(\mathscr{X}_{t-1}), \tag{31}$$

$$\mathscr{Y}_{t|t-1} = h(\mathscr{X}_{t|t-1}). \tag{32}$$

The posterior mean $a_{t|t-1}$, one-step forecast $\hat{y}_{t|t-1}$ and covariances are weighted sample mean / covariance of sigma points. The remaining steps in the UKF are similar to those in the EKF.

5 Results

In this section, we present the outcomes of our numerical experiments. In Sect. 5.1, we compare seven distinct numerical methods for the evaluation of a matrix exponential. Evaluating the matrix exponential of the matrix G is necessary for obtaining a closed-form of futures prices according to Theorem 1. With the increase in the degree of the polynomial, the dimension of G grows exponentially. In Sect. 5.2, we assess the performance of the polynomial diffusion model through a simulation study.

5.1 Matrix Exponential

Theorem 1 gives a direct way of deriving the futures price with the spot price given in a polynomial form. However, the computation of matrix exponential $e^{(T-t)G}$ is required. As the degree of polynomial increases, the dimension of the G matrix increases quickly. As a consequence, one must carefully choose the numerical methods to compute the matrix exponential. A discussion of different methods to compute matrix exponential is given in [14]. We compare different methods to compute e^A in the following three aspects:

Stability: If small changes in matrix A cause large changes in e^A. It is evaluated by

$$\phi = \frac{||e^{A+E} - e^A||}{||e^A||},$$

where $||\cdot||$ represents 2-norm and E is a matrix whose norm is small.

Accuracy: The difference between an approximation and the true value. As we know the vector of eigenvalues Λ and the matrix of corresponding eigenvectors V (as shown in the steps of generating matrix), the true value is calculated as $e^A = Ve^{\Lambda}V^{-1}$. Let $B = \{b_{ij}\}$ be the matrix exponential calculated by one method and $C = \{c_{ij}\}$ be the true value of the matrix exponential, then the accuracy is evaluated as

$$\psi = \sum_{i=1}^{n}\sum_{j=1}^{n}(b_{ij} - c_{ij})^2.$$

Efficiency: The computing time required for this methods.

Seven different methods are tested on 100 realisations of 10×10 random matrices. The matrix is generated by the following steps (where the dimension $n = 10$):

1. Generate a n-dimensional vector of eigenvalues $\Lambda = (\lambda_1, ..., \lambda_n)$, where $\lambda_i \sim N(0, 10^2)$.
2. Generate a diagonal matrix V, with diagonal elements Λ.
3. Generate a $n \times n$ matrix $U = \{u_{ij}\}$, where $u_{ij} \sim N(0,1)$. Then, divide each column of U by its L_2 norm. Therefore, each column of U forms an eigenvector.
4. Calculate $A = U * V * U^{-1}$.

Table 1 Stability, accuracy and efficiency of the listed methods.

Methods	Stability	Accuracy	Efficency
Taylor series	1.7184	10.0214	0.1038
Pade approximation	1.7182	7.1447e+22	0.2415
Scaling and squaring	1.7183	4.3527	0.0657
Lagrange	1.7183	0.5412	0.1030
Newton	1.7183	0.0938	0.2098
Vandermonde	1.7183	4.0583e+10	0.1153
Eigen-decomposition	1.7183	0.0094	0.0340

The mean stability, mean accuracy and efficiency (in second) of seven methods are given in Table 1. All methods have similar stability. Eigen-decomposition method has the best accuracy which is only 0.0094. The Eigen-decomposition and Scaling and Squaring methods have the fastest computing time in ascending order.

In conclusion, eigen-decomposition approximates the matrix exponential accurately and efficiently, and it is used in the following section to compute the matrix exponential.

5.2 Simulation Study

In this section, we assess the performance of the polynomial diffusion model in terms of futures estimation and parameter estimation. To evaluate the model, we conducted a simulation study on two datasets. The first data has 1000 observations and 13 contracts, with maturity time from T_1 = 1 month for the first available contract to T_{13} = 13 months for the last available contract. The second data also has 1000 observations but contains 20 contracts with maturity times up to 20 months. The time series of two data are given in Fig. 1 and Fig. 2, and the term structures are given in Fig. 3 and Fig. 4.

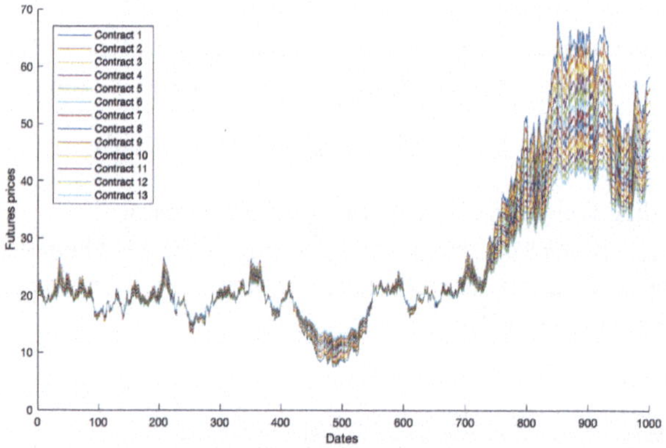

Fig. 1 The time series plots of 13 simulated futures contract prices with the maturity times $T_1 = 1$ month, $T_2 = 2$ months, ..., $T_{13} = 13$ months respectively.

Our target is to show that there exist some identification problems between model parameters and coordinate representation of spot price, but this would not affect the contract estimations. To achieve this, we first separate all parameters into two sets: model parameters $\theta = (\kappa, \gamma, \mu_\xi, \sigma_\chi, \sigma_\xi, \rho, \lambda_\chi, \lambda_\xi, \sigma_1, \sigma_2, \ldots, \sigma_{20})$, which includes all state parameters and measurement errors, and coordinate representation $\vec{p} = (\alpha_1, \alpha_2, \alpha_3, \alpha_4, \alpha_5, \alpha_6)$. Then, the polynomial diffusion model performances were evaluated through root mean square error (RMSE) in the following 4 cases:

Case 1. All model parameters and coordinate representations are known. Only the futures contracts and state variables are estimated.

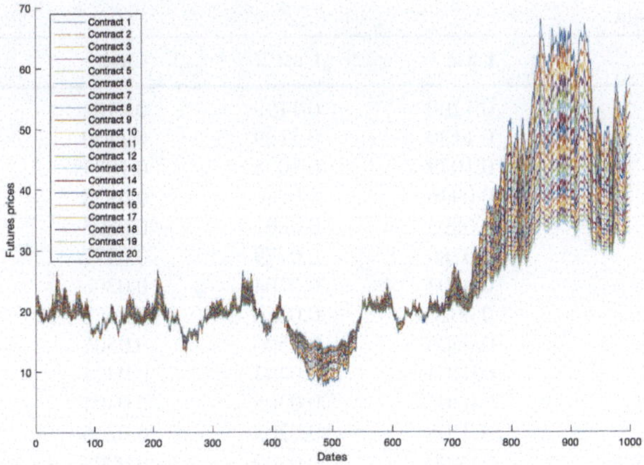

Fig. 2 The time series plots of 20 simulated futures contract prices with the maturity times $T_1 = 1$ month, $T_2 = 2$ months, ..., $T_{20} = 20$ months respectively.

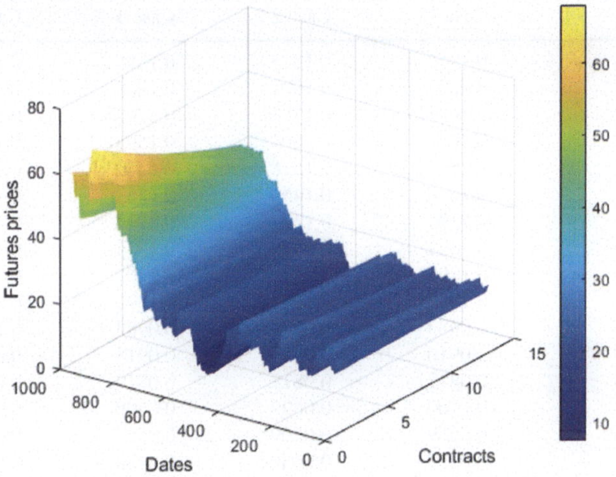

Fig. 3 Term structure plot of 13 futures contracts prices versus their maturities.

Case 2. Model parameters are fixed to the true values.

Case 3. Coordinate representations are fixed to the true values.

Case 4. All model parameters and coordinate representations are estimated.

The RMSE for each contract is presented in Table 2 and Table 4 for the case of EKF and UKF algorithms respectively, based on 13-contract data. Additionally,

Table 2 Root mean square errors (RMSEs) for futures contracts in the case of EKF: comparison of 13-contract data.

Contracts	Case 1	Case 2	Case 3	Case 4
Contract 1	0.1200	0.1199	0.1182	0.1173
Contract 2	0.1130	0.1129	0.1130	0.1150
Contract 3	0.1079	0.1078	0.1084	0.1075
Contract 4	0.0946	0.0946	0.0938	0.0910
Contract 5	0.0893	0.0892	0.0895	0.0872
Contract 6	0.0780	0.0779	0.0777	0.0769
Contract 7	0.0704	0.0704	0.0706	0.0706
Contract 8	0.0620	0.0620	0.0622	0.0621
Contract 9	0.0506	0.0506	0.0506	0.0506
Contract 10	0.0404	0.0404	0.0404	0.0412
Contract 11	0.0305	0.0305	0.0304	0.0317
Contract 12	0.0205	0.0205	0.0206	0.0239
Contract 13	0.0100	0.0100	0.0098	0.0066
Mean	0.0683	0.0682	0.0681	0.0678

Table 3 Root mean square errors (RMSEs) for futures contracts in the case of EKF: comparison of 20-contract data.

Contracts	Case 1	Case 2	Case 3	Case 4
Contract 1	0.1873	0.1873	0.1858	0.1939
Contract 2	0.1802	0.1801	0.1803	0.1864
Contract 3	0.1783	0.1783	0.1792	0.1827
Contract 4	0.1603	0.1604	0.1593	0.1585
Contract 5	0.1577	0.1578	0.1580	0.1600
Contract 6	0.1456	0.1456	0.1453	0.1459
Contract 7	0.1394	0.1394	0.1395	0.1410
Contract 8	0.1329	0.1329	0.1334	0.1346
Contract 9	0.1200	0.1199	0.1201	0.1217
Contract 10	0.1091	0.1091	0.1092	0.1099
Contract 11	0.0983	0.0984	0.0983	0.1007
Contract 12	0.0910	0.0910	0.0913	0.0925
Contract 13	0.0817	0.0817	0.0819	0.0832
Contract 14	0.0700	0.0699	0.0699	0.0720
Contract 15	0.0569	0.0569	0.0567	0.0579
Contract 16	0.0489	0.0489	0.0488	0.0503
Contract 17	0.0381	0.0381	0.0382	0.0391
Contract 18	0.0292	0.0292	0.0294	0.0309
Contract 19	0.0192	0.0192	0.0197	0.0206
Contract 20	0.0077	0.0077	0.0074	0.0112
Mean	0.1026	0.1026	0.1026	0.1047

Table 4 Root mean square errors (RMSEs) for futures contracts in the case of UKF: comparison of 13-contract data.

Contracts	Case 1	Case 2	Case 3	Case 4
Contract 1	0.1190	0.1219	0.1173	0.1173
Contract 2	0.1131	0.1155	0.1132	0.1132
Contract 3	0.1077	0.1097	0.1082	0.1082
Contract 4	0.0938	0.0950	0.0930	0.0930
Contract 5	0.0887	0.0898	0.0889	0.0889
Contract 6	0.0773	0.0778	0.0771	0.0771
Contract 7	0.0703	0.0711	0.0704	0.0705
Contract 8	0.0619	0.0628	0.0621	0.0621
Contract 9	0.0500	0.0506	0.0500	0.0500
Contract 10	0.0400	0.0406	0.0399	0.0400
Contract 11	0.0300	0.0311	0.0299	0.0300
Contract 12	0.0200	0.0211	0.0201	0.0202
Contract 13	0.0089	0.0111	0.0087	0.0087
Mean	0.0677	0.0691	0.0676	0.0676

Table 5 Root mean square errors (RMSEs) for futures contracts in the case of UKF: comparison of 20-contract data.

Contracts	Case 1	Case 2	Case 3	Case 4
Contract 1	0.2375	0.2133	0.1868	0.1886
Contract 2	0.2217	0.2085	0.1809	0.1820
Contract 3	0.2252	0.1952	0.1820	0.1839
Contract 4	0.1979	0.1804	0.1595	0.1607
Contract 5	0.2024	0.1687	0.1614	0.1633
Contract 6	0.1781	0.1631	0.1456	0.1466
Contract 7	0.1734	0.1531	0.1414	0.1425
Contract 8	0.1706	0.1426	0.1369	0.1384
Contract 9	0.1539	0.1306	0.1225	0.1237
Contract 10	0.1448	0.1179	0.1126	0.1139
Contract 11	0.1319	0.1081	0.1014	0.1026
Contract 12	0.1281	0.0975	0.0963	0.0977
Contract 13	0.1138	0.0899	0.0853	0.0864
Contract 14	0.1030	0.0788	0.0739	0.0750
Contract 15	0.0911	0.0657	0.0609	0.0621
Contract 16	0.0827	0.0584	0.0532	0.0544
Contract 17	0.0774	0.0461	0.0454	0.0467
Contract 18	0.0704	0.0373	0.0378	0.0392
Contract 19	0.0624	0.0303	0.0294	0.0308
Contract 20	0.0574	0.0223	0.0228	0.0244
Mean	0.1412	0.1154	0.1068	0.1081

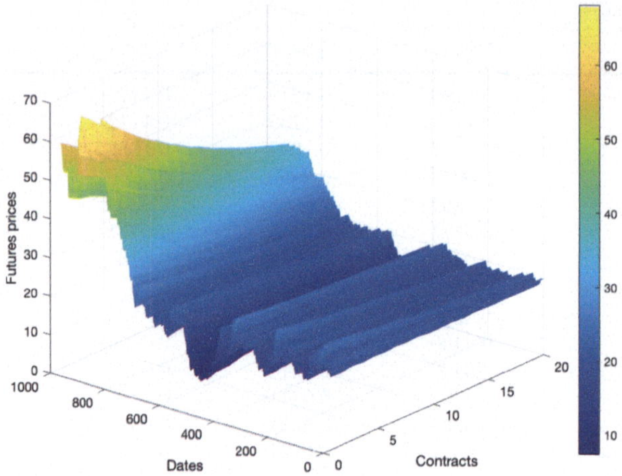

Fig. 4 Term structure plot of 20 futures contracts prices versus their maturities.

Table 3 and Table 5 display the RMSE for the 20-contract data. It is easy to see that the RMSE values are similar across all four cases, regardless of the contract considered. This indicates that all futures contracts are estimated accurately. Furthermore, by assigning a larger measurement error to the short-term contract and a smaller measurement error to the long-term contract, the RMSE values decrease as the maturity time increases. Moreover, there is minimal difference in contract estimation when using either the EKF or UKF. The performance of both algorithms in terms of contract estimation is comparable, demonstrating their effectiveness in the context of this study.

Table 6 Estimated state parameters for the 13-contract data for Case 3 and Case 4. For Case 1 and Case 2 the state parameters are fixed to the true values.

	True	Case 3 - EKF	Case 4 - EKF	Case 3 - UKF	Case 4 - UKF
κ	0.5	0.5166	0.9540	0.5169	0.5198
γ	0.3	0.3151	0.4311	0.3153	0.3200
μ_ξ	1	1.8929	-0.1651	1.8745	2.3371
σ_χ	1.5	1.4166	7.0949	1.4172	5.1781
σ_ξ	1.3	1.4068	5.6780	1.4054	1.3884
ρ	-0.3	-0.3025	-0.2937	-0.3019	-0.2896
λ_χ	0.5	0.8498	7.8781	0.8537	2.0501
λ_ξ	0.3	0.8423	4.5633	0.8195	1.2683
χ_0	0	-1.6138	3.5164	0.1731	-4.2461
ξ_0	3.33	5.0634	-9.3820	5.0790	3.6159

Table 7 Estimated state parameters for the 20-contract data for Case 3 and Case 4. For Case 1 and Case 2 the state parameters are fixed to the true values.

	True	Case 3 - EKF	Case 4 - EKF	Case 3 - UKF	Case 4 - UKF
κ	0.5	0.5115	0.4629	0.5122	0.5121
γ	0.3	0.3083	0.4101	0.3092	0.3072
μ_ξ	1	1.7924	-0.8217	1.8131	2.5670
σ_χ	1.5	1.4506	1.3957	1.4482	1.6945
σ_ξ	1.3	1.3273	1.3459	1.3307	1.1172
ρ	-0.3	-0.3046	0.3317	-0.3037	-0.2987
λ_χ	0.5	0.7988	-0.7015	0.8082	1.0558
λ_ξ	0.3	0.7986	-1.8602	0.8081	0.8872
χ_0	0	0.7635	2.0328	0.9607	-2.9934
ξ_0	3.33	3.0004	2.0159	3.9452	6.5649

Table 8 Estimated measurement errors for the 13-contract data for Case 3 and Case 4. For Case 1 and Case 2 the measurement errors are fixed to the true values.

	True	Case 3 - EKF	Case 4 - EKF	Case 3 - UKF	Case 4 - UKF
σ_1	0.13	0.1255	0.1302	0.1255	0.1255
σ_2	0.12	0.1207	0.1220	0.1207	0.1207
σ_3	0.11	0.1133	0.1135	0.1133	0.1133
σ_4	0.10	0.0966	0.0952	0.0967	0.0966
σ_5	0.09	0.0925	0.0904	0.0925	0.0925
σ_6	0.08	0.0790	0.0783	0.0790	0.0790
σ_7	0.07	0.0721	0.0719	0.0721	0.0721
σ_8	0.06	0.0630	0.0635	0.0630	0.0630
σ_9	0.05	0.0507	0.0515	0.0507	0.0507
σ_{10}	0.04	0.0404	0.0411	0.0404	0.0404
σ_{11}	0.03	0.0298	0.0313	0.0298	0.0298
σ_{12}	0.02	0.0207	0.0227	0.0207	0.0207
σ_{13}	0.01	0.0100	1.03E-05	0.0100	0.0100

However, when we look at the parameter estimations, the conclusion changes. Table 6 - Table 11 give the estimated state parameters, measurement errors and coordinate representations, respectively. For the state parameters, the estimations in Case 4 (coordinate representations are estimated) change a lot compared to the estimations in Case 3 (coordinate representations are fixed), which suggests the existence of parameter identification problems. The trends of state variables are somehow captured by the coordinate representations. Moreover, the estimations in Case 3 are closer to the true values.

For the measurement errors (Table 8 and Table 9), there are not many differences between Case 3 and Case 4. Measurement errors are estimated accurately. There are also not many differences between EKF and UKF.

The estimated coordinate representations are given in Table 10 and Table 11. Comparing Case 2 (state parameters and measurement errors are fixed) to Case 4 (state parameters and measurement errors are estimated), like state parameters, the

Table 9 Estimated measurement errors for the 20-contract data for Case 3 and Case 4. For Case 1 and Case 2 the measurement errors are fixed to the true values.

	True	Case 3 - EKF	Case 4 - EKF	Case 3 - UKF	Case 4 - UKF
σ_1	0.20	0.1953	0.2015	0.1953	0.1953
σ_2	0.19	0.1902	0.1962	0.1903	0.1903
σ_3	0.18	0.1869	0.1888	0.1869	0.1869
σ_4	0.17	0.1646	0.1647	0.1646	0.1646
σ_5	0.16	0.1636	0.1636	0.1636	0.1636
σ_6	0.15	0.1487	0.1486	0.1486	0.1486
σ_7	0.14	0.1433	0.1434	0.1433	0.1433
σ_8	0.13	0.1366	0.1365	0.1366	0.1366
σ_9	0.12	0.1225	0.1233	0.1225	0.1225
σ_{10}	0.11	0.1109	0.1112	0.1109	0.1109
σ_{11}	0.10	0.0995	0.1002	0.0995	0.0995
σ_{12}	0.09	0.0924	0.0926	0.0925	0.0925
σ_{13}	0.08	0.0826	0.0834	0.0826	0.0826
σ_{14}	0.07	0.0706	0.0708	0.0706	0.0706
σ_{15}	0.06	0.0570	0.0574	0.0570	0.0570
σ_{16}	0.05	0.0493	0.0498	0.0493	0.0493
σ_{17}	0.04	0.0387	0.0389	0.0388	0.0387
σ_{18}	0.03	0.0294	0.0294	0.0294	0.0294
σ_{19}	0.02	0.0208	0.0204	0.0208	0.0208
σ_{20}	0.01	0.0093	0.0113	0.0093	0.0093

Table 10 Estimated coordinate representations for the 13-contract data for Case 2 and Case 4. For Case 1 and Case 3 the coordinate representations are fixed to the true values.

	True	Case 2 - EKF	Case 4 - EKF	Case 2 - UKF	Case 4 - UKF
α_1	5	4.5149	-3.8661	10	3.8891
α_2	2	1.8340	-2.2637	8.5185	-0.2656
α_3	2	2.3138	-9.9986	-2.2797	0.3508
α_4	2	1.9874	0.0163	1.8713	0.1451
α_5	3	3.0488	0.0470	-3.6190	0.8563
α_6	1	0.9661	-0.7228	0.5497	1.0349

Table 11 Estimated coordinate representations for the 20-contract data for Case 2 and Case 4. For Case 1 and Case 3 the coordinate representations are fixed to the true values.

	True	Case 2 - EKF	Case 4 - EKF	Case 2 - UKF	Case 4 - UKF
α_1	5	5.3698	8.0156	6.7448	8.9205
α_2	2	2.5473	-4.2205	8.5185	-5.7008
α_3	2	1.9943	6.7163	-2.2797	-4.9896
α_4	2	2.0308	2.1835	1.8713	1.4698
α_5	3	2.8492	2.1582	-3.6190	2.9955
α_6	1	0.9686	-2.3905	0.5497	1.4239

estimations of coordinate representations change a lot. Moreover, in Case 2, the estimations of coordinate representations filtered by EKF are close to the true values, but not the estimations filtered by UKF are not. In that case, coordinate representations can be estimated through EKF. In Case 4, both EKF and UKF cannot estimate coordinate representations.

6 Conclusion

In the modelling of commodity futures, it is common to assume that the logarithm of the underlying spot price is expressed as a sum of various factors. However, this class of models is subject to two limitations. Firstly, for deriving a closed-form expression for the futures price, the spot price used to be a linear function of Gaussian distributed factors. Secondly, these models imply that the spot price always be positive. To overcome these two limitations, we introduced the polynomial diffusion model in this paper, which serves as a generalisation of the Schwartz-Smith two-factor model. This model allows for a more flexible and non-linear representation of the spot price. Specifically, we applied a polynomial diffusion model of degree 2 to the two-factor model. The estimation of parameters and hidden state variables was performed using the EKF and the UKF.

We conducted a study to assess the performance of the models in four different cases using the simulated data. Overall, we found that the futures contracts can be accurately estimated. However, parameter estimation remains challenging, even when we impose constrains on the model parameters and estimate coordinate representations of the polynomial of spot price. In other words, while the state variables χ_t and ξ_t can not be estimated correctly, however the futures prices were recovered reasonably well. This phenomenon suggests the existence of identification problems. Firstly, the trends of state variables are captured by the spot price polynomials. As a consequence, the estimates of both state parameters and coordinate representations of the polynomial are far away from the true values. Secondly, even when we fix one set of parameters (either the model parameters or coordinate representations), the estimation of the other set of parameters still fails to converge to the true values. The identification issue poses the challenge of finding the correct order for the polynomial diffusion process. Further research on the parameter identification problem is needed to address this issue.

References

1. Ames, M., Bagnarosa, G., Matsui, T., Peters, G.W., Shevchenko, P.V.: Which risk factors drive oil futures price curves? Energy Economics **87**, 104676 (2020)
2. Black, F.: The pricing of commodity contracts. Journal of Financial Economics **3**, 167–179 (1976)

3. Carr, P., Wu, L.: Time-changed lévy processes and option pricing. Journal of Financial economics **71**(1), 113–141 (2004)
4. Cortazar, G., Millard, C., Ortega, H., Schwartz, E.S.: Commodity price forecasts, futures prices, and pricing models. Management Science **65**(9), 4141–4155 (2019)
5. Cortazar, G., Naranjo, L.: An n-factor gaussian model of oil futures prices. Journal of Futures Markets **26**(3), 243–268 (2006)
6. Eydeland, A., Geman, H.: Fundamentals of electricity derivatives. Energy Modelling and the Management of Uncertainty pp. 35–43 (1999)
7. Fallahgoul, H., Hugonnier, J., Mancini, L.: Risk premia and lévy jumps: theory and evidence. Swiss Finance Institute Research Paper (19-49) (2020)
8. Favetto, B., Samson, A.: Parameter estimation for a bidimensional partially observed Ornstein–Uhlenbeck process with biological application. Scandinavian Journal of Statistics **37**(2), 200–220 (2010)
9. Filipovic, D., Larsson, M.: Polynomial diffusions and applications in finance. Finance and Stochastics **20**(4), 931–972 (2016)
10. Gibson, R., Schwartz, E.S.: Stochastic convenience yield and the pricing of oil contingent claims. Journal of Finance **45**(3), 959–976 (1990)
11. Huang, J.z., Wu, L.: Specification analysis of option pricing models based on time-changed lévy processes. The Journal of Finance **59**(3), 1405–1439 (2004)
12. Kiesel, R., Schindlmayr, G., Börger, R.H.: A two-factor model for the electricity forward market. Quantitative Finance **9**(3), 279–287 (2009)
13. Kleisinger-Yu, X., Komaric, V., Larsson, M., Regez, M.: A multi-factor polynomial framework for long-term electricity forwards with delivery period. arXiv: Mathematical Finance (2019)
14. Moler, C., Van Loan, C.: Nineteen dubious ways to compute the exponential of a matrix, twenty-five years later. SIAM review **45**(1), 3–49 (2003)
15. Peters, G.W., Briers, M., Shevchenko, P., Doucet, A.: Calibration and filtering for multi factor commodity models with seasonality: incorporating panel data from futures contracts. Methodology and Computing in Applied Probability **15**(4), 841–874 (2013)
16. Schwartz, E.S.: The stochastic behavior of commodity prices: Implications for valuation and hedging. Journal of Finance **52**(3), 923–973 (1997)
17. Schwartz, E.S., Smith, J.E.: Short-term variations and long-term dynamics in commodity prices. Management Science **46**(7), 893–911 (2000)
18. Sorensen, C.: Modeling seasonality in agricultural commodity futures. Journal of Futures Markets **22**(5), 393–426 (2002)
19. Wan, E.A., Van Der Merwe, R.: The unscented kalman filter for nonlinear estimation. In: Proceedings of the IEEE 2000 Adaptive Systems for Signal Processing, Communications, and Control Symposium (Cat. No. 00EX373), pp. 153–158. Ieee (2000)

On Autoregressive Measurement Errors in a Two-Factor Model

Jun S. Han, Nino Kordzakhia, Pavel V. Shevchenko, Stefan Trück

Abstract In this study, we consider the extended two-factor model, originally introduced by Schwartz and Smith (2000), which has been commonly used for the pricing of commodity derivatives. In this model setup, we assume the latent short and long-term factors represent correlated mean-reverting processes. We develop a Kalman filter for jointly estimating the state variables and unknown parameters. In the measurement equation system, we assume that the residuals are serially correlated and inter-dependent. We derive the Kalman filter under the assumption that these residuals are AR(p) processes. Then the Kalman filter is used for obtaining the marginalised likelihood estimators for the state variables and the model parameters. We provide an extensive, reproducible simulation study for examining the convergence of the estimators in our proposed model. The `MATLAB` code for this study is available via the link to GitHub[1].

1 Introduction

Commodity markets are a crucial part of the global economy, as they enable market participants and industries to manage price risk and ensure efficient allocation of available resources. Challenges arise in the pricing of commodity derivatives due

Jun S. Han
School of Mathematical and Physical Sciences, Macquarie University
e-mail: jun.han1@hdr.mq.edu.au

Nino Kordzakhia
School of Mathematical and Physical Sciences, Macquarie University

Pavel V. Shevchenko
Department of Actuarial Studies and Business Analytics, Macquarie University

Stefan Trück
Department of Actuarial Studies and Business Analytics, Macquarie University

[1] https://github.com/Junee1992/EUA-Futures-Pricing/tree/main/MATRIX

D. R. Wood et al. (eds.), *2021-2022 MATRIX Annals*, MATRIX Book Series 5,
https://doi.org/10.1007/978-3-031-47417-0_19

to the complex nature of the market, including supply and demand fluctuations, geopolitical risks, and climate risks. Over the years various pricing models have been developed for capturing the complexities of the market dynamics.

The factor models provided a convenient framework for pricing derivatives written on commodities with unobservable spot prices. Study in [4] proposed a reduced form model, a single factor model that models the spot price of a commodity using the geometric Brownian motion, with the convenience yield being a deterministic function of the commodity price and a constant interest rate. The model was extended in [10] and [14], allowing stochasticity in the convenience yield and incorporating volatilities of historical returns on futures contracts of different maturities. Comparative analysis was conducted in [25], assessing reduced-form models that are comprised of one to three factors, starting from the spot price following a mean-reverting process, and subsequently adding stochastic convenience yields and interest rates.

The Schwartz-Smith two-factor model was introduced in [24] and since then is widely used in the pricing of commodities. In fact, the authors acknowledged that their approach is formally equivalent to their pioneering pricing framework in [14], where the stochastic yield factor was considered. The spot price of a commodity in [24] is decomposed into two distinct unobservable factors, representing short-term deviations and a long-term equilibrium price level. The short-term factor is assumed to converge to zero over time, reflecting the temporary effects of fluctuations in supply, demand, and market conditions. Although the long-term factor was originally assumed to follow a Brownian motion with drift, often it is also assumed to follow a mean-reverting process, see [8] and the references therein.

In this paper, we will study the joint estimation problem of the two latent variables, as well as model parameters, by establishing the likelihood estimation procedure using the Kalman filter in the extended setting. In previous works the two variants of Schwartz-Smith (2000) model have been calibrated to various futures markets such as the copper market (see [15], [23]), the crude oil market (see [7], [3], [8]), and also the carbon emission allowance market (see [1], [16]).

As supply and demand for certain commodities are influenced by the time of the year, the deterministic functions representing seasonal components were introduced in the model framework. Some common functions include a step function to model seasonal effects for different calendar months, and the sum of sinusoids (see studies in [18], [21], [27]). Stochastic jumps were incorporated in a short-term factor in [29] and [26] for modelling sudden changes in the market in the data-driven setting. The generalisation of the Schwartz-Smith Two-Factor model has been designed in [18] and [9], allowing for more than two factors that follow joint mean-reverting processes as well as introducing time-varying economic factors in the pricing model. Study in [6] developed a pricing model for long-dated commodity derivatives, allowing for both stochastic volatilities and interest rates to be correlated. In [30], the risk premium followed a stochastic process, and its performance was empirically analysed using the European Union Allowance futures (EUA) and options prices.

The joint estimation of state variables and model parameters has several limitations. The study in [13] pointed out the identifiability problem due to the invertibility issue of the observed Fisher information matrix of model parameters. An additional constraint was proposed in [3] for rectifying the parameter identification problem within the maximum likelihood estimation procedure using Kalman filter approach. The filtering process itself was also limited to a set of standard assumptions on measurement errors, e.g. Gaussian white noise. The study in [17] relaxed the assumption of independence of the state and measurement error processes, and allowed the measurement errors to follow a GARCH(1,1) process to incorporate the heteroscedastic feature of commodity spot prices. For relaxing the assumption of independence between different futures contracts, [27] has taken an approach of estimating covariances, by defining each element as a parametrised function of time-to-maturity, but no substantial improvement was made in parameter estimation. The study in [11] has demonstrated through numerical simulation, that augmentation methods do not work well for the estimation of the covariance matrix of measurement errors. In [12], the effects on parameter estimation were studied in the framework when the option prices were used instead of futures prices; the extended Kalman filter was deployed for the estimation of state variables in a nonlinear system. Other than financial application, estimation of dependence component-wise is of high importance in geoscience, and the ensemble Kalman filter is usually adopted for developing various methods of estimation of the covariance matrix of measurement errors, (see [20], [28]).

In [25], the correlations of log prices of commodity contracts were modelled through the correlated bivariate latent state variables. The empirical studies have shown that often the observed measurement equation residuals exhibit significant correlations, necessitating a closer investigation of their sources. In a recent study, [16] proposed a two-step approach to joint estimation of model parameters and inter-correlations of the errors in measurement equations for the contracts with different maturities. For the incorporation of inter-correlated measurement errors and AR(1) serial correlation in each error series, a two-step approach was used, first estimating the model parameters via maximising the marginalised likelihood process and subsequently fitting an autoregressive model to the model residuals.

In difference from [16], here we propose a unified estimation procedure for the estimation of serially correlated series in the form of AR(p) of any order $p \geq 1$. The derivation of Kalman filter for AR(1) measurement errors can be found in [5] and [22]. The method suggested in this paper, can be used for improving out-of-sample commodity price forecasts over the method with any other reduced form of the setting of the measurement equations. In this paper, using a simulation study, we illustrate the convergence of the estimates of parameters in the Schwartz-Smith two-factor model by taking different numbers of contracts and sample sizes.

This paper is outlined as follows. In Section 2, we formulate the main model with a measurement equations system accommodating inter-dependent and serially correlated error terms. Under these assumptions on the error processes, we derive the Kalman filter. In Section 3, we present the results of the simulation study illustrating the convergence of the estimates to true values of model parameters for different practical scenarios. Section 4 concludes with a brief discussion of the results.

2 The Model

Consider the risk-neutral probability space denoted by $(\Omega, \mathfrak{I}, \mathbb{Q})$. We define the risk-neutral dynamics of the spot price S_t as $\ln S_t = \chi_t + \xi_t$, where χ_t and ξ_t are latent variables that are short-term and long-term factors, respectively, given through following stochastic differential equations

$$d\chi_t = (-\kappa\chi_t - \lambda_\chi)dt + \sigma_\chi dW_t^\chi, \tag{1}$$

$$d\xi_t = (\mu_\xi - \lambda_\xi - \gamma\xi_t)dt + \sigma_\xi dW_t^\xi, \tag{2}$$

$$\mathbb{E}_\mathbb{Q}[dW_t^\chi dW_t^\xi] = \rho_{\chi\xi} dt, \tag{3}$$

where $\kappa, \gamma \in \mathbb{R}^+$ are the rate of the mean-reversion for χ_t and ξ_t, respectively, λ_χ, $\lambda_\xi \in \mathbb{R}$ are deterministic unknown risk premia, σ_χ, $\sigma_\xi \in \mathbb{R}^+$ are the volatility parameters, and $\mu_\xi \in \mathbb{R}$ is the mean of the long-term factor. W_t^χ, W_t^ξ are correlated standard Brownian processes under $\mathbb{Q}$, where $\rho_{\chi\xi}$ is the correlation coefficient of the two processes. The futures price at time t, maturing in T years, is $F_{t,T} = \mathbb{E}[S_T|\mathfrak{I}_t]$.

In discrete time, we consider the following linear state space model

$$\mathbf{x}_t = \mathbf{c} + \mathbf{G}\mathbf{x}_{t-1} + \mathbf{w}_t, \quad \mathbf{w}_t \sim \mathcal{N}(0, \mathbf{W}), \tag{4}$$

$$\mathbf{y}_t = \mathbf{d}_t + \mathbf{B}_t\mathbf{x}_t + \mathbf{v}_t, \tag{5}$$

$$\mathbf{v}_t = \Phi_p(L)\mathbf{v}_t + \varepsilon_t, \quad \varepsilon_t \sim \mathcal{N}(0, \mathbf{V}_\varepsilon), \tag{6}$$

where, for $t = 1, 2, \cdots, n$, $\mathbf{x}_t$ and $\mathbf{y}_t$ are the vectors of latent and observable variables, respectively. Further, we assume that N futures contracts have the maturities $T_1, \cdots, T_N$, where $T_1 < T_2 < \cdots < T_N$, and

$$\mathbf{x}_t = \begin{pmatrix} \chi_t \\ \xi_t \end{pmatrix}, \ \mathbf{c} = \begin{pmatrix} 0 \\ \frac{\mu_\xi(1-e^{-\gamma\Delta t})}{\gamma} \end{pmatrix}, \ \mathbf{G} = \begin{pmatrix} e^{-\kappa\Delta t} & 0 \\ 0 & e^{-\gamma\Delta t} \end{pmatrix}, \tag{7}$$

$$\mathbf{y}_t = (\ln F_{t,T_1}, \ln F_{t,T_2}, \cdots, \ln F_{t,T_N})', \tag{8}$$

$$\mathbf{d}_t = (A(T_1 - t), A(T_2 - t), \cdots, A(T_N - t))', \tag{9}$$

$$\mathbf{B}_t = \begin{pmatrix} e^{-\kappa(T_1-t)} & \dots & e^{-\kappa(T_N-t)} \\ e^{-\gamma(T_1-t)} & \dots & e^{-\gamma(T_N-t)} \end{pmatrix}', \tag{10}$$

with $\Phi_p(L) = \sum_{r=1}^{p} \phi_r L^r$ being a function of the lag operator, Δt is the time difference in years between discretised time instances $t-1$ and t, and the function $A(T-t)$ is defined as

$$A(T-t) = -\frac{\lambda_\chi}{\kappa}\left(1-e^{-\kappa(T-t)}\right) + \frac{\mu_\xi - \lambda_\xi}{\gamma}\left(1-e^{-\gamma(T-t)}\right)$$
$$+ \quad \frac{1}{2}\left(\frac{1-e^{-2\kappa(T-t)}}{2\kappa}\sigma_\chi^2 + \frac{2(1-e^{-(\kappa+\gamma)(T-t)})}{\kappa+\gamma}\sigma_\chi\sigma_\xi\rho_{\chi\xi} + \frac{1-e^{-2\gamma(T-t)}}{2\gamma}\sigma_\xi^2\right). \tag{11}$$

The covariance matrices of error terms $\mathbf{w}_t$ and ε_t are defined as

$$\mathbf{W} = \begin{pmatrix} \frac{1-e^{-2\kappa\Delta t}}{2\kappa}\sigma_\chi^2 & \frac{1-e^{-(\kappa+\gamma)\Delta t}}{\kappa+\gamma}\sigma_\chi\sigma_\xi\rho_{\chi\xi} \\ \frac{1-e^{-(\kappa+\gamma)\Delta t}}{\kappa+\gamma}\sigma_\chi\sigma_\xi\rho_{\chi\xi} & \frac{1-e^{-2\gamma\Delta t}}{2\gamma}\sigma_\xi^2 \end{pmatrix}, \quad \mathbf{V}_\varepsilon = \begin{pmatrix} s_{11}^2 & s_{12} & \cdots & s_{1N} \\ s_{12} & s_{22}^2 & \cdots & s_{2N} \\ \vdots & \vdots & \ddots & \vdots \\ s_{1N} & s_{2N} & \cdots & s_{NN}^2 \end{pmatrix}. \tag{12}$$

Applying the lag polynomial $I - \Phi_p(L)$ to (5), where I is the identity matrix, the measurement equation is expressed as

$$\mathbf{y}_t = \mathbf{d}_t^*(p) + \mathbf{B}_t^*(p)\mathbf{x}_t + \mathbf{v}_t^*(p), \tag{13}$$

where

$$\mathbf{d}_t^*(p) = \sum_{r=1}^{p}\phi_r\mathbf{y}_{t-r} + \mathbf{d}_t - \sum_{r=1}^{p}\phi_r\mathbf{d}_{t-r} + \sum_{r=1}^{p}\phi_r\mathbf{B}_{t-r}\left[\sum_{k=1}^{r}(\mathbf{G}^{-1})^k\right]\mathbf{c}, \tag{14}$$

$$\mathbf{B}_t^*(p) = \mathbf{B}_t - \sum_{r=1}^{p}\phi_r\mathbf{B}_{t-r}(\mathbf{G}^{-1})^r, \tag{15}$$

$$\mathbf{v}_t^*(p) = \sum_{r=1}^{p}\phi_r\mathbf{B}_{t-r}\left[\sum_{k=1}^{r}(\mathbf{G}^{-1})^{r-k+1}\mathbf{w}_{t-k+1}\right] + \varepsilon_t. \tag{16}$$

Proof. We wish to prove that (14) - (16) are true for forming the measurement equation (13) when measurement errors marginally follow AR(p) processes. For $p = 1$, we consider the following linear state-space model

$$\mathbf{x}_t = \mathbf{c} + \mathbf{G}\mathbf{x}_{t-1} + \mathbf{w}_t, \ \ \mathbf{w}_t \sim \mathcal{N}(0, \mathbf{W}), \tag{17}$$

$$\mathbf{y}_t = \mathbf{d}_t + \mathbf{B}_t\mathbf{x}_t + \mathbf{v}_t, \tag{18}$$

$$\mathbf{v}_t = \phi\mathbf{v}_{t-1} + \varepsilon_t, \ \ \varepsilon_t \sim \mathcal{N}(0, \mathbf{V}_\varepsilon). \tag{19}$$

Let $\Phi(L) = \phi_1 L$. By applying $I - \phi_1 L$ to equation (18), and using equation (17), we obtain

$$(I - \Phi(L))\mathbf{y}_t = (I - \Phi(L))\mathbf{d}_t + (I - \Phi(L))\mathbf{B}_t\mathbf{x}_t + (I - \Phi(L))\mathbf{v}_t,$$

which can be extended to

$$\mathbf{y}_t = \phi\mathbf{y}_{t-1} + \mathbf{d}_t - \phi_1\mathbf{d}_{t-1} + (\mathbf{B}_t - \phi_1\mathbf{B}_{t-1}\mathbf{G}^{-1})\mathbf{x}_t + \phi_1\mathbf{B}_{t-1}\mathbf{G}^{-1}(\mathbf{c} + \mathbf{w}_t) + \varepsilon_t.$$

Hence,

$$\mathbf{y}_t = \mathbf{d}_t^*(1) + \mathbf{B}_t^*(1)\mathbf{x}_t + \mathbf{v}_t^*(1),$$

where

$$\begin{aligned}
\mathbf{d}_t^*(1) &= \phi_1\mathbf{y}_{t-1} + \mathbf{d}_t - \phi_1\mathbf{d}_{t-1} + \phi_1\mathbf{B}_{t-1}\mathbf{G}^{-1}\mathbf{c},\\
\mathbf{B}_t^*(1) &= \mathbf{B}_t - \phi_1\mathbf{B}_{t-1}\mathbf{G}^{-1},\\
\mathbf{v}_t^*(1) &= \phi_1\mathbf{B}_{t-1}\mathbf{G}^{-1}\mathbf{w}_t + \varepsilon_t.
\end{aligned}$$

Therefore, (14) - (16) are obvious for $p = 1$. Now we assume that (13) - (16) holds for $p = m$. Then for $p = m + 1$ we obtain

$$\begin{aligned}
\mathbf{y}_t &= \phi_1\mathbf{y}_{t-1} + \cdots + \phi_m\mathbf{y}_{t-m} + \phi_{m+1}\mathbf{y}_{t-(m+1)}\\
&\quad + \mathbf{d}_t - \phi_1\mathbf{d}_{t-1} - \cdots - \phi_m\mathbf{d}_{t-m} - \phi_{m+1}\mathbf{d}_{t-(m+1)}\\
&\quad + \mathbf{B}_t\mathbf{x}_t - \phi_1\mathbf{B}_{t-1}\mathbf{x}_{t-1} - \cdots - \phi_m\mathbf{B}_{t-m}\mathbf{x}_{t-m} - \phi_{m+1}\mathbf{B}_{t-(m+1)}\mathbf{x}_{t-(m+1)} + \varepsilon_t\\
&= \mathbf{d}_t^*(m) + \phi_{m+1}\mathbf{y}_{t-(m+1)} - \phi_{m+1}\mathbf{d}_{t-(m+1)} + \mathbf{B}_t^*(m)\mathbf{x}_t - \phi_{m+1}\mathbf{B}_{t-(m+1)}\mathbf{x}_{t-(m+1)}\\
&\quad + \mathbf{v}_t^*(m)\\
&= \mathbf{d}_t^*(m+1) + \mathbf{B}_t^*(m+1)\mathbf{x}_t + \mathbf{v}_t^*(m+1).
\end{aligned}$$

Hence, (14) - (16) are proven true for all $p \geq 1$. □

Given the first p observations, the prediction error for $t > p$ is defined as

$$\begin{aligned}
\mathbf{e}_{t|t-1}^* &= \mathbf{y}_t - \mathbb{E}[\mathbf{y}_t|\mathfrak{I}_{t-1}]\\
&= \mathbf{y}_t - \mathbf{d}_t^*(p) - \mathbf{B}_t^*(p)\mathbb{E}[\mathbf{x}_t|\mathfrak{I}_{t-1}] - \mathbb{E}[\mathbf{v}_t^*(p)|\mathfrak{I}_{t-1}],
\end{aligned} \tag{20}$$

and we assume $\mathbf{w}_t$ and ε_t are mutually independent, and serially uncorrelated. For $p = 1$, we have $\mathbb{E}[\mathbf{v}_t^*(1)|\mathfrak{I}_{t-1}] = 0$, since $\mathbb{E}[\mathbf{w}_t|\mathfrak{I}_{t-1}] = 0$. For $p > 1, \mathbb{E}[\mathbf{v}_t^*(p)|\mathfrak{I}_{t-1}]$ is defined as

$$\mathbb{E}[\mathbf{v}_t^*(p)|\Im_{t-1}] = \sum_{r=2}^{p} \phi_r \mathbf{B}_{t-r} \left[\sum_{k=2}^{r} (\mathbf{G}^{-1})^{r-k+1} \mathbb{E}[\mathbf{w}_{t-k+1}|\Im_{t-1}] \right],$$

and the conditional expectation of $\mathbf{w}_{t-k+1}|\Im_{t-1}$ can be estimated as

$$\mathbb{E}[\mathbf{w}_{t-k+1}|\Im_{t-1}] = \mathbb{E}[\mathbf{x}_{t-k+1}|\Im_{t-1}] - \mathbf{c} - \mathbf{G}\mathbb{E}[\mathbf{x}_{t-k}|\Im_{t-1}],$$

with $\mathbb{E}[\mathbf{x}_{t-k}|\Im_{t-1}]$ that can be estimated via the Kalman smoother for $k > 1$. We use the following recursive formulas to obtain estimates of smoother,

$$\begin{aligned}
\mathbb{E}[\mathbf{x}_{t-k}|\Im_{t-1}] &= \mathbb{E}[\mathbf{x}_{t-k}|\Im_{t-k}] + \mathbf{J}_{t-k}(\mathbb{E}[\mathbf{x}_{t-k+1}|\Im_{t-1}] - \mathbb{E}[\mathbf{x}_{t-k+1}|\Im_{t-k}]),\\
Var[\mathbf{x}_{t-k}|\Im_{t-1}] &= Var[\mathbf{x}_{t-k}|\Im_{t-k}] + \mathbf{J}_{t-k}(Var[\mathbf{x}_{t-k+1}|\Im_{t-1}] - Var[\mathbf{x}_{t-k+1}|\Im_{t-k}])\mathbf{J}'_{t-k},\\
\mathbf{J}_{t-k} &= Var[\mathbf{x}_{t-k}|\Im_{t-k}]\mathbf{G}' Var[\mathbf{x}_{t-k+1}|\Im_{t-k}]^{-1}.
\end{aligned}$$

Let $\mathbf{P}_{t|t-1} = Var[\mathbf{x}_t|\Im_{t-1}]$ be the variance of state prediction error for the next state. The variance of the prediction error is

$$\begin{aligned}
Var[\mathbf{e}^*_{t|t-1}] &= Var[\mathbf{y}_t|\Im_{t-1}] = \mathbf{L}^*_{t|t-1} = Var[\mathbf{B}_t^*(p)\mathbf{x}_t + \mathbf{v}_t^*(p)|\Im_{t-1}]\\
&= \mathbf{B}_t^*(p)\mathbf{P}_{t|t-1}[\mathbf{B}_t^*(p)]' + \mathbf{V}^*_{t|t-1}(p)\\
&\quad + \mathbf{B}_t^*(p)\mathbf{W}\left[\sum_{r=1}^{p}\phi_r\mathbf{B}_{t-r}(\mathbf{G}^{-1})^r\right] + \left[\sum_{r=1}^{p}\phi_r\mathbf{B}_{t-r}(\mathbf{G}^{-1})^r\right]\mathbf{W}'\mathbf{B}_t^*(p)\\
&= \mathbf{B}_t^*(p)\mathbf{P}_{t|t-1}[\mathbf{B}_t^*(p)]' + \mathbf{V}^*_{t|t-1}(p) + \mathbf{B}_t^*(p)\mathbf{C}_t^*(p) + [\mathbf{C}_t^*(p)]'[\mathbf{B}_t^*(p)]'.
\end{aligned} \tag{21}$$

where $\mathbf{C}_t^*(p)$ is the covariance of the two error processes $\mathbf{w}_t, \mathbf{v}_t^*(p)$, as $\mathbf{w}_t$ is a function of $\mathbf{v}_t^*(p)$, and is defined as

$$\mathbf{C}_t^*(p) = \mathbf{W}\left(\sum_{r=1}^{p}\phi_r\mathbf{B}_{t-r}(\mathbf{G}^{-1})^r\right)' \tag{22}$$

The conditional variance of $\mathbf{v}_t^*(p)$ is

$$Var[\mathbf{v}_t^*(p)|\Im_{t-1}] = \mathbf{V}^*_{t|t-1}(p) = \left[\sum_{r=1}^{p}\phi_r\mathbf{B}_{t-r}(\mathbf{G}^{-1})^r\right]\mathbf{W}\left[\sum_{r=1}^{p}\phi_r\mathbf{B}_{t-r}(\mathbf{G}^{-1})^r\right]' + \mathbf{V}_\varepsilon. \tag{23}$$

Proof. Assuming $\mathbf{w}_t, \varepsilon_t$ are mutually independent and serially uncorrelated, the variance of the prediction error is

$$\begin{aligned}
\mathbf{L}^*_{t|t-1} &= Var[\mathbf{e}^*_{t|t-1}] = Var[\mathbf{B}^*_t(p)\mathbf{x}_t + \mathbf{v}^*_t(p)|\mathfrak{I}_{t-1}] \\
&= \mathbb{E}[(\mathbf{B}^*_t(p)\mathbf{x}_t + \mathbf{v}^*_t(p))(\mathbf{B}^*_t(p)\mathbf{x}_t + \mathbf{v}^*_t(p))'|\mathfrak{I}_{t-1}] \\
&- \mathbb{E}[\mathbf{B}^*_t(p)\mathbf{x}_t + \mathbf{v}^*_t(p)|\mathfrak{I}_{t-1}]\mathbb{E}[(\mathbf{B}^*_t(p)\mathbf{x}_t + \mathbf{v}^*_t(p))'|\mathfrak{I}_{t-1}] \\
&= \mathbf{B}^*_t(p)\mathbf{P}_{t|t-1}\mathbf{B}^{*\prime}_t(p) + \mathbf{V}^*_{t|t-1}(p) + \mathbf{B}^*_t(\mathbb{E}[\mathbf{x}_t[\mathbf{v}^*_t(p)]'|\mathfrak{I}_{t-1}] \\
&- \mathbb{E}[\mathbf{x}_t|\mathfrak{I}_{t-1}]\mathbb{E}[\mathbf{v}^*_t(p)|\mathfrak{I}_{t-1}]) + (\mathbb{E}[\mathbf{v}^*_t(p)\mathbf{x}_t{}'|\mathfrak{I}_{t-1}] \\
&- \mathbb{E}[\mathbf{v}^*_t(p)|\mathfrak{I}_{t-1}]\mathbb{E}[\mathbf{x}_t|\mathfrak{I}_{t-1}])[\mathbf{B}^*_t(p)]'|\mathfrak{I}_{t-1}] \\
&= \mathbf{B}^*_t(p)\mathbf{P}_{t|t-1}\mathbf{B}^{*\prime}_t(p) + \mathbf{V}^*_{t|t-1}(p) + \mathbf{B}^*_t(p)Cov(\mathbf{x}_t, \mathbf{v}^*_t(p)|\mathfrak{I}_{t-1}) \\
&+ Cov(\mathbf{v}^*_t(p), \mathbf{x}^{*\prime}_t|\mathfrak{I}_{t-1})'[\mathbf{B}^*_t(p)]',
\end{aligned}$$

where $\mathbf{P}_{t|t-1} = Var[\mathbf{x}_t|\mathfrak{I}_{t-1}]$. Since the new measurement error $\mathbf{v}^*_t(p)$ is a function of $\mathbf{w}_t$,

$$\begin{aligned}
\mathbf{C}^*_t(p) = Cov(\mathbf{x}_t, \mathbf{v}^*_t(p)|\mathfrak{I}_{t-1}) &= Cov(\mathbf{w}_t, \sum_{r=1}^{p} \phi_r \mathbf{B}_{t-r}(\mathbf{G}^{-1})^r \mathbf{w}_t|\mathfrak{I}_{t-1}) \\
&= \mathbf{W}\left(\sum_{r=1}^{p} \phi_r \mathbf{B}_{t-r}(\mathbf{G}^{-1})^r\right)',
\end{aligned}$$

and

$$\begin{aligned}
\mathbf{V}^*_{t|t-1}(p) &= Var[\mathbf{v}^*_t(p)|\mathfrak{I}_{t-1}] \\
&= Var\left[\sum_{r=1}^{p} \phi_r \mathbf{B}_{t-r}\left[\sum_{k=1}^{r} (\mathbf{G}^{-1})^{r-k+1}\mathbf{w}_{t-k+1}\right] + \varepsilon_t \middle| \mathfrak{I}_{t-1}\right] \\
&= \sum_{r=1}^{p} \phi_r \mathbf{B}_{t-r}\left[\sum_{k=1}^{r} (\mathbf{G}^{-1})^k \mathbf{W}[(\mathbf{G}^{-1})^k]'\right]\mathbf{B}'_{t-r}\phi'_r + \mathbf{V}_\varepsilon \\
&= \left[\sum_{r=1}^{p} \phi_r \mathbf{B}_{t-r}(\mathbf{G}^{-1})^r\right]\mathbf{W}\left[\sum_{r=1}^{p} \phi_r \mathbf{B}_{t-r}(\mathbf{G}^{-1})^r\right]' + \mathbf{V}_\varepsilon,
\end{aligned}$$

with $Var[\mathbf{w}_{t-k+1}|\mathfrak{I}_{t-1}] = 0$ for any $k > 1$. □

Lastly, the optimal Kalman gain matrix $\mathbf{K}^*_t$ is

$$\mathbf{K}^*_t = (\mathbf{P}_{t|t-1}[\mathbf{B}^*_t(p)]' + \mathbf{C}^*_t(p))(\mathbf{L}^*_{t|t-1} + \mathbf{B}^*_t(p)\mathbf{C}^*_t(p) + [\mathbf{C}^*_t(p)]'[\mathbf{B}^*_t(p)]')^{-1}, \quad (24)$$

which is used to correct and forecast the distribution of the state vectors for $t > p$.

Proof. Defining $\mathbf{a}_{t|t-1} = \mathbb{E}[\mathbf{x}_t|\mathfrak{I}_{t-1}]$ and $\mathbf{v}^*_{t|t-1}(p) = \mathbb{E}[\mathbf{v}^*_t(p)|\mathfrak{I}_{t-1}]$,

$$\begin{aligned}\mathbf{a}_t &= \mathbb{E}[\mathbf{x}_t|\mathfrak{I}_t] = \mathbf{a}_{t|t-1} + \mathbf{K}^*_t\mathbf{e}^*_{t|t-1}\\ &= \mathbf{a}_{t|t-1} + \mathbf{K}^*_t[\mathbf{B}^*_t(p)\mathbf{x}_t + \mathbf{v}^*_t(p) - \mathbf{B}^*_t(p)\mathbf{a}_{t|t-1} - \mathbf{v}^*_{t|t-1}(p)].\end{aligned}$$

and the state estimation error η_t is

$$\begin{aligned}\eta_t &= \mathbf{x}_t - \mathbf{a}_t\\ &= \mathbf{x}_t - \mathbf{a}_{t|t-1} - \mathbf{K}^*_t[\mathbf{B}^*_t(p)\mathbf{x}_t + \mathbf{v}^*_t(p) - \mathbf{B}^*_t(p)\mathbf{a}_{t|t-1} - \mathbf{v}^*_{t|t-1}(p)]\\ &= (\mathbf{I} - \mathbf{K}^*_t\mathbf{B}^*_t(p))(\mathbf{x}_t - \mathbf{a}_{t|t-1}) - \mathbf{K}^*_t(\mathbf{v}^*_t(p) - \mathbf{v}^*_{t|t-1}(p)).\end{aligned}$$

The covariance matrix of the state estimation error will be

$$\begin{aligned}\mathbf{P}_t &= \mathbb{E}[\eta_t\eta_t']\\ &= \mathbb{E}[((\mathbf{I} - \mathbf{K}^*_t\mathbf{B}^*_t(p))(\mathbf{x}_t - \mathbf{a}_{t|t-1})(\mathbf{x}_t - \mathbf{a}_{t|t-1})'(\mathbf{I} - \mathbf{K}^*_t\mathbf{B}^*_t(p))'\\ &- (\mathbf{I} - \mathbf{K}^*_t\mathbf{B}^*_t(p))(\mathbf{x}_t - \mathbf{a}_{t|t-1})(\mathbf{K}^*_t(\mathbf{v}^*_t(p) - \mathbf{v}^*_{t|t-1}(p)))'\\ &- \mathbf{K}^*_t(\mathbf{v}^*_t(p) - \mathbf{v}^*_{t|t-1}(p))(\mathbf{x}_t - \mathbf{a}_{t|t-1})'(\mathbf{I} - \mathbf{K}^*_t\mathbf{B}^*_t(p))'\\ &+ \mathbf{K}^*_t(\mathbf{v}^*_t(p) - \mathbf{v}^*_{t|t-1}(p))(\mathbf{K}^*_t(\mathbf{v}^*_t(p) - \mathbf{v}^*_{t|t-1}(p))'']\\ &= \mathbf{P}_{t|t-1} - \mathbf{K}^*_t\mathbf{B}^*_t(p)\mathbf{P}_{t|t-1} - \mathbf{P}_{t|t-1}[\mathbf{B}^*_t(p)]'\mathbf{K}^{*\prime}_t + \mathbf{K}_t\mathbf{B}^*_t(p)\mathbf{P}_{t|t-1}[\mathbf{B}^*_t(p)]'\mathbf{K}^{*\prime}_t\\ &- (\mathbf{I} - \mathbf{K}^*_t\mathbf{B}^*_t(p))\mathbf{C}^*_t(p)\mathbf{K}^{*\prime}_t - \mathbf{K}^*_t[\mathbf{C}^*_t(p)]'(\mathbf{I} - \mathbf{K}^*_t\mathbf{B}^*_t(p))' + \mathbf{K}^*_t\mathbf{V}^*_{t|t-1}\mathbf{K}^{*\prime}_t\\ &= \mathbf{P}_{t|t-1} - \mathbf{K}^*_t\mathbf{B}^*_t(p)\mathbf{P}_{t|t-1} - \mathbf{P}_{t|t-1}[\mathbf{B}^*_t(p)]'\mathbf{K}^{*\prime}_t\\ &+ \mathbf{K}^*_t(\mathbf{B}^*_t(p)\mathbf{P}_{t|t-1}[\mathbf{B}^*_t(p)]' + \mathbf{V}^*_{t|t-1})\mathbf{K}^{*\prime}_t - \mathbf{C}^*_t(p)\mathbf{K}^{*\prime}_t + \mathbf{K}^*_t\mathbf{B}^*_t(p)\mathbf{C}^*_t(p)\mathbf{K}^{*\prime}_t\\ &- \mathbf{K}^*_t[\mathbf{C}^*_t(p)]' + \mathbf{K}^*_t[\mathbf{C}^*_t(p)]'[\mathbf{B}^*_t(p)]'\mathbf{K}^{*\prime}_t.\end{aligned}$$

Applying trace to the above equation, we minimise the following first derivative in order to find the optimal Kalman gain matrix.

$$\begin{aligned}tr(\mathbf{P}_t) &= tr(\mathbf{P}_{t|t-1}) - 2tr(\mathbf{K}^*_t\mathbf{B}^*_t(p)\mathbf{P}_{t|t-1}) + tr(\mathbf{K}^*_t\mathbf{L}^*_{t|t-1}\mathbf{K}^{*\prime}_t)\\ &+ tr(\mathbf{K}^*_t\mathbf{B}^*_t(p)\mathbf{C}^*_t(p)\mathbf{K}^{*\prime}_t) + tr(\mathbf{K}^*_t[\mathbf{C}^*_t(p)]'[\mathbf{B}^*_t(p)]'\mathbf{K}^{*\prime}_t)\\ &- 2tr(\mathbf{K}^*_t[\mathbf{C}^*_t(p)]'),\\ \frac{d(tr(\mathbf{P}_t))}{d\mathbf{K}^*_t} &= -2(\mathbf{B}^*_t(p)\mathbf{P}_{t|t-1})' + 2\mathbf{K}^*_t\mathbf{L}^*_{t|t-1} + 2\mathbf{K}^*_t\mathbf{B}^*_t(p)\mathbf{C}^*_t(p)\\ &+ 2\mathbf{K}^*_t[\mathbf{C}^*_t(p)]'[\mathbf{B}^*_t(p)]' - 2\mathbf{C}^*_t(p) = 0.\end{aligned}$$

Therefore, the optimal Kalman gain matrix $\mathbf{K}^*_t$ is

$$\mathbf{K}^*_t = (\mathbf{P}_{t|t-1}[\mathbf{B}^*_t(p)]' + \mathbf{C}^*_t(p))(\mathbf{L}^*_{t|t-1} + \mathbf{B}^*_t(p)\mathbf{C}^*_t(p) + [\mathbf{C}^*_t(p)]'[\mathbf{B}^*_t(p)]')^{-1}. \quad \square$$

Now, assuming the prediction error $\mathbf{e}^*_{t|t-1}$ follows a multivariate normal distribution, we construct the marginalised likelihood based on the prediction errors

$$\ell_t(\psi, \mathbf{V}_\varepsilon, \phi; \mathbf{y}_t) = -\frac{1}{2}\ln(\det(\mathbf{L}^*_{t|t-1})) - \frac{1}{2}(\mathbf{e}^*_{t|t-1})'(\mathbf{L}^*_{t|t-1})^{-1}\mathbf{e}^*_{t|t-1}, \tag{25}$$

where $\psi = (\kappa, \sigma_\chi, \lambda_\chi, \gamma, \mu_\xi, \sigma_\xi, \lambda_\xi, \rho_{\chi\xi})$ is the set of model parameters and $\det(\cdot)$ denotes the determinant of the matrix. The sum of the marginalised likelihoods for $t > p$ is the likelihood function to be maximised for the estimation of the model parameters. That is,

$$\ell(\psi, \mathbf{V}_\varepsilon, \phi; \mathbf{y}_t) = -\frac{1}{2}\sum_{t=p+1}^{n}\left[\ln(\det(\mathbf{L}^*_{t|t-1})) + (\mathbf{e}^*_{t|t-1})'(\mathbf{L}^*_{t|t-1})^{-1}\mathbf{e}^*_{t|t-1}\right], \tag{26}$$

where, (26) is to be maximised subject to the constraint $\kappa \geq \gamma$ from [3]. As discussed in [3], this constraint is used to overcome the parameter identification problem in the state process $\mathbf{x}_t$. In addition, the code specifications and structure of the covariance matrix of the measurement errors, $\mathbf{V}_\varepsilon$, used are similar to that provided in [16].

3 Simulation Study

In this section, we present the results of the simulation study designed to address the convergence of the estimates of the model parameters and the state variables in the two scenarios, where the error terms in the system of measurement equations are inter-correlated and follow:

Model 1: AR(1) process (the two-step approach is used for estimation, [16]);
Model 2: AR(p) process (the unified procedure presented in Section 2, is used for estimation).

In this study, the system of measurement equations is formulated for the logarithms of the prices of futures contracts with varying maturity times. The samples of the futures prices (with the number of the contracts $N = 5, 10$) were simulated using Model 2 scenario with $p = 1$. The futures contracts have maturities ranging from 1 to 5 years. Specifically, for $N = 5$, we simulate annual contracts, while for $N = 10$, each vector of futures contracts matures biannually. To align with real market conditions, we took the model parameters from an empirical study on the EUA futures prices in [1] as the true values in our simulation study. In addition, to initialise the serial and inter-correlations of the measurement errors in Model 2 we use the results from Model 1, [16]. The long-term factor ξ_t is simulated assuming it follows a Brownian motion, equivalent to setting $\gamma = 0$.

Figures 1 - 2 present visualisations illustrating the convergence of the model parameter estimates by plotting the differences between the estimates and their true values. The figures correspond to the cases of $N = 5$ and $N = 10$, respectively. For $N = 5$, we observe that the estimates tend to exhibit fluctuations for sample sizes $n \leq 1000$. However, they start stabilising thereafter. In most cases, Model 2 demonstrates superior performance compared to Model 1, particularly when the sample size is large. Although there are minimal differences between the two models, except for

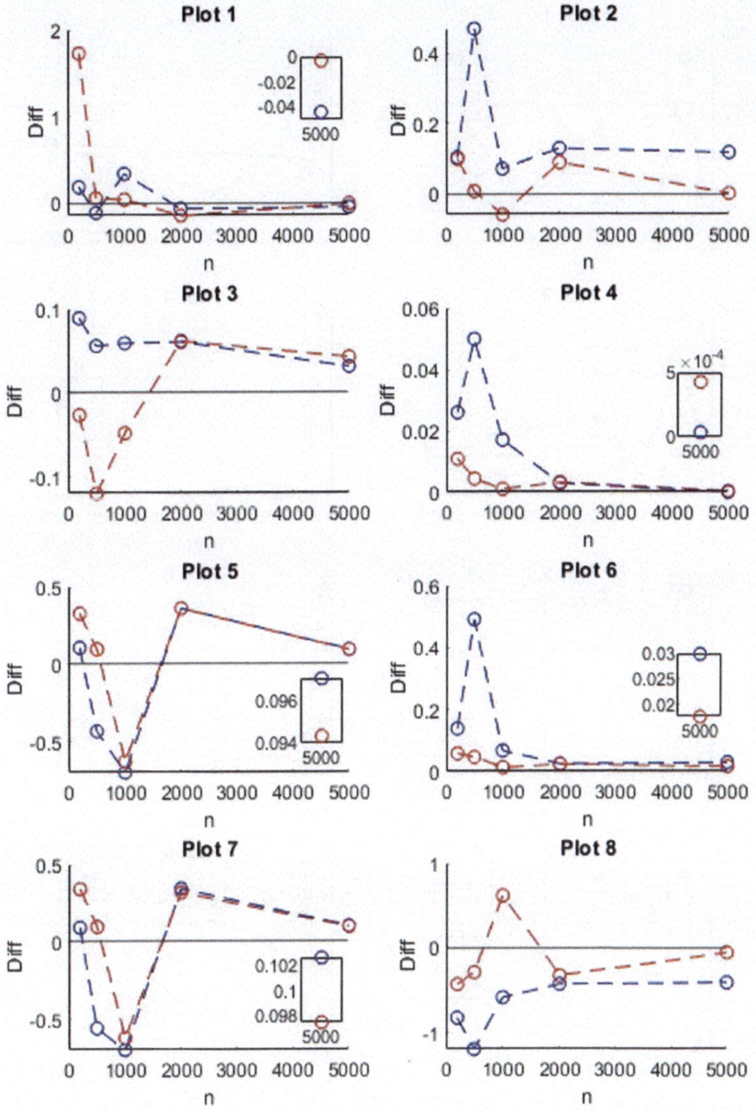

Fig. 1 Plots 1-8 show the difference of the estimated parameter to their true values in the following order - $\kappa, \sigma_\chi, \lambda_\chi, \gamma, \mu_\xi, \sigma_\xi, \lambda_\xi, \rho_{\chi\xi}$, when $N = 5$. The x-axis shows the sample size in the simulated dataset, while the y-axis shows the absolute difference. The blue line represents the result using Model 1, and the red line shows the result using Model 2. Insets are added at $n = 5{,}000$ for better comparison between two models.

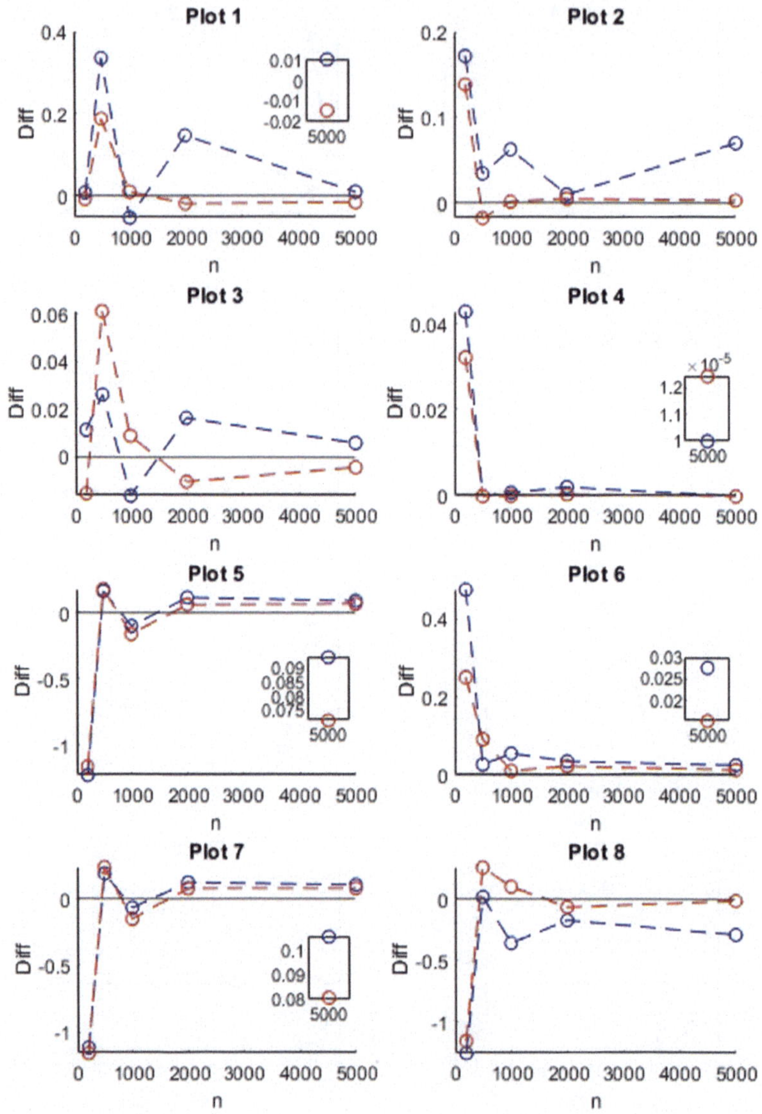

Fig. 2 Plots 1-8 show the difference of the estimated parameter to their true values in the following order - $\kappa, \sigma_\chi, \lambda_\chi, \gamma, \mu_\xi, \sigma_\xi, \lambda_\xi, \rho_{\chi\xi}$, when $N = 10$. The x-axis shows the sample size in the simulated dataset, while the y-axis shows the absolute difference. The blue line represents the result using Model 1, and the red line shows the result using Model 2. Insets are added at $n = 5,000$ for better comparison between two models.

$\hat{\lambda}_\chi$ and $\hat{\gamma}$, Model 2 generally demonstrates better performance. Both models exhibit similar behaviour, except for $\hat{\sigma}_\chi$ and $\hat{\rho}_{\chi\xi}$, where we see a significant difference in estimates at $n = 5{,}000$.

Table 1 Simulation Study - Parameter estimates with a full covariance matrix and AR(1) measurement errors, $n = 5{,}000$, $N = 10$

	Aspinall et al. [2]	Model 1	Model 2	True
$\hat{\kappa}$	0.2746	0.2571	0.2319	0.2462
$\hat{\sigma}_\chi$	0.1171	0.1454	0.0787	0.0748
$\hat{\lambda}_\chi$	-0.0327	0.0122	0.0021	0.0062
$\hat{\gamma}$	1.00×10^{-5}	1.00×10^{-5}	1.25×10^{-5}	0
$\hat{\mu}_\xi$	0.4956	0.5913	0.5692	0.4973
$\hat{\sigma}_\xi$	0.5212	0.5354	0.5231	0.5077
$\hat{\lambda}_\xi$	-0.1748	0.5590	0.5337	0.4527
$\hat{\rho}_{\chi\xi}$	0.2283	0.0094	0.2852	0.2960
$\hat{s}_1$	1.00×10^{-10}	0.0054	0.0046	0.01
$\hat{s}_2$	0.0180	0.0203	0.0046	0.01
$\hat{s}_3$	0.0145	0.0175	0.0047	0.01
$\hat{s}_4$	0.0138	0.0176	0.0043	0.01
$\hat{s}_5$	0.0148	0.0187	0.0042	0.01
$\hat{s}_6$	0.0141	0.0192	0.0047	0.01
$\hat{s}_7$	0.0132	0.0179	0.0047	0.01
$\hat{s}_8$	0.0140	0.0197	0.0048	0.01
$\hat{s}_9$	0.0136	0.0200	0.0045	0.01
$\hat{s}_{10}$	0.0127	0.0010	0.0046	0.01
$\hat{\rho}_1$	-	0.9666	0.3888	0.90
$\hat{\rho}_2$	-	0.5298	0.2730	0.90
$\hat{\rho}_3$	-	0.6047	0.3858	0.90
$\hat{\rho}_4$	-	0.6422	0.1978	0.90
$\hat{\rho}_5$	-	0.6300	0.1509	0.90
$\hat{\rho}_6$	-	0.6983	0.3523	0.90
$\hat{\rho}_7$	-	0.6419	0.3267	0.90
$\hat{\rho}_8$	-	0.6516	0.4005	0.90
$\hat{\rho}_9$	-	0.6737	0.2693	0.90
$\hat{\rho}_{10}$	-	0.3210	0.7079	0.90
$\hat{\phi}_{1,1}$	-	0.8929	0.9467	0.95
$\hat{\phi}_{1,2}$	-	0.9526	0.9558	0.95
$\hat{\phi}_{1,3}$	-	0.9401	0.9451	0.95
$\hat{\phi}_{1,4}$	-	0.9408	0.9456	0.95
$\hat{\phi}_{1,5}$	-	0.9495	0.9592	0.95
$\hat{\phi}_{1,6}$	-	0.9485	0.9446	0.95
$\hat{\phi}_{1,7}$	-	0.9392	0.9485	0.95
$\hat{\phi}_{1,8}$	-	0.9494	0.9498	0.95
$\hat{\phi}_{1,9}$	-	0.9501	0.9500	0.95
$\hat{\phi}_{1,10}$	-	0.9187	0.9468	0.95
AIC	-2.71×10^5	-4.01×10^5	-4.60×10^5	-

The estimation results for model parameters, encompassing volatilities, inter-correlations, and serial correlations, are presented in Table 1 for a sample size of $n = 5{,}000$ and $N = 10$. Additionally, we present the estimates using the model in [2] in Table 1 for testing on the available R package. This work involves a recently published R code for the estimation of model parameters under the Schwartz-Smith two-factor model framework, where model parameters are estimated based on the 'Genetic Optimization Using Derivatives' algorithm that does not require setting the values to initiate the estimation procedure, see [19]. However, this model setup assumes measurement errors follow white noise, and are independent component-wise. In contrast, in [16], they use the gradient-based algorithm to minimise the non-linear function with constraints, with the initial values determined based on the grid search method. The result shows the model in [2] is able to closely estimate $\hat{\gamma}$ and $\hat{\mu}_\xi$, but other model parameters deviates from their true values.

It is important to note that in Model 1, the estimated parameters $\hat{s}_j$ and $\hat{\rho}_j$ correspond to the volatilities and inter-correlation parameters of the measurement error process $\mathbf{v}_t$. Conversely, in Model 2, they represent parameters related to ε_t. The serial correlations, denoted as $\hat{\phi}$, for Model 1 are estimated by fitting an autoregressive (AR) model to the fitted residuals. Using the Akaike Information Criterion (AIC) Model 2 is selected over Model 1. In [2], $\mathbf{V}$ is assumed to be a diagonal matrix, hence s_j, $j = 1, \dots, N$ are to be estimated.

In addition, we evaluate the performance of state estimation for Models 1 and 2. The quality of the fit for the state variables is assessed using the mean squared error (MSE). Table 2 presents the MSE values for each model at various sample sizes when $N = 5, 10$. We observe that for smaller datasets ($n \leq 500$, $N = 5$), estimations of state variables are unstable, whereas when $n \geq 1{,}000$, we observe, especially with longer-dated contracts, that the fit is better in terms of MSE. Hence, to use this model for empirical applications, at least $n = 1{,}000$ may be required in order to achieve a better fit of state variables, and hence, closer parameter estimates to their true values.

Table 2 Mean Square Errors of state variables for Models 1 and 2

		Model 1		Model 2	
N	n	MSE(χ)	MSE(ξ)	MSE(χ)	MSE(ξ)
5	200	0.2365	0.2422	0.0691	0.0911
	500	0.7216	0.7230	0.2174	0.2180
	1,000	0.0019	0.0039	0.0343	0.0282
	2,000	0.2432	0.2380	0.1167	0.1118
	5,000	0.1291	0.1313	0.1693	0.1720
10	200	0.0464	0.0528	0.0200	0.0142
	500	0.0134	0.0174	0.0329	0.0418
	1,000	0.0013	0.0022	0.0005	0.0020
	2,000	0.0015	0.0022	0.0015	0.0025
	5,000	0.0016	0.0020	0.0005	0.0011

Our findings, backed by AIC, demonstrate that incorporating the AR(p) structure of the measurement errors in Model 2 and using the unified estimation procedure surpassed the two-step approach employed in Model 1. The noticeable superiority is also evident through MSE criterion in the fitting of the state variables.

In summary, given the assumption of independence of measurement errors cannot be validated, our study provides a new tool which can be used for Schwartz-Smith two factor model calibration and forecasting of futures prices.

4 Conclusion

This paper presents an extended two-factor model from [24] and [8], for which we propose the data-driven parameter estimation through modelling of the autoregressive structure of inter-correlated errors in the measurement equations' system. In this framework we derived the linear Kalman filter, which was used for joint estimation of the state variables and model parameters.

In the simulation study, we carried out a comparative analysis of the convergence of the estimates in Models 1 and 2 with respect to the sample size and number of contracts, as well as accuracy. Model 2 outperformed Model 1 in terms of MSE and AIC for state variables and parameter estimates, respectively. The simulation study offered valuable insights into the performance and suitability of the models, contributing to the advancement of data-driven estimation methodology.

The empirical results presented in Section 3 can be reproduced using `MATLAB` code developed by the first author, the link to the code is provided in the footnote on the title page of the paper.

References

1. Aspinall, T., Gepp, A., Harris, G., Kelly, S., Southam, C., Vanstone, B.: Estimation of a term structure model of carbon prices through state space methods: The European Union emissions trading scheme. Accounting & Finance **61**(2), 3797–3819 (2021)
2. Aspinall, T., Gepp, A., Harris, G., Kelly, S., Southam, C., Vanstone, B.: NFCP: N-Factor Commodity Pricing Through Term Structure Estimation (2022). URL https://CRAN.R-project.org/package=NFCP. R package version 1.2.1
3. Binkowski, K., He, P., Kordzakhia, N., Shevchenko, P.: On the parameter estimation in the Schwartz-Smith's two-factor model. In: Statistics and Data Science: Research School on Statistics and Data Science, RSSDS 2019, Melbourne, VIC, Australia, July 24–26, 2019, Proceedings 1, pp. 226–237. Springer (2019)
4. Brennan, M.J., Schwartz, E.S.: Evaluating natural resource investments. Journal of business pp. 135–157 (1985)
5. Brown, R.G., Hwang, P.Y.: Introduction to random signals and applied Kalman filtering. John Wiley and Sons, New York (1997)
6. Cheng, B., Nikitopoulos, C.S., Schlögl, E.: Pricing of long-dated commodity derivatives: Do stochastic interest rates matter? Journal of Banking & finance **95**, 148–166 (2018)

7. Cortazar, G., Milla, C., Severino, F.: A multicommodity model of futures prices: Using futures prices of one commodity to estimate the stochastic process of another. Journal of Futures Markets: Futures, Options, and Other Derivative Products **28**(6), 537–560 (2008)
8. Cortazar, G., Millard, C., Ortega, H., Schwartz, E.S.: Commodity price forecasts, futures prices, and pricing models. Management science **65**(9), 4141–4155 (2019)
9. Cortazar, G., Naranjo, L.: An N-factor Gaussian model of oil futures prices. Journal of Futures Markets: Futures, Options, and Other Derivative Products **26**(3), 243–268 (2006)
10. Cortazar, G., Schwartz, E.S.: The valuation of commodity contingent claims. Journal of derivatives **1**(4), 27–39 (1994)
11. DelSole, T., Yang, X.: State and parameter estimation in stochastic dynamical models. Physica D: Nonlinear Phenomena **239**(18), 1781–1788 (2010)
12. Ewald, C.O., Zhang, A., Zong, Z.: On the calibration of the Schwartz two-factor model to WTI crude oil options and the extended Kalman filter. Annals of Operations Research **282**, 119–130 (2019)
13. Favetto, B., Samson, A.: Parameter estimation for a bidimensional partially observed Ornstein–Uhlenbeck process with biological application. Scandinavian Journal of Statistics **37**(2), 200–220 (2010)
14. Gibson, R., Schwartz, E.S.: Stochastic convenience yield and the pricing of oil contingent claims. The Journal of Finance **45**(3), 959–976 (1990)
15. Goodwin, Dominice: Schwartz-Smith Two-Factor Model in the Copper Market: before and after the New Market Dynamics (2013). Student Paper
16. Han, J.S., Kordzakhia, N., Shevchenko, P.V., Trück, S.: On correlated measurement errors in the Schwartz-Smith two-factor model. Dependence Modeling **10**(1), 108–122 (2022)
17. Kuruppuarachchi, D., Lin, H., Premachandra, I.: Testing commodity futures market efficiency under time-varying risk premiums and heteroscedastic prices. Economic Modelling **77**, 92–112 (2019)
18. Manoliu, M., Tompaidis, S.: Energy futures prices: term structure models with Kalman filter estimation. Applied mathematical finance **9**(1), 21–43 (2002)
19. Mebane Jr, W.R., Sekhon, J.S.: Genetic optimization using derivatives: the rgenoud package for R. Journal of Statistical Software **42**, 1–26 (2011)
20. Moye, M.J., Diekman, C.O.: Data assimilation methods for neuronal state and parameter estimation. The Journal of Mathematical Neuroscience **8**, 1–38 (2018)
21. Peters, G.W., Briers, M., Shevchenko, P., Doucet, A.: Calibration and filtering for multi factor commodity models with seasonality: incorporating panel data from futures contracts. Methodology and Computing in Applied Probability **15**, 841–874 (2013)
22. Petovello, M.G., O'Keefe, K., Lachapelle, G., Cannon, M.E.: Consideration of time-correlated errors in a Kalman filter applicable to GNSS. Journal of Geodesy **83**, 51–56 (2009)
23. Sauvageau, M., Kumral, M.: Genetic algorithms for the optimisation of the Schwartz-Smith two-factor model: a case study on a copper deposit. International Journal of Mining, Reclamation and Environment **32**(3), 163–181 (2018)
24. Schwartz, E., Smith, J.E.: Short-term variations and long-term dynamics in commodity prices. *Management Science* **46**(7), 893–911 (2000)
25. Schwartz, E.S.: The stochastic behavior of commodity prices: Implications for valuation and hedging. The Journal of finance **52**(3), 923–973 (1997)
26. Seifert, J., Uhrig-Homburg, M.: Modelling jumps in electricity prices: theory and empirical evidence. Review of Derivatives Research **10**, 59–85 (2007)
27. Sørensen, C.: Modeling seasonality in agricultural commodity futures. Journal of Futures Markets: Futures, Options, and Other Derivative Products **22**(5), 393–426 (2002)
28. Tandeo, P., Ailliot, P., Bocquet, M., Carrassi, A., Miyoshi, T., Pulido, M., Zhen, Y.: A review of innovation-based methods to jointly estimate model and observation error covariance matrices in ensemble data assimilation. Monthly Weather Review **148**(10), 3973–3994 (2020)
29. Villaplana, P.: Pricing power derivatives: A two-factor jump-diffusion approach. Available at SSRN 493943 (2003)
30. Wen, Y., Kiesel, R.: Pricing options on EU ETS certificates with a time-varying market price of risk model. Stochastics of Environmental and Financial Economics **138**, 341–360 (2016)

On extension of the Markov chain approximation method for computing Feynman–Kac type expectations

Vincent Liang and Konstantin Borovkov

Abstract An efficient discrete time and space Markov chain approximation employing a Brownian bridge correction for computing curvilinear boundary crossing probabilities for general diffusion processes was recently proposed in Liang and Borovkov (2021). One of the advantages of that method over alternative approaches is that it can be readily extended to computing expectations of path-dependent functionals over the event of the process trajectory staying between two curvilinear boundaries. In the present paper, we extend the scheme to compute expectations of the Feynman–Kac type that frequently appear in option pricing. To illustrate our approximation scheme, we apply it in three special cases. For sufficiently smooth integrands, numerical experiments suggest that the proposed approximation converges at the rate $O(n^{-2})$, where n is the number of steps on the uniform time grid used.

1 Introduction

Barrier options are widely used path-dependent derivative securities, and there is extensive literature on their valuation (see e.g. [2], [7] and the references therein). Most of the literature is focused on the case where the barriers are "flat". However, curved boundaries also appear in the context of pricing options on underlying assets that have a volatility term structure. Using a deterministic time-change, the time dependence in the volatility can be transferred into the boundary, reducing the problem of pricing call and put options in the Black–Scholes type setting with deterministic

Vincent Liang
School of Mathematics and Statistics, The University of Melbourne, Parkville 3010, Australia e-mail: vliang@student.unimelb.edu.au

Konstantin Borovkov
School of Mathematics and Statistics, The University of Melbourne, Parkville 3010, Australia e-mail: borovkov@unimelb.edu.au

D. R. Wood et al. (eds.), *2021-2022 MATRIX Annals*, MATRIX Book Series 5,
https://doi.org/10.1007/978-3-031-47417-0_20

interest rates to computing the curvilinear boundary crossing probability of a diffusion process (see e.g. Section 3 in [6]). Methods for solving the latter problem are numerous, and we refer the interested reader to [14] for a literature review. To extend the aforementioned problem to the case of stochastic interest rates, one needs to introduce a numeraire process.

It is well-known that, under the no-arbitrage assumption, the fair price of a replicable derivative on an underlying asset X with maturity T and general payoff $\phi(X(T))$ is equal to $\mathbf{E}\phi(X(T))/N(T)$, where $\mathbf{E}$ denotes the expectation with respect to the martingale measure associated with the numeraire process N (for details, see e.g. Chapter 6 in [4]). If we introduce a barrier feature for the above option using two time-dependent barriers $g^- < g^+$, the payoff of the new option will be given by

$$\phi(X(T))\mathbf{1}\{g^-(s) < X(s) < g^+(s), s \in [0,T]\}.$$

Hence the fair price of the barrier option is equal to

$$\mathbf{E}\big[\phi(X(T))/N(T); g^-(s) < X(s) < g^+(s), s \in [0,T]\big]. \tag{1}$$

The purpose of this paper is to extend the Markov chain approximation method proposed in [14] to compute the above expectation in the case when the numeraire process takes the form $N(T) = \exp\{\int_0^T V(X(t))\,dt\}$ for some continuous function V, i.e. to evaluate

$$\mathbf{E}\Big[e^{-\int_0^T V(X(u))\,du}\phi(X(T)); g^-(s) < X(s) < g^+(s), s \in [0,T]\Big]. \tag{2}$$

The method suggested in [14] was a development of the approach from [10] for computing boundary non-crossing probabilities of the Brownian motion process. That approach suggested to replace the continuous dynamics of the underlying Brownian motion process with those of a denumerable discrete time Markov chain, by discretising both time and space using uniform grids. The transition probabilities for these Markov chains were specified to be proportional to the values of the Brownian motion transition densities for the respective time and space increments. The desired approximation for the boundary non-crossing probability was then computed in [10] by multiplying finite-dimensional substochastic matrices obtained by retaining the entries corresponding to the space nodes located between the given boundaries at the respective times.

Our modification of this method developed in [14] dramatically improved the convergence rate of the procedure by shifting and adjusting the uniform space grids so that they would have nodes lying exactly on the boundaries, and by implementing one-step Brownian bridge corrections taking into account the possibility that the process could cross a boundary at a time between the points on the discrete time grid. At the same time, the method was extended in [14] to a much broader class of diffusions.

Roughly speaking, one can think about the workings of this method as summing up the probabilities of all the Markov chain trajectories lying between the boundaries (and applying appropriate corrections). As, for each trajectory, its probability is given by the product of one-step transition probabilities, one basically "tracks" each trajectory of the chain in this computational process. Unlike alternative approaches to computing boundary crossing probabilities, such as the method of integral equations (see e.g. [11] and further references therein), this provides one with an opportunity to compute not only probabilities but also expectations of path-dependent functionals, finding the values of expressions of the form (2) being of particular convenience.

Turning back to applications related to (1), we note that when the underlying X is the spot interest rate, this includes the case when the numeraire is the bank account process, i.e. $V(x) = x$. This expression can also be used to price what we call a *hybrid step-barrier option*, an option that has a step option feature (see [15]) and a barrier above the step level (see Example 3 below).

It was reported in [14] that numerical experiments had showed that the proposed method produces approximations for boundary crossing probabilities with convergence rate $O(n^{-2})$, where n is the number of nodes on the regular time discretisation grid. It turns out that this convergence rate is preserved in the new extension of the method presented in this paper when V is sufficiently smooth. Note that proving the above-mentioned very fast convergence rate for our scheme is a very difficult task already in the basic case of computing boundary crossing probabilities. Establishing this rate in the one-sided boundary case with $g^+ \in C^2$ for the Brownian motion process is work in progress.

The paper is organised as follows. In Section 2 we describe our approximation procedure and demonstrate its convergence. In Section 3 we present the results of applying it in three special cases.

2 The method

We will begin by specifying the diffusion process model for the underlying asset X. We assume a unit diffusion coefficient since we can transform a large class of diffusion processes to this case using the well-known unit-diffusion transform (see e.g. Section 2 in [14]). Then, we will describe the sequence of Markov chain approximations and prove that the sequence converges weakly to the desired diffusion process. Using the weak convergence result, we will show that the corresponding approximations to the option price that are using the Brownian bridge correction techniques converge as well. Recall that employing this correction dramatically improves convergence rates in the basic problem of computing boundary crossing probabilities [14]. As we will see this applies in the present paper setup as well.

Without loss of generality, we set the terminal time $T := 1$ since a deterministic time change can be applied to get the results for a general fixed $T < \infty$.

2.1 The setup

Suppose that the underlying asset's price is modelled by the one-dimensional diffusion process

$$X(t) = x_0 + \int_0^t \mu(s, X(s))\,ds + W(t), \quad t \in [0,1],$$

where $\{W(t)\}_{t\in[0,1]}$ is a standard Brownian motion process, and x_0 is non-random. Assume the following condition is satisfied:

(C) For any fixed $x \in \mathbb{R}$, one has $\mu(\cdot, x) \in C^1([0,1])$, and for any fixed $t \in [0,1]$, one has $\mu(t,\cdot) \in C^2(\mathbb{R})$. Moreover, for any $r > 0$, there exists a $K_r < \infty$ such that one has

$$|\mu(t,x)| + |\partial_t \mu(t,x)| + |\partial_x \mu(t,x)| + |\partial_{xx}\mu(t,x)| \le K_r, \quad t \in [0,1],\ |x| \le r.$$

We will also need some notations and conditions related to the boundaries $g^{\pm}$. Denote by $C = C([0,1])$ the space of continuous functions $f : [0,1] \to \mathbb{R}$ equipped with the uniform norm $\|f\|_\infty := \sup_{t\in[0,1]}|f(t)|$. For a fixed $x_0 \in \mathbb{R}$, consider the class

$$\mathscr{G} := \Big\{(f^-, f^+) : f^{\pm} \in C,\ f^-(0) < x_0 < f^+(0),\ \min_{0\le t\le 1}(f^+(t) - f^-(t)) > 0\Big\}$$

of pairs of functions from C and introduce the notation

$$S(f^-, f^+) := \{v \in C : f^-(t) < v(t) < f^+(t),\ t \in [0,1]\}, \quad (f^-, f^+) \in \mathscr{G}$$

for the "bunch" of continuous functions whose graphs are entirely contained in the strip between the boundaries $f^{\pm}$.

Let $V : \mathbb{R} \to \mathbb{C}$ be a (possibly complex-valued) continuous function and $\phi : \mathbb{R} \to \mathbb{R}$ be some continuous function. The problem we deal with in this paper is how to compute

$$Q := \mathbf{E}\Big(e^{-\int_0^1 V(X(s))\,ds}\phi(X(1)); X \in G\Big), \quad G := S(g^-, g^+), \tag{3}$$

for some $(g^-, g^+) \in \mathscr{G}$. To explain our approach, set $\mathbf{E}^{s,x}(\cdot) := \mathbf{E}(\cdot\,|X(s) = x)$ and introduce Feynman–Kac type time-dependent semigroup $\{T_{s,t}\}_{0\le s<t\le 1}$ defined by

$$T_{s,t}f(x) = \mathbf{E}^{s,x}\Big(e^{-\int_s^t V(X(u))\,du} f(X(t)); g^-(u) < X(u) < g^+(u), u \in [s,t]\Big) \tag{4}$$

and represent (3) as

$$Q = (T_{0,t_{n,1}} T_{t_{n,2},t_{n,3}} \cdots T_{t_{n,n-1},t_{n,n}}\phi)(x_0), \tag{5}$$

for

$$t_{n,k} := k/n, \quad k = 0, 1, \ldots, n,$$

the uniform partition of $[0,1]$ of rank $\Delta_n := 1/n$, $n \geq 1$. The idea of our method is to approximate operators $T_{s,t}$ with their discrete versions. We will implement it in the next section, leading to approximation (8) to our Q.

2.2 Markov chain approximation

To specify our time-dependent Markov chain approximation, we first define the space grids $E_{n,k}$ and the transition probabilities $p_{n,k}$, and then introduce the corrective terms $\pi_{n,k}$ and $e_{n,k}$ which account for boundary correction and the presence of the term with V respectively.

The grids $E_{n,k}$ are constructed as follows. Set $g^{\pm}_{n,k} := g^{\pm}(t_{n,k})$, $k = 1,\dots,n$, and, for fixed $\delta \in (0,\frac{1}{2}]$ and $\gamma > 0$, put

$$w_{n,k} := \begin{cases} \frac{(g^{+}_{n,k}-g^{-}_{n,k})/\Delta_n^{1/2+\delta}}{\lfloor \gamma (g^{+}_{n,k}-g^{-}_{n,k})/\Delta_n^{1/2+\delta} \rfloor}, & 1 \leq k < n, \\ \frac{(g^{+}(1)-g^{-}(1))/\Delta_n}{\lfloor \gamma (g^{+}(1)-g^{-}(1))/\Delta_n \rfloor}, & k = n, \end{cases}$$

assuming that n is large enough such that the integer parts in all the denominators are positive. We set the time-dependent space lattice step sizes to be

$$h_{n,k} := \begin{cases} w_{n,k}\Delta_n^{1/2+\delta}, & 1 \leq k < n, \\ w_{n,n}\Delta_n, & k = n. \end{cases}$$

Next, we define

$$E_{n,k} := \{g^{+}_{n,k} - j h_{n,k} : j \in \mathbb{Z}\}, \quad k = 1,\dots,n.$$

We also put $E_{n,0} := \{x_0\}$ and define the corresponding boundary restricted lattices

$$E^{G}_{n,k} := \{x \in E_{n,k} : g^{-}_{n,k} < x < g^{+}_{n,k}\}, \quad k = 1,\dots,n.$$

Further, for $k = 1,\dots,n$, we introduce the discrete time drift and diffusion coefficients

$$\begin{aligned} \mu_{n,k}(x) &:= \big(\mu + \tfrac{1}{2}\Delta_n(\partial_t \mu + \mu\partial_x\mu + \tfrac{1}{2}\partial_{xx}\mu)\big)(t_{n,k-1},x)\Delta_n, \\ \sigma^2_{n,k}(x) &:= (1 + \tfrac{1}{2}\Delta_n \partial_x \mu(t_{n,k-1},x))^2 \Delta_n, \end{aligned} \tag{6}$$

and define one-step "pseudo-transition probabilities" (we use this expression because they do not sum to one, although the normalising constants converge to 1 so quickly as $n \to \infty$ that it makes no sense to normalise them, see Remark 7 in [14]) as

$$p_{n,k}(x,y) := \varphi(y\,|\,x + \mu_{n,k}(x), \sigma^2_{n,k}(x))h_{n,k}, \quad (x,y) \in E_{n,k-1} \times E_{n,k},$$

where $\varphi(x\,|\,\mu,\sigma^2) := (2\pi\sigma^2)^{-1/2}e^{-(x-\mu)^2/(2\sigma^2)}$, $x,\mu \in \mathbb{R}$, $\sigma^2 > 0$. We will implement the Brownian bridge correction for two-sided boundary crossing probabilities by using the factor

$$\pi_{n,k}(x,y) = 1 - \exp\{\tfrac{-2}{\Delta_n}(g^+_{n,k-1} - x)(g^+_{n,k} - y)\} - \exp\{\tfrac{-2}{\Delta_n}(g^-_{n,k-1} - x)(g^-_{n,k} - y)\}.$$

Note that we ignore here the highly unlikely event that the trajectory of the Brownian bridge process hits both the upper and lower boundaries in the small time interval $[t_{n,k-1}, t_{n,k}]$, see [14]. The following expression will be used as a trapezoidal approximation of the exponential term $\exp\{-\int_{(k-1)/n}^{k/n} V(X(s))\,ds\}$:

$$e_{n,k}(x,y) := \exp\{\tfrac{-\Delta_n}{2}(V(x)+V(y))\}. \tag{7}$$

Now we introduce the pseudo-transition matrices $S_{n,k} \in \mathbb{R}^{|E^G_{n,k-1}|\times|E^G_{n,k}|}$ involving both the boundary crossing correction terms $\pi_{n,k}$ and our adjustments $e_{n,k}$ for the presence of the exponential terms with V:

$$S_{n,k} := \big[p_{n,k}(x,y)\pi_{n,k}(x,y)e_{n,k}(x,y)\big]_{(x,y)\in E^G_{n,k-1}\times E^G_{n,k}}.$$

Denoting by $\phi_n := [\phi(x)]_{x\in E^G_{n,n}}$ the n-th payoff column vector, our approximation to (3) is given by the sequence of matrix products

$$Q_n := S_{n,1}S_{n,2}\cdots S_{n,n}\phi_n, \quad n \ge 1, \tag{8}$$

which are discrete analogues of (5).

2.3 *Weak convergence*

Theorem 1. *Let condition* (C) *be met,* $(g^-, g^+) \in \mathscr{G}$, *and* $V : \mathbb{R} \to \mathbb{C}$, $\phi : \mathbb{R} \to \mathbb{R}$ *be continuous functions. Then* $Q_n \to Q$ *as* $n \to \infty$.

Proof. We will use the method of weak convergence. First, we normalise the pseudo-transition probabilities $p_{n,k}$ and prove that the corresponding sequence of Markov chains converges weakly to the target diffusion process X using convergence results from [14]. Then we extend this convergence result to a sequence of auxiliary bivariate processes, where from the desired convergence $Q_n \to Q$ will follow as these quantities can be expressed as expectations of a suitable functional.

Let $\{\xi_{n,k}\}_{k=1}^n$ denote a Markov chain with transition probabilities $p_{n,k}(x,y)/C_{n,k}(x)$, $(x,y) \in E_{n,k-1} \times E_{n,k}$, where $C_{n,k}(x) := \sum_{y\in E_{n,k}} p_{n,k}(x,y)$. Furthermore, let $\widetilde{X}_n$ denote the Brownian bridge-interpolated version of $\{\xi_{n,k}\}_{k=1}^n$, i.e.

$$\widetilde{X}_n(t) := B_{n,k}^{\xi_{n,k-1},\xi_{n,k}}(t), \quad t \in [t_{n,k-1}, t_{n,k}], \quad k = 1,\dots,n,$$

where

$$B^{x,y}_{n,k}(t) := B^{\circ}_{n,k}(t) + x + n(t - t_{n,k-1})(y - x), \quad x, y \in \mathbb{R},$$

and $B^{\circ}_{n,k}(t), t \in [t_{n,k-1}, t_{n,k}]$, are independent Brownian motions "pinned" at the time-space points $(t_{n,k-1}, 0)$ and $(t_{n,k}, 0)$, these bridges being independent of our chain. From Corollary 1 in [14], we know that $\widetilde{X}_n \Rightarrow X$ as $n \to \infty$, where "$\Rightarrow$" denotes weak convergence of random elements in the respective functional space (in this case, in space $(C, \|\cdot\|_\infty)$).

Define the two-dimensional process $\mathbf{X}(t) := (X(t), Z(t))$ by letting

$$Z(t) := \int_0^t V(X(s))\,ds, \quad t \in [0,1].$$

Then we can rewrite Q from (3) as

$$Q = \mathbf{E}\big(\Psi(\mathbf{X}(1)); X \in G\big),$$

where $\Psi(\mathbf{X}(1)) := \phi(X(1))e^{Z(1)}$. Our approximation (8) can also be rewritten as

$$Q_n - \mathbf{E}\big(\Psi(\mathbf{X}_n(1)); \widetilde{X}_n \in G\big),$$

where $\mathbf{X}_n(t) := (\widetilde{X}_n(t), Z_{n,\lfloor nt \rfloor})$, $Z_{n,0} = 0$ and, for $k = 0, 1, \ldots, n$,

$$Z_{n,k} := \frac{\Delta_n}{2} \sum_{i=1}^{k} \big[V(\widetilde{X}_n(t_{n,i-1})) + V(\widetilde{X}_n(t_{n,i}))\big].$$

We will now show that $(\widetilde{X}_n, Z_n) \Rightarrow (X, Z)$ in the respective Skorokhod space by verifying the consistency of infinitesimal moments and tightness. We first define the auxiliary process $\widehat{Z}_n(t) := \widehat{Z}_{n,\lfloor nt \rfloor}, t \in [0,1]$, with $\widehat{Z}_{n,k} = \sum_{i=0}^{k-1} V(\widetilde{X}_n(t_{n,i}))\Delta_n$ and show that $(\widetilde{X}_n, \widehat{Z}_n) \Rightarrow (X, Z)$.

Define the joint process $\widetilde{\mathbf{X}}_n(t) := (t, \widetilde{X}_n(t), \widehat{Z}_{n,\lfloor nt \rfloor})$ and the stopping time

$$\tau^r_n := \inf\{t \ge 0 : \|\widetilde{\mathbf{X}}_n(t)\| \vee \|\widetilde{\mathbf{X}}_n(t-)\| > r\},$$

$\|\mathbf{u}\| := |u_1| \vee |u_2| \vee |u_3|$ being the Chebyshev norm of $\mathbf{u} = (u_1, u_2, u_3) \in \mathbb{R}^3$. Note that

$$\widehat{Z}_{n,k+1} = \widehat{Z}_{n,k} + V(\widetilde{X}_n(t_{n,k}))\Delta_n, \quad \widehat{Z}_{n,0} = 0.$$

Let $\Delta\widehat{Z}_{n,k+1} := \widehat{Z}_{n,k+1} - \widehat{Z}_{n,k}$. It follows immediately that, for all $r > 0$,

$$\lim_{n\to\infty} \max_{1 \le k \le \lfloor n\tau^r_n \rfloor} \Big|\Delta_n^{-1}\mathbf{E}\big(\Delta\widehat{Z}_{n,k+1} \,|\, \widehat{Z}_{n,k} = z, \widetilde{X}_n(t_{n,k}) = x\big) - V(x)\Big| = 0$$

and

$$\lim_{n\to\infty} \max_{1 \le k \le \lfloor n\tau^r_n \rfloor} \Delta_n^{-1}\mathrm{Var}\big(\Delta\widehat{Z}_{n,k+1} \,|\, \widehat{Z}_{n,k} = z, \widetilde{X}_n(t_{n,k}) = x\big) = 0.$$

Tightness of the sequence of the distributions of $\widetilde{\mathbf{X}}_n$ in the respective Skorokhod space is immediate since V is bounded on compact sets and hence it follows from Corollary 4.2 in [9] that $(\widetilde{X}_n, \widehat{Z}_n) \Rightarrow (X, Z)$. For all $r > 0$,

$$\max_{1 \le k \le \lfloor n\tau_n^r \rfloor} |\widehat{Z}_{n,k} - Z_{n,k}| \le \frac{\Delta_n}{2}|V(0) + V(r)| \xrightarrow{\text{a.s.}} 0, \quad n \to \infty,$$

and hence it follows that $(\widetilde{X}_n, Z_n) \Rightarrow (X, Z)$ as well. By the continuous mapping theorem applied to the function $(x, y) \mapsto \phi(x)e^y$, the Portmanteau theorem (see e.g. p. 24 in [3]), and Corollary 2 from [14], the claimed result follows from the established weak convergence of the processes and boundedness of the integrands in Q. □

3 Numerical examples

In this section, we will illustrate the efficiency of our approximation scheme (8). The run times for these computations using the programming language `Julia` run on a MacBook Pro 2020 laptop computer with an i5 processor (2 GHz, 16 GB RAM) are basically the same as the ones reported in [14] for computing boundary-crossing probabilities employing the same hardware and software. To evaluate the partial derivatives in (6), we used the package `HyperDualNumbers.jl`

We apply our algorithm in three different examples, where $V(x)$ is equal to $-ix^2$, x, and $\kappa\mathbf{1}\{x > r\}$ respectively, for some $x, \kappa, r \in \mathbb{R}$. Due to the discontinuity of the function V at $x = r$ in the last case, we modify the scheme slightly for that choice of V.

For each example, we compute Q_n and plot the observed convergence rate of $Q_{n+1} - Q_n$ to zero as n grows. Closed-form expressions for Q are not available in any of these examples. Due to the iterative nature of our scheme (8), as a byproduct, we obtain approximations to $v(t, x) := (T_{t,1}\phi)(x)$, where $T_{s,t}$ appeared in (4). We will plot the surfaces $v(t, x)$ for each example as well.

3.1 *The case* $V(x) = -ix^2$

To demonstrate general methods for obtaining distributional properties of Wiener functionals of the form $\int_0^1 V(W(s))\,ds$, the characteristic function of the random variable $\int_0^1 W(s)^2\,ds$ was computed in [12] and [8] yielding

$$\mathbf{E}\exp\left\{i\lambda \int_0^1 W(u)^2\,du\right\} = \sqrt{\operatorname{sech}\sqrt{2i\lambda}}, \quad \lambda \in \mathbb{R}.$$

Using our scheme, we compute approximations to the expressions of this form restricted to the event of non-crossing given boundaries:

$$v(t,x) = \mathbf{E}\left(\exp\left\{i\int_t^1 W(u)^2\,du\right\}; g^-(s) < W(s) < g^+(s),\, s\in[t,1]\,\middle|\, W(t)=x\right),$$

with time-dependent boundaries

$$g^{\pm}(t) = \pm 4 \mp t^2, \quad t\in[0,1].$$

In Fig 1, we plot our approximation of the form (8) to $v(t,x)$. We see from Fig 2, that $|\operatorname{Re} Q_{n+1} - \operatorname{Re} Q_n|$ converges to zero at the rate $O(n^{-3})$, suggesting that $|\operatorname{Re} Q_n - \operatorname{Re} Q| = O(n^{-2})$ as $n\to\infty$.

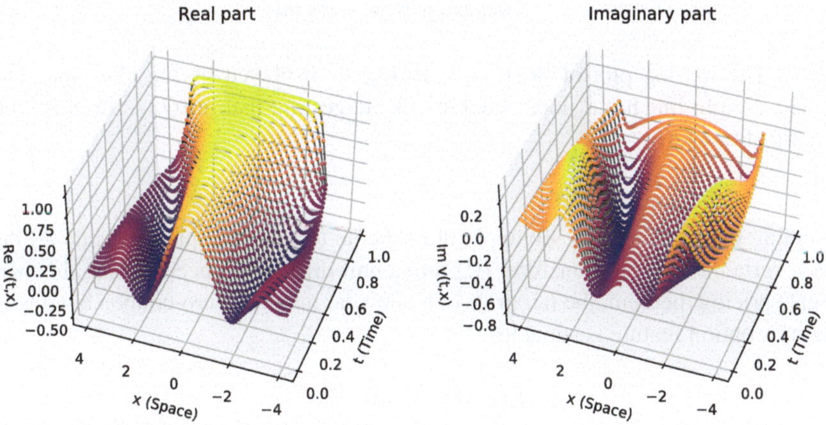

Fig. 1: Approximations to $\operatorname{Re} v(t,x)$ and $\operatorname{Im} v(t,x)$ using (8) with parameters: $n=30$, $\gamma=2$, $\delta=0$, $\mu(t,x)=0$, and $V(x)=-ix^2$.

3.2 *The case* $V(x) = x$

In this example, we price a zero-coupon bond with a two-sided time-dependent knock-out barrier, using a one-factor Hull–White model for spot interest rates. This problem was also explored in [13]. For $\alpha,\sigma>0$, the Hull–White model assumes that the spot interest rate r has the following dynamics:

$$dr(t) = (\theta(t) - \alpha r(t))\,dt + \sigma\,dW(t), \quad t\geq 0,$$

where $\theta(t)$ depends on α, σ and the currently observed instantaneous forward rate curve $f(t)$ (see e.g. Proposition 10.1.6 in [1]):

$$\theta(t) = f'(t) + \alpha f(t) + \frac{\sigma^2}{2\alpha}(1-e^{-2\alpha t}),$$

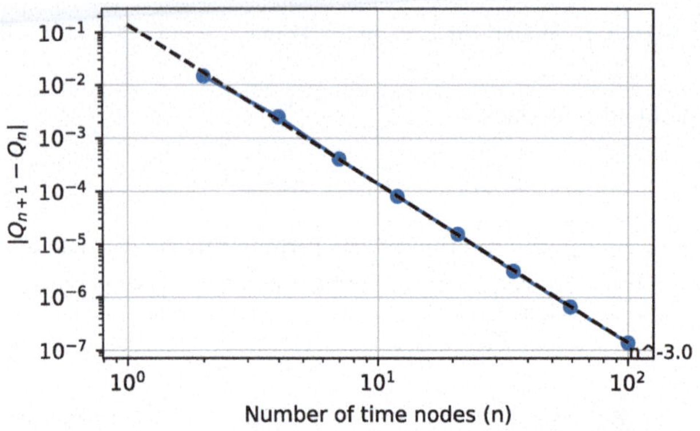

Fig. 2: The log-log plot of $|\mathrm{Re}\, Q_{n+1} - \mathrm{Re}\, Q_n|$ is displayed by the blue line. The dashed straight line has a slope equal to -3, suggesting that $|\mathrm{Re}\, Q_n - \mathrm{Re}\, Q|$ is converging at rate $O(n^{-2})$.

assuming that $r(0) = f(0)$. To apply our scheme to this example, we scale all the space variables by σ to transform $r(t)$ into a unit-diffusion process. Using the martingale pricing theorem, the fair price of a one-year maturity zero-coupon bond with a barrier option feature is equal to

$$\mathbf{E}\big(e^{-\int_0^1 r(s)\,ds}; r \in G\big).$$

We used the time-dependent boundaries given by

$$g^{\pm}(t) = \pm 0.04(1 \mp \tfrac{1}{2}\sin(3t)), \quad t \in [0,1].$$

To demonstrate the performance of our scheme, we set $\alpha = 0.01$, $\sigma = 0.01$ and a flat instantaneous forward curve $f(t) \equiv 0.03$. In the left pane in Fig 3, we plot our approximation of

$$v(t,x) = \mathbf{E}\left(\exp\left\{-\int_t^1 r(u)\,du\right\}; g^-(s) < r(s) < g^+(s),\, s \in [t,1] \,\middle|\, r(t) = x\right), \quad (9)$$

for different initial values using our approximation algorithm (8). We see from the right pane in Fig 3 that $Q_{n+1} - Q_n$ converges to zero at the rate $O(n^{-3})$, suggesting that $|Q_n - Q| = O(n^{-2})$ as $n \to \infty$.

3.3 *The case* $V(x) = \kappa \mathbf{1}\{x > r\}$

The standard barrier option is one of the most popular first-generation exotic derivatives, due to its cheaper price compared to their vanilla counterparts. If an investor

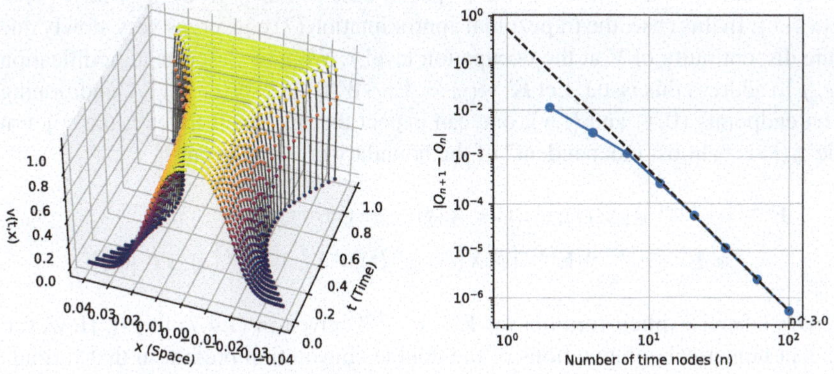

Fig. 3: The left pane shows an approximation to $v(t,x)$ from (9) for different initial values (t,x) using (8) with parameters $n = 30$, $\gamma = 2$, $\delta = 0$, $\mu(t,x) = (\theta(t) - \alpha x)$, $\sigma = 0.01$ and $V(x) = x$. On the right pane, we present a log-log plot of $|Q_{n+1} - Q_n|$ (the blue line). The dashed black straight line has a slope equal to -3.

believes that a certain price is unlikely to fall below a certain level, they can choose to include a knock-out feature in the option, which reduces the price. These cost reductions can be substantial when volatility is high. However, standard barrier options come with a few disadvantages. Option buyers can lose their entire investment due to short time price spikes once the price is near the barrier. The delta sensitivity of such a barrier option is discontinuous near the boundary, making it difficult to hedge for options dealers.

The so-called "step option" was introduced in [15] to address these issues. The payoff of an "up-and-out" step option with finite knock-out rate $\kappa \geq 0$ written on an option with payoff ϕ is given by the formula

$$\exp\{-\kappa \tau_r(T)\}\phi(X(T)),$$

where $\kappa > 0$ is a constant and $\tau_r(t)$ is the sojourn time of the underlying price process X above the level $r > 0$ until time t:

$$\tau_r(t) := \int_0^t \mathbf{1}\{X(s) > r\}\, ds, \quad t \in [0,T].$$

We will assume X to be the standard Brownian motion process W in this example.

This option can be modified by adding barrier features that will further reduce its price. We introduce a time-dependent knock-out barrier g^+ above the step option barrier r and a lower knock-out barrier g^-. We call this option a "hybrid step-barrier option". Its payoff is given by

$$\exp\{-\kappa \tau_r(T)\}\phi(X(T))\mathbf{1}\{g^-(s) < X(s) < g^+(s),\, s \in [0,T]\}.$$

The expectation of this functional is a special case of our Q from (3) with $V(x) = \kappa\mathbf{1}\{x > r\}$. In this case, the trapezoidal approximation (7) converges very slowly due to the discontinuity of V at the occupation level r. We now describe a modification of $e_{n,k}$ to address this issue. Let $\mathbf{E}^{t,x,y}(\cdot) := \mathbf{E}(\cdot\,|\,W(0) = x, W(t) = y)$. Conditioning W on endpoints $(0,x)$ and (t,y), one can expect that for small times t, the sojourn time $\tau_r(t)$ is "almost independent" of the boundary crossing event:

$$\begin{aligned}&\mathbf{E}^{t,x,y}\left[e^{-\kappa\tau_r(t)}\phi(X(t)); g^-(t) < X(s) < g^+(s),\, s \in [0,t]\right]\\&\qquad\approx \mathbf{E}^{t,x,y}e^{-\kappa\tau_r(t)}\mathbf{E}^{t,x,y}\left[\phi(X(t)); g^-(t) < X(s) < g^+(s),\, s \in [0,t]\right].\end{aligned}$$

There is a semi-explicit formula for $\mathbf{E}^{t,x,y}e^{-\kappa\tau_r(t)}$ given in (1.4.7) in [5]. However, efficient numerical computations of the double convolution present in that formula was challenging in the case when $x < r$ and $y > r$. Hence we decided to use the following simple approximation.

Lemma 1. *For $x, y, \kappa \in \mathbb{R}$, as $t \downarrow 0$,*

$$\mathbf{E}^{t,x,y}e^{-\kappa\tau_r(t)} = 1 - \kappa\int_0^t \overline{\Phi}\left(\frac{r - x - \frac{s}{t}(y-x)}{\sqrt{(t-s)s/t}}\right) ds + O(t^2),$$

where $\Phi(\cdot)$ is the standard normal distribution function and $\overline{\Phi}(x) := 1 - \Phi(x)$.

Proof. Since $\tau_r(t) \le t$, a Taylor series expansion of e^{-x} yields

$$\mathbf{E}^{t,x,y}e^{-\kappa\tau_r(t)} = 1 - \kappa\mathbf{E}^{t,x,y}\tau_r(t) + O(t^2).$$

Changing the order of integration using Fubini's theorem we get

$$\mathbf{E}^{t,x,y}\tau_r(t) = \int_0^t \mathbf{P}^{t,x,y}(W(s) > r)\,ds.$$

Since

$$(W(s)\,|\,W(0) = x, W(t) = y) \sim N\left(x + \frac{s}{t}(y-x), \frac{s}{t}(t-s)\right),$$

one has

$$\mathbf{E}^{t,x,y}\tau_r(t) = \int_0^t \overline{\Phi}\left(\frac{r - x - (y-x)s/t}{\sqrt{(t-s)s/t}}\right) ds.$$

□

Remark 1. One could actually compute the second-order term in the expansion for $\mathbf{E}^{t,x,y}e^{-\kappa\tau_r(t)}$ using the following formula for the second moment of $\tau_r(t)$:

$$\mathbf{E}^{t,x,y}\tau_r(t)^2 = \int_0^t\int_0^t \mathbf{P}^{t,x,y}(W(u) > r, W(s) > r)\,du\,ds.$$

Let $W^{t,x,y} := \{W^{t,x,y}(s) : s \in [0,t]\}$ denote the Brownian motion pinned at $(0,x)$ and (t,y). For brevity, we also set $W^t := W^{t,0,0}$. It is well-known that, for $s < t$,

$$\mathbf{E}W^{t,x,y}(s) = x + \frac{s}{t}(y-x), \quad \mathrm{Var}\, W^{t,x,y}(s) = \frac{s}{t}(t-s),$$

and hence

$$\mathbf{P}^{t,x,y}(W(u) > r, W(s) > r) = \mathbf{P}\left(\frac{W^t(u)}{\sqrt{\mathrm{Var}\, W^t(u)}} > r_u, \frac{W^t(s)}{\sqrt{\mathrm{Var}\, W^t(s)}} > r_s\right),$$

where

$$r_v := \frac{r - x - (y-x)v/t}{\sqrt{(t-v)v/t}}, \quad v \in [0,t].$$

Now since $\mathbf{E}W^t(s)W^t(u) = (t-u)s/t$, $s < u$, we get

$$\rho := \mathrm{Cov}\left(\frac{W^t(s)}{\sqrt{\mathrm{Var}\, W^t(s)}}, \frac{W^t(u)}{\sqrt{\mathrm{Var}\, W^t(u)}}\right)$$
$$= \frac{\mathbf{E}W^t(s)W^t(u)}{\sqrt{\mathrm{Var}\, W^t(s)\,\mathrm{Var}\, W^t(u)}} = \sqrt{\frac{s(t-u)}{u(t-s)}}. \quad (10)$$

Therefore

$$\mathbf{P}^{t,x,y}(W(u) > r, W(s) > r) = 1 - \Phi_\rho(-r_s, -r_u),$$

where Φ_ρ denotes the bivariate standard normal cumulative distribution function with correlation ρ given by (10). Hence we have

$$\mathbf{E}^{t,x,y}\tau_r(t)^2 = \int_0^t \int_0^t (1 - \Phi_\rho(-r_s, -r_u))\, du\, ds.$$

However, numerically computing this integral turned out to be too computationally expensive and did not improve the efficiency of the algorithm.

To summarise, our modification of the term $e_{n,k}$ in this special case when V is discontinuous amounts to replacing (7) with

$$\tilde{e}_{n,k}(x,y) = 1 - \kappa \int_0^t \overline{\Phi}\left(\frac{r - x - (y-x)s/t}{\sqrt{(t-s)s/t}}\right) ds. \quad (11)$$

In our numerical implementation, we computed the time integral appearing in (11) using Gaussian quadratures. To demonstrate the performance of our scheme, we set $r = 1/19$, $\kappa = 2$ and $g^{\pm}(t) = \pm 4 \mp t^2$. Without replacing $e_{n,k}(x,y)$ with (11), the convergence of $|Q_{n+1} - Q_n|$ to zero is highly non-smooth and slow. With the proposed modification, the scheme appears to converge at rate $O(n^{-1})$, as shown in the right pane of Fig 4. The left pane in Fig 4 shows our approximation for the values of

$$v(t,x) = \mathbf{E}\left(e^{-\kappa \int_t^1 \mathbf{1}\{W(s)>r\}\, du}; g^-(s) < W(s) < g^+(s), s \in [t,1] \,\middle|\, W(t) = x\right). \quad (12)$$

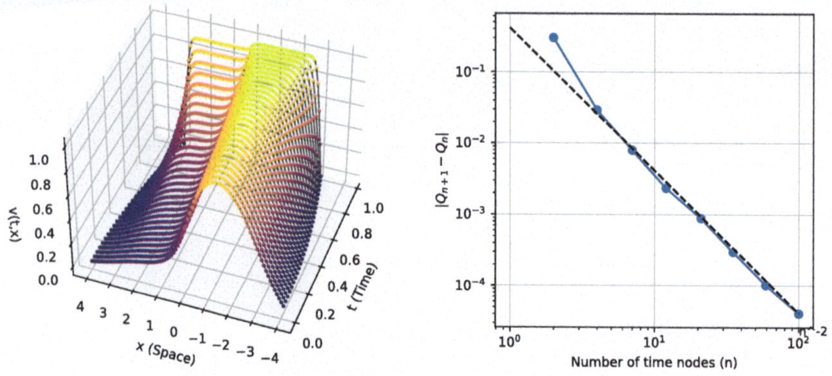

Fig. 4: The left pane shows our approximation to $v(t,x)$ from (12) using (8) with modification (11). In this example we used parameters $n = 30$, $r = 1/19$, $\kappa = 2$, $\delta = 0$, $\mu(t,x) = 0$, and $V(x) = \kappa \mathbf{1}\{x > r\}$. The right pane shows a log-log plot of $|Q_{n+1} - Q_n|$, displayed by the blue line. The dashed black straight line has a slope equal to -2, which indicates that $Q_n - Q = O(n^{-1})$.

References

1. Andersen, L. and Piterbarg, L.: Interest Rate Modelling, Vol 2: Term Structure Models. Atlantic Financial Press, London (2010).
2. Armstrong, G.: Valuation formulae for window barrier options. Appl. Math. Fin. **8**, 197–208 (2001)
3. Billingsley, P.: Convergence of Probability Measures. Wiley, New York (1968)
4. Bingham, N. and Kiesel, R.: Risk-Neutral Valuation, 2nd ed., Springer, London (2004)
5. Borodin, A. and Salminen, P.: Handbook of Brownian Motion: Facts and Formulae, 2nd ed., Birkhaüser, (2002)
6. Borovkov, K. and Novikov, A.: Explicit bounds for approximation rates of boundary crossing probabilities for the Wiener process. J. Appl. Probab. **42**, 82-92 (2005)
7. Carr, P.: Two extensions to barrier option valuation. Appl. Math. Fin. **2**, 173–209 (1995)
8. Cameron, R. and Martin, W.: Transformations of Wiener Integrals Under a General Class of Linear Transformations. Trans. Amer. Math. Soc. **58**, 184–219 (1945)
9. Ethier, S. and Kurtz, G.: Markov Processes. Wiley, New York, (1986)
10. Fu, J.C. and Wu, T.-L.: Linear and nonlinear boundary crossing probabilities for Brownian motion and related processes. J. Appl. Probab. **47**, 1058—1071 (2010)
11. Gutiérrez, R., Ricciardi, L.M., Román, P. and and Torres, F.: First-passage-time densities for time-non-homogeneous diffusion processes. J. Appl. Probab. **34**, 623–631 (1997)
12. Kac, M.: On distributions of certain Wiener functionals. Trans. Amer. Math. Soc. **65**, 1–13 (1949)
13. Kuan, G. and Webber, N.: Pricing barrier options with one-factor interest rate models. J. Deriv. **4**, 33–50 (2003)
14. Liang, V. and Borovkov, K.: On Markov chain approximations for computing boundary crossing probabilities of diffusion processes. arXiv:2112.05268 (2021). To appear in Adv. Appl. Prob.
15. Linetsky, V.: Step options. Math. Fin. **9**, 55–96 (1999)

On ruin probabilities in a Sparre Andersen type model in the presence of risky investments and random switching

Ying He and Konstantin Borovkov

Abstract We study the asymptotic behavior of ruin probabilities, as the initial reserve goes to infinity, for a reserve process model where claims arrive according to a renewal process, while between the claim times the process has the dynamics of geometric Brownian motion-type Itô processes with time-dependent random coefficients. These coefficients are "reset" after each claim time, switching to new values independent of the past history of the process. We use the implicit renewal theory to obtain power-function bounds for the eventual ruin probability. In the special case when the random drift and diffusion coefficients of the investment returns process remain unchanged between consecutive claim arrivals, we obtain conditions for the validity of the power function decay behaviour (as the initial reserve tends to infinity) for the ruin probability for our model.

1 Introduction and the main result

In the classical Cramér–Lundberg collective risk model (going back to a 1903 F. Lundberg's work), the insurance company reserve process X is assumed to have dynamics of the form

$$X(t) = u + ct - \sum_{j=1}^{N(t)} \xi_j, \qquad t \geq 0, \tag{1}$$

Ying He
School of Mathematics and Statistics, The University of Melbourne, Parkville 3010, Australia, e-mail: hey4@student.unimelb.edu.au

Konstantin Borovkov
School of Mathematics and Statistics, The University of Melbourne, Parkville 3010, Australia, e-mail: borovkov@unimelb.edu.au

D. R. Wood et al. (eds.), *2021-2022 MATRIX Annals*, MATRIX Book Series 5,
https://doi.org/10.1007/978-3-031-47417-0_21

where c is a constant premium payment rate, N is a Poisson process of claim epochs T_j, $j \geq 1$, and ξ_j, $j \geq 1$, are positive i.i.d. random variables modeling claim sizes made at the respective claim times, their sequence being independent of N.

The main question posed in the context of this model was on the behavior of the ultimate ruin probability

$$\psi(u) := \mathbf{P}\Big(\inf_{t>0} X(t) < 0\Big)$$

as the initial reserve u tends to infinity. Clearly, in model (1), the ruin (X turning negative) can only occur at a claim time. Hence one deals here with a question on the asymptotic behavior of the distribution tail of the global maximum of a random walk with jumps of the form $Z_j := \xi_j - c\tau_j$, where $\tau_j := T_j - T_{j-1}$, $j \geq 1$ (setting $T_0 := 0$). Hence $\psi(u) < 1$ for $u > 0$ and $\psi(u) \to 0$ as $u \to \infty$ once the safety loading condition

$$\mathbf{E}Z \equiv \mathbf{E}\xi - c\mathbf{E}\tau < 0 \tag{2}$$

is met (here and in what follows, we use the convention that $Z \stackrel{d}{=} Z_1$, $\xi \stackrel{d}{=} \xi_1$, *etc.*). But what can one say about the rate at which $\psi(u)$ vanishes at infinity? The most famous result in this classical setting is the celebrated Cramér–Lundberg approximation that holds in the case of exponentially light tails and can be stated as follows.

For a random variable V, denote by

$$\phi_V(q) := \mathbf{E}e^{qV}, \quad q \in \mathbb{R}, \qquad q_V := \sup\{q \in \mathbb{R} : \phi_V(q) < \infty\} \tag{3}$$

its moment generating function and the right end-point of the interval on which the latter is finite, respectively. If $\phi_Z(q_Z) \geq 1$ then, under condition (2), there exists a unique solution $\gamma > 0$ to the equation $\phi_Z(q) = 1$ and if, in addition, $\phi_Z(q_Z) > 1$ or $\phi_Z(q_Z) = 1$ and $\phi_Z'(q_Z-) < \infty$ then

$$\psi(u) = Ce^{-\gamma u}(1 + o(1)) \quad \text{as} \quad u \to \infty, \tag{4}$$

where the constant C admits a closed-form expression (see e.g. Section 22 in [3] or Section I.4d in [2]). Moreover, it turns out that approximation (4) is rather sharp: there is an $\varepsilon > 0$ such that the remainder term $o(1)$ in it can be replaced with $o(e^{-\varepsilon u})$. Furthermore, under the same moment assumptions on the distribution of Z, approximation (4) also holds for the Sparre Andresen model that differs from (1) only in that the process N is just a renewal process (so that the inter-claim times τ_j are general positive i.i.d. random variables). In this case, the remainder term will be decaying exponentially fast under the additional assumption that the distribution of Z contains an absolutely continuous component (p. 129 in [3]).

Note that in the case where $\phi_Z(q_Z) = 1$ and $\phi_Z'(q_Z-) = \infty$, the problem on the asymptotics of $\psi(u)$ is more difficult and the asymptotic behavior of this probability as $u \to \infty$ can have a different form, see e.g. p. 136 in [3] and Section 6.5 in [4].

Of course, the Cramér–Lundeberg model (1) and its Sparre Andersen extension are oversimplifications of real-life situations. These models assume that all the reserves of the insurance company are kept in a safe bank account. Over the last two decades, several authors turned their attention to more realistic models in which the reserve capital can be invested in a risky financial asset (considering a single risky asset is reasonable due to the common practice of investing in a market portfolio or an index). Models with surplus generating process and investments in risky asset modelled by Lévy processes were discussed, e.g., in [20, 21, 23]. In particular, it was noted in [23] that the ultimate ruin probability and the Laplace transform of the ruin time are solutions to suitable boundary value problems for the respective integro-differential equations.

A discrete time model with stochastic interest rates and returns was considered in [18], the main results (obtained using the "crude" large deviation theory) included power asymptotic behavior of the ruin probability as a function of the initial reserve. A power function ruin probability asymptotics behavior was also obtained in [22] for the Lévy processes-based models under suitable conditions, basing on the results from [21]. Assuming (1) (and also allowing a more general Lévy process model) and that the risky investment returns follow an independent geometric Lévy process, power function bounds for $\psi(u)$ were obtained in [15]. A power function asymptotic behavior was obtained in [9] for a modification of the classical model (1) with investments in a risky asset with price following an independent geometric Brownian motion (BM) process with mean return $\mu \in \mathbb{R}$ and volatility $\sigma > 0$ (as in (6) below, but with a constant $c(s) \equiv c$); the first steps in that direction were announced in [8]. Assuming that the claim sizes are exponentially distributed and setting $\beta := 2\mu/\sigma^2 - 1$, it was shown in [9] that

$$\psi(u) = Cu^{-\beta}(1+o(1)) \quad \text{as} \quad u \to \infty \tag{5}$$

for some constant $C > 0$ when $\beta > 0$ (and that $\psi(u) \equiv 1$ when $\beta < 0$). For claims with a general distribution such that $\mathbf{E}\xi_1^\beta < \infty$, were obtained upper and lover power bounds with the right-hand sides of the form $Cu^{-\beta}$ for come constants C.

Note that the presence of the moment condition on ξ_1 (here and in our Theorem 1 below) is quite natural as for heavy-tailed claim distributions, the asymptotics of the ruin will be governed by the distribution tail of the "integrated tail law" for ξ_1 when that tail dominates $u^{-\beta}$ (cf. [1] and Chapter X in [2]).

These results were extended in [24] to a modification of the above model with a variable premium payment rate $c(t)$ yielding the following dynamics:

$$X(t) = u + \int_0^t c(s)\,ds + \int_0^t \mu X(s)\,ds + \int_0^t \sigma X(s)\,dW(s) - \sum_{j=1}^{N(t)} \xi_j, \tag{6}$$

where W is a BM process independent of N and $\{\xi_j\}$, the coefficients μ and σ are constant, and $c(t) = c(t,X) \in [0,\overline{c}]$ (with a constant $\overline{c} < \infty$) is a bounded adapted function such that there exists a unique strong solution to the above equation. Upper and lover bounds with the right-hand sides of the form $Cu^{-\beta}$ were obtained under appropriate moment conditions on ξ_1, whereas exact asymptotics of the form (5) were established for generally distributed ξ_1 (satisfying $\mathbf{E}\xi_1^{\beta+\delta} < \infty$ for some $\delta > 0$) in the special case where $c(t) = c_1 e^{\gamma t}$ for some $\gamma \le 0$. The toolbox used in that paper, as in some other previous work as well, was based on the implicit renewal theory.

It may seem paradoxical at the first glance that, in all these papers establishing power asymptotics of the form (5), the distributions of the "main source of risk"—the claims made against the insurer—could have a finite exponential moment, as in the case leading to the much faster exponential decay (4). This means that investing in a risky asset (even with significant mean positive returns) dramatically increases the riskiness of the insurance business. In Remark 5 below we will provide an intuitive explanation of the emergence of the power behavior at infinity for ψ. Roughly speaking, it is due to the closeness of the dynamics of an embedded discrete time process (the values of the risk process $u - X$ at the claims times) to those of the exponential of a random walk with i.i.d. jumps and negative trend. The ruin occurs when the global supremum of that walk is "large", of the order of magnitude of $\ln u$, and the probability of this has the form of the right-hand side of (4), with u replaced by $\ln u$

Over the last few years, several authors turned their attention to versions of model (6) with random switching. In [25], the authors considered a Markov-modulated model, governed by an "external environment process" $\{J(t)\}_{t\ge 0}$, assumed to be a homogeneous, irreducible and recurrent continuous time Markov process, with a finite state space E. All the components of the reserve process depend on the state of $J(t)$: given that $J(t) = i \in E$, arrivals occur according to a Poisson process with rate λ_i, the (absolutely continuous) claim distribution depends on i, premiums are received at a rate $c_i > 0$, and the drift and diffusion coefficients of the risky return process also depend on i. Under these assumptions, a system of integro-differential equations was derived for the vector of ruin probabilities (indexed by the initial state of the modulating Markov chain). Switching to the corresponding Laplace transforms and using Karamata Tauberian theorems, the authors then derive the asymptotic behaviour of the ruin probailities, but only in the case when both the drift and diffusions coefficients for the risky asset price process do not depend on the state of the modulating process J.

In [7], it was assumed that the geometric BM process modelling the dynamics of the risky asset has stochastic drift and volatility coefficients: $\mu = \mu_{\theta(t)}$, $\sigma = \sigma_{\theta(t)}$, where $\{\theta(t)\}_{t\ge 0}$ is a time-homogeneous (hidden) Markov chain with state space $\{0,1\}$ independent of all the other stochastic ingredients of the model. Using implicit renewal theory, the authors derived two-sided power function bounds of the form

$$0 < \liminf_{u\to\infty} u^\beta \psi(u) \le \limsup_{u\to\infty} u^\beta \psi(u) < \infty \tag{7}$$

for the ruin probability. These results were extended in [12] to the case where $\{\theta(t)\}_{t\ge 0}$ has an arbitrary finite state space.

In [6] a Sparre Andersen type model was considered, where the dynamics of the risky asset used for investment was given by a general Lévy process $\{R(t)\}_{t\ge 0}$ (with the assumption that its jumps are always greater than -1):

$$X(t) = u + \int_0^t X(s-)dR(s) - \sum_{j=1}^{N(t)} \xi_j,$$

where $\{N(t)\}_{t\geq 0}$ is now a renewal process (all the components of the model were, as usual, assumed to be independent). Using recent results from the theory of distributional equations, the authors derived for this model two-sided power function bounds of the form (7). That paper was complemented by note [13], where the authors used semi-Markov processes techniques to extend the results to the case of annuities and models with two-sided jumps.

One can also mention here paper [14] that deals with the ruin problem when an insurance company that has two business branches, life insurance and non-life insurance (so that the reserve process would have both positive and negative jumps), and invests its reserve into a risky asset with the price dynamics given by a geometric Brownian motion. For the case of exponentially distributed jumps, it was shown that the non-ruin probability is a solution of an ordinary differential equation of the fourth order. Asymptotic analysis of the latter led to the conclusion that the ruin probability vanishes the same way as in the already studied cases of models with one-side jumps.

In the present note, we extend (6) to another version of the Sparre Andersen-type model with investment in a risky asset that involves random switching. To formally describe our model, in addition to the i.i.d. sequence $\{\xi_n\}_{n\geq 1}$ of claim sizes (as above), introduce an independent of it i.i.d. sequence of quadruples

$$(\mu_n(\cdot), \sigma_n(\cdot), \tau_n, W_n(\cdot)), \quad n \geq 1, \tag{8}$$

and (independent) filtrations $\{\mathbb{H}_n = \{\mathscr{H}_n(t), t \geq 0\}\}_{n\geq 1}$, where W_n is a standard Wiener process which is a martingale w.r.t. filtration $\mathbb{H}_n$, while the process μ_n is adapted to $\mathbb{H}_n$ and locally integrable a.s., σ_n is progressively measurable (w.r.t. $\mathbb{H}_n$) and locally square-integrable a.s., and $\tau_n > 0$ are stopping times w.r.t. $\mathbb{H}_n$ (in particular, they may be independent of W_n, assuming $\mathbb{H}_n$ large enough). About $c(t)$ we will assume, as in [24], that it is right-continuous and takes values in $[0, \overline{c}]$ with some $0 < \overline{c} < \infty$ and is adapted in an appropriate way (omitting technical details to avoid making exposition too cumbersome) such that there exist unique strong solutions to the equations describing our model.

Our reserve process follows the dynamics of (6), where the drift and diffusion coefficients μ and σ are random processes of the form

$$\mu(t) = \sum_{n=1}^{\infty} \mu_n(t - T_{n-1}) \mathbf{1}_{[T_{n-1}, T_n)}(t), \quad \sigma(t) = \sum_{n=1}^{\infty} \sigma_n(t - T_{n-1}) \mathbf{1}_{[T_{n-1}, T_n)}(t),$$

while $N(t) = \sum_{n=1}^{\infty} \mathbf{1}_{(0,t]}(T_n)$, $T_n := \sum_{i=1}^{n} \tau_i$, is the renewal process generated by the inter-arrival times $\tau_n > 0$. We assume that

$$\mu_n(t) \geq \underline{\mu} > -\infty, \quad 0 < \sigma_n(t) \leq \overline{\sigma} < \infty \quad \text{a.s.}$$

for some constant $\underline{\mu}, \overline{\sigma}$.

Thus, according to the suggested model, our insurance company commences at time $t = 0$ with an initial endowment u, faces a renewal-reward claims process with claim sizes ξ_n and inter-claim times τ_n, and receives premium inflow at a bounded non-negative random rate $c(t)$. During the time period (T_{n-1}, T_n), the company obtains a rate of return following a diffusion process with random time-dependent drift coefficient μ_n and volatility σ_n, which are "switched" to μ_{n+1} and σ_{n+1} at time T_n. The random regime switching for the investment component may be related to changing the investment policy or insurer's economic environment following claim payments. Considering the proposed model is also suggested by the inner logic of the mathematical problem per se.

To state our main results, we first need to introduce some notations. Following the standard approach used, in particular, in [15] and [24], we note that ruin for this model can only occur at one of the claim times T_n. Therefore, for the ruin probability analysis, it suffices to consider the embedded discrete time process $\{S_n := X(T_n)\}_{n\geq 0}$ (setting $T_0 := 0$) since

$$\psi(u) = \mathbf{P}\big(\inf_{n\geq 1} S_n < 0\big). \tag{9}$$

The dynamics of (6) inside intervals $[T_{n-1}, T_n)$ are those of solutions to linear stochastic differential equations with the respective initial values S_{n-1}. Using the available in closed form solutions to such problems (see e.g. Chapter 9 in [17]), noting that $S_n = X(T_n-) - \xi_n$, and introducing notations

$$K_n(s) := \int_s^{\tau_n} (\mu_n(u) - \sigma_n^2(u)/2)du, \quad Z_n(s) := \int_s^{\tau_n} \sigma_n(u)dW_n(u), \quad s \in [0, \tau_n],$$

$K_n := K_n(0)$, $Z_n := Z_n(0)$,

$$\nu_n := -K_n - Z_n, \quad \lambda_n := e^{-\nu_n}, \quad \zeta_n := \int_0^{\tau_n} e^{K_n(s)+Z_n(s)} c(T_{n-1}+s)ds - \xi_n, \tag{10}$$

we obtain that

$$S_n = \lambda_n S_{n-1} + \zeta_n, \quad n \geq 1, \quad S_0 = u, \tag{11}$$

Note that, due to our assumptions, $\{(K_n, \nu_n)\}_{n\geq 1}$ is an i.i.d. sequence, whereas $\{\zeta_n\}_{n\geq 1}$ does not need to be so.

Recall that, for sequences of random elements, we agreed to omit for brevity's sake the subscript n in the case where $n = 1$.

Referring to (3), we will use the following lemma to introduce one more notation.

Lemma 1. *If $\phi_\nu(q_\nu) \in (1, \infty]$, $\mathbf{E}\tau < \infty$ and $\mathbf{E}K \in (0, \infty)$ then $q_\nu > 0$ and there exists a $\beta \in (0, q_\nu)$ such that $\phi_\nu(\beta) = 1$.*

We will refer to the coefficient β from Lemma 1 as the *Frolova exponent* for our model, as the power decay behaviour of the form (5) (with $\beta := 2\mu/\sigma^2 - 1$ in the basic Cramér–Lundberg model with risky investments described by a geometric BM process) was apparently first conjectured in [8].

Remark 1. Note that as ϕ_ν is left-continuous on $(0, q_\nu)$, one has $\beta < q_\nu$. Therefore $\mathbf{E}\lambda^{-(\beta+\delta)} \equiv \phi_\nu(\beta+\delta) < \infty$ for any $\delta \in (0, q_\nu - \beta) \neq \varnothing$.

Our main result is stated in the following theorem.

Theorem 1. *Assume that* $\phi_\nu(q_\nu) \in (1, \infty]$, $\mathbf{E}\tau < \infty$, $\mathbf{E}K \in (0, \infty)$, *and* $\mathbf{E}\xi^\beta < \infty$ *for some* $\delta > 0$, *where* β *is the Frolova exponent for our model. Then*

$$\limsup_{u\to\infty} u^\beta \psi(u) \le C_+. \tag{12}$$

If, in addition, $(\mu(\cdot), \sigma(\cdot), \tau)$ *and* $W(\cdot)$ *are independent,* $\mathbf{E}\xi^{\beta+\delta} < \infty$ *for some* $\delta > 0$ *and*

$$q_\tau > \beta^2\overline{\sigma}^2/2 + \beta(\overline{\sigma}^2/2 - \underline{\mu})^+ \tag{13}$$

then

$$\liminf_{u\to\infty} u^\beta \psi(u) \ge C_-. \tag{14}$$

Here $0 < C_- \le C_+ < \infty$ *are some constants.*

Remark 2. The existence of the Frolova exponent $\beta > 0$ is ensured since the conditions of Lemma 1 are clearly met under the assumptions of Theorem 1. Without loss of generality, in what follows we will assume about the δ from the conditions of Theorem 1 that $\delta \in (0, q_\nu - \beta)$ (see Remark 1).

Remark 3. Condition $\mathbf{E}K = \mathbf{E}\int_0^\tau (\mu(u) - \sigma^2(u)/2)du > 0$ means that "volatility" $\sigma(t)$ cannot be "large" in some average sense. Recall that $\sigma^2/2 > \mu$ implies certain ruin in the models with constant μ and σ considered in [9] and [24].

Remark 4. Observe that if $\phi_\tau(q) < \infty$ for any $q > 0$ then condition (13) is clearly superfluous.

The proof of Theorem 1 is given in Section 2.

The existence of the Frolova exponent β is the key factor for establishing the power behaviour of the ruin probability. Given the structure of our random variable ν, verifying the existence of such a β in the general case is a complicated task. In Section 3, we will establish a sufficient condition for the existence of the Frolova exponent in the more tractable special case when

$$\mu_n(t) \equiv \mu_n(0) =: \mu_n \quad \text{and} \quad \sigma_n(t) \equiv \sigma_n(0) =: \sigma_n \tag{15}$$

do not depend on time (so that the random drift and diffusions coefficients for the return on investments process remain unchanged during each of the intervals

$[T_{n-1}, T_n)$, $n \geq 1$). Moreover, we will assume that the components (μ_n, σ_n), τ_n and $W_n(\cdot)$ of our quadruples (8) are jointly independent. The problem admits in this case an elegant solution: it turns out that the answer (given in Theorem 2 stated and proved in Section 3) basically depends on "concentration of probability" in vicinity of a certain straight line tangent to the support of the distribution of the random vector $(\mu, \sigma^2/2)$.

2 Proof of Theorem 1

Proof of Lemma 1. That $q_\nu > 0$ is clear since $\phi_\nu(q_\nu) > 1$. Further,

$$\mathbf{E}\nu = -\mathbf{E}K - \mathbf{E}Z = -\mathbf{E}K < 0 \tag{16}$$

as $\mathbf{E}Z = \mathbf{E}\int_0^\tau \sigma(u)dW(u) = 0$ by the optional stopping theorem (note that $\mathbf{E}|\nu| < \infty$). Since ϕ_ν is a convex function and $\phi_\nu(q_\nu) \in (1, \infty]$, the existence of the claimed β is equivalent to having $\phi_\nu'(0+) < 0$, which is an immediate consequence of (16). □

Proof of Theorem 1. Our line of argument follows the overall logic employed in [24]. Iterating (11) and setting $\Lambda_n := \prod_{k=1}^n \lambda_k$, $n \geq 1$, we get

$$S_n = \Lambda_n u + \Lambda_n \sum_{k=1}^n \Lambda_k^{-1} \zeta_k, \quad n \geq 1. \tag{17}$$

First we will prove the upper bound (12). Clearly,

$$(Q_k, M_k) := \{(\xi_k, 1)/\lambda_k\}_{k \geq 1}$$

is an i.i.d. sequence. Set

$$R_n := \sum_{k=1}^n Q_k \prod_{i=1}^{k-1} M_i, \quad n \geq 1, \tag{18}$$

with the usual convention that $\prod_{i=j}^k = 1$ when $j > k$. Since $Q, M > 0$, the sequence $\{R_n\}_{n \geq 1}$ is clearly increasing so that

$$R_n \uparrow R \quad \text{a.s.} \tag{19}$$

for some (possibly improper) random variable $R \leq \infty$.

In view of (10), one has $\zeta_k \geq -\xi_k$, $k \geq 1$, and hence we obtain from (17) that

$$S_n \geq \Lambda_n u - \Lambda_n \sum_{k=1}^n \Lambda_k^{-1} \xi_k = \Lambda_n(u - R_n) \geq \Lambda_n(u - R), \quad n \geq 1. \tag{20}$$

Hence it follows from (9) that

$$\psi(u) \le \mathbf{P}(R > u), \quad u > 0. \tag{21}$$

Remark 5. One can clarify the emergence of the power decay for ψ as follows. Clearly, $U_n := \sum_{k=1}^{n} v_k$, $n \ge 1$, is a random walk with i.i.d. jumps v_k with negative trend (see (16)) and $\phi_v(\beta) = 1$. Hence by the classical Cramér–Lundberg result (4) for $\overline{U} := \max_{n\ge 1} U_n$, one has $\mathbf{P}(\overline{U} > w) \sim Ce^{-\beta w}$ as $w \to \infty$.

Now in view of (17), ruin is equivalent to the event $\left\{ \sup_{n\ge 1} \sum_{k=1}^{n} (-\zeta_k) e^{U_k} > u \right\}$ which actually occurs "due" to a few terms in these sums, with k close to the point n' such that $\overline{U} = U_{n'}$ (cf. the argument in the proof of Theorem 4 in [5]). So one can expect that the probability of ruin behaves like $\mathbf{P}(\overline{U} > \ln u) \sim Ce^{-\beta \ln u} = Cu^{-\beta}$ as $u \to \infty$.

That R is a proper random variable follows immediately from the following lemma, which is a direct consequence of Theorem 1.6 in [26]:

Lemma 2. *Let $\{(A_n, B_n)\}_{n\ge 1}$ be an i.i.d. sequence of bivariate random vectors, and*

$$Z_n(x) := x \prod_{j=1}^{n} A_j + \sum_{k=1}^{n} B_k \prod_{j=1}^{k-1} A_j, \quad n \ge 1, \quad x \in \mathbb{R}. \tag{22}$$

Assume that $\mathbf{E} \ln|A| < 0$ *and* $\mathbf{E}(\ln|B|)^+ < 0$, *where* $z^+ := \max\{0, z\}$, $z \in \mathbb{R}$. *Then* $Z_n(x) \to Z$ *in distribution as* $n \to \infty$ *for all* $x \in \mathbb{R}$, *where the distribution of the proper random variable* Z *satisfies the random equation*

$$Z \stackrel{d}{=} B + AZ, \tag{23}$$

(A, B) *and* Z *on the right-had side being independent of each other.*

Indeed, our sequence (18) is of the form (22) with $x = 0$ and $(A_n, B_n) = (M_n, Q_n)$, $n \ge 1$, and $\mathbf{E} \ln|A| = \mathbf{E} \ln \lambda^{-1} = \mathbf{E} v < 0$ by (16), whereas

$$\mathbf{E}(\ln|B|)^+ = \mathbf{E}(\ln(\xi/\lambda))^+ = \mathbf{E}(\ln \xi + v)^+ \le \mathbf{E}(\ln \xi)^+ + \mathbf{E} v^+ < \infty$$

as $\mathbf{E}\xi^{\beta} < \infty$ and $\mathbf{E}|v| < \infty$ (cf. Lemma 1).

Hence, by Lemma 2, the sequence $\{R_n\}$ converges as $n \to \infty$ in distribution to a proper random variable, which implies that the a.s. limit R from (19) is proper as well and satisfies the random equation

$$R \stackrel{d}{=} Q + MR, \tag{24}$$

where R and (M, Q) on the right-hand side are independent of each other.

Now to complete the derivation of the desired upper bound using (21) it remains to turn to the implicit renewal theory. We will make use of the following lemma which is a direct consequence of Theorem 4.1 in [10].

Lemma 3. *Assume that the distribution of a bivariate random vector* (A, B) *with* $A \ge 0$ *a.s. is such that, for some* $\alpha > 0$,

$$\mathbf{E}A^{\alpha} = 1, \quad \mathbf{E}A^{\alpha}(\ln A)^{+} < \infty, \quad \mathbf{E}|B|^{\alpha} < \infty,$$

while the conditional distribution of $\ln A$ *given* $A \neq 0$ *is non-arithmetic. Then solution to* (23) *satisfies*

$$\lim_{u\to\infty} u^{\alpha}\mathbf{P}(Z > u) = C,$$

where $C := \mathbf{E}\big[((B+AZ)^{+})^{\alpha} - (AZ^{+})^{\alpha}\big]/(\alpha\mathbf{E}A^{\alpha}\ln A) \in (0,\infty)$.

To apply this lemma to our equation (24) with $(A,B) = (M,Q)$ and $\alpha = \beta$, it suffices to note that $\mathbf{E}M^{\beta} = \phi_{\nu}(\beta) = 1$, $\mathbf{E}Q^{\beta} = \mathbf{E}\xi^{\beta}M^{\beta} = \mathbf{E}\xi^{\beta}\mathbf{E}M^{\beta} = \mathbf{E}\xi^{\beta} < \infty$ due to independence, and $\mathbf{E}M^{\beta}(\ln M)^{+} = \mathbf{E}e^{\beta\nu}\nu^{+} < \infty$ since $\mathbf{E}e^{(\beta+\delta)\nu} < \infty$ for some $\delta > 0$ (see Remark 2). That $\ln M$ given $M \neq 0$ is non-arithmetic is obvious from the definition of $M = e^{\nu}$ and the presence of the Itô integral in ν. This completes the proof of the upper bound (12).

Now we will proceed to proving the lower bound (14). The main tool here is the following assertion from [16] (see also [10] and [19]).

Lemma 4. *Assume that* Y *satisfies the equation*

$$Y \stackrel{d}{=} B + AY^{+}, \tag{25}$$

where (A,B) *and* Y *on the right-hand side are independent of each other,* $A > 0$ *a.s., and the distribution of* (A,B) *is such that* $\mathbf{P}(A > 1, B > 0) > 0$. *If, for some* $\alpha, \delta > 0$,

$$\mathbf{E}A^{\alpha} = 1, \quad \mathbf{E}A^{\alpha+\delta} < \infty, \quad \mathbf{E}|B|^{\alpha+\delta} < \infty,$$

and $\ln A$ *is absolutely continuous, then*

$$\lim_{u\to\infty} u^{\alpha}\mathbf{P}(Y > u) = C + o(u^{-h})$$

for some positive constants C *and* h.

To apply this result, we turn to representation (17) and use the natural upper bound for ζ_n :

$$\zeta_n \leq \overline{\zeta}_n := \overline{c}\int_0^{\tau_n} \exp\{K_n(s) + Z_n(s)\}ds - \xi_n$$

to get the inequality

$$S_n \leq \overline{S}_n := \Lambda_n u + \Lambda_n \sum_{k=1}^{n} \Lambda_k^{-1}\overline{\zeta}_k = \Lambda_n(u - \overline{R}_n), \quad n \geq 1,$$

where

$$\overline{R}_n := \sum_{k=1}^{n} \overline{Q}_k \prod_{i=1}^{k-1} M_i, \quad \overline{Q}_n := -\overline{\zeta}_n/\lambda_n, \quad n \geq 1.$$

In view of (9), this implies the bound

$$\psi(u) \ge \mathbf{P}\big(\inf_{n\ge 1} \overline{S}_n < 0\big) \ge \mathbf{P}\big(\overline{R} > u\big), \quad \text{where } \overline{R} := \sup_{n\ge 1} \overline{R}_n.$$

Next we note that, since $\overline{R}_1 = \overline{Q}_1$ and $M_1 > 0$, one has

$$\begin{aligned}\overline{R} &= \overline{Q}_1 \vee \sup_{n\ge 2} \overline{R}_n = \overline{Q}_1 \vee \Big(\overline{Q}_1 + M_1 \sup_{n\ge 2} \sum_{k=2}^{n} \overline{Q}_k \prod_{i=2}^{k-1} M_i\Big)\\ &= \overline{Q}_1 \vee (\overline{Q}_1 + M_1 \overline{R}') = \overline{Q}_1 + M_1 (\overline{R}')^+,\end{aligned}$$

where $\overline{R}' := \sup_{n\ge 2} \sum_{k=2}^{n} \overline{Q}_k \prod_{i=2}^{k-1} M_i \overset{d}{=} \overline{R}$ is independent of (M_1, Q_1). Therefore our $\overline{R}$ satisfies the random equation

$$\overline{R} \overset{d}{=} \overline{Q} + M(\overline{R})^+,$$

where (M, Q) and $\overline{R}$ on the right-hand side are independent of each other. This relation is exactly of the form (25), and we will now verify whether the conditions of Lemma 4 are met when $(A, B) = (M, \overline{Q})$, $\alpha = \beta$.

First of all, it follows from Proposition 6.1 in [10] that $\overline{R}$ is a proper random variable provided that $\mathbf{E}\ln(1 \vee \overline{Q}) < \infty$. The latter will immediately follow from the condition $\mathbf{E}|\overline{Q}|^{\beta+\delta} < \infty$ of Lemma 4 that we need to verify. To demonstrate the latter relation, note that

$$\begin{aligned}\overline{Q} &= -\frac{\overline{\zeta}}{\lambda} = \frac{\xi}{\lambda} - \overline{c} e^{-K(0)-Z(0)} \int_0^{\tau} e^{K(s)+Z(s)} ds\\ &= \frac{\xi}{\lambda} - \overline{c} \int_0^{\tau} \exp\Big\{-\int_0^s (\mu(u) - \sigma^2(u)/2) du - \int_0^s \sigma(u) dW(u)\Big\} ds. \qquad (26)\end{aligned}$$

It is obvious from the elementary inequality $|x+y|^p \le (1 \vee 2^{p-1})(|x|^p + |y|^p)$, $x, y, p > 0$, that it suffices to show that the absolute moments of the order $\beta + \delta$ are finite for both terms on the right-hand side. By independence, one has

$$\mathbf{E}\left|\frac{\xi}{\lambda}\right|^{\beta+\delta} = \mathbf{E}\xi^{\beta+\delta} \mathbf{E}\lambda^{-(\beta+\delta)} = \mathbf{E}\xi^{\beta+\delta} \phi_\nu(\beta+\delta) < \infty$$

in view of Remark 2.

Next note that, due to our assumption about independence of $(\mu(\cdot), \sigma(\cdot), \tau)$ and $W(\cdot)$, one has $\{-\int_0^s \sigma(u) dW(u)\}_{s\ge 0} \overset{d}{=} \{W(\Sigma(s))\}_{s\ge 0}$, where we set $\Sigma(s) := \int_0^s \sigma^2(u) du$, $s \ge 0$. Therefore, putting $\overline{W}(t) := \max_{0\le s\le t} W(s)$, $t \ge 0$, we get

$$\max_{0\le s\le \tau} \Big(-\int_0^s \sigma(u) dW(u)\Big) \overset{d}{=} \max_{0\le s\le \tau} W(\Sigma(s)) = \overline{W}(\Sigma(\tau)) \le \overline{W}(\overline{\sigma}^2 \tau).$$

Now, setting $\kappa(u) := \sigma^2(u)/2 - \mu(u)$ and noting that $\kappa(u) \le \overline{\kappa} := \overline{\sigma}^2/2 - \underline{\mu}$ a.s., we get for the second term on the right-hand side of (26) that

$$\mathbf{E}\left(\int_0^\tau \cdots ds\right)^{\beta+\delta} \le \mathbf{E}\left(e^{\overline{W}(\overline{\sigma}^2\tau)}\int_0^\tau e^{\overline{\kappa}s}ds\right)^{\beta+\delta} \le \mathbf{E}\left(e^{\overline{W}(\overline{\sigma}^2\tau)}\tau e^{\overline{\kappa}^+\tau}\right)^{\beta+\delta}. \quad (27)$$

Due to the reflection principle, for any $a,t>0$,

$$\mathbf{E}e^{a\overline{W}(t)} = 2\mathbf{E}(e^{aW(t)}; W(t)>0) < 2\mathbf{E}e^{aW(t)} = 2e^{a^2t/2}, \quad t>0,$$

so conditioning the last expectation in (27) on τ and using independence, we obtain that it is less than

$$2\mathbf{E}e^{((\beta+\delta)^2\overline{\sigma}^2/2+(\beta+\delta)\overline{\kappa}^+)\tau}\tau^{\beta+\delta} < \infty,$$

using assumption (13) and choosing $\delta>0$ small enough. Thus we showed that $\mathbf{E}|\overline{Q}|^{\beta+\delta}<\infty$, which implies, in particular, that $\overline{R}$ is proper.

To verify the remaining assumptions of Lemma 4, we observe that condition $\mathbf{E}M^\beta = 1$ is met by Lemma 1 and that $\mathbf{E}M^{\beta+\delta}<\infty$ as explained in Remark 2. That $M = e^\nu$ is absolutely continuous follows from the presence of the Itô integral in ν and independence of W from the other participating random quantities. Thus it only remains to verify that $\mathbf{P}(M>1, \overline{Q}>0)>0$. Setting

$$V(t) := \int_0^t (\mu(u) - \sigma^2(u)/2)du + \int_0^t \sigma(u)dW(u), \quad t\ge 0,$$

and choosing $a>0$ such that $b := \mathbf{P}(\xi > a\overline{c})>0$, the previous probability is clearly equal to

$$\begin{aligned}\mathbf{P}(V(\tau)<0, \overline{\zeta}<0) &\ge \mathbf{P}\left(V(\tau)<0, \overline{c}\int_0^\tau e^{V(\tau)-V(s)}ds<\xi, a\overline{c}<\xi\right)\\ &\ge b\mathbf{P}\left(V(\tau)<0, \int_0^\tau e^{V(\tau)-V(s)}ds<a\right)\\ &\ge b\mathbf{P}\left(V(\tau)<0, \int_0^\tau e^{-V(s)}ds<a\right)\\ &\ge b\mathbf{P}(\tau e^{-\underline{V}(\tau)}<a \,|\, V(\tau)<0)\mathbf{P}(V(\tau)<0),\end{aligned}$$

where we put $\underline{V}(t) := \inf_{0\le s\le t} V(s)$. Obviously, $\mathbf{P}(V(\tau)<0)>0$, and as $-\underline{V}(\tau)>0$ on the event $\{V(\tau)<0\}$ while a can be chosen arbitrary small, the product in the last line of the displayed formula is positive, establishing that the last condition of Lemma 4 is met as well. This completes the proof of Theorem 1. □

3 Frolova's exponent when coefficients $\mu_n(t)$ and $\sigma_n(t)$ do not depend on time

In this section we assume satisfied condition (15) and also that the components (μ_n, σ_n), τ_n and $W_n(\cdot)$ of our quadruples (8) are jointly independent. Under these assumptions, one has $v = -(\mu - \sigma^2/2) - \sigma W(\tau)$. Introducing the random vector $\Theta := (\mu, \sigma^2/2)$, setting

$$u(q) := (-q, q(q+1)), \quad q \in \mathbb{R},$$

and conditioning, we get

$$\begin{aligned} \phi_v(q) &= \mathbf{E}e^{q(-(\mu-\sigma^2/2)\tau-\sigma W(\tau))} = \mathbf{E}e^{-q(\mu-\sigma^2/2)\tau+q^2\sigma^2\tau/2} \\ &= \mathbf{E}\phi_\tau(-q(\mu-\sigma^2/2)+q^2\sigma^2/2) = \mathbf{E}\phi_\tau(\langle u(q), \Theta\rangle), \end{aligned} \tag{28}$$

where $\langle\cdot,\cdot\rangle$ stands for the inner product in $\mathbb{R}^2$.

Note that our key condition $\mathbf{E}K > 0$ for the existence of the Frolova exponent is equivalent in the case under consideration to

$$\mathbf{E}(\mu - \sigma^2/2) > 0 \tag{29}$$

(assuming that $\mathbf{E}\tau < \infty$), which is a "mean version" of the condition $2\mu/\sigma^2 - 1 > 0$ under which the asymptotics (5) was established in the case of constant deterministic μ and σ in [24].

Assuming that the above condition is met, the case $q_\tau = \infty$ is trivial: it is clear from Lemma 1 and (28) that β will then always exist. So we will only consider the case where

$$q_\tau < \infty, \quad \phi_\tau(q_\tau) = \infty. \tag{30}$$

Note that the latter is a typical situation when $q_\tau < \infty$; this is so, for instance, for gamma-distributed τ. It turns out that, in this situation, the desired β may or may not exist depending on the distribution of Θ, .

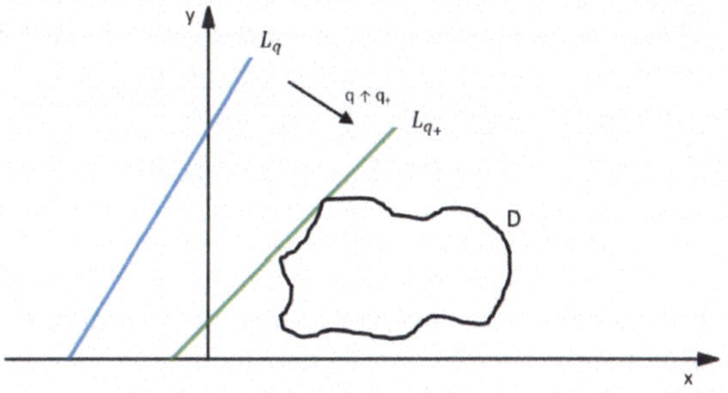

Fig. 1 As $q \uparrow q_+$, the line L_q approaches the set D.

Introduce rays $L_q := \{(x,y) \in \mathbb{R} \times \mathbb{R}^+ : \langle u(q),(x,y)\rangle = q\tau\}$, $q > 0$. Clearly, $(x,y) \in L_q$ iff

$$y = \frac{x}{q+1} + \frac{q\tau}{q(q+1)}, \quad x \geq -\frac{q\tau}{q} \tag{31}$$

(the last inequality is equivalent to $y \geq 0$). Denote by $D \subseteq [\underline{\mu},\infty) \times [0,\overline{\sigma}^2/2]$ the support of Θ and put

$$q_+ := \inf\{q > 0 : L_q \cap D \neq \varnothing\}.$$

Note that, as q increases, the ray L_q "moves" to the right and "rotates" in the clockwise direction, and as D is bounded from the left and from the top, q_+ is a finite positive number (see Fig. 1: q_+ is the value of q for which L_q first "touches" D).

Note that if $q > q_+$ then $\mathbf{P}(\langle u(q),\Theta\rangle > q\tau) > 0$, so that $\phi_\nu(q) = \infty$ by (28). As for $q < q_+$ one clearly has $\mathbf{P}(\langle u(q),\Theta\rangle < q\tau - \varepsilon) = 1$ for some $\varepsilon > 0$, we get $q < q_\nu$ (again by (28)). We conclude that $q_+ = q_\nu$.

Clearly, $\phi_\nu(q_\nu) = \infty$ is sufficient for the existence of the Frolova exponential under the condition that $\mathbf{E}\nu < 0$. In view of our assumption (30), representation (28) suggests that whether $\phi_\nu(q_\nu)$ is infinite or not depends on how strongly the distribution of Θ is concentrated in vicinity of the ray L_{q_+}. To capture this, we introduce the random variable H by setting, for any $\theta_+ \in L_{q_+}$,

$$H := \langle u(q_+),\theta_+ - \Theta\rangle = q\tau + q_+\mu - q_+(q_+ + 1)\sigma^2/2,$$

where the second equality was obtained choosing $\theta_+ = \theta_0 := (-q\tau/q_+,0) \in L_{q_+}$, and denote by F_H its distribution function. We see that $H \geq 0$ a.s. (as the point Θ is below the ray L_{q_+} given by (31)) with $q = q_+$, the value of H being equal to the Euclidean length of the vector $u(q_+)$ times the distance from Θ to L_{q_+}

Now, from (28),

$$\begin{aligned}\phi_\nu(q) &= \mathbf{E}\phi_\tau(\langle u(q) - u(q_+),\Theta\rangle + \langle u(q_+),\Theta - \theta_0\rangle + \langle u(q_+),\theta_0\rangle) \\ &= \mathbf{E}\phi_\tau(q\tau - H - \langle u(q_+) - u(q),\Theta\rangle) = \mathbf{E}\phi_\tau(q\tau - H - \varepsilon\langle a(\varepsilon),\Theta\rangle),\end{aligned} \tag{32}$$

where we first noted that $u(q_+) - u(q) = (q_+ - q)(-1,q_+ + q + 1)$ and then put $\varepsilon := q_+ - q$, $a(\varepsilon) := (-1,2q_+ + 1 - \varepsilon) \to (-1,2q_+ + 1)$ as $\varepsilon \downarrow 0$.

There is no monotone dependence on ε in the integrand on the right-hand side on (32), so we need an argument establishing convergence of these expectations as $\varepsilon \downarrow 0$. Let $D'' := \{\theta = (x,y) \in D : y < x/(q_+ + 1)\}$. Clearly, $\langle u(q_+),\theta\rangle = -q_+x + q_+(q_+ + 1)y < 0$ for $\theta \in D''$, so that $\phi_\tau(\langle u(q_+),\theta\rangle) \leq 1$ in that domain and hence

$$\mathbf{E}(\phi_\tau(\langle u(q),\Theta\rangle);\Theta \in D'') \to \mathbf{E}(\phi_\tau(q\tau - H);\Theta \in D''), \quad q \uparrow q_+,$$

by the dominated convergence theorem. Turning to $D' := D \setminus D''$, one can easily verify that there exist $r_\pm \in \mathbb{R}$ such that $r_- \leq \langle a(\varepsilon),\theta\rangle \leq r_+$ for all $\theta \in D'$, $\varepsilon \in (0,1)$. Since ϕ_τ is an increasing function, we get

$$\begin{aligned}\mathbf{E}(\phi_\tau(q_\tau - H - r_+\varepsilon);\Theta\in D') &\le \mathbf{E}(\phi_\tau(q_\tau - H - \varepsilon\langle a(\varepsilon),\Theta\rangle));\Theta\in D')\\ &\le \mathbf{E}(\phi_\tau(q_\tau - H - r_-\varepsilon);\Theta\in D').\end{aligned}$$

Now we can apply the monotone convergence theorem to both lower and upper bounds in the last displayed formula since the integrands in them have monotone dependence on ε. We conclude that

$$\phi_\nu(q) \to \int_0^\infty \phi_\tau(q_\tau - h)dF_H(h), \quad q\uparrow q_+.$$

Since clearly $\int_\delta^\infty \phi_\tau(q_\tau - h)dF_H(h) \le \phi_\tau(q_\tau - \delta) < \infty$ for any $\delta > 0$, we arrive at the following result.

Theorem 2. *Under the assumptions stated at the beginning of this section, assume that* (29) *and* (30) *hold true. Then* $q_\nu = q_+$ *and, moreover,* $\phi_\nu(q_\nu) = \infty$ *iff*

$$\int_0^\delta \phi_\tau(q_\tau - h)dF_H(h) = \infty$$

for some (and then for any) $\delta > 0$.

Thus there must be significant presence of probability mass in vicinity of the tangent to D line L_{q_+} to ensure that $\phi_\nu(q_\nu) = \infty$.

It is not hard to get closed-form expressions for ϕ_ν in several tractable examples in the special case of the Poisson arrival process with rate 1, which means that $\phi_\tau(q) = 1/(1-q)$, $q < q_\tau := 1$ (so that (30) is true). In one such example one has $\mathbf{P}(\Theta = (1/j, 1-1/j)) = j^{-p}/\zeta(p)$, $j \ge 1$, for a fixed $p \in \mathbb{N}$, where ζ is the Euler–Riemann zeta function. In this case, $q_+ = (\sqrt{5}-1)/2$ and $D \cap L_1 = \{(0,1)\}$, and it turns out that $\phi_\nu(q_+) = \infty$ iff $p = 2$, in obvious agreement with the claim of Theorem 2. If, further, one assumes that Θ is uniformly distributed in a unit square D with vertices at the points (i,j), $i,j \in \{0,1\}$, then again $q_+ = (\sqrt{5}-1)/2$, $D \cap L_1$ consists of the single point $(0,1)$ (the vertex of our D at which it touches the line L_{q_+}), and one can also derive a closed form expression for ϕ_ν yielding $\phi_\nu(q_+) < \infty$. If, however, we rotate the square in the anticlockwise direction around the vertex $(0,1)$ until its upper edge runs along the line L_{q_+} (with clearly the same value of q_+ as in the previous examples) then one would have $\phi_\nu(q_+) = \infty$, also in agreement with Theorem 2. In the latter case, there is "too much probability" in vicinity of L_{q_+} (the probability mass in the ε-neighbourhood of that line is $\asymp \varepsilon$ as $\varepsilon \downarrow 0$ compared to $\asymp \varepsilon^2$ in the former case).

Acknowledgements The authors are grateful to the anonymous referee for spotting several typos in the manuscript and providing a number of useful comments, including additional references relevant to the topic of the paper.

References

1. Albrecher, H., Constantinescu C., Thomann E.: Asymptotic results for renewal risk models with risky investments. Stoch. Proc. Appl. **122**, 3767–3789 (2012)
2. Asmussen, S., Albrecher, H.: Ruin Probabilities. 2nd edn. World Scientific, Singapore (2010)
3. Borovkov, A. A. Stochastic Processes in Queueing Theory. Springer, New York (1972)
4. Borovkov, A.A., Borovkov, K.A.: Asymptotic Analysis of Random Walks: Heavy-Tailed Distributions. Cambridge University Press, Cambridge (2008)
5. Borovkov, K.A.; Vatutin, V.A.: On the asymptotic behaviour of random recursive trees in random environments. Adv. Appl. Prob. **38**, 1047–1070 (2006)
6. Eberlain, E., Kabanov, Yu., Schmidt, T.: Ruin probabilities for a Sparre Andersen model with investments. Stoch. Proc. Appl. **144**, 72–84 (2022)
7. Ellanskaya, A., Kabanov, Yu.: On ruin probabilities with risky investments in a stock with stochastic volatility. Extremes. **24**, 687–697 (2021)
8. Frolova, A.G.: Some mathematical models of risk theory. In: Klimentova, S.B., Pavlova, I.V., Prokhorov, Yu.V., Khokhlov, V.I. (eds) All-Russian School–Colloquium on Stochastic Methods in Geometry and Analysis. 1994, 117–118. TVP, Moscow (1994)
9. Frolova, A.G., Kabanov, Yu.M., Pergamenshchikov, S.M.: In the insurance business risky investments are dangerous. Financ. Stoch. **6**, 227–235 (2002)
10. Goldie, C.M.: Implicit renewal theory and tails of solutions of random equations. Ann. Appl. Probab. **1**, 126–166 (1991)
11. Kabanov, Yu., Pergamenshchikov, S.: In the insurance business risky investments are dangerous: the case of negative risk sums. Financ. Stoch. **20**, 355–379 (2016)
12. Kabanov, Yu., Pergamenshchikov, S.: On ruin probabilities with investments in a risky asset with a regime-switching price. Financ. Stoch. **26**, 877–897 (2022)
13. Kabanov Yu., Promyslov P.: Ruin probabilities for a Sparre Andersen Model with Investments: the case of annuity payments. arXiv:2301.01966 (2023)
14. Kabanov, Yu., Pukhlyakov, N.: Ruin probabilities with investments: smoothness, IDE and ODE, asymptotic behavior. J. Appl. Probab. **59**, 556–570 (2020)
15. Kalshnikov, V, Norberg, R.: Power tailed ruin probabilities in the presence of risky investments. Stoch. Proc. Appl. **98**, 211–228 (2002)
16. Letac, G.: A contraction principle for certain Markov chains and its applications. In: Cohen, J.E., Kesten, H., Newman, C.M. (eds) Random Matrices and Their Applications. Proc. AMS–IMS–SIAM Joint Summer Research Conf. 1984. Contemp. Math. **50**, 263–273. Amer. Math. Soc., Providence, RI (1986)
17. Mackevicius, V.: Introduction to Stochastic Analysis: Integrals and Differential Equations. Wiley, New York (2013)
18. Nyrhinen, H.: On the ruin probabilities in a general economic environment. Stoch. Proc. Appl. **83**, 319–330 (1999)
19. Nyrhinen, H.: Finite and infinite time ruin probabilities in a stochastic economic environment. Stoch. Proc. Appl. **92**, 265–285 (2001)
20. Paulsen, J.: Risk theory in a stochastic economic environment. Stoch. Proc. Appl. **46**, 327–361 (1993)
21. Paulsen, J.: Sharp conditions for certain ruin in a risk process with stochastic return on investments. Stoch. Proc. Appl. **75**, 135–148 (1998)
22. Paulsen, J.: On Cramér-like asymptotics for risk processes with stochastic return on investments. Ann. Appl. Probab. **12**, 1247–1260 (2002)
23. Paulsen, J., Gjessing, H.K.: Ruin theory with stochastic return on investments. Adv. Appl. Prob. **29**, 965–985 (1997)
24. Pergamenshchikov, S., Zeitouny, O.: Ruin probability in the presence of risky investments. Stoch. Proc. Appl. **116**, 267–278 (2006)
25. Ramsden, L., Papaioannou, A.D.: Asymptotic results for a Markov-modulated risk process with stochastic investment. J. Comp. Appl. Math. **313**, 38–53 (2017)
26. Vervaat, W.: On a stochastic difference equation ond a representation of non-negative infinitely divisible random variables. Adv. Appl. Prob. **11**, 750–783 (1979)

On some asymptotic expansions of skew diffusions

Noufel Frikha and Arturo Kohatsu-Higa

Abstract Relying on a perturbation technique of Markov semigroup, also known as the parametrix method, we establish some asymptotic expansion for the Markov semigroup generated by a skew diffusion process with respect to a small parameter. Though the approach developed here is general, we focus on the case of short time and small skew heat kernel expansions.

1 Introduction

The parametrix method introduced by Levi [18] in 1907 has been extensively used to establish semi-explicit expansions of the fundamental solution of a variety of partial differential equations. For a review on these subjects, we refer the reader to [12]. It has also recently been revisited in a series of works [1, 7, 4] to provide some probabilistic representation and integration by parts formula of the marginal law of various processes as an alternative to standard Malliavin calculus techniques or to establish the well-posedness in the weak sense of some degenerate stochastic differential equations (SDEs for short) [8, 9].

The purpose of the present paper is to introduce the inexperienced reader to an application of this method to some asymptotic expansions of skew diffusions. For a general introduction to skew Brownian motion and its applications, we refer the reader to [17]. For an application of the parametrix technique to skew diffusion for the study of its transition density or a time discretization scheme, we refer the reader to [15] and [6] respectively.

Noufel Frikha
Université Paris 1 Panthéon-Sorbonne, Centre d'Economie de la Sorbonne, 106 Boulevard de l'Hôpital, 75642 Paris Cedex 13, France. e-mail: noufel.frikha@univ-paris1.fr

Arturo Kohatsu-Higa
Department of Mathematical Sciences Ritsumeikan University, 1-1-1 Nojihigashi, Kusatsu, Shiga, 525-8577, Japan. e-mail: khts00@fc.ritsumei.ac.jp

D. R. Wood et al. (eds.), *2021-2022 MATRIX Annals*, MATRIX Book Series 5,
https://doi.org/10.1007/978-3-031-47417-0_22

While there have been several important works on the problem of density expansions for diffusions with regular coefficients using different methods, see e.g. [21], [2] or [3], to the best of our knowledge, asymptotic expansions for the transition density of a skew diffusion have not been considered so far. For explicit asymptotic expansions using this method in the case of regular diffusions, we refer the reader to [14].

To motivate our results, we also present a financial application with a model representing regime switching which has been recently discussed in [10]. The parameters that are considered to be small in this model are not the same as in our main results but we will show the reader that the same technique applies. Clearly, these techniques may be useful in other models as well and should have broad applications in pricing and risk management.

Our approach in this article is more didactical than general so that we have chosen some simplified models in order for the reader to grasp the ideas quickly. In this sense, some technical details are deliberately omitted by referring the reader to the corresponding article.

2 Asymptotic expansions of skew diffusions

In this section we study a one-dimensional Markov semigroup with singular coefficients for which we derive an asymptotic expansion with respect to a small parameter ε. We consider a skew diffusion process, that is the unique strong solution of the following one dimensional SDE with dynamics

$$X_t^\varepsilon = x + \int_0^t b_\varepsilon(X_s^\varepsilon)ds + \int_0^t \sigma_\varepsilon(X_s^\varepsilon)dW_s + \alpha\varepsilon^\ell L_t^0(X^\varepsilon), \quad \ell \in \mathbb{R}_+, \tag{1}$$

where $(\alpha,\varepsilon) \in (-1,1)\times[0,1]$, $W = (W_t)_{t\geq 0}$ is a one-dimensional standard Brownian motion and $L_t^0(X^\varepsilon)$ is the symmetric local time at the origin accumulated by X^ε up to time t.

Note that for $\sigma \equiv 1$, $b \equiv 0$, $\varepsilon = 1$ and $\ell = 0$ the process X^ε is known as the skew Brownian motion. We refer the reader to [17] for a survey on this process (see also Section 5 for some basic information). When $b \equiv 0$ and $\sigma \equiv 1$, [20] constructed a semigroup, closely related to the process (1), for which the associated infinitesimal generator has a Dirac mass as drift part. Additionally, he proved that $P_t^\varepsilon f(x) = \mathbb{E}[f(X_t^\varepsilon)]$ generates a Feller semigroup. Actually, his approach is more general since he considers a multi-dimensional framework and the singular drift is the Dirac mass of a smooth surface.

In order to use the parametrix method for the expansion, we need to first gather some regularity properties for the law of X_t^ε.

Regularity properties: Let $T > 0$. Assuming that for any $\varepsilon \in [0,1]$, b_ε is continuous[1] and σ_ε is bounded, uniformly elliptic and Hölder continuous, the random variable X_t^ε given by the unique weak solution at time $t > 0$ of the SDE (1) admits a density $p_t^\varepsilon(x,y)$ satisfying some Gaussian upper bounds as well as the Markov property[2]. We refer the reader to Kohatsu-Higa *et al.* [15]. Notably, the authors proved that for all $t \in (0,T]$, $y \mapsto p_t^\varepsilon(x,y)$ is not continuous at 0 but admits left and right limits (as it is the case for the standard skew Brownian motion) and that there exist $C > 1$ and $c > 1$ that may depend on ε such that

$$\forall (x,y) \in \mathbb{R} \times \mathbb{R}\backslash\{0\},\ p_t^\varepsilon(x,y) \leq C g_{c\varepsilon^2 t}(y-x), \tag{2}$$

where $g_c(y) := 1/(2\pi c)^{1/2} \exp(-y^2/(2c))$.

In order to carry out the expansion, we introduce the time-homogeneous semigroup $(P_t^\varepsilon)_{t\geq 0}$ induced by the SDE (1) defined for any bounded and measurable function f by $P_t^\varepsilon f(x) = \mathbb{E}[f(X_t^\varepsilon)]$. We will frequently use the following boundary condition which is called the "transmission condition" due to its meaning as interaction at the boundary. Given a function $f : \mathbb{R} \to \mathbb{R}$ such that $f'(0+)$ and $f'(0-)$ exist, we say that f satisfies the transmission condition at 0 if

$$(1 + \alpha\varepsilon^\ell) f'(0+) = (1 - \alpha\varepsilon^\ell) f'(0-). \tag{3}$$

We are interested in obtaining the asymptotic expansion of $p_t^\varepsilon(x,y)$ in powers of ε. Importantly, by a weak uniqueness argument, we remark that the case $b_\varepsilon(x) = \varepsilon^2 b(x)$, $\sigma_\varepsilon(x) = \varepsilon\sigma(x)$ and $\ell = 0$ corresponds to the short time asymptotics of the density of X_t given by the unique weak solution taken at time t of the following SDE

$$X_t = x + \int_0^t b(X_s)ds + \int_0^t \sigma(X_s)dW_s + \alpha L_t^0(X). \tag{4}$$

Indeed, it is sufficient to remark that by weak uniqueness $\{X_{\varepsilon^2 t}; t \geq 0\}$ is equal in law to $\{X_t^\varepsilon; t \geq 0\}$. Hence, the case $\ell > 0$ corresponds to a skew parameter α close to zero and will be referred as to the small skew case.

Our main results are established in Sections 2.1 and 4. The first one is Theorem 2.1 where we establish an asymptotic expansion of the transition density $y \mapsto p_t^\varepsilon(x,y)$ with respect to the skew parameter. Not only such expansion appears to be new and of independent interest but also it is the first step towards our second main result, developed in Section 4, where the first order expansion of $y \mapsto p_t^\varepsilon(x,y)$ (in the case $b_\varepsilon \equiv \varepsilon^2 b$, $\sigma_\varepsilon = \varepsilon\sigma$), which notably includes the short time asymptotic, is established.

[1] In fact, a deeper analysis of the proofs in [15] results in the conclusion that this assumption is only used in Proposition 3.2 which is not required in order to obtain Proposition 5.3. In fact, this result can be obtained using an equivalent proof which uses the Itô-Tanaka formula instead of the generator L (see the proof of Lemma 1). Therefore one may assume that b is a measurable and bounded function. For a quick review of the argument, we refer to the Appendix.

[2] The proof of this fact in [15] can also be applied under the assumption that b is bounded and measurable.

2.1 Small skew expansion: a representation formula for the transition density

The first step consists in defining an approximation process that does not involve the local time component of X^{ε}. For some prescribed finite time horizon $T > 0$, we want to expand $(P^{\varepsilon}_t)_{t\in[0,T]}$ around the semigroup $(\bar{P}^{\varepsilon}_t)_{t\in[0,T]}$ associated to the unique weak solution of the following SDE with dynamics

$$\bar{X}^{\varepsilon}_t = x + \int_0^t b_{\varepsilon}(\bar{X}^{\varepsilon}_s)ds + \int_0^t \sigma_{\varepsilon}(\bar{X}^{\varepsilon}_s)dW_s. \tag{5}$$

We will work under the following mild assumptions on the coefficients:

(R) For any $\varepsilon \in [0,1]$, the measurable functions $b_{\varepsilon} : \mathbb{R} \to \mathbb{R}$ and $\sigma_{\varepsilon} : \mathbb{R} \to \mathbb{R}$ are bounded uniformly in ε. Moreover, $a_{\varepsilon} = \sigma^2_{\varepsilon}$ is η-Hölder continuous for some $\eta \in (0,1]$ uniformly in ε.

(UE) For any $\varepsilon \in [0,1]$, $\sigma_{\varepsilon}(x) = \sigma_0 + \varepsilon\sigma(x)$ for some constant $\sigma_0 \geq 0$ and where σ is a positive function such that there exist positive constants $\underline{a}, \overline{a}$ satisfying $\underline{a} \leq a(x) = \sigma^2(x) \leq \overline{a}$ for any $x \in \mathbb{R}$.

Since σ_{ε} is continuous and uniformly elliptic, without loss of generality, we assume that σ_{ε} is positive. Under **(R)** and **(UE)**, it is known that the transition density associated to $(\bar{P}^{\varepsilon}_t)_{t\geq 0}$ exists for $\varepsilon \in [0,1]$. We denote it by $(t,x,y) \mapsto \bar{p}^{\varepsilon}_t(x,y)$. According to classical results using the parametrix method (see e.g. Chapter 1, Section 6 in [5][3]), under **(R)** and **(UE)**, it is known that $\bar{p}^{\varepsilon}$ satisfies the following properties: for any $(t,x,y) \in (0,T) \times \mathbb{R} \times \mathbb{R}$,

$$\bar{p}^{\varepsilon}_t(x,y) = \bar{p}^{\varepsilon,1}_t(x,y) + \bar{p}^{\varepsilon,2}_t(x,y),$$

where

$$\bar{p}^{\varepsilon,1}_t(x,y) = \frac{1}{\sqrt{2\pi t}\,\sigma_{\varepsilon}(y)} \exp\left(-\frac{(y - b_{\varepsilon}(y)t - x)^2}{2\sigma^2_{\varepsilon}(y)t}\right), \tag{6}$$

and $\bar{p}^{\varepsilon,2}_t(x,y)$ satisfies the following Gaussian type estimates for $r = 0$ and $r = 1$

$$|\partial^r_x \bar{p}^{\varepsilon,2}_t(x,y)| \leq K_{\varepsilon} t^{-\frac{(1+r-\eta)}{2}} \exp\left(-C_{\varepsilon}\frac{(y-x)^2}{\varepsilon^2 t}\right), \tag{7}$$

for some positive constants C_{ε} and K_{ε}. Note that $\bar{p}^{\varepsilon,1}_t(x,y)$ and $\bar{p}^{\varepsilon,2}_t(x,y)$ are not density functions. Moreover, it holds

$$|\partial_x \bar{p}^{\varepsilon}_t(x,y)| \leq K_{\varepsilon} t^{-1} \exp\left(-C_{\varepsilon}\frac{(y-x)^2}{\varepsilon^2 t}\right), \tag{8}$$

[3] Although in this reference, the author assumes that the drift coefficient b is Hölder continuous, a careful analysis of the arguments reveals that this can be replaced by the weaker condition of being bounded and measurable. For a sketch of the argument in a particular case, see the Appendix.

and from (6) and (7), we easily obtain for $(y,t) \in \mathbb{R} \times (0,T]$,

$$|\partial_x \bar{p}_t^{\varepsilon,1}(y,y)| + |\partial_x \bar{p}_t^{\varepsilon,2}(y,y)| \le K_\varepsilon t^{-1+\frac{\eta}{2}}. \tag{9}$$

Note also that if $\sigma_0 > 0$ then there exist $M > 1$ such that for any $\varepsilon \in [0,1]$,

$$M^{-1} \le K_\varepsilon \vee \frac{C_\varepsilon}{\varepsilon^2} \le M. \tag{10}$$

An important point has to be noted at this step. Since $x \mapsto \bar{p}_t^\varepsilon(x,y)$ is continuously differentiable under **(R)** and **(UE)**, $\bar{P}_t^\varepsilon f$ does not satisfy the transmission condition (3). This is due to the fact that the dynamics (5) does not contain a local time term.

This important observation will characterize the first order expansion of P^ε around $\bar{P}^\varepsilon$ that represents a first step towards our main results. In order to simplify the writing of our next result, we introduce the following operators

$$\Delta_m^{\pm} f = \frac{f(m+) \pm f(m-)}{2}.$$

Lemma 1. *Assume that* **(UE)** *and* **(R)** *are satisfied. For all $f \in \mathscr{C}_c^\infty(\mathbb{R})$ and all $(t,x) \in [0,T] \times \mathbb{R}$, it holds*

$$P_t^\varepsilon f(x) = \bar{P}_t^\varepsilon f(x) + \alpha a_\varepsilon(0)\varepsilon^\ell \int_0^t \int_{\mathbb{R}} \Delta_0^+ p_{s_1}^\varepsilon(x,\cdot)\partial_x \bar{p}_{t-s_1}^\varepsilon(0,y) f(y)\, dy\, ds_1.$$

By a monotone class argument, one also has the following representation for $(t,x,y) \in (0,T] \times \mathbb{R} \times (\mathbb{R}\backslash\{0\})$,

$$p_t^\varepsilon(x,y) = \bar{p}_t^\varepsilon(x,y) + \alpha a_\varepsilon(0)\varepsilon^\ell \int_0^t \Delta_0^+ p_{s_1}^\varepsilon(x,\cdot)\partial_x \bar{p}_{t-s_1}^\varepsilon(0,y)\, ds_1. \tag{11}$$

Proof. We apply the Itô-Tanaka formula to the process $(X_s^\varepsilon)_{s\in[0,t]}$ with the map $[0,t] \times \mathbb{R} \ni (s,x) \mapsto \bar{P}_{t-s}^\varepsilon f(x)$ (see e.g. [11]) and take expectations to obtain:

$$\mathbb{E}[f(X_t^\varepsilon)] = \bar{P}_t^\varepsilon f(x) + \frac{1}{2}\alpha\varepsilon^\ell \mathbb{E}\left[\int_0^t \partial_x \bar{P}_{t-s}^\varepsilon f(0) dL_s^0(X^\varepsilon)\right].$$

Therefore the result follows from the space time formula for local times and the properties stated for the transition density of X^ε. For more details, we apply formula (7.3) in Section 3.7 of [13] and remark that the process $2(\Lambda_t(0,\omega) + \Lambda_t(0-,\omega))$ therein corresponds to $L_t^0(X^\varepsilon)$ in our situation.

For the density representation, one has to use the fact that the formula obtained can be extended to measurable and bounded functions whose support does not contain 0 and from here the density representation follows. □

In the sense of operators which generate the corresponding diffusions we see that the following expansion is satisfied in a *generalized sense*:

$$\mathscr{L}^{\varepsilon} = \bar{\mathscr{L}}^{\varepsilon} + \varepsilon^{\ell} \mathscr{L}^{1,\varepsilon},$$
$$\mathscr{L}^{1,\varepsilon} = \alpha a_{\varepsilon}(0)\delta_0 \partial_x.$$

where $\mathscr{L}^{\varepsilon}$ and $\bar{\mathscr{L}}^{\varepsilon}$ stand for the infinitesimal generators associated to P^{ε} and $\bar{P}^{\varepsilon}$.

Our next goal in order to obtain an expansion for p_t^{ε} is to iterate the above identity. This gives our first main result.

Theorem 2.1 *Let $T > 0$. Under* **(UE)** *and* **(R)**, *for all $(\varepsilon, \alpha) \in [0,1] \times (-1,1)$, the transition density associated to the unique weak solution to the SDE* (1) *admits the following representation for any $(t,x,y) \in (0,T] \times \mathbb{R} \times (\mathbb{R}\backslash\{0\})$,*

$$p_t^{\varepsilon}(x,y) = \bar{p}_t^{\varepsilon}(x,y) + \sum_{p \geq 1} (\alpha a_{\varepsilon}(0)\varepsilon^{\ell})^p I_p^{\varepsilon}(t,x,y), \tag{12}$$

with

$$I_p^{\varepsilon}(t,x,y) := \int_{\Delta_p(t)} \prod_{k=1}^{p-1} \partial_x \bar{p}^{\varepsilon}_{s_k - s_{k+1}}(0,0) \bar{p}^{\varepsilon}_{s_p}(x,0) \partial_x \bar{p}^{\varepsilon}_{t-s_1}(0,y)\, d\mathbf{s}_p, \quad p \geq 1,$$

and the convention $s_0 = t$ with $\Delta_p(t) := \{(s_1, ..., s_p); 0 < s_p < ... < s_1 < t\}$ and $d\mathbf{s}_p = ds_1 ... ds_p$.

Proof. In order to iterate formula (11), one needs to derive a similar formula for $\Delta_0^+ p^{\varepsilon}_{s_1}(x,\cdot)$. As we want to take limits in (11), we need to prove the uniform integrability of the integrands. We note here that the estimate in (9) can not be directly used here because formula (11) is not valid for $y = 0$.

In order to obtain the uniform integrability, we remark that (11) implies that for any $h \in (0,1]$

$$\begin{aligned}
\frac{p_t^{\varepsilon}(x,h) + p_t^{\varepsilon}(x,-h)}{2} &= \frac{\bar{p}_t^{\varepsilon}(x,h) + p_t^{0}(x,-h)}{2} \\
&+ \alpha a_{\varepsilon}(0)\varepsilon^{\ell} \int_0^t \Delta_0^+ p^{\varepsilon}_{s_1}(x,\cdot) \frac{\partial_x \bar{p}^{\varepsilon}_{t-s_1}(0,h) + \partial_x \bar{p}^{\varepsilon}_{t-s_1}(0,-h)}{2} ds_1.
\end{aligned}$$

Moreover, from (6) and (7), one has

$$\begin{aligned}
&\frac{\partial_x \bar{p}^{\varepsilon}_{t-s_1}(0,h) + \partial_x \bar{p}^{\varepsilon}_{t-s_1}(0,-h)}{2} \\
=& \frac{1}{2}\left\{ \frac{h - b_{\varepsilon}(h)(t-s_1)}{\sigma_{\varepsilon}^2(h)(t-s_1)} - \frac{h + b_{\varepsilon}(-h)(t-s_1)}{\sigma_{\varepsilon}^2(-h)(t-s_1)} \right\} \bar{p}^{\varepsilon,1}_{t-s_1}(0,h) \\
&+ \frac{h + b_{\varepsilon}(-h)(t-s_1)}{2\sigma_{\varepsilon}^2(-h)(t-s_1)} (\bar{p}^{\varepsilon,1}_{t-s_1}(0,h) - \bar{p}^{\varepsilon,1}_{t-s_1}(0,-h)) \\
&+ \frac{\partial_x \bar{p}^{\varepsilon,2}_{t-s_1}(0,h) + \partial_x \bar{p}^{\varepsilon,2}_{t-s_1}(0,-h)}{2} \\
=:& A_1 + A_2 + A_3.
\end{aligned}$$

From (7), it is clear that $|A_3| \le K_\varepsilon (t-s_1)^{-1+\frac{\eta}{2}}$ for some $K_\varepsilon > 0$ independent of h. Now, from **(UE)**, **(R)** and standard computations that we omit, on the set $\{s_1 \in [0,t] : h < (t-s_1)^\theta\}$, with θ chosen so that $\frac{1}{1+\eta} < 2\theta < 1$, we obtain

$$|A_1| + |A_2| \le K_\varepsilon \left\{ (t-s_1)^{-\frac{1}{2}} + (t-s_1)^{-\frac{3}{2}+\theta(1+\eta)} \right\}.$$

On the set $\{s_1 \in [0,t] : h \ge (t-s_1)^\theta\}$, from (8) and $\theta < 1/2$, one gets

$$\begin{aligned}|\partial_x \bar{p}^\varepsilon_{t-s_1}(0,h)| + |\partial_x \bar{p}^\varepsilon_{t-s_1}(0,-h)| \le & K_\varepsilon \frac{1}{(t-s_1)} \exp\left(-C_\varepsilon \frac{h^2}{\varepsilon^2 (t-s_1)}\right) \\ \le & K_\varepsilon \frac{1}{(t-s_1)} \exp\left(-C_\varepsilon \frac{1}{\varepsilon^2 (t-s_1)^{1-2\theta}}\right).\end{aligned}$$

Letting $h \downarrow 0$, it then follows from the dominated convergence and the continuity at 0 of $y \mapsto \bar{p}^\varepsilon_t(0,y)$ and $y \mapsto \partial_x \bar{p}^\varepsilon_t(0,y)$ that

$$\Delta_0^+ p^\varepsilon_t(x,.) = \bar{p}^\varepsilon_t(x,0) + \alpha a_\varepsilon(0)\varepsilon^\ell \int_0^t \Delta_0^+ p^\varepsilon_{s_1}(x,.)\partial_x \bar{p}^\varepsilon_{t-s_1}(0,0)\, ds_1. \quad (13)$$

Moreover, it follows from estimates (2) and (9) that

$$s_1 \mapsto \Delta_0^+ p^\varepsilon_{s_1}(x,.)\partial_x \bar{p}^\varepsilon_{t-s_1}(0,0) \in L^1([0,t]).$$

Hence, one may iterate (13) so that

$$\Delta_0^+ p^\varepsilon_t(x,.) = \bar{p}^\varepsilon_t(x,0) + \sum_{p \ge 1} (\alpha a_\varepsilon(0)\varepsilon^\ell)^p I^\varepsilon_p(t,x,0). \quad (14)$$

Indeed, it follows from (8) and (9) that there exists K_ε such that for any positive integer p

$$\begin{aligned}&(\alpha a_\varepsilon(0)\varepsilon^\ell)^p \left| I^\varepsilon_p(t,x,0) \right| \\ \le & K_\varepsilon (\alpha a_\varepsilon(0)\varepsilon^\ell K_\varepsilon)^p \int_{\Delta_p(t)} \prod_{k=1}^{p-1} (s_k - s_{k+1})^{-1+\frac{\eta}{2}} s_p^{-\frac{1}{2}} (t-s_1)^{-1+\frac{\eta}{2}}\, d\mathbf{s}_p \\ =: & \int_{\Delta_p(t)} G_p(\mathbf{s}_p)\, d\mathbf{s}_p.\end{aligned}$$

The function $G_p(\mathbf{s}_p)$ defined above satisfies for $C_1^\varepsilon := \alpha \varepsilon^\ell a_\varepsilon(0) K_\varepsilon t^{\frac{\eta}{2}}$ and any $r > 0$

$$
\begin{aligned}
&\sum_{p\geq 1}\int_{\Delta_p(t)} d\mathbf{s}_p G_p(\mathbf{s}_p) \qquad (15)\\
=&K_\varepsilon t^{-\frac{1+\eta}{2}}\sum_{p\geq 1}(C_1^\varepsilon)^p\prod_{k=0}^{p-1}B\left(\frac{1+k\eta}{2},\frac{\eta}{2}\right)\\
\leq&K_\varepsilon t^{-\frac{1+\eta}{2}}\sum_{p\geq 1}\left(C_1^\varepsilon\Gamma(\frac{\eta}{2})\right)^p\frac{1}{\Gamma(\frac{1+p\eta}{2})}<\infty.
\end{aligned}
$$

where we have used the inequality $x^r e^{-x}\leq C_r$, $x>0$, for the last inequality. Here, B and Γ denote the Beta and Gamma functions respectively. From this last estimate, we deduce that the expansion (14) is satisfied. We then plug (14) into (13) and use Fubini's theorem in order to derive the small skew expansion (12). □

As an illustration of the theorem, we consider the case of the skew Brownian motion that is we set $\sigma_0=0$, $\sigma\equiv 1$, $b\equiv 0$. Hence, $\bar{p}^\varepsilon$ is the fundamental solution associated to the operator $\frac{1}{2}\varepsilon^2\partial_x^2$ and is given by $\bar{p}_t^\varepsilon(x,y)=g_{\varepsilon^2 t}(y-x)$ so that Theorem 2.1 asserts that for all $(t,x,y)\in(0,T]\times\mathbb{R}\times(\mathbb{R}\backslash\{0\})$

$$
p_t^\varepsilon(x,y)=\bar{p}_t^\varepsilon(x,y)+\alpha\varepsilon^{2+\ell}\int_0^t ds\,\bar{p}_s^\varepsilon(x,0)\partial_x\bar{p}_{t-s}^\varepsilon(0,y). \qquad (16)
$$

This is one possible expression for the transition density of the skew Brownian motion, see e.g. Lejay [17]. Hence, Theorem 2.1 appears as an extension of the representation formula (16) to the more general class of skew diffusions.

Another application consists of measuring the distance between models.

Corollary 1. *For any $t\in(0,T]$, the total variation distance between the laws of X_t^ε and $\bar{X}_t^\varepsilon$ can be bounded by*

$$
\int|p_t^\varepsilon(x,y)-\bar{p}_t^\varepsilon(x,y)|dy\leq K_\varepsilon\alpha a_\varepsilon(0)\varepsilon^\ell,
$$

for some constant $K_\varepsilon:=K_\varepsilon(T)$, uniformly bounded in $\varepsilon\in[0,1]$ such that $T\mapsto K_\varepsilon(T)$ is non-decreasing.

Proof. Note that

$$
\begin{aligned}
&(\alpha a_\varepsilon(0)\varepsilon^\ell)^p\left|I_p^\varepsilon(t,x,y)\right|\\
\leq&K_\varepsilon(\alpha a_\varepsilon(0)\varepsilon^\ell K_\varepsilon)^p\int_{\Delta_p(t)}\prod_{k=1}^{p-1}(s_k-s_{k+1})^{-1+\frac{\eta}{2}}s_p^{-\frac{1}{2}}(t-s_1)^{-1}\exp\left(-C_\varepsilon\frac{y^2}{\varepsilon^2(t-s_1)}\right)d\mathbf{s}_p
\end{aligned}
$$

so that

$$
\begin{aligned}
&(\alpha a_\varepsilon(0)\varepsilon^\ell)^p\int_{\mathbb{R}}\left|I_p^\varepsilon(t,x,y)\right|dy\\
\leq&\varepsilon C_\varepsilon^{-1/2}(\alpha a_\varepsilon(0)\varepsilon^\ell)^p K_\varepsilon^{p+1}\int_{\Delta_p(t)}d\mathbf{s}_p\prod_{k=1}^{p-1}(s_k-s_{k+1})^{-1+\frac{\eta}{2}}s_p^{-\frac{1}{2}}(t-s_1)^{-1/2}.
\end{aligned}
$$

Now the same line of calculations using the Beta function and (10) yields the result.

3 Application to a financial model

We will consider the following stochastic differential equation as an extension of the Black-Scholes model where the local volatility is given by a simple regime switching in the following sense

$$S_t = S_0 + \int_0^t \sigma(S_s) S_s dB_s, \tag{17}$$

with

$$\sigma(s) = \sigma_1 1_{[0,\mathrm{s}_1)}(s) + \sigma_2 1_{[\mathrm{s}_1,\infty)}(s).$$

For more background on the financial meaning of the model, we refer the reader to [10]. In order to transform this model into a skew diffusion, we let $F : \mathbb{R} \to \mathbb{R}$ be the continuous increasing function defined by

$$\begin{aligned} F(y) &= \frac{y}{\sigma_1} 1_{(-\infty,\mathrm{s}_1^*)}(y) + \left(\frac{y}{\sigma_2} + c_2\right) 1_{[\mathrm{s}_1^*,\infty)}(y), \\ c_2 &= -\frac{\mathrm{s}_1^*}{\sigma_2} + \frac{\mathrm{s}_1^*}{\sigma_1}, \\ \mathrm{s}_1^* &= \exp(s_1). \end{aligned}$$

Using Itô's formula, we have

$$\ln(S_t) = \ln(S_0) + \int_0^t \left(\sigma(S_u) dB_u - \frac{\sigma^2(S_u)}{2} du\right).$$

Furthermore, applying the Itô-Tanaka formula for $X_t = F(\ln(S_t))$, we obtain

$$X_t = X_0 + B_t - \int_0^t \frac{\tilde{\sigma}(X_u)}{2} du + \Delta_{a_1}^-(\tilde{\sigma}^{-1}) L_t^{a_1}(\ln(S)).$$

Here, $X_0 = F(\ln(S_0))$, $a_1 := \mathrm{s}_1^*/\sigma_1$ and

$$\tilde{\sigma}(x) := \sigma(e^{F^{-1}(x)}) = \sigma_1 1_{(-\infty,a_1)}(x) + \sigma_2 1_{[a_1,\infty)}(x).$$

Similarly, we apply the Itô-Tanaka formula to $g(\ln(S_t)) := |F(\ln(S_t)) - a_1|$. This gives for $\Delta_{a_1}^+(\tilde{\sigma}^{-1}) := 2^{-1}(\sigma_2^{-1} + \sigma_1^{-1})$,

$$g(\ln(S_t)) = g(\ln(S_0)) + \int_0^t g'(\ln(S_u)) d\ln(S_u) + \Delta_{a_1}^+(\tilde{\sigma}^{-1}) L_t^{a_1}(\ln(S)). \tag{18}$$

Furthermore,

$$\int_0^t \mathrm{sgn}(X_u - a_1)dX_u = \int_0^t \mathrm{sgn}(F(\ln(S_u)) - a_1)\left(dB_u - \frac{\tilde{\sigma}(X_u)}{2}du\right)$$
$$= \int_0^t \mathrm{sgn}(F(\ln(S_u)) - a_1)\sigma^{-1}(X_u)d\ln(S_u)$$
$$= \int_0^t g'(\ln(S_u))d\ln(S_u). \tag{19}$$

Recall that an application of the Itô-Tanaka formula to $|X_t - a_1|$ gives

$$L_t^{a_1}(X) = |X_t - a_1| - |X_0 - a_1| - \int_0^t \mathrm{sgn}(X_u - a_1)dX_u.$$

Comparing the above results in (18) and (19), we get

$$\Delta_{a_1}^+(\tilde{\sigma}^{-1})L_t^{a_1}(\ln(S)) = L_t^{a_1}(X).$$

This gives that

$$X_t = X_0 + B_t - \int_0^t \frac{\tilde{\sigma}(X_u)}{2}du + \alpha_1 L_t^{a_1}(X),$$

with $\alpha_1 := \frac{\Delta_{a_1}^-(\tilde{\sigma}^{-1})}{\Delta_{a_1}^+(\tilde{\sigma}^{-1})} \in (-1,1)$ for $\Delta_{a_1}^-(\tilde{\sigma}^{-1}) := 2^{-1}(\sigma_2^{-1} - \sigma_1^{-1})$.

Using the technique exposed in the previous section, we can now study an expansion of this model with $\varepsilon = 2\frac{\sigma_2-\sigma_1}{\sigma_2+\sigma_1} > 0$, $\alpha = -\frac{1}{2}$, $\ell = 1$. In this setting, we can set-up

$$b_\varepsilon(x) := \tilde{\sigma}(x) = \sigma_1 + \frac{\varepsilon}{2}(\sigma_2 + \sigma_1)1_{[a_1,\infty)}(x)$$
$$a_\varepsilon(x) = \sigma_0 = 1.$$

There will be a slight difference in the definition of $\bar{X}^\varepsilon$ which will be given by

$$\bar{X}_t = X_0 + B_t - \frac{\sigma_1}{2}t.$$

Besides the above change the hypotheses **(R)** and **(UE)** are clearly satisfied. We now mention the main modifications to the proof of Theorem 2.1. First, in Lemma 1, we obtain instead that for all $f \in \mathscr{C}_c^\infty(\mathbb{R})$, for any $x \in \mathbb{R}$, one has

$$P_t^\varepsilon f(x) = \bar{P}_t f(x) + \int_0^t \int_{\mathbb{R}^2} p_s^\varepsilon(x,x_1)(b_\varepsilon - \sigma_1)(x_1)\partial_x \bar{p}_{t-s}(x_1,y)f(y)dx_1dyds$$
$$+ \alpha_1 \varepsilon \int_0^t \int_{\mathbb{R}} \Delta_{a_1}^+ p_s^\varepsilon(x,\cdot)\partial_x \bar{p}_{t-s}(a_1,y)f(y)dyds. \tag{20}$$

Here $\bar{p}_t(x,y) = g_t(y - x + \frac{\sigma_1}{2}t)$ is the density of $\bar{X}_t$. The above shows that the argument can be applied using the "error operator " $\mathscr{L}^{1,\varepsilon}$ given by

$$\mathscr{L}^{1,\varepsilon} f(x) = \varepsilon \left(\frac{\sigma_2 + \sigma_1}{2} 1_{[a_1,\infty)}(x) + \alpha_1 \delta_{a_1}(x) \right) f'(x).$$

The analysis of integrability of multiple integrals and the estimates in (6)-(9) are straightforward because the density of $\bar{X}_t$ can be explicitly written and in fact the bounds can be improved in comparison with (7)~(10). This is due to the fact that the diffusion coefficient is constant and previously it was a general Hölder continuous function.

Continuing with the argument, instead of $I_p^\varepsilon(t,x,y)$ in (12), we obtain

$$\begin{aligned}
&\bar{P}_{s_p} \mathscr{L}^{1,\varepsilon} \bar{P}_{s_{p-1}-s_p} \cdots \bar{P}_{s_1-s_2} \mathscr{L}^{1,\varepsilon} \bar{P}_{t-s_1}(\delta_y(.))(x) \\
&= \int_{\mathbb{R}^p} dx_1 ... dx_p \bar{p}_{s_p}(x,x_p)((b_\varepsilon(x_1) - \sigma_1)\partial_x \bar{p}_{t-s_1}(x_1,y) + \alpha_1 \varepsilon \delta_{a_1}(x_1)\partial_x \bar{p}_{t-s_1}(x_1,y)) \\
&\times \prod_{k=1}^{p-1} ((b_\varepsilon(x_{k+1}) - \sigma_1)\partial_x \bar{p}_{s_k - s_{k+1}}(x_{k+1},x_k) + \alpha_1 \varepsilon \delta_{a_1}(x_{k+1})\partial_x \bar{p}_{s_k - s_{k+1}}(x_{k+1},x_k)) .
\end{aligned}$$

In the present case the partial derivatives are explicit. In particular,

$$\begin{aligned}
|\partial_x \bar{p}_t(x,y)| \leq & \frac{C}{\sqrt{t}} \bar{p}_t(x,y), \\
|\partial_x \bar{p}_t(x,x)| \leq & \frac{C}{\sqrt{t}}.
\end{aligned}$$

Therefore one can easily conclude an equivalent result to Corollary 1 so that for any bounded payoff function f, it holds

$$\left| \mathbb{E}[f(S_t)] - \mathbb{E}[f(\tilde{S}_t)] \right| \leq C|\sigma_2 - \sigma_1|.$$

Here $\tilde{S}$ corresponds to the Black-Scholes model with volatility σ_1. By inspecting the proof, one can see that with a little bit more effort, one can achieve a similar result for linearly growing test functions f. Furthermore as $\bar{p}$ can be expressed explicitly in the present case higher order expansions can be obtained.

A similar argument could be achieved for a more general class of regime switching models such as the following extension of the Black-Scholes model where the volatility is subjected to $n+1$ regime switching models in the following sense

$$S_t = S_0 + \int_0^t \sigma(S_s) S_s dB_s, \tag{21}$$

where σ is defined by

$$\sigma(s) = \sum_{i=1}^{n+1} \sigma_i 1_{[s_{i-1},s_i)}(s).$$

Here $\sigma_1, \dots, \sigma_{n+1} \in \mathbb{R}_+$ and $0 = s_0 < s_1 < \cdots < s_n < s_{n+1} = \infty$.

Rather than continuing on this path, we would like to present how to compute the full expansion of the density p^ε. First note that the first term in (20) gives as the lowest order term of the expansion

$$\bar{P}_{s_p} \mathscr{L}^{1,\varepsilon} \bar{P}_{s_{p-1}-s_p} \cdots \bar{P}_{s_1-s_2} \mathscr{L}^{1,\varepsilon} \bar{P}_{t-s_1}(\delta_y(.))(x) = \varepsilon^p A_p(s_p,...,s_1).$$

Here, for $\tilde{b}(x) := \frac{\sigma_1+\sigma_2}{2} 1_{[a_1,\infty)}(x)$

$$\begin{aligned} &A_p(s_p,...,s_1) := \\ &\int_{\mathbb{R}^p} dx_1...dx_p \bar{p}_{s_p}(x,x_p) \left(\tilde{b}(x_1)\partial_x \bar{p}_{t-s_1}(x_1,y) + \alpha_1 \delta_{a_1}(x_1)\partial_x \bar{p}_{t-s_1}(x_1,y)\right) \\ &\times \prod_{k=1}^{p-1} \left(\tilde{b}(x_{k+1})\partial_x \bar{p}_{s_k-s_{k+1}}(x_{k+1},x_k) + \alpha_1 \delta_{a_1}(x_{k+1})\partial_x \bar{p}_{s_k-s_{k+1}}(x_{k+1},x_k)\right). \end{aligned}$$

From here one has the expansion

$$p_t^\varepsilon(x,y) = \bar{p}_t(x,y) + \sum_{p=1}^{\infty} \varepsilon^p \int_{\Delta_p(t)} d\mathbf{s}_p A_p(s_p,...,s_1).$$

The full expansion for the density of S_t can be easily achieved by a change of variables formula. In the next section, we will return to our main problem which is slightly more complex as the approximating densities also depend on ε.

4 Asymptotics of the density $\bar{p}^\varepsilon$

Now we return to the general case studied in Theorem 2.1 and note that we have reduced the problem of the expansion of p^ε to the expansion of the density of $\bar{X}_t^\varepsilon$ which is a one dimensional diffusion. In order to carry out this expansion, we can use classical techniques to expand its density in the case[4] $\sigma_0 = 0$. In fact, we will use the Lamperti transformation as follows:

$$F(x) := \int_0^x \frac{1}{\sigma(y)} dy$$

Note that this function is well defined due to hypothesis **(UE)**. Then a combination of Itô's formula and the Girsanov measure transformation under hypothesis **(R)**, gives the following expression with $Z_t^\varepsilon := F(x) + \varepsilon W_t$:

[4] The case $\sigma_0 > 0$ can also be achieved with further computational effort.

$$\begin{aligned}\mathbb{E}[f(\bar{X}_t^\varepsilon)] =& \mathbb{E}\left[f(F^{-1}(Z_t^\varepsilon))M_t^\varepsilon\right] \\ \bar{b}_\varepsilon(x) :=& -\frac{1}{2}\sigma'(F^{-1}(x)) + \varepsilon^{-2}\frac{b_\varepsilon}{\sigma}(F^{-1}(x)) \\ M_t^\varepsilon :=& \exp\left(\varepsilon\int_0^t \bar{b}_\varepsilon(Z_s^\varepsilon)dW_s - \frac{\varepsilon^2}{2}\int_0^t \bar{b}_\varepsilon^2(Z_s^\varepsilon)ds\right).\end{aligned} \tag{22}$$

Using the explicit density of $F(x)+\varepsilon W_t$ and the fact that $\{W_s; s\in[0,t]\}$ conditioned to $W_t = y$ has the law of a Brownian bridge, we obtain :

$$\bar{p}_t^\varepsilon(x,y) = \frac{g_{\varepsilon^2 t}(F(x),F(y))}{\sigma(y)}\mathbb{E}\left[\exp\left(\varepsilon\int_0^t \bar{b}_\varepsilon(B_s^{x,y})dB_s^{x,y} - \frac{\varepsilon^2}{2}\int_0^t \bar{b}_\varepsilon^2(B_s^{x,y})ds\right)\right].$$

Here $B_s^{x,y}$ represents a Brownian bridge with variance ε^2 at time s, starting at $F(x)$ and arriving at $F(y)$ at time t. From now on, we will assume the following hypothesis:

(R') The measurable functions $\bar{b}_\varepsilon : \mathbb{R} \to \mathbb{R}$ are bounded uniformly in ε.

Therefore in order to obtain an expansion for the above expectation, one may just use the Taylor expansion for the exponential function and the following representation for the Brownian bridge (see Section 5.6.B in [13])

$$\begin{aligned}B_s^{x,y} =& F_{s,t}(x,y) + \varepsilon(t-s)\int_0^s \frac{dW_u}{t-u}, \\ F_{s,t}(x,y) :=& F(x)\left(1-\frac{s}{t}\right) + F(y)\frac{s}{t}.\end{aligned}$$

As the proofs on the order of the residues are cumbersome, from now on we only give heuristic expansions so that the way calculations are performed is clear.

For example, the first term in the expansion requires the computation of

$$\mathbb{E}\left[\varepsilon\int_0^t \bar{b}_\varepsilon(B_s^{x,y})dB_s^{x,y}\right] = \mathbb{E}\left[\varepsilon\int_0^t \bar{b}_\varepsilon(B_s^{x,y})\left(\partial_s F_{s,t}(x,y) - \varepsilon\int_0^s \frac{dW_u}{t-u}\right)ds\right]. \tag{23}$$

Further calculations will require explicit values for the coefficient functions b_ε and σ but it should be clear that the only underlying law needed for this calculation is the one related to the Brownian motion W. For example, assume that one may expand $\bar{b}_\varepsilon(B_s^{x,y})$ as follows:

$$\begin{aligned}\bar{b}_\varepsilon(B_s^{x,y}) \approx& \bar{b}_\varepsilon(F_{s,t}(x,y)) + \bar{b}'_\varepsilon(F_{s,t}(x,y))(t-s)\varepsilon\int_0^s \frac{dW_u}{t-u} \\ &+ \frac{1}{2}\bar{b}''_\varepsilon(F_{s,t}(x,y))(t-s)^2\varepsilon^2\left(\int_0^s \frac{dW_u}{t-u}\right)^2.\end{aligned}$$

Using the above expansion in (23), one obtains:

$$\mathbb{E}\left[\varepsilon\int_0^t \bar{b}_\varepsilon(B_s^{x,y})dB_s^{x,y}\right] \approx \varepsilon\int_0^t \bar{b}_\varepsilon(F_{s,t}(x,y))\partial_s F_{s,t}(x,y)ds - \varepsilon^3\int_0^t \bar{b}'_\varepsilon(F_{s,t}(x,y))\frac{s}{t}ds$$
$$+\frac{\varepsilon^3}{2}\int_0^t \bar{b}_\varepsilon\bar{b}''_\varepsilon(F_{s,t}(x,y))(t-s)^2\partial_s F_{s,t}(x,y)ds.$$

Putting the above expansions together, one obtains:

$$\bar{p}_t^\varepsilon(x,y) \approx \frac{g_{\varepsilon^2 t}(F(x),F(y))}{\sigma(y)} \tag{24}$$
$$\times\left(1+\varepsilon\int_0^t \bar{b}_\varepsilon(F_{s,t}(x,y))\partial_s F_{s,t}(x,y)ds - \frac{\varepsilon^2}{2}\int_0^t \bar{b}_\varepsilon^2(F_{s,t}(x,y))ds\right)$$
$$=: B_0(\varepsilon). \tag{25}$$

This gives the expansion for the first term of the main result in Theorem 2.1. For the following term, we see that one also needs an expansion for the derivative $\partial_x \bar{p}_t^\varepsilon(x,y)$.

In order to do this, we take $f=\delta_y$ in (22) and differentiate with respect to x. This gives:

$$\partial_x \bar{p}_t^\varepsilon(x,y) = \frac{\sigma(y)}{\sigma(x)}\mathbb{E}\left[\delta'_y(\bar{X}_t^\varepsilon)M_t^\varepsilon\right] + \sigma(x)^{-1}\mathbb{E}\left[\delta_y(\bar{X}_t^\varepsilon)M_t^\varepsilon Y_t^\varepsilon\right] \tag{26}$$
$$Y_t^\varepsilon := \varepsilon\left(\int_0^t \bar{b}'_\varepsilon(F(x)+\varepsilon W_s)dW_s - \varepsilon\int_0^t \bar{b}_\varepsilon\bar{b}'_\varepsilon(F(x)+\varepsilon W_s)ds\right).$$

The second term in (26) can be treated as we did in the expansion of $\bar{p}^\varepsilon$. For the first term, we need to use the integration by parts formula of Malliavin Calculus. Without going into much detail into this theory[5], let us just state the result:

$$\mathbb{E}\left[\delta'_y(\bar{X}_t^\varepsilon)M_t^\varepsilon\right] = (\sigma(y)\varepsilon t)^{-1}\mathbb{E}\left[\delta_y(\bar{X}_t^\varepsilon)M_t^\varepsilon\left(W_t - \varepsilon\int_0^t(\bar{b}_\varepsilon(Z_s^\varepsilon)ds + s dY_s^\varepsilon)\right)\right]$$
$$=: (\sigma(y)^2\varepsilon t)^{-1} g_{\varepsilon^2 t}(F(x),F(y))A_1(t).$$

The above expression can now be expanded as we did previously using the Brownian bridge $B^{x,y}$. The explicit expansion of $A_1(t)$ is performed in Section 6.

These are the essential mathematical elements required in order to obtain the expansions. The rest of the calculations, which are briefly explained in the Appendix, require careful attention as there are many terms. We refer the reader to Appendix 2 for the explicit but long expressions. Taking into account the calculations in the Appendix 2, we obtain the following result:

Result Let X^ε be the unique solution of (1) with $\ell \geq 1$ and $(t,x,y)\in(0,T]\times\mathbb{R}\times(\mathbb{R}\backslash\{0\})$. Assume the hypotheses **(UE)**, **(R)** and **(R')**. Furthermore, assume that $\bar{b}_\varepsilon \in C^2$ and that $\|\bar{b}_\varepsilon^{(i)}\|_\infty \leq C$, $i=1,2$. Then, it holds

$$p_t^\varepsilon(x,y) \approx B_0(\varepsilon) + \alpha\varepsilon^\ell\left(\sigma(y)^{-1}g_{\varepsilon^2 t}(|F(x)|,|F(y)|) + \varepsilon B_1 + \varepsilon^2 B_2\right).$$

[5] We refer the interested reader to [19] for definitions and notation.

We quote the above statement as a result and not a theorem because proving error rates is a technical and cumbersome argument which we have avoided here. In particular, the error rate seems to be bounded by $\varepsilon^3 g_{\varepsilon^2 t}(F(x),F(y))$. In the case that $\bar{b}_\varepsilon$ is not differentiable but explicit as in Section 3, a careful analysis about the expansion of M_t^ε is needed. We do not develop these issues here.

5 Appendix 1: The parametrix method applied to a skew diffusion

In this section, we give a brief and sketchy argument to show what is claimed in footnote 3. In this sense, it may be easier for a beginner as we will not deal with technicalities.

Our goal is to prove the existence and some regularity estimates for the density of the random variable X_t^ε in (1). In order to simplify notation, we will just explain how to obtain properties of the law of

$$X_t = x + \int_0^t b(X_s)ds + \int_0^t \sigma(X_s)dW_s + \alpha L_t^m(X).$$

Recall that we are assuming that b is bounded and measurable and σ is Hölder continuous and bounded.

The approximation will be based on the skew Brownian motion with $\alpha \in (-1,1)$ and $m \in \mathbb{R}$ which is the unique solution of the equation

$$X_t^\alpha = x + \bar{\sigma} W_t + \alpha L_t^m(X^\alpha).$$

for some well-chosen $\bar{\sigma}$. Its density is given explicitly by

$$p_t^\alpha(x,y) := g_{t\bar{\sigma}^2}(y-x) + \operatorname{sgn}(y-m)\alpha g_{t\bar{\sigma}^2}(|y-m|+|x-m|)$$
$$g_t(y) := \frac{1}{\sqrt{2\pi t}} \exp\left(-\frac{y^2}{2t}\right).$$

Note that this density is in general discontinuous at $y = m$.

We will be using as the parametrix operator, the following linear operator which is well defined for any bounded measurable function f:

$$\bar{P}_t f(x) = \int_{\mathbb{R}} f(y)\bar{p}_t(x,y)dy$$
$$\bar{p}_t(x,y) := g_{t\sigma^2(y)}(y-x) + \operatorname{sgn}(y-m)\alpha g_{t\sigma^2(y)}(|y-m|+|x-m|).$$

Note that $y \mapsto \bar{p}_t(x,y)$ is not a density but $\bar{P}_t f$ is differentiable w.r.t. x with a second generalized derivative which satisfies the transmission condition (3) with $\varepsilon = 1$ for any bounded measurable function $f : \mathbb{R} \to \mathbb{R}$.

From these properties, we see that the Itô-Tanaka formula can be applied (see e.g. formula 2.5 in [11]) to $\bar{P}_{T-t}f(X_t)$, $t\in[0,T]$. This gives after explicit calculations [6]

$$\begin{aligned}
&\mathbb{E}[f(X_T)]-\bar{P}_T f(x)\\
&=\int_0^T \mathbb{E}\left[(\partial_s\bar{P}_{T-s}f)(X_s)+b(X_s)\partial_x\bar{P}_{T-s}f(X_s)+\frac{1}{2}\sigma^2(X_s)\partial_x^2\bar{P}_{T-s}f(X_s)1_{(X_s\neq m)}\right]ds\\
&=\int_0^T \mathbb{E}\left[b(X_s)\partial_x\bar{P}_{T-s}f(X_s)+\left(\partial_s\bar{P}_{T-s}f+\frac{1}{2}\sigma^2\partial_x^2\bar{P}_{T-s}f\right)(X_s)1_{(X_s\neq m)}\right]ds.
\end{aligned}$$

Iterating the above expression one can obtain a semi-explicit expansion of the form:

$$\begin{aligned}
\mathbb{E}[f(X_T)] &=\bar{P}_T f(x)+\sum_{p=1}^{\infty}\int_{\Delta_p}\bar{P}_{s_p}Q_{s_{p-1}-s_p}\cdots Q_{T-s_1}f(x)ds_1\ldots ds_p \qquad (27)\\
Q_t f(x) &:=b(x)\partial_x\bar{P}_t f(x)+\left(-\partial_t\bar{P}_t f+\frac{1}{2}\sigma^2\partial_x^2\bar{P}_t f\right)(x)1_{(x\neq m)}\\
\Delta_p(T) &:=\{(s_1,\ldots,s_p);0<s_p<\ldots<s_1<T\}.
\end{aligned}$$

Clearly, for the above expansion to be valid we need to prove convergence properties for the above infinite sum. Direct calculations using the definition of $\bar{P}$ and the assumptions on b and σ give (see (29) to guess why the estimate below is correct)

$$\|Q_t f\|_\infty\leq\frac{C}{t^{(1-\eta/2)\wedge(1/2)}},$$

where η is the Hölder exponent of σ^2. Therefore iterative integration over the simplex[7] Δ_p gives that the infinite sum in (27) is absolutely convergent.

As this formula is valid for all bounded and measurable functions $f:\mathbb{R}\to\mathbb{R}$, this result also gives an expression for the density of X_T which can then be differentiated term by term in order to obtain the claimed properties. In fact, the density $p_t(x,y)$ of X_t can be written as in the following expansion

$$p_t(x,y)=\bar{p}_t(x,y)+\sum_{p=1}^{\infty}\int_{\Delta_p}\bar{P}_{s_p}Q_{s_{p-1}-s_p}\cdots Q_{T-s_1}\delta_y(x)ds_1\ldots ds_p. \qquad (28)$$

We remark here that

$$Q_s\delta_y(x)=b(x)\partial_x\bar{p}_t(x,y)+\frac{1}{2}(\sigma^2(y)-\sigma^2(x))\partial_x^2\bar{p}_t(x,y)1_{(x\neq m)}. \qquad (29)$$

From (28), one can easily deduce that $p_t(x,m\pm)$ exist. Moreover, by direct differentiation one also obtains for $x\neq m$

[6] Note that $\int_0^T 1_{(X_s=m)}ds=0$ due to the time-space formula for semimartingales. For more details, see formula (7.3) in Section 3.7 of [13].

[7] This requires the use of the Beta function.

$$\partial_x p_t(x,y) = \partial_x \bar{p}_t(x,y) + \sum_{p=1}^{\infty} \int_{\Delta_p} \int \partial_x \bar{p}_{s_p}(x,z) Q_{s_{p-1}-s_p} \cdots Q_{T-s_1} \delta_y(z) dz ds_1 \ldots ds_p.$$

This gives the existence of the derivative $\partial_x p_t(x, \pm m)$ and Gaussian type upper estimates by using the inequality $x^p e^{-cx^2} \leq C(p,c)$ for $x \geq 0$. For a probabilistic interpretation of a variation of this methodology, we refer the reader to [16].

6 Appendix 2: Explicit calculations for the expansion of $\bar{p}^\varepsilon$ and $\partial_x \bar{p}^\varepsilon$

In this section, we compute the explicit expressions for the expansion of p^ε up to order $\varepsilon^2 g_{\varepsilon^2 t}$.

The argument has already been explained in Section 4. The application of that argument gives (24).

Now, we need to expand $\partial_x \bar{p}_t^\varepsilon(x,y)$. This is done through the expansion of the two terms appearing in the right-hand side of (26). Let us start with the simplest of the two which is the second term. To expand this term one uses the same technique as in the above expansion. That is,

$$\begin{aligned}
\mathbb{E}[\delta_y(\bar{X}_t^\varepsilon) M_t^\varepsilon Y_t^\varepsilon] \approx & \varepsilon \frac{g_{\varepsilon^2 t}(F(x),F(y))}{\sigma(y)} \left(\mathbb{E}\left[\int_0^t \bar{b}'_\varepsilon(B_s^{x,y}) dB_s^{x,y} - \varepsilon \int_0^t \bar{b}_\varepsilon \bar{b}'_\varepsilon(B_s^{x,y}) ds \right] \right. \\
& \left. + \varepsilon \mathbb{E}\left[\int_0^t \bar{b}'_\varepsilon(B_s^{x,y}) dB_s^{x,y} \int_0^t \bar{b}_\varepsilon(B_s^{x,y}) dB_s^{x,y} \right] \right) \\
\approx & \varepsilon \frac{g_{\varepsilon^2 t}(F(x),F(y))}{\sigma(y)} \left(\int_0^t \bar{b}'_\varepsilon(F_{s,t}(x,y)) \partial_s F_{s,t}(x,y) ds - \varepsilon \int_0^t \bar{b}_\varepsilon \bar{b}'_\varepsilon(F_{s,t}(x,y)) ds \right. \\
& \left. + \varepsilon \int_0^t \bar{b}'_\varepsilon(F_{s,t}(x,y)) \partial_s F_{s,t}(x,y) ds \int_0^t \bar{b}_\varepsilon(F_{s,t}(x,y)) \partial_s F_{s,t}(x,y) ds \right).
\end{aligned}$$

Finally, for the first term in (26), we have for $G_\varepsilon(s,z) := \bar{b}_\varepsilon(z)\left(1 - s\varepsilon \bar{b}'_\varepsilon(z)\right)$

$$\begin{aligned}
A_1(t) := \mathbb{E}\Bigg[& \exp\left(\varepsilon \int_0^t \bar{b}_\varepsilon(B_s^{x,y}) dB_s^{x,y} - \frac{\varepsilon^2}{2} \int_0^t \bar{b}_\varepsilon^2(B_s^{x,y}) ds \right) \\
& \times \left(\frac{F(y) - F(x)}{\varepsilon} - \varepsilon \int_0^t G_\varepsilon(s, B_s^{x,y}) ds + s \bar{b}'_\varepsilon(B_s^{x,y}) dB_s^{x,y} \right) \Bigg].
\end{aligned}$$

Using the fact that $\exp(x) = 1 + x + \frac{1}{2}x^2 + \frac{1}{6}x^3 + O(x^4)$ gives

$$
\begin{aligned}
&\exp\left(\varepsilon\int_0^t \bar{b}_\varepsilon(B_s^{x,y})dB_s^{x,y}-\frac{\varepsilon^2}{2}\int_0^t \bar{b}_\varepsilon^2(B_s^{x,y})ds\right)\\
&\overset{\mathbb{E}}{=}1+\varepsilon\int_0^t \bar{b}_\varepsilon(B_s^{x,y})dB_s^{x,y}-\frac{\varepsilon^2}{2}\int_0^t \bar{b}_\varepsilon^2(B_s^{x,y})ds\\
&+\frac{1}{2}\varepsilon^2(\int_0^t \bar{b}_\varepsilon(B_s^{x,y})dB_s^{x,y}-\frac{\varepsilon}{2}\int_0^t \bar{b}_\varepsilon^2(B_s^{x,y})ds)^2\\
&+\frac{1}{6}\varepsilon^3(\int_0^t \bar{b}_\varepsilon(B_s^{x,y})dB_s^{x,y}-\frac{\varepsilon}{2}\int_0^t \bar{b}_\varepsilon^2(B_s^{x,y})ds)^3.
\end{aligned}
$$

Using the previous expansion technique, we expand the above expression. This gives

$$
\begin{aligned}
A_1(t)\approx&\frac{F(y)-F(x)}{\varepsilon}\left(1+\varepsilon\int_0^t \bar{b}_\varepsilon(F_{s,t}(x,y))\partial_s F_{s,t}(x,y)ds-\frac{\varepsilon^2}{2}\int_0^t \bar{b}_\varepsilon^2(F_{s,t}(x,y))ds\right.\\
&\left.-\varepsilon^3\int_0^t \bar{b}'_\varepsilon(F_{s,t}(x,y))\frac{s}{t}ds+\frac{\varepsilon^3}{2}\int_0^t \bar{b}''_\varepsilon(F_{s,t}(x,y))\frac{s(t-s)}{t}\partial_s F_{s,t}(x,y)ds\right)\\
&-\varepsilon\left(1+\varepsilon\int_0^t \bar{b}_\varepsilon(F_{s,t}(x,y))\partial_s F_{s,t}(x,y)ds\right)\\
&\times\int_0^t \left(G_\varepsilon(s,F_{s,t}(x,y))-s\bar{b}'_\varepsilon(F_{s,t}(x,y))\partial_s F_{s,t}(x,y)\right)ds.
\end{aligned}
$$

Now we need to put all these estimates together into the first order expression in $I_1^\varepsilon(t,x,y)$ defined in Theorem 2.1. That is,

$$
\begin{aligned}
&\alpha\sigma(0)^2\varepsilon^{2+\ell}\int_0^t \bar{p}_s^\varepsilon(x,0)\partial_x\bar{p}_{t-s}^\varepsilon(0,y)ds\\
\approx&\alpha\varepsilon^\ell\left(\sigma(y)^{-1}g_{\varepsilon^2 t}(|F(x)|,|F(y)|)+\varepsilon B_1+\varepsilon^2 B_2\right).
\end{aligned}
$$

Here, we have used the identity (proven in Lemma 10 in [7])

$$
\int_0^t g_{\varepsilon^2 s}(F(x),0)g_{\varepsilon^2(t-s)}(0,F(y))\frac{F(y)}{\varepsilon^2(t-s)}ds=g_{\varepsilon^2 t}(|F(x)|,|F(y)|).
$$

The following terms in the expansions can be computed and give:

$$B_1 := \int_0^t g_{\varepsilon^2 s}(F(x),0) g_{\varepsilon^2(t-s)}(0,F(y)) \frac{F(y)}{\sigma(y)} \frac{B_1^1(s)}{t-s} ds$$

$$B_1^1(s) := \int_0^s \bar{b}_\varepsilon(F_{u,s}(x,0)) \partial_u F_{u,s}(x,0) du + \int_s^t \bar{b}_\varepsilon(F_{u,t}(0,y)) \partial_u F_{u,t}(0,y) du.$$

$$B_2 := \int_0^t g_{\varepsilon^2 s}(F(x),0) g_{\varepsilon^2(t-s)}(0,F(y)) \frac{F(y) B_2^1(s) + B_2^2(s)}{\sigma(y)(t-s)} ds$$

$$B_2^1(s) := -\frac{1}{2} \int_0^s \bar{b}_\varepsilon^2(F_{u,s}(x,0)) du - \frac{1}{2} \int_s^t \bar{b}_\varepsilon^2(F_{u,t}(0,y)) du$$

$$+ \int_0^s \bar{b}_\varepsilon(F_{u,s}(x,0)) \partial_u F_{u,s}(x,0) du \int_s^t \bar{b}_\varepsilon(F_{u,t}(0,y)) \partial_u F_{u,t}(0,y) du$$

$$B_2^2(s) := \int_s^t \bar{b}_\varepsilon(F_{u,t}(0,y) - (u-s) \bar{b}'_\varepsilon(F_{u,t}(0,y)) \partial_u F_{u,t}(0,y) du.$$

In fact, with the above formulas one can also compute B_3 but as the reader can guess the expressions are longer. Still, one can see some algebraic structure in these constants. This is inherent to the parametrix approach and it will be interesting to understand this structure which is clearly linked to the so called skeleton of Wiener functionals.

As we have previously explained, our goal here is to explain the technique rather than providing full technical details. For example, one may entertain the possibility that condition **(R')** is not satisfied and expand the density with the same technique if one has an expansion of $\bar{b}_\varepsilon$ in powers of ε.

Acknowledgements The second author would like to thank the organizers of the Matrix Workshop "Mathematics of Risk" for the invitation to the conference and the opportunity to present these ideas. The research of N. Frikha has benefited from the support of the Institut Europlace de Finance. The second author was supported by KAKENHI grant 20K03666.

References

1. Bally, V., Kohatsu-Higa, A.: A probabilistic interpretation of the parametrix method. Annals of Applied Probability **25**, 3095–3138 (2015)
2. Ben Arous, G.: Développement asymptotique du noyau de la chaleur hypoelliptique hors du cut-locus. Ann. Sci. École Norm. Sup. (4) **21**(3), 307–331 (1988)
3. Ben Arous, G., Léandre, R.: Décroissance exponentielle du noyau de la chaleur sur la diagonale. II. Prob. Th. Rel. Fields **90**(3), 377–402 (1991)
4. Chen, J., Frikha, N., Li, H.: Probabilistic representation of integration by parts formulae for some stochastic volatility models with unbounded drift. ESAIM: Probability and Statistics **26**, 304–351 (2022)
5. Friedman, A.: Partial differential equations of parabolic type. Prentice-Hall (1964)
6. Frikha, N.: On the weak approximation of a skew diffusion by an Euler-type scheme. Bernoulli **24**(3), 1653–1691 (2018)
7. Frikha, N., Kohatsu-Higa, A., Li, L.: Integration by parts formula for killed processes: a point of view from approximation theory. Electron. J. Probab. **24**(95), 1–44 (2019)
8. Frikha, N., Li, L.: Weak uniqueness and density estimates for sdes with coefficients depending on some path-functionals. Ann. Inst. H. Poincaré Probab. Statist. **56**, 1002–1040 (2020)

9. Frikha, N., Li, L.: Parametrix method for the first hitting time of an elliptic diffusion with irregular coefficients. Stochastics **93**(2), 167–195 (2021). DOI 10.1080/17442508.2019.1711092. URL https://doi.org/10.1080/17442508.2019.1711092
10. Gairat, A., Shcherbakov, V.: Density of skew Brownian motion and its functionals with application in finance. Mathematical Finance **27(4)**, 1069–1088 (2017)
11. Ghomrasni, R., Peskir, G.: Local time-space calculus and extensions of Itô's formula. In: In: Hoffmann-Jørgensen, J., Wellner, J.A., Marcus, M.B. (eds) High Dimensional Probability III., *Progress in Probability*, vol. 55, p. 177–192. Birkhäuser, Basel (2004)
12. Ivasyshen, S., Medynsky, I.P.: On applications of the Levi method in the theory of parabolic equations. F Mat. Stud. **47**, 33–46 (2017)
13. Karatzas, I., Shreve, S.: Brownian motion and stochastic calculus. Second Edition, Springer Verlag (1991)
14. Kohatsu-Higa, A., Ivanenko, D., Kulik, A.: Small time chaos approximations for heat kernels of multidimensional diffusions. BIT Numerical Mathematics, To appear
15. Kohatsu-Higa, A., Taguchi, D., Zhong, J.: The Parametrix method for skew diffusions. Potential Analysis **45**, 299–329 (2016)
16. Kohatsu-Higa, A., Yuki, G.: Stochastic formulations of the parametrix method. ESAIM: Probability and Statistics **22** (2018)
17. Lejay, A.: On the constructions of the skew Brownian motion. Probab. Surv. **3**, 413–466 (2006)
18. Levi, E.: Sulle equazioni lineari totalmente ellitiche alle derivate parziali. Rend. Circ. Matem. Palermo **24**, 275–317 (1907)
19. Nualart, D.: Malliavin calculus and related topics. Springer Verlag (1995)
20. Portenko, N.I.: Generalized diffusion processes, *Translations of Mathematical Monographs*, vol. 83. American Mathematical Society, Providence, RI (1990). Translated from the Russian by H. H. McFaden
21. Watanabe, S.: Analysis of Wiener functionals (Malliavin calculus) and its applications to heat kernels. Ann. Probab. **15**(1), 1–39 (1987)

Fractional Growth Portfolio Investment

A.E. Brockwell

Abstract We review some fundamental concepts of investment from a mathematical perspective, concentrating specifically on fractional-Kelly portfolios, which allocate a fraction of wealth to a growth-optimal portfolio, with the remaining fraction collecting (or paying) interest at a risk-free rate. Using stochastic differential equations, we lay out a coherent continuous-parameter time-series framework for analysis of these portfolios, explaining relationships between Sharpe ratios, growth rates, and leverage. We see how Kelly's criterion prescribes the same leverage as Markowitz mean-variance optimization. Furthermore, for fractional Kelly portfolios, we state a simple distributional relationship between portfolio Sharpe ratio, the fractional coefficient, and portfolio log-returns. These results provide critical insight into realistic expectations of growth for different classes of investors, from individuals to quantitative trading operations. We then illustrate application of the results by analyzing performance of various bond and equity mixes for an investor. We also demonstrate how the relationships can be exploited by a simple method-of-moments calculation to estimate portfolio Sharpe ratios and levels of risk deployment, given a fund's reported returns.

1 Introduction

This paper describes quantitative relationships which are important in investment, including the concept of leverage. In doing so, it provides a range of insights that could be useful to home investors, quantitative traders, or anyone who is managing liquid investments.

We are interested in managing a block of capital which can be invested into one or more risky *instruments*, or held in cash/debt at a "risk-free" interest rate. An instrument is simply a particular type of asset that can be bought and sold, and

e-mail: anthony.brockwell@gmail.com

D. R. Wood et al. (eds.), *2021-2022 MATRIX Annals*, MATRIX Book Series 5,
https://doi.org/10.1007/978-3-031-47417-0_23

exhibits some growth (or decay) in value over time. It is typical to aim for good long-term growth rate in total capital by investing in a mix of different instruments. Below we study the question of determining the "optimal" mix. By nature of the construction of the portfolio, we ensure theoretically, that if gains/losses on capital are re-invested continuously in time, then the pool of capital will retain non-negative value. In other words, we cannot lose more than our initial block of capital. [1]

Leverage is a critical tool in optimizing capital growth. Roughly speaking, leverage is a mechanism by which we can amplify the returns of an investment. While it might seem natural that we should amplify the returns of a "good" investment as much as possible, we will see that this is not a good idea. To give a concrete example, imagine that on two successive days we observe a 10% loss followed by a 10% gain. This leaves us with a cumulative 1% loss ($0.9 \times 1.1 = 0.99$). If we were to amplify the two returns by a factor of two, however, a 20% loss followed by a 20% gain leaves us with a 4% loss ($0.8 \times 1.2 = 0.96$). Thus, somewhat counter-intuitively, doubling the size of our returns causes our cumulative loss to be be *worse* than double the original loss. This asymmetry suggests to us a potential connection between volatility, expected growth, and optimal leverage.

In the remainder of this paper, we consider this problem from a continuous-parameter time series perspective, studying the impact of leverage analytically using stochastic calculus. It is beyond the scope of this document to provide an introduction to stochastic calculus, but for a useful starting point, see, e.g. [9], or [4] for a more comprehensive treatment. To the author's knowledge, the results stated in theorem form in this paper do not appear in existing financial literature, although components of the framework itself, along with associated topics of discussion, can be found in the literature. In particular, [1, 3, 10] give much detailed discussion of Kelly portfolios and their advantages and disadvantages. Useful related information can also be found in [7]. As in [2], we will adopt a continuous-time framework based on the use of stochastic differential equations, to analyze a portfolio of instruments whose prices follow a multivariate geometric Brownian motion. [6] (and references therein) point out the importance of avoiding "bad" outcomes in such portfolios, pointing out that fractional Kelly schemes (among other solutions) are useful in this context. In this paper, we also address this problem, obtaining a specific characterization of the trade-off between growth and safety.

Acknowledgements The author is grateful to Peter Brockwell, Jonathan Baxter, and Peter Dodds for their valuable comments and suggestions, as well as an anonymous reviewer. This work has benefited substantially from their feedback.

---

[1] However, it is worth noting that this is a theoretical result, since reinvestment of profits/losses continuously in time is a practical impossibility in most cases.

2 Problem Formulation

Suppose that, at time $t = 0$, an investor has a block of capital A_0 to invest, and assume that it can be allocated to cash, which earns interest continuously at the *risk-free rate* r per unit time, or to one or more different risky instruments which generate returns over time. Over time, we will denote the total capital owned by the investor as

$$\{A_t,\ t \geq 0\}. \tag{1}$$

This quantity should be thought of as the total value of all assets, including investments in particular instruments, and cash/debt. We will assume the existence of a set of m different investment instruments (not including cash), which have price time series

$$\{P_{t,j},\ t \geq 0\},\ j = 1,2,\ldots,m. \tag{2}$$

In what follows, we will often refer to the vector of prices by

$$\mathbf{P}_t = (P_{t,1},\ldots,P_{t,m})^T. \tag{3}$$

The collection of investments in the m instruments as well as cash will be referred to as the *portfolio*. As mentioned above, the value of the portfolio at time t is denoted by A_t. Any cash component of the portfolio receives continuously-compounded interest at the risk-free rate r. It will be useful to define a vector version of the risk-free rate

$$\mathbf{r} = (r, r, \ldots, r)^T \in \mathbb{R}^m. \tag{4}$$

We define the *leverage* vector to be

$$\mathbf{k} = (k_1,\ldots,k_m)^T. \tag{5}$$

and we define *total-leverage*

$$\kappa = \sum_{j=1}^{m} k_j. \tag{6}$$

We will say that our portfolio is *non-leveraged* if $\kappa \leq 1$.

At any given time t, the investor invests fractions k_j of his/her capital A_t into the instruments with price $P_{t,j}$, holding the remainder in cash if that remainder is positive, or maintaining the required debt otherwise. Cash is assumed to earn the interest-free rate r, while debt pays interest at the same rate. By construction, the amount of cash held at time t is clearly

$$A_t(1-\kappa). \tag{7}$$

The leverage vector determines the manner in which returns on the m investment instruments are related to returns in the total capital. To state this property formally, let us define the infinitesimal return of an investment instrument by

$$dP_{t,j}/P_{t,j}. \tag{8}$$

Then define the infinitesimal return of our total capital by

$$dA_t/A_t. \tag{9}$$

Given initial capital A_0, the price processes $\{P_{t,j}\}$, $j = 1,\ldots,m$, and a leverage vector $\mathbf{k}$, we can solve for $\{A_t,\ t \geq 0\}$ using the relationship

$$dA_t/A_t = (1-\kappa)rdt + \sum_{j=1}^{m} k_j dP_{t,j}/P_{t,j} \tag{10}$$

Equation (10) determines the distribution of the process $\{A_t\}$, but it can also be viewed as a mathematical definition of leverage. The first term on the right represents the return on capital from risk-free rate interest accrual/payment on the cash/debt portion of the portfolio, while the summation represents return contributed by the leveraged investments when the fractions of A_t invested in $P_{t,j}$ are k_j, for $j = 1,\ldots,m$.

The components of our leverage vector do not necessarily have to be less than one, or add up to one, or even be non-negative.

For example, suppose that we have $m = 2$ possible investments: an S&P500 mutual fund and a long-term bond fund. (We study a case like this in more detail later in the paper.) A leverage vector of $(0.2, 0.2)$ would indicate that at any given point in time, we keep one fifth of our total capital in the S&P500 fund, one fifth in the bond fund, and the remaining three fifths held in cash, accruing interest payment at the risk-free rate. Alternately, a leverage vector of $(0.2, 1.8)$ would indicate that we would invest one fifth of initial capital in the S&P500 fund, borrow an amount equal to the initial capital, and invest the four fifths remaining initial capital, along with the borrowed sum, in the bond fund.

2.1 Objective

There are many possible investment objectives. For example, one could attempt to minimize the probability of ultimately losing all their capital, attempt to maximize the expected return relative to some benchmark, maximize some utility function of wealth, etc. In this paper we aim for good long-term growth profiles for our capital $\{A_t\}$. To state this more precisely, we will be particularly interested in the *expected log-return per unit time*

$$L = \mathbf{E}\left[\log(A_{t+\delta}/A_t)\right]/\delta \tag{11}$$

and *log-return variance per unit time*

$$V = \mathbf{Var}(\log(A_{t+\delta}/A_t))/\delta. \tag{12}$$

As we will see below, this approach leads us to a convenient stochastic calculus-based derivation of Kelly's formula, and yields additional useful insights.

Direct maximization of L (with no regard for V) yields the so-called *growth-optimal portfolio*, also referred to as the *Kelly portfolio*. In the long-run, the growth-optimal portfolio almost-surely leads to more wealth than any competing portfolio. [2] The price to pay for this growth dominance however, is relatively large up and down-swings in $\{A_t\}$. *Fractional Kelly schemes*, which simply apply a fractional multiplier to the growth-optimal portfolio leverage vector, provide one way to address this problem.

2.2 Mathematical Analysis

We are now in a position to analyze the impact of leverage choice. For the sake of exposition, we address the univariate case before developing the multivariate case.

2.2.1 Univariate Case

In the univariate ($m = 1$) case, we can write down a formal stochastic differential equation to describe the price $\{P_t\}$ of an instrument over time $t \in \mathbb{R}, t \geq 0$, as

$$dP_t = \mu P_t dt + \sigma P_t dW_t, \tag{13}$$

$$P_0 = 1, \tag{14}$$

where $\{W_t\}$ is a standard Brownian motion.

The solution to (13,14) is a geometric Brownian motion, for which

$$d\log(P_t) = (\mu - \sigma^2/2)dt + \sigma dW_t. \tag{15}$$

(This follows directly from an application of Itô's formula to (13,14).) The process has several important properties, which we will simply state here without proof.

1. It provides a realistic description of many real-life instruments that can be bought as investments. The parameters μ and σ vary, however, over different investments and arguably also over time for a particular investment.
2. $P_t > 0$.
3. The log-return of the price satisfies

$$\log(P_{t+\delta}/P_t) \sim \mathrm{N}(\mu\delta - \sigma^2\delta/2, \sigma^2\delta). \tag{16}$$

---

[2] More formally, if $\{A_t\}$ represents the capital associated with the growth-optimal portfolio, and $\{A'_t\}$ represents the capital from another portfolio in the same instruments, chosen in a different manner, then under a modest set of regularity conditions, we can show that $(A'_t/A_t) \stackrel{a.s.}{\to} 0$ as $t \to \infty$.

Now let us consider the behavior of total capital $\{A_t,\ t \geq 0\}$ over time when we invest a fraction k (leverage) in the instrument whose price P_t is given by (13,14). It follows directly from (10) that

$$dA_t = (1-k)rA_t dt + kA_t dP_t/P_t \tag{17}$$
$$= [(1-k)r + k\mu]A_t dt + kA_t\sigma dW_t. \tag{18}$$

In other words, like the underlying instrument price, our amount of capital A_t also follows a geometric Brownian motion process, but with different parameters. Consequently,

$$d\log(A_t) = [(1-k)r + k\mu - k^2\sigma^2/2]dt + k\sigma dW_t. \tag{19}$$

It is very important to note[3] that

$$d\log(A_t) \neq k \cdot d\log(P_t). \tag{20}$$

It follows directly from (19) that

1. $A_t > 0$. This is our guarantee that we do not lose more than our initial capital. However, note that it relies on the unrealistic assumption that we are able to carry out continuous reinvestment of profits/losses.
2. The log of total amount of capital (including reinvested profits/losses) satisfies

$$\log(A_{t+\delta}/A_t) \sim \mathrm{N}(r\delta + k(\mu - r)\delta - k^2\sigma^2\delta/2, k^2\sigma^2\delta). \tag{21}$$

At any time t, equation (21) provides a predictive distribution for our amount of capital at a time point $(t+\delta)$ in the future. Specifically, it is log-normal.

Following on from (21), we see that the *expected log-return per unit time* of our capital is

$$L(k) = \mathbf{E}\left[\log(A_{t+\delta}/A_t)\right]/\delta = r + k(\mu - r) - k^2\sigma^2/2. \tag{22}$$

Differentiating this with respect to k and setting to zero gives us a formula for the value of leverage k that maximizes expected log-return per unit time. This is simply

$$k^* = (\mu - r)/\sigma^2, \tag{23}$$

and the corresponding expected log-return per unit time is

$$L(k^*) = r + \frac{1}{2}(\mu - r)^2/\sigma^2. \tag{24}$$

The expression (23) is well-known, and is typically referred to as *Kelly's formula*, for Kelly's analysis of a closely-related problem (see [5]).

---

[3] It is a common mistake to presume, for example, that a triple-leverage investment yields triple the log-returns over time. This is not the case.

2.2.2 Multivariate Case

It is straightforward to extend both (22) and (23) to the multivariate case when we have multiple investments ($m > 1$), and we allocate proportions $\mathbf{k} = (k_1, k_2, \ldots, k_m)^T$ of capital to the respective investments. In this case, we define $\mu = (\mu_1, \ldots, \mu_m)^T$, $\sigma = (\sigma_1, \ldots, \sigma_m)^T$, as well as a correlation matrix $R \in \mathbb{R}^{m \times m}$. The multivariate analog of (13) becomes

$$d\mathbf{P}_t = \text{diag}(\mu)\mathbf{P}_t dt + \text{diag}(\sigma)\text{diag}(\mathbf{P}_t)d\mathbf{U}_t, \tag{25}$$

where $\{\mathbf{U}_t\}$ is a multivariate Brownian motion with correlation matrix R, so that

$$\mathbf{E}[\mathbf{U}_t] = 0, \quad \mathbf{Var}(\mathbf{U_t}) = Rt, \tag{26}$$

and diag$(\cdot)$ represents a square matrix with diagonal elements given by the vector argument, and zeros in all off-diagonal positions. It will be convenient to define

$$\Sigma = [\text{diag}(\sigma)]R[\text{diag}(\sigma)]. \tag{27}$$

Also, recall the definition $\kappa = \sum k_j = \mathbf{k} \cdot \mathbf{e}_m$, where $\mathbf{e}_m = (1, 1, \ldots, 1)^T \in \mathbb{R}^m$.

Applying (10), it is straightforward to show that there is another standard Brownian motion $\{W_t\}$ such that the total capital $\{A_t\}$ satisfies

$$\begin{aligned} dA_t &= ((1 - \mathbf{k} \cdot \mathbf{e}_m)r + \mathbf{k} \cdot \mu)A_t dt + (\mathbf{k}^T \Sigma \mathbf{k})^{1/2} A_t dW_t & (28) \\ &= (r - \mathbf{k} \cdot \mathbf{r} + \mathbf{k} \cdot \mu)A_t dt + (\mathbf{k}^T \Sigma \mathbf{k})^{1/2} A_t dW_t, & (29) \end{aligned}$$

and it follows directly from application of Itô's formula that

$$d\log(A_t) = (r + \mathbf{k} \cdot (\mu - \mathbf{r}) - \mathbf{k}^T \Sigma \mathbf{k}/2)dt + (\mathbf{k}^T \Sigma \mathbf{k})^{1/2} dW_t, \tag{30}$$

so the expected log-return per unit time is given by

$$L(\mathbf{k}) = \mathbf{E}[\log(A_{t+\delta}/A_t)]/\delta = r + \mathbf{k} \cdot (\mu - \mathbf{r}) - \mathbf{k}^T \Sigma \mathbf{k}/2, \tag{31}$$

Consequently Kelly's formula becomes

$$\mathbf{k}^* = \Sigma^{-1}(\mu - \mathbf{r}), \tag{32}$$

with maximum expected log-return per unit time

$$L(\mathbf{k}^*) = r + \frac{1}{2}(\mu - \mathbf{r})^T \Sigma^{-1}(\mu - \mathbf{r}). \tag{33}$$

We will also be interested in the variance of the log-return per unit time. From (30), this is easily seen to be

$$V(\mathbf{k}^*) = \mathbf{Var}(\log(A_{t+\delta}/A_t))/\delta = \mathbf{k}^T \Sigma \mathbf{k}. \tag{34}$$

Readers may recognize (32) as the solution of a standard Markowitz mean-variance portfolio weight selection problem [8].

2.3 Sharpe Ratios and Fractional Kelly Investment

In the finance industry, in the univariate case, it is common practice to refer to the quantity $(\mu - r)/\sigma$, as the *Sharpe ratio* of the investment. [4] We will work with the closely-related *maximal Sharpe ratio* of a single investment or a set of investments, which we define by

$$S_M = [(\mu - \mathbf{r})^T \Sigma^{-1} (\mu - \mathbf{r})]^{1/2}, \tag{35}$$

where μ and Σ are the parameters defining the growth and volatility of the component investments, and $\mathbf{r}$ is the vector whose elements are all equal to the risk-free rate. This is "maximal" in the sense that it represents the maximum Sharpe ratio over all possible linear combinations of portfolio components. It is easily verified that in the univariate case, it reduces to $S_M = |(\mu - r)/\sigma|$, which is indeed the maximum Sharpe ratio one can obtain by scaling a single investment with Sharpe ratio $(\mu - r)/\sigma$.

The maximal Sharpe S_M is crucially important, because the maximum growth rates in (24, 33) increase quadratically with respect to S_M.

We have seen that Kelly's formula prescribes the leverage that maximizes expected log-return per unit time L of an investment. However, to limit up/down-swings in capital, it is desirable to operate with a lower total level of risk. To achieve this, we can apply a multiplier to (32),

$$\mathbf{k}_\alpha = \alpha \Sigma^{-1} (\mu - \mathbf{r}), \quad \alpha \in [0, 1]. \tag{36}$$

This quantity is often referred to as *fractional Kelly* leverage. In this light, fractional Kelly investment can be viewed as a particular means of leverage selection, with α specifying the level of risk relative to that required for the growth-optimal portfolio.

The following result makes a formal connection between leverage, log-returns and Sharpe ratios.

Theorem 1 (Sharpe-Leverage Performance Profile). *Suppose that we apply fractional Kelly leverage (36) with multiplier $\alpha \in [0, 1]$ to the portfolio with prices governed by (25), earning/paying the risk-free interest rate r on the cash/debt component. Then the resulting capital process $\{A_t\}$ is a geometric Brownian motion satisfying*

$$d\log(A_t) = [r + (\alpha - \alpha^2/2) S_M^2] dt + \alpha S_M dW_t, \tag{37}$$

where $S_M = [(\mu - \mathbf{r})^T \Sigma^{-1} (\mu - \mathbf{r})]^{1/2}$ denotes the maximal Sharpe ratio of the portfolio. Consequently, $\{A_t\}$ has expected log-return per unit time

---

[4] It is important to differentiate between this quantity, and estimators thereof. Unfortunately, people often use the same term to refer to both the quantity and its estimator(s).

$$L(\mathbf{k}_\alpha) = \mathbf{E}\left[\log(A_{t+\delta}/A_t)\right]/\delta = r + (\alpha - \alpha^2/2)S_M^2 \tag{38}$$

and log-return variance per unit time

$$V(\mathbf{k}_\alpha) = \mathbf{Var}(\log(A_{t+\delta}/A_t))/\delta = \alpha^2 S_M^2. \tag{39}$$

Proof. The derivation of (30) above establishes that $\{A_t\}$ is a geometric Brownian motion. Substituting (36) into (30) and making use of the definition (35) yields (37). The remaining results then follow directly.

When the conditions of Theorem 1 are met and $\alpha = 1$, we obtain the growth-optimal portfolio. More generally, equations (38, 39) show the trade-off we obtain with different values of α. Values closer to one yield higher expected log-return per unit time on capital, but values closer to zero give better (reduced) variance of the log-return around the mean.

3 Examples and Applications

We now consider some practical implications of the theory outlined above, beginning with analysis that would be of interest to a typical individual with some savings to invest.

3.1 Equity and Bonds Mix

Let us consider a portfolio consisting of only two instruments: $\{P_{t,1}\}$ will represent the price of Vanguard 500 Mutual Fund Index shares (VFIAX), and $\{P_{t,2}\}$ will represent the price of Vanguard Long-Term Treasury Fund shares (VUSUX).

Figure 1 shows differences of logarithms of adjusted daily closing prices for these two instruments from 2002 to 2023. This time period covers the dot-com bubble of 2002, the sub-prime crisis of 2008, the more recent market turbulence at the onset of the coronavirus pandemic in early 2020, and the interest-rate reset period beginning at the end of 2021.

In what follows, we will assume that the risk-free rate r is equal to zero. This is a fairly reasonable approximation for the post-sub-prime period from 2008 to 2022, although results here can be easily adapted if r is assumed to be non-zero.

3.1.1 Parameter Estimation

From the raw data shown in the plots in Figure 1, we can easily compute estimators of $\mu = (\mu_1, \mu_2)$,

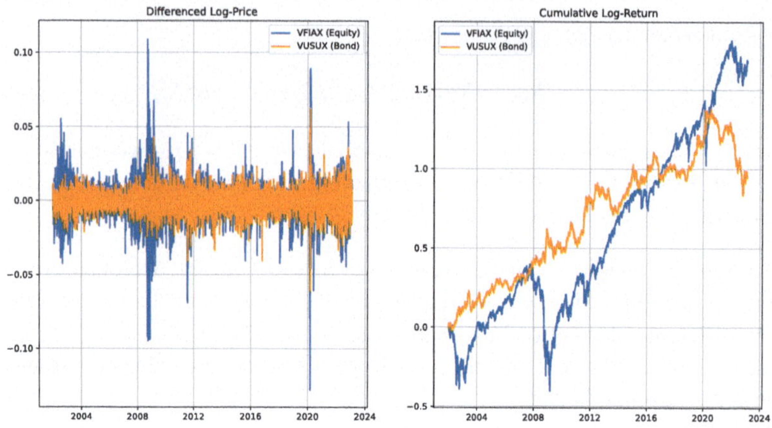

Fig. 1 Equity (VFIAX) and Bond (VUSUX) performance from Jan. 2nd, 2002 to Feb. 13th, 2023. Data source: Yahoo Finance.

Let the unit of time be one year, assume there are 260 trading days each year, and define the daily log-returns (shown in the left part of Figure 1) as

$$D_{t,j} = \log(P_{(t+1)/260,j}) - \log(P_{t/260,j}), \quad t = 1, 2, \ldots, n-1, \tag{40}$$

where n is the total number of trading days of data. We know from (16) that

$$D_{t,j} \sim \mathrm{N}((\mu_j - \sigma_j^2/2)/260, \sigma^2/260). \tag{41}$$

Furthermore,

$$\mathrm{Corr}(D_{t,1}, D_{t,2}) = R_{12}. \tag{42}$$

We can therefore take sample mean and variance from our two daily price series, compute sample correlation, and use them to construct method-of-moments estimators

$$\hat{\sigma}_j^2 = \frac{260}{n-2} \sum_{t=1}^{n-1} (D_{t,j} - \bar{D}_j)^2, \quad \bar{D}_j = \frac{1}{n-1} \sum_{t=1}^{n-1} D_{t,j}, \tag{43}$$

$$\hat{\sigma}_j = \sqrt{\hat{\sigma}_j^2}, \tag{44}$$

$$\hat{\mu}_j = 260\bar{D}_j + \hat{\sigma}_j^2/2, \tag{45}$$

$$\hat{R}_{jk} = \frac{260}{\hat{\sigma}_j \hat{\sigma}_k (n-2)} \sum_{t=1}^{n-1} (D_{t,j} - \bar{D}_j)(D_{t,k} - \bar{D}_k) \tag{46}$$

With the raw data described above, we obtain

$$\hat{\mu} = (0.102, 0.055)^T, \quad \hat{\sigma} = (0.200, 0.127)^T, \quad \hat{R}_{12} = -0.343. \tag{47}$$

From a practical perspective, we also want to consider the different impact of taxation on these two instruments. In the United States, tax on a mutual fund that tracks the S&P500 is typically close to the long-term capital gain rate of 20%, while tax on bond funds, although more complex to compute, is typically closer to standard income tax, which is often around 40%. To approximate the impact of taxation, we will therefore adjust the estimates (47) above by multiplying components of μ by one minus the corresponding tax rate, giving us final estimates

$$\hat{\mu} = (0.082, 0.033), \quad \hat{\Sigma} = \begin{bmatrix} 0.0399 & -0.0087 \\ -0.0087 & 0.0160 \end{bmatrix}. \tag{48}$$

From these estimates, in turn, we can determine the Kelly leverage for the growth-optimal portfolio by applying (32), obtaining

$$\mathbf{k}^* = \hat{\Sigma}^{-1}\hat{\mu} \simeq (2.83, 3.58)^T. \tag{49}$$

If μ and Σ were indeed the same as these estimates, and if we could borrow money to provide leverage, this says we could maximize growth over time of capital by borrowing 5.41 times our initial capital (assuming a no-interest loan) to obtain total leverage of $2.83 + 3.58 = 6.41$. We would then invest the original capital along with the borrowed capital into the two components of the portfolio, rebalancing to hold the fraction of capital in the two components constant. The resulting expected annual log-return on initial (non-borrowed) capital would be

$$\mathbf{k}^* \cdot \hat{\mu} - \mathbf{k}^{*T}\hat{\Sigma}\mathbf{k}^*/2 \simeq 0.175 \tag{50}$$

In this argument we have pretended that μ and Σ are known precisely. In fact, estimation error on these quantities can have a significant impact, but that analysis is beyond the scope of this paper.

3.1.2 Leverage Value Configurations

We now consider a range of different investment possibilities. Some of these will be applied to only one of our two instruments, and for these cases we will use $m = 1$ along with the corresponding (marginal) components of $\hat{\mu}$, $\hat{\Sigma}$ and the corresponding estimate $\hat{S}$ of Sharpe ratio. In the more interesting cases we will examine portfolios consisting of both instruments. We consider the following cases.

Cases are constructed as follows.

- *Non-leveraged/Double/Triple Equities/Bonds:* The non-leveraged approach invests all capital directly into the equities/bonds instrument. Double and triple

	Leverage	Drift/Diffusion			Kelly-Fraction
Name	k	$\hat{\mu}$	$\hat{\Sigma}$	$\hat{S}$	$\hat{\alpha}$
Single-instrument ($m = 1$) portfolios					
Non-Leveraged Equity	1.00	0.082	0.200^2	0.410	0.502
Double-Equity	2.00	0.082	0.200^2	0.410	1.003
Triple-Equity	3.00	0.082	0.200^2	0.410	1.505
Non-Leveraged Bonds	1.00	0.033	0.127^2	0.260	0.488
Double-Bonds	2.00	0.033	0.127^2	0.260	0.976
Two-instrument ($m = 2$) portfolios					
Fractional Kelly(0.31)	(0.88,1.12)	(0.082,0.033)	$\hat{\Sigma}$ (eqn 48)	0.592	0.31
Constr. Kelly(2.0)	(1.30,0.70)	(0.082,0.033)	$\hat{\Sigma}$ (eqn 48)	0.592	NA
Full-Kelly	(2.83,3.58)	(0.082,0.033)	$\hat{\Sigma}$ (eqn 48)	0.592	1.0

Table 1 Example portfolios.

versions invest all capital in the same instrument but with leverage 2 or 3, respectively.

- *Fractional Kelly (0.31):* This applies leverage $\alpha = 0.31$ times the full-Kelly leverage. In this example, it results in total leverage κ approximately equal to 2.0.
- *Constrained Kelly (2.0):* Here we choose $\mathbf{k}$ so as to maximize expected growth L subject to the constraint that $\kappa = 2$. This can be done using Lagrange multipliers, as described later in Section 5.1.
- *Full-Kelly:* This choice, given by (49) leads to maximum growth rate over time, at the cost of high variance and large draw-downs.

Table 1 also shows a term

$$\hat{\alpha} = \mathbf{k}/(\hat{\Sigma}^{-1}\hat{\mu}), \tag{51}$$

which is only defined when the leverage vector $\mathbf{k}$ is a multiple of $\hat{\Sigma}^{-1}\hat{\mu}$, and in that case represents an estimate of the ratio of portfolio leverage to growth-optimal leverage. Values less than 1.0 are desirable in this context, since it is obvious from (38) and (39) that $\alpha > 1$ implies we are taking unnecessary extra risk for a return that can also be achieved with $\alpha < 1$.

3.1.3 Simulation Results

It is straightforward to simulate hypothetical capital invested in the portfolios corresponding to the test cases. Using data ranging from Jan. 1st, 2002 to Feb. 13th, 2023, we compute the daily portfolio returns using the recorded data. In order to approximate the tax-effect on the two components, we also apply a negative drift by subtracting the tax adjustment used on μ in (48) on a daily basis from the log-returns of the data.

Figure 2 shows cumulative capital $\{A_t\}$ over time for a range of different leverage settings. The full-Kelly leverage case is omitted from the plot since it distorts

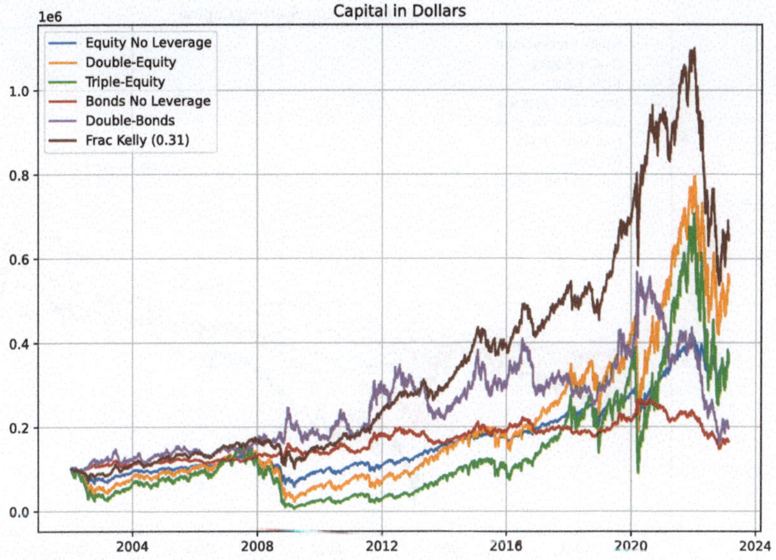

Fig. 2 Simulated post-tax capital of hypothetical portfolios built of VFIAX and VUSUX, assuming starting capital of $100000, for a range of different leverage vectors.

the scale, but the full-Kelly case is included in Figure 3, which shows the log of cumulative capital.

Table 2 shows corresponding summary statistics, including the sample (annualized) growth rate $\hat{L}$, the corresponding sample variance $\hat{V}$, as well as the final value and maximum "draw-down" over the time period, which is defined as the maximum peak-to-trough drop such that the trough occurs after the peak. From a psychological perspective, investors pay significant attention to difference between the historical maximum and the current value of an investment. Large values of draw-down typically induce great anxiety.

Several important observations can be made from the results. First, the full-Kelly case and the triple-equity case both exhibit very large maximum draw-downs over this time period - 92.4% and 95.6%, respectively. Few investors would hold onto an investment after losses that large. Both the fractional Kelly and the constrained Kelly options appear to give a fairly good compromise between greater growth and higher variance/draw-downs. Of course different investors may different preferences, but it is notable that of the choices in Table 2, apart from the non-leveraged bond case, the fractional Kelly option has the smallest value of $\hat{V}$ and still yields a respectable growth rate $\hat{L}$.

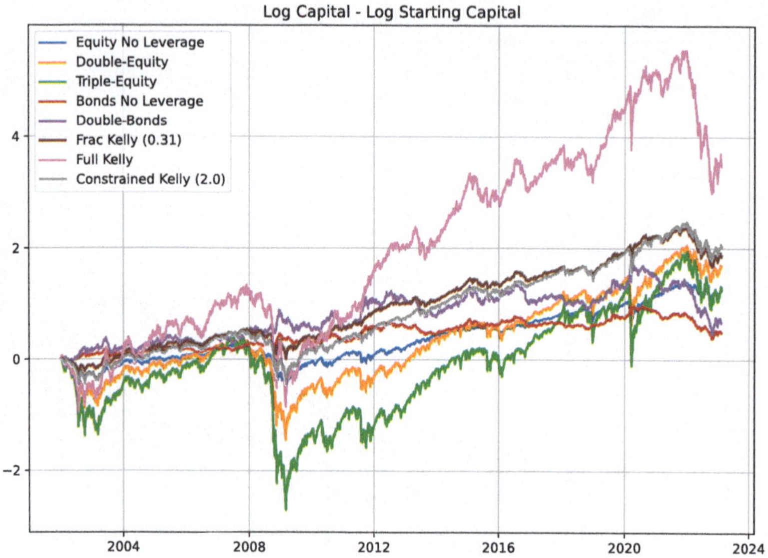

Fig. 3 Simulated post-tax log-return of hypothetical portfolios built of VFIAX and VUSUX, for a range of different leverage vectors. The full-Kelly case clearly dominates the others in terms of growth, but it also loses 92.4% of its capital value from Dec. 23rd, 2021 to Oct. 19th, 2022.

3.2 Application to Fund Evaluation

Hedge fund managers generally release very little information about the nature of their trading operations. However, they often release yearly or quarterly statments of actual returns on their hedge funds. We have already seen that Theorem 1 assists in choosing a desirable leverage vector. Here we see how it leads to a simple method-of-moments approach to estimate portfolio maximal Sharpe ratio and risk deployment of a fund, using only such publicly available return data. The approach relies on an implicit assumption that the fund is applying a fractional Kelly approach, or something similar. In light of the previous observation at the end of Subsection 2.2.2 that fractional Kelly leverage is equivalent to industry-standard Markowitz mean-variance optimization, this is not an unrealistic assumption. This "reverse-engineering" allows us to evaluate funds in a more nuanced way than simple inspection of past returns. For example, we can easily differentiate between funds that obtain high returns by taking excessive risk, and those that obtain high returns by deployment of high-Sharpe ratio portfolios.

	Growth		Draw-down			
Name	$\hat{L}$	$\hat{V}^{1/2}$	Max.	Start	End	Final Value
Single-instrument ($m = 1$) portfolios						
Non-Leveraged Equity	0.062	0.200	56.4%	Oct 8 2007	Mar 6 2009	$354,781
Double-Equity	0.083	0.401	84.6%	Oct 8 2007	Mar 6 2009	$553,393
Triple-Equity	0.063	0.605	95.6%	Jul 18 2007	Mar 6 2009	$372,336
Non-Leveraged Bonds	0.025	0.127	44.9%	Aug 3 2020	Oct 21 2022	$166,362
Double-Bonds	0.033	0.253	71.7%	Mar 6 2020	Oct 21 2022	$199,389
Two-instrument ($m = 2$) portfolios						
Fractional Kelly(0.31)	0.091	0.184	51.2%	Dec 23 2021	Oct 19 2022	$656,615
Constr. Kelly(2.0)	0.099	0.244	62.5%	Oct 8 2007	Mar 6 2009	$768,512
Full-Kelly	0.172	0.596	92.4%	Dec 23 2021	Oct 19 2022	$3,473,850

Table 2 Summary of simulated performance of example portfolios derived from the VFIAX and VUSUX instruments, simulated from Feb. 12th, 2001 to Aug. 24th, 2021, with initial capital $\$100,000$. Risk-free interest rate is assumed to be zero. $\hat{L}$ denotes the sample average of the annualized log-return, $\hat{V}^{1/2}$ denotes the annualized sample standard deviation of those log-returns, and the maximum *draw-down* is defined as the maximum drop from peak to trough with trough occurring after the peak.

Consider, for example, the annual returns of the well-known (and very strongly performing) Renaissance Medallion fund. These returns are publicly available in Appendix 1 of [11], and can easily be translated into log-returns and viewed in light of equations (38) and (39). Without re-printing the full table itself, we give summary statistics of their resulting (before-fee) log-returns in Table 3 below.

Date Range	1988-2018 inclusive
# Data Points	31
Average Log-Return	0.490
Std. Dev. (Log-Return)	0.187

Table 3 Summary statistics of Renaissance Medallion Fund returns as given in [11].

Re-arranging the pair of equations (38,39), we see that for a fractional Kelly investor,

$$\alpha = 2V(2L+V)^{-1} \tag{52}$$

$$S_M^2 = \alpha^{-1}(L+V/2). \tag{53}$$

By simply matching $L = 0.490$ and $V = 0.187^2$ from the table above, we see that the Medallion returns are consistent with deployment of a portfolio with maximal Sharpe ratio $S_M \simeq 2.72$ at approximately $\alpha = 0.068$ times the Kelly-optimal leverage. The (estimated) value of 0.068 places them comfortably below the obvious danger point. A value equal to 1.0 would indicate deployment of risk at the growth-optimal level, with its extreme volatility. Such a value is not palatable to a typical investor. Any value larger than 1.0 would indicate a sub-optimal deployment of risk.

A value larger than 2.0 would indicate that the fund will likely eventually collapse in the sense that its value will converge in probability to zero.

4 Concluding Remarks

Given a portfolio of instruments with risk, some capital A_0, and a risk-free interest rate r, we have seen that the Kelly portfolio maximizes long-term growth of capital, but with distressingly large draw-downs along the way. This undesirable quality can be mitigated by the use of fractional Kelly portfolios, at the cost of a reduction in long-term growth rate. In Theorem 1, we have directly quantified this distributional relationship, and given explicit expressions that help us to find a desirable balance between growth and variance of returns. We have also seen how, under a set of reasonable assumptions, fractional Kelly investment is the same as Markowitz mean-variance optimization.

The stochastic differential equation methodology used in this paper has been used by others in the same context, notably [2], but appears not to be widely appreciated. Within this framework, one could potentially generalize the geometric Brownian motion price models (13, 25) to jump-diffusion models, thereby allowing for the more realistic case of skewed and heavy-tailed log-returns. Using multivariate stochastic calculus, it would also possible to carry out analysis of situations where drift and/or diffusion coefficients μ and Σ are not known and must be estimated.

On the practical side, the relationships given in Theorem 1 have broad applicability. They apply equally well to simple investment of an individual's personal funds as to the more sophisticated operations of a quantitative trading organization or other fund. For example, individual investors can use the implied return distributions to assess the quality of leveraged mutual funds under various assumptions. At the more sophisticated end of the spectrum, a quantitative trading operation with a porfolio of known Sharpe ratio could use the result to determine how much capital/leverage to deploy, while remaining within distributional constraints imposed by their investors. More generally, the stochastic differential equation framework provides a solid foundation with which to address important yet-unsolved practical investment problems.

5 Supporting Results

5.1 Constrained Maximization of Expected Log-Return per Unit Time

Suppose we wish to choose the leverage vector $k \in \mathbb{R}^m$ to maximize expected log-return per unit time (restating equation (31))

$$L(\mathbf{k}) = \mathbf{k} \cdot \mu - \frac{1}{2} \mathbf{k}^T \Sigma \mathbf{k}, \tag{54}$$

subject to the constraint that total-leverage is

$$\kappa = \sum_{j=1}^{m} k_j = \mathbf{k} \cdot \mathbf{e}_m = \kappa_0, \tag{55}$$

where $\mathbf{e}_m = (1, \ldots, 1)^T \in \mathbb{R}^m$. We can express the constraint as

$$(\mathbf{k} \cdot \mathbf{e}_m - \kappa_0) = 0. \tag{56}$$

Thus the Lagrangian is

$$\mathscr{L}(\mathbf{k}) = \mathbf{k} \cdot \mu - \frac{1}{2} \mathbf{k}^T \Sigma \mathbf{k} - \lambda (\mathbf{k} \cdot \mathbf{e}_m - \kappa_0), \tag{57}$$

Differentiating with respect to the vector $\mathbf{k}$, and setting to zero, we find that the value maximizing (54) takes the form

$$\mathbf{k} = \Sigma^{-1} (\mu \quad \lambda \mathbf{e}_m). \tag{58}$$

Substituting (58) into the constraint (56), we can solve for λ, obtaining

$$\lambda = (\mathbf{e}_m^T \Sigma^{-1} \mu - \kappa_0)(\mathbf{e}_m^T \Sigma^{-1} \mathbf{e}_m)^{-1}. \tag{59}$$

Equations (58) and (59) together specify the leverage vector maximizing L subject to the required constraint.

References

1. Breiman, L.: Optimal gambling systems for favorable games. Fourth Berkeley Symposium on Probability and Statistics **1**, 65–78 (1961)
2. Browne, S.: Survival and growth with a liability: Optimal portfolio strategies in continuous time. Mathematics of Operations Research **22**(2), 468–498 (1997)
3. Hakansson, N.: Capital growth and the mean-variance approach to portfolio selection. Journal of Financial and Quantitative Analysis **26**, 857–884 (1971)
4. Karatzas, I., Shreve, S.: Brownian Motion and Stochastic Calculus. Springer (1991)
5. Kelly, J.L.: A new interpretation of information rate. Bell System Technical Journal **35**(4), 917–926 (1956)
6. MacLean, L., Sanegre, R., Zhao, Y., Ziemba, W.: Capital growth with security. Journal of Economic Dynamics and Control **28**, 937–954 (2004)
7. MacLean, L., Thorp, E., Ziemba, W. (eds.): The Kelly Capital Growth Investment Criterion, *World Scientific Handbook in Financial Economic Series*, vol. 3. World Scientific (2012)
8. Markowitz, H.: Portfolio selection. Journal of Finance **7**(1), 77–91 (1952)
9. Øksendal, B.: Stochastic Differential Equations, 6th edn. Springer (2003)
10. Thorpe, E.: Portfolio choice and the Kelly criterion. In: W. Ziemba, T. Vickson (eds.) Stochastic Optimization Models in Finance, pp. 599–619. Academic Press (1975)

11. Zuckerman, G.: The Man Who Solved the Market: How Jim Simons Launched the Quant Revolution. Penguin Random House LLC (2019)

Lévy bandits under Poissonian decision times

José-Luis Pérez and Kazutoshi Yamazaki

Abstract We consider a version of the continuous-time multi-armed bandit problem where decision opportunities arrive at Poisson arrival times, and study its Gittins index policy. When driven by spectrally one-sided Lévy processes, the Gittins index can be written explicitly in terms of the scale function, and is shown to converge to that in the classical Lévy bandit of Kaspi and Mandelbaum [8].

Key words: multi-armed bandits, Gittins index, Lévy processes, stochastic control
AMS 2020 Subject Classifications: 60G51, 93E20, 90B36

1 Introduction

The multi-armed bandit (MAB) problem is a stochastic dynamic optimization problem where at each decision time, one of J alternatives (called *arms*) is selected and a reward is then collected. In its infinite-time horizon formulation, the objective is to derive a policy (strategy) that maximizes the expected total discounted rewards. An important assumption of the MAB is that the state of each arm changes only when it is selected and the evolution of each arm is independent of others. By exploiting these features, the existing results have shown the optimality of the so-called *Gittins index* policy [6], which selects at each time the arm with the highest Gittins index. The Gittins index is defined in terms of an optimal stopping problem and can be computed separately for each arm. The state of each arm is usually assumed to evolve in a Markovian fashion, and thus the MAB can be seen as an instance of

José-Luis Pérez
Department of Probability and Statistics, Centro de Investigación en MatemáticasA.C. Calle Jalisco s/n. C.P. 36240, Guanajuato, Mexico, e-mail: jluis.garmendia@cimat.mx

Kazutoshi Yamazaki
School of Mathematics and Physics, The University of Queensland, St Lucia, Brisbane, QLD 4072, Australia e-mail: k.yamazaki@uq.edu.au

D. R. Wood et al. (eds.), *2021-2022 MATRIX Annals*, MATRIX Book Series 5,
https://doi.org/10.1007/978-3-031-47417-0_24

the Markov decision process (MDP) problem (see [22]). This optimality result is important in the MDP research in that the original J-dimensional MDP problem can be reduced to J independent 1-dimensional MDPs, providing a way to overcome the *curse of dimensionality*.

The continuous-time MAB theory has been developed in the 1990s by, for example, El Karoui and Karatzas [7] and Kaspi and Mandelbaum [8, 9]. For the setting where an arm is continuously selected, advanced stochastic analysis techniques, in particular local time analysis, are required to precisely model the problem. However, main results in the discrete-time model continue to hold in the continuous-time model. The Gittins index can be defined in a similar way, and it is optimal to choose the arm with the highest Gittins index at all times. The advantage of studying a continuous-time model is that the Gittins index can be sometimes obtained in an explicit form while numerical approaches are usually required in the discrete time case. In particular, Kaspi and Mandelbaum [8] considered the model driven by one-dimensional Lévy processes and expressed the Gittins index concisely using the Wiener-Hopf factorization and excursion theory of Lévy processes.

In this paper, we revisit the MAB problem, in particular the Lévy model of [8], and study a version where decision opportunities arrive only at independent Poisson arrival times. This problem is at the interface between the discrete-time and continuous-time models. The state of each arm evolves continuously whereas decision times are discrete. We consider two variations of the problem and study the corresponding Gittins index policy for each. We introduce cutting-edge results of the fluctuation theory of Lévy processes (e.g. [3, 11]), which had not been available when the results of [8] were developed.

The Poissonian observation model has recently drawn much attention in the stochastic control literature (see, e.g., [2, 5, 18, 19, 21, 24]), but it has not been considered in the MAB problem. With the assumption that decision opportunities arrive at exponential times, an analytical approach is still possible thanks to recently developed fluctuation-theory-based methods such as [1, 15, 20]. The Gittins index can be written in terms of the so-called *scale function* when the Lévy process has only one-sided jumps (spectrally one-sided). We further investigate its connection with [8] by showing the convergence of the Gittins index to that obtained in [8].

The rest of the paper is organized as follows. In Section 2, we review the classical discrete-time and continuous-time models of the MAB problem, as well as the Lévy model of [8]. In Section 3, we introduce the Poissonian decision time models and propose our Gittins index policy. In Section 4, we review the fluctuation theory of spectrally one-sided Lévy processes and the scale function, and then obtain the Gittins index explicitly in terms of the scale function. In Section 5, we show the convergence of our Gittins index to that in [8]. We conclude the paper in Section 6.

2 Preliminaries

2.1 Discrete-time Markovian bandits

A (multi-armed) Markovian bandit consists of $J \geq 1$ arms. For each arm labeled $j \in \mathscr{J} := \{1,\ldots,J\}$, we use an E_j-valued discrete-time Markov process $X^j = (X^j(k))_{k\geq 0}$ defined on a probability space $(\Omega^j, \mathscr{F}^j, \mathbb{P}^j)$ to model the state dynamics of arm j. These J processes are assumed mutually independent. At each time $k \geq 0$, one of the J arms is selected. A state-dependent reward is then collected and the state of the arm changes. On the other hand, the states of all the other arms remain the same.

A policy (strategy) $\pi = (\pi(k))_{k\geq 0}$ models the arm to be selected at each time $k \geq 0$. It is adapted in the sense that $\pi(k)$ must be determined based only on the information collected until time k. Let $\sigma_j^\pi(k)$ be the number of times arm $j \in \mathscr{J}$ has been selected before time $k \geq 0$ under policy π. At time k, immediately before making the decision, the state of the arms are $(X^j(\sigma_j^\pi(k)), j \in \mathscr{J})$. Using these, the decision maker needs to determine $\pi(k)$ and the reward from arm $\pi(k)$ is collected. The counters $(\sigma_j^\pi; j \in \mathscr{J})$ are updated as follows: $\sigma_j^\pi(k+1) = \sigma_j^\pi(k)+1$ if $j = \pi(k)$ while $\sigma_j^\pi(k+1) = \sigma_j^\pi(k)$ if $j \neq \pi(k)$.

The objective of the MAB problem is to obtain an optimal policy that maximizes the total expected reward over an infinite-time horizon. Given a discount factor $0 < \beta < 1$ and a reward function $R^j : E_j \to \mathbb{R}$ for each $j \in \mathscr{J}$, the problem is to compute the value function

$$v(\mathbf{x}) = \sup_\pi \mathbb{E}_\mathbf{x}\left[\sum_{k=0}^{\infty} \beta^k R^{\pi(k)}(X^{\pi(k)}(\sigma^\pi_{\pi(k)}(k)))\right], \tag{1}$$

for $\mathbf{x} = (x^1,\ldots,x^J) \in E_1 \times \cdots \times E_J$ and obtain an optimal strategy π^*. Here, $\mathbb{E}_\mathbf{x}$ is the conditional expectation when $X^j(0) = x^j$ for $j \in \mathscr{J}$.

In this problem, there exists a simple policy, known as the Gittins index policy, that works optimally in every instance of the MAB problem. It is optimal to choose the arm with the highest Gittins index, i.e. to select, at each $k \geq 0$,

$$\pi^*(k) = \arg\max_{j\in\mathscr{J}} \Gamma_j(X^j(\sigma_j^\pi(k))),$$

where the Gittins index of arm j at state x is a "normalized maximal reward"

$$\Gamma_j(x) := \sup_{\tau\geq 1} \frac{\mathbb{E}_x^j\left[\sum_{k=0}^{\tau-1}\beta^k R^j(X^j(k))\right]}{\mathbb{E}_x^j\left[\sum_{k=0}^{\tau-1}\beta^k\right]}, \tag{2}$$

which can be computed without any knowledge about the other arms. Here, τ is chosen over all stopping times greater than or equal to 1 with respect to the filtration generated by $(X^j(k))_{k\geq 0}$, and $\mathbb{E}_x^j$ is the expectation operator of the law of $\mathbb{P}_x^j$ under which $X^j(0) = x$ *assuming arm j is always selected.*

The Gittins index (2) can be derived by computing the unique value of γ such that the value of the optimal stopping problem

$$\sup_{\tau \geq 1} \mathbb{E}_x \left[\sum_{k=0}^{\tau-1} \beta^k \left(R^j(X^j(k)) - \gamma \right) \right] \tag{3}$$

becomes zero; see [9, Eqn. (20)].

2.2 Continuous-time bandits

In the continuous-time model, the state process $X^j = (X^j(t))_{t \geq 0}$, for $j \in \mathscr{J}$, is an E_j-valued continuous-time process defined on a probability space $(\Omega^j, \mathscr{F}^j, \mathbb{P}^j)$. Again, these processes are assumed mutually independent.

The evolution of the arms is given by

$$\mathbf{X} := (\mathbf{X}(\mathbf{s}) = (X^1(s_1), \ldots, X^J(s_J)), \mathbf{s} = (s_1, \ldots, s_J) \geq \mathbf{0}),$$

which is adapted to the multiparameter filtration $(\mathscr{F}(\mathbf{s}) = \mathscr{F}^1(s_1) \vee \cdots \vee \mathscr{F}^J(s_J))_{\mathbf{s} \geq 0}$ where $(\mathscr{F}^j(t))_{t \geq 0}$ is the natural filtration generated by X^j for $j \in \mathscr{J}$. For more details, we refer the reader to [8].

In the continuous-time model, a strategy is given by a J-dimensional process

$$\mathbf{T} = (\mathbf{T}(t) = (T^1(t), \ldots, T^J(t)); t \geq 0)$$

where $T^j(t)$ is the amount of time allocated to arm j until t. Naturally, $\mathbf{T}(t)$ is nondecreasing, and $T^1(t) + \cdots + T^J(t) = t$ for all $t \geq 0$. A strategy has to be adapted and hence

$$\{T^1(t) \leq s_1, \ldots, T^J(t) \leq s_J\} \in \mathscr{F}(\mathbf{s})$$

for all $t \geq 0$ and $\mathbf{s} = (s_1, \ldots, s_J) \in [0, \infty)^J$.

Given a discount factor $q > 0$ and a reward function $r^j : E_j \to \mathbb{R}$, $j \in \mathscr{J}$, the objective is to maximize the total discounted expected reward

$$v(\mathbf{x}) = \sup_{\mathbf{T}} \mathbb{E}_{\mathbf{x}} \left[\int_0^{\infty} e^{-qt} \sum_{j \in \mathscr{J}} r^j \left(X^j(T^j(t)) \right) \mathrm{d}T^j(t) \right]$$

over all allocation strategies $\mathbf{T}$. Regarding the reward function, for the problem to be well-defined, it is assumed that

$$\mathbb{E}_x^j \left[\int_0^{\infty} e^{-qt} |r^j(X^j(t))| \mathrm{d}t \right] < \infty, \quad j \in \mathscr{J}, x \in E_j \tag{4}$$

where again $\mathbb{E}_x^j$ is the expectation operator of the law $\mathbb{P}_x^j$ of X^j under which $X^j(0) = x$ assuming arm j is always selected.

The Gittins index in the continuous-time model is analogous to (2) where the summations are replaced by integrals:

$$\Gamma_j(x) := \sup_{\tau>0} \frac{\mathbb{E}_x^j\left[\int_0^\tau e^{-qt} r^j(X^j(t))\mathrm{d}t\right]}{\mathbb{E}_x^j\left[\int_0^\tau e^{-qt}\mathrm{d}t\right]}, \quad j \in \mathscr{J}, x \in E_j. \tag{5}$$

Similar to the discrete-time counterpart, it is optimal to choose the arm with the highest Gittins index, although some technical details are needed to deal with the times at which multiple Gittins indices coincide.

2.3 Lévy bandits

As a special case of the continuous-time model, in Kaspi and Mandelbaum [8], an explicit expression of the Gittins index (5) was obtained under the following assumption: for each $j \in \mathscr{J}$,

1. the state process X^j is an $\mathbb{R}$-valued Lévy process ($E_j = \mathbb{R}$);
2. the reward function r^j is increasing and continuous.

Under these assumptions, the Gittins index (5) for arm $j \in \mathscr{J}$ when its state is $x \in \mathbb{R}$ can be written concisely by

$$\Gamma_j(x) = \int_{[0,\infty)} r^j(x+y)\mu^j(\mathrm{d}y), \quad x \in \mathbb{R},$$

where μ^j is the probability measure determined by

$$\int_{[0,\infty)} e^{-\theta y}\mu^j(\mathrm{d}y) = \frac{\bar{\varphi}^j(q,0)}{\bar{\varphi}^j(q,\theta)}, \quad \theta \geq 0, \tag{6}$$

in terms of the Laplace exponent $\bar{\varphi}^j(q,\theta)$ of the inverse local time at maximum and the ascending ladder height process $(L^{-1}(t), H(t))_{t\geq 0}$ (see Section 5.1). The above characterization in terms of (6) is given in (4.5) of [8], which is derived from an alternative expression (3.3) of [8] via the Wiener-Hopf factorization.

The optimal value function can be written in terms of the Gittins index function. We have

$$v(\mathbf{x}) = \mathbb{E}_{\mathbf{x}}\left[\int_0^\infty e^{-qt} \bigvee_{j\in\mathscr{J}} \Gamma_j[\underline{X}^j(T^{*,j}(t))]\mathrm{d}t\right]$$

where $\underline{X}^j(t) := \inf_{0\leq u\leq t} X^j(u)$ is the running infimum process and $(T^{*,j}; j \in \mathscr{J})$ is the allocation time under the Gittins index policy.

3 Periodic observation models

We now consider a version of the continuous-time model where decision times arrive at exponential times. Let $N = (N(t))_{t\geq 0}$ be a Poisson process with rate $\lambda > 0$, independent of X^j for all $j \in \mathscr{J}$, with arrival times $\mathscr{T} := \{T_1, T_2, \ldots\}$ (hence $T_{k+1} - T_k$, $k \geq 0$, is an independent exponential time with mean λ^{-1} where $T_0 := 0$ for convenience). We assume the decision times are given by $\mathscr{T} \cup \{0\}$. For each T_k, $k \geq 0$, one arm is selected. If arm j is selected at T_k, its state changes as a Lévy process according to the law $\mathbb{P}^j$ until the next decision time T_{k+1}, while the states of the other arms remain the same during the period.

We consider two variations of the problem when X^j is a one-dimensional Lévy process for each $j \in \mathscr{J}$.

3.1 Problem 1

With a discount factor $q > 0$ and a reward function $R^j : \mathbb{R} \to \mathbb{R}$, we consider the maximization problem:

$$v(\mathbf{x}) = \sup_{\pi} \mathbb{E}_{\mathbf{x}} \left[\sum_{k=0}^{\infty} e^{-qT_k} R^{\pi(k)}(X^{\pi(k)}(\sigma^{\pi}_{\pi(k)}(k))) \right],$$

which can be seen as a version of the discrete-time model (1) where the deterministic decision times $\mathbb{N} \cup \{0\}$ are replaced by the Poisson arrival times (in addition to zero) $\mathscr{T} \cup \{0\}$. Here, analogously to (1), $\pi(k) \in \mathscr{J}$ denotes the arm selected at time T_k and

$$\sigma^{\pi}_j(k) := \sum_{i=0}^{k-1} (T_{i+1} - T_i) 1_{\{\pi(i)=j\}}$$

is the amount of times arm j has been active before T_k under policy π. The selection $\pi(k)$ must be determined based only on the information collected right before making decisions, i.e.

$$T_1, \ldots, T_k \quad \text{and} \quad X^j(t), t \leq \sigma^{\pi}_j(k) \text{ for all } j \in \mathscr{J}. \tag{7}$$

Analogous to what is assumed in [8], we assume the following.

Assumption 3.1 *For each $j \in \mathscr{J}$, we assume*

1. *the reward function R^j is strictly increasing,*
2. *We have $\mathbb{E}^j_x \left[\sum_{k=0}^{\infty} e^{-qT_k} |R^j(X^j(T_k))| \right] < \infty$ for $x \in \mathbb{R}$, where $\mathbb{E}^j_x$ is the same as that in the continuous-time model in Section 2.2.*

3.2 Problem 2

As a special case of Problem 1, we also consider a variant of the continuous-time model reviewed in Section 2.2, where the change of selection can be made only at $\mathcal{T} \cup \{0\}$. At each T_k, an arm is selected and it is locked in until T_{k+1}. The difference from Problem 1 is that the reward depends on the path between the decision times. This can be modeled as a maximization problem

$$v(\mathbf{x}) = \sup_{\pi} v_\pi(\mathbf{x}),$$

with

$$v_\pi(\mathbf{x}) := \mathbb{E}_{\mathbf{x}} \left[\int_0^\infty e^{-qt} \sum_{j \in \mathcal{J}} r^j \left(X^j(S_j^\pi(t)) \right) \mathrm{d}S_j^\pi(t) \right],$$

where, for each $j \in \mathcal{J}$, $r^j : \mathbb{R} \to \mathbb{R}$ and $S_j^\pi(t)$ is the amount of time at which j has been active until time t under π. For $T_k \le t < T_{k+1}$, $S_{\pi(k)}^\pi(t) = \sigma_{\pi(k)}^\pi(k) + (t - T_k)$, and hence

$$v_\pi(\mathbf{x}) = \mathbb{E}_{\mathbf{x}} \left[\sum_{k=0}^\infty \int_{T_k}^{T_{k+1}} e^{-qt} r^{\pi(k)} \left(X^{\pi(k)}(\sigma_{\pi(k)}^\pi(k) + (t - T_k)) \right) \mathrm{d}t \right].$$

With $\mathcal{G}(T_k)$ all information observed until T_k under π (i.e. (7)) and e_λ an independent exponential random variable with mean λ^{-1}, noting that $\pi(k) \in \mathcal{G}(T_k)$ and because $T_{k+1} - T_k \sim \mathrm{e}_\lambda$ and is independent of $\mathcal{G}(T_k)$, the strong Markov property gives

$$\begin{aligned} v_\pi(\mathbf{x}) &= \mathbb{E}_{\mathbf{x}} \left[\sum_{k=0}^\infty \mathbb{E}_{\mathbf{x}} \left(\int_{T_k}^{T_{k+1}} e^{-qt} r^{\pi(k)} \left(X^{\pi(k)}(\sigma_{\pi(k)}^\pi(k) + (t - T_k)) \right) \mathrm{d}t \middle| \mathcal{G}(T_k) \right) \right] \\ &= \mathbb{E}_{\mathbf{x}} \left[\sum_{k=0}^\infty e^{-qT_k} \mathbb{E}_{\mathbf{x}} \left(\int_0^{\mathrm{e}_\lambda} e^{-qt} r^{\pi(k)} \left(X^{(\pi(k))}(\sigma_{\pi(k)}^\pi(k) + t) \right) \mathrm{d}t \middle| \mathcal{G}(T_k) \right) \right] \\ &= \mathbb{E}_{\mathbf{x}} \left[\sum_{k=0}^\infty e^{-qT_k} R^{\pi(k)}(X^{\pi(k)}(\sigma_{\pi(k)}^\pi(k))) \right], \end{aligned} \tag{8}$$

where

$$R^j(x) := \mathbb{E}_x^j \left[\int_0^{\mathrm{e}_\lambda} e^{-qt} r^j(X^j(t)) \mathrm{d}t \right] = \mathbb{E}_x^j \left[\int_0^\infty e^{-(q+\lambda)t} r^j(X^j(t)) \mathrm{d}t \right], \quad j \in \mathcal{J} \text{ and } x \in \mathbb{R}. \tag{9}$$

Hence, this is a special case of Problem 1 with this choice of the reward function.

For Problem 2, we assume below, following what are assumed in [8].

Assumption 3.2 *For each $j \in \mathcal{J}$, we assume*

1. *the reward function r^j is strictly increasing and continuous.*
2. *the integrability condition* (4) *is satisfied.*

This assumption guarantees that (9) is finite and Assumption 3.1 is fulfilled.

3.3 Gittins index

We define the Gittins index for our problems, analogously to those in the classical case. For simplicity, we fix an arm $j \in \mathscr{J}$ and omit the index in an obvious manner (e.g. $R \equiv R^j$ and $X \equiv X^j$). Here, we follow similar steps as [9, Section 3.6] to derive our Gittins index.

Recall that the Gittins index in the discrete-time case (2) is derived by considering the optimal stopping problem (3) parameterized by γ. Thus, it is natural to consider an analogous index in the Poissonian decision time settings, by considering (3) with the deterministic decision times replaced by Poisson arrival times.

For each $\gamma \in \mathbb{R}$, consider the following auxiliary optimal stopping problem

$$v(x,\gamma) = \sup_{M \geq 1} \mathbb{E}_x \left[\sum_{k=0}^{M-1} e^{-qT_k} \left(R(X(T_k)) - \gamma\right) \right],$$

where M is selected from a set of stopping times greater than or equal to 1 with respect to the filtration $\mathbb{A} := (\mathscr{A}_k)_{k \geq 0}$ with $\mathscr{A}_k := \sigma(T_k, 1 \leq i \leq k, X(t), t \leq T_k)$, i.e. we stop at T_M for an $\mathbb{A}$-stopping time M.

Lemma 1. *For $x \in \mathbb{R}$, the mapping $\gamma \mapsto v(x,\gamma)$ is strictly decreasing and convex.*

Proof. For $\overline{\gamma} > \underline{\gamma}$ and an $\mathbb{A}$-stopping time $M \geq 1$,

$$\mathbb{E}_x\Big[\sum_{k=0}^{M-1} e^{-qT_k}\left(R(X(T_k)) - \underline{\gamma}\right)\Big] \geq (\overline{\gamma} - \underline{\gamma}) + \mathbb{E}_x\Big[\sum_{k=0}^{M-1} e^{-qT_k}\left(R(X(T_k)) - \overline{\gamma}\right)\Big].$$

After maximizing both sides over M, the decreasing property is immediate.

Regarding the convexity, for $\delta \in [0,1]$, $\gamma_1, \gamma_2 \in \mathbb{R}$, and $\mathbb{A}$-stopping time M,

$$\begin{aligned} v_M(x, \delta\gamma_1 + (1-\delta)\gamma_2) &:= \mathbb{E}_x\Big[\sum_{k=0}^{M-1} e^{-qT_k}\left[R(X(T_k)) - (\delta\gamma_1 + (1-\delta)\gamma_2)\right]\Big] \\ &= \delta \mathbb{E}_x\Big[\sum_{k=0}^{M-1} e^{-qT_k}\left[R(X(T_k)) - \gamma_1\right]\Big] + (1-\delta)\mathbb{E}_x\Big[\sum_{k=0}^{M-1} e^{-qT_k}\left[R(X(T_k)) - \gamma_2\right]\Big] \end{aligned}$$

where the decomposition makes sense from Assumption 3.1(2) together with the fact that $\mathbb{E}_x\left[\sum_{k=0}^{\infty} e^{-qT_k}\right] = \mathbb{E}_x[1 + \int_{(0,\infty)} e^{-qs} \mathrm{d}N(s)] = (\lambda + q)/q < \infty$. This gives $v_M(x, \delta\gamma_1 + (1-\delta)\gamma_2) \leq \delta v(x,\gamma_1) + (1-\delta)v(x,\gamma_2)$ and hence $v(x, \delta\gamma_1 + (1-\delta)\gamma_2) \leq \delta v(x,\gamma_1) + (1-\delta)v(x,\gamma_2)$, showing the convexity.

Fix $x \in \mathbb{R}$. By Lemma 1 we have that the mapping $\gamma \mapsto v(x,\gamma)$ is continuous, and monotone convergence gives $\lim_{\gamma\downarrow-\infty} v(x,\gamma) = \infty$ and $\lim_{\gamma\uparrow\infty} v(x,\gamma) = -\infty$. Hence, $v(x,\cdot)$ has a unique root, which we denote by $\Gamma(x)$ such that

$$\begin{aligned} v(x,\gamma) > 0 &\Longleftrightarrow \gamma < \Gamma(x), \\ v(x,\gamma) = 0 &\Longleftrightarrow \gamma = \Gamma(x), \\ v(x,\gamma) < 0 &\Longleftrightarrow \gamma > \Gamma(x). \end{aligned} \tag{10}$$

We call $\Gamma(x)$ the Gittins index for our problem.

By these obtained characteristics of the optimal stopping problem, following the same arguments in Section 3.6 of [9], we derive our Gittins index in the form analogous to (2).

By Snell's optimal-stopping theory (see, e.g., [23]) we have that the stopping rule that attains $v(x,\gamma)$ is given by

$$\begin{aligned} \tau(x,\gamma) &:= \inf\{m \geq 1 : v(X(T_m),\gamma) \leq 0\} \\ &= \inf\{m \geq 1 : \Gamma(X(T_m)) \leq \gamma\}, \end{aligned} \tag{11}$$

where the latter holds by (10).

Therefore, for any $\mathbb{A}$-stopping time $M' \geq 1$,

$$\begin{aligned} 0 = v(x,\Gamma(x)) &= \sup_{M\geq 1} \mathbb{E}_x\Big[\sum_{k=0}^{M-1} e^{-qT_k}\left[R(X(T_k)) - \Gamma(x)\right]\Big] \\ &\geq \mathbb{E}_x\Big[\sum_{k=0}^{M'-1} e^{-qT_k}\left[R(X(T_k)) - \Gamma(x)\right]\Big] \\ &= \mathbb{E}_x\Big[\sum_{k=0}^{M'-1} e^{-qT_k} R(X(T_k))\Big] - \Gamma(x)\mathbb{E}_x\Big[\sum_{k=0}^{M'-1} e^{-qT_k}\Big], \end{aligned} \tag{12}$$

and thus

$$\Gamma(x) \geq \frac{\mathbb{E}_x\left[\sum_{k=0}^{M'-1} e^{-qT_k} R(X(T_k))\right]}{\mathbb{E}_x\left[\sum_{k=0}^{M'-1} e^{-qT_k}\right]}.$$

On the other hand, by (11), the inequality (12) holds with equality when M' equals

$$\tau(x,\Gamma(x)) = \inf\{m \geq 1 : \Gamma(X(T_m)) \leq \Gamma(x)\}.$$

This implies that

$$\Gamma(x) = \sup_{M\geq 1} \frac{\mathbb{E}_x\left[\sum_{k=0}^{M-1} e^{-qT_k} R(X(T_k))\right]}{\mathbb{E}_x\left[\sum_{k=0}^{M-1} e^{-qT_k}\right]} = \frac{\mathbb{E}_x\left[\sum_{k=0}^{\tau(x,\Gamma(x))-1} e^{-qT_k} R(X(T_k))\right]}{\mathbb{E}_x\left[\sum_{k=0}^{\tau(x,\Gamma(x))-1} e^{-qT_k}\right]}. \tag{13}$$

Note that both the numerator and denominator are finite by Assumption 3.1(2).

We now write (13) more explicitly. By Assumption 3.1 and because X is spatially homogeneous, $\Gamma(x)$ is strictly increasing in x and thus

$$\tau(x,\Gamma(x)) = \inf\{m \geq 1 : X(T_m) \leq x\}. \tag{14}$$

Because N is a Poisson process with arrival times $\mathscr{T} := (T_i)_{i\in\mathbb{N}}$, using

$$T_x^- := T_{\tau(x,\Gamma(x))} = \inf\{S \in \mathscr{T} : X(S) \leq x\},$$

we have

$$\begin{aligned}
&\mathbb{E}_x\left[\sum_{k=0}^{\tau(x,\Gamma(x))-1} e^{-qT_k} R(X(T_k))\right] \\
&= R(x) + \mathbb{E}_x\left[\int_{(0,T_x^-]} e^{-qs} R(X(s)) \mathrm{d}N(s)\right] - \mathbb{E}_x\left[e^{-qT_x^-} R(X(T_x^-))\right] \\
&= R(x) + \mathbb{E}_x\left[\int_{(0,T_x^-]} e^{-qs} R(X(s)) 1_{\{X(s)>x\}} \mathrm{d}N(s)\right] \\
&= R(x) + \lambda \mathbb{E}_x\left[\int_0^{T_x^-} e^{-qs} R(X(s)) 1_{\{X(s)>x\}} \mathrm{d}s\right].
\end{aligned}$$

Here, the second equality holds because if $T_i < T_x^-$ then $X(T_i) > x$ for all $i \geq 1$ whereas $X(T_x^-) \leq x$ by the definition of T_x^-. The last equality holds by compensation formula.

We hence have the following.

Theorem 1. *The Gittins index Γ for Problem 1 is*

$$\Gamma(x) = \frac{R(x) + \lambda \mathbb{E}_x\left[\int_0^{T_x^-} e^{-qs} R(X(s)) 1_{\{X(s)>x\}} \mathrm{d}s\right]}{1 + \lambda \mathbb{E}_x\left[\int_0^{T_x^-} e^{-qs} 1_{\{X(s)>x\}} \mathrm{d}s\right]}, \quad x \in \mathbb{R}. \tag{15}$$

The Gittins index for Problem 2 can be written by substituting (9) in (15).

Remark 1. As $\lambda \to 0$, the Gittins index $\Gamma(x)$ converges to the instantaneous reward $R(x)$ for all $x \in \mathbb{R}$. In particular, it converges to the q-resolvent $\mathbb{E}_x\left[\int_0^\infty e^{-qt} r(X(t)) \mathrm{d}t\right]$ for Problem 2. These results are consistent because, as λ goes to zero, the future rewards vanish.

4 Spectrally one-sided cases

In this section, we focus on the spectrally negative and positive cases and compute the Gittins index (15) in terms of the scale function. Recall that a spectrally negative

Lévy process is a Lévy process without positive jumps that does not have monotone paths a.s. The spectrally positive Lévy process is its dual process.

4.1 Review of fluctuation theory for spectrally negative Lévy processes

We first review the fluctuation theory and scale function of the spectrally negative Lévy process.

Let $(Y(t))_{t\geq 0}$ be a spectrally negative Lévy process defined on a probability space $(\Omega, \mathscr{F}, \mathbb{P})$. We denote its Laplace exponent by

$$\psi(\theta) := \log \mathbb{E}\left[e^{\theta Y(1)}\right], \qquad \theta \geq 0, \tag{16}$$

which is known to be convex on $[0,\infty)$ and admits its right-inverse

$$\Phi(q) := \sup\{s \geq 0 : \psi(s) = q\}, \quad q \geq 0. \tag{17}$$

For $y \in \mathbb{R}$, let $\mathbb{P}_y$ be the conditional probability under which $Y(0) = y$ and $\mathbb{E}_y$ be its expectation operator. We omit the subscript when $y = 0$.

Fix $q \geq 0$. The q-scale function $W^{(q)}$ is a mapping from $\mathbb{R}$ to $[0,\infty)$ that takes value zero on $(-\infty, 0)$, while on $[0,\infty)$ it is a continuous and strictly increasing function with the Laplace transform

$$\int_0^\infty e^{-\theta x} W^{(q)}(x)\mathrm{d}x = \frac{1}{\psi(\theta) - q}, \quad \theta > \Phi(q). \tag{18}$$

Define also the *second scale function*

$$Z^{(q)}(x;\theta) := e^{\theta x}\left(1 + (q - \psi(\theta)) \int_0^x e^{-\theta z} W^{(q)}(z)\mathrm{d}z\right), \quad x \in \mathbb{R},\, \theta \geq 0.$$

In particular, for $x \in \mathbb{R}$ and $\lambda > 0$,

$$Z^{(q)}(x;\Phi(q+\lambda)) = e^{\Phi(q+\lambda)x}\left(1 - \lambda \int_0^x e^{-\Phi(q+\lambda)z} W^{(q)}(z)\mathrm{d}z\right),$$

which can also be written by (18),

$$\begin{aligned} Z^{(q)}(x;\Phi(q+\lambda)) &= e^{\Phi(q+\lambda)x} \lambda \int_x^\infty e^{-\Phi(q+\lambda)z} W^{(q)}(z)\mathrm{d}z \\ &= \lambda \int_0^\infty e^{-\Phi(q+\lambda)z} W^{(q)}(z+x)\mathrm{d}z. \end{aligned} \tag{19}$$

There are a number of applications of the scale function. For example, the q-resolvent can be written

$$\mathbb{E}_y\left[\int_0^\infty e^{-qs} 1_A(Y(s))\mathrm{d}s\right] = \int_{\mathbb{R}} 1_A(y+u) g^{(q)}(u)\mathrm{d}u, \tag{20}$$

for any Borel set A on $\mathbb{R}$ where

$$g^{(q)}(u) := \frac{e^{-\Phi(q)u}}{\psi'(\Phi(q))} - W^{(q)}(-u), \qquad u \in \mathbb{R}; \tag{21}$$

see [10, Theorem 2.7 (iv)].

The joint Laplace transform of the first passage time

$$\tau_0^- := \inf\{t > 0 : Y(t) < 0\} \tag{22}$$

and the overshoot $Y(\tau_0^-)$ is given by the following identity

$$\begin{aligned} H^{(q)}(y;\theta) &:= \mathbb{E}_y\left[e^{-q\tau_0^- + \theta Y(\tau_0^-)} \mathbf{1}_{\{\tau_0^- < \infty\}}\right] \\ &= Z^{(q)}(y;\theta) - \frac{\psi(\theta) - q}{\theta - \Phi(q)} W^{(q)}(y), \quad y \in \mathbb{R},\ \theta \geq 0; \end{aligned} \tag{23}$$

see, e.g., Eqn. (4.5) in [12]. In particular,

$$H^{(q)}(y;\Phi(q+\lambda)) = Z^{(q)}(y;\Phi(q+\lambda)) - \frac{\lambda}{\Phi(q+\lambda) - \Phi(q)} W^{(q)}(y), \quad y \in \mathbb{R}.$$

Similar results have been obtained for the Poisson observation case. Recall that $\mathscr{T} := (T_n)_{n \geq 1}$ is the set of jump times of an independent Poisson process with rate λ. We define for $z \in \mathbb{R}$

$$\tilde{T}_z^- := \inf\{S \in \mathscr{T} : Y(S) < z\} \quad \text{and} \quad \tilde{T}_z^+ := \inf\{S \in \mathscr{T} : Y(S) > z\}.$$

Using [19, Theorem B.1] and [13, Theorem 3.1], we have the following.

Lemma 2. *Let A be a Borel set on $\mathbb{R}$.*

(1) We have

$$\mathbb{E}\left[\int_0^{\tilde{T}_0^-} e^{-qt} 1_A(Y(t))\mathrm{d}t\right] = \frac{\Phi(q+\lambda) - \Phi(q)}{\lambda} \int_A H^{(q+\lambda)}(-u;\Phi(q))\mathrm{d}u.$$

(2) We have

$$\mathbb{E}\left[\int_0^{\tilde{T}_0^+} e^{-qt} 1_A(-Y(t))\mathrm{d}t\right] = \frac{\Phi(q+\lambda) - \Phi(q)}{\lambda} \int_A H^{(q)}(u;\Phi(q+\lambda))\mathrm{d}u.$$

4.2 Gittins index for spectrally one-sided Lévy processes

We now express the Gittins index (15) in terms of the scale function for both spectrally negative and positive cases.

Proposition 1. *Suppose X is a spectrally negative Lévy process with its Laplace exponent ψ and scale function $W^{(q)}$ as defined in Section 4.1.*

1. The Gittins index (15) *for Problem 1 can be written, for $x \in \mathbb{R}$,*

$$\Gamma(x) = \frac{\Phi(q)}{\Phi(q+\lambda)} \left(R(x) + (\Phi(q+\lambda) - \Phi(q)) \int_0^\infty R(x+y) e^{-\Phi(q)y} \mathrm{d}y \right). \quad (24)$$

2. The Gittins index (15) *for Problem 2 can be written, for $x \in \mathbb{R}$,*

$$\Gamma(x) = \frac{\Phi(q)}{\Phi(q+\lambda)} \left[\int_{\mathbb{R}} r(x+u) g^{(q+\lambda)}(u) \mathrm{d}u \right.$$
$$\left. + (\Phi(q+\lambda) - \Phi(q)) \int_0^\infty e^{-\Phi(q)y} \int_{\mathbb{R}} r(x+y+u) g^{(q+\lambda)}(u) \mathrm{d}u \mathrm{d}y \right], \quad (25)$$

where $g^{(q+\lambda)}$ is defined as in (21) *(with q replaced by $q+\lambda$).*

Proof. (1) By Lemma 2(1) with $Y = X$ and because $H^{(q+\lambda)}(-y; \Phi(q)) = e^{-\Phi(q)y}$ for all $y \geq 0$,

$$\mathbb{E}_x \left[\int_0^{T_x^-} e^{-qs} R(X(s)) 1_{\{X(s)>x\}} \mathrm{d}s \right] = \mathbb{E} \left[\int_0^{T_0^-} e^{-qs} R(X(s)+x) 1_{\{X(s)>0\}} \mathrm{d}s \right]$$
$$= \frac{\Phi(q+\lambda) - \Phi(q)}{\lambda} \int_0^\infty R(x+y) e^{-\Phi(q)y} \mathrm{d}y.$$

By replacing $R(\cdot)$ with 1, we also have $\mathbb{E}_x \left[\int_0^{T_x^-} e^{-qs} 1_{\{X(s)>x\}} \mathrm{d}s \right] = (\Phi(q+\lambda) - \Phi(q))/(\lambda \Phi(q))$. Substituting these in (15), we have (24).

(2) By (9) and (20), we have $R(x) = \int_{\mathbb{R}} r(x+u) g^{(q+\lambda)}(u) \mathrm{d}u$, which is finite by Assumption 3.2. Substiuting this in (24), we have (25).

Proposition 2. *Suppose X is a spectrally positive Lévy process whose dual process $-X$ is a spectrally negative Lévy process with its Laplace exponent ψ and scale function $W^{(q)}$.*

1. The Gittins index for Problem 1 is given by, for $x \in \mathbb{R}$,

$$\Gamma(x) = \frac{q\Phi(q+\lambda)}{(\lambda+q)\Phi(q)} \left(R(x) + (\Phi(q+\lambda) - \Phi(q)) \int_0^\infty R(x+y) H^{(q)}(y; \Phi(q+\lambda)) \mathrm{d}y \right). \quad (26)$$

2. *The Gittins index for Problem 2 is given by, for* $x \in \mathbb{R}$,

$$\Gamma(x) = \frac{q\Phi(q+\lambda)}{(\lambda+q)\Phi(q)} \Big(\int_{\mathbb{R}} r(x-u) g^{(q+\lambda)}(u) \mathrm{d}u \\ +(\Phi(q+\lambda)-\Phi(q)) \int_0^\infty H^{(q)}(y;\Phi(q+\lambda)) \int_{\mathbb{R}} r(x+y-u) g^{(q+\lambda)}(u) \mathrm{d}u \mathrm{d}y \Big). \tag{27}$$

Proof. (1) By Lemma 2(2) with $Y = -X$,

$$\begin{aligned} &\mathbb{E}_x \left[\int_0^{T_x^-} e^{-qs} R(X(s)) 1_{\{X(s)>x\}} \mathrm{d}s \right] \\ &= \mathbb{E} \left[\int_0^{\tilde{T}_0^+} e^{-qs} R(-Y(s)+x) 1_{\{-Y(s)>0\}} \mathrm{d}s \right] \\ &= \frac{\Phi(q+\lambda)-\Phi(q)}{\lambda} \int_0^\infty R(x+y) H^{(q)}(y;\Phi(q+\lambda)) \mathrm{d}y, \end{aligned}$$

which is finite by Assumption 3.1(2). On the other hand,

$$\mathbb{E}_x \left[\int_0^{T_x^-} e^{-qs} 1_{\{X(s)>x\}} \mathrm{d}s \right] = \mathbb{E}_x \left[\int_0^{T_x^-} e^{-qs} \mathrm{d}s \right] - \mathbb{E}_x \left[\int_0^{T_x^-} e^{-qs} 1_{\{X(s)\le x\}} \mathrm{d}s \right]. \tag{28}$$

Here, by identity (62) in [16],

$$\begin{aligned} \mathbb{E}_x \left[\int_0^{T_x^-} e^{-qs} \mathrm{d}s \right] \frac{1}{q} \left(1 - \mathbb{E}_x \left[e^{-qT_x^-} \right] \right) &= \frac{1}{q} \left(1 - \frac{\Phi(q+\lambda)-\Phi(q)}{\Phi(q+\lambda)} \right) \\ &= \frac{1}{q} \frac{\Phi(q)}{\Phi(q+\lambda)}, \end{aligned}$$

and, by Lemma 2(2) with $Y = -X$ and $A = (-\infty, 0]$,

$$\begin{aligned} \mathbb{E}_x \left[\int_0^{T_x^-} e^{-qs} 1_{\{X(s)\le x\}} \mathrm{d}s \right] &= \mathbb{E} \left[\int_0^{\tilde{T}_0^+} e^{-qs} 1_{\{-Y(s)\le 0\}} \mathrm{d}s \right] \\ &= \int_{-\infty}^0 \frac{\Phi(q+\lambda)-\Phi(q)}{\lambda} H^{(q)}(y,\Phi(q+\lambda)) \mathrm{d}y \\ &= \int_0^\infty \frac{\Phi(q+\lambda)-\Phi(q)}{\lambda} e^{-\Phi(q+\lambda)y} \mathrm{d}y \\ &= \frac{\Phi(q+\lambda)-\Phi(q)}{\Phi(q+\lambda)\lambda}. \end{aligned}$$

Substituting these in (28) gives

$$1+\lambda\mathbb{E}_x\left[\int_0^{T_x^-} e^{-qs}1_{\{X(s)>x\}}\mathrm{d}s\right] = 1+\lambda\left(\frac{1}{q}\frac{\Phi(q)}{\Phi(q+\lambda)} - \frac{\Phi(q+\lambda)-\Phi(q)}{\Phi(q+\lambda)\lambda}\right)$$
$$= \frac{(\lambda+q)\Phi(q)}{q\Phi(q+\lambda)}.$$

Substituting these we have (26).

(2) By (9) and (20),

$$R(x) = \mathbb{E}\left[\int_0^\infty e^{-(q+\lambda)s} r(X(s)+x)\mathrm{d}s\right] = \mathbb{E}\left[\int_0^\infty e^{-(q+\lambda)s} r(-Y(s)+x)\mathrm{d}s\right]$$
$$= \int_{\mathbb{R}} r(x-u) g^{(q+\lambda)}(u)\mathrm{d}u,$$

which is finite by Assumption 3.2(2). Substiuting this in (26), we have (27).

5 Convergence of the Gittins index as $\lambda \to \infty$

In this section, we show the convergence of the Gittins indices obtained in Propositions 1 and 2 to those in the classical Lévy case [9].

5.1 Classical Lévy bandit case

We first review and obtain a characterization of the Gittins index of [9]. Let Y be a spectrally negative Lévy process and use the same notations used in Section 4.1.

For the spectrally negative case, the local time at maximum can be selected to be $L(t) = \overline{Y}(t) := \sup_{0\le u\le t} Y(u)$ for $t \geq 0$. Then, the ascending ladder height process becomes $(L^{-1}(t), H(t))_{t\geq 0}$ where

$$L^{-1}(t) := \inf\{s > 0 : \overline{Y}(s) > t\} = \inf\{s > 0 : Y(s) > t\},$$

and because Y does not jump upwards,

$$H(t) := Y(L^{-1}(t)) = t,$$

(see [11, Section 6.5.2]). As in [11, (6.34)], the pair $(L^{-1}(t), H(t))_{t\geq 0}$ becomes a two-dimensional subordinator with its Laplace exponent

$$\mathbb{E}[e^{-qL^{-1}(1)-\theta H(1)} 1_{\{1<L(\infty)\}}] = e^{-\kappa(q,\theta)} \tag{29}$$

where

$$\kappa(q,\theta) := \Phi(q) + \theta, \quad q,\theta \geq 0.$$

By the Wiener-Hopf factorization, with $\widehat{L}$ and $(\widehat{L}^{-1},\widehat{H})$ those for the dual process $-Y$ (called the descending ladder height process)

$$\mathbb{E}[e^{-q\widehat{L}^{-1}(1)-\theta\widehat{H}(1)}1_{\{1<\widehat{L}(\infty)\}}] = e^{-\widehat{\kappa}(q,\theta)} \tag{30}$$

with

$$\widehat{\kappa}(q,\theta) := \frac{q-\psi(\theta)}{\Phi(q)-\theta}, \quad q,\theta \geq 0;$$

see [11, Eqn. (6.35)].

Using these, the Laplace transform (6) that characterizes the Gittins index in the classical Lévy model can be written explicitly.

(1) Suppose X is a spectrally negative Lévy process. By (29) for $Y = X$,

$$\bar{\varphi}(q,\theta) = \kappa(q,\theta) = \Phi(q) + \theta, \quad q,\theta \geq 0,$$

and thus (6) becomes

$$\int_{[0,\infty)} e^{-\theta y}\mu(\mathrm{d}y) = \frac{\Phi(q)}{\Phi(q)+\theta}, \quad \theta \geq 0. \tag{31}$$

(2) Suppose X is a spectrally positive Lévy process. Let ψ be the Laplace exponent of its dual (spectrally negative) Lévy process and Φ its right inverse. By (30) for $Y = -X$,

$$\bar{\varphi}(q,\theta) = \widehat{\kappa}(q,\theta) = \frac{q-\psi(\theta)}{\Phi(q)-\theta}, \quad \theta \geq 0,$$

and (6) becomes

$$\int_{[0,\infty)} e^{-\theta y}\mu(\mathrm{d}y) = \frac{q}{\Phi(q)}\frac{\theta-\Phi(q)}{\psi(\theta)-q}, \quad \theta \geq 0. \tag{32}$$

5.2 Convergence of the Gittins index

Suppose X is spectrally negative. For Problem 1, the Gittins index (24) can be written

$$\Gamma(x) = \int_{[0,\infty)} R(x+y)\mu_1^\lambda(\mathrm{d}y), \quad x \in \mathbb{R}, \tag{33}$$

with

$$\mu_1^\lambda(\mathrm{d}y) := \frac{\Phi(q)}{\Phi(q+\lambda)}\Big(\delta_0(\mathrm{d}y) + 1_{\{y>0\}}(\Phi(q+\lambda)-\Phi(q))e^{-\Phi(q)y}\mathrm{d}y\Big), \quad y \geq 0,$$

where δ_0 is the Dirac measure at zero.

For Problem 2, the (normalized) Gittins index (25) can be written

$$(q+\lambda)\Gamma(x) = \frac{(q+\lambda)\Phi(q)}{\Phi(q+\lambda)} \left[\int_{\mathbb{R}} r(x+u) g^{(q+\lambda)}(u) \mathrm{d}u \right. \tag{34}$$

$$\left. + (\Phi(q+\lambda) - \Phi(q)) \int_0^\infty \int_{\mathbb{R}} r(x+w) e^{-\Phi(q)y} g^{(q+\lambda)}(w-y) \mathrm{d}w \mathrm{d}y \right] \tag{35}$$

$$= \int_{\mathbb{R}} r(x+u) \mu_2^\lambda(\mathrm{d}u), \quad x \in \mathbb{R}, \tag{36}$$

where

$$\mu_2^\lambda(\mathrm{d}u) := \frac{(q+\lambda)\Phi(q)}{\Phi(q+\lambda)} \left[g^{(q+\lambda)}(u) + (\Phi(q+\lambda) - \Phi(q))(f^{(q)} \star g^{(q+\lambda)})(u) \right] \mathrm{d}u, \quad u \in \mathbb{R}.$$

Here,

$$(f^{(q)} \star g^{(q+\lambda)})(u) := \int_0^\infty e^{-\Phi(q)y} g^{(q+\lambda)}(u-y) \mathrm{d}y$$

which is a convolution of $f^{(q)}(u) := e^{-\Phi(q)u} 1_{\{u>0\}}$ and $g^{(q+\lambda)}$ as defined in (21).

In (36), we consider a normalized version because $R(\cdot)$ in (9) depends on λ. This normalization is appropriate in view of Remark 2(2) below.

Remark 2. 1. By setting $R \equiv 1$ in (33), we have $1 = \Gamma(x) = \mu_1^\lambda([0,\infty))$ and hence μ_1^λ is a probability measure.
2. When $r \equiv 1$, $R = (q+\lambda)^{-1}$ (see (9)) and thus $\Gamma(x) \equiv (q+\lambda)^{-1}$. Substituting these in (36), we have $\mu_2^\lambda(\mathbb{R}) = 1$, and thus μ_2^λ is a probability measure.

Proposition 3. *When X is spectrally negative, the measures μ_1^λ and μ_2^λ converge weakly to μ (defined by (31)) as $\lambda \to \infty$.*

Proof. Proof for Problem 1. For $\theta \geq 0$, because

$$\int_{[0,\infty)} e^{-\theta y} [\delta_0(\mathrm{d}y) + (\Phi(q+\lambda) - \Phi(q)) 1_{\{y>0\}} e^{-\Phi(q)y} \mathrm{d}y]$$
$$= 1 + \frac{\Phi(q+\lambda) - \Phi(q)}{\Phi(q)+\theta} = \frac{\theta + \Phi(q+\lambda)}{\Phi(q)+\theta},$$

the Laplace transform of μ_1^λ becomes

$$\int_{[0,\infty)} e^{-\theta y} \mu_1^\lambda(\mathrm{d}y) = \frac{\Phi(q)}{\Phi(q+\lambda)} \frac{\theta + \Phi(q+\lambda)}{\Phi(q)+\theta}, \tag{37}$$

which converges to (31) as $\lambda \to \infty$. By the the continuity theorem for Laplace transforms, the weak convergence holds.

Proof for Problem 2. Because μ_2^λ has a support $\mathbb{R}$, we consider Fourier transforms. To this end, define the characteristic exponent of $X = Y$ by $\Psi(\theta) = -\psi(\mathrm{i}\theta)$ such that

$$\mathbb{E}[e^{\mathrm{i}\theta X(s)}] = e^{-\Psi(\theta)s}, \quad \theta \in \mathbb{R}, s \geq 0. \tag{38}$$

Fix $\theta \in \mathbb{R}$. Using (20) and (38), the Fourier transform of $g^{(q+\lambda)}$ becomes

$$\int_{\mathbb{R}} e^{\mathrm{i}\theta y} g^{(q+\lambda)}(y)\mathrm{d}y = \mathbb{E}\left[\int_0^\infty e^{-(q+\lambda)s} e^{\mathrm{i}\theta X(s)}\mathrm{d}s\right] \\ = \int_0^\infty e^{-(q+\lambda)s} e^{-\Psi(\theta)s}\mathrm{d}s = \frac{1}{\Psi(\theta)+(q+\lambda)}. \tag{39}$$

On the other hand, the Fourier transform of $f^{(q)}$ becomes

$$\int_{\mathbb{R}} e^{\mathrm{i}\theta y} f^{(q)}(y)\mathrm{d}y = \int_0^\infty e^{\mathrm{i}\theta y} e^{-\Phi(q)y}\mathrm{d}y = (\Phi(q)-\mathrm{i}\theta)^{-1}.$$

By these and the convolution theorem,

$$\int_{\mathbb{R}} e^{\mathrm{i}\theta y}(f^{(q)} \star g^{(q+\lambda)})(y)\mathrm{d}y = \frac{1}{(\Psi(\theta)+(q+\lambda))(\Phi(q)-\mathrm{i}\theta)}.$$

This together with (39) gives

$$\int_{\mathbb{R}} e^{\mathrm{i}\theta y}\mu_2^\lambda(\mathrm{d}y) = \frac{\Phi(q)}{\Phi(q+\lambda)}\frac{(q+\lambda)}{\Psi(\theta)+(q+\lambda)}\left(1+\frac{\Phi(q+\lambda)-\Phi(q)}{\Phi(q)-\mathrm{i}\theta}\right) \\ = \frac{\Phi(q+\lambda)-\mathrm{i}\theta}{\Phi(q+\lambda)}\frac{(q+\lambda)}{\Psi(\theta)+(q+\lambda)}\frac{\Phi(q)}{\Phi(q)-\mathrm{i}\theta}, \tag{40}$$

which converges as $\lambda \to \infty$ to $\int_{[0,\infty)} e^{\mathrm{i}\theta y}\mu(\mathrm{d}y) = \Phi(q)/(\Phi(q)-\mathrm{i}\theta)$, matching (31) (with θ replaced by $-\mathrm{i}\theta$). By the continuity theorem, the weak convergence holds.

Remark 3. (1) For Problem 1, the Laplace transform (37) can be written

$$\int_{[0,\infty)} e^{-\theta y}\mu_1^\lambda(\mathrm{d}y) = \frac{\varphi_1(0,q,\lambda)}{\varphi_1(\theta,q,\lambda)} \quad \text{where } \varphi_1(\theta,q,\lambda) := \frac{\Phi(q)+\theta}{\theta+\Phi(q+\lambda)}.$$

(2) For Problem 2, the Fourier transform (40) can be written

$$\int_{\mathbb{R}} e^{\mathrm{i}\theta y}\mu_2^\lambda(\mathrm{d}y) = \frac{\varphi_2(0,q,\lambda)}{\varphi_2(\theta,q,\lambda)} \quad \text{where } \varphi_2(\theta,q,\lambda) := \frac{(\Psi(\theta)+(q+\lambda))(\Phi(q)-\mathrm{i}\theta)}{\Phi(q+\lambda)-\mathrm{i}\theta}.$$

These expressions can be seen as generalizations of the classical case (6).

We now consider the spectrally positive case. For Problem 1, the Gittins index (26) can be written

$$\Gamma(x) = \int_{[0,\infty)} R(x+y)\mu_1^\lambda(\mathrm{d}y), \quad x \in \mathbb{R},$$

with

$$\mu_1^\lambda(\mathrm{d}y) := \frac{q\Phi(q+\lambda)}{(\lambda+q)\Phi(q)}\Big(\delta_0(\mathrm{d}y) + 1_{\{y>0\}}(\Phi(q+\lambda)-\Phi(q))H^{(q)}(y,\Phi(q+\lambda))\mathrm{d}y\Big), \quad y \geq 0.$$

For Problem 2, the Gittins index (27) can be written with $\widehat{g}(u) := g(-u)$, $u \in \mathbb{R}$,

$$
\begin{aligned}
&(q+\lambda)\Gamma(x)\\
&= \frac{q\Phi(q+\lambda)}{\Phi(q)}\Bigg[\int_{\mathbb{R}} r(x-u)g^{(q+\lambda)}(u)\mathrm{d}u\\
&\quad + (\Phi(q+\lambda)-\Phi(q))\int_0^\infty\int_{\mathbb{R}} H^{(q)}(y;\Phi(q+\lambda))r(x+y-u)g^{(q+\lambda)}(u)\mathrm{d}u\mathrm{d}y\Bigg]\\
&= \frac{q\Phi(q+\lambda)}{\Phi(q)}\Bigg[\int_{\mathbb{R}} r(x+u)\widehat{g}^{(q+\lambda)}(u)\mathrm{d}u\\
&\quad + (\Phi(q+\lambda)-\Phi(q))\int_{\mathbb{R}}\int_0^\infty H^{(q)}(y;\Phi(q+\lambda))r(x+w)\widehat{g}^{(q+\lambda)}(w-y)\mathrm{d}y\mathrm{d}w\Bigg]\\
&= \int_{\mathbb{R}} r(x+u)\mu_2^\lambda(\mathrm{d}u)
\end{aligned}
$$

where

$$
\mu_2^\lambda(\mathrm{d}u) := \frac{q\Phi(q+\lambda)}{\Phi(q)}\left(\widehat{g}^{(q+\lambda)}(u) + (\Phi(q+\lambda)-\Phi(q))(f^{(q)}\star\widehat{g}^{(q+\lambda)})(u)\right)\mathrm{d}u, \quad u\in\mathbb{R},
$$

with $f^{(q)}(u) := H^{(q)}(u,\Phi(q+\lambda))1_{\{u>0\}}$.

Remark 4. Similar to Remark 2, both μ_1^λ and μ_2^λ are probability measures.

Proposition 4. *When X is spectrally positive, the measures μ_1^λ and μ_2^λ converge weakly to μ (defined by* (32)*) as $\lambda\to\infty$.*

Proof. Proof for Problem 1. In order to characterize the measure μ^λ we compute its Laplace transform. To this end, for $\theta > \Phi(q)$, using (19),

$$
\begin{aligned}
&\int_0^\infty e^{-\theta y} Z^{(q)}(y;\Phi(q+\lambda))\mathrm{d}y\\
&= \lambda\int_0^\infty e^{-\theta y}\int_y^\infty e^{-\Phi(q+\lambda)(u-y)}W^{(q)}(u)\mathrm{d}u\mathrm{d}y\\
&= \lambda\int_0^\infty e^{-\Phi(q+\lambda)u}W^{(q)}(u)\int_0^u e^{-(\theta-\Phi(q+\lambda))y}\mathrm{d}y\mathrm{d}u\\
&= \frac{\lambda}{\Phi(q+\lambda)-\theta}\int_0^\infty e^{-\Phi(q+\lambda)u}W^{(q)}(u)\left(e^{(\Phi(q+\lambda)-\theta)u}-1\right)\mathrm{d}u\\
&= \frac{\lambda}{\Phi(q+\lambda)-\theta}\left(\int_0^\infty e^{-\theta u}W^{(q)}(u)\mathrm{d}u - \int_0^\infty e^{-\Phi(q+\lambda)u}W^{(q)}(u)\mathrm{d}u\right)\\
&= \frac{\lambda+q-\psi(\theta)}{(\Phi(q+\lambda)-\theta)(\psi(\theta)-q)},
\end{aligned}
\tag{41}
$$

and hence, again using (18),

$$\begin{aligned}&(\Phi(q+\lambda)-\Phi(q))\int_0^{\infty} e^{-\theta y} H^{(q)}(y,\Phi(q+\lambda))\mathrm{d}y\\&=(\Phi(q+\lambda)-\Phi(q))\frac{\lambda+q-\psi(\theta)}{(\Phi(q+\lambda)-\theta)(\psi(\theta)-q)}-\frac{\lambda}{\psi(\theta)-q}\\&=\frac{(q-\psi(\theta))\Phi(q+\lambda)-(\lambda+q-\psi(\theta))\Phi(q)+\lambda\theta}{(\psi(\theta)-q)(\Phi(q+\lambda)-\theta)},\end{aligned} \tag{42}$$

which holds for any $\theta \geq 0$ by analytic continuation and is finite in view of Lemma 2(2). Thus,

$$\begin{aligned}&\int_{[0,\infty)} e^{-\theta y}\left[\delta_0(\mathrm{d}y)+1_{\{y>0\}}(\Phi(q+\lambda)-\Phi(q))H^{(q)}(y,\Phi(q+\lambda))\mathrm{d}y\right]\\&=1+\frac{(q-\psi(\theta))\Phi(q+\lambda)-(\lambda+q-\psi(\theta))\Phi(q)+\lambda\theta}{(\psi(\theta)-q)(\Phi(q+\lambda)-\theta)}\\&=\frac{(\lambda+q-\psi(\theta))(\theta-\Phi(q))}{(\psi(\theta)-q)(\Phi(q+\lambda)-\theta)}.\end{aligned}$$

Therefore,

$$\int_{[0,\infty)} e^{-\theta y}\mu_1^{\lambda}(\mathrm{d}y)=\frac{\lambda+q-\psi(\theta)}{\lambda+q}\frac{\Phi(q+\lambda)}{\Phi(q+\lambda)-\theta}\frac{\theta-\Phi(q)}{\psi(\theta)-q}\frac{q}{\Phi(q)}, \quad \theta\geq 0, \tag{43}$$

which converges to (32) as $\lambda \to \infty$. By the continuity theorem, the weak convergence holds.

Proof for Problem 2. Again let $\Psi(\theta) = -\psi(\mathrm{i}\theta)$ (where we recall ψ is for the dual spectrally negative Lévy process $Y = -X$) and hence $\mathbb{E}[e^{\mathrm{i}\theta Y(s)}] = e^{-\Psi(\theta)s}$, $\theta \in \mathbb{R}, s \geq 0$.

We have

$$\begin{aligned}\int_{\mathbb{R}} e^{\mathrm{i}\theta y}\hat{g}^{(q+\lambda)}(y)\mathrm{d}y&=\mathbb{E}\left[\int_0^{\infty} e^{-(q+\lambda)s}e^{-\mathrm{i}\theta Y(s)}\mathrm{d}s\right]\\&=\int_0^{\infty} e^{-(q+\lambda)s}e^{-\Psi(-\theta)s}\mathrm{d}s=\frac{1}{\Psi(-\theta)+(q+\lambda)}.\end{aligned} \tag{44}$$

Because (42) also holds for $\theta = 0$ and it is finite, by analytic continuation, we obtain

$$\begin{aligned}&(\Phi(q+\lambda)-\Phi(q))\int_0^{\infty} e^{\mathrm{i}\theta y} H^{(q)}(y,\Phi(q+\lambda))\mathrm{d}y\\&=\frac{(q+\Psi(-\theta))\Phi(q+\lambda)-(\lambda+q+\Psi(-\theta))\Phi(q)-i\lambda\theta}{(-\Psi(-\theta)-q)(\Phi(q+\lambda)+\mathrm{i}\theta)}.\end{aligned}$$

Together with (44) and the convolution theory for Fourier transforms,

$$\begin{aligned}
&\int_{\mathbb{R}} e^{i\theta y}\mu_2^\lambda(\mathrm{d}y) \\
&= \frac{q\Phi(q+\lambda)}{\Phi(q)}\frac{1}{\Psi(-\theta)+(q+\lambda)} \\
&\left(1+\frac{(q+\Psi(-\theta))\Phi(q+\lambda)-(\lambda+q+\Psi(-\theta))\Phi(q)-i\lambda\theta}{(-\Psi(-\theta)-q)(\Phi(q+\lambda)+\mathrm{i}\theta)}\right) \\
&= \frac{\mathrm{i}\theta+\Phi(q)}{\Phi(q)}\frac{q}{\Psi(-\theta)+q}\frac{\Phi(q+\lambda)}{\Phi(q+\lambda)+\mathrm{i}\theta},
\end{aligned} \tag{45}$$

which converges as $\lambda \to \infty$ to

$$\int_{[0,\infty)} e^{\mathrm{i}\theta y}\mu(\mathrm{d}y) = \frac{q}{\Phi(q)}\frac{\mathrm{i}\theta+\Phi(q)}{\Psi(-\theta)+q},$$

matching (32) (with θ replaced by $-\mathrm{i}\theta$). Therefore the continuity theorem shows the weak convergence.

Remark 5. (1) For Problem 1, the Laplace transform (43) can be written

$$\int_{[0,\infty)} e^{-\theta y}\mu_1^\lambda(\mathrm{d}y) = \frac{\varphi_1(0,q,\lambda)}{\varphi_1(\theta,q,\lambda)} \quad \text{where } \varphi_1(\theta,q,\lambda) := \frac{(\psi(\theta)-q)(\Phi(q+\lambda)-\theta)}{(\lambda+q-\psi(\theta))(\theta-\Phi(q))}.$$

(2) For Problem 2, the Fourier transform (45) can be written

$$\int_{\mathbb{R}} e^{\mathrm{i}\theta y}\mu_2^\lambda(\mathrm{d}y) = \frac{\varphi_2(0,q,\lambda)}{\varphi_2(\theta,q,\lambda)} \quad \text{where } \varphi_2(\theta,q,\lambda) := \frac{(\Psi(-\theta)+q)(\Phi(q+\lambda)+\mathrm{i}\theta)}{\mathrm{i}\theta+\Phi(q)}.$$

These can be seen as generalizations of (6).

Remark 6. The weak convergence as in Propositions 3 and 4 implies the convergence of the Gittins index, if $R(\cdot)$ and $r(\cdot)$ are bounded and continuous in Problems 1 and 2, respectively.

6 Concluding remarks

In this paper, we studied an extension of the MAB driven by Lévy processes [8] where decision opportunities arrive at Poisson arrival times. The Gittins index can be written analogously to the classical case (15). In particular, we obtained explicitly its form in terms of the scale function for the case of spectrally one-sided Lévy processes and showed its convergence to those in [8] as the arrival rate increases to infinity.

The studied problem is new to the best of our knowledge and there are several venues for future research.

First, in the current paper, we did not discuss the optimality of the Gittins index policy, and it is of great importance to analyze its near or exact optimality. The convergence results obtained in Section 5 suggest that it is near optimal at least when λ is high. It is a non-trivial and crucial problem to investigate whether the optimality remains to hold generally under the Poissonian decision times.

Second, one potential application of the considered problem is to approximate the optimal solutions in the classical discrete-time model (in Section 2.1) in terms of the Gittins index policy in the considered problem by approximating deterministic decision times in terms of Poisson arrival times. This is related to Carr's randomization [4], where constant discrete decision times are approximated by Erlang random variables, or the sum of independent exponential random variables. This approximation method is conjectured to be practical and accurate in view of the existing related results in stochastic control such as [14].

Third, the obtained results are conjectured to hold for more general processes. The convergence of our Gittins index to that in (15) is easily conjectured to hold for a general Lévy process with two-sided jumps. In addition, the Gittins index may admit an analytical expression written in terms of the Wiener-Hopf factorization, even when the spectrally one-sided assumption is relaxed. It is also of great interest to generalize the results to a more general Markov process with jumps by taking advantage of the recent development of excursion theory such as [17].

References

1. Albrecher, H., Ivanovs, J., Zhou, X.: Exit identities for Lévy processes observed at Poisson arrival times. Bernoulli. **22**, 1364–1382 (2016)
2. Avanzi, B., Cheung, E. C., Wong, B., Woo, J.K.: On a periodic dividend barrier strategy in the dual model with continuous monitoring of solvency. Insurance Math. Econom. **52**, 98–113 (2013)
3. Bertoin, J.: Lévy processes. Cambridge University Press, Cambridge (1996)
4. Carr, P.: Randomization and the American put. Rev. Financ. Stud. **11(3)**, 597-626 (1998)
5. Czarna, I., Palmowski, Z.: Dividend problem with Parisian delay for a spectrally negative Lévy risk process. J. Optim. Theory Appl. **161(1)**, 239-256 (2014)
6. Gittins, J.: A dynamic allocation index for the sequential design of experiments. In Progress in Statistics, 241-266 (1974)
7. El Karoui, N., Karatzas, I.: Dynamic allocation problems in continuous time. Ann. Appl. Probab. **4(2)**, 255-286 (1994)
8. Kaspi, H., Mandelbaum, A.: Multi armed bandits driven by Lévy processes. Ann. Appl. Probab. **5(2)**, 541–565 (1995)
9. Kaspi, H., Mandelbaum, A.: Multi armed bandits in discrete and continuous time. Ann. Appl. Probab. **8(4)**, 1270–1290 (1998)
10. Kuznetsov, A., Kyprianou, A.E., Rivero, V.: The theory of scale functions for spectrally negative Lévy processes. In Lévy Matters II, pp. 97–186, Springer Lecture Notes in Mathematics (2013)
11. Kyprianou, A.E.: Fluctuations of Lévy processes with applications. Springer, Berlin (2014)
12. Kyprianou, A.E., Surya, B.A.: Principles of smooth and continuous fit in the determination of endogenous bankruptcy levels. Finance Stoch. **11**, 131–152 (2007)
13. Landriault, D., Li, B., Wong, J. T., Xu D.: Poissonian potential measures for Lévy risk models, Insurance Math. Econom. **82**, 152–166 (2018)
14. Leung, T., Yamazaki, K., Zhang, H.: An analytic recursive method for optimal multiple stopping: Canadization and phase-type fitting. Int. J. Theor. Appl. Finance. **18(05)**, 1550032 (2015)

15. Lkabous, M. A.: Poissonian occupation times of spectrally negative Lévy processes with applications. Scand. Actuar. J. **2021(10)**, 916-935 (2021)
16. Mata, D., Pérez, J.L. and Yamazaki, K.: Effects of positive jumps of assets on endogenous bankruptcy and optimal capital structure: continuous- and periodic-observation models. SIAM J Financial Math. **12(3)** 1112–1149 (2021)
17. Noba, K.: Generalized scale functions of standard processes with no positive jumps. Electron. Commun. Probab. **25**, 1-12 (2020)
18. Palmowski, Z., Pérez, J. L., Yamazaki, K.: Double continuation regions for American options under Poisson exercise opportunities. Math. Financ., **31**(2), 722-771 (2021)
19. Palmowski, Z., Pérez, J. L., Surya, B. A., Yamazaki, K.: The Leland-Toft optimal capital structure model under Poisson observations, Finance Stoch., **24**, 1035–1082 (2020)
20. Pérez, J.L. and Yamazaki, K.: Mixed periodic-classical barrier strategies for Lévy risk processes. Risks **6(2)**, 33 (2018)
21. Pérez, J.L., Yamazaki, K., Bensoussan, A.: Optimal periodic replenishment policies for spectrally positive Lévy demand processes. SIAM J. Control Optim. **58(6)**, 3428–3456 (2020)
22. Puterman, M. L.: Markov decision processes: discrete stochastic dynamic programming. John Wiley & Sons (2014)
23. Snell, L.: Applications of martingale systems theorems. Trans. Amer. Math. Soc. **73**, 293-312 (1952)
24. Zhao, Y., Chen, P., Yang, H.: Optimal periodic dividend and capital injection problem for spectrally positive Lévy processes. Insurance Math. Econom., **74**, 135-146 (2017)

Entrance laws for continuous-state nonlinear branching processes coming down from infinity

Bo Li and Xiaowen Zhou

Abstract We consider a class of non-negative valued, time-changed spectrally positive Lévy processes stopped whenever hitting 0, which can be identified as continuous-state branching processes with population dependent branching rates. Given the process comes down from infinity, we find expressions for Laplace transform of the first passage time and for the potential measure for the process started from infinity. Those expressions are in terms of the generalized scale functions for the corresponding spectrally positive Lévy processes.

1 Introduction

Continuous-state branching processes (CSBPs for short) are non-negative valued Markov processes satisfying the so called branching property, i.e. given two independent CSBPs with identical branching mechanism and initial values x and y, respectively, the sum of the two processes is also a CSBP with initial value $x+y$. They are the continuous-state counterparts of the Bienaymé-Galton-Watson processes. We refer to Kyprianou [9] and Li [14] for introductions on CSBPs.

In Li [13] a class of nonlinear CSBPs with power branching rate functions is introduced as solutions to the associated SDEs driven by spectrally positive Lévy processes (SPLPs for short). Nonlinear CSBPs with more general rate functions are introduced in Foucart et al. [7] by time-changing spectrally positive Lévy processes via Lamperti type transforms. The criteria for coming down from infinity and the speed of coming down from infinity for such nonlinear CSBPs are obtained in [7].

Bo Li
School of Mathematics and LPMC, Nankai University, Tianjin, China, e-mail: libo@nankai.edu.cn

Xiaowen Zhou
Department of Mathematics and Statistics, Concordia University, Montreal, Canada e-mail: xiaowen.zhou@concordia.ca

D. R. Wood et al. (eds.), *2021-2022 MATRIX Annals*, MATRIX Book Series 5,
https://doi.org/10.1007/978-3-031-47417-0_25

Foucart et al. [6] further show that such a process can be extended to a $[0,\infty]$-values Markov process started from ∞. We refer to Bansaye et al. [1] for previous work on the speed of coming down from infinity for birth and death processes.

The Lamperti type transform allows to express the interested quantities of nonlinear CSBP in terms of the associated SPLP that often involves functional integral of the SPLP. As a result, one can study the nonlinear CSBPs using fluctuation theory for spectrally one-sided Lévy processes. In this note, we adopt the above approach to further derive some expressions for nonlinear CSBPs that come down from infinity. More precisely, for the process started at ∞ and coming down from ∞, we find expressions for the Laplace transform of the first passage time and for the potential measure where those expressions are in terms of generalized scale functions for SPLPs. These quantities also characterize the entrance law for the associated $[0,\infty)$-valued nonlinear CSBP. We also obtain expressions on the moments of the total progeny for the nonlinear CSBP.

In Section 2 we introduce the needed fluctuation results for SPLP, the definition of nonlinear CSBP using Lamperti transform on SPLP and some previous results on nonlinear CSBPs and the coming down from infinity. The main results and their proofs are presented in Section 3.

2 Spectrally positive Lévy processes and continuous-state nonlinear branching processes

2.1 Spectrally positive Lévy processes

Let ξ be a spectrally positive Lévy process, that is, a stochastic process with stationary and independent increments and without negative jumps defined on a filtered probability space $(\Omega,\mathscr{F},(\mathscr{F}_t)_{t\geq 0},\mathbb{P})$. The law of ξ for $\xi_0=x$ is denoted by $\mathbb{P}_x$ and the corresponding expectation by $\mathbb{E}_x$. We write $\mathbb{P}$ and $\mathbb{E}$ when $x=0$. The Laplace component of ξ is well defined and takes the Lévy-Khintchine form, for every $\theta\geq 0$

$$\psi(\theta):=\log\mathbb{E}\big[e^{-\theta\xi_1}\big]=\gamma\theta+\frac{\sigma^2}{2}\theta^2+\int_0^\infty(e^{-\theta z}-1+\theta z)\pi(dz), \tag{1}$$

where $\gamma\in\mathbb{R},\sigma\geq 0$ are constants and π is a σ-finite measure on $(0,\infty)$ with $\int_0^\infty z\wedge z^2\nu(dz)<\infty$. We exclude the case of ξ being a subordinator. It is well-known that ψ in (1) is continuous and strictly convex on $[0,\infty)$ with $-\psi'(0)=\mathbb{E}[\xi(1)]=-\gamma<\infty$. Its right inverse is defined by

$$\phi(s):=\sup\{\theta\geq 0:\psi(\theta)=s\}\quad\text{for}\quad s\geq 0.$$

In the fluctuation theory of spectrally negative Lévy processes, the following scale functions $W^{(q)}$ plays a key role, which is defined as the unique continuous and increasing function on $[0,\infty)$ satisfying

$$\int_0^\infty e^{-sy} W^{(q)}(y)\,dy = \frac{1}{\psi(s)-q} \quad \text{for } s > \phi(q) \text{ and } q \geq 0, \tag{2}$$

with extension $W^{(q)}(x) = 0$ for $x < 0$. We write $W(x) \equiv W^{(0)}(x)$ for $q = 0$ and refer the readers to [2] and [9] for more detailed discussions on the spectrally negative Lévy process and its scale functions. We also call $W^{(q)}$ the scale function of SPLP ξ for the obvious reason.

Define the first passage times of ξ by

$$\tau_b^+ := \inf\{t \geq 0 : \xi_t > b\} \quad \text{and} \quad \tau_a^- := \inf\{t \geq 0 : \xi_t < a\}$$

with the convention $\inf \emptyset = \infty$. Then the potential measure of ξ killed upon leaving $[a,b]$ is given by

$$\begin{aligned} \Theta^{(q)}(x,dy) &:= \int_0^\infty e^{-qt} \mathbb{P}_x(\xi_t \in dy; t < \tau_a^- \wedge \tau_b^+)\,dt \\ &= \Big(\frac{W^{(q)}(b-x)}{W^{(q)}(b-a)} W^{(q)}(y-a) - W^{(q)}(y-x)\Big)\,dy \end{aligned} \tag{3}$$

for $x, y \in (a,b)$.

Let ω be a nonnegative locally bounded Borel function and define the functional

$$\eta(t) := \int_0^t \omega(\xi_s)\,ds$$

which is called ω-weighted occupation time in Li and Palmowski [10]. We will need the following result on potential measure for process ξ.

Lemma 1 ([10] Theorem.2.2). *For $b > x, y > a$, we have*

$$\begin{aligned} \Theta^{(\omega)}(x,dy) &= \int_0^\infty \mathbb{E}_x\Big[e^{-\eta(t)}; t \leq \tau_a^- \wedge \tau_b^+, \xi_t \in dy\Big]\,dt \\ &= \Big(\frac{W^{(\omega)}(b,x)}{W^{(\omega)}(b,a)} W^{(\omega)}(y,a) - W^{(\omega)}(y,x)\Big)\,dy \end{aligned}$$

where $W^{(\omega)}(x,y)$ is defined as the unique locally bounded function satisfying

$$W^{(\omega)}(x,y) = W(x-y) + \int_y^x W^{(\omega)}(x,z)\omega(z)W(z-x)\,dz. \tag{4}$$

Remark 1. The ω-weighted scale function $W^{(\omega)}$ defined above extends the classical scale function in (3) with $W^{(\omega)}(x,x) = W(0)$ and $W^{(\omega)}(x,y) = 0$ if $x < y$. One can show that if $\omega \equiv q$, then $W^{(\omega)}(x,y) = W^{(q)}(x-y)$ for all $x, y \in \mathbb{R}$ and $\Theta^{(\omega)} = \Theta^{(q)}$ as expected.

2.2 Time-changed SPLP and nonlinear CSBP

In this paper, we are interested in the following time-changed SPLP taking values on $[0,\infty)$

$$X(t) := \xi(\eta^{-1}(t) \wedge \tau_0^-), \tag{5}$$

stopped at time $\zeta = \eta(\tau_0^-) \in [0,\infty]$ with absorbing states 0 and ∞, where

$$\eta(t) := \int_0^t \omega(\xi_s)\,ds = \int_0^t \frac{ds}{R(\xi_s)} \quad \text{for } t < \tau_0^- \text{ and } \eta^{-1}(s) := \inf\{t \geq 0 : \eta(t) > s\} \tag{6}$$

and $R(z) := \frac{1}{\omega(z)}$ is a positive and continuous function on $(0,\infty)$. The Laplace exponent ψ for ξ is called the branching mechanism for X and R is called the branching rate function. Since X is obtained by time-changing SPLP ξ, it is defined on the same probability space as ξ with $X(0) = \xi(0)$, and $\mathbb{P}_x$ can also be understood as the law of X with $X(0) = x$.

The time-changed SPLP defined above is also referred to as the Lamperti-type transformed process. It is well known that, when the function R is the identity function, X reduces to the CSBP with Ψ being the branching mechanism, and if R is an exponential function, X is related to a positive self-similar Markov process. The two transforms are called the classical Lamperti transformations in the literatures; see e.g. [9, 4].

The above positive Markov process X in (5) can also be characterized as the solution to SDE

$$X(t) = x - \gamma \int_0^t R(X_s)\,ds + \sigma \int_0^t \sqrt{R(X_s)}dB_s + \int_0^t \int_0^\infty \int_0^{R(X_{s-})} z\tilde{N}(ds,dz,du), \tag{7}$$

where B is a standard Brownian motion and $\{\tilde{N}(ds,dz,du)\}_{s,z,u>0}$ is an independent compensated Poisson random measure on $(0,\infty)^3$ with intensity $ds\nu(dz)du$, where ν is a nonnegative σ-finite measure on $(0,\infty)$ such that $\int_0^\infty (z \wedge z^2)\nu(dz) < \infty$, comparing with ψ in (1). For the above stochastic differential equation to have a unique solution, we need additional technical assumptions such as $R(\cdot)$ is locally Lipschitz on $(0,\infty)$.

Processes of this type are first introduced in [13], for power functions R, as nonlinear CSBPs that describe the evolution of populations whose reproduction rates are power functions of the population sizes instead of the identity function for the classical CSBP; also see [7] for nonlinear CSBPs with more general rate functions.

For the main result in the note we need the following Laplace transforms on downward first passage times, also see [7] for related results, where the first passage time of X is defined by

$$T_a^- := \inf\{t \geq 0, X_t < a\}$$

with the convention $\inf \emptyset = \infty$.

Lemma 2 ([12] Lemma.4.4). *If* $\int^\infty \frac{W_{\phi(0)}(x)}{R(x)}dx < \infty$, *then for* $x > a > 0$

$$\mathbb{E}_x\left[e^{-T_a^-}\right] = \frac{H^{(\omega)}(x)}{H^{(\omega)}(a)} \tag{8}$$

where $H^{(\omega)}(x)$ is defined as the unique decreasing function satisfying

$$H^{(\omega)}(x) := \lim_{b\to\infty} \frac{W^{(\omega)}(b,x)}{W(b)} = e^{-\phi(0)x} + \int_x^\infty H^{(\omega)}(z)\omega(z)W(z-x)\,dz. \tag{9}$$

Remark 2. By the definition of X in (5), the locally boundedness assumption on ω in Lemma 2 ensures the finiteness of T_a^- on the set $\{\tau_a^- < \infty\}$. By the absence of negative jumps, there always exists some function $H^{(\omega)}$ so that the ratio identity (8) holds for all $x > a > 0$.

The condition $\int^\infty W_{\phi(0)}(x)\omega(x)\,dx < \infty$ enables us to derive a proper integral equation for $H^{(\omega)}$ in (9), and whose solution can be expressed as the following infinite sum of integrals

$$H^{(\omega)}(x) = e^{-\phi(0)x}\Big(1 + \int_{z>x} \omega(z)W_{\phi(0)}(z-x)dz$$
$$+ \iint_{z_2>z_1>x} \omega(z_2)\omega(z_1)W_{\phi(0)}(z_2-z_1)W_{\phi(0)}(z_1-x)dz_1dz_2 + \cdots\Big).$$

On the other hand, if $\int^\infty W_{\phi(0)}(z)\omega(z)dz = \infty$, then $\lim_{b\to\infty} \frac{W^{(\omega)}(b,x)}{W(b)} = \infty$. Therefore, the introduction of appropriate $H^{(\omega)}$ is more involved and the associated integral equation can be different. We thus consider the following two examples to illustrate this.

If $\omega(z) = \frac{1}{z+a}$ for some $a \geq 0$, then $H^{(\omega)}$ in (9) solves a singular integral equation

$$H^{(\omega)}(x) := \lim_{b\to\infty} \frac{W^{(\omega)}(b,x)}{W(b)} = \int_x^\infty H^{(\omega)}(z)\frac{W(z-x)}{z+a}dz,$$

whose solution can be expressed, similar to [5], as

$$H^{(\omega)}(x) := \int_{\phi(0)}^\infty e^{-(x+a)y} \exp\left(\int_{\phi(0)+1}^y \frac{dz}{\psi(z)}\right)\frac{dy}{\psi(y)}, \quad \forall x > 0.$$

However, if $\int^\infty W_{\phi(q)}(x)\left|\omega(x)-q\right|dx < \infty$ for some ω and $q > 0$, then one would expect the following equation from [10, equation (2.3)] and [11, Lem.4.3]

$$W^{(\omega)}(x,y) = W^{(q)}(x-y) + \int_y^x W^{(\omega)}(x,z)\big(\omega(z)-q\big)W^{(q)}(z-y)dz,$$

and $H^{(\omega)}$ can be defined as a limit that solves equation

$$H^{(\omega)}(x) := \lim_{b\to\infty} \frac{W^{(\omega)}(b,x)}{W^{(q)}(b)} = e^{-\phi(q)x} + \int_x^\infty H^{(\omega)}(z)(\omega(z)-q)W^{(q)}(z-x)dz,$$

which can be further expressed as an infinite sum of integrals.

Remark 3. If we assume $\omega(z) = \int_0^\infty e^{-zx}\mu(dx)$ for some positive measure μ on $(0,\infty)$ such that

$$\int^\infty \frac{W_{\phi(0)}(x)}{R(x)}dx \asymp \int_a^\infty \frac{W_{\phi(0)}(x)}{R(x+a)}dx = \int_{0+}^\infty e^{-at}\frac{\mu(dt)}{\psi_{\phi(0)}(t)} \asymp \int_{0+} \frac{\mu(dt)}{\psi_{\phi(0)}(t)} < \infty,$$

then $H^{(\omega)}(x) = \int_0^\infty e^{-xy}\nu(dy)$ for some positive measure ν on $[\phi(0),\infty)$ so that

$$\nu(ds) = \delta_{\{\phi(0)\}}(ds) + \frac{\nu * \mu(ds)}{\psi(s)} \quad \text{for } s \geq \phi(0),$$

where δ denotes the delta measure and $\nu * \mu$ denotes the convolution of ν and μ.

2.3 Nonlinear CSBPs coming down from infinity

By saying a stochastic process *comes down from infinity*, we mean intuitively that the process drops dramatically within arbitrarily short time when initiated from a very high level. The following definitions can be found in [8] and in recent papers [13, 6].

Definition 1. Let $(Y_t, t \geq 0)$ be a positive Markov process. The boundary ∞ is an instantaneous entrance boundary for the process Y if the process does not explode and

$$\forall t > 0, \quad \lim_{a\to\infty} \liminf_{x\to\infty} \mathbb{P}_x(S_a^- \leq t) = 1,$$

where $S_x^- := \inf\{t > 0, Y(t) < x\}$ denotes the first downcrossing time of Y.

If a strong Markov process comes down from infinity in the above sense, it is natural to extend the state space and define the process starting at ∞. Kallenberg [8] considers such an extension for regular diffusions. For positive-valued strong Markov processes with no negative jumps, besides some equivalent conditions [6, Lem.1.2.] for the entrance boundary, the following results for Markov process coming down from infinity are also shown in the paper, see also [13] and references there in. Recalling that for a positive Markov process Y, its Markov semigroup $(P_t, t \geq 0)$ is a family of linear operators on $L^\infty(\mathbb{R}^+)$ indexed by $t \geq 0$ and given by, c.f. [2, I.2]

$$P_t f(x) = \mathbb{E}_x[f(Y_t)] = \int_0^\infty f(y)\mathbb{P}_x(Y_t \in dy).$$

Definition 2 (C_b-Feller property). For $E = [0,\infty)$ or $[0,\infty]$, a semigroup $(P_t, t \geq 0)$ satisfies the *Feller property* if

(i) for any $f \in C_b(E)$, $x \to P_t f(x) \in C_b(E)$ for every $t \geq 0$,
(ii) for any $f \in C_b(E)$ and $x \in E$, $P_t f(x) \to f(x)$, as $t \to 0$,

where $C_b(E)$ denotes the space of bounded continuous functions on E. A Markov process Y on E is called a Feller process if its transition group satisfies the Feller property.

Lemma 3 ([6] Theorem 2.2). *Let Y be a non-explosive Markov process taking values in $[0,\infty)$ with no negative jumps, satisfying the Feller property and with an entrance boundary at ∞, in the sense of Definitions 1 and 2, respectively. Then Y can be extended to a Feller process valued in $[0,\infty]$ such that under $\mathbb{P}_\infty$, it starts from ∞, leaves from it instantaneously and stays finite almost-surely:*

$$\mathbb{P}_\infty\big(Y_0 = \infty\big) = 1 \quad \textit{and} \quad \mathbb{P}_\infty\big(\forall t > 0, Y_t < \infty\big) = 1.$$

Lemma 4 ([6] Theorem 2.5). *Assume that Y is a Feller process on $[0,\infty]$. Let $Y^{(x)}$ be the Markov process started from $x \in [0,\infty]$ with càdlàg sample paths. Then the family of processes $\big(Y^{(x)}\big)_{x\in[0,\infty)}$ converges weakly, in the Skorohod topology, as $x \to \infty$ towards $Y^{(\infty)}$.*

Lemma 3 shows the convergence of semigroups, that is,

$$P_t f(x) \to P_t f(\infty) = \mathbb{E}_\infty\big[f(X_t)\big]$$

as $x \to \infty$ for every $f \in C_b[0,\infty)$ to some probability measures $\big(\mathbb{P}_\infty(t,dy)\big)_{t>0}$ on $[0,\infty)$. Lemma 4 deals with the convergence of processes as elements in cádlág space. As far as the convergence of stopping times is concerned, the following useful regularity property of the first passage times under the probability measures $\big(\mathbb{P}_x\big)_{x\geq 0}$ is also proved in the above mentioned papers.

Lemma 5 ([6] Proposition 2.4). *Suppose that the conditions in Lemma 3 hold. Let $h \in C_b[0,\infty)$ be bounded or nonnegative increasing. Then*

(a) for any $\theta > 0$, there exists $b_\theta > 0$ such that for all $b \geq b_\theta, \mathbb{E}_\infty\big[e^{\theta S_b^-}\big] < \infty$;
(b) for any $b > 0$, if $S_b^- < \infty$, $\mathbb{P}_\infty$-a.s. and $\mathbb{E}_\infty\big[h(S_b^-)\big] < \infty$, then

$$\mathbb{E}_x\big[h(S_b^-)\big] \to \mathbb{E}_\infty\big[h(S_b^-)\big], \quad \textit{as } x \to \infty.$$

For the time-changed SPLP X in (5), the Feller property as well as the necessary and sufficient condition for the process to come down from infinity has been shown in [7].

Lemma 6 ([7] Proposition 2.1). *Assume that function R is strictly positive and continuous function on $(0,\infty)$, and ξ is a spectrally positive Lévy process. Then for any $x > 0$ and $t \in [0, \tau_0^-)$, we have $\eta(t) < \infty$ $\mathbb{P}_x$-a.s. The process X is well-defined, strong Markov and with cádlág paths, whose semigroup $(P_t, t \geq 0)$ satisfies the Feller property.*

Lemma 7 ([7] Theorem 3.1). *Assume that $\mathbb{E}[-\xi_1] \geq 0$. The boundary ∞ is an instantaneous entrance boundary for the process X in the sense of Definition 1 if and only if*

$$\int^{\infty} \frac{1}{x\psi(1/x)R(x)}\,dx < \infty.$$

Moreover, $\mathbb{E}_\infty\left[T_b\right] = \int_b^\infty \frac{W(x-b)}{R(x)}\,dx$ for the measure $\mathbb{P}_\infty$ in Lemma 3.

Since $W(x) \asymp \frac{1}{x\psi(1/x)}$ by Proposition VII.10 in [2], the above integral condition is equivalent to

$$\int^{\infty} \frac{W(x)}{R(x)}\,dx < \infty.$$

3 Main results

In this note we take use of the results from [6, 7] to compute quantities related to X that starts from infinity and comes down from infinity. Therefore, for the rest of the note we always assume that the conditions in Lemma 7 hold, that is, the following assumption holds.

Assumption $\mathbb{H}$: $-\mathbb{E}[\xi_1] \geq 0$, R is a strictly positive, continuous function on $(0,\infty)$ and

$$\int^{\infty} \frac{1}{x\psi(1/x)R(x)}\,dx \asymp \int^{\infty} \frac{W(x)}{R(x)}\,dx = \int^{\infty} \omega(z)W(z)\,dz < \infty.$$

Under the above assumption, $\phi(0) = 0$ and $H^{(\omega)}$ in Lemma 2 is the unique decreasing solution to equation

$$H^{(\omega)}(x) = 1 + \int_x^{\infty} H^{(\omega)}(z)\omega(z)W(z-x)\,dz,$$

which converges to 1 as $x \to \infty$.

Theorem 1. *Under Assumption $\mathbb{H}$ we have for any $q, a > 0$, $T_a^- < \infty$ $\mathbb{P}_\infty$-a.s. and*

$$\mathbb{E}_\infty\left[e^{-qT_a^-}\right] = \left(H^{(q\omega)}(a)\right)^{-1}.$$

In addition, the resolvent measure of X is

$$U_a^{(q)}(\infty, dy) := \int_0^{\infty} e^{-qt}\mathbb{P}_\infty(X_t \in dy, t \leq T_a^-)\,dt = \omega(y)\frac{W^{(q\omega)}(y,a)}{H^{(q\omega)}(a)}\,dy.$$

Proof. It follows from Assumption $\mathbb{H}$ that $T_a^- < \infty$ $\mathbb{P}_\infty$-a.s. Applying Lemma 5 (b) to Lemma 2 directly with weight function $q\omega$ and noticing that $\phi(0) = 0$, we obtain the Laplace transform for T_a^-.

For the resolvent measure of X, notice that for every $f \in C_b[0,\infty)$ we have from Lemma 1

$$\begin{aligned}
&\int_0^\infty e^{-qt}\mathbb{E}_x\left[f(X(t)); t \le T_a^- \wedge T_b^+\right] dt \\
&= \int_0^\infty \mathbb{E}_x\left[e^{-q\eta(s)}\omega(\xi_s) f(\xi(s)); s \le \tau_a^- \wedge \tau_b^+\right] dt \\
&= \int_a^b f(y)\omega(y)\Big(\frac{W^{(q\omega)}(b,x)}{W^{(q\omega)}(b,a)} W^{(q\omega)}(y,a) - W^{(q\omega)}(y,x)\Big)\, dy.
\end{aligned}$$

Letting $b \to \infty$ first and then $x \to \infty$, we have

$$\begin{aligned}
U_a^{(q)} f(x) &:= \int_0^\infty e^{-qt}\mathbb{E}_x\left[f(X_t); t \le T_a^-\right] dt \\
&= \int_a^\infty f(y)\omega(y)\Big(\frac{H^{(q\omega)}(x)}{H^{(q\omega)}(a)} W^{(q\omega)}(y,a) - W^{(q\omega)}(y,x)\Big)\, dy \\
&\to \int_a^\infty f(y)\omega(y)\frac{W^{(q\omega)}(y,a)}{H^{(q\omega)}(a)}\, dy.
\end{aligned}$$

Thus, as a function of x, $U_a^{(q)} f(x)$ has a limit as $x \to \infty$.

On the other hand, to give a probabilistic interpretation of the limit of $U_a^{(q)} f(x)$. Let e_q be an exponential random variable with parameter $q > 0$ independent of X. For every fixed $t > 0$, applying the Markov property of X and the memoryless property of e_q at t under $\mathbb{P}_x$, we have for $x \in (0,\infty)$

$$\begin{aligned}
qU_a^{(q)} f(x) =& \mathbb{E}_x\Big[f\big(X(e_q)\big); e_q \le T_a^-\Big] \\
=& e^{-qt}\mathbb{E}_x\Big[U_a^{(q)} f(X_t); t \le T_a^-\Big] + \mathbb{E}_x\Big[f\big(X(e_q)\big); e_q < T_a^- \wedge t\Big].
\end{aligned} \tag{10}$$

Applying Lemma 3 to the first term above, for the Feller semigroup of $\big(\mathbb{P}_x(X_s \in dy, s < T_a^-), s \ge 0\big)$ and function $U_a^{(q)} f \in C_b[0,\infty)$, we have as $x \to \infty$

$$\begin{aligned}
e^{-qt}\mathbb{E}_x\Big[U_a^{(q)} f(X_t); t \le T_a^-\Big] &\to e^{-qt}\mathbb{E}_\infty\Big[U_a^{(q)} f(X_t); t \le T_a^-\Big] \\
&= \mathbb{E}_\infty\Big[f\big(X(e_q)\big); e_q \le T_a^-\Big] - \mathbb{E}_\infty\Big[f\big(X(e_q)\big); e_q < T_a^- \wedge t\Big].
\end{aligned}$$

For the second term in (10), by the boundedness of f and Lemma 5 we have

$$\begin{aligned}
&\limsup_{x\to\infty}\mathbb{E}_x\Big[f\big(X(e_q)\big); e_q < T_a^- \wedge t\Big] \le ||f||_\infty \cdot \limsup_{x\to\infty}\mathbb{P}_x\big(e_q < T_a^- \wedge t\big) \\
&= ||f||_\infty \cdot \lim_{x\to\infty}\mathbb{E}_x\left[1 - e^{-q(T_a^- \wedge t)}\right] = ||f||_\infty \cdot \mathbb{E}_\infty\left[1 - e^{-q(T_a^- \wedge t)}\right].
\end{aligned}$$

Finally, letting $t \to 0$ gives

$$\lim_{x\to\infty} \mathbb{E}_x\left[f(X(e_q)); e_q \le T_a^-\right] = \mathbb{E}_\infty\left[f(X(e_q)); e_q \le T_a^-\right]$$

and this finishes the proof. □

Recall the *total progeny* until time $t \ge 0$

$$J(t) := \int_0^t X(s)\,ds$$

defined in [9, Sec.12.2.3.]. Note that $T_a^- = \eta(\tau_a^-) < \infty$. By a change of variable, we have

$$J(T_a^-) = \int_0^{\eta(\tau_a^-)} \xi(\eta^{-1}(t))\,dt = \int_0^{\tau_a^-} \xi(s)\eta(ds) = \int_0^{\tau_a^-} \frac{\xi(s)}{R(\xi(s))}\,ds.$$

Applying Theorem 1, we have the following Laplace transform.

Corollary 1. *If Assumption* $\mathbb{H}$ *holds with* $R(x)$ *replaced by* $R(x)/x$, *then we have* $J(T_a^-) < \infty$ $\mathbb{P}_\infty$*-a.s. and for* $\omega^*(x) := x\omega(x)$,

$$\mathbb{E}_\infty\left[e^{-qJ(T_a^-)}\right] = \left(H^{(q\omega^*)}(a)\right)^{-1}.$$

We can also derive the moments for $J(T_a^-)$.

Theorem 2. *Under Assumption* $\mathbb{H}$, *for every* $a > 0$ *denote by*

$$m_{a,0} = 1 \text{ and } m_{a,n} := \mathbb{E}_\infty\left[(T_a^-)^n\right]$$

the nth *moment of* T_a^- *under* $\mathbb{P}_\infty$. *Then* $\{m_{a,n}\}_{n\ge 1}$ *satisfies equation*

$$m_{a,n} = n! \sum_{k=1}^{n} (-1)^{k-1} a_k m_{a,n-k} = n! \sum_{k=0}^{n-1} (-1)^{n-k-1} m_{a,k} a_{n-k}, \tag{11}$$

where the sequence $(a_n)_{n\ge 1}$ *is defined by* $a_1 := \int_a^\infty \frac{W(x-a)}{R(x)}\,dx$ *and*

$$a_n := \iint_{x_n > x_{n-1} > \cdots > x_1 > a} \frac{W(x_n - x_{n-1})W(x_{n-1}-x_{n-2})\cdots W(x_1 - a)}{R(x_n)R(x_{n-1})\cdots R(x_1)}\,dx_n \cdots dx_1$$

that are finite for every $a > 0$ *and* $n \ge 1$.

Proof. Fix $a > 0$. Under Assumption $\mathbb{H}$, the following functions are well-defined for $n \in \mathbb{N}$.

$$f_0(x) = 1 \quad \text{and} \quad f_n(x) = \int_x^\infty f_{n-1}(z)\frac{W(z-x)}{R(z)}\,dz. \tag{12}$$

Then we have from [3, Lem.8.11.1.] (see also [12, Prop.4.7.]) that

$$\begin{aligned}m_{a,n}(x) &:= \mathbb{E}_x\Big[\big(T_a^-\big)^n; T_a^- < \infty\Big] = n\int_a^\infty \frac{m_{a,n-1}(y)}{R(y)}\big(W(y-a)-W(y-x)\big)\,dy \\ &= n\Big(\int_a^\infty m_{a,n-1}(y)\frac{W(y-a)}{R(y)}\,dy - \int_x^\infty m_{a,n-1}(y)\frac{W(y-x)}{R(y)}\,dy\Big)\end{aligned} \tag{13}$$

for the moments of T_a^+ under $\mathbb{P}_x$ and again the finiteness assumption from $\mathbb{H}$. One can check from the identity that for $n \in \mathbb{N}$ and $x > a$

$$m_{a,n}(x) \le n!\Big(\int_a^\infty \frac{W(y-a)}{R(y)}\,dy\Big)^n.$$

Given the identity (13) and $f_n(x)$ defined in (12), one can also prove by induction on n that the moments $m_{a,n}$ can be written as

$$m_{a,n}(x) = n!\sum_{k=0}^n (-1)^k f_k(x)\cdot c_{n,k} \tag{14}$$

for some constant array $\{c_{n,k}, k = 0, \cdots, n\}_{n\ge 0}$. Plugging (14) into (13) and then applying (12), we have

$$\begin{aligned}\frac{m_{a,n+1}(x)}{(n+1)!} &= \sum_{k=0}^n (-1)^k c_{n,k}\Big(\int_a^\infty f_k(z)\frac{W(z-a)}{R(z)}\,dz - \int_x^\infty f_k(z)\frac{W(z-x)}{R(z)}\,dz\Big) \\ &= \sum_{k=0}^n (-1)^k c_{n,k}\big(f_{k+1}(a) - f_{k+1}(x)\big).\end{aligned} \tag{15}$$

Comparing (14) with (15) we have $c_{0,0} = 1$ for $n \in \mathbb{N}$

$$c_{n,0} = \sum_{k=1}^n (-1)^{k-1} f_k(a) c_{n-1,k-1} \quad \text{and} \quad c_{n,k} = c_{n-1,k-1}.$$

Therefore, letting $x \to \infty$ in (13) and applying Lemma 5 (b), we have

$$m_{a,n} = \lim_{x\to\infty} m_{a,n}(x) = n!\cdot c_{n,0} = n!\sum_{k=1}^n (-1)^{k-1} a_k m_{a,n-k}.$$

This finishes the proof. □

Remark 4. If $\omega(z) = R^{-1}(z) = \hat{\mu}(z)$ for some measure μ on $(0,\infty)$ as described in Remark 3, one can have from (12) in the proof that

$$f_1(x) = \int_x^\infty dz\int_0^\infty e^{-zy}W(z-x)\mu(dy) = \int_0^\infty \frac{e^{-xy}}{\psi(y)}\mu(dy)$$

and then by induction on n,

$$f_n(x) = \iint_{y_1<y_2<\cdots<y_n} e^{-xy_n}\frac{\mu(dy_1)\mu(dy_2-y_1)\cdots\mu(dy_n-y_{n-1})}{\psi(y_1)\psi(y_2)\cdots\psi(y_n)}.$$

In particular, if $R(x) = e^{-\lambda x}$ for some $\lambda > 0$, then $\mu(dz) = \delta_\lambda(dz)$ for delta measure δ_λ. We have

$$a_n = f_n(a) = e^{-na\lambda} \prod_{k=1}^{n} \frac{1}{\psi(k\lambda)} \quad \text{for } n \geq 1.$$

If $\psi(x) = x^\alpha$ and $R(x) = x^\beta$ for some $\beta > \alpha > 1$, then $\mu(dz) = \frac{z^{\beta-1}}{\Gamma(\beta)} dz$ and

$$a_n = f_n(a) = a^{n(\alpha-\beta)} \prod_{k=1}^{n} \frac{\Gamma(k(\beta-\alpha))}{\Gamma(k(\beta-\alpha)+\alpha)} \quad \text{for } n \geq 1.$$

Acknowledgements Bo Li and Xiaowen Zhou are supported by Natural Sciences and Engineering Research Council of Canada (RGPIN-2021-04100).

References

1. Bansaye, V., Méléard, S., Richard, M.: Speed of coming down from infinity for birth-and-death processes. Advances in Applied Probability **48**(4), 1183–1210 (2016). DOI 10.1017/apr.2016.70
2. Bertoin, J.: Lévy Processes. Cambridge University Press (1996). URL `https://www.ebook.de/de/product/3779806/jean_bertoin_bertoin_jean_levy_processes.html`
3. Bingham, N.H., Goldie, C.M., Teugels, J.L.: Regular Variation. No. volume 27 in Encyclopedia of mathematics and its applications. Cambridge University Press, Cambridge (1987). DOI 10.1017/cbo9780511721434
4. Caballero, M.E., Lambert, A., Bravo, G.U.: Proof(s) of the Lamperti representation of continuous-state branching processes. Probability Surveys **6**, 62–89 (2009). DOI 10.1214/09-ps154
5. Duhalde, X., Foucart, C., Ma, C.: On the hitting times of continuous-state branching processes with immigration. Stochastic Processes and their Applications **124**(12), 4182–4201 (2014). DOI 10.1016/j.spa.2014.07.019
6. Foucart, C., Li, P.S., Zhou, X.: On the entrance at infinity of Feller processes with no negative jumps. Statistics & Probability Letters **165**, 108,859 (2020). DOI 10.1016/j.spl.2020.108859
7. Foucart, C., Li, P.S., Zhou, X.: Time-changed spectrally positive Lévy processes started from infinity. Bernoulli **27**(2), 1291–1318 (2021). DOI 10.3150/20-bej1274
8. Kallenberg, O.: Foundations of Modern Probability. Springer International Publishing (2021). DOI 10.1007/978-3-030-61871-1
9. Kyprianou, A.E.: Fluctuations of Lévy Processes with Applications. Springer Berlin Heidelberg (2014). DOI 10.1007/978-3-642-37632-0
10. Li, B., Palmowski, Z.: Fluctuations of Omega-killed spectrally negative Lévy processes. Stochastic Processes and their Applications **128**(10), 3273–3299 (2018). DOI 10.1016/j.spa.2017.10.018
11. Li, B., Zhou, X.: Local Times for Spectrally Negative Lévy Processes. Potential Analysis **52**(4), 689–711 (2019). DOI 10.1007/s11118-018-09756-6
12. Li, B., Zhou, X.: On the explosion of a class of continuous-state nonlinear branching processes. Electronic Journal of Probability **26**, 1–25 (2021). DOI 10.1214/21-ejp715
13. Li, P.S.: A continuous-state polynomial branching process. Stochastic Processes and their Applications **129**(8), 2941–2967 (2019). DOI 10.1016/j.spa.2018.08.013

14. Li, Z.: Measure-Valued Branching Markov Processes. Springer Berlin Heidelberg (2011). DOI 10.1007/978-3-642-15004-3

The matrix sequential probability ratio test and multivariate ruin theory

Hansjoerg Albrecher and Oscar Peralta

Abstract The matrix sequential probability ratio test (MSPRT) is a statistical method to decide which law governs a collection of independent and identically distributed data amongst a finite set of possibilities. By focusing on the case where the possible laws are exponentially tilted versions of each other, in this note we exhibit novel links between the MSPRT and multivariate risk processes with common shocks, as well as with one-dimensional renewal theory.

1 Introduction

The celebrated *sequential probability ratio test* (SPRT) developed in [8] is an efficient tool to identify the nature of a collection of i.i.d. observations employing only a finite subset of data points whose cardinality is random. More specifically, for an independent and identically distributed (i.i.d.) sequence $\{X_\ell\}_{\ell\geq 1}$ with density f, one tests the null hypothesis $H_0 = \{f = f_0\}$ against the alternative hypothesis $H_1 = \{f = f_1\}$, where f_0 and f_1 are given density functions. The decision rule associated to the SPRT is based on employing the likelihood ratio of the first T observations, say $\Lambda^{(T)}$ for $\Lambda^{(k)} := \prod_{\ell=1}^{k} f_0(X_\ell)/f_1(X_\ell)$, where T corresponds to the number of observations it takes $\{\Lambda^{(k)}\}_{k\geq 1}$ to exit the strip (a,b), $0 < a < 1 < b < \infty$. In particular, the null hypothesis H_0 is rejected if and only if $\Lambda^{(T)} \leq a$, with corresponding errors

$$\alpha_0 := \mathbb{P}(\text{reject } H_0 \mid H_0) \quad \text{and} \quad \alpha_1 := \mathbb{P}(\text{do not reject } H_0 \mid H_1)$$

Hansjoerg Albrecher
University of Lausanne and Swiss Finance Institute, Faculty of Business and Economics, Quartier de Chambronne, 1015 Lausanne, Switzerland, e-mail: hansjoerg.albrecher@unil.ch

Oscar Peralta
Cornell University, School of Operations Research and Information Engineering, Rhodes Hall, Ithaca, New York 14853, United States of America, e-mail: op65@cornell.edu

D. R. Wood et al. (eds.), *2021-2022 MATRIX Annals*, MATRIX Book Series 5,
https://doi.org/10.1007/978-3-031-47417-0_26

that depend on a and b.

Although the SPRT is a well-established method which in practice produces efficient results in terms of the number T of observations needed, in the general case it is difficult to obtain constants a, b that precisely yield prescribed errors α_0 and α_1; in most cases only bounds exist (see e.g. [9]). In [1], the authors established exact boundaries (a, b) for the prescribed errors for the case f_0 follows a phase-type distribution and f_1 is an exponentially-tilted transform of f_0. To do so, the authors follow a probabilistic approach based on establishing a link between the SPRT and certain Markov additive risk processes and their exit probabilities. The theory of scale functions developed in [7] was then used to compute α_0 and α_1, as well as the expected amount of observations needed, $\mathbb{E}(T)$.

In this paper, we are interested in establishing novel geometric considerations for the matrix sequential probability ratio test (MSPRT) as defined in [5], an extension of the SPRT that allows for multiple hypotheses. Similarly to [1], we focus on the case where the candidate densities are exponentially-tilted transforms of some fixed function. This allows us to redefine the decision rule associated to the MSPRT in terms of exit times of a multivariate risk process with common shocks. An additional transformation resets the problem in terms of a one-dimensional renewal process and its exit time from a region that varies in time. While both problems are, at the moment, not tractable in the ruin theory and applied probability literature, our geometric approach elucidates avenues of research within these disciplines that can yield powerful results for the MSPRT.

The structure of this paper is as follows. In Section 2, we provide a brief introduction of the MSPRT and its properties. In Section 3, we specialize onto the case of multiple hypothesis testing for exponentially-tilted versions of each other, which will allow to draw novel risk- and renewal-theoretical perpectives on the MSPRT. In Section 4 we provide some synthetic examples that illustrate the performance of the MSPRT in the case of exponentially-tilted hypotheses. We finalize in Section 5 by summarizing our findings and pointing out some potential avenues for further resarch.

2 The Matrix Sequential Probability Ratio Test

In a probability space $(\Omega, \mathbb{P}, \mathscr{F})$, suppose that we have a sequence of identically distributed random variables $\{X_\ell\}_{\ell\geq 1}$ and a collection of events that partition Ω, say $\{H_i\}_{i\in\Theta}$ with $i \in \Theta = \{1, \ldots, M\}$, where

$$H_i = \{X_\ell \sim F_i \text{ for all } \ell \geq 1\},$$

and each F_i corresponding to a pre-specified distribution function. Furthermore, suppose that conditional on each event H_i, the random variables $\{X_\ell\}_{\ell\geq 1}$ are independent. The previous corresponds to the classical Bayesian setting for *multiple hypothesis testing*. Given a realization of the random variables $\{X_\ell\}_{\ell\geq 1}$, the main goal of multiple hypothesis testing is to choose the event H_i which is the most likely, given the observed sequence.

Let $\mathbb{P}_i$ be the probability measure conditioned on the hypothesis H_i. Here we assume that $\mathbb{P}_i$ and $\mathbb{P}_j$ are equivalent measures for all $i, j \in \Theta$, in the sense that the Radon-Nikodym derivative $\mathrm{d}\mathbb{P}_i/\mathrm{d}\mathbb{P}$ exists and is $\mathbb{P}$-a.e. positive. This is equivalent to assuming the existence of a reference measure ν on $(-\infty, \infty)$ and ν-integrable positive functions $f_1, f_2, \ldots, f_M$, which share the same support, such that

$$F_i(x) = \int_{(-\infty,x]} f_i(y)\mathrm{d}\nu(y), \qquad x \in (-\infty,\infty), i \in \Theta.$$

For the set of observations $\{X_\ell\}$, here we are interested in a method that, within some prespecified error bounds, allows us to reach a consensus w.r.t. the possible events $\{H_i\}_{i\in\Theta}$, using a finite amount of observations *only*. The *matrix sequential probability ratio test* (MSPRT), constructed in [5], produces a pair (d,T) where $T \geq 1$ is an $\mathscr{F}_n$-stopping time associated to the number of observations needed, and d is an $\mathscr{F}_T$-measurable function taking values in Θ associated to the hypothesis that is the most likely. In the previous, $\mathscr{F}_n$ denotes the σ-algebra generated by $X_1, X_2, \ldots, X_n$, while $\mathscr{F}_T$ is the σ-algebra generated by $X_1, \ldots, X_T$.

For the MSPRT, one of the aims is to get small error probabilities

$$\alpha_{ij} = \mathbb{P}_i(d = j), \quad i \neq j, \tag{1}$$

as well as a "small" stopping time T. To properly specify what we mean by this, define the *n-th step likelihood function*

$$\Lambda_i^{(n)} := \left.\frac{\mathrm{d}\mathbb{P}_i}{\mathrm{d}\nu}\right|_{\mathscr{F}_n} (\{X_\ell\}_{\ell\geq 1}) = \prod_{\ell=1}^{n} f_i(X_\ell). \tag{2}$$

Then (d,T) corresponding to the MSPRT in [5] takes the form

$$T = \min_{i\in\Theta}\{T_i\} \quad \text{where} \quad T_i = \inf\left\{n \geq 1 : \Lambda_i^{(n)} \geq A_i\left(\Lambda_1^{(n)}, \cdots, \Lambda_{i-1}^{(n)}, \Lambda_{i+1}^{(n)}, \ldots, \Lambda_M^{(n)}\right)\right\}, \tag{3}$$

$$d = \arg\min_{i\in\Theta}\{T_i\}, \tag{4}$$

for some collection of deterministic functions $A_i : \mathbb{R}^{M-1} \mapsto \mathbb{R}$, $i \in \Theta$. The choice for the collection of functions A_i will impact the *risk* associated to making the decision i, here defined by

$$R_i(d,T) = \sum_{j\neq i} \pi_j \alpha_{ji} \quad \text{with} \quad \pi_j = \mathbb{P}(H_j),$$

as well as on the expected number of observations until a decision is met, $\mathbb{E}(T)$. In fact, in [5] it is shown that $A_i(\cdot)$ can be chosen to be of the form

$$A_i(x_1,\dots,x_{i-1},x_{i+1},\dots,x_M) = \frac{M}{\rho_i}\max_{j\neq i}\pi_j x_j \text{ for some fixed } \rho_i > 0, \tag{5}$$

and such a choice yields a risk for the decision i which is lower than ρ_i, i.e. $R_i(d,T) \le \rho_i$.

While the choice (5) does not yield the lowest expected number of observations for a predetermined level of risk ρ_i, it is a tractable option which still produces an optimal decision in an asymptotic sense. More specifically, let $\boldsymbol{\rho} = (\rho_1,\rho_2,\dots,\rho_M)$ be a vector with positive entries. We then consider the *class of tests* $\boldsymbol{\Delta}(\boldsymbol{\rho})$ with

$$\boldsymbol{\Delta}(\boldsymbol{\rho}) = \{(c,S) : R_i(c,S) \le \rho_i \text{ for all } i \in \Theta\},$$

where the set on the r.h.s. spans over all pairs (c,S) defined analogously as (d,T) was defined in (3) and (4) for arbitrary measurable functions A_i.

Theorem 1 ([5]). *Let $\boldsymbol{\rho} \downarrow 0$ denote the instance when the elements of $\boldsymbol{\rho}$ uniformly tend to 0 while mantaining the asymptotic relationship*

$$\frac{\log\rho_i}{\log\rho_j} \to c_{ij}, \quad 0 < c_{ij} < \infty.$$

Then,

$$\inf_{(c,S)\in\boldsymbol{\Delta}(\boldsymbol{\rho})} \mathbb{E}_i(S^m) = \mathbb{E}_i(T^m)(1+o(1)),$$

where $o(1)$ denotes some generic real function which converges to 0 as $\boldsymbol{\rho} \downarrow 0$.

Theorem 1 essentially implies that, amongst all the tests which yield risks which are below a prespecified vector $\boldsymbol{\rho}$, (d,T) with A_i defined as in (5) enjoys asymptotic optimality w.r.t. the moments of T as the entries of $\boldsymbol{\rho}$ tend to 0. Due to this and its simplicity, in practice it is common to employ the MSRPT with the test (d,T), even if it is not optimal in a non-asymptotic sense. From now on, whenever we refer to the MSPRT test (d,T), it will be understood that it is the one defined by (5).

3 MSPRT for exponential tilting

In this section, we look at a special type of MSPRT, where the collection of hypotheses corresponds to distributions which are all exponentially tilted versions of some reference probability measure. More specifically, suppose that there exists a distribution function taking the form

$$F(x) = \int_{(-\infty,x]} f(y)\mathrm{d}\nu(y), \quad x \in (-\infty,\infty) \tag{6}$$

for some nonnegative measurable function f, and a collection of real numbers $\{\gamma_i\}_{i\in\Theta}$ such that $f_i(x) = \frac{e^{\gamma_i x} f(x)}{L(\gamma_i)}$, where $L(\gamma_i) = \int e^{\gamma_i s} f(s)\mathrm{d}\nu(s)$ is assumed to be finite for all $i \in \Theta$. Under these particular circumstances, the n-th step likelihood function takes the form

$$\Lambda_i^{(n)} := \prod_{\ell=1}^{n} \frac{e^{\gamma_i X_\ell}}{L(\gamma_i)}. \tag{7}$$

Transforming (7) by taking logarithms leads us to an equivalent definition to (3),

$$T_i = \inf\left\{ n \geq 1 : \lambda_i^{(n)} \geq \log M/\rho_i + \max_{j \neq i} \left\{ \log \pi_j + \lambda_j^{(n)} \right\} \right\}$$

where

$$\lambda_i^{(n)} := \log \Lambda_i^{(n)} = \gamma_i X_\ell - \log L(\gamma_i).$$

3.1 Links to multivariate risk processes

From now on, let us assume that f has support in $(0, \infty)$, i.e. $X_\ell > 0$ a.s. for all $\ell \geq 1$, and w.l.o.g. suppose that $\gamma_1 < \gamma_2 < \cdots < \gamma_M$. Let $\{N(t)\}_{t \geq 0}$ be the *renewal process* whose interarrival times are $\{X_\ell\}_{\ell \geq 1}$. A key aspect of this paper is to notice that the stopping time T_i corresponds to the minimum number of arrivals of $\{N(t)\}_{t \geq 0}$ needed so that the i-th row of the multidimensional array process $\boldsymbol{Y}(t) = \{Y_{ij}(t)\}_{i,j \in \Theta}$ of the form

$$Y_{ij}(t) = (\gamma_i - \gamma_j)t - \sum_{\ell=1}^{N(t)} (\log L(\gamma_i) - \log L(\gamma_j))$$

is simultaneously above certain boundaries. More precisely, T_i can be expressed as

$$T_i = \inf\left\{ \ell \geq 1 : Y_{ij}(W_\ell) \geq b_{ij} \text{ for all } j \neq i \right\},$$

where $W_n = \sum_{\ell=1}^{n} X_\ell$ and $b_{ij} = \log M/\rho_i + \log \pi_j$. Examining each of the processes $\{Y_{ij}(t)\}_{t \geq 0}$ for all $i, j \in \Theta$, we get:

- In the case $i > j$, $\{Y_{ij}(t)\}_{t \geq 0}$ corresponds to a process that increases linearly with slope $\gamma_i - \gamma_j$ and has negative jumps of size $(\log L(\gamma_i) - \log L(\gamma_j))$ at the arrival times of $\{N(t)\}$; this can be regarded as a Sparre-Andersen model with deterministic jump sizes (cf. [1]).
- In the case $i < j$, $\{Y_{ij}(t)\}_{t \geq 0}$ decreases linearly with slope $\gamma_i - \gamma_j$ and has positive jumps of size $(\log L(\gamma_i) - \log L(\gamma_j))$ instead; this can be also be regarded as a Sparre-Andersen model, but flipped along the x-axis.
- The case $i = j$ simply yields $\{Y_{ij}(t)\}_{t \geq 0}$ to be the zero process.

This elucidates a novel risk-theoretic understanding of the MSPRT, for the exponentially tilted case, in terms of risk processes (classical and flipped) and their upcrossing probabilities at the epochs $\{S_\ell\}_{\ell \geq 0}$. In case we want to link the MSPRT to the continuous upcrossing probabilities (as opposed to the discretely observed ones), we can define

$$b_{ij}^* = \begin{cases} b_{ij} & \text{if } i < j, \\ b_{ij} + (\log L(\gamma_i) - \log L(\gamma_j)) & \text{if } i > j, \\ 0 & \text{if } i = j. \end{cases}$$

In this case, it can be readily verified that

$$T_i = \inf\left\{\ell \geq 1 : \text{for some } s \in [0, W_\ell],\, Z_{ij}(s) \geq (\gamma_i - \gamma_j)^{-1} b_{ij}^* \text{ for all } j \in \theta\right\}, \quad i \in \Theta.$$

For geometric considerations, it is convenient to normalize the processes Y_{ij} to move linearly at a positive unit rate. Thus, we define the Sparre-Andersen processes

$$Z_{ij}(t) = \frac{Y_{ij}}{\gamma_i - \gamma_j} = t - \sum_{\ell}^{N(t)} \frac{\log L(\gamma_i) - \log L(\gamma_j)}{\gamma_i - \gamma_j}, \quad i \neq j. \tag{8}$$

Given that $\gamma_i - \gamma_j > 0$ if and only if $i > j$, then the stopping rule for the i-th hypothesis becomes

$$\begin{aligned} T_i &= \inf\left\{\ell \geq 1 : Z_{ij}(W_\ell) \geq c_{ij} \text{ for all } j < i \text{ and } Z_{ij}(W_\ell) \leq c_{ij} \text{ for all } j > i\right\} \qquad (9) \\ &= \inf\left\{\ell \geq 1 : \begin{array}{c} \text{for some } s \in [0, S_\ell],\, Z_{ij}(s) \geq c_{ij}^* \text{ for all } j < i \\ \text{and } Z_{ij}(s) \leq c_{ij}^* \text{ for all } j > i \end{array}\right\}, \end{aligned}$$

where $c_{ij} = (\gamma_i - \gamma_j)^{-1} b_{ij}$ and $c_{ij}^* = (\gamma_i - \gamma_j)^{-1} b_{ij}^*$. Then T_i can be seen to correspond to the simultaneous hitting time of a multivariate Sparre-Andersen process with common shocks.

Remark 1. In the case that $f_1, \ldots, f_M$ are all exponential density functions, the processes will be linked to Cramér-Lundberg processes with common shocks, and moreover, with jumps that are constant. In fact, the resulting multivariate process is a subclass of the risk models considered in [4], where the authors present a multidimensional Cramér-Lundberg process with simultaneous arrivals and ordered claim sizes. Unfortunately, their results yield ruin probabilities only, while in our case, simultaneous downcrossing and upcrossing probabilities are needed. Nevertheless, this presents a promising avenue for further research.

3.2 Links to one-dimensional renewal theory

Similar to the transformation of the bivariate problem in [2] and [3], an additional understanding of T_i can be obtained in terms of the *one-dimensional* process $\{N(t)\}_{t \geq 0}$ and certain hitting times of barriers that are time-varying. More specifically, dividing (8) by $(\log L(\gamma_i) - \log L(\gamma_j))/(\gamma_i - \gamma_j)$, we get

$$N(t) = (t - Z(t)) \frac{\gamma_i - \gamma_j}{\log L(\gamma_i) - \log L(\gamma_j)}.$$

By plugging the previous in (9),

$$T_i = \inf\left\{\ell \geq 1 : N(W_\ell) \leq s_{ij}W_\ell - d_{ij} \text{ for all } j < i \text{ and } N(W_\ell) \geq s_{ij}W_\ell - d_{ij} \text{ for all } j > i\right\} \tag{10}$$

$$= \inf\left\{\ell \geq 1 : \begin{array}{c} \text{for some } s \in [0, S_\ell], N(s) \leq s_{ij}t - d^*_{ij} \text{ for all } j < i \\ \text{and } N(s) \geq s_{ij}t - d^*_{ij} \text{ for all } j > i \end{array}\right\},$$

where $s_{ij} = \frac{\gamma_i - \gamma_j}{\log L(\gamma_i) - \log L(\gamma_j)}$, $d_{ij} = (\log L(\gamma_i) - \log L(\gamma_j))^{-1} b_{ij}$ and $d^*_{ij} = (\log L(\gamma_i) - \log L(\gamma_j))^{-1} b^*_{ij}$.

Now, note that due to the monotonicity of $L(\cdot)$, $s_{ij} = s_{ji} > 0$, $d^*_{ij} > 0$ for all $j < i$, and $d^*_{ij} < 0$ for all $j > i$. The values $\{s_{ij}\}_{i \neq j}$ are, in general, not ordered. Nevertheless, it is possible to order a subset of it. To showcase this, let us first prove the following technical result.

Lemma 1. *Define* $\Sigma(\lambda) = \log(L(\gamma))$ *for all* $\gamma \in (-\infty, c)$, *where* $c = \sup\{z \in \mathbb{R} : L(\gamma) < \infty\}$. *Then, the function* $\Sigma(\cdot)$ *is convex in* $(-\infty, c)$.

Proof. Fix $-\infty < \gamma < \gamma_* < c$ and $s \in (0,1)$. Then, for a random variable $Z \sim f$ and constants $p, q \geq 0$ such that $1/p + 1/q = 1$, by Hölder's inequality we get

$$\begin{aligned}
\Sigma(s\gamma + (1-s)\gamma_*) &= \log\left(\mathbb{E}\left(e^{s\gamma Z} e^{(1-s)\gamma_* Z}\right)\right) \\
&\leq \log\left(\mathbb{E}\left(e^{ps\gamma Z}\right)^{1/p} \mathbb{E}\left(e^{q(1-s)\gamma_* Z}\right)^{1/q}\right) \\
&= \frac{1}{p}\Sigma(ps\gamma) + \frac{1}{q}\Sigma(q(1-s)\gamma_*) \\
&= s\Sigma(\gamma) + (1-s)\Sigma(\gamma_*),
\end{aligned}$$

where we subsitute $p = 1/s$ and $q = 1/(1-s)$ in the last equality, verifying the convexity of $\Sigma(\cdot)$.

Corollary 1. *Define* $s_{M,M+1} = 0$. *Then, for all* $0 \leq i \leq j \leq M$, *we have* $s_{i,i+1} \geq s_{j,j+1}$.

Proof. If $j = M$, then the inequality $s_{i,i+1} \geq s_{j,j+1}$ trivially holds. If $j < M$, employing the convexity of Σ we get

$$\frac{1}{s_{i,i+1}} = \frac{\Sigma(\gamma_{i+1}) - \Sigma(\gamma_i)}{\gamma_{i+1} - \gamma_i} \leq \frac{\Sigma(\gamma_{j+1}) - \Sigma(\gamma_j)}{\gamma_{j+1} - \gamma_j} = \frac{1}{s_{j,j+1}},$$

which in turn implies that $s_{i,i+1} \geq s_{j,j+1}$.

Given the order of the collection $\{s_{i,i+1}\}$ that Corollary 1 yields, the decision rule takes the following simple geometric form:

1. In the Euclidian space $\mathbb{R}^2_+$, draw lines that cross 0 with slopes $\{s_{ij}\}_{i<j}$.

2. For the line corresponding to the slope s_{ij} draw a barrier d_{ij}-units above and another barrier $|d_{ji}|$-units below. Let $B_{ij} \subset \mathbb{R}_+^2$ be the set of points lying strictly between those barriers (see Figure 1).
3. Let C_i be the set of points lying strictly between the lines with slope $s_{(i-1),i}$ and $s_{i,(i+1)}$ (defining $s_{0,1} = +\infty$ and $s_{M,M+1} = 0$).
4. Accept H_i as the correct hypothesis for the i for which $\{(W_\ell, N(W_\ell))\}_{\ell \geq 1}$ enters $C_i \setminus ((\cup_{\ell<i} B_{\ell i}) \cup (\cup_{\ell>i} B_{i\ell}))$ for the first time (see Figure 1).

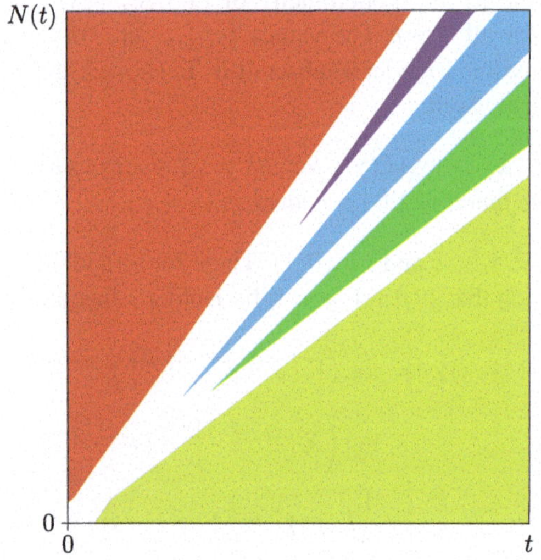

Fig. 1: Acceptance areas associated to $C_i \setminus ((\cup_{\ell<i} B_{\ell i}) \cup (\cup_{\ell>i} B_{i\ell}))$ for $i \in \Theta = \{1,2,3,4,5\}$: Acceptance area for H_1 (red), for H_2 (violet), for H_3 (blue), for H_4 (green), and for H_5 (mustard).

Remark 2. Analogously, we can also define the process in terms of the hitting events of the continuously observed process $\{(t, N(t))\}_{t \geq 0}$. More specifically, we declare i as the correct hypothesis if $\{(t, N(t))\}_{t \geq 0}$ enters $C_i \setminus \left(\left(\cup_{\ell<i} B^*_{\ell i}\right) \cup \left(\cup_{\ell>i} B^*_{i\ell}\right)\right)$ first, where the definition of B^*_{ij} is akin to that of B_{ij} with d_{ij} and d_{ji} being replaced by d^*_{ij} and d^*_{ji}.

Within this framework, there are some non-trivial quantities of interest to compute. First, note that

$$\mathbb{P}_i(d \neq i) \leq \mathbb{P}_i(\sigma_i < \infty), \quad \text{where} \tag{11}$$

$$\sigma_i = \inf\{\ell \geq 1 : (W_\ell, N(W_\ell)) \notin C_i \cup B_{i-1,i} \cup B_{i,i+1}\} \tag{12}$$

The previous simply states that no *wrong* decision can be taken if the set of points $\{(W_\ell, N(W_\ell))\}_{\ell\geq 1}$ never leaves an area which contains the entire acceptance area for H_i. Interestingly, the r.h.s in (11) corresponds to the exit probability of a two-dimensional random walk on a cone. This is investigated in [6]. However, the authors there investigate the exponential decay of the non-exit probability. Their work relies on identifying the dual cone K_* of C_i, defined by

$$K_* = \{z \in \mathbb{R}^2 : \langle x, z\rangle \geq 0\, \forall\, x \in C_i\}. \tag{13}$$

Then, they show that

$$\limsup_{n\to\infty} \mathbb{P}(\sigma_i > n)^{1/n} = \inf_{z\in K_*} \tilde{L}(z), \quad \text{where} \tag{14}$$

$$\tilde{L}(z) = \int_{\mathbb{R}^2} e^{\langle z,y\rangle} \left(\delta_1(y_1) \times f_i(y_2) \mathrm{d}\nu(y_2)\right). \tag{15}$$

In our case, K_* is the cone between the lines $z_2 = (-1/s_{i,i+1})z_1$ and $z_2 = (-1/s_{i-1,i})z_1$; it is readily verified that the infimum of $\tilde{L}(z)$ over K^* is actually reached at these lines. Thus, we need to find the infimum of the functions

$$g_1(z_2) = \int_{\mathbb{R}^2} e^{-s_{i,i+1}z_2y_1 + z_2y_2} \left(\delta_1(y_1) \times f_i(y_2) \mathrm{d}\nu(y_2)\right)$$
$$g_2(z_2) = \int_{\mathbb{R}^2} e^{-s_{i-1,i}z_2y_1 + z_2y_2} \left(\delta_1(y_1) \times f_i(y_2) \mathrm{d}\nu(y_2)\right).$$

In the case of g_1,

$$\begin{aligned}\log(g_1(z_1)) &= -s_{i,i+1}z_2 + \log\left(\int_{\mathbb{R}^2} e^{z_2y_2} f_i(y_2)\mathrm{d}\nu(y_2)\right)\\ &= -s_{i,i+1}z_2 + \log\left(L(z_2+\gamma_i)\right).\end{aligned}$$

Given that the function Σ is convex and increasing, then

$$\frac{\log\left(L(z_2+\gamma_i)\right)}{\mathrm{d}z_2} = \Sigma'(\gamma_i) \geq \frac{\Sigma(\gamma_i) - \Sigma(\gamma_{i-1})}{\gamma_i - \gamma_{i-1}} = s_{i-1,i},$$

meaning that the minumum of g_1 is reached at $z_2 = 0$. Analogous arguments follow for g_2, which in turn implies that the infimum of $\tilde{L}$ over K^* is reached at $(z_1, z_2) = (0,0)$. In short, the results of [6] confirm that exiting from the cone C_i is not a certain event.

4 Numerical examples

In the following, we perform a simulation study of the MSPRT in order to measure its efficiency from an empirical perspective; we do this following the implementation laid out in Subsection 3.2, though we stress that the one in [5] or Subsection 3.1 produces the same results by alternative means. Suppose that ν in (6) is the Lebesgue measure and $f(x) = e^{-x}$, from which we sample an i.i.d. sequence $\{X_\ell\}_{\ell\geq 1}$ with $X_\ell \sim f$. In this analysis, we desire to test if the data $\{X_\ell\}_{\ell\geq 1}$ identifies via the MSPRT the correct hypohesis or not, with alternative hypotheses being "close" to the original one. More specifically, we take $\Theta = \{1,2,3,4,5\}$, and let the tilting values be $\gamma_1 = -2\beta$, $\gamma_2 = -\beta$, $\gamma_3 = 0$, $\gamma_4 = \beta$ and $\gamma_5 = 2\beta$ for some $\beta \in (0,1/2)$. Furthermore, let $\boldsymbol{\rho}$ take the form

$$\boldsymbol{\rho} = (\rho/2, \rho, \rho, \rho, \rho/2), \quad \rho \in (0,1);$$

given the geometric considerations of Figure 1, it makes sense to consider a stricter level of risk for H_1 and H_5, as they have only one neighbouring region each.

Now, take $\beta = 0.2$ and generate 200 samples of sequences $\{X_\ell\}_{\ell\geq 1}$ which follow the density function f; note that the MSPRT choosing H_3 for a given sample would correspond to a correct labeling, and any other case is a mislabel; see Figure 2 for a sample which was labelled correctly. Out of the 200 samples, we

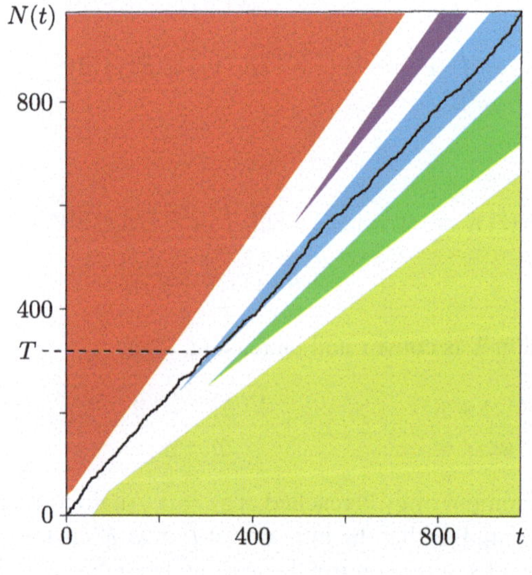

Fig. 2: Renewal process $N(t)$, shown in black, associated to one sample of simulated data $\{X_\ell\}_{\ell\geq 1}$. The process enters the acceptance region of H_3 at $T = 321$ observations.

then count for how many of them the MSPRT chooses H_i, $i \in \Theta$, for different $\rho = 0.5, 0.25, 0.1, 0.05, 0.025$, and provide the empirical average of samples needed to reach a conclusion, T. These results are shown in Table 1 below.

ρ	H_1	H_2	H_3	H_4	H_5	Empirical mean of T
0.5	0	23	159	18	0	123.97
0.25	0	1	199	0	0	373.15
0.1	0	0	200	0	0	981.975
0.05	0	0	200	0	0	1904.245
0.025	0	0	200	0	0	3720.985

Table 1: Counts of 200 sample runs resulting in H_i, $i \in \Theta$, for $\beta = 0.2$ and $\rho = 0.5, 0.25, 0.1, 0.05, 0.025$, along with empirical average of T.

As we can observe, the MSPRT is remarkably accurate at identifying the correct hypothesis, even when ρ is moderately large. In fact, the discrepancy between the risk ρ and the proportion of instances the algorithm chooses the wrong hypothesis is considerable: this is explained by the fact that ρ is proven to be an upper bound to the actual risk, which heuristically seems to be much lower. As expected, the mean number of samples until a decision is taken increases as ρ decreases.

Repeating the aforementioned procedure for $\beta = 0.1$ yields the results presented in Table 2 below.

ρ	H_1	H_2	H_3	H_4	H_5	Empirical mean of T.
0.5	0	23	151	26	0	467.95
0.25	0	0	199	1	0	1469.19
0.1	0	0	200	0	0	3741.455
0.05	0	0	200	0	0	7433.485
0.025	0	0	200	0	0	14445.49

Table 2: Counts of 200 sample runs resulting in H_i, $i \in \Theta$, for $\beta = 0.1$ and $\rho = 0.5, 0.25, 0.1, 0.05, 0.025$, along with empirical average of T

We can notice that in this case, the proportion of times the MSPRT chooses the right hypothesis is comparable to the case $\beta = 0.2$, however, the mean number of samples until a decision is taken is considerably higher. This is consistent with the fact that we are comparing hypotheses that are "closer" than those in the case $\beta = 0.2$, and the algorithm needs more data to reach a proper conclusion.

To stress the MSPRT further, we consider the case where $\beta = 0.05$ and collect the results in Table 3 below.

Once again, the frequency with which the MSPRT chooses the right hypothesis in this instance is comparable to the cases $\beta = 0.2, 0.1$, while the mean number of needed samples until a decision is taken is the largest amongst all cases.

ρ	H_1	H_2	H_3	H_4	H_5	Empirical mean of T
0.5	0	24	156	20	0	1857.165
0.25	0	1	197	2	0	5738.82
0.1	0	0	200	0	0	15084.21
0.05	0	0	200	0	0	29105.39
0.025	0	0	200	0	0	56722.98

Table 3: Counts of 200 sample runs resulting in H_i, $i \in \Theta$, for $\beta = 0.05$ and $\rho = 0.5, 0.25, 0.1, 0.05, 0.025$, along with empirical average of T.

5 Conclusion

We provided a novel understanding of the MSPRT for exponential tilting in terms of the first passage times of a multivariate risk process and a renewal process with time-varying barriers. Both proposed alternative approaches indicate interesting future avenues of research: the former is connected to recent research on queues with simultaneous arrivals [4], and the latter with first passage probabilities of multidimensional random walks in cones [6]. Furthermore, our simulated numerical experiments suggest the need to pursue tighter bounds for the risks guaranteed for a given MSPRT, via both proposed alternative interpretations provided in this paper.

Acknowledgements The authors would like to acknowledge financial support from the Swiss National Science Foundation Project 200021_191984.
Declaration HA and OP declare no conflict of interest related to the current manuscript.

References

1. Albrecher, H., Asadi, P., Ivanovs, J.: Exact boundaries in sequential testing for phase-type distributions. Journal of Applied Probability **51**(A), 347–358 (2014)
2. Avram, F., Palmowski, Z., Pistorius, M.: A two-dimensional ruin problem on the positive quadrant. Insurance: Mathematics and Economics **42**(1), 227–234 (2008)
3. Badescu, A.L., Cheung, E.C., Rabehasaina, L.: A two-dimensional risk model with proportional reinsurance. Journal of Applied Probability **48**(3), 749–765 (2011)
4. Badila, E., Boxma, O., Resing, J., Winands, E.: Queues and risk models with simultaneous arrivals. Advances in Applied Probability **46**(3), 812–831 (2014)
5. Draglia, V., Tartakovsky, A.G., Veeravalli, V.V.: Multihypothesis sequential probability ratio tests. I. Asymptotic optimality. IEEE Transactions on Information Theory **45**(7), 2448–2461 (1999)
6. Garbit, R., Raschel, K.: On the exit time from a cone for random walks with drift. Revista Matematica Iberoamericana **32**(2), 511–532 (2016)
7. Ivanovs, J., Palmowski, Z.: Occupation densities in solving exit problems for Markov additive processes and their reflections. Stochastic Processes and their Applications **122**(9), 3342–3360 (2012)
8. Wald, A.: Sequential Tests of Statistical Hypotheses. The Annals of Mathematical Statistics **16**(2), 117 – 186 (1945)
9. Wald, A., Wolfowitz, J.: Optimum character of the sequential probability ratio test. The Annals of Mathematical Statistics pp. 326–339 (1948)

Bayesian Parameter Estimation for Poisson AR Model

Vladimira Seckarova and Kamil Dedecius

Abstract A popular model in time series analysis is the autoregression model. It explains the current value of a modeled process by means of a weighted average of its past values. In this paper, we focus on two important aspects. Firstly, the modeled variable stands for counts, i.e., it takes values in the set of nonnegative integers. Secondly, we focus on *online* modeling, where the arriving observations sequentially update estimates of the parameters. The emphasis is put on low computational requirements, opening the way towards high-rate real world applications. The paper describes our initial results in this domain. The solution is based on a Poisson autoregression model, where the linear predictor and the modeled variable are canonically linked by the logarithmic function. The adopted Bayesian estimation framework relies on an analytical approximation of the Poisson model by a Gaussian density. A Gaussian prior then provides analytically tractable posterior estimator. Two examples demonstrate the feasibility of the solution. We demonstrate a potential of our approach on the simulated data as well as COVID-19 data and notice advantageous performance of considered methods especially for the use in machine learning.

1 Introduction

Traditional statistical analyses of time series rely on techniques of exponential smoothing and ARMA modeling. In our work, we focus on the autoregression models, where the prediction of the variable of interest is based on a weighted linear combination of its past values. Based on their structure, they allow for modeling

Vladimira Seckarova
School of Mathematics, Monash University, 9 Rainforest Walk, Clayton, Victoria 3800 Australia, e-mail: vladka.seckarova@monash.edu

Kamil Dedecius
Czech Academy of Sciences, Institute of Information Theory and Automation, Pod Vodárenskou věží 4, 182 00 Prague, Czech Republic, e-mail: dedecius@utia.cas.cz

D. R. Wood et al. (eds.), *2021-2022 MATRIX Annals*, MATRIX Book Series 5,
https://doi.org/10.1007/978-3-031-47417-0_27

of stationary processes, random walks (potentially with a drift), or oscillating processes. For this substantial flexibility, the autoregression models have found an immense number of applications in virtually all domains of human activity, from humanities over science to technology [4, 15, 21]. Their flexibility has also led to development of numerous modifications and improvements of the basic autoregression model. To name only a few: the autoregression model with exogeneous inputs (ARX) is popular in the control theory [17], the seasonal autoregression (SAR) model used in domains with fixed periodicity of the observed variable, the nonlinear autoregression and its variants [3], or the vector autoregression (VAR) [22]. The autoregression can even represent the temporal evolution of a random process variance, which gives rise to the autoregressive conditional heteroskedasticity (ARCH) models [9]. The combination of various AR models with moving average models then yields the phenomenal class of ARMA models [4, 21].

We focus on autoregression models of count variables. Recall, that the 'classical' AR models assume Gaussian distribution of the error component. Violation of this assumption need not be critical if the counts are high enough. However, if the counts are low, the 'classical' models are prone to failure. In particular, the forecasted values may take negative values. One possible solution is to consider the variable at time t to be connected with its (transformed) past values through a convenient link function. This gives rise to the generalized linear models (GLMs) [20]. If the link function is the natural logarithm, the resulting model is Poisson. For completeness we mention, that the 'classical' AR model is a GLM too – the link function is the identity function.

A common feature of the inferential methods for the Poisson AR model is their offline character. This means, that they are able to estimate the model parameters from a batch of observed data. However, if new observations are acquired, there is no computationally cheap way to update the previous estimates. In combination with the iterative character of the inferential methods, this may be a significant obstacle to the use of the Poisson AR model in quick online applications.

The present paper reports our initial results in the domain of *sequential* modeling of time series of counts. We exploit the results of the second author in Bayesian estimation of the Poisson GLM [7]. However, they are not fully transferable, as the time series domain faces specific issues, e.g., those connected with the stationarity of AR processes, or the invertibility of MA processes. These topics and their impact on estimation are postponed to future research.

2 Poisson autoregression

Let us assume a probability space $(\Omega, \mathscr{A}, P)$ and $\mathscr{T} \equiv \mathscr{N}_0$ a set of discrete time indices. Furthermore, let $(Y_t)_{t\in\mathscr{T}}$ be a discrete-time stochastic process with nonnegative values $y_t \in \mathbb{N}_0$. Let the variable X_t be measurable with respect to the σ-algebra

$$\mathscr{F}_{t-1} = \sigma(Y_\tau; \tau \leq t-1).$$

If it follows from the generative model

$$Y_t|\mathscr{F}_{t-1} \sim \text{Poisson}(\lambda_t) \tag{1}$$

with $\lambda_t = \exp(\nu_t)$ where

$$\nu_t = \beta_0 + \sum_{i=1}^{p} \beta_i \log(L^i Y_t) \tag{2}$$

$$= \boldsymbol{\beta} \boldsymbol{X}_t \tag{3}$$

$$= \begin{bmatrix} \beta_0 \\ \beta_1 \\ \vdots \\ \beta_p \end{bmatrix}^\mathsf{T} \begin{bmatrix} 1 \\ \log(Y_{t-1}) \\ \vdots \\ \log(Y_{t-p}) \end{bmatrix}, \qquad p \geq 1, \tag{4}$$

we talk about the Poisson autoregressive process of order p. The operator L is the lag (or backshift) operator. The resulting series $(Y_1, \ldots, Y_t)$ is the *time series of counts*. $\boldsymbol{\beta} \in \mathbb{R}^{p+1}$ is the vector of regression coefficients, and $\boldsymbol{X_t} \in \mathbb{R}^{p+1}$ is the regressor. From the properties of the underlying Poisson distribution, it holds

$$\mathbb{E}[Y_t|\mathscr{F}_{t-1}] = \lambda_t = \exp(\boldsymbol{\beta} \boldsymbol{X}_t), \tag{5}$$

$$\text{var}[Y_t|\mathscr{F}_{t-1}] = \lambda_t = \exp(\boldsymbol{\beta} \boldsymbol{X}_t). \tag{6}$$

This property is known as the equivariance.

For obvious reasons, the given model admits zero values of Y_t only in the limit case if $\nu_t \to -\infty$, which makes it inappropriate for most real time series. Several modifications have been proposed in the literature to deal with this issue. For instance, the vector $\boldsymbol{X_t}$ is replaced with

$$\boldsymbol{X}_t' = \begin{bmatrix} 1 \\ \log(Y_{t-1} + c) \\ \vdots \\ \log(Y_{t-p} + c) \end{bmatrix}, \qquad \text{or} \qquad \boldsymbol{X}_t' = \begin{bmatrix} 1 \\ \log(\max(Y_{t-1}, c)) \\ \vdots \\ \log(\max(Y_{t-p}, c)) \end{bmatrix}. \tag{7}$$

Both models have been used to model counts, see [23]. In the former case, a positive constant c ("immigration" rate) is added to the measurements in order to suppress zero values. In the latter case, c in the argument $\max(Y_{t-1}, c)$ relates to the probability of $Y_t > 0$ given that $y_{t-1} = 0$ for Poisson distributed random variable.

Mostly, the value $c = 1$, but it is reported that a reasonable values of c do not have a gross impact on modeling performance [10].

The presented model is only one of many possible choices. For further details see book of Kedem and Fokianos [14, Chapter 4], [16], [6].

3 Sequential estimation of model parameters

The Poisson model (1) is a standard generalized linear model with a *canonical* logarithmic link function [18]. It is characterized by the probability density function

$$p(y_t|\mathscr{F}_{t-1}) = \frac{\lambda_t^{y_t}\exp(-\lambda_t)}{y_t!} \tag{8}$$

$$= \frac{\exp(\boldsymbol{\beta}\boldsymbol{X}_t y_t)\exp(-\exp(\boldsymbol{\beta}\boldsymbol{X}_t))}{y_t!}. \tag{9}$$

The estimation of the vector $\boldsymbol{\beta}$ traditionally relies on the maximum likelihood method [10]. As the score equation is nonlinear in respect of the model parameters, the estimator is not analytically tractable and will have to be found using iterative optimization. Typically, iterative reweighted least squares (IRLS) relying on Newton-Raphson procedure are used. This and other MLE-oriented approaches are discussed, e.g., in [19] and [5, Chapter 3.1]. Alternatively, it is possible to infer $\boldsymbol{\beta}$ by means of the Bayesian framework. This domain clearly relies on expensive Markov chain Monte Carlo (MCMC) methods [19]. In either case, the estimation is

- *offline*, i.e., it requires the batch of data and does not allow for sequential updating of estimates by new observations;
- intrinsically *computationally expensive* due the iterative character of existing methods.

In order to perform a true sequential inference of $\boldsymbol{\beta}$ from continually acquired observations, we may take advantage of the method suggested by the second author in [7] and originally inspired by [8]. The transformation, ascribed to Bartlett and Kendall [2], is that any probability density function of the form

$$\theta^{y-1}\exp(-\theta) \tag{10}$$

is approximately proportional to

$$\exp\left(\frac{y}{2}(\log(\theta)-\log(y))^2\right), \tag{11}$$

see [8]. This allows to rewrite the densities proportional to (10) as the densities of $N(logy, y^{-1})$. This approximation works relatively well for y large enough. The publication [7] of the second author additionally considers a calibration suppressing the approximation error for low values of y. Since the approximation does not admit zero values of y_t, we adopt the principle presented in Section 2 and add a small constant to it. The resulting variable $\tilde{y}_t = y_t + c$ now allows us to rewrite the Poisson GLM as

$$p(y_t|\mathscr{F}_{t-1}) \propto \exp(\boldsymbol{\beta}\boldsymbol{X}_t\tilde{y}_t)\exp(-\exp(\boldsymbol{\beta}\boldsymbol{X}_t)) \tag{12}$$

$$\approx \sqrt{\frac{\tilde{y}_t}{2\pi}}\exp\left(-\frac{\tilde{y}_t}{2}(\log(\tilde{y}_t)-\boldsymbol{\beta}\boldsymbol{X}_t)^2\right). \tag{13}$$

It is well known that a convenient prior distribution for $\boldsymbol{\beta}$ obeying the Gaussian likelihood (13) with known variance is a $p+1$-dimensional Gaussian distribution,

$$\pi(\boldsymbol{\beta}|\mathbb{F}_{t-1}) = \frac{1}{\sqrt{(2\pi)^{p+1}|\boldsymbol{\Sigma}_{t-1}|}} \exp\left(-\frac{1}{2}(\boldsymbol{\beta}-\hat{\boldsymbol{\beta}}_{t-1})\boldsymbol{\Sigma}_{t-1}^{-1}(\boldsymbol{\beta}-\hat{\boldsymbol{\beta}}_{t-1})^{\mathsf{T}}\right), \tag{14}$$

where $\hat{\boldsymbol{\beta}}_{t-1} \in \mathbb{R}^{p+1}$ is the prior point estimator of $\boldsymbol{\beta}$ and $\boldsymbol{\Sigma} \in \mathbb{R}^{(p+1)\times(p+1)}$ is the covariance matrix. The filtration

$$\mathbb{F}_{t-1} = (\mathscr{F}_{\tau})_{\tau=0,\dots,t-1} \tag{15}$$

encompasses all information about the process up to time $t-1$.

The posterior distribution resulting from the Bayes' theorem reads

$$\pi(\boldsymbol{\beta}|\mathbb{F}_t) \propto p(y_t|\mathscr{F}_{t-1})\pi(\boldsymbol{\beta}|\mathbb{F}_{t-1}) \tag{16}$$

$$\propto \sqrt{\frac{y_t}{2\pi}} \exp\left(-\frac{y_t}{2}(\log(y_t)-\boldsymbol{\beta}\boldsymbol{X}_t)^2\right)$$
$$\times \frac{1}{\sqrt{(2\pi)^{p+1}|\boldsymbol{\Sigma}_{t-1}|}} \exp\left(-\frac{1}{2}(\boldsymbol{\beta}-\hat{\boldsymbol{\beta}}_{t-1})\boldsymbol{\Sigma}_{t-1}^{-1}(\boldsymbol{\beta}-\hat{\boldsymbol{\beta}}_{t-1})^{\mathsf{T}}\right) \tag{17}$$

$$\propto \frac{1}{\sqrt{(2\pi)^{p+1}|\boldsymbol{\Sigma}_t|}} \exp\left(-\frac{1}{2}(\boldsymbol{\beta}-\hat{\boldsymbol{\beta}}_t)\boldsymbol{\Sigma}_t^{-1}(\boldsymbol{\beta}-\hat{\boldsymbol{\beta}}_t)^{\mathsf{T}}\right), \tag{18}$$

where the updated hyperparameters $\hat{\boldsymbol{\beta}}_t$ and $\boldsymbol{\Sigma}_t$ follow from a simple algebra. They are given by the formulas

$$\boldsymbol{\Sigma}_t = \left(\tilde{y}_t\boldsymbol{X}_t\boldsymbol{X}_t^{\mathsf{T}} + \boldsymbol{\Sigma}_{t-1}^{-1}\right)^{-1}, \tag{19}$$

$$\hat{\boldsymbol{\beta}}_t = \boldsymbol{\Sigma}_t\left(\boldsymbol{\Sigma}_{t-1}^{-1}\hat{\boldsymbol{\beta}}_{t-1} + \tilde{y}_t\log(\tilde{y}_t)\boldsymbol{X}_t\right). \tag{20}$$

The following sections contain application of the above theory to simulated and real data to see whether choice of modifications (7) and/or constant c affects estimates of hyperparameters $\boldsymbol{\beta}$ and predictions.

4 Example I: Simulated data

In this chapter we focus on performance of Poisson autoregressive model (3) of order $p=2$ for 3 modifications given in (7) and the following values of constant c:

$$\text{Model 1 } [c=1]: \ \nu_t = \beta_0 + \beta_1\log(y_{t-1}+1) + \beta_2\log(y_{t-2}+1) \tag{21}$$

$$\text{Model 2 } [c=1]: \ \nu_t = \beta_0 + \beta_1\log(\max(y_{t-1},1)) + \beta_2\log(\max(y_{t-2},1)) \tag{22}$$

	β_0	β_1	β_2
Model 1	0.233	0.083	0.073
Model 2	0.103	0.05	0.02
Model 3	0.071	0.039	0.022

Table 1 Simulated data example: RMSE of estimates of unknown hyperparameters $\boldsymbol{\beta}$ against pre-set values $\boldsymbol{\beta} = (\beta_0, \beta_1, \beta_2) = (1.1, 0.5, 0.1)$.

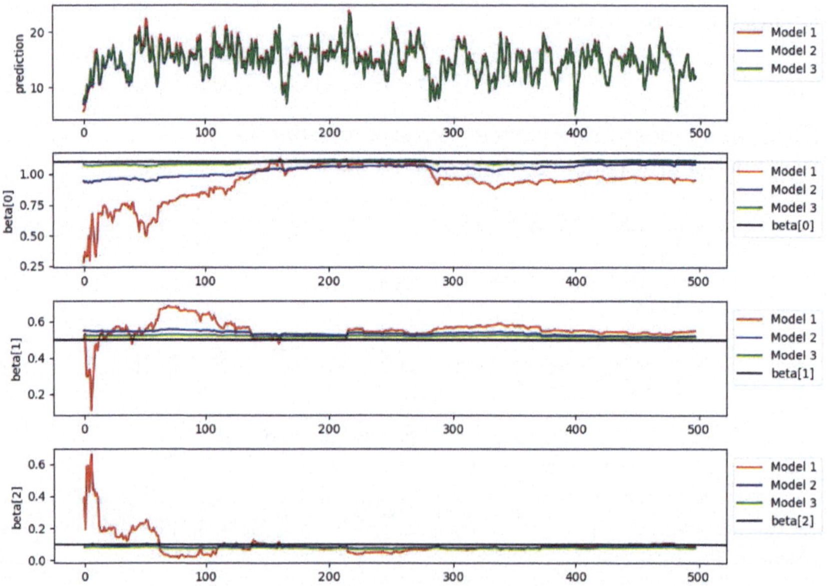

Fig. 1 Simulated data example. Top: Predictions given by models (21), (22), (23). Bottom three plots depict the evolution of estimates of $(\beta_0, \beta_1, \beta_2)$ plotted against values $(1.1, 0.5, 0.1)$ used to generate observations from autoregressive Poisson process of 2nd order.

$$\text{Model 3 } [c = e^{-10}]: \; v_t = \beta_0 + \beta_1 \log(\max(y_{t-1}, e^{-10})) + \beta_2 \log(\max(y_{t-2}, e^{-10})) \tag{23}$$

To demonstrate effectiveness of model (3) with modifications (7) we generated first two observations from Poisson distribution with $\lambda = 5$ and the rest of the data from Poisson distribution with updated value of the rate parameter according to modifications (21), (22), (23).

Figure 1 shows the predictions and comparison of estimated values $\hat{\boldsymbol{\beta}}$ to the original values of hyperparameters $\boldsymbol{\beta}$ based on models (21), (22), (23). The prediction trendline as well as the root mean square errors (RMSE) for estimates of $\boldsymbol{\beta}$ are similar for all three models with Model 2 and Model 3 having slightly lower values of RMSE, see Table 1. However, Model 1 estimates only stabilise around time step 150 while the estimates based on Model 2 and Model 3 stabilise early on. Although all three models performed well in terms of predictions, the behaviour of regression co-

	Model 1	Model 2	Model 3
RMSE (whole dataset)	45.26	45.86	45.36
RMSE (train/test)	3.98	3.26	4.36

Table 2 Real data example: RMSE of predictions based on COVID-19 data from the whole year and on testing set when splitting into training set and testing set.

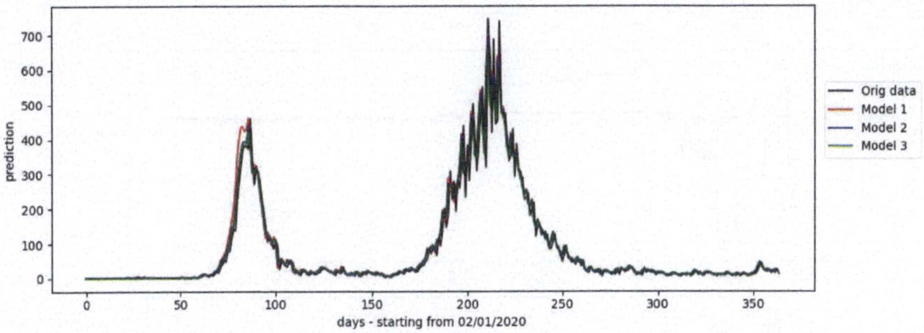

Fig. 2 Black line: numbers of new confirmed COVID-19 cases - raw data. Coloured lines: predictions using models (21), (22), (23) for the whole year.

efficients estimates especially for Model 1 (with possible constraints on the values) is of interest in future research.

5 Example II: Real time series

To demonstrate effectiveness of models (21), (22), (23) on real time series we use data from COVID-19 Google Open Dataset (COD) https://health.google.com/covid-19/open-data/, i.e., we use daily numbers of new confirmed cases in Australia from the first year (365 days) starting from 01/01/2020. Example of another approach dealing with COVID-19 data using Poisson AR model can be found in [1].

5.1 Whole dataset

In this section we let our models run on the whole set of 365 data. Figure 2 shows the original data and trendlines obtained – produced trendlines are very similar and resemble the raw data trendline. RMSE for all three models are very similar, see Table 2.

Figure 3 shows the estimates of unknown $\boldsymbol{\beta}$ at each time step. While the estimates based on Model 2 and Model 3 stabilise early, estimates based on Model 1 only stabilise after $2/3$ of the data has been processed.

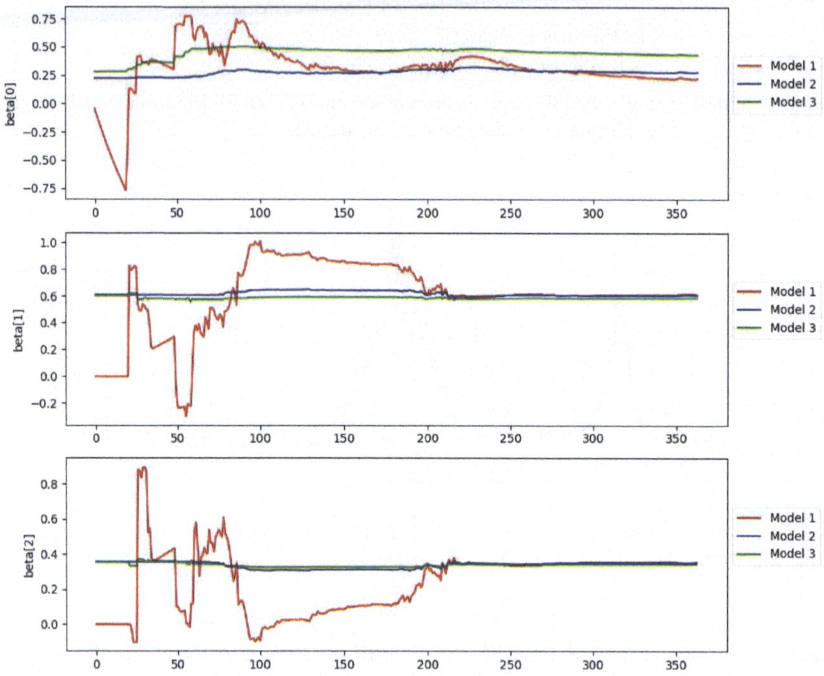

Fig. 3 Estimates of hyperparameters $\boldsymbol{\beta} = (\beta_0, \beta_1, \beta_2)$ based on modifications (21), (22), (23) for COVID-19 data from Australia (whole year starting from 01/01/2020).

This is a potential issue especially if we were to use machine learning techniques for $\boldsymbol{\beta}$ estimation in future, e.g., splitting dataset into training part and testing part - we investigate this in the next subsection.

5.2 Training and testing set

In this section we divide COVID-19 dataset into two disjoint subsets - a training set and a testing set. We let all three models run on the first subset of data (training set); last estimated value of hyperpameters $\boldsymbol{\beta}$ is then taken and used to obtain predictions. We then compare the predictions to appropriate subset of raw data (testing set).

Firstly, we use first 80% of data as a training set to obtain an estimate for unknown $\boldsymbol{\beta}$. Since the estimates, especially in Model 1, are stabilised by this time point the predictions for all three considered models are very similar to the values of raw data, see Figure 4.

All three considered modifications perform well in terms of predictions; their RMSE are quite small and not significantly different, see Table 2; however the behaviour of hyperparameter (regression coefficients) estimates deserves further in-

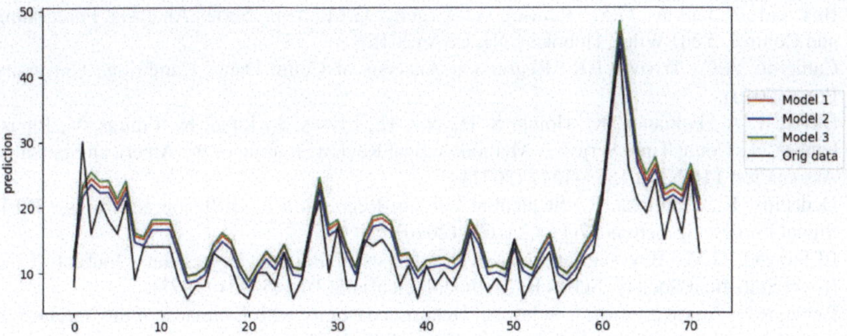

Fig. 4 Predictions based on splitting of the real data into: first 80% training data, the rest 20% testing data.

vestigation with introduction of possible constraints on their values and perhaps inclusion of new modifications.

6 Conclusion and future work

We have presented our initial results in the domain of sequential modeling of count data. This domain becomes increasingly popular due to the omnipresence of count data, and the rapid development of cheap smart appliances, Internet of Things etc. The examples demonstrate that the proposed method is able to reliably estimate the true model parameters. The example focused on a real time series then shows that the method is applicable to real-world problems.

There is a huge research potential in the domain. In particular, we will focus on two issues. First, the real-world count data often contain more zeros than prescribed by the Poisson model. This issue is known as zero-inflation [11]. Second, the equivariance assumption is rather limiting in practical applications. In the GLM domain, there exist more apt models for under- and overdispersion, e.g., the negative binomial or the generalized Poisson models [12, 13].

References

1. Agosto, A., Campmas, A., Giudici, P., Renda, A.: Monitoring COVID-19 contagion growth. Statistics in Medicine **40**, 4150–4160 (2021).
2. Bartlett, M.S., Kendall, D.G.: The Statistical Analysis of Variance-Heterogeneity and the Logarithmic Transformation. Supplement to the Journal of the Royal Statistical Society **8**(1), 128–138 (1946).
3. Billings, S.A.: Nonlinear System Identification: NARMAX Methods in the Time, Frequency, and Spatio-Temporal Domains. John Wiley & Sons (2013)

4. Box, G.E.P., Jenkins, G.M., Reinsel, G.C., Ljung, G.M.: Time Series Analysis: Forecasting and Control, 5 ed. Wiley, Hoboken, NJ, USA (2015)
5. Cameron, A.C., Trivedi, P.K.: Regression Analysis of Count Data. Cambridge University Press (2013).
6. Davis, R. A., Fokianos, K., Holan, S. H., Joe, H., Livsey, J., Lund, R., Pipiras, V., Ravishanker, N.: Count Time Series: A Methodological Review, Journal of the American Statistical Association **116**(535), 1533–1547 (2021).
7. Dedecius, K., Zemlicka, R.: Sequential Poisson Regression in Diffusion Networks. IEEE Signal Processing Letters **27**(1), 625–629 (2020).
8. El-Sayyad, G.M.: Bayesian and Classical Analysis of Poisson Regression. Journal of the Royal Statistical Society: Series B (Methodological) **35**(3), 445–451 (1973).
9. Engle, R.F.: Autoregressive Conditional Heteroscedasticity with Estimates of the Variance of United Kingdom Inflation. Econometrica **50**(4), 987–1007 (1982).
10. Fokianos, K., Tjøstheim, D.: Log-Linear Poisson Autoregression. Journal of Multivariate Analysis **102**(3), 563–578 (2011).
11. Haslett, J., Parnell, A.C., Hinde, J., Andrade Moral, R.: Modelling Excess Zeros in Count Data: A New Perspective on Modelling Approaches. International Statistical Review pp. 1–21 (2021).
12. Hilbe, J.M.: Negative Binomial Regression. Cambridge University Press, Cambridge (2007).
13. Hilbe, J.M.: Modeling Count Data. Cambridge University Press (2014).
14. Kedem, B., Fokianos, K.: Regression Models for Time Series Analysis. Wiley (2005).
15. Koop, G.: Bayesian Econometrics. Wiley (2003).
16. Liboschik, T., Fokianos, K., Fried, R.: tscount: An R Package for Analysis of Count Time Series Following Generalized Linear Models. Journal of Statistical Software **82**(5), 1–51 (2017).
17. Ljung, L., Glad, T.: Modeling and Identification of Dynamic Systems. Prentice Hall (1994).
18. McCullagh, P., Nelder, J.A.: Generalized Linear Models, Second Edition. Chapman and Hall/CRC Monographs on Statistics and Applied Probability Series. Chapman & Hall (1989).
19. Myers, R.H., Montgomery, D.C., Vining, G.G., Robinson, T.J.: Generalized Linear Models. Wiley Series in Probability and Statistics. John Wiley & Sons, Inc., Hoboken, NJ, USA (2010).
20. Nelder, J.A., Wedderburn, R.W.M.: Generalized Linear Models. Journal of the Royal Statistical Society. Series A (General) **135**(3), 370 (1972).
21. Pollock, D.S.G.: Handbook of Time-Series Analysis, Signal Processing and Dynamics. Academic Press (1999)
22. Reinsel, G.C.: Elements of Multivariate Time Series Analysis. Springer (2003).
23. Zeger, S.L., Qaqish, B.: Markov Regression Models for Time Series: A Quasi-Likelihood Approach. Biometrics **44**(4), 1019–1031 (1988).

Nonparametric Bayesian inference for stochastic processes with piecewise constant priors

Denis Belomestny, Frank van der Meulen, and Peter Spreij

Abstract We present a survey of some of our recent results on Bayesian nonparametric inference for a multitude of stochastic processes. The common feature is that the prior distribution in the cases considered is on suitable sets of piecewise constant or piecewise linear functions, that differ for the specific situations at hand. Posterior consistency and in most cases contraction rates for the estimators are presented. Numerical studies on simulated and real data accompany the theoretical results.

Key words: Diffusion coefficient; Dispersion coefficient; Gibbs sampler; (Inverse) Gamma Markov chain prior; Nonparametric Bayesian estimation; Stochastic differential equation; Volatility; Point process; Microstructure noise; Lévy density, Gamma-type subordinators.

2000 Mathematics Subject Classification. Primary: 62G20, Secondary: 62M05.

1 Introduction

This paper is a survey of our recent results, obtained with collaborators, in Bayesian nonparametric function estimation. Although the settings and corresponding mod-

Denis Belomestny
Faculty of Mathematics, Duisburg-Essen University, Thea-Leymann-Str. 9, D-45127 Essen, Germany and Faculty of Computer Sciences, HSE University, Moscow, Russian Federation. e-mail: denis.belomestny@uni-due.de

Frank van der Meulen
Department of Mathematics, Vrije Universiteit, De Boelelaan 1111, 1081HV Amsterdam, The Netherlands. e-mail: f.h.van.der.meulen@vu.nl

Peter Spreij
Korteweg-de Vries Institute for Mathematics, University of Amsterdam, P.O. Box 94248, 1090 GE Amsterdam, The Netherlands and Institute for Mathematics, Astrophysics and Particle Physics, Radboud University, Nijmegen, The Netherlands. e-mail: spreij@uva.nl

D. R. Wood et al. (eds.), *2021-2022 MATRIX Annals*, MATRIX Book Series 5,
https://doi.org/10.1007/978-3-031-47417-0_28

els that we consider exhibit a large variety (volatility estimation in diffusion models, Poisson intensity estimation, estimation for gamma-driven stochastic differential equations, estimation in the presence of microstructure noise, and Lévy density estimation for Gamma-type subordinators), they all share the following common property. The function to be estimated, which can be a function of time or space, is approximated by piecewise constant functions (with one exception that deals with piecewise linear functions). This assumption is one of the possibilities to reduce the estimation of an infinite dimensional parameter to the estimation of only finitely many parameters, this being inevitable if one has to *compute* estimators.

We adopt the Bayesian approach by specifying a prior distribution supported on piecewise constant functions. The goal is then to compute the posterior distribution and show it has good properties, when considered from a frequentist point of view. In this work, the first goal is either accomplished by using conjugate priors or using Markov Chain Monte Carlo to obtain samples from the posterior. The second goal is achieved by showing posterior consistency and quantification of posterior contraction rates. A thorough introduction to these concepts is given in [13]. Informally, posterior consistency means that the posterior mass of a fixed ε-ball around the "true" function tends to 1 as the sample size increases. The contraction rates quantifies the rate at which we can shrink the radius of these balls with increasing sample size while still asymptotically having posterior mass 1.

It turns out that this procedure is feasible in all cases that we consider. It is intuitively clear that rougher (continuous) functions are more difficult to estimate than smoother ones. The degree of smoothness is measured by the Hölder exponent in our approach. We assume in all cases that the functions to be estimated are Hölder continuous. We obtain contraction rates that depend on the Hölder exponent, and the smoother the function, the better the estimation procedure's contraction rate. As we will demonstrate, despite the entirely different settings, the contraction rates are identical in all cases.

For theoretical and numerical implementations, one has to specify priors on the approximating piecewise continuous functions. We will assume a fixed set of bins, reflecting the resolution at which we aim to learn the unknown function. In a first approach, we will a priori take the (finitely many) values on the bins as *independent* random variables with distributions from an appropriate chosen class (mostly gamma or inverse gamma distributions), such that the prior is (partially) conjugate. These specifications lead to the theoretical results on the posterior distributions mentioned above.

It also appeared that better numerical results, clearly visible in the figures we provide and less sensitive to the chosen set of bins, can be obtained by relaxing the independence assumption on the prior distribution. We accomplish this observation by using appropriate Markov chain priors on the levels of the piecewise constant functions. A principal motivation is that such a prior introduces dependence between the adjacent levels of the piecewise constant approximating functions, which is a rather natural property if one realizes that values of continuous functions tend to be close to each other on adjacent intervals. The theoretical results on the contraction rates we show in several cases coincide with the minimax convergence rates in frequentist estimation. From that asymptotic point of view, the estimators with the Markov chain prior cannot be better than the ones with the independent priors. Nevertheless, in practical situations, they turn out to be more accurate.

A great variety of numerical examples in all settings are presented. They include results from simulated examples to illustrate the theoretical properties and from real data to derive some conclusions based on our methods.

The paper is organized as follows. Section 2 details the results on volatility estimation for stochastic differential equations driven by a Brownian motion, Section 3 is devoted to estimation for partial observations of diffusion processes in presence of microstructure noise, Section 4 treats intensity estimation for Poisson processes, Section 5 gives results on estimation of the volatility function in gamma-driven stochastic differential equations, Section 6 is on estimation of the Lévy density for Gamma-type subordinators. Finally Section 7 summarizes and concludes.

A remark on the notation throughout the paper: for two sequences (a_n) and (b_n) of real numbers we write $a_n \asymp b_n$ if there are constants c and C in $(0,\infty)$ such that eventually $cb_n \leq a_n \leq Cb_n$.

2 Volatility estimation

In this section, we consider the nonparametric Bayesian estimation of a volatility function, also synonymously referred to as the dispersion coefficient, that arises in diffusion models. This function may be a smooth function of the time parameter, but it may also happen that it is a realization of another diffusion process, like a Brownian motion. Hence the function to be estimated may have various orders of regularity, which we characterize by their degree of Hölder continuity. This section is based on [15], [17], which also contain many references to related literature.

2.1 Problem formulation

Stochastic differential equations (SDEs) driven by a Brownian motion have been widely used as models in numerous applications ranging from physics to engineering and to finance. Their solutions, weak or strong, are diffusion processes. In this section we consider a one-dimensional diffusion process X satisfying the stochastic differential equation (SDE)

$$\mathrm{d}X_t = a(t,X_t)\,\mathrm{d}t + s(t)\,\mathrm{d}W_t, \quad X_0 = x_0, \quad t \in [0,1]. \tag{1}$$

Here W is a standard Wiener process, and a and s are the drift function and volatility function, respectively. Note that s is a function of time only. It is further assumed that b and s are such that (1) admits a weak solution that is unique in law. We assume that observations

$$\mathscr{X}_n = \{X_{t_{0,n}}, \ldots, X_{t_{n,n}}\}$$

from the solution X to (1) are available, where $t_{i,n} = \frac{i}{n}$, $i = 0, \ldots, n$. In the statistical set-up, we denote the true drift and diffusion coefficient by a_0 and s_0, Our aim is to estimate s_0 nonparametrically within the Bayesian setup. We note that nonparametric Bayesian estimation of volatility is very different from the drift coefficient estimation: the latter fundamentally relies on the equivalence of laws of continuously observed diffusion processes that have the same diffusion coefficient. This is not applicable when the diffusion coefficient itself is unknown and is a parameter to be estimated when a continuous record of observations is available. In the current set-up on the contrary, with *discrete time* observations, there is equivalence of laws and it is therefore possible to derive a likelihood, at least theoretically, a fundamental tool for Bayesian inference. An implementable version of such a likelihood is obtained by *intentionally* misspecifying the drift coefficient a by taking it equal to zero, and employing a (conjugate) histogram-type prior on the diffusion coefficient, that has piecewise constant realisations on bins forming a partition of $[0,1]$. Due to this, our nonparametric Bayesian method to estimate the dispersion coefficient s_0 in (1) is easily implemented. Theoretical asymptotic results for estimating s will be obtained in the case when $n \to \infty$ (high frequency observations), keeping the observation interval $[0,1]$ fixed. For simplicity, we take $x_0 = 0$. Because of the fixed interval $[0,1]$, consistent estimation of the drift is impossible, but the high frequency observations make it possible to ignore the drift if the aim is to (consistently) estimate the volatility. We will make this precise in Proposition 2.3.

2.2 Likelihood, prior and posterior

Under the assumption that the drift in (1) is zero, the (pseudo-)likelihood is that of a Gaussian sample and is given by

$$L_n(s) = \prod_{i=1}^{n} \left\{ \frac{1}{\sqrt{2\pi \int_{t_{i-1,n}}^{t_{i,n}} s^2(u)\,\mathrm{d}u}} \phi_0 \left(\frac{Y_{i,n}}{\sqrt{\int_{t_{i-1,n}}^{t_{i,n}} s^2(u)\,\mathrm{d}u}} \right) \right\}, \tag{2}$$

where $Y_{i,n} = X_{t_{i,n}} - X_{t_{i-1,n}}$ and $\phi_0(u) = \exp(-u^2/2)$. With Π_n denoting a prior on the dispersion coefficient, provided all the involved quantities are suitably measurable, Bayes' theorem gives that the posterior probability of any measurable subset $S \subset \mathscr{S}$ of dispersion coefficients is given by

$$\Pi_n(S \mid X_{t_{0,n}} \ldots, X_{t_{n,n}}) = \frac{\int_S L_n(s) \Pi_n(\mathrm{d}s)}{\int_{\mathscr{S}} L_n(s) \Pi_n(\mathrm{d}s)},$$

where $\mathscr{S}$ denotes a space on which the prior Π_n is defined.

It follows from (2) that the likelihood depends on the parameter of interest only through the integrals $\int_{t_{i-1,n}}^{t_{i,n}} s^2(u)\,\mathrm{d}u$. Consequently, it appears natural to a priori model the diffusion coefficient as piecewise constant on intervals $[t_{i-1,n}, t_{i,n}]$. However, some smoothing should also be performed, and this can be achieved by aggregated several neighbouring intervals $[t_{i-1,n}, t_{i,n}]$. Thus, to construct the prior, we proceed as follows: Let m be an integer smaller than n. Then we can uniquely write $n = mN + r$ with $0 \leq r < m$, and in fact $N = \lfloor \frac{n}{m} \rfloor$. Both m and N will depend on n (and we also write m_n and N_n to emphasize this when appropriate). With this assumption we have bins $B_k = [t_{m(k-1),n}, t_{mk,n})$, $k = 1, \ldots, N-1$ and $B_N = [t_{m(N-1),n}, 1]$.

Let $\mathscr{S}_n$ be the set of functions s that can be represented as $s = \sum_{k=1}^{N_n} \xi_k \mathbf{1}_{B_k}$. The prior Π_n on the dispersion coefficient $s \in \mathscr{S}_n$ is defined by putting a prior on the coefficients ξ_k's. Since

$$s^2 = \sum_{k=1}^{N_n} \xi_k^2 \mathbf{1}_{B_k} = \sum_{k=1}^{N_n} \theta_k \mathbf{1}_{B_k}, \tag{3}$$

where we have put $\theta_k = \xi_k^2$, equivalently one can place the prior on the coefficients θ_k's of the diffusion coefficient s^2. We call the prior Π_n a histogram-type prior.

As a result of choosing Π_n such that the θ_i are independent having an inverse gamma distribution, we can explicitly compute the conjugate posterior of the θ_i, again of inverse gamma type.

Lemma 2.1. *Assume $\theta_1, \ldots, \theta_N$ are independent with the inverse gamma* $\mathrm{IG}(\alpha, \beta)$ *distribution. Then $\theta_1, \ldots, \theta_N$ are a posteriori independent and, for $k = 1, \ldots, N-1$,*

$$\theta_k \mid \mathscr{X}_n \sim \mathrm{IG}\left(\alpha + m/2, \beta + nZ_k/2\right).$$

with

$$Z_k = \sum_{i=(k-1)m+1}^{km} Y_{i,n}^2,$$

whereas

$$\theta_N \mid \mathscr{X}_n \sim \mathrm{IG}\left(\alpha + (m+r)/2, \beta + nZ_N/2\right),$$

with

$$Z_N = \sum_{i=(N-1)m+1}^{n} Y_{i,n}^2.$$

2.3 Asymptotics

We let $\mathbb{P}_{a_0,s_0}$ denote the law of the path $(X_t \colon t \in [0,1])$ from (1) under the true parameter values (a_0, s_0). In particular the notation $\mathbb{P}_{0,s_0}$ is used for such a law when the drift coefficient is equal to zero. For expectation we use the same subscripts. We denote the prior distribution on the dispersion coefficient by Π_n (with n the number of observations) and write the posterior as $\Pi_n(\cdot \mid \mathscr{X}_n)$. By $\|\cdot\|_2$ we denote the L_2-norm with respect to the Lebesgue measure on the Borel sets of $[0,1]$.

The standing assumption on the volatility in this section is the following.

Assumption 2.2 Assume that the true volatility function s_0 is Hölder continuous on $[0,1]$ with Hölder constant L and Hölder exponent $\lambda \in (0,1]$, $|s_0(u) - s_0(v)| \leq L|u-v|^\lambda$ for all $u, v \in [0,1]$, and is bounded away from zero.

The first result formalises that under our asymptotic regime, we can safely ignore the drift by, erroneously in general, putting it equal to zero.

Proposition 2.3. *Let, for* $\varepsilon > 0$,

$$U_{s_0,\varepsilon} = \{s \in \mathscr{S}_n \colon \|s - s_0\|_2 < \varepsilon\}$$

be the L_2*-neighbourhood of* s_0 *of radius* ε. *Let Assumption 2.2 hold and assume that there exists a sequence* $\varepsilon_n \to 0$

$$\mathbb{E}_{0,s_0}[\Pi_n(U^c_{s_0,\varepsilon_n} \mid \mathscr{X}_n)] \to 0$$

as $n \to \infty$. *Then also*

$$\mathbb{E}_{a_0,s_0}[\Pi_n(U^c_{s_0,\varepsilon_n} \mid \mathscr{X}_n)] \to 0.$$

As we consider the asymptotics $n \to \infty$, we take the number of bins N to depend on the sample size n, and indicate this in our notation by writing N_n. The next result gives the posterior contraction rate $n^{-\lambda/(2\lambda+1)}$ for the L_2-norm. Establishing posterior contraction in the L_2-metric is rather natural, as $\int_0^1 s_0^2(t)\mathrm{d}t$ is the quadratic variation of the process $(X_t : t \in [0,1])$ over the interval $[0,1]$.

Theorem 2.4. *Assume that Assumption 2.2 holds. Assume* $N_n \asymp n^{1/(2\lambda+1)}$. *If we let* $\varepsilon_n \asymp n^{-\lambda/(2\lambda+1)}$, *then for any sequence* h_n *tending to infinity (as* $n \to \infty$*) we have*

$$\mathbb{E}_{a_0,s_0}\left[\Pi_n(\|s^2 - s_0^2\|_2 \geq h_n \varepsilon_n \mid \mathscr{X}_n)\right] \to 0$$

as $n \to \infty$.

A comparison with the frequentist minimax convergence rate in [20] shows that the posterior for the diffusion coefficient contracts at the optimal rate in the L_2-metric. With histogram-type priors considered in this work no further improvement in the posterior contraction rate, when $\lambda = 1$, is possible beyond $n^{-1/3}$, even if the function s_0 is smoother than a Lipschitz function. An intuitive reason for this is that realisations of our histogram-type priors are too rough for this; cf. p. 629 in [37].

2.4 Numerical results

In this section we illustrate the theoretical results of Section 2.3 with a number of numerical examples. For an implementation of our Bayesian analysis a judicious choice of the number of bins N is crucial. For obtaining such a choice there are various methodologies. For instance one can use the Deviance Information Criterion (DIC) of [38] to select N, or choose N as the maximizer of the marginal likelihood

$n \mapsto \int_{\mathscr{S}} L_n(s)\Pi_n(\mathrm{d}s)$. We refer to [15] for details. We use simulations of diffusion processes with known drift and diffusion coefficient to gain insight into the numerical performance of our method. We are particularly interested in the practical consequence of using a pseudo-likelihood ignoring the drift.

First we simulate realisations from the model for different dispersion and drift coefficients. Given subsamples of those realisations sampled at different rates we compute the posterior distribution of the dispersion function using the piecewise constant prior (histogram-type prior) of Lemma 2.1 with varying numbers of bins. As an illustration, plots of marginal posterior bands are compared with the true dispersion function. The marginal posterior bands are obtained by computing $1-\alpha$ central posterior intervals (see [11], p. 33) separately for the coefficients θ_k's using Lemma 2.1. This way of computing the posterior intervals will also be used in later sections. By Proposition 2.3, assuming that there is no drift still leads to consistent Bayesian estimation of the dispersion coefficient, even if the data are from a diffusion process with nonzero drift. This is illustrated by our simulation results of this section.

We simulated sample paths of the diffusion $(X_t : t \in [0,1])$ where the true dispersion coefficient is given by one of

$$s_1(t) = 3/2 + \sin(2(4t-2)) + 2\exp(-16(4t-2)^2),$$
$$s_2(t) = B_t(\omega_0) + 1,$$

and the drift is given by one of

$$a_1(x) = 0,$$
$$a_2(x) = -10x + 20.$$

Note the change of notation in these examples, true dispersion and drift are not denoted s_0 and a_0 here, but differently to distinguish these special choices from the general case. The function s_1 is a benchmark function used in [10] in the context of nonparametric regression, up to a vertical shift to ensure positivity. To define $s_2(t)$, we took a fixed realisation of a Brownian motion B, indicated by ω_0, starting at 1 with $B_t(\omega_0) > -1$ for $t \in [0,1]$. The function s_1 is Lipschitz continuous, while s_2 is Hölder continuous with exponent essentially $\frac{1}{2}$. As the prior on the coefficients on the individual bins we used $\mathrm{IG}(0.1, 0.1)$ distributions and the coefficients were assumed to be independent.

Figures 1 and 3 show marginal 98 % posterior bands for different combinations of bin number and observation regime for both dispersion coefficients when the drift is zero. Figure 2 shows the marginal posterior bands for s_1 which are obtained if an affine drift term $a_2(x) = -10x + 20$ is present, but neglected in the estimation procedure. Comparison with Figure 1 shows that presence of a strong nonzero drift hardly affects the obtained credible bands. Note that credible bands successfully recover the overall shape of the functions s_i, although the recovery is not too refined. The recovery is somewhat less accurate for function s_2, see Figure 3, than for function s_1, see Figure 1. This is in perfect agreement with our theoretical results from Theorem 2.4, that give a slower posterior contraction rate for less smooth functions.

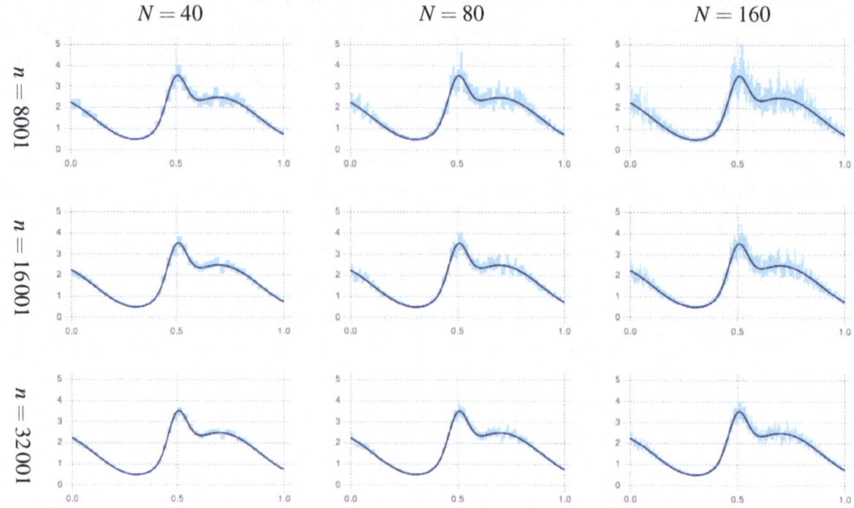

Fig. 1 Estimation results for s_1 with varying number of observations and bins, no drift. The solid curve is the true dispersion coefficient while the light blue areas are 98 % marginal posterior bands. n is the number of observations, N is the number of bins.

2.5 *Inverse Gamma Chain Markov prior*

In this section we postulate another prior on the values of the bins. Rather than assuming independent inverse Gamma distributions as in Section 2.2, we take an *inverse Gamma Markov chain (IGMC)* prior. Posterior inference is still straightforward to implement via Gibbs sampling, as the full conditional distributions are available explicitly and turn out to be inverse Gamma again. We will compare the performance of under both priors. The results of this section are based on [17].

We still consider the diffusion process that solves the SDE (1), with the observations $\mathscr{X}_n = \{X_{t_{0,n}}, \dots, X_{t_{n,n}}\}$. We also keep on taking the drift equal to zero in our statistical analysis. A main reason to depart from the independent prior setting is that this has only limited possibility for adaptation to the local structure of the volatility coefficient, which may become an issue if the volatility has a wildly varying curvature on the time interval $[0, 1]$.

The alternative, fundamentally different, approach, we undertake here, is inspired by ideas in [4] in the context of audio signal modelling, see also [7] and [30], different from the SDE setting that we consider. Instead of using a prior on the (squared)

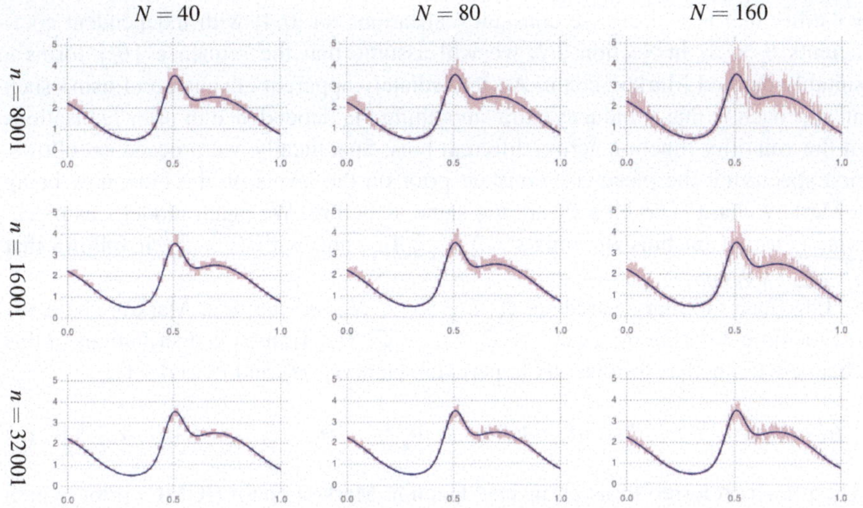

Fig. 2 Estimation for s_1 with varying number of observations and bins and drift $a_2(x) = -10x + 20$. The solid curve is the true dispersion coefficient while the light red areas are 98 % marginal posterior bands. n is the number of observations, N is the number of bins.

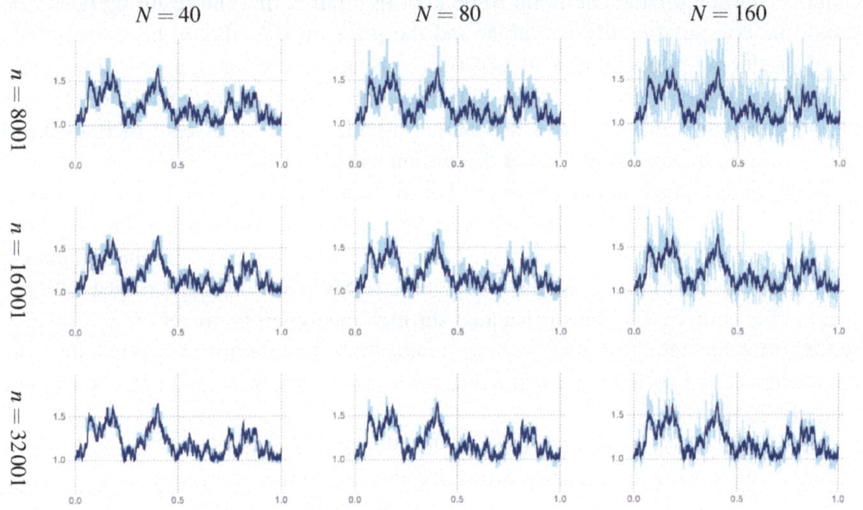

Fig. 3 Estimation results for s_2 with varying number of observations and bins, no drift. The solid curve is the true dispersion coefficient while the light blue areas are 98 % marginal posterior bands. n is the number of observations, N is the number of bins.

volatility that has piecewise constant realisations on $[0,1]$ with independent coefficients θ_k's, as in Section 2.2, we will assume that the sequence $\{\theta_k\}$ forms a suitably defined Markov chain. An immediately apparent advantage of using such an approach is that it induces extra smoothing via dependence in prior realisations of the volatility function across different bins. Specifically, we proceed as follows, first specifying the piecewise constant prior on the levels on the bins, now being a Markov chain. The bins B_k are the same as before. We again model s as piecewise constant on bins B_k, thus $s = \sum_{k=1}^N \xi_k \mathbf{1}_{B_k}$, and with $\theta_k = \xi_k^2$ it follows that $s^2 = \sum_{k=1}^N \theta_k \mathbf{1}_{B_k}$.

Introduce auxiliary variables ζ_k, $k = 2,\ldots,N$, and define a Markov chain using the time ordering $\theta_1, \zeta_2, \theta_2, \ldots, \zeta_k, \theta_k, \ldots, \zeta_N, \theta_N$. Transition distributions of this chain are defined as follows: fix hyperparameters α_1, α_ζ and α, and set

$$\theta_1 \sim \mathrm{IG}(\alpha_1, \alpha_1), \quad \zeta_{k+1}|\theta_k \sim \mathrm{IG}(\alpha_\zeta, \alpha_\zeta \theta_k^{-1}), \quad \theta_{k+1}|\zeta_{k+1} \sim \mathrm{IG}(\alpha, \alpha \zeta_{k+1}^{-1}). \quad (4)$$

The prior is referred to as an inverse Gamma Markov chain (IGMC) prior, which reflects the fact that the distributions in (4) are inverse Gamma. In fact, it can even be verified by direct computations employing (4) that the full conditional distributions of θ_k's and ζ_k's are inverse gamma too, see [17] for details, which facilitates Gibbs sampling. As observed in [4], there are various ways of defining an inverse Gamma Markov chain. The point to be kept in mind is that the resulting posterior should be computationally tractable, and the prior on θ_k's should have a capability of producing realisations with positive correlation structures, as this introduces smoothing among the θ_k's in adjacent bins. The definition of the IGMC prior in the present work, that employs the latent variables ζ_k's, takes care of both these important points. For an additional discussion see [4].

In the practical examples presented below, both the number of bins, N, and the parametes $\alpha, \alpha_1, \alpha_\zeta$ are considered as hyperparameters. Assigning values to them can be done in multiple ways, see [17], but it has been observed in our simulation examples that inferential conclusions with the IGMC prior are quite robust with respect to the choice of N. This is because through the hyperparameters α and α_ζ, the IGMC prior has an additional layer for controlling the amount of applied smoothness; when α and α_ζ are equipped with a prior (as we do), they can in fact be learned from the data.

The test function to be estimated is the blocks function from [8] visualized in Figure 4. With a vertical shift for positivity, this is defined as follows:

$$s(t) = 10 + 3.655606 \times \sum_{j=1}^{11} h_j K(t - t_j), \quad t \in [0,1], \quad (5)$$

where $K(t) = (1 + \mathrm{sgn}(t))/2$, and

$$\{t_j\} = (0.1, 0.13, 0.15, 0.23, 0.25, 0.4, 0.44, 0.65, 0.76, 0.78, 0.81),$$
$$\{h_j\} = (4, -5, 3, -4, 5, -4.2, 2.1, 4.3, -3.1, 2.1, -4.2).$$

Our main goal here was to compare the performance of the IGMC prior-based approach to the IIG prior-based one from Section 2.2. To complete the SDE specification, the true drift coefficient was chosen to be a rather strong linear drift $a_2(x) = -10x + 20$.

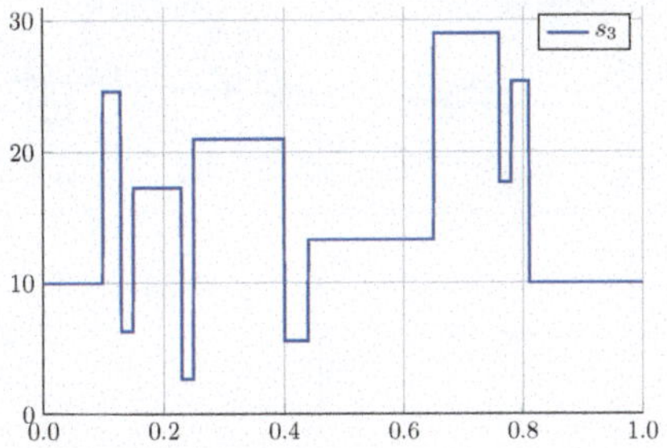

Fig. 4 The blocks function s from Equation (5).

The top left and top right panels of Figurc 5 give estimation results obtained with the IIG prior-based approach from Section 2.2. The number of bins was $N = 160$ or $N = 320$, and in both these cases we used diffuse independent $\mathrm{IG}(0.1, 0.1)$ priors on the coefficients of the (squared) volatility function (see [15] for details). These results have to be contrasted to those obtained with the IGMC prior, plotted in the bottom left and bottom right panels of Figure 5, where we assumed $\alpha_1 = 0.1$ and $\alpha = \alpha_\zeta \sim \mathrm{IG}(0.3, 0.3)$. The following conclusions emerge from Figure 5:

- Although both the IGMC and IIG approaches recover globally the shape of the volatility function, the IIG approach results in much greater uncertainty in inferential conclusions, as reflected in wider marginal confidence bands. The effect is especially pronounced in the case $N = 320$, where the width of the band for the IIG prior renders it almost useless for inference.
- The bands based on the IGMC prior look more 'regular' than the ones for the IIG prior.
- The method based on the IIG prior is sensitive to the bin number selection: compare the top left panel of Figure 5 using $N = 160$ bins to the top right panel using $N = 320$ bins, where the credible band is much wider. On the other hand, the method based on the IGMC prior automatically rebalances the amount of smoothing it uses with different numbers of bins N, thanks to the hyperprior on the parameters α, α_ζ; in fact, the bottom two plots in Figure 5 look similar to each other.

We have shown more pronounced results when using the IGMC prior than by using the IIG prior, although the latter already shows optimal asymptotic properties, as reported in Section 2.3. We will see similar behaviour in Section 4.6.

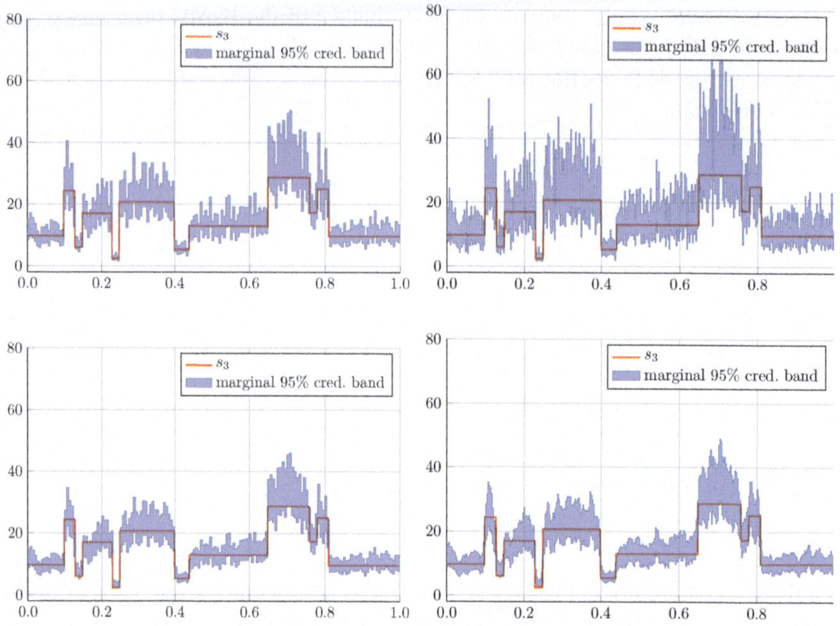

Fig. 5 Volatility function s with superimposed 95% marginal credible band for the IIG prior IG(0.1,0.1), using $N = 160$ (top left panel) and $N = 320$ bins (top right panel). Volatility function s from (5) with superimposed 95% marginal credible band for the IGMC prior, using $N = 160$ (bottom left panel) and $N = 320$ bins (bottom right panel); in both cases, $\alpha_1 = 0.1$ and $\alpha = \alpha_\zeta \sim \mathrm{IG}(0.3, 0.3)$.

2.6 Closing remarks

We have studied nonparametric Bayesian estimation of a volatility function, and presented posterior contraction results when the prior was of independent inverse gamma type. We have given numerical results and have also illustrated that the performance of the statistical procedure had increased when switching from independent priors on the coefficients to a Markov chain prior. The theoretical results can also be applied to real data, see [15] for an example with exchange rate data and Section 3 for a follow up with partial observations.

3 Volatility estimation in presence of microstructure noise

The goal of this section is nonparametric Bayesian estimation of the volatility in presence of market microstructure noise. The exposition is based on [16], which in turn uses some ideas from [17], whose findings have also been used in Section 2.5.

3.1 Problem formulation

As in Section 2, let X solve the one-dimensional stochastic differential equation (SDE) (1). Assume (not necessarily uniformly spaced) observation times $\{t_i : 1 \leq i \leq n\}$ and suppose our observations are $\mathscr{Y}_n = \{Y_1, \dots, Y_n\}$. Here

$$Y_i = X_{t_i} + V_i, \quad 0 < t_1 < \cdots < t_n = 1, \tag{6}$$

and $\{V_i\}$ is a sequence of independent and identically distributed random variables, independent of the Wiener process W driving (1). The measurement errors $\{V_i\}$ are referred to as microstructure noise. They model such features of financial time series as their discreteness or approximations due to market friction. For widely spaced data in time, microstructure noise can be ignored without great consequences for inferring the volatility function. However, for densely spaced data in time (the so-called high-frequency data regime), this ceases to be true and even small amount of the microstructure noise adversely affects the volatility estimates.

As referenced in [16], there exists a large body of literature on nonparametric volatility estimation under microstructure noise. The existing work mainly focusses on estimation of the integrated volatility and is frequentist in its nature. Here we take a Bayesian approach instead. General advantages of the latter have been sufficiently highlighted in the preceding sections.

The key steps of our approach are as follows: we first reduce our model to the Gaussian linear state space model, which is computationally convenient. Next we specify an inverse Gamma Markov chain prior on the volatility function s. This allows us to employ for posterior inference the Gibbs sampler. The latter incorporates a classic Forward Filtering Backward Simulation (FFBS) step.

After detailing the algorithm, we demonstrate the performance of our method on simulated and real data examples, and also provide a theoretical result on the limiting behaviour of the Gamma Markov chain prior.

3.2 Linear state space model

We rigorously derive our Bayesian algorithm in the case when the "true", data-generating volatility s is a deterministic function of time t. Next, if the "true" s is in fact a stochastic process, we simply apply the algorithm without further changes, as if s were deterministic. The intuition is as follows: first the stochastic volatility is generated, and given a realisation of it, the observations are generated by an independent Brownian motion W.

Denote $t_0 = 0$. From equation (1),

$$X_{t_i} = X_{t_{i-1}} + \int_{t_{i-1}}^{t_i} a(t, X_t) \mathrm{d}t + \int_{t_{i-1}}^{t_i} s(t) \mathrm{d}W_t. \tag{7}$$

Due to the properties of the Wiener process paths, over short time intervals $[t_{i-1},t_i]$ the term $\int_{t_{i-1}}^{t_i} s(t)\mathrm{d}W_t$ in (7) dominates the term $\int_{t_{i-1}}^{t_i} a(t,X_t)\mathrm{d}t$. Indeed, the latter scales as Δt_i, whereas the former as $\sqrt{\Delta t_i}$. Following [15], as in Section 2, we act as if the process X had a zero drift, $a \equiv 0$.

We thus *assume* $X_{t_i} = X_{t_{i-1}} + U_i$, where $U_i = \int_{t_{i-1}}^{t_i} s(t)\mathrm{d}W_t$. This implies that

$$U_i \sim N(0, w_i) \quad \text{with} \quad w_i = \int_{t_{i-1}}^{t_i} s^2(t)\mathrm{d}t. \tag{8}$$

Furthermore, $\{U_i\}$ is a sequence of independent random variables, for so are the increments of a Wiener process over disjoint time intervals. To lighten the notation, in the sequel we will write $x_i = X_{t_i}$, $y_i = Y_i$, $u_i = U_i$, $v_i = V_i$. Our model reduces to the *linear state space model*

$$\begin{aligned} x_i &= x_{i-1} + u_i, \\ y_i &= x_i + v_i, \end{aligned} \tag{9}$$

where $i = 1,\ldots,n$. In the time series literature, the first equation in (9) is called the *state equation*, whereas the second equation is referred to as the *observation equation*. We assume that $\{v_i\}$ is a sequence of independent $N(0,\eta_v)$ distributed random variables, also independent of the Wiener process W in (1). As a result, $\{v_i\}$ is independent of $\{u_i\}$. We equip the *initial state* x_0 with the $N(\mu_0, C_0)$ prior distribution. Then (9) turns into a *Gaussian* linear state space model. The latter is most convenient computationally, as one does not have to deal with an *intractable likelihood*. The latter constitutes the main *computational bottleneck* for Bayesian inference in SDE models.

3.3 Prior

For the measurement error variance η_v, we assume a priori an inverse Gamma distribution, $\eta_v \sim \mathrm{IG}(\alpha_v, \beta_v)$. This choice is dictated by conjugacy considerations. In general, the inverse gamma prior on the variance parameter in Gaussian models is common practice in Bayesian statistics.

The construction of the prior for s is more involved. Fix an integer $m < n$. Then we can write $n = mN + r$ with $0 \le r < m$, where $N = \lfloor n/m \rfloor$. Define bins $B_k = [t_{m(k-1)}, t_{mk})$, $k = 1,\ldots,N-1$, and $B_N = [t_{m(N-1)}, 1]$. Assume a priori

$$s = \sum_{k=1}^{N} \xi_k \mathbf{1}_{B_k}, \tag{10}$$

where N (the number of bins) is a hyperparameter. Then $s^2 = \sum_{k=1}^{N} \theta_k \mathbf{1}_{B_k}$, where $\theta_k = \xi_k^2$. The prior specification for s is completed by assigning a prior distribution to the coefficients $\theta_{1:N}$ as in Section 2.5, which we repeat here. We introduce auxiliary variables $\zeta_{2:N}$, and suppose the sequence $\theta_1, \zeta_2, \theta_2, \ldots, \zeta_k, \theta_k, \ldots, \zeta_N, \theta_N$ forms a Markov chain (in this order of variables). The transition distributions of the chain are defined by

$$\theta_1 \sim \mathrm{IG}(\alpha_1, \beta_1), \quad \zeta_{k+1} | \theta_k \sim \mathrm{IG}(\alpha, \alpha\theta_k^{-1}), \quad \theta_{k+1} | \zeta_{k+1} \sim \mathrm{IG}(\alpha, \alpha\zeta_{k+1}^{-1}), \tag{11}$$

where $\alpha_1, \beta_1, \alpha$ are hyperparameters. We refer to this chain as an inverse Gamma Markov chain, see [4]. The variables $\theta_1, \ldots, \theta_N$ are positively correlated, which imposes *smoothing* across different bins B_k: under the prior, the small or large values of θ's on adjacent bins tend to go together. At the same time, we attain partial conjugacy in the Gibbs sampler that we derive below. This leads to simple and tractable MCMC inference.

An uninformative choice $\alpha_1, \beta_1 \to 0$ performs well in practice. The parameter α plays the role of the smoothing parameter. We equip it with a log-normal prior $\log \alpha \sim N(a, b)$. The hyperparameters $a \in \mathbb{R}, b > 0$ are chosen so as to render the hyperprior on α diffuse. The interpretation of hyperparameter N (or equivalently m) is somewhat different than merely a smoothing parameter. It can also be considered as the resolution at which one wants to infer the volatility function. Under finite amounts of data, this resolution cannot be made arbitrarily fine. On the other hand, inference with the IGMC prior is robust with respect to a wide range of values of N due to the fact that the corresponding Bayesian procedure has an additional *regularisation parameter* α that is estimated from the data. See also the earlier Section 2.5 and Section 4 for related phenomena.

3.4 Gibbs sampler for sampling from the posterior

An expression for the posterior of s can be written down in closed form, but does not scale up computationally. The problem is circumvented by following a data augmentation approach: we treat $x_{0:n}$ as *missing data*, whereas the $y_{1:n}$ are the observed data; cf. [39]. An expression for the joint density of all the quantities involved in the model is easily obtained as

$$\begin{aligned} &p(y_{1:n}, x_{0:n}, \theta_{1:N}, \zeta_{2:N}, \alpha, \eta_v) = \\ &\quad \left(\prod_{k=1}^{n} p(y_k \mid x_k, \eta_v) \right) p(x_{0:n} \mid \theta_{1:N}) \\ &\quad \times p(\theta_1) \prod_{k=1}^{N-1} \left[p(\zeta_{k+1} \mid \theta_k, \alpha) p(\theta_{k+1} \mid \zeta_{k+1}, \alpha) \right] p(\alpha) p(\eta_v). \end{aligned}$$

Except for $p(x_{0:n} \mid \theta_{1:N})$, all the densities have been given in the previous subsection. To obtain an expression for the latter, define (with $\Delta_i \equiv \Delta t_i$)

$$Z_k = \sum_{i=(k-1)m+1}^{km} \frac{(x_i - x_{i-1})^2}{\Delta_i}, \quad k = 1, \ldots, N-1,$$

$$Z_N = \sum_{i=(N-1)m+1}^{n} \frac{(x_i - x_{i-1})^2}{\Delta_i}.$$

Next set $m_k = m$ for $k = 1, \ldots, N-1$, and let $m_N = m + r$. It follows that

$$p(x_{0:N} \mid \theta_{1:N}) \propto e^{(x_0 - \mu_0)^2/(2C_0)} \prod_{k=1}^{N} \theta_k^{-m_k/2} \exp\left(-\frac{Z_k}{2\theta_k}\right).$$

Now we can use the Gibbs sampler to sample from the joint conditional distribution of $(x_{0:n}, \theta_{1:N}, \zeta_{2:N}, \eta_v, \alpha)$ given observations $y_{1:n}$. The full conditionals of $\theta_{1:N}$, $\zeta_{2:N}$, η_v required in the Gibbs sampler are easy to derive and are of the inverse Gamma type. Furthermore, the parameter α is updated via a Metropolis-Hastings step; see [16] for all the details. This is quite similar conceptually to Section 4.5 below. On the other hand, for updating $x_{0:N}$, conditional on all other quantities in the model, we use the classical Forward Filtering Backward Simulation (FFBS) algorithm for Gaussian state space models (cf. Section 4.4.3 in [31]). The Gibbs sampler that results from these steps is summarized in Algorithm 1. For details, see [16].

Algorithm 1: Gibbs sampler for volatility learning

Data: Observations $y_{1:n}$
Hyperparameters α_1, β_1, α_v, β_v, a, b, N;
Result: Posterior samples $\theta^i_{1:N} : i = 1, \ldots, M$
Initialization $\theta^0_{1:N}$, $\zeta^0_{1:N}$, η^0_v, α^0;
while $i \le M$ **do**
    sample $x^i_{0:n}$ via FFBS;
    sample $\theta^i_{1:N}$ from the inverse Gamma full conditionals;
    sample $\zeta^i_{2:N}$ from the inverse Gamma full conditionals;
    sample η^i_v from the inverse Gamma full conditional;
    sample α^i via a Metropolis-Hastings step;
    set $i = i + 1$.
end

3.5 Asymptotics

In this section we provide some results on the limit behaviour of the Gamma Markov Chain prior. They shed additional light on the prior and are expected to be of use when establishing frequentist asymptotic properties of our nonparametric Bayesian volatility estimation procedure.

We start with results on the prior conditional mean and variance of the θ_k. These results will be exploited to find an asymptotic regime for the θ_k when the number of bins tends to infinity. To simplify the exposition, in this section we depart from the setting with α as a random hyperparameter, but instead we assume it to be a deterministic sequence. The proof of the following proposition is in [16].

Proposition 3.1. *If $\alpha > 2$ is a fixed parameter then the IGMC prior satisfies*

$$\mathbb{E}_k[\theta_{k+1} - \theta_k] = \frac{1}{\alpha - 1}\theta_k, \tag{12}$$

$$\mathbb{V}_k(\theta_{k+1} - \theta_k) = \frac{\alpha(2\alpha - 1)}{(\alpha - 1)^2(\alpha - 2)}\theta_k^2, \tag{13}$$

where $\mathbb{E}_k$ and $\mathbb{V}_k$ respectively denote expectation and variance, conditional on θ_k. Hence, the conditional mean squared error $\mathbb{E}_k(\theta_{k+1} - \theta_k)^2$ equals $\frac{2(\alpha+1)\theta_k^2}{(\alpha-1)(\alpha-2)}$.

As seen from Proposition 3.1, the conditional mean squared error in Proposition 3.1 decreases in α. This, in turn, illustrates the regularising property of this parameter.

We are interested in the behaviour of the prior distribution on the θ_k for large values of α. We will suitably scale α with the number of bins N to obtain a limit result by applying Donsker's theorem (cf. Corollary VII.3.11 in [23]). Take $\alpha = \gamma N$, where γ is a positive scaling factor. We are thus interested in the law of θ_k for $N \to \infty$. This entails simultaneously increasing the number of bins, as well as the dependence of the values on the bins. Then, as shown in [16], with θ_1 fixed, the process $t \mapsto s^2(t)$, with $s^2(t) = \sum_{k=1}^{N} \theta_k \mathbf{1}_{B_k}(t)$, converges weakly to the continuous time process $t \mapsto \theta_1 Z_t$, where

$$\log Z_t = \sqrt{\frac{2}{\gamma}} W_t - \frac{1}{\gamma} t. \tag{14}$$

As a result, the continuous time approximation Z of the process θ_k/θ_1, with Z as in (14), is the Doléans exponential $\mathscr{E}(\sqrt{\frac{2}{\gamma}} W)$, which satisfies the stochastic differential equation

$$\mathrm{d}Z_t = \sqrt{\frac{2}{\gamma}} Z_t \, \mathrm{d}W_t. \tag{15}$$

3.6 Numerical results for the Heston model

The Heston model (see [19]) is a popular stochastic volatility model. According to Heston's model, the asset price process, denoted by S, is governed by the SDE

$$\mathrm{d}S_t = \mu S_t \mathrm{d}t + \sqrt{Z_t} S_t \mathrm{d}W_t,$$

where the process Z follows the CIR or square root process (see [6]),

$$dZ_t = \kappa(\theta - Z_t)dt + \sigma\sqrt{Z_t}dB_t. \tag{16}$$

The driving Wiener processes W and B are correlated with correlation ρ. A common approach in quantitative finance is to log transform S to $X_t = \log S_t$. Then X obeys a diffusion equation with volatility $s(t) = \sqrt{Z_t}$,

$$dX_t = (\mu - \frac{1}{2}Z_t)dt + \sqrt{Z_t}dW_t,$$

which can verified by Itô's formula.

Assume high-frequency observations on the log-price process X with additive noise $V_i \sim N(0, \eta_v)$ are available. Note that the volatility s is a random function. No further knowledge of the data generation mechanism is assumed. The Heston model setting is challenging for any nonparametric volatility estimation method and as such serves as a good test example. To generate synthetic data, we used the parameter values $\mu = 0.05$, $\kappa = 7$, $\theta = 0.04$, $\sigma = 0.6$, $\rho = -0.6$. These are similar to the ones obtained via fitting the Heston model to real data (see Table 5.2 in [32]). Furthermore, the noise variance was $\eta_v = 10^{-6}$. This guarantees a large enough signal-to-noise ratio in the model (9), which is a prerequisite for successful estimation of volatility. Finally, the parameter choice $\mu = 0.04$ corresponds to a realistic 4% log-return rate.

Inference results with $N = 80$ bins are given in Figure 6. These are surprisingly accurate, despite the general difficulty of the problem and the amount of assumptions that went into our estimation procedure. Indeed, the Heston model does not formally satisfy assumptions under which our Bayesian method was derived; this concerns specifically the non-deterministic character of the volatility function s.

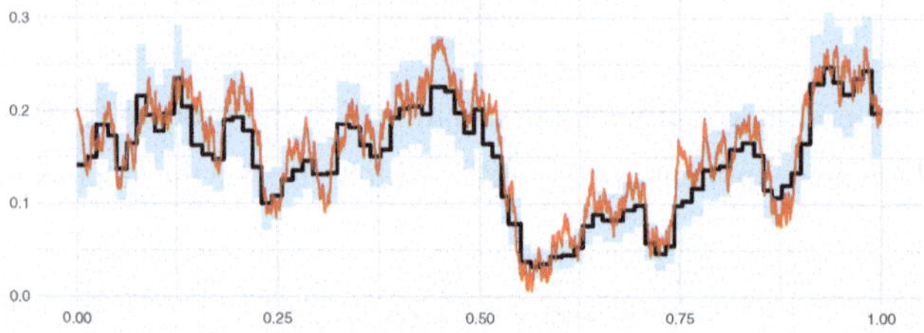

Fig. 6 Posterior mean and pointwise 95% credible band for the Heston model. The true volatility function is plotted in red, the black step function gives the posterior mean, and the credible band is shaded in blue.

We also performed experiments for a variation of the above Heston model, where we replaced the squared volatility process with the limit process $\theta_1 Z$, where Z is as in (14), or, which is the same, as in (15), with $\sqrt{\frac{2}{\gamma}} = 0.6$. These experiments illustrate the influence of the number of bins N and the starting values, see Figures 7 and 8. Here the results with a smaller number of bins $N = 40$ combined with a low initial value are less satisfactory than with $N = 80$ and higher starting value.

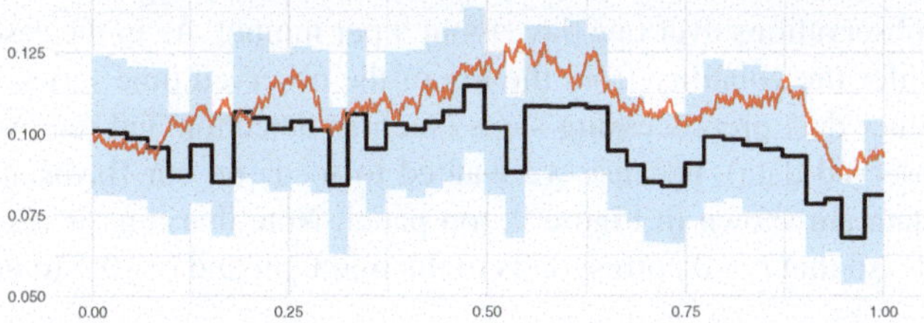

Fig. 7 Posterior results for the Heston model where the CIR volatility is replaced by the root of the continuous time limit of the prior, $N = 40$, starting value of the volatility is 0.1. The colors have the same meaning as in Figure 6.

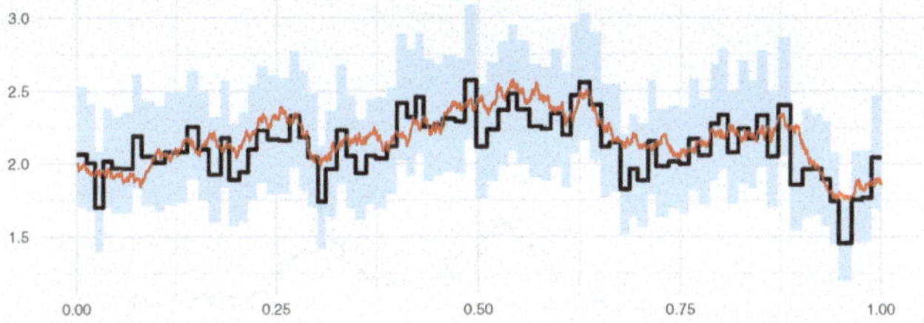

Fig. 8 Posterior results for the Heston model where the CIR volatility is replaced by the root of the continuous time limit of the prior, $N = 80$, starting value of the volatility is 2.0.
f

3.7 Exchange rate data

In this section, we infer volatility of the high frequency foreign exchange rate data made available by Pepperstone Limited, the London based forex broker.[1] As we shall show shortly, the inferred volatility looks realistic, however there is substantial uncertainty surrounding the inferential results (as demonstrated by the wide marginal credible band).

---

[1] Note that the data are not available from the Pepperstone website any longer, but can be obtained directly from the present authors.

In greater detail, we use the EUR/USD tick data (bid prices) for 2 March 2015. Note that the independent additive measurement error model (6) becomes harder to justify for highly densely spaced data in time. Therefore we opted to subsample the data, retaining every 10th observation. Subsampling is a common strategy in this context; see, e.g., Section 5 in [28]. In our case subsampling results in a total of $n = 13025$ observations over one day, about 9 per minute. As in the case of Heston model, we infer the volatility from the log of the observed time series. We did not apply any other data preprocessing steps (such as detection and removal of jumps from the observed data), because we wanted to focus on our Bayesian procedure itself. The data are shown in Figure 9, top panel. Note that time is rescaled to the interval $[0,1]$, so that $t = 0$ corresponds to the midnight and $t = 0.5$ to noon.

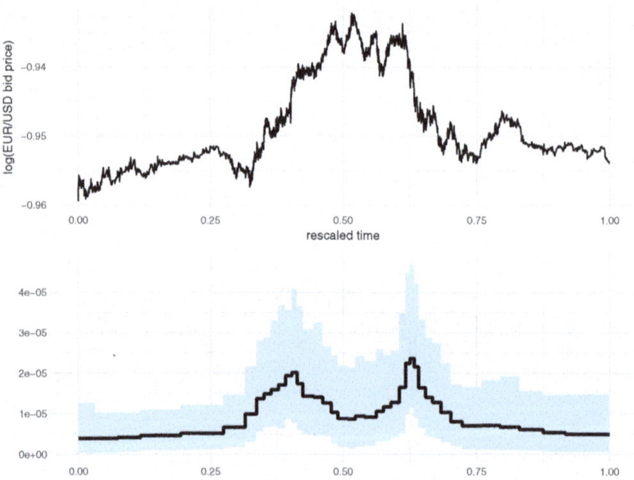

Fig. 9 Top: Natural logarithm of the EUR/USD exchange rate data for 2 March 2015 analysed in Section 3.7. Bottom: Posterior mean (black curve) and pointwise 95% credible band (blue band) for the volatility function. The time axis is rescaled to $[0,1]$.

The inference results are given in the bottom panel of Figure 9. We used $N = 96$ bins, which corresponds to updating volatility every 15 minutes. There is considerable uncertainty in the inferred volatility. Reassuringly, the volatility is lower during the night hours. It also displays two peaks corresponding to the early morning and late afternoon hours of the day. In Figure 10 we give inference results obtained via further subsampling of the data, retaining 50% of the already subsampled observations. The posterior mean is quite similar to that in Figure 9, whereas the wider credible band reflects greater inferential uncertainty due to a smaller sample size. The stability of the inferred volatility across different observation frequency regimes partially validates the model and the approach that we use.

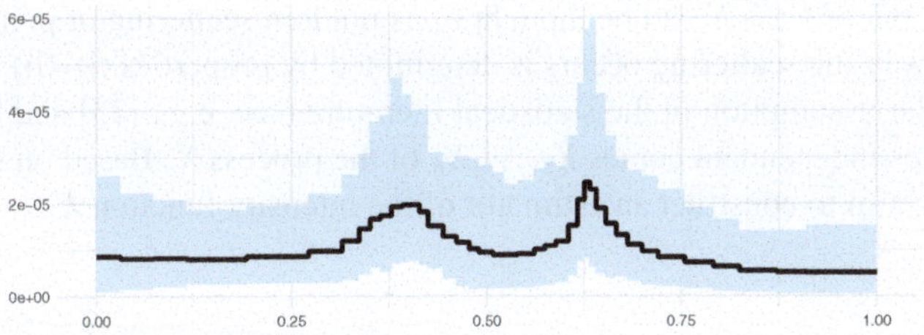

Fig. 10 Posterior mean (black curve) and pointwise 95% credible band (blue band) for the volatility of the further subsampled EUR/USD exchange rate data analysed in Section 3.7. The time axis is rescaled to $[0,1]$.

3.8 Conclusions

Volatility estimation under the presence of microstructure noise is much more challenging than the same task under noiseless observations. As such, accurate inference on volatility under microstructure noise requires large amounts of data. These, however, are readily available in financial applications. In turn, this necessitates finding a volatility estimation method that scales well with data.

The prior that we employ together with and a deliberate, but asymptotically harmless misspecification of the drift (by taking $b \equiv 0$) enable us to combine our earlier work in [17], partially reproduced here in Section 2.5, with the FFBS algorithm for Gaussian linear state space models. This results in an intuitive and fast algorithm to obtain samples from the posterior.

An interesting future research topic in this context is to explicitly account for the possible presence of jumps in financial time series.

4 Poisson intensity estimation

This section is based on [14]. Consider a Poisson point process X on an interval $[0,T]$ of the real line $\mathbb{R}$, which we equip with the Borel σ-field $\mathscr{B}([0,T])$. Such Poisson point processes are also often called non-homogeneous Poisson processes. The process X is a random integer-valued measure on $[0,T]$ (we assume the underlying (complete) probability space $(\Omega,\mathscr{F},\mathbb{Q})$ in the background), such that

1. for any disjoint subsets $B_1,B_2,\ldots,B_m \in \mathscr{B}([0,T])$, the random variables $X(B_1)$, $X(B_2),\ldots,X(B_m)$ are independent, and
2. for any $B \in \mathscr{B}([0,T])$, the random variable $X(B)$ is Poisson distributed with parameter $\Lambda(B)$. Here Λ is a finite measure on $([0,T],\mathscr{B}([0,T]))$, called the intensity measure of the process X. Moreover, it is assumed that Λ admits a density λ with respect to the Lebesgue measure on $\mathscr{B}([0,T])$.

Intuitively, the process X can be thought of as random scattering of points in $[0,T]$, where the way the scattering occurs is determined by properties (i)–(ii).

A popular assumption in the statistical literature (see, e.g., [25] and [26]) is that one observes independent copies $X_1,\dots,X_n$ of the process X. Based on these observations, we aim to construct an estimator of the intensity function λ.

4.1 Likelihood

By Proposition 6.11 in [25] or Theorem 1.3 in [26], the law of X under the parameter value λ admits a density $p(\cdot;\lambda)$ with respect to the measure induced by a standard Poisson point process of rate 1. This density is given by

$$p(\xi;\lambda) = \exp\left(\int_{[0,T]} \log\lambda(x)\mathrm{d}\xi(x) - \int_{[0,T]} (\lambda(x)-1)\mathrm{d}x\right),$$

where $\xi = \sum_{i=1}^{m} \delta_{x_i}$ is a realisation of X (here δ_{x_i} denotes the Dirac measure at x_i). We assume independent observations $X_1,\dots,X_n$ with the same distribution as X and write $X_j = \sum_{i=1}^{m_j} \delta_{X_{ij}}$ for $j = 1,\dots,n$. If we define $X^{(n)} = (X_1,X_2,\dots,X_n)$, then it follows that the likelihood $L(X^{(n)};\lambda)$ can be written as

$$L(X^{(n)};\lambda) = \prod_{j=1}^{n} \exp\left(\int_{[0,T]} \log\lambda(x)\mathrm{d}X_j(x) - \int_{[0,T]} (\lambda(x)-1)\,\mathrm{d}x\right). \tag{17}$$

4.2 Prior

Fix a positive integer N. Let $0 = b_0 < b_1 < \cdots < b_{N-1} < b_N = T$ be a grid of points on the interval $\mathscr{X} = [b_0,b_N]$, for instance a uniform grid. Using this grid, define the bins $B_k = [b_{k-1},b_k)$, $k = 1,\dots,N-1$, with the last bin $B_N = [b_{N-1},b_N]$. In order to define a prior on λ, introduce a collection of piecewise constant functions λ on bins B_k,

$$\lambda(x) = \sum_{k=1}^{N} \psi_k \mathbf{1}_{B_k}(x), \quad x \in [0,T]. \tag{18}$$

We assume $\psi_k \overset{\text{i.i.d.}}{\sim} \mathrm{G}(\alpha,\beta)$ and define the prior Π_N on λ to be the law of the random function (18). We will refer to the prior Π_N as the independent gamma prior. If the grid on $\mathscr{X} = [0,T]$ is taken to be uniform, the prior has three hyperparameters: α, β, and N. Depending on the amount of available data and the degree of prior knowledge on λ, the hyperparameters α,β can for instance be chosen to render the $\mathrm{G}(\alpha,\beta)$ prior diffuse (non-informative). This corresponds to the case when one has little information on the magnitude and shape of λ. The hyperparameter N can be viewed as a smoothing parameter and tuned using an empirical Bayes procedure.

Using a prior with piecewise constant realisations as in (18) is not unnatural. In fact, in practical applications it is often the case that Poisson point processes are only partially observed through aggregate counts of points over successive intervals (cf. [18]). The intensity function cannot be learned beyond the resolution of these intervals, which lends support to using priors with piecewise constant realisations to model the intensity function.

4.3 Posterior

Since the function λ in our case is parametrised by the coefficients $\psi_1, \ldots, \psi_N$, the posterior distribution of the intensity function λ given the data $X^{(n)}$ can be equivalently described through the posterior distribution of $\psi_1, \ldots, \psi_N$ given $X^{(n)}$. Transition from the prior to the posterior can be thought of as updating our prior opinion on λ upon seeing the data $X^{(n)}$.

Lemma 4.1. *Define*

$$H_k = \sum_{j=1}^{n} \sum_{i=1}^{m_j} \mathbf{1}_{\{X_{ij} \in B_k\}} \tag{19}$$

and $\Delta_k = b_k - b_{k-1}$ *for* $k = 1, \ldots, N$. *Then* $\psi_1, \ldots, \psi_N$ *are a posteriori independent and*

$$\psi_k \mid X^{(n)} \sim \mathrm{G}(\alpha + H_k, \beta + n\Delta_k), \quad k = 1, \ldots, N. \tag{20}$$

Thus, the posterior for λ is known in closed form. The posterior mean is given by $\hat{\lambda}(x) = \sum_{k=1}^{N} \hat{\psi}_k \mathbf{1}_{B_k}(x)$, $x \in [0, T]$, with

$$\hat{\psi}_k = \frac{H_k + \alpha}{n\Delta_k + \beta}, \quad k = 1, \ldots, N. \tag{21}$$

Marginal $1 - \gamma$ credible bands for λ can be obtained by producing $1 - \gamma$ credible intervals for ψ_k, using the lower and upper $\gamma/2$-quantiles of the gamma distribution. This gives Bayesian uncertainty quantification for estimation of λ.

4.4 Asymptotics

In this subsection we perform the asymptotic frequentist analysis of the Bayesian procedure we introduced above. This concerns the study of the asymptotic properties of the posterior measure as the sample size $n \to \infty$.

For simplicity in the proofs, we assume that the grid $\{b_k\}$ defining the bins B_k is uniform: $b_k = Tk/N$, $k = 0, \ldots, N$, so that the bins are of equal width $\Delta_k = \Delta = T/N$. Our first result shows that the posterior mean $\hat{\lambda}$ is a consistent estimator of λ_0, and establishes its convergence rate. The expectation $\mathbb{E}[\cdot]$ here and in the sequel is always under the law of the observations with the "true" parameter value λ_0.

Theorem 4.2. *Assume the intensity function $\lambda_0\colon [0,T] \to (0,\infty)$ is Hölder continuous: there exist constants $L > 0$ and $0 < h \le 1$, such that*

$$|\lambda_0(x) - \lambda_0(y)| \le L|x-y|^h, \quad \forall x,y \in [0,T].$$

If $N \asymp n^{1/(2h+1)}$, then

$$\mathbb{E}[\|\hat{\lambda} - \lambda_0\|_2^2] \lesssim n^{-2h/(2h+1)}.$$

The right-hand side is the optimal rate for estimating an h-Hölder-regular intensity function, see [26].

The next result gives a posterior contraction rate in the L_2-metric. Whereas Theorem 4.2 dealt with the 'centre' of the posterior distribution, the theorem below deals with the entire posterior distribution.

Theorem 4.3. *Let the assumptions of Theorem 4.2 hold, and let $\varepsilon_n \asymp n^{-h/(2h+1)}$. Then, for any sequence $M_n \to \infty$,*

$$\mathbb{E}[\Pi_N(\|\lambda - \lambda_0\|_2 \ge M_n \varepsilon_n \mid X^{(n)})] \to 0$$

as $n \to \infty$.

The first conclusion that follows from Theorem 4.3 is that the proposed Bayesian procedure is consistent: as the sample size $n \to \infty$, the posterior puts most of its mass on (shrinking) L_2-neighbourhoods around the true parameter λ_0. Furthermore, the rate $\varepsilon_n \asymp n^{-h/(2h+1)}$ in Theorem 4.3 is the optimal posterior contraction rate for h-Hölder-regular intensity functions.

4.5 Gamma Markov chain prior

The assumption that the coefficients ψ_k of the intensity function λ are a priori independent and gamma distributed can be relaxed. For this we use ideas that have appeared in various works in the audio signal modelling literature (see, e.g., [4], [7] and [30]). This is similar to the approach taken in Section 2.5. Specifically, we postulate that the $\{\psi_k\}$ form a gamma Markov chain (GMC). This chain is defined as follows: introduce auxiliary variables $\zeta_k, k = 2,\ldots,N$, use the time ordering $\psi_1, \zeta_2, \psi_2, \ldots, \zeta_N, \psi_N$, and set

$$\psi_1 \sim \mathrm{G}(\alpha_1, \beta_1), \quad \zeta_k \mid \psi_{k-1} \sim \mathrm{IG}(\alpha_\zeta, \alpha_\zeta \psi_{k-1}), \quad \psi_k \mid \zeta_k \sim \mathrm{G}\left(\alpha_\psi, \frac{\alpha_\psi}{\zeta_k}\right), \tag{22}$$

where $k \in \{2,\ldots,N\}$. The name of the chain reflects the fact that its transition distributions are (inverse) gamma. The parameters α_1, β_1, α_ζ and α_ψ are the hyperparameters of the GMC prior. The hyperparameters α_1, β_1 allow one to 'release' the chain at the origin. This is important to avoid possible edge effects in nonparametric estimation due to a strong specification of the prior at the time origin $t = 0$. A principal aim in using latent variables ζ_k's in (22) is to attain positive correlation between ψ_k's. In the intensity function modelling context this induces smoothing across different bins.

The posterior distribution with the GMC prior is not available in closed form. However, "full" conditional distributions are known in closed form which facilitates Gibbs sampling. The full conditional distributions of ψ_k, ζ_k are

$$\zeta_k \mid \psi_k, \psi_{k-1} \sim \mathrm{IG}\left(\alpha_\zeta + \alpha_\psi, \alpha_\zeta \psi_{k-1} + \alpha_\psi \psi_k\right), \quad k = 2, \ldots, N, \tag{23}$$

$$\psi_k \mid \zeta_{k+1}, \zeta_k, X^{(n)} \sim \mathrm{G}\left(\alpha_\psi + \alpha_\zeta + H_k, \frac{\alpha_\psi}{\zeta_k} + \frac{\alpha_\zeta}{\zeta_{k+1}} + n\Delta_k\right), \quad k = 2, \ldots, N-1, \tag{24}$$

$$\psi_1 \mid \zeta_2, X^{(n)} \sim \mathrm{G}\left(\alpha_1 + \alpha_\zeta + H_1, \beta_1 + \frac{\alpha_\zeta}{\zeta_2} + n\Delta_1\right), \tag{25}$$

$$\psi_N \mid \zeta_N, X^{(n)} \sim \mathrm{G}\left(\alpha_\psi + H_N, \frac{\alpha_\psi}{\zeta_N} + n\Delta_N\right). \tag{26}$$

The Gibbs sampler cycles through formulae (23)–(26) to generate samples from the posterior distribution of $\{\psi_k\}$, $\{\zeta_k\}$ given the data $X^{(n)}$. One can initialize the sampler e.g. by providing values for $\psi_1, \ldots, \psi_N$.

4.6 Numerical results

Consider

$$\lambda_0(x) = 2e^{-x/5}(5 + 4\cos(x)), \quad x \in [0, 10].$$

A principal challenge in inferring this function consists in the fact that it takes small values in the middle part of its domain and has a slope of changing sign. In Figure 11 the posterior mean and marginal 95% posterior credible bands are shown. From the lower panels one can clearly see the beneficial smoothing effect of the GMC-prior.

We refer to [35] for an application of the proposed methods in estimating heart rate variability.

4.7 Closing remarks

The presented approach is computationally appealing, as it scales with the number of bins and not the sample size. One can think of the chosen number of bins as the resolution at which one estimates the intensity function. If this resolution is too fine, the independent Gamma prior results in overfitting. This can be avoided by using the GMC-prior, which induces a smoothing effect on neighbouring bins.

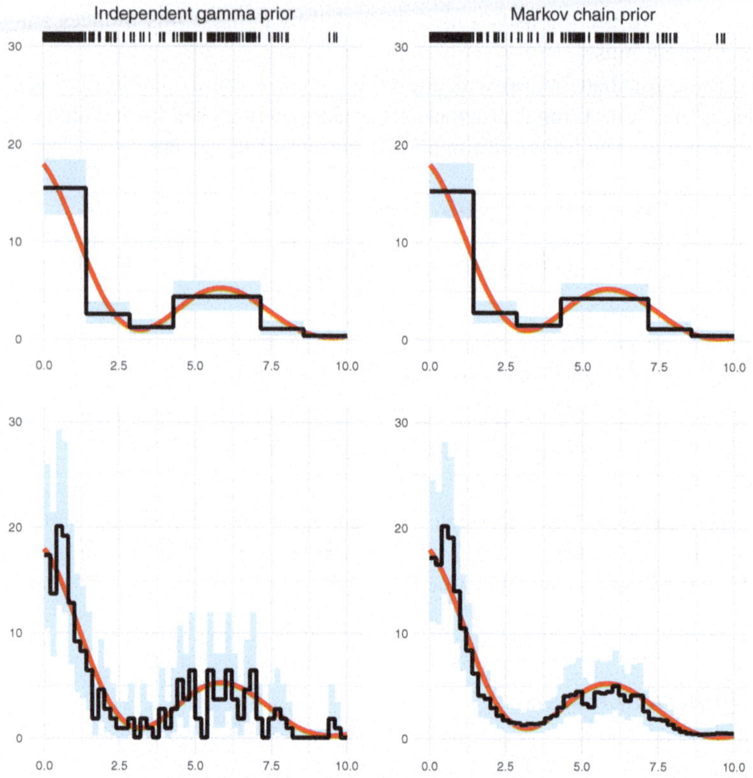

Fig. 11 Estimation results for the oscillating exponential function $\lambda_0(x) = 2e^{-x/5}(5+4\cos(x))$ with $n = 5$ replicates. The top plot corresponds to the method based on the independent gamma prior, with $N = 7$ chosen via the empirical Bayes method (details are in Section 3.3 of [14]). The bottom plot corresponds to the method based on the GMC prior with $N = 50$ bins.

5 Estimation for gamma-driven stochastic differential equations

The goal of the present section, based on [2], is Bayesian nonparametric estimation of the positive local scale function σ that appears in the Lévy-driven stochastic differential equation

$$dX_t = \sigma(X_{t-})\,dL_t,\ X_0 = 0, \tag{27}$$

from observations of the solution X. Here L is a gamma process with $L_0 = 0$ and therefore L is a subordinator, i.e. a stochastic process with monotonous sample paths. Furthermore, L has a Lévy measure ν admitting the Lévy density

$$\nu(x) = \alpha x^{-1}\exp(-\beta x),\ x > 0, \tag{28}$$

where α and β are two positive constants. The process L has independent increments, and $L_t - L_s$ has the Gamma$(\alpha(t-s),\beta)$ distribution for $t > s$, i.e. the distribution with density $p(x) = \frac{\beta^{\alpha(t-s)}}{\Gamma(\alpha(t-s))} x^{\alpha(t-s)-1} \exp(-\beta x)$, $x \geq 0$.

5.1 Preliminaries

Under the assumption that the function σ (in view of financial applications we refer to it as volatility function) is measurable and satisfies a linear growth condition, it has been shown in [3] that Equation (27) admits a weak solution that is unique in law. Note that X is a Markov process. Under the stronger condition that σ is Lipschitz continuous, Equation (27) even has a unique strong solution, see [33, Theorem V.6].

We are ultimately interested in asymptotic properties of our Bayesian procedure, detailed below, for estimating the volatility function σ. To that end, we need a sufficiently rich set of observations and this can be accomplished by the scaled observation process in (29) below, instead of X satisfying (27). So we consider the process X^n given as the solution to

$$\mathrm{d}X_t^n = \frac{1}{n}\sigma(X_{t-}^n)\,\mathrm{d}L_t,\ X_0^n = 0. \tag{29}$$

The scaling factor $\frac{1}{n}$ causes, for large values of n, a 'slow growth' of the process X^n and 'long times' to stay around certain levels. We will thus assume that X^n is observed on a long time interval $[0, T^n]$, where $T^n \to \infty$. In Section 5.4 we will specify T^n and derive asymptotic results for $n \to \infty$.

5.2 Likelihood

For a Bayesian analysis we need to convert a prior distribution into a posterior one, and for this a likelihood ratio is needed. Let us first sketch the set-up.

Let $(\Omega, \mathscr{F}, \mathbb{F}, \mathbb{P})$ be a filtered probability space and let $(L_t)_{t\geq 0}$ be a gamma process adapted to $\mathbb{F}$, whose Lévy measure admits the density ν given by (28). Assume that X is a (weak) solution to (27), and assume that X is observed on an interval $[0,T]$. We denote by $\mathbb{P}_T^\sigma$ its law, a probability measure on the Skorohod space $D[0,T]$ with the Skorohod J_1-topology and the σ-algebra generated by it. In agreement with this notation we let $\mathbb{P}_T^1$ be the law of X when $\sigma \equiv 1$, in which case $X_t = L_t$, $t \in [0,T]$. The measure $\mathbb{P}_T^1$ will serve as a reference measure. Recall (28) and let

$$Y(t,x) := \frac{1}{\sigma(X_{t-})}\nu\Big(\frac{x}{\sigma(X_{t-})}\Big)/\nu(x).$$

In [3] the following proposition has been proven.

Proposition 5.1. *Let σ be a positive locally bounded measurable function on $[0,\infty)$ such that* (27) *admits a weak solution that is unique in law. Assume moreover that σ is lower bounded by a strictly positive constant. Let T be a finite (stopping) time $T>0$. Then the laws $\mathbb{P}_T^\sigma$ and $\mathbb{P}_T^1$ are equivalent on $\mathscr{F}_T^X$ and the corresponding Radon-Nikodym derivative Z_T has the explicit representation:*

$$Z_T = \exp\left(\int_0^T \int_0^\infty \log Y(t,x)\,\mu^X(\mathrm{d}x,\mathrm{d}t) - \int_0^T \int_0^\infty (Y(t,x)-1)\nu(x)\,\mathrm{d}x\,\mathrm{d}t\right), \quad (30)$$

where μ^X is the jump measure of X and both double integrals are a.s. finite.

In the next section we will present, under additional assumptions, an explicit expression for Z_T.

5.3 Prior and posterior

We model σ as a piecewise constant function:

$$\sigma(x) = \sum_{k=1}^K \xi_k \mathbf{1}_{B_k}(x) \quad (31)$$

for bins $B_1 = [0,b_1]$, $B_k = (b_{k-1},b_k], k=2,\dots,K-1$, and $B_K = (b_{K-1},b_K)$, with appropriately chosen increasing sequence of bin endpoints $\{b_k\}$ and the bin number K. The $\{\xi_k\}$ are positive numbers (later on positive random variables). Although we use (31) for our model, we emphasize that the 'true' σ does not need to be piecewise constant. As a final remark we note that when σ is given by (31), (29) still has a unique solution, obtained as a concatenation of stopped gamma processes. Under the specification (31), the general likelihood ratio Z_T of (30) gets an explicit form.

Proposition 5.2. *Suppose that X^n is given by* (29) *with σ given by* (31)*. Let $\tau_k^n = \inf\{t \geq 0 : X_t^n \geq b_k\}$, $k=1,\dots,K$, and write $T=\tau_K^n$. Then*

$$\frac{\mathrm{d}\mathbb{P}_T^\sigma}{\mathrm{d}\mathbb{P}_T^1} = \exp\left\{\beta \sum_{k=1}^K \left[1-n\xi_k^{-1}\right](X_{\tau_k^n}^n - X_{\tau_{k-1}^n}^n) - \alpha \sum_{k=1}^K \left(\tau_k^n - \tau_{k-1}^n\right)\log(\xi_k/n)\right\}.$$

To assign a prior distribution, we take the ξ_k to be independent random variables with an inverse gamma distributions for each of them, that is $\xi_k \sim \mathrm{IG}(\alpha_k,\beta_k)$, and thus have marginal densities $p_k(x) = \beta_k^{\alpha_k}\Gamma(\alpha_k)^{-1}x^{-\alpha_k-1}\exp(-\beta_k/x)$. Therefore, we possibly have to extend the original probability space to carry the ξ_k as well, taking into account that L and the ξ_k have to be independent. We can then use Proposition 5.2, where $\mathbb{P}_T^\sigma$ is to be interpreted as the conditional law of X^n on $[0,T]$ given the ξ_k. It turns out that the posterior distributions of the ξ_k are again of inverse gamma type.

Lemma 5.3. *Let $x_1,\dots,x_K$ denote realisations of $\xi_1,\dots\xi_K$ and let X^n stand for the path $X^n_t, t\in[0,T]$. With $T=\tau^n_K$, we then have from Proposition 5.2 that the posterior joint density of $(\xi_1,\dots,\xi_K)$ is given by*

$$\Pi(x_1,\dots,x_K \mid X^n) \propto \prod_{k=1}^{K} \exp\left(-(n\beta(X^n_{\tau^n_k}-X^n_{\tau^n_{k-1}})+\beta_k)x_k^{-1}\right) x_k^{-\alpha(\tau^n_k-\tau^n_{k-1})-\alpha_k-1}.$$

It follows that the ξ_k are independent under the posterior distribution, and

$$\xi_k \mid X^n \sim \mathrm{IG}(\alpha\Delta\tau^n_k+\alpha_k, n\beta\Delta X^n_{\tau^n_k}+\beta_k), \tag{32}$$

where $\Delta\tau^n_k=\tau^n_k-\tau^n_{k-1}$, $\Delta X^n_{\tau^n_k}=X^n_{\tau^n_k}-X^n_{\tau^n_{k-1}}$.

5.4 Asymptotics

We consider the process X^n satisfying (29), but the standing assumption from here on is that $\sigma:[0,\infty)\to[0,\infty)$ is Hölder continuous, that is, there are constants $H\geq 0$ and $0<\lambda\leq 1$ such that for all $x,y>0$ it holds that $|\sigma(x)-\sigma(y)|\leq H|x-y|^\lambda$. Moreover, σ is assumed to be bounded from below by a positive constant $\underline{\sigma}$. Here are some further assumptions.

- The number of bins and their width depend on n. So we write $B^n_k=(b^n_{k-1},b^n_k]$, $k=1,\dots,K$, $K=K_n$. We assume equidistant bins. Let b_K be the endpoint of the last bin, assumed to be fixed. We take the other bin boundaries b^n_k as $b^n_k=\frac{b_K k}{K}$, $k=1,\dots,K$. A given $x\in(0,b_K]$ then belongs to bin B^n_k with $k=k_n(x)=\lceil\frac{Kx}{b_K}\rceil$.
- We assume the number of bins $K=K_n\asymp n^\kappa$ for $0<\kappa<1$. Then, given also the above assumption on the b^n_k, one has $\Delta b^n_k:=b^n_k-b^n_{k-1}\asymp n^{-\kappa}$ for all k and $x\in B_k$. Note that the bin widths decrease to zero and that the number of bins increases to infinity, with a rate depending on the parameter κ (to be specified below). This behaviour is needed to have better and better approximations of the true function σ with piecewise constant functions.
- We observe the process X^n until it crosses the last bin, i.e. until time $T^n=\tau^n_K$. In [2] it has been shown that the random time T^n is roughly proportional to n.

Although σ is continuous, we assume it to be apriori piecewise constant. That is, parallel to (31) but note that the bins now depend on n, as

$$\xi^n(x)=\sum_{k=0}^{K}\xi_k\mathbf{1}_{B^n_k}(x), \tag{33}$$

where the ξ_k are assigned the inverse gamma prior distributions as in Section 5.3. Consequently, the posterior distributions of the ξ_k are as in (32).

Here is the main result, it says that the contraction rate of the posterior distribution is $n^{-\lambda/(2\lambda+1)}$, λ being the Hölder exponent of σ. We use the notation Π_n

to denote posterior probabilities given the observations of X^n on the time interval $[0,T^n]$.

Theorem 5.4. *Assume the model with volatility* (33) *whereas the true volatility function $x \mapsto \sigma(x)$ is Hölder continuous of order $\lambda \leq 1$ and bounded from below. Let the bin sizes shrink proportional to $n^{-\frac{1}{2\lambda+1}}$ (i.e. $\kappa = \frac{1}{2\lambda+1}$), and let (m_n) be any sequence of real numbers diverging to infinity. Then, for $n \to \infty$*

$$\sup_{x\in[0,b_K]} \Pi_n\left(|\xi^n(x) - \sigma(x)| > m_n n^{-\frac{\lambda}{2\lambda+1}}\right) \to 0 \textit{ in } \mathbb{P}^{\sigma}_{T^n}\textit{-probability.}$$

Moreover, this contraction rate is minimax optimal.

5.5 Numerical illustration

Here we graphically illustrate our inferential results. In agreement with the notation pertaining to the asymptotic regime detailed in (29) we took X^n as the solution to the Lévy SDE (27) with the volatility function

$$\sigma_0(x) = \frac{3}{2} + \sin(2\pi x) \tag{34}$$

and scaled it with $1/n$, $n = 500$. We have partitioned the unit interval into 10 bins setting $b_k = k\delta x$, $\delta x = 0.1$, $k = 1,\ldots,10$, and used the piecewise constant prior of the form $\sigma(x) = \sum_k \xi_k \mathbf{1}_{B_k}(x)$, where $\xi_1,\ldots,\xi_{10}$ are i.i.d. random variables with the inverse gamma IG(α_k,β_k) distribution with $\alpha_k = \beta_k = 0.1$. The posterior is given in closed form by (32). Figure 12 contrasts the corresponding marginal 90%-posterior credible bands for σ with the true volatility function σ_0 from (34).

5.6 Closing remarks

There are many extensions conceivable. The inclusion of a drift in Equation (27) is one possibility, another one is to move from a continuous record of observations to a discrete time setting. See [2] for short discussions of such possible extensions and also for an extensive treatment of an application to real data obtained from the North Greenland Ice Core Project. The present section concentrated on the theoretical side.

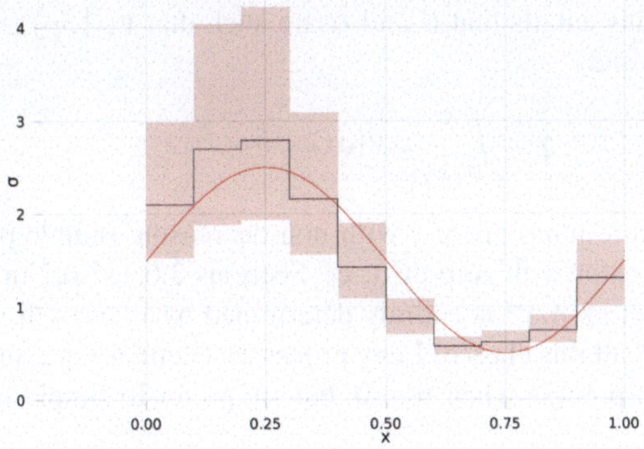

Fig. 12 Red: true volatility σ_0 as a function of x. Shaded: marginal 90%-posterior credible band for the piecewise constant posterior σ. Black: marginal posterior median.

6 Gamma-type Lévy subordinators

Given discrete time observations over a growing time interval, we consider in this section a nonparametric Bayesian approach to estimation of the Lévy density of a Lévy process belonging to a flexible class of infinite activity subordinators. We establish that our nonparametric Bayesian procedure is consistent: in the low frequency data setting, with equispaced in time observations and intervals between successive observations remaining fixed, the posterior asymptotically, as the sample size $n \to \infty$, concentrates around the Lévy density under which the data have been generated. Finally, we test our method on a classical insurance dataset. This section is based on [1].

6.1 Problem formulation

Consider a univariate Lévy process $X = (X_t : t \geq 0)$ with Lévy triplet (γ, A, ν). By the Lévy-Khintchine formula, see Theorem 8.1 in [36], the characteristic function ϕ_{X_1} of X_1 admits the unique representation of the type

$$\phi_{X_1}(z) = \exp\left(\mathrm{i}\gamma z - \frac{1}{2}Az^2 + \int_{\mathbb{R}} \left(e^{\mathrm{i}zx} - 1 - \mathrm{i}zx\mathbf{1}_{|x|\leq 1}\right)\nu(\mathrm{d}x)\right),$$

and the triplet (γ, A, ν) uniquely determines the law of the process X. We assume that the Lévy measure ν admits the representation

$$\nu(\mathrm{d}x) = \frac{\beta}{x}e^{-\alpha x - \theta(x)}\mathrm{d}x, \quad x > 0, \tag{35}$$

where α and $\theta\colon [0,\infty) \to \mathbb{R}$ are parameters to be estimated, while β is a known or unknown parameter. This entails that θ and α are such that $\nu([1,\infty))$ is finite. We also assume that $A = 0$ and

$$\gamma = \int_{0\le x\le 1} x\nu(\mathrm{d}x) < \infty. \tag{36}$$

It follows that X is a pure jump process with non-decreasing sample paths, or put another way, a subordinator with zero drift, cf. Sections 2.6.1–2.6.2 in [27]. As in our case the Lévy triplet (γ, A, ν) is entirely determined by ν, we will write $\mathbb{P}_\nu$ for the law of X. One may call this class of Lévy processes Gamma-type subordinators, because X is a Gamma process when $\theta \equiv 0$, but we prefer to simply refer to it as θ-subordinators.

Assume that the process X is observed at discrete time instances $0 = t_0 < t_1 < \cdots < t_n = T$, so our observations are $X^{(n)} = (X_{t_i} : i \in \{0,\ldots,n\})$. Our aim is nonparametric Bayesian estimation for the parameter triple (α, β, θ). This requires specification of the likelihood and the prior in our model, that are next combined via Bayes' formula to form the posterior distribution. This latter encodes all the necessary inferential information within the Bayesian setup.

6.2 Likelihood

We consider the problem with β known and fixed in the present treatment. This choice has been made because of two reasons. First, the parameter β plays a role similar to the diffusion coefficient σ in a stochastic differential equation driven by a Wiener process; different values of σ there result in mutually singular laws of a continuous record of observations, and similar singularity is valid for θ-subordinators. Derivation of nonparametric Bayesian asymptotics for the stochastic differential equations (all of which is a recent work) historically proceeded from the assumption of a known σ to the one where σ is unknown and has to be estimated. In that sense, the fact that we assume β is known at this stage does not appear unexpected or unnatural. There is also a technical reason to assume β to be known. It assists in deriving useful bounds on the Kullback-Leibler and Hellinger distances between marginals of θ-subordinators under different Lévy triplets, which in general, is the key to establishing consistency properties of nonparametric Bayesian procedures. For a treatment of the statistical problem with observations in discrete time when also θ is unknown, we refer to [1, Section 6], and for now, we maintain the assumption of known β.

All processes and their laws in this section are restricted to the time interval $[0,T]$ for a fixed $T > 0$. Note that for any two Lévy measures ν and ν_0 given by (35) with parameters β, α, θ and $\beta, \alpha_0, \theta_0$, respectively, provided $\theta(0) = \theta_0(0) = 0$ and both functions θ and θ_0 are Lipschitz continuous in some neighbourhood of zero, we have ν and ν_0 are equivalent. Moreover,

$$d^2_{\mathscr{H}}(\nu,\nu_0) = \frac{1}{2}\int\limits_{(0,\infty)} (\sqrt{\mathrm{d}\nu} - \sqrt{\mathrm{d}\nu_0})^2 < \infty, \tag{37}$$

where $d_{\mathscr{H}}(\cdot,\cdot)$ is the Hellinger distance between two (infinite) measures. By assumption (36) and property (37), together with Theorem 33.1 in [36], it follows that the laws $\mathbb{P}_\nu$ and $\mathbb{P}_{\nu_0}$ of $X = (X_t : t \in [0,T])$ are equivalent. Furthermore, Theorem 33.2 in [36] implies that a.s.

$$U_T = \log\left(\frac{\mathrm{d}\mathbb{P}_\nu}{\mathrm{d}\mathbb{P}_{\nu_0}}(X)\right) = \sum_{(s,\Delta X_s)\in(0,T]\times\{\Delta X_s>0\}} \phi(\Delta X_s) - T\int\limits_{(0,\infty)} (e^{\phi(x)} - 1)\nu_0(\mathrm{d}x),$$

where $\Delta X_s = X_s - X_{s-}$, and

$$\phi(x) = \log\left(\frac{\mathrm{d}\nu}{\mathrm{d}\nu_0}(x)\right) = -(\alpha x + \theta(x) - \alpha_0 x - \theta_0(x)), \quad x > 0.$$

We can also write the log-likelihood ratio U_T as

$$U_T = \int_{(0,T]}\int_{(0,\infty)} \phi(x)\mu(\mathrm{d}s,\mathrm{d}x) - T\int\limits_{(0,\infty)} (\nu - \nu_0)(\mathrm{d}x),$$

where the jump measure μ is defined by

$$\mu((0,t]\times B) = \#\{s\colon (s,\Delta X_s) \in (0,t]\times B\}$$

for any Borel subset B of $(0,\infty)$. We can view $\mathbb{P}_{\nu_0}$ as the dominating measure for $\mathbb{P}_\nu$. From an inferential point of view the specific choice of the dominating measure is immaterial. A convenient choice of ν_0 for the theoretical development in Section 6.4 is to actually take ν_0 to be the 'true' Lévy measure ν_0 with parameters α_0 and θ_0.

6.3 Prior

To define the prior, we consider a subclass of processes defined in (35), where the parameter θ in the Lévy measure ν has the following form. Fix a sequence $0 < b_1 < \cdots < b_N < \infty$, set for convenience $b_0 = 0$ and $b_{N+1} = \infty$, and define bins B_k by

$$B_k = [b_k, b_{k+1}), \quad k = 0,\ldots,N.$$

Given bins B_k, assume the function θ is, unlike those in previous sections, piecewise linear, i.e.,

$$\theta(x) = \sum_{k=1}^{N} (\rho_k + \theta_k x)\mathbf{1}_{B_k}, \tag{38}$$

where $\rho_k \in \mathbb{R}$, $k = 1,\ldots,N$, $\theta_k \in \mathbb{R}$, $k = 1,\ldots,N$, and $\theta_N > -\alpha$. Together with α, the parameter θ_k determines the slope of the function $\theta(x) + \alpha x$ on the bin B_k, while ρ_k gives the intercept. The process X with the law $\mathbb{P}_\nu$ can be viewed as a Gamma process with rate parameter α and shape parameter β, subjected to local deviations in the behaviour of jumps of sizes falling in bins B_k compared to that of a Gamma process. The parameters θ_k, ρ_k quantify the extent of these local deviations on the bin B_k.

We model α, θ_k, ρ_k as independent under the prior distribution. Note such a prior on α, θ_k, ρ_k implicitly defines a prior on the Lévy measure ν as well. The specific form of the prior is not crucial for many arguments that follow, but is convenient computationally. In fact, theoretical results in Section 6.4 can be derived for other series priors as well. However, the local linear structure in (38) (which also means that the prior could be rewritten as a series prior with basis functions having compact support) is important to derive some simple update formulae below. For a realisation ν from the implicit prior on ν as above in the present section, let us work out the integral

$$\nu(B_k) = \int_{b_k}^{b_{k+1}} \frac{\beta}{x} e^{-(\alpha+\theta_k)x-\rho_k} \mathrm{d}x,$$

which enters the expression for the likelihood. To that end, remember the definition of the exponential integral, $E_1(z) = \int_z^\infty t^{-1}e^{-t}\mathrm{d}t$, see, e.g., §15.09 in [24] for its basic properties. Then a change of the integration variable gives

$$\nu(B_k) = \beta e^{-\rho_k}\{E_1((\theta_k+\alpha)b_k) - E_1((\theta_k+\alpha)b_{k+1})\}, \quad k = 1,\ldots,N. \tag{39}$$

Observe that $\nu(B_N) = \beta e^{-\rho_N} E_1((\theta_k+\alpha)b_N)$. Similar to the case of ν,

$$\nu_0(B_k) = \beta\{E_1(\alpha_0 b_k) - E_1(\alpha_0 b_{k+1})\}, \quad k = 1,\ldots,N.$$

Also here remark that $\nu_0(B_N) = \beta E_1(\alpha_0 b_N)$. For future reference, note that for any α, α',

$$\lim_{x\to 0}\{E_1(\alpha x) - E_1(\alpha' x)\} = \log\left(\frac{\alpha'}{\alpha}\right), \tag{40}$$

which follows from the formula for Frullani's integral, see §12.16 in [24].

6.4 Asymptotics

Let the process X be observed at equidistant times t_i, $i = 1,\ldots,n$. Without loss of generality we assume that the observations are $X_1,\ldots,X_n$. We pass to the increments $Z_i = X_i - X_{i-1}$, $i = 1,\ldots,n$, and denote our observations by $\mathscr{Z}_n = \{Z_1,\ldots,Z_n\}$. We assume that under the true Lévy density v_0, $Z_1 \sim \mathbb{Q}_{v_0}$. In general, $\mathbb{Q}_v$ will stand for the law of the increment Z_1 under the Lévy density v. Furthermore, we introduce the law $(X_t : t \in [0,1]) \sim \mathbb{R}_{v_0}$ under the true Lévy density v_0. The law of this path under the Lévy density v will be denoted by $\mathbb{R}_v$. For asymptotic results, we will let the number of bins N depend on the sample size n, and write N_n instead. We postulate a condition on the underlying model and on the chosen prior distribution.

Condition 1 Let the function θ_0 have a compact support on the interval $[\underline{b}, \overline{b}]$ where the boundary points $0 < \underline{b} < \overline{b} < \infty$ are known, $\|\theta_0\|_\infty < \bar{\theta}$, and suppose θ_0 is λ-Hölder continuous, that is, $|\theta_0(x) - \theta_0(y)| \leq L|x-y|^\lambda$ ($\lambda \in (0,1]$, $L > 0$). Suppose also that $\alpha_0 \in [\underline{\alpha}, \overline{\alpha}]$ with known boundary points $0 < \underline{\alpha} < \overline{\alpha} < \infty$. Finally, assume that the parameter β_0 is known (and, without loss of generality, equal to 1.

Condition 2 The prior Π_n is defined on a special class of Lévy densities, V_n. These are the densities that on the bins $B_k = (b_{k-1}, b_k]$, $k = 1, \ldots, N$, $b_0 = 0$, $b_1 = \underline{b}$, $b_N = \overline{b}$, have the form $v(x) = \frac{\beta_0}{x} \exp(-\alpha x - \theta_k(x))$, with $\theta_k(x) = \rho_k + \theta_k x$, with the special choice $\rho_0 = \theta_0 = 0$ and $\beta_0 = 1$. So, with the above notation,

$$V_n = \left\{ v\colon v_{|B_k}(x) = \frac{\beta_0}{x} \exp(-\alpha x - \theta_k(x)), k = 1, \ldots, N \right\}.$$

Under the prior distribution, the coefficients θ_i, $i = 1, \ldots N-1$, are independent and uniformly distributed on the known interval $[-\overline{\theta}, \overline{\theta}]$, $\overline{\theta} > 0$. Likewise, the coefficients ρ_i, $i = 1, \ldots, N-1$, are independent uniform on the interval $[-\overline{\theta}, \overline{\theta}]$, whereas α is uniform on $[\underline{\alpha}, \overline{\alpha}]$, $\overline{\alpha} > 0$. We assume a priori independence of all ρ_i, θ_i and α. Implicitly, this defines a prior Π_n on the class of Lévy densities V_n, which are realisations from the prior.

Theorem 6.1 below implies that our Bayesian procedure is consistent in probability; this in turn implies the existence of consistent Bayesian point estimates, see, e.g., [12], pp. 506–507. We use the notation $\Pi_n(\mathrm{d}v \mid \mathscr{Z}_n)$ for the posterior measure. Also, $\mathbb{Q}^n_{v_0}$ denotes the law of the sample $\mathscr{Z}_n$ under the true Lévy density v_0.

Theorem 6.1. *Assume that Conditions 1 and 2 hold and that $N_n \to \infty$ and $N_n/n \to 0$ as $n \to \infty$. Let $d_{\mathscr{H}}$ be the Hellinger metric. Then, for any fixed $\varepsilon, \varepsilon' > 0$,*

$$\mathbb{Q}^n_{v_0}\left(\Pi_n(v\colon d_{\mathscr{H}}(\mathbb{Q}_{v_0}, \mathbb{Q}_v) > \varepsilon \mid \mathscr{Z}_n) > \varepsilon'\right) \to 0$$

as $n \to \infty$.

6.5 Numerical results

Over the last two decades there has been an increasing interest in applying Bayesian methods to insurance problems, see, e.g., [21] and references therein. [22] apply a Dirichlet process mixture prior to model the density of insurance claim sizes, and provide motivation for using a nonparametric Bayesian approach in the actuarial science. In this section we will apply our Bayesian approach to the Danish data on large fire losses. This dataset is a standard test example in extreme value theory, and from that point of view it has been a subject of several deep studies, such as [29] and [34]. Our goals here are more modest, and aim at demonstrating the facts that firstly, θ-subordinators can be potentially used to capture some aggregate features of the Danish data on large fire losses, and secondly, statistical inference for real data modelled through such processes can be successfully performed using the Bayesian methodology developed here.

Data description and visualisation

A succinct description of the Danish data on large fire losses can be found on p. 298 in [9]. The dataset (scaled for privacy reasons) comprises 2167 fire losses (adjusted suitably for inflation to reflect the 1985 values) in Denmark over the 10 year period starting on 6 January 1980 and ending on 30 December 1990, that exceed in size one million DKK, and that were registered by Copenhagen Reinsurance. The rationale for thresholding losses at one million DKK is given in [29], pp. 119–120, and consists in the fact that in practice it is virtually impossible to collect exhaustive data on small losses: insurance is typically provided against significant losses, while small losses are dealt with by insured parties directly. The data can be accessed through the **QRM** package in **R** under the name `danish`. The time plot of the data is given in the left panel of Figure 13.

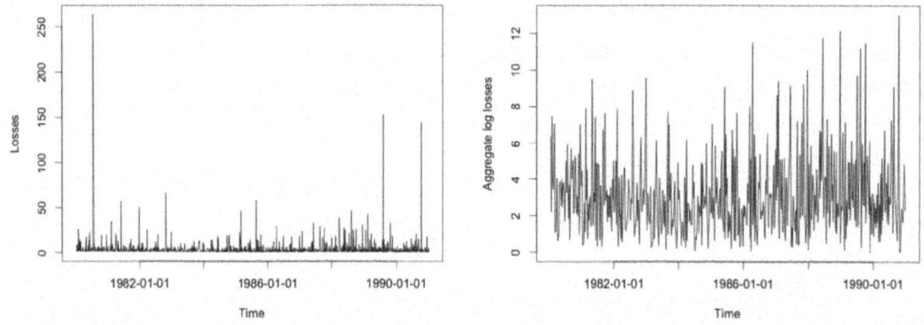

Fig. 13 Danish data on large fire losses. *Left*: original daily data (the unit is one million DKK). *Right*: logarithmically transformed and aggregated data.

According to the exploratory analysis of the transformed data, the data can be modelled as an i.i.d. sequence that follows a Gamma-like distribution, but perhaps is not genuinely Gamma. This suggests a possibility of using a θ-subordinator to model the data.

Modelling fire losses with a θ-subordinator

Because the sample size is much smaller compared to our simulation examples, we chose $N = 1$ corresponding to a single grid point $b_1 = 2$ and four parameters

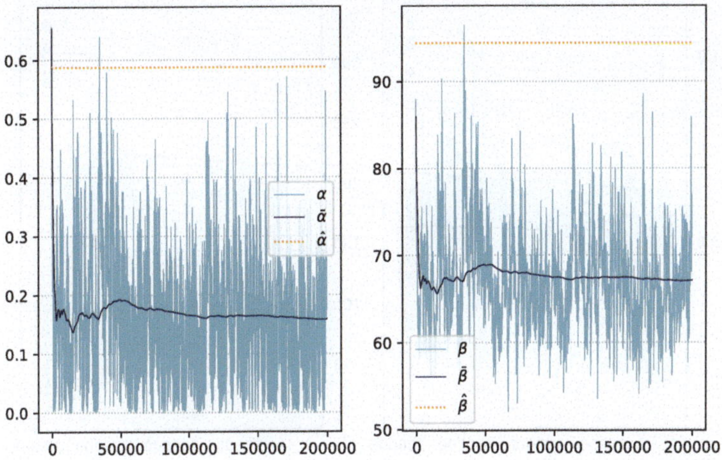

Fig. 14 Trace plots of the parameters α and β for the fire loss data. Left: trace and running average of samples of α. (The latter indicated by $\bar{\alpha}$.) The maximum likelihood estimate $\hat{\alpha}$ of α using a Gamma process model is marked as the dotted yellow line. Right: trace and running average of samples of β. (The latter indicated by $\bar{\beta}$.) The maximum likelihood estimate $\hat{\beta}$ of β using a Gamma process model is marked as the yellow dotted line.

α, β, θ_1, ρ_1. In order to improve mixing of the chain, we use a reparametrization $\alpha_1 = \alpha + \theta_1$, $\beta_1 = \beta \exp(-\rho)$, and work with four parameters α, β, α_1, β_1, so that

$$v(x) = \begin{cases} \frac{\beta}{x} \exp(-\alpha x) & x \leq b_1, \\ \frac{\beta_1}{x} \exp(-\alpha_1 x) & x > b_1. \end{cases}$$

A priori we equip these four parameters with Gamma distributions, with mean 0.75 and variance 0.36 for the parameters α, α_1, and mean 90 and variance 2500 for the parameters β, β_1, and in such a way that they become independent random variables. In the data augmentation step we take intermediate points at distance 0.0192, corresponding to $m = 1000$.

For the parameter updates we took independent Gaussian innovations with standard deviations $\sigma_\alpha = \sigma_{\alpha_1} = 0.03$, $\sigma_\beta = 1$ and $\sigma_{\beta_1} = 6$, respectively. In the Gibbs sampler in each step new Gamma bridges are proposed in the data augmentation step, followed by a parameter update Metropolis-Hastings step cycling through updates of β in the first and second and the other parameters jointly in each of the remaining three of in total 5 stages. With these choices, the chains mix sufficiently well. The MCMC algorithm was ran for 200 000 iterations. Figure 14 shows trace plots and running averages of the posterior samples of the parameters α and β, whereas Figure 15 shows similar plots for the parameters α_1 and β_1.

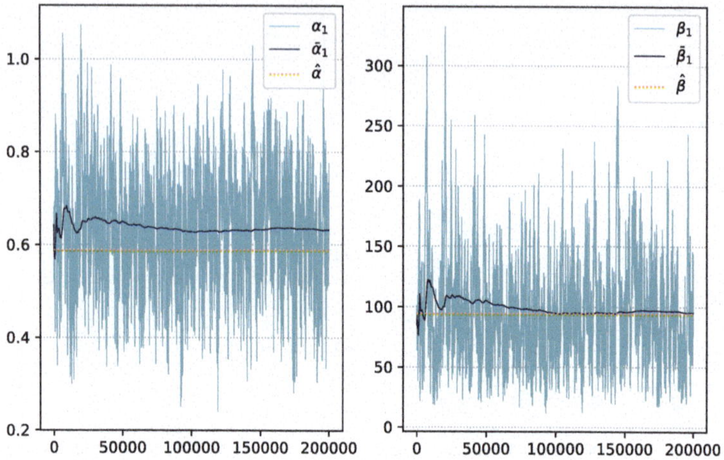

Fig. 15 Trace plots of the parameters used for the bin (b_1, ∞) for the fire loss data. Left: trace and running average of the samples of α_1. The maximum likelihood estimate of α using a Gamma process model is marked as yellow line. Right: trace and running average of the samples of β_1. The maximum likelihood estimate of β using a Gamma process model is marked as yellow line.

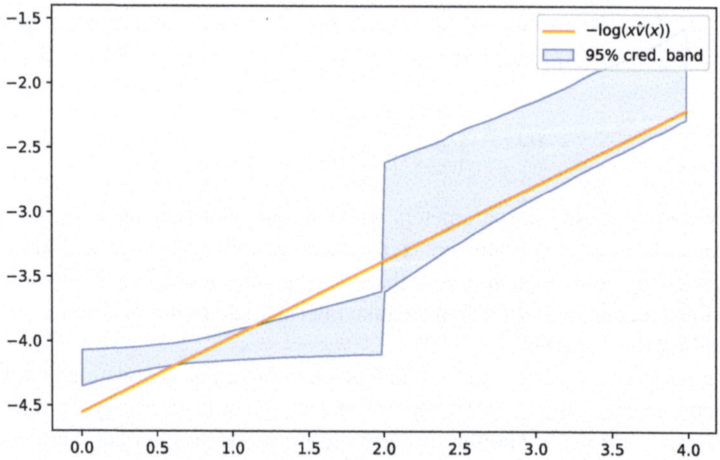

Fig. 16 Marginal Bayesian credible bands for the fire loss data for the function $-\log(xv(x))$ based on all samples. Yellow: maximum likelihood estimate $-\log(x\hat{v}(x))$ assuming a Gamma process.

Figure 16 shows the 95% marginal Bayesian credible band for the function $\theta(x) + \alpha x$ contrasted to the maximum likelihood estimate that assumes the observations come from a Gamma process. This plot suggests that modelling the losses with a Gamma process leads to overestimation of the number of small jumps and possibly of large jumps too; however, more data is necessary to make a definitive statement (unfortunately, as observed in [5], it is difficult for academia to gain access to the insurance data). In connection to this, we note that a difference in the estimates of the rate of decay of the Lévy density (value of α_1 in the model) has serious implications of practical relevance for the assessment of the risk of very large fire losses.

6.6 Conclusions and possible extensions

We have considered in this section a nonparametric Bayesian approach to estimation of the Lévy density of a Lévy process belonging to a flexible class of infinite activity subordinators and established that this procedure is consistent. We have also tested our method on a classical insurance dataset.

There are various directions for extensions of the model studied in this section. One may consider a class of increasing, infinite activity Lévy processes X for which the characteristic function of X_1 is given by

$$\phi(z) = E[e^{\mathrm{i}zX_1}] = \int (e^{\mathrm{i}zx} - 1)\nu(dx),$$

where $\theta(\cdot)$ is nonnegative and the Lévy measure ν is given by

$$\nu(\mathrm{d}x) = \frac{1}{x^{1+a}} e^{-\theta(x)} \mathbf{1}_{(0,\infty)}(x)\mathrm{d}x.$$

This model generalizes the Gamma process with $a = 0$ and $\theta(x) \equiv \lambda$ and also covers the class of one-sided tempered stable processes. Here one could think of estimating the stability index α from a Bayesian point of view. This is difficult due to singularity of the measures induced by two Lévy processes with different stability indices, and so this poses an interesting challenge that deserves to be explored.

7 Summary and concluding remarks

In this paper, we have reviewed some of our contributions, obtained with our collaborators, to Bayesian nonparametric function estimation. We have considered a large variety of models (volatility estimation in diffusion models, Poisson intensity estimation, estimation for gamma-driven stochastic differential equations, estimation in the presence of microstructure noise, and Lévy density estimation for Gamma-type subordinators), which all had a common denominator, the function to be estimated, is well approximated by piecewise constant functions (with one exception that deals with piecewise linear functions) giving rise to tractable prior distributions for Bayesian inference.

The goal of inference in all models under consideration was to obtain posterior consistency of the proposed estimators or even to quantify posterior contraction rates. We have seen that rougher continuous functions are more difficult to estimate than smoother ones. The degree of smoothness is in our approach measured by the Hölder exponent; the smoother the function is, the better the contraction rate of the estimation procedure is. The precise results are formulated as Theorems 2.4, 4.3, and 5.4. The conclusion that emerges from these theorems is that despite the entirely different settings and the different kinds of functions (their arguments being time or space variables), the contraction rates, all depending on the Hölder exponent, are the same across all cases. A deeper analysis in terms of equivalence of statistical experiments is likely to explain the similarity of the rates. These rates were obtained under an independence assumption on the prior, i.e., the approximating functions with constant levels on each of their bins are such that these levels are a priori independent. The setting of Section 6 was somewhat different, and we had to content ourselves with posterior consistency only, Theorem 6.1.

It also appeared that better numerical results, visible in the plots, can be obtained by relaxing the independence assumption on the prior distribution. This observation has been accomplished by using a specific Markov chain prior on the levels of the piecewise constant functions, introducing a priori dependence between the levels on adjacent bins of the piecewise constant approximating functions. However, the contraction rates induced by such a prior cannot improve upon the one for the prior that renders the levels independent, as the latter has been seen to be optimal.

Throughout the paper, we have given ample numerical illustrations that support the theoretical results, ranging from simulated experiments to real-world cases.

Acknowledgements The authors gratefully acknowledge the contributions to this paper by Shota Gugushvili. The work of D. Belomestny was done within the framework of the HSE University Basic Research Program.

References

1. Belomestny, D., Gugushvili, S., Schauer, M., Spreij, P.: Nonparametric Bayesian inference for Gamma-type Lévy subordinators. Commun. Math. Sci. **17**(3), 781–816 (2019). DOI 10.4310/CMS.2019.v17.n3.a8
2. Belomestny, D., Gugushvili, S., Schauer, M., Spreij, P.: Nonparametric Bayesian volatility estimation for gamma-driven stochastic differential equations. Bernoulli **28**(4), 2151–2180 (2022). DOI 10.3150/21-BEJ1413. URL https://doi.org/10.3150/21-BEJ1413
3. Belomestny, D., Gugushvili, S., Schauer, M., Spreij, P.: Weak solutions to gamma-driven stochastic differential equations. Indagationes Mathematicae **34**(4), 820–829 (2023). DOI 10.1016/j.indag.2023.03.004. URL https://doi.org/10.1016/j.indag.2023.03.004

4. Cemgil, A.T., Dikmen, O.: Conjugate gamma Markov random fields for modelling nonstationary sources. In: M.E. Davies, C.J. James, S.A. Abdallah, M.D. Plumbley (eds.) Independent Component Analysis and Signal Separation: 7th International Conference, ICA 2007, London, UK, September 9–12, 2007. Proceedings, pp. 697–705. Springer Berlin Heidelberg, Berlin, Heidelberg (2007)
5. Chavez-Demoulin, V., Embrechts, P., Hofert, M.: An extreme value approach for modeling operational risk losses depending on covariates. Journal of Risk and Insurance **83**(3), 735–776 (2016)
6. Cox, J.C., Ingersoll Jr., J.E., Ross, S.A.: A theory of the term structure of interest rates. Econometrica **53**(2), 385–407 (1985)
7. Dikmen, O., Cemgil, A.T.: Gamma Markov random fields for audio source modeling. IEEE Trans. Audio, Speech, Language Process. **18**(3), 589–601 (2010)
8. Donoho, D.L., Johnstone, I.M.: Adapting to unknown smoothness via wavelet shrinkage. J. Amer. Statist. Assoc. **90**(432), 1200–1224 (1995)
9. Embrechts, P., Klüppelberg, C., Mikosch, T.: Modelling extremal events: for insurance and finance, vol. 33. Springer Science & Business Media (2013)
10. Fan, J., Gijbels, I.: Local polynomial modelling and its applications. Routledge (2018)
11. Gelman, A., Carlin, J.B., Stern, H.S., Dunson, D.B., Vehtari, A., Rubin, D.B.: Bayesian data analysis, third edn. Texts in Statistical Science Series. CRC Press, Boca Raton, FL (2014)
12. Ghosal, S., Ghosh, J.K., van der Vaart, A.W.: Convergence rates of posterior distributions. Ann. Statist. **28**(2), 500–531 (2000). DOI 10.1214/aos/1016218228. URL `https://doi.org/10.1214/aos/1016218228`
13. Ghosal, S., van der Vaart, A.: Fundamentals of nonparametric Bayesian inference, *Cambridge Series in Statistical and Probabilistic Mathematics*, vol. 44. Cambridge University Press, Cambridge (2017)
14. Gugushvili, S., van der Meulen, F., Schauer, M., Spreij, P.: Fast and scalable non-parametric bayesian inference for poisson point processes. Researchers.One (2020). URL `https://researchers.one/articles/19.06.00001v1`
15. Gugushvili, S., van der Meulen, F., Schauer, M., Spreij, P.: Nonparametric Bayesian estimation of a Hölder continuous diffusion coefficient. Braz. J. Probab. Stat. **34**(3), 537–579 (2020)
16. Gugushvili, S., van der Meulen, F., Schauer, M., Spreij, P.: Nonparametric bayesian volatility learning under microstructure noise. Japanese Journal of Statistics and Data Science **online** (2022). URL `https://doi.org/10.1007/s42081-022-00185-9`
17. Gugushvili, S., van der Meulen, F., Schauer, M., Spreij, P.: Nonparametric Bayesian volatility estimation. In: J. de Gier, C.E. Praeger, T. Tao (eds.) 2017 MATRIX Annals, pp. 279–302. Springer International Publishing, Cham (2019). DOI 10.1007/978-3-030-04161-8_19
18. Henderson, S.G.: Estimation for nonhomogeneous Poisson processes from aggregated data. Oper. Res. Lett. **31**(5), 375–382 (2003). DOI 10.1016/S0167-6377(03)00027-0
19. Heston, S.L.: A closed-form solution for options with stochastic volatility with applications to bond and currency options. Rev. Financial Stud. **6**(2), 327–343 (1993)
20. Hoffmann, M.: Minimax estimation of the diffusion coefficient through irregular samplings. Statistics & probability letters **32**(1), 11–24 (1997)
21. Hong, L., Martin, R.: A review of bayesian asymptotics in general insurance applications. European Actuarial Journal **7**, 231–255 (2017)
22. Hong, L., Martin, R.: Dirichlet process mixture models for insurance loss data. Scandinavian Actuarial Journal **2018**(6), 545–554 (2018)
23. Jacod, J., Shiryaev, A.: Limit theorems for stochastic processes, vol. 288. Springer Science & Business Media (2013)
24. Jeffreys, H., Jeffreys, B., Swirles, B.: Methods of mathematical physics. Cambridge university press (1999)
25. Karr, A.F.: Point processes and their statistical inference. Probability: Pure and Applied, 2. New York-Basel: Marcel Dekker, Inc. (1986)
26. Kutoyants, Y.A.: Statistical inference for spatial Poisson processes, *Lecture Notes in Statistics*, vol. 134. Springer-Verlag, New York (1998). DOI 10.1007/978-1-4612-1706-0

27. Kyprianou, A.E.: Fluctuations of Lévy processes with applications. Introductory lectures, 2nd edn. Universitext. Springer, Heidelberg (2014)
28. Mancini, C., Mattiussi, V., Renò, R.: Spot volatility estimation using delta sequences. Finance and Stochastics **19**(2), 261–293 (2015)
29. McNeil, A.J.: Estimating the tails of loss severity distributions using extreme value theory. ASTIN Bulletin: The Journal of the IAA **27**(1), 117–137 (1997)
30. Peeling, P., Cemgil, A.T., Godsill, S.: Bayesian hierarchical models and inference for musical audio processing. In: 2008 3rd International Symposium on Wireless Pervasive Computing, pp. 278–282 (2008). DOI 10.1109/ISWPC.2008.4556214
31. Petris, G., Petrone, S., Campagnoli, P.: Dynamic linear models with R. Use R! Springer, New York (2009)
32. van der Ploeg, A.P.C.: Stochastic volatility and the pricing of financial derivatives. Ph.D. thesis, University of Amsterdam (2006)
33. Protter, P.E.: Stochastic integration and differential equations, *Applications of Mathematics (New York)*, vol. 21, 2nd edn. Springer-Verlag, Berlin (2004). Stochastic Modelling and Applied Probability
34. Resnick, S.I.: Discussion of the danish data on large fire insurance losses. ASTIN Bulletin: The Journal of the IAA **27**(1), 139–151 (1997)
35. Rosas, F.E., Candia-Rivera, D., Luppi, A.I., Guo, Y., Mediano, P.A.: Bayesian at heart: Towards autonomic outflow estimation via generative state-space modelling of heart rate dynamics. arXiv preprint arXiv:2303.04863 (2023)
36. Sato, K.i.: Lévy processes and infinitely divisible distributions, *Cambridge Studies in Advanced Mathematics*, vol. 68. Cambridge University Press, Cambridge (2013). Translated from the 1990 Japanese original, Revised edition of the 1999 English translation
37. Scricciolo, C.: On rates of convergence for bayesian density estimation. Scandinavian Journal of Statistics **34**(3), 626–642 (2007)
38. Spiegelhalter, D.J., Best, N.G., Carlin, B.P., van der Linde, A.: Bayesian measures of model complexity and fit. J. R. Stat. Soc. Ser. B Stat. Methodol. **64**(4), 583–639 (2002)
39. Tanner, M.A., Wong, W.H.: The calculation of posterior distributions by data augmentation. J. Amer. Statist. Assoc. **82**(398), 528–550 (1987). With discussion and with a reply by the authors

Quickest Changepoint Detection in General Multistream Stochastic Models: Recent Results, Applications and Future Challenges

Alexander G. Tartakovsky and Valentin Spivak

Abstract Modern information systems generate large volumes of data with anomalies that occur at unknown points in time and have to be detected quickly and reliably with low false alarm rates. The paper develops a general theory of quickest multistream detection in non-i.i.d. stochastic models when a change may occur in a set of multiple data streams. The first part of the paper focuses on the asymptotic quickest detection theory. Nearly optimal pointwise detection strategies that minimize the expected detection delay are proposed and analyzed when the false alarm rate is low. The general theory is illustrated in several examples. In the second part, we discuss challenging applications associated with the rapid detection of new COVID waves and the appearance of near-Earth space objects. Finally, we discuss certain open problems and future challenges.

1 Introduction

The problem of changepoint detection in multiple data streams (sensors, populations, or multichannel systems) arises in numerous applications that include but are not limited to the medical sphere (detection of an epidemic present in only a fraction of hospitals [5, 8, 22, 34]); environmental monitoring (detection of the presence of hazardous materials or intruders [7, 20]); military defense (detection of an unknown number of targets by multichannel sensor systems [1, 30]); near-Earth space informatics (detection of debris and satellites with telescopes [3, 11, 24]); cyber security (detection of attacks in computer networks [23, 26, 32, 33]); detection of malicious activity in social networks [18, 19], to name a few.

Alexander G. Tartakovsky
AGT StatConsult, Los Angeles, California, USA, e-mail: alexg.tartakovsky@gmail.com

Valentin Spivak
MIPT, Russia, e-mail: valyas.spivak@gmail.com

D. R. Wood et al. (eds.), *2021-2022 MATRIX Annals*, MATRIX Book Series 5,
https://doi.org/10.1007/978-3-031-47417-0_29

In many change detection applications, the pre-change (the baseline or in-control) distribution of observed data is known, but the post-change (out-of-control) distribution is not completely known. As discussed in [29, Ch 3, 6] there are three conventional approaches in this case: (i) to select a representative value of the post-change parameter and apply efficient detection rules tuned to this value such as the Shiryaev rule, the Shiryaev–Roberts rule or CUSUM, (ii) to select a mixing measure over the parameter space and apply mixture-type rules, (iii) to estimate the parameter and apply adaptive schemes. In Chapters 4 and 6 of [29], a single stream case was considered. In this paper, we consider a more general case where the change occurs in multiple data streams and the number and location of affected data streams are unknown.

To be more specific, suppose there are N data streams observed sequentially in time subject to a change at an unknown point in time $\nu \geq 0$, so that the data up to the time ν are generated by one stochastic model and after $\nu + 1$ by another model. The change in distributions may occur at a subset of streams of a size $1 \leq K \leq N$, where K is an assumed maximal number of streams that can be affected, which can be substantially smaller than N. A sequential detection procedure is a stopping time T with respect to an observed sequence. A false alarm is raised when the detection is declared before the change occurs. One wants to detect the change with as small a delay as possible while controlling the false alarm rate.

We consider a fairly general stochastic model assuming that the observations may be dependent and non-identically distributed (non-i.i.d.) before and after the change and that streams may be mutually dependent.

In the case of i.i.d. observations (in pre-change and post-change modes with different distributions), this problem was considered in [6, 15, 25, 31, 32, 35]. Specifically, in the case of a known post-change parameter and $K = 1$ (i.e., when only one stream can be affected but it is unknown which one), Tartakovsky [25] proposed to use a multi-chart CUSUM procedure that raises an alarm when one of the partial CUSUM statistics exceeds a threshold. This procedure is very simple, but it is not optimal and performs poorly when many data streams are affected. To avoid this drawback, Mei [15] suggested a SUM-CUSUM rule based on the sum of CUSUM statistics in streams and evaluated its first-order performance, which shows that this detection scheme is first-order asymptotically minimax minimizing the maximal expected delay to detection when the average run length to false alarm approaches infinity. Fellouris and Sokolov [6] suggested more efficient generalized and mixture-based CUSUM rules that are second-order minimax. Xie and Siegmund [35] considered a particular Gaussian model with an unknown post-change mean. They suggested a rule that combines mixture likelihood ratios that incorporate an assumption about the proportion of affected data streams with the generalized CUSUM statistics in streams and then add up the resulting local statistics. They also performed a detailed asymptotic analysis of the proposed detection procedure in terms of the average run length to a false alarm and the expected delay as well as MC simulations. Chan [4] studied a version of the mixture likelihood ratio rule for detecting a change in the mean of the normal population assuming independence of data streams and establishing its asymptotic optimality in a minimax setting as well as dependence of operating characteristics on the fraction of affected streams.

In the present paper, we consider a Bayesian problem with a general prior distribution of the change point and multiple data streams with an unknown pattern, i.e., when the size and location of the affected streams are unknown. It is assumed that the observations can be dependent and non-identically distributed in data streams and even across the streams. Furthermore, in contrast to most previous publications where asymptotically stationary models were considered, we consider substantially non-stationary models, even asymptotically. We address two scenarios when the pre- and post-change distributions are completely known and also a parametric uncertainty when the post-change distribution is known up to an unknown parameter. We introduce mixture detection procedures that mix the Shiryaev–Robers-type statistic over the distributions of the unknown pattern and unknown post-change parameter (in the case of prior uncertainty). The resulting statistics are then compared to appropriate thresholds.

The paper is organized as follows. In Section 2, we present a general theory for very general stochastic models, providing sufficient conditions under which the suggested detection procedures are first-order asymptotically optimal. In Section 3, we provide illustrative examples. In Section 4, we evaluate the performance of proposed mixture detection procedures using Monte Carlo simulations for the non-stationary Gaussian model. In Section 5, theoretical results are applied to rapid detection of the COVID-19 outbreak in Australia based on monitoring the percentage of infections in the total population as well as to rapid detection and extraction of faint near-Earth space objects with telescopes. Section 6 concludes the paper with several remarks, including future research challenges.

2 Asymptotic Theory of Multistream Quickest Change Detection for General Non-i.i.d. Models

In this section, we develop the quickest detection theory in the general multistream non-i.i.d. scenario. We design mixture-based change detection procedures which are nearly optimal in the class of change detection procedures with the prespecified average (weighted) probability of false alarm when this probability is small, assuming that the change point is random with a given prior distribution.

2.1 A General Multistream Model and Basic Notations

We begin with a preliminary description of the scenario of interest and general notation.

Consider the multistream scenario where the observations $\mathbf{X} = (X(1), \ldots, X(N))$ are sequentially acquired in N streams, i.e., in the i-th stream one observes a sequence $X(i) = \{X_n(i)\}_{n\in\mathbb{N}}$, where $i \in \mathscr{N} := \{1,\dots,N\}$. The observations are subject to a change at an unknown time $\nu \in \{0,1,2,\dots\}$, so that $X_1(i),\dots,X_\nu(i)$ are generated by one stochastic model and $X_{\nu+1}(i), X_{\nu+2}(i),\dots$ by another model when the change occurs in the i-th stream. The change in distributions happens at a subset of streams $\mathsf{B} \subseteq \{1,\dots,N\}$ with cardinality $1 \le |\mathsf{B}| \le K \le N$, where K is an assumed maximal number of streams that can be affected, which can be and often is substantially smaller than N. A sequential detection rule is a stopping time T with respect to an observed sequence $\{\mathbf{X}_n\}_{n\in\mathbb{N}}$, $\mathbf{X}_n = (X_n(1),\dots,X_n(N))$, i.e., T is an integer-valued random variable, such that the event $\{T = n\}$ belongs to the sigma-algebra $\mathscr{F}_n = \sigma(\mathbf{X}_1,\dots,\mathbf{X}_n)$ generated by observations $\mathbf{X}_1,\dots,\mathbf{X}_n$.

The observations may have a very general stochastic structure. Specifically, if we let $\mathbf{X}^n(i) = (X_1(i),\dots,X_n(i))$ denote the sample of size n in the i-th stream and if $\{f_{\theta_i,n}(X_n(i)|\mathbf{X}^{n-1}(i))\}_{n\ge 1}$, $\theta_i \in \Theta_i$ is a parametric family of conditional densities of $X_n(i)$ given $\mathbf{X}^{n-1}(i)$, then when $\nu = \infty$ (there is no change) the parameter θ_i is equal to the known value $\theta_{i,0}$, i.e., $f_{\theta_i,n}(X_n(i)|\mathbf{X}^{n-1}(i)) = f_{\theta_{i,0},n}(X_n(i)|\mathbf{X}^{n-1}(i))$ for all $n \in \mathbb{N}$ and when $\nu = k < \infty$, then $\theta_i = \theta_{i,1} \neq \theta_{i,0}$, i.e., $f_{\theta_i,n}(X_n(i)|\mathbf{X}^{n-1}(i)) = f_{\theta_{i,0},n}(X_n(i)|\mathbf{X}^{n-1}(i))$ for $n \le k$ and $f_{\theta_i,n}(X_n(i)|\mathbf{X}^{n-1}(i)) = f_{\theta_{i,1},n}(X_n(i)|\mathbf{X}^{n-1}(i))$ for $n > k$. Not only the point of change ν, but also the subset B, its size $|\mathsf{B}|$, and the post-change parameters $\theta_{i,1}$ are unknown. For further details with a certain change of notation see below.

Let P_∞ denote the probability measure corresponding to the sequence of observations $\{\mathbf{X}_n\}_{n\in\mathbb{N}}$ from all N streams when there is never a change ($\nu = \infty$) in any of the components and, for $k = 0,1,\dots$ and $\mathsf{B} \subset \mathscr{N}$, let $\mathsf{P}_{k,\mathsf{B}}$ denote the measure corresponding to the sequence $\{\mathbf{X}_n\}_{n\in\mathbb{N}}$ when $\nu = k < \infty$ and the change occurs in a subset B of the set $\mathscr{P}$ (i.e., $X_{\nu+1}(i)$, $i \in \mathsf{B}$ is the first post-change observation). By $\mathsf{H}_\infty : \nu = \infty$ we denote the hypothesis that the change never occurs and by $\mathsf{H}_{k,\mathsf{B}}$ – the hypothesis that the change occurs at time $0 \le k < \infty$ is the subset of streams $\mathsf{B} \subset \mathscr{P}$. The set $\mathscr{P}$ is a class of subsets of $\mathscr{N}$ that incorporates available prior information regarding the subset B where the change may occur. For example, in applications frequently it is known that at most K streams can be affected, in which case $\mathscr{P} = \mathscr{P}_K = \{\mathsf{B} \subset \mathscr{N} : 1 \le |\mathsf{B}| \le K\}$. Hereafter $|\mathsf{B}|$ denotes the size of a subset B (the number of affected streams under $\mathsf{H}_{k,\mathsf{B}}$) and $|\mathscr{P}|$ denotes the size of class $\mathscr{P}$ (the number of possible alternatives in $\mathscr{P}$). Note that $|\mathscr{P}|$ takes maximum value when there is no prior information regarding the subset of affected streams, i.e., when $\mathscr{P} = \mathscr{P}_N$, in which case $|\mathscr{P}| = 2^N - 1$.

The problem is to detect the change as soon as possible after it occurs regardless of the subset B, i.e., we are interested in detecting the event $\cup_{\mathsf{B}\in\mathscr{P}} \mathsf{H}_{k,\mathsf{B}}$ that the change has occurred in some subset but not in identifying the subset of streams where it occurs.

Write $\mathbf{X}^n(i) = (X_1(i),\dots,X_n(i))$ for the concatenation of the first n observations from the i-th data stream and $\mathbf{X}^n = (\mathbf{X}_1,\dots,\mathbf{X}_n)$ for the concatenation of the first n observations from all N data streams. Let $\{g(\mathbf{X}_n|\mathbf{X}^{n-1})\}_{n\in\mathbb{N}}$ and $\{f_{\mathsf{B}}(\mathbf{X}_n|\mathbf{X}^{n-1})\}_{n\in\mathbb{N}}$ be sequences of conditional densities of $\mathbf{X}_n$ given $\mathbf{X}^{n-1}$, which may depend on n, i.e., $g = g_n$ and $f_{\mathsf{B}} = f_{\mathsf{B},n}$. We omit the subscript n for the sake of brevity. For the general non-i.i.d. changepoint model the joint density $p(\mathbf{X}^n|\mathsf{H}_{k,\mathsf{B}})$ under hypothesis

$\mathsf{H}_{k,\mathsf{B}}$ can be written as follows

$$p(\mathbf{X}^n|\mathsf{H}_{k,\mathsf{B}}) = \begin{cases} \prod_{t=1}^n g(\mathbf{X}_t|\mathbf{X}^{t-1}) & \text{for } \nu = k \geq n, \\ \prod_{t=1}^k g(\mathbf{X}_t|\mathbf{X}^{t-1}) \times \prod_{t=k+1}^n f_{\mathsf{B}}(\mathbf{X}_t|\mathbf{X}^{t-1}) & \text{for } \nu = k < n, \end{cases} \quad (1)$$

where $\mathsf{B} \subset \mathscr{P}$. Therefore, $g(\mathbf{X}_n|\mathbf{X}^{n-1})$ is the pre-change conditional density and $f_{\mathsf{B}}(\mathbf{X}_n|\mathbf{X}^{n-1})$ is the post-change conditional density given that the change occurs in the subset B.

In most practical applications, the post-change distribution is not completely known – it depends on an unknown (generally multidimensional) parameter $\theta \in \Theta$, so that the model (1) may be treated only as a benchmark for a more practical case where the post-change densities $f_{\mathsf{B}}(\mathbf{X}_t|\mathbf{X}^{t-1})$ are replaced by $f_{\mathsf{B},\theta}(\mathbf{X}_t|\mathbf{X}^{t-1})$, i.e.,

$$p(\mathbf{X}^n|H_{k,\mathsf{B}},\theta) = \prod_{t=1}^{k} g(\mathbf{X}_t|\mathbf{X}^{t-1}) \times \prod_{t=k+1}^{n} f_{\mathsf{B},\theta}(\mathbf{X}_t|\mathbf{X}^{t-1}) \quad \text{for } \nu = k < n. \quad (2)$$

Notice that the probabilistic models given by (1) and (2) are very general and do not assume that the data streams are mutually independent.

2.2 Optimality Criterion

In the sequel, we assume that the change point ν is a random variable independent of the observations with a prior distribution $\pi_k = \mathsf{P}(\nu = k)$, $k = 0,1,2,\dots$ with $\pi_k > 0$ for $k \in \{0,1,2,\dots\} = \mathbb{Z}_+$ and that a change point may take negative values, i.e., $k \in \mathbb{Z}$, but the detailed structure of the distribution $\mathsf{P}(\nu = k)$ for $k \in \mathbb{Z}^- = \{-1,-2,\dots,-\infty\}$ is not important. Only the total probability $\pi_- = \mathsf{P}(\nu \leq -1)$ of the change being in effect before the observations become available matters.

Let $\mathsf{E}_{k,\mathsf{B},\theta}$ and E_∞ denote expectations under $\mathsf{P}_{k,\mathsf{B},\theta}$ and P_∞, respectively, where $\mathsf{P}_{k,\mathsf{B},\theta}$ corresponds to model (2) with an unknown parameter $\theta \in \Theta$. Define the probability measure on the Borel σ-algebra $\mathscr{B}$ in $\mathbb{R}^\infty \times \mathbb{Z}$ as

$$\mathsf{P}^\pi_{\mathsf{B},\theta}(\mathscr{A} \times \mathscr{K}) = \sum_{k \in \mathscr{K}} \pi_k \mathsf{P}_{k,\mathsf{B},\theta}(\mathscr{A}), \quad \mathscr{A} \in \mathscr{B}(\mathbb{R}^\infty), \quad \mathscr{K} \in \mathbb{Z}.$$

Under measure $\mathsf{P}^\pi_{\mathsf{B},\theta}$ the change point ν has distribution $\pi = \{\pi_k\}$ and the model for the observations is of the form (2), i.e., $\mathbf{X}(t)$ has conditional density $g(\mathbf{X}(t)|\mathbf{X}^{t-1})$ if $\nu \leq k$ and conditional density $f_{\mathsf{B},\theta}(\mathbf{X}(t)|\mathbf{X}^{t-1})$ if $\nu > k$ and the change occurs in the subset B with the parameter θ. Let $\mathsf{E}^\pi_{\mathsf{B},\theta}$ denote the expectation under $\mathsf{P}^\pi_{\mathsf{B},\theta}$.

For the prior distribution of the change point $\pi = \{\pi_k\}_{k \in \mathbb{Z}}$, introduce the average (weighted) probability of false alarm associated with the change detection procedure T

$$\mathsf{PFA}_\pi(T) = \mathsf{P}^\pi_{\mathsf{B},\theta}(T \leq \nu) = \sum_{k=0}^{\infty} \pi_k \mathsf{P}_\infty(T \leq k) \quad (3)$$

that corresponds to the risk due to a false alarm. Note that here we took into account that $\mathsf{P}_{k,\mathsf{B},\theta}(T \le k) = \mathsf{P}_\infty(T \le k)$ since the event $\{T \le k\}$ depends on the observations $\mathbf{X}_1,\dots,\mathbf{X}_k$ generated by the pre-change probability measure P_∞ (recall that by our convention $\mathbf{X}_k$ is the last pre-change observation if $\nu = k$).

For $\nu = k \in \mathbb{Z}_+$, $\mathsf{B} \in \mathscr{P}$, and $\theta \in \Theta$ the risk associated with the detection delay is measured by the conditional expected delay to detection

$$\mathsf{EDD}_{k,\mathsf{B},\theta}(T) = \mathsf{E}_{k,\mathsf{B},\theta}\left[T - k \,|\, T > k\right]. \tag{4}$$

Note that if the change occurs before the observations become available, i.e., $k \in \mathbb{Z}^- = \{-1,-2,\dots\}$, then $\mathsf{EDD}_{k,\mathsf{B},\theta}(T) = \mathsf{E}_{k,\mathsf{B},\theta}[T] \equiv \mathsf{E}_{0,\mathsf{B},\theta}[T]$ since $T \ge 0$ with probability one.

Next, for the prior distribution π and $\alpha \in (0,1)$, define the Bayesian class of changepoint detection procedures with the weighed probability of false alarm $\mathsf{PFA}_\pi(T) = \mathsf{P}^\pi_{\mathsf{B},\theta}(T \le \nu)$ not greater than a prescribed number α:

$$\mathbb{C}_\pi(\alpha) = \{T \in \mathscr{M} : \mathsf{PFA}_\pi(T) \le \alpha\} = \left\{T \in \mathscr{M} : \sum_{k=0}^{\infty} \pi_k \mathsf{P}_\infty(T \le k) \le \alpha\right\}, \tag{5}$$

where $\mathscr{M}$ stands for the totality of Markov times.

In this section, we are interested in the uniform Bayesian constrained optimization criterion

$$\inf_{\{T:\mathsf{PFA}_\pi(T)\le\alpha\}} \mathsf{EDD}_{k,\mathsf{B},\theta}(T) \quad \text{for all } k \in \mathbb{Z}_+,\ \mathsf{B} \in \mathscr{P} \text{ and } \theta \in \Theta.$$

However, this problem is intractable for arbitrary values of $\alpha \in (0,1)$. For this reason, we will consider the following first-order asymptotic problem assuming that the given PFA α approaches zero: Find a change detection procedure T^* such that it minimizes the expected detection delay $\mathsf{EDD}_{k,\mathsf{B},\theta}(T)$ asymptotically to first order as $\alpha \to 0$ uniformly for all possible values of $k \in \mathbb{Z}_+$, subsets $\mathsf{B} \in \mathscr{P}$, and $\theta \in \Theta$. That is, our goal it to design such detection procedure T^* that, as $\alpha \to 0$,

$$\inf_{T\in\mathbb{C}_\pi(\alpha)} \mathsf{EDD}_{k,\mathsf{B},\theta}(T) = \mathsf{EDD}_{k,\mathsf{B},\theta}(T^*)(1+o(1)) \ \forall\, \mathsf{B} \in \mathscr{P},\ \theta \in \Theta,\ k \in \mathbb{Z}_+, \tag{6}$$

where $\mathbb{C}_\pi(\alpha)$ is the class of detection procedures for which the PFA does not exceed a prescribed number $\alpha \in (0,1)$ defined in (5) and $o(1) \to 0$ as $\alpha \to 0$.

2.3 Multistream Changepoint Detection Procedures

We begin by considering the most general scenario where the observations across streams are dependent.

2.3.1 Parametric Prior Uncertainty

Let $\mathscr{L}_{\mathsf{B},\theta}(n) = f_{\mathsf{B},\theta}(\mathbf{X}_n|\mathbf{X}^{n-1})/g(\mathbf{X}_n|\mathbf{X}^{n-1})$. Note that in the general non-i.i.d. case the statistic $\mathscr{L}_{\mathsf{B},\theta}(n) = \mathscr{L}_{\mathsf{B},\theta}^{(k)}(n)$ depends on the change point $\nu = k$ since the post-change density $f_{\mathsf{B},\theta}(\mathbf{X}_n|\mathbf{X}^{n-1}) = f_{\mathsf{B},\theta}^{(k)}(\mathbf{X}_n|\mathbf{X}^{n-1})$ may depend on k. The likelihood ratio (LR) of the hypothesis $\mathsf{H}_{k,\mathsf{B}}$ that the change occurs at $\nu = k$ in the subset of streams B against the no-change hypothesis H_∞ based on the sample $\mathbf{X}^n = (\mathbf{X}_1, \dots, \mathbf{X}_n)$ is given by the product

$$LR_{\mathsf{B},\theta}(k,n) = \prod_{t=k+1}^{n} \mathscr{L}_{\mathsf{B},\theta}(t), \quad n > k$$

and we set $LR_{\mathsf{B},\theta}(k,n) = 1$ for $n \le k$.

For $\mathsf{B} \in \mathscr{P}$ and $\theta \in \Theta$, define the generalized Shiryaev–Roberts (SR) statistic

$$R_{\mathsf{B},\theta}(n) = r\, LR_{\mathsf{B},\theta}(0,n) + \sum_{k=0}^{n-1} LR_{\mathsf{B},\theta}(k,n), \quad n \ge 1 \tag{7}$$

with the initial condition (non-negative head-start) $R_{\mathsf{B},\theta}(0) = r$, $r \ge 0$.

Now, introduce the probability mass function (mixing measure)

$$\mathbf{p} = \{p_{\mathsf{B}}, \mathsf{B} \in \mathscr{P}\}, \quad p_{\mathsf{B}} > 0\ \forall\, \mathsf{B} \in \mathscr{P}, \quad \sum_{\mathsf{B} \in \mathscr{P}} p_{\mathsf{B}} = 1, \tag{8}$$

where p_{B} is the prior probability of the change being in effect on the set of streams B, and also the mixing probability measure

$$\mathbf{W} = \{W(\theta), \theta \in \Theta\}, \quad \int_{\Theta} \mathrm{d}W(\theta) = 1. \tag{9}$$

For a fixed value of θ, introduce the mixture statistic

$$\begin{aligned} R_{\mathbf{p},\theta}(n) &= \sum_{\mathsf{B} \in \mathscr{P}} p_{\mathsf{B}} R_{\mathsf{B},\theta}(n) \\ &= r\Lambda_{\mathbf{p},\theta}(0,n) + \sum_{k=0}^{n-1} \Lambda_{\mathbf{p},\theta}(k,n), \quad n \ge 1, \quad R_{\mathbf{p},\theta}(0) = r, \end{aligned} \tag{10}$$

where

$$\Lambda_{\mathbf{p},\theta}(k,n) = \sum_{\mathsf{B} \in \mathscr{P}} p_{\mathsf{B}} LR_{\mathsf{B},\theta}(k,n) \tag{11}$$

is the mixture LR.

As discussed in [29, 31], when the parameter θ is unknown there are two main conventional approaches – either to estimate θ (say maximize) or average (mix) over θ. Using the mixing measure (prior distribution) $\mathbf{W} = \{W(\theta), \theta \in \Theta\}$ given in (9), define the double LR-mixture

$$\Lambda_{\mathbf{p},W}(k,n) = \int_{\Theta} \sum_{\mathsf{B}\in\mathscr{P}} p_{\mathsf{B}} LR_{\mathsf{B},\theta}(k,n)\,\mathrm{d}W(\theta) = \int_{\Theta} \Lambda_{\mathbf{p},\theta}(k,n)\,\mathrm{d}W(\theta), \quad k<n \tag{12}$$

and the double-mixture statistic

$$\begin{aligned} R_{\mathbf{p},W}(n) &= \int_{\Theta} \sum_{\mathsf{B}\in\mathscr{P}} p_{\mathsf{B}} R_{\mathsf{B},\theta}(n)\,\mathrm{d}W(\theta) \\ &= r\Lambda_{\mathbf{p},W}(0,n) + \sum_{k=0}^{n-1} \Lambda_{\mathbf{p},W}(k,n), \quad n \geq 1, \quad R_{\mathbf{p},W}(0) = r \end{aligned} \tag{13}$$

(with a non-negative head-start r).

The corresponding double-mixture LR-based detection procedure is given by the stopping rule which is the first time $n \geq 1$ such that the statistic $R_{\mathbf{p},W}(n)$ exceeds the level $A > 0$:

$$T_A^{\mathbf{p},W} = \inf\{n \geq 1 : R_{\mathbf{p},W}(n) \geq A\}. \tag{14}$$

The main result of our theory is that this changepoint detection procedure is first-order asymptotically optimal under certain very general conditions if threshold $A = A_\alpha$ is adequately selected, i.e., asymptotic equality (6) holds with $T^* = T_A^{\mathbf{p},W}$.

2.3.2 Known Parameters of the Post-Change Distribution

If the value of the post-change parameter θ is known or its putative value is of special interest, representing a nominal change, then it is reasonable to turn the double-mixture procedure $T_A^{\mathbf{p},W}$ in single-mixture procedure $T_A^{\mathbf{p},\theta}$ by taking the degenerate weight function $W(\vartheta)$ concentrated at $\vartheta = \theta$, i.e.,

$$T_A^{\mathbf{p},\theta} = \inf\{n \geq 1 : R_{\mathbf{p},\theta}(n) \geq A\}, \tag{15}$$

and ask whether or not it has first-order asymptotic optimality properties at the point θ, i.e., that under certain conditions asymptotic formula (6) holds for $T^* = T_A^{\mathbf{p},\theta}$.

2.4 Asymptotic Optimality of Mixture-Based Detection Procedures

2.4.1 Basic Conditions

While we consider a general prior and a very general stochastic model for the observations in streams and between streams, to study asymptotic optimality properties we still need to impose certain constraints on the prior distribution $\pi = \{\pi_k\}$ and on the general stochastic model (1) that guarantee asymptotic stability of the detection statistics as the sample size increases.

Regarding the prior distribution, we will assume that the following condition holds:

$$\lim_{n\to\infty} \frac{1}{n} |\log \mathsf{P}(\nu > n)| = \beta \ \text{ for some } \beta \geq 0.$$

However, this condition being quite general does not cover the case where β is positive but may go to zero. Indeed, the distributions with an exponential right tail that satisfy this condition with $\beta > 0$ do not converge as $\beta \to 0$ to heavy-tailed distributions with $\beta = 0$. For this reason, any assertions for heavy-tailed distributions with $\beta = 0$ do not hold if $\beta \to 0$ with an arbitrary rate; the rate must be matched with the PFA probability α, $\alpha \to 0$. Hence, in what follows we consider the scenario with the prior distribution $\pi^\alpha = \{\pi_k^\alpha\}$ that depends on the PFA constraint α and modify the above condition as

CP$_1$. For some $\beta_\alpha \geq 0$ such that $\beta_\alpha \to 0$ as $\alpha \to 0$

$$\lim_{n\to\infty} \frac{1}{n} \left| \log \sum_{k=n+1}^{\infty} \pi_k^\alpha \right| = \beta_\alpha.$$

For $\mathsf{B} \in \mathscr{P}$ and $\theta \in \Theta$, introduce the log-likelihood ratio (LLR) process between the hypotheses $\mathsf{H}_{k,\mathsf{B},\theta}$ $(k = 0, 1, \dots)$ and H_∞:

$$\lambda_{\mathsf{B},\theta}(k,n) = \sum_{t=k+1}^{n} \log \frac{f_{\mathsf{B},\theta}(\mathbf{X}_t|\mathbf{X}^{t-1})}{g(\mathbf{X}_t|\mathbf{X}^{t-1})}, \qquad n > k$$

$(\lambda_{\mathsf{B},\theta}(k,n) = 0$ for $n \leq k)$.

Let $\psi : \mathbb{R}_+ \to \mathbb{R}_+$ be an increasing and continuous function and by Ψ denote the inverse of ψ, which is also increasing and continuous. We also assume that $\lim_{x\to\infty} \psi(x) = \infty$, and thus, $\Psi(x)$ is properly defined on the entire positive real line.

In the rest of the paper, we assume that the strong law of large numbers (SLLN) holds for the LLR with the rate $\psi(n)$, that is, $\lambda_{\mathsf{B},\theta}(k, k+n)/\psi(n)$ converges almost surely (a.s.) under probability $\mathsf{P}_{k,\mathsf{B},\theta}$ to a finite and positive number $I_{\mathsf{B},\theta}$:

$$\frac{1}{\psi(n)} \lambda_{\mathsf{B},\theta}(k, k+n) \xrightarrow[n\to\infty]{\mathsf{P}_{k,\mathsf{B},\theta}-\text{a.s.}} I_{\mathsf{B},\theta} \ \text{ for all } k \in \mathbb{Z}_+,\ \mathsf{B} \in \mathscr{P},\ \theta \in \Theta. \tag{16}$$

Condition (16) is the first main assumption regarding the general stochastic model for observed data.

Notice that if the data in streams and across the streams are independent (but non-stationary), then

$$I_{\mathsf{B},\theta} = \lim_{n\to\infty} \frac{1}{\psi(n)} \sum_{i\in\mathsf{B}} \sum_{t=k+1}^{k+n} \mathsf{E}_{k,i,\theta_i} \left[\frac{f_{i,\theta_{i,t}}(X_t(i))}{g_{i,t}(X_t(i))} \right],$$

where $g_{i,t}(X_t(i))$ and $f_{i,\theta_{i,t}}(X_t(i))$ are pre- and post-change densities of the t-th observation in the i-th stream, respectively. In other words, the number $I_{\mathsf{B},\theta}$ can be interpreted as the limiting local Kullback-Leibler divergence.

If the function $\psi(n)$ is non-linear we will say that the model is *non-stationary (even asymptotically)*. However, in many applications the observations are non-identically distributed but $\psi(n) = n$, in which case we will say that the non-i.i.d. model is *asymptotically stationary*.

In Subsection 2.4.3, we establish the asymptotic lower bound (as $\alpha \to 0$) for the minimal value of the expected detection delay $\mathsf{EDD}_{k,\mathsf{B},\theta}(T)$ in class $\mathbb{C}_\pi(\alpha)$ whenever the almost sure convergence condition (16) holds. To obtain the lower bound it suffices to require the following right-tail condition

$$\lim_{L\to\infty} \mathsf{P}_{k,\mathsf{B},\theta}\left\{\frac{1}{\psi(L)} \max_{1\le n\le L} \lambda_{\mathsf{B},\theta}(k,k+n) \ge (1+\varepsilon)I_{\mathsf{B},\theta}\right\} = 0 \quad \text{for all } \varepsilon > 0.$$

This condition holds whenever the SLLN (16) takes place. However, the SLLN is not sufficient to show that this lower bound is attained for the mixture detection procedures (14) and (15). To this end, the SLLN should be strengthened into a complete convergence version.

Definition 1. We say that the process $\{Y_{k,n}\}_{n\in\mathbb{N}}$, $k \in \mathbb{Z}_+$ converges to a random variable Y uniformly completely under the measure P_k if

$$\sum_{n=1}^{\infty} \sup_{k\in\mathbb{Z}_+} \mathsf{P}_k\left\{\left|Y_{k,n} - Y\right| > \varepsilon\right\} < \infty \quad \text{for all } \varepsilon > 0.$$

The proof of the upper bound presented in Subsection 2.4.5 shows that if along with the SLLN (16) the following left-tail condition is satisfied

$$\sum_{n=1}^{\infty} \sup_{k\in\mathbb{Z}_+} \mathsf{P}_{k,\mathsf{B},\theta}\left\{\frac{1}{\psi(n)}\lambda_{\mathsf{B},\theta}(k,k+n) < I_{\mathsf{B},\theta} - \varepsilon\right\} < \infty \quad \text{for all } \varepsilon > 0, \tag{17}$$

where the "information number" $I_{\mathsf{B},\theta}$ is positive and finite, then the detection procedure (15) is asymptotically optimal for the fixed (prespecified) post-change parameter θ.

Obviously, both conditions (16) and (17) are satisfied whenever there exist positive and finite numbers $I_{\mathsf{B},\theta}$ ($\mathsf{B} \in \mathscr{P}$, $\theta \in \Theta$) such that the normalized LLR $\lambda_{\mathsf{B},\theta}(k,k+n)/\psi(n) \to I_{\mathsf{B},\theta}$ uniformly completely under $\mathsf{P}_{k,\mathsf{B},\theta}$-probability. Therefore, this condition turns out to be sufficient for asymptotic optimality of the detection procedure (15) when the post-change parameter θ is known. In the case of the unknown post-change parameter, the left-tail condition is similar to but more sophisticated than condition (17). The details will be given in Subsection 2.4.6.

2.4.2 Heuristic Argument

We begin with a heuristic argument that allows us to obtain approximations for the expected detection delay when the threshold in the mixture detection procedure $T_A^{\mathbf{p},W}$ is large. For the sake of simplicity, the head-start in the detection statistic $R_{\mathbf{p},W}(n)$ is set to zero, $R_{\mathbf{p},W}(n) = r = 0$. This argument also explains the reason why the condition $\mathbf{CP}_1$ on the prior distribution is imposed.

Assume first that the change occurs at $\nu = k \le 0$. It is easy to see that the logarithm of the statistic $R_{\mathbf{p},W}(n)$ can be written as

$$\log R_{\mathbf{p},W}(n) = \lambda_{\mathbf{p},W}(0,n) + Y_n,$$

where

$$Y_n = \log\left[\sum_{t=1}^{n-1} \frac{\Lambda_{\mathbf{p},W}(t,n)}{\Lambda_{\mathbf{p},W}(0,n)}\right].$$

Due to the SLLN (16) $\lambda_{\mathbf{p},W}(n)/\psi(n)$ converges almost surely as $n \to \infty$ under $\mathsf{P}_{0,\mathsf{B},\theta}$ to $I_{\mathsf{B},\theta}$, so we can expect that for a large n

$$\lambda_{\mathbf{p},W}(0,n) \approx I_{\mathsf{B},\theta}\, \psi(n) + \xi_n, \tag{18}$$

where $\xi_n/\psi(n)$ converges to 0. Also, Y_n, $n \ge 1$ are "slowly changing" and converge to a random variable Y_∞. Thus, ignoring the overshoot of $\log R_{\mathbf{p},W}(T_A^{\mathbf{p},W})$ over $\log A$, we obtain

$$\log A \approx \log R_{\mathbf{p},W}(T_A^{\mathbf{p},W}) \approx I_{\mathsf{B},\theta}\, \psi\left(T_A^{\mathbf{p},W}\right) + \xi_{T_A^{\mathbf{p},W}} + Y_\infty. \tag{19}$$

If $\psi(n)$ increases sufficiently fast, at least not slower than n, then the last two terms can be ignored, and hence,

$$\psi(T_A^{\mathbf{p},W}) \approx \frac{\log A}{I_{\mathsf{B},\theta}},$$

which implies

$$T_A^{\mathbf{p},W} \approx \Psi\left(\frac{\log A}{I_{\mathsf{B},\theta}}\right).$$

Taking expectation yields

$$\mathsf{E}_{0,\mathsf{B},\theta}[T_A^{\mathbf{p},W}] \approx \Psi\left(\frac{\log A}{I_{\mathsf{B},\theta}}\right).$$

A similar argument leads to the following approximate formula (for a large A) for the expected delay $\mathsf{E}_{k,\mathsf{B},\theta}[(T_A^{\mathbf{p},W} - k)^+]$ when the change occurs at any $\nu = k \in \mathbb{Z}_+$:

$$\mathsf{E}_{k,\mathsf{B},\theta}[(T_A^{\mathbf{p},W} - k)^+] \approx \Psi\left(\frac{\log A}{I_{\mathsf{B},\theta}}\right), \quad k \in \mathbb{Z}_+. \tag{20}$$

Next, in Lemma 1 (see Subsection 2.4.4) it is established that the probability of a false alarm of the procedure $T_A^{\mathbf{p},W}$ satisfies the inequality

$$\mathsf{PFA}_\pi(T_A^{\mathbf{p},W}) \le \bar{\nu}_A/A,$$

where $\bar{\nu}_A = \sum_{k=1}^\infty k\pi_k^A$ is the mean of the prior distribution π^A. If we assume that $\bar{\nu}_A/A \to 0$ as $A \to \infty$, then this inequality along with approximate equality (20) yields the following approximation for the expected delay to detection:

$$\mathsf{E}_{k,\mathsf{B},\theta}[T_A^{\mathbf{p},W} - k | T_A^{\mathbf{p},W} > k] = \frac{\mathsf{E}_{k,\theta}[(T_A^W - k)^+]}{1 - \mathsf{PFA}_\pi(T_A^{\mathbf{p},W})} \approx \Psi\left(\frac{\log A}{I_{\mathsf{B},\theta}}\right). \tag{21}$$

In subsequent sections, this approximation is justified rigorously.

There are two key points we would like to address. The first one is that if we impose condition $\mathbf{CP}_1$ on the prior distribution of the change point with $\beta > 0$ that does not depend on α and does not converge to 0, then the lower bound for the expected delay to detection in class $\mathbb{C}_\pi(\alpha)$ has the form

$$\inf_{T \in \mathbb{C}_\pi(\alpha)} \mathsf{EDD}_{k,\mathsf{B},\theta} \ge \Psi\left(\frac{|\log \alpha|}{I_{\mathsf{B},\theta} + \beta}\right)(1 + o(1)) \quad \text{as } \alpha \to 0. \tag{22}$$

In this case, this lower bound is not attained by the procedure $T_A^{\mathbf{p},W}$ since if we take $A = A_\alpha \sim |\log \alpha|$, then it follows from (21) that

$$\mathsf{E}_{k,\mathsf{B},\theta}[T_A^{\mathbf{p},W} - k | T_A^{\mathbf{p},W} > k] \sim \Psi\left(\frac{|\log \alpha|}{I_{\mathsf{B},\theta}}\right) > \Psi\left(\frac{|\log \alpha|}{I_{\mathsf{B},\theta} + \beta}\right).$$

Hereafter we use a conventional notation $y_a \sim z_a$ as $a \to a_0$ if $\lim_{a \to a_0}(y_a/z_a) = 1$. This can be expected since the detection statistic $R_{\mathbf{p},W}(n)$ is based on the uniform prior on a positive half line. However, if $\beta = \beta_\alpha \to 0$, as we assumed in condition $\mathbf{CP}_1$, then the lower bound is

$$\Psi\left(\frac{|\log \alpha|}{I_{\mathsf{B},\theta}}\right)(1 + o(1))$$

and it is attained by $T_A^{\mathbf{p},W}$.

Yet another key point is that to obtain the lower bound (22) with $\beta = \beta_\alpha \to 0$ but $\beta_\alpha > 0$, as shown in the proof of Theorem 1, we need the function $\psi(x)$ to be either linear or super-linear. Indeed, if now we define the statistic

$$S_{\mathbf{p},W}^\pi(n) = \frac{1}{\mathsf{P}(\nu \ge n)}\left[\pi_{-} \Lambda_{\mathbf{p},W}(0,n) + \sum_{k=0}^{n-1} \pi_k \Lambda_{\mathbf{p},W}(k,n)\right]$$

and the corresponding detection procedure

$$\widetilde{T}_A^{\mathbf{p},W} = \inf\left\{n \ge 1 : S_{\mathbf{p},W}^\pi(n) \ge A\right\},$$

then instead of (19) we have

$$\log A \approx \log S_{\mathbf{p},W}(\widetilde{T}_A^{\mathbf{p},W}) \approx \beta \widetilde{T}_A^{\mathbf{p},W} + I_{\mathsf{B},\theta}\psi\left(\widetilde{T}_A^{\mathbf{p},W}\right) + \xi_{\widetilde{T}_A^{\mathbf{p},W}} + Y_\infty.$$

So if $\psi(x) < x$ is sub-linear, i.e., $\Psi(n) \gg n$ for large n, we expect that the prior distribution gives much more contribution than the observations and

$$\mathsf{E}_{k,\mathsf{B},\theta}[\widetilde{T}_A^{\mathbf{p},W} - k | \widetilde{T}_A^{\mathbf{p},W} > k] \approx \Psi\left(\frac{\log A}{\beta}\right)$$

as long as $\beta > 0$, i.e., the prior distribution has an exponential right tail.

Despite the simplicity of the basic ideas and the approximate calculations, the rigorous argument in proving these results is rather tedious.

2.4.3 Asymptotic Lower Bound for Expected Detection Delay

For establishing asymptotic optimality of changepoint detection procedures, we first obtain, under the a.s. convergence condition (16), the asymptotic lower bound for expected detection delay $\mathsf{EDD}_{k,\mathsf{B},\theta}(T) = \mathsf{E}_{k,\mathsf{B},\theta}[T - k | T > k]$ of any detection rule T from class $\mathbb{C}_\pi(\alpha)$. In the following subsections, we show that under certain additional conditions associated with complete convergence of LLR these bounds are attained for the mixture procedures $T_A^{\mathbf{p},W}$ and $T_A^{\mathbf{p},\theta}$.

For $\varepsilon \in (0,1)$ and $\delta > 0$, define

$$N_\alpha = N_\alpha(\varepsilon, \delta, \mathsf{B}, \theta) = \Psi\left(\frac{(1-\varepsilon)|\log \alpha|}{I_{\mathsf{B},\theta} + \beta_\alpha + \delta}\right). \tag{23}$$

The following theorem specifies the asymptotic lower bound for the expected detection delay.

Theorem 1. *Let the prior distribution of the change point satisfy condition* $\mathbf{CP}_1$ *and assume that for some positive and finite numbers* $I_{\mathsf{B},\theta}$ $(\mathsf{B} \in \mathscr{P},\ \theta \in \Theta)$ *the a.s. convergence condition* (16) *holds. Suppose in addition that the function* $\psi(x)$ *increases not slower than x, i.e.,*

$$\psi(x) \ge x, \quad x > 0. \tag{24}$$

Then for all $0 < \varepsilon < 1$ *and* $\delta > 0$

$$\lim_{\alpha \to 0} \sup_{T \in \mathbb{C}_\pi(\alpha)} \mathsf{P}_{k,\mathsf{B},\theta}\{k < T \le k + N_\alpha(\varepsilon, \delta, \mathsf{B}, \theta)\} = 0 \text{ for all } \mathsf{B} \in \mathscr{P}, \theta \in \Theta, \tag{25}$$

and, as a result, for all $\nu = k \in \mathbb{Z}_+$, $\mathsf{B} \in \mathscr{P}$, *and* $\theta \in \Theta$

$$\inf_{T \in \mathbb{C}_\pi(\alpha)} \mathsf{EDD}_{k,\mathsf{B},\theta}(T) \ge \Psi\left(\frac{|\log \alpha|}{I_{\mathsf{B},\theta}}\right)(1 + o(1)) \text{ as } \alpha \to 0. \tag{26}$$

Proof. In the asymptotically stationary case where $\psi(n) = n$, the methodology of the proof is analogous to that used by Tartakovsky [27, 28] in the proofs of the lower bounds in a single stream change detection problem with slightly different assumptions on the prior distribution. A generalization of the proof in the substantially non-stationary case has several technical details. We present complete proof.

To begin, consider an arbitrary stopping time $T \in \mathbb{C}_\pi(\alpha)$ and note that by Markov's inequality

$$\begin{aligned}\mathsf{EDD}_{k,\mathsf{B},\theta}(T) &\geq \mathsf{E}_{k,\mathsf{B},\theta}\left[(T-k)^+\right] \geq N_\alpha \, \mathsf{P}_{k,\mathsf{B},\theta}\left\{(T-k)^+ > N_\alpha\right\} \\ &= N_\alpha \, \mathsf{P}_{k,\mathsf{B},\theta}\left(T > k + N_\alpha\right).\end{aligned}$$

If assertion (25) holds, then

$$\lim_{\alpha \to 0} \inf_{T \in \mathbb{C}_\pi(\alpha)} \mathsf{P}_{k,\mathsf{B},\theta}\left(T > k + N_\alpha\right) = 1. \tag{27}$$

Indeed,

$$\mathsf{P}_{k,\mathsf{B},\theta}\left(T > k + N_\alpha\right) = \mathsf{P}_\infty\left(T > k\right) - \mathsf{P}_{k,\mathsf{B},\theta}\left(k < T \leq k + N_\alpha\right),$$

where we used the identity $\mathsf{P}_{k,\mathsf{B},\theta}(T > k) = \mathsf{P}_\infty(T > k)$. Next, since for any change detection procedure $T \in \mathbb{C}_\pi(\alpha)$,

$$\alpha \geq \sum_{i=k}^{\infty} \pi_i^\alpha \mathsf{P}_\infty(T \leq i) \geq \mathsf{P}_\infty(T \leq k) \sum_{i=k}^{\infty} \pi_i^\alpha,$$

it follows that

$$\inf_{T \in \mathbb{C}_\pi(\alpha)} \mathsf{P}_\infty(T > k) \geq 1 - \alpha / \Pi_{k-1}^\alpha, \quad k \in \mathbb{Z}_+, \tag{28}$$

where $\Pi_\ell^\alpha = \mathsf{P}(\nu > \ell) = \sum_{k=\ell+1}^\infty \pi_k^\alpha$. Hence, we obtain

$$\inf_{T \in \mathbb{C}_\pi(\alpha)} \mathsf{P}_{k,\mathsf{B},\theta}\left(T > k + N_\alpha\right) \geq 1 - \alpha / \Pi_{k-1}^\alpha - \sup_{T \in \mathbb{C}_\pi(\alpha)} \mathsf{P}_{k,\mathsf{B},\theta}(k < T \leq k + N_\alpha).$$

This inequality yields (27), and we obtain the asymptotic inequality

$$\mathsf{EDD}_{k,\mathsf{B},\theta}(T) \geq N_\alpha(1 + o(1)) \text{ as } \alpha \to 0,$$

which holds for arbitrary values of $\varepsilon \in (0,1)$ and $\delta > 0$. By our assumption, the function $N_\alpha = N_\alpha(\varepsilon, \delta, \mathsf{B}, \theta)$ is continuous, so we may take a limit $\delta, \varepsilon \to 0$, which implies inequality (26). Consequently, to show the validity of asymptotic inequality (26), it suffices to prove the equality (25).

Let

$$\lambda_{\mathsf{B},\theta}^*(t) = \log \frac{f_{\mathsf{B},\theta}(\mathbf{X}_t | \mathbf{X}^{t-1})}{g(\mathbf{X}_t | \mathbf{X}^{t-1})}$$

and notice that for $n > k$

$$\frac{d\mathsf{P}_\infty^{(n)}}{d\mathsf{P}_{k,\mathsf{B},\theta}^{(n)}} = \frac{1}{LR_{\mathsf{B},\theta}(k,n)} = \exp\left\{-\sum_{t=k+1}^{n} \lambda_{\mathsf{B},\theta}^*(t)\right\},$$

where hereafter $\mathsf{P}^{(n)}$ denotes a restriction of the measure P to the sigma-algebra $\mathscr{F}_n$. Therefore, changing the measure $\mathsf{P}_\infty \to \mathsf{P}_{k,\mathsf{B},\theta}$ and using Wald's likelihood ratio identity, we obtain for any $C > 0$:

$$\begin{aligned}
\mathsf{P}_\infty(k < T \le k+N_\alpha) &= \mathsf{E}_{k,\mathsf{B},\theta}\left[\mathbb{1}_{\{0<T-k\le N_\alpha\}}\frac{d\mathsf{P}_\infty^{(T)}}{d\mathsf{P}_{k,\mathsf{B},\theta}^{(T)}}\right] \\
&= \mathsf{E}_{k,\mathsf{B},\theta}\left[\mathbb{1}_{\{0<T-k\le N_\alpha\}}e^{-\sum_{t=k+1}^{T}\lambda_{\mathsf{B},\theta}^*(t)}\right] \\
&\ge \mathsf{E}_{k,\mathsf{B},\theta}\left[\mathbb{1}_{\{0<T-k\le N_\alpha,\sum_{t=k+1}^{T}\lambda_{\mathsf{B},\theta}^*(t)<C\}}e^{-\sum_{t=k+1}^{T}\lambda_{\mathsf{B},\theta}^*(t)}\right] \\
&\ge e^{-C}\mathsf{P}_{k,\mathsf{B},\theta}\left\{0<T-k\le N_\alpha, \max_{0<n-k\le N_\alpha}\sum_{t=k+1}^{n}\lambda_{\mathsf{B},\theta}^*(t)<C\right\} \\
&\ge e^{-C}\left[\mathsf{P}_{k,\mathsf{B},\theta}\{0<T-k\le N_\alpha\}-\mathsf{P}_{k,\mathsf{B},\theta}\left\{\max_{0\le n<N_\alpha}\sum_{t=k+1}^{k+n+1}\lambda_{\mathsf{B},\theta}^*(t)\ge C\right\}\right],
\end{aligned}$$

where the last inequality follows from the fact that $\mathsf{P}(\mathscr{A}\cap\mathscr{B}) \ge \mathsf{P}(\mathscr{A}) - \mathsf{P}(\mathscr{B}^c)$ for any events $\mathscr{A}$ and $\mathscr{B}$, where $\mathscr{B}^c$ is the complement event of $\mathscr{B}$. Setting

$$C = \psi(N_\alpha)I_{\mathsf{B},\theta}(1+\varepsilon) = (1+\varepsilon)K_\alpha, \quad K_\alpha = \frac{(1-\varepsilon)|\log\alpha| I_{\mathsf{B},\theta}}{I_{\mathsf{B},\theta}+\beta_\alpha+\delta}$$

yields

$$\mathsf{P}_{k,\mathsf{B},\theta}(k < T \le k+N_\alpha) \le \kappa_{k,\alpha}(T) + \sup_{k\ge 0}\gamma_{k,\alpha}, \tag{29}$$

where

$$\kappa_{k,\alpha}(T) = \kappa_{k,\alpha}(\varepsilon,\delta,\theta,\mathsf{B},T) = e^{(1+\varepsilon)K_\alpha}\mathsf{P}_\infty\{0<T-k\le N_\alpha\}$$

and

$$\gamma_{k,\alpha} = \gamma_{k,\alpha}(\varepsilon,\mathsf{B},\theta) = \mathsf{P}_{k,\mathsf{B},\theta}\left\{\max_{0\le n<N_\alpha}\sum_{t=k+1}^{k+n+1}\lambda_{\mathsf{B},\theta}^*(t) \ge (1+\varepsilon)I_{\mathsf{B},\theta}\psi(N_\alpha)\right\}.$$

By the a.s. convergence condition (16),

$$\lim_{\alpha\to 0}\sup_{k\ge 0}\gamma_{k,\alpha} = 0. \tag{30}$$

To evaluate $\kappa_{k,\alpha}(T)$ it suffices to note that, by condition $\mathbf{CP}_1$, for all sufficiently small α, there exists a small δ such that

$$\frac{|\log \Pi_{k-1+N_\alpha}^\alpha|}{k-1+N_\alpha} \le \beta_\alpha + \delta,$$

and to use inequality (28), which yields that for all sufficiently small α

$$\kappa_{k,\alpha}(T) \le e^{(1+\varepsilon)K_\alpha}\mathsf{P}_\infty(T \le k+N_\alpha) \le \alpha\, e^{(1+\varepsilon)K_\alpha}/\Pi_{k-1+N_\alpha}^\alpha$$

$$\le \exp\left\{(1+\varepsilon)K_\alpha - |\log\alpha| + (k-1+N_\alpha)\frac{|\log \Pi^\alpha_{k-1+N_\alpha}|}{k-1+N_\alpha}\right\}$$
$$\le \exp\{(1+\varepsilon)K_\alpha - |\log\alpha| + (k-1+N_\alpha)(\beta_\alpha+\delta)\}.$$

Since by condition (24),

$$N_\alpha \le K_\alpha / I_{\mathsf{B},\theta} = \frac{(1-\varepsilon)|\log\alpha|}{I_{\mathsf{B},\theta}+\beta_\alpha+\delta}$$

it follows that

$$\kappa_{k,\alpha}(T) \le \exp\left\{-\frac{I_{\mathsf{B},\theta}\varepsilon^2 + (\beta_\alpha+\delta)\varepsilon}{I_{\mathsf{B},\theta}+\beta_\alpha+\delta}|\log\alpha| + (\beta_\alpha+\delta)k\right\}$$
$$\le \exp\{-\varepsilon^2|\log\alpha| + (\beta_\alpha+\delta)k\} := \overline{\kappa}_{\alpha,k}(\varepsilon,\delta).$$

for all $\varepsilon \in (0,1)$ and every small δ. Hence, for all $\varepsilon \in (0,1)$ and sufficiently small δ,

$$\kappa_{k,\alpha}(T) \le \overline{\kappa}_{k,\alpha}(\varepsilon,\delta), \tag{31}$$

where $\overline{\kappa}_{k,\alpha}(\varepsilon,\delta)$ does not depend on the stopping time T and goes to 0 as $\alpha \to 0$ for any fixed $k \in \mathbb{Z}_+$, which implies that

$$\sup_{T\in\mathbb{C}_\pi(\alpha)} \kappa_{k,\alpha}(T) \to 0 \quad \text{as } \alpha \to 0 \ \text{ for every fixed } k \in \mathbb{Z}_+.$$

The proof of the lower bound (26) is complete. □

Remark 1. Super-linear condition (24) for $\psi(x)$ can be generalized to asymptotic super-linearity, i.e., it suffices to require that the growth rate of $\psi(x)$ is not slower than a linear function for sufficiently large x.

2.4.4 Probabilities of False Alarm of Mixture Change Detection Procedures

An important question is how to select threshold A in change detection procedures $T_A^{\mathbf{p},\theta}$ and $T_A^{\mathbf{p},W}$ defined in (15) and (14), respectively, to imbed them into class $\mathbb{C}_\pi(\alpha)$.

The following lemma provides the upper bounds for the PFA of these procedures.

Lemma 1. *For all $A > 0$ and any prior distribution $\pi = \{\pi_k\}$ of the change point ν with finite mean $\bar\nu = \sum_{k=1}^\infty k\pi_k$, the weighted probabilities of false alarms of the detection procedures $T_A^{\mathbf{p},\theta}$ and $T_A^{\mathbf{p},W}$ satisfy the inequalities*

$$\mathsf{PFA}_\pi(T_A^{\mathbf{p},\theta}) \le \frac{rb+\bar\nu}{A}, \tag{32}$$

$$\mathsf{PFA}_\pi(T_A^{\mathbf{p},W}) \le \frac{rb+\bar\nu}{A}, \tag{33}$$

where $b = \sum_{k=1}^{\infty} \pi_k$. *Hence, if* $A = A_\alpha = (rb + \bar{\nu})/\alpha$, *then* $\mathsf{PFA}_\pi(T_{A_\alpha}^{\mathbf{p},\theta}) \le \alpha$ *and* $\mathsf{PFA}_\pi(T_{A_\alpha}^{\mathbf{p},W}) \le \alpha$, *i.e.,* $T_{A_\alpha}^{\mathbf{p},\theta} \in \mathbb{C}_\pi(\alpha)$ *and* $T_{A_\alpha}^{\mathbf{p},W} \in \mathbb{C}_\pi(\alpha)$.

Proof. We have

$$\mathsf{E}_\infty[R_{\mathbf{p},\theta}(n)|\mathscr{F}_{n-1}] = \sum_{\mathsf{B}\in\mathscr{P}} p_{\mathsf{B}} + \sum_{\mathsf{B}\in\mathscr{P}} p_{\mathsf{B}} R_{\mathsf{B},\theta}(n-1) = 1 + R_{\mathbf{p},\theta}(n-1),$$

and hence, the statistic $\{R_{\mathbf{p},\theta}(n)\}_{n\in\mathbb{N}}$ is a non-negative $(\mathsf{P}_\infty, \mathscr{F}_n)$-submartingale with mean

$$\mathsf{E}_\infty[R_{\mathbf{p},\theta}(n)] = r + n, \quad n \in \mathbb{N}.$$

By Doob's maximal submartingale inequality, for any $k \ge 1$,

$$\mathsf{P}_\infty(T_A^{\mathbf{p},\theta} \le k) = \mathsf{P}_\infty\left\{\max_{0\le n\le k} R_{\mathbf{p},\theta}(n) \ge A\right\} \le \frac{\mathsf{E}_\infty[R_{\mathbf{p},\theta}(k)]}{A} = \frac{r+k}{A}, \tag{34}$$

which implies

$$\sum_{k=1}^{\infty} \pi_k \mathsf{P}_\infty(T_A^{\mathbf{p},\theta} \le k) \le \frac{r\sum_{k=1}^{\infty}\pi_k + \sum_{k=1}^{\infty} k\pi_k}{A},$$

and inequality (32) follows.

To prove inequality (33) it suffices to note that the double mixture statistic $\{R_{\mathbf{p},W}(n)\}_{n\in\mathbb{N}}$ is also a $(\mathsf{P}_\infty, \mathscr{F}_n)$-submartingale with mean $\mathsf{E}_\infty[R_{\mathbf{p},W}(n) = r+n$ since

$$\begin{aligned}\mathsf{E}_\infty[R_{\mathbf{p},W}(n)|\mathscr{F}_{n-1}] &= \mathsf{E}_\infty\left[\int_\Theta R_{\mathbf{p},\theta}(n)\mathrm{d}W(\theta)|\mathscr{F}_{n-1}\right] = \int_\Theta \mathsf{E}_\infty\left[R_{\mathbf{p},\theta}(n)|\mathscr{F}_{n-1}\right]\mathrm{d}W(\theta)\\ &= 1 + \int_\Theta R_{\mathbf{p},\theta}(n-1)\mathrm{d}W(\theta) = 1 + R_{\mathbf{p},W}(n-1).\end{aligned}$$

Applying Doob's submartingale inequality yields

$$\mathsf{P}_\infty(T_A^{\mathbf{p},W} \le k) \le (r+k)/A, \quad k = 1, 2, \ldots, \tag{35}$$

which implies (33) and the proof is complete. □

2.4.5 Asymptotic Optimality of Mixture Detection Procedure $T_A^{\mathbf{p},\theta}$ for Known Post-change Parameter

We now proceed with establishing asymptotic optimality properties of the mixture detection procedure $T_A^{\mathbf{p},\theta}$ in class $\mathbb{C}_\pi(\alpha)$ as $\alpha \to 0$ assuming that the post-change parameter θ is prespecified (known or selected).

Throughout this subsection, we assume that there exist positive and finite numbers $I_{\mathsf{B},\theta}$ ($\mathsf{B} \in \mathscr{P}$, $\theta \in \Theta$) such that the normalized LLR $\lambda_{\mathsf{B},\theta}(k, k+n)/\psi(n)$ converges as $n \to \infty$ to $I_{\mathsf{B},\theta}$ uniformly completely under probability measure $\mathsf{P}_{k,\mathsf{B},\theta}$, i.e.,

$$\sum_{n=1}^{\infty} \sup_{k\in\mathbb{Z}_+} \mathsf{P}_{k,\mathsf{B},\theta}\left\{\left|\frac{\lambda_{\mathsf{B},\theta}(k,k+n)}{\psi(n)} - I_{\mathsf{B},\theta}\right| > \varepsilon\right\} < \infty \quad \text{for all } \varepsilon > 0. \tag{36}$$

2.4.5(a) Asymptotic Operating Characteristics of Detection Procedure $T_A^{\mathbf{p},\theta}$ for Large Threshold Values

The following theorem provides asymptotic operating characteristics of the mixture detection procedure $T_A^{\mathbf{p},\theta}$ for large values of threshold A. Since no constraints are imposed on the false alarm rate (FAR) – neither on the PFA nor any other FAR measure – this result is universal and can be used in a variety of optimization problems. We need to impose some constraints on the behavior of the head-start r_A, which may depend on A and approach infinity as $A \to \infty$ in such a way that

$$\lim_{A\to\infty} (r_A A^{-\varepsilon}) = 0 \quad \text{for any } \varepsilon > 0. \tag{37}$$

We also need to assume that the function $\psi(x)$ increases not too slowly, at least faster than the logarithmic function. Specifically, we assume the following condition on the inverse function $\Psi(x)$

$$\lim_{x\to\infty} \frac{\log \Psi(x)}{x} = 0. \tag{38}$$

Theorem 2. *Suppose that conditions* (37) *and* (38) *hold and there exist positive and finite numbers* $I_{\mathsf{B},\theta}$, $\mathsf{B} \in \mathscr{P}$, $\theta \in \Theta$, *such that the uniform complete convergence condition* (36) *is satisfied. Then the following asymptotic as* $A \to \infty$ *approximation holds*

$$\mathsf{EDD}_{k,\mathsf{B},\theta}(T_A^{\mathbf{p},\theta}) = \Psi\left(\frac{\log A}{I_{\mathsf{B},\theta}}\right)(1+o(1)) \quad \textit{for all } k \in \mathbb{Z}_+,\ \mathsf{B} \in \mathscr{P}. \tag{39}$$

Proof. For $\varepsilon \in (0,1)$, let $N_A = N_A(\varepsilon,\mathsf{B},\theta) = \Psi\left((1-\varepsilon)(\log A)/I_{\mathsf{B},\theta}\right)$. Inequality (34) implies that for all $k \in \mathbb{Z}_+$

$$\mathsf{P}_{k,\mathsf{B},\theta}(T_A^{\mathbf{p},\theta} > k) = \mathsf{P}_\infty(T_A^{\mathbf{p},\theta} > k) \ge 1 - \frac{k+r_A}{A}.$$

Hence, using Markov's inequality, we obtain

$$\begin{aligned}
\mathsf{EDD}_{k,\mathsf{B},\theta}(T_A^{\mathbf{p},\theta}) &\ge \mathsf{E}_{k,\mathsf{B},\theta}\left[(T_A^{\mathbf{p},\theta} - k)^+\right] \\
&\ge N_A\, \mathsf{P}_{k,\mathsf{B},\theta}(T_A^{\mathbf{p},\theta} - k > N_A) \\
&\ge N_A\left[\mathsf{P}_{k,\mathsf{B},\theta}(T_A^{\mathbf{p},\theta} > k) - \mathsf{P}_{k,\mathsf{B},\theta}(k < T_A^{\mathbf{p},\theta} < k+N_A)\right] \\
&\ge N_A\left[1 - \frac{r_A+k}{A} - \mathsf{P}_{k,\mathsf{B},\theta}(k < T_A^{\mathbf{p},\theta} < k+N_A)\right].
\end{aligned} \tag{40}$$

Similarly to (29)

$$\mathsf{P}_{k,\mathsf{B},\theta}\left(k < T_A^{\mathbf{p},\theta} \le k+N_A\right) \le \kappa_{k,A}(T_A^{\mathbf{p},\theta}) + \gamma_{k,A}, \tag{41}$$

where

$$\kappa_{k,A}(T_A^{\mathbf{p},\theta}) = e^{(1-\varepsilon^2)\log A}\mathsf{P}_\infty\left(0 < T_A^{\mathbf{p},\theta} - k \le N_A\right)$$

and

$$\gamma_{k,A} = \mathsf{P}_{k,\mathsf{B},\theta}\left\{\max_{0\le n<N_A} \lambda_{\mathsf{B},\theta}(k,k+n) \ge (1+\varepsilon)I_{\mathsf{B},\theta}\,\psi(N_A)\right\}.$$

Using inequality (34), we obtain

$$\mathsf{P}_\infty\left(0 < T_A^{\mathbf{p},\theta} - k < N_A\right) \le \mathsf{P}_\infty\left(T_A^{\mathbf{p},\theta} < k+N_A\right) \le (k+r_A+N_A)/A,$$

and consequently,

$$\kappa_{k,A}(T_A^{\mathbf{p},\theta}) \le \frac{k+r_A+\Psi((1-\varepsilon)I_{\mathsf{B},\theta}^{-1}\log A)}{A^{\varepsilon^2}}. \tag{42}$$

Since, by conditions (37) and (38), for any $\varepsilon > 0$

$$\frac{r_A+\Psi((1-\varepsilon)I_{\mathsf{B},\theta}^{-1}\log A)}{A^{\varepsilon^2}} \to 0 \quad \text{as } A\to\infty.$$

it follows from (42) that $\kappa_{k,A}(T_A^{\mathbf{p},\theta}) \to 0$ as $A\to\infty$ for any $k\in\mathbb{Z}_+$. Also, $\gamma_{k,A}\to 0$ as $A\to\infty$ by condition (36). Therefore,

$$\mathsf{P}_{k,\mathsf{B},\theta}\left(0 < T_A^{\mathbf{p},\theta} - k < N_A\right) \to 0 \quad \text{as } A\to\infty$$

for any fixed $k\in\mathbb{Z}_+$. It follows from (40) that as $A\to\infty$

$$\mathsf{EDD}_{k,\mathsf{B},\theta}(T_A^{\mathbf{p},\theta}) \ge \Psi\left(\frac{(1-\varepsilon)\log A}{I_{\mathsf{B},\theta}}\right)(1+o(1)),$$

where ε can be arbitrarily small. This yields the asymptotic lower bound (for any fixed $k\in\mathbb{Z}_+$, $\mathsf{B}\in\mathscr{P}$)

$$\liminf_{A\to\infty}\frac{\mathsf{EDD}_{k,\mathsf{B},\theta}(T_A^{\mathbf{p},\theta})}{\log A} \ge \frac{1}{I_{\mathsf{B},\theta}}. \tag{43}$$

To prove (39) it suffices to show that this bound is attained by $T_A^{\mathbf{p},\theta}$, i.e.,

$$\limsup_{A\to\infty}\frac{\mathsf{EDD}_{k,\mathsf{B},\theta}(T_A^{\mathbf{p},\theta})}{\log A} \le \frac{1}{I_{\mathsf{B},\theta}}. \tag{44}$$

For $0 < \varepsilon < I_{\mathsf{B},\theta}$, define

$$M_A = M_A(\varepsilon, \mathsf{B}, \theta) = 1 + \left\lfloor \Psi\left(\frac{\log A}{I_{\mathsf{B},\theta} - \varepsilon}\right)\right\rfloor. \tag{45}$$

Recall the definitions of the mixture SR statistic $R_{\mathbf{p},\theta}(n)$ and mixture LR $\Lambda_{\mathbf{p},\theta}(n)$ given in (10) and (11), respectively. Obviously, for any $n \in \mathbb{N}$ and $\mathsf{B} \in \mathscr{P}$,

$$\log R_{\mathbf{p},\theta}(k+n) \geq \log \Lambda_{\mathbf{p},\theta}(k,k+n) \geq \lambda_{\mathsf{B},\vartheta}(k,k+n) + \log p_{\mathsf{B}},$$

and we have

$$\begin{aligned} &\mathsf{P}_{k,\mathsf{B},\theta}\left(T_A^{\mathbf{p},\theta} - k > n\right) \leq \mathsf{P}_{k,\mathsf{B},\theta}\left\{\frac{1}{\psi(n)} \log R_{\mathbf{p},\theta}(k+n) < \frac{1}{\psi(n)} \log A\right\} \\ &\leq \mathsf{P}_{k,\mathsf{B},\theta}\left\{\frac{1}{\psi(n)} \lambda_{\mathsf{B},\theta}(k,k+n) < \frac{1}{\psi(n)} (\log A + |\log p_{\mathsf{B}}|)\right\}, \end{aligned}$$

where for any $n \geq M_A$ the last probability does not exceed the probability

$$\mathsf{P}_{k,\mathsf{B},\theta}\left\{\frac{1}{\psi(n)} \lambda_{\mathsf{B},\theta}(k,k+n) < I_{\mathsf{B},\theta} - \varepsilon + |\log p_{\mathsf{B}}|/\psi(M_A)\right\}.$$

Therefore, for $I_{\mathsf{B},\theta} > \varepsilon > 0$, all $n \geq M_A$ and all sufficiently large A such that $|\log p_{\mathsf{B}}|/\psi(M_A) < \varepsilon/2$, we have

$$\mathsf{P}_{k,\mathsf{B},\theta}\left(T_A^{\mathbf{p},\theta} - k > n\right) \leq \mathsf{P}_{k,\mathsf{B},\theta}\left\{\frac{1}{\psi(n)} \lambda_{\mathsf{B},\theta}(k,k+n) < I_{\mathsf{B},\theta} - \varepsilon/2\right\}. \tag{46}$$

Now, by Lemma A1 in Tartakovsky [29, page 239], for any $\ell \geq 1$, $k \in \mathbb{Z}_+$, and $\mathsf{B} \in \mathscr{P}$,

$$\mathsf{E}_{k,\mathsf{B},\theta}\left[(T_A^{\mathbf{p},\theta} - k)^+\right]^\ell \leq M_A^\ell + \ell 2^{\ell-1} \sum_{n=M_A}^{\infty} n^{\ell-1} \mathsf{P}_{k,\mathsf{B},\theta}\left(T_A^{\mathbf{p},\theta} > n\right), \tag{47}$$

which along with inequality (46) yields

$$\begin{aligned} \mathsf{E}_{k,\mathsf{B},\theta}\left[\left(T_A^{\mathbf{p},\theta} - k\right)^+\right] &\leq 1 + \left\lfloor \Psi\left(\frac{\log A}{I_{\mathsf{B},\theta} - \varepsilon}\right)\right\rfloor \\ &\quad + \sum_{n=M_A}^{\infty} \mathsf{P}_{k,\mathsf{B},\theta}\left\{\frac{1}{\psi(n)} \lambda_{\mathsf{B},\theta}(k,k+n) < I_{\mathsf{B},\theta} - \varepsilon/2\right\}. \end{aligned} \tag{48}$$

Write

$$\Upsilon_A(\varepsilon, \mathsf{B}, \theta) = \sum_{n=M_A}^{\infty} \sup_{k \in \mathbb{Z}_+} \mathsf{P}_{k,\mathsf{B},\theta}\left\{\frac{1}{\psi(n)} \lambda_{\mathsf{B},\theta}(k,k+n) < I_{\mathsf{B},\theta} - \varepsilon\right\}.$$

Using (48) and the inequality $\mathsf{P}_\infty(T_A^{\mathbf{p},\theta} > k) > 1 - (r_A + k)/A$ (see (34)), we obtain

$$\mathsf{EDD}_{k,\mathsf{B},\theta}(T_A^{\mathbf{p},\theta}) = \frac{\mathsf{E}_{k,\mathsf{B},\theta}\left[\left(T_A^{\mathbf{p},\theta}-k\right)^+\right]}{\mathsf{P}_\infty(T_A^{\mathbf{p},\theta}>k)} \le \frac{1+\left\lfloor\Psi\left(\frac{\log A}{I_{\mathsf{B},\theta}-\varepsilon}\right)\right\rfloor+\Upsilon_A(\varepsilon/2,\mathsf{B},\theta)}{1-(r_A+k)/A}. \tag{49}$$

Since due to condition (37) $r_A/A \to 0$ and, by complete convergence condition (36), $\lim_{A\to\infty}\Upsilon_A(\varepsilon/2,\mathsf{B},\theta)=0$ for all $\varepsilon>0$ and $\mathsf{B}\in\mathscr{P}$, inequality (49) implies the asymptotic inequality

$$\mathsf{EDD}_{k,\mathsf{B},\theta}(T_A^{\mathbf{p},\theta}) \le \left(\frac{\log A}{I_{\mathsf{B},\theta}-\varepsilon}\right)(1+o(1)) \quad \text{as } A\to\infty.$$

Since ε can be arbitrarily small the asymptotic upper bound (44) follows and the proof of the asymptotic approximation (39) is complete. □

2.4.5(b) Asymptotic Optimality

Since we assume that the prior distribution may depend on the prescribed PFA α the mean $\bar{\nu}_\alpha=\sum_{k=0}^\infty k\pi_k^\alpha$ depends on α. We also suppose that the head-start $r=r_\alpha$ may depend on α and may go to infinity as $\alpha\to 0$. We further assume that r_α and $\bar{\nu}_\alpha$ approach infinity with such a rate that

$$\lim_{\alpha\to 0}\frac{\log(r_\alpha+\bar{\nu}_\alpha)}{|\log\alpha|}=0. \tag{50}$$

The following theorem establishes the first-order asymptotic (as $\alpha\to 0$) optimality of the mixture detection procedure $T_A^{\mathbf{p},\theta}$ for the fixed value of θ in the general non-i.i.d. case. In this theorem, we assume that $\psi(x)$ satisfies condition (24) in contrast to Theorem 2 where this function can be practically arbitrary.

Theorem 3. *Let the prior distribution of the change point satisfy condition* $\mathbf{CP}_1$ *and let the mean of the prior distribution* ν_α *and the head-start* r_α *of the statistic* $R_{\mathbf{p},\theta}(n)$ *go to infinity with such a rate that condition* (50) *hold. Assume that for some* $0<I_{\mathsf{B},\theta}<\infty$ ($\mathsf{B}\in\mathscr{P}$) *the uniform complete convergence condition* (36) *holds. If threshold* $A=A_\alpha$ *is so selected that* $\mathsf{PFA}_\pi(T_{A_\alpha}^{\mathbf{p},\theta})\le\alpha$ *and* $\log A_\alpha\sim|\log\alpha|$ *as* $\alpha\to 0$, *in particular as* $A_\alpha=(r_\alpha+\bar{\nu}_\alpha)/\alpha$, *then for all* $k\in\mathbb{Z}_+$ *and* $\mathsf{B}\in\mathscr{P}$ *as* $\alpha\to 0$

$$\mathsf{EDD}_{k,\mathsf{B},\theta}(T_{A_\alpha}^{\mathbf{p},\theta})=\Psi\left(\frac{|\log\alpha|}{I_{\mathsf{B},\theta}}\right)(1+o(1)); \tag{51}$$

$$\inf_{T\in\mathbb{C}_\pi(\alpha)}\mathsf{EDD}_{k,\mathsf{B},\theta}(T)=\mathsf{EDD}_{k,\mathsf{B},\theta}(T_{A_\alpha}^{\mathbf{p},\theta})(1+o(1)). \tag{52}$$

That is, the detection procedure $T_{A_\alpha}^{\mathbf{p},\theta}$ *is first-order asymptotically optimal as* $\alpha \to 0$ *in class* $\mathbb{C}_\pi(\alpha)$, *minimizing average detection delay* $\mathsf{EDD}_{k,\mathsf{B},\theta}(T)$ *uniformly for all* $k \in \mathbb{Z}_+$ *and* $\mathsf{B} \in \mathscr{P}$.

Proof. If threshold A_α is so selected that $\log A_\alpha \sim |\log \alpha|$ as $\alpha \to 0$, then asymptotic approximation (51) follows from Theorem 2. This asymptotic approximation coincides with the asymptotic lower bound (26) in Theorem 1 if condition (24) on $\psi(x)$ holds. Hence, the lower bound is attained by $T_{A_\alpha}^{\mathbf{p},\theta}$, proving (52) and the assertion of the theorem.

In particular, if threshold is selected as $A_\alpha = (r_\alpha + \bar{v}_\alpha)/\alpha$, then by Lemma 1 $\mathsf{PFA}_\pi(T_{A_\alpha}^{\mathbf{p},\theta}) \le \alpha$ and $\log A_\alpha \sim -\log \alpha$ due to condition (50). □

Remark 2. If the prior distribution π of the change point does not depend on α and condition $\mathbf{CP}_1$ is satisfied with $\beta \ge 0$, then the assertion of Theorem 3 holds if, and only if, $\beta = 0$, i.e., for heavy-tailed prior distributions, but not for priors with exponential right tail with $\beta > 0$. This follows from the fact that for $\beta > 0$ the lower bound has the form

$$\inf_{T \in \mathbb{C}_\pi(\alpha)} \mathsf{EDD}_{k,\mathsf{B},\theta}(T) \ge \Psi\left(\frac{|\log \alpha|}{I_{\mathsf{B},\theta} + \beta}\right)(1 + o(1) \quad \text{as } \alpha \to 0.$$

2.4.6 Asymptotic Optimality of the Double-Mixture Procedure $T_A^{\mathbf{p},W}$ for Unknown Post-change Parameter

Consider now the case where the post-change parameter $\theta \in \Theta$ is unknown. The goal is to show that the double-mixture detection procedure $T_A^{\mathbf{p},W}$ defined in (14) is asymptotically optimal to first order.

Recall that in the case of the known parameter θ, the sufficient condition for asymptotic optimality (imposed on the data model) is the uniform complete version of the SLLN (36). The proofs of Theorem 1 and Theorem 2 show that to establish asymptotic optimality of procedure $T_A^{\mathbf{p},\theta}$ it suffices to require the following two (right-tail and left-tail) conditions: for all $\varepsilon > 0$, $k \in \mathbb{Z}_+$ and $\mathsf{B} \in \mathscr{P}$

$$\lim_{L \to \infty} \mathsf{P}_{k,\mathsf{B},\theta}\left\{\frac{1}{\psi(L)} \max_{0 \le n < L} \lambda_{\mathsf{B},\theta}(k, k+n) \ge (1+\varepsilon) I_{\mathsf{B},\theta}\right\} = 0 \tag{53}$$

and

$$\sum_{n=1}^{\infty} \mathsf{P}_{k,\mathsf{B},\theta}\left\{\frac{1}{\psi(n)} \lambda_{\mathsf{B},\theta}(k, k+n) < I_{\mathsf{B},\theta} - \varepsilon\right\} < \infty. \tag{54}$$

Note that the SLLN (16) guarantees right-tail condition (53) and complete version (36) guarantees both conditions (53) and (54).

If the post-change parameter θ is unknown to obtain the upper bound for the expected detection delay the left-trail condition (54) has to be modified as follows. For $\delta > 0$, define $\Gamma_{\delta,\theta} = \{\vartheta \in \Theta : |\vartheta - \theta| < \delta\}$ and assume that there exist positive and finite numbers $I_{B,\theta}$ ($B \in \mathscr{P}$, $\theta \in \Theta$) such that for any $\varepsilon > 0$ and for all $B \in \mathscr{P}$ and $\theta \in \Theta$

$$\lim_{\delta \to 0} \sum_{n=1}^{\infty} \sup_{k \in \mathbb{Z}_+} \mathsf{P}_{k,B,\theta} \left\{ \frac{1}{\psi(n)} \inf_{\vartheta \in \Gamma_{\delta,\theta}} \lambda_{B,\vartheta}(k,k+n) < I_{B,\theta} - \varepsilon \right\} < \infty. \tag{55}$$

We ignore parameter values with W-measure null, i.e., without special emphasis we will always assume that

$$W(\Gamma_{\delta,\theta}) > 0 \text{ for all } \theta \in \Theta \text{ and } \delta > 0.$$

The following theorem generalizes Theorem 2 to the case of the unknown post-change parameter.

Theorem 4. *Suppose that conditions* (37) *and* (38) *hold and there exist positive and finite numbers* $I_{B,\theta}$, $B \in \mathscr{P}$, $\theta \in \Theta$, *such that right-tail and left-tail conditions* (53) *and* (55) *are satisfied. Then the following asymptotic as* $A \to \infty$ *approximation holds*

$$\mathsf{EDD}_{k,B,\theta}(T_A^{\mathbf{p},W}) = \Psi\left(\frac{\log A}{I_{B,\theta}}\right)(1+o(1)) \quad \textit{for all } k \in \mathbb{Z}_+,\ B \in \mathscr{P},\ \theta \in \Theta. \tag{56}$$

Proof. The proof of the asymptotic lower bound (for any fixed $k \in \mathbb{Z}_+$, $B \in \mathscr{P}$ and $\theta \in \Theta$)

$$\mathsf{EDD}_{k,B,\theta}(T_A^{\mathbf{p},W}) \geq \Psi\left(\frac{\log A}{I_{B,\theta}}\right)(1+o(1)) \quad \text{as } A \to \infty \tag{57}$$

is essentially identical to that used to establish the lower bound (43) for the expected delay to detection $\mathsf{EDD}_{k,B,\theta}(T_A^{\mathbf{p},\theta})$ in the proof of Theorem 2. So it is omitted.

To prove asymptotic approximation (56) it suffices to show that the lower bound (57) is attained by $T_A^{\mathbf{p},W}$, i.e.,

$$\limsup_{A \to \infty} \frac{\mathsf{EDD}_{k,B,\theta}(T_A^{\mathbf{p},W})}{\log A} \leq \frac{1}{I_{B,\theta}}. \tag{58}$$

Let M_A be as in (45). Since for any $n \geq 1$,

$$\log R_{\mathbf{p},W}(k+n) \geq \log \Lambda_{\mathbf{p},W}(k,k+n) \geq \inf_{\vartheta \in \Gamma_{\delta,\theta}} \lambda_{B,\vartheta}(k,k+n) + \log W(\Gamma_{\delta,\theta}) + \log p_B,$$

using essentially the same argument as in the proof of Theorem 2 that have led to inequality (46) we obtain that for all $n \geq M_A$

$$\mathsf{P}_{k,B,\theta}\left(T_A^{\mathbf{p},W} > n\right) \leq \mathsf{P}_{k,B,\theta}\left\{\frac{1}{\psi(n)} \inf_{\vartheta \in \Gamma_{\delta,\theta}} \lambda_{B,\vartheta}(k,k+n) < I_{B,\theta} - \varepsilon \right.$$
$$\left. - \frac{1}{\psi(M_A)} \left(\log p_B + \log W(\Gamma_{\delta,\theta})\right) \right\}.$$

Hence, for all $n \ge M_A$ and all sufficiently large A such that $|\log[p_{\mathsf{B}} W(\Gamma_{\delta,\theta})]|/\psi(M_A) < \varepsilon/2$,

$$\mathsf{P}_{k,\mathsf{B},\theta}\left(T_A^{\mathbf{p},W} - k > n\right) \le \mathsf{P}_{k,\mathsf{B},\theta}\left(\frac{1}{\psi(n)} \inf_{\vartheta \in \Gamma_{\delta,\theta}} \lambda_{\mathsf{B},\vartheta}(k,k+n) < I_{\mathsf{B},\theta} - \frac{\varepsilon}{2}\right). \quad (59)$$

Lemma A1 in Tartakovsky [29, Page 239] yields the inequality (for any $k \in \mathbb{Z}_+$, $\mathsf{B} \in \mathscr{P}$, and $\theta \in \Theta$)

$$\mathsf{E}_{k,\mathsf{B},\theta}\left[(T_A^{\mathbf{p},W} - k)^+\right] \le M_A + \sum_{n=M_A}^{\infty} \mathsf{P}_{k,\mathsf{B},\theta}\left(T_A^{\mathbf{p},W} > n\right). \quad (60)$$

Using (60) and (59), we obtain

$$\mathsf{E}_{k,\mathsf{B},\theta}\left[\left(T_A^{\mathbf{p},W} - k\right)^+\right] \le 1 + \Psi\left(\left\lfloor \frac{\log A}{I_{\mathsf{B},\theta} - \varepsilon} \right\rfloor\right) + \Upsilon_A(\varepsilon/2, \mathsf{B}, \theta), \quad (61)$$

where

$$\Upsilon_A(\varepsilon, \mathsf{B}, \theta) = \sum_{n=M_A}^{\infty} \sup_{k \in \mathbb{Z}_+} \mathsf{P}_{k,\mathsf{B},\theta}\left\{\frac{1}{\psi(n)} \inf_{\vartheta \in \Gamma_{\delta,\theta}} \lambda_{\mathsf{B},\vartheta}(k,k+n) < I_{\mathsf{B},\theta} - \varepsilon\right\}.$$

Next, inequality (61) along with the inequality $\mathsf{P}_\infty(T_A^{\mathbf{p},W} > k) > 1 - (r_A + k)/A$ (see (35)) implies the inequality

$$\mathsf{EDD}_{k,\mathsf{B},\theta}(T_A^{\mathbf{p},W}) \le \frac{1 + \Psi\left(\left\lfloor \frac{\log A}{I_{\mathsf{B},\theta} - \varepsilon} \right\rfloor\right) + \Upsilon_A(\varepsilon/2, \mathsf{B}, \theta)}{1 - (r_A + k)/A}. \quad (62)$$

Since $r_A/A \to 0$ (see condition (37)) and, by condition (55), $\Upsilon_A(\varepsilon/2, \mathsf{B}, \theta) \to 0$ as $A \to \infty$ for all $\varepsilon > 0$, $\mathsf{B} \in \mathscr{P}$, and $\theta \in \Theta$ inequality (62) implies the asymptotic inequality

$$\mathsf{EDD}_{k,\mathsf{B},\theta}(T_A^{\mathbf{p},W}) \le \left(\frac{\log A}{I_{\mathsf{B},\theta} - \varepsilon}\right)(1 + o(1)) \quad \text{as } A \to \infty,$$

where ε can be taken arbitrarily small so that the asymptotic upper bound (58) follows and the proof of the asymptotic approximation (56) is complete. □

Using asymptotic approximation (56) in Theorem 4 and the lower bound (26) in Theorem 1, it is easy to prove that the mixture procedure $T_A^{\mathbf{p},W}$ is asymptotically optimal to first order as $\alpha \to 0$ in class $\mathbb{C}_\pi(\alpha)$. We silently assume that $\psi(x)$ is either a linear or super-linear function (see (24)).

Theorem 5. *Let the prior distribution of the change point satisfy condition* $\mathbf{CP}_1$. *Assume also that the mean value* $\bar{\nu} = \bar{\nu}_\alpha$ *of the prior distribution and the head-start* $r = r_\alpha$ *of the statistic* $R_{\mathbf{p},W}(n)$ *approach infinity as* $\alpha \to 0$ *with such a rate* *that condition* (50) *holds. Suppose further that there exist numbers* $0 < I_{\mathsf{B},\theta} < \infty$ $(\mathsf{B} \in \mathscr{P}, \theta \in \Theta)$ *such that conditions* (53) *and* (55) *are satisfied. If threshold* A_α *is so selected that* $\mathsf{PFA}_\pi(T_{A_\alpha}^{\mathbf{p},W}) \le \alpha$ *and* $\log A_\alpha \sim |\log \alpha|$ *as* $\alpha \to 0$, *in particular as* $A_\alpha = (r_\alpha + \bar{\nu}_\alpha)/\alpha$, *then for all* $k \in \mathbb{Z}_+$, $\mathsf{B} \in \mathscr{P}$ *and* $\theta \in \Theta$ *as* $\alpha \to 0$

$$\mathsf{EDD}_{k,\mathsf{B},\theta}(T_{A_\alpha}^{\mathbf{p},W}) = \Psi\left(\frac{|\log \alpha|}{I_{\mathsf{B},\theta}}\right)(1+o(1)); \tag{63}$$

$$\inf_{T \in \mathbb{C}_\pi(\alpha)} \mathsf{EDD}_{k,\mathsf{B},\theta}(T) = \mathsf{EDD}_{k,\mathsf{B},\theta}(T_{A_\alpha}^{\mathbf{p},W})(1+o(1)), \tag{64}$$

that is, the procedure $T_{A_\alpha}^{\mathbf{p},W}$ *is first-order asymptotically optimal as* $\alpha \to 0$ *in class* $\mathbb{C}_\pi(\alpha)$.

Proof. If A_α is so selected that $\log A_\alpha \sim |\log \alpha|$ as $\alpha \to 0$, then asymptotic approximation (63) follows from asymptotic approximation (56) in Theorem 4. Since this approximation is the same as the asymptotic lower bound (26) in Theorem 1, this shows that the lower bound is attained by the detection rule $T_{A_\alpha}^{\mathbf{p},W}$, so (64) follows and the proof is complete.

If, in particular, $A_\alpha = (r_\alpha + \bar{\nu}_\alpha)/\alpha$, then $\log A_\alpha \sim |\log \alpha|$ and by Lemma 1 $T_{A_\alpha}^{\mathbf{p},W} \in \mathbb{C}_\pi(\alpha)$. □

Remark 3. The assertion of Theorem 5 holds when $\bar{\nu}$ and r do not depend on α and the condition $\mathbf{CP}_1$ is satisfied with $\beta \equiv 0$, i.e., for priors π with heavy tails. This could be expected since the detection statistic $R_{\mathbf{p},W}(n)$ uses an improper uniform prior distribution of the change point on the whole positive line.

2.4.7 The Case of Independent Streams

Notice that so far we considered a very general stochastic model where not only the observations in streams may be dependent and non-identically distributed, but also the streams may be mutually dependent. In this very general case, computing statistic $R_{\mathbf{p},W}(n)$ is problematic even when the statistics in data streams can be computed. Consider now still a very general scenario where the data streams are mutually independent (but have a general statistical structure), which is of special interest for many applications. The computational problem becomes more manageable when the data between data streams are independent.

2.4.7(a) Computational Issues

In the case where the data across streams are independent, the model has the form

$$p(\mathbf{X}^n|H_{k,\mathsf{B}},\boldsymbol{\theta}_\mathsf{B}) = \prod_{t=1}^{n}\prod_{i=1}^{N} g_i(X_t(i)|\mathbf{X}^{t-1}(i)) \quad \text{for } \nu = k \ge n,$$

$$p(\mathbf{X}^n|H_{k,\mathsf{B}},\boldsymbol{\theta}_\mathsf{B}) = \prod_{t=1}^{k}\prod_{i=1}^{N} g_i(X_t(i)|\mathbf{X}^{t-1}(i)) \times \tag{65}$$

$$\prod_{t=k+1}^{n}\prod_{i\in\mathsf{B}} f_{i,\theta_i}(X_t(i)|\mathbf{X}^{t-1}(i)) \prod_{i\notin\mathsf{B}} g_i(X_t(i)|\mathbf{X}^{t-1}(i)) \quad \text{for } \nu = k < n,$$

where $g_i(X_t(i)|\mathbf{X}^{t-1}(i))$ and $f_{i,\theta_i}(X_t(i)|\mathbf{X}^{t-1}(i))$ are conditional pre- and post-change densities in the i-th data stream, respectively, $\theta_i \in \Theta_i$ is the unknown post-change parameter (generally multidimensional) in the i-th stream ($i \in \{1,\dots,N\} = \mathcal{N}$), and $\boldsymbol{\theta}_\mathsf{B} = (\theta_i, i \in \mathsf{B})$ is the vector of parameters in the set B. So the LR processes are

$$LR_{\mathsf{B},\boldsymbol{\theta}_\mathsf{B}}(k,n) = \prod_{i\in\mathsf{B}} LR_{i,\theta_i}(k,n), \quad LR_{i,\theta_i}(k,n) = \prod_{t=k+1}^{n} \mathscr{L}_{i,\theta_i}(t), \quad n > k, \tag{66}$$

where $\mathscr{L}_{i,\theta_i}(t) = f_{i,\theta_i}(X_t(i)|\mathbf{X}^{t-1}(i))/g_i(X_t(i)|\mathbf{X}^{t-1}(i))$.

Recall that $\mathscr{P}_K = \{\mathsf{B} : 1 \le |\mathsf{B}| \le K\}$ is the subclass of $\mathscr{P}$ for which the cardinality $|\mathsf{B}|$ of the sets where the change may occur does not exceed $K \le N$ streams, and by $\mathscr{P}_K^\star = \{\mathsf{B} : |\mathsf{B}| = K\}$ denote the subclass of $\mathscr{P}$ for which the change occurs in exactly K streams.

Assume, in addition, that the mixing measure is such that

$$p_\mathsf{B} = C(\mathscr{P}_K)\prod_{i\in\mathsf{B}} p_i, \quad C(\mathscr{P}_K) = \left(\sum_{\mathsf{B}\in\mathscr{P}_K}\prod_{i\in\mathsf{B}} p_i\right)^{-1}.$$

Then the mixture LR is

$$\Lambda_{\mathbf{p},\boldsymbol{\theta}}(k,n) = C(\mathscr{P}_K)\sum_{i=1}^{K}\sum_{\mathsf{B}\in\mathscr{P}_i^\star}\prod_{j\in\mathsf{B}} p_j LR_{j,\theta_j}(k,n),$$

and its computational complexity is polynomial in the number of data streams. Moreover, in the special, most difficult case of $K = N$ and $p_j = p$, we obtain

$$\Lambda_{\mathbf{p},\boldsymbol{\theta}}(k,n) = C(\mathscr{P}_N)\left[\prod_{i=1}^{N}(1 + pLR_{i,\theta_i}(k,n)) - 1\right], \tag{67}$$

so its computational complexity is only $O(N)$. The representation (67) corresponds to the case when each stream is affected independently with probability $p/(1+p)$, the assumption that was made in [35].

2.4.7(b) Asymptotic Optimality of Detection Procedures

Since the data are independent across streams, for an assumed value of the change point $\nu = k$, stream $i \in \mathscr{N}$, and the post-change parameter value in the i-th stream θ_i, the LLR of observations accumulated by time $k+n$ is given by

$$\lambda_{i,\theta_i}(k,k+n) = \sum_{t=k+1}^{k+n} \log \frac{f_{i,\theta_i}(X_t(i)|\mathbf{X}^{t-1}(i))}{g_i(X_t(i))|\mathbf{X}^{t-1}(i))}, \quad k \in \mathbb{Z}_+, \ n \in \mathbb{N}.$$

Define $\varGamma_{\delta,\theta_i} = \{\vartheta \in \Theta_i : |\vartheta - \theta_i| < \delta\}$,

$$\gamma_{k,L}^{(i)}(\varepsilon,\theta_i) = \mathsf{P}_{k,i,\theta_i}\left\{\frac{1}{\psi(L)} \max_{1\le n\le L} \lambda_{i,\theta_i}(k,k+n) \ge (1+\varepsilon)I_{i,\theta_i}\right\},$$

$$\Upsilon^{(i)}(\varepsilon,\theta_i) = \lim_{\delta\to 0} \sum_{n=1}^{\infty} \sup_{k\in\mathbb{Z}_+} \mathsf{P}_{k,i,\theta_i}\left\{\frac{1}{\psi(n)} \inf_{\vartheta\in\varGamma_{\delta,\theta_i}} \lambda_{i,\vartheta}(k,k+n) < I_{i,\theta_i} - \varepsilon\right\},$$

and assume that the following right-tail and left-tail conditions are satisfied for local LLR statistics in data streams: There exist positive and finite numbers I_{i,θ_i}, $\theta_i \in \Theta_i$, $i \in \mathscr{N}$, such that for any $\varepsilon > 0$

$$\lim_{L\to\infty} \gamma_{k,L}^{(i)}(\varepsilon,\theta_i) = 0 \quad \text{for all } k \in \mathbb{Z}_+, \ \theta_i \in \Theta_i, \ i \in \mathscr{N}; \tag{68}$$

and

$$\Upsilon^{(i)}(\varepsilon,\theta_i) < \infty \quad \text{for all } \theta_i \in \Theta_i, \ i \in \mathscr{N}. \tag{69}$$

We also assume that $W(\varGamma_{\delta,\theta_i}) > 0$ for all $\delta > 0$ and $i \in \mathscr{N}$.

Let $I_{\mathsf{B},\boldsymbol{\theta}_\mathsf{B}} = \sum_{i\in\mathsf{B}} I_{i,\theta_i}$. Recall that

$$\gamma_{k,L}(\varepsilon,\mathsf{B},\boldsymbol{\theta}_\mathsf{B}) = \mathsf{P}_{k,\mathsf{B},\boldsymbol{\theta}_\mathsf{B}}\left\{\frac{1}{\psi(L)} \max_{1\le n\le L} \lambda_{\mathsf{B},\boldsymbol{\theta}_\mathsf{B}}(k,k+n) \ge (1+\varepsilon)I_{\mathsf{B},\boldsymbol{\theta}_\mathsf{B}}\right\}.$$

Since the LLR process $\lambda_{\mathsf{B},\boldsymbol{\theta}_\mathsf{B}}(k,k+n)$ is the sum of independent local LLRs, $\lambda_{\mathsf{B},\boldsymbol{\theta}_\mathsf{B}}(k,k+n) = \sum_{i\in\mathsf{B}} \lambda_{i,\theta_i}(k,k+n)$ (see (66)), it is easy to see that

$$\gamma_{k,L}(\varepsilon,\mathsf{B},\boldsymbol{\theta}_\mathsf{B}) \le \sum_{i\in\mathsf{B}} \gamma_{k,L}^{(i)}(\varepsilon,\theta_i),$$

so that local conditions (68) imply global right-tail condition (53). This is true, in particular, if $\lambda_{i,\theta_i}(k,k+n)/\psi(n)$ converge $\mathsf{P}_{k,i,\theta_i}$-a.s. to I_{i,θ_i}, $i = 1,\dots,N$, in which case the SLLN for the global LLR (16) holds with $I_{\mathsf{B},\boldsymbol{\theta}_\mathsf{B}} = \sum_{i\in\mathsf{B}} I_{i,\theta_i}$. Also,

$$\Upsilon(\varepsilon,\mathsf{B},\boldsymbol{\theta}_\mathsf{B}) \le \sum_{i\in\mathsf{B}} \Upsilon^{(i)}(\varepsilon,\mathsf{B},\theta_i),$$

which shows that local left-tail conditions (69) imply global left-tail condition (55).

Theorem 5 implies the following results on asymptotic properties of the mixture procedure $T_A^{\mathbf{p},W}$.

Corollary 1. *Assume that for some positive and finite numbers I_{i,θ_i}, $\theta_i \in \Theta_i$, $i = 1,\dots,N$, right-tail and left-tail conditions* (68) *and* (69) *for local data streams are satisfied. If threshold A_α is so selected that $\mathsf{PFA}_\pi(T_{A_\alpha}^{\mathbf{p},W}) \le \alpha$ and $\log A_\alpha \sim |\log \alpha|$ as $\alpha \to 0$, in particular as $A_\alpha = (r_\alpha + \bar{v}_\alpha)/\alpha$, and if conditions* (24) *and* (50) *are satisfied, then asymptotic formulas* (63) *and* (64) *hold with $I_{\mathrm{B},\theta} = I_{\mathrm{B},\boldsymbol{\theta}_\mathrm{B}} = \sum_{i\in\mathrm{B}} I_{i,\theta_i}$. Therefore, detection procedure $T_{A_\alpha}^{\mathbf{p},W}$ is first-order asymptotically optimal as $\alpha \to 0$ in class $\mathbb{C}_\pi(\alpha)$.*

Remark 4. The assertions of Corollary 1 also hold for different distribution functions W_i, $i \in \mathcal{N}$ in streams if we assume that $W_i(\Gamma_{\delta,\theta_i}) > 0$ for all $\delta > 0$ and $i \in \mathcal{N}$. A modification in the proof is trivial.

Remark 5. In the case where the post-change parameters $\theta_1,\dots,\theta_N$ are known, essentially similar corollary follows from Theorem 3 if we require the uniform complete version of the SLLN for all streams $i \in \mathcal{N}$:

$$\sum_{n=1}^{\infty} \sup_{k\in\mathbb{Z}_+} \mathsf{P}_{k,i,\theta_i}\left\{\left|\frac{\lambda_{i,\theta_i}(k,k+n)}{\psi(n)} - I_{i,\theta_i}\right| > \varepsilon\right\} < \infty \quad \text{for all } \varepsilon > 0. \tag{70}$$

3 Examples

3.1 *Detection of Changes in the Mean Values of Multistream Autoregressive Non-stationary Processes*

This example related to detecting changes in unknown means of multistream non-stationary AR(p) processes has a specific real application in addition to several other applications in mathematical statistics. Specifically, it arises in multi-channel sensor systems (such as radars and electro-optic imaging systems) where it is required to detect an unknown number of randomly appearing signals from objects in clutter and sensor noise (cf., e.g., [1, 30, 31]).

Observations in the i-th channel have the form

$$X_n(i) = \theta_i S_n(i) \mathbb{1}_{\{n>\nu\}} + \xi_n(i), \quad n \ge 1,\ i = 1,\dots,N, \tag{71}$$

where $\theta_i S_n(i)$ are deterministic signals with unknown "amplitudes" $\theta_i > 0$ that appear at an unknown time ν in additive noises $\xi_n(i)$ in an N-channel system. Assume that noise processes $\{\xi_n(i)\}_{n\in\mathbb{N}}$ ($i = 1,\dots,N$) are mutually independent p-th order Gaussian autoregressive processes AR(p) that obey recursions

$$\xi_n(i) = \sum_{j=1}^{p} \rho_{i,j} \xi_{n-j}(i) + w_n(i), \quad n \ge 1, \tag{72}$$

where $\{w_n(i)\}_{n\ge 1}, i=1,\dots,N$, are mutually independent i.i.d. normal $\mathcal{N}(0,\sigma_i^2)$ sequences ($\sigma_i>0$), so the observations in channels $X_n(1),\dots,X_N(n)$ are independent of each other. For simplicity, let us set zero initial conditions $\xi_{1-p}(i)=\xi_{2-p}(i)=\dots=\xi_0(i)=0$. The coefficients $\rho_{i,1},\dots,\rho_{i,p}$ and variances σ_i^2 are known and all roots of the equation $z^p-\rho_{i,1}z^{p-1}-\dots-\rho_{i,p}=0$ are in the interior of the unit circle, so that the AR(p) processes are stable.

For $n\in\mathbb{N}$, define the p_n-th order residuals

$$\widetilde{S}_n(i)=S_n(i)-\sum_{j=1}^{p_n}\rho_{i,j}S_{n-j}(i),\quad \widetilde{X}_n(i)=X_n(i)-\sum_{j=1}^{p_n}\rho_{i,j}X_{n-j}(i),$$

where $p_n=p$ if $n>p$ and $p_n=n$ if $n\le p$. It is easily shown that the conditional pre-change and post-change densities in the i-th channel are

$$g_i(X_n(i)|\mathbf{X}^{n-1}(i))=f_{0,i}(X_n(i)|\mathbf{X}^{n-1}(i))=\frac{1}{\sqrt{2\pi\sigma_i^2}}\exp\left\{-\frac{\widetilde{X}_n(i)^2}{2\sigma_i^2}\right\},$$

$$f_{i,\theta_i}(X_n(i)|\mathbf{X}^{n-1}(i))=\frac{1}{\sqrt{2\pi\sigma_i^2}}\exp\left\{-\frac{(\widetilde{X}_n(i)-\theta_i\widetilde{S}_n(i))^2}{2\sigma_i^2}\right\},\quad \theta_i\in(0,\infty),$$

and that for all $k\in\mathbb{Z}_+$ and $n\in\mathbb{N}$ the LLR in the i-th channel has the form

$$\lambda_{i,\theta_i}(k,k+n)=\frac{\theta_i}{\sigma_i^2}\sum_{t=k+1}^{k+n}\widetilde{S}_t(i)\widetilde{X}_t(i)-\frac{\theta_i^2\sum_{t=k+1}^{k+n}\widetilde{S}_t(i)^2}{2\sigma_i^2}.$$

Since under measure $\mathsf{P}_{k,i,\vartheta_i}$ the random variables $\{\widetilde{X}_n(i)\}_{n\ge k+1}$ are independent Gaussian random variables $\mathcal{N}(\vartheta_i\widetilde{S}_{n(i)},\sigma_i^2)$, under $\mathsf{P}_{k,i,\vartheta_i}$ the LLR $\{\lambda_{i,\theta_i}(k,k+n)\}_{n\in\mathbb{N}}$ is a Gaussian process (with independent but non-identically distributed increments) with mean and variance

$$\begin{aligned}\mathsf{E}_{k,i,\vartheta_i}[\lambda_{i,\theta_i}(k,k+n)]&=\frac{2\theta_i\vartheta_i-\theta_i^2}{2\sigma_i^2}\sum_{t=k+1}^{k+n}\widetilde{S}_t(i)^2,\\ \mathsf{Var}_{k,i,\vartheta_i}[\lambda_{i,\theta_i}(k,k+n)]&=\frac{\theta_i^2}{\sigma_i^2}\sum_{t=k+1}^{k+n}\widetilde{S}_t(i)^2.\end{aligned}\qquad(73)$$

Assume that

$$\lim_{n\to\infty}\frac{1}{\psi(n)}\sup_{k\in\mathbb{Z}_+}\sum_{t=k+1}^{k+n}\widetilde{S}_t(i)^2=Q_i,$$

where $0<Q_i<\infty$. In a variety of signal processing applications this condition holds with $\psi(n)=n$, e.g., in radar applications where the signals $\theta_i S_{i,n}$ are the sequences of harmonic pulses. In some applications such as detection, recognition, and tracking of objects on ballistic trajectories that can be approximated by polynomials of order $m=2-3$, the function $\psi(n)=n^m$, $m>1$. Then for all $k\in\mathbb{Z}_+$ and $\theta_i\in(0,\infty)$

$$\frac{1}{\psi(n)}\lambda_{i,\theta_i}(k,k+n)\xrightarrow[n\to\infty]{\mathsf{P}_{k,i,\theta_i}-\text{a.s.}}\frac{\theta_i^2 Q_i}{2\sigma_i^2}=I_{i,\theta_i},$$

so that the SLLN (16) takes place, and therefore, the right-tail condition (68) holds. Furthermore, since the second moment of the LLR is finite it can be shown that uniform complete convergence conditions (70) as well as the left-tail condition (69) also hold. Thus, by Corollary 1, the mixture detection procedure $T_{A_\alpha}^{\mathbf{p},W}$ minimizes as $\alpha\to 0$ the expected delay to detection and asymptotic formulas (63) and (64) hold with

$$I_{\mathsf{B},\theta}=I_{\mathsf{B},\boldsymbol{\theta}_\mathsf{B}}=\sum_{i\in\mathsf{B}}\frac{\theta_i^2 Q_i}{2\sigma_i^2}.$$

3.2 Change Detection in the Spectrum of the AR(p) Multistream Process

Consider the problem of detecting the change of the correlation coefficient in the p-th order AR process which in the i-th stream satisfies the recursion

$$X_n(i)=\sum_{j=1}^{p}\rho_{i,j}^{(n)}X_{n-j}(i)+w_n(i),\quad n\geq 1,\tag{74}$$

where

$$\rho_{i,\ell}^{(n)}=\theta_{i,\ell}^*\mathbb{1}_{\{n\leq\nu\}}+\theta_{i,\ell}\mathbb{1}_{\{n>\nu\}}$$

and $\{w_n(i)\}_{n\geq 1}$ are i.i.d. (mutually independent) Gaussian random variables with $\mathsf{E}[w_1(i)]=0$, $\mathsf{E}[w_1^2(i)]=1$. Additional notation:

$$\theta_i^*=(\theta_{i,1}^*,\dots,\theta_{i,p}^*)^\top,\ \ \theta_i=(\theta_{i,1},\dots,\theta_{i,p})^\top,\ \ \mathbf{X}_i^{n-1,n-p}=(X_{n-1}(i),\dots,X_{n-p}(i)).$$

($\top$ denotes transpose).

It is easy to see that the pre-change and post-change conditional densities $g_i(X_n(i)\,|\mathbf{X}_i^{n-1})=g_i(X_n(i)|\mathbf{X}_i^{n-1,n-p})$ and $f_{i,\theta_i}(X_n(i)|\mathbf{X}_i^{n-1}(i))=f_{i,\theta_i}(X_n(i)|\mathbf{X}_i^{n-1,n-p})$ are given by

$$\begin{aligned} g_i(X_n(i)|\mathbf{X}_i^{n-1,n-p}) &= \frac{1}{(2\pi)^{p/2}}\exp\left\{-\frac{(\eta_i^*(X_n(i),\mathbf{X}_i^{n-1,n-p}))^2}{2}\right\},\\ f_{i,\theta_i}(X_n(i)|\mathbf{X}_i^{n-1,n-p}) &= \frac{1}{(2\pi)^{p/2}}\exp\left\{-\frac{(\eta_{\theta_i,i}(X_n(i),\mathbf{X}_i^{n-1,n-p}))^2}{2}\right\},\end{aligned}\tag{75}$$

where $\eta_i^*(y,x)=y-(\theta_i^*)^\top x$ and $\eta_{\theta_i,i}(y,x)=y-(\theta_i)^\top x$ ($x\in\mathbb{R}$, $y\in\mathbb{R}^p$). Therefore, for any $\theta_i\in\mathbb{R}^p$, the LLR is

$$\lambda_i(k,k+n) = \sum_{t=k+1}^{k+n} \lambda_i^*(t)$$

where

$$\lambda_i^*(t) = \log \frac{f_{i,\theta_i}(X_t(i)|\mathbf{X}_i^{t-1,t-p})}{g_i(X_t(i)|\mathbf{X}_i^{t-1,t-p})} \\ = X_t(i)(\theta_i - \theta_i^*)^\top \mathbf{X}_i^{t-1,t-p} + \frac{((\theta_i^*)^\top \mathbf{X}_i^{t-1,t-p})^2 - (\theta_i^\top \mathbf{X}_i^{t-1,t-p})^2}{2}. \tag{76}$$

The process (74) is not Markov, but the p-dimensional processes

$$\Phi_{i,n} = (X_{i,n}, \dots, X_{i,n-p+1})^\top \in \mathbb{R}^p, \; i = 1, \dots, N$$

are Markov. Now, for any $\vartheta = (\vartheta_1, \dots, \vartheta_p) \in \mathbb{R}^p$, define the matrix

$$\Lambda(\vartheta) = \begin{pmatrix} \vartheta_1 & \vartheta_2 & \dots & \vartheta_p \\ 1 & 0 & \dots & 0 \\ \vdots & \vdots & \ddots & \vdots \\ 0 & 0 & \dots 1 & 0 \end{pmatrix}.$$

Note that

$$\Phi_{i,n} = \begin{cases} \Lambda(\theta_i^*)\Phi_{i,n-1} + \widetilde{w}_{i,n} & \text{for } n \le \nu, \\ \Lambda(\theta_i)\Phi_{i,n-1} + \widetilde{w}_{i,n} & \text{for } n > \nu, \end{cases} \tag{77}$$

where $\widetilde{w}_{i,n} = (w_{i,n}, 0, \dots, 0)^\top \in \mathbb{R}^p$. Obviously,

$$\mathsf{E}[\widetilde{w}_n \widetilde{w}_n^\top] = B = \begin{pmatrix} 1 & \dots & 0 \\ \vdots & \ddots & \vdots \\ 0 & \dots & 0 \end{pmatrix}.$$

Assume that all eigenvalues of the matrices $\Lambda(\theta_i^*)$ in modules are less than 1 and that θ_i belongs to the set

$$\Theta_i^{st} = \{\theta_i \in \mathbb{R}^p : \max_{1 \le j \le p} |\mathbf{e}_j(\Lambda(\theta_i))| < 1\} \setminus \{\theta_i^*\}, \tag{78}$$

where $\mathbf{e}_j(\Lambda)$ is the j-th eigenvalue of the matrix Λ. Using (77) it can be shown that in this case the processes $\{\Phi_{i,n}\}_{n>\nu}$ $(i = 1, \dots, N)$ are ergodic with stationary normal distributions $\mathcal{N}(0, \mathrm{F}_i(\theta_i))$, where

$$\mathrm{F}_i(\theta_i) = \sum_{n=0}^{\infty} (\Lambda(\theta_i))^n B (\Lambda^\top(\theta_i))^n.$$

Taking into account that $\max_{1 \le i \le N} \sup_{t \ge 1} \mathsf{E}_\infty |X_t(i)|^m < \infty$ for any $m > 0$ and using techniques developed in [16] it can be shown that the uniform complete convergence conditions (70) as well as the left-tail conditions (69) hold. Therefore, Corollary 1 implies that the mixture detection procedure $T_{A_\alpha}^{\mathbf{p},W}$ minimizes as $\alpha \to 0$ the expected delay to detection for any compact subset of $\Theta = \Theta_1^{st} \times \cdots \times \Theta_N^{st}$ and asymptotic formulas (63) and (64) hold with

$$I_{\mathsf{B},\theta} = I_{\mathsf{B},\boldsymbol{\theta}_\mathsf{B}} = \frac{1}{2}\sum_{i\in\mathsf{B}}(\theta_i - \theta_i^*)^\top \mathrm{F}_i(\theta_i)(\theta_i - \theta_i^*).$$

In particular, in the purely Markov scalar case where $p = 1$ in (74),

$$I_{\mathsf{B},\theta} = I_{\mathsf{B},\boldsymbol{\theta}_\mathsf{B}} = \sum_{i\in\mathsf{B}} \frac{(\theta_i - \theta_i^*)^2}{2(1-\theta_i^2)}.$$

4 Monte Carlo

In this section, we perform MC simulations for the example considered in Subsection 3.1, assuming for simplicity (and with a very minor loss of generality) that the noise processes $\xi_n(i) = w_n(i)$ in (71) are i.i.d. Gaussian with mean zero and variances σ_i^2, which is equivalent setting $p = 0$ in (72). We also assume that $S_n(i) = n^{1.1}$, $\sigma_i^2 = 4$, $\theta_i = 0$ pre-change and $\theta_i = \theta = 0.1$ in the post-change mode when the change occurs in the i-th stream.

We suppose that the change in each channel occurs independently with probability q, so the probability of the event $\{M = m\}$ that m streams (out of N) are affected is

$$\mathsf{P}(M = m) = \frac{N!}{m!(N-m)!}\, q^m\, (1-q)^{N-m}.$$

The change occurs at time ν according to the geometric prior distribution

$$\mathsf{P}(\nu = k) = \rho(1-\rho)^k, \quad k = 0, 1, 2, \dots$$

with a parameter $\rho \in (0,1)$.

In MC simulations, we set $N = 10$ for the total number of streams, $\rho = 0.1$, $q = 1/N = 0.1$, and consider three cases: (a) the change occurs in a single stream with $\theta = 0.1$, (b) the change occurs in two streams with $\theta = 0.1$ in each, and (c) the change occurs in three streams with $\theta = 0.1$ in each.

For each MC run, using equations (66) and (67), we compute the statistics (10) and (13) and get the stopping times for the cases where the parameter of the post-change distribution is known (15) and when the parameter of the post-change distribution is unknown (14). We take uniform prior $W(\theta)$ on $[0.1, 0.3]$ with the step 0.01. The number of Monte Carlo runs is 10^6, which ensures very high accuracy of the estimated characteristics.

The results are shown in Table 1 and Fig. 1.

In Fig. 1, PFA (x-axis) is presented in a logarithmic scale. Marks on the figure $2, 3, \dots, 9$ mean how many times the PFA value is greater than the previous decimal

Table 1 Operating characteristics of the mixture detection procedures for the number of streams $N = 10$ with parameters $S_n(i) = n^{1.1}$ and $\theta_1 = \theta_2 = \theta_3 = \theta = 0.1$. The number of MC runs is 10^6.

$\mathsf{PFA}_\pi(T)$	10^{-1}	$5\cdot10^{-2}$	10^{-2}	$5\cdot10^{-3}$	10^{-3}	$5\cdot10^{-4}$
$\mathsf{EDD}_{\mathsf{B}_1,\theta}(T_A^{\mathbf{p},\theta})$	10.77	11.28	12.76	13.34	14.68	15.22
$\mathsf{EDD}_{\mathsf{B}_1,\theta}(T_A^{\mathbf{p},W})$	11.33	12.07	13.32	13.93	15.55	16.02
$\overline{\mathsf{EDD}}_{\mathsf{B}_1,\theta}$	15.99	16.92	18.84	19.50	21.05	21.63
$\mathsf{EDD}_{\mathsf{B}_2,\theta}(T_A^{\mathbf{p},\theta})$	8.77	9.33	10.50	10.88	12.00	12.46
$\mathsf{EDD}_{\mathsf{B}_2,\theta}(T_A^{\mathbf{p},W})$	8.90	9.52	10.82	11.30	12.34	12.81
$\overline{\mathsf{EDD}}_{\mathsf{B}_2,\theta}$	11.78	12.66	14.43	15.02	16.38	16.88
$\mathsf{EDD}_{\mathsf{B}_3,\theta}(T_A^{\mathbf{p},\theta})$	8.01	8.34	9.33	9.79	10.77	11.12
$\mathsf{EDD}_{\mathsf{B}_3,\theta}(T_A^{\mathbf{p},W})$	8.05	8.47	9.55	9.99	10.96	11.39
$\overline{\mathsf{EDD}}_{\mathsf{B}_3,\theta}$	8.90	9.20	11.35	12.00	13.45	13.96

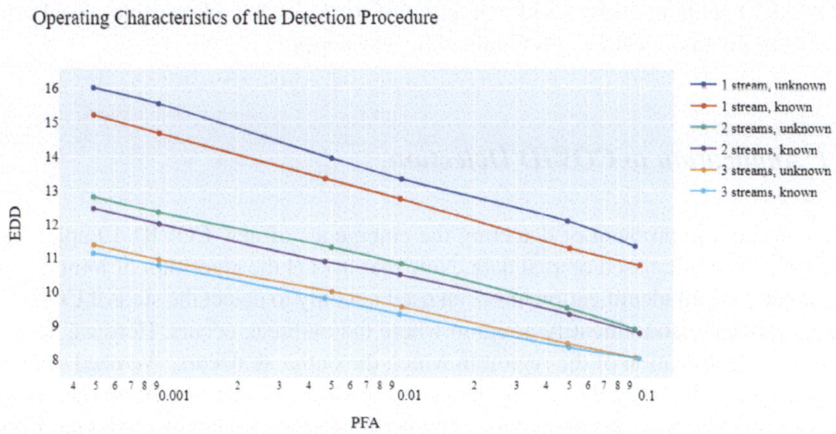

Fig. 1 Operating characteristics of the detection procedures for the total number of streams $N = 10$ and for the number of affected streams $m = 1,2,3$: $\mathsf{EDD}_{\mathsf{B}_m,\theta}(T_A^{\mathbf{p},\theta})$ vs $\log \mathsf{PFA}_\pi$ and $\mathsf{EDD}_{\mathsf{B}_m,\theta}(T_A^{\mathbf{p},W})$ vs $\log \mathsf{PFA}_\pi$. Number of MC runs 10^6. x-axis in logarithmic scale

mark (0.0001, 0.001, or 0.01). It is seen that the detection algorithm has good performance even with such a low signal-to-noise ratio. As expected, in the case when the parameter of the post-change distribution is not known, the algorithm works worse than when it is known, but only slightly – the difference is small. When the change occurs in multiple streams, the expected detection delays $\mathsf{EDD}_{\mathsf{B}_2,\theta}(T_A^{\mathbf{p},W})$ and $\mathsf{EDD}_{\mathsf{B}_3,\theta}(T_A^{\mathbf{p},W})$ decrease compared to $\mathsf{EDD}_{\mathsf{B}_1,\theta}(T_A^{\mathbf{p},W})$ in the case (a), and also $\mathsf{EDD}_{\mathsf{B}_3,\theta}(T_A^{\mathbf{p},W})$ becomes smaller than $\mathsf{EDD}_{\mathsf{B}_2,\theta}(T_A^{\mathbf{p},W})$, as expected from theoretical results (see, e.g., (56)). Here $\mathsf{EDD}_{\mathsf{B}_m,\theta}(T_A^{\mathbf{p},W})$ denotes the expected detection delay when the change occurs in m streams ($m = 1,2,3$).

It is also interesting to compare the MC estimates of the expected detection delay with theoretical asymptotic approximations given by (56). In our example, these approximations reduce to

$$\overline{\mathsf{EDD}}_{\mathsf{B}_m,\theta} \approx \left(\frac{\log A_m}{I_{\mathsf{B}_m,\theta}}\right)^{1/3.2}$$

with $I_{\mathsf{B}_m,\theta} = m(\theta^2/6.4\sigma^2)$, $m = 1,2,3$. Note that thresholds A_m are different for different m to guarantee the same PFA. The data in Table 1 show that the first-order asymptotic approximation is not too bad but not especially accurate.

5 Applications

In this section, we consider two different applications – rapid detection of new COVID-19 epidemic waves and extraction of tracks of low-observable near-Earth space objects in optical images obtained by telescopes.

5.1 Application to COVID Detection

We consider the problem of detecting the emergence of new COVID-19 epidemic waves in Australia based on real data. Note that in [17] the algorithm of joint disorder detection and identification has been used not only to detect the start of COVID-19 in Italy but also to identify a region where the outbreak occurs. Here we do not consider identification of the region in which the outbreak occurs. We need only to decide on the occurrence of an epidemic in the whole country based on data from various regions. Also, in contrast to [17] where a stationary Markov model has been used, we now propose a substantially different non-stationary model taking into account that a new wave of COVID typically spreads faster than a linear law. This fact has been recently noticed in [13].

We selected data on COVID in Australia, which has 8 regions: Australian Capital Territory, New South Wales, Northern Territory, Queensland, South Australia, Tasmania, Victoria, Western Australia. We propose to use the non-stationary model (71) considered in Subsection 3.1 with super-linear functions $S_n(i) = C_i n^\gamma$, $\gamma > 1$. As an observation, we use the percentage of infections in the total population. We study the COVID outbreak in Australia, which was recorded in the winter of 2021-2022. The data are taken from the World Health Organization and presented in Fig. 2.

A performed statistical analysis shows that the model with parameters $\theta_1 = 0.1$ and $\theta_2 = 0.08$ for New South Wales and Victoria, respectively, describes the beginning of the pandemic outbreak well. We also selected $S_n(i) = n^{1.127}$ and $\xi_n(i) = w_n(i) \sim \mathcal{N}(0, \sigma_i^2)$ with $\sigma_i = 2.4$ in (71). We apply the 8-stream double mixture detection algorithm discussed above when the parameter of the post-change distri-

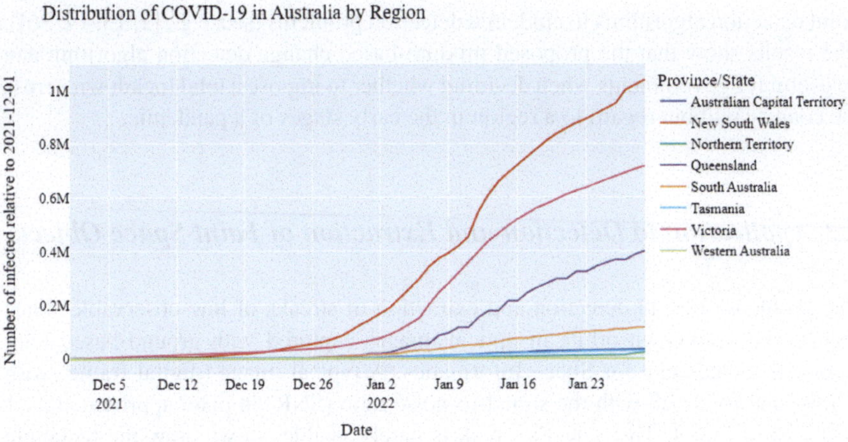

Fig. 2 Distribution of COVID-19 in Australia by region.

bution is not known. We used the discrete uniform prior $W(\theta)$ on $[0.08, \ldots, 0.11]$ with step 0.01. The proposed change detection algorithm with a threshold selected so that the probability of false detection does not exceed 0.01 (average detection delay is about 10) decides on the outbreak of the epidemic on January 9, 2022 (see Fig. 3).

The plots in Fig. 3 illustrate the behavior of the change detection statistic $R_{\mathbf{p},W}(n)$ defined in (13). The COVID wave is detected at the moment when the statistic $R_{\mathbf{p},W}(n)$ crosses the threshold.

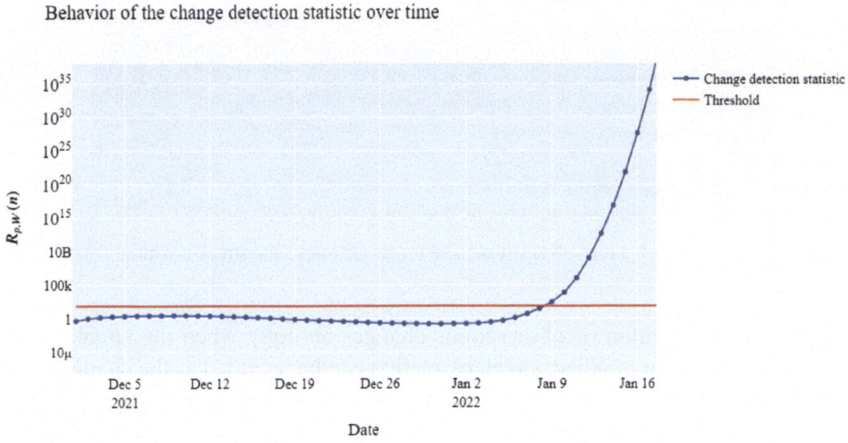

Fig. 3 Behavior of the change detection statistic $R_{\mathbf{p},W}(n)$ over time.

The obtained results complement the existing work on the application of changepoint detection algorithms to epidemic detection problems (see, e.g., [2, 13, 17, 36]). The results show that the proposed mixture-based change detection algorithm can be useful for governments when deciding whether to impose a total lockdown across the country without regard to a region in the early stages of a pandemic.

5.2 Application to Detection and Extraction of Faint Space Objects

The problem of rapid detection and extraction of streaks of low-observable space objects with unknown orbits in optical images captured with ground-based telescopes is a challenge for Space Informatics. A typical image (digital frame) with a low-contrast streak with the signal-to-noise ratio (SNR) in pixel approximately 1 is shown in Fig. 4. The object's streak is barely visible, so we drew the rectangle around the streak that marks its position.

Fig. 4 Digital image with a low-SNR streak. The rectangle marks the streak position.

Since the distribution of observations changes abruptly when the streak starts and ends the problem of object streak extraction can be regarded as the changepoint detection problem in 2-D space (but not in time since we consider a single image).

In realistic situation, the problem is aggravated by the presence of stars and background that produce strong clutter, and special image preprocessing for clutter removal has to be implemented (see, e.g., [30]). We assume that after appropriate preprocessing, the clutter-removed input frame contains Gaussian noise independent from pixel to pixel. For the sake of simplicity, we also consider a scenario where only one streak may be present in the image. We further assume that the satellite has a linear and uniform motion in the frame. The satellite streak is given by the vector $\mathbf{Y} = (x_0, y_0, x_1, y_1)$, where (x_0, y_0) corresponds to the start point and (x_1, y_1) corresponds to the endpoint. We consider the following model of the observation $X_{i,j}$ in pixel (i, j) of the 2-D frame:

$$X_{i,j} = \theta S_{i,j}(\mathbf{Y}) + \xi_{i,j}, \tag{79}$$

where θ is an unknown signal intensity from the object, $\{S_{i,j}(\mathbf{Y})\}$ are values of the model profile of the streak that are calculated beforehand assuming the point spread function (PSF) is Gaussian with a certain effective width, which is shown in Fig. 5; and $\xi_{i,j} \sim \mathcal{N}(0, \sigma^2)$ is Gaussian noise after preprocessing with zero mean and known (estimated empirically) local variance σ^2. Thus, the observation $X_{i,j}$ has normal pre-change distribution $g(X_{i,j}) \sim \mathcal{N}(0, \sigma^2)$ when the streak does not cover pixel (i, j) and normal post-change distribution $f(X_{i,j}) \sim \mathcal{N}(\theta S_{i,j}(\mathbf{Y}), \sigma^2)$ when the streak covers the pixel (i, j).

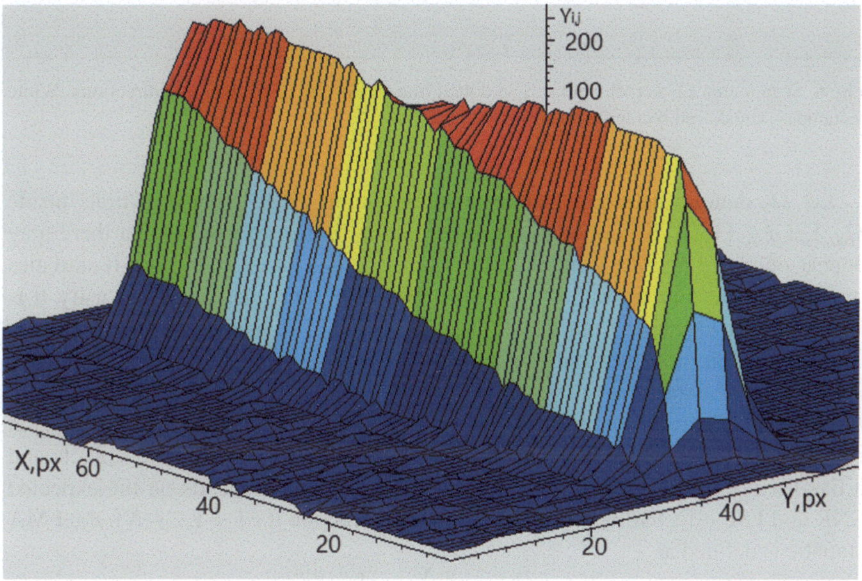

Fig. 5 Model profile of the streak.

The problem is to detect the streak with minimal delay or to make a decision that there is no streak in the frame.

We consider detection of faint streaks of subequatorial satellites with unknown orbits with telescopes mounted at the equator. In this case, a signal from a satellite (with unknown intensity, start, and end points) is located almost vertically in a small area at the center of the frame, as shown in Fig. 6.

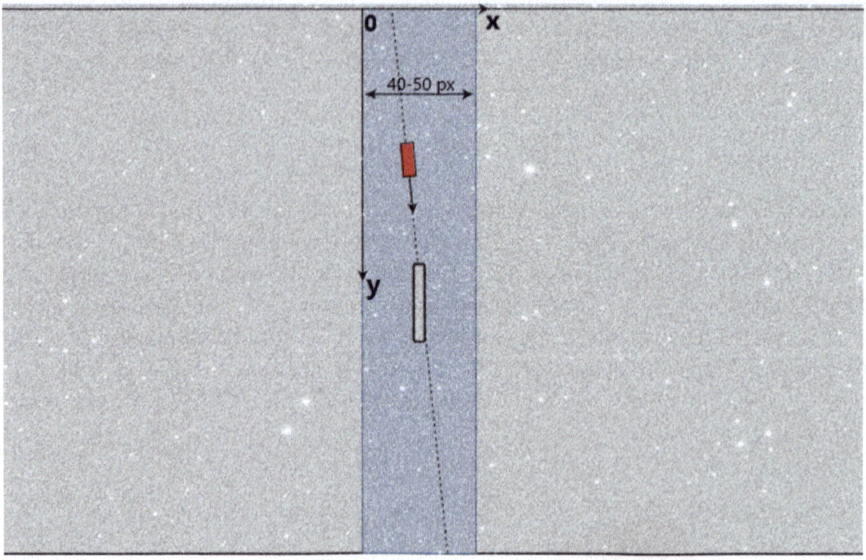

Fig. 6 Search area Ω_S is rectangular. The dotted line shows one of the possible directions. White rectangle – streak, red rectangle – sliding window.

Let Ω_S denote the streak search area. We define N different directions inside Ω_S. Let $d \in \{1,\dots,N\}$ denote a certain direction. Since we assume that there may appear only one satellite's streak in the frame there is no need to mix SR statistics over streak location but rather maximize. However, since the signal intensity θ is not known we still need to mix over the distribution of θ, which is a drawback.

A reasonable way to avoid this averaging is to use the so-called Finite Moving Average (FMA) statistics, as suggested in [24]. Specifically, let $M_d(n), n \geq 1$ stand for a 2-D sliding rectangular window which contains certain pixel numbers (i,j) at each step n in the direction d. Window $M_d(n)$ has a fixed length of L_d pixels and a fixed width of K_d pixels (the choice of the parameters depends on the expected SNR and PSF effective width). For the certain direction d ($d = 1,\dots,N$), the FMA statistic is defined as

$$V_{M_d(n)}(n) = \sum_{(i,j)\in M_d(n)} S_{i,j}^{(d)} X_{i,j}^{(d)},$$

where $\{S_{i,j}^{(d)}\}$ are values of the Gaussian model profile in the direction d and $X_{i,j}^{(d)}$ are observed data in the direction d. Profile location is given by the vector $\mathbf{Y}_d$ in the direction d. Then the multistream[1] FMA detection procedure is defined as

[1] Here "streams" are not streams per se but rather data $X_{i,j}^{(d)}$ in different directions $d \in \{1,\dots,N\}$ in the search area Ω_S.

$$T_h = \inf\left\{n \geq 1 : \max_{1\leq d\leq N} V_{M_d(n)}(n) \geq h\right\}.$$

For the Gaussian model, the FMA procedure is invariant to the unknown signal intensity θ, which is a big advantage over the SR-type versions.

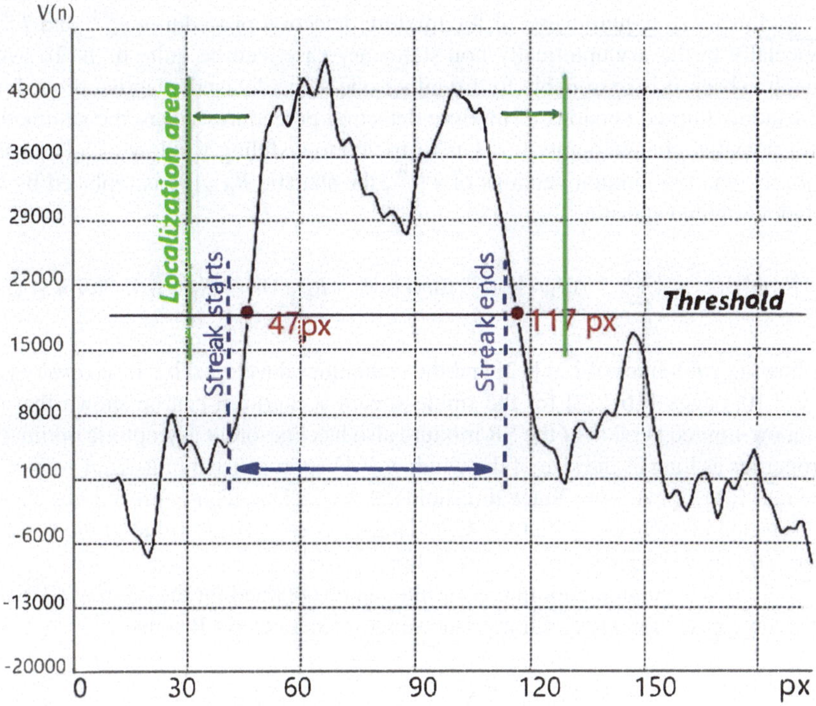

Fig. 7 The behavior of the FMA $V_{M_d(n)}(n)$ statistic along the correct direction.

Fig. 7 shows a typical behavior of the $V_{M_d(n)}(n)$ statistic along the direction containing the streak in the case of a very low SNR $= 0.9$. In this case, the streak is detected with coordinates of start and end at points 47 and 117, respectively, while the true values are 40 and 110 so that the precision is 7 pixels. Experiments show that when sliding the 2-D window in various directions inside Ω_S and then comparing the largest value $\max_d V_{M_d(n)}(n)$ to a threshold we typically determine the approximate position of the streak with an accuracy of 5-10 pixels. Therefore, the proposed FMA version of the change detection algorithm turns out to be efficient – it allows us to rapidly determine a localization area, which with a high probability contains the streak.

6 Concluding Remarks and Future Challenges

6.1 Remarks

1. As we already discussed, for general non-i.i.d. models SR-type statistics cannot be computed recursively even in separate streams, so the computational complexity and memory requirements of the mixture detection procedures $T_A^{\mathbf{p},\theta}$ and $T_A^{\mathbf{p},W}$, especially in the asymptotically non-stationary case, can be quite high. To avoid this drawback it is reasonable to use either one-stage delayed adaptive procedures or window-limited versions of mixture detection procedures where the summation over potential change points $\nu = k$ is restricted to a sliding window of a fixed size ℓ. In the window-limited versions of $T_A^{\mathbf{p},W}$, the statistic $R_{\mathbf{p},W}(n)$ is replaced by the window-limited statistic

$$\widehat{R}_{\mathbf{p},W}(n) = \sum_{k=n-(\ell+1)}^{n-1} \Lambda_{\mathbf{p},W}(k,n) \quad \text{for } n > \ell; \quad \widehat{R}_{\mathbf{p},W}(n) = R_{\mathbf{p},W}(n) \quad \text{for } n \le \ell.$$

Following guidelines of Lai [12] and the techniques developed by Tartakovsky [29, Sec 3.10, pages 116-122] for the single-stream scenario, it can be shown that the window-limited version of the SR mixture also has first-order asymptotic optimality properties as long as the size of the window $\ell(A)$ approaches infinity as $A \to \infty$ with the rate $\ell(A)/\log A \to \infty$. Since thresholds, $A = A_\alpha$, in detection procedures should be selected in such a way that $\log A_\alpha \sim |\log \alpha|$ as $\alpha \to 0$, the value of the window size $\ell(\alpha)$ should satisfy $\lim_{\alpha\to 0}[\ell(\alpha)/|\log \alpha|] = \infty$.

2. Similar asymptotic optimality results can be obtained for the mixture CUSUM procedure based on thresholding of the sum of generalized LR statistic

$$W(n) = \sum_{i=1}^{N} \max_{0\le k<n} \sup_{\theta_i\in\Theta_i} \lambda_{i,\theta_i}(k,n)$$

and for the corresponding window-limited version.

3. We also conjecture that asymptotic optimality properties hold for the multi-stream Finite Moving Average (FMA) detection procedure given by the stopping time

$$T_h = \inf\left\{ n \ge 1 : \sum_{i=1}^{N} \sup_{\theta_i\in\Theta_i} \sum_{k=\max(1,n-\ell+1)}^{n} \lambda^*_{i,\theta_i}(k) \ge h \right\},$$

where $\lambda^*_{i,\theta_i}(k) = \log[f_{i,\theta_i}(X_k(i)|\mathbf{X}^{k-1}(i))/g_i(X_k(i)|\mathbf{X}^{k-1}(i))]$. In cases where the LLR $\lambda^*_{i,\theta_i}(k)$ is a monotone function of some statistic, the FMA procedure can be appropriately modified in such a way that it is invariant to the unknown parameters θ_i. This is a great advantage over corresponding SR-based and CUSUM-based procedures. See also discussion in Subsection 5.2.

4. The results can be easily generalized to the case where the change points ν_i are different for different streams $i = 1,\ldots,N$.

5. The Bayesian-type results of this paper can be used to establish asymptotic optimality properties of the detection procedures $T_A^{\mathbf{p},\theta}$ and $T_A^{\mathbf{p},W}$ in a non-Bayesian setting. In particular, the optimization problem can be solved in the class of procedures with the upper-bounded maximal local probability of false alarms

$$\sup_{m\geq 0} \mathsf{P}_\infty(T \leq m+k|T > m) \leq \alpha,$$

both in pointwise and minimax settings, by embedding into the Bayesian class, similar to what was done by Pergamenchtchikov and Tartakovsky [16] for the case of a single stream.

6.2 Future Challenges

1. The results of MC simulations in Section 4 show that first-order approximations to EDD and PFA are typically not especially accurate, so higher-order approximations are in order. However, it is not feasible to obtain such approximations in the general non-i.i.d. case considered in this paper. Higher order approximations to the expected detection delay and the probability of false alarm for the i.i.d. models, assuming that the observations in streams are independent and also independent across streams, can be derived based on the renewal and nonlinear renewal theories. This important problem will be considered in the future.

2. The results of this paper cover the scenario where the number of streams N is fixed so that $\log(N/\alpha) \sim \log(1/\alpha)$ for small α. In Big Data problems, N may be very large and go to infinity. These problems require different approaches and different detection procedures. This challenging problem will be considered in the future.

3. For detecting transient (or intermittent) changes of unknown duration (like object streaks in Subsection 5.2) it is often more reasonable to consider not quickest detection criteria but reliable detection criteria that require minimization of the detection probability in a fixed time (or space) window (see, e.g., [3, 24, 29] and references therein). While certain interesting asymptotic results for single-stream scenarios and i.i.d. data models exist [9, 10, 14, 21, 24, 29], to the best of our knowledge this problem has never been considered in the multistream setting and for non-i.i.d. models.

Acknowledgements We are grateful to an anonymous referee and Prof Kais Hamza for a comprehensive review and useful comments.

References

1. Bakut, P.A., Bolshakov, I.A., Gerasimov, B.M., Kuriksha, A.A., Repin, V.G., Tartakovsky, G.P., Shirokov, V.V.: Statistical Radar Theory, vol. 1 (G. P. Tartakovsky, Editor). Sovetskoe Radio, Moscow, USSR (1963). In Russian
2. Baron, M.: Early detection of epidemics as a sequential change-point problem. In: V. Antonov, C. Huber, M. Nikulin, V. Polischook (eds.) Longevity, Aging and Degradation Models in Reliability, Public Health, Medicine and Biology, vol. 2, pp. 31–43. St. Petersburg (2004)
3. Berenkov, N.R., Tartakovsky, A.G., Kolessa, A.E.: Reliable detection of dynamic anomalies with application to extracting faint space object streaks from digital frames. In: 2020 International Conference on Engineering and Telecommunication (EnT-MIPT 2020). Dolgoprudny, Russia (2020)
4. Chan, H.P.: Optimal sequential detection in multi-stream data. Annals of Statistics **45**(6), 2736–2763 (2017)
5. Chang, F.K.: Structural health monitoring: Promises and challenges. In: Proceedings of the 30th Annual Review of Progress in Quantitative NDE (QNDE), Green Bay, WI, USA. American Institute of Physics (2003)
6. Fellouris, G., Sokolov, G.: Second-order asymptotic optimality in multichannel sequential detection. IEEE Transactions on Information Theory **62**(6), 3662–3675 (2016). DOI 10.1109/TIT.2016.2549042
7. Fienberg, S.E., Shmueli, G.: Statistical issues and challenges associated with rapid detection of bio-terrorist attacks. Statistics in Medicine **24**(4), 513–529 (2005)
8. Frisén, M.: Optimal sequential surveillance for finance, public health, and other areas (with discussion). Sequential Analysis **28**(3), 310–393 (2009)
9. Guépié, B.K., Fillatre, L., Nikiforov, I.: Sequential detection of transient changes. Sequential Analysis **31**(4), 528–547 (2012). DOI 10.1080/07474946.2012.719443
10. Guépié, B.K., Fillatre, L., Nikiforov, I.: Detecting a suddenly arriving dynamic profile of finite duration. IEEE Transactions on Information Theory **63**(5), 3039–3052 (2017). DOI 10.1109/TIT.2017.2679057
11. Kolessa, A., Tartakovsky, A., Ivanov, A., Radchenko, V.: Nonlinear estimation and decision-making methods in short track identification and orbit determination problem. IEEE Transactions on Aerospace and Electronic Systems **56**(1), 301–312 (2020). DOI 10.1109/TAES.2019.2911760
12. Lai, T.L.: Information bounds and quick detection of parameter changes in stochastic systems. IEEE Transactions on Information Theory **44**(7), 2917–2929 (1998)
13. Liang, Y., Tartakovsky, A.G., Veeravalli, V.V.: Quickest change detection with non-stationary post-change observations. IEEE Transactions on Information Theory **69**(5), 3400–3414 (2023). DOI 0.1109/TIT.2022.3230583
14. Mana, F.E., Guépié, B.K., Nikiforov, I.: Sequential detection of an arbitrary transient change profile by the FMA test. Sequential Analysis **42**(2), 91–111 (2023). DOI 10.1080/07474946.2023.2171056
15. Mei, Y.: Efficient scalable schemes for monitoring a large number of data streams. Biometrika **97**(2), 419–433 (2010)
16. Pergamenchtchikov, S., Tartakovsky, A.G.: Asymptotically optimal pointwise and minimax change-point detection for general stochastic models with a composite post-change hypothesis. Journal of Multivariate Analysis **174**(11), 1–20 (2019)
17. Pergamenchtchikov, S.M., Tartakovsky, A.G., Spivak, V.S.: Minimax and pointwise sequential changepoint detection and identification for general stochastic models. Journal of Multivariate Analysis **190**, 1–22 (2022)
18. Raghavan, V., Galstyan, A., Tartakovsky, A.G.: Hidden Markov models for the activity profile of terrorist groups. Annals of Applied Statistics **7**(4), 2402–2430 (2013)
19. Raghavan, V., Steeg, G.V., Galstyan, A., Tartakovsky, A.G.: Modeling temporal activity patterns in dynamic social networks. IEEE Transactions on Computational Social Systems **1**(1), 89–107 (2013)

20. Rolka, H., Burkom, H., Cooper, G.F., Kulldorff, M., Madigan, D., Wong, W.K.: Issues in applied statistics for public health bioterrorism surveillance using multiple data streams: research needs. Statistics in Medicine **26**(8), 1834–1856 (2007)
21. Sokolov, G., Spivak, V.S., Tartakovsky, A.G.: Detecting an intermittent change of unknown duration. Sequential Analysis Vol. 42, Issue 3, pp. 269–302 **42**(3), 269–302. DOI: 10.1080/07474946.2023.2211126
22. Sonesson, C., Bock, D.: A review and discussion of prospective statistical surveillance in public health. Journal of the Royal Statistical Society A **166**, 5–21 (2003)
23. Szor, P.: The Art of Computer Virus Research and Defense. Addison-Wesley Professional, Upper Saddle River, NJ, USA (2005)
24. Tartakovsky, A., Berenkov, N., Kolessa, A., Nikiforov, I.: Optimal sequential detection of signals with unknown appearance and disappearance points in time. IEEE Transactions on Signal Processing **69**, 2653–2662 (2021). DOI 10.1109/TSP.2021.3071016
25. Tartakovsky, A.G.: Asymptotic performance of a multichart CUSUM test under false alarm probability constraint. In: Proceedings of the 44th IEEE Conference Decision and Control and European Control Conference (CDC-ECC'05), Seville, SP, pp. 320–325. IEEE, Omnipress CD-ROM (2005)
26. Tartakovsky, A.G.: Rapid detection of attacks in computer networks by quickest changepoint detection methods. In: N. Adams, N. Heard (eds.) Data Analysis for Network Cyber-Security, pp. 33–70. Imperial College Press, London, UK (2014)
27. Tartakovsky, A.G.: On asymptotic optimality in sequential changepoint detection: Non-iid case. IEEE Transactions on Information Theory **63**(6), 3433–3450 (2017). DOI 10.1109/TIT.2017.2683496
28. Tartakovsky, A.G.: Asymptotic optimality of mixture rules for detecting changes in general stochastic models. IEEE Transactions on Information Theory **65**(3), 1413–1429 (2019). DOI 10.1109/TIT.2018.2876863
29. Tartakovsky, A.G.: Sequential Change Detection and Hypothesis Testing: General Non-i.i.d. Stochastic Models and Asymptotically Optimal Rules. Monographs on Statistics and Applied Probability 165. Chapman & Hall/CRC Press, Taylor & Francis Group, Boca Raton, London, New York (2020)
30. Tartakovsky, A.G., Brown, J.: Adaptive spatial-temporal filtering methods for clutter removal and target tracking. IEEE Transactions on Aerospace and Electronic Systems **44**(4), 1522–1537 (2008)
31. Tartakovsky, A.G., Nikiforov, I.V., Basseville, M.: Sequential Analysis: Hypothesis Testing and Changepoint Detection. Monographs on Statistics and Applied Probability 136. Chapman & Hall/CRC Press, Taylor & Francis Group, Boca Raton, London, New York (2015)
32. Tartakovsky, A.G., Rozovskii, B.L., Blaźek, R.B., Kim, H.: Detection of intrusions in information systems by sequential change-point methods. Statistical Methodology **3**(3), 252–293 (2006)
33. Tartakovsky, A.G., Rozovskii, B.L., Blaźek, R.B., Kim, H.: A novel approach to detection of intrusions in computer networks via adaptive sequential and batch-sequential change-point detection methods. IEEE Transactions on Signal Processing **54**(9), 3372–3382 (2006)
34. Tsui, K., Chiu, W., Gierlich, P., Goldsman, D., Liu, X., Maschek, T.: A review of healthcare, public health, and syndromic surveillance. Quality Engineering **20**(4), 435–450 (2008)
35. Xie, Y., Siegmund, D.: Sequential multi-sensor change-point detection. Annals of Statistics **41**(2), 670–692 (2013)
36. Yu, X., Baron, M., Choudhary, P.K.: Change-point detection in binomial thinning processes, with applications in epidemiology. Sequential Analysis **32**(3), 350–367 (2013)

Chapter 9

MATRIX Program: Mathematics of the Interactions between Brain Structure and Brain Functions

Mathematical Foundations for Measurement of Communication Efficiency in the Human Brain

Alex Hocking, Kerri Morgan, Nicholas Parsons, Govinda Poudel, Sergiy Shelyag, Julien Ugon

Abstract In this paper we provide a detailed mathematical (graph-theoretical) foundation for our newly developed metric for measurement of efficiency of communication in human brain. The metric is based on the assumption that the structural connectome paths facilitate the functional connectome. Our metric demonstrates that majority of the functional connections in the human brain can be explained by indirect functional paths. This further indicates that information transport in the human brain is a complex process, where multiple regions can participate as intermediary communication nodes.

1 Introduction

The human brain is a complex and dynamic network comprised of structurally connected and functionally interactive elements. These elements are critical for the brain to seamlessly manifest cognition and behaviour and can be measured with

Alex Hocking
Deakin University, Melbourne, Australia, e-mail: alexhocking23@gmail.com

Kerri Morgan
RMIT, Melbourne, Australia e-mail: kerri.morgan@rmit.edu.au

Nicholas Parsons
Monash University, Melbourne, Australia e-mail: nicholas.parsons@monash.edu

Govinda Poudel
Australian Catholic University, Melbourne, Australia e-mail: govinda.poudel@acu.edu.au

Sergiy Shelyag
Flinders University, Adelaide, Australia e-mail: sergiy.shelyag@flinders.edu.au

Julien Ugon
Deakin University, Melbourne, Australia e-mail: julien.ugon@deakin.edu.au

D. R. Wood et al. (eds.), *2021-2022 MATRIX Annals*, MATRIX Book Series 5,
https://doi.org/10.1007/978-3-031-47417-0_30

magnetic resonance imaging (MRI). Functional connectivity (FC) can be viewed as which brain regions utilise oxygenated blood at the same time while performing a specific function. Structural connectivity (SC) can be viewed as how physically connected brain regions are. Current methodologies are yet to mathematically elucidate how and where SC relates to FC within a suitable framework.

A recently introduced model [11] explains how the information is communicated in the brain to facilitate the synchronous activity in distant brain regions. That model relies on mathematical concepts and ideas introduced in [9]. The purpose of this paper is to elaborate on these ideas to provide a full explanation of the mathematical tools that we developed to underpin this model.

Classically, both the physical structure and the functional activity of the brain are represented in the form of graphs in a multilayer network. Each layer is a graph with the same set of nodes, one layer has structural edges and the other functional edges.

Most of the existing research approaches in the application of graph theory to neuroscience adapt existing techniques from other application fields, particularly social networks analysis. These early investigations have yielded interesting insights on how the brain is organised [3, 4]. It was suggested the human brain has small world properties [12], is modularised into distinct areas that perform specific roles [10] and maintains cost-efficient transmission of information [5, 8, 2].

Recent developments consider the structural layer as a transportation or communication network [1]. This makes sense as the general agreement is that communication occurs through neurons (and neuronal populations) from which complex brain functions arise [1]. To extend this idea, we ask: "how does structural connectivity permit the synchronicity of brain regions?" In [11] we proposed that functional synchronicity of brain regions must be facilitated by communication across structural paths.

In this paper, we provide a detailed description of the graph-theoretical machinery we utilised for analysis of SC-FC interaction. The paper is organised as follows. In Section 2 we present the mathematical method and equations for SC-FC interaction analysis. In Section 3 we apply the developed method to the human structural and functional connectome data provided by Human Connectome Project (HCP, `http://www.humanconnectome.org/`). Section 4 provides a brief summary and interpretation for our findings.

2 Mathematical method and equations

The structural and functional brain connectomes, after processing, are represented by two real-valued matrices. The elements of these matrices correspond to either the number of streamlines (in the case of structural connectome) or the synchronicity (in the case of functional connectome) between any two regions of the human brain. The matrices are both symmetric, have the same order, but not the same elements. Each node in the connectome is represented by the same row (or column) index in both matrices. These matrices represent undirected weighted graphs. After applying an appropriate (arbitrary) threshold θ to the matrix values, the matrices are converted into binary adjacency matrices with the element values of 0 and 1 if the original elements were less or equal, or greater than θ, respectively. The adjacency matrices represent undirected unweighted graphs.

We will denote by $A^{[s]}$ and $A^{[f]}$, and $w^{[s]}$ and $w^{[f]}$ the (binary) adjacency matrices for the structural and the functional connectome, and the edge weight matrices for the structural and the functional connectome, respectively. The set of nodes in the connectome is denoted by V, and for a given node in the connectome (that is, a vertex in the graphs), we denote by $d^{[s]}$ or $d^{[f]}$ its degree in the structural or functional layer respectively.

The matrix E represents a matrix whose elements are all 1, of appropriate order, and $\circ$ is the Hadamard (or element-wise) product of matrices.

In the following we present the mathematical formulations for the measures developed in [11].

2.1 Node-clustering (nodewise measure)

First, we define the matrix of direct SC-FC connections as

$$N = \frac{A^{[f]} \circ A^{[s]}}{2} \tag{1}$$

We start by recalling coefficients proposed in [6]. The number of triangles is given by

$$T(k) = \frac{\left(A^{[s]}(A^{[f]} \circ (E - A^{[s]}))A^{[s]}\right)_{kk}}{2}, \tag{2}$$

where the numerator is the number of structural tuples centred on node k closed by an FC edge ($A^{[f]}$), divided by 2 in order to control for the undirected layer. The term $E - A^{[s]}$ limits the FC edges to only those that are not present in the SC layer ($A^{[s]}$). In order to count a three cycle (i.e. triangle), we take only the element kk from the resulting matrix.

Again, we restrict induced paths to where a FC edge $A^{[f]}$ exists without a possible SC edge (i, j) reflected by $(E - A^{[s]})$.

2.2 Weighted edgewise equation

The *weight* of a path ikj is calculated as the minimum weight across its structural edges: $\min(w_{ik}^{[s]}, w_{kj}^{[s]})$. Given two vertices i and j, we propose to consider the path with the largest of the minimum of these path weights (i.e., $\max_{k \in V} \min(w_{ik}^{[s]}, w_{kj}^{[s]})$)

as the one most likely to facilitate communication between the brain regions i and j.

The weight of the maximum throughput triangle and the total throughput of the weighted triangles are then given by:

$$WT_{ij}^{\max} = A_{ij}^{[f]}(1 - A_{ij}^{[s]}) \max_{k \in V} \left(\min(w_{ik}^{[s]}, w_{kj}^{[s]}) \right), \tag{3}$$

$$WT_{ij}^{total} = A_{ij}^{[f]}(1 - A_{ij}^{[s]}) \sum_{k \in V} \left(\min(w_{ik}^{[s]}, w_{kj}^{[s]}) \right). \tag{4}$$

Any triangle containing a functional edge (i, j) and structural edges (i,k) and (k, j) has a structural edge of minimum weight. In line with our analogy of the structural layer being akin to a transportation network, this minimum weight provides the maximum possible throughput from i to j through k on the structural layer. We consider each vertex k in our set of nodes V, i.e. $k \in V$, as a possible intermediary node. Among all possible two-paths, the maximum throughput on paths from i to j considering intermediary nodes k is given by $\max_{k \in V} \left(\min(w_{ik}^{[s]}, w_{kj}^{[s]}) \right)$, while the total throughput is given by $\sum_{k \in V} \left(\min(w_{ik}^{[s]}, w_{kj}^{[s]}) \right)$. Finally, we are only interested in tuples that form a functional edge (i, j) which we count using $A_{ij}^{[f]}$, and excluding cases where there is also a structural edge (i, j) which are removed using the $(1 - A_{ij}^{[s]})$ term.

Note 1. If no intermediary node between vertex i and vertex j exists, then $\min(w_{ik}^{[s]}, w_{kj}^{[s]}) = 0$ for any $k \in V$, therefore $WT_{ij}^{\max} = 0$ and $WT_{ij}^{total} = 0$ (see Fig. 1).

Note 2. The values of $WT_{ij}^{\max}$ and WT_{ij}^{total} are small if every ikj path has low throughput. The value WT_{ij}^{total} is also small if there are few nodes structurally adjacent to both i and j. The values of $WT_{ij}^{\max}$ and WT_{ij}^{total} are large if some ikj path has large throughput (see Fig. 2). For the value WT_{ij}^{total} to be large, there also can be many nodes structurally adjacent to both i and j.

In order to count weighted subgraphs, we first define the summed minimum weight of subgraphs. For a given functional edge (i, j), the summed minimum weight of subgraphs that can be obtained using a structural neighbour k_1 incident to node i and a structural neighbour k_2 incident to node j, is given as follows:

$$S_{ij} = A_{ij}^{[f]}(1 - A_{ij}^{[s]}) \sum_{k_1, k_2 \in V} \min(w_{ik_1}^{s}, w_{k_2 j}^{s}). \tag{5}$$

The quantity S_{ij} can be thought of the total throughput that can be transmitted through both i and j (not necessarily on the same path).

A triangle exists if $k_1 = k_2$. The weighted proportion of subgraphs that are triangles is given by

$$S_{ij}^{prop} = \frac{A_{ij}^{[f]}(1 - A_{ij}^{[s]}) \sum_{k \in V} \min(w_{ik}^{[s]}, w_{kj}^{[s]})}{A_{ij}^{[f]}(1 - A_{ij}^{[s]}) \sum_{k_1, k_2 \in V} \min(w_{ik_1}^{[s]}, w_{k_2 j}^{[s]})} = \frac{WT_{ij}^{total}}{S_{ij}} \tag{6}$$

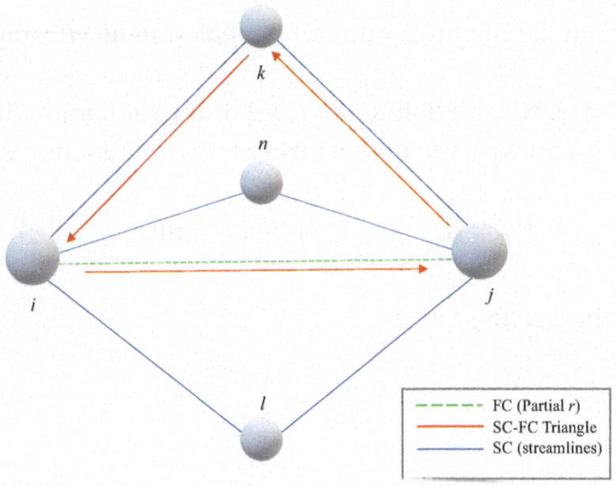

Figure 1: Unweighted triangles cannot consider structural connectivity edge weights that close functionally connected nodes *i* and *j*.

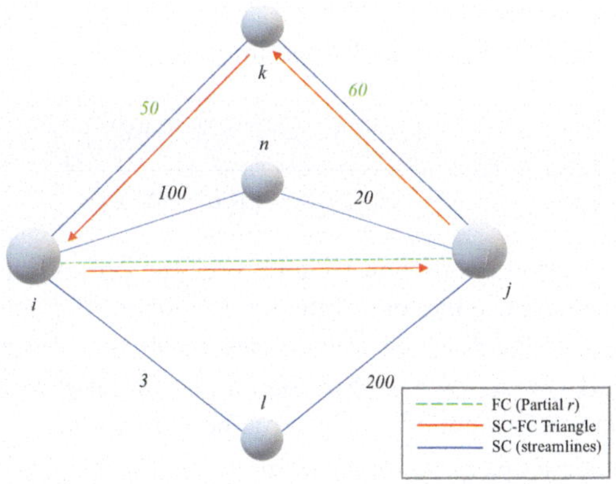

Figure 2: Our method weights structural tuples with the largest *minimum* edge weight facilitating communication between nodes *i* and *j* (the edge *ik* with weight 50).

Generally, if $S_{ij} = 0$, there are no subgraphs of this type, so we set $S_{ij}^{prop} = 0$. In our particular application, since the connectomes form connected graphs (there is no region in the brain that is disconnected from the rest of the brain), $S_{ij} \neq 0$.

Note 3. It is clear that $0 \leq S_{ij}^{prop} \leq 1$. $S_{ij}^{prop} = 0$, if there are no weighted triangles (regardless of whether there are any weighted subgraphs). $S_{ij}^{prop} = 1$, if every weighted subgraph is a triangle.

We now look at the the proportion of weighted induced subgraphs that are triangles.

In a similar way, we can define the ratio of the maximum throughput triangle to the induced subgraph with the highest weight on the function edge (i, j). We set

$$P_{ij} = A_{ij}^{[f]}(1 - A_{ij}^{[s]}) \max_{k_1,k_2 \in V} \min(w_{ik_1}^{[s]}, w_{k_2 j}^{[s]}).$$

The ratio is then defined to be

$$I_{ij}^{prop} = \frac{WT_{ij}^{\max}}{P_{ij}}. \tag{7}$$

P_{ij} can be thought as the maximum throughput that can be "sent" or "received" by both i and j. When $I_{ij}^{prop} = 1$, this means that this maximum throughput is achieved through an existing 2-path between i and j.

As before, we set $I_{ij}^{prop} = 0$ if $P_{ij} = 0$. In our experiments this does not happen.

2.3 Functional edge clustering (edgewise measure)

The *local clustering coefficient* is defined in [6] as:

$$C_k = \frac{\left(A^{[s]}(A^{[f]} \circ (E - A^{[s]}))A^{[s]}\right)_{kk}}{d_k^{[s]}(d_k^{[s]} - 1)(1 - c^{[s]}(k))}. \tag{8}$$

The local clustering coefficient C_k measures the proportion of structural tuples that are closed by an FC edge, out of the total number of structural tuples.

The functional connections between nodes that are not structurally connected are calculated with $A^{[f]} \circ (E - A^{[s]})$. Then the number of such connections where both vertices are adjacent to a node k is given by the numerator.

The number of two-paths centered around node k is given by $d_k^{[s]}(d_k^{[s]} - 1)$. We exclude those that are part of a triangle, using $(1 - c^{[s]}(k))$ where $c^{[s]}(k)$ is the directed clustering coefficient (the number of triangles containing the vertex k) as defined in [7].

While in [6] the main focus is given on a given structural node, here we propose a new measure that focuses on understanding how the synchronicity between two functional nodes is facilitated using structural information.

We start by definition the following quantity:

$$T_{ij} = ((A^{[s]})^2 \circ A^{[f]} \circ (E - A^{[s]}))_{ij}. \tag{9}$$

Here T_{ij} is the number of triangles (i, j, k) in the multiplex, satisfying the conditions:

1. The nodes i and j are functionally (counted by $A^{[f]}$) but not *directly* structurally connected (counted by $(E - A^{[s]})$). That is, we aim to “explain” the functional link between these nodes by looking for possible communication channels in the structural layer.
2. We then count the number of vertices k which are structurally connected to both i and j, indicating that there is a functional route between i and j through k. In other words, we count the number of 2-paths from i to j (counted by $(A^{[s]})^2$). See Figure 1.

To calculate coefficients (proportions) we consider the possibility that a longer structural path may facilitate communication between the functionally connected vertices i and j. An upper bound for the count of such paths is given by

$$(d_i^{[s]} - T_{ij})(d_j^{[s]} - T_{ij})A_{ij}^{[f]}(1 - A_{ij}^{[s]}),$$

that is, the number of neighbours of i that are not adjacent to j multiplied by the number of neighbours of j that are not adjacent to i.

The proportion of structural two-paths that facilitate a functional connection ij is therefore given by:

$$J_{ij}^{prop} = \frac{T_{ij}}{T_{ij} + (d_i^{[s]} - T_{ij})(d_j^{[s]} - T_{ij})A_{ij}^{[f]}(1 - A_{ij}^{[s]})}. \tag{10}$$

Alternatively, we can consider *all* potential structural paths between i and j, regardless of whether the edges could be part of a shorter structural 2-path:

$$R_{ij}^{prop} = \frac{T_{ij}}{d_i^{[s]} d_j^{[s]} A_{ij}^{[f]} (E - A_{ij}^{[s]})}. \tag{11}$$

The proportion of subgraph triangles R_{ij}^{prop} can be calculated by dividing T_{ij} by the total number of possible *triplets* (i.e. a subgraph that contains one neighbour incident to each edge end of a functional edge) reflected by counting all the neighbours of i and j. Again, we restrict this to where a FC edge $A^{[f]}$ exists without a possible SC edge (i, j) reflected by $(E - A^{[s]})$.

3 Application to the brain connectomes

We applied the measures described above (see [11] for details) to the publicly available brain connectome data obtained from Human Connectome Project (HCP, `http://www.humanconnectome.org/`). 484 subjects were included in the dataset. The data are available at three different resolutions, namely 68×68, 114×114 and 219×219 parcels, corresponding to the graph nodes in our analysis. We also generated 484 Watts-Strogatz random graphs of the densities equal to the densities of the structural connectome graphs. Then, we applied the calculation we proposed in the section above to true SC-FC pairs and the simulated connectomes where the structural connectome was replaced by its corresponding Watts-Strogatz random graph. The results of the calculations for 219×219 resolution are shown in Figures 3 and 4. Both figures show that there are nearly 4 times more functional connections, which are facilitated by indirect 2-paths.

The true connectomes had a mean of 1240.61 functional connections (standard deviation 84.66) explained by 1-paths and 4467.39 functional connections (standard deviation 353.68) explained by 2-paths. The Watts-Strogatz simulated connectomes have a mean of 1091.9 functional connections (standard deviation 77.4) explained by 1-paths, and 3393.49 functional connections (standard deviation 402.7) explained by 2-paths.

We also note that the distributions produced by synthetic Watts-Strogatz connectomes are much more symmetric with the real FC-SC connectomes skewed to the higher numbers of the FC edges explained by the given number of paths in the structural connectome.

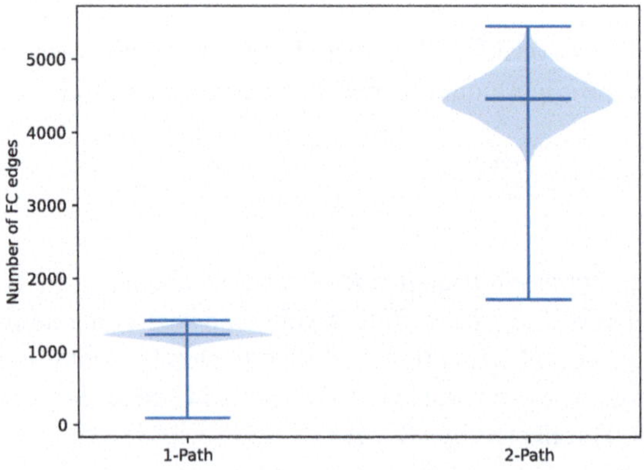

Figure 3: A box plot showing the number of FC edges that are closed by 1- and 2-path lengths for all 484 subjects at 219×219 resolution.

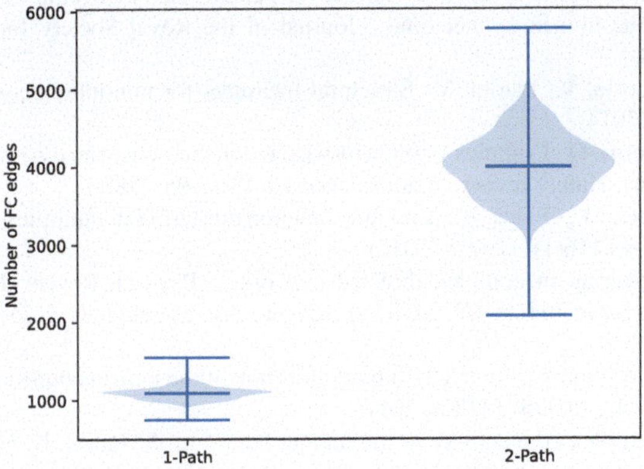

Figure 4: A box plot showing the number of FC edges that are closed by 1- and 2-path lengths in 484 219-node Watts-Strogatz graphs of the same density as the human connectomes considered in this study.

4 Conclusions

In this short paper, we describe a new graph-theoretic method for analysis of the relationship between structural and functional connectomes of the human brain.

The method is based on counting the paths of various lengths in the structural connectome, which are able to provide a physical communication links, which are described by the functional connectome.

We applied our method for the analysis of real human brain and synthetic (Watts-Strogatz) structural connectomes and real functional connectomes. As is evident from the comparison between the true connectomes and the randomly generated structural paths with true functional paths, there are more functional connections in the real brain, which may be facilitated by indirect multiple-node connections in the structural connectome.

We also demonstrate that, according to our introduced measure, Watts-Strogatz graphs exhibit somewhat different statistical properties in comparison to the real brain connectomes.

References

1. Avena-Koenigsberger, A., Mišić, B., Sporns, O.: Communication dynamics in complex brain networks. Nature Reviews Neuroscience **19**, 17–33 (2017)

2. Bassett, D.S., Bullmore, E.: Small-world brain networks. The neuroscientist **12**(6), 512 (2006)
3. Battiston, F., Guillon, J., Chavez, M., Latora, V., Fallani, F.D.V.: Multiplex core–periphery organization of the human connectome. Journal of the Royal Society Interface **15**(146), 20180,514 (2018)
4. Battiston, F., Nicosia, V., Latora, V.: Structural measures for multiplex networks. Physical Review E **89**(3) (2014)
5. Bullmore, E., Sporns, O.: Complex brain networks: graph theoretical analysis of structural and functional systems. Nature reviews neuroscience **10**, 186–198 (2009)
6. Crofts, J., Forrester, M., O'Dea, R.: Structure-function clustering in mulitplex brain networks. Europhysics Letters **116**(1), 18,003 (2016)
7. Fagiolo, G.: Clustering in complex directed networks. Physical Review E **76**(2) (2007). DOI 10.1103/physreve.76.026107. URL `https://doi.org/10.1103/physreve.76.026107`
8. van den Heuvel, M., Sporns, O.: Rich-club organization of the human connectome. Journal of neuroscience **31**(44), 15,775–15,786 (2011)
9. Hocking, A.: Multiplex connectivity in the human brain. In: S. James, K. Morgan, J. Ugon (eds.) Deakin Mathematics Yearbook, chap. 8, pp. 85–90. Deakin University (2021)
10. Meunier, D., Lambiotte, R., Bullmore, E.T.: Modular and hierarchically modular organization of brain networks. Frontiers in Neuroscience **4** (2010). DOI 10.3389/fnins.2010.00200
11. Parsons, N., Ugon, J., Morgan, K., Shelyag, S., Hocking, A., Chan, S.Y., Poudel, G., Domìnguez D, J.F., Caeyenberghs, K.: Structural-functional connectivity bandwidth of the human brain. NeuroImage **263**, 119,659 (2022). DOI https://doi.org/10.1016/j.neuroimage.2022.119659. URL `https://www.sciencedirect.com/science/article/pii/S1053811922007807`
12. Watts, D.J., Strogatz, S.H.: Collective dynamics of 'small-world' networks. Nature **393**(6684), 440–442 (1998). DOI 10.1038/30918

k-Means clustering in EEG (brain waves) timeseries

Nadezda Sukhorukova and Liam Kelly

Abstract Clustering is a process of assigning data points into groups (clusters). The goal is to assign "similar" points to the same cluster, while "different" points belong to distinct clusters. k-means and its modification form a very important and commonly used group of clustering methods. k-means is fast, simple and can be easily implemented in any programming package. Classical k-means leads to a local minimum of a function that represents the sum of the squares of the Euclidean distances between all the points and the centres of the clusters where the points were assigned. Therefore, the mathematical interpretation of k-means is straigthforward. In this paper, we use k-means methods and their modifications to produce clusters in timeseries data and use these clusters to create classification rules for distinguishing between the classes. The results of numerical experiments demonstrate that these fast and simple approaches provide high classification accuracy.

1 Introduction

Clustering is a process of grouping similar points. This technique allows one to identify the structure of data. In signal processing, there are two main approaches for representing signals (timeseries): by treating each time moment as an independent dimension or by constructing a prototype curve for each group of similar timeseries. In this study the former approach is used: each timeseries of n time moments is simply a point in $\mathbf{R}^n$. At first glance, this approach completely ignores the fact that the timeseries values at the nearby time moment strongly relate to each other, but

Nadezda Sukhorukova
Swinburne University of Technology, John Street, Hawthorn, Victoria, Australia 3122, e-mail: nsukhorukova@swin.edu.au

Liam Kelly
Swinburne University of Technology, John Street, Hawthorn, Victoria, Australia 3122, e-mail: lkelly@student.swin.edu.au

D. R. Wood et al. (eds.), *2021-2022 MATRIX Annals*, MATRIX Book Series 5,
https://doi.org/10.1007/978-3-031-47417-0_31

the corresponding optimisation problems are simple and can be handled even for high dimensional timeseries. Our numerical experiments demonstrate that, despite its simplicity, this approach is very efficient for the data we use in this study.

The k-means algorithm is a fast and well-studied clustering method developed for grouping points in $\mathbf{R}^n$. The name "k-means" was first proposed in 1967 by James MacQueen [6]. Mathematically, this method is based on the minimisation of the so-called total dissimilarity function, which is the sum of squares of the Euclidean distances between the points and the cluster centres to which these points are assigned. There are a number of modifications of k-means: k-medians, where the dissimilarity is computed in the Manhattan-distance formulation, some combinations of k-means and k-medians. For theoretical and computational properties of this group of methods and their applicability to data analysis and signal processing, we refer to [2, 10].

Another clustering method that we use in this study is k-medoids. The k-medoids problems are known to be NP-complete [8]. A comprehensive review of such methods can be found in [4].

In general, the k-medoids methods are slower than k-means, but in some cases they may be more suitable. For example, when the dissimilarity measure is not the square of the Euclidean distance and/or it is beneficial to assign one of the sample point from the data to the cluster centres. Examples of such problems can be found in [3].

2 Mathematical background of k-means

2.1 k-means and k-means with L_1 norm

The classical k-means method contains two steps.

1. Assign each point to the cluster with the nearest centre. The initial location of the cluster centres can be chosen in a number of ways. One of the most common ways is to take k random points from the data.
2. For each cluster, we recompute the centres by minimising the total sum of dissimilarities within each cluster.

We repeat these two steps until none of the points move from one cluster to another. In the case when the square of the Euclidean distance is chosen as the dissimilarity function, the minimal total dissimilarity within each cluster is simply the barycenter of the cluster and therefore the algorithm is very fast. This implementation of k-means terminates at a local minimum of the function

$$\min \sum_{i=1}^{N} \min_{c_1,\dots,c_k} \|x_i - c_j\|, \tag{1}$$

where N is the total number of points in the dataset and $c_1,\dots,c_k$ are the cluster centres (each point is assigned to the cluster with the nearest centre).

k-medians has a similar structure, but the dissimilarity function is based on L_1 norm (step 1) and at step 2 one computes the median point rather than the barycentre. Finally, k-means with L_1 is a combination of k-means and k-medians: step 1 is the same as in k-medians, while the updated centre is the barycentre.

2.1.1 k-medoids

As mentioned in section 1, the k-medoids problem (in its classical formulation) is NP-complete [8], but there are a number of approaches for k-medoids that deal with some reformulations of the classical k-medoids problem, for example, the Partitioning Around Medoids (PAM) algorithm [5]. PAM relies on iterative clustering rather than optimisation and uses a number of heuristics.

In its classical form, the k-medoids problem can be formulated as an integer linear programming problem (ILPP) [7]. Assume that there are n data points in total and the distance matrix

$$\mathbf{D} = \{d_{ij}\},\ i = 1,\dots,n,\ j = 1,\dots,n.$$

The goal is to select k points as centres. The decision variables are binary: $x_{ij} \in \{0,1\},\ i = 1,\dots,n,\ j = 1,\dots,n$ and $y_i \in \{0,1\},\ i = 1,\dots,n$. Variable y_i is 1 if point i is treated as a centre, otherwise, this variable is zero. Variable x_{ij} is 1 if point i was assigned to centre j. The corresponding optimisation problem is as follows:

$$\min \sum_{i=1}^{n} \sum_{j=1}^{n} d_{ij} x_{ij} \tag{2}$$

subject to

$$\sum_{j=1}^{n} x_{ij} = 1,\ i = 1,\dots,n; \tag{3}$$

$$x_{ij} \leq y_j,\ i,j = 1,\dots,n; \tag{4}$$

$$\sum_{i=1}^{n} y_i = k; \tag{5}$$

$$x_{ij},\ y_i \in \{0,1\},\ i,j = 1,\dots,n. \tag{6}$$

Coefficients d_{ij} are the distances between points i and j (based on chosen similarity). Constraints (3) ensure that each point is assigned to a single centre. Constraints (4) ensure that point i can only be assigned to point j if this point is also a centre. Finally, constraint (5) ensures that exactly k points are selected as centres. It is clear that problem (2)-(6) is an ILPP. It can be solved, for example, by using a Branch and Bound method.

3 Numerical experiments

In our study, we use EEG signals (brain waves), from a publicly available dataset, collected by the epileptic center at the University of Bonn, Germany [1]. This dataset consists of five classes: 1, 2, 3, 4 and 5. Each class contains 100 signal segments recorded during 23.6 seconds with a sampling frequency of 173.61 Hz, which means that every signal has 4097 recordings.

The same dataset (class 1 and class 2 only) was used in [9], where the authors applied a more advanced and time-consuming optimisation based approach for data classification. The models in [9] are more accurate (for separating class 1 and class 2), but our approach is simple and fast. Moreover, it provides a reasonable level of classification accuracy for separating the remaining classes within this dataset and this topic was not touched at all in [9].

Classes 1 and 2 were recorded from five healthy volunteers. Volunteers were relaxed and awake with eyes open (class 1) and eyes closed (class 2). Classes 3 and 4 correspond to the patients who are seizure free during the recording, but had a seizure in the past and it occurred in the opposite hemisphere of the brain (class 3) or in the same (class 4). Finally, class 5 segments were recorded during an active stage of seizure.

In our experiments, we take 80% of the points of each class for the training set, while the remaining 20% are the test set. All the experiments were performed on all possible pairs of classes (pair-wise separation). We produced up to six clusters in each class (training sets), assigned the points from the test sets to the clusters with the nearest centres and calculated the percentage of correctly classified points (test set classification accuracy). We use four types of clustering: k-means, k-means with L_1, k-medoids with squared Euclidean distance as the similarity measure (we call it k-medoids for simplicity) and k-medoids with L_1 as the similarity measure.

The k-means group of clustering is very sensitive to the choice of the initial centres. To be able to compare results obtained for different clustering settings, we always choose the first k points from the data as initial centres rather than choosing them randomly.

Tables 1-4 represent the test set classification results. In these tables, we only present the highest classification accuracy for each pair-wise combination of classes. The conclusions are as follows.

1. The k-means group of methods is significantly faster and more accurate than the k-medoids group.
2. Theoretically, the increase in the number of clusters leads to more accurate approximations of classes and higher classification accuracy. In practice, this is not always the case. One possible explanation for this observation is that the increase in the number of clusters leads to the increase in the dimension of the corresponding optimisation problems. Since we can only guarantee local optimality, the approximation quality for a higher number of clusters may be questionable.
3. Despite their simplicity, methods from the k-means group demonstrated their classification efficiency even for clustering very complex and long timeseries.

Table 1 *k*-means clustering and classification results

Classes to separate	Optimal number of clusters per class (highest test accuracy)	Test classification accuracy	Time (seconds)
1,2	2,3	52.5%	1.016071
1,3	5,5	82.5%	0.961105
1,4	6,6	72.5%	1.019396
1,5	6,6	62.5%	0.99491
2,3	4,4	77.5%	1.01654
2,4	5,6	77.5%	1.050314
2,5	5,6	65%	0.974254
3,4	4,6	72.5%	1.051843
3,5	1,2	65%	0.943939
4,5	6,6	65%	1.01614

Table 2 *k*-means with L_1 clustering and classification results

Classes to separate	Optimal number of clusters per class (highest test accuracy)	Test classification accuracy	Time (seconds)
1,2	5,6	57.5%	1.496592
1,3	4,4	80%	1.617608
1,4	5,5	77.5%	1.428125
1,5	1,3	72.5%	1.354719
2,3	2,2	72.5%	1.714098
2,4	5,5	77.5%	1.341092
2,5	4,6	77.5%	1.338823
3,4	1,1	72.5%	1.606779
3,5	1,2	70%	1.623985
4,5	1,1	67.5%	1.29161

Table 3 *k*-medoids clustering and classification results

Classes to separate	Optimal number of clusters per class (highest test accuracy)	Test classification accuracy	Time (seconds)
1,2	1,2	55%	33.510218
1,3	2,6	50%	33.318993
1,4	1,1	52.5%	33.941688
1,5	1,2	52.5%	33.284459
2,3	2,1	57.5%	33.552932
2,4	2,1	55%	33.776406
2,5	1,2	52.5%	33.459145
3,4	5,1	60%	33.954362
3,5	1,2	52.5%	33.355816
4,5	1,2	52.5%	33.840175

Table 4 k-mediods with L_1 clustering and classification results

Classes to separate	Optimal number of clusters per class (highest test accuracy)	Test classification accuracy	Time (seconds)
1,2	1,3	57.5%	31.80871
1,3	1,3	52.5%	31.541411
1,4	2,5	62.5%	31.948276
1,5	1,2	52.5%	31.609486
2,3	3,1	60%	31.762258
2,4	1,1	52.5%	31.880342
2,5	1,2	52.5%	31.729709
3,4	3,5	70%	31.871092
3,5	1,2	52.5%	31.58272
4,5	1,2	52.5%	31.682192

4 Conclusions and further research directions

Our numerical experiments demonstrated that k-means group of methods can be used for distinguishing classes of timeseries recorded from epileptic patients. For pair-wise separation of classes, the test set classification accuracy is around 70%-80%.

For our future studies, we identify two main directions.

1. The k-means group of methods is very sensitive to the initial choice of cluster centres and therefore the study on the choice of efficient initial centres is needed.
2. The k-medoids group of methods did not perform well in our experiments. The only potential direction here is the separation between class 3 and class 4, which essentially means the prediction which brain hemisphere one can expect a seizure. Despite being slow and not very accurate, this group of methods is less sensitive to the initial choice of cluster centres and therefore still has some potential if its efficiency is improved.

Acknowledgements This research was supported by the Australian Research Council (ARC), Solving hard Chebyshev approximation problems through nonsmooth analysis (Discovery Project DP180100602).

References

1. Ralph G Andrzejak, Klaus Lehnertz, Florian Mormann, Christoph Rieke, Peter David, and Christian E Elger. Indications of nonlinear deterministic and finite-dimensional structures in time series of brain electrical activity: Dependence on recording region and brain state. *Physical Review E*, 64(6):061907, 2001.
2. Adil Bagirov and Julien Ugon. Nonsmooth dc programming approach to clusterwise linear regression: optimality conditions and algorithms. *Optimization methods and software*, 33(1):194–219, 2018.

3. Behrooz Bodaghi and Nadezda Sukhorukova. Solving resource allocation problems in disaster management using linear programming relaxations. *Communications in Optimization Theory*, 2022:1–11, 2022.
4. Trevor Hastie, Robert Tibshirani, and Jerome Friedman. *The Elements of Statistical Learning Data Mining, Inference, and Prediction*. Springer Series in Satistics, 2008.
5. L. Kaufman and P.J. Rousseeuw. *Finding Groups in Data: an introduction to cluster analysis*. Wiley, 1990.
6. James MacQueen. Some methods for classification and analysis of multivariate observations. In *Proceedings of 5th Berkeley Symposium on Mathematical Statistics and Probability*, pages 281–297. University of California Press, 1967.
7. Hrishikesh D. Vinod Mathematica. Integer programming and the theory of grouping. *Journal of the American Statistical Association*, 64:506–519, 1969.
8. C. Papadimitriou. Worst-case and probabilistic analysis of a geometric location problem. *SIAM Journal on Computing*, 10(3):542–557, 1981.
9. V. Peiris, N. Sharon, N. Sukhorukova, and J. Ugon. Generalised rational approximation and its application to improve deep learning classifiers. *Applied Mathematics and Computation*, 389:125560, 2021.
10. H. Späth. *Cluster Analysis Algorithms for Data Reduction and Classification of Objects*. Ellis Horwood Limited, Chichester, 1980.

Chapter 10

IBS-CGP & MATRIX Workshop: Symplectic Topology

Analysis of pseudoholomorphic curves on symplectization: Revisit via contact instantons

Yong-Geun Oh and Taesu Kim

Abstract In this survey article, we present the analysis of pseudoholomorphic curves $u : (\dot{\Sigma}, j) \to (Q \times \mathbb{R}, \widetilde{J})$ on the symplectization of contact manifold (Q, λ) as a subcase of the analysis of contact instantons $w : \dot{\Sigma} \to Q$, i.e., of the maps w satisfying the equation

$$\overline{\partial}^\pi w = 0, d(w^* \lambda \circ j) = 0$$

on the contact manifold (Q, λ), which has been carried out by a coordinate-free covariant tensorial calculus. The latter was initiated by Wang and the first author of the present survey in [OW18a, OW18b] for the closed string case. More recently the first author has extended the machinery to the open string case and applied it to the problems of quantitative contact topology and contact dynamics [Oh21a], [Oh22a], [OY23]. When the analysis is applied to that of pseudoholomorphic curves $u = (w, f)$ with $w = \pi \circ u$, $f = s \circ u$ on symplectization, the outcome is generally stronger and more accurate than the common results on the regularity presented in the literature in that all of our a priori estimates can be written purely in terms w not involving f. The a priori elliptic estimates for w, especially $W^{2,2}$-estimate, are largely consequences of various Weitzenböck-type formulae *with respect to the contact triad connection* introduced by Wang and the first author in [OW14], and the estimate for f is a consequence thereof by simple integration of the equation $df = w^* \lambda \circ j$.
We also derive a simple precise tensorial formulae for the linearized operator and for the asymptotic operator that admit a perturbation theory of the operators with

Yong-Geun Oh
Center for Geometry and Physics, Institute for Basic Science (IBS)
77 Cheongam-ro, Nam-gu, Pohang-si, Gyeongsangbuk-do, Korea 790-784
& POSTECH, Gyeongsangbuk-do, Korea
e-mail: yongoh1@postech.ac.kr

Taesu Kim
Center for Geometry and Physics, Institute for Basic Science (IBS)
77 Cheongam-ro, Nam-gu, Pohang-si, Gyeongsangbuk-do, Korea 790-784
e-mail: tkim@ibs.re.kr

D. R. Wood et al. (eds.), *2021-2022 MATRIX Annals*, MATRIX Book Series 5,
https://doi.org/10.1007/978-3-031-47417-0_32

respect to (adapted) almost complex structures: The latter has been missing in the analysis of pseudoholomorphic curves on symplectization in the existing literature.

Key words: Contact manifolds, Legendrian submanifolds, contact triad metric, contact triad connection, contact instantons, symplectization, locally conformal symplectic manifold, almost Hermitian manifold, canonical connection, pseudo-holomorphic curves on symplectization, Weitzenböck formula, asymptotic operator

Contents

Part I
Introduction and overview

Let (Q,ξ) be a contact manifold. Assume ξ is coorientable. Then we can choose a contact form λ with $\ker\lambda = \xi$. With λ given, we have the Reeb vector field R_λ uniquely determined by the equation $R_\lambda \rfloor d\lambda = 0$, $R_\lambda \rfloor \lambda = 1$. Then we have decomposition $TQ = \xi \oplus \mathbb{R}\{R_\lambda\}$. We denote by $\pi : TQ \to \xi$ the associated projection and $\Pi = \Pi_\lambda : TQ \to TQ$ the associated idempotent whose image is ξ.

A *contact triad* is a triple (Q,λ,J) where λ is a contact form of ξ, i.e., $\ker\lambda = \xi$ and J is an endomorphism $J : TQ \to TQ$ that satisfies the following.

Definition 0.1 (λ-adapted CR almost complex structure). A *CR almost complex structure* J is an endomorphism $J : TQ \to TQ$ satisfying $J^2 = -\Pi$, or more explicitly

$$(J|_\xi)^2 = -id|_\xi, \quad J(R_\lambda) = 0.$$

We say J is *adapted to* λ if $d\lambda(Y,JY) \geq 0$ for all $Y \in \xi$ with equality only when $Y = 0$. In this case, we all the pair (λ,J) an *adapted pair* of the contact manifold (Q,ξ).

The associated contact triad metric is given by

$$g_\lambda := d\lambda(\cdot,J\cdot) + \lambda \otimes \lambda. \tag{1}$$

The symplectization of (Q,λ) is the symplectic manifold $(Q \times \mathbb{R}, d(e^s\lambda))$ with $\mathbb{R}$-coordinate s also called the radial coordinate. We equip the symplectization with the s-translation invariant almost complex structure

$$\widetilde{J} = J \oplus J_0$$

where J_0 is the almost complex structure on the plane $\mathbb{R}\{\frac{\partial}{\partial s}, R_\lambda\}$ satisfying $J_0(\frac{\partial}{\partial s}) = R_\lambda$.

The main purpose of the present survey is to advertise the *covariant* tensorial approach to the analysis of pseudoholomorphic curves on symplectization via the study of contact instantons which was initiated by Wang and the first author of the present survey in [OW18a, OW18b], further developed by the first author in [Oha]–[Oh22a] and also by the present authors in [KO23].

Remark 0.2. When we say 'covariant tensorial', it means that we follow the way how Riemannian geometers and physicists do their tensor calculations. More specifically, we fix a 'best' connection ∇ on Q once and for all that optimizes tensor calculations. In our case, the contact triad connection is such a connection as illustrated by [OW18a, OW18b], [Oha] and [OY22]. Then for a given smooth map $w : \dot\Sigma \to Q$ or $u : \dot\Sigma \to Q \times \mathbb{R}$, we take covariant derivatives of any induced tensorial quantities only in terms of the induced connection of ∇ and the Levi-Civita connection of the domain $\dot\Sigma$.

Also the relevant Fredholm theory and compactification results are developed by the first author in [Oha] for the closed string case. More recently the theory of contact instantons has been extended in two different directions. On the one hand, in the joint work by Savelyev and the first author [OSar], they lifted the theory of contact instantons to the theory of pseudoholomorphic curves on the $\mathfrak{lcs}$-fication $(Q \times S^1_\rho, \omega_\lambda)$, Banyaga's locally conformal symplectification (which they call the $\mathfrak{lcs}$-fication) of contact manifold (Q,λ) [Ban02] on

$$(Q \times S^1_\rho, d\lambda + d\theta \wedge \lambda)$$

with the canonical angular form $d\theta$ satisfying $\int_{S^1_\rho} d\theta = 1$. According to the terminology adopted in [OSar], the authors call them the $\mathfrak{lcs}$-fication 'of nonzero temperature' on which the theory of pseudoholomorphic curves is developed. Here 'lcs' stands for the standard abbreviation of 'locally conformal symplectic'. The authors of ibid. call the relevant pseudoholomorphic curves *lcs instantons*. This family can be augmented by including the case of the product $Q \times \mathbb{R}$ as the 'zero temperature limit' with $1/\rho \to 0$,

$$(Q \times \mathbb{R}, \omega_\lambda), \quad \omega_\lambda := d\lambda + ds \otimes \lambda \tag{2}$$

(Here ρ represents the radius of the circle S^1.)

On the other hand, in [Oh21a], the first author of the present paper also extended the theory to the open-string case. It is further developed in [OY22] and applied for the construction of Legendrian contact instanton cohomology and the associated spectral invariants on the one-jet bundle in [OY23] jointly with Seungook Yu. Then the present authors have carried out precise asymptotic analyses near the punctures of finite energy contact instantons and of finite energy pseudoholomorphic curves in [KO23] in preparation by developing a generic perturbation theory of asymptotic operators over the change of adapted pairs (λ, J). The tensorial approach also clarifies the relationship between the background geometries of the contact triad (Q,λ,J), the symplectization

$$(M, d(e^s\lambda)) = (Q \times \mathbb{R}, e^s\omega_\lambda)$$

and the lcs manifold (4). Consider the decomposition

$$TM \cong \xi \oplus \mathbb{R}\left\{\frac{\partial}{\partial s}, R_\lambda\right\}. \tag{3}$$

Let $\widetilde{\nabla} := \nabla^{\text{can}}$ be the canonical connection of this almost Hermitian manifold

$$(Q \times \mathbb{R}, \widetilde{g}_\lambda, \widetilde{J}), \quad \widetilde{g}_\lambda := g_\lambda + ds \otimes ds \tag{4}$$

i.e., the unique Riemannian connection whose torsion T satisfies $T(X, \widetilde{J}X) = 0$ for all $X \in T(Q \times \mathbb{R})$ (See [Gau97, Kob03, Oh15] for its definition and basic properties.) We note that this almost Hermitian structure on $Q \times \mathbb{R}$ is translational invariant in the s-direction.

The *first upshot* of our tensorial approach utilizing the contact triad connection lies in our point of view of regarding the product

$$(Q\times\mathbb{R},\omega_\lambda), \quad \omega_\lambda := d\lambda + ds\wedge\lambda$$

as an $\mathfrak{lcs}$-fication of contact manifold (Q,λ), and Hofer's analysis of pseudoholomorphic curves as the analysis of pseudoholomorphic curves on the $\mathfrak{lcs}$*-fication (in zero temperature)*

$$(Q\times\mathbb{R},\omega_\lambda,\widetilde{J})$$

of contact triad (Q,λ,J). It is quite apparent that in Hofer's global analysis of pseudoholomorphic curves on symplectization *the symplectic form $d(e^s\lambda)$ plays little role* but $d\lambda$ and Hofer's ingenious way of considering translational invariant *λ-energy* does.

The $\mathfrak{lcs}$-fication carries a canonical translational invariant metric

$$\widetilde{g} := \omega_\lambda(\cdot, J\cdot)$$

so that the triple $(Q\times\mathbb{R},\widetilde{g},\widetilde{J})$ becomes an almost Hermitian (but not almost Kähler) manifold. The following geometric relationship between the contact triad connection and its lcs lifting has been implicitly exploited.

Proposition 0.3 (Canonical connection on $\mathfrak{lcs}$-fication). *Let $\widetilde{g} = \widetilde{g}_\lambda$ be the almost Hermitan metric given above. Let $\widetilde{\nabla}$ be the canonical connection of the almost Hermitian manifold* (4)*, and ∇ be the contact triad connection for the triad (Q,λ,J). Then $\widetilde{\nabla}$ preserves the splitting* (3) *and satisfies $\widetilde{\nabla}|_\xi = \nabla|_\xi$.*

The *second upshot* is our utilization of an important property of the Levi-Civita connection proved by Blair [Bla10] for the triad metric on Q. This important property has been completely unnoticed (as far as the authors are aware) in the symplectic community around pseudoholomorphic curves, including the first author until very recently at the time of preparing the paper [KO23] and this survey. (Indeed this property had been already mentioned in the paper [OW14, Proposition 4] by Wang and the first author himself!)

Proposition 0.4 (Lemma 6.1 [Bla10]). *Let ∇^{LC} be the Levi-Civita connection of the triad metric associated to the triad (Q,λ,J). Then*

$$\nabla^{\mathrm{LC}}_{R_\lambda} J = 0. \tag{5}$$

This property of the Levi-Civita connection together with the usage of canonical connection on the $\mathfrak{lcs}$-fication $(Q\times\mathbb{R},\widetilde{J},\omega_\lambda)$ (or equivalently via contact triad connection on the triad (Q,λ,J)) plays an important role in the present authors' analysis of the asymptotic operators of finite energy contact instantons in [KO23] and hence of finite energy pseudoholomorphic curves too.

Remark 0.5. In fact if we let $\nabla^{\mathrm{LC}}_{R_\lambda}$ or ∇_{R_λ} acted upon $\Gamma(\xi)\subset\Gamma(TQ)$, then two covariant derivatives coincide, i.e.,

$$\pi\nabla^{\mathrm{LC}}_{R_\lambda}|_\xi = \pi\nabla_{R_\lambda}|_\xi,$$

even though the full connections are not the same, i.e., $\pi\nabla^{\mathrm{LC}} \neq \pi\nabla$. See [OW14, Section 6] or more clearly see its arXiv version 1212.4817(v2) Theorem 1.4 therein where ∇ is expressed as $\nabla = \nabla^{\mathrm{LC}} + B$ for some $(2,1)$ tensor B. From the expression of B there, we have $B(R_\lambda, Y) = 0$ for any $Y \in \xi$. However, while we have $\nabla^{\mathrm{LC}}_{R_\lambda} J = 0$, $\nabla_{R_\lambda} J \neq 0$ on the full tangent bundle TQ.

Roughly speaking Proposition 0.4 enables us to compute the asymptotic operator

$$A^{\pi}_{(\lambda,J,\nabla)} : \Gamma(\gamma^*\xi) \to \Gamma(\gamma^*\xi)$$

in the covariant tensorial way *uniformly in terms of the pull-back connection of the Levi-Civita connection of the triad metric* of the given compatible pair (λ, J).

Remark 0.6. 1. In the literature, the notion of 'asymptotic operator' of a pseudoholomorphic curve on symplectization has been used. For example, Hofer-Wysocki-Zehnder use special coordinates followed by some adjustment of the given almost complex structures along the Reeb orbit in the analysis of asymptotic operators of finite energy planes, while Siefring [Sie08] used a symmetric connection, Wendl [Wen] and Pardon [Par19] a connection obtained by declaring the Lie derivative $\mathscr{L}_{R_\lambda}$ to be the covariant differential along the Reeb orbit γ. (Compare these practices with those given in [KO23] a summary of which is given in Section 21 of the present survey.)

2. Such an important property (5) of the Levi-Civita connection has not attracted any attention from the symplectic community around pseudoholomorphic curves, because contact Hamiltonian dynamics has not attracted much attention from the researchers around pseudoholomorphic curves, and there has been no serious investigation thereof up to the level of symplectic Hamiltonian dynamics. As a consequence, there might not have been enough motivation for them to redo the analysis of pseudoholomorphic curves on symplectization *from scratch starting from its starting place, the contact triad* (Q, λ, J), especially when there is already the well-established Gromov's theory of pseudoholomorphic curves around.
3. However in relation to thermodynamics, contact completely integrable systems and new constructions of Sasaki-Einstein manifolds which is motivated by AdS/CFT correspondence and black-hole dynamics, there has been a systematic development of contact Hamiltonian calculus by a group of geometers and physicists. (See [BCT17], [dLLV19] and [Ler04] and [MS05, MS06], [Boy11] and references therein). Through their study, it has been becoming increasingly clearer that contact Hamiltonian dynamics deserves much more attention than now in many respects. We believe that the analysis and geometry of (perturbed) contact instantons will provide a flexible geometro-analytical package for the study of contact Hamiltonian dynamics and quantitative contact topology as the symplectic Floer theory does. These are illustrated by the first author and his collaborators' recent applications of the package to the problems of quantitative contact topology and contact dynamics. (See [Oh21c, Oh22a] and [OY23].)
4. The common folklore practice of lifting contact dynamics to the homogeneous Hamiltonian dynamics to the symplectization, attempting to exploit the machin-

ery of Floer's theory on the symplectization and then extracting information on the original contact dynamics therefrom does not produce optimal results because the lifting process is not reversible. Because of this, the optimal results proved in [Oh21c, Oh22a] or the Floer theoretic construction of Legendrian spectral invariants on the one-jet bundle given in [OY23] have not been obtainable by the existing machinery of pseudoholomorphic curves via symplectization at least by now. It is an interesting open problem, if possible, to recover those results belonging to purely contact realm by the symplectic machinery of pseudoholomorphic curves.

1 Pseudoholomorphic curves on symplectization

In the seminal work [Hof93], Hofer initiated the study of pseudoholomorphic curves on the symplectization $Q \times \mathbb{R} =: M$ for the study of contact topology and developed the analysis thereof in a series of papers [HWZ96b]-[HWZ02] with applications to contact dynamics in 3 dimensions. Bourgeois [Bou02] and Wendl [Wen] extend their analysis to higher dimensions, and Cant to the relative case [Can22].

Throughout the paper, we adopt the following notations.

Notation 1.1. We denote by (Σ, j) a closed Riemann surface, $\dot{\Sigma}$ the associated punctured Riemann surface and $\overline{\Sigma}$ the real blow-up of $\dot{\Sigma}$ along the punctures.

Note that in the presence of contact form λ, any smooth map $u : \dot{\Sigma} \to Q \times \mathbb{R}$ has the form $u = (w, f)$ with

$$f = s \circ u, \, w = \pi \circ u. \tag{1}$$

We have the splitting

$$TM \cong \xi \oplus \mathbb{R}\{R_\lambda\} \oplus \mathbb{R}\left\{\frac{\partial}{\partial s}\right\}.$$

We have a canonical almost complex structure

$$J_0 : \mathbb{R}\left\{\frac{\partial}{\partial s}, R_\lambda\right\} \to \mathbb{R}\left\{\frac{\partial}{\partial s}, R_\lambda\right\}$$

defined by $J_0 \frac{\partial}{\partial s} = R_\lambda$. We equip (Q, ξ) with a triad (Q, λ, J) and the cylindrical almost complex structure $\widetilde{J} = J \oplus J_0$ which is s-translation invariant.

Remark 1.2. As mentioned above, Hofer's analysis of pseudoholomorphic curves on symplectization should be regarded as a special case for the analysis of pseudoholomorphic curves on lcs manifolds $(Q \times \mathbb{R}, d\lambda + ds \wedge \lambda)$ equipped with the above mentioned almost Hermitian structure arising from the contact triad (Q, λ, J) as illustrated by Savelyev and the first author in [OSar]. See [Oha], [OSar] for further explanation on this point of view. We would like to reiterate that the symplectic form $d(e^s\lambda)$ itself plays very little role in the compactification of punctured pseudoholomorphic curves but Hofer's translational invariant energy $E(w) = E^\pi(w) + E^\perp(w)$ does.

For $M = Q \times \mathbb{R}$, we have the decomposition of the derivative

$$du = dw \oplus \left(df \otimes \frac{\partial}{\partial s} \right)$$

viewed as a TM-valued one-form which can be further decomposed to

$$du(z) = d^\pi w \oplus (w^*\lambda \otimes R_\lambda) \oplus \left(df \otimes \frac{\partial}{\partial s} \right) \tag{2}$$

with respect to the splitting

$$\mathrm{Hom}(T_z\dot{\Sigma}, T_{u(z)}M) = \mathrm{Hom}(T_z\dot{\Sigma}, HT_{u(z)}M) \oplus \mathrm{Hom}(T_z\dot{\Sigma}, VT_{u(z)}M).$$

(For the notational simplicity, we often omit '$\otimes$' except the situation that could cause confusion to the readers without it.)

By definition, we have $d\pi du = dw$. It was observed by Hofer [Hof93] that u is $\widetilde{J}$-holomorphic if and only if (w, f) satisfies

$$\begin{cases} \overline{\partial}^\pi w = 0 \\ w^*\lambda \circ j = df. \end{cases} \tag{3}$$

2 Definitions of contact instantons and bordered contact instantons

Let $\dot{\Sigma}$ a boundary punctured Riemann surface associated a bordered compact Riemann surface (Σ, j). Then for a given map $w : \dot{\Sigma} \to Q$, we can decompose its derivative du, regarded as a w^*TQ-valued one-form on $\dot{\Sigma}$, into

$$dw = d^\pi w + w^*\lambda \otimes R_\lambda \tag{1}$$

where $d^\pi w := \pi dw$. Furthermore $d^\pi w$ is decomposed into

$$d^\pi w = \overline{\partial}^\pi w + \partial^\pi w \tag{2}$$

where $\overline{\partial}^\pi w := (dw^\pi)_J^{(0,1)}$ (resp. $\partial^\pi w := (dw^\pi)_J^{(1,0)}$) is the anti-complex linear part (resp. the complex linear part) of $d^\pi w : (T\dot{\Sigma}, j) \to (\xi, J|_\xi)$. (For the simplicity of notation, we will abuse our notation by often denoting $J|_\xi$ by J. We also simply write $((\cdot)^\pi)_J^{(0,1)} = (\cdot)^{\pi(0,1)}$ and $((\cdot)^\pi)_J^{(1,0)} = (\cdot)^{\pi(1,0)}$ in general.)

A contact instanton is a map $w : \dot{\Sigma} \to Q$ that satisfies the system of nonlinear partial differential equation

$$\overline{\partial}^\pi w = 0, \quad d(w^*\lambda \circ j) = 0 \tag{3}$$

on a contact triad (Q,λ,J). The equation itself had been first mentioned by Hofer [Hof00, p.698], and some attempt to utilize the equation to attack the Weinstein conjecture for dimension 3 was made by Abbas [Abb11], Abbas-Cieliebak-Hofer [ACH05] as well as by Bergmann [Bera, Berb].

In a series of papers, [OW18a, OW18b] jointed with Wang and in [Oha], the first named author systematically developed analysis of contact instantons (for the closed string case) *without taking symplectization* by the global covariant tensorial calculations using the notion of *contact triad connection* which was introduced in [OW14]

More recently he also studied its open string counterpart of the boundary value problem of (3) under the Legendrian boundary condition whose explanation is now in order. For the simplicity and for the main purpose of the present paper, we focus on the genus zero case so that $\dot\Sigma$ is conformally the unit disc with boundary punctures $z_0,\ldots,z_k \in \partial D^2$ ordered counterclockwise, i.e.,

$$\dot\Sigma \cong D^2 \setminus \{z_0,\ldots,z_k\}$$

Then, for a $(k+1)$-tuple $\vec R = (R_0,R_1,\cdots,R_k)$ of Legendrian submanifolds, which we call an (ordered) Legendrian link, we consider the boundary value problem

$$\begin{cases} \overline{\partial}^\pi w = 0, \quad d(w^*\lambda \circ j) = 0, \\ w(\overline{z_i z_{i+1}}) \subset R_i \end{cases} \tag{4}$$

as an elliptic boundary value problem for a map $w : \dot\Sigma \to Q$ by deriving the a priori coercive elliptic estimates. Here $\overline{z_i z_{i+1}} \subset \partial D^2$ is the open arc between z_i and z_{i+1}.

3 $W^{2,2}$-estimates, Weitzenböck formulae and contact triad connection

Let us start with stating the general Weitzenböck formula in differential geometry. A good exposition of general Weitzenböck formula is given in [FU84, Appendix C].

Let (P,h) be a Riemannian manifold and $E \to P$ is a Euclidean vector bundle with inner product $\langle\cdot,\cdot\rangle$ and assume ∇ is a connection compatible with $\langle\cdot,\cdot\rangle$. We denote by d^∇ the covariant differential and δ^∇ its Hodge dual. We also denote by

$$\Delta^\nabla := \delta^\nabla d^\nabla + d^\nabla \delta^\nabla, \quad Tr\nabla^2 := \sum_i \nabla^2_{e_i,e_i} = \nabla^*\nabla$$

the covariant Hodge Laplacian and the trace Laplacian respectively, both of which are *second-order differential operators* acted upon the section space $\Gamma(E)$. A general Weitzenböck formula provides an explicit formula for the difference between the two Laplacians in terms of the action of the Ricci curvature operator of the underlying Riemannian manifold (P,h): *An upshot of the formula is that the difference is a zero-order differential operator.*

It is well established in the analysis of geometric PDE of the types, harmonic maps, minimal surface equation and Yang-Mills equations and so on that all a priori elliptic regularity results are based on suitable applications of the following general Weitzenböck Formula and the integration by parts one way or the other. (See [SY76], [SU81], [Uhl82], [SU83], [Sch84], [PW93], [RT95], to name a few.)

Theorem 3.1 (Weitzenböck Formula). *Let $E \to P$ be a vector bundle equipped with inner product $\langle \cdot, \cdot \rangle$ and an Euclidean connection ∇. Assume $\{e_i\}$ is an orthonormal frame of P, and $\{\alpha^i\}$ is the dual frame. We denote by R the curvature tensor of the bundle E with respect to the connection ∇. Then, when applied to E-valued differential forms, we have*

$$\begin{aligned} \Delta^\nabla &= -Tr\nabla^2 + \sum_{i,j} \alpha^j \wedge (e_i \rfloor R(e_i, e_j)(\cdot)) \\ &= -\nabla^*\nabla + \sum_{i,j} \alpha^j \wedge (e_i \rfloor R(e_i, e_j)(\cdot)). \end{aligned}$$

(For readers' convenience, we will provide a summary of the exterior calculus of E-valued differential forms and a derivation of the above Weitzenböck formula in Appendix 22.)

For each given equation, to get the optimal regularity estimates, it is important to use the 'best' connection compatible with the given geometry such as the Chern connection or the canonical connection in the harmonic theory of holomorphic vector bundles in complex geometry [Che67]. (See [Wel73] for a nice exposition on the harmonic theory of holomorphic vector bundles on complex manifolds. One may view that our calculations are largely the almost complex counterpart thereof on the Riemann surface $\dot\Sigma$.)

We specialize this general Weitzenböck formula to our purpose of tensorial study of contact instantons. For each given contact triad (Q, λ, J), we consider the vector bundles E such as

$$w^*TQ, \quad w^*\xi, \quad \Lambda^1(w^*TQ), \quad \Lambda^{(0,1)}(w^*TQ)$$

on punctured Riemann surface $(\dot\Sigma, j, h)$ equipped with Kähler metric h that is cylindrical near each puncture. As the aforementioned 'best' connection in this setting of contact triads, Wang and the first author introduced the notion of *contact triad connection* which is unique for each given contact triad (Q, λ, J). (See [OW14] for its construction and full properties. See also Section 9 for a summary.)

In the geometric PDEs of minimal surface-type map $w : S \to Q$, such as pseudo-holomorphic curves or contact instantons, the regularity estimates usually starts with computing the formula for the Laplacian of the harmonic energy density function

$$\Delta |dw|^2.$$

A priori, this quantity involves degree 3 derivatives for general smooth maps (*in the off-shell*), but which is hoped to be expressed as a sum of the terms of degree less than 3 for maps satisfying the equation (*on shell*). It will then enable one to develop a priori boot-strap arguments. We always regard dw as a w^*TQ-valued one-form on the domain $\dot\Sigma$.

Following this general practice and using the decomposition (1) and the properties $\xi \perp R_\lambda$ and $|R_\lambda| = 1$ for the contact triad metric, we can decompose the (full) harmonic energy density into the sum

$$|dw|^2 = |d^\pi w|^2 + |w^*\lambda|^2 \tag{1}$$

where $|d^\pi w|^2$ is the contribution from the ξ-direction and $|w^*\lambda|^2$ is the energy density in the Reeb direction: Recall the decomposition $dw = d^\pi w + w^*\lambda \otimes R_\lambda$ is orthogonal with respect to the induced metric from the triad metric of the target and the Kähler metric h of the domain Riemann surface $(\dot\Sigma, j)$.

The following differential identity for contact Cauchy-Riemann map plays a fundamental role for all the estimates needed for the contact instantons.

Theorem 3.2 (Fundamental Equation; Theorem 4.2 [OW18a]). *Let w be any contact Cauchy–Riemann map, i.e., a solution of $\overline\partial^\pi w = 0$. Then*

$$d^{\nabla^\pi}(d^\pi w) = -w^*\lambda \circ j \wedge \left(\frac{1}{2}(\mathcal{L}_{R_\lambda} J)\, d^\pi w\right). \tag{2}$$

Here d^∇ is the skew-symmetrization of the covariant derivative ∇ on $\dot\Sigma$.

Remark 3.3. When we apply similar calculation to a J-holomorphic map u with respect to the canonical connection on the almost Hermitian manifold (M, ω, J), then the corresponding equation is the simple harmonic map equation $d^\nabla(du) = 0$ for u. (See [Oh15, Corollary 7.3.3].) Together with the conformality of any J-holomorphic map, this provides a computational proof with the well-known fact that the image of a J-holomorphic curve is a minimal surface with respect to the compatible metric. This equation was the basis for the $W^{2,2}$-estimate for J-holomorphic map equation $\overline\partial_J u = 0$ on symplectic manifolds. (See [Oh15, Proposition 7.4.5].)

Then we can derive the following differential identity for the ξ-component $d^\pi w$ of the derivative dw, utilizing the Weitzenböck formula associated to the contact triad connection.

Proposition 3.4 (Equation (4.11) [OW18a]). *Let w be a contact Cauchy-Riemann map. Then*

$$\begin{aligned}-\frac{1}{2}\Delta e^\pi(w) &= |\nabla^\pi(d^\pi w)|^2 + K|d^\pi w|^2 + \langle \mathrm{Ric}^{\nabla^\pi}(d^\pi w), d^\pi w\rangle \\ &\quad + \langle \delta^{\nabla^\pi}[(w^*\lambda \circ j)\wedge(\mathcal{L}_{R_\lambda} J)d^\pi w], d^\pi w\rangle. \end{aligned} \tag{3}$$

If w is a contact instanton, i.e., if it satisfies $d(w^*\lambda \circ j) = 0$ in addition, we also have the following identity for the Laplacian of the energy density along the Reeb direction.

Proposition 3.5 (Equation (5.4) [OW18a]; Proposition 11.8). *Let w be a contact instanton. Then*

$$-\frac{1}{2}\Delta|w^*\lambda|^2 = |\nabla w^*\lambda|^2 + K|w^*\lambda|^2 + \langle *\langle \nabla^\pi d^\pi w, d^\pi w\rangle, w^*\lambda\rangle. \tag{4}$$

By adding the two identities, we obtain a formula for the Laplacian $\Delta|dw|^2$ of the full harmonic energy density function $|dw|^2$ on shell. Once these are established, the following local $W^{2,2}$-estimate is obtained by a standard trick of multiplying cut-off function and integrating by parts.

Theorem 3.6 (Theorem 1.6 [OW18a]). *Let $w : \dot\Sigma \to Q$ satisfy* (4). *Then for any relatively compact domains D_1 and D_2 in $\dot\Sigma$ such that $\overline{D_1} \subset D_2$, we have*

$$\|dw\|^2_{W^{1,2}(D_1)} \leq C_4\|dw\|^4_{L^4(D_2)}$$

where C_4 is a constant depending only on D_1, D_2 and (Q, λ, J) and R_i's.

The same estimates is also proved in the similar spirit by incorporating the Legendrian boundary condition which is a (nonlinear) elliptic boundary value problem. (See [Oh21a, Theorem 1.4] for the statement and [OY22] for the same statement with corrected proof.) This boundary estimate is rather nontrivial unlike the closed string case.

As the first step towards the analytic study of the above boundary value problem (4), we first show that the Legendrian boundary condition for the contact instanton is a free boundary value problem, i.e., it satisfies

$$\frac{\partial w}{\partial \nu} \perp TR$$

for any Legendrian submanifold. (See [Jos86] for the importance of the free boundary value condition for a general study of elliptic estimates of the minimal surface type equations.) Then we prove the elliptic $W^{2,2}$-estimate as an application of Stokes' formula combined with the Legendrian boundary condition. The global tensorial calculation deriving the a priori estimate in [OY22] illustrates how well the Legendrian boundary condition interacts with triad connection and the contact instanton equation.

3.0.1 Higher $C^{k,\delta}$ Hölder estimates

Once this $W^{2,2}$ estimates is established, we proceed with the higher boundary regularity estimates. Obviously the same estimates hold for the closed string case (3) the corresponding statement of which had been established in [OW18a]. Since this case is easier, we focus on the statements for the boundary estimates below.

Starting from Theorem 3.6 and using the embedding $W^{2,2} \hookrightarrow C^{0,\delta}$ with $0 < \delta < 1/2$, we also establish the following higher local $C^{k,\delta}$-estimates on punctured surfaces $\dot\Sigma$ in terms of the $W^{2,2}$-norms.

Theorem 3.7 (Theorem 1.4 [OY22]). *Let w satisfy* (4). *Then for any pair of disk $D_1 \subset D_2 \subset \dot\Sigma$ of semi-disk domains $(D_1, \partial D_1) \subset (D_2, \partial D_2) \subset (\Sigma, \partial\Sigma)$ such that $\overline{D_1} \subset D_2$,*

$$\|dw\|_{C^{k,\delta}(D_1)} \leq C_\delta\left(\|dw\|_{W^{1,2}(D_2)}\right)$$

where $C_\delta = C_\delta(r) > 0$ is a function continuous at $r = 0$ and depends only on J, λ and D_1, D_2 but independent of w.

With some adjustment of the function C_δ, combining the two theorems, we obtain

Corollary 3.8. *Assume $k \geq 1$ and $0 < \delta < 1/2$. Let w satisfy* (4). *Then for any pair of domains $D_1 \subset D_2 \subset \dot\Sigma$ such that $\overline{D_1} \subset D_2$,*

$$\|dw\|_{C^{k,\delta}(D_1)} \leq C_\delta\left(\|dw\|_{L^4(D_2)}\right)$$

where $C_\delta = C_\delta(r) > 0$ is a function continuous at $r = 0$ and depends only on J, λ and D_1, D_2 but independent of w.

In particular, we prove that any weak solution of (4) in $W^{1,4}_{\mathrm{loc}}$ automatically becomes a classical solution. (Compare [Oh15, Theorem 8.5.5] for a similar theorem for the Lagrangian boundary condition in symplectic geometry.)

The proof of Theorem 3.7 is carried out by an *alternating boot strap argument* by decomposing

$$dw = d^\pi w + w^*\lambda \otimes R_\lambda$$

as follows. Let $z = x + iy$ be any isothermal coordinates on $(D_2, \partial D_2) \subset (\dot\Sigma, \partial\dot\Sigma)$ adapted to the boundary, i.e., satisfying that $\frac{\partial}{\partial x}$ is tangent to $\partial\dot\Sigma$. We set

$$\zeta := d^\pi w(\partial_x),$$
$$\alpha := \lambda\left(\frac{\partial w}{\partial y}\right) + \sqrt{-1}\lambda\left(\frac{\partial w}{\partial x}\right)$$

Then we show that the fundamental equation (2) is transformed into the following system of equations for the pair (ζ, α)

$$\begin{cases} \nabla_x^\pi \zeta + J\nabla_y^\pi \zeta + \frac{1}{2}\lambda(\frac{\partial w}{\partial y})(\mathcal{L}_{R_\lambda}J)\zeta - \frac{1}{2}\lambda(\frac{\partial w}{\partial x})(\mathcal{L}_{R_\lambda}J)J\zeta = 0 \\ \zeta(z) \in TR_i \quad \text{for } z \in \partial D_2 \end{cases} \tag{5}$$

and

$$\begin{cases} \bar\partial\alpha = \frac{1}{2}|\zeta|^2 \\ \alpha(z) \in \mathbb{R} \quad \text{for } z \in \partial D_2 \end{cases} \tag{6}$$

for some $i = 0,\ldots,k$. With this coupled system of equations for (ζ,α) at our disposal, the proof of higher regularity results is carried out by the alternating boot strap argument between ζ and α in [Oh21a, OY22].

4 Asymptotic convergence and vanishing of asymptotic charge

Next we study the asymptotic convergence result of contact instantons of finite energy $E(w) = E^\pi(w) + E^\perp(w) < \infty$ for the closed string case (resp. with Legendrian boundary condition of pair (R_0,R_1) for the open string case) near the punctures of a Riemann surface $\dot\Sigma$. (We refer to [Oha, Oh21c, OY23] for the precise definition of total energy.)

Let $\dot\Sigma$ be a punctured Riemann surface with punctures

$$\{p_i^+\}_{i=1,\cdots,l^+} \cup \{p_j^-\}_{j=1,\cdots,l^-}$$

equipped with a metric h with cylinder-like ends (resp. *strip-like ends* for the open string case) outside a compact subset K_Σ. Let $w : \dot\Sigma \to Q$ be any such smooth map.

As in [OW18a], we define the total π-harmonic energy $E^\pi(w)$ is easy to define as

$$E^\pi(w) = E^\pi_{(\lambda,J;\dot\Sigma,h)}(w) = \frac{1}{2}\int_{\dot\Sigma} |d^\pi w|^2\, dA \tag{1}$$

where dA is the associated area form and the norm is taken in terms of the given metric h on $\dot\Sigma$ and the triad metric on Q.

4.0.1 The case of closed strings

Under the hypotheses of nondegeneracy λ (resp. of the pair $(\lambda,(R_0,R_1))$ for the open string case) and of asymptotic convergence at the punctures, we can associate two natural asymptotic invariants at each puncture defined as

$$T_w := \lim_{r\to\infty}\int_{\{r\}\times S^1} (w|_{\{r\}\times S^1})^*\lambda \tag{2}$$

$$Q_w := \lim_{r\to\infty}\int_{\{r\}\times S^1} ((w|_{\{r\}\times S^1})^*\lambda\circ j) \tag{3}$$

at each puncture. (Here we only look at positive punctures. The case of negative punctures is similar.) As in [OW18a], we call $T = T_w$ the *asymptotic contact action* and $Q = Q_w$ the *asymptotic contact charge* of the contact instanton w at the given puncture.

Remark 4.1. It is unfortunate that we ended up using the same letter Q for both contact manifold and the asymptotic charge but this practice should not confuse readers.

The proof of the following subsequence convergence result is given in [OW18a, Theorem 6.4]. A similar asymptotic convergence result in 3 dimension in the setting of pseudoholomorphic curves on symplectization, i.e., the case of $Q = 0$ is proved in [HWZ02]. (See also [HWZ96b, HWZ96a] and compare their proofs with the proof given in [OW18a].)

Theorem 4.2 (Subsequence Convergence, Theorem 6.4 [OW18a]). *Let $w : [0, \infty) \times S^1 \to Q$ satisfy the contact instanton equations* (4) *and Hypothesis* (2). *Then for any sequence $s_k \to \infty$, there exists a subsequence, still denoted by s_k, and a massless instanton $w_\infty(\tau, t)$ (i.e., $E^\pi(w_\infty) = 0$) on the cylinder $\mathbb{R} \times S^1$ that satisfies the following:*

1. *$\overline{\partial}^\pi w_\infty = 0$ and*
$$\lim_{k \to \infty} w(s_k + \tau, t) = w_\infty(\tau, t)$$
in the $C^l(K \times S^1, Q)$ sense for any l, where $K \subset [0, \infty)$ is an arbitrary compact set.
2. *$w_\infty^* \lambda = -Q\, d\tau + T\, dt$*

In general $Q = 0$ does not necessarily hold for the closed string case. When $Q \neq 0$ combined with $T = 0$ happens, we say w has the bad limit of *appearance of spiraling instantons along the Reeb core*. It is also proven in [Oha] that If $Q = 0 = T$, then the puncture is removable.

When $Q = 0$, which is always the case when contact instanton is exact such as those arising from the symplectization case, we have the following asymptotic convergence result.

Corollary 4.3. *Assume that λ is nondegenerate. Let w be as above and assume $Q = 0, T \neq 0$ and that $w_\tau : S^1 \to Q$ converges as $|\tau| \to \infty$. Then w_τ converges to a Reeb orbit of period $|T|$ exponentially fast.*

We would like to emphasize that the asymptotic limit could be a constant path, i.e., the Reeb action $T := \int \gamma^* \lambda$ is zero, which is normally not regarded as a Reeb orbit. To unify the cases for $T \neq 0$ and $T = 0$, it is useful to utilize the following Moore-type notion of paths introduced and utilized in [Oh21c, Oh22a] for the study of Sandon-Shelukhin type conjectures.

Definition 4.4 (Isospeed Reeb trajectories). A pair (γ, T) of a parameterized curve $\gamma : [0, 1] \to Q$ with $T = \int \gamma^* \lambda$ is called an *isospeed Reeb trajectory* if it satisfies the equation
$$\dot{\gamma}(t) = T R_\lambda(\gamma(t)). \tag{4}$$

Note that when $T \neq 0$, each such pair gives rise to a Reeb trajectory on the domain $[0, |T|]$, while when $T = 0$ the resulting path is a constant path. From now on, we will say *$w(\tau, \cdot)$ converges to an isospeed Reeb trajectory (γ, T)* if $w(\tau, \cdot) \to \gamma$ as $|\tau| \to \infty$ with $\int \gamma^* \lambda = T$.

4.0.2 The case of open strings

Now we make the corresponding statement for the open string case proved in [Oh21a].

Theorem 4.5 (Subsequence Convergence; the case of open strings). *Let $w : [0,\infty) \times [0,1] \to Q$ satisfy the contact instanton equations* (4). *Then for any sequence $s_k \to \infty$, there exists a subsequence, still denoted by s_k, and a massless instanton $w_\infty(\tau,t)$ (i.e., $E^\pi(w_\infty) = 0$) on the cylinder $\mathbb{R} \times [0,1]$ such that*

$$\lim_{k\to\infty} w(s_k + \tau, t) = w_\infty(\tau,t)$$

in the $C^l(K \times [0,1], Q)$ sense for any l, where $K \subset [0,\infty)$ is an arbitrary compact set. Furthermore, w_∞ has $Q = 0$ and the formula $w_\infty(\tau,t) = \gamma(t)$ with asymptotic action T, where (γ,t) is an isospeed Reeb chord with its action $T = \int \gamma^*\lambda$ joining R_0 and R_1 of period $|T|$.

Corollary 4.6 (Vanishing Charge). *Assume the pair $(\lambda,\vec{R})$ is nondegenerate. Let w be as above with finite energy. Suppose that $w(\tau,\cdot)$ converges as $\tau \to \infty$ in the strip-like coordinate at a puncture $p \in \partial\Sigma$ with associated Legendrian pair (R,R'). Then its asymptotic charge Q vanishes at p.*

5 Asymptotic operators and their analysis

We first mention a few differences between the way how we study the asymptotic operators and those of [HWZ96b] and of other literature such as [RS01, Appendix C], [Sie08, Sie11], [Wen], [Can22].

Remark 5.1. In [RS01, Appendix E], [Sie08, Sie11], [Wen, Section 3.3], there have been attempts to give a coordinate-free definition of the asymptotic operator along the associated asymptotic Reeb orbit for a pseudoholomorphic curve $u = (w,f)$ on symplectization. However both fall short of a seamless definition of the 'asymptotic operator' of the Reeb orbits because the Reeb orbit lives on Q while the pseudoholomorphic curves live on the product $Q \times \mathbb{R}$ and the asymptotic limit of pseudoholomorphic curve live at infinity $Q \times \{\pm\infty\}$ *where only the contact structure makes sense, i.e., is canonically defined, but not the contact form itself.* Cant studies the asymptotic operator for the relative context in [Can22, Section 6.3] by adapting Wendl's. What these literature (e.g.[Wen, Section 3.3]) are describing is actually the asymptotic operator of the contact instanton w but trying to describe it in terms of the pseudoholomorphic curves which prevents them from being able to give a seamless definition. (See our definition of the asymptotic operator of contact instantons given in Definition 21.2 and compare it therewith. See also [OW18b, Section 11.2 & 11.5] for the precursor of our definition.)

In their series of works [HWZ95] – [HWZ02] in 3 dimension, Hofer-Wysocki-Zehnder carried out fundamental analytic study of pseudoholomorphic curves on symplectization utilizing *special coordinates followed by some local adjustment of given almost complex structure along the Reeb orbit of interest.* This practice has been propagated to other literature, such as [Bou02], [Hut02, Section 1.3] giving rise to some unnecessary restrictions on the choice of almost complex structure beyond the natural λ-adaptedness. [1]

Largely, thanks to the property $\nabla^{\mathrm{LC}}_{R_\lambda} J = 0$ from Proposition 0.4 combined with our usage of contact triad connection in the derivation of the formula for the asymptotic operator, our asymptotic analysis provided in [KO23] does not need any of those special coordinates and so Hutchings' assumption (or similar ansatz in other literature) is really not needed.

Theorem 5.2 (Corollary 21.7). *Let (λ, J) be any adapted pair and let ∇^{LC} be the Levi-Civita connection of the triad metric of (Q, λ, J). For given contact instanton w with its action $\int \gamma^* \lambda = T$ at a puncture, let $A_{(\lambda,J,\nabla)}$ be the asymptotic operator of w written in cylindrical coordinate (τ, t). Then*

1. $[\nabla^{\mathrm{LC}}_t, J] (= \nabla^{\mathrm{LC}}_t J) = 0,$
2. *We have*

$$\begin{aligned} A^\pi_{(\lambda,J,\nabla)} &= -J\nabla_t + \frac{T}{2}\mathscr{L}_{R_\lambda} J J \\ &= -J\nabla^{\mathrm{LC}}_t - \frac{T}{2} Id + \frac{T}{2}\mathscr{L}_{R_\lambda} J J. \end{aligned}$$

We also have the following formula of $A^\pi_{(\lambda,J,\nabla)}$ that is independent of the choice of connections.

Proposition 5.3 (Proposition 1.7 [KO23]). *We have*

$$A^\pi_{(\lambda,J,\nabla)} = T\left(-\frac{1}{2}\mathscr{L}_{R_\lambda} J - \mathscr{L}_{R_\lambda} + \frac{1}{2}(\mathscr{L}_{R_\lambda} J) J\right). \tag{1}$$

Because the existing literature on the pseudoholomorphic curves on symplectization lack this kind of explicit formula of the asymptotic operator (together with the commuting property $[\nabla^{\mathrm{LC}}_t, J] = 0$), it has been the case that the general abstract perturbation theory of linear operators Kato [Kat95] is just quoted in their study of asymptotic operators which prevents one from making any statement on specific dependence on the adapted almost complex structures. (See [Wen, Lemma 3.17 & Theorem 3.35], for example, which in turn follows the statements and arguments given in [HWZ95].)

Remark 5.4. 1. The formula of an asymptotic operator canonically applied to every contact instanton w of the adapted pair (λ, J) (as well as that of a pseudoholomorphic curve $u = (w, f)$ on symplectization) depends not only on the adapted

[1] Hutchings has informed the first author that this part of his paper is now obsolete in that Siefring proved the asymptotics he needed without the extra assumptions in [Sie08] quoted above. Siefring's paper is now quoted for this in Hutching's later paper [Hut09] which is a kind of update to [Hut02].

pair (λ,J) but also on the connection that is used to compute the linearization operator of w (or (u,f)). (See Definition 21.2 and Remark 21.3 for the explanation for why.) In this regard, we denote the asymptotic operator by

$$A^{\pi}_{(\lambda,J,\nabla)}(u) := A^{\pi}_{(\lambda,J,\nabla)}(w)$$

So it is conceivable to expect that a good choice of connection will give rise to a formula of the asymptotic operator that is easier to analyze.

2. In fact, the lack of precise definition together with the non-commuting property $[\nabla_t, J] \neq 0$ in the literature combined with the practice of *using special coordinates followed by adjusting the almost complex structure along the Reeb orbits* is bound to make the analysis of asymptotic operators very complicated as seen from [HWZ02], [Sie08], and prevents one from developing any perturbation theory of asymptotic operators under the perturbation of J's such as those developed by the present authors in [KO23]. The latter is summarized in Section 21 of the present survey.

On the other hand, our explicit formula of the asymptotic operator given in Theorem 5.2, which simultaneously applies to all closed Reeb orbits, enables us to prove the following natural generic perturbation result [KO23].

Theorem 5.5 (Generic simpleness of eigenvalues; [KO23]). *Let (Q,ξ) be a contact manifold. Assume that λ is nondegenerate. For a generic choice of λ-adapted CR almost complex structures J, all eigenvalues μ_i of the asymptotic operator are simple for all closed Reeb orbits of λ.*

See Section 21 for our derivation of the formula of the asymptotic operator and a summary of our analysis in [KO23] of the asymptotic operators. We believe that this kind of perturbation result will play some role in the construction of Kuranishi structures on the moduli space of finite energy contact instantons so that certain natural functor can be defined in our Fukaya-type category of contact manifolds [KO], [Ohc] (See Remark 9.10 (2) for the relevant remark.)

6 Comparison of compactifications of two moduli spaces

(The materials in this subsection is borrowed from [KO], [Ohc] which are in preparation.)

Now let us consider contact instantons w arising from a pseudoholomorphic curves on symplectization (w,f). In particular all such w has its the *charge class* vanishes $[w^*\lambda \circ j] = 0$ in $H^1(\dot\Sigma,\mathbb{Z})$. (See [OSar] or Subsection 19.2 for its definition.)

Remark 6.1. For the boundary punctured case, the charge class can be lifted to $H^1(\overline{\Sigma},\partial_\infty\dot\Sigma)$ where $\overline{\Sigma}$ is the real blow-up of $\dot\Sigma$ along the punctures. See Subsection 19.2 for its definition.

Let Σ be a closed Riemann surface of genus g and $\dot\Sigma$ be the associated punctured Riemann surface $\dot\Sigma = \Sigma \setminus \{z_1, \cdots, z_\ell\}$. We denote the moduli space of such contact instantons $w : \dot\Sigma \to Q$ of finite energy by

$$\widetilde{\mathcal{M}}^{\text{exact}}_{g,\ell}(Q, J; \vec\gamma^-, \vec\gamma^+)$$

and

$$\mathcal{M}^{\text{exact}}_{g,k,\ell}(Q, J; \vec\gamma^-, \vec\gamma^+) := \widetilde{\mathcal{M}}^{\text{exact}}_{g,\ell}(Q, J; \vec\gamma^-, \vec\gamma^+)/\mathrm{Aut}(\dot\Sigma)$$

the set of isomorphism classes thereof. We have the natural forgetful map $(w, f) \mapsto w$ which descends to

$$\mathfrak{forget} : \mathcal{M}_{g,\ell}(M, \widetilde J; \vec\gamma^-, \vec\gamma^+) \to \mathcal{M}^{\text{exact}}_{g,\ell}(Q, J; \vec\gamma^-, \vec\gamma^+). \tag{1}$$

By definition of the equivalence relation on $\widetilde{\mathcal{M}}_{g,\ell}(M, \widetilde J; \vec\gamma^-, \vec\gamma^+)$ defined in [EGH00], [BEHZ03], it follows that this forgetful map is a bijective correspondence, *provided $\dot\Sigma$ is connected.*

However when one considers the SFT compactification, one needs to also consider the case of pseudoholomorphic curves with disconnected domains. So let us consider such cases. Suppose that the Riemann surface $\dot\Sigma$ is the union

$$\dot\Sigma = \bigsqcup_{i=1}^{k} \dot\Sigma_i$$

of connected components with $k \geq 2$. We denote by

$$\overline{\mathcal{M}}_{g,\ell}(M, \widetilde J; \vec\gamma^-, \vec\gamma^+)$$

and

$$\overline{\mathcal{M}}^{\text{exact}}_{g,\ell}(Q, J; \vec\gamma^-, \vec\gamma^+)$$

the relevant stable map compactifications, respectively. The following proposition shows the precise relationship between the two. We know that each story carries at least one non-cylindrical component.

Proposition 6.2 (Proposition 19.7)**.** *Let $1 \leq \ell \leq k$ be the number of connected components which are not cylinderical. The forgetful map* $\mathfrak{forget}$ (1) *extends to the compactified moduli spaces, and is a principle $\mathbb{R}^{\ell-1}$ fibration.*

So the above mentioned bijective correspondence still holds when there is exactly one non-trivial component. (See [BEHZ03, p. 835] for a relevant remark.)

This proposition clearly shows that the compactification proposed in [EGH00], [BEHZ03] have spurious strata that contract in the compactification of (exact) contact instantons under the forgetful map. In this regard, our compactification of exact contact instanton moduli spaces is closely related to Pardon's compactification of moduli spaces of pseudoholomorphic curves on symplectization [Par19], which is slightly different from that of [EGH00], [BEHZ03]. We will make precise their relationship in [KO] and [Ohc]. Pardon [Par19] used his compactification for his construction of Kuranishi structure on the moduli space of pseudoholomorphic curves on symplectization to define contact homology.

7 Fredholm theory and the index formula

Next, we study another crucial component, the relevant Fredholm theory and the index formula for the equation (4) by adapting the one from [Oha], [OSar] to the current case of contact instantons *with boundary*. The relevant Fredholm theory in general has been developed by the first named author in [Oha] for the closed string case and [Ohb] for the case with boundary.

In the present paper, we state the index formula for general disk instanton w with finite number of boundary punctures.

We recall the following Fredholm property of the linearized operator that is proved in [Ohb].

Proposition 7.1 (Proposition 3.18 & 3.20 [Ohb]). *Suppose that w is a solution to* (4). *Consider the completion of $D\Upsilon(w)$, which we still denote by $D\Upsilon(w)$, as a bounded linear map from $\Omega^0_{k,p}(w^*TQ,(\partial w)^*T\vec{R})$ to $\Omega^{(0,1)}(w^*\xi)\oplus\Omega^2(\Sigma)$ for $k\geq 2$ and $p\geq 2$. Then*

1. *The off-diagonal terms of $D\Upsilon(w)$ are relatively compact operators against the diagonal operator.*
2. *The operator $D\Upsilon(w)$ is homotopic to the operator*

$$\begin{pmatrix} \bar{\partial}^{\nabla^\pi} + T^{\pi,(0,1)}_{dw} + B^{(0,1)} & 0 \\ 0 & -\Delta(\lambda(\cdot))\,dA \end{pmatrix} \tag{1}$$

via the homotopy

$$s\in[0,1]\mapsto \begin{pmatrix} \bar{\partial}^{\nabla^\pi} + T^{\pi,(0,1)}_{dw} + B^{(0,1)} & \frac{s}{2}\lambda(\cdot)(\mathscr{L}_{R_\lambda}J)J(\pi dw)^{(1,0)} \\ s\,d((\cdot)\rfloor d\lambda)\circ j) & -\Delta(\lambda(\cdot))\,dA \end{pmatrix} =: L_s \tag{2}$$

which is a continuous family of Fredholm operators.

3. *And the principal symbol*

$$\sigma(z,\eta): w^*TQ|_z \to w^*\xi|_z\oplus\Lambda^2(T_z\Sigma), \quad 0\neq\eta\in T^*_z\Sigma$$

of (1) *is given by the matrix*

$$\begin{pmatrix} \frac{\eta+i\eta\circ j}{2}Id & 0 \\ 0 & |\eta|^2 \end{pmatrix}.$$

Then we have the Fredholm index of $D\Upsilon(w)$ is given by

$$\operatorname{Index} D\Upsilon(w) = \operatorname{Index}(\bar{\partial}^{\nabla^\pi} + T^{\pi,(0,1)}_{dw} + B^{(0,1)}) + \operatorname{Index}(-\Delta_0).$$

For this purpose of computing the Fredholm index of the linearized operator in terms of a topological index, especially in terms of those who will equip the relevant Floer-type complex with the absolute grading, we need the notion of anchored Legendrian submanifolds [OY22] which is an adaptation of that of Lagrangian submanifolds studied in [FOOO10] in symplectic geometry.

Definition 7.2. Fix a base point y of ambient contact manifold (Q,ξ). Let R be a Legendrian submanifold of (Q,ξ). We define an *anchor* of R to y is a path $\ell : [0,1] \to Q$ such that $\ell(0) = y$, $\ell(1) \in R$. We call a pair (R,ℓ) an *anchored* Legendrian submanifold. A chain $\mathscr{E} = ((R_0,\ell_0),\dots,(R_k,\ell_k))$ is called an *anchored Legendrian chain*.

We refer readers to [OY22] for the details of derivation of the relevant index formula which expresses the analytical index of the linearized problem of (4) in terms of a topological index of the Maslov-type. This index is made more explicit in [OY23] for the case of Hamiltonian isotopes of the zero section of the one-jet bundle.

Remark 7.3. While we are preparing this survey, Dylan Cant informed the first author of his thesis work [Can22] in which he studied the Fredholm theory and index formula *for the open string case* of pseudoholomorphic curves for the symplectization of *general pair* $(Q \times \mathbb{R}, R \times \mathbb{R})$ with Legendrian submanifold R in general dimension. As mentioned in [OY22, Section 1.5], this case is included as a part of the study of pseudoholomorphic curves on the $\mathfrak{lcs}$-fication of contact manifold as the exact case or as the 'zero-temperature limit' thereof.

8 Finer asymptotics

Finally we describe our work on the precise fine asymptotics of contact instantons (and so of pseudoholomorphic curves on symplectization) through the coordinate-free analysis of asymptotic operators and the study of their eigenvalues under the perturbation of adapted pairs (λ, J) in our work [KO23]. We refer readers to Section 21.3 for more detailed summary of the materials below.

Remark 8.1. A study of precise asymptotic formula of pseudholomorphic curves on symplectization in the more general context of stable Hamiltonian structures is given by Siefring in [Sie08, Sie11] which is applied to develop local intersection theory and embedding controls (in 4 dimension) of two embedded pseudoholomorphic curves with the same asymptotic limits.

Our study provides a precise generic description of the spectral behavior of asymptotic operators under the change $d\lambda$-compatible CR almost complex structures J when λ is fixed. For example, we prove the following in [KO23].

An upshot of our study of asymptotic behavior of finite contact instantons (and hence that of pseudoholomorphic curves on symplectization) is our usage of the following variable

$$\zeta(\tau,t) := \left(\frac{\partial w}{\partial \tau}\right)^{\pi}$$

on the cylindrical coordinates of $\dot\Sigma$ near a puncture in our various study of contact instantons which enables us to avoid any usage of special target coordinates that has been essential in Hofer-Wyosocki-Zehnder's approach, when we take the covariant tensorial approach using the contact triad connection.

We will derive in (4) that ζ satisfies

$$\nabla^{\pi}_{\tau}\zeta + J\nabla^{\pi}_{t} + S\zeta = 0 \tag{1}$$

where S is the zero operator

$$S\zeta = \frac{1}{2}w^*\lambda(\partial_\tau)(\mathscr{L}_{R_\lambda}J)\zeta + \frac{1}{2}w^*\lambda(\partial_t)(\mathscr{L}_{R_\lambda}J)J\zeta.$$

We write the loop $w_\tau := w(\tau,\cdot)$ and

$$A^{\tau} = (-J\nabla^{\pi}_{t} - S)|_{w^*_\tau\xi}$$

which we know converges to the asymptotic operator $A^{\pi}_{(\lambda,J,\nabla)}$. We define

$$v_\tau := \frac{\zeta_\tau}{\|\zeta_\tau\|_{L^2}}$$

so that it has unit L^2-norm on S^1 for all $\tau \geq \tau_0$. With this definition, one can following essentially verbatim the proof of [HWZ01, Theorem 2.8] using the tensorial way considering parallel transport of sections of $w^*\xi$ with respect to the canonical connection.

Then the following theorem describes the asymptotic behavior of w as $|\tau| \to \infty$, which is the analog to [HWZ96b, Theorem 1.4]. Our proof is the covariant tensorial version of the Hofer-Wysocki-Zehnder's proof of [HWZ01, Theorem 2.8] given in Section 3 therein via coordinate calculations in 3 dimension.

Theorem 8.2 (Asymptotic behavior). *Assume that (γ,T) is nondegenerate. Consider the unit vectors.*

$$v(\tau,t) := \frac{\zeta(\tau,t)}{\|\zeta(\tau)\|_{L^2(w^*_\tau\xi)}}, \quad \alpha(\tau) := \frac{d}{d\tau}\log\|\zeta(\tau)\|_{L^2(w^*_\tau\xi)}.$$

Then we have the following:

1. *Either $\zeta(\tau,t) = 0$ for all $(\tau,t) \in [\tau_0,\infty)$ or*
2. *the following hold:*
   a. *There exists an eigenvector e of $A^{\pi}_{(\lambda,J,\nabla)}(\gamma_T)$ of eigenvalue μ such that $v_\tau \to e$ as $\tau \to \infty$ and*

$$\zeta(\tau,t) = e^{\int_{\tau_0}^{\tau}\alpha(s)\,ds}(e(t) + \tilde r(\tau,t))$$

where $\tilde{r}(\tau,t) \to 0$ *in* C^∞ *topology.*

b. *For any* $0 < r < \mu$, *there exist constants* $\delta > 0$, $\tau_0 > 0$ *and* C_β *such that for all multi-indices* $\beta = (\beta_1, \beta_2)) \in \mathbb{N}^2$ *such that*

$$\sup_{(\tau,t)} |(\nabla^\beta \zeta)(\tau,t)| \le C_\beta e^{-r\tau}$$

for all $\tau \ge \tau_0$ *and* $t \in S^1$. *The same exponential decay also holds for the function* $\tilde{r}$ *appearing in Statement (a) above.*

The exponential convergence here appearing can be derived by a boot-strap argument using the exponential convergence result on ζ as $\tau \to \infty$ given in Section 17. (See [OW18b], [OY22] for the details of this exponential convergence and the boot-strap argument.)

Note that the above representation formula of ζ implies the following convergence of the tangent plane.

Corollary 8.3 (Convergence of tangent plane)**.** *Assume the second alternative in Theorem 8.2 and denote*

$$P(\tau,t) := \mathrm{Image}\, dw(\tau,t) \in \mathrm{Gr}_2(\xi_{w(\tau,t)})$$

where $\mathrm{Gr}_2(\xi_x)$ *is the set of 2 dimensional subspaces of the contact hyperplane* $\xi_x \subset T_xQ$. *Then* $P(\tau,t) \to \mathrm{span}_{\mathbb{R}}\{e(t), Je(t) + TR_\lambda(\gamma(t))\}$ *exponentially fast in* C^∞ *topology uniformly in* $t \in S^1$.

This convergence statement is a trivial vacuous statement in dimension 3 since $\dim \xi = 2$ but is a nontrivial statement for higher dimensions. We refer readers to [Sie08, Sie11] for a precise local intersection theory of two pseudoholomorphic curves at the punctures and topological controls on the intersection number of two curves with the same 'fine' asymptotic limit.

Finally we would like to just mention that the same asymptotic study can be made in a straightforward way by incorporating the boundary condition by now as done in [OY22], [Ohb], [Oh22b].

This article is largely a survey of the first author and his collaborators' series of works on the analysis of contact instantons, focusing mainly on the *unperturbed ones* [OW14], [OW18a, OW18b], [Oha] and [Oh21a].

This survey does not touch upon the case of *Hamiltonian-perturbed contact instantons* studied in [Oh21b] for the elliptic regularity theory generalizing that of [Oh21a, OY22] in which does lie the real power of our approach through the interplay between geometric analysis of perturbed contact instantons and the calculus of Hamiltonian geometry and dynamics. We refer readers to [Oh21c], [Oh22a] in which the interplay has been exhibited by the proof of Sandon-Shelukhin type conjectures.

The case of unperturbed contact instantons corresponds to the case of Gromov's original pseudoholomorphic curves [Gro85] while the perturbed ones correspond to solutions of Floer's Hamiltonian-perturbed trajectory equations [Flo89, SZ92] in symplectic geometry. We refer interested readers to [Oh21b] for the case of perturbed equation. We also refer readers to [OW18b] for the tensorial proof of exponential convergence result in the Morse-Bott nondegenerate case of contact instantons again utilizing Weitezenböck-type formulae with respect to contact triad connections.

Throughout the paper, we freely use the (covariant) exterior calculus of E-valued differential forms and Weitzenböck formula with respect to the contact triad connection ∇. For the convenience of the prospective readers of this survey article, we duplicate [OW18a, Appendix A & B] here which summarizes the exterior calculus of vector-valued forms and the derivation of Weitzenböck formula in Appendix of the present paper.

There is one exception of the usage of contact triad connection: This is for the study of asymptotic operators along the Reeb orbits for the closed case (or along the Reeb chords in the open string case) where the derivation of asymptotic operators is done using the triad connection but converted the formula in the final conclusion to one involving the Levi-Civita connection of the triad metric *along the Reeb orbits (or along the Reeb chords*) because of our desire to more widely advertize the wonderful property of the Levi-Civita connection given in Proposition 0.4.

Acknowledgement: We thank MATRIX for providing an excellent research environment and Brett Parker for his great effort for smoothly running the IBSCGP-MATRIX Symplectic Topology Workshop. We also thank all participants of the workshop for making the workshop a big success. The first author also thanks Givental for useful communication on SFT compactification, and Hutchings for brining our attention to Siefring's paper [Sie08]. We also thank the unknown referee for many suggestions to improve the presentation throughout the paper.

Conventions: All the conventions regarding the definition of Hamiltonian vector fields, canonical symplectic forms on the cotangent bundle and definition of contact Hamiltonians and others are the same as those adopted and listed in [Oh21a, Conventions]. These also coincide with the conventions used in [dLLV19].

Part II
Contact triads and their $\mathfrak{lcs}$-fications

9 Contact triad connection and canonical connection

Let (Q,ξ) be a given contact manifold. When a contact form λ is given, we have the projection $\pi = \pi_\lambda$ from TQ to ξ associated to the decomposition

$$TQ = \xi \oplus \mathbb{R}\langle R_\lambda \rangle.$$

We denote by $\Pi = \Pi_\lambda : TQ \to TQ$ the corresponding idempotent, i.e., the endomorphism of TQ satisfying $\Pi^2 = \Pi$, $\operatorname{Im}\Pi = \xi$, $\ker\Pi = \mathbb{R}\{R_\lambda\}$.

9.1 Contact triads and triad connections

Let (Q,λ,J) be a contact triad of dimension $2n+1$ for the contact manifold (Q,ξ), and equip with it the contact triad metric $g = g_\xi + \lambda \otimes \lambda$. In [OW14], Wang and the first author introduced the *contact triad connection* associated to every contact triad (Q,λ,J) with the contact triad metric and proved its existence and uniqueness and naturality.

Theorem 9.1 (Contact Triad Connection [OW14]). *For every contact triad (Q,λ,J), there exists a unique affine connection ∇, called the contact triad connection, satisfying the following properties:*

1. *The connection ∇ is metric with respect to the contact triad metric, i.e., $\nabla g = 0$;*
2. *The torsion tensor T of ∇ satisfies $T(R_\lambda,\cdot) = 0$;*
3. *The covariant derivatives satisfy $\nabla_{R_\lambda} R_\lambda = 0$, and $\nabla_Y R_\lambda \in \xi$ for any $Y \in \xi$;*
4. *The projection $\nabla^\pi := \pi\nabla|_\xi$ defines a Hermitian connection of the vector bundle $\xi \to Q$ with Hermitian structure $(d\lambda|_\xi, J)$;*
5. *The ξ-projection of the torsion T, denoted by $T^\pi := \pi T$ satisfies the following property:*
$$T^\pi(Y,JY) = 0 \tag{1}$$
for all X tangent to ξ;
6. *For $X \in \xi$, we have the following*
$$\partial^\nabla_Y R_\lambda := \frac{1}{2}(\nabla_Y R_\lambda - J\nabla_{JY} R_\lambda) = 0.$$

From this theorem, we see that the contact triad connection ∇ canonically induces a Hermitian connection ∇^π for the Hermitian vector bundle (ξ, J, g_ξ), and we call it the *contact Hermitian connection*. This connection will be used to study estimates for the π-energy in later sections.

Moreover, the following fundamental properties of the contact triad connection was proved in [OW14]

Corollary 9.2 (Naturality). *1. Let ∇ be the contact triad connection of the triad (Q,λ,J). Then for any diffeomorphism $\phi : Q \to Q$, the pull-back connection $\phi^*\nabla$*

is the triad connection associated to the triad $(Q, \phi^*\lambda, \phi^*J)$ associated to the pull-back contact structure $\phi^*\xi$.

2. *In particular if ϕ is contact, i.e., $d\phi(\xi) \subset \xi$, then $(Q, \phi^*\lambda, \phi^*J)$ is a contact triad of ξ and $\phi^*\nabla$ the contact triad connection (Q, ξ).*

The following identities are also very useful to perform tensorial calculations in the study of a priori elliptic estimates and in the derivation of the linearization formula.

Corollary 9.3. *Let ∇ be the contact triad connection. Then*

1. *For any vector field X on Q,*

$$\nabla_X R_\lambda = \frac{1}{2}(\mathcal{L}_{R_\lambda} J)JX; \tag{2}$$

2. *For the Reeb component of the torsion T, we have $\lambda(T) = d\lambda$.*

We refer readers to [OW14] for more discussion on the contact triad connection and its relation with other related canonical type connections.

9.2 Canonical connection on almost Hermitian manifold

In this subsection, we give the definition of canonical connection on general almost Hermitian manifolds and apply it to the case of $\mathfrak{lcs}$-fication of contact triad (Q, λ, J). See [Oh15, Chapter 7] for the exposition of canonical connection for almost Hermitian manifolds and its relationship with the Levi-Civita connection, and in relation to the study of pseudoholomorphic curves on general symplectic manifolds emphasizing the Weitzenböck formulae in the study of elliptic regularity in the same spirit of the present survey.

Let (M, J) be any almost complex manifold.

Definition 9.4. A metric g on (M, J) is called Hermitian, if g satisfies

$$g(Ju, Jv) = g(u, v), \quad u, v \in T_xM, x \in M.$$

We call the triple (M, J, g) an almost Hermitian manifold.

For any given almost Hermitian manifold (M, J, g), the bilinear form

$$\Phi := g(J\cdot, \cdot)$$

is called the fundamental two-form in [KN96], which is nondegenerate.

Definition 9.5. An almost Hermitian manifold (M, J, g) is an *almost Kähler manifold* if the two-form Φ above is closed.

Definition 9.6. A (almost) Hermitian connection ∇ is an affine connection satisfying

$$\nabla g = 0 = \nabla J.$$

Existence of such a connection is easy to check. In general the torsion $T = T_\nabla$ of the almost Hermitian connection ∇ is not zero, even when J is integrable. The following is the almost complex version of the Chern connection in complex geometry.

Theorem 9.7 ([Gau97], [Kob03]). *On any almost Hermitian manifold (M, J, g), there exists a unique Hermitian connection ∇ on TM satisfying*

$$T(X, JX) = 0 \tag{3}$$

for all $X \in TM$.

In complex geometry [Che67] where J is integrable, a Hermitian connection satisfying (3) is called the Chern connection.

Definition 9.8. A *canonical connection* of an almost Hermitian connection is defined to be one that has the torsion property (3).

The triple (4)

$$(Q \times \mathbb{R}, \widetilde{J}, \widetilde{g}_\lambda)$$

is a natural example of an almost Hermitian manifold associated to the contact triad (Q, λ, J).

Let $\widetilde{\nabla}$ be the canonical connection thereof. Then we have the following which also provides a natural relationship between the contact triad connection and the canonical connection.

Proposition 9.9 (Canonical connection versus contact triad connection). *Let $\widetilde{g} = \widetilde{g}_\lambda$ be the almost Hermitan metric given above. Let $\widetilde{\nabla}$ be the canonical connection of the almost Hermitian manifold* (4)*, and ∇ be the contact triad connection for the triad (Q, λ, J). Then $\widetilde{\nabla}$ preserves the splitting* (3) *and satisfies $\widetilde{\nabla}|_\xi = \nabla|_\xi$.*

Remark 9.10. 1. In fact it was shown in [OW14] that for each real constant c, there is the unique connection that satisfies all properties (1)–(5) and (6) replaced by (6;c)

$$\nabla_{JY} R_\lambda + J \nabla_Y R_\lambda = cY, \text{or equivalently } \partial_Y^\nabla R_\lambda = \frac{c}{2} Y \tag{4}$$

for all $Y \in \xi$. Our canonical connection corresponds to $c = 0$. In particular all of these connections, temporarily denoted by ∇^c and called c-triad connection, induce the same Hermitian connection on ξ, i.e.,

$$\nabla^{c;\pi} = \nabla^\pi$$

for all c. With this choice of connection ∇^c, (2) is replaced by

$$\nabla_Y^c R_\lambda = -\frac{c}{2} JY + \frac{1}{2} (\mathscr{L}_{R_\lambda} J) JY.$$

It seems to us that this constant c is related to the way how one lifts triad connection to a canonical connection on the symplectization. Recall if one starts from a Liouville manifold M with cylindrical end, only the contact structure ξ is naturally induced from the Liouville structure, but not the contact form, on its ideal boundary $\partial_\infty M$.

2. We suspect that Proposition 9.9 will be important to make the Kuranishi structure constructed in [KO] compatible with Kuranishi structure on the moduli spaces of pseudoholomorphic curves on noncompact symplectic manifolds M of contact-type boundary such as on Liouville manifolds. In this way, we conjecture existence of an A_∞-type functor

$$D^\pi \mathscr{W}(M)/D^\pi \mathscr{F}(M) \to D^\pi \mathfrak{Leg}(\partial_\infty M):$$

Here the codomain $D^\pi \mathfrak{Leg}(Q,\xi)$ is the derived category of the Fukaya-type category $\mathfrak{Leg}(Q) = \mathfrak{Leg}(Q,\xi)$ generated by Legendrian submanifolds that will be constructed in [KO] and [Ohc]. For the domain, $D^\pi \mathscr{W}(M)$ and $D^\pi \mathscr{F}(M)$ are the derived wrapped Fukaya category and the derived compact Fukaya category of the Liouville manifold M respectively. See [BJK] for the description of the quotient category associated to a pluming space in terms of a cluster category associated. According thereto, Ganatra-Gao-Venkatesh relate this quotient to a derived Rabinowitz Fukaya category.

10 Contact instantons and pseudoholomorphic curves on symplectization

Denote by $(\dot\Sigma, j)$ a punctured Riemann surface (including the case of closed Riemann surfaces without punctures).

10.1 Analysis of pseudoholomorphic curves on symplectization in the literature

We would like to make it clear that the analytical results themselves we describe on the symplectization in the present article are mostly known and established by Hofer-Wysocki-Zehnder's in a series of papers in [HWZ96a] - [HWZ99] in 3 dimension and some in higher dimensions by Bourgeois [Bou02] and Siefring [Sie08, Sie11]. Wendl's book manuscript [Wen] also describes the analysis in general dimension and Cant's thesis [Can22] explains the relative case in general dimension.

To highlight the main differences between the above and Wang and the first author's analysis of contact instantons [OW18a, OW18b], [Oh21a], we start here with quoting a few sample statements of the main results from [HWZ96a]-[HWZ99], which are propagated to other later literature.

The following is the prototype of the statements made in [HWZ96a] - [HWZ99] on the description of finer asymptotics of pseudoholomorphic curves near punctures, and repeated in other later literature on the pseudoholomorphic curves on symplectization.

10.1.1 Choice of special coordinates

Let γ be a closed Reeb orbit with action $T \neq 0$.

Hofer-Wysocki-Zehnder first take special coordinates (a, ϑ, z) in an S^1*-invariant neighborhood* $W \subset Q$ of the given Reeb orbit $x(T\cdot) \in C^\infty(S^1, Q)$ in such a way that

1. Image $v \subset W$ for all s large enough,
2. They choose the coordinates $(\theta, x, y) \in S^1 \times \mathbb{R}^2$ by considering contact diffeomorphism onto its image contained in $(S^1 \times \mathbb{R}^2, f \cdot \lambda_0)$ to (Q, λ) where
   - the periodic solution corresponds to $S^1 \times \{0\}$,
   - f a positive function,
   - $f \cdot \lambda_0$ is a contact form with $\lambda_0 = d\theta + x dy$ the standard contact form on $S^1 \times \mathbb{R}^2$.
   - They lift the map v to $\widetilde{v}$ defined on the universal covering space $[s_0, \infty) \times \mathbb{R}$ of $[s_0, \infty) \times S^1$.

With these preparations, the authors therefrom start with a finite energy cylinder $\widetilde{v} : \mathbb{R} \times S^1 \to Q \times \mathbb{R}$ with $\widetilde{v} = (v, a)$ for a given contact manifold (Q, λ), and derive the following estimates.

Theorem 10.1 (Theorem 2.8 [HWZ96b], Asymptotic behavior of nondegenerate finite energy planes). *Assume the functions* $(a, u); [s_0, \infty) \times \mathbb{R} \to \mathbb{R}^4$ *meet the above conditions. Then either*

1. There exists $c \in \mathbb{R}$ *such that*

$$(a(s,t), \vartheta(s,t), z(s,t)) = (Ts + c, kt, 0)$$

or

2. There are constants $c \in \mathbb{R}$, $d > 0$ *and* $M_\alpha > 0$ *for all* $\alpha = (\alpha_1, \alpha_2) \in \mathbb{N} \times \mathbb{N}$ *such that*

$$|\partial^\alpha [a(s,t) - Ts - c]| \leq M_\alpha e^{-d \cdot s}$$
$$|\partial^\alpha [\vartheta(s,t) - kt]| \leq M_\alpha e^{-d \cdot s}$$

for all $s \geq s_0$, $t \in \mathbb{R}$. *Moreover*

$$z(s,t) = e^{\int_{s_0}^s \gamma(\tau)\, d\tau} [e(t) + r(s,t)].$$

Here $e \neq 0$ is an eigenvector of the self-adjoint operator A_∞ corresponding to a negative eigenvalue $\lambda < 0$ and $\gamma : [s_0, \infty) \to \mathbb{R}$ is a smooth function satisfying $\gamma(s) \to \lambda$ as $\lambda \to \infty$. In particular $e(t) \neq 0$ pointwise and the remainder $r(s,t)$ satisfies $\partial^\alpha r(s,t) \to 0$ for all derivatives $\alpha = (\alpha_1, \alpha_2)$, uniformly in $t \in \mathbb{R}$.

10.1.2 Local adjustment of J along the Reeb orbits

Next, they put some constraints on the almost complex structure in $W \times \mathbb{R} \subset Q \times \mathbb{R}$ which has troubled the first author much: Along the way, they *change the given almost complex structure* on W *in the way that depends on the given Reeb orbit* as follows. Consider the symplectic inner product $d\lambda$ on the $S^1 \times \mathbb{R}^2$ family of 2-dimensional contact plane $\xi_m = \ker \lambda_m$ given by

$$\xi_m \subset \mathbb{R}^3 \cong T_m(S^1 \times \mathbb{R}^2)$$

where $\xi_m = \text{span}\langle e_1, e_2 \rangle$ with

$$e_1 = (0,1,0), \quad e_2 = (-x_1, 0, 1).$$

Then the 2×2 matrix $\Omega = \Omega(\theta, x, y)$ associated to $d\lambda$ on ξ_m is given by

$$\Omega = fJ_0, \quad J_0 = \begin{pmatrix} 0 & -1 \\ 1 & 0 \end{pmatrix}$$

where J_0 is the standard complex structure on $\mathbb{R}^2$. Then the almost complex structure $j_m : \xi_m \to \xi_m$ is the pull-back of a standard λ-adapted CR almost complex structure on Q restricted to its contact distributions. Then they put a 'very special' almost complex structure [HWZ96b, In the middle of p.351] denoted by $\widetilde{J}$.

10.1.3 Weak points of the aforementioned coordinate approach

This usage of special coordinates and adjustment on J along the Reeb orbits *depending on the Reeb orbits* prevents one from enabling to study the J-dependence on the generic properties of asymptotic behaviors of pseudoholomorphic curves, e.g., about the generic properties of eigenvalues of the asymptotic operators in terms of the choice of J. In this regard, Siefring [Sie08, Sie11] studied an operator that he calls the asymptotic operator in the coordinate-free way but did not develop its spectral perturbation theory under the change of CR almost complex structures J.

The first author's attempt to find a more canonical way of doing the aforementioned asymptotic study of Reeb orbits, especially that of doing those practiced in the higher dimensional Morse-Bott case given in [Bou02], had motivated Wang and him to pursue the current tensorial approach at the time of writing [OW14, OW18a, OW18b] in the beginning of 2010's. Furthermore our tensorial approach also helps the understanding of the background geometry of contact triads

(Q,λ,J) and the symplectization irrespective of the analytic study of pseudoholomorphic curves as described in Section 9.

10.2 Contact Cauchy-Riemann maps

The following definition is introduced in [OW18a].

Definition 10.2 (Contact Cauchy-Riemann map). A smooth map $w : \dot\Sigma \to Q$ is called a *contact Cauchy-Riemann map* (with respect to the contact triad (Q,λ,J), if w satisfies the following Cauchy–Riemann equation

$$\overline{\partial}^\pi w := \overline{\partial}^\pi_{j,J} w := \frac{1}{2}(\pi dw + J\pi dw \circ j) = 0.$$

Recall that for a fixed smooth map $w : \dot\Sigma \to Q$, the triple

$$(w^*\xi, w^*J, w^*g_\xi)$$

becomes a Hermitian vector bundle over the punctured Riemann surface $\dot\Sigma$. This introduces a Hermitian bundle structure on $Hom(T\dot\Sigma, w^*\xi) \cong T^*\dot\Sigma \otimes w^*\xi$ over $\dot\Sigma$, with inner product given by

$$\langle \alpha \otimes \zeta, \beta \otimes \eta \rangle = h(\alpha,\beta) g_\xi(\zeta,\eta),$$

where $\alpha,\beta \in \Omega^1(\dot\Sigma)$, $\zeta,\eta \in \Gamma(w^*\xi)$, and h is the Kähler metric on the punctured Riemann surface $(\dot\Sigma, j)$.

Let ∇^π be the contact Hermitian connection. Combining the pulling-back of this connection and the Levi–Civita connection of the Riemann surface, we get a Hermitian connection for the bundle $T^*\dot\Sigma \otimes w^*\xi \to \dot\Sigma$, which we still denote by ∇^π by a slight abuse of notation. *This is the setting where we apply the harmonic theory and Weitzenböck formulae to study the global a priori $W^{1,2}$-estimate of $d^\pi w$ mentioned in the introduction of the present survey:* The smooth map w has an associated π-harmonic energy density, the function $e^\pi(w) : \dot\Sigma \to \mathbb{R}$, defined by

$$e^\pi(w)(z) := |d^\pi w|^2(z).$$

(Here we use $|\cdot|$ to denote the norm from $\langle\cdot,\cdot\rangle$ which should be clear from the context.)

Similar to standard Cauchy–Riemann maps for almost Hermitian manifolds (i.e., pseudo-holomorphic curves), we have the following whose proofs are straightforward and so omitted. (See [OW18a, Lemma 3.2] for the proofs.)

Lemma 10.3. *Fix a Kähler metric h on $(\dot\Sigma, j)$, and consider a smooth map $w : \dot\Sigma \to Q$. Then we have the following equations*

1. $e^\pi(w) := |d^\pi w|^2 = |\partial^\pi w|^2 + |\overline{\partial}^\pi w|^2$;

2. $2w^*d\lambda = (-|\overline{\partial}^\pi w|^2 + |\partial^\pi w|^2)\, dA$ *where* dA *is the area form of the metric* h *on* $\dot\Sigma$;
3. $w^*\lambda \wedge w^*\lambda \circ j = -|w^*\lambda|^2 dA$.

As a consequence, if w *is a contact Cauchy-Riemann map, i.e.,* $\overline{\partial}^\pi w = 0$, *then*

$$|d^\pi w|^2 = |\partial^\pi w|^2 \quad \textit{and} \quad w^*d\lambda = \frac{1}{2}|d^\pi w|^2 dA. \tag{1}$$

However the contact Cauchy-Riemann equation itself $\overline{\partial}^\pi w = 0$ does not form an elliptic system because it is degenerate along the Reeb direction: Note that the rank of w^*TQ has $2n+1$ while that of $w^*\xi \otimes \Lambda^{0,1}(\Sigma)$ is $2n$. Therefore to develop suitable deformation theory and a priori estimates, one needs to lift the equation to an elliptic system by incorporating the data of the Reeb direction. In hindsight, the pseudoholomorphic curve system of the pair (a, w) is one of many possible such liftings by introducing an auxiliary variable a, *when the one-form* $w^*\lambda \circ j$ *is exact.* Hofer [Hof93] did this by lifting the equation to the symplectization $Q \times \mathbb{R}$ and considering the pull-back function $f := s \circ w$ of the $\mathbb{R}$-coordinate function s of $Q \times \mathbb{R}$. By doing so, he *added one more variable* to the equation $\overline{\partial}^\pi w = 0$ while *adding 2 more equations* $w^*\lambda \circ j = df$ which becomes Gromov's pseudoholomorphic curve system on the product $Q \times \mathbb{R}$.

We now introduce two other possible elliptic liftings of Cauchy-Riemann maps.

10.3 Gauged sigma model lifting of contact Cauchy-Riemann map

We first recall that any contact manifold (Q, ξ), whether it is coorientable or not, carries a natural real line bundle $TQ/\xi \to Q$. This bundle is trivial when ξ is coorientable. By complexifying the line bundle, we look for a lifting of the Cauchy-Riemann map equation by coupling a section of complex line bundle over Q

$$\mathscr{L}_\lambda \to Q \tag{2}$$

whose fiber at $q \in Q$ is given by

$$\mathscr{L}_{\lambda,q} = \mathbb{R}_{\lambda,q} \otimes \mathbb{C}$$

where $\mathbb{R}_\lambda \to Q$ is the trivial real line bundle whose fiber at q is given by

$$\mathbb{R}_{\lambda,q} = \mathbb{R}\{R_\lambda(q)\}.$$

Now let $w : \Sigma \to Q$ be a smooth map where $\dot\Sigma$ is either closed or a punctured Riemann surface, and χ be a section of the pull-back bundle $w^*\mathscr{L}_\lambda$.

Definition 10.4. We call a triple (w, j, χ) consisting of a complex structure j on Σ, $w : \Sigma \to Q$ and a $\mathbb{C}$-valued one-form χ a *gauged contact instanton* if they satisfy

$$\begin{cases} \overline{\partial}^\pi w = 0 \\ \overline{\partial}\chi = 0, \quad \operatorname{Im}\chi = w^*\lambda. \end{cases} \tag{3}$$

This system is a coupled system of the contact Cauchy-Riemann map equation and the well-known Riemann-Hilbert problem of the type which solves the real part in terms of the imaginary part of holomorphic functions in complex variable theory.

10.4 Contact instanton lifting of contact Cauchy-Riemann map

By augmenting the closedness condition $d(w^*\lambda \circ j) = 0$ to Contact Cauchy-Riemann map equation $\overline{\partial}^\pi w = 0$, we arrive at an elliptic system (4) which is of our main interest in the present survey. The current contact instanton map system

$$\overline{\partial}^\pi w = 0, \quad d(w^*\lambda \circ j) = 0 \tag{4}$$

is another such an elliptic lifting which is more natural in some respect in that it does not introduce any additional variable and keeps the original 'bulk', the contact manifold Q.

To illustrate the effect of the closedness condition on the behavior of contact instantons, we look at them on *closed* Riemann surface and prove the following classification result. The following proposition is stated by Abbas as a part of [Abb11, Proposition 1.4]. We refer readers [OW18a, Proposition 3.4] for another proof which is somewhat different from Abbas' in [Abb11].

Proposition 10.5 (Proposition 1.4, [Abb11])**.** *Assume $w : \Sigma \to Q$ is a smooth contact instanton from a closed Riemann surface. Then*

1. *If $g(\Sigma) = 0$, w is a constant map;*
2. *If $g(\Sigma) \geq 1$, w is either a constant or the locus of its image is a* closed *Reeb orbit.*

Part III
A priori estimates

11 Weitzenböck formulae

In this section, we use the contact triad connection, first to derive Weitzenböck-type formula associated to the π-harmonic energy *for contact Cauchy–Riemann maps*, and then to derive another formula associated to the full harmonic energy density *for the contact instantons*. The contact triad connection fits well for this purpose which will be seen clearly in this section.

11.1 Weitzenböck formulae for contact Cauchy-Riemann maps

We start with by looking at the (Hodge) Laplacian of the π-harmonic energy density of an arbitrary smooth map $w : \dot\Sigma \to M$, i.e., in the *off-shell level* in physics terminology. As the first step, we apply the standard Weitzenböck formula to the connection ∇^π on $T^*\dot\Sigma \otimes w^*\xi$ that is induced by the pull-back connection on bundle $w^*\xi$ and the Levi–Civita connection on $T\dot\Sigma$, and obtain the following formula

$$-\frac{1}{2}\Delta e^\pi(w) = |\nabla^\pi(d^\pi w)|^2 - \langle \Delta^{\nabla^\pi} d^\pi w, d^\pi w\rangle + K \cdot |d^\pi w|^2 + \langle \mathrm{Ric}^{\nabla^\pi}(d^\pi w), d^\pi w\rangle. \tag{1}$$

Here $e^\pi(w) = |d^\pi w|^2$ is the π-harmonic energy density, K the Gaussian curvature of the Kähler manifold $(\dot\Sigma, h)$, and $\mathrm{Ric}^{\nabla^\pi}$ is the Ricci tensor of the connection ∇^π on the vector bundle $w^*\xi$. (We refer to [OW18a, Appendix A] for the he proof of (1).)

The following fundamental identity is derived in [OW18a, Lemma 4.1] to which we refer readers for its derivation. This is an analog to a similar formula [Oh15, Lemma 7.3.2] in the symplectic context.

Lemma 11.1. *Let $w : \dot\Sigma \to M$ be any smooth map. Denote by T^π the torsion tensor of ∇^π. Then as a two form with values in $w^*\xi$, $d^{\nabla^\pi}(d^\pi w)$ has the expression*

$$d^{\nabla^\pi}(d^\pi w) = T^\pi(\Pi dw, \Pi dw) + w^*\lambda \wedge \left(\frac{1}{2}(\mathscr{L}_{R_\lambda} J) J d^\pi w\right). \tag{2}$$

We now restrict the above lemma to the case of contact Cauchy–Riemann map, i.e., maps satisfying $\overline\partial^\pi w = 0$. In this case, by the property $T(Y, JY) = 0$ of the torsion T of the contact triad connection, we derive the following formula as an immediate corollary of the previous lemma.

Theorem 11.2 (Fundamental Equation; Theorem 4.2 [OW18a]). *Let w be a contact Cauchy–Riemann map, i.e., a solution of $\overline\partial^\pi w = 0$. Then*

$$d^{\nabla^\pi}(d^\pi w) = -w^*\lambda \circ j \wedge \left(\frac{1}{2}(\mathscr{L}_{R_\lambda} J) d^\pi w\right). \tag{3}$$

The following elegant expression of Fundamental Equation in any *isothermal coordinates* (x,y), i.e., one such that $z = x+iy$ provides a complex coordinate of $(\dot\Sigma, j)$ such that $h = dx^2+dy^2$, will be useful for the study of higher a priori $C^{k,\alpha}$ Hölder estimates.

Corollary 11.3 (Fundamental Equation in Isothermal Coordinates). *Let (x,y) be an isothermal coordinates. Write $\zeta := \pi\frac{\partial w}{\partial \tau}$ as a section of $w^*\xi \to Q$. Then*

$$\nabla_x^\pi \zeta + J\nabla_y^\pi \zeta - \frac{1}{2}\lambda\left(\frac{\partial w}{\partial x}\right)(\mathcal{L}_{R_\lambda}J)\zeta + \frac{1}{2}\lambda\left(\frac{\partial w}{\partial y}\right)(\mathcal{L}_{R_\lambda}J)J\zeta = 0. \tag{4}$$

Proof. We denote $\pi\frac{\partial w}{\partial x}$ by ζ and $\pi\frac{\partial w}{\partial y}$ by η. By the isothermality of the coordinate (x,y), we have $j\frac{\partial}{\partial x} = \frac{\partial}{\partial y}$. Using the (j,J)-linearity of $d^\pi w$, we derive

$$\eta = dw^\pi\left(\frac{\partial}{\partial y}\right) = dw^\pi\left(j\frac{\partial}{\partial x}\right) = Jdw^\pi\left(\frac{\partial}{\partial x}\right) = J\zeta.$$

Now we evaluate each side of (3) against $(\frac{\partial}{\partial x}, \frac{\partial}{\partial y})$. For the left hand side, we get

$$\nabla_x^\pi \eta - \nabla_y^\pi \zeta = \nabla_x^\pi J\zeta - \nabla_y^\pi \zeta = J\nabla_x^\pi \zeta - \nabla_y^\pi \zeta.$$

For the right hand side, we get

$$\begin{aligned}
&\frac{1}{2}\lambda\left(\frac{\partial w}{\partial x}\right)(\mathcal{L}_{R_\lambda}J)J\eta - \frac{1}{2}\lambda\left(\frac{\partial w}{\partial y}\right)(\mathcal{L}_{R_\lambda}J)J\zeta \\
&= -\frac{1}{2}\lambda\left(\frac{\partial w}{\partial x}\right)(\mathcal{L}_{R_\lambda}J)\zeta - \frac{1}{2}\lambda\left(\frac{\partial w}{\partial y}\right)(\mathcal{L}_{R_\lambda}J)J\zeta
\end{aligned}$$

where we use the equation $\eta = J\zeta$ for the equality. By setting them equal and applying J to the resulting equation using the fact that $\mathcal{L}_{R_\lambda}J$ anti-commutes with J, we obtain the equation. □

Remark 11.4. The fundamental equation in cylindrical (or strip-like) coordinates is nothing but the linearization equation of the contact Cauchy-Riemann equation in the direction $\frac{\partial}{\partial \tau}$. This plays an important role in the derivations of the exponential decay of the derivatives at cylindrical ends and of finer asymptotic study of the asymptotic operators. (See [OW18b, Part II].)

Now we can convert the general Weitzenböck formula (1) into the following in our case.

Proposition 11.5 (Equation (4.11) [OW18a]). *Let w be a contact Cauchy-Riemann map. Then*

$$\begin{aligned}
-\frac{1}{2}\Delta e^\pi(w) &= |\nabla^\pi(\partial^\pi w)|^2 + K|\partial^\pi w|^2 + \langle \mathrm{Ric}^{\nabla^\pi}(\partial^\pi w), \partial^\pi w\rangle \\
&\quad + \langle \delta^{\nabla^\pi}[(w^*\lambda\circ j)\wedge(\mathcal{L}_{R_\lambda}J)\partial^\pi w], \partial^\pi w\rangle.
\end{aligned} \tag{5}$$

The upshot of the equation is that the *a priori the third derivatives involving the LHS for general smooth maps can be written in terms of those involving at most the second derivatives for the contact Cauchy-Riemann maps*. This enables us to perform the bootstrap arguments to obtain the higher regularity results.

Outline of the proof. The following formula expresses $\Delta^{\nabla^\pi} d^\pi w$, which involves the third derivatives of w, in terms of the terms involving derivatives of order at most two. Here Δ^{∇^π} is the covariant harmonic Laplacian

$$\Delta^{\nabla^\pi} = \delta^{\nabla^\pi} d^{\nabla^\pi} + d^{\nabla^\pi} \delta^{\nabla^\pi}.$$

Lemma 11.6. *For any contact Cauchy–Riemann map w,*

$$\begin{aligned} -\Delta^{\nabla^\pi} d^\pi w &= \delta^{\nabla^\pi}[(w^*\lambda \circ j) \wedge (\mathscr{L}_{R_\lambda} J)\partial^\pi w] \\ &= -*\langle (\nabla^\pi(\mathscr{L}_{R_\lambda} J))\partial^\pi w, w^*\lambda\rangle \\ &\quad -*\langle (\mathscr{L}_{R_\lambda} J)\nabla^\pi \partial^\pi w, w^*\lambda\rangle - *\langle (\mathscr{L}_{R_\lambda} J)\partial^\pi w, \nabla w^*\lambda\rangle. \end{aligned}$$

Proof. The first equality immediately follows from the fundamental equation, Theorem 11.2, for contact Cauchy–Riemann maps. For the second equality, we calculate by writing

$$\delta^{\nabla^\pi}[(w^*\lambda \circ j) \wedge (\mathscr{L}_{R_\lambda} J)\partial^\pi w] = -*d^{\nabla^\pi} *[(\mathscr{L}_{R_\lambda} J)\partial^\pi w \wedge (*w^*\lambda)],$$

and then by applying the definition of the Hodge $*$ to the expression $*[(\mathscr{L}_{R_\lambda} J)\partial^\pi w \wedge (*w^*\lambda)]$, we further get

$$\begin{aligned} &\delta^{\nabla^\pi}[(w^*\lambda \circ j) \wedge (\mathscr{L}_{R_\lambda} J)\partial^\pi w] \\ &= -*d^{\nabla^\pi}\langle (\mathscr{L}_{R_\lambda} J)\partial^\pi w, w^*\lambda\rangle \\ &= -*\langle (\nabla^\pi(\mathscr{L}_{R_\lambda} J))\partial^\pi w, w^*\lambda\rangle - *\langle (\mathscr{L}_{R_\lambda} J)\nabla^\pi \partial^\pi w, w^*\lambda\rangle - *\langle (\mathscr{L}_{R_\lambda} J)\partial^\pi w, \nabla w^*\lambda\rangle. \end{aligned}$$

□

This leads us to the following formula

Corollary 11.7. *For any contact Cauchy-Riemann map w, we have*

$$\begin{aligned} -\langle \Delta^{\nabla^\pi} d^\pi w, d^\pi w\rangle &= \langle \delta^{\nabla^\pi}[(w^*\lambda \circ j) \wedge (\mathscr{L}_{R_\lambda} J)d^\pi w], d^\pi w\rangle \\ &= -\langle *\langle (\nabla^\pi(\mathscr{L}_{R_\lambda} J))\partial^\pi w, w^*\lambda\rangle, d^\pi w\rangle \\ &\quad -\langle *\langle (\mathscr{L}_{R_\lambda} J)\nabla^\pi \partial^\pi w, w^*\lambda\rangle, d^\pi w\rangle \\ &\quad -\langle *\langle (\mathscr{L}_{R_\lambda} J)\partial^\pi w, \nabla w^*\lambda\rangle, d^\pi w\rangle. \end{aligned}$$

Here in the above lemma $\langle \cdot, \cdot \rangle$ denotes the inner product induced from h, i.e., $\langle \alpha_1 \otimes \zeta, \alpha_2 \rangle := h(\alpha_1, \alpha_2)\zeta$, for any $\alpha_1, \alpha_2 \in \Omega^k(P)$ and ζ a section of E. This inner product should not be confused with the inner product of the vector bundles.

By applying δ^{∇^π} to (2) and the resulting expression of $\Delta^{\nabla^\pi}(d^\pi w)$ thereinto, we can convert the Weitzenböck formula (1) to (5). □

11.2 The case of contact instantons

Now we consider contact instantons which are Cauchy–Riemann maps satisfying $d(w^*\lambda \circ j) = 0$ in addition.

Proposition 11.8. *Let w be a contact instanton. Then*

$$-\frac{1}{2}\Delta|w^*\lambda|^2 = |\nabla w^*\lambda|^2 + K|w^*\lambda|^2 + \langle *\langle \nabla^\pi \partial^\pi w, \partial^\pi w\rangle, w^*\lambda\rangle. \tag{6}$$

Proof. In this case again by using the Bochner–Weitzenböck formula (for forms on Riemann surface), we get the following inequality

$$-\frac{1}{2}\Delta|w^*\lambda|^2 = |\nabla w^*\lambda|^2 + K|w^*\lambda|^2 - \langle \Delta(w^*\lambda), w^*\lambda\rangle. \tag{7}$$

Write

$$\Delta(w^*\lambda) = d\delta(w^*\lambda) + \delta d(w^*\lambda),$$

in which the first term vanishes since w satisfies the contact instanton equation which includes $0 = d(w^*\lambda \circ j) = -\delta(w^*\lambda)$ in addition. Then a straightforward calculation gives rise to

$$\begin{aligned}
\langle \Delta(w^*\lambda), w^*\lambda\rangle &= \langle \delta d(w^*\lambda), w^*\lambda\rangle \\
&= -\frac{1}{2}\langle *d|\partial^\pi w|^2, w^*\lambda\rangle \\
&= -\langle *\langle \nabla^\pi \partial^\pi w, \partial^\pi w\rangle, w^*\lambda\rangle.
\end{aligned}$$

Substituting this into (7) we have finished the proof. □

12 A priori $W^{2,2}$-estimates for contact instantons

In this section, we derive basic estimates for the harmonic energy density $|dw|^2$ of contact instantons w. These estimates are important for the derivation of local regularity and the ε-regularity needed for the compactification of moduli spaces in general.

We first remark that the total harmonic energy density is decomposed into

$$e(w) := |dw|^2 = e^\pi(w) + |w^*\lambda|^2 :$$

This follows from the decomposition $dw = d^\pi w + w^*\lambda \otimes R_\lambda$ and the orthogonality of the two summands with respect to the triad metric and $|R_\lambda| \equiv 1$.

Therefore we will derive the Laplacian of each summand, $|d^\pi w|^2$ and $|w^*\lambda|^2$.

12.1 Computation of $\Delta|dw|^2$ and Weitzenböck formulae

Recall the formula (3) of the Laplacian $\Delta e^\pi(w)$ from the last section. The following estimate is proved in [OW18a, Equation (5.3)].

Lemma 12.1. *For any constant $c > 0$, we have*

$$|\langle \delta^{\nabla^\pi}[(w^*\lambda \circ j) \wedge (\mathscr{L}_{R_\lambda} J)\partial^\pi w], \partial^\pi w\rangle| \\ \leq \frac{1}{2c}\left(|\nabla^\pi(\partial^\pi w)|^2 + |\nabla w^*\lambda|^2\right) + \left(c\|\mathscr{L}_{R_\lambda} J\|^2_{C^0(Q)} + \|\nabla^\pi(\mathscr{L}_{R_\lambda} J)\|_{C^0(Q)}\right)|dw|^4 \tag{1}$$

This gives rise to the following differential inequality

$$-\frac{1}{2}\Delta e^\pi(w) \leq \left(1 + \frac{1}{2c}|\nabla^\pi(d^\pi w)|^2 + \frac{1}{2c}\right) + \frac{1}{2c}\left(|\nabla w^*\lambda|^2\right) \\ +(\|K\|_{C^0(Q)} + \|\operatorname{Ric}^{\nabla^\pi}\|_{C^0(Q)})|d^\pi w|^2 \\ + \left(c\|\mathscr{L}_{R_\lambda} J\|^2_{C^0(Q)} + \|\nabla^\pi(\mathscr{L}_{R_\lambda} J)\|_{C^0(Q)}\right)|dw|^4 \tag{2}$$

Similarly as in the previous estimates for the Laplacian term of $\partial^\pi w$, we can bound

$$|-\langle \Delta(w^*\lambda), w^*\lambda\rangle| = |\langle *\langle \nabla^\pi\partial^\pi w, \partial^\pi w\rangle, w^*\lambda\rangle| \\ \leq |\nabla^\pi\partial^\pi w||dw|^2 \\ \leq \frac{1}{2c}|\nabla^\pi\partial^\pi w|^2 + \frac{c}{2}|dw|^4. \tag{3}$$

We add (2) and (7), and then apply the estimates (1) and (3) respectively. This yields the following differential inequality for the total energy density

$$-\frac{1}{2}\Delta e(w) \\ \geq \left(1 - \frac{1}{c}\right)|\nabla^\pi(\partial^\pi w)|^2 + \left(1 - \frac{1}{2c}\right)|\nabla w^*\lambda|^2 \\ - \left(c\|\mathscr{L}_{R_\lambda} J\|^2_{C^0(Q)} + \|\nabla^\pi(\mathscr{L}_{R_\lambda} J)\|_{C^0(Q)} + \frac{c}{2} + \|\operatorname{Ric}\|_{C^0(Q)}\right) e(w)^2 + Ke(w) \tag{4}$$

$$\geq -\left(c\|\mathscr{L}_{R_\lambda} J\|^2_{C^0(Q)} + \|\nabla^\pi(\mathscr{L}_{R_\lambda} J)\|_{C^0(Q)} + \frac{c}{2} + \|\operatorname{Ric}\|_{C^0(Q)}\right) e(w)^2 + Ke(w),$$

for any $c > 1$. We fix $c = 2$ and get the following

Theorem 12.2 (Theorem 5.1 [OW18a]). *For a contact instanton w, we have the following total energy density estimate*

$$\Delta e(w) \leq Ce(w)^2 + \|K\|_{L^\infty(\Sigma)} e(w), \tag{5}$$

where

$$C = 2\|\mathscr{L}_{R_\lambda} J\|^2_{C^0(Q)} + \|\nabla^\pi(\mathscr{L}_{R_\lambda} J)\|_{C^0(Q)} + \|\,\mathrm{Ric}\,\|_{C^0(Q)} + 1$$

which is a positive constant independent of w.

An immediate corollary of (5) is the following density estimate which is derived by the standard argument from [Sch84]. (Also see the proof of [Oh15, Theorem 8.1.3] given in the context of pseudoholomorphic curves.) It is in turn a consequence of an application of the mean value inequality of Morrey (See [GT70, Problems 4.5 in p.67] for the relevant extension of the mean-value inequality for the Poisson equation.)

Corollary 12.3 (ε-regularity and interior density estimate; Corollary 5.2 [OW18a]). *There exist constants* C, ε_0 *and* $r_0 > 0$, *depending only on* J *and the Hermitian metric* h *on* $\dot\Sigma$, *such that for any* C^1 *contact instanton* $w : \dot\Sigma \to Q$ *with*

$$E(r_0) := \frac{1}{2}\int_{D(r_0)} |dw|^2 \le \varepsilon_0,$$

and discs $D(2r) \subset \mathrm{Int}\,\Sigma$ *with* $0 < 2r \le r_0$, w *satisfies*

$$\max_{\sigma\in(0,r]} \left(\sigma^2 \sup_{D(r-\sigma)} e(w)\right) \le CE(r) \tag{6}$$

for all $0 < r \le r_0$. *In particular, letting* $\sigma = r/2$, *we obtain*

$$\sup_{D(r/2)} |dw|^2 \le \frac{4CE(r)}{r^2} \tag{7}$$

for all $r \le r_0$.

Now we rewrite (4) into

$$\begin{aligned} &\left(1-\frac{1}{c}\right)|\nabla^\pi(\partial^\pi w)|^2 + \left(1-\frac{1}{2c}\right)|\nabla w^*\lambda|^2 \\ &\le -\frac{1}{2}\Delta e(w) - Ke(w) \\ &\quad + \left(c\|\mathscr{L}_{R_\lambda} J\|^2_{C^0(Q)} + \|\nabla^\pi(\mathscr{L}_{R_\lambda} J)\|_{C^0(Q)} + \frac{c}{2} + \|\,\mathrm{Ric}\,\|_{C^0(Q)}\right) e^2 \end{aligned} \tag{8}$$

Taking $c = 2$, we obtain the following differential inequality, which is to used derive coercive elliptic estimates for contact instantons.

Proposition 12.4 (Equation (5.13) [OW18a]).

$$|\nabla(dw)|^2 \le C_1|dw|^4 - 4K|dw|^2 - 2\Delta e(w) \tag{9}$$

where

$$C_1 := 9\|\mathscr{L}_{R_\lambda} J\|^2_{C^0(Q)} + 4\|\nabla^\pi(\mathscr{L}_{R_\lambda} J)\|_{C^0(Q)} + 4\|\,\mathrm{Ric}\,\|_{C^0(Q)} + 4$$

denotes a constant.

Then by multiplying a cut-off function and doing integration by parts, we obtain the following local $W^{2,2}$ estimate.

Proposition 12.5 (Proposition 5.3 [OW18a]). *For any pair of open domains D_1 and D_2 in $\dot{\Sigma}$ such that $\overline{D}_1 \subset \mathrm{Int}(D_2)$,*

$$\|\nabla(dw)\|^2_{L^2(D_1)} \leq C_1(D_1,D_2)\|dw\|^2_{L^2(D_2)} + C_2(D_1,D_2)\|dw\|^4_{L^4(D_2)}$$

for any contact instanton w, where $C_1(D_1,D_2)$, $C_2(D_1,D_2)$ are some constants which depend on D_1, D_2 and (Q,λ,J), but are independent of w.

12.2 Local boundary $W^{2,2}$-estimate

Now let us consider the contact instantons with Legendrian boundary condition. Let $\vec{R} = (R_0,\cdots,R_k)$ be a $(k+1)$-component Legendrian link.

Consider the equation

$$\begin{cases} \overline{\partial}^\pi w = 0, \; d(w^*\lambda \circ j) = 0 \\ w(\overline{z_i z_{i+1}}) \subset R_i \end{cases} \tag{10}$$

for a smooth map $w : (\dot{\Sigma}, \partial\dot{\Sigma}) \to (Q, \vec{R})$. We will simplify writing of the boundary condition as

$$w(\partial\dot{\Sigma}) \subset \vec{R}$$

in a single equation.

The following local boundary a priori estimate is established in [Oh21a], [OY22].

Theorem 12.6. *Let $w : \mathbb{R}\times[0,1] \to Q$ satisfy* (4). *Then for any relatively compact domains D_1 and D_2 of the semi-disc type in $\dot{\Sigma}$ such that $\overline{D_1} \subset D_2$, we have*

$$\|dw\|^2_{W^{1,2}(D_1)} \leq C_1\|dw\|^2_{L^2(D_2)} + C_2\|dw\|^4_{L^4(D_2)} \tag{11}$$

where C_1, C_2 are some constants which depend only on D_1, D_2 and (Q,λ,J) and C_3 is a constant which also depends on R_i with $w(\partial D_2) \subset R_i$ as well.

Leaving the details of the proof to [OY22], we just outline the strategy of the proof here (See [Oh15, Section 8.2 & 8.3] for the same strategy used for pseudo-holomorphic curves with Lagrangian boundary condition.):

1. As the first step, we utilize the contact triad connection ∇ for the study of boundary value problem to derive the following differential inequality

$$\|dw\|^2_{W^{1,2}(D_1)} \leq C_1\|dw\|^2_{L^2(D_2)} + C_2\|dw\|^4_{L^4(D_2)} + \int_{\partial D_2} |C(\partial D_2)| \tag{12}$$

where we have

$$C(\partial D_2) := -8\left\langle B\left(\frac{\partial w}{\partial x},\frac{\partial w}{\partial x}\right),\frac{\partial w}{\partial y}\right\rangle \tag{13}$$

for the second fundamental form $B = B^\nabla$ of ∇ in the isothermal coordinate $z = x+iy$ adapted to $\partial\dot{\Sigma}\cap D_2$.

2. Once we derive (12), noting that Legendrian boundary condition $\vec{R} = (R_0,\cdots,R_k)$ for the contact instanton is automatically a free boundary value problem, i.e., a solution w of

$$\frac{\partial w}{\partial \nu} \perp TR_i,$$

one can use the Levi-Civita connection $\nabla^{\rm LC}$ of a metric for which each component of Legendrian link $\vec{R}$ becomes *totally geodesic* (i.e., $B^{\nabla^{\rm LC}} = 0$) which will eliminate the boundary contribution appearing above in (12). Then recalling the standard fact that $\nabla = \nabla^{\rm LC} + P$ for a $(2,1)$ tensor P, we can convert the inequality (12) into (11) with some adjustment of constants C_1, C_2. (See [Oh15, Section 8.3] for such detail.)

The following lemma is an important ingredient entering in the proof.

Lemma 12.7. *Let $e = |dw|^2$ be the total harmonic energy density function. Then we have*

$$*de = -4\left\langle B\left(\frac{\partial w}{\partial x},\frac{\partial w}{\partial x}\right),\frac{\partial w}{\partial y}\right\rangle \tag{14}$$

on ∂D.

13 $C^{k,\delta}$ coercive estimates for $k \geq 1$: alternating boot-strap

Once we have established $W^{2,2}$ estimate, we can proceed with the $W^{k+2,2}$ estimate $k \geq 1$ inductively as in [OW18a, Section 5.2]. Because of the effect of the Legendrian boundary condition on the higher derivative estimate, it is not quite straightforward to boot-strap using the Sobolev norms but is better to work with $C^{k,\delta}$ Hölder norms as in [Oh21a], [OY22].

Here we provide an outline of the main steps of the *alternating boot-strap arguments* from [OY22, Section 4] to establish higher $C^{k,\delta}$ regularity results.

We start with the fundamental equation in isothermal coordinates $z = x+iy$

$$\nabla_x^\pi \zeta + J\nabla_y^\pi \zeta + \frac{1}{2}\lambda\left(\frac{\partial w}{\partial y}\right)(\mathscr{L}_{R_\lambda}J)\zeta - \frac{1}{2}\lambda\left(\frac{\partial w}{\partial x}\right)(\mathscr{L}_{R_\lambda}J)J\zeta = 0.$$

(See Corollary 11.3.) By writing

$$\overline{\nabla}^\pi := \nabla^{\pi(0,1)} = \frac{\nabla^\pi + J\nabla^\pi_{j(\cdot)}}{2}$$

which is the anti-complex linear part of ∇^π, and the linear operator

$$P_{w^*\lambda}(\zeta) := \frac{1}{4}\lambda\left(\frac{\partial w}{\partial y}\right)(\mathscr{L}_{R_\lambda}J)\zeta - \frac{1}{4}\lambda\left(\frac{\partial w}{\partial x}\right)(\mathscr{L}_{R_\lambda}J)J\zeta,$$

the equation becomes

$$\overline{\nabla}^\pi \zeta + P_{w^*\lambda}(\zeta) = 0 \tag{1}$$

which is a linear first-order PDE for ζ of Cauchy-Riemann type once $w^*\lambda$ is given. By the Sobolev embedding, we have $W^{2,2} \subset C^{0,\delta}$ for $0 \leq \delta < 1/2$. Therefore we start from $C^{0,\delta}$ bound with $0 < \delta < 1/2$ and will inductively bootstrap it to get $C^{k,\delta}$ bounds for $k \geq 1$.

WLOG, we assume that $D_2 \subset \dot{\Sigma}$ is a semi-disc with $\partial D_2 \subset \partial\dot{\Sigma}$ and equipped with an isothermal coordinates (x,y) such that

$$D_2 = \{(x,y) \mid |x|^2 + |y|^2 < \varepsilon, y \geq 0\}$$

for some $\varepsilon > 0$ and so $\partial D_2 \subset \{(x,y) \in D_2 \mid y = 0\}$. Assume $D_1 \subset D_2$ is the semi-disc with radius $\varepsilon/2$. We denote $\zeta = \pi\frac{\partial w}{\partial x}$, $\eta = \pi\frac{\partial w}{\partial y}$ as in [OW18a], and consider the complex-valued function

$$\alpha(x,y) = \lambda\left(\frac{\partial w}{\partial y}\right) + \sqrt{-1}\left(\lambda\left(\frac{\partial w}{\partial x}\right)\right) \tag{2}$$

as in [OW18b, Subsection 11.5].

Remark 13.1. In [OW18b, Subsection 11.5], the global isothermal coordinate (τ,t) of $[0,\infty) \times S^1$ with circle-valued flat coordinate t is used and the function α defined by

$$\alpha(x,y) = \lambda\left(\frac{\partial w}{\partial y}\right) - T + \sqrt{-1}\left(\lambda\left(\frac{\partial w}{\partial x}\right)\right)$$

is used for the exponential convergence result. See the displayed formula right above in Lemma 11.19 of [OW18b, Subsection 11.5].

We note that since w satisfies the Legendrian boundary condition $w(\partial\dot{\Sigma}) \subset \vec{R}$, we have

$$\lambda\left(\frac{\partial w}{\partial x}\right) = 0 \tag{3}$$

on ∂D_2. The following formula is crucially used in [OW18b, Subsection 11.5] for the exponential decay result, and in [Oh21a, OY22] for the alternating boot-strap argument for the higher regularity but without detailed proof. Because the proof well demonstrates how important closedness of $w^*\lambda \circ j$ and the equality (1) are in the study of a priori elliptic estimates of contact instanton w and also because how the Legendrian boundary condition interacts with the equation, we give the full details of its proof here.

Proposition 13.2. *The complex-valued function α satisfies the equations*

$$\begin{cases} \overline{\partial}\alpha = \nu, & \nu = \frac{1}{2}|\zeta|^2 \\ \alpha(z) \in \mathbb{R} & \text{for } z \in \partial D_2 \end{cases} \tag{4}$$

Proof. For the equation, we first recall

$$d(w^*\lambda \circ j) = 0, \quad dw^*\lambda = \frac{1}{2}|d^\pi w|^2\, dA$$

where the second equality is from (1). In isothermal coordinate (x,y), we have $dA = dx \wedge dy$ and hence we have

$$*(dw^*\lambda) = \frac{1}{2}|d^\pi w|^2.$$

By the isothermality of (x,y), $\{\frac{\partial}{\partial x}, \frac{\partial}{\partial y}\}$ is an orthonormal frame of $T\dot{\Sigma}$ and hence

$$|d^\pi w|^2 = \left|\pi\frac{\partial w}{\partial x}\right|^2 + \left|\pi\frac{\partial w}{\partial y}\right|^2 = 2\left|\pi\frac{\partial w}{\partial x}\right|^2$$

where the last equality follows since $0 = \overline{\partial}^\pi w = (\overline{\partial} w)^\pi$ for the Cauchy-Riemann operator $\overline{\partial}$ for the standard complex structure $J_0 = \sqrt{-1}$.

Therefore if we write $\zeta = \pi\frac{\partial w}{\partial x}$,

$$*d(w^*\lambda) = |\zeta|^2.$$

On the other hand, using this identity, the isothermality of the coordinate again (x,y) and the equation $d(w^*\lambda \circ j) = 0$, we derive

$$\begin{aligned} \overline{\partial}\alpha &= \frac{1}{2}\left(\frac{\partial}{\partial x} + i\frac{\partial}{\partial y}\right)\left(\lambda\left(\frac{\partial w}{\partial y}\right) + i\left(\lambda\left(\frac{\partial w}{\partial x}\right)\right)\right) \\ &= \frac{1}{2}\left(\frac{\partial}{\partial x}\left(\lambda\left(\frac{\partial w}{\partial y}\right)\right) - \frac{1}{2}\frac{\partial}{\partial y}\left(\lambda\left(\frac{\partial w}{\partial x}\right)\right)\right) \\ &\quad + \frac{i}{2}\left(\frac{\partial}{\partial x}\left(\lambda\left(\frac{\partial w}{\partial x}\right)\right) + \frac{1}{2}\frac{\partial}{\partial y}\left(\lambda\left(\frac{\partial w}{\partial y}\right)\right)\right) \\ &= \frac{1}{2}(d(w^*\lambda) - id(w^*\lambda \circ j))\left(\frac{\partial}{\partial x}, \frac{\partial}{\partial y}\right) \\ &= \frac{1}{2}d(w^*\lambda)\left(\frac{\partial}{\partial x}, \frac{\partial}{\partial y}\right) = \frac{1}{2} * d(w^*\lambda). \end{aligned}$$

Combining the two, we have derived $\overline{\partial}\alpha = \frac{1}{2}|\zeta|^2$. This finishes the proof of the equation.

The boundary condition $\alpha(z) \in \mathbb{R}$ for $z \in \partial D_2$ follows from the Legendrian boundary condition: We have

$$\operatorname{Im} \alpha(z) = \lambda\left(\frac{\partial w}{\partial x}\right) = 0$$

since the vector $\frac{\partial w}{\partial x}$ is tangent to the given Legendrian submanifold by the adaptedness $\frac{\partial}{\partial x} \in \partial D_2$ of the isothermal coordinate (x, y) to the boundary ∂D_2. This finishes the proof. □

Then we arrive at the following system of equations for the pair (ζ, α)

$$\begin{cases} \nabla_x^\pi \zeta + J\nabla_y^\pi \zeta + \frac{1}{2}\lambda(\frac{\partial w}{\partial y})(\mathscr{L}_{R_\lambda} J)\zeta - \frac{1}{2}\lambda(\frac{\partial w}{\partial x})(\mathscr{L}_{R_\lambda} J)J\zeta = 0 \\ \zeta(z) \in TR_i \quad \text{for } z \in \partial D_2 \end{cases} \tag{5}$$

for some $i = 0, \ldots, k$, and

$$\begin{cases} \bar{\partial}\alpha = \frac{1}{2}|\zeta|^2 \\ \alpha(z) \in \mathbb{R} \quad \text{for } z \in \partial D_2. \end{cases} \tag{6}$$

These two equations form a nonlinear elliptic system for (ζ, α) which are coupled: α is fed into (5) through its coefficients and then ζ provides the input for the right hand side of the equation (6) and then back and forth. Using this structure of coupling, we can derive the higher derivative estimates by alternating boot strap arguments between ζ and α which is now in order.

Theorem 13.3 (Theorem 1.4, [OY22]). *Assume $k \geq 1$ and $0 < \delta < 1/2$. Let w be a contact instanton satisfying* (4). *Then for any pair of domains $D_1 \subset D_2 \subset \dot{\Sigma}$ such that $\overline{D_1} \subset D_2$, we have*

$$\|dw\|_{C^{k,\delta}(D_1)} \leq C_\delta(\|dw\|_{W^{1,2}(D_2)})$$

for some positive function $C_\delta = C_\delta(r)$ that is continuous at $r = 0$ which depends on J, λ and D_1, D_2 but independent of w.

The rest of the proof of this theorem given in [OY22] consist of the following steps:

1. Start of alternating boot-strap: $W^{1,2}$-estimate for dw.
2. $C^{1,\delta}$-estimate for $w^*\lambda = f\,dx + g\,dy$.
3. $C^{1,\delta}$-estimate for $d^\pi w$.
4. $C^{2,\delta}$-estimate for $w^*\lambda$.
5. $C^{2,\delta}$-estimate for $d^\pi w$.
6. Wrap-up of the alternating boot-strap argument: We repeat the above alternating boot strap arguments between ζ and α back and forth by taking the differential with respect to ∇_x^{LC} to inductively derive the $C^{k,\delta}$-estimates both for ζ and α in terms of $\|\zeta\|_{L^4(D_2)}$ and $\|\alpha\|_{L^4(D_2)}$ which is equivalent to considering the full $\|dw\|_{L^4}$. This completes the proof of Theorem 13.3.

We refer readers to [OY22] for complete details of this for the alternating boot strap arguments which go back and forth between ζ and α.

Part IV
Asymptotic convergence and charge vanishing

In this part, we study the most basic asymptotic behavior of finite energy contact instantons near the punctures. We divide the study into two cases separately, one the case of closed strings, i.e., on the cylinderical region $[0,\infty)\times S^1$ near interiors punctures, and the other the case of open strings, i.e., on the strip-like region $[0,\infty)\times[0,1]$ with Legendrian boundary condition of the pair (R,R').

14 Generic nondegeneracy of Reeb orbits and of Reeb chords

Nondegeneracy of closed Reeb orbits or of Reeb chords is fundamental in the Fredholm property of the linearized operator of contact instanton equations as well as of pseudoholomorphic curves on symplectization. The main conclusion of the present subsection will be the statement on the generic nondegeneracy under the perturbation of contact forms or of Legendrian boundary conditions. The study of the case of closed Reeb orbits is standard (see [ABW10]), and our exposition on the results for the case of open strings is based on [Ohb, Appendix B].

14.1 *The case of closed Reeb orbits*

Let γ be a closed Reeb orbit of period $T>0$. In other words, $\gamma:\mathbb{R}\to Q$ is a solution of $\dot x = R_\lambda(x)$ satisfying $\gamma(T)=\gamma(0)$. By definition, we can write $\gamma(T)=\phi^T_{R_\lambda}(\gamma(0))$ for the Reeb flow $\phi^T=\phi^T_{R_\lambda}$ of the Reeb vector field R_λ. Therefore if γ is a closed orbit, then we have

$$\phi^T_{R_\lambda}(\gamma(0))=\gamma(0)$$

i.e., $p=\gamma(0)$ is a fixed point of the diffeomorphism ϕ^T. Since $\mathscr{L}_{R_\lambda}\lambda=0$, $\phi^T_{R_\lambda}$ is a (strict) contact diffeomorphism and so induces an isomorphism

$$d\phi^T(p)|_{\xi_p}:\xi_p\to\xi_p$$

which is the linearization restricted to ξ_p of the Poincaré return map.

Definition 14.1. We say a T-closed Reeb orbit (T,λ) is *nondegenerate* if $d\phi^T(p)|_{\xi_p}:\xi_p\to\xi_p$ with $p=\gamma(0)$ has not eigenvalue 1.

Denote $\mathscr{L}(Q)=C^\infty(S^1,Q)$ the space of loops $z:S^1=\mathbb{R}/\mathbb{Z}\to Q$. We consider the assignment

$$\Phi:(T,\gamma,\lambda)\mapsto\dot\gamma-T\,R_\lambda(\gamma)$$

which we would like to consider a section of some Banach vector bundle over $(0,\infty)\times\mathscr{L}^{1,2}(Q)\times\mathscr{C}(Q,\xi)$ where $\mathscr{L}^{1,2}(Q)$ is the $W^{1,2}$-completion of $\mathscr{L}(Q)$. We note the value

$$\dot\gamma-T\,R_\lambda(\gamma)\in\Gamma(\gamma^*TQ).$$

We denote by $L^2(\gamma^*TQ)$ the space of L^2-sections of the vector bundle γ^*TQ. Then we define the vector bundle

$$\mathscr{L}^2 \to (0,\infty) \times \mathscr{L}^{1,2}(Q) \times \mathrm{Cont}(Q,\xi)$$

whose fiber at (T,γ,λ) is $L^2(\gamma^*TQ)$. We denote by π_i, $i = 1, 2, 3$ the corresponding projections.

We denote $\mathfrak{Reeb}(Q,\xi) = \Phi^{-1}(0)$. Then the set $Reeb(\lambda)$ of λ-Reeb orbits (γ,T) is nothing but $\mathfrak{Reeb}(Q,\xi) \cap \pi_3^{-1}(\lambda)$.

Proposition 14.2. *A T-closed Reeb orbit (T,γ) is nondegenerate if and only if the linearization*

$$d_{(T,\gamma)}\Phi : \mathbb{R} \times W^{1,2}(\gamma^*TQ) \to L^2(\gamma^*TQ)$$

is surjective.

The following generic nondegeneracy result is proved by Albers-Bramham-Wendl in [ABW10]. We denote by

$$\mathscr{C}(Q,\xi)$$

the set of contact forms of (Q,ξ) equipped with C^∞-topology.

Theorem 14.3 (Albers-Bramham-Wendl)**.** *Let (Q,ξ) be a contact manifold. Then there exists a residual subset $\mathscr{C}^{\text{reg}}(Q,\xi) \subset \mathscr{C}(Q,\xi)$ such that for any contact form $\lambda \in \mathscr{C}^{\text{reg}}(Q,\xi)$ all Reeb orbits are nondegenerate for $T > 0$.*

(The case $T = 0$ can be included as the Morse-Bott nondegenerate case if we allow the action $T = 0$ by extending the definition of Reeb trajectory to *isospeed Reeb trajectories* of the pairs (γ,T) with $\gamma: [0,1] \to Q$ with $T = \int \gamma^*\lambda$ as done in [Oh21c, Ohb].)

14.2 The case of Reeb chords

We first recall the notion of iso-speed Reeb trajectories used in [Oh21c] and recall the definition of nondegeneracy of thereof.

Consider contact triads (Q,λ,J) and the boundary value problem for (γ,T) with $\gamma: [0,1] \to Q$

$$\begin{cases} \dot\gamma(t) = TR_\lambda(\gamma(t)), \\ \gamma(0) \in R_0, \quad \gamma(1) \in R_1. \end{cases} \tag{1}$$

Definition 14.4 (Isospeed Reeb trajectory; Definition 2.1 [Ohb])**.** We call a pair (γ,T) of a smooth curve $\gamma: [0,1] \to Q$ and $T \in \mathbb{R}$ an *iso-speed Reeb trajectory* if they satisfy

$$\dot\gamma(t) = TR_\lambda(\gamma(t)), \quad \int \gamma^*\lambda = T \tag{2}$$

for all $t \in [0,1]$. We call (γ, T) an iso-speed closed Reeb orbit if $\gamma(0) = \gamma(1)$, and an iso-speed Reeb chord of (R_0, R_1) it $\gamma(0) \in R_0$ and $\gamma(1) \in R_1$ from R_0 to R_1.

With this definition, we state the corresponding notion of nondegeneracy

Definition 14.5. We say a Reeb chord (γ, T) of (R_0, R_1) is nondegenerate if the linearization map $d\phi^T(p) : \xi_p \to \xi_p$ satisfies

$$d\phi^T(p)(T_{\gamma(0)}R_0) \pitchfork T_{\gamma(1)}R_1 \quad \text{in } \xi_{\gamma(1)}$$

or equivalently

$$d\phi^T(p)(T_{\gamma(0)}R_0) \pitchfork T_{\gamma(1)}Z_{R_1} \quad \text{in}\, T_{\gamma(1)}Q.$$

Here $\phi^t_{R_\lambda}$ is the flow generated by the Reeb vector field R_λ.

More generally, we consider the following situation. We recall the definition of *Reeb trace* Z_R of a Legendrian submanifold R, which is defined to be

$$Z_R := \bigcup_{t \in \mathbb{R}} \phi^t_{R_\lambda}(R).$$

(See [Ohb, Appendix B] for detailed discussion on its genericity.)

Definition 14.6 (Nondegeneracy of Legendrian links). Let $\vec{R} = (R_1, \cdots, R_k)$ be a chain of Legendrian submanifolds, which we call a (ordered) *Legendrian link*. We assume that we have

$$Z_{R_i} \pitchfork R_j$$

for all i, $j = 1, \ldots, k$ and $i \neq j$.

We denote by

$$\mathscr{L}eg(Q, \xi)$$

the set of Legendrian submanifold and by $\mathscr{L}eg(Q, \xi; R)$ its connected component containing $R \in \mathscr{L}eg(Q, \xi)$, i.e, the set of Legendrian submanifolds Legendrian isotopic to R. We denote by

$$\mathscr{P}(\mathscr{L}eg(Q, \xi))$$

the monoid of Legendrian isotopies $[0,1] \to \mathscr{L}eg(Q, \xi)$. We have natural evaluation maps

$$\mathrm{ev}_0,\, \mathrm{ev}_1 : \mathscr{P}(\mathscr{L}eg(Q, \xi)) \to \mathscr{L}eg(Q, \xi)$$

and denote by

$$\mathscr{P}(\mathscr{L}eg(Q, \xi), R) = \mathrm{ev}_0^{-1}(R) \subset \mathscr{P}(\mathscr{L}eg(Q, \xi))$$

and

$$\mathscr{P}(\mathscr{L}eg(Q, \xi), (R_0, R_1)) = (\mathrm{ev}_0 \times \mathrm{ev}_1)^{-1}(R_0, R_1) \subset \mathscr{P}(\mathscr{L}eg(Q, \xi)).$$

We now provide the off-shell framework for the proof of nondegeneracy in general. Denote by $\mathscr{P}(Q; R_0, R_1)$ the space of paths

$$\gamma : ([0,1],\{0,1\}) \to (Q;R_0,R_1).$$

We consider the assignment

$$\Phi : (T,\gamma,\lambda) \mapsto \dot{\gamma} - T R_\lambda(\gamma) \tag{3}$$

as a section of the Banach vector bundle over

$$(0,\infty) \times \mathscr{P}^{1,2}(Q;R_0,R_1) \times \mathscr{C}(Q,\xi)$$

where $\mathscr{P}^{1,2}(Q;R_0,R_1)$ is the $W^{1,2}$-completion of $\mathscr{P}(Q;R_0,R_1)$. We have

$$\dot{\gamma} - T R_\lambda(\gamma) \in \Gamma(\gamma^* TQ; T_{\gamma(0)}R_0, T_{\gamma(1)}R_1).$$

We define the vector bundle

$$\mathscr{L}^2(Q;R_0,R_1) \to (0,\infty) \times \mathscr{P}^{1,2}(Q;R_0,R_1) \times \mathscr{C}(Q,\xi)$$

whose fiber at (T,γ,λ) is $L^2(\gamma^* TQ)$. We denote by π_i, $i = 1, 2, 3$ the corresponding projections.

We denote $\mathfrak{Reeb}(M,\lambda;R_0,R_1) = \Phi_\lambda^{-1}(0)$, where

$$\Phi_\lambda := \Phi|_{(0,\infty)\times\mathscr{L}^{1,2}(Q;R_0,R_1)\times\{\lambda\}}.$$

Then we have

$$\mathfrak{Reeb}(\lambda;R_0,R_1) = \Phi_\lambda^{-1}(0) = \mathfrak{Reeb}(Q,\xi) \cap \pi_3^{-1}(\lambda).$$

The following relative version of Theorem 14.3 is proved in [Ohb, Appendix B].

Theorem 14.7 (Perturbation of contact forms; Theorem B.3 [Ohb]). *Let (Q,ξ) be a contact manifold. Let (R_0,R_1) be a pair of Legendrian submanifolds allowing the case $R_0 = R_1$. There exists a residual subset $\mathscr{C}_1^{\text{reg}}(Q,\xi) \subset \mathscr{C}(Q,\xi)$ such that for any $\lambda \in \mathscr{C}_1^{\text{reg}}(Q,\xi)$ all Reeb chords from R_0 to R_1 are nondegenerate for $T > 0$ and Bott-Morse nondegenerate when $T = 0$.*

The following theorem is also proved in [Ohb].

Theorem 14.8 (Perturbation of boundaries; Theorem B.10 [Ohb]). *Let (Q,ξ) be a contact manifold. Let (R_0,R_1) be a pair of Legendrian submanifolds allowing the case $R_0 = R_1$. For a given contact form λ and R_1, there exists a residual subset*

$$R_0 \in \mathscr{L}eg^{\text{reg}}(Q,\xi) \subset \mathscr{L}eg(Q,\xi)$$

of Legendrian submanifolds such that for all $R_0 \in \mathscr{L}eg(Q,\xi)$ all Reeb chords from R_0 to R_1 are nondegenerate for $T > 0$ and Morse-Bott nondegenerate when $T = 0$.

We refer readers to [Ohb, Appendix B] for the proofs of these results.

15 Subsequence convergence

In this section, we study the asymptotic behavior of contact instantons on the Riemann surface $(\dot{\Sigma}, j)$ associated with a metric h with *cylinder-like ends* for the closed string context and with *strip-like ends* for the open string context.

15.1 Closed string case

We assume there exists a compact set $K_\Sigma \subset \dot{\Sigma}$, such that $\dot{\Sigma} - \mathrm{Int}(K_\Sigma)$ is a disjoint union of punctured disks each of which is isometric to the half cylinder $[0,\infty) \times S^1$ or the half strip $(-\infty, 0] \times [0,1]$, where the choice of positive or negative strips depends on the choice of analytic coordinates at the punctures. We denote by $\{p_i^+\}_{i=1,\cdots,l^+}$ the positive punctures, and by $\{p_j^-\}_{j=1,\cdots,l^-}$ the negative punctures. Here $l = l^+ + l^-$. Denote by $\phi_i^\pm$ such cylinder-like coordinates. We first state our assumptions for the study of the behavior of boundary punctures. (The case of interior punctures is treated in [OW18a, Section 6].)

Definition 15.1. Let $\dot{\Sigma}$ be a boundary-punctured Riemann surface of genus zero with punctures $\{p_i^+\}_{i=1,\cdots,l^+} \cup \{p_j^-\}_{j=1,\cdots,l^-}$ equipped with a metric h with *strip-like ends* outside a compact subset K_Σ. Let $w : \dot{\Sigma} \to Q$ be any smooth map with Legendrian boundary condition. We define the total π-harmonic energy $E^\pi(w)$ by

$$E^\pi(w) = E^\pi_{(\lambda,J;\dot{\Sigma},h)}(w) = \frac{1}{2}\int_{\dot{\Sigma}} |d^\pi w|^2 \tag{1}$$

where the norm is taken in terms of the given metric h on $\dot{\Sigma}$ and the triad metric on M.

Throughout this section, we work locally near one interior puncture p, i.e., on a punctured semi-disc $D^\delta(p) \setminus \{p\}$. By taking the associated conformal coordinates $\phi^+ = (\tau, t) : D^\delta(p) \setminus \{p\} \to [0,\infty) \times [0,1]$ such that $h = d\tau^2 + dt^2$, we need only look at a map w defined on the half cylinder $[0,\infty) \times S^1 \to Q$ without loss of generality.

We put the following hypotheses in our asymptotic study of the finite energy contact instanton maps w as in [OW18a]:

Hypothesis 15.2. Let h be the metric on $\dot{\Sigma}$ given above. Assume $w : \dot{\Sigma} \to Q$ satisfies the contact instanton equations (4), and

1. $E^\pi_{(\lambda,J;\dot{\Sigma},h)}(w) < \infty$ (finite π-energy);
2. $\|dw\|_{C^0(\dot{\Sigma})} < \infty$.
3. Image $w \subset \mathsf{K} \subset Q$ for some compact subset K.

The above finite π-energy and C^0 bound hypotheses imply

$$\int_{[0,\infty)\times S^1} |d^\pi w|^2 \, d\tau \, dt < \infty, \quad \|dw\|_{C^0([0,\infty)\times S^1)} < \infty \tag{2}$$

in these coordinates.

Definition 15.3 (Asymptotic action and charge)**.** Assume that the limit of $w(\tau, \dot{)}$ as $\tau \to \infty$ exists. Then we can associate two natural asymptotic invariants at each puncture defined as

$$T := \lim_{r\to\infty} \int_{\{r\}\times S^1} (w|_{\{0\}\times S^1})^*\lambda$$

$$Q := \lim_{r\to\infty} \int_{\{r\}\times S^1} ((w|_{\{0\}\times[0,1]})^*\lambda \circ j).$$

(Here we only look at positive punctures. The case of negative punctures is similar.) We call T the *asymptotic contact action* and Q the *asymptotic contact charge* of the contact instanton w at the given puncture.

The proof of the following subsequence convergence result is proved in [OW18a, Theorem 6.4].

Theorem 15.4 (Subsequence Convergence, Theorem 6.4 [OW18a])**.** *Let* $w : [0,\infty) \times S^1 \to Q$ *satisfy the contact instanton equations* (4) *and Hypothesis* (2)*. Then for any sequence* $s_k \to \infty$*, there exists a subsequence, still denoted by* s_k*, and a* massless instanton $w_\infty(\tau,t)$ *(i.e.,* $E^\pi(w_\infty) = 0$*) on the cylinder* $\mathbb{R} \times S^1$ *that satisfies the following:*

1. $\overline{\partial}^\pi w_\infty = 0$ *and*
$$\lim_{k\to\infty} w(s_k + \tau, t) = w_\infty(\tau,t)$$
in the $C^l(K \times S^1, Q)$ *sense for any* l*, where* $K \subset [0,\infty)$ *is an arbitrary compact set.*
2. $w_\infty^*\lambda = -Q\,d\tau + T\,dt$

In general $Q = 0$ does not necessarily hold for the closed string case. When this $Q \neq 0$ combined with $T = 0$, we say w has the bad limit of *appearance of spiraling instantons along the Reeb core*. It is also proven in [Oha] that If $Q = 0 = T$, then the puncture is removable.

When $Q = 0$, which is always the case when contact instanton is exact such as those arising from the symplectization case, we have the following asymptotic convergence result.

Corollary 15.5. *Assume that* λ *is nondegenerate. Suppose that* w_τ *converges as* $|\tau| \to \infty$ *and its massless limit instanton has* $Q = 0$ *but* $T \neq 0$*, then the* w_τ *converges to a Reeb orbit of period* $|T|$ *exponentially fast.*

15.2 Open string case

In this section, we study the asymptotic behavior of contact instantons on bordered Riemann surface $(\dot\Sigma, j)$ associated with a metric h with *strip-like ends*. To be precise, we assume there exists a compact set $K_\Sigma \subset \dot\Sigma$, such that $\dot\Sigma - \mathrm{Int}(K_\Sigma)$ is a disjoint union of punctured semi-disks each of which is isometric to the half strip $[0,\infty) \times [0,1]$ or $(-\infty,0] \times [0,1]$, where the choice of positive or negative strips depends on the choice of analytic coordinates at the punctures.

Again under the assumption that the limit $\lim_{\tau\to\infty} w(\tau,\cdot)$ exists, we can define the asymptotic action T and charge Q at each puncture in the way as in the closed string case by replacing S^1 by $[0,1]$.

The following subsequence convergence and charge vanishing results are proved in [Oh21a], [OY22]. One may say that *the presence of Legendrian barrier prevents the instanton from spiraling.*

Theorem 15.6 (Subsequence convergence and Charge vanishing). *Let* $w : [0,\infty) \times [0,1] \to Q$ *satisfy the contact instanton equations* (4) *and converges to an isospeed Reeb chord* (γ, T) *as* $|\tau| \to \infty$. *Then for any sequence* $s_k \to \infty$, *there exists a subsequence, still denoted by* s_k, *and a massless instanton* $w_\infty(\tau,t)$ *(i.e.,* $E^\pi(w_\infty) = 0$*) on the cylinder* $\mathbb{R} \times [0,1]$ *that satisfies the following:*

1. $\overline{\partial}^\pi w_\infty = 0$ *and*
$$\lim_{k\to\infty} w(s_k+\tau,t) = w_\infty(\tau,t)$$
in the $C^l(K \times [0,1], Q)$ *sense for any* l, *where* $K \subset [0,\infty)$ *is an arbitrary compact set.*
2. w_∞ *has vanishing asymptotic charge* $Q = 0$ *and satisfies* $w_\infty(\tau,t) \equiv \gamma(t)$ *for some Reeb chord* γ *is some Reeb chord joining* R_0 *and* R_1 *with period* T *at each puncture.*
3. $T \neq 0$ *at each puncture with the associated pair* (R,R') *of boundary condition with* $R \cap R' = \emptyset$.

Remark 15.7. This charge vanishing $Q = 0$ of the massless instanton is a huge advantage over the closed string case studied in [OW18a, OW18b, Oha]. We refer to [Oha, Section 8.1] for the classification of massless instantons on the cylinder $\mathbb{R} \times S^1$ for which the kind of massless instantons with $Q \neq 0$ but $T_w = 0$ appear for the closed string case. The *appearance of spiraling instantons along the Reeb core* is the only obstacle towards the Fredholm theory and the compactification of the moduli space of finite energy contact instantons for the general closed string context. (See [OSar] for a way of taking care of this phenomenton by 'quantizing' the charge through the $\mathfrak{lcs}$*-fication*.)

The above theorem automatically removes this obstacle for the open string case of Legendrian boundary condition. One could say that the presence of the Legendrian obstacle blocks this spiraling phenomenon of the contact instantons.

From the previous theorem, we immediately get the following corollary as in [OW18a, Section 8].

Corollary 15.8 (Corollary 5.11 [OY22]). *Assume that the pair* $(\lambda, \vec{R})$ *is nondegenerate in the sense of Definition 14.6. Let* $w : \dot\Sigma \to Q$ *satisfy the contact instanton equation* (4) *and Hypothesis* (2). *Then on each strip-like end with strip-like coordinates* $(\tau,t) \in [0,\infty) \times [0,1]$ *near a puncture*

$$\lim_{s\to\infty}\left|\pi\frac{\partial w}{\partial\tau}(s+\tau,t)\right|=0,\quad \lim_{s\to\infty}\left|\pi\frac{\partial w}{\partial t}(s+\tau,t)\right|=0$$

$$\lim_{s\to\infty}\lambda(\frac{\partial w}{\partial\tau})(s+\tau,t)=0,\quad \lim_{s\to\infty}\lambda(\frac{\partial w}{\partial t})(s+\tau,t)=T$$

and

$$\lim_{s\to\infty}|\nabla^l dw(s+\tau,t)|=0 \quad \textit{for any} \quad l\geq 1.$$

All the limits are uniform for (τ,t) *in* $K\times[0,1]$ *with compact* $K\subset\mathbb{R}$.

16 Off-shell energy of contact instantons

Now following [Oha], [Oh21c], we explain the definition of off-shell energy that governs the global convergence behavior of finite energy contact instantons.

Assume that λ for the closed string case (or $(\lambda,\vec{R})$ for the open string case) is nondegenerate. We also assume that the asymptotic charges vanish so that in cylindrical (or in strip-like) coordinates $w^*\lambda\to T\,dt$ and $w^*\lambda\circ j\to T\,d\tau$ exponentially fast.

We start with the π-energy

Definition 16.1 (The π-energy of contact instanton)**.** Let $w:\mathbb{R}\times[0,1]\to J^1B$ be any smooth map, where J^1B is the 1-jet bundle of B. We define

$$E^\pi(w):=\frac{1}{2}\int|d^\pi w|_J^2.$$

Next we borrow the presentation of λ-energy from [Oha, Section 5], [Oh21c, Section 11]. We follow the procedure exercised in [Oha] for the closed string case. We introduce the following class of test functions. Especially the automatic charge vanishing in our current circumstance also enables us to define the vertical part of energy, called the λ-energy whose definition is in order. We will focus on the more nontrivial open-string case below.

Definition 16.2. We define

$$\mathscr{C}=\left\{\varphi:\mathbb{R}\to\mathbb{R}_{\geq 0}\,\middle|\,\operatorname{supp}\varphi \text{ is compact}, \int_{\mathbb{R}}\varphi=1\right\} \tag{1}$$

Definition 16.3 (Contact instanton potential)**.** We call the above function f the *contact instanton potential* of the contact instanton charge form $w^*\lambda\circ j=df$ on $\dot\Sigma$.

Such a function exists modulo addition by a constant. *Using the assumption of vanishing of asymptotic charge*, we can explicitly write the potential as

$$f(z)=\int_{+\infty}^{z} w^*\lambda\circ j \tag{2}$$

on the strip-like regions of $\dot\Sigma$ where the integral is over any path from any puncture denote by ∞ to z along a path $\dot\Sigma$. By the closedness of $w^*\lambda\circ j$ on $\dot\Sigma$, the function f is well-defined and satisfies $w^*\lambda\circ j = df$ on each of the strip-like regions. (Compare this with [Oha, Formula above (5.5)] where the general case with nontrivial charge is considered.) By the vanishing theorem, Theorem 15.6 of the asymptotic charge of bordered contact instantons, the local version of potential function is always well-defined locally near punctures for the bordered contact instantons

We denote by ψ the function determined by

$$\psi' = \varphi, \quad \psi(-\infty) = 0, \psi(\infty) = 1. \tag{3}$$

Definition 16.4. Let w satisfy $d(w^*\lambda\circ j) = 0$. Then we define

$$\begin{aligned} E_{\mathscr{C}}(j,w;p) &= \sup_{\varphi\in\mathscr{C}} \int_{D_\delta(p)\setminus\{p\}} df\circ j \wedge d(\psi(f)) \\ &= \sup_{\varphi\in\mathscr{C}} \int_{D_\delta(p)\setminus\{p\}} (-w^*\lambda)\wedge d(\psi(f)). \end{aligned}$$

We note that

$$df\circ j\wedge d(\psi(f)) = \psi'(f)df\circ j\wedge df = \varphi(f)df\circ j\wedge df \geq 0$$

since

$$df\circ j\wedge df = |df|^2\, d\tau\wedge dt.$$

Therefore we can rewrite $E_{\mathscr{C}}(j,w;p)$ into

$$E_{\mathscr{C}}(j,w;p) = \sup_{\varphi\in\mathscr{C}} \int_{D_\delta(p)\setminus\{p\}} \varphi(f)df\circ j\wedge df.$$

The following proposition shows that the definition of $E_{\mathscr{C}}(j,w;p)$ does not depend on the constant shift in the choice of f.

Proposition 16.5 (Proposition 11.6 [Oh21c])**.** *For a given smooth map w satisfying $d(w^*\lambda\circ j) = 0$, we have $E_{\mathscr{C};f}(w) = E_{\mathscr{C},g}(w)$ for any pair (f,g) with*

$$df = w^*\lambda\circ j = dg$$

on $D^2_\delta(p)\setminus\{p\}$.

This proposition enables us to introduce the following vertical energy on each strip-like region, where we write $E^\lambda_\pm := E^\lambda_{\pm\infty}$ on $\mathbb{R}\times[0,1]\cong D^2\setminus\{\pm 1\}$.

Definition 16.6 (Vertical energy)**.** We define the *vertical energy*, denoted by $E^\perp(w)$, to be the sum

$$E^\perp(w) = E^\lambda_+(w) + E^\lambda_-(w)$$

Now we define the final form of the off-shell energy.

Definition 16.7 (Total energy). Let $w : \dot{\Sigma} \to Q$ be any smooth map. We define the *total energy* to be the sum

$$E(w) = E^{\pi}(w) + E^{\perp}(w). \tag{4}$$

Remark 16.8 (Uniform C^1 bound). The upshot is that the Sachs-Uhlenbeck [SU81], Gromov [Gro85] and Hofer [Hof93] style bubbling-off analysis can be carried out with this choice of energy. (See [Oha, Oh21c] for the details of this bubbling-off analysis.) In particular *all moduli spaces of finite energy perturbed contact instantons we consider in the present paper will have uniform C^1-bounds inside each given moduli spaces.*

17 Exponential C^∞ convergence

In this section, we outline the main steps for proving the exponential convergence result from [OW18b, Section 11] for the closed string case and from [OY22, Section 6] for the open string case, respectively.

Under the nondegeneracy hypothesis from Definition 14.6 we can improve the subsequence convergence to the exponential C^∞ convergence under the transversality hypothesis. Suppose that the tuple $\vec{R} = (R_0, \ldots, R_k)$ are transversal in the sense all pairwise Reeb chords are nondegenerate. In particular we assume that the tuples are pairwise disjoint. The proof is divided into several steps.

17.1 L^2-exponential decay of the Reeb component of dw

We will prove the exponential decay of the Reeb component $w^*\lambda$. We focus on a punctured neighborhood around a puncture $z_i \in \partial\Sigma$ equipped with strip-like coordinates $(\tau, t) \in [0, \infty) \times [0, 1]$.

We again consider a complex-valued function α given in (2). Then by the Legendrian boundary condition, we know $\alpha(\tau, i) \in \mathbb{R}$, i.e.

$$\operatorname{Im} \alpha = 0$$

for $i = 0, 1$.

The following lemma was proved in the closed string case in [OW18b] (in the more general context of Morse-Bott case). For readers' convenience, we provide some details by indicating how we adapt the argument with the presence of boundary condition.

Lemma 17.1 (Lemma 6.1 [OY22]; Compare with Lemma 11.20 [OW18b]). *Suppose the complex-valued functions α and ν defined on $[0, \infty) \times [0, 1]$ satisfy*

$$\begin{cases} \overline{\partial}\alpha = \nu, \\ \alpha(\tau,i) \in \mathbb{R} \text{ for } i = 0, 1, \\ \|\nu\|_{L^2([0,1])} + \|\nabla \nu\|_{L^2([0,1])} \leq Ce^{-\delta\tau} \quad \text{for some constants } C, \delta > 0 \\ \lim_{\tau\to+\infty} \alpha(\tau,t) = T \end{cases}$$

then $\|\alpha - T\|_{L^2(S^1)} \leq \overline{C}e^{-\delta\tau}$ *for some constant* $\overline{C}$.

Outline of proof. Notice that from previous section we have already established the $W^{1,2}$-exponential decay of $\nu = \frac{1}{2}|\zeta|^2$. Once this is established, the proof of this L^2-exponential decay result is proved by the standard three-interval method. See Theorem 24.11 for a general abstract framework which is in Appendix 24 of the present paper. We also refer to the Appendix of the arXiv version of [OW18a] for friendly details for the current nondegenerate case. □

(See Remark 13.1 for the difference of the definitions of α here and in Lemma 11.20 [OW18b].)

17.2 C^0 exponential convergence

Now the C^0-exponential convergence of $w(\tau,\cdot)$ to some Reeb chord as $\tau \to \infty$ can be proved from the L^2-exponential estimates presented in previous sections by the verbatim same argument as the proof of [OW18b, Proposition 11.21] with S^1 replaced by $[0,1]$ here. Therefore we omit its proof.

Proposition 17.2 (Proposition 11.21 [OW18b], Proposition 6.3 [OY22]). *Under Hypothesis 15.2, for any contact instanton w satisfying the Legendrian boundary condition, there exists a unique Reeb orbit γ such that the curve $z(\cdot) = \gamma(T\cdot)$: $[0,1] \to Q$ satisfies*

$$\|d(w(\tau,\cdot),z(\cdot))\|_{C^0([0,1])} \to 0,$$

as $\tau \to +\infty$, where d denotes the distance on Q defined by the triad metric. Here $T = T_\gamma$ is action of γ given by $T_\gamma = \int \gamma^*\lambda$.

Then the following C^0-exponential convergence is also proved.

Proposition 17.3 (Proposition 11.23 [OW18b], Proposition 6.5 [OY22]). *There exist some constants $C > 0$, $\delta > 0$ and τ_0 large such that for any $\tau > \tau_0$,*

$$\|d\,(w(\tau,\cdot),z(\cdot))\,\|_{C^0([0,1])} \leq Ce^{-\delta\tau}$$

17.3 C^∞-exponential decay of $dw - R_\lambda(w)\,d\tau$

So far, we have established the following:

- $W^{1,2}$-exponential decay of w,
- C^0-exponential convergence of $w(\tau,\cdot) \to z(\cdot)$ as $\tau \to \infty$ for some Reeb chord z between two Legendrians R, R'.

Now we are ready to complete the proof of C^∞-exponential convergence $w(\tau,\cdot) \to z$ by establishing the C^∞-exponential decay of $dw - R_\lambda(w)\,dt$. The proof of the latter decay is now in order which will be carried out by the bootstrapping arguments applied to the system (4).

Combining the above three, we have obtained L^2-exponential estimates of the full derivative dw. By the bootstrapping argument using the local uniform a priori estimates on the strip-like region as in the proof of Lemma 17.1, we obtain higher order $C^{k,\alpha}$-exponential decays of the term

$$\frac{\partial w}{\partial t} - TR_\lambda(z), \quad \frac{\partial w}{\partial \tau}$$

for all $k \geq 0$, where $w(\tau,\cdot)$ converges to z as $\tau \to \infty$ in C^0 sense. Combining this, Lemma 17.1 and elliptic $C^{k,\alpha}$-estimates given in Theorem 13.3, we complete the proof of C^∞-convergence of $w(\tau,\cdot) \to z$ as $\tau \to \infty$.

Part V
Compactification, Fredholm theory and asymptotic analysis

In this section, we develop the Fredholm theory of moduli space of pseudoholomorphic curves on the symplectization as a special case of the theory of pseudoholomorphic curves on the $\mathfrak{lcs}$-fication of contact manifolds as done in by the first author with Savelyev [OSar]. The symplectization is just the zero-temperature limit of the $\mathfrak{lcs}$-fications. Contact triad connection and its symplectification interact very well with the decomposition

$$TQ = \xi \oplus \mathbb{R}\langle R_\lambda \rangle$$
$$TM \cong \xi \oplus \mathbb{R}\langle R_\lambda \rangle \oplus \mathbb{R}\left\langle \frac{\partial}{\partial s} \right\rangle$$

which enable us to derive a tensorial formula for the linearized operator in a precise matrix form all whose summands carry natural geometric meaning. This enables us to do completely coordinate free asymptotic analysis. On the other hand, [HWZ96b, HWZ96a, HWZ02] rely on the choice of special coordinates in 3 dimension. Such a coordinate approach, especially in higher dimensions, leads to many complicated tedious expressions in the asymptotic convergence results. (See [Bou02] for example.) This was the starting point of Wang and the first author of the present survey for them to develop the notion of contact instantons and its tensorial study [OW18a, OW18b], and to have discovered the notion of *contact triad connection* [OW14] which well suits the purpose. This approach especially leads to an simple transparent formulae for the linearized operators both for the contact instantons and for the pseudoholomorphic curves in symplectization.

Leaving the case of contact instantons to [Oha] for the closed string case and to [Ohb] for the open string case, we now focus on the case of pseudoholomorphic curves in symplectization.

We start with the following completion of exponential convergence result for the finite energy punctured pseudoholomorphic curves.

18 Exponential convergence in symplectization

In this section, we consider the symplectization

$$M = Q \times \mathbb{R}, \quad \omega = d(e^s \pi^* \lambda) = e^s (ds \wedge \pi^* \lambda + d\pi^* \lambda)$$

of the contact manifold (Q, ξ) equipped with contact form λ.

On Q, the Reeb vector field R_λ associated to the contact form λ is the unique vector field satisfying

$$R_\lambda \rfloor \lambda = 1, \quad R_\lambda \rfloor d\lambda = 0. \tag{1}$$

We call (y, s) the cylindrical coordinates. On the cylinder $[0, \infty) \times Q \subset (-\infty, \infty) \times Q$, we have the natural splitting

$$TM \cong TQ \oplus \mathbb{R} \cdot \frac{\partial}{\partial s} \cong \xi \oplus \text{span}\left\{\widetilde{R}_\lambda, \frac{\partial}{\partial s}\right\} \cong \xi \oplus \mathbb{R}^2.$$

We denote by $\widetilde{R}_\lambda$ the unique vector field on $[0,\infty) \times Q$ which is invariant under the translation, tangent to the level sets of s and projected to R_λ. When there is no danger of confusion, we will sometimes just denote it by R_λ.

Now we describe a special family of almost complex structure adapted to the given cylindrical structure of M.

Definition 18.1. An almost complex structure $\widetilde{J}$ on $Q \times \mathbb{R}$ is called *λ-adapted* if it is split into

$$\widetilde{J} = J \oplus J_0 : TM \cong \xi \oplus \mathbb{R}^2 \to TM \cong \xi \oplus \mathbb{R}^2$$

where J is compatible to $d\lambda|_\xi$ and $J_0 : \mathbb{R}^2 \to \mathbb{R}^2$ maps $\frac{\partial}{\partial s}$ to R_λ.

For our purpose, we will need to consider a family of symplectic forms to which the given $\widetilde{J}$ is compatible and their associated metrics. For any λ-adapted $\widetilde{J}$, the $\widetilde{J}$-compatible metric associated to ω is expressed as

$$g_{(\omega,\widetilde{J})} = e^s(ds^2 + g_Y) \tag{2}$$

on $Q \times \mathbb{R}$.

Now we regard the triple $(\omega, \widetilde{J}, g_{(\omega,\widetilde{J})})$ be an almost Kähler manifold near the level surface $s = 1$. We then fix the canonical connection ∇ associated to $(\omega, \widetilde{J}, g_{(\omega,\widetilde{J})})$. The following is a general property of the canonical connection.

Proposition 18.2. *Let $(W, \omega, \widetilde{J})$ be an almost Kähler manifold and ∇ be the canonical connection. Denote by T be its torsion tensor. Then*

$$T(\widetilde{J}Y, Y) = 0 \tag{3}$$

for all vector fields Y on W.

Consider the decomposition

$$TM \cong \xi \oplus \mathbb{R}\{R_\lambda\} \oplus \mathbb{R}\left\{\frac{\partial}{\partial r}\right\}$$

and the canonical connection $\widetilde{\nabla}$ on $Q \times \mathbb{R}$, which in particular is $\widetilde{J}$-linear.

Now denote $f = s \circ u$ and $w = \pi \circ u$, i.e., $u = (w, f)$ in $Q \times \mathbb{R}$ be $\widetilde{J}$-holomorphic. Then w is automatically an exact contact instanton for which we have already shown the exponential convergence of $w(\tau, \cdot)$ to an isospeed Reeb orbit (γ, T) such that $w(\infty, t) = \gamma(t)$. For the convergence, we use the equation

$$w^*\lambda \circ j = df$$

which implies

$$w^*\lambda = -df \circ j$$

By taking differential, we obtain

$$-d(df \circ j) = w^* d\lambda = \frac{1}{2}|d^\pi w|^2 \, d\tau \wedge dt.$$

This first shows that f is a subharmonic function and satisfies

$$\frac{\partial^2 f}{\partial \tau^2} + \frac{\partial^2 f}{\partial t^2} - \frac{1}{2}|d^\pi w|^2 = 0 \tag{4}$$

where we know from Theorem 15.4 that the convergence $\frac{1}{2}|d^\pi w|^2 \to T^2$ is exponentially fast. This immediately gives rise to the following exponential convergence of the radial component which will complete the study of asymptotic convergence property of finite energy pseudoholomorphic planes in symplectization by combining that of w. (See Theorem 24.11 for the general abstract framework of establishing exponential convergence.)

Proposition 18.3 (Exponential convergence of radial component). *Let $u = (w, f)$ be a finite energy $\widetilde{J}$-holomorphic plane in $Q \times \mathbb{R}$. Then we have convergence*

$$df \to T \, d\tau$$

exponentially fast.

Proof. We have already established $w^*\lambda \to T\,dt$ before. By composing by j, the statement follows. □

19 The moduli spaces of contact instantons and of pseudoholomorphic curves

In this section, we recall the definitions of the moduli spaces of contact instantons and of pseudoholomorphic curves and compare their compactifications.

19.1 Moudli space of pseudoholomorphic curves on symplectization

Let $(\dot{\Sigma}, j)$ be a punctured Riemann surface and let

$$p_1, \cdots, p_{s^+}, q_1, \cdots, q_{s^-}$$

be the positive and negative punctures. For each p_i (resp. q_j), we associate the isothermal coordinates $(\tau, t) \in [0, \infty) \times S^1$ (resp. $(\tau, t) \in (-\infty, 0] \times S^1$) on the punctured disc $D_{e^{-2\pi R_0}}(p_i)\setminus\{p_i\}$ (resp. on $D_{e^{-2\pi R_0}}(q_i)\setminus\{q_i\}$) for some sufficiently large $R_0 > 0$.

Following [Hof93], [BEHZ03], we define the associated energy

$$E_\eta(u) = E^\pi(u) + E^\perp_\eta(u)$$

for each smooth map $u = (w, f)$ in class η, i.e., $[u]_{S^1} := f^* d\theta = \eta$.

Definition 19.1. We define

$$\widetilde{\mathcal{M}}_{k,\ell}(\dot\Sigma, M; \widetilde{J}), \quad k, \ell \geq 0, k+\ell \geq 1$$

to be the moduli space of pseudoholomorphic curve $u = (w, f)$ with $E_\eta(u) < \infty$.

Then we have a decomposition

$$\widetilde{\mathcal{M}}_{k,\ell}(\dot\Sigma, M; \widetilde{J}) = \bigcup_{\vec\gamma^+, \vec\gamma^-} \widetilde{\mathcal{M}}_{k,\ell}\left(\dot\Sigma, M; \widetilde{J}; (\vec\gamma^+, \vec\gamma^-)\right)$$

by Theorem 15.4 where

$$\widetilde{\mathcal{M}}_{k,\ell}(\dot\Sigma, M; \widetilde{J}; (\vec\gamma^+, \vec\gamma^-)) = \{u = (w, f) \mid u \text{ is an lcs instanton with } E_\eta(u) < \infty, w(-\infty_j) = \gamma_j^-, w(\infty_i) = \gamma_i\}.$$

Here we have the collections of Reeb orbits γ_i^+ and γ_j^- and of points p_i, q_j for $i = 1, \cdots, s^+$ and for $j = 1, \cdots, s^-$ respectively such that

$$\lim_{\tau\to\infty} w((\tau,t)_i) = \gamma_i^+(T_i(t+t_i)), \quad \lim_{\tau\to-\infty} w((\tau,t)_j) = \gamma_j^-(T_j(t-t_j)) \qquad (1)$$

for some $t_i, t_j \in S^1$, where

$$T_i = \int_{S^1} (\gamma_i^+)^* \lambda, \; T_j = \int_{S^1} (\gamma_j^-)^* \lambda.$$

Here t_i, t_j depends on the given analytic coordinate and the parameterization of the Reeb orbits.

Due to the $\mathbb{R}$-equivariance of the equation (3) under the $\mathbb{R}$ action of translations, this action induces a free action on $\widetilde{\mathcal{M}}_{k,\ell}(\dot\Sigma, M; \widetilde{J})$. Then we denote

$$\mathcal{M}_{k,\ell}(\dot\Sigma, M; \widetilde{J}) = \widetilde{\mathcal{M}}_{k,\ell}(\dot\Sigma, M; \widetilde{J})/\mathbb{R}. \qquad (2)$$

We also have the decomposition

$$\widetilde{\mathcal{M}}_{k,\ell}(\dot\Sigma, M; \widetilde{J}) = \bigcup_{\vec\gamma^\pm} \widetilde{\mathcal{M}}_{k,\ell}(\dot\Sigma, M; \widetilde{J}; (\vec\gamma^+, \vec\gamma^-)).$$

Here we denote $\vec\gamma^+ = (\gamma_i^+)$, $\vec\gamma^- = (\gamma_j^-)$ By quotienting the above out by the $\mathbb{R}$-action of $Q \times \mathbb{R}$, we obtain the moduli space of J-holomorphic curves

$$\mathscr{M}_{k,\ell}(\dot{\Sigma},M;\widetilde{J}) = \bigcup_{\vec{\gamma}^{\pm}} \mathscr{M}_{k,\ell}(\dot{\Sigma},M;\widetilde{J};(\vec{\gamma}^{+},\vec{\gamma}^{-})).$$

We now introduce a uniform energy bound for $u = (w, f)$ with given asymptotic condition at its punctures. Recall that they satisfy $w^*\lambda \circ j = df$.

The following proposition is the analog to [BEHZ03, Lemma 5.15] and [Oha, Proposition 9.2] whose proof is also similar.

Proposition 19.2. *Let $u = (w, f) \in \mathscr{M}_{k,\ell}(\dot{\Sigma},M;\widetilde{J};(\vec{\gamma}^{+},\vec{\gamma}^{-}))$. Suppose that $E^{\pi}(w) < \infty$ and the function $f : \dot{\Sigma} \to \mathbb{R}$ is proper. Then $E(w) < \infty$.*

The following a priori energy bounds for $u = (w, f)$ is proved in [BEHZ03].

Proposition 19.3 (Lemma 5.15, [BEHZ03]). *Let $u = (w, f) \in \mathscr{M}_{k,\ell}(\dot{\Sigma},M;\widetilde{J};(\vec{\gamma}^{+},\vec{\gamma}^{-}))$. Let $\widetilde{f} : \dot{\Sigma} \to \mathbb{R}$ be the function satisfying $w^*\lambda \circ j = d\widetilde{f}$. Suppose the function $\widetilde{f}$ is proper. Then we have*

$$E^{\pi}(u) = E^{\pi}(w) = \sum_{i=1}^{k} \mathscr{A}_{\lambda}(\gamma_i^{+}) - \sum_{j=1}^{\ell} \mathscr{A}_{\lambda}(\gamma_j^{-})$$

$$E_{\eta}^{\perp}(u) = \sum_{j=1}^{k} \mathscr{A}_{\lambda}(\gamma_i^{+})$$

$$E(u) = 2\sum_{i=1}^{k} \mathscr{A}_{\lambda}(\gamma_i^{+}) - \sum_{j=1}^{\ell} \mathscr{A}_{\lambda}(\gamma_j^{-}).$$

19.2 Moduli space of contact instantons with prescribed charge

We first consider the closed string case. In this case, the asymptotic charge $Q(p)$ at some interior puncture p may not vanish in general although they always vanish at the boundary punctures. In [OSar], the authors introduced the notion of *charge class* $\eta \in H^1(\dot{\Sigma},\mathbb{Z})$ and defined the moduli space of pseudoholomorphic curves on the lcs-fication

$$(Q \times S^1, \omega_\lambda), \quad \omega_\lambda = d\lambda + d\theta \wedge \lambda$$

for the canonical angular form $d\theta$ on S^1 which is the generator of $H^1(S^1,\mathbb{Z})$ similarly as in the case of symplectization or more precisely in the zero-temperature limit of lcs-fication

$$(Q \times \mathbb{R}, \omega_\lambda), \quad \omega_\lambda = d\lambda + ds \wedge \lambda.$$

They lift a λ-adapted CR almost complex structure to an almost complex structure on $Q \times S^1$ by requiring

$$\widetilde{J}|_{\xi} = J|_{\xi}, \quad \widetilde{J}\left(\frac{\partial}{\partial \theta}\right) = R_\lambda.$$

Similarly as in the case of symplectization, they showed that a $\widetilde{J}$-holomorphic curve $u = (w, f)$ satisfies

$$\bar{\partial}^{\pi} w = 0, \quad w^*\lambda \circ j = df \tag{3}$$

where $f : \dot{\Sigma} \to S^1$ given by $f := \theta \circ u$. By definition, each lcs instanton carries a cohomology class $\eta_w := [w^*\lambda \circ j] \in H^1(\dot{\Sigma}, \mathbb{Z})$.

Recalling the isomorphism

$$[\dot{\Sigma}, S^1] \cong H^1(\dot{\Sigma}, \mathbb{Z})$$

we may also regard the cohomology class $[u]_{S^1}$ as an element in $[\dot{\Sigma}, S^1]$. This enables us to define an element in the set of homotopy classes $[\dot{\Sigma}, S^1]$, which we also denote by $\eta = \eta_u$. In fact, the isomorphism $[\dot{\Sigma}, S^1] \cong H^1(\dot{\Sigma}; \mathbb{Z})$ is directly induced by the period map

$$[f] \mapsto [f^* d\theta].$$

Definition 19.4 (Period map and the charge class; Definition 5.5 [OSar]). Let $u = (w, f) : \dot{\Sigma} \to Q \times S^1$ be a smooth map.

1. We call the map

$$C^\infty(\dot{\Sigma}, S^1) \to H^1(\dot{\Sigma}, \mathbb{Z}); \quad f \mapsto [f^* d\theta]$$

the *period map* and call the cohomology class $[f^* d\theta]$ the *charge class* of the map f.

2. For an lcs instanton $u = (w, f) : \dot{\Sigma} \to Q \times S^1$, we call the cohomology class

$$[f^* d\theta] \in H^1(\dot{\Sigma}, \mathbb{Z})$$

the *charge class* of u and write

$$[u]_{S^1} := [f^* d\theta].$$

Now we consider the maps $u = (w, f)$ with a fixed charge class $\eta = \eta_u$ and define the moduli space

$$\widetilde{\mathcal{M}}^{\eta}_{k,\ell}(\dot{\Sigma}, Q; J; (\vec{\gamma}^+, \vec{\gamma}^-)) = \{u = (w, f) \mid u \text{ is an lcs instanton with } E_\eta(u) < \infty, w(-\infty_j) = \gamma_j^-, w(\infty_i) = \gamma_i\}.$$

For the open string case, as proven in Theorem 15.6 the (local) charge near every boundary puncture vanishes. This shows that the closed one-form $w^*\lambda \circ j$ defines a cohomology class

$$[w^*\lambda \circ j] \in H^1((\overline{\Sigma}, \partial_\infty \dot{\Sigma}); \mathbb{Z}). \tag{4}$$

Definition 19.5 (The charge class; open string case). Let $w : (\dot{\Sigma}, \partial \dot{\Sigma}) \to (Q, \vec{R})$ be a bordered contact instanton. We call the cohomology class $[w^*\lambda \circ j]$ given in (4) the charge class of w.

The following is an immediate consequence of this definition and Theorem 15.6.

Corollary 19.6. *Suppose Σ is an open Riemann surface of genus 0. Then for any bordered contact instanton on $\dot\Sigma = \Sigma \setminus \{z_0, \cdots z_k\}$, the charge class vanishes.*

19.3 Comparison of compactifications of the two moduli spaces

Now we consider contact instantons w arising from a pseudoholomorphic curves on symplectization (w, f). In particular all such w has its charge class $[w^*\lambda \circ j] = 0$ in $H^1(\dot\Sigma, \mathbb{Z})$ but can be lifted to $H^1(\overline{\Sigma}, \partial_\infty \dot\Sigma)$ where $\overline{\Sigma}$ is the real blow-up of $\dot\Sigma$ along the punctures. We denote the moduli space of such contact instantons of finite energy by

$$\widetilde{\mathscr{M}}^{\text{exact}}(\dot\Sigma, Q, J; \vec\gamma^-, \vec\gamma^+)$$

and

$$\mathscr{M}^{\text{exact}}(\dot\Sigma, Q, J; \vec\gamma^-, \vec\gamma^+) := \widetilde{\mathscr{M}}^{\text{exact}}(\dot\Sigma, Q, J; \vec\gamma^-, \vec\gamma^+)/\text{Aut}(\dot\Sigma)$$

the set of isomorphism classes thereof. We have natural forgetful map $(w, f) \mapsto w$ which descends to

$$\mathfrak{forget} : \mathscr{M}(\dot\Sigma, M, \widetilde{J}; \vec\gamma^-, \vec\gamma^+) \to \mathscr{M}^{\text{exact}}(\dot\Sigma, Q, J; \vec\gamma^-, \vec\gamma^+).$$

By definition of the equivalence relation on $\widetilde{\mathscr{M}}(\dot\Sigma, M, \widetilde{J}; \vec\gamma^-, \vec\gamma^+)$, it follows that this forgetful map is a bijective correspondence, *provided $\dot\Sigma$ is connected.*

However when one considers the SFT compactification as in [EGH00], [BEHZ03], one needs to consider the case of pseudoholomorphic curves with disconnected domain. So let us consider such cases. Suppose we have the union

$$\dot\Sigma = \bigsqcup_{i=1}^{k} \dot\Sigma_i$$

of connected components with $k \geq 2$. We denote by

$$\overline{\mathscr{M}}(\dot\Sigma, M, \widetilde{J}; \vec\gamma^-, \vec\gamma^+)$$

and

$$\overline{\mathscr{M}}^{\text{exact}}(\dot\Sigma, Q, J; \vec\gamma^-, \vec\gamma^+)$$

the stable map compactification respectively. The following proposition shows the precise relationship between the two. We know that each story carries at least one non-cylindrical component.

Proposition 19.7. *Let $1 \leq \ell \leq k$ be the number of connected components which are not cylinderical. The forgetful map $\mathfrak{forget}$ is a principle $\mathbb{R}^{\ell-1}$ fibration.*

Proof. Recall the equivalence relation on $\widetilde{\mathscr{M}}(\dot\Sigma, M, \widetilde{J}; \vec\gamma^-, \vec\gamma^+)$: We say two elements $(u_1, \cdots, u_k) \sim (u_1', \cdots, u_k')$ if there is a $s_0 \in \mathbb{R}$ and a reparameterization $\varphi = (\varphi_1, \cdots, \varphi_k)$ of $\dot\Sigma$ such that

$$(w_i', f_i') = (w \circ \varphi_i, f_i \circ \varphi_i - s_0)$$

for all $i = 1, \cdots, k$. We define the map

$$T_{\vec{s}} : \widetilde{\mathcal{M}}(\dot{\Sigma}, M, \widetilde{J}; \vec{\gamma}^-, \vec{\gamma}^+) \to \widetilde{\mathcal{M}}(\dot{\Sigma}, M, \widetilde{J}; \vec{\gamma}^-, \vec{\gamma}^+)$$

by

$$T_{\vec{s}}(u_1, \cdots, u_k) = (T_{s_1} u_1, \cdots, T_{s_k} u_k)$$

where $T_{s_i} u_i = T_{s_i}(w_i, f_i) := (w_i, f_i - s_i)$. Note that this action is trivial for each trivial component.

Then all the following maps

$$T_{\vec{s}}(u) = (T_{s_1} u_1, \cdots, T_{s_k} u_k)$$

project down to the same element $[(w_1, \cdots, w_k)] \in \mathcal{M}^{\text{exact}}(\dot{\Sigma}, Q, J; \vec{\gamma}^-, \vec{\gamma}^+)$. The abelian group $\mathbb{R}^{\ell-1}$ admits a free transitive action of $\mathbb{R}^{\ell-1}$ on each fiber

$$\mathfrak{forget}^{-1}([(w_1, \cdots, w_k)])$$

given by

$$([(u_1, \cdots, u_k)], (s_1, \cdots s_k)) \mapsto ([(T_{s_1} u_1, \cdots, T_{s_k} u_k)]$$

where we realize

$$\mathbb{R}^{\ell-1} \cong \{(s_1, \cdots, s_\ell) \mid s_1 + \cdots + s_\ell = 0\}$$

This finishes the proof. □

20 Fredholm theory and index calculations

In this section, we work out the Fredholm theories of pseudoholomorphic curves on symplectization. We will adapt the exposition given in [Oha] [OSar] for the case of contact instantons to that of pseudoholomorphic curves thereon as the zero-temperature limit lcs instantons considered in [OSar] just by incorporating the presence of the $\mathbb{R}$-factor in the product $M = Q^{2n-1} \times \mathbb{R}$.

We divide our discussion into the closed case and the punctured case.

20.1 Calculation of the linearization map

Let Σ be a closed Riemann surface and $\dot{\Sigma}$ be its associated punctured Riemann surface. We allow the set of whose punctures to be empty, i.e., $\dot{\Sigma} = \Sigma$. We would like to regard the assignment $u \mapsto \overline{\partial}_J u$ which can be decomposed into

$$u = (w, f) \mapsto \left(\overline{\partial}^{\pi} w, w^*\lambda \circ j - f^* ds\right)$$

for a map $w : \dot{\Sigma} \to Q$ as a section of the (infinite dimensional) vector bundle over the space of maps of w. In this section, we lay out the precise relevant off-shell framework of functional analysis.

Let $(\dot{\Sigma}, j)$ be a punctured Riemann surface, the set of whose punctures may be empty, i.e., $\dot{\Sigma} = \Sigma$ is either a closed or a punctured Riemann surface. We will fix j and its associated Kähler metric h.

We consider the map

$$\Upsilon(w, f) = \left(\overline{\partial}^{\pi} w, w^*\lambda \circ j - f^* ds\right)$$

which defines a section of the vector bundle

$$\mathcal{H} \to \mathcal{F} = C^{\infty}(\Sigma, Q)$$

whose fiber at $u \in C^{\infty}(\Sigma, Q \times \mathbb{R})$ is given by

$$\mathcal{H}_u := \Omega^{(0,1)}(u^*\xi) \oplus \Omega^{(0,1)}(u^*\mathcal{V}).$$

Recalling $\mathcal{V}_{(q,s)} = \mathrm{span}_{\mathbb{R}}\{R_\lambda, \frac{\partial}{\partial s}\}$, we have a natural isomorphism

$$\Omega^{(0,1)}(u^*\mathcal{V}) \cong \Omega^1(\Sigma) = \Gamma(T^*\mathbb{R})$$

via the map

$$\alpha \in \Gamma(T^*\mathbb{R}) \mapsto \frac{1}{2}\left(\alpha \otimes \frac{\partial}{\partial s} + \alpha \circ j \otimes R_\lambda\right).$$

Utilizing this isomorphism, we decompose $\Upsilon = (\Upsilon_1, \Upsilon_2)$ where

$$\Upsilon_1 : \Omega^0(w^*TQ) \to \Omega^{(0,1)}(w^*\xi); \quad \Upsilon_1(w) = \overline{\partial}^{\pi}(w) \tag{1}$$

and

$$\Upsilon_2 : \Omega^0(w^*TQ) \to \Omega^1(\Sigma); \quad \Upsilon_2(w) = w^*\lambda \circ j - f^* ds. \tag{2}$$

We first compute the linearization map which defines a linear map

$$D\Upsilon(u) : \Omega^0(w^*TQ) \oplus \Omega^0(f^*T\mathbb{R}) \to \Omega^{(0,1)}(w^*\xi) \oplus \Omega^1(\dot{\Sigma})$$

where we have

$$T_u\mathcal{F} = \Omega^0(w^*TQ) \oplus \Omega^0(f^*T\mathbb{R}).$$

For the optimal expression of the linearization map and its relevant calculations, we use the lcs-fication connection ∇ of $(Q \times \mathbb{R}, \lambda, J)$ which is the lcs-lifting of contact triad connection introduced in [OW18a]. We refer readers to [OW18a], [Oha], [OSar] for the unexplained notations appearing in our tensor calculations during the proof.

We define the covariant differential

$$\overline{\partial}^{\nabla^\pi} := \frac{1}{2}(\nabla^\pi + J\nabla^\pi \circ j).$$

Theorem 20.1 (Theorem 10.1 [OSar]). *We decompose $dw = d^\pi w + w^*\lambda \otimes R_\lambda$ and $Y = Y^\pi + \lambda(Y)R_\lambda$, and $X = (Y,v) \in \Omega^0(w^*T(Q\times\mathbb{R}))$. Denote $\kappa = \lambda(Y)$ and $\upsilon = ds(v)$. Then we have*

$$D\Upsilon_1(w)(Y,v) = \overline{\partial}^{\nabla^\pi} Y^\pi + B^{(0,1)}(Y^\pi) + T^{\pi,(0,1)}_{dw}(Y^\pi) + \frac{1}{2}\kappa\cdot((\mathscr{L}_{R_\lambda}J)J(\partial^\pi w)) \tag{3}$$

$$D\Upsilon_2(u)(Y,v) = w^*(\mathscr{L}_Y\lambda)\circ j - \mathscr{L}_v ds = d\kappa\circ j - d\upsilon + w^*(Y\rfloor d\lambda)\circ j \tag{4}$$

where $B^{(0,1)}$ and $T^{\pi,(0,1)}_{dw}$ are the $(0,1)$-components of B and of T^π_{dw}, where B, T^π_{dw} : $\Omega^0(w^*TQ) \to \Omega^1(w^*\xi)$ are zero-order differential operators given by

$$B(Y) = -\frac{1}{2}w^*\lambda\left((\mathscr{L}_{R_\lambda}J)JY\right)$$

and

$$T^\pi_{dw}(Y) = \pi T(Y,dw)$$

respectively.

We often omit w^* from the Lie derivative $\mathscr{L}_Y\lambda$ and from the interior product $Y\rfloor d\lambda$ regarding them 'over the map w' when Y is already a vector field along the map w below.

We can also express the operator $D\Upsilon(u)$ in the following matrix form

$$D\Upsilon(u) = \begin{pmatrix} \overline{\partial}^{\nabla^\pi} + B^{(0,1)} + T^{\pi,(0,1)}_{dw} & , \frac{1}{2}(\cdot)\cdot((\mathscr{L}_{R_\lambda}J)J(\partial^\pi w)) \\ ((\cdot)^\pi\rfloor d\lambda)\circ j & , \overline{\partial} \end{pmatrix} \tag{5}$$

with respect to the decomposition

$$\left(Y, \upsilon\frac{\partial}{\partial s}\right) = \left(Y^\pi + \kappa R_\lambda, \upsilon\frac{\partial}{\partial s}\right) \cong (Y^\pi, \upsilon + i\kappa)$$

in terms of the splitting

$$T(Q\times\mathbb{R}) = \xi \oplus (\mathrm{span}\{R_\lambda\}\oplus T\mathbb{R}) \cong \xi \oplus \mathbb{C}.$$

Now we evaluate $D\Upsilon_1(w)$ more explicitly. We have the expression of $B^{(0,1)}(Y)$

$$B^{(0,1)}(Y) = -\frac{1}{4}\left(w^*\lambda\otimes\pi((\mathscr{L}_{R_\lambda}J)JY) + (w^*\lambda\circ j)\otimes\pi((\mathscr{L}_{R_\lambda}J)Y)\right).$$

Remark 20.2. Abstractly the linearization of the equation is well-defined on shell which does not depend on the choice of connections. The upshot of making a good choice of connection is to have a good formula that makes it easier to extract its consequence from it.

20.2 The punctured case

We consider some choice of weighted Sobolov spaces

$$\mathscr{W}^{k,p}_{\delta;\eta}(\dot\Sigma, Q \times \mathbb{R}; \vec\gamma^+, \vec\gamma^-)$$

as the off-shell function space and linearize the map

$$(w, \widetilde{f}) \mapsto \left(\overline{\partial}^\pi w, d\widetilde{f}\right).$$

This linearization operator then becomes cylindrical in cylindrical coordinates near the punctures.

The local model of the tangent space of $\mathscr{W}^{k,p}_\delta(\dot\Sigma, Q; J; \gamma^+, \gamma^-)$ at $w \in C^\infty_\delta(\dot\Sigma, Q) \subset W^{k,p}_\delta(\dot\Sigma, Q)$ is given by the sum

$$\Gamma_{s^+,s^-} \bigoplus W^{k,p}_\delta(w^* TQ) \tag{6}$$

where $W^{k,p}_\delta(w^*TQ)$ is the Banach space

$$\begin{aligned}&\{Y = (Y^\pi, \lambda(Y) R_\lambda) \mid e^{\frac{\delta}{p}|\tau|} Y^\pi \in W^{k,p}(\dot\Sigma, w^*\xi), \lambda(Y) \in W^{k,p}(\dot\Sigma, \mathbb{R})\}\\ &\cong W^{k,p}(\dot\Sigma, \mathbb{R}) \otimes R_\lambda(w) \bigoplus W^{k,p}(\dot\Sigma, w^*\xi).\end{aligned}$$

Here we measure the various norms in terms of the triad metric of the triad (Q, λ, J).

To describe the choice of $\delta > 0$, we need to recall the *covariant linearization* of the map $D\Phi_{\lambda,T} : W^{1,2}(z^*\xi) \to L^2(z^*\xi)$ of the map

$$\Phi_{\lambda,T} : z \mapsto \dot z - T R_\lambda(z)$$

for a given T-periodic Reeb orbit (T, z). The operator has the expression

$$D\Phi_{\lambda,T} = \nabla^\pi_t - \frac{T}{2}(\mathscr{L}_{R_\lambda} J) J \tag{7}$$

where ∇^π_t is the covariant derivative with respect to the pull-back connection $z^*\nabla^\pi$ along the Reeb orbit z and $(\mathscr{L}_{R_\lambda} J)J$ is a (pointwise) symmetric operator with respect to the triad metric. (See Lemma 3.4 [OW14].)

Remark 20.3. Again this *covariant* linearization map can be defined along any smooth curve and does not depend on the choice of connection *along the Reeb chords.*

We choose $\delta > 0$ so that $0 < \delta/p < 1$ is smaller than the spectral gap

$$\mathrm{gap}(\gamma^+,\gamma^-) := \min_{i,j}\{d_{\mathrm{H}}(,Q,\mathrm{Spec}A_{(T_i,z_i)},0),\, d_{\mathrm{H}}(,Q,\mathrm{Spec}A_{(T_j,z_j)},0)\}. \tag{8}$$

We now provide details of the Fredholm theory and the index calculation. Fix an elongation function $\rho : \mathbb{R} \to [0,1]$ so that

$$\rho(\tau) = \begin{cases} 1 & \tau \geq 1 \\ 0 & \tau \leq 0 \end{cases}$$
$$0 \leq \rho'(\tau) \leq 2.$$

Then we consider sections of w^*TQ by

$$\overline{Y}_i = \rho(\tau - R_0)R_\lambda(\gamma_k^+(t)), \quad \underline{Y}_j = \rho(\tau + R_0)R_\lambda(\gamma_k^+(t)) \tag{9}$$

and denote by $\Gamma_{s^+,s^-} \subset \Gamma(w^*TQ)$ the subspace defined by

$$\Gamma_{s^+,s^-} = \bigoplus_{i=1}^{s^+}\mathbb{R}\{\overline{Y}_i\} \oplus \bigoplus_{j=1}^{s^-}\mathbb{R}\{\underline{Y}_j\}.$$

Let $k \geq 2$ and $p > 2$. We denote by

$$\mathscr{W}_\delta^{k,p}(\dot{\Sigma},Q;J;\gamma^+,\gamma^-), \quad k \geq 2$$

the Banach manifold such that

$$\lim_{\tau\to\infty} w((\tau,t)_i) = \gamma_i^+(T_i(t+t_i)), \quad \lim_{\tau\to-\infty} w((\tau,t)_j) = \gamma_j^-(T_j(t-t_j))$$

for some $t_i, t_j \in S^1$, where

$$T_i = \int_{S^1}(\gamma_i^+)^*\lambda,\, T_j = \int_{S^1}(\gamma_j^-)^*\lambda.$$

Here t_i, t_j depends on the given analytic coordinate and the parameterization of the Reeb orbits.

Now for each given $w \in \mathscr{W}_\delta^{k,p} := \mathscr{W}_\delta^{k,p}(\dot{\Sigma},Q;J;\gamma^+,\gamma^-)$, we consider the Banach space

$$\Omega_{k-1,p;\delta}^{(0,1)}(w^*\xi)$$

the $W_\delta^{k-1,p}$-completion of $\Omega^{(0,1)}(w^*\xi)$ and form the bundle

$$\mathscr{H}^{(0,1)}_{k-1,p;\delta}(\xi) = \bigcup_{w \in \mathscr{W}^{k,p}_\delta} \Omega^{(0,1)}_{k-1,p;\delta}(w^*\xi)$$

over $\mathscr{W}^{k,p}_\delta$. Then we can regard the assignment

$$\Upsilon_1 : (w, f) \mapsto \overline{\partial}^\pi w$$

as a smooth section of the bundle $\mathscr{H}^{(0,1)}_{k-1,p;\delta}(\xi) \to \mathscr{W}^{k,p}_\delta$.

Furthermore the assignment

$$\Upsilon_2 : (w, f) \mapsto w^*\lambda \circ j - f^*ds$$

defines a smooth section of the bundle

$$\Omega^1_{k-1,p}(u^*\mathscr{V}) \to \mathscr{W}^{k,p}_\delta.$$

We have already computed the linearization of each of these maps in the previous section.

With these preparations, the following is a corollary of exponential estimates established in [OW18a].

Proposition 20.4 (Corollary 6.5 [OW18a]). *Assume λ is nondegenerate. Let $w : \dot{\Sigma} \to Q$ be a contact instanton and let $w^*\lambda = a_1\, d\tau + a_2\, dt$. Suppose*

$$\begin{aligned} \lim_{\tau \to \infty} a_{1,i} &= -Q(p_i), \quad \lim_{\tau \to \infty} a_{2,i} = T(p_i) \\ \lim_{\tau \to -\infty} a_{1,j} &= -Q(q_j), \quad \lim_{\tau \to -\infty} a_{2,j} = T(q_j) \end{aligned} \tag{10}$$

at each puncture p_i and q_j. Then $w \in \mathscr{W}^{k,p}_\delta(\dot{\Sigma}, Q; J; \gamma^+, \gamma^-)$.

Now we are ready to describe the moduli space of lcs instantons with prescribed asymptotic condition as the zero set

$$\mathscr{M}(\dot{\Sigma}, Q; J; \gamma^+, \gamma^-) = \left(\mathscr{W}^{k,p}_\delta(\dot{\Sigma}, Q; J; \gamma^+, \gamma^-) \bigoplus \mathscr{W}^{k,p}_\delta(\dot{\Sigma}, \mathbb{R})\right) \cap \Upsilon^{-1}(0) \tag{11}$$

whose definition does not depend on the choice of k, p or δ as long as $k \geq 2$, $p > 2$ and $\delta > 0$ is sufficiently small. One can also vary λ and J and define the universal moduli space whose detailed discussion is postponed.

In the rest of this section, we establish the Fredholm property of the linearization map

$$D\Upsilon_{(\lambda,T)}(u) : \Omega^0_{k,p;\delta}(u^*T(Q \times \mathbb{R}); J; \gamma^+, \gamma^-) \to \Omega^{(0,1)}_{k-1,p;\delta}(w^*\xi) \oplus \Omega^{(0,1)}_{k-1,p}(f^*T\mathbb{R})$$

and compute its index. Here we also denote

$$\Omega^0_{k-1,p;\delta}(u^*T(Q \times \mathbb{R}); J; \gamma^+, \gamma^-) = W^{k-1,p}_\delta(u^*T(Q \times \mathbb{R}); J; \gamma^+, \gamma^-)$$

for the semantic reason.

For this purpose, we remark that as long as the set of punctures is non-empty, the symplectic vector bundle $w^*\xi \to \dot{\Sigma}$ is trivial. We recall that $\overline{\Sigma}$ stands for the real blow-up of the boundary punctured Riemann surface $\dot{\Sigma}$. We denote by $\Phi : E \to \overline{\Sigma} \times \mathbb{R}^{2n}$ a trivialization of $E \to \overline{\Sigma}$ and by

$$\Phi_i^+ := \Phi|_{\partial_i^+ \overline{\Sigma}}, \quad \Phi_j^- = \Phi|_{\partial_j^- \overline{\Sigma}}$$

its restrictions on the corresponding boundary components of $\partial\overline{\Sigma}$. Using the cylindrical structure near the punctures, we can extend the bundle to the bundle $E \to \overline{\Sigma}$. We then consider the following set

$$\begin{aligned}\mathscr{S} := \{A : [0,1] \to Sp(2n,\mathbb{R}) \mid 1 \notin \operatorname{Spec}(A(1)),\\ A(0) = id, \dot{A}(0)A(0)^{-1} = \dot{A}(1)A(1)^{-1}\}\end{aligned}$$

of regular paths in $Sp(2n,\mathbb{R})$ and denote by $\mu_{CZ}(A)$ the Conley-Zehnder index of the paths following [RS93]. Recall that for each closed Reeb orbit γ with a fixed trivialization of ξ, the covariant linearization $A_{(T,z)}$ of the Reeb flow along γ determines an element $A_\gamma \in \mathscr{S}$. We denote by Ψ_i^+ and Ψ_j^- the corresponding paths induced from the trivializations Φ_i^+ and Φ_j^- respectively.

We have the decomposition

$$\Omega^0_{k,p;\delta}(w^*T(Q\times\mathbb{R}); J; \gamma^+, \gamma^-) = \Omega^0_{k,p;\delta}(w^*\xi) \oplus \Omega^0_{k,p;\delta}(u^*\mathscr{V}),$$

and decomposition of the operator

$$D\Upsilon_{(\lambda,T)}(u) : \Omega^0_{k,p;\delta}(w^*T(Q\times\mathbb{R}); J; \gamma^+, \gamma^-) \to \Omega^{(0,1)}_{k-1,p;\delta}(w^*\xi) \oplus \Omega^{(0,1)}_{k-1,p;\delta}(u^*\mathscr{V}) \tag{12}$$

into

$$D\Upsilon_1(u)(Y,v) \oplus D\Upsilon_2(u)(\kappa)$$

where the summands are given as in (3) and (4) respectively. We see therefrom that $D\Upsilon_{(\lambda,T)}$ is the first-order differential operator whose first-order part is given by the direct sum operator

$$(Y^\pi, (\kappa, \upsilon)) \mapsto \overline{\partial}^{\nabla^\pi} Y^\pi \oplus (d\kappa \circ j - d\upsilon)$$

where we write $(Y,v) = \left(Y^\pi + \kappa R_\lambda, \upsilon \frac{\partial}{\partial s}\right)$ for $\kappa = \lambda(Y)$, $\upsilon = ds(v)$. Here we have

$$\overline{\partial}^{\nabla^\pi} : \Omega^0_{k,p;\delta}(w^*\xi; J; \gamma^+, \gamma^-) \to \Omega^{(0,1)}_{k-1,p;\delta}(w^*\xi)$$

and the second summand can be written as the standard Cauchy-Riemann operator

$$\overline{\partial} : W^{k,p}(\dot{\Sigma}; \mathbb{C}) \to \Omega^{(0,1)}_{k-1,p}(\dot{\Sigma}, \mathbb{C}); \quad \upsilon + i\kappa =: \varphi \mapsto \overline{\partial}\varphi.$$

The following proposition can be derived from the arguments used by Lockhart and McOwen [LM85]. However before applying their general theory, one needs to pay some preliminary measure to handle the fact that the order of the operators $D\Upsilon(w)$ are different depending on the direction of ξ or on that of R_λ.

Proposition 20.5. *Suppose $\delta > 0$ satisfies the inequality*

$$0 < \delta < \min\left\{\frac{\mathrm{gap}(\gamma^+,\gamma^-)}{p}, \frac{2}{p}\right\}$$

where $\mathrm{gap}(\gamma^+,\gamma^-)$ *is the spectral gap, given in* (8), *of the asymptotic operators* $A_{(T_j,z_j)}$ *or* $A_{(T_i,z_i)}$ *associated to the corresponding punctures. Then the operator* (12) *is Fredholm.*

Proof. We first note that the operators $\overline{\partial}^{\nabla^\pi} + T_{dw}^{\pi,(0,1)} + B^{(0,1)}$ and $\overline{\partial}$ are Fredholm: The relevant a priori coercive $W^{k,2}$-estimates for any integer $k \geq 1$ for the derivative dw on the punctured Riemann surface $\dot{\Sigma}$ with cylindrical metric near the punctures are established in [OW18a] for the operator $\overline{\partial}^{\nabla^\pi} + T_{dw}^{\pi,(0,1)} + B^{(0,1)}$ and the one for $\overline{\partial}$ is standard. From this, the standard interpolation inequality establishes the $W^{k,p}$-estimates for $D\Upsilon(w)$ for all $k \geq 2$ and $p \geq 2$.

Proposition 20.6. *The off-diagonal terms decay exponentially fast as $|\tau| \to \infty$.*

Proof. For the $(1,2)$-term, we derive from $\overline{\partial}^\pi w = 0$

$$\left(\frac{\partial w}{\partial \tau}\right)^\pi + J\left(\frac{\partial w}{\partial t}\right)^\pi = 0.$$

Therefore we have

$$\partial^\pi w\left(\frac{\partial}{\partial \tau}\right) = \frac{1}{2}\left(\left(\frac{\partial w}{\partial \tau}\right)^\pi - J\left(\frac{\partial w}{\partial t}\right)^\pi\right) = -J\left(\frac{\partial w}{\partial t}\right)^\pi.$$

By the exponential convergence $\frac{\partial w}{\partial t} \to TR_\lambda(\gamma_\infty(t))$, we derive

$$J\partial^\pi w\left(\frac{\partial}{\partial \tau}\right) = \left(\frac{\partial w}{\partial t}\right)^\pi \to 0$$

since $\frac{\partial w}{\partial t} \to TR_\lambda$. Therefore the off-diagonal term converges to the zero operator exponentially fast.

For the $(2,1)$-term, we evaluate

$$(Y^\pi \rfloor d\lambda) \circ j\left(\frac{\partial}{\partial \tau}\right) = d\lambda\left(Y, \frac{\partial w}{\partial t}\right)$$

$$(Y^\pi \rfloor d\lambda) \circ j\left(\frac{\partial}{\partial t}\right) = -d\lambda\left(Y, \frac{\partial w}{\partial \tau}\right).$$

Therefore we have derived

$$(Y^\pi \rfloor d\lambda) \circ j = d\lambda \left(Y, \frac{\partial w}{\partial t}\right) d\tau - d\lambda \left(Y, \frac{\partial w}{\partial \tau}\right) dt.$$

Therefore we have shown

$$((\cdot)^\pi \rfloor d\lambda) \circ j \to d\lambda \left(\cdot, \frac{\partial w}{\partial t}\right) d\tau - d\lambda \left(\cdot, \frac{\partial w}{\partial \tau}\right) dt.$$

Since $\frac{\partial w}{\partial t} \to TR_\lambda$, the first term converges to zero, and the second term converges to

$$-d\lambda(\cdot, JTR_\lambda) = Td\lambda\left(\cdot, \frac{\partial}{\partial s}\right) = 0.$$

This finishes the proof. □

Therefore it can be homotoped to the block-diagonal form, i.e., into the direct sum operator

$$\left(\overline{\partial}^{\nabla^\pi} + T_{dw}^{\pi,(0,1)} + B^{(0,1)}\right) \oplus \overline{\partial}$$

via a continuous path of Fredholm operators given by

$$s \in [0,1] \mapsto \begin{pmatrix} \overline{\partial}^{\nabla^\pi} + B^{(0,1)} + T_{dw}^{\pi,(0,1)} & , \frac{s}{2}(\cdot) \cdot ((\mathscr{L}_{R_\lambda} J) J(\partial^\pi w)) \\ s((\cdot)^\pi \rfloor d\lambda) \circ j & , \qquad \underline{\partial} \end{pmatrix}$$

from $s = 1$ to $s = 0$. The Fredholm property of this path follows from the fact that the off-diagonal terms are 0-th order linear operators. □

Then by the continuous invariance of the Fredholm index, we obtain

$$\operatorname{Index} D\Upsilon_{(\lambda,T)}(w) = \operatorname{Index}\left(\overline{\partial}^{\nabla^\pi} + T_{dw}^{\pi,(0,1)} + B^{(0,1)}\right) + \operatorname{Index}(\overline{\partial}). \tag{13}$$

Therefore it remains to compute the latter two indices.

We denote by $m(\gamma)$ the multiplicity of the Reeb orbit in general. Then we have the following index formula.

Theorem 20.7. *We fix a trivialization $\Phi : E \to \overline{\Sigma}$ and denote by Ψ_i^+ (resp. Ψ_j^-) the induced symplectic paths associated to the trivializations Φ_i^+ (resp. Φ_j^-) along the Reeb orbits γ_i^+ (resp. γ_j^-) at the punctures p_i (resp. q_j) respectively. Then we have*

$$\begin{aligned} &\operatorname{Index}(\overline{\partial}^{\nabla^\pi} + T_{dw}^{\pi,(0,1)} + B^{(0,1)}) \\ &= n(2 - 2g - s^+ - s^-) + 2c_1(w^*\xi) + (s^+ + s^-) \\ &\quad + \sum_{i=1}^{s^+} \mu_{CZ}(\Psi_i^+) - \sum_{j=1}^{s^-} \mu_{CZ}(\Psi_j^-) \end{aligned} \tag{14}$$

$$\text{Index}(\overline{\partial}) = 2\sum_{i=1}^{s^+} m(\gamma_i^+) + 2\sum_{j=1}^{s^-} m(\gamma_j^-) - 2g. \tag{15}$$

In particular,

$$\begin{aligned}
&\text{Index}\, D\Upsilon_{(\lambda,T)}(u) \\
&= n(2-2g-s^+-s^-) + 2c_1(w^*\xi) \\
&\quad + \sum_{i=1}^{s^+} \mu_{CZ}(\Psi_i^+) - \sum_{j=1}^{s^-} \mu_{CZ}(\Psi_j^-) \\
&\quad + \sum_{i=1}^{s^+} (2m(\gamma_i^+)+1) + \sum_{j=1}^{s^-} (2m(\gamma_j^-)+1) - 2g.
\end{aligned} \tag{16}$$

Proof. The formula (14) can be immediately derived from the general formula given in the top of p. 52 of Bourgeois's thesis [Bou02]: The summand $(s^+ + s^-)$ comes from the factor Γ_{s^+,s^-} in the decomposition (6) which has dimension $s^+ + s^-$.

So it remains to compute the index (15). To compute the (real) index of $\overline{\partial}$, we consider the Dolbeault complex

$$0 \to \Omega^0(\Sigma; D) \to \Omega^1(\Sigma; D) \to 0$$

where $D = D^+ + D^-$ is the divisor associated to the set of punctures

$$D^+ = \sum_{i=1}^{s^+} m(\gamma_i^+) p_i, \quad D^- = \sum_{j=1}^{s^-} m(\gamma_j^-) q_j$$

where $m(\gamma_i^+)$ (resp. $m(\gamma_j^-)$) is the multiplicity of the Reeb orbit γ_i^+ (resp. γ_j^-). The standard Riemann-Roch formula then gives rise to the formula for the Euler characteristic

$$\begin{aligned}
\chi(D) &= 2\dim_{\mathbb{C}} H^0(D) - 2\dim_{\mathbb{C}} H^1(D) = 2\deg(D) - 2g \\
&= \sum_{i=1}^{s^+} 2m(\gamma_i^+) + \sum_{j=1}^{s^-} 2m(\gamma_j^-) - 2g.
\end{aligned}$$

This finishes the proof. □

Remark 20.8. We can also symplectify the Fredholm theory from [Ohb] and the index calculation given in [OY22] in the similar way to give rise to the relevant theory for the pseudoholomorphic curves on the symplectization which we leave to the readers as an exercise.

21 Exponential asymptotic analysis

Recalling that for any $\widetilde{J}$-holomorphic curve (w,f), w is a contact instanton for J on Q. Furthermore we have

$$(w,f)^*T(Q\times\mathbb{R}) = w^*TQ \oplus f^*T\mathbb{R} = w^*\xi \oplus \mathrm{span}_{\mathbb{R}}\left\{\frac{\partial}{\partial s}, R_\lambda\right\}.$$

The last splitting is respected to the *canonical connection* of the almost Hermitian manifold

$$(Q\times\mathbb{R}, d\lambda + ds\wedge\lambda, \widetilde{J}).$$

Indeed, we have

$$\nabla^{\mathrm{can}} = \nabla^\pi \oplus \nabla_0$$

where $\nabla^\pi = \nabla|_\xi$ and ∇_0 is the trivial connection on $\mathrm{span}_{\mathbb{R}}\{\frac{\partial}{\partial s}, R_\lambda\}$. (Recall Proposition 0.3.)

Remark 21.1. Here again we would like to emphasize that the usage of the canonical connection of the almost Hermitian manifold, not the Levi-Civita connection, admits this splitting.

21.1 Definition of asymptotic operators and their formulae

Now we study a finer analysis of the asymptotic behavior along the Reeb orbit. Our discussion thereof is close to the one given in [OW18b, Section 11.2 &11.5] where the more general Morse-Bott case is studied.

For this purpose, we evaluate the linearization operator $D\Upsilon$ against $\frac{\partial}{\partial\tau}$. We have already checked in Proposition 20.6 that the off-diagonal terms of the matrix representation of $D\Upsilon(w)$ decays exponentially fast in the direction τ in the previous section and so we have only to examine the diagonal terms $D\Upsilon_1(w)$ and $D\Upsilon_2(w)$.

First we consider $D\Upsilon_2$ and rewrite

$$D\Upsilon_2 = \bar\partial = \frac{1}{2}(\partial_\tau + i\partial_t).$$

Therefore we have the asymptotic operator

$$A^\perp_{(\lambda,J,\nabla)} := i\partial_t \tag{1}$$

which does not depend on the choice of $J\in\mathscr{J}_\lambda(Q,\xi)$. The eigenfunction expansions for this operator is nothing but the standard Fourier series for $f\in L^2(S^1,Q)$.

This being said, we now focus on the Q-component $D\Upsilon_1$ of the asymptotic operator and compute

$$D\Upsilon_1(w)\left(\frac{\partial}{\partial\tau}\right) = \frac{1}{2}(\nabla^\pi_\tau + J\nabla^\pi_t) + T^{\pi,(0,1)}_{dw}\left(\frac{\partial}{\partial\tau}\right) + B^{(0,1)}\left(\frac{\partial}{\partial\tau}\right). \tag{2}$$

In fact, this is nothing but the left hand side of (4). We write

$$2D\Upsilon(w)\left(\frac{\partial}{\partial\tau}\right) = \nabla^\pi_\tau + A^\pi_{(\lambda,J,\nabla)}.$$

and define the family of operators

$$A^\tau_{(\lambda,J,\nabla)} : \Gamma(w^*_\tau\xi) \to \Gamma(w^*_\tau\xi)$$

Thanks to the exponential convergence of $w_\tau \to \gamma_\pm$ as $\tau \to \pm\infty$, we can take the limit of the conjugate operators

$$\Pi^\infty_\tau A^\tau_{(\lambda,J,\nabla)}(\Pi^\infty_\tau)^{-1} : \Gamma(\gamma^*_\pm\xi) \to \Gamma(\gamma^*_\pm\xi) \tag{3}$$

as $\tau \to \pm\infty$ respectively, where Π^∞_τ is the parallel transport along the short geodesics from $w(\tau,t)$ to $w(\infty,t)$. This conjugate is defined for all sufficiently large $|\tau|$.

Since the discussion at $\tau = -\infty$ will be the same, we will focus our discussion on the case at $\tau = +\infty$ from now on.

Definition 21.2 (Asymptotic operator)**.** Let (τ,t) be the cylindrical (or strip-like) coordinate, and let ∇^π be the almost Hermitian connection on $w^*\xi$ induced by the contact triad connection ∇ of (Q,λ,J). We define the *asymptotic operator* of a contact instanton w to be the limit operator

$$A^\pi_{(\lambda,J,\nabla)} := \lim_{\tau\to+\infty} \Pi^\infty_\tau A^\tau_{(\lambda,J,\nabla)}(\Pi^\infty_\tau)^{-1}. \tag{4}$$

Obviously we can define the asymptotic operator at negative punctures in the similar way.

Remark 21.3. The upshot of our definition lies in its naturality depending only on the given adapted pair (λ,J) and its associated triad connection. Other literature definition of the asymptotic operators is given differently in the way how the dependence on the given J of the final formula is hard to analyze. This definition can be given any connection ∇ on Q that has the property that $\nabla(\xi) \subset \xi$ and $\nabla^\pi(J|_\xi) = 0$. Using the exponential convergence, one can check that the definition does not depend on the choice of such connections. (See Proposition 5.3 for such a formula.)

Therefore it is conceivable that a good choice of connection will facilitate the study asymptotic operators, which is precisely what is happening by our choice of contact triad connection.

Now we find the formula for this limit operator *with respect to the contact triad connection*. Since $T(R_\lambda,\cdot) = 0$, $\left(\frac{\partial w}{\partial\tau}\right)^\pi = -J\left(\frac{\partial w}{\partial t}\right)^\pi$ and $\frac{\partial w}{\partial t}(\tau,\cdot) \mapsto TR_\lambda$ exponentially fact, we obtain

$$2T^{\pi,(0,1)}_{dw}(\frac{\partial}{\partial \tau}) = T^\pi\left(\frac{\partial w}{\partial \tau},\cdot\right) \to 0.$$

On the other hand, we have

$$2B^{(0,1)}\left(\frac{\partial}{\partial \tau}\right) = -\frac{1}{2}\lambda\left(\frac{\partial w}{\partial \tau}\right)\mathscr{L}_{R_\lambda}J - \frac{1}{2}\lambda\left(\frac{\partial w}{\partial t}\right)J\mathscr{L}_{R_\lambda}J.$$

We note that the right hand side converges to $\frac{T}{2}J\nabla_{R_\lambda}$ since $w^*\lambda \to T\,dt$ as $\tau \to \infty$. This convergence proves

Proposition 21.4. *Let ∇ be the contact triad connection associated to any adapted pair (λ, J). Then the asymptotic operator $A^\pi_{(\lambda,J,\nabla)}$ is given by*

$$A^\pi_{(\lambda,J,\nabla)} = -J\nabla_t + \frac{T}{2}\mathscr{L}_{R_\lambda}JJ.$$

Proof. We recall $T(R_\lambda,\cdot) = 0$ and the identity (2). On the other hand we have $\frac{\partial w}{\partial t}(\infty,t) = TR_\lambda$. Combining the two and above convergence of $B^{(0,1)}(\partial_\tau)$, we have finished the proof. □

Now one may define the full asymptotic operator $A^\pi_{(\lambda,J,\nabla)}$ of the pseudoholomorphic curves to be the operator

$$A^\pi_{(\lambda,J,\nabla)} : \gamma^*\xi \oplus \mathbb{C} \to \gamma^*\xi \oplus \mathbb{C}$$

defined by

$$A_{(\lambda,J,\nabla)} = A^\pi_{(\lambda,J,\nabla)} \oplus A^\perp_{(\lambda,J,\nabla)}.$$

Here $\mathbb{C}$ stands for the pull-back bundle

$$(\pi \circ u_\tau)^*\left(\mathbb{R}\left\{\frac{\partial}{\partial s}, R_\lambda\right\}\right) = w_\tau^*\left(\mathbb{R}\left\{\frac{\partial}{\partial s}, R_\lambda\right\}\right) \cong \mathbb{R}\left\{\frac{\partial}{\partial s}, R_\lambda\right\}$$

which is canonically trivialized, and hence may be regarded as a vector bundle over the curves w_τ on Q. (Compare this with [Par19, Definition 2.28].)

21.2 Asymptotic operator and the Levi-Civita connection

Up until now, we have emphasized the usage of triad connection which give rise to an optimal form of tensorial expression. The main reason behind this switch is the following surprising property of Levi-Civita connection of the triad metric.

We recall that while $\nabla_Y J = 0$ for all $Y \in \xi$, $\nabla_{R_\lambda} J \neq 0$ for the contact triad connection in general. (In fact, the latter holds if and only if R_λ is a Killing vector field, i.e., $\nabla R_\lambda = 0$ with respect to ∇. See [OW18a, Remark 2.4].) However while $\nabla^{LC}_Y J \neq 0$

for the Levi-Civita connection in general, the Levi-Civita connection carries the following useful property for the study of asymptotic operators.

Proposition 21.5 (Lemma. 6.1 [Bla10], Proposition 4 [OW14]). $\nabla^{\text{LC}}_{R_\lambda} J = 0.$

To utilize this property in our study of eigenfunction analysis for the linearized operator $A_{(\lambda,K,\nabla)} = -\frac{1}{2}J\nabla_t$, we convert it in terms of the Levi-Civita connection. For this purpose, the following lemma is crucial.

Lemma 21.6 (Lemma 6.2 [Bla10], Lemma 9 [OW14]). *For any $Y \in \xi$, we have*

$$\nabla^{\text{LC}}_Y R_\lambda = \frac{1}{2}JY + \frac{1}{2}\mathscr{L}_{R_\lambda} JJY.$$

Therefore combining Proposition 21.4 and this lemma, we can rewrite the linearized operator in terms of the Levi-Civita connection as follows. It reveals a rather remarkable property of the asymptotic operator *computed in terms of the triad connection* that it becomes a Hermitian operator for any choice of adapted pair (λ, J) for the contact manifold (Q, ξ).

Corollary 21.7. *Let (λ, J) be any adapted pair and let ∇^{LC} be the Levi-Civita connection of the triad metric of (Q, λ, J). For given contact instanton w with its action $\int \gamma^* \lambda = T$ at a puncture, let $A_{(\lambda,J,\nabla)}$ be the asymptotic operator of w at the with cylindrical coordinate (τ, t). Then*

1. $[\nabla^{\text{LC}}_t, J] (= \nabla^{\text{LC}}_t J) = 0,$
2. $A^{\pi}_{(\lambda,J,\nabla)} = -J\nabla_t + \frac{T}{2}\mathscr{L}_{R_\lambda} JJ = -J\nabla^{\text{LC}}_t - \frac{T}{2} Id + \frac{T}{2}\mathscr{L}_{R_\lambda} JJ.$

In particular, it induces a J-Hermitian operator on $(\xi, g|_\xi)$ with respect to the triad metric g.

The same kind of property also holds for the open string case for the Legendrian pair (R_0, R_1). This explicit formula for the asymptotic operator enables us to prove the following series of perturbation results in [KO23] on the eigenfunctions and eigenvalues of the asymptotic operators *under the perturbation of J's* inside the set $\mathscr{J}_\lambda$ of λ-adapted CR almost complex structures J. The above explicit form of asymptotic operator on J and λ enable us to prove the following genericity in terms of the choice of contact triads (Q, λ, J) of contact manifold (Q, ξ).

Theorem 21.8 (Simpleness of eigenvalues; [KO23]). *Let (Q, ξ) be a contact manifold. Assume that λ is nondegenerate. For a generic choice of adapted pair (λ, J), all eigenvalues μ_i of the asymptotic operator are simple for all closed Reeb orbits.*

21.3 Finer asymptotic behavior

Once these theorems at our disposal, we can further proceed with a finer asymptotic convergence result for the pseudoholomorphic curves on symplectization in a canonical covariant tensorial way. Similar study is given in [HWZ96b, HWZ96a,

HWZ02] in 3 dimension using special coordinates followed by adjusting the almost complex structure along the given Reeb orbit.

Let $u = (w, f)$ be any finite energy $\widetilde{J}$-pseudoholomorphic curve from a punctured Riemann surface $\dot{\Sigma}$ to the symplectization of $M = Q \times \mathbb{R}$. Then w is a contact instanton of the triad (Q, λ, J). Consider a puncture and let

$$w : [\tau_0, \infty) \times S^1 \to Q$$

be the representation of w in the cylindrical coordinate for some large $\tau_0 > 0$ and denote $w_\tau := w(\tau, \cdot)$.

We have shown that there exists a isospeed Reeb trajectory (γ, T) such that

$$w(\tau,t) = w_\tau(t) \to w_\infty = \gamma(T(\cdot)) \text{ as } \tau \to \infty.$$

Denote by $\gamma_T : S^1 \to Q$ the curve given by $\gamma_T(t) := \gamma(Tt)$.

At each τ we have the operator

$$A^{\tau,\pi}_{(\lambda,J,\nabla)} : \Gamma(w_\tau^*\xi) \to \Gamma(w_\tau^*\xi)$$

defined by

$$A^{\tau,\pi}_{(\lambda,J,\nabla)} := J\nabla^\pi_t + T^{\pi,(0,1)}_{dw}\left(\frac{\partial}{\partial \tau}\right) + B^{(0,1)}\left(\frac{\partial}{\partial \tau}\right). \tag{5}$$

We have shown that as $\tau \to \infty$ the operator

$$\Pi_\tau^\infty \circ A^{\tau,\pi}_{(\lambda,J,\nabla)} \circ (\Pi_\tau^\infty)^{-1} : \Gamma(\gamma_T^*\xi) \to \Gamma(\gamma_T^*\xi),$$

converges to the asymptotic operator

$$A^{\pi}_{(\lambda,J,\nabla)} : \Gamma(\gamma_T^*\xi) \to \Gamma(\gamma_T^*\xi), \tag{6}$$

We have shown that there exists an iso-speed Reeb trajectory (γ, T) such that

$$w(\tau,t) = w_\tau(t) \to w_\infty = \gamma(\cdot) \text{ as } \tau \to \infty.$$

At each τ we have the operator

$$A^{\tau,\pi}_{(\lambda,J,\nabla)} : \Gamma(w_\tau^*\xi) \to \Gamma(w_\tau^*\xi)$$

defined by

$$A^{\tau,\pi}_{(\lambda,J,\nabla)} := J\nabla^\pi_t + T^{\pi,(0,1)}_{dw}\left(\frac{\partial}{\partial \tau}\right) + B^{(0,1)}\left(\frac{\partial}{\partial \tau}\right). \tag{7}$$

We have shown that as $\tau \to \infty$ the operator

$$\Pi_\tau^\infty \circ A^{\tau,\pi}_{(\lambda,J,\nabla)} \circ (\Pi_\tau^\infty)^{-1} : \Gamma(\gamma_T^*\xi) \to \Gamma(\gamma_T^*\xi),$$

converges to the asymptotic operator

$$A^{\pi}_{(\lambda,J,\nabla)} := A^{\pi}_{(\lambda,J,\nabla)}(\gamma) : \Gamma(\gamma_T^*\xi) \to \Gamma(\gamma_T^*\xi), \tag{8}$$

given by $A^{\pi}_{(\lambda,J,\nabla)}(\gamma)$.

Then the following theorem describes the asymptotic behavior of finite energy contact instanon w as $|\tau| \to \infty$, which is the analog to [HWZ96b, Theorem 1.4].

Theorem 21.9 (Asymptotic behavior; Theorem 1.10 [KO23]). *Assume that* (γ, T) *is nondegenerate. Consider t*

$$v(\tau,t) := \frac{\zeta(\tau,t)}{\|\zeta(\tau)\|_{L^2(w_\tau^*\xi}}, \quad \alpha(\tau) = \frac{d}{d\tau}\log\|\zeta\|_{L^2(w_\tau^*\xi)}$$

Then

1. *Either* $\zeta(\tau,t) = 0$ *for all* $(\tau,t) \in [\tau_0,\infty)$
2. *or otherwise we have the following:*

    a. *There exists an eigenvector e of* $A^{\pi}_{(\lambda,J,\nabla)}$ *of eigenvalue* μ *such that* $v_\tau \to e$ *as* $\tau \to \infty$ *and*

    $$\zeta(\tau,t) = e^{\int_{\tau_0}^{\tau}\alpha(s)\,ds}(e(t) + \widetilde{r}(\tau,t))$$

    b. *There exist constants* $\delta > 0$, *and* C_β *for all multi-indices* $\beta = (\beta_1,\beta_2) \in \mathbb{N}\times\mathbb{N}$ *such that*

    $$\sup_{(\tau,t)} |(\nabla^\beta v)(\tau,t)| \le C_\beta$$

    for all $\tau \ge \tau_0$ *and* $t \in S^1$.

Note that the above representation formula of ζ implies the following convergence of the tangent plane.

Corollary 21.10 (Convergence of tangent plane; Corollary 1.11 [KO23]). *Assume the second alternative in Theorem 8.2 and denote*

$$P(\tau,t) := \mathrm{Image}\, dw(\tau,t) \in \mathrm{Gr}_2(\xi_{w(\tau,t)})$$

where $\mathrm{Gr}_2(\xi_x)$ *is the set of 2 dimensional subspaces of the contact hyperplane* $\xi_x \subset T_xQ$. *Then* $P(\tau,t) \to \mathrm{span}_{\mathbb{R}}\{e(t), Je(t) + TR_\lambda(\gamma(t))\}$ *exponentially fast in* C^∞ *topology uniformly in* $t \in S^1$.

Finally we would like to just mention that the same asymptotic study can be made by now in a straightforward way by incorporating the boundary condition as done in [OY22], [Ohb], [Oh22b]. The exponential convergence statement can be derived by a boot-strap argument using the exponential convergence result on ζ as $\tau \to \infty$ given in Section 17. (See [OW18b], [OY22] for the details of this exponential convergence and the boot-strap argument.)

Combining all the above, we have finished the proof of Theorem 21.9.

Now we prove the following two theorems which are analogs to Theorem 5.1 and Theorem 5.2 respectively whose proofs follow those in [HWZ01, Section 5] after translating their coordinate exposition into that of covariant tensorial exposition, and so omitted.

Theorem 21.11 (Compare with Theorem 5.1 [HWZ01]). *Consider a nondegenerate finite energy contact instanton w on cylindrical coordinates* $(\tau,t) \in [0,\infty) \times S^1$. *Then there exists some* $\tau_0 > 0$ *such that the sets*

$$\{(\tau,t) \mid w(\tau,z) \in \operatorname{Image}\gamma,\, \tau \geq \tau_0\}$$

$$\{(\tau,z) \mid \zeta(z) = 0,\, \tau \geq \tau_0\}$$

consist of finitely many points.

The following naturally occurs in the study of asymptotic behavior of bubbles which provides a strong embdding control of such planes in 3 dimension.

Theorem 21.12 (Compare with Theorem 5.2 [HWZ01]). *Consider a nondegenerate finite energy contact instanton w on* $\mathbb{C} \cong S^2 \setminus \{pt\}$. *Then each of the sets*

$$\{z \in \mathbb{C} \mid w(z) \in \operatorname{Image}\gamma\}, \quad \{z \in \mathbb{C} \mid d^\pi w(z) = 0\}$$

consists of finitely many points.

Significance of this theorem is not as large as in 3 dimension in higher dimension because they are expected to be empty for a generic choice of J by the transversality argument by the dimensional reason.

22 Wedge products of vector-valued forms

In this section, we continue with the setting from Appendix 23. To be specific, we assume (P,h) is a Riemannian manifold of dimension n with metric h, and denote by D the Levi–Civita connection. $E \to P$ is a vector bundle with inner product $\langle \cdot,\cdot \rangle$ and ∇ is a connection of E which is compatible with $\langle \cdot,\cdot \rangle$.

We remark that we include this section for the sake of completeness of our treatment of vector valued forms, and the content of this appendix is not used in any section of this article. Actually one can derive exponential decay using the differential inequality method from the formulas we provide here. We leave the proof to interested reader or to the earlier arXiv version of [OW18a].

The wedge product of forms can be extended to E-valued forms by defining

$$\wedge : \Omega^{k_1}(E) \times \Omega^{k_2}(E) \to \Omega^{k_1+k_2}(E)$$
$$\beta_1 \wedge \beta_2 = \langle \zeta_1, \zeta_2 \rangle \alpha_1 \wedge \alpha_2,$$

where $\beta_1 = \alpha_1 \otimes \zeta_1 \in \Omega^{k_1}(E)$ and $\beta_2 = \alpha_2 \otimes \zeta_2 \in \Omega^{k_2}(E)$ are E-valued forms.

Lemma 22.1. *For* $\beta_1, \beta_2 \in \Omega^k(E)$,

$$\langle \beta_1, \beta_2 \rangle = *(\beta_1 \wedge *\beta_2).$$

Proof. Write $\beta_1 = \alpha_1 \otimes \zeta_1$ and $\beta_2 = \alpha_2 \otimes \zeta_2$. Then

$$\begin{aligned} *(\beta_1 \wedge *\beta_2) &= *((\alpha_1 \otimes \zeta_1) \wedge ((*\alpha_2) \otimes \zeta_2)) \\ &= *(\langle \zeta_1, \zeta_2 \rangle \, \alpha_1 \wedge *\alpha_2) \\ &= \langle \zeta_1, \zeta_2 \rangle * (\alpha_1 \wedge *\alpha_2) \\ &= \langle \zeta_1, \zeta_2 \rangle \, h(\alpha_1, \alpha_2) \\ &= \langle \beta_1, \beta_2 \rangle. \end{aligned}$$

□

The following lemmas exploit the compatibility of ∇ with the inner product $\langle \cdot, \cdot \rangle$.

Lemma 22.2.

$$d(\beta_1 \wedge \beta_2) = d^\nabla \beta_1 \wedge \beta_2 + (-1)^{k_1} \beta_1 \wedge d^\nabla \beta_2,$$

where $\beta_1 \in \Omega^{k_1}(E)$ and $\beta_2 \in \Omega^{k_2}(E)$ are E-valued forms.

Proof. We write $\beta_1 = \alpha_1 \otimes \zeta_1$ and $\beta_2 = \alpha_2 \otimes \zeta_2$ and calculate

$$\begin{aligned} d(\beta_1 \wedge \beta_2) &= d(\langle \zeta_1, \zeta_2 \rangle \, \alpha_1 \wedge \alpha_2) \\ &= d\langle \zeta_1, \zeta_2 \rangle \wedge \alpha_1 \wedge \alpha_2 + \langle \zeta_1, \zeta_2 \rangle \, d(\alpha_1 \wedge \alpha_2) \\ &= \langle \nabla \zeta_1, \zeta_2 \rangle \wedge \alpha_1 \wedge \alpha_2 + \langle \zeta_1, \nabla \zeta_2 \rangle \wedge \alpha_1 \wedge \alpha_2 \\ &\quad + \langle \zeta_1, \zeta_2 \rangle \, d\alpha_1 \wedge \alpha_2 + (-1)^{k_1} \langle \zeta_1, \zeta_2 \rangle \, \alpha_1 \wedge d\alpha_2, \end{aligned}$$

while

$$\begin{aligned} d^\nabla \beta_1 \wedge \beta_2 &= d^\nabla (\alpha_1 \otimes \zeta_1) \wedge (\alpha_2 \otimes \zeta_2) \\ &= (d\alpha_1 \otimes \zeta_1 + (-1)^{k_1} \alpha_1 \wedge \nabla \zeta_1) \wedge (\alpha_2 \otimes \zeta_2) \\ &= \langle \zeta_1, \zeta_2 \rangle \, d\alpha_1 \wedge \alpha_2 + \langle \nabla \zeta_1, \zeta_2 \rangle \wedge \alpha_1 \wedge \alpha_2. \end{aligned}$$

A similar calculation shows that

$$(-1)^{k_1} \beta_1 \wedge d^\nabla \beta_2 = (-1)^{k_1} \langle \zeta_1, \zeta_2 \rangle \, \alpha_1 \wedge d\alpha_2 + \langle \zeta_1, \nabla \zeta_2 \rangle \wedge \alpha_1 \wedge \alpha_2.$$

Summing these up, we get the equality we want. □

Lemma 22.3. *Assume $\beta_0 \in \Omega^k(E)$ and $\beta_1 \in \Omega^{k+1}(E)$, then we have*

$$\langle d^\nabla \beta_0, \beta_1 \rangle - (-1)^{n(k+1)} \langle \beta_0, \delta^\nabla \beta_1 \rangle = *d(\beta_0 \wedge *\beta_1).$$

Proof. We calculate

$$\begin{aligned} *d(\beta_0 \wedge *\beta_1) &= *(d^\nabla \beta_0 \wedge *\beta_1 + (-1)^k \beta_0 \wedge (d^\nabla * \beta_1)) \\ &= \langle d^\nabla \beta_0, \beta_1 \rangle + (-1)^n * (\beta_0 \wedge *(*d^\nabla * \beta_1) \\ &= \langle d^\nabla \beta_0, \beta_1 \rangle - (-1)^{n(k+1)} \langle \beta_0, \delta^\nabla \beta_1 \rangle. \end{aligned}$$

□

23 The Weitzenböck formula for vector-valued forms

In this appendix, we recall the standard Weitzenböck formulas applied to our current circumstance. A good exposition on the general Weitzenböck formula is provided in the appendix of [FU84].

Assume (P,h) is a Riemannian manifold of dimension n with metric h, and D is the Levi–Civita connection. Let $E \to P$ be any vector bundle with inner product $\langle \cdot,\cdot \rangle$, and assume ∇ is a connection on E which is compatible with $\langle \cdot,\cdot \rangle$.

For any E-valued form s, calculating the (Hodge) Laplacian of the energy density of s, we get

$$-\frac{1}{2}\Delta|s|^2 = |\nabla s|^2 + \langle Tr\nabla^2 s, s\rangle,$$

where for $|\nabla s|$ we mean the induced norm in the vector bundle $T^*P\otimes E$, i.e., $|\nabla s|^2 = \sum_i |\nabla_{E_i} s|^2$ with $\{E_i\}$ an orthonormal frame of TP. $Tr\nabla^2$ denotes the connection Laplacian, which is defined as $Tr\nabla^2 = \sum_i \nabla^2_{E_i,E_i} s$, where $\nabla^2_{X,Y} := \nabla_X\nabla_Y - \nabla_{\nabla_X Y}$.

Denote by $\Omega^k(E)$ the space of E-valued k-forms on P. The connection ∇ induces an exterior derivative by

$$\begin{aligned} d^\nabla &: \Omega^k(E) \to \Omega^{k+1}(E) \\ d^\nabla(\alpha\otimes\zeta) &= d\alpha\otimes\zeta + (-1)^k\alpha\wedge\nabla\zeta. \end{aligned}$$

It is not hard to check that for any 1-forms, equivalently one can write

$$d^\nabla\beta(v_1,v_2) = (\nabla_{v_1}\beta)(v_2) - (\nabla_{v_2}\beta)(v_1),$$

where $v_1, v_2 \in TP$.

We extend the Hodge star operator to E-valued forms by

$$\begin{aligned} * &: \Omega^k(E) \to \Omega^{n-k}(E) \\ *\beta &= *(\alpha\otimes\zeta) = (*\alpha)\otimes\zeta \end{aligned}$$

for $\beta = \alpha\otimes\zeta \in \Omega^k(E)$.

Define the Hodge Laplacian of the connection ∇ by

$$\Delta^\nabla := d^\nabla\delta^\nabla + \delta^\nabla d^\nabla,$$

where δ^∇ is defined by

$$\delta^\nabla := (-1)^{nk+n+1} * d^\nabla * .$$

The following lemma is important for the derivation of the Weitzenböck formula.

Lemma 23.1. *Assume $\{e_i\}$ is an orthonormal frame of P, and $\{\alpha^i\}$ is the dual frame. Then we have*

$$d^\nabla = \sum_i \alpha^i \wedge \nabla_{e_i}$$
$$\delta^\nabla = -\sum_i e_i \rfloor \nabla_{e_i}.$$

Proof. Assume $\beta = \alpha \otimes \zeta \in \Omega^k(E)$. Then

$$\begin{aligned} d^\nabla(\alpha \otimes \zeta) &= (d\alpha) \otimes \zeta + (-1)^k \alpha \wedge \nabla \zeta \\ &= \sum_i \alpha^i \wedge \nabla_{e_i} \alpha \otimes \zeta + (-1)^k \alpha \wedge \nabla \zeta. \end{aligned}$$

On the other hand,

$$\begin{aligned} \sum_i \alpha^i \wedge \nabla_{e_i}(\alpha \otimes \zeta) &= \sum_i \alpha^i \wedge \nabla_{e_i} \alpha \otimes \zeta + \alpha^i \wedge \alpha \otimes \nabla_{e_i} \zeta \\ &= \sum_i \alpha^i \wedge \nabla_{e_i} \alpha \otimes \zeta + (-1)^k \alpha \wedge \nabla \zeta, \end{aligned}$$

so we have proved the first statement.

For the second equality, we compute

$$\begin{aligned} \delta^\nabla(\alpha \otimes \zeta) &= (-1)^{nk+n+1} * d^\nabla * (\alpha \otimes \zeta) \\ &= (\delta \alpha) \otimes \zeta + (-1)^{nk+n+1} * (-1)^{n-k} (*\alpha) \wedge \nabla \zeta \\ &= -\sum_i e_i \rfloor \nabla_{e_i} \alpha \otimes \zeta + \sum_i (-1)^{nk-k+1} * ((*\alpha) \wedge \alpha^i) \otimes \nabla_{e_i} \zeta \\ &= -\sum_i e_i \rfloor \nabla_{e_i} \alpha \otimes \zeta - \sum_i e_i \rfloor \alpha \otimes \nabla_{e_i} \zeta \\ &= -\sum_i e_i \rfloor \nabla_{e_i} (\alpha \otimes \zeta). \end{aligned}$$

□

Theorem 23.2 (Weitzenböck Formula). *Assume $\{e_i\}$ is an orthonormal frame of P, and $\{\alpha^i\}$ is the dual frame. Then when applied to E-valued forms*

$$\Delta^\nabla = -Tr\nabla^2 + \sum_{i,j} \alpha^j \wedge (e_i \rfloor R(e_i, e_j)(\cdot))$$

where R is the curvature tensor of the bundle E with respect to the connection ∇.

Proof. Since the right hand side of the equality is independent of the choice of orthonormal basis, and it is a pointwise formula, we can take the normal coordinates $\{e_i\}$ at a point $p \in P$ (and $\{\alpha^i\}$ the dual basis), i.e., $h_{ij} := h(e_i, e_j)(p) = \delta_{ij}$ and $dh_{i,j}(p) = 0$, and prove that the given formula holds at p for such coordinates. For the Levi–Civita connection, the condition $dh_{i,j}(p) = 0$ of the normal coordinate is equivalent to letting $\Gamma^k_{i,j}(p) := \alpha^k(D_{e_i}e_j)(p)$ be 0.

For $\beta \in \Omega^k(E)$, using Lemma 23.1 we calculate

$$\begin{aligned}\delta^\nabla d^\nabla \beta &= -\sum_{i,j} e_i \rfloor \nabla_{e_i}(\alpha^j \wedge \nabla_{e_j}\beta) \\ &= -\sum_{i,j} e_i \rfloor (D_{e_i}\alpha^j \wedge \nabla_{e_j}\beta + \alpha^j \wedge \nabla_{e_i}\nabla_{e_j}\beta).\end{aligned}$$

At the point p, the first term vanishes, and we get

$$\begin{aligned}\delta^\nabla d^\nabla \beta(p) &= -\sum_{i,j} e_i \rfloor (\alpha^j \wedge \nabla_{e_i}\nabla_{e_j}\beta)(p) \\ &= -\sum_i \nabla_{e_i}\nabla_{e_i}\beta(p) + \sum_{i,j} \alpha^j \wedge (e_i \rfloor \nabla_{e_i}\nabla_{e_j}\beta)(p) \\ &= -\sum_i \nabla^2_{e_i,e_i}\beta(p) + \sum_{i,j} \alpha^j \wedge (e_i \rfloor \nabla_{e_i}\nabla_{e_j}\beta)(p).\end{aligned}$$

Also,

$$\begin{aligned}d^\nabla \delta^\nabla \beta &= -\sum_{i,j} \alpha^i \wedge \nabla_{e_i}(e_j \rfloor \nabla_{e_j}\beta) \\ &= -\sum_{i,j} \alpha^i \wedge (e_j \rfloor \nabla_{e_i}\nabla_{e_j}\beta) - \sum_{i,j} \alpha^i \wedge ((D_{e_i}e_j)) \rfloor \nabla_{e_j}\beta),\end{aligned}$$

As before, at the point p, the second term vanishes.

Now we sum the two parts $d^\nabla \delta^\nabla$ and $\delta^\nabla d^\nabla$ and get

$$\Delta^\nabla \beta(p) = -\sum_i \nabla^2_{e_i,e_i}\beta(p) + \sum_{i,j} \alpha^j \wedge (e_i \rfloor R(e_i,e_j)\beta)(p).$$

□

In particular, when acting on zero forms, i.e., sections of E, the second term on the right hand side vanishes, and there is

$$\Delta^\nabla = -Tr\nabla^2.$$

When acting on full rank forms, the above also holds by easy checking.

When $\beta \in \Omega^1(E)$, which is the case we use in this article, there is the following

Corollary 23.3. *For $\beta = \alpha \otimes \zeta \in \Omega^1(E)$, the Weizenböck formula can be written as*

$$\Delta^\nabla \beta = -\sum_i \nabla^2_{e_i,e_i}\beta + \mathrm{Ric}^{D*}(\alpha) \otimes \zeta + \mathrm{Ric}^\nabla \beta,$$

where Ric^{D*} denotes the adjoint of Ric^D, which acts on 1-forms.

In particular, when P is a surface, we have

$$\Delta^\nabla \beta = -\sum_i \nabla^2_{e_i,e_i}\beta + K\cdot\beta + \mathrm{Ric}^\nabla(\beta)$$

$$-\frac{1}{2}\Delta|\beta|^2 = |\nabla\beta|^2 - \langle \Delta^\nabla\beta,\beta\rangle + K\cdot|\beta|^2 + \langle \mathrm{Ric}^\nabla(\beta),\beta\rangle. \quad (1)$$

where K is the Gaussian curvature of the surface P.

24 Abstract framework of the three-interval method

In this section, we provide the method in proving exponential decay using the abstract framework of the three-interval method from [OW18b] referring interested readers [OW18b, Section 11]. We remark that the method can deal with the case with any exponentially decaying perturbation too (see Theorem 24.11).

The three-interval method is based on the following analytic lemma.

Lemma 24.1 ([MiRT09] Lemma 9.4). *For a sequence of nonnegative numbers* $\{x_k\}_{k=0,1,\cdots,N}$, *if there exists some constant* $0<\gamma<\frac{1}{2}$ *such that*

$$x_k \le \gamma(x_{k-1}+x_{k+1})$$

for every $1\le k\le N-1$, *then it follows*

$$x_k \le x_0\xi^{-k} + x_N\xi^{-(N-k)}, \quad k=0,1,\cdots,N,$$

where $\xi := \frac{1+\sqrt{1-4\gamma^2}}{2\gamma}$.

Remark 24.2. 1. If we write $\gamma=\gamma(c) := \frac{1}{e^c+e^{-c}}$ where $c>0$ is uniquely determined by γ, then the conclusion can be written into the exponential form

$$x_k \le x_0 e^{-ck} + x_N e^{-c(N-k)}.$$

2. For an infinite nonnegative sequence $\{x_k\}_{k=0,1,\cdots}$, if we have a uniform bound of in addition, then the exponential decay follows as

$$x_k \le x_0 e^{-ck}.$$

The analysis of proving the exponential decay will be carried on a Banach bundle $\mathscr{E}\to[0,\infty)$ modelled by the Banach space $\mathbb{E}$, for which we mean every fiber $\mathbb{E}_\tau$ is identified with the Banach space $\mathbb{E}$ smoothly depending on τ. We omit this identification if there is no way of confusion.

First we emphasize the base $[0,\infty)$ is non-compact and carries a natural translation map for any positive number r, which is $\sigma_r : \tau\mapsto\tau+r$. We introduce the following definition which ensures us to study the sections in local trivialization after taking a subsequence.

Definition 24.3. Let $\mathscr{E}$ be a Banach bundle modelled with a Banach space $\mathbb{E}$ over $[0,\infty)$. Let $[a,b] \subset [0,\infty)$ be any given bounded interval and let $s_k \to \infty$ be any given sequence. A *tame family of trivialization* over $[a,b]$ relative to the sequence s_k is defined to be a sequence of trivializations $\{\Phi_k\} : \mathscr{E}|_{[a,b]} \to [a,b] \times \mathbb{E}$

$$\Phi_k : \sigma_{s.}^* \mathscr{E}|_{[a+s_k, b+s_k]} \to [a,b] \times \mathbb{E}$$

for $k \geq 0$ satisfying the following: There exists a sufficiently large $k_0 > 0$ such that for any $k \geq k_0$ the bundle map

$$\Phi_{k_0+k} \circ \Phi_{k_0}^{-1} : [a,b] \times \mathbb{E} \to [a,b] \times \mathbb{E}$$

satisfies

$$\|\nabla_\tau^l (\Phi_{k_0+k} \circ \Phi_{k_0}^{-1})\|_{\mathscr{L}(\mathbb{E},\mathbb{E})} \leq C_l < \infty \tag{1}$$

for constants $C_l = C_l(|b-a|)$ depending only on $|b-a|$, $l = 0, 1, \cdots$.

We call $\mathscr{E}$ *uniformly locally tame*, if it carries a tame family of trivializations over $[a,b]$ relative to the sequence s_k for any given bounded interval $[a,b] \subset [0,\infty)$ and a sequence $s_k \to \infty$.

Definition 24.4. Suppose $\mathscr{E}$ is uniformly locally tame. We say a connection ∇ on $\mathscr{E}$ is *uniformly locally tame* if the push-forward $(\Phi_k)_* \nabla_\tau$ can be written as

$$(\Phi_k)_* \nabla_\tau = \frac{d}{d\tau} + \Gamma_k(\tau)$$

for any tame family $\{\Phi_k\}$ so that $\sup_{\tau \in [a,b]} \|\Gamma_k(\tau)\|_{\mathscr{L}(\mathbb{E},\mathbb{E})} < C$ for some $C > 0$ independent of k's.

Definition 24.5. Consider a pair $\mathscr{E}_2 \subset \mathscr{E}_1$ of uniformly locally tame bundles, and a bundle map $B : \mathscr{E}_2 \to \mathscr{E}_1$. We say B is *uniformly locally bounded*, if for any compact set $[a,b] \subset [0,\infty)$ and any sequence $s_k \to \infty$, there exists a subsequence, still denoted by s_k, a sufficiently large $k_0 > 0$ and tame families $\Phi_{1,k}$, $\Phi_{2,k}$ such that for any $k \geq 0$

$$\sup_{\tau \in [a,b]} \|\Phi_{i,k_0+k} \circ B \circ \Phi_{i,k_0}^{-1}\|_{\mathscr{L}(\mathbb{E}_2,\mathbb{E}_1)} \leq C \tag{2}$$

where C is independent of k.

For a given locally tame pair $\mathscr{E}_2 \subset \mathscr{E}_1$, we denote by $\mathscr{L}(\mathscr{E}_2, \mathscr{E}_1)$ the set of bundle homomorphisms which are uniformly locally bounded.

Lemma 24.6. *If $\mathscr{E}_1, \mathscr{E}_2$ are uniformly locally tame, then so is $\mathscr{L}(\mathscr{E}_2, \mathscr{E}_1)$.*

Definition 24.7. Let $\mathscr{E}_2 \subset \mathscr{E}_1$ be as above and let $B \in \mathscr{L}(\mathscr{E}_2, \mathscr{E}_1)$. We say B is *precompact* on $[0,\infty)$ if for any locally tame families Φ_1, Φ_2, there exists a further subsequence such that $\Phi_{1,k_0+k} \circ B \circ \Phi_{1,k_0}^{-1}$ converges to some $B_{\Phi_1 \Phi_2;\infty} \in \mathscr{L}(\Gamma([a,b] \times \mathbb{E}_2), \Gamma([a,b] \times \mathbb{E}_1))$.

Assume B is a bundle map from $\mathscr{E}_2$ to $\mathscr{E}_1$ which is uniformly locally bounded, where $\mathscr{E}_1 \supset \mathscr{E}_2$ are uniformly locally tame with tame families $\Phi_{1,k}$, $\Phi_{2,k}$. We can write

$$\Phi_{2,k_0+k} \circ (\nabla_\tau + B) \circ \Phi_{1,k_0}^{-1} = \frac{\partial}{\partial \tau} + B_{\Phi_1 \Phi_2,k}$$

as a linear map from $\Gamma([a,b] \times \mathbb{E}_2)$ to $\Gamma([a,b] \times \mathbb{E}_1)$, since ∇ is uniformly locally tame.

Next we introduce the following notion of coerciveness.

Definition 24.8. Let $\mathscr{E}_1, \mathscr{E}_2$ be as above and $B : \mathscr{E}_2 \to \mathscr{E}_1$ be a uniformly locally bounded bundle map. We say the operator

$$\nabla_\tau + B : \Gamma(\mathscr{E}_2) \to \Gamma(\mathscr{E}_1)$$

is *uniformly locally coercive*, if the following holds:

1. For any pair of bounded closed intervals I, I' with $I \subset \mathrm{Int} I'$,

$$\|\zeta\|_{L^2(I,\mathscr{E}_2)} \leq C(I,I') \|\nabla_\tau \zeta + B\zeta\|_{L^2(I,\mathscr{E}_1)}. \tag{3}$$

2. if for given bounded sequence $\zeta_k \in \Gamma(\mathscr{E}_2)$ satisfying

$$\nabla_\tau \zeta_k + B\zeta_k = L_k$$

with $\|L_k\|_{\mathscr{E}_1}$ bounded on a given compact subset $K \subset [0,\infty)$, there exists a subsequence, still denoted by ζ_k, that uniformly converges in $\mathscr{E}_2$.

Remark 24.9. Let $E \to [0,\infty) \times S$ be a (finite dimensional) vector bundle and denote by $W^{k,2}(E)$ the set of $W^{k,2}$-section of E and $L^2(E)$ the set of L^2-sections. Let $D : L^2(E) \to L^2(E)$ be a first order elliptic operator with cylindrical end. Denote by $i_\tau : S \to [0,\infty) \times S$ the natural inclusion map. Then there is a natural pair of Banach bundles $\mathscr{E}_2 \subset \mathscr{E}_1$ over $[0,\infty)$ associated to E, whose fiber is given by $\mathscr{E}_{1,\tau} = L^2(i_\tau^* E)$, $\mathscr{E}_{2,\tau} = W^{1,2}(i_\tau^* E)$. Furthermore assume $\mathscr{E}_i$ for $i = 1,2$ is uniformly local tame if S is a compact manifold (without boundary). Then D is uniformly locally coercive, which follows from the elliptic bootstrapping and the Sobolev's embedding.

Finally we introduce the notion of asymptotically cylindrical operator B.

Definition 24.10. We call B *locally asymptotically cylindrical* if the following holds: Any subsequence limit $B_{\Phi_1 \Phi_2;\infty}$ appearing in Definition 24.7 is a *constant* section, and $\|B_{\Phi_1 \Phi_2,k} - \Phi_{2,k_0+k} \circ B \circ \Phi_{1,k_0}^{-1}\|_{\mathscr{L}(\mathbb{E}_i,\mathbb{E}_i)}$ converges to zero as $k \to \infty$ for both $i = 1,2$.

Now we specialize to the case of Hilbert bundles $\mathscr{E}_2 \subset \mathscr{E}_1$ over $[0,\infty)$ and assume that $\mathscr{E}_1$ carries a connection which is compatible with the Hilbert inner product of $\mathscr{E}_1$. We denote by ∇_τ the associated covariant derivative. We assume that ∇_τ is uniformly locally tame.

Denote by $L^2([a,b];\mathscr{E}_i)$ the space of L^2-sections ζ of $\mathscr{E}_i$ over $[a,b]$, i.e., those satisfying

$$\int_a^b |\zeta(\tau)|^2_{\mathscr{E}_i}\, dt < \infty.$$

where $|\zeta(\tau)|_{\mathscr{E}_i}$ is the norm with respect to the given Hilbert bundle structure of $\mathscr{E}_i$.

Theorem 24.11 (Three-Interval Method). *Assume $\mathscr{E}_2 \subset \mathscr{E}_1$ is a pair of Hilbert bundles over $[0,\infty)$ with fibers $\mathbb{E}_2$ and $\mathbb{E}_1$, and $\mathbb{E}_2 \subset \mathbb{E}_1$ is dense. Let B be a section of the associated bundle $\mathscr{L}(\mathscr{E}_2, \mathscr{E}_1)$ and $L \in \Gamma(\mathscr{E}_1)$. We assume the following:*

1. *There exists a covariant derivative ∇_τ that preserves the Hilbert structure;*
2. *$\mathscr{E}_i$ for $i = 1, 2$ are uniformly locally tame;*
3. *B is precompact, uniformly locally coercive and asymptotically cylindrical;*
4. *Every subsequence limit B_∞ is a self-adjoint unbounded operator on $\mathbb{E}_1$ with its domain $\mathbb{E}_2$, and satisfies $\ker B_\infty = \{0\}$;*
5. *There exists some positive number δ such that any subsequence limiting operator B_∞ of the above mentioned pre-compact family has their first non-zero eigenvalues $|\lambda_1| > \delta$;*
6. *There exists some $R_0 > 0$, $C_0 > 0$ and $\delta_0 > \delta$ such that*

$$\|L(\tau)\|_{\mathscr{E}_{1,\tau}} \leq C_0 e^{-\delta_0 \tau}$$

   for all $\tau \geq R_0$.

Then for any (smooth) section $\zeta \in \Gamma(\mathscr{E}_2)$ with

$$\sup_{\tau \in [R_0,\infty)} \|\zeta(\tau,\cdot)\|_{\mathscr{E}_{2,\tau}} < \infty \tag{4}$$

and satisfying the equation

$$\nabla_\tau \zeta + B(\tau)\zeta(\tau) = L(\tau), \tag{5}$$

there exist some constants R, $C > 0$ such that for any $\tau > R$,

$$\|\zeta(\tau)\|_{\mathscr{E}_{1,\tau}} \leq C e^{-\delta \tau}.$$

Acknowledgements This work is supported by the IBS project # IBS-R003-D1. Both authors would like to also acknowledge MATRIX and the Simons Foundation for their support and funding through the MATRIX-Simons Collaborative Fund of the IBS-CGP and MATRIX workshop on Symplectic Topology.

References

[Abb11] C. Abbas, *Holomorphic open book decompositions*, Duke Math. J. **158** (2011), 29–82.

[ABW10] P. Albers, B. Bramham, and C. Wendl, *On non-separating contact hypersurfaces in symplectic 4-manifolds*, Algebraic & Geometric Topology (2010), 697–737.

[ACH05] C. Abbas, K. Cieliebak, and H. Hofer, *The Weinstein conjecture for planar contact structures in dimension three*, Comment. Math. Helv. **80** (2005), 771–793.

[Ban02] A. Banyaga, *Some properties of locally conformal symplectic structures*, Comm. Math. Helv. **77** (2002), 383–398.

[BCT17] A. Bravetti, H. Cruz, and D. Tapias, *Contact Hamiltonian mechanics*, Ann. Physics **376** (2017), 17–39.

[BEH$^+$03] F. Bourgeois, Y. Eliashberg, H. Hofer, K. Wysocki, and E. Zehnder, *Compactness results in symplectic field theory*, Geom. Topol. **7** (2003), 799–888.

[Bera] J. Bergmann, *Compactness resutls for H-holomorphic maps*, preprint 2009, arXiv:0904.1603.

[Berb] ______, *Embedded H-holomorphic maps and open book decompositions*, preprint 2009, arXiv:0907.3939.

[BJK] Hanwool Bae, Wonbo Jeong, and Jongmyeong Kim, *Cluster categories from Fukaya categories*, prepring 2022, arXiv:2209.09442.

[Bla10] David E. Blair, *Riemannian geometry of contact and symplectic manifolds*, Progress in Mathematics, vol. 203, Birkhäuser Boston, Ltd., Boston, MA, 2010, Second edition.

[Bou02] F. Bourgeois, *A Morse-Bott approach to contact homology*, Ph D Dissertation, Stanford University, 2002.

[Boy11] C. Boyer, *Completely integrable contact Hamiltonian systems and toric contact structures on* $S^2 \times S^3$, Sigma **7** (2011), 058, 22 pages.

[Can22] Dylan Cant, *A dimension formula for relative symplectic field theory*, 2022, thesis, Stanford University.

[Che67] S.S. Chern, *Complex manifolds without potential theory*, Van Nostrand Mathematical Studies, no. 15D, VAN Nostrand Co. Inc., Princeton, 1967.

[dLLV19] Manuel de León and Manuel Lainz Valcázar, *Contact Hamiltonian systems*, J. Math. Phys. **60** (2019), no. 10, 102902, 18 pp.

[EGH00] Y. Eliashberg, A. Givental, and H. Hofer, *Introduction to symplectic field theory*, Geom. and Funct. Anal. (2000), 560–673.

[Flo89] Andreas Floer, *Symplectic fixed points and holomorphic spheres*, Comm. Math. Phys. **120** (1989), no. 4, 575–611.

[FOOO10] Kenji Fukaya, Y.-G. Oh, Hiroshi Ohta, and Kaoru Ono, *Anchored Lagrangian submanifolds and their Floer theory*, Mirror symmetry and tropical geometry (Providence, RI), Contemp. Math., vol. 527, Amer. Math. Soc., 2010, pp. 15–54.

[FU84] D. Freed and K. Uhlenbeck, *Instantons and Four-Manifolds*, MSRI Publ., vol. 1, Srpringer-Verlag, New York, 1984.

[Gau97] P. Gauduchon, *Hermitian connection and Dirac operators*, Boll. Un. Math. Ital. B (7) **11** (1997), no. 2 suppl., 2587 –288.

[Gro85] M. Gromov, *Pseudoholomorphic curves in symplectic manifolds*, Invent. Math. **82** (1985), no. 2, 307–347. MR 809718 (87j:53053)

[GT70] D. Gilbarg and N. S. Trudinger, *Elliptic partial differential equations of second order*, Comprehensive Studies in Math., vol. 224, Springer-Verlag, 1970.

[Hof93] H. Hofer, *Pseudoholomorphic curves in symplectizations with applications to the Weinstein conjecture in dimension three*, Invent. Math. **114** (1993), 515–563.

[Hof00] ______, *Holomorphic curves and real three-dimensional dynamics*, Geom. Func. Anal. (2000), 674–704, Special Volume, Part II.

[Hut02] M. Hutchings, *An index inequality for embedded pseudoholomorphic curves in symplectizations*, J. Eur. Math. Soc. (JEMS) **4** (2002), no. 4, 313–361.

[Hut09] ______, *The embedded contact homology index revisited*, vol. 49, 2009, pp. 263–297.

[HWZ95] H. Hofer, K. Wysocki, and E. Zehnder, *Properties of pseudoholomorphic curves in symplectizations, II. Embedding control and algebraic invariants*, Geom. Funct. Anal. **5** (1995), no. 2, 270–328.

[HWZ96a] ______, *Correction to: "properties of pseudoholomorphic curves in symplectisations. I. Asymptotics"*, Ann. Inst. H. Poincaré (C) Anal. Non Linéaire **15** (1996), no. 4, 535–538.

[HWZ96b] ______, *Properties of pseudoholomorphic curves in symplectizations, I: asymptotics*, Annales de l'insitut Henri Poincaré, (C) Analyse Non Ninéaire **13** (1996), 337 – 379.

[HWZ99] ______, *Properties of pseudoholomorphic curves in symplectizations. III. Fredholm theory.*, Progr. Nonlinear Differential Equations Appl., vol. 35, pp. 381–475, Birkhäuser, Basel, 1999.

[HWZ01] ______, *The asymptotic behavior of a finite energy plane*, FIM preprint, ETH, Zürich, 2001.

[HWZ02] ______, *Finite energy cylinders of small area*, Ergodic Theory Dynam. Systems **22** (2002), no. 5, 1451–1486.

[Jos86] J. Jost, *On the regularity of minimal surfaces with free boundaries in Riemannian manifolds*, Manuscripta Math. **56** (1986), no. 3, 279–291.

[Kat95] T. Kato, *Perturbation theory for linear operators*, Classics in Mathematics, Springer-Verlag, Berlin, 1995, Reprint of the 1980 edition.

[KN96] S. Kobayashi and K. Nomizu, *Foundations of Differential Geometry, vol.2*, John Wiley & Sons, New York, 1996, Wiley Classics Library Edition.

[KO] Taesu Kim and Y.-G. Oh, *Kuranishi structures on contact instanton moduli spaces*, in preparation.

[KO23] ______, *Perturbation theory of asymptotic operators of contact instantons and of pseudoholomorphic curves on symplectization*, preprint, arXiv:2303.01011, 2023.

[Kob03] S. Kobayashi, *Natural connections in almost complex manifolds, Explorations in Complex and Riemannian Geometry*, Contemp. Math., vol. 332, pp. 153–169, Amer. Math. Soc., Providence, RI, 2003.

[Ler04] E. Lerman, *Contact fiber bundles*, J. Geom. Phys. **49** (2004), no. 1, 52–66.

[LM85] R. Lockhart and R. McOwen, *Elliptic differential operators on noncompact manifolds*, Ann. Scuola Norm. Sup. Pisa Cl. Sci. (4) **12** (1985), no. 3, 409–447.

[MiRT09] I. Mundet i Riera and G. Tian, *A compactification of the moduli space of twisted holomorphic maps*, Adv. Math. **222** (2009), no. 4, 1117–1196.

[MS05] D. Martelli and J. Sparks, *Toric Sasaki–Einstein metrics on $S^2 \times S^3$*, Phys. Lett. B **621** (2005), 208–212.

[MS06] ______, *Toric geometry, Sasaki–Einstein manifolds and a new infinite class of AdS/CFT duals*, Comm. Math. Phys. **262** (2006), 51–89.

[Oha] Y.-G. Oh, *Analysis of contact Cauchy-Riemann maps III: energy, bubbling and Fredholm theory*, Bulletin of Math. Sci.

[Ohb] ______, *Bordered contact instantons and their Fredholm theory and generic transversalities*, preprint, submitted for the proceedigns of Bumsig Kim's Memorial Conference, October 2021, KIAS, arXiv:2209.03548(v2).

[Ohc] ______, *Geometry and analysis of contact instantons and entangement of Legendrian links II*, in preparation.

[Oh15] ______, *Symplectic Topology and Floer Homology. vol. 1.*, New Mathematical Monographs, 28., Cambridge University Press, Cambridge., 2015.

[Oh21a] ______, *Contact Hamiltonian dynamics and perturbed contact instantons with Legendrian boundary condition*, preprint, arXiv:2103.15390(v2), 2021.

[Oh21b] ______, *Geometric analysis of perturbed contact instantons with Legendrian boundary conditions*, preprint, arXiv:2205.12351, 2021.

[Oh21c] ______, *Geometry and analysis of contact instantons and entangement of Legendrian links I*, preprint, arXiv:2111.02597, 2021.

[Oh22a] ______, *Contact instantons, anti-contact involution and proof of Shelukhin's conjecture*, preprint, arXiv:2212.03557, 2022.

[Oh22b] ______, *Gluing theories of contact instantons and of pseudoholomorphic curves in SFT*, preprint, arXiv:2205.00370, 2022.

[OSar] Y.-G. Oh and Y. Savelyev, *Pseudoholomoprhic curves on the $\mathfrak{LCS}$-fication of contact manifolds*, Advances in Geometry (to appear), arXiv:2107.03551.

[OW14] Y.-G. Oh and R. Wang, *Canonical connection on contact manifolds*, Real and Complex Submanifolds, Springer Proceedings in Mathematics & Statistics, vol. 106, 2014, (arXiv:1212.4817 in its full version), pp. 43–63.

[OW18a] ______, *Analysis of contact Cauchy-Riemann maps I: A priori C^k estimates and asymptotic convergence*, Osaka J. Math. **55** (2018), no. 4, 647–679.

[OW18b] ______, *Analysis of contact Cauchy-Riemann maps II: Canonical neighborhoods and exponential convergence for the Morse-Bott case*, Nagoya Math. J. **231** (2018), 128–223.

[OY22] Y.-G. Oh and Seungook Yu, *Contact instantons with Legendrian boundary condition: a priori estimates, asymptotic convergence and index formula*, preprint, arXiv:2301.06023, 2022.

[OY23] ______, *Legendrian spectral invariants on the one jet bundle via perturbed contact instantons*, preprint, arXiv:2301.06704, 2023.

[Par19] John Pardon, *Contact homology and virtual fundamental cycles*, J. Amer. Math. Soc. **32** (2019), no. 3, 825–919.

[PW93] Thomas H. Parker and Jon G. Wolfson, *Pseudo-holomorphic maps and bubble trees*, J. Geom. Anal. **3** (1993), no. 1, 63–98.

[RS93] J. Robbin and D. Salamon, *The Maslov index for paths*, Topology **32** (1993), 827–844.

[RS01] ______, *Asymptotic behavior of holomorphic strips*, Ann. I. H. Poincareé–AN **18** (2001), 573–612.

[RT95] Yongbin Ruan and Gang Tian, *A mathematical theory of quantum cohomology*, J. Differential Geom. **42** (1995), no. 2, 259–367.

[Sch84] R. Schoen, *Analytic aspects of the harmonic map problem*, Math. Sci. Res. Inst. Publ. 2, Springer, New York, 1984, S.S. Chern, ed., pp. 321–358.

[Sie08] Richard Siefring, *Relative asymptotic behavior of pseudoholomorphic half-cylinders*, Comm. Pure Appl. Math. **61** (2008), no. 12, 1631–1684.

[Sie11] ______, *Intersection theory of punctured pseudoholomorphic curves*, Geom. Topol. **15** (2011), no. 4, 2351–2457.

[SU81] J. Sacks and K. Uhlenbeck, *The existence of minimal immersions of 2 spheres*, Ann. Math. **113** (1981), 1–24.

[SU83] R. Schoen and K. Uhlenbeck, *Boundary regularity and the Dirichlet problem for harmonic maps*, J. Differential Geom. **18** (1983), no. 2, 253–268.

[SY76] R. Schoen and S.-T. Yau, *Harmonic maps and the topology of stable hypersurfaces and manifolds with non-negative Ricci curvature*, Comment. Math. Helv. **51** (1976), no. 3, 333–341.

[SZ92] Dietmar Salamon and Eduard Zehnder, *Morse theory for periodic solutions of Hamiltonian systems and the Maslov index*, Comm. Pure Appl. Math. **45** (1992), no. 10, 1303–1360.

[Uhl82] K. Uhlenbeck, *Removable singularities in Yang-Mills fields*, Comm. Math. Phys. **83** (1982), no. 1, 11–29.

[Wel73] Raymond O. Wells, *Differential analysis on complex manifolds*, Graduate Texts in Mathematics, vol. 65, Springer, New York, 1973.

[Wen] C. Wendl, *Lectures on Symplectic Field Theory*, a book manuscript, arXiv:1612.01009.

High-jet Relations of the Heat Kernel: Embedding Map and Applications

Ke Zhu

Abstract For any compact Riemannian manifold (M, g) and its heat kernel embedding map $\psi_t : M \to l^2$ constructed in [BBG], we study the higher derivatives of ψ_t with respect to an orthonormal basis at x on M. As the heat flow time $t \to 0_+$, it turns out the limiting angles between these derivative vectors are universal constants independent on g, x and the choice of orthonormal basis. Geometric applications to the mean curvature and the Riemannian curvature are given. Some algebraic structures of the ∞-jet space of ψ_t are explored.

1 Introduction

Let (M, g) be a n-dimensional compact Riemannian manifold with smooth metric g. In [BBG], Bérard, Besson and Gallot considered the *heat kernel embedding*

$$\Phi_t : x \in M \to \left\{ e^{-\lambda_j t/2} \phi_j (x) \right\}_{j \geq 1} \in l^2 \tag{1}$$

where λ_j is the j-th eigenvalue of the Laplacian Δ_g of (M, g), $\{\phi_j\}_{j\geq 0}$ is the L^2 orthonormal eigenbasis of Δ_g, and l^2 is the Hilbert space of real-valued, square summable series. Let $\langle , \rangle$ be the standard inner product of l^2, then $\langle \Phi_t (x), \Phi_t (y) \rangle$ is the heat kernel

$$H (t, x, y) = \sum_{j=1}^{\infty} e^{-\lambda_j t} \phi_j (x) \phi_j (y)$$

on M. It was proved in [BBG] that the *normalized heat kernel embedding*

Ke Zhu
Department of Mathematics and Statistics, Minnesota State University
Mankato, MN 56001, USA
e-mail: ke.zhu@mnsu.edu

D. R. Wood et al. (eds.), *2021-2022 MATRIX Annals*, MATRIX Book Series 5,
https://doi.org/10.1007/978-3-031-47417-0_33

$$\psi_t : x \in M \to \sqrt{2}\,(4\pi)^{n/4}\, t^{\frac{n+2}{4}} \cdot \left\{ e^{-\lambda_j t/2} \phi_j(x) \right\}_{j \geq 1} \in l^2 \tag{2}$$

is *asymptotically isometric*, i.e. the induced metric of the embedded image $\psi_t(M)$ in l^2 gets closer and closer to the original metric g as the heat flow time $t \to 0_+$. More precisely, as $t \to 0_+$

$$\psi_t^* g_0 = g + \frac{t}{3}\left(\frac{1}{2} S_g \cdot g - \mathrm{Ric}_g\right) + O\left(t^2\right), \tag{3}$$

where g_0 is the standard metric in l^2, S_g is the scalar curvature, and Ric_g is the Ricci curvature of (M, g). The embedding ψ_t is *canonical*, in the sense that it is constructed by the eigenfunctions of the Laplacian of (M, g).

Given any $x \in M$, we choose the normal coordinates $\left(x^1, \cdots x^n\right)$ near x such that $\left\{\frac{\partial}{\partial x^j}\right\}_{1 \leq j \leq n}$ is *orthonormal* at x. Let $\overrightarrow{\alpha}$ be the multi-index of the mixed derivative operator $D^{\overrightarrow{\alpha}}$ on this chart. As $t \to 0_+$, it turns out these (l^2-valued) coefficients $\left\{D^{\overrightarrow{\alpha}} \Phi_t(x)\right\}$ have a special property: the angles between any two $D^{\overrightarrow{\alpha}} \Phi_t(x)$ and $D^{\overrightarrow{\beta}} \Phi_t(x)$ converge to universal constants *independent* on the metric g, the point x on M, and the orthonormal basis $\{V_i\}_{1 \leq i \leq n} = \left\{\frac{\partial}{\partial x^j}\right\}_{1 \leq j \leq n}$ at x, but only on $\overrightarrow{\alpha}$ and $\overrightarrow{\beta}$. The asymptote of the length of $D^{\overrightarrow{\alpha}} \Phi_t(x)$ only depends on $\overrightarrow{\alpha}$ and n too (Theorem 2). These *asymptotic relations* in the k-jet space of Φ_t (or ψ_t) have interesting applications:

1. When $k = 1$, [BBG] used them to construct *asymptotically isometric embeddings* $\psi_t = \sqrt{2}\,(4\pi)^{n/4}\, t^{\frac{n+2}{4}} \Phi_t : M \to l^2$ as in (3). They also used the embedding ψ_t to define the *spectral distance* between different metrics on M. When the metrics have a uniform upper bound of the diameter and a uniform lower bound of the Ricci curvature, they established a *precompactness* theorem on the space of such metrics;
2. When $k = 2$, by perturbing the almost isometric embedding ψ_t, [WZ] constructed *isometric embeddings* of compact Riemannian manifolds M into $\mathbb{R}^{q(t)}$ with controlled second fundamental form and $q(t) \sim t^{-n/2}$, where the 2-jet bundle of ψ_t appeared in the linearized operator of the isometric embedding problem. It was also showed that for any compact Riemannian manifolds, the mean curvature vectors of the image $\frac{1}{\sqrt{t}} \psi_t(M)$ converge to constant length $\sqrt{\frac{n+2}{2n}}$ as $t \to 0_+$;
3. When $k = 3$, we use them to show the embedded image $\frac{1}{\sqrt{t}} \psi_t(M) \subset l^2$ is *asymptotically umbilical* in the mean curvature vector direction as $t \to 0_+$ (Proposition 4). We also propose to construct submanifolds in $\mathbb{R}^q$ with constant mean curvature length by truncating and perturbing $\frac{1}{\sqrt{t}} \psi_t(M) \subset l^2$;
4. When $k = 2$, but also considering the secondary leading terms in the asymptote of the limiting angles, we prove the "asymptotic" Gauss formula (Proposition 5) to express the *Riemannian curvature* tensor of (M, g) as

$$R(X, Y, Z, W) = \lim_{t \to 0+} \left[\langle \nabla_X \nabla_W \psi_t, \nabla_Y \nabla_Z \psi_t \rangle - \langle \nabla_X \nabla_Z \psi_t, \nabla_Y \nabla_W \psi_t \rangle \right].$$

From this formula we can easily see the symmetry of the Riemannian curvature tensor, including the second Bianchi identity (Lemma 3). We can also express the *Levi-Civita connection* in terms of ψ_t (Lemma 2).

For example, the following 2-jet relations were proved in [WZ] and played a crucial role to construct isometric embeddings of M into $\mathbb{R}^q$. Let $\langle , \rangle$ be the standard inner product in l^2 and $|\cdot|$ be the standard metric in l^2. We have

Theorem 1. *([WZ] Theorem 2) For any $x \in M$, let $(x^1, \cdots x^n)$ be the normal coordinates near x such that $\left\{\frac{\partial}{\partial x^j}\right\}_{1\le j\le n}$ is* orthonormal *at x. Then for the n first derivatives vectors $\nabla_i \Phi_t$ $(1 \le i \le n)$ and the $n(n+1)/2$ second derivatives vectors $\nabla_i \nabla_j \Phi_t$ $(1 \le i \le j \le n)$ of Φ_t with respect to $\left\{\frac{\partial}{\partial x^j}\right\}_{1\le j\le n}$, as $t \to 0_+$ we have*

$$\frac{\left\langle \nabla_i \Phi_t(x), \nabla_j \Phi_t(x)\right\rangle}{\left|\nabla_i \Phi_t(x)\right|\left|\nabla_j \Phi_t(x)\right|} \to \delta_{ij}, \text{ and } \frac{\left\langle \nabla_i \nabla_j \Phi_t(x), \nabla_k \Phi_t(x)\right\rangle}{\left|\nabla_i \nabla_j \Phi_t(x)\right|\left|\nabla_k \Phi_t(x)\right|} \to 0,$$

and for $i \neq j$ or $k \neq l$,

$$\frac{\left\langle \nabla_i \nabla_j \Phi_t(x), \nabla_k \nabla_l \Phi_t(x)\right\rangle}{\left|\nabla_i \nabla_j \Phi_t(x)\right|\left|\nabla_k \nabla_l \Phi_t(x)\right|} \to 0,$$

except when $\{i, j\} = \{k, l\}$ as sets. Furthermore, for $i \neq j$,

$$\frac{\left\langle \nabla_i \nabla_i \Phi_t(x), \nabla_j \nabla_j \Phi_t(x)\right\rangle}{\left|\nabla_i \nabla_i \Phi_t(x)\right|\left|\nabla_j \nabla_j \Phi_t(x)\right|} \to \frac{1}{3}. \tag{4}$$

The convergence is uniform for all x on M.

In this paper, we generalize Theorem 1 to obtain the limiting angles between any two derivative vectors $D^{\vec{\alpha}}\Phi_t(x)$ and $D^{\vec{\beta}}\Phi_t(x)$ as $t \to 0_+$. It is a refinement of Proposition 15 in [WZ]; The new observation is that the coefficients there can be translated into certain enumeration problems of graphs and can be solve explicitly (Proposition 2).

In the following we denote the disjoint union of two sets by $\amalg$, and let $\left|\vec{\alpha}\right|$ be the total degree of the derivative operator $D^{\vec{\alpha}}$.

Definition 1. For any two derivative vectors $D^{\vec{\alpha}}\Phi_t(x)$ and $D^{\vec{\beta}}\Phi_t(x)$, suppose there are s distinct indices $\{j_1, j_2, \cdots, j_s\}$ in $\{\vec{\alpha}\} \amalg \left\{\vec{\beta}\right\}$, such that each j_r has multiplicity a_r in $\vec{\alpha}$, multiplicity b_r in $\vec{\beta}$, and let the average multiplicity $\sigma_r := \frac{a_r+b_r}{2}$ for $r = 1, 2, \cdots, s$. If all $a_r + b_r$ are even, we define the constants

$$A\left(\vec{\alpha},\vec{\beta}\right) = (-1)^{\frac{|\vec{\alpha}|-|\vec{\beta}|}{2}} \prod_{r=1}^{s} \left[\frac{(2\sigma_r)!}{2^{\sigma_r}(\sigma_r)!}\right], \tag{5}$$

$$\begin{aligned} B\left(\vec{\alpha},\vec{\beta}\right) &= A\left(\vec{\alpha},\vec{\beta}\right) \Big/ \left[A\left(\vec{\alpha},\vec{\alpha}\right) A\left(\vec{\beta},\vec{\beta}\right)\right]^{1/2} \\ &= (-1)^{\frac{|\vec{\alpha}|-|\vec{\beta}|}{2}} \prod_{r=1}^{s} \frac{(2\sigma_r)!}{\sigma_r!} \Big/ \left[\frac{(2a_r)!}{a_r!}\frac{(2b_r)!}{b_r!}\right]^{1/2}. \end{aligned} \tag{6}$$

If some $a_r + b_r$ is odd, we simply let

$$A\left(\vec{\alpha},\vec{\beta}\right) = 0 = B\left(\vec{\alpha},\vec{\beta}\right). \tag{7}$$

We have our main theorem:

Theorem 2. *(Asymptotic high-jet relations) For any two derivative vectors $D^{\vec{\alpha}}\Phi_t(x)$ and $D^{\vec{\beta}}\Phi_t(x)$, as $t \to 0_+$,*

1. *If there is an index appeared with odd total multiplicity in $\{\vec{\alpha}\} \amalg \{\vec{\beta}\}$, then*

$$\frac{\left\langle D^{\vec{\alpha}}\Phi_t(x), D^{\vec{\beta}}\Phi_t(x)\right\rangle}{\left|D^{\vec{\alpha}}\Phi_t(x)\right|\left|D^{\vec{\beta}}\Phi_t(x)\right|} \to 0. \tag{8}$$

Especially this is the case when $|\vec{\alpha}| + |\vec{\beta}|$ is odd;

2. *If each index appears with even total multiplicity in $\{\vec{\alpha}\} \amalg \{\vec{\beta}\}$, then*

$$\frac{\left\langle D^{\vec{\alpha}}\Phi_t(x), D^{\vec{\beta}}\Phi_t(x)\right\rangle}{\left|D^{\vec{\alpha}}\Phi_t(x)\right|\left|D^{\vec{\beta}}\Phi_t(x)\right|} \to B\left(\vec{\alpha},\vec{\beta}\right) \neq 0; \tag{9}$$

3. *We have*

$$\left|D^{\vec{\alpha}}\Phi_t(x)\right|^2 \to \frac{1}{(4\pi t)^{n/2}}\left(\frac{1}{2t}\right)^{|\vec{\alpha}|} A\left(\vec{\alpha},\vec{\alpha}\right). \tag{10}$$

The above convergences are uniform for all x on M.

For example, for $n \geq 2$, $i \neq j$, $\vec{\alpha} = (i,i)$ and $\vec{\beta} = (j,j)$, we recover Theorem 2 in [WZ], because

$$A\left(\vec{\alpha},\vec{\beta}\right)=\prod_{r=1}^{2}\left[(-1)^0\frac{2!}{2^1\cdot 1!}\right]=1,$$

$$A\left(\vec{\alpha},\vec{\alpha}\right)=\prod_{r=1}^{1}\left[(-1)^0\frac{4!}{2^2\cdot 2!}\right]=3=A\left(\vec{\beta},\vec{\beta}\right),$$

$$B\left(\vec{\alpha},\vec{\beta}\right)=A\left(\vec{\alpha},\vec{\beta}\right)\Big/\left[A\left(\vec{\alpha},\vec{\alpha}\right)A\left(\vec{\beta},\vec{\beta}\right)\right]^{1/2}=\frac{1}{3}.$$

Theorem 2 is a consequence of the more general result in Section 2 (Proposition 3), stated as follows:

Proposition 1. *As $t\to 0_+$, we have*

$$(4\pi t)^{n/2}\,(2t)^{\left\lfloor\frac{|\vec{\alpha}|+|\vec{\beta}|}{2}\right\rfloor}\cdot D_y^{\vec{\beta}}D_x^{\vec{\alpha}}H(t,x,y)\Big|_{x=y}=A\left(\vec{\alpha},\vec{\beta}\right)+O(t),\tag{11}$$

where $[c]$ is the largest integer less or equal to the given real number c. The convergence is uniform for all x on M.

After the completion of this paper, the author received the paper [Ni], which is related to our results. The paper [Ni], motivated by probabilistic questions, describes the small ε-asymptotics of the jets along the diagonal of the integral kernel of the smoothing operator $w\left(\varepsilon\sqrt{\Delta_g}\right)$ for an arbitrary nonnegative even Schwartz function w. For $w(x)=e^{-x^2}$, these reduce to the short time asymptotics of the jets along the diagonal of the heat kernel.

The paper is organized as follows: In Section 2 we first review the heat kernel and its near-diagonal expansion, then introduce certain graphs to aid the computation of the leading terms in (11), and then prove our main Theorem 2. In Section 3 we give applications of the high-jet relations of ψ_t to the mean curvature of $\psi_t(M)\subset l^2$ and the Riemannian curvature of M. In Section 4 we explore more algebraic structures of the high-jet relations, and reformulate them by the lattice geometry on $\mathbb{Z}_+^n$, with the motivation for future applications.

Acknowledgements The author thanks Xiaowei Wang for many enlightening discussions during the collaboration of [WZ], where some ideas of the current work emerged. The author thanks Liviu I. Nicolaescu for pointing out the probabilistic point of view on our results. He also thanks Clifford Taubes for interest and support. The author thanks the IBS-CGP & MATRIX workshop on Symplectic Topology in 2022, where he learned some related asymptotic limits in tropical geometry.

2 High-jet relations

2.1 Heat kernel and derivatives of the distance function

Let $H(t,x,y)$ be the heat kernel of the Laplacian Δ_g on (M,g),

$$H(t,x,y) := \sum_{j=1}^{\infty} e^{-\lambda_j t}\phi_j(x)\phi_j(y). \tag{1}$$

It is well known that $H(t,x,y)$ has the *Minakshisundaram-Pleijel expansion*

$$H(t,x,y) = \frac{1}{(4\pi t)^{n/2}} e^{-\frac{r^2}{4t}} U(t,x,y)$$

(e.g. [BeGaM], p.213, or [Ch], p.154), where $r = r(x,y)$ is the distance function for points x and y on M,

$$U(t,x,y) = u_0(x,y) + tu_1(x,y) + \cdots + t^p u_p(x,y) + O\left(t^{p+1}\right) \tag{2}$$

as $t \to 0_+$, where the convergence is in C^l for any $l \geq 0$, and

$$\begin{aligned} u_0(x,y) &= [\theta(x,y)]^{-1/2} \text{ (for } x \text{ and } y \text{ close enough)}, \\ u_1(x,x) &= \frac{S_g(x)}{6}, \end{aligned} \tag{3}$$

where

$$\begin{aligned} \theta(x,y) &= \frac{\text{volume density at } y \text{ read in the normal coordinates around } x}{r^{n-1}} \\ &= \left|\det\left(d_y \exp_x\right)\right| = \left|\det\left(g_{ij}(y)\right)\right|^{1/2}, \end{aligned} \tag{4}$$

and in the second line of (4), y is regarded as a point on T_xM, and $d_y \exp_x$ is the differential of the exponential map $\exp_x : T_xM \to M$ ([BeGaM], p.208, [BGV], p.36). Given any x on M, let $\left(x^1, \cdots, x^n\right)$ be the normal coordinates around x and $\left\{V_j\right\}_{j=1}^n = \left\{\frac{\partial}{\partial x^j}\right\}_{j=1}^n$ be an orthonormal basis at x. For any unit vector $V \in T_xM$, it was derived in [BBG] that for $x_s = \exp_x(sV)$ and $x_\tau = \exp_x(\tau V)$ with small s and τ,

$$\theta(x_s, x_\tau) = 1 - \mathrm{Ric}_g(\dot{x}(s), \dot{x}(s)) \frac{(s-\tau)^2}{3!} + O\left(|s-\tau|^3\right),$$

so

$$u_0(x_s, x_\tau) = 1 + \frac{1}{12}\mathrm{Ric}_g(\dot{x}(s), \dot{x}(s))(s-\tau)^2 + O\left(|s-\tau|^3\right).$$

Hence

$$u_0(x,x) = 1, \tag{5}$$
$$\nabla_i u_0(x,y)|_{x=y} = 0, \tag{6}$$
$$\nabla_i U(t,x,y)|_{x=y} = O(t), \tag{7}$$
$$\nabla_i^x \nabla_j^y u_0(x,y)|_{x=y} = -\frac{1}{6}\mathrm{Ric}_g(V_i, V_j), \tag{8}$$

For later computations, we will need the following facts on the derivatives of the squared distance function $r^2(x,y)$ on the diagonal $x = y$.

Lemma 1. *For $r = r(x,y)$ on $M \times M$, let $\nabla_i = \nabla^x_{V_j}$ be the partial derivative with respect to V_j in the x variables, $\nabla_{\bar{j}} = \nabla^y_{V_j}$ be the partial derivative with respect to V_j in the y variables, $\nabla_{ij} = \nabla^x_{V_i}\nabla^x_{V_j}$, $\nabla_{i\bar{j}} = \nabla^x_{V_i}\nabla^y_{V_j}$, $\nabla_{ij\bar{k}} = \nabla^x_{V_i}\nabla^x_{V_j}\nabla^y_{V_k}$ and so on. Then we have the following identities*

$$\nabla_i r^2(x,y)|_{x=y} = 0 = \nabla_{\bar{i}} r^2(x,y)|_{x=y}, \tag{9}$$
$$\nabla_{ij} r^2(x,y)|_{x=y} = 2\delta_{ij} = -\nabla_{i\bar{j}} r^2(x,y)|_{x=y}, \tag{10}$$
$$\nabla_{ijk} r^2(x,y)|_{x=y} = \nabla_{ij\bar{k}} r^2|_{x=y} = \nabla_{i\bar{j}\bar{k}} r^2|_{x=y} = \nabla_{\bar{i}\bar{j}\bar{k}} r^2|_{x=y} = 0, \tag{11}$$
$$\nabla_{ijkl} r^2(x,y)|_{x=y} = -\nabla_{ij\bar{k}\bar{l}} r^2|_{x=y}(x,y) = -\frac{2}{3}\left(R_{ikjl}(x) + R_{iljk}(x)\right), \tag{12}$$

where $R_{ijkl}(x) = R(V_i, V_j, V_k, V_l)(x)$ is the Riemannian curvature.

The identities (9) and (10) are well-known (e.g. [BBG]). The identities (11) and (12) can be found in Chapter 16 (p.282) of [De], where $\frac{1}{2}r^2(x,y)$ is called the *world function*, and the limit at $x = y$ is called the coincidence limit.

2.2 Admissible graphs

To compute $\left\langle D^{\vec{\alpha}}\Phi_t(x), D^{\vec{\beta}}\Phi_t(x)\right\rangle$, it is useful to introduce certain graphs.

Definition 2. (Vertex set) Let $\mathcal{S} = \{\vec{\alpha}\} \amalg \{\vec{\beta}\}$ be the set with $|\vec{\alpha}| + |\vec{\beta}|$ vertices, where each vertex is an index $i \in \mathcal{S}$, with the sign "+" if it is in $\{\vec{\alpha}\}$ and sign "−" if it is in $\{\vec{\beta}\}$. We say the vertex has color i if we do not distinguish its sign.

Definition 3. (Admissible graph) A graph G on the vertex set $\mathcal{S} = \{\vec{\alpha}\} \amalg \{\vec{\beta}\}$ is called *admissible* if (i) Each vertex is connected to only one another vertex; (ii) Each edge only connects two vertices with the same color; (iii) Each edge has a sign ± being the negative of the product of the signs of its two end-points. Let Sign(G) be the product of the signs of all edges of G.

Remark 1. The necessary and sufficient condition for $\mathcal{S} = \{\vec{\alpha}\} \amalg \{\vec{\beta}\}$ to have admissible graph(s) is that each index appears in $\mathcal{S}$ with even multiplicity. Especially $|\vec{\alpha}| + |\vec{\beta}|$ must be even. If G is not admissible, we simply let $\mathrm{Sign}(G) = 0$.

Proposition 2. *Suppose there are s distinct indices* $\{j_1, j_2, \cdots, j_s\} \subset \{1, 2, \cdots, n\}$ *in* $\{\vec{\alpha}\} \amalg \{\vec{\beta}\}$, *such that each* j_r *has multiplicity* a_r *in* $\vec{\alpha}$, *and multiplicity* b_r *in* $\vec{\beta}$. *Let* $\sigma_r = \frac{a_r + b_r}{2}$. *Then we have*

1. *For each admissible graph G of* $\mathcal{S} = \{\vec{\alpha}\} \amalg \{\vec{\beta}\}$, $sign(G) = (-1)^{\frac{|\vec{\alpha}| - |\vec{\beta}|}{2}}$;
2. $\#\{$*admissible graphs of* $\mathcal{S}\} = \prod_{r=1}^{s} \left[\frac{(2\sigma_r)!}{2^{\sigma_r}(\sigma_r)!}\right] = \left|A\left(\vec{\alpha}, \vec{\beta}\right)\right|$.

Proof. By the definition of $\mathrm{sign}(G)$, it is the product of the signs of its edges. There are $\frac{|\vec{\alpha}| + |\vec{\beta}|}{2}$ edges, each edge connecting vertices i and i' has the sign $(-1) \cdot \mathrm{sign}(i)\mathrm{sign}(i')$, and the disjoint union of edges covers all vertices of G, so

$$
\begin{aligned}
\mathrm{sign}\,(G) &= \prod_{\text{edge } e \text{ of } G} \mathrm{sign}(e) = (-1)^{\frac{|\vec{\alpha}| + |\vec{\beta}|}{2}} \cdot \prod_{\text{vertex } i \text{ of } G} \mathrm{sign}(i) \\
&= (-1)^{\frac{|\vec{\alpha}| + |\vec{\beta}|}{2}} \cdot (-1)^{|\vec{\beta}|} = (-1)^{\frac{|\vec{\alpha}| - |\vec{\beta}|}{2}},
\end{aligned}
$$

where the second line is because only vertices in $\vec{\beta}$ have "–" sign. For each index j_r, it has total multiplicity $2\sigma_r$ in $\mathcal{S} = \{\vec{\alpha}\} \amalg \{\vec{\beta}\}$, i.e. there are $2\sigma_r$ vertices with color j_r in $\mathcal{S}$, so the ways to draw disjoint edges on this subset of vertices in $\mathcal{S}$ is

$$
\begin{aligned}
&\binom{2\sigma_r}{2\sigma_r - 2}\binom{2\sigma_r - 2}{2\sigma_r - 4} \cdots \binom{2}{2} \Big/ \sigma_r! \\
&= \frac{(2\sigma_r)!}{(2\sigma_r - 2)!2!} \frac{(2\sigma_r - 2)!}{(2\sigma_r - 4)!2!} \cdots \frac{2!}{2!} \Big/ \sigma_r! \\
&= \frac{(2\sigma_r)!}{2^{\sigma_r}(\sigma_r)!}
\end{aligned}
$$

Considering all distinct indices $j_1, j_2, \cdots$ and j_s in $\mathcal{S}$, we see the number of admissible graphs on $\mathcal{S}$ is $\prod_{r=1}^{s} \left[\frac{(2\sigma_r)!}{2^{\sigma_r}(\sigma_r)!}\right]$. □

Remark 2. The number $A\left(\vec{\alpha}, \vec{\beta}\right)$ is related to the moments of the Gaussian $e^{-|x|^2}$ by

$$
A\left(\vec{\alpha}, \vec{\beta}\right) = 2^{\left[\frac{|\vec{\alpha}| + |\vec{\beta}|}{2}\right]} \pi^{-n/2} \int_{\mathbb{R}^n} e^{-|x|^2} x^{\vec{\alpha} + \vec{\beta}} dx.
$$

Actually, the above combinatorics of the numbers $A\left(\overrightarrow{\alpha}, \overrightarrow{\beta}\right)$ is a manifestation of a classical *Wick formula* (or *Isserlis' theorem*) of the Gauss integrals, see e.g. [AT], Section 11.6. The formula is frequently used in the Feynman integral.

2.3 High-jet relations from the heat kernel expansion

Given any x on M, let $\{V_i\}_{i=1}^n$ be an orthonormal basis at x, $\nabla_i = \nabla_{V_i}^x$ be the partial derivative with respect to V_i in the x variables, and $\nabla_{\bar{j}} = \nabla_{V_j}^y$ be the partial derivative with respect to V_j in the y variables, where "–" indicates the derivative is with respect to the y-variables. We also denote $\nabla_{ij} = \nabla_{V_i}^x \nabla_{V_j}^x$, $\nabla_{i\bar{j}} = \nabla_{V_i}^x \nabla_{V_j}^y$ and so on. Since

$$\langle \Phi_t(x), \Phi_t(y)\rangle = \sum_{j=1}^{\infty} e^{-\lambda_j t}\phi_j(x)\phi_j(y) = H(t,x,y),$$

applying the derivative operators $D_x^{\overrightarrow{\alpha}}$ and $D_y^{\overrightarrow{\beta}}$ to the above identity and letting $x = y$, we get

$$\left\langle D^{\overrightarrow{\alpha}}\Phi_t(x), D^{\overrightarrow{\beta}}\Phi_t(x)\right\rangle = D_y^{\overrightarrow{\beta}} D_x^{\overrightarrow{\alpha}} H(t,x,y)\,|_{x=y}. \tag{13}$$

Proposition 3. *Let $\overrightarrow{\alpha}$ and $\overrightarrow{\beta}$ be two multi-indices in the set $\{1, 2, \cdots n\}$. Then as $t \to 0_+$, we have*

$$(4\pi t)^{n/2}(2t)^{\left[\frac{|\overrightarrow{\alpha}|+|\overrightarrow{\beta}|}{2}\right]} \cdot D_y^{\overrightarrow{\beta}} D_x^{\overrightarrow{\alpha}} H(t,x,y)\Big|_{x=y} \to A\left(\overrightarrow{\alpha}, \overrightarrow{\beta}\right), \tag{14}$$

where $[c]$ is the largest integer less or equal to the given real number c. If $A\left(\overrightarrow{\alpha}, \overrightarrow{\beta}\right) = 0$, we further have

$$(4\pi t)^{n/2}(2t)^{\left[\frac{|\overrightarrow{\alpha}|+|\overrightarrow{\beta}|}{2}\right]} \cdot D_y^{\overrightarrow{\beta}} D_x^{\overrightarrow{\alpha}} H(t,x,y)\Big|_{x=y} = O(t). \tag{15}$$

The convergence is uniform for all x on M.

Proof. For any multi-index $\vec{\gamma}$, applying the Leibniz rule to

$$H(t,x,y) = \frac{1}{(4\pi t)^{n/2}} e^{-\frac{r^2}{4t}} U(t,x,y)$$

we can write

$$D^{\overrightarrow{\gamma}} H(t,x,y) = \frac{1}{(4\pi t)^{n/2}} e^{-\frac{r^2}{4t}} P_{\overrightarrow{\gamma}}(t,x,y),$$

where $P_{\overrightarrow{\gamma}}(t,x,y)$ is a polynomial in $D^{\overrightarrow{\mu_j}}\left(-\frac{r^2(x,y)}{4t}\right)$ and $D^{\overrightarrow{\eta}}U(t,x,y)\Big|_{x=y}$, i.e. each summand is of the form

$$\left(\prod_{j=1}^{l} D^{\overrightarrow{\mu_j}}\left(-\frac{r^2(x,y)}{4t}\right)\right)\Bigg|_{x=y} \cdot D^{\overrightarrow{\eta}}U(t,x,y)\Big|_{x=y} \tag{16}$$

for some multi-indices $\overrightarrow{\mu_j}$ and $\overrightarrow{\eta}$ with

$$\sum_{j=1}^{l}\left|\overrightarrow{\mu_j}\right| + \left|\overrightarrow{\eta}\right| = \left|\overrightarrow{\gamma}\right|. \tag{17}$$

For example, when $\overrightarrow{\gamma} = \partial_{x_i}$,

$$P_{\overrightarrow{i}}(t,x,y) = \partial_i\left(-\frac{r^2(x,y)}{4t}\right)U(t,x,y) + \partial_i U(t,x,y).$$

(more examples of $P_{\overrightarrow{\gamma}}(t,x,y)$ with $\left|\overrightarrow{\gamma}\right| \le 4$ are in [WZ]). Now we take

$$\vec{\gamma} = \left(\overrightarrow{\alpha}, \overrightarrow{\overline{\beta}}\right), \left|\overrightarrow{\gamma}\right| = \left|\overrightarrow{\alpha}\right| + \left|\overrightarrow{\beta}\right|, D^{\vec{\gamma}} = D_y^{\overrightarrow{\beta}} D_x^{\overrightarrow{\alpha}},$$

where $\overrightarrow{\overline{\beta}}$ indicates the derivative $D_y^{\overrightarrow{\beta}}$ is with respect to the y-variables. We have the following claims:

1. As $t \to 0_+$, the nonzero summands of $P_{\vec{\gamma}}(t,x,y)$ involving the highest power of $\frac{1}{t}$ must have $\overrightarrow{\mu_j}'s$ with $\left|\overrightarrow{\mu_j}\right| = 2$ as many as possible, such that each $\overrightarrow{\mu_j}$ is of the form (i,i), $\left(i,\overline{i}\right)$ or $\left(\overline{i},\overline{i}\right)$ for some $i \in \{1,2,\cdots,n\}$, and $\left|\overrightarrow{\eta}\right| = 0$ or 1. We also have
$$\left|D^{\overrightarrow{\gamma}}H(t,x,y)\,|_{x=y}\right| \le C\frac{1}{(4\pi t)^{n/2}} \cdot t^{-\left[\frac{|\overrightarrow{\gamma}|}{2}\right]},$$
where the constant C depends on n and $\left|\overrightarrow{\gamma}\right|$.
2. If the total multiplicity of each index in $\{\overrightarrow{\alpha}\} \amalg \left\{\overrightarrow{\beta}\right\}$ is even, then we further have $l = \frac{|\overrightarrow{\gamma}|}{2}$, and $\left|\overrightarrow{\eta}\right| = 0$ for the the nonzero summands of $P_{\vec{\gamma}}(t,x,y)$ involving the highest power of $\frac{1}{t}$. These terms are of the same sign $(-1)^{\frac{|\overrightarrow{\alpha}|-|\overrightarrow{\beta}|}{2}}$, and the total number of such terms is $\left|A\left(\overrightarrow{\alpha},\overrightarrow{\beta}\right)\right|$.

For Claim 1, from Lemma 1 we have

$$\partial_i \left(-\frac{r^2(x,y)}{4t} \right) |_{x=y} = 0,$$
$$\partial_{ij} \left(-\frac{r^2(x,y)}{4t} \right) |_{x=y} = -\frac{\delta_{ij}}{2t} = -\partial_{i\bar{j}} \left(-\frac{r^2(x,y)}{4t} \right) |_{x=y},$$
$$\partial_{ijk} \left(-\frac{r^2(x,y)}{4t} \right) |_{x=y} = 0 = \partial_{ij\bar{k}} \left(-\frac{r^2(x,y)}{4t} \right) |_{x=y},$$
$$\left| D^{\overrightarrow{\mu_j}} \left(-\frac{r^2(x,y)}{4t} \right) |_{x=y} \right| \leq C \cdot \frac{1}{t}, \text{ for any } \vec{\mu}_j. \tag{18}$$

So from the equations in (18) and the total degree condition (17), the summand

$$\left(\prod_{j=1}^{l} D^{\overrightarrow{\mu_j}} \left(-\frac{r^2(x,y)}{4t} \right) \right) \Bigg|_{x=y} \cdot D^{\vec{\eta}} U$$

can not exceed $\left[\frac{|\vec{\gamma}|}{2} \right]$ copies of factor $\frac{1}{t}$. The $\left[\frac{|\vec{\gamma}|}{2} \right]$ copies of factor $\frac{1}{t}$ can be achieved if $|\overrightarrow{\mu_j}| = 2$ for all j, and $|\vec{\eta}| = 0$ or 1 depending on $|\vec{\gamma}|$ is even or odd, so

$$\left| D^{\vec{\eta}} U \right| \leq |u_0(x,x)| + \left| \nabla_i u_0(x,y)|_{x=y} \right| \leq 2$$

by (5) and (6) as $t \to 0_+$. Such summand is nonzero if and only if each $\overrightarrow{\mu_j}$ is of the form (i,i), $\left(i,\bar{i}\right)$ or $\left(\bar{i},\bar{i}\right)$ for some $i \in \{1,2,\cdots,n\}$ by Lemma 1, so there are at most $3n$ choices of each $\overrightarrow{\mu_j}$. Therefore as $t \to 0_+$,

$$\begin{aligned} \left| D^{\vec{\gamma}} H(t,x,y)|_{x=y} \right| &= \left| \frac{1}{(4\pi t)^{n/2}} e^{-\frac{r^2}{4t}} P_{\vec{\gamma}}(t,x,y) \Big|_{x=y} \right| \\ &\leq \frac{1}{(4\pi t)^{n/2}} \cdot \left(\frac{1}{2t} \right)^{\left[\frac{|\vec{\gamma}|}{2} \right]} \cdot (3n)^{\left[\frac{|\vec{\gamma}|}{2} \right]} \cdot 2. \end{aligned}$$

For Claim 2, since each index appears with even multiplicity in $\{\vec{\alpha}\} \amalg \{\vec{\beta}\}$, it is easy to see the summands (16) with $l = \frac{|\vec{\gamma}|}{2}$ and $|\vec{\eta}| = 0$ can have maximal number of $\overrightarrow{\mu_j}$'s satisfying $|\overrightarrow{\mu_j}| = 2$, with each $\overrightarrow{\mu_j}$ is of the form (i,i), $\left(i,\bar{i}\right)$ or $\left(\bar{i},\bar{i}\right)$ for some $i \in \{1,2,\cdots,n\}$. It is of the form

$$\left(\prod_{j=1}^{l} D^{\overrightarrow{\mu_j}} \left(-\frac{r^2(x,y)}{4t} \right) \right) \Bigg|_{x=y} \cdot U(t,x,x).$$

Since

$$\partial_{ij}\left(-\frac{r^2(x,y)}{4t}\right)|_{x=y}=-\frac{\delta_{ij}}{2t}=-\partial_{\bar{i}\bar{j}}\left(-\frac{r^2(x,y)}{4t}\right)|_{x=y}, \tag{19}$$

the product

$$\left(\prod_{j=1}^{l}D^{\overrightarrow{\mu_j}}\left(-\frac{r^2(x,y)}{4t}\right)\right)\Bigg|_{x=y}=\prod_{j=1}^{l}\operatorname{sign}(\vec{\mu}_j)\cdot\left(\frac{1}{2t}\right)^{l}$$

gives the highest power of $\frac{1}{t}$, where we let $\operatorname{sign}(\vec{\mu}_j)=1$ if $\overrightarrow{\mu_j}=(i,i)$ or $\left(\bar{i},\bar{i}\right)$, and $\operatorname{sign}(\vec{\mu}_j)=-1$ if $\overrightarrow{\mu_j}=\left(i,\bar{i}\right)$. If these summands do not cancel each other, they give the leading term in $D^{\vec{\gamma}}H(t,x,y)|_{x=y}$ as $t\to 0_+$, of the form

$$\frac{1}{(4\pi t)^{n/2}}\cdot\left(\frac{1}{2t}\right)^{\frac{|\vec{\gamma}|}{2}}\cdot(-1)^a A,$$

where A is the number of such summands, and $(-1)^a$ is their common sign.

To determine the number A, we build a one-one correspondence between such summands and admissible graphs in Definition 3. For each $\overrightarrow{\mu_j}=(i,i)$, $\left(i,\bar{i}\right)$ or $\left(\bar{i},\bar{i}\right)$ of the summand, we draw an edge between the corresponding vertices in $\mathcal{S}=\{\vec{\alpha}\}\amalg\{\vec{\beta}\}$. Drawing l edges on $\mathcal{S}$ with no intersecting vertex, we obtain an admissible graph G. The sign on each edge of G reflects the sign in (19). Taking product of them we see $\operatorname{sign}(G)$ is exactly the sign of the summand. Conversely, given any admissible graph G, we can read $\overrightarrow{\mu_j}$ from its edges, and then can write down the corresponding summand $\left(\prod_{j=1}^{l}D^{\overrightarrow{\mu_j}}\left(-\frac{r^2(x,y)}{4t}\right)\right)\Big|_{x=y}\cdot U(t,x,x)$ in $D_y^{\vec{\beta}}D_x^{\vec{\alpha}}H(t,x,y)\Big|_{x=y}$. From Proposition 2 we know $\operatorname{sign}(G)=(-1)^{\frac{|\vec{\alpha}|-|\vec{\beta}|}{2}}$, so the summands are of the same sign $(-1)^{\frac{|\vec{\alpha}|-|\vec{\beta}|}{2}}$. Proposition 2 also gives the signed count of admissible graphs, which is $A\left(\vec{\alpha},\vec{\beta}\right)$. Claim 2 is proved.

If each index appears with even multiplicity in $\{\vec{\alpha}\}\amalg\{\vec{\beta}\}$, then $|\vec{\gamma}|$ is even, $\left[\frac{|\vec{\gamma}|}{2}\right]=\frac{|\vec{\gamma}|}{2}$. From the above argument we have as $t\to 0_+$,

$$(4\pi t)^{n/2}(2t)^{\left[\frac{|\vec{\alpha}|+|\vec{\beta}|}{2}\right]}\cdot D_y^{\vec{\beta}}D_x^{\vec{\alpha}}H(t,x,y)\Big|_{x=y}$$

$$\to(4\pi t)^{n/2}(2t)^{\frac{|\vec{\gamma}|}{2}}\cdot\frac{1}{(4\pi t)^{n/2}}\cdot\left(\frac{1}{2t}\right)^{\frac{|\vec{\gamma}|}{2}}A\left(\vec{\alpha},\vec{\beta}\right)=A\left(\vec{\alpha},\vec{\beta}\right).$$

If some index i appears with odd multiplicity in $\{\vec{\alpha}\} \amalg \{\vec{\beta}\}$, then by our definition $A\left(\vec{\alpha}, \vec{\beta}\right) = 0$. For a nonzero summand of the form

$$\left(\prod_{j=1}^{l} D^{\vec{\mu_j}}\left(-\frac{r^2(x,y)}{4t}\right)\right)\Bigg|_{x=y} \cdot D^{\vec{\eta}} U,$$

by the equations in (18) and the total degree condition (17), it can have the $\frac{1}{t}$ power at most $\left[\frac{|\vec{\gamma}|-1}{2}\right]$. If the power $\left[\frac{|\vec{\gamma}|-1}{2}\right]$ is achieved, then $|\vec{\eta}| \le 2$ by the total degree condition (17), and $|\vec{\eta}| = 2$ is only possible when $\left|\vec{\gamma}\right|$ is even. If $|\vec{\eta}| = 0$, there will be a term $D^{\vec{\sigma}}\left(r^2\right)|_{x=y}$ in the summand, such that $|\vec{\sigma}| \le 3$ and i appears with odd multiplicity in $\vec{\sigma}$, making the summand zero by Lemma 1. So we must have $|\vec{\eta}| = 1$ when $\left|\vec{\gamma}\right|$ is odd.

Similar to the argument in Claim 2, we have

$$\left|D^{\vec{\gamma}} H(t,x,y)\Big|_{x=y}\right| \le C\frac{1}{(4\pi t)^{n/2}} \cdot \left(\frac{1}{t}\right)^{\left[\frac{|\vec{\gamma}|-1}{2}\right]} \left|D^{\vec{\eta}} U\right|. \tag{20}$$

When $|\vec{\eta}| = 1$, by (7) we have $\left|D^{\vec{\eta}} U\right| \le Ct$. Therefore

$$(4\pi t)^{n/2}(2t)^{\left[\frac{|\vec{\alpha}|+|\vec{\beta}|}{2}\right]} \cdot D_y^{\vec{\beta}} D_x^{\vec{\alpha}} H(t,x,y)\Big|_{x=y}$$

$$\to \left\{\begin{array}{l} t^{\left[\frac{|\vec{\gamma}|}{2}\right]} \cdot O\left(t^{-\left[\frac{|\vec{\gamma}|-1}{2}\right]}\right) \cdot O(t) = O(t) \to 0 \text{ (if } \left|\vec{\gamma}\right| \text{ is odd),} \\ t^{\left[\frac{|\vec{\gamma}|}{2}\right]} \cdot O\left(t^{-\left[\frac{|\vec{\gamma}|-1}{2}\right]}\right) \cdot O(1) = O(t) \to 0 \text{ (if } \left|\vec{\gamma}\right| \text{ is even).} \end{array}\right\} = A\left(\vec{\alpha}, \vec{\beta}\right).$$

Proposition 3 is proved. The convergence is uniform for all x on M, because in *Minakshisundaram-Pleijel expansion*

$$H(t,x,y) = \frac{1}{(4\pi t)^{n/2}} e^{-\frac{r^2}{4t}} U(t,x,y)$$

the term

$$U(t,x,y) = u_0(x,y) + tu_1(x,y) + \cdots + t^p u_p(x,y) + O\left(t^{p+1}\right) \tag{21}$$

as $t \to 0_+$, where the convergence is in C^l for any $l \ge 0$ (Theorem 2.30 in [BGV]). □

Now we are ready to give the proof of Theorem 2.

Proof. Letting $\overrightarrow{\alpha} = \overrightarrow{\beta}$ in Proposition 3, and noticing

$$\left\langle D^{\overrightarrow{\alpha}}\Phi_t(x), D^{\overrightarrow{\beta}}\Phi_t(x)\right\rangle = D_y^{\overrightarrow{\beta}} D_x^{\overrightarrow{\alpha}} H(t,x,y)\,|_{x=y},$$

we have

$$\begin{aligned}&(4\pi t)^{n/2}(2t)^{\left|\overrightarrow{\alpha}\right|} \cdot \left|D^{\overrightarrow{\alpha}}\Phi_t(x)\right|^2\\ &= (4\pi t)^{n/2}(2t)^{\left\lfloor\frac{\left|\overrightarrow{\alpha}\right|+\left|\overrightarrow{\alpha}\right|}{2}\right\rfloor} \cdot D_y^{\overrightarrow{\alpha}} D_x^{\overrightarrow{\alpha}} H(t,x,y)\Big|_{x=y} \to A\left(\overrightarrow{\alpha},\overrightarrow{\alpha}\right),\end{aligned}$$

as $t \to 0_+$, i.e.

$$\left|D^{\overrightarrow{\alpha}}\Phi_t(x)\right|^2 \to \frac{1}{(4\pi t)^{n/2}}\left(\frac{1}{2t}\right)^{\left|\overrightarrow{\alpha}\right|} A\left(\overrightarrow{\alpha},\overrightarrow{\alpha}\right). \tag{22}$$

If $\left|\overrightarrow{\alpha}\right| + \left|\overrightarrow{\beta}\right|$ is even, by Proposition 3 we have

$$\begin{aligned}&\lim_{t\to 0_+} \frac{\left\langle D^{\overrightarrow{\alpha}}\Phi_t(x), D^{\overrightarrow{\beta}}\Phi_t(x)\right\rangle}{\left|D^{\overrightarrow{\alpha}}\Phi_t(x)\right|\left|D^{\overrightarrow{\beta}}\Phi_t(x)\right|}\\ &= \lim_{t\to 0_+} \frac{(4\pi t)^{n/2}(2t)^{\frac{\left|\overrightarrow{\alpha}\right|+\left|\overrightarrow{\beta}\right|}{2}} \cdot \left\langle D^{\overrightarrow{\alpha}}\Phi_t(x), D^{\overrightarrow{\beta}}\Phi_t(x)\right\rangle}{\left((4\pi t)^{n/4}(2t)^{\frac{\left|\overrightarrow{\alpha}\right|}{2}}\left|D^{\overrightarrow{\alpha}}\Phi_t(x)\right|\right)\cdot\left((4\pi t)^{n/4}(2t)^{\frac{\left|\overrightarrow{\beta}\right|}{2}}\left|D^{\overrightarrow{\beta}}\Phi_t(x)\right|\right)}\\ &= \left[A\left(\overrightarrow{\alpha},\overrightarrow{\beta}\right)\right] \Big/ \left[A\left(\overrightarrow{\alpha},\overrightarrow{\alpha}\right)A\left(\overrightarrow{\beta},\overrightarrow{\beta}\right)\right]^{1/2} = B\left(\overrightarrow{\alpha},\overrightarrow{\beta}\right).\end{aligned}$$

If $\left|\overrightarrow{\alpha}\right| + \left|\overrightarrow{\beta}\right|$ is odd, then some index must appear with odd multiplicity in $\left\{\overrightarrow{\alpha}\right\} \amalg \left\{\overrightarrow{\beta}\right\}$. By (20) and (22),

$$\begin{aligned}&\lim_{t\to 0_+} \frac{\left|\left\langle D^{\overrightarrow{\alpha}}\Phi_t(x), D^{\overrightarrow{\beta}}\Phi_t(x)\right\rangle\right|}{\left|D^{\overrightarrow{\alpha}}\Phi_t(x)\right|\left|D^{\overrightarrow{\beta}}\Phi_t(x)\right|}\\ &\le \lim_{t\to 0_+} \frac{C\frac{1}{(4\pi t)^{n/2}}\left(\frac{1}{t}\right)^{\left\lfloor\frac{\left|\overrightarrow{\alpha}\right|+\left|\overrightarrow{\beta}\right|-1}{2}\right\rfloor}}{\frac{1}{(4\pi t)^{n/4}}\left(\frac{1}{2t}\right)^{\frac{\left|\overrightarrow{\alpha}\right|}{2}} A\left(\overrightarrow{\alpha},\overrightarrow{\alpha}\right)\cdot\frac{1}{(4\pi t)^{n/4}}\left(\frac{1}{2t}\right)^{\frac{\left|\overrightarrow{\beta}\right|}{2}} A\left(\overrightarrow{\beta},\overrightarrow{\beta}\right)}\\ &= O(t) \to 0 = B\left(\overrightarrow{\alpha},\overrightarrow{\beta}\right).\end{aligned}$$

By Proposition 3, the convergence is uniform for all x on M. Theorem 2 is proved. □

Remark 3. If we only care about the leading order terms in $\left\langle D^{\vec{\alpha}}\Phi_t(x), D^{\vec{\beta}}\Phi_t(x)\right\rangle$ or $D_y^{\vec{\beta}} D_x^{\vec{\alpha}} H(t,x,y)\,|_{x=y}$ as $t \to 0_+$, the ordering of derivatives within $\vec{\alpha}$ and $\vec{\beta}$ are not important, and we can regard $\vec{\alpha}$ and $\vec{\beta}$ as *sets* of their indices, respectively. This is because when we switch the ordering of derivatives, say from $D_x^i D_x^j$ to $D_x^j D_x^i$, their difference is a lower order differential operator, and for lower order differential operator $D^{\vec{\alpha'}}$ with $\left|\vec{\alpha'}\right| < \left|\vec{\alpha}\right|$, $\left|D^{\vec{\alpha'}}\Phi_t(x)\right| = o\left(\left|D^{\vec{\alpha}}\Phi_t(x)\right|\right)$ as $t \to 0_+$ for all x on M by (10).

Remark 4. We remark that except the constant term $A\left(\vec{\alpha}, \vec{\beta}\right)$, the next order terms in $t^{\frac{n}{2}+\left\lceil\frac{|\vec{\alpha}|+|\vec{\beta}|}{2}\right\rceil} \cdot D_y^{\vec{\beta}} D_x^{\vec{\alpha}} H(t,x,y)\Big|_{x=y}$ do depend on g (or curvature), and are important for some geometric applications (e.g. Proposition 5), but to compute them is significantly harder.

3 Applications

In this section, we give several applications of the high-jet relations of ψ_t in Theorem 2 and Proposition 3. The applications to the mean curvature and the Riemannian curvature tensor are related to the *extrinsic geometry* and *intrinsic geometry* of $\psi_t(M) \subset l^2$ as $t \to 0_+$ respectively. We also discuss the *approximation* of the geometry of (M, g) by finitely many eigenfunctions.

3.1 Mean curvature of $\psi_t(M)$ in l^2

The first application is about the mean curvature of the embedded image $\psi_t(M) \subset l^2$. Intuitively speaking, as $t \to 0_+$, the extrinsic geometry of $\psi_t(M) \subset l^2$ locally looks the same at any point, and is evenly bumpy everywhere. More precisely we have

Proposition 4. *For the heat kernel embedding $\psi_t : M \to l^2$ and $x \in M$, let*

$$A(x,t) : T_{\psi_t(x)}\psi_t(M) \times T_{\psi_t(x)}\psi_t(M) \to T_{\psi_t(x)}\psi_t(M)^{\perp}$$

be the second fundamental form of the embedded image $\psi_t(M) \subset l^2$ and $H(x,t)$ be the mean curvature vector at $\psi_t(x)$. We have

1. *([WZ] Corollary 22) $\sqrt{t}A(x,t)$ converges to certain normal form for all x on M as $t \to 0_+$. The mean curvature vector $H(x,t)$, after scaled by a factor $\sqrt{t}$, converges to constant length:*

$$\lim_{t\to 0_+} \sqrt{t}\,|H(x,t)| = \sqrt{\frac{n+2}{2n}}. \tag{1}$$

2. *$\psi_t(M)$ is asymptotically umbilical in the mean curvature vector direction as $t \to 0_+$, in the following sense:*

$$t \cdot \lim_{t\to 0_+} S(x,t)\langle W, H\rangle = -\frac{3}{2}W, \textit{ for any } W \in T_{\psi_t(x)}\psi_t(M),$$

where

$$S(x,t) : T_{\psi_t(x)}\psi_t(M) \times T_{\psi_t(x)}\psi_t(M)^{\perp} \to T_{\psi_t(x)}\psi_t(M)$$

is the second fundamental tensor on $\psi_t(M)$ defined by $S(X,\upsilon) = (\nabla_X \upsilon)^T$ for any $X \in T_{\psi_t(x)}\psi_t(M)$ and $\upsilon \in T_{\psi_t(x)}\psi_t(M)^{\perp}$, $\top : T_{\psi_t(x)}l^2 \to T_{\psi_t(x)}\psi_t(M)$ is the projection, and "$\perp$" means the orthogonal complement in l^2.

The convergence is uniform for all x on M.

Proof. For any $x \in M$, let $(x_1, \cdots, x_n)$ be a local coordinates near x such that the coordinate vectors $\left\{\frac{\partial}{\partial x^i}\right\}_{i=1}^n := \{V_i\}_{i=1}^n$ are orthonormal at x. The *second fundamental form* of the submanifold $\psi_t(M) \subset l^2$ can be written as

$$A(x,t) = \sum_{1\le i,j\le n} h_{ij}(x,t)\,dx^i dx^j$$

where $h_{jk}(x,t)$ $(1 \le j,k \le n)$ are vectors in l^2. Then as $t \to 0_+$, we have

$$\lim_{t\to 0_+}\left(h_{jk} - \nabla_{V_j}\nabla_{V_k}\psi_t\right)(x,t) = 0, \text{ for } 1 \le j,k \le n. \tag{2}$$

This is due to the 2-jet relation

$$\begin{aligned}
&\lim_{t\to 0_+}\left\langle \nabla_{V_i}\psi_t, \nabla_{V_j}\nabla_{V_k}\psi_t\right\rangle(x,t)\\
&= \lim_{t\to 0_+}(4\pi t)^{n/2}(2t)\left\langle \nabla_{V_i}\Phi_t, \nabla_{V_j}\nabla_{V_k}\Phi_t\right\rangle(x,t) = 0
\end{aligned} \tag{3}$$

from (15), so the second order terms in the Taylor expansion of $\psi_t : M \to l^2$ approximate the second fundamental form of $\psi_t(M) \subset l^2$ as $t \to 0_+$. Therefore, for the *mean curvature* $H(x,t) = \frac{1}{n}\Sigma_{k=1}^n h_{kk}(x)$, we have

$$\lim_{t\to 0_+}\left(H(x,t) - \frac{\Sigma_{k=1}^n \nabla_{V_k}\nabla_{V_k}\psi_t}{n}\right)(x,t) = 0. \tag{4}$$

Item 1 was proved in [WZ] using the 2-jet relations of ψ_t in Theorem 1. For Item 2, using the 3-jet relations in Theorem 2 for $\overrightarrow{\alpha} = (i,k,k)$ and $\overrightarrow{\beta} = (j)$, we have

$$\frac{\left\langle \nabla_{V_i}\nabla_{V_k}\nabla_{V_k}\psi_t, \nabla_{V_j}\psi_t \right\rangle (x,t)}{\left|\nabla_{V_i}\nabla_{V_k}\nabla_{V_k}\psi\right| \left|\nabla_{V_j}\psi_t\right| (x,t)} \to B\left(\overrightarrow{\alpha}, \overrightarrow{\beta}\right)$$

$$= \begin{cases} (-1)^{\frac{2}{2}} \cdot \frac{(4!/2!)}{[(6!/3!)(2!/1!)]^{1/2}} = -\sqrt{\frac{3}{5}}, & \text{if } i = j = k, \\ (-1)^{\frac{2}{2}} \cdot \frac{(2!/1!)}{[(2!/1!)(2!/1!)]^{1/2}} \cdot \frac{(2!/1!)}{[(4!/2!)(0!/0!)]^{1/2}} = -\frac{1}{\sqrt{3}}, & \text{if } i = j \neq k, \\ 0, & \text{if } i \neq j, \end{cases}$$

and

$$\left|\nabla_{V_i}\nabla_{V_k}\nabla_{V_k}\psi_t\,(x,t)\right| \to \sqrt{2}\,(4\pi)^{n/4}\, t^{\frac{n+2}{4}} \cdot \left(\frac{1}{4\pi t}\right)^{n/4} \left(\frac{1}{2t}\right)^{3/2} A\left(\overrightarrow{\alpha}, \overrightarrow{\alpha}\right)^{1/2}$$

$$= \begin{cases} \frac{1}{2t}\sqrt{\frac{6!}{2^3 3!}} = \frac{1}{2t}\sqrt{15}, \text{ if } i = k \\ \frac{1}{2t}\sqrt{\frac{4!}{2^2 2!}\cdot\frac{2!}{2^1 2!}} = \frac{1}{2t}\sqrt{3}, \text{ if } i \neq k \end{cases},$$

$$\left|\nabla_{V_j}\psi_t\,(x,t)\right| \to \sqrt{2}\,(4\pi)^{n/4}\, t^{\frac{n+2}{4}} \cdot \left(\frac{1}{4\pi t}\right)^{n/4} \left(\frac{1}{2t}\right)^{1/2} A\left(\overrightarrow{\beta}, \overrightarrow{\beta}\right)^{1/2} = 1.$$

Therefore as $t \to 0_+$,

$$\left\langle \nabla_{V_i}\nabla_{V_k}\nabla_{V_k}\psi_t, \nabla_{V_j}\psi_t \right\rangle (x,t) \to \begin{Bmatrix} -\sqrt{\frac{3}{5}} \cdot \frac{1}{2t}\sqrt{15} = -\frac{3}{2t}, \text{ if } i = j = k, \\ -\frac{1}{\sqrt{3}} \cdot \frac{1}{2t}\sqrt{3} = -\frac{3}{2t}, \quad \text{if } i = j \neq k. \end{Bmatrix} = -\frac{3}{2t}.$$

By the definition of the second fundamental tensor,

$$\begin{aligned} & 2t \cdot \lim_{t\to 0_+} \left\langle S(x,t)\left(\nabla_{V_i}\psi_t, H\right), \nabla_{V_j}\psi_t \right\rangle (x,t) \\ &= 2t \cdot \lim_{t\to 0_+} \left\langle \nabla_{\nabla_{V_i}\psi_t} H\left(\psi_t(x), t\right), \nabla_{V_j}\psi_t \right\rangle \left(\psi_t(x), t\right) \\ &= 2t \cdot \lim_{t\to 0_+} \left\langle \nabla_{V_i} H(x,t), \nabla_{V_j}\psi_t \right\rangle (x,t) \\ &= 2t \cdot \lim_{t\to 0_+} \left\langle \nabla_{V_i}\left(\frac{1}{n}\sum_{k=1}^{n} \nabla_{V_k}\nabla_{V_k}\psi_t\right), \nabla_{V_j}\psi_t \right\rangle (x,t) \\ &= \frac{1}{n} 2t \cdot \sum_{k=1}^{n} \lim_{t\to 0_+} \left\langle \nabla_{V_i}\nabla_{V_k}\nabla_{V_k}\psi_t, \nabla_{V_j}\psi_t \right\rangle (x,t) \\ &= \begin{cases} \frac{1}{n} 2t \cdot \sum_{k=1}^{n} \left(-\frac{3}{2t}\right), & \text{if } i = j, \\ 0, & \text{if } i \neq j, \end{cases} \\ &= -3\delta_{ij}. \end{aligned}$$

Since $\left\{\nabla_{V_j}\psi_t\right\}_{1\le j\le n}$ span $T_{\psi_t(x)}\psi_t(M)$, this means for any $W \in T_{\psi_t(x)}\psi_t(M)$,

$$t \cdot \lim_{t\to 0_+} S(x,t)\langle W, H\rangle = -\frac{3}{2}W.$$

So $\psi_t(M)$ is asymptotically umbilical in the mean curvature vector direction as $t \to 0_+$. □

Remark 5. In other words, for the image $\frac{1}{\sqrt{t}}\psi_t(M) \subset l^2$, as $t \to 0_+$ its mean curvature vector converges to constant length $\sqrt{\frac{n+2}{2n}}$, and $\frac{1}{\sqrt{t}}\psi_t(M)$ is asymptotically umbilical in the mean curvature vector direction. If we want to construct submanifolds in $\mathbb{R}^q$ with constant mean curvature length, perhaps we can truncate $\frac{1}{\sqrt{t}}\psi_t(M) \subset l^2$ to $\mathbb{R}^q$ by taking the first q components for large q, and then perturb it by the implicit function theorem.

3.2 Riemannian curvature tensor

In the second application we give a *heat kernel embedding* interpretation of the Levi-Civita connection and the Riemannian curvature tensor. The idea is that $\psi_t : M \to l^2$ is an almost isometric embedding, so the "extrinsic" constructions of the Levi-Civita connection and Gaussian curvature of surfaces in $\mathbb{R}^3$ using the ambient space can be mimicked here, with $\mathbb{R}^3$ replaced by l^2. The difficulty is that we need to take the limits as $t \to 0_+$ in our interpretation, and to show that these limits exist, we need the high-jet relations in Proposition 3 and some refinement.

As pointed to the author by L. I. Nicolaescu, there exists a *random function* interpretation of the Levi-Civita connection and the Riemannian curvature tensor in Section 12 of [AT] (equation (12.2.6) and Lemma 12.2.1, respectively). If the random function f in [AT] is taken as $f_t = \sum_{j\geq 1} u_j\phi_j(x)$ with the coefficients u_j being independent Gaussian random variables with the expectation $E(u_j) = 0$ and variance $Var(u_j) = e^{-\lambda_j t}$, then Lemma 2 and Proposition 5 follows from the results in [AT] by taking limits as $t \to 0$. (For this approach, see [Ni], appendix B.)

As before, for any $p \in M$, we choose the normal coordinates $(x^1, \cdots x^n)$ at p, and choose an orthonormal frame field $\{V_j\}_{j=1}^n$ near p such that $\{V_j\}_{i=1}^n |_p = \{\frac{\partial}{\partial x^j}\}$. Let ∇ be the Levi-Civita connection of (M, g), and $\nabla_{V_j} := \nabla_j$ for $1 \leq j \leq n$.

Lemma 2. *(Levi-Civita Connection) For any $X, Y \in T_pM$, we have*

$$\begin{aligned}\nabla_X Y(p) &= \lim_{t\to 0_+} \sum_{j=1}^n \langle \nabla_X\nabla_Y\psi_t, \nabla_{V_j}\psi_t\rangle(p)\,V_j(p) \\ &= \lim_{t\to 0_+} d\psi_t^{-1}\left[(\nabla_X\nabla_Y\psi_t(p))^\top\right],\end{aligned} \tag{5}$$

where $\top : T_{\psi_t(p)}l^2 \to T_{\psi_t(p)}\psi_t(M)$ is the orthogonal projection with respect to the standard metric on l^2, and on the right hand side ∇_X and ∇_Y are the derivatives for l^2-valued functions $f : M \to l^2$.

Proof. Since all quantities in (5) are linear in X, we may assume $X = V_i$. In a neighborhood of p, we write $Y(x) = \sum_{k=1}^n y^k(x)V_k(x)$. Since $\nabla_{V_i}V_j(p) = \delta_{ij}$, we have

$$\nabla_{V_i} Y(p) = \sum_{k=1}^{n} \nabla_i y^k(p) V_k(p),$$

where $\partial_i y^k(x) = \nabla_{V_i} y^k(x)$. On the other hand, by (3) and the 2-jet relation (3), we have

$$\begin{aligned}
&\lim_{t\to 0_+} \sum_{i=1}^{n} \langle \nabla_i \nabla_Y \psi_t, \nabla_j \psi_t \rangle (p) V_j(p) \\
&= \lim_{t\to 0_+} \sum_{i=1}^{n} \left\langle \nabla_i \left(\sum_{k=1}^{n} y^k \nabla_k \psi_t \right), \nabla_j \psi_t \right\rangle (p) V_j(p) \\
&= \sum_{i=1}^{n} \sum_{k=1}^{n} \lim_{t\to 0_+} \nabla_i y^k(p) \langle \nabla_k \psi_t, \nabla_j \psi_t \rangle (p) V_j(p) \\
&\quad + y^k(p) \langle \nabla_i \nabla_k \psi_t, \nabla_j \psi_t \rangle (p) V_j(p) \\
&= \sum_{i=1}^{n} \sum_{k=1}^{n} \left(\partial_i y^k(p) \delta_{kj} V_j(p) + 0 \right) \\
&= \nabla_i y^k(p) V_k(p).
\end{aligned}$$

Therefore the first identity

$$\nabla_X Y(p) = \lim_{t\to 0_+} \sum_{j=1}^{n} \langle \nabla_X \nabla_Y \psi_t, \nabla_j \psi_t \rangle (p) V_j(p)$$

is proved. Since $\{V_j\}_{j=1}^{n}$ is an orthonormal basis of $T_p M$, and ψ_t is an almost isometric embedding in the sense of (3), $\{\nabla_{V_j} \psi_t(p)\}_{j=1}^{n}$ is asymptotically an orthonormal basis of $T_{\psi_t(p)} \psi_t(M)$ as $t \to 0_+$, and we have

$$\lim_{t\to 0_+} \left[(\nabla_X \nabla_Y \psi_t(p))^\top - \sum_{j=1}^{n} \langle \nabla_X \nabla_Y \psi_t, \nabla_{V_j} \psi_t \rangle (p) \nabla_{V_j} \psi_t(p) \right] = 0.$$

Applying $d\psi_t^{-1}$ on both sides, noting $d\psi_t^{-1} \left(\nabla_{V_j} \psi_t(p) \right) = V_j$, the second identity is proved. □

Suppose $f : (M, g) \to \mathbb{R}^q$ is an isometric embedding, and h is the second fundamental form of M. It is well-known that for vectors $X, Y, Z, W \in TM$, the Riemannian curvature tensor R of M satisfies

$$R(X, Y, Z, W) = \langle h(X, W), h(Y, Z) \rangle - \langle h(X, Z), h(Y, W) \rangle.$$

This is the *Gauss formula*, which relates the intrinsic curvature R of (M, g) with its extrinsic geometry in the ambient space. In our case $\psi_t : M \to l^2$ is asymptotically isometric as $t \to 0_+$, and by (2) the second fundamental form $h(X,Y)$ of $\psi_t(M) \subset l^2$ is approximated by $\nabla_X \nabla_Y \psi_t$ as $t \to 0_+$, so it is natural to expect the following "asymptotic" Gauss formula

Proposition 5. *(Riemannian curvature) For any X, Y, Z, W of TM, we have*

$$R(X,Y,Z,W) = \lim_{t \to 0+} \left[\langle \nabla_X \nabla_W \psi_t, \nabla_Y \nabla_Z \psi_t \rangle - \langle \nabla_X \nabla_Z \psi_t, \nabla_Y \nabla_W \psi_t \rangle \right]. \qquad (6)$$

Proof. We only need to prove the above identity when X, Y, Z and W are taken from the orthonormal basis $\{V_j\}_{i=1}^n$ near x as above, say they are V_i, V_j, V_k and V_l respectively. For general vector fields, we can express $X = \sum_{j=1}^n a^j(x) V_j(x)$ etc., and plug them in (6). Then by the high-jet relations (3) and (3), as $t \to 0_+$ the terms of the type $\langle \nabla_k a^i \nabla_i \psi_t, \nabla_l b^j \nabla_j \psi_t \rangle$ will vanish or cancel each other, and the terms of the type $\langle \nabla_k a^i \nabla_i \psi_t, \nabla_j \psi_t \nabla_k \psi_t \rangle$ will converge to 0. For the same reason, switching the ordering of the covariant derivatives in (6), say from $\nabla_X \nabla_W \psi_t$ to $\nabla_W \nabla_X \psi_t$ etc., will not affect the limit in (6).

We first refine our estimate of $\lim_{t \to 0_+} \nabla^x_{V_i} \nabla^x_{V_l} \nabla^y_{V_j} \nabla^y_{V_k} H(t,x,y)|_{x=y}$ in Proposition 3. We have proved that

$$\lim_{t \to 0_+} (4\pi t)^{\frac{n}{2}} (2t)^2 D_x^{\vec{\alpha}} D_y^{\vec{\beta}} H(t,x,y) \Big|_{x=y} = A\left(\vec{\alpha}, \vec{\beta}\right)$$

for $\vec{\alpha} = (i,l)$ and $\vec{\beta} = (j,k)$, where $A\left(\vec{\alpha}, \vec{\beta}\right)$ is the leading (constant) term of $(4\pi t)^{\frac{n}{2}} (2t)^2 D_x^{\vec{\alpha}} D_y^{\vec{\beta}} H(t,x,y) \Big|_{x=y}$ as $t \to 0_+$, but to prove our theorem we will need the next order term, which is linear in t. Recall in Proposition 3, for $\vec{\gamma} = \left(\vec{\alpha}, \vec{\vec{\beta}}\right)$,

(where the notation $\vec{\vec{\beta}}$ means the derivative $D^{\vec{\beta}}$ is with respect to the y-variables), we write $D^{\vec{\gamma}} H(t,x,y) = \frac{1}{(4\pi t)^{n/2}} e^{-\frac{r^2}{4t}} P_{\vec{\gamma}}(t,x,y)$, where $P_{\vec{\gamma}}(t,x,y)$ is a polynomial in $D^{\vec{\mu_s}}\left(-\frac{r^2(x,y)}{4t}\right)$ and $D^{\vec{\eta}} U(t,x,y)\Big|_{x=y}$, i.e. each summand is of the form

$$\left(\prod_{s=1}^h D^{\vec{\mu_j}} \left(-\frac{r^2(x,y)}{4t} \right) \right) \Bigg|_{x=y} \cdot D^{\vec{\eta}} U(t,x,y) \Big|_{x=y}.$$

Using Lemma 1 and the total degree condition $|\vec{\gamma}| = 4$, by the same method in Proposition 3 to find the leading terms in $P_{\vec{\gamma}}(t,x,y)$ of order $\left(\frac{1}{t}\right)^2$, we see the secondary terms of order $\left(\frac{1}{t}\right)$ must have $h = 1$, $|\vec{\mu_1}| = 2$, and $|\vec{\eta}| = 2$, or $h = 1$, $|\vec{\mu_1}| = 4$, and $\vec{\eta} = 0$, namely of the form

$$D^{\overrightarrow{\mu_1}}\left(-\frac{r^2(x,y)}{4t}\right)\Big|_{x=y}\cdot D^{\overrightarrow{\eta}}U(t,x,y)\Big|_{x=y},\ \text{with } \overrightarrow{\mu_1}\cup\overrightarrow{\eta}=\overrightarrow{\gamma},\ \left|\overrightarrow{\mu_1}\right|=\left|\overrightarrow{\eta}\right|=2,$$

or

$$D^{\overrightarrow{\gamma}}\left(-\frac{r^2(x,y)}{4t}\right)\Big|_{x=y}\cdot U(t,x,y)|_{x=y}=-\frac{1}{6t}\left(R_{ijkl}(x)+R_{ilkj}(x)\right),$$

where we have used (12). So as $t\to 0_+$, we have the refined convergence

$$\begin{aligned}&(4\pi t)^{n/2}(2t)^2\,\nabla^x_{V_i}\nabla^x_{V_l}\nabla^y_{V_j}\nabla^y_{V_k}H(t,x,y)\,|_{x=y}\\ &=A\left(\overrightarrow{(i,l)},\overrightarrow{(j,k)}\right)\\ &-t\sum_{\left|\overrightarrow{\mu_1}\right|=\left|\overrightarrow{\eta}\right|=2,\ \overrightarrow{\mu_1}\cup\overrightarrow{\eta}=\left(\overrightarrow{(i,l)},\overrightarrow{(j,k)}\right)}\left(D^{\overrightarrow{\mu_1}}\left(r^2(x,y)\right)\Big|_{x=y}\cdot D^{\overrightarrow{\eta}}U(t,x,y)\Big|_{x=y}\right)\\ &-\frac{2}{3}t\left(R_{ijkl}(x)+R_{ilkj}(x)\right)+O\left(t^2\right).\end{aligned}\tag{7}$$

For $\overrightarrow{\gamma}=\left(\overrightarrow{(i,l)},\overrightarrow{(j,k)}\right)$ and $\overrightarrow{\gamma}=\left(\overrightarrow{(i,k)},\overrightarrow{(j,l)}\right)$, by considering all partitions of $\overrightarrow{\gamma}$ into $\overrightarrow{\mu_1}\cup\overrightarrow{\eta}$ with $\left|\overrightarrow{\mu_1}\right|=2$, and $\left|\overrightarrow{\eta}\right|=2$, and noticing $\nabla^a_x\nabla^b_y f(x,y)\,|_{x=y}=\nabla^a_y\nabla^b_x f(x,y)\,|_{x=y}$ for any indices a,b and functions f symmetric to x and y (here $r^2(x,y)$ and $U(t,x,y)$), we have

$$\begin{aligned}&\sum_{\left|\overrightarrow{\mu_1}\right|=\left|\overrightarrow{\eta}\right|=2,\ \overrightarrow{\mu_1}\cup\overrightarrow{\eta}=\left(\overrightarrow{(i,l)},\overrightarrow{(j,k)}\right)}\left(D^{\overrightarrow{\mu_1}}\left(r^2(x,y)\right)\Big|_{x=y}\cdot D^{\overrightarrow{\eta}}U(t,x,y)\Big|_{x=y}\right)\\ &=\sum_{\left|\overrightarrow{\mu_1}\right|=\left|\overrightarrow{\eta}\right|=2,\ \overrightarrow{\mu_1}\cup\overrightarrow{\eta}=\left(\overrightarrow{(i,k)},\overrightarrow{(j,l)}\right)}\left(D^{\overrightarrow{\mu_1}}\left(r^2(x,y)\right)\Big|_{x=y}\cdot D^{\overrightarrow{\eta}}U(t,x,y)\Big|_{x=y}\right).\end{aligned}$$

We also have $A\left(\overrightarrow{(i,l)},\overrightarrow{(j,k)}\right)=A\left(\overrightarrow{(i,k)},\overrightarrow{(j,l)}\right)$ by the definition of $A\left(\overrightarrow{\alpha},\overrightarrow{\beta}\right)$. Therefore, from (7) we have

$$
\begin{aligned}
&\lim_{t\to 0_+}\left[\left\langle \nabla_{V_i}\nabla_{V_l}\psi_t, \nabla_{V_j}\nabla_{V_k}\psi_t\right\rangle(x) - \left\langle \nabla_{V_i}\nabla_{V_k}\psi_t, \nabla_{V_j}\nabla_{V_l}\psi_t\right\rangle(x)\right]\\
&= \lim_{t\to 0_+}\frac{1}{2t}\left[(4\pi t)^{n/2}(2t)^2\left(\nabla^x_{V_i}\nabla^x_{V_l}\nabla^y_{V_j}\nabla^y_{V_k}H(t,x,y) - \nabla^x_{V_i}\nabla^x_{V_k}\nabla^y_{V_j}\nabla^y_{V_l}H(t,x,y)\right)|_{x=y}\right]\\
&= \lim_{t\to 0_+}\frac{1}{2t}\frac{2t}{3}\left[-R_{ijlk}(x) - R_{iklj}(x) + R_{ijkl}(x) + R_{ilkj}(x) + O\left(t^2\right)\right]\\
&= \frac{1}{3}\lim_{t\to 0_+}\left[R_{ijkl}(x) + R_{ikjl}(x) + R_{ijkl}(x) + R_{kjil}(x) + O\left(t^2\right)\right]\\
&= \frac{1}{3}\lim_{t\to 0_+}\left[2R_{ijkl}(x) + \left(R_{ikjl}(x) + R_{kjil}(x)\right) + O\left(t^2\right)\right]\\
&= \frac{1}{3}\lim_{t\to 0_+}\left[2R_{ijkl}(x) + R_{ijkl}(x) + O\left(t^2\right)\right]\\
&= R_{ijkl}(x),
\end{aligned}
$$

where in the last four identities we have used the symmetries of the Riemannian curvature tensor. □

Remark 6. There are several ways to express the curvature tensors by the eigenfunctions and the heat kernel on M, e.g. the Ricci curvature (8) and scalar curvature (3) in [BBG] and [Gil], but the expression (6) has a clear geometric meaning: the Riemannian curvature tensor on (M, g) is computed from the Gauss map of $\psi_t(M) \to S^\infty$ in l^2 as $t \to 0_+$. One more interesting point is that although our construction of the Riemannian curvature tensor via the embedding $\psi_t : M \to l^2$ looks "extrinsic", it is actually "intrinsic", since the embedding ψ_t is constructed by the eigenfunctions of Δ_g.

If we only want to prove the *existence* of the limit in (6), we only need the cancellation of the leading terms $A\left(\overrightarrow{(i,l)}, \overrightarrow{(j,k)}\right)$ and $A\left(\overrightarrow{(i,k)}, \overrightarrow{(j,l)}\right)$ in (7), but do not need the symmetries of the Riemannian curvature tensor. On the other hand, (6) *manifests the symmetries* of the Riemannian curvature tensor in a rather direct way. From (6), it is easy to see $R(X,Y)Z = -R(Y,X)Z$, $\langle R(X,Y)Z, W\rangle = -\langle R(X,Y)W, Z\rangle$, $\langle R(X,Y)Z, W\rangle = \langle R(Z,W)X, Y\rangle$, and $R(X,Y)Z + R(Y,Z)X + R(Z,X)Y = 0$ (the first Bianchi identity). (Note that switching the order of differentiation in the second derivatives, say from $\nabla_X\nabla_Z\psi_t$ to $\nabla_Z\nabla_X\psi_t$, will not affect the limit in (6), as remarked in the beginning of the proof of Proposition 5.) With a little computation the second Bianchi identity also follows from our formula of R, as follows:

Lemma 3. *(Second Bianchi Identity) For any coordinates $\left(x^1, \cdots, x^n\right)$ near x, the Riemannian curvature tensor R satisfies*

$$
\frac{\partial}{\partial x^h}R_{klij} + \frac{\partial}{\partial x^k}R_{lhij} + \frac{\partial}{\partial x^l}R_{hkij} = 0. \tag{8}
$$

Proof. Since all terms on the left side are tensors, we only need to prove the identity when $\left(x^1, \cdots, x^n\right)$ is a normal coordinate near x. From (6) we have

$$
\begin{aligned}
\frac{\partial}{\partial x^h} R_{klij}(x) &= \nabla_h \lim_{t\to 0+} \left[\left\langle \nabla_k \nabla_j \psi_t, \nabla_l \nabla_i \psi_t \right\rangle - \left\langle \nabla_k \nabla_i \psi_t, \nabla_l \nabla_j \psi_t \right\rangle \right](x) \\
&= \lim_{t\to 0+} \left[\left\langle \nabla_{hkj} \psi_t, \nabla_{li} \psi_t \right\rangle + \left\langle \nabla_{hli} \psi_t, \nabla_{kj} \psi_t \right\rangle \right. \\
&\quad \left. - \left\langle \nabla_{hki} \psi_t, \nabla_{lj} \psi_t \right\rangle - \left\langle \nabla_{hlj} \psi_t, \nabla_{ki} \psi_t \right\rangle \right](x),
\end{aligned}
$$

where $\nabla_{hkj} := \nabla_h \nabla_k \nabla_j$. Similarly

$$
\begin{aligned}
\frac{\partial}{\partial x^k} R_{lhij}(x) &= \lim_{t\to 0+} \left[\left\langle \nabla_{klj} \psi_t, \nabla_{hi} \psi_t \right\rangle + \left\langle \nabla_{khi} \psi_t, \nabla_{lj} \psi_t \right\rangle \right. \\
&\quad \left. - \left\langle \nabla_{kli} \psi_t, \nabla_{hj} \psi_t \right\rangle - \left\langle \nabla_{khj} \psi_t, \nabla_{li} \psi_t \right\rangle \right](x), \\
\frac{\partial}{\partial x^l} R_{hkij}(x) &= \lim_{t\to 0+} \left[\left\langle \nabla_{lhj} \psi_t, \nabla_{ki} \psi_t \right\rangle + \left\langle \nabla_{lki} \psi_t, \nabla_{hj} \psi_t \right\rangle \right. \\
&\quad \left. - \left\langle \nabla_{lhi} \psi_t, \nabla_{kj} \psi_t \right\rangle - \left\langle \nabla_{lkj} \psi_t, \nabla_{hi} \psi_t \right\rangle \right](x).
\end{aligned}
$$

Adding the 3 identities together, (8) follows. There is a convergence issue of limits like $\lim_{t\to 0+} \left\langle \nabla_{hkj} \psi_t, \nabla_{li} \psi_t \right\rangle = O\left(\frac{1}{t}\right)$ from Proposition 3, but writing

$$
\frac{\partial}{\partial x^h} R_{klij} + \frac{\partial}{\partial x^k} R_{lhij} + \frac{\partial}{\partial x^l} R_{hkij} = \frac{1}{2t} \left[2t \left(\frac{\partial}{\partial x^h} R_{klij} + \frac{\partial}{\partial x^k} R_{lhij} + \frac{\partial}{\partial x^l} R_{hkij} \right) \right]
$$

and applying Proposition 3 easily fix that. □

3.3 Approximation of Riemannian geometry by eigenfunctions

For the embedding map $\psi_t : (M, g) \to l^2$, we can truncate the infinite dimensional l^2 to $\mathbb{R}^q$ by taking its first q components, and denote the resulted map by ψ_t^q. The following truncation lemma in [WZ] gives the bound of q that can preserve most good properties of ψ_t. It was proved by the Weyl's asymptotic formula of eigenvalues (see [Ch]) and C^0-estimates of eigenfunctions (and their derivatives) on compact Riemannian manifolds.

Lemma 4. *([WZ] Theorem 6 and Remark 7) Let $\rho > 0$ be a fixed small constant. For integers $q = q(t)$ of the order $t^{-\left(\frac{n}{2}+\rho\right)}$ as $t \to 0_+$, the truncated heat kernel mapping*

$$
\psi_t^{q(t)} : (M, g) \to \mathbb{R}^{q(t)}
$$

still satisfies the asymptote

$$
\left(\psi_t^{q(t)}\right)^* g_0 = g + \frac{t}{3}\left(\frac{1}{2} S_g \cdot g - Ric_g\right) + O\left(t^2\right), \tag{9}
$$

where g_0 is the standard metric in l^2. The truncation order $q(t) \sim t^{-\left(\frac{n}{2}+\rho\right)}$ is also valid to preserve all high-jet relations in Theorem 2 for higher derivatives of ψ_t.

Using the above truncation lemma, we can approximate the metric g, the Levi-Civita connection ∇, and the Riemannian curvature R by $q(t) \sim t^{-\frac{n}{2}-\rho}$ eigenfunctions by the expressions (3), (5), and (6), respectively, with the error of certain orders of t. This suggests the possibility of approximating the Riemannian geometry of (M, g) by finitely many eigenfunctions. Indeed, if the Schwartz function w is chosen with compact support, Theorem 1.6 of [Ni] gives one of such approximations.

It will be useful to get a precise control of the coefficients in the above remainder $O(t^2)$ and their higher derivative counterparts, so we know how large $q(t)$ (i.e. how small t) should be in order to yield a given accuracy. For recent progress on truncation dimension of heat kernel embedding maps, and for Riemannian metrics of lower regularity (only $C^{2,\alpha}$), see [CW] and [Po].

4 Algebraic structures in the ∞-jet space of $\Phi_t(x)$

In Section 2 we have obtained the angle and length relations of all derivative vectors $D^{\vec{\alpha}}\Phi_t(x)$ as $t \to 0_+$. These relations are independent on g, x and the choice of orthonormal basis, so it is worthwhile to investigate them more closely. The relations are expressed by the constants $A\left(\vec{\alpha}, \vec{\beta}\right)$ and $B\left(\vec{\alpha}, \vec{\beta}\right)$. In this section we explore more relations between these constants, or in other words study the ∞-jet space of $\Phi_t(x)$ as $t \to 0_+$.

4.1 Inductive relations of $A\left(\vec{\alpha}, \vec{\beta}\right)$ and $B\left(\vec{\alpha}, \vec{\beta}\right)$

It turns out the constants $A\left(\vec{\alpha}, \vec{\beta}\right)$ and $B\left(\vec{\alpha}, \vec{\beta}\right)$ involved in the high-jet relations in Theorem 2 can be defined inductively.

Definition 4. Suppose $j_r \in \{1, 2, \cdots n\}$, but not necessarily in $S = \{\vec{\alpha}\} \amalg \{\vec{\beta}\}$. Let $\vec{\alpha}_+ = \left(\vec{\alpha}, j_r\right)$, $\vec{\beta}_+ = \left(\vec{\beta}, j_r\right)$, and $\overrightarrow{\alpha_{+k}} = \left(\vec{\alpha}, \underbrace{j_r, \cdots, j_r}_{k}\right)$. If $j_r \in \vec{\beta}$, we let $\vec{\beta}_- = \vec{\beta} \backslash \{j_r\}$.

By definition it is easy to check $A\left(\vec{\alpha}, \vec{\beta}\right)$ satisfies the following four properties:

$$\text{Normalization:}\quad A\left(\overrightarrow{\varnothing},\overrightarrow{\varnothing}\right)=1,\ \text{(where } \varnothing \text{ is the empty set)} \tag{1}$$

$$\text{Symmetry:}\quad A\left(\overrightarrow{\alpha},\overrightarrow{\beta}\right)=A\left(\overrightarrow{\beta},\overrightarrow{\alpha}\right), \tag{2}$$

$$\text{Leibniz rule:}\quad A\left(\overrightarrow{\alpha}_{+},\overrightarrow{\beta}_{-}\right)=-A\left(\overrightarrow{\alpha},\overrightarrow{\beta}\right), \tag{3}$$

$$\text{Adding index:}\quad A\left(\overrightarrow{\alpha}_{+},\overrightarrow{\beta}_{+}\right)=A\left(\overrightarrow{\alpha},\overrightarrow{\beta}\right)(a_r+b_r+1). \tag{4}$$

Proposition 6. *The above four properties completely characterize* $A\left(\overrightarrow{\alpha},\overrightarrow{\beta}\right)$.

Proof. To obtain a general $A\left(\overrightarrow{\alpha},\overrightarrow{\beta}\right)$, we can start from $\overrightarrow{\alpha}=\overrightarrow{\beta}=\overrightarrow{\varnothing}$, use (4) to obtain $A\left(\overrightarrow{\varnothing_{+k}},\overrightarrow{\varnothing_{+k}}\right)$, and then use (3) to obtain $A\left(\overrightarrow{\varnothing_{+(k+l)}},\overrightarrow{\varnothing_{+(k-l)}}\right)$ for $0\le l\le k$. Then we can use (4) to add different indices j' to $\left(\overrightarrow{\alpha},\overrightarrow{\beta}\right)$ to get all $A\left(\overrightarrow{\alpha},\overrightarrow{\beta}\right)$ (note if $j'\notin\left\{\overrightarrow{\alpha}\right\}\amalg\left\{\overrightarrow{\beta}\right\}$, then $A\left(\overrightarrow{\alpha_{+}},\overrightarrow{\beta_{+}}\right)=A\left(\overrightarrow{\alpha},\overrightarrow{\beta}\right)$ by (4)). □

Similarly, $B\left(\overrightarrow{\alpha},\overrightarrow{\beta}\right)$ is characterized by the following four properties:

$$B\left(\overrightarrow{\varnothing},\overrightarrow{\varnothing}\right)=1,\ B\left(\overrightarrow{\alpha},\overrightarrow{\beta}\right)=B\left(\overrightarrow{\beta},\overrightarrow{\alpha}\right),$$
$$B\left(\overrightarrow{\alpha}_{+},\overrightarrow{\beta}_{-}\right)=-B\left(\overrightarrow{\alpha},\overrightarrow{\beta}\right)\left(\frac{2b_r-1}{2a_r+1}\right)^{1/2},$$
$$B\left(\overrightarrow{\alpha}_{+},\overrightarrow{\beta}_{+}\right)=B\left(\overrightarrow{\alpha},\overrightarrow{\beta}\right)\frac{a_r+b_r+1}{\left[(2a_r+1)(2b_r+1)\right]^{1/2}}. \tag{5}$$

4.2 Stabilization property from repeated differentials

For some index $j_r\in\{1,2,\cdots n\}$ (but not necessarily in $\left\{\overrightarrow{\alpha}\right\}\amalg\left\{\overrightarrow{\beta}\right\}$), if we apply ∂_{j_r} repeatedly to $D^{\overrightarrow{\alpha}}\Phi_t(x)$ and $D^{\overrightarrow{\beta}}\Phi_t(x)$, then the the resulted derivative vectors become "more and more collinear", and the angle between them has certain stabilization property. More precisely, using the conventions in Definition 4, we have the following

Proposition 7. $B\left(\overrightarrow{\alpha}_{+},\overrightarrow{\beta}_{+}\right)$ *and* $B\left(\overrightarrow{\alpha},\overrightarrow{\beta}\right)$ *always have the same sign, and*

$$\left|B\left(\overrightarrow{\alpha}_{+},\overrightarrow{\beta}_{+}\right)\right|\ge\left|B\left(\overrightarrow{\alpha},\overrightarrow{\beta}\right)\right|, \tag{6}$$

where the equality holds if and only if j_r *appears with the same multiplicity in* $\overrightarrow{\alpha}$ *and* $\overrightarrow{\beta}$. *Further more we have the following limit: if we let the multi-index*

$\overrightarrow{\alpha_{+k}} = \left(\overrightarrow{\alpha}, \underbrace{j_r, \cdots, j_r}_{k}\right)$, *similarly for* $\overrightarrow{\beta_{+k}}$, *then*

$$\lim_{k\to\infty} B\left(\overrightarrow{\alpha_{+k}}, \overrightarrow{\beta_{+k}}\right) := B\left(\overrightarrow{\alpha}_*, \overrightarrow{\beta}_*\right), \tag{7}$$

where the multi-indices $\overrightarrow{\alpha_*}$ *and* $\overrightarrow{\beta}_*$ *are obtained from deleting all* j_r *indices in* $\overrightarrow{\alpha}$ *and* $\overrightarrow{\beta}$ *respectively. We also have*

$$\lim_{k\to\infty} B\left(\overrightarrow{\alpha}, \overrightarrow{\beta_{+k}}\right) := 0. \tag{8}$$

Proof. Since $\left|\overrightarrow{\alpha}_+\right| - \left|\overrightarrow{\beta}_+\right| = \left|\overrightarrow{\alpha}\right| - \left|\overrightarrow{\beta}\right|$, by definition $B\left(\overrightarrow{\alpha}_+, \overrightarrow{\beta}_+\right)$ and $B\left(\overrightarrow{\alpha}, \overrightarrow{\beta}\right)$ always have the same sign. Let us assume $B\left(\overrightarrow{\alpha}, \overrightarrow{\beta}\right) \geq 0$ (the $B\left(\overrightarrow{\alpha}, \overrightarrow{\beta}\right) \leq 0$ case is similar). From (5) we have

$$\begin{aligned} B\left(\overrightarrow{\alpha}_+, \overrightarrow{\beta}_+\right) &= B\left(\overrightarrow{\alpha}, \overrightarrow{\beta}\right) \cdot \left(\frac{(a_r + b_r + 1)^2}{(2a_r + 1)(2b_r + 1)}\right)^{1/2} \\ &= B\left(\overrightarrow{\alpha}, \overrightarrow{\beta}\right) \cdot \left(1 + \frac{(a_r - b_r)^2}{(2a_r + 1)(2b_r + 1)}\right)^{1/2} \\ &\geq B\left(\overrightarrow{\alpha}, \overrightarrow{\beta}\right), \end{aligned}$$

and the equality holds if and only if $a_r = b_r$. Recall

$$B\left(\overrightarrow{\alpha}, \overrightarrow{\beta}\right) = (-1)^{\frac{|\overrightarrow{\alpha}| - |\overrightarrow{\beta}|}{2}} \prod_{r=1}^{s} \frac{(2\sigma_r)!}{\sigma_r!} \Big/ \left[\frac{(2a_r)!}{a_r!}\frac{(2b_r)!}{b_r!}\right]^{1/2}.$$

Since $\sigma_r = \frac{a_r + b_r}{2}$, if we let $l_r = \frac{b_r - a_r}{2} \geq 0$ (assuming $b_r \geq a_r$), then

$$\frac{a_r!}{\sigma_r!}\frac{b_r!}{\sigma_r!} = \frac{(\sigma_r + 1)\cdots(\sigma_r + l_r)}{(\sigma_r - 1)\cdots(\sigma_r - l_r)}, \text{ and } \frac{(2a_r)!}{(2\sigma_r)!}\frac{(2b_r)!}{(2\sigma_r)!} = \frac{(2\sigma_r + 1)\cdots(2\sigma_r + 2l_r)}{(2\sigma_r - 1)\cdots(2\sigma_r - 2l_r)}.$$

So the r-th factor in $B\left(\overrightarrow{\alpha}, \overrightarrow{\beta}\right)$ is

$$\begin{aligned} &\frac{(2\sigma_r)!}{\sigma_r!} \Big/ \left[\frac{(2a_r)!}{a_r!}\frac{(2b_r)!}{b_r!}\right]^{1/2} = \left[\frac{(2\sigma_r)!}{(2a_r)!}\frac{(2\sigma_r)!}{(2b_r)!}\frac{a_r!}{\sigma_r!}\frac{b_r!}{\sigma_r!}\right]^{1/2} \\ &= \left[\frac{(2\sigma_r - 1)(2\sigma_r - 3)\cdots(2\sigma_r - 2l_r + 1)}{(2\sigma_r + 1)(2\sigma_r + 3)\cdots(2\sigma_r + 2l_r - 1)}\right]^{1/2} \leq 1. \end{aligned} \tag{9}$$

As $k \to \infty$, in the multi-index $\left(\overrightarrow{\alpha_{+k}}, \overrightarrow{\beta_{+k}}\right)$, σ_r is changed to $\sigma_r + k$, but l_r is unchanged, so the factor involving j_r in $B\left(\overrightarrow{\alpha_{+k}}, \overrightarrow{\beta_{+k}}\right)$ becomes

$$\left[\frac{(2\sigma_r + 2k - 1)(2\sigma_r + 2k - 3)\cdots(2\sigma_r + 2k - 2l_r + 1)}{(2\sigma_r + 2k + 1)(2\sigma_r + 2k + 3)\cdots(2\sigma_r + 2k + 2l_r - 1)}\right]^{1/2} \to 1$$

as $k \to \infty$, and the other factors not involving j_r are unchanged. This gives the limit $B\left(\overrightarrow{\alpha_*}, \overrightarrow{\beta_*}\right)$.

For (8), w.l.o.g we can assume $\left|\overrightarrow{\alpha}\right| + \left|\overrightarrow{\beta}\right|$ is even (otherwise we can replace $\overrightarrow{\beta}$ by $\overrightarrow{\beta}_+$ before we take limit). In the multi-index $\left(\overrightarrow{\alpha}, \overrightarrow{\beta_{+2k}}\right)$, σ_r and l_r are changed to $\sigma_r + k$ and $l_r + k$ respectively, so the factor involving j_r in $B\left(\overrightarrow{\alpha}, \overrightarrow{\beta_{+2k}}\right)$ becomes

$$\left[\frac{(2\sigma_r + 2k - 1)(2\sigma_r + 2k - 3)\cdots(2\sigma_r - 2l_r + 1)}{(2\sigma_r + 2k + 1)(2\sigma_r + 2k + 3)\cdots(2\sigma_r + 4k + 2l_r - 1)}\right]^{1/2} \le \left[\frac{2\sigma_r - 2l_r + 1}{2\sigma_r + 4k + 2l_r - 1}\right]^{1/2} \to 0$$

as $k \to \infty$. Therefore $B\left(\overrightarrow{\alpha}, \overrightarrow{\beta_{+2k}}\right) \to 0$. By our definition $B\left(\overrightarrow{\alpha}, \overrightarrow{\beta_{+(2k+1)}}\right) = 0$. Combining the two cases, $\lim_{k\to\infty} B\left(\overrightarrow{\alpha}, \overrightarrow{\beta_{+k}}\right) = 0$. □

Corollary 1. $-1 < B\left(\overrightarrow{\alpha}, \overrightarrow{\beta}\right) \le 1$, *and* $B\left(\overrightarrow{\alpha}, \overrightarrow{\beta}\right) = 1$ *if and only if* $\overrightarrow{\alpha} = \overrightarrow{\beta}$.

Proof. From (9) we see for any multi-indices $\overrightarrow{\alpha}$ and $\overrightarrow{\beta}$,

$$\left|B\left(\overrightarrow{\alpha}, \overrightarrow{\beta}\right)\right| \le 1. \tag{10}$$

If $B\left(\overrightarrow{\alpha}, \overrightarrow{\beta}\right) = 1$ but $\overrightarrow{\alpha} \ne \overrightarrow{\beta}$, then there must be some index j_r appearing with *different* multiplicities in $\overrightarrow{\alpha}$ and $\overrightarrow{\beta}$. We can construct $\overrightarrow{\alpha_+}$ and $\overrightarrow{\beta_+}$ by adding the index j_r. By (6) (notice the equality can *not* hold by the condition on j_r), we have

$$\left|B\left(\overrightarrow{\alpha}_+, \overrightarrow{\beta}_+\right)\right| > \left|B\left(\overrightarrow{\alpha}, \overrightarrow{\beta}\right)\right| = 1,$$

contradicting with (10). Therefore $\overrightarrow{\alpha} = \overrightarrow{\beta}$.

We claim $B\left(\overrightarrow{\alpha}, \overrightarrow{\beta}\right) \ne -1$. Otherwise, by the sign of $B\left(\overrightarrow{\alpha}, \overrightarrow{\beta}\right)$, we see $\frac{\left|\overrightarrow{\alpha}\right| - \left|\overrightarrow{\beta}\right|}{2}$ must be odd, especially $\overrightarrow{\alpha} \ne \overrightarrow{\beta}$. Therefore there is some $l_r \ne 0$. By (9), this makes the r-th factor in $B\left(\overrightarrow{\alpha}, \overrightarrow{\beta}\right)$ have absolute value strictly less than 1, so $\left|B\left(\overrightarrow{\alpha}, \overrightarrow{\beta}\right)\right| < 1$. But this contradicts with $\left|B\left(\overrightarrow{\alpha}, \overrightarrow{\beta}\right)\right| = |-1| = 1$. □

4.3 Lattice geometry on $\mathbb{Z}_+^n$

In this subsection, we use the high-jet relations of Φ_t in Theorem 2 to put a metric d on the lattice space

$$\mathbb{Z}_+^n = \left\{(m_1, \cdots, m_n)\, |m_j \in \mathbb{Z}, m_j \geq 0 \text{ for } 1 \leq j \leq n\right\}.$$

The motivation is to visualize these relations by the geometry of $\left(\mathbb{Z}_+^n, d\right)$.

For any multi-index $\overrightarrow{\alpha}$ with indices in $\{1, 2, \cdots, n\}$, taking the multiplicity of its each index, we obtain a point $z \in \mathbb{Z}_+^n$; conversely we can use z to recover $\overrightarrow{\alpha}$. For example, if we let z_1, z_2, z_3 be the three axes of $\mathbb{Z}_+^3$, then the multi-index $\overrightarrow{\alpha} = (1, 1, 2, 3)$ has 1 with multiplicity 2, 2 with multiplicity 1, and 3 with multiplicity 1. So $\overrightarrow{\alpha}$ correspond to $z = (2, 1, 1) \in \mathbb{Z}_+^3$. With this identification, we have

Definition 5. (Angle distance) For any two multi-indices $\overrightarrow{\alpha}$ and $\overrightarrow{\beta}$ in $\mathbb{Z}_+^n$, we define their *angle distance* to be

$$d\left(\overrightarrow{\alpha}, \overrightarrow{\beta}\right) = \cos^{-1}\left(B\left(\overrightarrow{\alpha}, \overrightarrow{\beta}\right)\right). \tag{11}$$

By Theorem 2, $d\left(\overrightarrow{\alpha}, \overrightarrow{\beta}\right)$ is the *limiting distance* of $\frac{D^{\overrightarrow{\alpha}}\Phi_t(x)}{\left|D^{\overrightarrow{\alpha}}\Phi_t(x)\right|}$ and $\frac{D^{\overrightarrow{\beta}}\Phi_t(x)}{\left|D^{\overrightarrow{\beta}}\Phi_t(x)\right|}$ on the unit sphere $S^\infty \subset l^2$ as $t \to 0_+$. Let $\theta_t\left(\overrightarrow{\alpha}, \overrightarrow{\beta}\right)$ be the minimal distance of $\frac{D^{\overrightarrow{\alpha}}\Phi_t(x)}{\left|D^{\overrightarrow{\alpha}}\Phi_t(x)\right|}$ and $\frac{D^{\overrightarrow{\beta}}\Phi_t(x)}{\left|D^{\overrightarrow{\beta}}\Phi_t(x)\right|}$ on S^∞, then $0 \leq \theta_t\left(\overrightarrow{\alpha}, \overrightarrow{\beta}\right) \leq \pi$, and

$$\lim_{t \to 0_+} \theta_t\left(\overrightarrow{\alpha}, \overrightarrow{\beta}\right) = d\left(\overrightarrow{\alpha}, \overrightarrow{\beta}\right). \tag{12}$$

Lemma 5. *The angle distance d defined in* (11) *is a metric on* $\mathbb{Z}_+^n$. *For any* $\overrightarrow{\alpha}$ *and* $\overrightarrow{\beta} \in \mathbb{Z}_+^n$, $0 \leq d\left(\overrightarrow{\alpha}, \overrightarrow{\beta}\right) < \pi$.

Proof. To prove d is a metric, we only need to verify that it is *nondegenerate*, since the triangle inequality of $\theta_t\left(\overrightarrow{\alpha}, \overrightarrow{\beta}\right)$ on S^∞ is preserved when we take the limit in (12) as $t \to 0_+$, and $d\left(\overrightarrow{\alpha}, \overrightarrow{\beta}\right) = d\left(\overrightarrow{\beta}, \overrightarrow{\alpha}\right)$ follows from the definition of $B\left(\overrightarrow{\alpha}, \overrightarrow{\beta}\right)$. Suppose $d\left(\overrightarrow{\alpha}, \overrightarrow{\beta}\right) = 0$, then $B\left(\overrightarrow{\alpha}, \overrightarrow{\beta}\right) = 1$. From Corollary 1, we have $\overrightarrow{\alpha} = \overrightarrow{\beta}$. By Corollary 1, $B\left(\overrightarrow{\alpha}, \overrightarrow{\beta}\right) \in (-1, 1]$, so $0 \leq d\left(\overrightarrow{\alpha}, \overrightarrow{\beta}\right) < \pi$. □

We can interpret the stabilization property in (7) of Proposition 7 in terms of the distance d on $\mathbb{Z}_+^n$:

Lemma 6. *(Stabilization) For any coordinate vector* $v = \frac{\partial}{\partial z^j}$, *and any two vectors* $\overrightarrow{\alpha}, \overrightarrow{\beta} \in \mathbb{Z}_+^n$, *we have*

$$\left|d\left(\vec{\alpha}+v,\vec{\beta}+v\right)-\frac{\pi}{2}\right|\geq\left|d\left(\vec{\alpha},\vec{\beta}\right)-\frac{\pi}{2}\right|,$$
$$\lim_{k\to\infty}d\left(\vec{\alpha}+kv,\vec{\beta}+kv\right)=d\left(\vec{\alpha}_*,\vec{\beta}_*\right),$$
$$\lim_{k\to\infty}d\left(\vec{\alpha},\vec{\beta}+kv\right)=\frac{\pi}{2},$$

where $\vec{\alpha}_*$ and $\vec{\beta}_*$ are obtained by setting the j-th component of $\vec{\alpha}$ and $\vec{\beta}$ to be zero.

The stabilization properties in the above lemma make each axis of $(\mathbb{Z}_+^n, d)$ somehow looks like a real projective line. It will be interesting to say more about the geometry of $(\mathbb{Z}_+^n, d)$ as $\left|\vec{\alpha}\right|\to\infty$.

In the following we are interested in the orthogonal relations of $D^{\vec{\alpha}}\Phi_t(x)$ as $t\to 0_+$. From our definition, $B\left(\vec{\alpha},\vec{\beta}\right)=0$ if and only if some index appears with odd multiplicity in $\{\vec{\alpha}\}\amalg\left\{\vec{\beta}\right\}$. This can be reformulated as

Lemma 7. *(Orthogonal relations) $d\left(\vec{\alpha},\vec{\beta}\right)=\frac{\pi}{2}$ if and only if $\vec{\alpha}-\vec{\beta}\notin(2\mathbb{Z})^n$. Consequently,*

1. *Any two adjacent lattice points in $\mathbb{Z}_+^n$ have the angle distance $\frac{\pi}{2}$;*
2. *$\mathbb{Z}_+^n/(2\mathbb{Z}_+)^n$ has 2^n "orthogonal" cosets $A_1,\cdots,A_{2^n}\subset\mathbb{Z}_+^n$, such that for any two elements in different cosets, their angle distance is $\frac{\pi}{2}$.*

This $\mathbb{Z}_+^n/(2\mathbb{Z}_+)^n$-grading of the ∞-jet space of Φ_t as $t\to 0$ may have future applications. Locally, in an orthonormal coordinates $(x_1,\cdots,x_n)$ around $x\in M$, the Weyl algebra spanned by $\left\{x_1,\cdots,x_n,\frac{\partial}{\partial x_1},\cdots,\frac{\partial}{\partial x_n}\right\}$ acts on the ∞-jet space $j^\infty(x)$ of $\Phi_t(x)$, and when $t\to 0_+$, by the above lemma, $j^\infty(x)$ splits into 2^n asymptotically orthogonal summands that are permuted by the Weyl algebra action.

In the following we give some estimate of $d\left(\vec{\alpha},\vec{\beta}\right)$. Let d_0 be the "*set distance*" function on $\mathbb{Z}_+^n$: for $\vec{\alpha}=(a_1,\cdots,a_n)$ and $\vec{\beta}=(b_1,\cdots,b_n)$,

$$d_0\left(\vec{\alpha},\vec{\beta}\right)=\sum_{j=1}^n\left|a_j-b_j\right|=\left|\vec{\alpha}\vartriangle\vec{\beta}\right|,$$

where $\vartriangle$ is the difference of two sets A and B: $A\vartriangle B=(A\backslash B)\cup(B\backslash A)$.

Intuitively, if $\left|\vec{\alpha}\vartriangle\vec{\beta}\right|$ is very large relative to $\left|\vec{\alpha}\right|+\left|\vec{\beta}\right|$, then $\vec{\alpha}$ and $\vec{\beta}$ are very different, and we expect the angle $d\left(\vec{\alpha},\vec{\beta}\right)$ between the derivative vectors $D^{\vec{\alpha}}\Phi_t(x)$ and $D^{\vec{\beta}}\Phi_t(x)$ to be not too close to 0 or π as $t\to 0_+$. Indeed, we have

Proposition 8. *(Distance comparison) There exists a constant $\delta>0$ independent on n, such that for any multi-indices $\vec{\alpha}$ and $\vec{\beta}\in\mathbb{Z}_+^n$,*

$$\left|\cos d\left(\vec{\alpha},\vec{\beta}\right)\right| \leq 1-\delta\frac{\left|\vec{\alpha}\vartriangle\vec{\beta}\right|}{\left|\vec{\alpha}\right|+\left|\vec{\beta}\right|}. \tag{13}$$

Proof. Inequality (13) is equivalent to the following combinatorial inequality

$$\prod_{r=1}^{s}\frac{(2\sigma_r-1)(2\sigma_r-3)\cdots(2\sigma_r-2l_r+1)}{(2\sigma_r+1)(2\sigma_r+3)\cdots(2\sigma_r+2l_r-1)} \leq 1-\delta\frac{\Sigma_{r=1}^{s}l_r}{\Sigma_{r=1}^{s}\sigma_r}, \tag{14}$$

where $\sigma_r \geq l_r \geq 0$ are integers. We first prove (14) when $s = 1$. Let $f(x) = \ln\left(\frac{1-x}{1+x}\right)$, then $f''(x) \leq 0$ on $[0, 1)$. Therefore

$$\frac{f\left(\frac{1}{2\sigma_r}\right)+\cdots+f\left(\frac{2l_r-1}{2\sigma_r}\right)}{l_r} \leq f\left(\frac{\frac{1}{2\sigma_r}+\cdots+\frac{2l_r-1}{2\sigma_r}}{l_r}\right) = f\left(\frac{l_r}{2\sigma_r}\right), \text{ i.e.}$$

$$\left(\frac{1-\frac{1}{2\sigma_r}}{1+\frac{1}{2\sigma_r}}\right)\cdots\left(\frac{1-\frac{2l_r-1}{2\sigma_r}}{1+\frac{2l_r-1}{2\sigma_r}}\right) \leq \left(\frac{1-\frac{l_r}{2\sigma_r}}{1+\frac{l_r}{2\sigma_r}}\right)^{l_r} \leq \frac{1-\frac{l_r}{2\sigma_r}}{1+\frac{l_r}{2\sigma_r}} \leq 1-\frac{l_r}{2\sigma_r}. \tag{15}$$

For $s > 1$, using (15) we have

$$\prod_{r=1}^{s}\frac{(2\sigma_r-1)(2\sigma_r-3)\cdots(2\sigma_r-2l_r+1)}{(2\sigma_r+1)(2\sigma_r+3)\cdots(2\sigma_r+2l_r-1)}$$

$$\leq \prod_{r=1}^{s}\left(1-\frac{l_s}{2\sigma_s}\right) \leq \left(\frac{\Sigma_{r=1}^{s}\left(1-\frac{l_r}{2\sigma_r}\right)}{s}\right)^{s} \quad \text{(arithmetic-geometric inequality)}$$

$$= \left(1-\frac{1}{2s}\Sigma_{r=1}^{s}\frac{l_r}{\sigma_r}\right)^{s} \leq \left(1-\frac{1}{2s}\left(\frac{\Sigma_{r=1}^{s}l_r}{\Sigma_{r=1}^{s}\sigma_r}\right)\right)^{s} \quad \left(\text{since } \frac{a}{b}+\frac{c}{d} \geq \frac{a+c}{b+d} \text{ for } a,b,c,d>0\right)$$

$$\leq 1-\delta\frac{\Sigma_{r=1}^{s}l_r}{\Sigma_{r=1}^{s}\sigma_r},$$

where the last inequality is because there exists a $\delta > 0$ such that

$$\left(1-\frac{1}{2s}x\right)^{s} \leq 1-\delta x \tag{16}$$

on $[0, 1]$ for any integer $s \geq 1$. This can be proved as follows: For each integer s, let $h_s(x) = \frac{1-\left(1-\frac{1}{2s}x\right)^{s}}{x}$. Defining $h_s(0) = \frac{1}{2}$ we see $h_s(x)$ is a continuous, positive function on $[0, 1]$, thus h_s has a minimum $\delta_s > 0$ on $[0, 1]$. Thus for all $x \in [0, 1]$,

$$\left(1-\frac{1}{2s}x\right)^{s} \leq 1-\delta_s x.$$

Next we improve δ_s to a uniform constant independent on s. Note when $s \longrightarrow \infty$, $h_s(x)$ tends to a constant function:

$$\lim_{s\to\infty} h_s(x) = \lim_{s\to\infty} \frac{1-\left(1-\frac{1}{2s}x\right)^s}{x} = \lim_{s\to\infty} \frac{1-e^{-\frac{1}{2}x}}{x} = \frac{1}{2}.$$

Thus $\lim_{s\to\infty}\delta_s = \frac{1}{2}$, and the constant $\delta := \inf_{s\geq 1}\delta_s > 0$ makes the inequality (16) hold for all $x \in [0, 1]$. □

Inequality (13) gives a quantitative control of the linear independence of the derivative vectors $D^{\vec{\alpha}}\Phi_t(x)$ and $D^{\vec{\beta}}\Phi_t(x)$ as $t \longrightarrow 0_+$. When $|\vec{\alpha}| = |\vec{\beta}| = 2$, the *uniform* linear independence of $D^{\vec{\alpha}}\Phi_t(x)$ and $D^{\vec{\beta}}\Phi_t(x)$ as $t \longrightarrow 0_+$ played an important role in [WZ] to construct isometric embeddings by the implicit function theorem. It will be interesting to see if there are other geometric problems where the higher-jet linear independence property can play role.

References

[AT] R. Adler, R.J.E. Taylor, *Random Fields and Geometry*, Springer Monographs in Mathematics, Springer Verlag, 2007

[BBG] P. Bérard, G. Besson, S. Gallot, *Embedding Riemannian Manifolds by Their Heat kernel*, Geom. and Func. Analysis, vol. 4, No. 4, (1994)

[BeGaM] M. Berger, P. Gauduchon, E. Mazet, *Le spectre d'une variété Riemannianne*, Springer Lecture Notes in Math. 194, 1971

[BGV] N. Berline , E. Getzler , M. Vergne, *Heat Kernels and Dirac Operators*, SPRINGER VERLAG; 199, 2004

[Ch] I. Chavel, *Eigenvalues in Riemannian Geometry*, Vol 115, Pure and Applied Mathematics, Academic Press, 1984

[CW] C. Lin, H. Wu, *Embeddings of Riemannian manifolds with finite eigenvector fields of connection Laplacian,* Calculus of Variations and Partial Differential Equations volume 57, Article number: 126 (2018)

[De] B. DeWitt, *The Global Approach to Quantum Field Theory*, Vol 114, The International Series of Monographs on Physics, Oxford University Press, 2003

[Gil] P. Gilkey, *Asymptotic Formulae in Spectral Geometry,* (Studies in Advanced Mathematics), Chapman and Hall/CRC; 1st edition (December 17, 2003)

[Ni] L. I. Nicolaescu, *Random Morse functions and spectral geometry*, arXiv:1209.0639

[Po] J. Portegies, *Embeddings of Riemannian manifolds with heat kernels and eigenfunctions*, Communications on Pure and Applied Mathematics 69 (3), 478-518, 2016

[WZ] X. Wang, K. Zhu, *Isometric embeddings via heat kernel,* J. Diff. Geom., (99), 497-538, 2015

Part II

Other Contributed Articles

Chapter 11

MATRIX-MFO Tandem Workshop: Invariants and Structures in Low-Dimensional Topology

Pachner's Theorem

Stefan Friedl and Julian Hannes

Abstract Pachner's Theorem is a purely combinatorial theorem which plays an important role in low-dimensional topology. We give a topology-free self-contained proof of Pachner's Theorem and we outline why Pachner's Theorem is such an essential tool in low-dimensional topology.

1 Introduction

We start out with our definition of an abstract simplicial complex.

Definition 1.

1. An *abstract simplicial complex* is a pair $K = (V,S)$, where V is a finite set, and S is a subset of the power set $\mathscr{P}(V)$ of V satisfying the following three conditions:

   (a) Each element $s \in S$ is a non-empty subset of V.
   (b) Given $s \in S$ and $t \subseteq s$ with $t \neq \varnothing$, we have $t \in S$.
   (c) For all $v \in V$, the singleton $\{v\}$ lies in S.

   We call elements in V *vertices of* K, elements in S *simplices of* K, and we refer to V as the *vertex set of* K and to S as the *simplex set of* K.
2. Let $K = (V,S)$ and $L = (W,T)$ be abstract simplicial complexes. A *simplicial map from* K *to* L is a map $f\colon V \to W$ on the vertex sets such that for all $s \in S$

Stefan Friedl
Department of Mathematics, University of Regensburg, Germany.
e-mail: `sfriedl@gmail.com`

Julian Hannes
Department of Mathematics, University of Regensburg, Germany.
e-mail: `jkmh.vadis@gmx.de`

D. R. Wood et al. (eds.), *2021-2022 MATRIX Annals*, MATRIX Book Series 5,
https://doi.org/10.1007/978-3-031-47417-0_34

we have $f(s) := \{f(v) | v \in s\} \in T$. Such a simplicial map is called a *simplicial isomorphism* if it is a bijection on the vertex sets and on the simplex sets. (One way to view simplicial isomorphisms is simply as renaming the vertices, of course in a way that fully preserves all the simplices.)

As an example we introduce two families of abstract simplicial complexes which play an important role throughout. Let $n \in \mathbb{N}_0$. We consider the set $V_n := \{0, \ldots, n\}$. We set

$$D_n := (V_n, \mathscr{P}(V_n) \setminus \{\varnothing\}) \qquad \text{(standard simplicial } n\text{-disk)},$$
$$S_n := (V_{n+1}, \mathscr{P}(V_{n+1}) \setminus \{\varnothing, V_{n+1}\}) \qquad \text{(standard simplicial } n\text{-sphere)}.$$

We furthermore define D_{-1} and S_{-1} to be the empty abstract simplicial complex $(\varnothing, \varnothing)$.

Fig. 1: A visualisation of the abstract simplicial complexes D_2 and S_1

In Chapter 4 we introduce the purely combinatorial notion of a stellar subdivision σ_s of an abstract simplicial complex K along a simplex s. Loosely speaking this is defined as follows: we add the barycenter of s to the vertex set and we subdivide all simplices that have s as a face accordingly. In Figure 2 we illustrate the stellar subdivision along a 1-dimensional simplex.

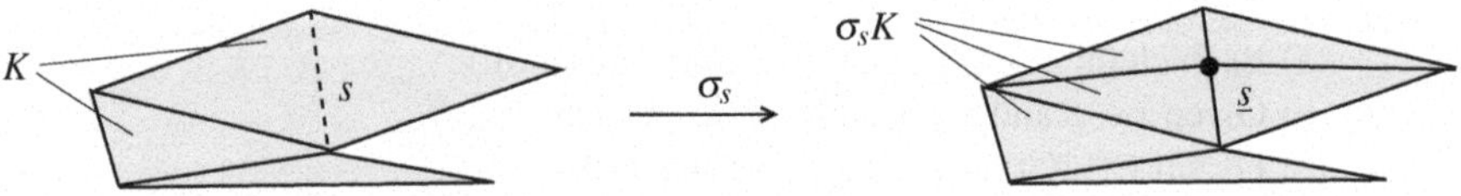

Fig. 2: Stellar subdivision at a 1-simplex

Two abstract simplicial complexes are called *stellar equivalent* if one can go from one to the other through a finite sequence of stellar subdivisions, inverses of stellar subdivisions and simplicial isomorphisms.

To any abstract simplicial complex K one can associate its topological realization $|K|$ and given two such topological realizations there is a natural notion of a PL-homeomorphism between them, which is now a topological notion. We do not give the precise definitions, instead we refer to [Gla70, Lic99, Fri23] for details. The following theorem, which we do not use otherwise, now gives one reason why the notion of stellar equivalence is so important.

Theorem 1. (Alexander-Newman Theorem) *Let K and L be two abstract simplicial complexes. The following two statements are equivalent:*

(1) *The abstract simplicial complexes K and L are stellar equivalent.*
(2) *The topological realizations $|K|$ and $|L|$ are* PL*-homeomorphic.*

Proof. The "(1)⇒(2)"-direction follows easily from the definitions. The much harder "(2)⇒(1)"-direction was first proved by James Alexander [Ale30, Theorem 15:1] in 1930, building on work of Max Newman [New26]. More modern expositions of the proof are given in [Gla70, Theorem II.17] and in [Lic99, Theorem 4.5].

In Chapter 4 we introduce the purely combinatorial notion of a closed combinatorial n-manifold. In a nutshell a closed combinatorial n-manifold is an abstract simplicial complex K such that for every vertex v the link $\mathrm{Lk}(K,v)$ is stellar equivalent to the standard simplicial n-sphere S_n.

Next, let $K=(V,S)$ be a closed combinatorial n-manifold. Furthermore let s be a k-simplex such the link $\mathrm{Lk}(K,s)$ is simplicially isomorphic to the standard simplicial $(n-k-1)$-sphere S_{n-k-1} and such that the vertices of $\mathrm{Lk}(K,s)$ do not form a k-simplex of K. Given such s we introduce in Chapter 6 the *bistellar move* along s to obtain a new closed combinatorial n-manifold $\tau_s K$. At this point we do not give the precise definition, but we hope that Figure 3 gives the flavor of the definition.

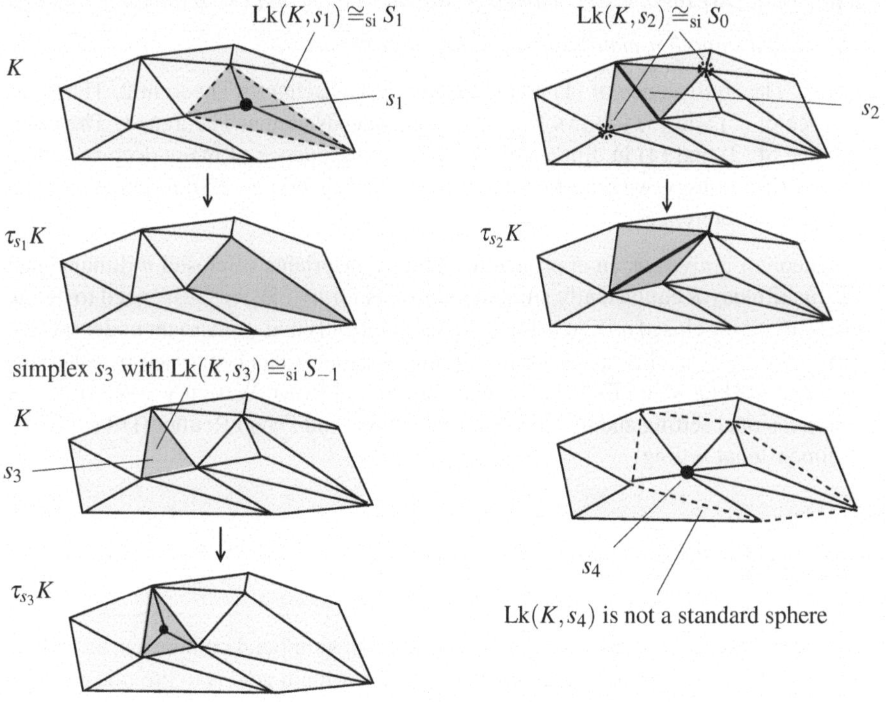

Fig. 3: Examples for three bistellar moves and one impossible move

We say that two closed combinatorial n-manifolds are *bistellar equivalent* if one can go from one to the other through a finite sequence of bistellar moves and simplicial isomorphisms. It follows immediately from the definitions that if two closed combinatorial n-manifolds are bistellar equivalent, then they are also stellar equivalent. Amazingly the converse to this statement also holds:

Theorem 2. (Pachner's Theorem) *Let $K = (V,S)$ and $K' = (V',S')$ be closed combinatorial n-manifolds. Then K and K' are bistellar equivalent if and only if they are stellar equivalent.*

The theorem was first proved by Udo Pachner [Pac91] in a seminal paper. Pachner's original paper is a highly cited and extremely influential paper in combinatorial topology. The significance of Pachner's theorem lies in the fact that given an n-dimensional combinatorial manifold there are infinitely many different types of stellar moves, whereas there are only $n+1$ different type of bistellar moves.

The following combines some of the previous results with results on the topology of smooth manifolds.

Theorem 3. *Let $n \in \mathbb{N}$. There exist natural bijections between the following sets:*

(1) *Closed combinatorial n-manifolds up to bistellar equivalence.*
(2) *Closed combinatorial n-manifolds up to stellar equivalence.*
(3) *Closed PL-manifolds up to PL-homeomorphisms.*

If $n \in \{1,\ldots,6\}$ there exists also a bijection of (1), (2) *and* (3) *to the following set:*

(4) *Closed smooth n-manifolds up to diffeomorphism.*

Proof. The equivalence of (1) and (2) is precisely Pachner's Theorem 2. The equivalence of (2) and (3) follows from the Alexander-Newman Theorem 1. The equivalence of (3) and (4) in dimensions ≤ 6 is a consequence of many deep results in geometric topology, we refer to [Cer68, p. IX] and [Sco05, p. 220] for an exposition and precise references.

Theorem 3 gives us an approach to defining invariants of closed n-dimensional PL-manifolds (or equivalently smooth manifolds for $n \leq 6$): we "just" need to define an invariant for closed n-dimensional stellar manifolds that is invariant under the $n+1$ bistellar moves. This approach to defining invariants has been used by Vladimir Turaev and Oleg Viro [TV92] and John Barrett and Bruce Westbury [BW96] in the 3-dimensional setting and by Christopher Douglas and David Reutter [DR18] in the 4-dimensional setting.

Organization

Chapters 2 and 3 introduce the concept of abstract simplicial complexes, as well as some of the standard notions related to them. Their main goal is to provide an overview for people who are unfamiliar with these concepts. An effort will be made to keep everything free of topology, save for the pictures that illustrate the definitions.

In Chapter 4, we turn our attention to introducing stellar equivalence, and follow it up with an introduction to combinatorial balls, spheres, and combinatorial manifolds. A lot of the properties whose proofs are commonly omitted in literature are proved in great detail.

We will use the results of Chapter 4 in Chapter 5 to introduce the concept of starrability, and subsequently prove that all combinatorial balls are starrable.

At this point, the last two key ingredients required for the proof of Pachner's theorem, namely bistellar moves and elementary shellings, will be introduced in Chapter 6. We also provide proof for some statements that will play a vital role in proving Pachner's theorem in Chapter 7.

Remark

These notes are based on the master thesis of the second author and on lecture notes by the first author. Our goal was to write a detailed topology-free completely combinatorial proof of Pachner's theorem. In the near future we hope to implement this proof in the proof assistant `lean`.

Acknowledgements We wish to thank Lars Munser for crucial help in proving a particular lemma. We also wish to thank the referee for helpful feedback. The first author was supported by the SFB 1085 "higher invariants" funded by the DFG.

2 Preliminaries

In order to formulate and consequently prove the main results, these first two chapters aim to introduce some of the basic concepts about simplicial complexes. Most of the definitions and lemmas in this chapter are included for rigour's sake, and can therefore be skipped by readers who are already familiar with the notion of (abstract) simplicial complexes. The less well-versed reader is invited to come up with their own examples in order to acquaint themselves with the definitions.

The following definitions and conventions play a major role in the remainder of these notes. It may look like a lot, but all of them are pretty much what one would naively define them as.

Definition 2. Let $K = (V, S)$ be an abstract simplicial complex.

(1) The *dimension of a simplex* $s \in S$ is defined to be

$$\dim s := \#s - 1 \in \mathbb{N}_0.$$

The *dimension of* K is defined as

$$\dim K := \sup_{s \in S} \dim s \in \mathbb{N}_0$$

if K is not the empty abstract simplicial complex $(\varnothing, \varnothing)$, and $\dim K := -1$ if it is.

(2) If $t \in S$ is a simplex and $s \subseteq t$ is a non-empty subset, then we call s a *face* of t.

(3) An *abstract subcomplex of K* is an abstract simplicial complex $L = (W, T)$ such that we have inclusions $W \subseteq V$ and $T \subseteq S$.

(4) Let $n \in \mathbb{N}_0$. The *n-skeleton of K,* defined as

$$K^n := \Big(V, \{s \in S\colon\ \dim s \leq n\}\Big).$$

This is an abstract subcomplex of K. For readers unfamiliar with abstract simplicial complexes, it can be a worthwhile exercise to prove this fact.

The following is another important example of abstract subcomplexes.

Remark 1. Let $K = (V, S)$ be an abstract simplicial complex, and let $s \in S$ be a k-simplex. Then we often view $(s, \mathscr{P}(s) \setminus \{\varnothing\})$ as an abstract subcomplex, and thus as an abstract simplicial complex in its own right, which we also refer to as s by an abuse of notation. It is furthermore simplicially isomorphic[1] to D_k.

3 Some More Constructions on Abstract Simplicial Complexes

Before we can start with the actual topics, we need more methods of constructing new abstract simplicial complexes from existing ones. We start out with an operation that at first may seem rather "out of the blue" but will shortly become one of the core concepts of these notes.

Definition 3.

1. Let V and W be two sets. We define their disjoint union as

$$V \sqcup W \ := \ (V \times \{1\}) \cup (W \times \{2\})$$

and we view V and W as subsets of $V \sqcup W$ in the obvious way.

2. Let $K = (V, S)$ and $L = (W, T)$ be abstract simplicial complexes. We define their *join* to be the abstract simplicial complex

$$K * L := \left(V \sqcup W, \left\{ s \sqcup t \,\middle|\, \begin{matrix} s \in S \cup \{\varnothing\}, t \in T \cup \{\varnothing\}, \\ \text{at least one of } s \text{ and } t \text{ non-empty} \end{matrix} \right\} \right).$$

Pictorially, the join operation means "connecting every point of K to every point of L". However, we are technically talking about an operation on **abstract** simplicial

[1] This concept will be defined in the subsequent definition.

complexes so the reader should take that intuition as well as the following suggestive pictures with the usual grain of salt.

Fig. 4: What one might imagine the joins $D_0 * D_1$ and $D_1 * D_1$ to look like.

Remark 2. Many an author will move on to say that the join operation is clearly associative and commutative. However, one has to be careful when taking disjoint unions as it is only commutative up to a bijection. Therefore, it is only reasonable to claim that the join operation is commutative up to a simplicial isomorphism. However, since this would be quite a hassle, we will ignore this problem for the remainder of these notes, and only take joins of abstract simplicial complexes whose vertex sets are disjoint already. In that case, commutativity and associativity follow quite easily from the definition.
It also follows immediately from the definition that the empty complex $(\varnothing,\varnothing)$ is a neutral element with respect to taking joins, i.e., that for any abstract simplicial complex K, we have $K * (\varnothing,\varnothing) = K$.

The following elementary statement will be used a lot, often without even explicitly mentioning it. Its proof is an easy yet refreshing exercise in set theory.

Lemma 1. *Let $K = (V,S)$ and $L = (W,T)$ be abstract simplicial complexes. Then*

$$\dim(K * L) = \dim K + \dim L + 1.$$

Recall that we had defined the dimension of the empty complex $(\varnothing,\varnothing)$ to be -1.

Proof. Let $K = (V,S)$ and $L = (W,T)$ be abstract simplicial complexes. If one of K or L is the empty complex, then the equation is a tautology. Now assume neither K nor L are empty. Denote by k and l the dimensions of K and L, respectively. By definition, that means there is a simplex $s \in S$ with $\#s = k+1$, and a simplex $t \in T$ with $\#t = l+1$ and no simplices in S or T that are of higher dimension than s or t, respectively. But then

$$\#(s \sqcup t) = \#s + \#t = k + l + 2,$$

and $\dim(s \sqcup t) = \#(s \sqcup t) - 1 = k + l + 1$, and there are no simplices in $S \sqcup T$ that are of higher dimension. Therefore, $\dim(K * L) = \dim K + \dim L + 1$.

Remark 3. There are natural injective simplicial maps $K \to K * L$ and $L \to K * L$. We will use these maps to view K and L as abstract subcomplexes of $K * L$. Looking at Figure 4 above, it should become clear what we mean by that.

One important example for joins is the cone of an abstract simplicial complex.

Example 1. Let $K = (V,S)$ be an abstract simplicial complex. We define the *cone of* K to be

$$\mathrm{Cone}(K) := D_0 * K,$$

where D_0 is the abstract simplicial complex $(\{0\},\{\{0\}\})$ that we had defined back in Chapter 2. From the remark above, it follows that we can view K as an abstract subcomplex of $\mathrm{Cone}(K)$. Furthermore, it follows from Lemma 1 that $\dim \mathrm{Cone}(K) = \dim K + 1$.

Definition 4. Let $K = (V,S)$ be an abstract simplicial complex, let $s \in S$ be a simplex. We define the following abstract subcomplexes:

$$\mathrm{St}(K,s) := \Bigg(\bigcup_{t\in S,\, s\subseteq t} t,\ \{t \in S | s\cup t \in S\} \Bigg)$$

$$\mathrm{Lk}(K,s) := \Bigg(\bigcup_{t\in S\, s\subseteq t} t\setminus s,\ \{t \in S | s\cup t \in S \text{ and } s\cap t = \varnothing\} \Bigg)$$

$$K \setminus \mathring{\mathrm{St}}(K,s) := \begin{cases} (V\setminus\{v\}, \{t\in S | s \not\subseteq t\}), & \text{if } s = \{v\} \text{ is a 0-simplex} \\ (V, \{t\in S | s\not\subseteq t\}), & \text{else} \end{cases}$$

$$\partial s := \begin{cases} (\varnothing,\varnothing), & \text{if } s \text{ is a 0-simplex} \\ (s, \mathscr{P}(s)\setminus\{\varnothing, s\}), & \text{else} \end{cases}$$

They are called the *(closed) star of s in K,* the *link of s in K,* the *complement*[2] *of the star of s in K,* and the *boundary of the simplex s,* respectively.

The sketches below show some of the definitions. For better understanding, the reader could go ahead and draw their own abstract simplicial complexes. Note that[3]

$$\mathrm{St}(K,s) \cap \big(K\setminus \mathring{\mathrm{St}}(K,s)\big) = \mathrm{Lk}(K,s).$$

This next lemma is of technical nature and is used every so often throughout these notes.

Lemma 2. *Let $K = (V,S)$ and $L = (W,T)$ be abstract simplicial complexes, and let $s \in S$ be a simplex. Then we have*

(1) $L * \mathrm{Lk}(K,s) = \mathrm{Lk}(K*L,s)$,
(2) $L * \mathrm{St}(K,s) = \mathrm{St}(K*L,s)$, *and*
(3) $L * \big(K\setminus \mathring{\mathrm{St}}(K,s)\big) = (K*L)\setminus \mathring{\mathrm{St}}(K*L,s)$,

---

[2] It is important to remember that $K\setminus \mathring{\mathrm{St}}(K,s)$ is simply a notation for what one might pictorially draw as the closure of the complement. By itself, "$\mathring{\mathrm{St}}(K,s)$" is in no way an abstract simplicial complex!

[3] The intersection of abstract simplicial complexes will be defined shortly.

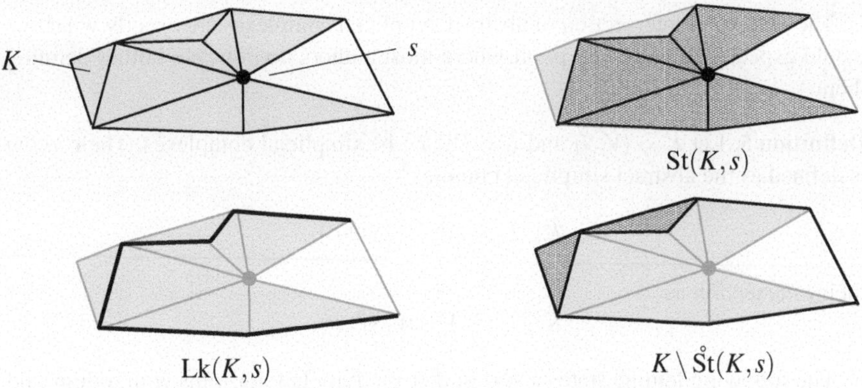

Fig. 5: Star, link, and complement of the star of a 0-simplex

Proof. For the equality involving links, we calculate on the simplex sets

$$\mathrm{Lk}(K*L,s) = \left\{ t \sqcup u \,\middle|\, \begin{array}{l} t \in S \cup \{\varnothing\}, u \in T \cup \{\varnothing\}, (t \sqcup u) \cap s = \varnothing, (t \sqcup u) \cup s \in S \sqcup T, \\ \text{at least one of } t \text{ and } u \text{ non-empty} \end{array} \right\}$$

Now, we use the fact that s is a simplex of K to simplify the conditions, namely we can simplify $(t \sqcup u) \cap s = \varnothing$ to $t \cap s = \varnothing$, and $(t \sqcup u) \cup s \in S \sqcup T$ to $t \cup s \in S$. Doing so yields

$$\mathrm{Lk}(K*L,s) = \left\{ t \sqcup u \,\middle|\, \begin{array}{l} t \in S \cup \{\varnothing\}, u \in T \cup \{\varnothing\}, t \cap s = \varnothing, t \cup s \in S, \\ \text{at least one of } t \text{ and } u \text{ non-empty} \end{array} \right\} = L * \mathrm{Lk}(K,s).$$

The proof of the other two equalities follows essentially the same strategy.

Since working with links in joins is so much fun, we also formulate the following statement, which will be very useful later on.

Lemma 3. *Let $K = (V,S)$ and $L = (W,T)$ be abstract simplicial complexes, and let $s \in S$ and $t \in T$ be simplices. Then we have*

$$\mathrm{Lk}(K*L, s*t) = \mathrm{Lk}(K,s) * \mathrm{Lk}(L,t).$$

Proof. We perform the following elementary calculation

$$\begin{aligned} \mathrm{Lk}(K*L, s*t) &= \{u \in K*L | u \cap (s \sqcup t) = \varnothing, u \cup (s \sqcup t) \in S \sqcup T\} = \\ &= \{u_1 \sqcup u_2 \in K*L | (u_1 \sqcup u_2) \cap (s \sqcup t) = \varnothing, (u_1 \sqcup u_2) \cup (s \sqcup t) \in S \sqcup T\} = \\ &= \{u_1 \in S | u_1 \cap s = \varnothing, u_1 \cup s \in S\} * \{u_2 \in T | u_2 \cap t = \varnothing, u_2 \cup t \in T\} = \\ &= \mathrm{Lk}(K,s) * \mathrm{Lk}(L,t). \square \end{aligned}$$

The union and intersection of abstract simplicial complexes are exactly what one would expect it to be, to the point where most authors do not even bother defining them.

Definition 5. Let $K=(V,S)$ and $L=(W,T)$ be simplicial complexes. Their *union* is defined as the abstract simplicial complex[4]

$$K\cup L := (V\cup W, S\cup T),$$

their *intersection* as

$$K\cap L := (V\cap W, S\cap T).$$

The following lemma states a sort of distributivity law for joins with unions and intersections. We will not mention it every time we use it.

Lemma 4. *Let $K=(V,S)$, $L_1=(W_1,T_1)$, and $L_2=(W_2,T_2)$ be abstract simplicial complexes. Then we have*

$$K*(L_1\cup L_2) = K*L_1\cup K*L_2$$

and

$$K*(L_1\cap L_2) = K*L_1\cap K*L_2.$$

Proof. We leave this proof as an exercise to the reader.

The following definition is our way to create vertices that are not in a given abstract simplicial complex.

Definition 6. Let $K=(V,S)$ be an abstract simplicial complex, let $s\in S$ be a simplex. We define the *barycentre* of s to be $\underline{s} := \frac{1}{\#s}\sum_{v\in s} v \in \mathbb{R}^V$.

For the remainder of these notes, the following slogan works: "If something is underlined, it is a barycentre and therefore a 0-simplex not in the vertex set of whatever abstract simplicial complex we are currently working with." Pictorially, we draw the barycentre of a simplex at the actual "physical" barycentre of the simplex.

4 Stellar Moves and Combinatorial Manifolds

Having finished two chapters that were riddled with definitions and conventions, we now start actually proving statements. This chapter loosely follows the approach pursued by Lickorish (cf. [Lic99, Chapter 3]), but in much more detail, and with an added focus on mathematical rigour.

We start out with the following definition, the second half of which is mostly about semantics.

---

[4] Note that here, we explicitly do not ask for the vertex (or simplex) sets to be disjoint.

Definition 7. Let $K = (V,S)$ be an abstract simplicial complex, let $s \in S$ be a simplex. We call the abstract simplicial complex[5]

$$\sigma_s K := \left(K \setminus \mathring{\mathrm{St}}(K,s)\right) \cup \underline{s} * \partial s * \mathrm{Lk}(K,s)$$

the *abstract simplicial complex obtained from K by stellar subdivision at s.* We also refer to the operation of going from K to $\sigma_s K$ as a *stellar subdivision.*
If an abstract simplicial complex K' is obtained from K via a stellar subdivision, then we say *K is obtained from K' via a stellar weld,* and the operation of going from K' to K is called likewise.
A *stellar move* is either a stellar subdivision or a stellar weld.

The intuition one should have about stellar subdivision is, at least for simplices of dimension at least 1, to add a vertex at the centre of that simplex, and subsequently split its entire star along that point in order for it to "become a simplicial complex again." If that description seems confusing at first, the reader is invited to refer to Figures 2 in the introduction and Figure 6 below, or come up with and draw their own examples.

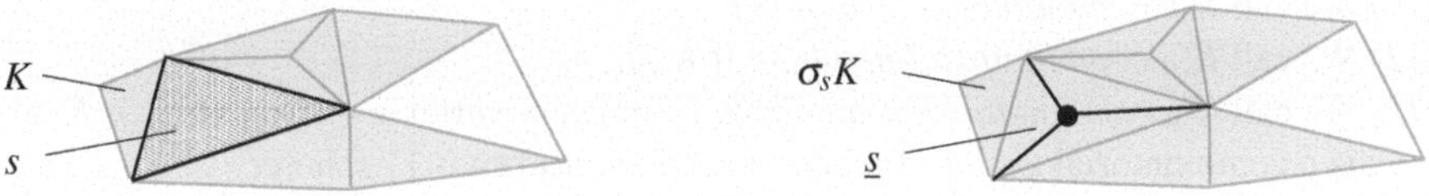

Fig. 6: Stellar subdivision at a 2-simplex

It follows immediately from the definition that stellar subdivision at a 0-simplex creates an abstract simplicial complex that is simplicially isomorphic to the original complex.

Definition 8. Let $K = (V,S)$ and $K' = (V',S')$ be abstract simplicial complexes. We call K and K' *stellar equivalent* if there exists a sequence $K_i = (V_i,S_i)$, $i = 1,\dots,n$, such that $K_1 = K$, $K_n = K'$, and for each $i \in \{1,\dots,n-1\}$ we have that K_{i+1} is a stellar subdivision of, a stellar weld of, or simplicially isomorphic to, K_i. This is clearly an equivalence relation, and we write $K \approx_{\mathrm{st}} K'$ if K and K' are stellar equivalent.

Now that we have some shiny new definitions, we want to put them to use. The following lemma probably comes as no big surprise, and might be used in the future without explicit mention.

Lemma 5. *Let K, K', L, and L' be abstract simplicial complexes. If $K \approx_{\mathrm{st}} K'$ and $L \approx_{\mathrm{st}} L'$, then $K * L \approx_{\mathrm{st}} K' * L'$.*

---

[5] Note that while the barycentre $\underline{s}$ is inspired by the topological setting, it is ultimately just a **natural** way to choose a vertex that is not already in V. Most authors just define it the latter way.

Proof. From the commutativity of the join and the fact that joins clearly commute with simplicial isomorphisms it follows that it suffices to show that for abstract simplicial complexes $K = (V,S)$ and $L = (W,T)$, and a simplex $s \in S$, we have the equality

$$(\sigma_s K) * L = \sigma_s(K * L).$$

By definition and using Lemma 4 and Lemma 2 (1) and (3), one calculates

$$\begin{aligned}\sigma_s(K*L) &= \underbrace{(K*L)\setminus \mathring{\mathrm{St}}(K*L,s)}_{=(K\setminus \mathring{\mathrm{St}}(K,s))*L} \cup \underline{s} * \partial s * \underbrace{\mathrm{Lk}(K*L,s)}_{=\mathrm{Lk}(K,s)*L} = \\ &= \left(K\setminus \mathring{\mathrm{St}}(K,s) \cup \underline{s} * \partial s * \mathrm{Lk}(K,s)\right) * L = \\ &= (\sigma_s K) * L.\end{aligned}$$

This concludes the proof.

Next, we want to take a look at abstract simplicial complexes that have an especially nice relationship with stellar equivalences.

Definition 9. Let $K = (V,S)$ be an abstract simplicial complex, let $n \in \mathbb{N}_0$.

(1) We call K a *combinatorial n-ball* if $K \approx_{\mathrm{st}} D_n$.
(2) We call K a *combinatorial n-sphere* if $K \approx_{\mathrm{st}} S_n$.
(3) We call K a *combinatorial n-manifold* if for every vertex $v \in V$, its link $\mathrm{Lk}(K,v)$ is a combinatorial $(n-1)$-ball or a combinatorial $(n-1)$-sphere.

The reader is invited to refer to Figure 7 below in order to acquaint themselves with the definition of a combinatorial manifold. We leave the formal proof that this abstract simplicial complex is a combinatorial manifold to the reader.

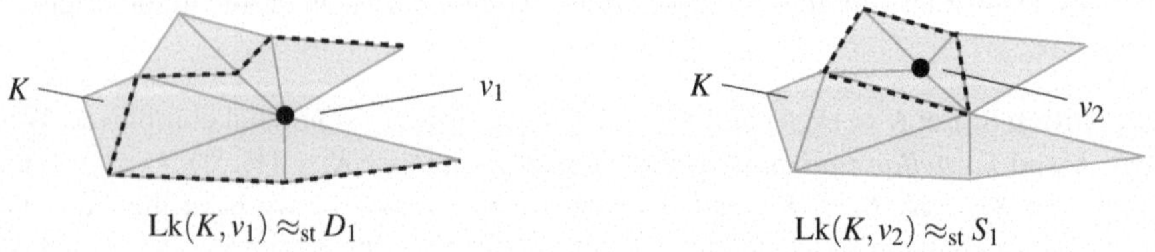

Fig. 7: A combinatorial 2-manifold with the different types of links

Lemma 6. *Let $n \in \mathbb{N}_0$, then D_n and S_n are combinatorial n-manifolds.*

In order to allow the reader to grow used to this topic, we shall be especially thorough in proving this statement.

Proof. Let v be a vertex of D_n. Its link is by definition the subcomplex of D_n that consists of all simplices whose union with $\{v\}$ is again a simplex and whose intersection with $\{v\}$ is empty. Since the simplex set of D_n is just $\mathscr{P}(\{0,\dots,n\}) \setminus \{\varnothing\}$, it follows that

$$\mathrm{Lk}(D_n, v) = \Big(V_n \setminus \{v\}, \mathscr{P}(V_n \setminus \{v\}) \setminus \{\varnothing\}\Big),$$

which is simplicially isomorphic to D_{n-1}. But by definition, a simplicial isomorphism is in particular also a stellar equivalence, so the link of v is in fact a combinatorial $(n-1)$-ball.
Now let v be a vertex of S_n. Using the same logic as before, we arrive at the conclusion that the link of v in S_n is given by

$$\mathrm{Lk}(S_n, v) = \Big(V_{n+1} \setminus \{v\}, \mathscr{P}(V_{n+1} \setminus \{v\}) \setminus \{\varnothing, V_{n+1} \setminus \{v\}\}\Big),$$

where we had to exclude $V_{n+1} \setminus \{v\}$ as a simplex because $V_{n+1} = \{v\} \cup (V_{n+1} \setminus \{v\})$ is not a simplex of S_n. But now we can, analogously to the case of D_n, write down a simplicial isomorphism from $\mathrm{Lk}(S_n, v)$ to S_{n-1}, therefore the link of v is in fact a combinatorial $(n-1)$-sphere.

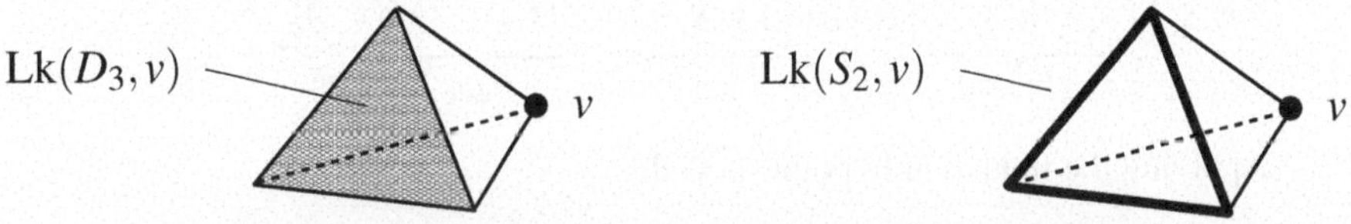

Fig. 8: The links of a vertex in D_3 and S_2 look a lot like D_2 and S_1

In order to get used to the concept of combinatorial balls and spheres, we start out by proving the following lemma about their behaviour under the join operation.

Lemma 7. *Let $m, n \in \mathbb{N}_0$, and let K and L be abstract simplicial complexes. Then the following are true.*

(1) *If K is a combinatorial m-ball and L is a combinatorial n-ball, then $K * L$ is a combinatorial $(m+n+1)$-ball.*
(2) *If K is a combinatorial m-sphere and L is a combinatorial n-sphere, then $K * L$ is a combinatorial $(m+n+1)$-sphere.*
(3) *If K is a combinatorial m-ball and L is a combinatorial n-sphere, then $K * L$ is a combinatorial $(m+n+1)$-ball.*

Proof. Let $m, n \in \mathbb{N}_0$, and let K and L be abstract simplicial complexes.

(1) Assume $K \approx_{\mathrm{st}} D_m$ and $L \approx_{\mathrm{st}} D_n$. From Lemma 5, it follows that

$$K * L \approx_{\mathrm{st}} D_m * D_n.$$

We can give an explicit simplicial isomorphism from $D_m * D_n$ to D_{m+n+1} as follows[6]

[6] Remember that the disjoint union of two sets A and B is defined as $(\{0\} \times A) \cup (\{1\} \times B)$.

$$\begin{aligned}\{0\}\times\{0,\dots,m\}\cup\{1\}\times\{0,\dots,n\} &\to \{0,\dots,m+n+1\}\\ (0,i) &\mapsto i\\ (1,i) &\mapsto m+i+1\end{aligned}$$

Since simplicial isomorphisms are stellar equivalences, this implies that $K*L$ is a combinatorial $(m+n+1)$-ball.

(2) Assume $K\approx_{\mathrm{st}} S_m$ and $L\approx_{\mathrm{st}} S_n$. As in the proof of (1), it follows from Lemma 5 that $K*L\approx_{\mathrm{st}} S_m*S_n$. It remains to show that $S_m*S_n\approx_{\mathrm{st}} S_{m+n+1}$. For this, we inductively prove that for any $k\in\mathbb{N}_0$, we have

$$S_k\approx_{\mathrm{st}} \underbrace{S_0*\cdots*S_0}_{(k+1)\text{-times}}.$$

The case $k=0$ is a tautology. Now let $k\in\mathbb{N}$. Let $s=\{0,\dots,k\}$ be one of the k-simplices of S_k. Now consider

$$\sigma_s S_k=\underbrace{S_k\setminus \mathring{\mathrm{St}}(S_k,s)}_{=\{k+1\}*\partial s}\cup\underline{s}*\partial s*\underbrace{\mathrm{Lk}(S_k,s)}_{=(\varnothing,\varnothing)}=\underbrace{(\{k+1\}\cup\underline{s})}_{\cong_{\mathrm{si}}S_0}*\underbrace{\partial s}_{=S_{k-1}}\cong_{\mathrm{si}} S_0*S_{k-1}.$$

Applying the induction hypothesis yields

$$S_k\approx_{\mathrm{st}} S_0*S_{k-1}\approx_{\mathrm{st}}\underbrace{S_0*\cdots*S_0}_{(k+1)\text{-times}}.$$

This result implies that S_m*S_n is stellar equivalent to the join of

$$(m+1)+(n+1)=m+n+2$$

copies of S_0, but so is S_{m+n+1}. In Figure 9, the idea behind the induction step is shown for $k=2$. It may not look like it, but it works the same for all dimensions, as the calculation above shows.

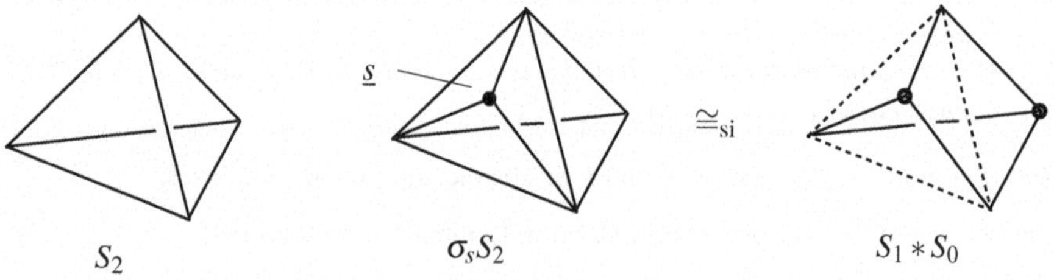

Fig. 9: The subdivision of S_2 is simplicially isomorphic to the join S_1*S_0

(3) Assume $K\approx_{\mathrm{st}} D_m$ and $L\approx_{\mathrm{st}} S_n$. As above, this implies that $K*L\approx_{\mathrm{st}} D_m*S_n$, and it suffices to show $D_m*S_n\approx_{\mathrm{st}} D_{m+n+1}$. We first show that $D_0*S_n\approx_{\mathrm{st}} D_{n+1}$.

For this, let $s := V_{n+1} = \{0,\dots,n+1\}$ be the maximal simplex of D_{n+1}. Then[7]

$$\sigma_s D_{n+1} = \underbrace{D_{n+1} \setminus \mathring{\mathrm{St}}(D_{n+1},s)}_{=(\varnothing,\varnothing)} \cup \underline{s} * \underbrace{\partial s}_{=S_n} * \underbrace{\mathrm{Lk}(D_{n+1},s)}_{=(\varnothing,\varnothing)} \cong_{\mathrm{si}} D_0 * S_n.$$

Now we can use the result from part (1) to calculate

$$D_m * S_n \cong_{\mathrm{si}} D_{m-1} * \underbrace{D_0 * S_n}_{\approx_{\mathrm{st}} D_{n+1}} \approx_{\mathrm{st}} D_{(m-1)+(n+1)+1} = D_{m+n+1},$$

which concludes the proof of the third statement.

Thus, all three statements have been proved.

The following proposition aims to summarise two nice properties of combinatorial manifolds. Both the statement and its proof are based on [Lic99, Lemma 3.2], however steps were taken to fill in some of the gaps.

Proposition 1. *Let $n \in \mathbb{N}_0$, and let $K = (V,S)$ be a combinatorial n-manifold. Then the following two statements hold true:*

(1) *For any k-simplex $s \in S$, its link $\mathrm{Lk}(K,s)$ is a combinatorial $(n-k-1)$-ball or a combinatorial $(n-k-1)$-sphere.*
(2) *If $K' = (V',S')$ is an abstract simplicial complex that is stellar equivalent to K, then K' is a combinatorial n-manifold.*

Remark 4. It shall be stressed that somewhat counter-intuitively, the second statement does *not* immediately follow from the definitions. In the future, however, we use part (2) without explicitly mentioning it.

We need the following technical lemma.

Lemma 8. *Let $K = (V,S)$ be an abstract simplicial complex, let $s \in S$ be a simplex. If $t \subseteq s$ is a face of s, then we have the equality*

$$\mathrm{Lk}(K,s) = \mathrm{Lk}\left(\mathrm{Lk}(K,t), s \setminus t\right).$$

Proof. If $s = t$, then the equality holds since we had set the link of the empty set in an abstract simplicial complex to be that complex. Therefore, we now assume that t is a proper face of s.
We prove this claim set-theoretically for the simplex sets

$$\begin{aligned}\mathrm{Lk}\left(\mathrm{Lk}(K,t), s\setminus t\right) &= \{u \in \mathrm{Lk}(K,t) \,|\, u \cap (s\setminus t) = \varnothing \wedge u \cup (s \setminus t) \in \mathrm{Lk}(K,t)\} = \\ &= \{u \in S | u \cap t = \varnothing \wedge u \cap (s\setminus t) = \varnothing \wedge u \cup s \in S\} = \\ &= \{u \in S | u \cap s = \varnothing \wedge u \cup s \in S\} = \\ &= \mathrm{Lk}(K,s). \square\end{aligned}$$

[7] Both $D_{n+1} \setminus \mathring{\mathrm{St}}(D_{n+1},s)$ and $\mathrm{Lk}(D_{n+1},s)$ are empty because the star of s is already all of D_{n+1}.

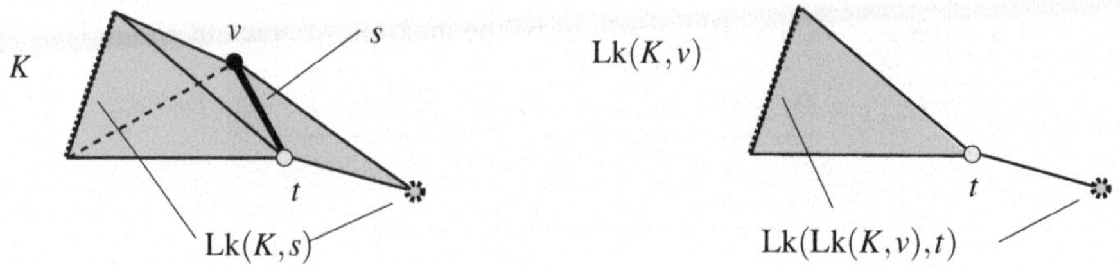

Fig. 10: Sketch of Lemma 8

Now that this is out of the way, we can turn to the proof of Proposition 1.

Proof. We perform induction on the dimension n. The case $n = 0$ is rather trivial: a combinatorial 0-manifold is a finite collection of 0-simplices, because every vertex $v \in V$ has an empty link $\mathrm{Lk}(K,v)$, so K cannot have any 1-simplices. Since every simplex is a 0-simplex, the first property follows immediately from the definition of a combinatorial manifold. The second property follows from the fact that any stellar move on a 0-simplex is trivial, therefore the stellar equivalence between K and K' is essentially a simplicial isomorphism, which restricts to the links and therefore turns any link in K' into a combinatorial (-1)-ball.
Now assume that K is a combinatorial n-manifold for some $n \in \mathbb{N}$, and both (1) and (2) are true for dimensions less than n.

(1) Let $s \in S$ be a k-simplex of K. If $k = 0$, then (1) follows immediately from the definition since any 0-simplex is a vertex. If $k > 0$, we write $s = \{v\} \sqcup t$, for v a vertex of s and $t := s \setminus \{v\}$. Note that t is a $(k-1)$-simplex of K. From Lemma 8, we obtain that
$$\mathrm{Lk}(K,s) = \mathrm{Lk}\left(\mathrm{Lk}(K,v),t\right).$$
Since K is a combinatorial n-manifold, the link $\mathrm{Lk}(K,v)$ is a combinatorial $(n-1)$-ball or a combinatorial $(n-1)$-sphere. By Lemma 6 and property (2), in dimension $n-1$, this implies that $\mathrm{Lk}(K,v)$ is a combinatorial $(n-1)$-manifold. Now we can use property (1), in dimension $n-1$, to obtain that $\mathrm{Lk}\left(\mathrm{Lk}(K,v),t\right)$ is a combinatorial ball or a combinatorial sphere of dimension
$$\Big(\underbrace{(n-1)}_{\dim \mathrm{Lk}(v)} - \underbrace{(k-1)}_{\dim(t)} - 1\Big) = (n-k-1).$$
Therefore, it follows from the claim that $\mathrm{Lk}(K,s)$ is a combinatorial $(n-k-1)$-ball or a combinatorial $(n-k-1)$-sphere. Since s was chosen arbitrarily, this concludes the proof of (1) in dimension n.

(2) Clearly, an abstract simplicial complex that is simplicially isomorphic to a combinatorial n-manifold is itself a combinatorial manifold. Thus, it suffices to show that if $K' = \sigma_s K$ for some k-simplex $s \in S$ and K is a combinatorial mani-

fold, then K' is a combinatorial manifold. For this, let $v \in V'$ be a vertex of K'. We consider the following cases[8]:

Case 1: $v \notin \mathrm{St}(K,s)$. In this case, it follows from the definition of stellar subdivision that we have the equality $\mathrm{Lk}(K',v) = \mathrm{Lk}(K,v)$, because the stellar subdivision σ_s does not change the abstract simplicial complex outside of the star of s. Since K is a combinatorial n-manifold, the link $\mathrm{Lk}(K,v)$ is a combinatorial $(n-1)$-ball or a combinatorial $(n-1)$-sphere, and hence so is $\mathrm{Lk}(K',v)$.

Case 2: $v \in \mathrm{Lk}(K,s)$. In this case, we know that $s \in \mathrm{Lk}(K,v)$. It follows from set theory, similar to the arguments used at the end of chapter 3, that

$$\mathrm{Lk}(K',v) \cong_{\mathrm{si}} \sigma_s \mathrm{Lk}(K,v).$$

In other words, a stellar move connects the links of v in K and K', respectively. Therefore, $\mathrm{Lk}(K,v)$ is a combinatorial $(n-1)$-ball or a combinatorial $(n-1)$-sphere if and only if $\mathrm{Lk}(K',v)$ is.

Case 3: $v \in s$. We let $t := s \setminus \{v\}$, thus $s = \{v\} * t$. Just as in the second case, it follows that $\mathrm{Lk}(K',v) \cong_{\mathrm{si}} \sigma_t \mathrm{Lk}(K,v)$. Therefore, the links again differ by a stellar move.

Case 4: $v = \underline{s}$. By definition of $K' = K \setminus \mathring{\mathrm{St}}(K,s) \cup \underline{s} * \partial s * \mathrm{Lk}(K,s)$, we have

$$\mathrm{Lk}(K',\underline{s}) = \partial s * \mathrm{Lk}(K,s).$$

Again, $\mathrm{Lk}(K,s)$ is a combinatorial $(n-k-1)$-ball or a combinatorial $(n-k-1)$-sphere by property (1), in dimension n. But since ∂s is the boundary of a k-simplex, it is simplicially isomorphic to S_{k-1}, and thus in particular a combinatorial $(k-1)$-sphere. By Lemma 7 (2) and (3), we obtain that $\mathrm{Lk}(K',\underline{s})$ is a combinatorial ball or a combinatorial sphere of dimension $(k-1)+(n-k-1)+1 = n-1$.

In each of the cases, $\mathrm{Lk}(K',v)$ is a combinatorial $(n-1)$-ball or a combinatorial $(n-1)$-sphere, and therefore, K' is a combinatorial manifold.

This concludes the proof by induction.

Next, we shall define the boundary of a combinatorial manifold.

Definition 10. Let $K = (V,S)$ be a combinatorial n-manifold. By Proposition 1, we know that the link of every simplex is a combinatorial ball or a combinatorial sphere. We define the *boundary* ∂K *of* K as the abstract subcomplex consisting of all the vertices and simplices of K whose link in K is not a combinatorial sphere[9]. Often, we slightly abuse notation and also denote the simplex set of ∂K by ∂K.
We say a combinatorial manifold K is *closed* if its boundary is the empty simplicial complex, in other words if the links of all simplices are combinatorial spheres.

---

[8] It can be a helpful exercise to think about why these four cases are the ones one has to consider.

[9] It is important to define it this way, because a (-1)-ball is the same as a (-1)-sphere.

Example 2. As one would expect, we have $\partial D_n = S_{n-1}$. This becomes evident as follows: If s is a k-simplex of D_n, then by definition of the link we have

$$\mathrm{Lk}(D_n, s) = \Big(V_n \setminus s, \mathscr{P}(V_n \setminus s) \setminus \{\varnothing\}\Big).$$

This is a fancy way of writing down what one would intuitively call "the opposite face of s." Due to the nature of this abstract simplicial complex, it is easy to write down a simplicial isomorphism to D_{n-k-1}. However, this is a combinatorial sphere if and only if $k = n$, thus the abstract subcomplex ∂D_n consists of all the simplices of D_n that are of dimension strictly less than n, which just happens to be S_{n-1}.

We use the following technical lemma on multiple occasions.

Lemma 9. *Let $K = (V,S)$ be a combinatorial n-manifold, let $s \in S$ be a k-simplex with $s \in \partial K$. Then we have*[10] $\mathrm{Lk}(\partial K, s) = \partial \mathrm{Lk}(K, s)$.

Proof. Using Lemma 8, we have[11]

$$\begin{aligned}\partial \mathrm{Lk}(K,s) &= \{u \in \mathrm{Lk}(K,s) | \mathrm{Lk}(\mathrm{Lk}(K,s),u) \approx_{\mathrm{st}} D_l\} = \\ &= \{u \in K | u \cap s = \varnothing, u \cup s \in K, \mathrm{Lk}(K, u \cup s) \approx_{\mathrm{st}} D_l, \mathrm{Lk}(K,u) \approx_{\mathrm{st}} D_m\} = \\ &= \{u \in \partial K | u \cap s = \varnothing, u \cup s \in \partial K\} = \\ &= \mathrm{Lk}(\partial K, s).\end{aligned}$$

Note that for legibility's sake, we did notationally not distinguish between simplicial complexes and their simplex sets. The dimensions of the combinatorial balls follow immediately from Lemma 7 (1) and (3).

For rigour's sake, it is also important to prove the following unsurprising statement.

Lemma 10. *Let $K = (V,S)$ be a combinatorial n-manifold. Then its boundary is in fact an abstract subcomplex, and it is furthermore a closed combinatorial $(n-1)$-manifold.*

Proof. Let $K = (V,S)$ be a combinatorial n-manifold, and let $s \in S$ be a k-simplex of K whose link in K is a combinatorial $(n-k-1)$-ball but not a combinatorial $(n-k-1)$-sphere. Since for $k = n$, this combinatorial (-1)-ball would also be a combinatorial (-1)-sphere, we can assume for the remainder of the proof that $k < n$.
For the first claim, we have to show that the link in K of any face of s is also a combinatorial ball. So let $t \subsetneq s$ be a proper face of s. From Proposition 1, we know that $\mathrm{Lk}(K,t)$ is a combinatorial ball or a combinatorial sphere, so it suffices to

[10] Here, we implicitly use that the link of s in K is a combinatorial ball, hence in particular a combinatorial manifold by Lemma 6 and Proposition 1 (1).

[11] Here, we set $l = l(u) := n-k-2-\dim u$ and $m = m(u) := n-1-\dim u$. The diligent reader can check for themselves that this is in fact the correct dimension for those links.

exclude the possibility of it being a sphere. Let us assume that it is a combinatorial sphere. Note that it follows from Lemma 8 that

$$\mathrm{Lk}(K,s) = \mathrm{Lk}\left(\mathrm{Lk}(K,t), s \setminus t\right).$$

Since we had assumed $\mathrm{Lk}(K,t)$ to be a combinatorial sphere, this is a combinatorial sphere of dimension[12]

$$\dim \mathrm{Lk}(K,t) - \dim(s \setminus t) - 1 = n - l - 1 - (k - l - 1) - 1 = n - k - 1 \geq 0,$$

in contradiction to $\mathrm{Lk}(K,s)$ being a combinatorial ball.
For the second claim, we have to show that the link of s in ∂K is a combinatorial $(n-k-2)$-sphere. Since by Corollary 1[13], the boundary $\partial\,\mathrm{Lk}(K,s)$ of the combinatorial $(n-k-1)$-ball $\mathrm{Lk}(K,s)$ is a combinatorial $(n-k-2)$-sphere, it follows from Lemma 9 that the links of all simplices of ∂K in ∂K are combinatorial $(n-k-2)$-spheres, which implies that the boundary of ∂K is empty.

The following lemma may not come as a surprise, but many authors fail to acknowledge that it is something that needs proof.

Lemma 11. *Let $K = (V,S)$ be a combinatorial ball, and let $s \in S$ be a simplex. If $s \notin \partial K$, then $\partial(\sigma_s K) = \partial K$.*

Proof. Let $K = (V,S)$ be a combinatorial ball, and let $s \in S$ be a k-simplex not in ∂K. This implies that $\mathrm{Lk}(K,s) \approx_{\mathrm{st}} S_{n-k-1}$. Using the definition of a stellar subdivision, we have

$$\mathrm{Lk}(\sigma_s K, \underline{s} * \partial s) = \mathrm{Lk}(K,s) \approx_{\mathrm{st}} S_{n-k-1},$$

thus $\underline{s} * \partial s \notin \partial(\sigma_s K)$. But since the only way that σ_s could potentially change ∂K is by changing $\underline{s} * \partial s$, it follows that the boundary remains unchanged under σ_s, i.e., that $\partial(\sigma_s K) = \partial K$.

In the following we state two nice corollaries, which we will use occasionally.

Corollary 1. *Let $K = (V,S)$ be a combinatorial n-ball. Then ∂K is a combinatorial $(n-1)$-sphere.*

Proof. Let $K = (V,S)$ be a combinatorial n-ball, i.e., there exists a sequence of stellar moves (and simplicial isomorphisms) from K to D_n. Our goal is to observe how a stellar move affects ∂K. If for a stellar move $\sigma_s^{\pm 1}$ the simplex s does not lie in the boundary, then it follows from Lemma 11 that the stellar move does not change the boundary. If on the other hand, the simplex s lies in the boundary, then $\partial(\sigma_s^{\pm 1} K) = \sigma_s^{\pm 1}(\partial K)$, so the boundaries of K and $\sigma_s^{\pm 1} K$ are related by a stellar move. Either way, we have $\partial K \approx_{\mathrm{st}} \partial D_n = S_{n-1}$, where the last equality was shown in the example on page 18.

[12] Here, we denote the dimension of t by l.

[13] A closer look shows that we indeed do not use the statement that the boundary has no boundary anywhere leading up to Corollary 1, so we can, in fact, use it here.

This next corollary is another well-known statement in the topological setting, but it does require some thought.

Corollary 2. *Let $K = (V,S)$ be a combinatorial n-sphere. Then K is closed, i.e., ∂K is the empty abstract simplicial complex.*

Proof. So let $K = (V,S)$ be a combinatorial n-sphere. Recall that it follows from the example on page 18 and Lemma 10 that ∂S_n is the empty abstract simplicial complex. Therefore, we can apply Lemma 11 for a sequence of stellar moves (and simplicial isomorphisms) from K to S_n in order to obtain that $\partial K = \partial S_n = (\varnothing,\varnothing)$.

We will also often use the following statement, which may look similar to statements from other fields of mathematics.

Proposition 2.

(1) *Let $K = (V,S)$ and $L = (W,T)$ be combinatorial balls. Then*

$$\partial(K * L) = (K * \partial L) \cup (\partial K * L).$$

(2) *Let $K = (V,S)$ be a combinatorial sphere and $L = (W,T)$ be a combinatorial ball. Then*

$$\partial(K * L) = K * \partial L.$$

Proof. We prove both statements individually on a set-theoretic level. Unsurprisingly, one of the main ingredients for the proof is Lemma 7. Note that by a slight abuse of notation, we will sometimes not differentiate between simplicial complexes and their simplex sets.

(1) Let K and L be combinatorial balls as above. We prove this equation by proving both that both sides are included in one another. For the "$\subseteq$" inclusion we let $s * t \in \partial(K * L)$, i.e., we have $s \in S \cup \{\varnothing\}$ and $t \in T \cup \{\varnothing\}$ not both empty, in such a way that[14]

$$\mathrm{Lk}(K * L, s * t) \approx_{\mathrm{st}} D.$$

Our goal is to show that $s * t$ lies in $(K * \partial L) \cup (\partial K * L)$.
If $s = \varnothing$, then

$$s * t = t \in T \subseteq \partial K * L,$$

and if $t = \varnothing$, then

$$s * t = s \in S \subseteq K * \partial L.$$

The final case is that neither s nor t are empty. From Lemma 3, we obtain that

$$\mathrm{Lk}(K,s) * \mathrm{Lk}(L,t) = \mathrm{Lk}(K * L, s * t) \approx_{\mathrm{st}} D,$$

which then implies by Lemmas 7, 6, and 1 (2) that at least one of the links of s and t is a combinatorial ball; in other words, we have $s \in \partial K$ or $t \in \partial L$, which

[14] In the remainder of this proof, we omit the dimensions of our combinatorial balls and spheres, and briefly write "$\approx_{\mathrm{st}} D$" and "$\approx_{\mathrm{st}} S$" for combinatorial balls and spheres, respectively.

concludes this inclusion.
For the opposite direction, it suffices to show that $\partial K * L \subseteq \partial(K * L)$, because the expression on the right-hand side is symmetric in K and L. Thus, let $s \in \partial K \cup \{\varnothing\}$ and $t \in L \cup \{\varnothing\}$ not both empty. We want to show that $s * t \in \partial(K * L)$.
If $s = \varnothing$, then $s * t = t \in L \subseteq K * L$, and we have

$$\mathrm{Lk}(K * L, s * t) = \mathrm{Lk}(K * L, t) \underset{2\,(1)}{=} \underbrace{\mathrm{Lk}(L,t)}_{\approx_{\mathrm{st}} D \text{ or } \approx_{\mathrm{st}} S} * \underbrace{K}_{\approx_{\mathrm{st}} D} \approx_{\mathrm{st}} D$$

by Lemma 7 (1) or (3). But by definition, this means that $s * t \in \partial(K * L)$.
If $t = \varnothing$, then $s * t = s \in \partial K \subset K * L$, and we have

$$\mathrm{Lk}(K * L, s * t) = \mathrm{Lk}(K * L, s) \underset{2\,(1)}{=} \underbrace{\mathrm{Lk}(K,s)}_{\approx_{\mathrm{st}} D} * \underbrace{L}_{\approx_{\mathrm{st}} D} \approx_{\mathrm{st}} D$$

by Lemma 7 (1). But by definition, this means that $s * t \in \partial(K * L)$.
If neither s nor t are empty, we can apply Lemma 3 in order to obtain

$$\mathrm{Lk}(K * L, s * t) = \underbrace{\mathrm{Lk}(K,s)}_{\approx_{\mathrm{st}} D} * \underbrace{\mathrm{Lk}(L,t)}_{\approx_{\mathrm{st}} D \text{ or } \approx_{\mathrm{st}} S} \approx_{\mathrm{st}} D$$

by Lemma 7 (1) or (3). But by definition, this means that $s * t \in \partial(K * L)$.
As mentioned above, the inclusion $K * \partial L \subseteq \partial(K * L)$ follows analogously, which concludes the proof of the statement.

(2) Now assume that K is a combinatorial sphere and L is a combinatorial ball. We again prove both inclusions, starting with "$\subseteq$". So let $s * t \in \partial(K * L)$, i.e., $s \in S \cup \{\varnothing\}$ and $t \in T \cup \{\varnothing\}$ not both empty, with

$$\mathrm{Lk}(K * L, s * t) \approx_{\mathrm{st}} D.$$

We want to show that $s * t \in K * \partial L$.
If $s = \varnothing$, then $s * t = t$, and from

$$D \approx_{\mathrm{st}} \mathrm{Lk}(K * L, s * t) = \mathrm{Lk}(K * L, t) = \underbrace{\mathrm{Lk}(L,t)}_{\approx_{\mathrm{st}} D \text{ or } \approx_{\mathrm{st}} S} * \underbrace{K}_{\approx_{\mathrm{st}} S}$$

and Lemma 7 (2) and (3), we obtain that $\mathrm{Lk}(L,t)$ must be a combinatorial ball; in other words, $s * t = t \in \partial L \subseteq K * \partial L$.
If $t = \varnothing$, then we immediately obtain that $s * t = s \in K \subseteq K * \partial L$.
If neither s nor t are empty, then we calculate using Lemma 3 that

$$D \approx_{\mathrm{st}} \mathrm{Lk}(K * L, s * t) = \underbrace{\mathrm{Lk}(K,s)}_{\approx_{\mathrm{st}} S \text{ by } 2} * \underbrace{\mathrm{Lk}(L,t)}_{\approx_{\mathrm{st}} D \text{ or } \approx_{\mathrm{st}} S}.$$

But by Lemma 7 (2) and (3), this again implies that $\mathrm{Lk}(L,t)$ must be a combinatorial ball; in other words, we have $s * t \in K * \partial L$.

For the converse inclusion, we let $s*t \in K*\partial L$, i.e., $s \in S \cup \{\varnothing\}$ and $t \in \partial L \cup \{\varnothing\}$ not both empty. Our goal is to show that we have $s*t \in \partial(K*L)$.
If $s = \varnothing$, then $s*t = t$, and we obtain from Lemma 7 (3) that

$$\mathrm{Lk}(K*L, s*t) = \mathrm{Lk}(K*L, t) = \underbrace{\mathrm{Lk}(L,t)}_{\approx_{\mathrm{st}} D} * \underbrace{K}_{\approx_{\mathrm{st}} S} \approx_{\mathrm{st}} D.$$

But this just means that $s*t \in \partial(K*L)$.
If $t = \varnothing$, then $s*t = s$, and we obtain from Lemma 7 (3) that

$$\mathrm{Lk}(K*L, s*t) = \mathrm{Lk}(K*L, s) = \underbrace{\mathrm{Lk}(K,s)}_{\approx_{\mathrm{st}} S \text{ by } 2} * \underbrace{L}_{\approx_{\mathrm{st}} D} \approx_{\mathrm{st}} D.$$

But this again just means that $s*t \in \partial(K*L)$.
The final case is that neither s nor t are empty. We use Lemmas 3 and 7 (3) in order to obtain that

$$\mathrm{Lk}(K*L, s*t) = \underbrace{\mathrm{Lk}(K,s)}_{\approx_{\mathrm{st}} S \text{ by } 2} * \underbrace{\mathrm{Lk}(L,t)}_{\approx_{\mathrm{st}} D} \approx_{\mathrm{st}} D.$$

This shows that in all cases, we have $s*t \in \partial(K*L)$. We have therefore seen both inclusions, which concludes the proof. □

The next Corollary, which is a direct consequence of Proposition 2 (1), will be used in a lot of the proofs of the main statements of the next chapter.

Corollary 3. *Let K be a combinatorial ball. Then $\partial\mathrm{Cone}(K) = \mathrm{Cone}(\partial K) \cup K$.*

Proof. Let K be a combinatorial ball. Remember that $\mathrm{Cone}(K) = D_0 * K$. (In particular, the cone of K is in fact a combinatorial manifold by Lemmas 7 (1) and 6.) It follows from Proposition 2 (1) that

$$\partial\mathrm{Cone}(K) = (D_0 * \partial K) \cup (\underbrace{\partial D_0}_{=(\varnothing,\varnothing)} * K) = \mathrm{Cone}(\partial K) \cup K,$$

where we used that taking the join with the empty abstract simplicial complex preserves any abstract simplicial complex.

5 Starrability of Stellar Balls

Now that we know a fair bit about some of the properties of stellar moves, balls, spheres, and manifolds, we want to introduce a concept of "especially nice" combinatorial balls, only to then prove that all combinatorial balls have this nice property. Just like the previous one, this chapter is also modelled after the approach made by Lickorish (cf. [Lic99, Chapter 3]), again however with a greater emphasis on details.

Definition 11. Let $K = (V,S)$ be a combinatorial n-ball. Denote by $s := \{0,\ldots,n\}$ the maximal simplex of D_n. We have the following chain of stellar equivalences

$$K \approx_{\text{st}} D_n \approx_{\text{st}} \sigma_s D_n = \underline{s} * S_{n-1} \approx_{\text{st}} \text{Cone}(\partial K),$$

where the last stellar equivalence is a consequence of Corollary 1 and Lemma 5. By transitivity, this implies that there exists a sequence of stellar moves and simplicial isomorphisms from K to $\text{Cone}(\partial K)$.

(1) A stellar subdivision σ_t is called *internal* if $t \notin \partial K$.
(2) A stellar weld σ_t^{-1} is called *internal* if the stellar subdivision that is its inverse is an internal stellar subdivision.
(3) A stellar move $\sigma_t^{\pm 1}$ is called an *internal move* if it is an internal stellar subdivision or an internal stellar weld.
(4) We call K *starrable in r moves* if there exists a sequence of r internal moves (and any number of simplicial isomorphisms) that transforms K into $\text{Cone}(\partial K)$. Such a sequence is called a *starring of K*.

Remark 5. One of the main goals of this chapter is to prove that *all* combinatorial balls are starrable. Put another way, this means that any abstract simplicial complex K that is stellar equivalent to D_n can be transformed into $\text{Cone}(\partial K)$ without "changing the boundary."

The following lemma may look harmless, but as mathematicians we know that looks can be deceiving. It can be found in [Lic99, Lemma 3.4].

Lemma 12. *Let $K = (V,S)$ be a starrable n-ball. If L is a stellar subdivision of K, then L is also starrable.*

Proof. Let $K = (V,S)$ be a starrable n-ball, i.e., there is a sequence of internal moves (and simplicial isomorphisms) $K \approx_{\text{st}} \underline{v} * \partial K$, and let $L := \sigma_s K$ for some k-simplex $s \in S$. If $s \notin \partial K$, then $\partial L = \partial K$ by Lemma 11, and therefore

$$L = \sigma_s K \approx_{\text{st}} K \approx_{\text{st}} \underline{v} * \partial K = \underline{v} * \partial L,$$

where all the stellar moves are internal since $s \notin \partial K$. Hence, L is also starrable. For the remainder of the proof, we can therefore assume that $s \in \partial K$.
We shall perform induction on r, which we define to be the minimum number of stellar moves in a sequence $K \approx_{\text{st}} \underline{v} * \partial K$ of internal moves and simplicial isomorphisms.
The case $r = 0$ is relatively straightforward: up to simplicial isomorphism, we have the equation $K = \underline{v} * \partial K$, and therefore

$$L = \sigma_s K \cong_{\text{si}} \sigma_s(\underline{v} * \partial K) = \underline{v} * \sigma_s(\partial K),$$

where the last equality follows from the fact that $s \in \partial K$. But since again $s \in \partial K$, we have $\sigma_s(\partial K) = \partial L$, thus it follows in fact that $L \cong_{\text{si}} \underline{v} * \partial L$, and L is (trivially) starrable.

Now assume that stellar subdivisions of combinatorial n-balls that are starrable in less than r moves are themselves starrable. We have to consider two cases:

Case 1: The first of the r internal moves is a stellar subdivision (as opposed to a stellar weld). We denote the result of this first subdivision by $K_1 := \sigma_t K$ for some $t \in S$ with $t \notin \partial K$. Remember that K_1 is, by construction, starrable in $r-1$ moves. Since σ_t is an internal move, we have $\partial K_1 = \partial K \ni s$. Hence, we can perform a stellar subdivision of K_1 at s, and we denote the resulting abstract simplicial complex by $L_1 := \sigma_s K_1$. Since L_1 is a stellar subdivision of K_1, which is starrable in $r-1$ moves, it follows from the induction hypothesis that L_1 is starrable. For clarity's sake, it makes sense to consider the following three cases:

Case 1a: If $s \cap t = \varnothing$, then up to a simplicial isomorphism, σ_s and σ_t commute, hence

$$L_1 = \sigma_s K_1 = \sigma_s \sigma_t K \cong_{\text{si}} \sigma_t \sigma_s K = \sigma_t L.$$

But L_1 is starrable and σ_t is an internal move, so L is starrable.

Case 1b: If there does not exist a simplex in K that has both s and t as its faces, then the open stars of s and t in K do not intersect, and therefore σ_s and σ_t still commute. The claim follows as in case 1a.

If one wanted to summarise cases 1a and 1b in a succinct way, the commutativity of the following diagram would be the key element.

$$\begin{array}{ccc} K & \xrightarrow{\sigma_t} & K_1 \\ \downarrow{\scriptstyle \sigma_s} & & \downarrow{\scriptstyle \sigma_s} \\ L & \xrightarrow{\sigma_t} & L_1 \end{array}$$

Case 1c: It remains to show the claim for the case that s and t are faces of a common simplex, and that $s \cap t = u$ for some simplex $u \in S$. If we define $s_0 := s \setminus u$ and $t_0 := t \setminus u$, we have $s = s_0 * u$ and $t = t_0 * u$, as well as $s_0 * t_0 * u \in S$. The situation is illustrated at the very left of Figure 11. We want to investigate the relationship between $\sigma_s \sigma_t K$ and $\sigma_t \sigma_s K$. Note that by definition of stellar subdivisions, the difference between these two abstract simplicial complexes has to lie entirely in what happens on the star $\operatorname{St}(K, s_0 * t_0 * u)$. First we shall focus on the simplex $s_0 * t_0 * u$ itself. Using associativity and commutativity of the simplicial join operation, one obtains

$$\begin{aligned}
\sigma_t \sigma_s (\underbrace{s_0 * u}_{=s} * t_0) &= \sigma_t (\underline{s} * \partial s * t_0) \overset{2\,(1)}{=} \\
&= \sigma_t ((\underline{s} * s_0 * \partial u * t_0) \cup (\underline{s} * \partial s_0 * \underbrace{u * t_0}_{=t})) = \\
&= (\underline{s} * s_0 * \partial u * t_0) \cup (\underline{s} * \partial s_0 * \underline{t} * \partial t) \overset{2\,(1)}{=} \\
&= (\underline{s} * (s_0 \cup \partial s_0 * \underline{t}) * t_0 * \partial u) \cup (\underline{s} * \partial s_0 * \underline{t} * \partial t_0 * u) \overset{2\,(1)}{=} \\
&= (\underline{s} * \partial (\underline{t} * s_0) * t_0 * \partial u) \cup (\underline{s} * \underline{t} * \partial s_0 * \partial t_0 * u)
\end{aligned}$$

where in the third step, the stellar subdivision σ_t does not change the first part because the dimension of the second part is precisely the dimension of $s_0 * t_0 * u$ and t is entirely contained in the second part[15]. In the last step, we used the fact that $\partial \underline{t} = (\varnothing, \varnothing)$ since it is a 0-simplex.
Figure 11 attempts to summarise the effects of the composition of two stellar subdivisions $\sigma_t \sigma_s$.

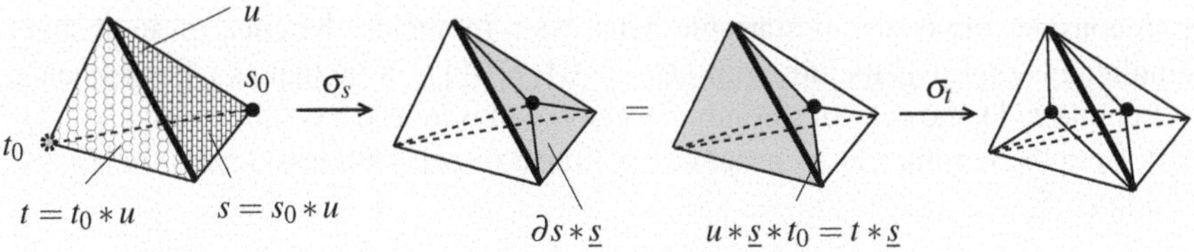

Fig. 11: The effects of $\sigma_t \sigma_s$ in case 1c

If we define $v := \underline{s} * t_0$, then stellar subdivision yields

$$\begin{aligned}\sigma_v \sigma_t \sigma_s (s_0 * u * t_0) &= \sigma_v((\partial(\underline{t} * s_0) * v * \partial u) \cup (\underline{s} * \underline{t} * \partial s_0 * \partial t_0 * u)) = \\ &= (\partial(\underline{t} * s_0) * \underline{v} * \partial v * \partial u) \cup (\underline{s} * \underline{t} * \partial s_0 * \partial t_0 * u) = \\ &= (\partial(\underline{t} * s_0) * \underline{v} * \partial(\underline{s} * t_0) * \partial u) \cup (\underline{s} * \underline{t} * \partial s_0 * \partial t_0 * u).\end{aligned}$$

However, since this expression is symmetric in s and t, we can use the same logic in order to perform another three subdivisions on the original $s_0 * u * t_0$ to obtain a simplicially isomorphic abstract simplicial complex[16]:

$$\sigma_{v'} \sigma_s \sigma_t (s_0 * u * t_0) = (\partial(\underline{t} * s_0) * \underline{v'} * \partial(\underline{s} * t_0) * \partial u) \cup (\underline{s} * \underline{t} * \partial s_0 * \partial t_0 * u).$$

Here, we set $v' := \underline{t} * s_0$ after performing $\sigma_s \sigma_t$.
We now take the join in K with $\mathrm{Lk}(K, s_0 * u * t_0)$ and obtain that

$$\sigma_v \sigma_t L = \sigma_v \sigma_t \sigma_s K \cong_{\mathrm{si}} \sigma_{v'} \sigma_s \sigma_t K = \sigma_{v'} L_1.$$

It now suffices to show that both v and v' are internal moves. Since $u \subseteq s \in \partial K$ implies $u \in \partial K$, and $t_0 * u = t \notin \partial K$, the simplex t_0 cannot lie in ∂K. Therefore, $v = \underline{s} * t_0 \notin \partial K$. Since $t \notin \partial K$, the 0-simplex $\underline{t}$ cannot lie in ∂K, thus also $v' = \underline{t} * s_0 \notin \partial K$.
Therefore, since σ_t is also still an internal move, L and L_1 differ by internal moves, proving that L is starrable.

---

[15] In Figure 11, this argument effectively means that only the magenta portion experiences a change from the stellar subdivision σ_t.

[16] If one were to draw $\sigma_s \sigma_t$, the result would essentially be Figure 11, but mirrored along a vertical axis.

Case 2: The first of the r internal moves is a stellar weld. If we again denote the result of this stellar weld by $K_1 := \sigma_t^{-1}K$, this time for some simplex t of K_1 with $t \notin \partial K_1$, we can use the same logic as before to conclude that $s \in \partial K_1$, and again define $L_1 := \sigma_s K_1$. Again, since L_1 is a subdivision of K_1, which is starrable in $r-1$ moves, L_1 is starrable. But we have already shown in Case 1 that L and L_1 are related by internal stellar moves, so L is also starrable. □

Our next step on the way to proving that all combinatorial balls are starrable is to consider the cones of starrable balls. As a reminder, the cone of an abstract simplicial complex is its join with $D_0 = (\{0\},\{\{0\}\})$. Note that as a consequence of Lemma 7 (1), the cone of a combinatorial n-ball is indeed a combinatorial $(n+1)$-ball. Both the lemma and its proof can be found (albeit with less details) in [Lic99, Lemma 3.5].

Lemma 13. *Let $K = (V,S)$ be a starrable n-ball. Then* Cone(K) *is starrable.*

Proof. So let $K = (V,S)$ be a starrable n-ball, that means $K \approx_{\text{st}} \underline{v} * \partial K$ by internal moves. As in the previous proofs, we shall perform induction on r, the minimum number of internal moves in such a sequence.
The case $r = 0$ means that $K \cong_{\text{si}} \underline{v} * \partial K$. Using associativity of the join operation, we obtain

$$\text{Cone}(K) = D_0 * K \cong_{\text{si}} D_0 * \underline{v} * \partial K.$$

We now define a 1-simplex $s := D_0 * \underline{v}$. Note that since by construction $\underline{v} \notin \partial K$, we have $s = D_0 * \underline{v} \notin \partial\text{Cone}(K) = \text{Cone}(\partial K) \cup K$, where the last equality is just Corollary 3, so performing a stellar subdivision along s in Cone(K) is an internal move. By a slight abuse of notation, we also refer to the preimage of s under the simplicial isomorphism as s. We then have

$$\sigma_s\text{Cone}(K) \cong_{\text{si}} \sigma_s(s * \partial K) = \underline{s} * \underbrace{\partial s}_{=D_0 \cup \underline{v}} * \underbrace{\text{Lk}(\text{Cone}(K), s)}_{=\partial K}.$$

By Corollary 3, we still have the identity $\partial\text{Cone}(K) = \text{Cone}(\partial K) \cup K$. Using this, we can resume the above calculation to obtain

$$\sigma_s\text{Cone}(K) \cong_{\text{si}} \underline{s} * \Big(\underbrace{D_0 * \partial K}_{=\text{Cone}(\partial K)} \cup \underbrace{\underline{v} * \partial K}_{\cong_{\text{si}} K} \Big) \cong_{\text{si}} \underline{s} * \partial\text{Cone}(K).$$

We have thus transformed Cone(K) into $\underline{s} * \partial\text{Cone}(K)$ using the internal move σ_s (together with some simplicial isomorphisms), so Cone(K) is starrable. This concludes the case that $r = 0$. Figure 12 hopefully gives some insight on what happened. Note that we intentionally avoided filling in the 2- and 3-simplices in order to keep it legible.
Now let us assume that $K \approx_{\text{st}} \underline{v} * \partial K$ in r moves, and that the cones of combinatorial n-balls that are starrable in less than r moves are themselves starrable. Based on the nature of the first of the r internal moves, we distinguish two cases:

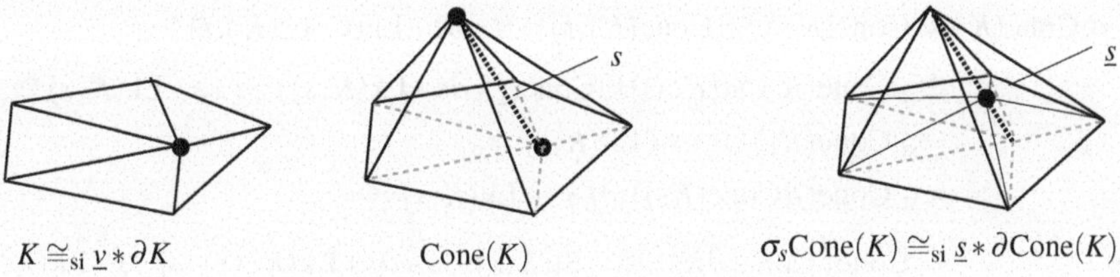

Fig. 12: The case $r = 0$ in the proof of Lemma 13

Case 1: If the first move is a stellar weld $K_1 = \sigma_s^{-1} K$ for some internal simplex s of K_1, then $K = \sigma_s K_1$ and $\mathrm{Cone}(K) \cong_{\mathrm{si}} \sigma_{\mathrm{Cone}(s)} \mathrm{Cone}(K_1)$. Since K_1 is starrable in $r-1$ moves, $\mathrm{Cone}(K_1)$ is starrable. By Lemma 12, $\mathrm{Cone}(K)$ is also starrable as it is a stellar subdivision of $\mathrm{Cone}(K_1)$.

Case 2: Now assume that the first of the r moves is a stellar subdivision $K_1 = \sigma_s K$ for some simplex $s \in S$ with $s \notin \partial K$. We have

$$K = (K \setminus \mathring{\mathrm{St}}(K,s)) \cup \mathrm{St}(K,s) = (K \setminus \mathring{\mathrm{St}}(K,s)) \cup s * \mathrm{Lk}(K,s),$$

so taking the cone according to Lemma 4 yields

$$\mathrm{Cone}(K) = \mathrm{Cone}(K \setminus \mathring{\mathrm{St}}(K,s)) \cup \mathrm{Cone}(s * \mathrm{Lk}(K,s)).$$

Now we want to perform stellar subdivision at the simplex $t := \mathrm{Cone}(s) = D_0 * s$ in the subcomplex $\mathrm{Cone}(\mathrm{St}(K,s)) = \mathrm{St}(\mathrm{Cone}(K),t)$ as follows:

$$\begin{aligned} \sigma_t \mathrm{Cone}(s * \mathrm{Lk}(K,s)) &= \underline{t} * \partial t * \mathrm{Lk}(K,s) = \\ &= \underline{t} * \mathrm{Cone}(\partial s) * \mathrm{Lk}(K,s) \cup \underline{t} * s * \mathrm{Lk}(K,s) = \\ &= \mathrm{Cone}(\underline{t} * \partial s * \mathrm{Lk}(K,s)) \cup \underline{t} * s * \mathrm{Lk}(K,s) \end{aligned}$$

where we used that $\partial t = \partial \mathrm{Cone}(s) = \mathrm{Cone}(\partial s) \cup s$ by Corollary 3, and the commutativity of the join operation, which allows us to "pull out the cone." Note that by definition of $K_1 = \sigma_s K$, and by using Lemma 4, we have

$$\mathrm{Cone}(K_1) = \mathrm{Cone}(K \setminus \mathring{\mathrm{St}}(K,s)) \cup \mathrm{Cone}(\underline{s} * \partial s * \mathrm{Lk}(K,s)).$$

Now recall that, by our induction assumption, the cone on K_1 is starrable as K_1 is starrable in $r-1$ moves, so $\mathrm{Cone}(K_1) \approx_{\mathrm{st}} \mathrm{Cone}(\partial \mathrm{Cone}(K_1))$ by internal moves. Using Corollary 3 to rewrite the boundary of the cone on K_1, we are now able to put everything together[17]:

[17] From the induction hypothesis and the fact that $s \notin \partial K$, thus $t \notin \partial \mathrm{Cone}(K)$, it follows that all the stellar moves are internal.

$$\begin{aligned}
\sigma_t\mathrm{Cone}(K) &= \mathrm{Cone}(K)\setminus \mathring{\mathrm{St}}(\mathrm{Cone}(K),t))\cup \underline{t}*\partial t*\mathrm{Lk}(\mathrm{Cone}(K),t)\cong_{\mathrm{si}}\\
&\cong_{\mathrm{si}} \mathrm{Cone}(K\setminus \mathring{\mathrm{St}}(K,s))\cup \mathrm{Cone}(\underline{t}*\partial s*\mathrm{Lk}(K,s))\cup \underline{t}*s*\mathrm{Lk}(K,s)\cong_{\mathrm{si}}\\
&\cong_{\mathrm{si}} \mathrm{Cone}(K_1)\cup \underline{t}*s*\mathrm{Lk}(K,s)\approx_{\mathrm{st}}\\
&\approx_{\mathrm{st}} \mathrm{Cone}(\partial\mathrm{Cone}(K_1))\cup \underline{t}*s*\mathrm{Lk}(K,s)=\\
&= \mathrm{Cone}\Big(\mathrm{Cone}(\partial K)\cup K\setminus \mathring{\mathrm{St}}(K,s)\cup \underline{s}*\partial s*\mathrm{Lk}(K,s)\Big)\cup \underline{t}*s*\mathrm{Lk}(K,s)\cong_{\mathrm{si}}\\
&\cong_{\mathrm{si}} \mathrm{Cone}\Big(\underbrace{\mathrm{Cone}(\partial K)\cup K}_{=\partial\mathrm{Cone}(K)}\cup \underline{s}*\partial s*\mathrm{Lk}(K,s)\Big)\approx_{\mathrm{st}}\\
&\approx_{\mathrm{st}} \mathrm{Cone}(\partial\mathrm{Cone}(K)).
\end{aligned}$$

In the fifth step, we again used Lemma 11 to replace ∂K_1 by ∂K, as well as the definition of $K_1 = \sigma_s K = K\setminus \mathring{\mathrm{St}}(K,s)\cup \underline{s}*\partial s*\mathrm{Lk}(K,s)$. We have thus shown that $\mathrm{Cone}(K)$ is starrable by internal moves.

In both cases, we have thus performed the induction step, which concludes the proof.

After this extremely technical proof, we can reap the fruit of our labour in form of the following theorem, which is essentially [Lic99, Theorem 3.6].

Theorem 4. *Let $n\in\mathbb{N}_0$, and let $K=(V,S)$ be a combinatorial n-ball. Then K is starrable.*

Proof. Unsurprisingly, we perform induction on n. The case $n=0$ is rather trivial: if K is a combinatorial 0-ball, it is by definition an abstract simplicial complex stellar equivalent to D_0. But stellar subdivisions and welds on D_0 are just simplicial isomorphisms, and therefore K is starrable.
Now suppose that combinatorial balls of dimension less than n are starrable. We perform another induction, this time on r, the number of stellar moves[18] in $K\approx_{\mathrm{st}} D_n$. If $r=0$, then there exists a simplicial isomorphism $f\colon D_n\to K$. Let $s:=\{0,\dots,n\}$ be the maximal simplex of D_n. Since $\mathrm{Lk}(K,f(s))\cong_{\mathrm{si}}\mathrm{Lk}(D_n,s)=(\varnothing,\varnothing)$, we have $f(s)\notin\partial K$, so

$$\sigma_{f(s)}K\cong_{\mathrm{si}}\sigma_s D_n=\underline{s}*S_{n-1}\cong_{\mathrm{si}}\underline{f(s)}*\partial K,$$

using an internal move, and K is starrable.
Now suppose $K\approx_{st} D_n$ in r stellar moves, and that combinatorial n-balls that are stellar equivalent to D_n in less than r moves are starrable. We denote the result of the first move by K_1. By our induction hypothesis, K_1 is starrable.
If the first move is internal, then K is starrable. If the first move is a weld $K_1=\sigma_s^{-1}K$, then $K=\sigma_s K_1$ is a stellar subdivision of a starrable n-ball, and therefore itself starrable by Lemma 12. We shall therefore assume for the remainder of the proof that $K_1=\sigma_s K$ for some k-simplex $s\in\partial K$.
Our plan now is as follows:

Step 1: Show that the star of s in K is starrable, and perform said starring.

[18] Remember that these stellar moves are in general not internal.

Step 2: From the resulting abstract simplicial complex, isolate a copy of
Step 3: Use the induction hypothesis to obtain a starring of this copy of K_1.
Step 4: Perform the starring from Step 1 backwards to recover a starring of K.

We choose a vertex $v \in s$ and call the "opposite face" $t := s \setminus \{v\}$. For the star of s in K, the following equation holds:

$$\mathrm{St}(K,s) = s * \mathrm{Lk}(K,s) = \{v\} * t * \mathrm{Lk}(K,s) \cong_{\mathrm{si}} \mathrm{Cone}\big(t * \mathrm{Lk}(K,s)\big).$$

We make the following observations:

1) Since t is a $(k-1)$-simplex, it is in particular a combinatorial $(k-1)$-ball.
2) The link $\mathrm{Lk}(K,s)$ is the link of a k-simplex from the boundary of a combinatorial manifold, therefore it is a combinatorial $(n-k-1)$-ball.
3) By Lemma 7 (1), the join of two combinatorial balls is a combinatorial ball, of dimension

$$\dim\big(t * \mathrm{Lk}(K,s)\big) = (k-1) + (n-k-1) + 1 = n-1.$$

By our induction hypothesis, this combinatorial $(n-1)$-ball is starrable.
4) The star $\mathrm{St}(K,s) \cong_{\mathrm{si}} \mathrm{Cone}\big(t * \mathrm{Lk}(K,s)\big)$ is therefore starrable by Lemma 13.

In order to show that, in fact, the entire combinatorial n-ball K is starrable, we perform the following calculation[19]

$$\begin{aligned}
K &= \mathrm{St}(K,s) \cup \big(K \setminus \mathring{\mathrm{St}}(K,s)\big) \overset{4)}{\approx_{\mathrm{st}}} \\
&\approx_{\mathrm{st}} \underline{w} * \partial(\mathrm{St}(K,s)) \cup \big(K \setminus \mathring{\mathrm{St}}(K,s)\big) \overset{2\,(1)}{=} \\
&= \underline{w} * s * \mathrm{Lk}(\partial K,s) \cup \underbrace{\underline{w} * \partial s * \mathrm{Lk}(K,s) \cup \big(K \setminus \mathring{\mathrm{St}}(K,s)\big)}_{\cong_{\mathrm{si}} \sigma_s K = K_1 \approx_{\mathrm{st}} \underline{u} * \partial K_1} \approx_{\mathrm{st}} \\
&\approx_{\mathrm{st}} \underbrace{\underline{u} * \partial\Big(\underline{w} * \partial s * \mathrm{Lk}(K,s) \cup \big(K \setminus \mathring{\mathrm{St}}(K,s)\big)\Big)}_{= \underline{w} * \partial s * \mathrm{Lk}(\partial K,s) \cup (\partial K \setminus \mathring{\mathrm{St}}(\partial K,s))} \cup \underline{w} * s * \mathrm{Lk}(\partial K,s) = \\
&= \underline{u} * \Big(\underline{w} * \partial s * \mathrm{Lk}(\partial K,s) \cup \big(\partial K \setminus \mathring{\mathrm{St}}(\partial K,s)\big)\Big) \cup \underline{w} * s * \mathrm{Lk}(\partial K,s) = \\
&= \underline{u} * \underline{w} * \partial s * \mathrm{Lk}(\partial K,s) \cup \underline{u} * \big(\partial K \setminus \mathring{\mathrm{St}}(\partial K,s)\big) \cup \underline{w} * s * \mathrm{Lk}(\partial K,s).
\end{aligned}$$

In the second to last step, we used the fact that $s \in \partial K$ to calculate the boundary of K_1.
Now we perform a stellar subdivision on $\underline{u} * \mathrm{St}(\partial K,s) = \underline{u} * s * \mathrm{Lk}(\partial K,s)$ as follows:

$$\begin{aligned}
\sigma_{\underline{u}*s}\big(\underline{u} * \mathrm{St}(\partial K,s)\big) &\cong_{\mathrm{si}} \underline{w} * \partial(\underline{u} * s) * \underbrace{\mathrm{Lk}(\underline{u} * \mathrm{St}(\partial K,s),s)}_{= \mathrm{Lk}(\partial K,s)} \overset{2\,(1)}{=} \\
&= \underline{u} * \underline{w} * \partial s * \mathrm{Lk}(\partial K,s) \cup \underline{w} * s * \mathrm{Lk}(\partial K,s).
\end{aligned}$$

---

[19] We leave it to the reader to verify that all the stellar moves are, in fact, internal.

Since $\underline{u} \notin \partial K$, and thus $\underline{u} * s \notin \partial K$, we can continue our sequence of internal stellar equivalences from above by adding

$$\begin{aligned} K &\approx_{\text{st}} \underline{u} * \underline{w} * \partial s * \text{Lk}(\partial K, s) \cup \underline{u} * \left(\partial K \setminus \mathring{\text{St}}(\partial K, s)\right) \cup \underline{w} * s * \text{Lk}(\partial K, s) \approx_{\text{st}} \\ &\approx_{\text{st}} \underline{u} * \text{St}(\partial K, s) \cup \underline{u} * \left(\partial K \setminus \mathring{\text{St}}(\partial K, s)\right) = \\ &= \underline{u} * \partial K. \end{aligned}$$

This finally completes our induction argument. The sketch below hopefully does a good job at illustrating what happened, as much as our dimensionally limited view allows.

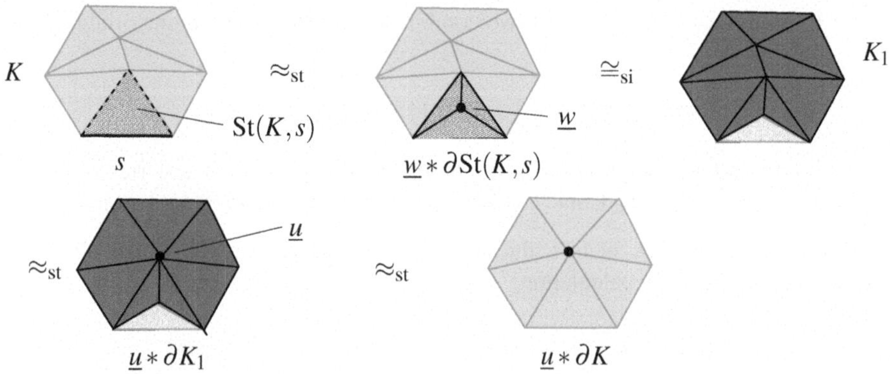

Fig. 13: Sketch of the induction step in the proof of Theorem 4

We want to continue with a technical lemma about "gluing on the cone of a nice piece of the boundary". It can be found in [Lic99, Lemma 3.7].

Lemma 14. *Let $n \in \mathbb{N}$. Let $K = (V, S)$ be a combinatorial n-manifold. If $L = (W, T)$ is an abstract subcomplex of ∂K that is a combinatorial $(n-1)$-ball, then there is a stellar equivalence*[20]

$$K \approx_{st} K \cup \text{Cone}(L) = K \cup \underline{u} * L.$$

The idea for the proof is to use Theorem 4 in order to find a combinatorial n-manifold that is stellar equivalent to K and that contains a "second copy of Cone(L)". Then, these two copies of Cone(L), which intersect in L, are a stellar subdivision of a single Cone(L) if we also show that L is a cone in itself.

Proof. Let $n \in \mathbb{N}$. Let $K = (V, S)$ be a combinatorial n-manifold, and let $L = (W, T)$ be an abstract subcomplex of ∂K with $L \approx_{\text{st}} D_{n-1}$. Then L is starrable by Theorem 4, so we can choose a starring sequence for L, i.e., a sequence of internal stellar

[20] Here, we assume that the cone point $\underline{u}$ does not lie in V.

moves (and simplicial isomorphisms, which we ignore as usual) from L to $\underline{v} * \partial L$. This leads us to the following claim:

Claim. We can extend this starring sequence to all of $K \cup \mathrm{Cone}(L)$, i.e., there is a sequence of stellar moves on $K \cup \mathrm{Cone}(L)$ that changes the abstract subcomplex L to $\underline{v} * \partial L$.

Proof. Each stellar move in the sequence is either an internal stellar subdivision or an internal stellar weld of L.
The stellar subdivisions in the sequence can easily be extended to all of K and to $\mathrm{Cone}(L)$ since L is a subcomplex of both K and $\mathrm{Cone}(L)$, and thus any simplex of L is also a simplex of these two abstract simplicial complexes.
If there is an internal stellar weld in the sequence, then without loss of generality we may assume it is the first of the stellar moves. So we can write

$$L = \sigma_s L_1 = K_1 \setminus \mathring{\mathrm{St}}(L_1, s) \cup \underline{s} * \partial s * \mathrm{Lk}(L_1, s),$$

for some combinatorial $(n-1)$-ball L_1 and some k-simplex s of L_1 with $s \notin \partial L_1$. Clearly, the weld extends to $\mathrm{Cone}(L)$, the difficult problem is proving it also extends to K. By construction, the star of $\underline{s}$ in K is given by

$$\mathrm{St}(K, \underline{s}) = \underline{s} * \mathrm{Lk}(K, \underline{s}).$$

Since $\underline{s}$ is a combinatorial 0-ball and $\mathrm{Lk}(K, \underline{s})$ is a combinatorial $(n-1)$-sphere we can conclude from Lemma 7 (3) that $\mathrm{St}(K, \underline{s})$ is a combinatorial n-ball, hence starrable by Theorem 4. That means there is a stellar equivalence

$$K \approx_{\mathrm{st}} K' := \left(K \setminus \mathring{\mathrm{St}}(K, \underline{s})\right) \cup \underline{s} * \partial\left(\mathrm{St}(K, \underline{s})\right) \cong_{\mathrm{si}} \left(K \setminus \mathring{\mathrm{St}}(K, \underline{s})\right) \cup \mathrm{Cone}\Big(\partial\left(\mathrm{St}(K, \underline{s})\right)\Big).$$

This stellar equivalence to the abstract simplicial complex K', which contains the cone on the boundary of $\mathrm{St}(K, \underline{s})$, allows us to extend the stellar weld to K', and thus also to the stellar equivalent K. This concludes the proof of the claim.

Since the stellar equivalence $L \approx_{\mathrm{st}} \underline{v} * \partial L$ extends to $K \cup \mathrm{Cone}(L)$, we may, without loss of generality, assume that $L = \underline{v} * \partial L$.
Just as before, the star $\mathrm{St}(K, \underline{v})$ is a combinatorial n-ball, which we star according to Theorem 4 to obtain the result that K contains

$$\underline{w} * \partial\left(\mathrm{St}(K, \underline{v})\right) = \underline{w} * \mathrm{Lk}(K, \underline{v})$$

as an abstract subcomplex. Since ∂L is an abstract subcomplex of $\mathrm{Lk}(K, \underline{v})$ by construction, this implies that $\underline{w} * L = \underline{w} * \underline{v} * \partial L$ is an abstract subcomplex of K.
Now, consider the following subcomplex of $K \cup \mathrm{Cone}(L)$:

$$\mathrm{Cone}(L) \cup \underline{w} * \underline{v} * \partial L = \mathrm{Cone}(\underline{v} * \partial L) \cup \underline{w} * \underline{v} * \partial L = \underline{u} * \underline{v} * \partial L \cup \underline{w} * \underline{v} * \partial L.$$

These two copies of $\mathrm{Cone}(\underline{v} * \partial L)$ are stellar equivalent to a single copy by welding at $\underline{v} * (\underline{u} \cup \underline{w})$. Hence,

$$K \cup \mathrm{Cone}(L) \approx_{\mathrm{st}} K \setminus \mathring{\mathrm{St}}(K,s) \cup \underline{w} * \underline{v} * \partial L \cup \mathrm{Cone}(L) \approx_{\mathrm{st}}$$
$$\approx_{\mathrm{st}} K \setminus \mathring{\mathrm{St}}(K,s) \cup \underline{w} * \underline{v} * \partial L \approx_{\mathrm{st}} K.$$

This concludes the proof. A sketch of the strategy for the proof when L consists of a single simplex can be found below.

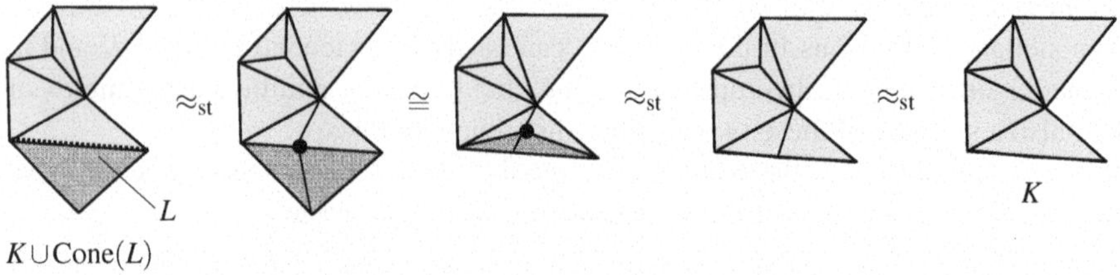

Fig. 14: Strategy for the proof of Lemma 14

The next theorem, which is due to Newman (cf. [New26]), has a very intuitive topological meaning, but is surprisingly tricky to prove in the abstract setting. It states that "removing" a combinatorial n-ball from a combinatorial n-sphere leaves behind another combinatorial n-ball. First, we have to define what we even mean by that. Most authors would give the more topologically inspired definition of "closure of $K \setminus L$", but we shall try and avoid topology at all costs in this chapter.

Definition 12. Let $K = (V,S)$ be a combinatorial n-sphere, and let $L = (W,T)$ be an abstract subcomplex that is a combinatorial n-ball. We define

$$K \setminus \mathring{L} := \left(\left\{ v \in V \,\middle|\, \begin{array}{l} \text{there is an } n\text{-simplex} \\ s \in S \setminus T \text{ with } v \in s \end{array} \right\}, \left\{ t \in S \,\middle|\, \begin{array}{l} \text{there is an } n\text{-simplex} \\ s \in S \setminus T \text{ with } t \subseteq s \end{array} \right\} \right).$$

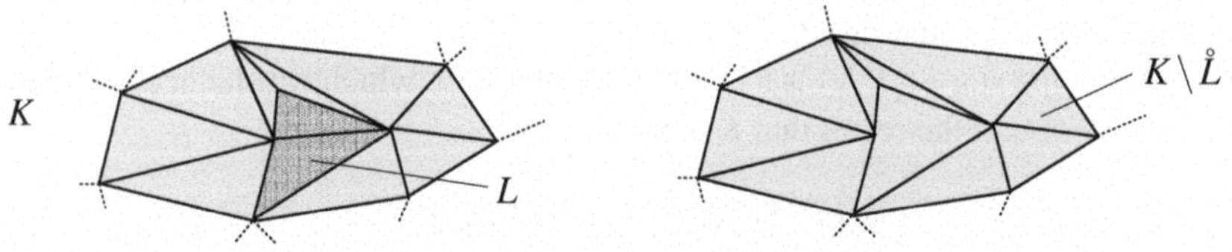

Fig. 15: Removing a 2-ball from a 2-sphere (indicated by the dotted lines)

This is a fancy way of saying to "keep only the vertices and simplices that actually contribute to something outside of L". Using the transitivity of the subset relation, it is an easy exercise to prove that $K \setminus \mathring{L}$ is in fact an abstract simplicial complex, and hence it is also a subcomplex of K by definition of subcomplexes.

Theorem 5. *Let $n \in \mathbb{N}_0$, let $K = (V,S)$ be a combinatorial n-sphere, and let $L = (W,T)$ be an abstract subcomplex of K that is a combinatorial n-ball. Then $K \setminus \mathring{L}$ is a combinatorial n-ball.*

Proof. Let $n \in \mathbb{N}_0$, let $K = (V,S)$ be a combinatorial n-sphere, and let $L = (W,T)$ be an abstract subcomplex of K that is a combinatorial n-ball. We start with an induction on n.

The case $n = 0$ is quite simple: As a combinatorial 0-sphere, K is simplicially isomorphic to S_0 since stellar moves at 0-simplices are just simplicial isomorphisms. On the other hand, L, being a combinatorial 0-ball, is simplicially isomorphic to D_0 for the same reason. So in the end, the statement for $n = 0$ boils down to the fact that if one takes two 0-simplices and removes one of them, they end up with a single 0-simplex. But that just means that $K \setminus \mathring{L}$ is a combinatorial 0-ball.

Now let $n \in \mathbb{N}$, and assume the claim is true for combinatorial spheres (and combinatorial balls) of dimension less than n.

In order to use some of the previous results, we first check that $K \setminus \mathring{L}$ is a combinatorial n-manifold[21]. Let v be a vertex of $K \setminus \mathring{L}$, i.e., there exists an n-simplex $s \in S \setminus T$ with $v \in s$. If for all n-simplices $t \in T$ we have $v \notin t$, then all n-simplices of K that v is a part of lie in $K \setminus \mathring{L}$, thus

$$\mathrm{Lk}(K \setminus \mathring{L}, v) = \mathrm{Lk}(K, v) \approx_{\mathrm{st}} S_{n-1},$$

because K is a combinatorial n-manifold. If, on the other hand, there does exist an n-simplex $t \in T$ with $v \in t$, then we have[22]

$$\mathrm{Lk}(K \setminus \mathring{L}, v) = \mathrm{Lk}(K, v) \setminus \mathrm{Lk}(\mathring{L}, v).$$

Here, $\mathrm{Lk}(K, v)$ is a combinatorial $(n-1)$-sphere as v is a vertex of the combinatorial n-sphere K, and $\mathrm{Lk}(L, v)$ is a combinatorial $(n-1)$-ball since v is a vertex in the boundary of the combinatorial n-ball L. By our induction hypothesis for the case $n-1$, the link $\mathrm{Lk}(K \setminus \mathring{L}, v)$ is a combinatorial $(n-1)$-ball. Therefore, $K \setminus \mathring{L}$ is a combinatorial n-manifold.

Now we shall start proving the actual claim. Since L is starrable by Theorem 4, we will, without loss of generality, assume for the remainder of the proof that $L = \mathrm{St}(K, v)$ for some vertex $v \in V$. This also has the convenient side effect that the definition of $K \setminus \mathring{L}$ coincides with the definition of $K \setminus \mathring{\mathrm{St}}(K, v)$ from page 8.[23]

We start another induction, on the number r of stellar moves in $K \approx_{\mathrm{st}} S_n$.

In the case $r = 0$, we have $K \cong_{\mathrm{si}} S_n$, so $L \cong_{\mathrm{si}} \mathrm{St}(S_n, v)$ is the star of a vertex of S_n, so $K \setminus \mathring{L}$ is the "n-simplex opposite to that vertex". But this is clearly simplicially

---

21 This is something we actually have to prove because we defined $K \setminus \mathring{L}$ in by "removing vertices and simplices" from a combinatorial sphere, which does not always result in a combinatorial manifold.

22 Note that the little circle on the right-hand side applies to all of $\mathrm{Lk}(L, v)$.

23 The proof that these two definitions coincide is a simple exercise in set theory that shall be left to the reader. Intuitively however, there is no doubt about it as they are both defined in terms of removing simplices that are in $\mathrm{St}(K, v)$.

isomorphic to D_n, so $K \setminus \mathring{L}$ is a combinatorial n-ball.

Now let $K \approx_{\text{st}} S_n$ in r moves, and assume that the claim is true for n-spheres that are stellar equivalent to S_n in less than r stellar moves. As so often, there are two cases for us to distinguish:

Case 1: If the first of the r moves is a stellar subdivision, denote its result by $K_1 = \sigma_s K$, for some k-simplex $s \in S$. We are interested in the relationship between $K \setminus \mathring{\text{St}}(K,v)$ and $K_1 \setminus \mathring{\text{St}}(K_1,v)$. Depending on v and s, we have to distinguish another two cases:

Case 1a: If $v \notin s$, then it follows from the definition of stellar moves that

$$K_1 \setminus \mathring{\text{St}}(K_1,v) = \sigma_s\big(K \setminus \mathring{\text{St}}(K,v)\big).$$

The expression on the left-hand side is a combinatorial n-ball by induction on r, thus $K \setminus \mathring{\text{St}}(K,v)$ is as well, because it is related to the left-hand side via a stellar move.

Case 1b: If $v \in s$, we set $t := s \setminus \{v\}$ and obtain

$$K_1 \setminus \mathring{\text{St}}(K_1,v) = K \setminus \mathring{\text{St}}(K,v) \cup \underline{s} * t * \text{Lk}(K,s).$$

Note that on the right-hand side, $K \setminus \mathring{\text{St}}(K,v)$ is a combinatorial n-manifold, and $t * \text{Lk}(K,s)$ is a combinatorial ball of dimension $(k-1)+(n-k-1)+1 = n-1$ that also is a subcomplex of $K \setminus \mathring{\text{St}}(K,v)$. Furthermore, the cone point $\underline{s}$ does not lie in K, so we can apply Lemma 14 to obtain

$$K_1 \setminus \mathring{\text{St}}(K_1,v) \cong_{\text{si}} K \setminus \mathring{\text{St}}(K,v) \cup \text{Cone}\big(t * \text{Lk}(K,s)\big) \overset{14}{\approx_{\text{st}}} K \setminus \mathring{\text{St}}(K,v)$$

Just as in Case 1a, we can therefore conclude that $K \setminus \mathring{\text{St}}(K,v)$ is a combinatorial n-ball.

Case 2: If the first of the r moves is a stellar weld, denote its result by $K_1 = \sigma_s^{-1} K$, i.e., $K = \sigma_s K_1$ for some k-simplex s of K_1. We are still interested in showing that $K \setminus \mathring{\text{St}}(K,v)$ is a combinatorial n-ball.

If $v \neq \underline{s}$, then the same arguments as in Case 1 apply, and $K \setminus \mathring{\text{St}}(K,v)$ is a combinatorial n-ball.

Now let us consider the case that $v = \underline{s}$. By construction, we have

$$K \setminus \mathring{\text{St}}(K,v) = K_1 \setminus \mathring{\text{St}}(K_1,s).$$

We pick a vertex $w \in s$, and again set $t := s \setminus \{w\}$. Since $K_1 \approx_{\text{st}} S_n$ in $r-1$ moves, we know from our induction hypothesis that $K_1 \setminus \mathring{\text{St}}(K_1,w)$ is a combinatorial n-ball. Thus, its boundary

$$\partial\big(K_1 \setminus \mathring{\text{St}}(K_1,w)\big) = \text{Lk}(K_1,w)$$

is a combinatorial $(n-1)$-sphere by Corollary 1. Since by construction $w \notin t * \text{Lk}(K_1,s)$, and $\{w\} * t * \text{Lk}(K_1,s) = \text{St}(K_1,s)$ is a subcomplex of K_1, it follows that $t * \mathrm{Lk}(K_1, s)$ is a subcomplex of $\partial\big(K_1 \setminus \mathring{\mathrm{St}}(K_1, w)\big)$. However, we also have

$$t * \mathrm{Lk}(K_1, s) \approx_{\mathrm{st}} D_{k-1} * S_{n-k-1} \overset{7\,(3)}{\approx}_{\mathrm{st}} D_{n-1},$$

and therefore we can again apply the induction hypothesis for the case $n-1$ to conclude that

$$M := \partial\big(K_1 \setminus \mathring{\mathrm{St}}(K_1, w)\big) \setminus \big(t * \mathring{\mathrm{Lk}}(K_1, s)\big)$$

is a combinatorial $(n-1)$-ball. Connecting all these deliberations, we can now finally calculate

$$\begin{aligned} K \setminus \mathring{\mathrm{St}}(K, v) &= K_1 \setminus \mathring{\mathrm{St}}(K_1, s) = \\ &= \big(K_1 \setminus \mathring{\mathrm{St}}(K_1, w)\big) \cup \big(\mathrm{St}(K_1, w) \setminus \mathring{\mathrm{St}}(K_1, s)\big) = \\ &= \big(K_1 \setminus \mathring{\mathrm{St}}(K_1, w)\big) \cup \{w\} * \Big(\mathrm{Lk}(K_1, w) \setminus \big(t * \mathring{\mathrm{Lk}}(K_1, s)\big)\Big) = \\ &= \big(K_1 \setminus \mathring{\mathrm{St}}(K_1, w)\big) \cup \{w\} * M \cong_{\mathrm{si}} \\ &\cong_{\mathrm{si}} \big(K_1 \setminus \mathring{\mathrm{St}}(K_1, w)\big) \cup \mathrm{Cone}(M) \overset{14}{\approx}_{\mathrm{st}} \\ &\approx_{\mathrm{st}} \big(K_1 \setminus \mathring{\mathrm{St}}(K_1, w)\big). \end{aligned}$$

As mentioned before, the very right-hand side is a combinatorial n-ball, and thus, so is $K \setminus \mathring{\mathrm{St}}(K, v)$.

In all cases, we showed that $K \setminus \mathring{\mathrm{St}}(K, v)$ is a combinatorial n-ball. This concludes both the induction on r and the one on n, and thus the proof.

The final theorem of this chapter, which is due to Alexander (cf. [Ale30]), is another gluing statement. It is, however, more of a corollary to the main statements of this chapter.

Theorem 6. *Let $K = (V, S)$ be a combinatorial n-manifold, and let $L = (W, T)$ be a combinatorial n-ball. We assume that the intersection $K \cap L$ satisfies the following two properties:*

(1) *We have $K \cap L = \partial K \cap \partial L$.*
(2) *The intersection $K \cap L$ is a combinatorial $(n-1)$-ball.*

Then $K \cup L \approx_{st} K$.

Proof. By Theorem 4, the combinatorial n-ball L is starrable. Therefore, without loss of generality, we can assume that $L = \underline{v} * \partial L$. Furthermore, ∂L is a combinatorial $(n-1)$-sphere by Corollary 1, and $K \cap L$ is a combinatorial $(n-1)$-ball that is an abstract subcomplex of ∂L by property (1). It now follows from Theorem 5 that

$$M := \partial L \setminus (K \mathring{\cap} L)$$

is a combinatorial $(n-1)$-ball, which is starrable by Theorem 4, i.e., $M \approx_{\mathrm{st}} \underline{w} * \partial M$. By Lemma 10, we have $\partial M = \partial(K \cap L)$, hence

$$M \approx_{\text{st}} \underline{w} * \partial(K \cap L).$$

This starring can be extended onto $\text{Cone}(M) \cong_{\text{si}} \underline{v} * M$, yielding

$$\underline{v} * M \approx_{\text{st}} \underline{v} * \underline{w} * \partial M = \underline{w} * \underline{v} * \partial(K \cap L).$$

Therefore, one obtains

$$\begin{aligned} K \cup L &= K \cup \underline{v} * \partial L = \\ &= K \cup \underline{v} * \big((K \cap L) \cup M\big) \stackrel{4}{=} \\ &= K \cup \underline{v} * (K \cap L) \cup \underline{v} * M. \end{aligned}$$

Since both $K \cap L$ and M are combinatorial $(n-1)$-balls that are abstract subcomplexes of ∂K, we can apply Lemma 14 twice in order to find a stellar equivalence from that final abstract simplicial complex to K, finishing the proof.

We end this chapter with the following lemma, which we will actually use in the proof of Pachner's Theorem 7. It can be found in [Lic99, Lemma 4.6].

Lemma 15. *Let $k \in \mathbb{N}$, and let $K = (V,S)$ be an abstract simplicial complex. If $K * S_{k-1}$ is a combinatorial n-sphere or a combinatorial n-ball, then K is a combinatorial $(n-k)$-sphere or a combinatorial $(n-k)$-ball.*

Proof. Let $k \in \mathbb{N}$, and let K be an abstract simplicial complex such that $K * S_{k-1}$ is a combinatorial n-sphere or a combinatorial n-ball. Consider the abstract simplicial complex

$$\text{Cone}(K * S_{k-1}).$$

It follows from Lemma 7 (1) or (3) that this cone is a combinatorial $(n+1)$-ball. It follows from the associativity and commutativity of the join operation that $K * D_k \cong_{\text{si}} K * \text{Cone}(S_{k-1})$ is also a combinatorial $(n+1)$-ball. Using Proposition 1 (1), we conclude that

$$K = \text{Lk}(K * D_k, D_k)$$

is a combinatorial sphere or a combinatorial ball of dimension

$$\dim(K * D_k) - \dim D_k - 1 = (n+1) - k - 1 = n - k. \square$$

6 Bistellar Moves and Elementary Shellings

Having now spent a lot of time on stellar moves, we want to turn to a second type of move on combinatorial manifolds. Despite initially sounding more complicated, there will be an extremely nice finiteness property to them, which is one of the reasons why Pachner's theorem is so important in the first place. This chapter follows the approach taken in [Lic99, Chapter 5].

Definition 13. Let $K = (V,S)$ be an abstract simplicial complex, let $s \in S$ be a simplex, and let $m \in \mathbb{N}_0 \cup \{-1\}$. We denote the link of s in K by $(U,R) := \mathrm{Lk}(K,s)$. We call (U,R) a *non-trivial m-sphere in K* if $(U,R) \cong_{\mathrm{si}} S_m$ and U itself is not a simplex of K.
Note that this definition makes sense for $m = -1$ since we had defined $S_{-1} := (\varnothing, \varnothing)$, and the empty set is not a simplex by definition of abstract simplicial complexes.
In this case, we also define

$$B(K,s) := \begin{cases} (U, \mathscr{P}(U) \setminus \{\varnothing\}), & \text{if } m \in \mathbb{N}_0, \\ (\{\underline{s}\}, \{\{\underline{s}\}\}), & \text{if } m = -1. \end{cases}$$

This abstract simplicial complex $B(K,s)$ is simplicially isomorphic to D_{m+1} by construction, and we call it the *dual simplex of s*.[24]

Figure 16 hopefully does a decent job at illustrating the situation. Essentially, we want the link of s to be the boundary of a simplex that is not in K.

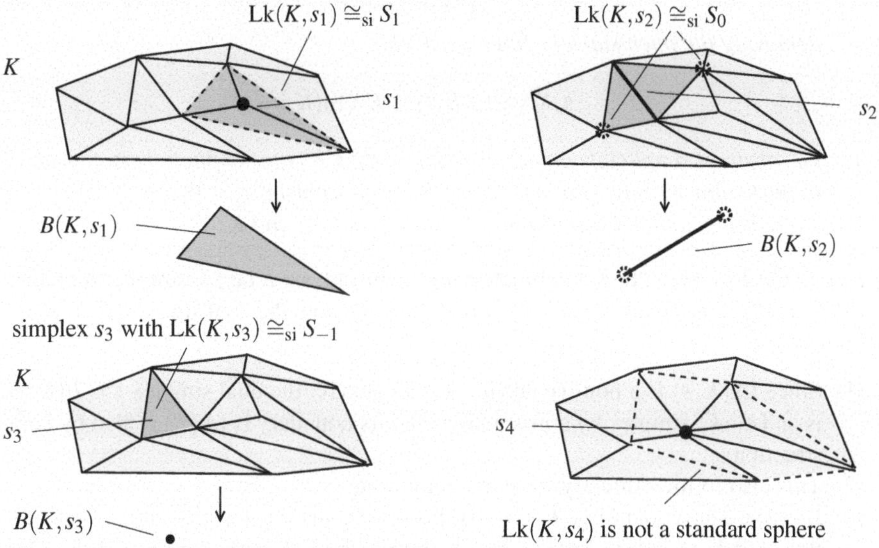

Fig. 16: From left to right, examples for a non-trivial 1-sphere, a non-trivial 0-sphere, a non-trivial (-1)-sphere, and a link that is not a non-trivial sphere

Definition 14. Let $K = (V,S)$ be a closed combinatorial n-manifold, $s \in S$ be a k-simplex such that $\mathrm{Lk}(K,s)$ is a non-trivial $(n-k-1)$-sphere in K. We call

$$\tau_s K := (K \setminus \mathring{\mathrm{St}}(K,s)) \cup \partial s * B(K,s)$$

[24] We will shortly see a justification for the name "dual simplex", more specifically in Lemma 16.

(the result of) a *bistellar move along s*.

Essentially, a bistellar move replaces the star $\mathrm{St}(K,s) = s * \mathrm{Lk}(K,s)$ by $\partial s * B(K,s)$. We refer to Figure 3 in the introduction for an illustration of the definition.

The following lemma summarises some of the main properties of bistellar moves. Another reason to include it is that proving it allows the reader to grow used to this new concept.

Lemma 16. *Let $K = (V,S)$ be a combinatorial n-manifold, $s \in S$ be a k-simplex such that $\mathrm{Lk}(K,s)$ is a non-trivial $(n-k-1)$-sphere in K. If we denote the dual simplex $B(K,s)$ by t, then we have the following useful properties:*

(1) *The dual simplex t is an $(n-k)$-simplex of $\tau_s K$, and its boundary is given by $\partial t = \mathrm{Lk}(K,s)$.*
(2) *The link of t in $\tau_s K$ is given by $\mathrm{Lk}(\tau_s K, t) = \partial s$.*
(3) *We have*

$$B(\tau_s K, t) = \begin{cases} s, & \text{if } k > 0, \\ (\{\underline{t}\}, \{\{\underline{t}\}\}), & \text{if } k = 0. \end{cases}$$

(4) *Removing the open star of t from $\tau_s K$ yields*

$$\tau_s K \setminus \mathring{\mathrm{St}}(\tau_s K, t) = K \setminus \mathring{\mathrm{St}}(K,s).$$

(5) *The abstract simplicial complexes $\sigma_s K$ and $\sigma_t(\tau_s K)$ are simplicially isomorphic. In particular we have, up to a simplicial isomorphism, $\tau_s = \sigma_t^{-1}\sigma_s$.*
(6) *There is a simplicial isomorphism between $\tau_t(\tau_s K)$ and K.*

Proof. Let $K = (V,S)$ be a combinatorial n-manifold, $s \in S$ be a k-simplex such that $\mathrm{Lk}(K,s)$ is a non-trivial $(n-k-1)$-sphere in K. Denote the dual simplex $B(K,s)$ by t.

(1) Since $\mathrm{Lk}(K,s)$ is a non-trivial $(n-k-1)$-sphere, the dual simplex $t = B(K,s)$ is an $(n-k)$-simplex by construction, and its boundary is $\mathrm{Lk}(K,s)$ also by construction.
(2) This also follows directly from the definition.
(3) By (2) we have that $\mathrm{Lk}(\tau_s K, t) = \partial s$. For $k = 0$, this is a non-trivial (-1)-sphere, thus $B(\tau_s K, t) = (\{\underline{t}\}, \{\{\underline{t}\}\})$. If $k > 0$, then the link is the non-trivial $(k-1)$-sphere ∂s, so in this case, the dual simplex of t in $\tau_s K$ is given by $B(\tau_s K, t) = s$.
(4) Using (2), we calculate

$$\mathrm{St}(\tau_s K, t) = \mathrm{Lk}(\tau_s K, t) * t = \partial s * B(K,s),$$

therefore

$$\begin{aligned} \tau_s K \setminus \mathring{\mathrm{St}}(\tau_s K, t) &= ((K \setminus \mathring{\mathrm{St}}(K,s)) \cup \partial s * B(K,s)) \setminus \mathring{\mathrm{St}}(\tau_s K, t) = \\ &= K \setminus \mathring{\mathrm{St}}(K,s). \end{aligned}$$

(5) Using (1), (2), and (4), one calculates

$$\begin{aligned}\sigma_t(\tau_s K) &= \tau_s K \setminus \mathring{\mathrm{St}}(\tau_s K, t) \cup \underline{t} * \partial t * \mathrm{Lk}(\tau_s K, t) = \\ &= K \setminus \mathring{\mathrm{St}}(K, s) \cup \underline{t} * \mathrm{Lk}(K, s) * \partial s \cong_{\mathrm{si}} \\ &\cong_{\mathrm{si}} K \setminus \mathring{\mathrm{St}}(K, s) \cup \underline{s} * \partial s * \mathrm{Lk}(K, s) = \\ &= \sigma_s K\end{aligned}$$

(6) This follows directly from (3) and (5) if we believe that a stellar subdivision, followed by a stellar weld on "the same" simplex, amount to a simplicial isomorphism.

All six claims have thus been proved.

Lemma 16 (6) allows us to give the following definition. To be more precise, it ensures that bistellar equivalence is, in fact, an equivalence relation.

Definition 15. Let $K = (V, S)$ and $L = (W, T)$ be combinatorial manifolds. We call K and L *bistellar equivalent* if they are connected by a finite sequence of bistellar moves and simplicial isomorphisms. If K and L are bistellar equivalent, we write $K \approx_{\mathrm{bst}} L$.

Now we have all the tools to formulate Pachner's theorem, which we already stated in the introduction, but without giving precise definitions.

Theorem 7. (Pachner's Theorem) *Let $K = (V, S)$ and $K' = (V', S')$ be closed combinatorial n-manifolds. Then*

$$K \approx_{\mathrm{bst}} K' \iff K \approx_{\mathrm{st}} K'.$$

This theorem is deceptively easy to state. To put it briefly, it says that if two combinatorial manifolds are stellar equivalent, then one can find a sequence of bistellar moves between them (and vice versa). At this point, there is one more tool we need to add to our toolbox before we can move onto the path toward proving Pachner's theorem.

Definition 16. Let $K = (V, S)$ be a combinatorial n-manifold, and let $s, t \in S$ be simplices such that the following three conditions are satisfied:

(1) the join $s * t$ is an n-simplex of K,
(2) we have $s \cap \partial K = \partial s$, and
(3) we have $t * \partial s \subseteq \partial K$ is a subcomplex.

Let $k := \dim t$. The *elementary k-shelling of K along t* is the combinatorial manifold[25]

$$K' := \mathrm{sh}_t K := K \setminus \mathring{\mathrm{St}}(K, s * t).$$

[25] Note that the star of the n-simplex $s * t$ is just the abstract simplicial complex consisting of all the faces of $s * t$. Therefore, a shelling can be imagined as "removing" the simplex $s * t$ from K.

Remark 6.

(1) Consider the situation from the above definition. As the notation "sh_t" suggests, the elementary k-shelling is uniquely determined by the k-simplex t. This follows from properties (2) and (3).
(2) In the sketches below, the two different "types" of elementary shellings on a combinatorial 2-manifold are illustrated. Note that for dimensional reasons[26], there cannot be an elementary n-shelling on an elementary n-manifold.

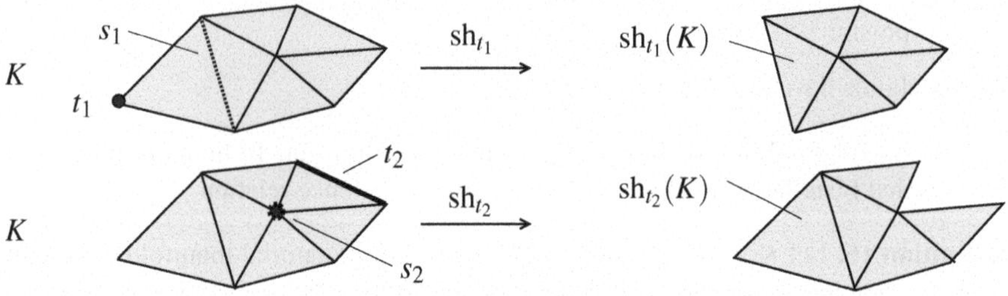

Fig. 17: An elementary 0-shelling (top) and an elementary 1-shelling (bottom) in a 2-manifold

We make the following observation about the boundary of a combinatorial manifold. We refer to the sketch above for some examples.

Lemma 17. *Let $K = (V, S)$ be a combinatorial manifold, and let $s, t \in S$ be as above. Then* $\mathrm{Lk}(\partial K, t) = \partial s$, *and*

$$\partial(\mathrm{sh}_t K) = \tau_t(\partial K).$$

Proof. From property (3) and Lemma 10, we can conclude that $\mathrm{Lk}(\partial K, t) = \partial s$. Now, property (2) ensures that $s \notin \partial K$, and together with property (1), this implies that the link of t in ∂K is a non-trivial $(n-k-2)$-sphere[27]. Therefore, the right-hand side is well-defined, and we have $B(\partial K, t) = s$.
Using Lemma 16, we calculate

$$\begin{aligned}\partial(\mathrm{sh}_t K) &= \partial K \setminus \mathring{\mathrm{St}}(\partial K, t) \cup \underbrace{\partial(s*t) \setminus (t * \mathring{\partial} s)}_{= s*\partial t} = \\ &= \partial K \setminus \mathring{\mathrm{St}}(\partial K, t) \cup \partial t * B(\partial K, t) = \\ &= \tau_t(\partial K).\end{aligned}$$

This proves the claim.

As the sketches above show, elementary shellings "make the abstract simplicial complex smaller." This intuitive concept shall be expanded upon in the following definition.

[26] More precisely, it follows from $n \geq \dim(s*t) = \dim s + \dim t + 1 > \dim t$.
[27] In detail, this follows from $n = \dim s + k + 1$ and $\dim \partial s = \dim s - 1$.

Definition 17. Let $K = (V, S)$ be a combinatorial n-manifold.

(1) If $L = (W, T)$ is another combinatorial n-manifold, and L can be obtained from K by means of a finite sequence of elementary shellings and simplicial isomorphisms, we write $K \xrightarrow{\text{sh}} L$.
(2) If K is a combinatorial n-ball, then we call K *shellable* if $K \xrightarrow{\text{sh}} D_n$.
(3) If K is a combinatorial n-sphere, then we call K *shellable* if there exists an n-simplex $s \in S$ such that $K \setminus \mathring{\text{St}}(K, s)$ is a shellable combinatorial n-ball[28].

The sketch below illustrates the concept of shellability. In each step, the dark grey simplex is the one that is "being removed" next. It is also evident that there exist many different sequences of elementary shellings that transform K into something simplicially isomorphic to D_2.

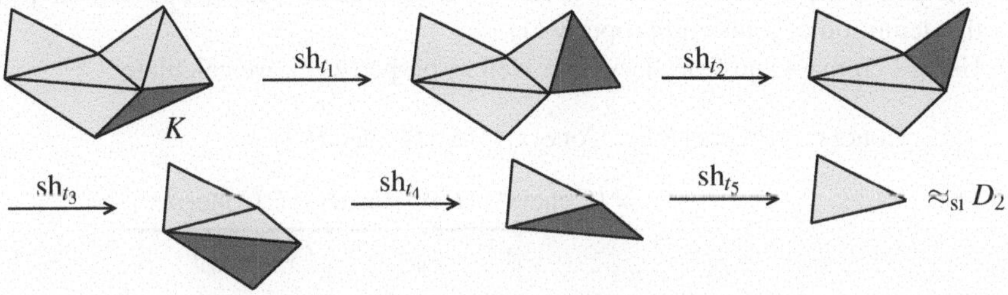

Fig. 18: A combinatorial 2-ball K that is shellable in 5 moves

Remark 7. While the picture makes it seem like every combinatorial n-ball is shellable, there are, in fact, examples of combinatorial n-balls that are not shellable. For instance, Rudin (cf. [Rud58]) found an unshellable combinatorial 3-ball with 41 simplices of dimension 3. Furthermore, for any $n \geq 3$, Lickorish (cf. [Lic91]) gives a combinatorial n-sphere that is not shellable. However, the intuition we have that all combinatorial 2-balls and -spheres are shellable is, in fact, correct. A proof for this statement can be found in Bing (cf. [Bin64]).

For us, the more interesting question is how we can use the concept of shellability in order to prove Pachner's theorem. We start by proving a lemma that deals with cones of shellable combinatorial balls and spheres. It is essentially [Lic99, Lemma 5.4], but as we have many times before, we fill in the details in order to make it more rigorous.

Lemma 18. *Let $K = (V, S)$ be a shellable combinatorial n-ball or a shellable combinatorial n-sphere. Then* Cone(K) *is shellable*[29].

[28] Note that it follows from Theorem 5 that $K \setminus \mathring{\text{St}}(K, s)$ is in fact a combinatorial n-ball.
[29] Remember that by Lemma 7 (1) or (3), the cone of K is a combinatorial $(n+1)$-ball.

Proof. The idea for the proof is to just "lift" the sequence of elementary shellings (and simplicial isomorphisms) that we have for K to $\mathrm{Cone}(K)$. Of course, there are some technical conditions we will have to check, but this is one of the cases where the intuitive approach actually works.

Let us first consider the case that $K = (V,S)$ is a shellable combinatorial n-ball. For each elementary shelling sh_t in a sequence $K \xrightarrow{\mathrm{sh}} D_n$, we want to lift sh_t to $\mathrm{Cone}(K)$. So let $s,t \in S$ satisfy the three conditions we had required[30]:

(1) The join $s * t$ is an n-simplex of K,
(2) we have $s \cap \partial K = \partial s$, and
(3) we have $t * \partial s \subseteq \partial K$.

Our goal is to show that these conditions still hold if we replace s by $\mathrm{Cone}(s)$, K by $\mathrm{Cone}(K)$, and n by $n+1$. We do so in great detail.

(1') The join $\mathrm{Cone}(s) * t = D_0 * s * t$ is an $(n+1)$-simplex of $\mathrm{Cone}(K) = D_0 * K$ by the definition of joins and property (1).
(2') Using Lemma 4 and Corollary 3, as well as property (2), we calculate

$$\begin{aligned}\mathrm{Cone}(s) \cap \partial\mathrm{Cone}(K) &= \mathrm{Cone}(s) \cap (K \cup \mathrm{Cone}(\partial K)) = \\ &= \underbrace{(\mathrm{Cone}(s) \cap K)}_{=s} \cup \underbrace{(\mathrm{Cone}(s) \cap \mathrm{Cone}(\partial K))}_{=\mathrm{Cone}(s \cap \partial K)} \overset{(2)}{=} \\ &= s \cup \mathrm{Cone}(\partial s) = \partial\mathrm{Cone}(s).\end{aligned}$$

(3') Using Lemma 4 and Corollary 3, as well as properties (1) and (3) and the commutativity of the join, we calculate

$$\begin{aligned}t * \partial\mathrm{Cone}(s) &= t * (s \cup \mathrm{Cone}(\partial s)) = \\ &= t * s \cup \underbrace{t * \mathrm{Cone}(\partial s)}_{=\mathrm{Cone}(t * \partial s)} \overset{(1),(3)}{\subseteq} \\ &\subseteq K \cup \mathrm{Cone}(\partial K) = \partial\mathrm{Cone}(K).\end{aligned}$$

Furthermore, we have

$$\begin{aligned}\mathrm{sh}_t\mathrm{Cone}(K) &= \mathrm{Cone}(K) \setminus \mathring{\mathrm{St}}(\mathrm{Cone}(K), \mathrm{Cone}(s) * t) = \\ &= \mathrm{Cone}(K \setminus \mathring{\mathrm{St}}(K, s * t)) = \mathrm{Cone}(\mathrm{sh}_t K).\end{aligned}$$

Therefore, we can "lift" an entire sequence $K \xrightarrow{\mathrm{sh}} D_n$ to one of the form

$$\mathrm{Cone}(K) \xrightarrow{\mathrm{sh}} \mathrm{Cone}(D_n) \cong_{\mathrm{si}} D_{n+1},$$

and $\mathrm{Cone}(K)$ is a shellable combinatorial $(n+1)$-ball.

Now consider the case that $K = (V,S)$ is a shellable combinatorial n-sphere. Let

[30] Here, we slightly abuse notation in order to write K instead of something like $\mathrm{sh}_{t_{k-1}} \dots \mathrm{sh}_{t_1} K$.

$s \in S$ be an n-simplex such that $K \setminus \mathring{\mathrm{St}}(K,s)$ is a shellable combinatorial n-ball. Note that $\mathrm{Cone}(s)$ is an $(n+1)$-simplex of $\mathrm{Cone}(K)$, so we can perform the following elementary n-shelling:

$$\mathrm{sh}_s\mathrm{Cone}(K) = \mathrm{Cone}(K) \setminus \mathring{\mathrm{St}}(\mathrm{Cone}(K), \mathrm{Cone}(s)) = \mathrm{Cone}\big(K \setminus \mathring{\mathrm{St}}(K,s)\big).$$

Since $K \setminus \mathring{\mathrm{St}}(K,s)$ is a shellable combinatorial n-ball, $\mathrm{Cone}\big(K \setminus \mathring{\mathrm{St}}(K,s)\big)$ is a shellable combinatorial $(n+1)$-ball by the first case. But since there is an elementary shelling from $\mathrm{Cone}(K)$ onto this shellable combinatorial $(n+1)$-ball, $\mathrm{Cone}(K)$ is a shellable combinatorial $(n+1)$-ball as well.

Next, we want to look into the join of shellable combinatorial balls or spheres with the simplicial complex S_m. It can be found in [Lic99, Lemma 5.5].

Lemma 19. *Let $K = (V,S)$ be a shellable combinatorial n-ball or a shellable combinatorial n-sphere, and let $m \in \mathbb{N}_0 \cup \{-1\}$. Then $K * S_m$ is shellable*[31].

Proof. We start with the case that $K = (V,S)$ is a shellable combinatorial n-ball. We perform induction on $m \in \mathbb{N}_0 \cup \{-1\}$.
The case $m = -1$ is trivial since $S_{-1} = (\varnothing, \varnothing)$, and joining with the empty complex preserves any abstract simplicial complex.
Now let $m \in \mathbb{N}_0$, and suppose the claim is true for balls of dimension less than m. Let v be a vertex of S_m, denote by $s := \{0,\ldots,m+1\} \setminus \{v\}$ the "opposite face". Since K is shellable, we can pick a sequence $\mathrm{sh}_{t_1},\ldots,\mathrm{sh}_{t_k}$ of elementary shellings for $K \overset{\mathrm{sh}}{\longrightarrow} D_n$. Then, the join simplices $t_1 * s,\ldots,t_k * s$ are part of the boundary of the iterated elementary shellings of $K * S_m$, and we can now perform the sequence $\mathrm{sh}_{t_1 * s},\ldots,\mathrm{sh}_{t_k * s},\mathrm{sh}_s$ on $K * S_m$ as follows:

$$\mathrm{sh}_s\mathrm{sh}_{t_k * s}\cdots\mathrm{sh}_{t_1 * s}(K * S_m) \cong_{\mathrm{si}} \mathrm{sh}_s(\underbrace{\{v\}}_{\cong_{\mathrm{si}}\{0\}} * K * \underbrace{\partial s}_{\cong_{\mathrm{si}} S_{m-1}} \cup s * D_n) \cong_{\mathrm{si}} \mathrm{Cone}(K * S_{m-1})$$

By our induction hypothesis, $K * S_{m-1}$ is a shellable combinatorial $(n+m)$-ball. Therefore, $\mathrm{Cone}(K * S_{m-1})$ is a shellable combinatorial $(n+m+1)$-ball by Lemma 18. Since there is a sequence of elementary shellings from $K * S_m$ to $\mathrm{Cone}(K * S_{m-1})$, we obtain that $K * S_m$ is a shellable combinatorial $(n+m+1)$-ball. The claim now follows from our induction argument.
The sketch in Figure **??** illustrates what has happened, even if it is a little unsatisfying because it only covers the case $m = 0$. It shall be mentioned that one could draw the case $m = 1$, but then K would have to be a 1-manifold, which is also not a very enlightening sketch.

Now we shall turn to the case that $K = (V,S)$ is a shellable combinatorial n-sphere, and we let $r \in S$ be an n-simplex such that $K \setminus \mathring{\mathrm{St}}(K,r)$ is a shellable combinatorial n-ball. As before, we let v be a vertex of S_m, and we denote by $s := \{0,\ldots,m+1\} \setminus \{v\}$ its "opposite face". Since $\partial s \cong_{\mathrm{si}} S_{m-1}$, we can remove

[31] Remember that by Lemma 7 (2) or (3), the join $K * S_m$ is a combinatorial $(n+m+1)$-ball or a combinatorial $(n+m+1)$-sphere.

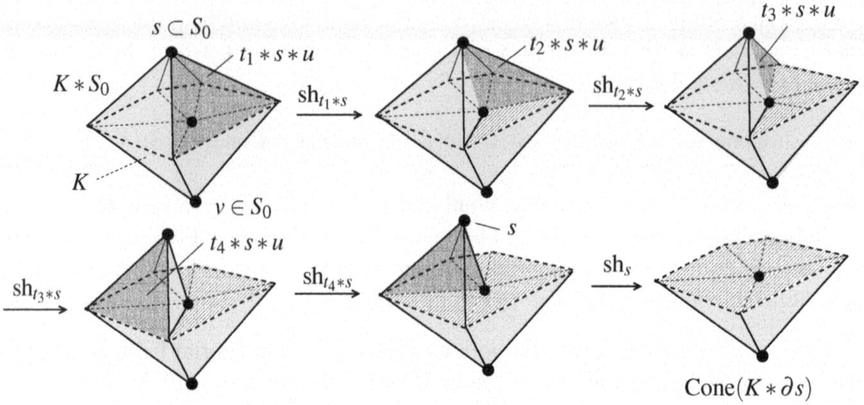

Fig. 19: Proof of Lemma 19 in the case of K being a 2-ball and $m = 0$

an $(m-1)$-simplex from ∂s, and then perform a total of $m-2$ elementary shellings $\mathrm{sh}_{s_1}, \ldots, \mathrm{sh}_{s_{m-2}}$ in order to obtain[32]

$$\mathrm{sh}_{s_{m-2}} \cdots \mathrm{sh}_{s_1} \left(\partial s \setminus \mathring{\mathrm{St}}(\partial s, s_0)\right) \cong_{\mathrm{si}} D_{m-1}.$$

Similarly to how we did in the previous case, we now perform the following sequence of elementary shellings:

$$\mathrm{sh}_r \mathrm{sh}_{r*s_{m-2}} \cdots \mathrm{sh}_{r*s_1} \left(K * S_m \setminus \mathring{\mathrm{St}}(K * S_m, r * s)\right) \cong_{\mathrm{si}} \left(K \setminus \mathring{\mathrm{St}}(K, r)\right) * S_m.$$

Since $K \setminus \mathring{\mathrm{St}}(K, r)$ is a shellable combinatorial n-ball, it follows from the first case that $\left(K \setminus \mathring{\mathrm{St}}(K, r)\right) * S_m$ is a shellable combinatorial $n+m+1$-ball. Due to the existence of the sequence of shellings

$$\left(K * S_m \setminus \mathring{\mathrm{St}}(K * S_m, r * s)\right) \xrightarrow{\mathrm{sh}} \left(K \setminus \mathring{\mathrm{St}}(K, r)\right) * S_m,$$

this implies that $\left(K * S_m \setminus \mathring{\mathrm{St}}(K * S_m, r * s)\right)$ also is a shellable combinatorial $n+m+1$-ball. But by definition, this means that $K * S_m$ is a shellable combinatorial $n+m+1$-sphere, which finishes the proof.

This next Lemma, which is just [Lic99, Lemma 5.7], has a certain similarity to our old friend Theorem 4, albeit with an additional condition and a stronger result.

Lemma 20. *If $K = (V, S)$ is a shellable combinatorial n-ball, then* $\mathrm{Cone}(\partial K) \approx_{\mathrm{bst}} K$.

Proof. Let $K = (V, S)$ be a shellable combinatorial n-ball. We perform induction on the number $r \in \mathbb{N}$ of n-simplices in K.

[32] This might seem like came completely out of the blue, but remember that an abstract simplicial complex simplicially isomorphic to S_{m-1} is just "S_{m-1} with its vertices renamed".

The case $r = 1$ is rather straightforward: If we denote this single n-simplex by s, then we make the following claim:

Claim. In this case, we have $(V,S) = (s, \mathscr{P}(s) \setminus \{\varnothing\})$.

Proof. The inclusion "$\supseteq$" follows from $s \in S$ and property (2) of abstract simplicial complexes. For the other inclusion, assume there was a vertex $v \in V \setminus s$. Then it follows from the definition of a combinatorial manifold[33] that $\mathrm{St}(K,v) = \{v\} * \mathrm{Lk}(K,v)$ is a combinatorial n-ball, and thus contains an n-simplex. But the only n-simplex of K is s, so since by assumption $v \notin s$, we have that s must be an n-simplex in the combinatorial $(n-1)$-ball $\mathrm{Lk}(K,v)$, which is clearly a contradiction.

Since K is just given by s and its faces, we have that $\mathrm{Lk}(K,s) = (\varnothing,\varnothing)$ is a non-trivial (-1)-sphere in K, so we can perform

$$\tau_s K = \underbrace{K \setminus \mathring{\mathrm{St}}(K,s)}_{=(\varnothing,\varnothing)} \cup \partial s * \underbrace{B(K,s)}_{=\underline{s}} = \underline{s} * \partial s \cong_{\mathrm{si}} \mathrm{Cone}(\partial K).$$

Now assume that $r \geq 2$, and that the claim is true for shellable combinatorial n-balls of dimension less than r. If we denote by $K_1 := \mathrm{sh}_t K$ the result of the first elementary shelling in a sequence $K \xrightarrow{\mathrm{sh}} D_n$, where $s * t$ is an n-simplex of K that fulfills $s \cap \partial K = \partial s$ and $t * \partial s \subseteq \partial K$. We also denote by k the dimension of t. By construction, K_1 has $r-1$ simplices of dimension n, so it follows from the induction hypothesis that $K_1 \approx_{\mathrm{bst}} \mathrm{Cone}(\partial K_1)$. We use this to rewrite

$$K = K_1 \cup s * t \approx_{\mathrm{bst}} \mathrm{Cone}(\partial K_1) \cup s * t.$$

Now, we have the following claim:

Claim. The link of s in this new combinatorial manifold is given by

$$\mathrm{Lk}\left(\mathrm{Cone}(\partial K_1) \cup s * t, s\right) = \partial \mathrm{Cone}(t),$$

and it is a non-trivial k-sphere in $\mathrm{Cone}(\partial K_1) \cup s * t$.

Proof. The second claim follows immediately from the first one since $\mathrm{Cone}(t) \cong_{\mathrm{si}} D_{k+1}$. For the first claim, we perform the following calculation:

$$\partial \mathrm{Cone}(t) = t \cup \mathrm{Cone}(\partial t) = t \cup \mathrm{Cone}\big(\mathrm{Lk}(\partial K_1, s)\big) = \mathrm{Lk}\left(\mathrm{Cone}(\partial K_1) \cup s * t, s\right)$$

The reader is invited to draw some pictures in order to verify that chain of equations.

Thanks to the claim, we can now perform[34]

$$\begin{aligned}
\tau_s\big(\mathrm{Cone}(\partial K_1) \cup s * t\big) &= \big(\mathrm{Cone}(\partial K_1) \cup s * t\big) \setminus \mathring{\mathrm{St}}\big(\mathrm{Cone}(\partial K_1) \cup s * t, s\big) \cup \partial s * \mathrm{Cone}(t) = \\
&= \big(\mathrm{Cone}(\partial K_1)\big) \setminus \mathring{\mathrm{St}}\big(\mathrm{Cone}(\partial K_1), \mathrm{Cone}(s)\big) \cup \mathrm{Cone}(t * \partial s) = \\
&= \mathrm{Cone}\big(\partial K_1 \setminus \mathring{\mathrm{St}}(\partial K_1, s) \cup t * \partial s\big).
\end{aligned}$$

[33] Recall that combinatorial balls are combinatorial manifolds by Lemma 6.

[34] The dual simplex of s in $\mathrm{Cone}(\partial K_1) \cup s * t$ is given by $\mathrm{Cone}(t)$.

In the last step, we used commutativity of the join and the fact that $s*t$ is entirely part of $\mathrm{St}\big(\mathrm{Cone}(\partial K_1)\cup s*t, s\big)$. Now, we use Lemma 17 to calculate the boundary of K_1 as follows:

$$\partial K_1 = \tau_t \partial K = \partial K \setminus \mathring{\mathrm{St}}(\partial K, t*\partial s) \cup s*\partial t.$$

This allows us to simplify the above expression.

$$\begin{aligned} K &\approx_{\mathrm{bst}} \mathrm{Cone}(\partial K_1)\cup s*t \approx_{\mathrm{bst}} \\ &\approx_{\mathrm{bst}} \mathrm{Cone}\big(\partial K_1 \setminus \mathring{\mathrm{St}}(\partial K_1, s)\big) \cup \mathrm{Cone}(t*\partial s) = \\ &= \mathrm{Cone}\Big(\big(\partial K \setminus \mathring{\mathrm{St}}(\partial K, t*\partial s)\cup s*\partial t\big)\setminus \mathring{\mathrm{St}}(\partial K, s*\partial t)\Big) \cup \mathrm{Cone}(t*\partial s) = \\ &= \mathrm{Cone}\big(\partial K \setminus \mathring{\mathrm{St}}(\partial K, t*\partial s)\big)\cup \mathrm{Cone}(t*\partial s) = \\ &= \mathrm{Cone}\big(\partial K \setminus \mathring{\mathrm{St}}(\partial K, t*\partial s)\cup (t*\partial s)\big) = \\ &= \mathrm{Cone}(\partial K). \end{aligned}$$

This concludes the induction step and the proof.

The induction step actually has a somewhat nice picture to back it up, see below. We remove a simplex, turn the remaining complex into a cone on its boundary, and are just one bistellar move away from a cone on the boundary of the original complex.

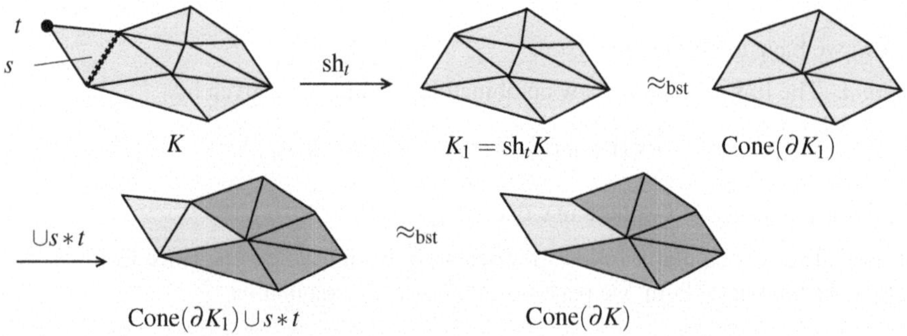

Fig. 20: The induction step in the proof of Lemma 20

The next corollary is beautiful not only because it is the first time we are able to connect stellar moves, bistellar moves, and shellability in a meaningful yet concise statement, but also because it will play a crucial role in proving Pachner's theorem.

Corollary 4. *Let $K = (V,S)$ be a combinatorial n-manifold, and let $s \in S$ be a k-simplex with $s \notin \partial K$. If $\mathrm{Lk}(K,s)$ is shellable, then $K \approx_{\mathrm{bst}} \sigma_s K$.*

Both the statement and the proof of Corollary 4 can be found in [Lic99, Corollary 5.8], but as per usual, we have attempted to fill in the details.

Proof. Since $s \notin \partial K$, its link $\mathrm{Lk}(K,s)$ is a shellable combinatorial $(n-k-1)$-sphere. We can apply Lemma 18 $(k+1)$-times to conclude that

$$\mathrm{St}(K,s) = s * \mathrm{Lk}(K,s) \cong_{\mathrm{si}} \underbrace{D_0 * \cdots * D_0}_{(k+1)\text{-times}} * \mathrm{Lk}(K,s)$$

is a shellable combinatorial n-ball. Thus, it follows from Lemma 20 that

$$\mathrm{St}(K,s) \approx_{\mathrm{bst}} \mathrm{Cone}\big(\partial \mathrm{St}(K,s)\big).$$

Since s is a combinatorial k-ball and $\mathrm{Lk}(K,s)$ is a combinatorial $(n-k-1)$-sphere, it follows from Proposition 2 (2) that we have $\partial \mathrm{St}(K,s) = \partial s * \mathrm{Lk}(K,s)$. Using this, we calculate

$$\begin{aligned} \sigma_s K &= K \setminus \mathring{\mathrm{St}}(K,s) \cup \underline{s} * \partial s * \mathrm{Lk}(K,s) \cong_{\mathrm{si}} \\ &\cong_{\mathrm{si}} K \setminus \mathring{\mathrm{St}}(K,s) \cup \mathrm{Cone}\big(\partial s * \mathrm{Lk}(K,s)\big) = \\ &= K \setminus \mathring{\mathrm{St}}(K,s) \cup \mathrm{Cone}\big(\partial \mathrm{St}(K,s)\big) \approx_{\mathrm{bst}} \\ &\approx_{\mathrm{bst}} K \setminus \mathring{\mathrm{St}}(K,s) \cup \mathrm{St}(K,s) = \\ &= K. \end{aligned}$$

Since any simplicial isomorphism is also a bistellar equivalence, this finishes the proof.

Apart from the fact that they will prove to be extremely useful in proving Pachner's theorem, reading and proving the statements in this chapter might have had the added benefit of making the reader grow used to bistellar moves.

7 Proof of Pachner's theorem

This penultimate chapter usues a lot of the results from the previous chapters in order to prove Pachner's theorem. For the reader's convenience, we shall quickly restate it in all its glory.

Theorem 7. (Pachner's Theorem) *Let $K = (V,S)$ and $K' = (V',S')$ be closed combinatorial n-manifolds. Then*

$$K \approx_{\mathrm{bst}} K' \iff K \approx_{\mathrm{st}} K'.$$

The proof of Pachner's theorem closely follows the one given in [Lic99, Theorem 5.9], which itself follows the original proof by Pachner (cf. [Pac91, Theorem 5.5]). However, Lickorish and Pachner make use of a concept called "stellar exchanges" that is essentially a generalisation of both stellar and bistellar moves. We stick to the (bi)stellar moves that we are familiar with in order to minimise confusion, but the interested reader can find the proof using stellar exchanges in the

aforementioned papers.
Another thing to note is that the proof uses a lot of notation and looks rather involved, but in the end boils down to the effects of stellar moves, Corollary 4, Lemma 15, and a glorified exercise in mathematical induction. For the reader just trying to skim over the proof, there are diagrams scattered throughout that attempt to capture a lot of what happens.

Proof. Let $K = (V,S)$ and $K' = (V',S')$ be closed combinatorial n-manifolds.
We start with the significantly easier direction. Assume that $K \approx_{\text{bst}} K'$. Recall that by Lemma 16 (5), any bistellar move in a sequence $K \approx_{\text{bst}} K'$ is (potentially up to a simplicial isomorphism) the composition of two stellar moves. Therefore, we have $K \approx_{\text{st}} K'$, and applying the Alexander-Newman Theorem concludes the proof of this implication.
For the converse, assume that $|K| \cong_{\text{PL}} |K'|$, or equivalently that $K \approx_{\text{st}} K'$. Due to the transitivity of bistellar equivalence, it suffices to prove the claim in the case that

$$K' = \sigma_s K = K \setminus \mathring{\mathrm{St}}(K,s) \cup \underline{s} * \partial s * \mathrm{Lk}(K,s)$$

for some k-simplex $s \in S$. For the remainder of the proof, we also use the notation $L := \mathrm{Lk}(K,s)$. As a first step, we split up L into

$$L = L' * \Sigma, \text{ where } \Sigma \cong_{\text{si}} S_{m_1} * \cdots * S_{m_p}, \text{ for suitable values of } p, m_1, \ldots, m_p \in \mathbb{N}_0.$$

Essentially, we are splitting off a part that is simplicially isomorphic to a join of spheres, but we do allow for Σ to be empty (in that case, $p = 0$). Note that since K is closed, the link L is a combinatorial $(n-k-1)$-sphere by Proposition 1 (1).
By applying Lemma 15 p-times, we obtain that L' is also a combinatorial sphere, the dimension of which we denote by[35]

$$m := \dim L' \in \mathbb{N}_0 \cup \{-1\}.$$

Furthermore, since L' is a combinatorial sphere, we can denote the minimum number of stellar moves in a sequence $L' \approx_{\text{st}} S_m$ by $r \in \mathbb{N}_0$.
We perform induction on m. The case $m = -1$ is non-trivial already: If $\dim L' = -1$, then L' is the empty abstract simplicial complex, and

$$L = \Sigma \cong_{\text{si}} S_{m_1} * \cdots * S_{m_p}.$$

Since the sphere S_{m_1} is a shellable combinatorial sphere, it follows from $(p-1)$-fold application of Lemma 19 that L is a shellable combinatorial $(n-k-1)$-sphere. Therefore, since $\partial s \cong_{\text{si}} S_{k-1}$, we obtain from applying Lemma 19 again that $\mathrm{Lk}(K',\underline{s}) = \partial s * \mathrm{Lk}(K,s)$ is a shellable combinatorial $(n-1)$-sphere. Since K is closed, we can apply Corollary 4 in order to conclude that

$$K \approx_{\text{bst}} \sigma_s K = K'.$$

[35] Recall that we had defined the dimension of the empty abstract simplicial complex as -1.

The case $m = 0$ is also of importance: If L' is a combinatorial 0-sphere, then it is simplicially isomorphic to S_0 since stellar moves along 0-simplices are just simplicial isomorphisms. But then L' can be absorbed into Σ, allowing us to reduce this case to the case that $m = -1$.
Now let $m \in \mathbb{N}$, and assume the claim is true for stellar moves along simplices whose links have a part "outside of Σ" of dimension less than m.
We shall now also perform induction on r. The case $r = 0$ is rather straightforward: If we have $L' \cong_{\text{si}} S_m$, then L' can again be absorbed into Σ, which also puts us back to the case that $m = -1$.
Now let $r \in \mathbb{N}$, and assume the claim is true for stellar moves along simplices whose links have a part "outside of Σ" that are stellar equivalent to S_m in less than r stellar moves.
Idea for the proof: *Cleverly subdivide both K and K' and use the induction hypotheses in order to find bistellar equivalent combinatorial manifolds.*
As we have done before, we distinguish two cases:

Case 1: The first of the r moves in $L' \approx_{\text{st}} S_m$ is a stellar subdivision σ_t for some simplex t of L'. We denote its link in L' by $L'' := \text{Lk}(L', t)$. We consider the following combinatorial n-manifolds: Since t is a simplex in $\text{Lk}(K, s)$, the simplex $s * t$ lies in K, and we can define

$$K_1 := \sigma_{s*t} K.$$

Furthermore, we use the fact that s is a simplex of

$$K_1 = K \setminus \mathring{\text{St}}(K, s*t) \cup \underline{s*t} * \underbrace{\partial(s*t)}_{=\partial s * t \cup s * \partial t} * \text{Lk}(K, s*t)$$

in order to be able to define

$$K_2 := \sigma_s K_1 = \sigma_s(\sigma_{s*t} K).$$

We are now interested in certain links that allow us to actually use the induction hypotheses. First, we calculate

$$\text{St}(K, s) = s * L = s * L' * \Sigma = s * t * L'' * \Sigma,$$

and therefore have $\text{Lk}(K, s*t) = L'' * \Sigma$. But since

$$m = \dim L' \geq \dim(L'' * t) = \dim L'' + \dim t + 1 > \dim L'',$$

we can conclude from the induction hypothesis on m that $K \approx_{\text{bst}} K_1$.
Now we use the above description of K_1 in order to calculate

$$\text{St}(K_1, s) = \underline{s*t} * s * \partial t * \underbrace{\text{Lk}(K, s*t)}_{=L'' * \Sigma} = s * \Sigma * \underbrace{\underline{s*t} * \partial t * L''}_{=\sigma_t L'} = s * \Sigma * \sigma_t L',$$

which yields us $\text{Lk}(K_1, s) = \Sigma * \sigma_t L'$. However, by construction, $\sigma_t L' \approx_{\text{st}} S_m$ in $r - 1$ stellar moves, thus it follows from the induction hypothesis on r that

$K_1 \approx_{\text{bst}} K_2$.
Note that $\underline{s} * t$ is a simplex of K' because t is a simplex in the link $\text{Lk}(K,s)$. Therefore, it makes sense to consider the following claim:

Claim. We have $\sigma_{\underline{s}*t} K' = K_2$.

Proof. It suffices to calculate the effects of[36] $\sigma_{\underline{s}*t}\sigma_s$ and $\sigma_s \sigma_{s*t}$ on $\text{St}(K, s*t)$ because on $K \setminus \mathring{\text{St}}(K, s*t)$, they certainly produce identical results. Using the results from above, we calculate[37]

$$\begin{aligned}
\sigma_{\underline{s}*t}\big(\sigma_s\big(\text{St}(K,s*t)\big)\big) &= \sigma_{\underline{s}*t}\big(\sigma_s(s*t*\Sigma*L'')\big) = \\
&= \sigma_{\underline{s}*t}\big(\underline{s}*\partial s*t*\Sigma*L''\big) = \\
&= \underline{\underline{s}*t}*\partial(\underline{s}*t)*\partial s*\Sigma*L'' = \\
&= \underline{\underline{s}*t}*(t\cup\underline{s}*\partial t)*\partial s*\Sigma*L'',
\end{aligned}$$

and

$$\begin{aligned}
\sigma_s\big(\sigma_{s*t}\big(\text{St}(K,s*t)\big)\big) &= \sigma_s\big(\sigma_{s*t}(s*t*\Sigma*L'')\big) = \\
&= \sigma_s\big(\underline{s*t}*\partial(s*t)*\Sigma*L''\big) = \\
&= \sigma_s\big(\underline{s*t}*(\partial s*t\cup s*\partial t)*\Sigma*L''\big) = \\
&= \underline{s*t}*\partial s*t*\Sigma*L''\cup\underline{s*t}*\underline{s}*\partial s*\partial t*\Sigma*L'' = \\
&= \underline{s*t}*\partial s*\Sigma*L''*(t\cup\underline{s}*\partial t),
\end{aligned}$$

which are identical by commutativity of the join operation.

There is an attempt at sketching the star of $s*t$ below, for disappointingly low dimensions of s, t, and Σ, the last of which is just a copy of S_0.

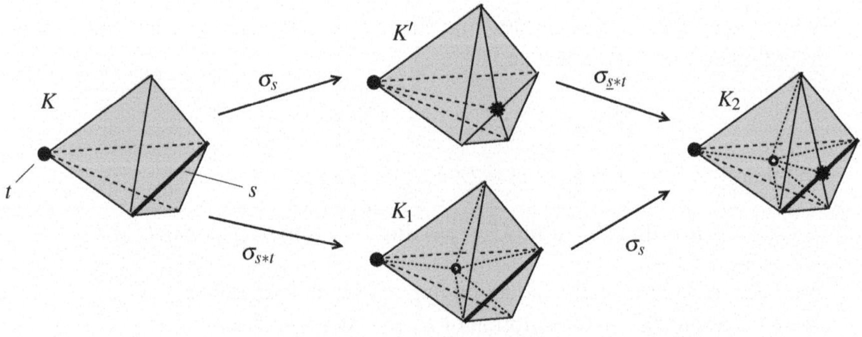

Fig. 21: Two ways to get from K to K_2 using stellar moves

[36] We care about these pairs of stellar subdivisions since they were used to define K' and K_2.

[37] These two calculations essentially utilise the description of stellar moves, and are probably easier to understand by looking at Figure 21.

The following diagram illustrates the situation. Note that the arrows are not maps, but instead indicate stellar subdivisions.

$$\begin{array}{ccc} K & \xrightarrow{\sigma_{s*t}} & K_1 \\ \downarrow{\scriptstyle \sigma_s} & & \downarrow{\scriptstyle \sigma_s} \\ K' & \xrightarrow{\sigma_{\underline{s}*t}} & K_2. \end{array}$$

We have already seen that the stellar subdivisions to the top and right induce bistellar equivalences. Our plan is now to show that the bottom horizontal stellar subdivision induces a bistellar equivalence, because then transitivity yields us a bistellar equivalence between K and K'.
Using the definition of $K' = \sigma_s K$, we calculate

$$\mathrm{St}(K', \underline{s} * t) = \underline{s} * \partial s * L = \underline{s} * \partial s * \Sigma * L' = \underline{s} * \partial s * \Sigma * t * L''.$$

Therefore, we have $\mathrm{Lk}(K', \underline{s} * t) = \partial s * \Sigma * L''$, and after absorbing the sphere ∂s into Σ, we can use the claim as well as our hypothesis on m to obtain that

$$K' \approx_{\mathrm{bst}} \sigma_{\underline{s}*t} K' = K_2.$$

But now we have shown the three bistellar equivalences

$$K \approx_{\mathrm{bst}} K_1 \approx_{\mathrm{bst}} K_2 \approx_{\mathrm{bst}} K',$$

so we have $K \approx_{\mathrm{bst}} K'$ by transitivity.

Case 2: The first of the r moves in $L' \approx_{\mathrm{st}} S_m$ is a stellar weld σ_t^{-1} for some simplex t of $\sigma_t^{-1} L'$, and we have[38]

$$\mathrm{Lk}(L', \underline{t}) = \partial t * \mathrm{Lk}(\sigma_t^{-1} L', t).$$

In Case 2, we denote that last link by $L'' := \mathrm{Lk}(\sigma_t^{-1} L', t)$. Because in this case we are dealing with a stellar weld, we have to distinguish another two cases:

Case 2a: The simplex t does not lie in K. We set $K_1 := \sigma_t^{-1} \sigma_{s*\underline{t}} K$ and $K_2 := \sigma_s K_1$. This is possible since $\underline{t}$ is a vertex in L' and hence also in $\mathrm{Lk}(K, s)$. Since for the star of s in K we have

$$\mathrm{St}(K, s) = s * L = s * L' * \Sigma = s * \underline{t} * \mathrm{Lk}(L', \underline{t}) * \Sigma = s * \underline{t} * \partial t * L'' * \Sigma,$$

we obtain $\mathrm{Lk}(K, s * \underline{t}) = \partial t * L'' * \Sigma = L'' * \Sigma'$, for $\Sigma' = \partial t * \Sigma$ still a link of spheres. Furthermore, we again have

$$m = \dim L' \geq \dim(L'' * \underline{t}) = \dim L'' + 0 + 1 > \dim L'',$$

so we can use the induction hypothesis on m in order to obtain that

[38] This is an immediate consequence of the definition of stellar moves.

$$K \approx_{\text{bst}} \sigma_{s*\underline{t}} K = \sigma_t K_1.$$

But $\text{St}(K_1,t) = \partial(s*\underline{t})*t*\Sigma*L''$, so $\text{Lk}(K_1,t) = \partial(s*\underline{t})*\Sigma*L''$, and we can absorb the sphere $\partial(s*\underline{t})$ into Σ in order to obtain from our induction hypothesis on m that $K_1 \approx_{\text{bst}} \sigma_t K_1$.
Similarly, we consider

$$\begin{aligned}\text{St}(K_1,s) &= \sigma_t^{-1}\left(\underline{s*t}*\partial(s*\underline{t})*\text{Lk}(K,s*\underline{t})\right) = \\ &= s*\sigma_t^{-1}L'*\Sigma \cup \sigma_t^{-1}s*\sigma_t^{-1}L'*\Sigma,\end{aligned}$$

and therefore

$$\text{Lk}(K_1,s) = \sigma_t^{-1}L'*\Sigma.$$

Note that $\sigma_t^{-1}L' \approx_{\text{st}} S_m$ in $r-1$ stellar moves, hence our induction hypothesis on r yields $K_1 \approx_{\text{bst}} K_2$.
Recall that since $\underline{t}$ is a vertex in $\text{Lk}(K,s) = \text{Lk}(K',\underline{s}*\partial s)$, the 1-simplex $\underline{s}*\underline{t}$ lies in K'. This allows us to make the following claim:

Claim. We have $\sigma_t^{-1}\left(\sigma_{\underline{s}*\underline{t}}K'\right) = K_2$.

Proof. It suffices to calculate the effects of both $\sigma_t^{-1}\sigma_{\underline{s}*\underline{t}}\sigma_s$ and $\sigma_s\sigma_t^{-1}\sigma_{s*\underline{t}}$ on $\text{St}(K,s*\underline{t})$ because on $K\setminus\mathring{\text{St}}(K,s*\underline{t})$, they certainly produce identical results. Using the results from above, we calculate

$$\begin{aligned}\sigma_t^{-1}\sigma_{\underline{s}*\underline{t}}\sigma_s\left(\text{St}(K,s*\underline{t})\right) &= \sigma_t^{-1}\sigma_{\underline{s}*\underline{t}}\sigma_s\left(s*\underline{t}*\partial t*\Sigma*L''\right) = \\ &= \sigma_t^{-1}\sigma_{\underline{s}*\underline{t}}\left(\underline{s}*\partial s*\underline{t}*\partial t*\Sigma*L''\right) = \\ &= \sigma_t^{-1}\left(\underline{s*t}*\partial(\underline{s}*\underline{t})*\partial s*\partial t*\Sigma*L''\right) = \\ &= (\underline{s}\cup\underline{t})*\partial s*t*\Sigma*L'',\end{aligned}$$

and

$$\begin{aligned}\sigma_s\sigma_t^{-1}\sigma_{s*\underline{t}}\left(\text{St}(K,s*\underline{t})\right) &= \sigma_s\sigma_t^{-1}\sigma_{s*\underline{t}}\left(s*\underline{t}*\partial t*\Sigma*L''\right) = \\ &= \sigma_s\sigma_t^{-1}\left(\underline{s*t}*\partial(s*\underline{t})*\partial t*\Sigma*L''\right) = \\ &= \sigma_s\sigma_t^{-1}\left(\underline{s*t}*(\partial s*\underline{t}\cup s)*\partial t*\Sigma*L''\right) = \\ &= \sigma_s\left((\partial s*\underline{t}\cup s)*t*\Sigma*L''\right) = \\ &= (\partial s*\underline{t}\cup\underline{s}*\partial s)*t*\Sigma*L'' = \\ &= (\underline{s}\cup\underline{t})*\partial s*t*\Sigma*L'',\end{aligned}$$

which are identical.

The following diagram illustrates the situation. Note that again, the arrows are not maps, but instead indicate stellar subdivisions.

$$\begin{array}{ccccc} K & \xrightarrow{\sigma_{s*\underline{t}}} & \sigma_t K_1 & \xleftarrow{\sigma_t} & K_1 \\ \big\downarrow{\scriptstyle\sigma_s} & & & & \big\downarrow{\scriptstyle\sigma_s} \\ K' & \xrightarrow{\sigma_{\underline{s}*\underline{t}}} & \sigma_t K_2 & \xleftarrow{\sigma_t} & K_2. \end{array}$$

We have already seen that the two top horizontal stellar subdivisions, as well as the one to right, all induce bistellar equivalences. Our plan is now to show that the two bottom horizontal stellar subdivisions induce bistellar equivalences, because then transitivity yields us a bistellar equivalence between K and K'.

As we had already seen in the proof of the claim, we have

$$\mathrm{St}(K', \underline{s} * \underline{t}) = \underline{s} * \partial s * \underline{t} * \partial t * \Sigma * L'',$$

and therefore

$$\mathrm{Lk}(K', \underline{s} * \underline{t}) = \partial s * \partial t * \Sigma * L''.$$

After absorbing the spheres ∂s and ∂t into Σ, we can use the claim as well as our hypothesis on m to obtain that

$$K' \approx_{\mathrm{bst}} (\sigma_{\underline{s}*\underline{t}} K') = \sigma_t K_2.$$

So it remains to show that $\sigma_t K_2 \approx_{\mathrm{bst}} K_2$. But this follows from the above result that

$$\mathrm{St}(K_2, t) = (\underline{s} \cup \underline{t}) * \partial s * t * \Sigma * L'',$$

hence

$$\mathrm{Lk}(K_2, t) = (\underline{s} \cup \underline{t}) * \partial s * \Sigma * L''.$$

Again, we absorb the spheres ∂s and $\underline{s} \cup \underline{t}$ into Σ, and can use the induction hypothesis on m to obtain $K_2 \approx_{\mathrm{bst}} \sigma_t K_2$.

Now we have shown the five stellar equivalences

$$K \approx_{\mathrm{bst}} \sigma_t K_1 \approx_{\mathrm{bst}} K_1 \approx_{\mathrm{bst}} K_2 \approx_{\mathrm{bst}} \sigma_t K_2 \approx_{\mathrm{bst}} K',$$

so we have $K \approx_{\mathrm{bst}} K'$ by transitivity.

Case 2b: If the simplex t happens to lie in K, we reduce the situation to Case 2a as follows: If t is a 0-simplex, then we can just use an alternative new vertex v_0 instead, and obtain the result from Case 2a up to a simplicial isomorphism. If $\dim t \geq 1$, we pick a vertex $u \in t$ and write $t' := t \setminus \{u\}$ for its opposite face in t. Now choose a vertex v that is not in K. Let φ be a simplicial isomorphism that sends the vertex $\underline{s*u}$ to v. We now define a series of combinatorial manifolds as follows:

$$\hat{K} := \varphi\sigma_{s*u} K, \qquad \hat{K}' := \sigma_s \hat{K}, \qquad K' := \sigma_s K.$$

Claim. We have $\varphi\sigma_{\underline{s}*u} K' = \hat{K}'$.

Proof. As in the previous cases, we calculate how $\varphi\sigma_{\underline{s}*u}\sigma_s$ and $\sigma_s\varphi\sigma_{s*u}$ affect $\mathrm{St}(K,s*t)$, because both clearly do the same on $K\setminus\mathring{\mathrm{St}}(K,s*t)$.

$$\begin{aligned}
\varphi\sigma_{\underline{s}*u}\sigma_s\big(\mathrm{St}(K,s*t)\big) &= \varphi\sigma_{\underline{s}*u}\sigma_s\big(s*u*t'*L''*\Sigma\big) = \\
&= \varphi\sigma_{\underline{s}*u}\big(\underline{s}*\partial s*u*t'*L''*\Sigma\big) = \\
&= v*\partial(\underline{s}*u)*\partial s*t'*L''*\Sigma = \\
&= v*(u\cup\underline{s})*\partial s*t'*L''*\Sigma,
\end{aligned}$$

and

$$\begin{aligned}
\sigma_s\varphi\sigma_{s*u}\big(\mathrm{St}(K,s*t)\big) &= \sigma_s\varphi\sigma_{s*u}\big(s*u*t'*L''*\Sigma\big) = \\
&= \sigma_s\big(v*\partial(s*u)*t'*L''*\Sigma\big) = \\
&= \sigma_s\big(v*(\partial s*u\cup s)*t'*L''*\Sigma\big) = \\
&= v*(\partial s*u\cup\underline{s}*\partial s)*t'*L''*\Sigma = \\
&= v*(u\cup\underline{s})*\partial s*t'*L''*\Sigma,
\end{aligned}$$

which are clearly identical.

The situation is now as follows:

$$\begin{array}{ccc}
K & \xrightarrow{\varphi\sigma_{s*u}} & \hat{K} \\
\big\downarrow{\scriptstyle\sigma_s} & & \big\downarrow{\scriptstyle\sigma_s} \\
K' & \xrightarrow{\varphi\sigma_{\underline{s}*u}} & \hat{K}'
\end{array}$$

In order to show that there is a bistellar equivalence between the combinatorial manifolds on the left-hand side, we show that the three stellar subdivisions at the bottom, top, and to the right induce bistellar equivalences between the respective combinatorial manifolds. First, we use that the link of s in $\hat{K}$ is given by[39]

$$\mathrm{Lk}(\hat{K},s) = v*t'*L''*\Sigma = \varphi(t)*L''*\Sigma = \varphi(L')*\Sigma,$$

and that $\varphi(L')$ is just another copy of L', but with one vertex renamed. Therefore, we can use the calculation from Case 2a in order to obtain that $\hat{K}\approx_{\mathrm{bst}}\hat{K}'$.

For the top horizontal map, we use that

$$\mathrm{Lk}(K,s*u) = t'*L''*\Sigma = \mathrm{Lk}(L',u)*\Sigma,$$

and since L' is a combinatorial m-sphere, we have

$$\dim\mathrm{Lk}(L',u) = m-1 < m,$$

[39] As before, we can read it off in the third line of the calculation of $\sigma_s\varphi\sigma_{s*u}\big(\mathrm{St}(K,s*t)\big)$.

so it follows from the induction hypothesis on m that $K \approx_{\mathrm{bst}} \hat{K}$.
The argument that the bottom horizontal map induces a bistellar equivalence $K' \approx_{\mathrm{bst}} \hat{K}'$ is essentially the same. Therefore, we can again conclude that

$$K \approx_{\mathrm{bst}} \hat{K} \approx_{\mathrm{bst}} \hat{K}' \approx_{\mathrm{bst}} K'.$$

In all cases, we have now shown the induction step, therefore Pachner's theorem follows by induction.

Remark 8. One of the downsides of the proof is that regrettably, drawing sketches that are both meaningful and non-trivial is pretty much impossible. This can be seen in the fact that at times in the proof, we had to consider joins of up to 5 simplicial complexes, which, according to the dimension formula from Lemma 1, automatically puts us at a dimension of at least 4, and significantly more if we ask for some of the links or simplices to be of higher dimensions. So unless the reader has an easy time visualising those dimensions, the beauty has to be found in the way everything ultimately ends up falling into place, as well as sketches of tiny parts of the complexes involved, similar to the way we did in Figure 21.

For the sake of completeness, we also state another result that could be described as "Pachner for manifolds with boundary". For this, we need a notion of connectedness for abstract simplicial complexes. Luckily, connectedness for abstract simplicial complexes is a rather straightforward concept to define.

Definition 18. Let $K = (V,S)$ be an abstract simplicial complex. We call K connected if for any two vertices $u, v \in V$, there exists a sequence $w_1, \dots, w_k \in V$ with $w_1 = u$, $w_k = v$, and the simplex $\{w_i, w_{i+1}\} \in S$ for all $i \in \{1, \dots, k-1\}$.

Theorem 8. (Pachner for Bounded Manifolds) *Let K and K' be connected combinatorial n-manifolds with non-empty boundary. Then $|K| \cong_{\mathrm{PL}} |K'|$ if and only if they are related by a sequence of elementary shellings, inverse shellings*[40]*, and simplicial isomorphisms.*

Proof. Proofs for this Theorem can be found in [Lic99, Theorem 5.10], or alternatively in [Pac91, Theorem 6.3].

References

Ale30. Alexander, J. W.: *The Combinatorial Theory of Complexes*. Annuals of Mathematics 31 (1930), pp. 292-320

BW96. Barrett, J. W. and Westbury B. W.: *Invariants of piecewise-linear 3-manifolds*, Trans. Amer. Math. Soc. 348 (1996), pp. 3997-4022.

Bin64. Bing, R. H.: *Some Aspects of the Topology of 3-Manifolds Related to the Poincaré Conjecture*. Lectures on Modern Mathematics, Volume II. Wiley, New York (1964), pp. 93-128

[40] An *inverse shelling* is the opposite operation of an elementary shelling, that is, if $K' = \mathrm{sh}_t K$ is an elementary shelling, then $K = \mathrm{sh}_t^{-1} K'$ is an inverse shelling.

Cer68. Cerf, J.: *Sur les difféomorphismes de la sphère de dimension trois* ($\Gamma_4 = 0$), Lecture Notes in Mathematics 53, Springer Verlag (1968)

Cer70. Cerf, J.: *La stratification naturelle des espaces de fonctions différentiables réelles et le théorème de la pseudo-isotopie*, Publ. Math., Inst. Hautes Étud. Sci. 39 (1970), pp. 5-173

DR18. Douglas C. L. and Reutter, D. J.: *Fusion 2-categories and a state-sum invariant for 4-manifolds*, Preprint, arXiv: 1812.11933 (2018)

Ewa86. Ewald, G.: *Über stellare Unterteilung von Simplizialkomplexen*. Archiv der Mathematik 46 (1986), pp. 153-158

Fri23. Friedl, S.: *Topology*, lectures notes, University of Regensburg (2023) `https://friedl.app.uni-regensburg.de/`

Gla70. Glaser, L.: *Geometrical combinatorial topology*, Vol. I. Van Nostrand Reinhold Mathematics Studies, 27. Van Nostrand Reinhold Co. (1970)

Hat01. Hatcher, A.: *Algebraic Topology*. First Edition Cambridge, Cambridge University Press (2001)

Hud69. Hudson, J. F. P.: *Piecewise Linear Topology*. New York, W. A. Benjamin, Inc. (1969)

Kle17. Kleinschmidt, P.: Udo Pachner (1947-2002) – A "Hidden Champion" in Mathematics University of Passau (2017)

Koz08. Kozlov, D.: *Combinatorial Algebraic Topology*. Algorithms and Computation in Mathematics 21. First Edition Heidelberg, Springer (2008)

Lic91. Lickorish, W. B. R.: *Unshellable Triangulations of Spheres*. European Journal of Combinatorics 12 (1991), pp. 527-530

Lic99. Lickorish, W. B. R.: *Simplicial Moves on Complexes and Manifolds*. Geometry and Topology Monographs 2: Proceedings of the Kirbyfest (1999), pp. 299-320

New26. Newman, M. H. A.: *On the Foundations of Combinatorial Analysis Situs*. Proceedings of the Royal Academy of Amsterdam 29 (1926), pp. 610-641

Pac91. Pachner, U.: *P.L. Homeomorphic Manifolds are Equivalent by Elementary Shellings*. European Journal of Combinatorics 12 (1991)

Rud58. Rudin, M. E.: *An Unshellable Triangulation of a Tetrahedron*. Bulletin of the American Mathematical Society 64 (1958), pp. 90-91

Sco05. Scorpan, A: *The wild world of 4-manifolds*, Providence, RI: American Mathematical Society (2005).

TV92. Turaev, V. G., & Viro, O. Y.: *State Sum Invariants of 3-Manifolds and Quantum 6j-Symbols*. Topology 31 (1992), pp. 865-902

Chapter 12

MATRIX Program: 2D Supersymmetric Theories and Related Topics

Vacuum Structures Revisited

Wei Gu

Abstract We consider the relationship between the higher symmetry and the dynamical decomposition in supersymmetric gauge theory in various dimensions by studying the semi-classical potential energy. We observe that besides the scalar moduli we shall also include the field strength $F_{0\cdots d}$ in the vacuum moduli in the 1+d dimensional theory along with a $\mathbb{Z}_p$ d-form symmetry. In gauge theory for charge-p matters with this symmetry, we find that the vacua decompose into p different universes at an intermediate scale, which means no dynamical domain wall can interpolate between them. In our setup, we re-derive the existing results on the decomposition in various dimensions. In four dimensions, we propose a UV gauge theory for the generalized super Yang-Mills theory, whose instanton sectors are restricted to the topological number with integer multiples of p.

1 Introduction

This note discusses the vacuum structure of supersymmetric gauge theory in various dimensions, and we only focus on the theory with a Lagrangian description.

Our analysis was motivated by the consideration of the relationship between the higher symmetry [1] and the emergent dynamical decomposition in field theory, which was first observed in two-dimensions [2–5], and then in other dimensions [6,7]. Certainly, they can be defined independently. The definition of a q-form symmetry in $1+d$ dimensions can be made abstractly, which is implemented by an operator associated with a codimension $q+1$ closed manifold M^{d-q}. The conserved current is a $q+1$-form, and in this note, we are primarily interested in the d-form symmetry. One of the fundamental new insights of [1] is the explanation of the Higgs mechanism in those "generalized global symmetries." While the de-

Wei Gu
Department of Physics, Virginia Tech, 850 West Campus Dr., Blacksburg, VA 24061, USA.
e-mail: weig8@vt.edu

D. R. Wood et al. (eds.), *2021-2022 MATRIX Annals*, MATRIX Book Series 5,
https://doi.org/10.1007/978-3-031-47417-0_35

composition can be made even in quantum mechanics, for example, consider the double-well potential energy,

$$V(x) = \lambda \left(x^2 - a^2\right)^2.$$

There are two interesting limits we can think about. The first one is to take a goes to infinity while λ is a finite value, one can, then, compute the fluctuation from the configuration at $x = -a$ to $x = a$ is zero because the energy barrier is infinity. The other limit is to let λ goes to infinity while a is a non-vanishing small value. One can easily observe that the wave function must be localized at $x = -a$ or $x = a$. Furthermore, the fluctuation between them is zero because of the same reason.

In quantum field theory, we can have a similar story, if we replace x and $V(x)$ by a scalar field ϕ and the semi-classical potential energy density function $U(\phi)$ respectively. It is well-known that parameters in quantum field theory depend on the physical scale. The first limit in the quantum mechanics example has a cousin in the quantum field theory called the "cluster decomposition." It is a far-infrared behavior. The second one is much more interesting in quantum field theory, if it started at an intermediate physical scale. We denote it as the "dynamical decomposition." This decomposition is not only physically nontrivial [1] but also could lead to interesting mathematical statements reviewed in section 2.3.

As mentioned, the higher symmetry, in general, does not have to be related to the dynamical decomposition directly [2]. However, there is a special one called d-form symmetry that enters our story. In previous studies of the vacuum configuration from the semi-classical potential, people only used scalar fields to parameterize the vacuum because of the Lorentz symmetry in quantum field theory. Now, if we have the d-form gauge field, we could also have an extra Lorentz invariant field to label the vacuum configuration. This note includes this in a region where the semi-classical analysis can be trusted. More specifically, we consider the semi-classical potential energy as a function in terms of both scalar fields and field strength $F_{0\cdots d}$ of the d-form gauge field,

$$U\left(\phi, F_{0\cdots d}\right).$$

The expectation value of $F_{0\cdots d}$ could teach us the nontrivial information of vacuum configuration, even though it is vanishing [3]. In this context, we will find the dynamical decomposition phenomenon in the effective theory with a $\mathbb{Z}_p$ d-form symmetry.

We begin in section 2 by re-discussing the known observations in two dimensions in our context. We try to clarify two points. The first one is where the dynamical decomposition can happen, and the second is why it is different from the "cluster decomposition" associated with the super-selection rule. Our motivation is to focus

---

[1] In [7, 8], they used the terminology "universe" in describing this structure to distinguish it from the usual vacua.

[2] If the quantum theory has symmetry, one can use it to label the Hilbert space and observables by this symmetry. However, this is a trivial "decomposition" of the vector space, which is not our interest in this note.

[3] In contrast, the vanishing of non-scalars is simply because of the Lorentz symmetry.

on the explanation rather than to produce new results in this section. It can be regarded as a warmup to the more interesting four-dimensional quantum field theory with a three-form $\mathbb{Z}_p$ symmetry.

In section 3, we discuss the three-dimensional $\mathcal{N}$=2 Chern-Simons matter theories. We first re-analyze the dynamical decomposition in the KK-reduction from the three-dimensional gauge theory to the two-dimensional one discussed in [6] via semi-classical potential energy. Then we comment on why there was no decomposition in the three-dimensional gauge theory. However, we do not claim we have physical proof.

In section 4, we briefly review the four-dimensional generalized super Yang-Mills theory. Then we propose a UV-fundamental theory for the generalized SYM theory with a $\mathbb{Z}_p$ 3-form symmetry. Finally, we observe the dynamical decomposition from the semi-classical approximation of the UV theory at an intermediate scale. It is similar to the two dimensions.

Our last section is devoted to conclusions and future directions.

2 Linear Sigma Models Revisited

This section investigates the dynamical decomposition in two-dimensional quantum field theory. We do not claim that our results are new, rather, we focus on the physical explanation of decomposition in our setup. The observations found in [2–5] relied on the computation of massless spectrums and other RG-protected quantities, see also [9–11] for more. The interested reader can refer to them for the calculation. Our approach is different, and it supports most observations found in them.

2.1 Charge-p $\mathbb{C}P^{N-1}$ Model

The $\mathcal{N} = (2,2)$ linear sigma model for charge-p $\mathbb{C}P^{N-1}$ is a $U(1)$ gauge theory with N gauge charge p chiral superfields, see [2, 13] for more details. As mentioned in the introduction, if we have a nontrivial vacuum configuration of F_{01}, besides the scalar fields considered in the previous studies, the flowing relevant terms are not necessary vanishing in a vacuum, and they are

$$\mathscr{L} = \frac{1}{2e^2}F_{01}^2 + (\theta + 2n\pi)\,F_{01} - \sum_{i=1}^{N} \mid \phi_i \mid^2 \left(\partial_\mu \varphi + pA_\mu\right)^2, \tag{1}$$

where we have written each ϕ_i in the polar variables $(\mid \phi_i \mid, \varphi_i)$ defined by $\phi_i = \mid \phi_i \mid e^{i\varphi_i}$. We keep the term $2n\pi F_{01}$ in our Lagrangian because of the Dirac quantization of the abelian group $U(1)$, so n is an integer. Focus on

$$\mathscr{L}_{\varphi} = -\sum_{i} |\phi_i|^2 \left(\partial_\mu \varphi_i + pA_\mu\right)^2, \tag{2}$$

unfortunately, the factor $\partial_\mu \varphi_i + pA_\mu$ in equation (1) does not simply depend on the scalar F_{01}, although it is gauge-invariant. To see how it relates to the F_{01}, we first introduce the auxiliary one-form variables $\lambda_{i,\mu}$, and then we have

$$-\sum_{i} \frac{1}{4|\phi_i|^2} \left(\lambda_{i,\mu}\right)^2 + \varepsilon^{\mu\nu} \lambda_{i,\mu} \left(\partial_\nu \varphi_i + pA_\nu\right). \tag{3}$$

It is easy to see that integrating out $\lambda_{i,\mu}$ in the above formula will give us the Lagrangian $\mathscr{L}_{\varphi}$. Now, instead, we integrate out φ_i first which gives the constraint $\lambda_{i,\mu} = \partial_\mu \vartheta_i$, where each ϑ_i is a 2π periodic variable. Plugging this back into the formula (3), we obtain the new Lagrangian[4]

$$\mathscr{L}_{\vartheta} = -\sum_{i} \frac{1}{4|\phi_i|^2} \left(\partial_\mu \vartheta_i\right)^2 - p\vartheta_i F_{01},$$

where the dual variable ϑ is coupled to the gauge field A_μ as a dynamical theta angle. The procedure we performed is called the abelian duality [14]. So we replace the Lagrangian in equation (1) by a new one

$$\widetilde{\mathscr{L}} = \frac{1}{2e^2} F_{01}^2 + \left(\theta + 2n\pi - p\sum_{i=1}^{N} \vartheta_i\right) F_{01} - \sum_{i} \frac{1}{4|\phi_i|^2} \left(\partial_\mu \vartheta_i\right)^2. \tag{4}$$

The Lorentz symmetry does not require the first two terms to vanish in a vacuum. Furthermore, the vacuum configuration, of course, does not depend on the field variables we use. The 2π periodicity of each ϑ_i is beautiful connected to the pair creation of heavy particles in the two-dimensional gauge theory [12] that changes θ by $2lp\pi$ for some integer l, so we can restrict $n \in [0, p-1]$. Now, we can write out the semi-classical potential energy

$$U(\phi_i, F_{01}) = \frac{e_{\text{eff}}^2}{2} \left(\sum_{i=1}^{N} p|\phi_i|^2 - r\right)^2 + \sum_{i=1}^{N} p^2 |\phi_i|^2 |\sigma|^2 + \frac{1}{2e_{\text{eff}}^2} F_{01}^2, \tag{5}$$

where we read off the potential energy density in the Hamiltonian formulation. The complex scalar σ is the lowest component of the super-field-strength[5], and r is the FI parameter at the physical scale. The last term in equation (5) was usually omitted in the previous papers because the gauge field is not a scalar field. In order to find the ground state, i.e. the vanishing configuration of potential energy, the quadratic terms in the potential energy must be vanishing separately because $|\phi_i|^2$ and F_{01}

[4] We assume all of the $|\phi_i|^2$ are nonzero, and if some of them vanished, we only take a dual expression for the remaining variables.

[5] The complex scalars σ and $\bar{\sigma}$ can be understood from the KK-reduction of the four-dimensional gauge field in $4d$ $\mathscr{N}=1$ gauge theory, see [13, sec. 2] for more details.

are all gauge-invariant Hermitian operators. This suggests that

$$\langle F_{01} \rangle = 0. \tag{6}$$

It looks like we were doing something trivial. However, this is too quick. Let us first look at what is F_{01} in the vacuum. From equation (4), we can observe that

$$F_{01} = e^2 \left(p \sum_{i=1}^{N} \vartheta_i - \theta - 2n\pi \right).$$

Then the semi-classical potential energy can be rewritten as

$$U(\phi_i, \vartheta_i) = \frac{e_{\text{eff}}^2}{2} \left(\sum_{i=1}^{N} p \mid \phi_i \mid^2 - r \right)^2 + \sum_{i=1}^{N} p^2 \mid \phi_i \mid^2 \mid \sigma \mid^2 + \frac{e_{\text{eff}}^2}{2} \left(p \sum_{i=1}^{N} \vartheta_i - \theta - 2n\pi \right)^2, \tag{7}$$

The parameters e^2 and r are both running under the RG-flow. Although we do not know the exact expression of e_{eff}^2, it becomes the strong coupling in the low energy physics because its mass dimension is two. While the RG-flow of parameter r can be computed exactly which is

$$r = pN \log \frac{\mu}{\Lambda}, \tag{8}$$

where μ is a physical scale and Λ is the dynamical scale. In order to make sense of the perturbative low energy effective theory NLSM, we restrict our physical scale

$$\Lambda \ll \mu \ll e_{\text{eff}} \sqrt{r}. \tag{9}$$

We call this the intermediate scale in our note. At this scale, the parameter r is very large suggesting the couplings in NLSM are very small such that the perturbative computation can be trusted. Furthermore, the procedure in integrating out massive modes in the linear sigma model around the vacuum configuration is also reasonable because their masses are proportional to $e_{\text{eff}} \sqrt{r}$ which are heavy objects compared to the physical scale. Finally, from equation (9), we can find that e_{eff}^2 approaches infinity at the intermediate scale. This would introduce some nontrivial physics defined in the introduction section called the dynamical decomposition. Let us investigate this by studying the vacuum configuration which can be read off from equation (7)

$$\left\langle \sum_{i=1}^{N} \mid \phi_i \mid^2 \right\rangle = \frac{r}{p}, \qquad \left\langle e^{i \sum_{i=1}^{N} \vartheta_i} \right\rangle = e^{i\frac{\theta}{p} + i\frac{2\pi n}{p}}, \tag{10}$$

for $n = 1, 2, \ldots p-1$. We have used the fact that all of the variables ϑ_i commute to each other, and only a single-valued field can be the physical observable. We see that different n label different *universes*, and the energy barrier between them is infinity because the parameter e_{eff}^2 goes to infinity. Each universe is a $\mathbb{C}P^{N-1}$, and there is no dynamics that connects one universe to others. Now consider the observable

$$e^{il*\sum_{i=1}^{N}\vartheta_i},$$

with an integer l. If $l \neq 0 \mod p$, then it is charged by a finite $\mathbb{Z}_p$ one-form symmetry with the charge l. The operation of this symmetry is to shift $\sum_{i=1}^{N}\vartheta_i$ by a 2π. We also call these observables as external probes that connect different universes. If $l = 0 \mod p$, then it is neutral under the one-form symmetry and only acts on a particular universe. The vacua in each universe can be connected by dynamical domain walls or called solitons associated with the superselection rule in a local quantum field theory.

We would like to make several comments:

- The vacuum configuration obtained in equation (10) is not in the far-infrared. Therefore, we do not expect any massless Goldstone particle appeared in the usual Higgs mechanism[6]. However, the expectation value of $\sum_{i=1}^{N} p \mid \phi_i \mid^2$ tells us that two bosons, σ, and A_μ, in the vector multiples are massive. This is because there are terms such as $e^2 \mid \phi \mid^2 \mid \sigma \mid^2$ and $e^2 \mid \phi \mid^2 A_\mu A^\mu$ in the Lagrangian, so their masses are proportional to $e_{\text{eff}}\sqrt{r}$. A similar argument can apply to gauginos.
- The non-vanishing expectation value of $e^{i\sum_{i=1}^{N}\vartheta_i}$ does not suggest that the flavor symmetry is broken. We still have $N-1$ free field variables which are generators of the group $U(1)^{N-1}$, and this group is the max-torus of the flavor symmetry of the target.
- We can also consider the linear sigma model with a superpotential, and the dynamical decomposition would not be changed. Because this phenomenon is due to the existence of the $\mathbb{Z}_p$ one-form symmetry and the strong coupling limit of the gauge coupling e_{eff}^2. Furthermore, the superpotential only introduces the usual F-term which, of course, does not affect the story that happened in the D-term.
- One may also worry that the full nonperturbative quantum correction would affect our statement on dynamical decomposition found in the semi-classical potential energy. However, one can easily estimate that the contribution of the degree one instanton is proportional to $q = e^{-r+i\theta}$, which is close to zero, and the higher degrees instantons are higher-order power of q. So one can expect that the summation of nonperturbative correction gives a finite convergence series, and this is what Hori and Vafa [14] originally performed in deriving the abelian mirror symmetry. An expert would already notice that our dual variables ϑ_i are the usual field variables in the mirror. A beautiful story of mirror symmetry is that the potential energy in the mirror is exact up to a renormalization of the gauge coupling. So it is better to discuss the decomposition in the mirror. See section 2.2.
- Our last comment in this section is whether we have a dynamical decomposition in the linear sigma model. Our answer would be *no*. This follows our definition of dynamical decomposition because, in the UV, the gauge coupling is a small parameter. One can still have an extra label on the physical observables and state spaces because of the one-form symmetry, and we call them sectors. However,

[6] In fact, there is no Goldstone particle even in the infinity volume limit in two-dimensional quantum field theory [15, 16].

this is a trivial vector space decomposition in a quantum theory[7]. Furthermore, the physical operators in the gauge theory do not depend on the gauge coupling, which means we can compute the NLSM correlation functions in the linear sigma model. The decomposition of correlation functions only follows the decomposition of the nonlinear sigma model that does not imply we have a decomposition in the linear sigma model. Finally, we want to mention two different field configurations which connect sectors. The first one, in the UV, is the non-BPS domain wall configuration that fluctuates from one vacuum in one sector to another vacuum in a different sector[8]. The tension is approximately proportional to $2\pi^2 e_{\text{eff}}^2$, and one can notice that they are infinity heavy in the infrared. The second field configuration that appeared in the UV gauge theory is the *point-like* instanton [13], and one can show that they can not be decomposed into different sectors. However, because they are positive codimension objects [17], they do not affect the correlation functions of NLSM on charge-p projective space. Furthermore, the masses of point-like instantons are $\sim e_{\text{eff}}\sqrt{r}$, which means they are also infinity massive in the low energy physics. All these objects suggest that we do not have a decomposition in the extreme UV, but we have a perfect decomposition at the intermediate scale.

2.2 Mirror of NLSM on Charge-p $\mathbb{C}P^{N-1}$ Model

As commented in the previous section, it is better to discuss the dynamical decomposition in the mirror. Following [14], we first define the mirror Landau-Ginzburg model: the target space $(\mathbb{C}^*)^N \times \mathbb{C}$ with the superpotential

$$W = \Sigma\left(p\sum_{i=1}^{N} Y_i - t\right) + \sum_{i=1}^{N} e^{-Y_i}.$$

The Kähler potential is

$$K\left(Y_i, \Sigma; \bar{Y}_i, \bar{\Sigma}\right) = -\frac{1}{2e^2}\bar{\Sigma}\Sigma - \sum_{i=1}^{N}\frac{1}{2}\left(Y_i + \bar{Y}_i\right)\log\left(Y_i + \bar{Y}_i\right).$$

Writing the Lagrangian in terms of the component fields of chiral super fields Y and Σ, there is a term like

$$p\vartheta_i F_{01},$$

[7] In quantum mechanics, we usually find a complete set of commuting operators to label the Hilbert space.

[8] The origin of scalar field spaces is the bridge in connecting different universes because there is no order parameter here to label the different "universe."

in the Lagrangian.[9] Because of the 2π periodicity of ϑ_i, we have the Dirac quantization condition

$$\frac{p}{2\pi}\int F \in \mathbb{Z}.$$

Now, we can write out the exact potential energy

$$U(y_i,\sigma) = \frac{e^2}{2}\left|p\sum_{i=1}^{N} y_i - t + 2i\pi n\right|^2 + \sum_{i=1}^{N}\left|p\sigma - e^{-y_i}\right|^2,$$

where σ is the lowest component of the chiral superfield Σ, while y_i is the lowest component of the chiral superfield Y_i,

$$y_i = \rho_i - i\vartheta_i.$$

One may already notice that ρ_i is equal to the dual variable $\mid \phi_i \mid^2$ in the linear sigma model. Then, from the exact potential, we can observe the same dynamical decomposition found in the gauge theory. Although we only have F-terms in the mirror, however, the Kähler potential of Σ-field provides the coefficient e^2 in the potential energy. The full nonperturbative correction only changes the expectation value of σ field and does not affect the decomposition conclusion. See also [18–20] for the dynamical decomposition in the nonabelian mirrors and [7,8,21–26] for the dynamical decomposition in the charge-p Schwinger model. Finally, we want to comment that if e^2 is very large, the Σ-field's Kähler potential is suppressed, which can be treated as an auxiliary field. Then we can integrate out it to introduce delta functions in the Y-field space, there are p different sectors. They can not talk to each other, because they localize at different delta functions. In the next section, we will give a brief review of the application of decomposition to math.

2.3 Mathematical Applications

In the previous two sections, we have reviewed a finer vacuum structure called the dynamical decomposition. Certainly, it is a nontrivial physical statement, which relates to several quantum field theoretic aspects such as higher symmetry, 't Hooft anomaly, etc. Furthermore, it can be applied to the string compactification [5]. However, in this section, we mainly focus on its applications to math.

The dynamical decomposition is a statement of the physical two-dimensional nonlinear sigma model. So the Witten-type topological quantum field theories, A-twisted or B-twisted of the physical NLSM, shall share the same conclusion. On the one hand, the topological A-model in math is related to the Gromov-Witten theory and Fukaya category. The dynamical decomposition nontrivially suggests that Gromov-Witten invariants [27,28] and Fukaya categories have decomposition

[9] In the mirror, the component field F_{01} in the chiral superfield Σ is not a gauge field strength, although we use the same notation.

properties if the target space manifold has a $\mathbb{Z}_p$ one-form symmetry. The Gromov-Witten theory is concerned with intersection numbers on the moduli space of stable map (from the genus-zero curve to the variety), and these invariants can be encoded into a Q series

$$\sum_{d=0} n_d Q^d, \tag{11}$$

where the rational number n_d is related to the Gromov-Witten invariants. The decomposition, in this context, means the moduli space of stable map can be decomposed into the stable map from the genus-zero curve to p same target manifold but with p different Novikov variables, and they are $Q^{\frac{1}{p}} e^{\frac{i2n\pi}{p}}$, for $n \in [0, \ldots, p-1]$. Therefore, only the map with a degree pd (d is a non-negative integer) can contribute to the invariants. While the categories, in mathematics, consists of a collection of "objects" that are linked by "morphisms." The decomposition of a category means that objects in this category can be decomposed into p sets, and the "morphisms" only link the objects in the same set. On the other hand, the topological B-model in math is related to the Picard-Fuchs equation and derived category of coherent sheaves. The decomposition of a Picard-Fuchs system says the solution space of this PF-equation with an order pn can be decomposed into p sectors, and each sector consists of the n-dimensional solution space of an order n Picard-Fuchs equation. Some studies of derived categories of coherent sheaves of weighted projective space can be found in the math literature [29]. However, a complete survey of the decomposition in categories is still worth pursuing in math.

We close this section by mentioning that we can also construct the A-twisted/B-twisted linear sigma model. The observables in the A-model do not depend on e^2, so the Gromov-Witten invariant can be computed in the A-twisted linear sigma model, which could have a decomposition phenomenon. There is no definition of Fukaya categories of linear sigma model in the literature yet, however, one may expect they do not depend on the parameter e^2, so they could have a dynamical decomposition if the theory has a nontrivial one-form symmetry. The topological B-model of linear sigma model is also interesting. First, we still do not have a definition of B-twisted closed string linear sigma model yet, however, the B-twisted boundary in the linear sigma model has been studied in [30–34] and its math theory, the derived category of a GIT quotient and its quotient stack[10], has been built in [35]. Furthermore, one lesson from these studies is that the derived category is not RG-protected, see also [36]. Thus from the physical reason, we should not expect the derived category of gauge theory has the decomposition feature [11], although its stable part can be decomposed because it corresponds to B-branes dynamics of the nonlinear sigma model. We leave this proof to the interested mathematician.

---

[10] We thank Ming Zhang for pointing out the correct terminology we should use here.

[11] Consider, for example, the linear sigma model for a projective space. The derived category of this GLSM, (in math terminology it is the derived category of stack quotient $[\mathbb{C}^N/U(1)]$), consists of, besides the data from the projective space, an extra part from the unstable locus where all of the coordinates are vanishing.

3 3d $\mathcal{N}=2$ Chern-Simons Matter Theories

In this section, we discuss the vacuum structure in the three-dimensional gauge theory, and we mainly focus on the $\mathcal{N}=2$ Chern-Simons matter theories. However, we expect that there should be a similar story in other 3d gauge theories.

As discussed in the introduction, the vacuum configuration is, not only, labeled by the scalar moduli, but also the field strength due to the two-form symmetry. So in 3d $\mathcal{N}=2$ gauge theory for charge-p $\mathbb{C}P^{N-1}$, we do not expect the $\mathbb{Z}_p$ one-form symmetry causes any nontrivial dynamical decomposition of vacua. However, as first discussed in [6], the 3d gauge theory for charge-p $\mathbb{C}P^{N-1}$ will decompose into p 2d linear sigma models for the same target via the study of the KK-reduction of the exact twisted superpotential. In this note, we review this fact by using the semi-classical potential energy.

Following [37] we relate the 3d data to the 2d's by KK-reduction. We first compactified the spacetime on $\mathrm{R}^2 \times S^1$, and the circle S^1 has the $2\pi\mathrm{R}$ periodic. Then the 2d matter field can be defined from the 3d one's as

$$\Phi_{2d} = \sqrt{2\pi R}\Phi_{3d},$$

with twisted masses given by $\frac{in}{R}$, for integer n. While the complex scalar field $\sigma = \sigma_1 + i\sigma_2$ in the two-dimensional vector multiplet can be descended from the scalar $\widetilde{\phi}$ and the circle component of gauge field v^μ in the three-dimensional abelian vector multiplet:

$$\sigma_1 = \frac{1}{2\pi R}\int_{S^1}\widetilde{\phi}, \qquad \sigma_2 = \frac{1}{2\pi R}\int_{S^1} v^2 \equiv \sigma_2 + \frac{1}{R},$$

where the periodicity of the Wilson line σ_2 arises from large gauge transformations. The 2d bare couplings also can be reduced from 3d's by

$$\frac{2\pi R}{e_{3\mathrm{d}}^2} = \frac{1}{e_{2\mathrm{d}}^2}, \qquad r_{2d} = 2\pi R r_{3d}.$$

Now, we can write out the potential energy of the three-dimensional gauge theory

$$U(\phi_i, \vartheta_i) = \frac{e_{3d}^2}{4\pi R}\left(\sum_{i=1}^{N} p \mid \phi_i \mid^2 - r\right)^2 + \frac{e_{3d}^2}{4\pi R}\left(p\sum_{i=1}^{N}\vartheta_i - \theta - 2n\pi\right)^2 \tag{1}$$
$$+ \sum_{i=1}^{N} \mid \phi_i \mid^2 \left| p\sigma_1 + i\left(p\sigma_2 + \frac{n}{R}\right)\right|^2,$$

where we have suppressed the information of the Chern-Simons level, but we have already chosen a proper one such that there is no topological vacuum in the Higgs phase [6, 38]. From equation (1), we can find the vacuum configuration

$$\left\langle \sum_{i=1}^{N} p \mid \phi_i \mid^2 \right\rangle = r, \quad \langle \sigma_1 \rangle = 0, \quad \langle e^{2i\pi R\sigma_2} \rangle = e^{2i\pi\frac{n}{p}}.$$

Besides the charged-p projective space moduli, we have an extra label due to the Wilson loop $e^{2i\pi R\sigma_2}$ [39, 40]. Furthermore, one can show that

$$e^{l*2i\pi R\sigma_2}$$

is a representation of the $\mathbb{Z}_p$ one-form symmetry with the charge l mod p [1, 39].

Now, take the vanishing limit of radius R and keep the gauge coupling e_{2d}^2 small, then we can observe that the 3d gauge theory decomposes into p universes, and each of them is a 2d linear sigma model for charge-p $\mathbb{C}P^{N-1}$. Furthermore, following section 2, each universe will further decompose into p universes when the gauge coupling becomes large.

Now, come back to the question mentioned at the beginning: do we have a dynamical decomposition in the 3d gauge theory? We suspect there is no decomposition in the three-dimensional gauge theory, however, we do not claim we have physical proof. The reason we expect this is we do not have theta angle in three-dimensional gauge theory[12]. Mathematically, it is because the second homotopy group of any gauge group is trivial.

4 Four-dimensional $\mathcal{N} = 1$ Generalized Super Yang-Mills Theory

The generalized four-dimensional Yang-Mills theory with 3-form symmetry was first proposed in [21], where Seiberg modified the instanton sectors to be a multiple of p by adding to the Lagrangian

$$i\vartheta\left(\frac{1}{8\pi^2}\mathrm{tr}\left[F(a)\wedge F(a)\right]-\frac{p}{2\pi}F^{(4)}\right),$$

where ϑ is a Lagrange multiplier and $F^{(4)} = dA^{(3)}$. This theory has not only the $\mathbb{Z}_N$ one-form symmetry but also a $\mathbb{Z}_p$ 3-form symmetry. The systematical study of this generalization can be found in [25]. Furthermore, the authors in [25] extended the story to the generalized $\mathcal{N} = 1$ SYM theory. They uncover the dynamical decomposition in this system by computing 't Hooft anomaly incorporated with the intriguing higher-group structure. They observed that the Np vacua split into p universes with N vacua in each. Furthermore, they found there are dynamical domain walls in between two vacua of a universe, but no dynamical domain wall connects two different universes. The remaining question is whether the auxiliary field ϑ can be descended from a dynamical field. Following the same idea used in two dimensions in section 2, we propose a 4d UV $\mathcal{N} = 1$ generalized super Yang-Mills theory that all of the fields are dynamical fields.

[12] One can certainly turn on the two-form symmetry in the three dimensions. However, even it has the decomposition feature due to this two-form symmetry, we expect that it would not talk to our Chern-Simons matter theories.

We first write out the known Lagrangian of $\mathcal{N}=1$ super Yang-Mills theory

$$S_{\rm SYM} = \frac{1}{2g^2}\int \mathrm{tr}\left[F(a)\wedge\star F(a)\right] + \frac{i\theta_{\rm YM}}{8\pi^2}\int \mathrm{tr}\left[F(a)\wedge F(a)\right] \tag{1}$$
$$+\,\overline{\lambda}\overline{\sigma}^{\mu}\left(\partial_\mu\lambda + i\left[a_\mu,\lambda\right]\right).$$

The system has a $\mathbb{Z}_N$ one-form symmetry. The $\mathbb{Z}_N$ two-form gauge field can be realized as a pair of $U(1)$ one-form and two-form gauge fields, $B^{(1)}$, $B^{(2)}$, with the constraint,

$$NB^{(2)} = dB^{(1)},$$

where the one-form symmetric transformation is

$$B^{(2)} \mapsto B^{(2)} + d\Lambda^{(1)}, B^{(1)} \mapsto B^{(1)} + N\Lambda^{(1)}.$$

In order to see this symmetry from the Lagrangian, following [1,41], we introduce the $U(N)$ gauge field, $\widetilde{a}$, and relate it to the dynamical $SU(N)$ gauge field locally as

$$\widetilde{a} = a + \frac{1}{N}B^{(1)}.$$

Then we replace the $SU(N)$ field strength with the gauge-invariant combination of the $U(N)$ field strength $F(\widetilde{a})$ and $B^{(2)}$:

$$F(a) \equiv F(\widetilde{a}) - B^{(2)}.$$

One can show that $F(a)$ is manifestly invariant under the one-form symmetric transformation. However, this system has a mixed anomaly between θ-angle periodicity and $\mathbb{Z}_N$ one-form symmetry. More specifically, the partition function is not a single-valued function under the shifting of $\theta_{\rm YM}$ by a 2π, i.e,

$$Z_{\rm SYM}(\theta_{\rm YM}+2\pi) = \exp\left(i\frac{N}{4\pi}\int B^{(2)}\wedge B^{(2)}\right) Z_{\rm SYM}(\theta_{\rm YM}),$$

where we have the non-trivial 't Hooft flux

$$\frac{N}{8\pi^2}\int B^{(2)}\wedge B^{(2)} \in \frac{1}{N}\mathbb{Z}. \tag{2}$$

It was observed in [21] that in the generalized Yang-Mills theory, the instanton sectors have been modified to

$$\left(\frac{1}{8\pi^2}\mathrm{tr}\left[F(a)\wedge F(a)\right] - \frac{p}{2\pi}F^{(4)}\right) = 0,$$

where $F^{(4)}$ is a four-form field strength with the normalization condition $\int F^{(4)} \in 2\pi\mathbb{Z}$. This suggests that

$$\frac{N}{8\pi^2}\int B^{(2)}\wedge B^{(2)} \in \mathbb{Z}, \tag{3}$$

which does not agree with the quantization condition in (2). Hence, we can not gauge the $\mathbb{Z}_N$ one-form symmetry solely in the generalized super Yang-Mills theory. This puzzle has been resolved in [25] by gauging the $\mathbb{Z}_p$ three-form symmetry at the same time, and a general statement for the $d+1$ group structure in $1+d$-dim QFT has been proved in [42]. The interested reader can read them for more details. In the remaining section, we propose a UV model that reduces to the generalized super Yang-Mills at the intermediate scale.

We first make the three-form gauge field dynamics by introducing a kinetic term

$$\frac{1}{2e_{4d}^2} F^{(4)} \wedge \star F^{(4)}. \tag{4}$$

The mass dimension of e_{4d}^2 is 4, so it is a weak coupling in the UV and becomes the strong coupling in the IR. Following the discussion of the one-form symmetry in the four-dimensional Maxwell theory in [41], one can observe that our generalized Maxwell equation has a three-forms $U(1)$ global symmetry. In order to have the correct degrees of freedom at the low energy physics, we also need to introduce a matter field ϕ charged by the two-forms $U(1)$ symmetry with the charge p [13], such that the symmetry in low energies is a three-forms $\mathbb{Z}_p$ global symmetry. The Lagrangian of the charged matter is

$$|D_\mu \phi|^2. \tag{5}$$

Since this "magnetic" charged matter has not appeared in the previous studies compared to the usual "electronic" charge matters, we would like to give more details about the gauging process.

To start, let us consider a free complex scalar ϕ in 1+d dimensions. We can write the field in polar coordinates $\phi = |\phi| e^{i\varphi}$ except at the origin. Without losing the generality, we restrict to the Lagrangian of the phase factor φ:

$$-d\varphi \wedge \star d\varphi. \tag{6}$$

This has a manifest global zero-form symmetry $\varphi \mapsto \varphi + p\alpha$ for a constant α. Then perform a local variation, we can find the conserved current of this zero-form symmetry

$$-2p \star d\varphi. \tag{7}$$

To gauge this zero-form symmetry, we couple to a one-form $U(1)$ gauge field A with the gauge-invariant action

$$-(d\varphi + pA) \wedge \star (d\varphi + pA). \tag{8}$$

In 1+1 dimensions, the above procedure gives the same expression as equation (2) up to an overall radial variable.

[13] It is consistent with the fact the instantons are classified by degree 4 cohomology.

The free theory actually has an extra (d-1)-form global symmetry with the conserved current:

$$(-)^d 2p d\varphi. \tag{9}$$

It is difficult to see this symmetry manifest in the "electronic" description; instead, it shifts the dual $(d-1)$-form variable by a $(d-1)$-form flat "connection": $C^{(d-1)}$. However, we still can gauge this symmetry as usual in the electronic variable. We first couple the current to a d-form field $A^{(d)}$:

$$(-)^d 2pA^{(d)} \wedge d\varphi. \tag{10}$$

This action is not yet gauge-invariant for a local gauge transformation. To get the gauge invariant one, we should introduce an extra term $p^2 A^{(d)} \wedge \star A^{(d)}$ in the Lagrangian. If we put it all together we obtain

$$-\left(d\varphi + p(-)^d \star A^{(d)}\right) \wedge \star \left(d\varphi + p(-)^d \star A^{(d)}\right). \tag{11}$$

This section mainly focuses on the $d = 3$ case, which means we are gauging the two-form $U(1)$ global symmetry. Furthermore, we are interested in a theory with the nontrivial instanton sector which means the mass dimension of our gauge field strength F_{0123} is four. Therefore, we expect the mass dimension of the radial part, $|\phi|$, is vanishing from equation (10). This also indicates, from the current (9), that the non-vanishing two-form conserved charge is infinity heavy in the far infrared. Thus, it does affect our usual infrared scattering amplitudes. Our gauge theory is, of course, a UV-fundamental theory. However, the matter has an unusual kinetic term

$$d|\phi| \wedge \star d|\phi|,$$

which has mass dimension two. Therefore, the coefficient of this kinetic term is not actually a dimensionless constant, rather, it has the mass dimension two. This means the only finite energy fluctuation in the far-infrared is the constant mode. One can compute the one-point correlation function of $|\phi|^2$ at the physical scale μ:

$$\left\langle |\phi|^2 \right\rangle \cong \int_{\mu \leq |k| \leq \Lambda_{\mathrm{UV}}} \frac{d^4k}{(2\pi)^4} \frac{1}{k^4} \cong \log\left(\frac{\Lambda_{\mathrm{UV}}}{\mu}\right). \tag{12}$$

The propagator k^{-4} in momentum space is an unusual one. It can be understood by replacing our variable $|\phi|$ with $\left|\partial_\mu^{-1}\widetilde{\phi}\right|$, where the field $\widetilde{|\phi|}$ has the usual propagator k^{-2}. There could be an overall constant off. However, the sign of this constant can be absorbed into the sign of charge p, while the absolute value of this constant can be absorbed into the scales.

We can, furthermore, turn on the FI parameter of this higher $U(1)$ symmetry

$$-r_{4d} \cdot D, \tag{13}$$

where D is an auxiliary field. The D-term coupling is

$$D\left(p|\phi|^2_{\text{Bare}} - r_{4d}^{\text{UV}}\right). \tag{14}$$

Because of SUSY, we claim the RG-flow of r_{4d} only receive a one-loop correction:

$$r_{4d}(\mu) = r_{4d}^{\text{UV}} + p\log\left(\frac{\mu}{\Lambda_{\text{UV}}}\right). \tag{15}$$

This correction can be understood from the calculation in equation (12).

The last term we introduce is

$$\left(\frac{\widetilde{\theta}}{2\pi} + n\right)\int F^{(4)}. \tag{16}$$

The integer n follows the Dirac quantization of the flux $F^{(4)}$. To affect the instanton sectors of the super Yang-Mills theory, we must impose a condition on theta angles:

$$\widetilde{\theta} \equiv p\theta_{\text{YM}}. \tag{17}$$

Therefore, besides the dynamical scale in Yang-Mills theory Λ, here we have the second RG-invariant dynamical scale defined by

$$\widetilde{\Lambda} \equiv \Lambda_{\text{UV}} \exp\left(-r_{4d}^{\text{UV}} + i\widetilde{\theta}\right) \gg \Lambda. \tag{18}$$

Hence, it is possible that the theory still stays at the perturbative region of Yang-Mills theory even though the coupling e_{4d}^2 is large. After gauging the one-form $\mathbb{Z}_N$ symmetry, the theory shall have nontrivial 't Hooft flux (2). In order to make this possible, we require that $A^{(3)}$ transforms under the one-form gauge transformation as [1, 25, 41]

$$pA^{(3)} \mapsto pA^{(3)} - i\left(\frac{N}{2\pi}B^{(2)} \wedge \Lambda^{(1)} + \frac{N}{4\pi}\Lambda^{(1)} \wedge d\Lambda^{(1)}\right). \tag{19}$$

One can find that $\frac{i}{4\pi}\int \text{tr}F(\widetilde{a}) \wedge F(\widetilde{a}) + p\int F^{(4)}$ is gauge-invariant under the one-form symmetry. This means the nontrivial 't Hooft flux is physically reasonable.

The vacuum structure, besides the D-term, also depends on the following three terms

$$\frac{1}{2e_{4d}^2}F^{(4)} \wedge \star F^{(4)} + \left(\frac{\widetilde{\theta}}{2\pi} + n\right)F^{(4)} + \left|D_\mu\phi\right|^2. \tag{20}$$

Following section 2, we use a dual variable to rewrite the formula as

$$\frac{1}{2e_{4d}^2}F^{(4)} \wedge \star F^{(4)} + \left(\frac{\widetilde{\theta} - p\vartheta}{2\pi} + n\right)F^{(4)} - \frac{1}{4|\phi|^2}\star d\vartheta \wedge d\vartheta. \tag{21}$$

From the equation of motion, we have

$$\star F^{(4)} = e_{4d}^2 \left(\frac{p\vartheta - \widetilde{\theta}}{2\pi} - n \right). \tag{22}$$

Finally, the semi-classical potential energy of our UV gauge theory is

$$U\left(\star F^{(4)}, \phi\right) = \frac{e_{4d,\text{eff}}^2}{2} \left(p \mid \phi \mid^2 - r_{4d}\right)^2 + \frac{2\pi^2}{e_{4d,\text{eff}}^2} \left(\star F^{(4)}\right)^2, \tag{23}$$

where $r_{4d} = p \log \frac{\mu}{\Lambda}$. Now, consider the intermediate physical scale that

$$e_{4d}\sqrt{r_{4d}} \gg \mu \gg \widetilde{\Lambda} \gg \Lambda, \tag{24}$$

where our perturbative theory is physics reasonable. The vacuum configuration from (23) is

$$\langle \mid \phi \mid^2 \rangle = \frac{r_{4d}}{p}, \qquad \left\langle e^{i\vartheta} \right\rangle = e^{i\frac{2n\pi + \widetilde{\theta}}{p}}, \tag{25}$$

for $n = 0, \ldots, p-1$. One can easily find that the theory has p different universes labeled by the integer n. The operator

$$e^{il*\vartheta}$$

is charged under the $\mathbb{Z}_p^{(3)}$ symmetry with the charge l mod p. The dynamical decomposition can be easily seen from our semi-classical analysis of the UV gauge theory we proposed. However, one drawback is that we could not see all Np vacua explicitly in the semi-classical analysis. For this purpose, we must include the full nonperturbative quantum correction in the potential energy to see the entire vacuum configuration. However, we do not perform this calculation in this note as this would not bring any new insight into the dynamical decomposition. Rather, we see that the radial direction of ϕ is heavy that can be integrated out. Furthermore, the kinetic term of ϑ is suppressed, which means it is an auxiliary field at the intermediate scale. Finally, we can estimate the tension of the non-BPS domain wall between different sectors is proportional to e_{4d}^2 which suggests that the domain wall becomes infinity heavy even at the intermediate scale. Therefore, our 4d UV gauge theory is correctly reduced to the effective theory proposed in [21] and studied in [25]. The vacua of the UV gauge theory can, then, be described explicitly in terms of the low energy degrees of freedom.

Although it is not crucial for our decomposition story; however, we end this section by proposing the full supersymmetric Lagrangian of our UV gauge theory. We first start with the chiral superfield defined as usual

$$\Phi = \phi(y) + \Theta\sqrt{2}\psi(y) + \Theta\Theta F(y), \tag{26}$$

where Θ is the Grassmann odd coordinate in the superspace, and $y^\mu = x^\mu + i\Theta\sigma^\mu\bar{\Theta}$. Our notation follows the book by Wess and Bagger [43]. It turns out that the most difficult technical part of defining our Lagrangian is the super p-form vector multiplet; however, it has been already systematically studied in the context of $4d$ $\mathcal{N}=1$ theory four decades ago [44]. We only focus on the super three-form vector multiplet in this note, let us denote the scalar vector multiplet to be V, then following [44, equation(3.12)], we define the super-field strength

$$\bar{\Sigma} = \frac{1}{4}D^2V = \bar{\sigma} + i\sqrt{2}\bar{\Theta}\bar{\lambda} + \bar{\Theta}\bar{\Theta}\left(D + iF_{0123}\right), \tag{27}$$

where we use a different notation and possibly a different normalization of each component field. Furthermore, the mass dimensions of fields are also different. For example, we require the σ-field to have the mass dimension three. The kinetic term of gauge field is

$$\begin{aligned}\mathscr{L}_{\text{vector}} &= \frac{1}{2e_{4d}^2}\int d^2\Theta d^2\bar{\Theta}\bar{\Sigma}\Sigma \\ &= \frac{1}{2e_{4d}^2}\left(\partial^\mu\bar{\sigma}\partial_\mu\sigma + i\partial_\mu\bar{\lambda}\bar{\sigma}^\mu\lambda + D^2 + F^{(4)}\wedge\star F^{(4)}\right),\end{aligned} \tag{28}$$

where σ^μ is the Pauli matrix. The kinetic term of the charged matter is also straightforward

$$\begin{aligned}\mathscr{L}_{\text{matter}} &= \int d^2\Theta d^2\bar{\Theta}\bar{\Phi}e^{pV}\Phi \\ &= |D_\mu\phi|^2 - i\bar{\psi}\bar{\sigma}^\mu D_\mu\psi \\ &\quad + Dp|\phi|^2 + |p\sigma|^2|\phi|^2 + |F|^2 + p\bar{\sigma}\bar{F}\phi + pF\sigma\bar{\phi} + ip\bar{\psi}\bar{\lambda}\phi + ip\bar{\phi}\lambda\psi.\end{aligned} \tag{29}$$

The last term we can add is the complexified FI-parameter ($t = r - i\widetilde{\theta}$) term:

$$\mathscr{L}_{\text{FI}} = -\frac{1}{2}\int d^2\Theta t\Sigma + h.c. \tag{30}$$

Finally, we would like to comment that the two-forms charged matters and the three-forms super gauge field are both heavy in low energies, thus there is no gauge anomaly to worry about. Although our UV "magnetic" matter theory only affects the topological sectors of super Yang-Mills theory; however, its UV dynamics are still very interesting, and we leave the complete investigation of this theory and its applications to future work [45].

5 Conclusions and Outlooks

The main observation in this note is to include the field strength $F_{0\cdots d}$, apart from the scalars, to parameterize the vacuum configuration in a $1+d$ gauge theory. The field strength $F_{0\cdots d}$ is Lorentz-invariant, so even the vanishing expectation value of this field could teach us the nontrivial information of vacua. For example, when a gauge theory has a d-form $\mathbb{Z}_p$ symmetry. One can observe, by the flux, its effective theory decomposes into p different universes at the intermediate scale. This is a stronger selection rule than we usually think of in the local quantum field theory. Following this setup, we reviewed the existing results in the two-dimensional quantum field theory from our perspective. Furthermore, we re-studied the dynamical decomposition in the KK-reduction of three-dimensional Chern-Simons matter theory and conjectured there is no decomposition in three dimensions. Finally, we propose a four-dimensional UV gauge theory which reduces to the generalized Yang-Mills theory in the low energy. This generalized Yang-Mills was first proposed in [21] and systematically investigated in [25], where Tanizaki and Ünsal observed the decomposition feature in the generalized Yang-Mills theory. Our proposed UV gauge theory gives a dynamical understanding.

So far, our discussions are limited to the theory with a Lagrangian description. However, it is known that there are many non-Lagrangian theories. Thus, it would be interesting to study how the d-form symmetry affects the vacuum configuration of these theories. Furthermore, we mainly focus on the continuous gauge theories in this note. However, we know many orbifolded theories can not be embedded into continuous gauge theories, so the decomposition of these orbifolded theories should be understood in a different approach. Finally, quantum field theories in more than four dimensions might also have the dynamical decomposition, which we missed in this note. We leave all these to future work.

Acknowledgements We would like to thank Ryan Thorngren and Yifan Wang for the debate on distinguishing between the topological current and the symmetric current in quantum field theory. It pushed the author to read the reference [1] and was finally convinced by the Higgs mechanism of the "generalized global symmetry" observed in [1]. We also thank Eric Sharpe for the discussion and Mithat Ünsal for his talk on the generalized Yang-Mills theory.

References

1. D. Gaiotto, A. Kapustin, N. Seiberg and B. Willett, "Generalized Global Symmetries," JHEP **02**, 172 (2015) [arXiv:1412.5148 [hep-th]].
2. T. Pantev and E. Sharpe, "GLSM's for Gerbes (and other toric stacks)," Adv. Theor. Math. Phys. **10**, no.1, 77-121 (2006) [arXiv:hep-th/0502053 [hep-th]].
3. T. Pantev and E. Sharpe, "Notes on gauging noneffective group actions," [arXiv:hep-th/0502027 [hep-th]].
4. T. Pantev and E. Sharpe, "String compactifications on Calabi-Yau stacks," Nucl. Phys. B **733**, 233-296 (2006) [arXiv:hep-th/0502044 [hep-th]].
5. S. Hellerman, A. Henriques, T. Pantev, E. Sharpe and M. Ando, "Cluster decomposition, T-duality, and gerby CFT's," Adv. Theor. Math. Phys. **11**, no.5, 751-818 (2007) [arXiv:hep-th/0606034 [hep-th]].
6. W. Gu, D. Pei and M. Zhang, "On Phases of 3d $\mathcal{N}=2$ Chern-Simons-Matter Theories," [arXiv:2105.02247 [hep-th]].
7. Y. Tanizaki and M. Ünsal, "Modified instanton sum in QCD and higher-groups," JHEP **03**, 123 (2020) doi:10.1007/JHEP03(2020)123 [arXiv:1912.01033 [hep-th]].
8. Z. Komargodski, K. Ohmori, K. Roumpedakis and S. Seifnashri, "Symmetries and strings of adjoint QCD_2," JHEP **03**, 103 (2021) [arXiv:2008.07567 [hep-th]].

9. E. Sharpe, "Decomposition in diverse dimensions," Phys. Rev. D **90**, no.2, 025030 (2014) [arXiv:1404.3986 [hep-th]].
10. E. Sharpe, "Notes on generalized global symmetries in QFT," Fortsch. Phys. **63**, 659-682 (2015) [arXiv:1508.04770 [hep-th]].
11. E. Sharpe, "Topological operators, noninvertible symmetries and decomposition," [arXiv:2108.13423 [hep-th]].
12. S. R. Coleman, "More About the Massive Schwinger Model," Annals Phys. **101**, 239 (1976)
13. E. Witten, "Phases of N=2 theories in two-dimensions," Nucl. Phys. B **403**, 159-222 (1993) [arXiv:hep-th/9301042 [hep-th]].
14. K. Hori and C. Vafa, "Mirror symmetry," [arXiv:hep-th/0002222 [hep-th]].
15. S. R. Coleman, "There are no Goldstone bosons in two-dimensions," Commun. Math. Phys. **31**, 259-264 (1973)
16. N. D. Mermin and H. Wagner, "Absence of ferromagnetism or antiferromagnetism in one-dimensional or two-dimensional isotropic Heisenberg models," Phys. Rev. Lett. **17**, 1133-1136 (1966)
17. D. R. Morrison and M. R. Plesser,"Summing the instantons: Quantum cohomology and mirror symmetry in toric varieties," Nucl. Phys. B **440**, 279-354 (1995) [arXiv:hep-th/9412236 [hep-th]].
18. W. Gu and E. Sharpe, "A proposal for nonabelian mirrors," [arXiv:1806.04678 [hep-th]].
19. Z. Chen, W. Gu, H. Parsian and E. Sharpe, "Two-dimensional supersymmetric gauge theories with exceptional gauge groups," Adv. Theor. Math. Phys. **24**, no.1, 67-123 (2020) [arXiv:1808.04070 [hep-th]].
20. W. Gu, E. Sharpe and H. Zou, "Notes on two-dimensional pure supersymmetric gauge theories," JHEP **04**, 261 (2021) [arXiv:2005.10845 [hep-th]].
21. N. Seiberg, "Modifying the Sum Over Topological Sectors and Constraints on Supergravity," JHEP **07**, 070 (2010) [arXiv:1005.0002 [hep-th]].
22. Z. Komargodski, A. Sharon, R. Thorngren and X. Zhou, "Comments on Abelian Higgs Models and Persistent Order," SciPost Phys. **6**, no.1, 003 (2019) [arXiv:1705.04786 [hep-th]].
23. M. M. Anber and E. Poppitz, "Anomaly matching, (axial) Schwinger models, and high-T super Yang-Mills domain walls," JHEP **09**, 076 (2018) [arXiv:1807.00093 [hep-th]].
24. A. Armoni and S. Sugimoto, "Vacuum structure of charge k two-dimensional QED and dynamics of an anti D-string near an O1-plane," JHEP **03**, 175 (2019) [arXiv:1812.10064 [hep-th]].
25. T. Misumi, Y. Tanizaki and M. Ünsal, "Fractional θ angle, 't Hooft anomaly, and quantum instantons in charge-q multi-flavor Schwinger model," JHEP **07**, 018 (2019) [arXiv:1905.05781 [hep-th]].
26. A. Cherman and T. Jacobson, "Lifetimes of near eternal false vacua," Phys. Rev. D **103**, no.10, 105012 (2021) [arXiv:2012.10555 [hep-th]].
27. E. Andreini, YF, Jiang, and HH. Tseng. "On Gromov-Witten theory of root gerbes." [arXiv:0812.4477 (2008)].
28. E. Andreini, YF, Jiang, and HH. Tseng. "Gromov-Witten theory of product stacks." Communications in Analysis and Geometry Volume 24, Number 2, 223–277, 2016 [arXiv:0905.2258 (2009)].
29. D. Auroux, L. Katzarkov, and D. Orlov. "Mirror symmetry for weighted projective planes and their noncommutative deformations." Annals of Mathematics 167 (2004): 867-943. [arXiv:math/0404281v1].
30. S. Govindarajan, T. Jayaraman and T. Sarkar, "On D-branes from gauged linear sigma models," Nucl. Phys. B **593**, 155-182 (2001) doi:10.1016/S0550-3213(00)00611-8 [arXiv:hep-th/0007075 [hep-th]].
31. K. Hori, "Linear models of supersymmetric D-branes," [arXiv:hep-th/0012179 [hep-th]].
32. S. Hellerman and J. McGreevy, "Linear sigma model toolshed for D-brane physics," JHEP **10**, 002 (2001) [arXiv:hep-th/0104100 [hep-th]].
33. M. Herbst, K. Hori and D. Page, "Phases Of N=2 Theories In 1+1 Dimensions With Boundary," [arXiv:0803.2045 [hep-th]].

34. K. Hori and M. Romo, “Exact Results In Two-Dimensional (2,2) Supersymmetric Gauge Theories With Boundary,” [arXiv:1308.2438 [hep-th]].
35. D. Halpern-Leistner. "The derived category of a GIT quotient.” Journal of the American Mathematical Society 28.3 (2015): 871-912.
36. A. Kapustin and Y. Li, “D branes in Landau-Ginzburg models and algebraic geometry,” JHEP **12**, 005 (2003) [arXiv:hep-th/0210296 [hep-th]].
37. M. Aganagic, K. Hori, A. Karch and D. Tong, “Mirror symmetry in (2+1)-dimensions and (1+1)-dimensions,” JHEP **07**, 022 (2001) [arXiv:hep-th/0105075 [hep-th]].
38. K. Intriligator and N. Seiberg, “Aspects of 3d N=2 Chern-Simons-Matter Theories,” JHEP **07**, 079 (2013) [arXiv:1305.1633 [hep-th]].
39. E. Witten, “Supersymmetric index of three-dimensional gauge theory,” [arXiv:hep-th/9903005 [hep-th]].
40. W. Gu, L. Mihalcea, E. Sharpe and H. Zou, “Quantum K theory of symplectic Grassmannians,” [arXiv:2008.04909 [hep-th]].
41. A. Kapustin and N. Seiberg, “Coupling a QFT to a TQFT and Duality,” JHEP **04**, 001 (2014) [arXiv:1401.0740 [hep-th]].
42. Y. Tachikawa, “On gauging finite subgroups,” SciPost Phys. **8**, no.1, 015 (2020) [arXiv:1712.09542 [hep-th]].
43. J. Wess and J. Bagger, Supersymmetry and Supergravity, Princeton University Press (second edition, 1992).
44. S. J. Gates, Jr., “Super *p*-form gauge superfields,” Nucl. Phys. B **184**, 381-390 (1981)
45. In preparation.

Chapter 13

MATRIX Program: Theory and Applications of Stable Polynomials

A large class of conjecturally stable chromatic symmetric functions

Jacob P. Matherne *

Abstract The theory of stable and Lorentzian polynomials has recently found a number of successes in a variety of research areas including combinatorics, engineering, and computer science; in particular, they have played a key role in solving long-standing open problems such as the Kadison–Singer problem and Mason's log-concavity conjecture. More recently, the classes of stable polynomials and Lorentzian polynomials have appeared in representation theory, algebraic combinatorics, and even knot theory. We further highlight their ubiquity by introducing a large class of chromatic symmetric functions related to Hessenberg varieties and the Stanley–Stembridge conjecture that are conjecturally Lorentzian and stable.

Graph coloring is a well-studied topic in combinatorics, computer science, and scheduling problems. On the other hand, there has been a recent explosion in the study of stable polynomials: these are a multivariate analogue of real-rooted polynomials that have led the way to the resolution of a number of open problems in fields as diverse as matroid theory [2, 5], knot theory [9], quantum mechanics, functional analysis, and engineering [11]. This note offers a large class of polynomials constructed via graph coloring that are conjecturally stable.

We write $G = (V,E)$ for an arbitrary graph with finite vertex set V and finite edge set E. The *chromatic symmetric function (CSF)* $X_G(x_1,\ldots,x_m)$ of G in m variables is

Jacob P. Matherne
Department of Mathematics, North Carolina State University, Raleigh, NC, USA,
e-mail: jpmather@ncsu.edu

* The author was supported by the Deutsche Forschungsgemeinschaft (DFG) under Germany's Excellence Strategy - GZ 2047/1, Projekt-ID 390685813.

D. R. Wood et al. (eds.), *2021-2022 MATRIX Annals*, MATRIX Book Series 5,
https://doi.org/10.1007/978-3-031-47417-0_36

$$X_G(x_1,\ldots,x_m) := \sum_{\substack{f\colon V\to[m]\\ \text{proper coloring}}} \prod_{v\in V} x_{f(v)},$$

where the sum is over all proper (vertex) colorings of G. A proper coloring $f\colon V \to [m]$ is a coloring of the vertices of G such that if $\{v,w\} \in E$, then $f(v) \neq f(w)$; that is, the two vertices of each edge are required to get different colors. This refinement of the chromatic polynomial χ_G of G was introduced by Stanley in [15]; indeed, they are related by the identity $X_G(\underbrace{1,\ldots,1}_{q\text{ times}},0,\ldots,0) = \chi_G(q)$. Note that we may define X_G using any number of variables; see Figure 1 for an example of the CSF of the path graph P_3 in both two and three variables. Much utility can be gained by considering CSFs $X_G(x_1,x_2,\ldots)$ in infinitely-many variables, where they lie in the ring of symmetric functions (see, for example, [7] and [13]); for simplicity, we will not go this route.

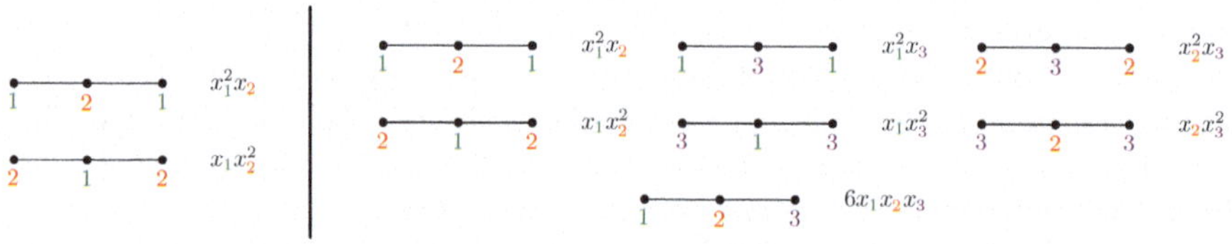

Fig. 1 On the left, the summands in $X_{P_3}(x_1,x_2)$, and on the right, the summands in $X_{P_3}(x_1,x_2,x_3)$.

A polynomial $f \in \mathbb{R}[x_1,\ldots,x_m]$ is called *stable* if it is either identically zero, or is nonvanishing on $\mathscr{H}^m$, where $\mathscr{H}$ is the open upper half-plane in $\mathbb{C}$. See [17, 3, 4] for more about the theory of stable polynomials. The CSF of an arbitrary graph need not be stable: for example, one may check that the CSF $X_{C_4}(x_1,x_2,x_3,x_4)$ in four variables of the four-cycle graph C_4 is not stable. However, the purpose of this note is to point out a large and interesting class where stability conjecturally does hold.

For the remainder of this note, we assume that our graphs G have vertex set $V = [n] = \{1,2,\ldots,n\}$. Define a set of graphs $\mathscr{D}$ by

$$\mathscr{D} = \{\text{graphs } G = ([n],E) \mid \text{if } \{i,j\} \in E \text{ and } i \leq i' < j' \leq j, \text{ then } \{i',j'\} \in E\}.$$

See Figure 2 for both an example of a graph in $\mathscr{D}$ and a non-example of a graph in $\mathscr{D}$ (for the case $n = 4$).

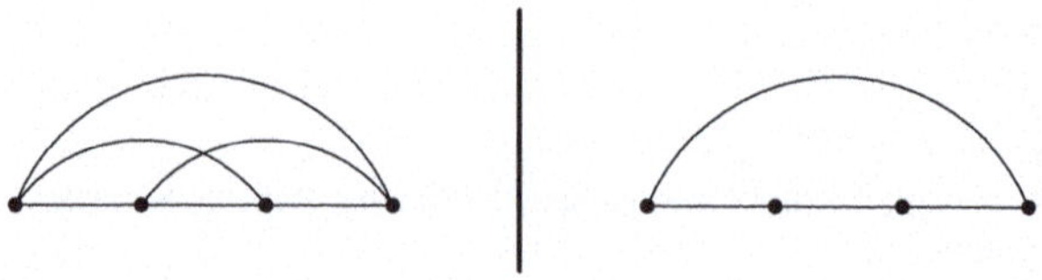

Fig. 2 The graph K_4 on the left is in $\mathscr{D}$, and the graph C_4 on the right is not in $\mathscr{D}$.

Graphs in the class $\mathscr{D}$ are enumerated by the n-th Catalan number. The graphs in $\mathscr{D}$ are also known as indifference graphs of Dyck paths, as unit interval orders, or as incomparability graphs of (2+2)- and (3+1)-free posets. The elusive Stanley–Stembridge conjecture [16] that has guided many developments in algebraic combinatorics involves exactly this class of graphs (due to a reduction in [8]); and on the other hand, the class $\mathscr{D}$ also naturally occurs in the study of Hessenberg varieties [14].

Conjecture 1 ([12, Conjecture 6.5]). If $G \in \mathscr{D}$, then $X_G(x_1,\ldots,x_m)$ is stable for every m.

Checking that a polynomial is stable generally involves verifying that an infinite number of univariate specializations are real-rooted [17, Lemma 2.3]. The Lorentzian property introduced in [6], and independently in [1], is a weaker notion than stability (e.g. Conjecture 1 implies Conjecture 2 below [6, Proposition 2.2]), but only involves a finite number of checks and has had much success uniting discrete and continuous log-concavity phenomena.

Conjecture 2 ([12, Conjecture 6.3]). If $G \in \mathscr{D}$, then $X_G(x_1,\ldots,x_m)$ is Lorentzian for every m.

Conjecture 2 has been checked for $n \leq 7$ and $m \leq 8$; and in the special case when $G \in \mathscr{D}$ has a bipartite complement graph, Conjecture 2 has been settled in [12, Theorem 6.8]. (Normalized) Schur polynomials were recently shown to be Lorentzian [10], and it is expected that many related polynomials in algebraic combinatorics should also have this property [10, Section 3]. On the other hand, (normalized) Schur polynomials are not stable [10, Example 9]. It is interesting to ponder where the line is: is there a satisfying reason why (normalized) Schur polynomials are not stable, but CSFs of graphs in $\mathscr{D}$ conjecturally are?

Acknowledgements The author is grateful to the MATRIX Institute and to the organizers of the "Theory and Applications of Stable Polynomials" research program, where the author thought about this problem with the participants. He is also thankful to Alejandro Morales and Jesse Selover for helpful comments on a draft of this note.

References

1. Anari, N., Gharan, S.O., Vinzant, C.: Log-concave polynomials, I: entropy and a deterministic approximation algorithm for counting bases of matroids. Duke Math. J. **170**(16), 3459–3504 (2021). DOI 10.1215/00127094-2020-0091. URL https://doi.org/10.1215/00127094-2020-0091
2. Anari, N., Lui, K., Gharan, S.O., Vinzant, C.: Log-Concave Polynomials III: Mason's Ultra-Log-Concavity Conjecture for Independent Sets of Matroids (2018). ArXiv:1811.01600

3. Borcea, J., Brändén, P.: The Lee-Yang and Pólya-Schur programs. I. Linear operators preserving stability. Invent. Math. **177**(3), 541–569 (2009). DOI 10.1007/s00222-009-0189-3. URL https://doi.org/10.1007/s00222-009-0189-3
4. Borcea, J., Brändén, P.: The Lee-Yang and Pólya-Schur programs. II. Theory of stable polynomials and applications. Comm. Pure Appl. Math. **62**(12), 1595–1631 (2009). DOI 10.1002/cpa.20295. URL https://doi.org/10.1002/cpa.20295
5. Brändén, P., Huh, J.: Hodge–Riemann relations for Potts model partition functions (2018). ArXiv:1811.01696
6. Brändén, P., Huh, J.: Lorentzian polynomials. Ann. of Math. (2) **192**(3), 821–891 (2020). DOI 10.4007/annals.2020.192.3.4. URL https://doi.org/10.4007/annals.2020.192.3.4
7. Carlsson, E., Mellit, A.: A proof of the shuffle conjecture. J. Amer. Math. Soc. **31**(3), 661–697 (2018). DOI 10.1090/jams/893. URL https://doi.org/10.1090/jams/893
8. Guay-Paquet, M.: A modular relation for the chromatic symmetric functions of (3+1)-free posets (2013). ArXiv:1306.2400
9. Hafner, E.S., Mészáros, K., Vidinas, A.: Log-concavity of the Alexander polynomial (2023). ArXiv:2303.04733
10. Huh, J., Matherne, J.P., Mészáros, K., St. Dizier, A.: Logarithmic concavity of Schur and related polynomials. Trans. Amer. Math. Soc. **375**(6), 4411–4427 (2022). DOI 10.1090/tran/8606. URL https://doi.org/10.1090/tran/8606
11. Marcus, A.W., Spielman, D.A., Srivastava, N.: Interlacing families II: Mixed characteristic polynomials and the Kadison-Singer problem. Ann. of Math. (2) **182**(1), 327–350 (2015). DOI 10.4007/annals.2015.182.1.8. URL https://doi.org/10.4007/annals.2015.182.1.8
12. Matherne, J.P., Morales, A.H., Selover, J.: The Newton polytope and Lorentzian property of chromatic symmetric functions (2022). to appear in Sel. Math. New Ser.
13. Orellana, R., Scott, G.: Graphs with equal chromatic symmetric functions. Discrete Math. **320**, 1–14 (2014). DOI 10.1016/j.disc.2013.12.006. URL https://doi.org/10.1016/j.disc.2013.12.006
14. Shareshian, J., Wachs, M.L.: Chromatic quasisymmetric functions. Adv. Math. **295**, 497–551 (2016). DOI 10.1016/j.aim.2015.12.018. URL https://doi.org/10.1016/j.aim.2015.12.018
15. Stanley, R.P.: A symmetric function generalization of the chromatic polynomial of a graph. Adv. Math. **111**(1), 166–194 (1995). DOI 10.1006/aima.1995.1020. URL https://doi.org/10.1006/aima.1995.1020
16. Stanley, R.P., Stembridge, J.R.: On immanants of Jacobi-Trudi matrices and permutations with restricted position. J. Combin. Theory Ser. A **62**(2), 261–279 (1993). DOI 10.1016/0097-3165(93)90048-D. URL https://doi.org/10.1016/0097-3165(93)90048-D
17. Wagner, D.G.: Multivariate stable polynomials: theory and applications. Bull. Amer. Math. Soc. (N.S.) **48**(1), 53–84 (2011). DOI 10.1090/S0273-0979-2010-01321-5. URL https://doi.org/10.1090/S0273-0979-2010-01321-5

A problem about virtual polytopes

Jidong Wang

Abstract This is a short summary on the problem of finding canonical representatives for elements in the Grothendieck group of polytopes.

1 Background

Let $\mathscr{K}$ be the set of polytopes in $\mathbb{R}^d$. Recall that $\mathscr{K}$ is equipped with the following two operations. For $P, Q \in \mathscr{K}$, $\lambda \geq 0$

$$\begin{aligned} &\text{Minkowski addition:} && P+Q=\{x+y \mid x \in P, y \in Q\} \\ &\text{scalar multiplication:} && \lambda P=\{\lambda x \mid x \in P\} \end{aligned}$$

The Minkowski addition turns $\mathscr{K}$ into a semigroup, with the identity element $\{0\}$. The semigroup $(\mathscr{K},+)$ satisfies the following *cancellation property*. For $P, Q, R \in \mathscr{K}$

$$P+R=Q+R \Leftrightarrow P=Q$$

This allows us to construct the *Grothendieck group* of $\mathscr{K}$. Concretely, let $\mathscr{P} = \mathscr{K} \times \mathscr{K} / \sim$, where

$$(P_1, Q_1) \sim (P_2, Q_2) \text{ if and only if } P_1+Q_2=P_2+Q_1$$

The additive structure on $\mathscr{P}$ is given by $(P_1, Q_1)+(P_2, Q_2)=(P_1+Q_1, P_2+Q_2)$. There is a the canonical embedding

$$\mathscr{K} \hookrightarrow \mathscr{P}, P \mapsto (P, \{0\})$$

Jidong Wang
University of Texas at Austin, 2515 Speedway, Austin, Texas, 78712, United States, e-mail: jidongw@utexas.edu

D. R. Wood et al. (eds.), *2021-2022 MATRIX Annals*, MATRIX Book Series 5,
https://doi.org/10.1007/978-3-031-47417-0_37

In other words, there is a canonical way to make sense of "Minkowski subtraction" in $\mathscr{K}$, and we can also write the equivalence class represented by (P,Q) as $P-Q$.

This construction seems to be known already in the early 20th century, first officially appearing in [7]. Elements in $\mathscr{P}$ are called *virtual polytopes* by Pukhlikov and Khovanskii [6]. It appears in McMullen's work on the polytope algebra [3]. Recently it also attracts attention from topologists in the context of Thurston norm [1]. For the history of virtual polytopes, see [5].

If one thinks of polytopes as they are, this construction seems a little bit surprising. However, it is very natural from a convex geometric perspective, since a convex body P is determined by its support function h_P and $h_{P+Q} = h_P + h_Q$. Clearly, one can subtract two functions. The difference of two functions is a function, but the difference of two convex bodies is an equivalence class. The natural question is, can we find a canonical representative for that equivalence class?

2 Question statement

The word "canonical" can mean different things. One natural way to construct canonical representatives is to define some minimality condition, as is shown in the following example.

Example 1. The nonzero rational numbers $\mathbb{Q}^\times$ are the Grothendieck group of the nonzero integers, with respect to its multiplicative structure. Each class in $\mathbb{Q}^\times$ has a canonical representative (up to multiplication by a unit) whose numerator and denominator both have the smallest absolute value. Since the integers have unique factorization, $\frac{a}{b}$ is a canonical representative when a and b are coprime. A more interesting example is the ring of quadratic integers $\mathbb{Z}[\sqrt{-5}]$. Unlike $\mathbb{Z}$, $\mathbb{Z}[\sqrt{-5}]$ doesn't have unique factorization. Let R be its Grothendieck group of the nonzero elements. An element in R may have two representatives $\frac{a_1}{b_1}$ and $\frac{a_2}{b_2}$ such that a_1 and b_1 are coprime, a_2 and b_2 are coprime, but a_1 and a_2 are not multiples of each other. For instance,

$$2 \cdot 3 = (1+\sqrt{-5})(1-\sqrt{-5}) \Rightarrow \frac{2}{1-\sqrt{-5}} = \frac{1+\sqrt{-5}}{3}$$

However, since $\mathbb{Z}[-5] \subset \mathbb{C}$, each element in R has a canonical representative up to a unit whose numerator and denominator both have the smallest modulus.

Question of such flavor first appeared in the context of quasidifferential calculus [4]. Lately, it was considered by Tran and the first author in the context of finding minimal representation of piecewise linear functions [9]. In summary, there are several ways to define what "minimal" means. They can be put into the following framework. Let $(S, \prec)$ be an partially ordered set and $f : \mathscr{K} \to S$ be a map. For two representatives of the same class $P_1 - Q_1 = P_2 - Q_2$, we say

$$P_1 - Q_1 \prec P_2 - Q_2 \text{ if } f(P_1) \prec f(P_2) \text{ and } f(Q_1) \prec f(Q_2)$$

- Take S to be $\mathscr{K}$, f to be the identity map, and $\prec$ to be inclusion, one gets the minimality condition defined by Pallaschke and Urbanski [4]. We call this *strong minimality*. It was shown that strongly minimal representatives are unique for $d = 2$, but not for higher dimensions. [2, 8]
- Take S to be the real numbers, f to be the d-dimensional volume, and $\prec$ to be the usual $\leq$, we get a minimality condition that minimizes the volume. We call this *volume minimality*. Volume-minimal representatives are unique when $d = 2$ (this is a consequence of the uniqueness of strongly minimal representatives, although direct elementary proofs can be found). The answer is unknown for higher dimensions.
  This minimality condition resembles Example 1 most. It is probably the most intriguing one since the volume function extends to virtual polytopes, and there is a strong connection with algebraic geometry where the (mixed) volume of convex bodies correspond to intersections of divisors. To our knowledge, this notion of minimality hasn't appeared in any literature.
- Take S to be $\mathbb{N}^{d+1}$, f to be the map extracting the f-vectors, and $\prec$ to be the product order induced by $\leq$ on $\mathbb{N}^{d+1}$, one gets a minimality condition that minimizes the combinatorial complexity, so we may call this *combinatorial minimality*. A version of this where f extracts only the vertices is considered in [9]. Note that this only makes sense for polytopes, while the above two minimality conditions apply for all compact convex bodies.

Question 1. In any of the above minimality criteria, what does the space of all minimal representatives look like?

References

1. Funke, F.: The integral polytope group. Advances in Geometry, **21(1)**, 45-62 (2021)
2. Grzybowski, J.: Minimal pairs of convex compact sets. Archiv der Mathematik, **63**, 173-181 (1994)
3. McMullen, P.: The polytope algebra. Advances in Mathematics, **78(1)**, 76-130 (1989)
4. Pallaschke, D., Urbanski, R.: Minimal pairs of compact convex sets, with application to quasidifferential calculus. Nonconvex Optimization and Its Applications, **43**, 173-214 (2000)
5. Panina, G. Y., Streinu, I.: Virtual polytopes. Russian Mathematical Surveys, **70(6)**, 1105-1165 (2015)
6. Pukhlikov, A. V., Khovanskii, A. G.: Finitely additive measures of virtual polytopes. St. Petersburg Math. J, **4(2)**, 337-356 (1993)
7. Rådström, H.: An embedding theorem for spaces of convex sets. Proceedings of the American mathematical society, **3(1)**, 165-169 (1952)
8. Scholtes, S.: Minimal pairs of convex bodies in two dimensions. Mathematika, **39(2)**, 267-273 (1992)
9. Tran, N. M., Wang, J.: Minimal Representations of Tropical Rational Functions. (2022) arXiv preprint arXiv:2205.05647.

A conjecture on spectral sparsification with respect to hyperbolicity cones

James Saunderson

Abstract Spectral sparsification of sums of positive semidefinite matrices is both a remarkable fact about positive semidefinite matrices and a key ingredient in fast algorithms for graph problems. This note presents a conjectural generalization of spectral sparsification to the setting of hyperbolicity cones.

1 Spectral sparsification

Let $G = (\mathscr{V}, \mathscr{E}, \mu)$ be an undirected simple graph with vertex set $\mathscr{V}$, edge set $\mathscr{E}$, and positive edge weights μ_e for $e \in \mathscr{E}$. We can associate, with G, a matrix L_G called the *weighted graph Laplacian*. This is a $|\mathscr{V}| \times |\mathscr{V}|$ matrix of the form

$$L_G = \sum_{\{i,j\} \in \mathscr{E}} \mu_{\{i,j\}}(e_i - e_j)(e_i - e_j)^T \tag{1}$$

where e_i is the ith standard basis vector in $\mathbb{R}^{|\mathscr{V}|}$. Note that L_G is positive semidefinite, as it is the sum of positive semidefinite (rank one) matrices. Many graph parameters, such as the maximum weight of a cut in the graph, can be expressed in terms of the quadratic form associated with L_G, i.e.,

$$x^T L_G x = \sum_{\{i,j\} \in \mathscr{E}} \mu_{\{i,j\}}(x_i - x_j)^2.$$

Suppose that $0 < \varepsilon < 1$ and $H = (\mathscr{V}, \mathscr{E}', \mu')$ is a weighted graph with the same vertex set as G and with $\mathscr{E}' \subseteq \mathscr{E}$. We say that H is an ε-*spectral sparsifier* of G if

$$(1-\varepsilon)L_G \preceq L_H \preceq (1+\varepsilon)L_G$$

James Saunderson
Department of Electrical and Computer Systems Engineering, Monash University, Clayton, VIC 3800, Australia, e-mail: james.saunderson@monash.edu

D. R. Wood et al. (eds.), *2021-2022 MATRIX Annals*, MATRIX Book Series 5,
https://doi.org/10.1007/978-3-031-47417-0_38

where $X \preceq Y$ indicates that $Y - X$ is a positive semidefinite matrix. This notion of approximation ensures that the quadratic form defined by L_H is uniformly close, in the relative error sense, to that of L_G. The idea behind this definition is that if we can find a sparsifier H with many fewer non-zero edges than G, it would likely be much more efficient to work with algorithmically.

The main existence result in this direction is the following.

Theorem 1 (Spectral sparsification [1]). *Let X be a $d \times d$ positive definite matrix with a decomposition as $X = \sum_{i=1}^{m} X_i$ where each X_i is positive semidefinite. Let $0 < \varepsilon < 1$. Then there exists $S \subseteq \{1, 2, \ldots, m\}$ with $|S| \leq \lceil 4d/\varepsilon^2 \rceil$, and positive scalars w_i for $i \in S$, such that*

$$(1-\varepsilon)X \preceq \sum_{i \in S} w_i X_i \preceq (1+\varepsilon)X.$$

This result is remarkable because the number of terms in the 'sparsified' sum is not only independent of m, the number of terms in the original decomposition, but is also linear in d. Applying this to the decomposition of L_G given in (1) (or rather of the restriction of L_G to the orthogonal complement of $(1, 1, \ldots, 1)$), shows that it is possible to find ε-spectral sparsifiers of dense graphs with $O(|\mathscr{V}|/\varepsilon^2)$ edges. The initial proof of this result (appearing in [1]) focused on the case where the X_i are all of rank one. A generalization to the case of X_i having any rank appears, for instance, in [4].

2 Hyperbolic polynomials and hyperbolicity cones

Let V denote a finite dimensional real vector space. We denote by $\mathbb{R}[V]_d$, polynomials on V with real coefficients that are homogeneous of degree d.

Definition 1. Let V be a finite dimensional real vector space and let $e \in V$. We say that $p \in \mathbb{R}[V]_d$ is *hyperbolic with respect to e* if $p(e) > 0$ and for all $x \in V$, the univariate polynomial $t \mapsto p(te - x)$ has only real zeros.

Associated with a hyperbolic polynomial is the following geometric object.

Definition 2. Let V be a finite dimensional vector space, and let $p \in \mathbb{R}[V]_d$ be hyperbolic with respect to $e \in V$. The *(open) hyperbolicity cone* associated with (p, e) is

$$\Lambda_{++}(p,e) = \{x \in V \ : \ \text{all zeros of } t \mapsto p(te - x) \text{ are positive}\}.$$

We denote by $\Lambda_{+}(p,e)$ the closure of $\Lambda_{++}(p,e)$. A foundational result of Gårding [2] is that $\Lambda_{++}(p,e)$ is a convex cone (and so $\Lambda_{+}(p,e)$ is a closed convex cone).

The prototypical example of a hyperbolic polynomial is $p(X) = \det(X)$, the determinant restricted to real symmetric matrices. This polynomial is hyperbolic with respect to $e = I$, the identity matrix, because $\det(tI - X)$, the characteristic polynomial of X, has only real zeros. The associated hyperbolicity cone is the cone of symmetric positive semidefinite matrices.

Generalizing the notation for the positive semidefinite order, given $x, y \in V$ we write $x \succeq_{\Lambda_+} y$ to mean that $x - y \in \Lambda_+(p,e)$.

3 Hyperbolic sparsification conjecture

The natural analogue of spectral sparsification in the setting of hyperbolicity cones, would be the following.

Conjecture 1 (Hyperbolic spectral sparsification). Let V be a finite dimensional real vector space and let $p \in \mathbb{R}[V]_d$ be hyperbolic with respect to $e \in V$. Let $x \in \Lambda_{++}(p,e)$ have a decomposition as $x = \sum_{i=1}^m x_i$ where $x_i \in \Lambda_+(p,e)$ for $i = 1,2,\ldots,m$. Let $0 < \varepsilon < 1$. Then there exists $S \subseteq \{1,2,\ldots,m\}$ with $|S| \leq \lceil 4d/\varepsilon^2 \rceil$, and positive scalars w_i for $i \in S$, such that

$$(1-\varepsilon)x \preceq_{\Lambda_+} \sum_{i \in S} w_i x_i \preceq_{\Lambda_+} (1+\varepsilon)x.$$

If true, this would directly generalize Theorem 1, and would imply that spectral sparsification depends only on the degree of the hyperbolic polynomial, and is completely independent of the number of terms in the decomposition of x and the dimension of the cone $\Lambda_+(p,e)$.

Spectral sparsification of sums of positive semidefinite matrices has a number of interesting applications to the simplification of semidefinite optimization problems, as discussed in [4]. It would be interesting to investigate what implications Conjecture 1 (if it were true) would have for hyperbolic optimization [3].

References

1. Batson, J., Spielman, D.A., Srivastava, N.: Twice-Ramanujan sparsifiers. SIAM Review **56**(2), 315–334 (2014)
2. Gårding, L.: An inequality for hyperbolic polynomials. Journal of Mathematics and Mechanics pp. 957–965 (1959)
3. Renegar, J.: Hyperbolic programs, and their derivative relaxations. Foundations of Computational Mathematics **6**(1), 59–79 (2006)
4. Silva, M.K.D.C., Harvey, N.J.A., Sato, C.M.: Sparse sums of positive semidefinite matrices. ACM Transactions on Algorithms (TALG) **12**(1), 1–17 (2015)

Topological Narayana polynomials and interlacing conjectures

Xavier Coulter and Norman Do and Ellena Moskovsky

1 From monotone Hurwitz numbers to topological Narayana polynomials

Monotone Hurwitz numbers enumerate factorisations of permutations into transpositions, with a certain monotonicity constraint [3]. They arise naturally as coefficients in the large N topological expansion of the well-studied HCIZ integral over the space of $N \times N$ unitary matrices. We introduce here a deformation of the monotone Hurwitz numbers that produces a generalisation of the Narayana polynomials [1]. Numerical evidence leads to conjectures concerning real-rootedness and interlacing of these so-called topological Narayana polynomials. It is then natural to investigate whether these polynomials admit multivariate generalisations that exhibit some form of stability.

Definition 1. Let g be a non-negative integer and let $\mu_1, \mu_2, \ldots, \mu_n$ be positive integers, whose sum we denote by $|\mu|$. The *monotone Hurwitz number* $\vec{H}_{g,n}(\mu_1, \mu_2, \ldots, \mu_n)$ is defined to be $\frac{1}{|\mu|!}$ times the number of tuples $(\tau_1, \tau_2, \ldots, \tau_m)$ of transpositions in the symmetric group $S_{|\mu|}$ such that

- $m = |\mu| + 2g - 2 + n$;
- $\tau_1 \circ \tau_2 \circ \cdots \circ \tau_m$ has n cycles labelled $1, 2, \ldots, n$, such that the length of cycle i is μ_i;
- $\langle \tau_1, \tau_2, \ldots, \tau_m \rangle$ is a transitive subgroup of $S_{|\mu|}$; and

Xavier Coulter
Department of Mathematics, The University of Auckland, Auckland 1010, New Zealand,
e-mail: xcou810@aucklanduni.ac.nz

Norman Do
School of Mathematics, Monash University, VIC 3800, Australia, e-mail: norm.do@monash.edu

Ellena Moskovsky
School of Mathematics, Monash University, VIC 3800, Australia,
e-mail: ellena.moskovsky@gmail.com

D. R. Wood et al. (eds.), *2021-2022 MATRIX Annals*, MATRIX Book Series 5,
https://doi.org/10.1007/978-3-031-47417-0_39

- if we write $\tau_i = (a_i\; b_i)$ with $a_i < b_i$, then $b_1 \leq b_2 \leq \cdots \leq b_m$.

The *t-monotone Hurwitz number* $\vec{H}^t_{g,n}(\mu_1, \mu_2, \ldots, \mu_n)$ is defined analogously, although each tuple is now counted with weight t to the power of its *hive number*, which is defined to be the number of distinct values in the tuple $(b_1, b_2, \ldots, b_m)$.

In the case $(g,n) = (0,1)$, one obtains the Narayana polynomials via the equation

$$\mu\, \vec{H}^t_{0,1}(\mu) = \text{Narayana}_{\mu-1}(t).$$

Thus, we refer to the t-monotone Hurwitz numbers collectively as *topological Narayana polynomials*. The word "topological" here refers to the fact that the t-monotone Hurwitz numbers fit into a large family of enumerative problems governed by the general theory of topological recursion [2]. In that context, the pair (g,n) can be thought of as recording the genus and number of boundary components of a surface.

Example 1. The t-monotone Hurwitz number $\vec{H}^t_{0,2}(2,1)$ counts certain triples (τ_1, τ_2, τ_3) of transpositions in the symmetric group S_3. All such triples appear in the table below and we observe that their compositions always result in a permutation of cycle type $(2,1)$, as required.

~~$(1\,2)\circ(1\,2)\circ(1\,2)$~~	~~$(1\,2)\circ(1\,3)\circ(1\,2)$~~	~~$(1\,2)\circ(2\,3)\circ(1\,2)$~~
$(1\,2)\circ(1\,2)\circ(1\,3)$	$(1\,2)\circ(1\,3)\circ(1\,3)$	$(1\,2)\circ(2\,3)\circ(1\,3)$
$(1\,2)\circ(1\,2)\circ(2\,3)$	$(1\,2)\circ(1\,3)\circ(2\,3)$	$(1\,2)\circ(2\,3)\circ(2\,3)$
~~$(1\,3)\circ(1\,2)\circ(1\,2)$~~	~~$(1\,3)\circ(1\,3)\circ(1\,2)$~~	~~$(1\,3)\circ(2\,3)\circ(1\,2)$~~
~~$(1\,3)\circ(1\,2)\circ(1\,3)$~~	~~$(1\,3)\circ(1\,3)\circ(1\,3)$~~	$(1\,3)\circ(2\,3)\circ(1\,3)$
~~$(1\,3)\circ(1\,2)\circ(2\,3)$~~	$(1\,3)\circ(1\,3)\circ(2\,3)$	$(1\,3)\circ(2\,3)\circ(2\,3)$
~~$(2\,3)\circ(1\,2)\circ(1\,2)$~~	~~$(2\,3)\circ(1\,3)\circ(1\,2)$~~	~~$(2\,3)\circ(2\,3)\circ(1\,2)$~~
~~$(2\,3)\circ(1\,2)\circ(1\,3)$~~	$(2\,3)\circ(1\,3)\circ(1\,3)$	$(2\,3)\circ(2\,3)\circ(1\,3)$
~~$(2\,3)\circ(1\,2)\circ(2\,3)$~~	$(2\,3)\circ(1\,3)\circ(2\,3)$	~~$(2\,3)\circ(2\,3)\circ(2\,3)$~~

The triples violating the transitivity and monotonicity conditions of Definition 1 are crossed out. Of the remaining twelve triples, the six in the top third of the table have $(b_1, b_2, b_3) = (2,2,3)$ or $(2,3,3)$ and hence hive number 2, while the remaining six in the bottom two-thirds of the table have $(b_1, b_2, b_3) = (3,3,3)$ and hence hive number 1. So with our normalisation, we have $\vec{H}^t_{0,2}(2,1) = \frac{1}{3!}(6t^2 + 6t) = t^2 + t$.

2 Recursions

It is natural to ask whether properties of the usual monotone Hurwitz numbers or of the usual Narayana polynomials can be generalised to the setting of topological Narayana polynomials. For example, the known linear recursions for monotone Hurwitz numbers and for Narayana polynomials have the following common generalisation, for which the original results can be recovered by specialising to $t = 1$ and $g = 0$, respectively.

Proposition 1 (Linear recursion). *For $g \geq 0$ and $\mu \geq 1$ such that $(g,\mu) \neq (0,1)$ or $(0,2)$, the "one-point" t-monotone Hurwitz numbers satisfy the recursion*

$$\mu^2 \vec{H}^t_{g,1}(\mu) = (\mu-1)(2\mu-3)(t+1)\vec{H}^t_{g,1}(\mu-1) \\ - (\mu-2)(\mu-3)(t-1)^2 \vec{H}^t_{g,1}(\mu-2) + \mu^2(\mu-1)^2 \vec{H}^t_{g-1,1}(\mu).$$

In this equation, we set $\vec{H}^t_{g,1}(\mu) = 0$ if $g < 0$ or $\mu < 1$.

Similarly, the known quadratic recursions for monotone Hurwitz numbers and for Narayana polynomials have the following common generalisation, for which the original results can be recovered by specialising to $t = 1$ and $g = 0$, respectively.

Proposition 2 (Quadratic recursion). *Let $S = \{2,3,\ldots,n\}$. For all $g \geq 0$ and $\mu_1 + \mu_2 + \cdots + \mu_n > 1$, the t-monotone Hurwitz numbers satisfy the recursion*

$$\mu_1 \vec{H}^t_{g,n}(\mu_1,\mu_S) = \sum_{i=2}^{n} (\mu_1+\mu_i)\vec{H}^t_{g,n-1}(\mu_1+\mu_i,\mu_{S\setminus\{i\}}) + \sum_{\alpha+\beta=\mu_1} \alpha\beta\,\vec{H}^t_{g-1,n+1}(\alpha,\beta,\mu_S) \\ + \sum_{\alpha+\beta=\mu_1} \sum_{\substack{g_1+g_2=g \\ I\sqcup J=S}} \alpha\beta\,\vec{H}^t_{g_1,|I|+1}(\alpha,\mu_I)\vec{H}^t_{g_2,|J|+1}(\beta,\mu_J) + (t-1)(\mu_1-1)\vec{H}^t_{g,n}(\mu_1-1,\mu_S).$$

In this equation, we set $\vec{H}^t_{g,n}(\mu_1,\mu_2,\ldots,\mu_n) = 0$ if $g < 0$ or $\mu_i < 1$ for some i and, for $I = \{i_1,i_2,\ldots,i_k\}$, we use μ_I to denote the sequence $\mu_{i_1},\mu_{i_2},\ldots,\mu_{i_k}$.

Along with the base cases $\vec{H}^t_{0,1}(1) = 1$ and $\vec{H}^t_{g,1}(1) = 0$ for $g \geq 1$, the quadratic recursion uniquely recovers all t-monotone Hurwitz numbers.

Observe that although $\vec{H}^t_{g,n}(\mu_1,\ldots,\mu_n)$ is inherently symmetric in its arguments, the quadratic recursion is manifestly asymmetric. This provides an instance of a much more general phenomenon exhibited by problems governed by the topological recursion. Although we have already hinted at the fact that t-monotone Hurwitz numbers are indeed governed by the topological recursion, we will not state the result explicitly nor consider the consequences thereof in the present note.

3 Real-rootedness, interlacing, and stability?

The usual Narayana polynomials are known to have only real roots and we conjecture that the same is true for topological Narayana polynomials more generally.

Conjecture 1 (Real-rootedness). Each topological Narayana polynomial $\vec{H}^t_{g,n}(\mu_1,\ldots,\mu_n)$ only has real roots.

One approach to proving the real-rootedness of Narayana polynomials is to show the stronger fact that any two consecutive Narayana polynomials interlace. A polynomial of degree d is said to *interlace* with a polynomial of degree $d+1$ if the roots $\alpha_1 \leq \alpha_2 \leq \cdots \leq \alpha_d$ of the first and the roots $\beta_1 \leq \beta_2 \leq \cdots \leq \beta_{d+1}$ of the second are real and satisfy

$$\beta_1 \leq \alpha_1 \leq \beta_2 \leq \alpha_2 \leq \cdots \leq \beta_d \leq \alpha_d \leq \beta_{d+1}.$$

One observes a similar phenomenon for topological Narayana polynomials.

Conjecture 2 (Interlacing). The topological Narayana polynomial $\vec{H}^t_{g,n}(\mu_1,\ldots,\mu_n)$ interlaces with each of the topological Narayana polynomials

$$\vec{H}^t_{g,n}(\mu_1+1,\mu_2,\ldots,\mu_n),\ \vec{H}^t_{g,n}(\mu_1,\mu_2+1,\ldots,\mu_n),\ \ldots,\ \vec{H}^t_{g,n}(\mu_1,\mu_2,\ldots,\mu_n+1).$$

The quadratic recursion gives an effective way to compute topological Narayana polynomials, from which one can then test the above predictions. Real-rootedness and interlacing have been confirmed for all $g+n \leq 5$ and $|\mu| \leq 10$. This amounts to over 300 checks, thus providing strong numerical evidence for the conjectures.

Given the close connection between real-rootedness for polynomials of one variable and stability for polynomials of many, one is led to consider whether the topological Narayana polynomials admit a natural multivariate generalisation that exhibits stability. Recall that a multivariate polynomial is said to be *stable* if it is non-zero whenever its arguments are evaluated at complex numbers with positive imaginary part.

An obvious way to define a multivariate version of the t-monotone Hurwitz numbers is to mimic the enumeration of Definition 1, associating to a tuple of transpositions $((a_1\ b_1),(a_2\ b_2),\ldots,(a_m\ b_m))$ the weight

$$\prod_{k\in(b_1,b_2,\ldots,b_m)} t_k,$$

where we ignore multiplicities in the tuple $(b_1,b_2,\ldots,b_m)$. One then obtains multiaffine polynomials in $\mathbb{Q}[t_2,t_3,\ldots]$ for which the corresponding t-monotone Hurwitz numbers are recovered by setting $t_k \mapsto t$ for all k. It is natural to wonder whether such polynomials are stable.

4 Data

The following table gives some examples of t-monotone Hurwitz numbers.

$(\mu_1,\ldots,\mu_n)$	$\mu_1\cdots\mu_n \vec{H}^t_{0,n}(\mu_1,\ldots,\mu_n)$	$\mu_1\cdots\mu_n \vec{H}^t_{1,n}(\mu_1,\ldots,\mu_n)$	$\mu_1\cdots\mu_n \vec{H}^t_{2,n}(\mu_1,\ldots,\mu_n)$
(1)	1	0	0
(2)	t	t	t
(3)	t^2+t	$5t^2+5t$	$21t^2+21t$
(4)	t^3+3t^2+t	$15t^3+40t^2+15t$	$161t^3+413t^2+161t$
(5)	$t^4+6t^3+6t^2+t$	$35t^4+175t^3+175t^2+35t$	$777t^4+3612t^3+3612t^2+777t$
$(1,1)$	t	t	t
$(2,1)$	$2t^2+2t$	$10t^2+10t$	$42t^2+42t$
$(3,1)$	$3t^3+9t^2+3t$	$45t^3+120t^2+45t$	$483t^3+1239t^2+483t$
$(2,2)$	$4t^3+10t^2+4t$	$50t^3+128t^2+50t$	$504t^3+1278t^2+504t$
$(4,1)$	$4t^4+24t^3+24t^2+4t$	$140t^4+700t^3+700t^2+140t$	$3108t^4+14448t^3+14448t^2+3108t$
$(3,2)$	$6t^4+30t^3+30t^2+6t$	$168t^4+792t^3+792t^2+168t$	$3402t^4+15450t^3+15450t^2+3402t$
$(1,1,1)$	$4t^2+4t$	$20t^2+20t$	$84t^2+84t$
$(2,1,1)$	$10t^3+28t^2+10t$	$140t^3+368t^2+140t$	$1470t^3+3756t^2+1470t$
$(3,1,1)$	$18t^4+102t^3+102t^2+18t$	$588t^4+2892t^3+2892t^2+588t$	$12726t^4+58794t^3+58794t^2+12726t$
$(2,2,1)$	$24t^4+120t^3+120t^2+24t$	$672t^4+3168t^3+3168t^2+672t$	$13608t^4+61800t^3+61800t^2+13608t$

References

1. Coulter, X., Do, N., Moskovsky, E.: Deformed monotone hurwitz numbers, topological narayana polynomials, and topological recursion. In preparation
2. Eynard, B., Orantin, N.: Invariants of algebraic curves and topological expansion. Commun. Number Theory Phys. **1**(2), 347–452 (2007)
3. Goulden, I.P., Guay-Paquet, M., Novak, J.: Monotone Hurwitz numbers and the HCIZ integral. Ann. Math. Blaise Pascal **21**(1), 71–89 (2014)

SPRINGER NATURE

GPSR Compliance

The European Union's (EU) General Product Safety Regulation (GPSR) is a set of rules that requires consumer products to be safe and our obligations to ensure this.

If you have any concerns about our products, you can contact us on ProductSafety@springernature.com

In case Publisher is established outside the EU, the EU authorized representative is:

Springer Nature Customer Service Center GmbH
Europaplatz 3
69115 Heidelberg, Germany

The manufacturer's authorised representative in the EU is Springer Nature Customer Service Centre GmbH, Europaplatz 3, 69115 Heidelberg, Germany. If you have any concerns regarding our products, please contact ProductSafety@springernature.com

Printed and bound by CPI Group (UK) Ltd, Croydon, CR0 4YY
01/10/2026
02225874-0001